P9-CDJ-445
3 2109 00820 4639 V.9

Comprehensive Supramolecular Chemistry

Comprehensive Supramolecular Chemistry

Executive Editors

Jerry L. Atwood
University of Missouri, Columbia, MO, USA

J. Eric D. Davies
University of Lancaster, UK

David D. MacNicol
University of Glasgow, UK

Fritz Vögtle
Institut für Organische Chemie und Biochemie der Rheinischen Friedrich-Wilhelms-Universität Bonn, Germany

Chairman of the Editorial Board

Jean-Marie Lehn
Université Louis Pasteur, Strasbourg, France & Collège de Paris, France

Volume 9
TEMPLATING, SELF-ASSEMBLY, AND SELF-ORGANIZATION

Volume Editors

Jean-Pierre Sauvage

M. Wais Hosseini
Université Louis Pasteur, Strasbourg, France

PERGAMON

UK	Elsevier Science Ltd., The Boulevard, Langford Lane, Kidlington, Oxford, OX5 1GB, UK
USA	Elsevier Science Inc., 660 White Plains Road, Tarrytown, New York, NY 10591-5153, USA
JAPAN	Elsevier Science Japan, Tsunashima Building Annex, 3-20-12 Yushima, Bunkyo-ku, Tokyo 113, Japan

---

First edition 1996.

Library of Congress Cataloging in Publication Data
A catalog record for this book is available from the Library of Congress.

British Library Cataloguing in Publication Data
A catalogue record for this book is available from the British Library.

ISBN 0–08–040610–6 (set : alk. paper)
ISBN 0–08–042721–9 (Volume 9)

∞™ The paper used in this publication meets the minimum requirements of the American National Standard for Information Sciences—Permanence of Paper for Printed Library Materials, ANSI Z39.48–1984.

Typeset by Variorum Publishing Ltd., Rugby, UK.
Printed and bound in Great Britain by BPC Wheatons Ltd., Exeter, UK.

Contents

Foreword

Since the middle of the nineteenth century, molecular chemistry, particularly synthetic chemistry, has resulted in an increasing mastery in formation of the covalent bond. A parallel evolution is now being encountered for noncovalent intermolecular forces. Beyond molecular chemistry, based on the covalent bond, lies the field of supramolecular chemistry, the aim of which is to gain control over the intermolecular bond.

Thus, supramolecular chemistry has been defined as chemistry beyond the molecule, referring to the organized entities of higher complexity that result from the association of two or more chemical species held together by intermolecular forces.

The field grew out of studies in the mid-1960s of molecular recognition of alkali metal ions using natural antibiotics and synthetic macro(poly)cyclic polyethers. Although it was not conceived as such at that time, its roots can be traced back to Paul Ehrlich's receptor idea, Alfred Werner's coordination theory, and Emil Fischer's lock-and-key image. In addition, there had been early investigations already in the mid-1930s of associations in solution, which were even termed "Übermolekeln" (supermolecules)!

The general concept was recognized and formulated only in the late 1970s. Its breadth and unifying power became progressively more and more apparent, so that recent years have seen an explosive growth in the number of laboratories that are working in this field and whose work has been reported in a vast range of publications, books, journals, meetings, and symposia.

Supramolecular chemistry has developed into a coherent and extremely lively body of concepts and objects, progressively generating and incorporating novel areas of investigation. A whole vocabulary, still incomplete, however, has been produced and is becoming more and more widely accepted and used.

Thus, supramolecular chemistry is a highly interdisciplinary field that has rapidly expanded at the frontiers of chemical science with physical and biological phenomena. Its roots extend over organic chemistry and the synthetic procedures for molecular construction; coordination chemistry and metal ion ligand complexes; physical chemistry and the experimental and theoretical studies of interactions; biochemistry and the biological processes that all start with substrate binding and recognition; and materials science and the mechanical properties of solids. A major feature is the range of perspectives offered by the cross-fertilization of supramolecular chemical research due to its location at the intersection of chemistry, biology, and physics. Drawing on the physics of organized condensed matter and expanding over the biology of large molecular assemblies, supramolecular chemistry expands into a supramolecular science. Such wide horizons are a challenge and a stimulus to the creative imagination of chemists.

For this reason, preparing "Comprehensive Supramolecular Chemistry" has also been a real challenge. Considering the breadth and the rapid expansion of the field, the Executive Editors, the Volume Editors, and of course the authors have performed an excellent task in bringing together such a vast and varied amount of information, results, and ideas. They deserve the warmest thanks of the whole community of chemists, who will find in this set of volumes not only the facts they need but also stimulation for further exploration of this most inspiring frontier of science.

Jean-Marie Lehn
Strasbourg

Preface

Among the many challenges still to be conquered by synthetic chemists, the construction of molecular objects presenting unusual structures (shape or topology) may still be considered to be an art. The preparation of elaborate large molecular assemblies composed of molecular units is certainly another largely unexplored area of research. This volume, based on contributions from the most research-active groups, gives an overview of both topics.

In terms of structure, topology, and function, an elegant and efficient way of building simple to extremely elaborate molecules involves template strategy. This is based on the preassembly and organization of several molecular fragments around a templating core using various construction principles. Covalent connection of the individual components affords a molecular entity, the high degree of complexity of which would have discouraged any nontemplate classical approach. An elegant example of this is template-directed synthesis of catenanes and rotaxanes. Using such a strategy, the size of the molecular objects lies typically in the nanoscale range.

Another viable approach to the construction of highly complex molecular edifices is based on noncovalent links, as opposed to the above-mentioned strategy using covalent bonds. Indeed, the construction of large size molecules (10^{-6}–10^{-4} cm in scale) with predicted and programmed structures can hardly be envisaged through stepwise classical synthesis using covalent linkages. However, the preparation of such higher-order materials may be attained through iterative processes based on self-assembly of individual modules. The synthesis of solids based on iterative assembling of individual complementary units still remains a challenge for chemists. A strict control of the self-assembly of molecular modules in the solid state should lead to structurally strictly controlled assemblies.

Another important area of research under active investigation deals with organometallic polymers. In recent years much attention has been focused on polymeric species (coordination polymers) in which the metal plays a structural role. This approach presents two interesting features. The metal, a structural constituent bridging other modules such as organic moieties, presents a wide variety of coordination geometries as well as specific electronic, redox, photonic, and magnetic properties. Therefore, coordination polymers may also exhibit preprogrammed or unexpected functional features. The preparation of such materials may be also envisaged through an iterative assembly of bridging ligands and metals.

The ultimate goal of the above-mentioned approaches is to convert molecular functions, built in within the framework of the individual modules, into macroscopic properties expressed at the level of the molecular assemblies and thus leading to the production of addressable devices.

Jean-Pierre Sauvage
Strasbourg

M. Wais Hosseini
Strasbourg

Contributors to Volume 9

Dr. D. B. Amabilino
Institut de Ciència de Materials de Barcelona, CSIC, Campus de la Universidad Autonoma Barcelaona, E-08193 Bellaterra, Spain

Professor W. Baumeister
Abteilung für Molekulare Strukturbiologie, Max-Planck-Institut für Biochemie, Am Klopferspitz, D-82152 Martinsried, Germany

Dr. P. N. W. Baxter
Laboratoire de Chimie Supramoléculaire, Institut le Bel, Université Louis Pasteur, 4 rue Blaise Pascal, F-67000 Strasbourg Cédex, France

Dr. D. H. Busch
Department of Chemistry, University of Kansas, Lawrence, KS 66045-0046, USA

Dr. J.-C. Chambron
Laboratoire de Chimie Organo-Minérale, Institut de Chimie, Université Louis Pasteur, 1 rue Blaise Pascal, F-67070 Strasbourg Cédex, France

Dr. I. S. Choi
Department of Chemistry, Harvard University, Cambridge, MA 02138, USA

Dr. E. C. Constable
Institut für Anorganische Chemie der Universität Basel, Spitalstrasse 51, CH-4056 Basel, Switzerland

Professor G. Decher
Université Louis Pasteur and CNRS, Institut Charles Sadron, 6 rue Boussingault, F-67083 Strasbourg Cédex, France

Dr. C. Dietrich-Buchecker
Laboratoire de Chimie Organo-Minérale, Institut de Chimie, Université Louis Pasteur, 1 rue Blaise Pascal, F-67070 Strasbourg Cédex, France

Mr. U. Drechsler
Lehrstuhl für Organische Chemie II, Universität Tübingen, Auf der Morgenstelle 18, D-72076 Tübingen, Germany

Dr. J. R. Fredericks
Department of Chemistry, University of Pittsburgh, 1201 Chevron Science Center, Pittsburgh, PA 15260, USA

Professor J.-H. Fuhrhop
Institut für Organische Chemie, Freie Universität Berlin, Takustrasse 3, D-14195 Berlin, Germany

Dr. M. Fujita
Department of Applied Chemistry, Faculty of Engineering, Chiba University, 1-33 Yayoicho, Inageku, Chiba 263, Japan

Dr. A. Garbesi
C.N.R.-I.Co.C.E.A., Area di Ricerca, Via P. Gobetti 101, I-40129 Bologna, Italy

Dr. M. R. Ghadiri
Department of Chemistry, The Scripps Research Institute, 10666 North Torrey Pines Road, La Jolla, CA 92037, USA

Dr. G. Gottarelli
Dipartimento di Chimica Organica, Università di Bologna, Via S. Donato 15, I-40127 Bologna, Italy

Professor A. D. Hamilton
Department of Chemistry, University of Pittsburgh, 1201 Chevron Science Center, Pittsburgh, PA 15260, USA

Professor Dr. M. Hanack
Lehrstuhl für Organische Chemie II, Universität Tübingen, Auf der Morgenstelle 18, D-72076 Tübingen, Germany

Dr. A. G. Kolchinksii
Department of Chemistry, University of Kansas, Lawrence, KS 66045-0046, USA

Professor T. Kunitake
Department of Chemical Science and Technology, Faculty of Engineering, Kyushu University, 6-10-1 Hakozaki, Higashi-ku, Fukuoka 812, Japan

Dr. D. H. Lee
Department of Chemistry, The Scripps Research Institute, 10666 North Torrey Pines Road, La Jolla, CA 92037, USA

Dr. X. Li
Department of Chemistry, Harvard University, Cambridge, MA 02138, USA

Professor S. Mann
School of Chemistry, University of Bath, Bath, BA2 7AY, UK

Dr. F. M. Raymo
School of Chemistry, University of Birmingham, Edgbaston, Birmingham, B15 2TT, UK

Dr. A. Reichert
Institut für Organische Chemie, Johannes-Gutenberg-Universität Mainz, Becherweg 18-20, D-55099 Mainz, Germany

Professor H. Ringsdorf
Institut für Organische Chemie, Johannes-Gutenberg-Universität Mainz, Becherweg 18-20, D-55099 Mainz, Germany

Dr. T. Scheybani
Abteilung für Molekulare Strukturbiologie, Max-Planck-Institut für Biochemie, Am Klopferspitz, D-82152 Martinsried, Germany

Dr. P. Schuhmacher
Institut für Organische Chemie, Johannes-Gutenberg-Universität Mainz, Becherweg 18-20, D-55099 Mainz, Germany

Dr. J. K. M. Sanders
Department of Chemistry, University Chemical Laboratory, University of Cambridge, Lensfield Road, Cambridge, CB2 1EW, UK

Professor J.-P. Sauvage
Laboratoire de Chimie Organo-Minérale, Institut de Chimie, Université Louis Pasteur, 1 rue Blaise Pascal, F-67070 Strasbourg Cédex, France

Dr. E. E. Simanek
Department of Chemistry, Harvard University, Cambridge, MA 02138, USA

Dr. G. P. Spada
Dipartimento di Chimica Organica, Università di Bologna, Via S. Donato 15, I-40127 Bologna, Italy

Dr. J. F. Stoddart
School of Chemistry, University of Birmingham, Edgbaston, Birmingham, B15 2TT, UK

Dr. A. L. Vance
Department of Chemistry, University of Kansas, Lawrence, KS 66045-0046, USA

Professor G. M. Whitesides
Department of Chemistry, Harvard University, Cambridge, MA 02138, USA

Abbreviations

The most commonly used abbreviations in "Comprehensive Supramolecular Chemistry" are listed below. Please note that in some instances these may differ from those used in other branches of chemistry.

Techniques and theories

AOM	angular overlap model
aq.	aqueous
at.%	atomic %
b.c.c.	body-centered-cubic
BM	Bohr magneton
b.p.	boiling point
c.c.p.	cubic-close-packed
c.d.	circular dichroism
CFSE	crystal field stabilization energy
CFT	crystal field theory
CIDNP	chemically induced dynamic nuclear polarization
CNDO	complete neglect of differential overlap
conc.	concentrated
c.p.	chemically pure
CP	cross-polarization
CPK	Corey–Pauling–Koltun
CT	charge transfer
cu.	cubic
dil.	dilute
DSC	differential scanning calorimetry
DTA	differential thermal analysis
EHMO	extended Hückel molecular orbital
ENDOR	external nuclear double resonance
equiv.	equivalent
ESR (or EPR)	electron spin (or paramagnetic) resonance
EXAFS	extended x-ray absorption fine structure
f.c.c.	face-centered-cubic
f.p.	freezing point
FT	Fourier transform
g.	gaseous
GC	gas chromatography
GLC	gas–liquid chromatography
GVB	generalized valence bond
h.c.p.	hexagonal-close-packed
HOMO	highest occupied molecular orbital
HPLC	high-performance liquid chromatography
HREELS	high-resolution electron energy loss spectroscopy
ICR	ion cyclotron resonance
INDO	incomplete neglect of differential overlap
IR	infrared
IUPAC	International Union of Pure and Applied Chemistry
l.	liquid
LAXS	large-angle x-ray scattering
LB	Langmuir–Blodgett
LCAO	linear combination of atomic orbitals
LFSE	ligand field stabilization energy
LFT	ligand field theory
LUMO	lowest unoccupied molecular orbital
MASNMR	magic angle spinning nuclear magnetic resonance
m.c.d.	magnetic circular dichroism
MD	molecular dynamics
MLCT	metal-to-ligand charge transfer
MM	molecular mechanics
MNDO	modified neglect of diatomic overlap
MO	molecular orbital
mol.%	molecular %
m.p.	melting point
MS	mass spectrometry
MW	molecular weight
NMR	nuclear magnetic resonance

NQR nuclear quadrupole resonance

ORD optical rotatory dispersion

PE photoelectron
PIO paired interacting orbitals
PRDDO partial retention of diatomic differential overlap
PSEPT polyhedral skeletal electron pair theory

RT room temperature

s. solid
SAXS small-angle x-ray scattering
SCE saturated calomel electrode
SCF self-consistent field
SET single-electron transfer
SHE standard hydrogen electrode

S_N1 substitution, nucleophilic, monomolecular
S_N2 substitution, nucleophilic, bimolecular

TGA thermogravimetric analysis
TLC thin-layer chromatography

UV ultraviolet

VB valence bond
vol.% volume %

WAXS wide-angle x-ray scattering
wt.% weight %

XANES x-ray absorption near-edge structure
XRD x-ray diffraction

Groups, reagents, and solvents

Ac acetyl
acac acetylacetonate
AIBN 2,2'-azobisisobutyronitrile
Ar aryl
arphos 1-(diphenylphosphino)-2-(diphenylarsino)ethane
ATP adenosine triphosphate
Azb azobenzene

9-BBN 9-borabicyclo[3.3.1]nonyl
9-BBN-H 9-borabicyclo[3.3.1]nonane
BHT 2,6-di-*t*-butyl-4-methylphenol (butylated hydroxytoluene)
bipy 2,2'-bipyridyl
Boc *t*-butoxycarbonyl
bsa *N,O*-bis(trimethylsilyl)-acetamide
bstfa *N,O*-bis(trimethylsilyl)-trifluoroacetamide
btaf benzyltrimethylammonium fluoride
Bn benzyl
Bz benzoyl

can ceric ammonium nitrate
cbd cyclobutadiene
cbz benzyloxycarbonyl
CD cyclodextrin
1,5,9-cdt cyclododeca-1,5,9-triene
1,3- or 1,4-chd 1,3- or 1,4-cyclohexadiene
chpt cycloheptatriene
[Co] cobalamin
(Co) cobaloxime ($Co(DMG)_2$) derivative
cod 1,5-cyclooctadiene
cot cyclooctatetraene

Cp η-cyclopentadienyl
Cp* pentamethylcyclopentadienyl
18-crown-6 1,4,7,10,13,16-hexaoxacyclo-octadecane
CSA camphorsulfonic acid
csi chlorosulfonyl isocyanate
Cy cyclohexyl

dabco 1,4-diazabicyclo[2.2.2]octane
dba dibenzylideneacetone
DBN 1,5-diazabicyclo[4.3.0]non-5-ene
DBU 1,8-diazabicyclo[5.4.0]undec-7-ene
dcc dicyclohexylcarbodiimide
dcpe 1,2-bis(dicyclohexyl-phosphino)ethane
DDQ 2,3-dichloro-5,6-dicyano-1,4-benzoquinone
deac diethylaluminum chloride
dead diethyl azodicarboxylate
depe 1,2-bis(diethylphosphino)-ethane
depm 1,2-bis(diethylphosphino)-methane
det diethyl tartrate (+ or –)
DHP dihydropyran
diars 1,2-bis(dimethylarsino)-benzene
DIBAL-H diisobutylaluminum hydride
dien diethylenetriamine
diglyme bis(2-methoxyethyl)ether
dimsyl Na sodium methylsulfinylmethide
DIOP 2,3-*O*-isopropylidene-2,3-dihydroxy-1,4-bis(diphenyl-phosphino)butane
dipt diisopropyl tartrate (+ or –)

dma	dimethylacetamide
dmac	dimethylaluminum chloride
DMAD	dimethyl acetylene-dicarboxylate
DMAP	4-dimethylaminopyridine
DME	dimethoxyethane
DMF	*N,N*-dimethylformamide
DMG	dimethylglyoximate
DMI	*N,N'*-dimethylimidazalone
dmpe	1,2-bis(dimethylphosphino)-ethane
dmpm	bis(dimethylphosphino)-methane
DMSO	dimethyl sulfoxide
dmtsf	dimethyl(methylthio)sulfonium fluoroborate
dpam	bis(diphenylarsino)methane
dppb	1,4-bis(diphenylphosphino)-butane
dppe	2-bis(diphenylphosphino)-ethane
dppf	1,1'-bis(diphenylphosphino)-ferrocene
dpph	1,6-bis(diphenylphosphino)-hexane
dppm	bis(diphenylphosphino)-methane
dppp	1,3-bis(diphenylphosphino)-propane
E^+	electrophile
eadc	ethylaluminum dichloride
EDG	electron-donating group
edta	ethylenediaminetetraacetate
eedq	*N*-ethoxycarbonyl-2-ethoxy-1,2-dihydroquinoline
en	ethylene-1,2-diamine (1,2-diamino-ethane)
Et_2O	diethyl ether
EWG	electron-withdrawing group
Fc	ferrocenyl
Fp	$Fe(CO)_2Cp$
HFA	hexafluoroacetone
hfacac	hexafluoroacetylacetonate
hfb	hexafluorobut-2-yne
HMPA	hexamethylphosphoramide
hobt	hydroxybenzotriazole
$IpcBH_2$	isopinocampheylborane
Ipc_2BH	diisopinocampheylborane
kapa	potassium 3-aminopropyl-amide
K-selectride	potassium tri-*s*-butylboro-hydride
L	ligand
LAH	lithium aluminum hydride
LDA	lithium diisopropylamide
LICA	lithium isopropylcyclohexyl-amide
LiTMP	lithium tetramethylpiperidide
L-selectride	lithium tri-*s*-butylborohydride
LTA	lead tetraacetate
M	metal
MCPBA	*m*-chloroperbenzoic acid
MEM	methoxyethoxymethyl
MEM-Cl	β-methoxyethoxymethyl chloride
Mes	mesityl
mma	methyl methacrylate
mmc	methylmagnesium carbonate
MOM	methoxymethyl
Ms	methanesulfonyl
MSA	methanesulfonic acid
MsCl	methanesulfonyl chloride
mvk	methyl vinyl ketone
nap	1-naphthyl
nbd	norbornadiene
NBS	*N*-bromosuccinimide
NCS	*N*-chlorosuccinimide
NMO	*N*-methylmorpholine *N*-oxide
NMP	*N*-methyl-2-pyrrolidone
Nu^-	nucleophile
ox	oxalate
pcc	pyridinium chlorochromate
pdc	pyridinium dichromate
phen	1,10-phenanthroline
phth	phthaloyl
ppa	polyphosphoric acid
ppe	polyphosphate ester
$[PPN]^+$	$[(Ph_3P)_2N]^+$
ppts	pyridinium *p*-toluenesulfonate$^-$
py	pyridine
pz	pyrazolyl
Red-Al	sodium bis(methoxyethoxy)-aluminum dihydride
sal	salicylaldehyde
salen	*N,N'*-bis(salicylaldehydo)-ethylenediamine (*N,N'*-bis(salicylidene)-1,2-diaminoethane)
SEM	β-trimethylsilylethoxymethyl
Sia_2BH	disiamylborane
tas	tris(diethylamino)sulfonium
tasf	tris(diethylamino)sulfonium difluorotrimethylsilicate
tbaf	tetra-*n*-butylammonium fluoride
TBDMS	*t*-butyldimethylsilyl
TBDPS	*t*-butyldiphenylsilyl
tbhp	*t*-butyl hydroperoxide

TCE	2,2,2-trichloroethanol
TCNE	tetracyanoethene
TCNQ	7,7,8,8-tetracyanoquino-dimethane
terpy	2,2':6',2"-terpyridyl
TES	triethylsilyl
Tf	triflyl (trifluoromethane-sulfonyl)
TFA	trifluoracetic acid
TFAA	trifluoroacetic anhydride
tfacac	trifluoroacetylacetonate
THF	tetrahydrofuran
THP	tetrahydropyranyl
tipbs-Cl	2,4,6-triisopropylbenzene-sulfonyl chloride
tips-Cl	1,3-dichloro-1,1,3,3-tetraisopropyldisiloxane
TMEDA	tetramethylethylenediamine (1,2-bis(dimethylamino)-ethane)
TMS	trimethylsilyl
Tol	tolyl
tpp	*meso*-tetraphenylporphyrin
Tr	trityl (triphenylmethyl)
tren	2,2',2"-triaminotriethylamine
trien	triethylenetetraamine
triphos	1,1,1-tris(diphenylphosphino-methyl)ethane
Ts	tosyl
TsMIC	tosylmethyl isocyanide
ttfa	thallium trifluoroacetate
ttn	thallium(III) nitrate
X	halogen

Contents of All Volumes

Volume 3 Cyclodextrins

Volume 4 Supramolecular Reactivity and Transport: Bioorganic Systems

Volume 5 Supramolecular Reactivity and Transport: Bioinorganic Systems

Volume 6 Solid-state Supramolecular Chemistry: Crystal Engineering

Volume 7 Solid-state Supramolecular Chemistry: Two- and Three-dimensional Inorganic Networks

Volume 8 Physical Methods in Supramolecular Chemistry

Volume 9 Templating, Self-assembly, and Self-organization

Volume 10 Supramolecular Technology

1

Molecular Template Effect: Historical View, Principles, and Perspectives

DARYLE H. BUSCH, ANDREW L. VANCE, and ALEXANDER G. KOLCHINSKI

University of Kansas, Lawrence, KS, USA

1.1 INTRODUCTION

1.1.1 Definition of Molecular Templates

For the rapidly moving realm of molecular design and chemical synthesis, a general definition of a chemical template has been suggested: "a chemical template organizes an assembly of atoms, with respect to one or more geometric loci, in order to achieve a particular linking of atoms."[1] This definition captures the essential elements of the concept: (i) the role of the templating agent is an active one in that it organizes an assembly of atoms; and (ii) the synthetic goal is directed toward compelling spatial, geometric, and/or topological relationships. The concept of template control of synthetic reactions was developed in conjunction with the application of metal ions in the design and synthesis of macrocyclic ligands, especially those involving nitrogen and sulfur donor atoms. It is important to distinguish between a templating phenomenon and the many other ways in which a metal ion can influence the progress and products of a chemical reaction. Consistent use of the term comes only with difficulty and this is especially true in those cases where a new ligand is synthesized, found bound to the metal ion, and is unstable in the absence of the metal ion. To accommodate this scenario, two kinds of metal template processes were recognized early on: the kinetic template and the thermodynamic template.[2] They may be defined as follows: in a kinetic template reaction, the binding moiety holds the reactive groups in a proper array to facilitate a stereochemically and/or topologically selective multistep reaction. In contrast, in a thermodynamic template reaction, the ligand-forming processes are reversible and may involve equilibration with more than one product; the complex formation is thermodynamically favored between the templating agent and one of those products, leading to a displacement of the equilibrium in favor of that product. Clearly, the definition given at the beginning of this paragraph is focused on the kinetic template effect, for it is most obvious that geometric and topological relationships are being served when the templating agent intervenes during the course of the reaction. In a lesser but nonetheless real way, the thermodynamic template effect is responsive to the same primary elements, but essentially because metal ions tend, at equilibrium, to select from mixtures of ligands the specific ones that bind most strongly and because, given time to equilibrate, ligands with relatively more constraining topologies bind most strongly. These ligands may be more difficult to isolate in pure form in the absence of the metal ion because their more intricate topologies/geometries may be entropically disfavored. Further aggravating the problem is the difficulty of adequately demonstrating the metal ion influence in the case of reactions suspected of involving thermodynamic templates. Operationally, the kinetic template effect has been used with great success in the carefully planned design of distinctive molecular motifs (cycles, bicycles, knots, interlocking rings, and more), whereas thermodynamic templates have commonly been invoked in systems where similar product motifs have been observed even though the planned course of the reaction might not have been subject to control. Two tests must be met in order to recognize the contribution of a template effect to a chemical reaction in the context of this chapter: (i) the templating agent must organize other molecular components, and (ii) the result of an ensuing or accompanying chemical reaction must include significant spatial, topological, and/or geometric control.

1.1.2 Scope of this Chapter

It is immediately necessary to recognize the fact that the word "template" has been used in many ways in the chemical and chemically related literature and that the use of the word "template" has grown phenomenally since the 1960s, as shown by the bar graph in Figure 1. The word "template" has been incorporated into the vocabularies of biochemistry, materials science, and other areas. This makes it necessary to define the scope of the present chapter in order to proceed with an account of the template effect in molecular design and synthetic chemistry. Some useful reviews and texts are cited.[3–5]

As indicated in the discussion of definitions of chemical templates, the primary goal of template synthesis is control of the overall topology of the product. Thus, syntheses in which the template serves to enhance the control of stereochemistry in more traditional ways are not discussed here, for example, chiral centers in natural product synthesis and, similarly, selective organometallic reactions and organometallic catalysis.[6] Equally fascinating are the syntheses exemplified by Martin and co-workers in which very small but macroscopic fibrils and tubules are formed within molds provided by the pores of a porous membrane.[7] These are certainly outside the scope of this

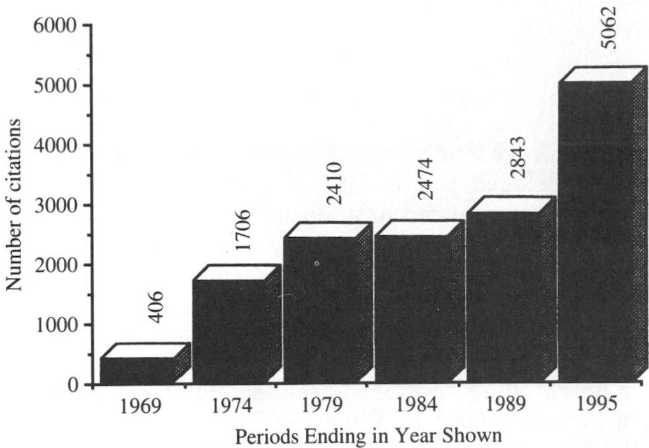

Figure 1 The appearance of the word "template" in *Chemical Abstracts.*

discussion because those templates serve to control the shape of the bulk materials produced within the membrane pores, not to facilitate a particular linking of atoms. This seems to be the application of the lost wax process of sculpture to materials science. A similar, but perhaps inverted, case is seen in the preparation of molecular sieves, as illustrated by the work of Beck *et al.*[8] They have achieved astounding control over the sizes of the pores in a family of silicate/aluminosilicate molecular sieves through the use of liquid crystals and have proposed a liquid crystal templating mechanism in which surfactant molecules that exhibit liquid crystalline behavior are surrounded by silicate material, resulting in highly ordered molecular sieves. This elegant sieve synthesis and the exciting preparations of fibrils and tubules both seem to utilize molds; thus, the former appears to use mechanical, not molecular, templates, while the latter shrinks this kind of template to the molecular scale. In the case of the fibrils and tubules, an external mechanical template operates as a mold, while in the case of the molecular sieves an internal molecular template acts to fill voids around which a porous material is produced. Certain polymerization processes have also been described as occurring on a template.[9] In these syntheses, the basic principles determining the construction of biological molecular chains are followed, and the focus is primarily on the composition of the polymer. In the biological sciences, the term "template" has been applied to the biosynthesis of DNA and other related substances. A review with information of interest to the synthetic chemist has been published by Lindsey.[10] Organometallic chemistry represents a vast field of study which is not discussed in this chapter, even though it does contain examples of template reactions, and readers are referred to the cited publications for further information.[11] While the present authors recognize these topics to be equally important and interesting, the present text is confined to template phenomena occurring on the molecular level for the design of molecules. It is for this reason that the phrase "molecular template" is contained in the title. A number of specific topics illustrating molecular template effects that are treated in more detail elsewhere in this volume are discussed only to elucidate the principles involved.

This presentation treats the fundamental science of molecular template reactions applied to the design of molecules. The main purpose is to provide a concise source of the principles associated with molecular template processes in the context of the development of the present state of understanding. The kinds of molecular entities that serve as templates and the molecular architectures their study has generated are of major interest. Since macrocycles, macrobicycles, catenanes, and knots figured heavily in the history of the subject, pertinent aspects of those subjects are reviewed. Because the potential of the subject of template processes has not been fully realized, attention is given to the systematics that may lead to further contributions from the field.

1.1.3 Historical Notes

The significance of molecular template processes is rapidly becoming understood, but the full potential of the phenomenon will not be realized until wide ranges of as yet unknown materials are produced through extensive expansion of the array of molecular motifs they are and will be generating. In the light of these expectations, it is a lesson in the development of knowledge to reflect on the beginnings, and near beginnings, of the subject. The "lock and key" of Fisher is sometimes recalled and labeled a template.[12] Insofar as this analogy relates to the substrate–enzyme complex, it reflects the concept of complementarity, as used in the complete coordination chemistry,[13] rather than a template relationship. The key is the obvious complement of the lock. A similar semantic misuse of the word "template" involves associating it with a foundational structure which might more reasonably be described as a "platform." In general, if the associated derivatization, addition, substitution, and so on is reasonably viewed as an addition to the appendages of the platform, then one is really adding superstructure,[14] not conducting a template reaction.

The unfolding of new scientific relationships depends on both mental and experimental advances. Success is achieved when experimental confirmation is found for a mental construct, and success also occurs when the significance of an unexpected experimental result is recognized. The initial period of discovery leading to the molecular template concept illustrates both processes. Furthermore, the beginnings of macrocyclic chemistry are closely tied to the appearance of the concept of molecular templates. In quest of the first intentionally synthesized new families of macrocyclic ligands, the Busch group developed both the molecular template concept and the first well-defined examples of both the kinetic[15] (Structure (**1**)) and thermodynamic[2,16] (Structures (**2**)–(**6**)) template effects. Publications of the kinetic template example and certain of the thermodynamic template examples were delayed by magnetic anomalies that were soon understood.[17] Some years and many experiments preceded the experimental confirmation of the mental construct.[18] In this earliest example of the kinetic template effect (Structures (**1**) and (**2**)), the metal ion (i) brings the two reactive centers (mercaptide donors) into adjacent *cis* positions to facilitate ring closure, and (ii) prevents reaction by the two additional reactive atoms (nitrogens) that occupy positions in between the terminal sulfur atoms. The alkylation of the nickel(II) thiolate complexes was developed specifically for the purposes of producing metal templates. Sound support for the kinetic template hypothesis was found in kinetic studies.[15c] When followed spectrophotometrically, the reaction between (**2**) and α,α′-dibromo-*o*-xylene gave no evidence for accumulation of detectable amounts of an intermediate, but proceeded directly from the complex of the linear tetradentate ligand to that of the macrocycle. In contrast, the reaction of (**2**) with a very similar monofunctional alkyl halide, benzyl bromide, proceeded smoothly in two steps with the rate of formation of the monoalkylated intermediate exceeding that of the final dialkylated product. The implicit acceleration is analogous to the kinetic chelate effect and characterizes the advantage of a kinetic template reaction.

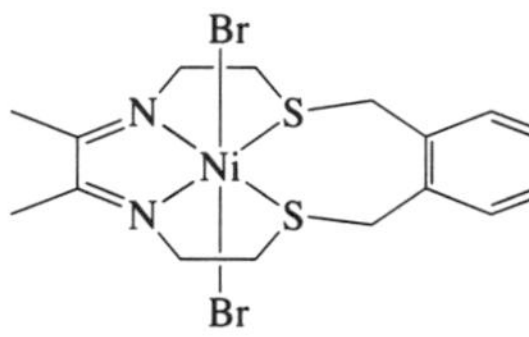

(**1**) 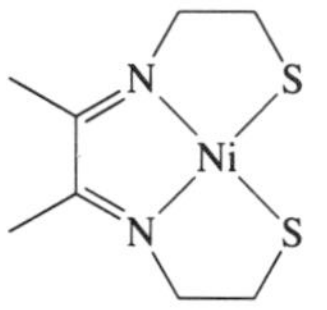 (**2**) 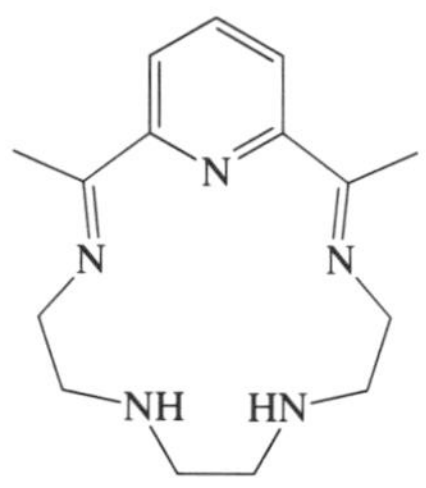 (**3**)

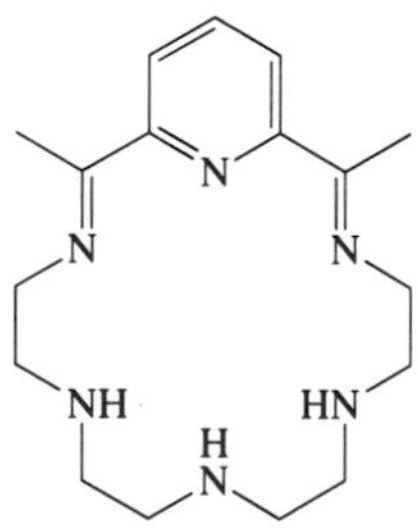

(**4**) 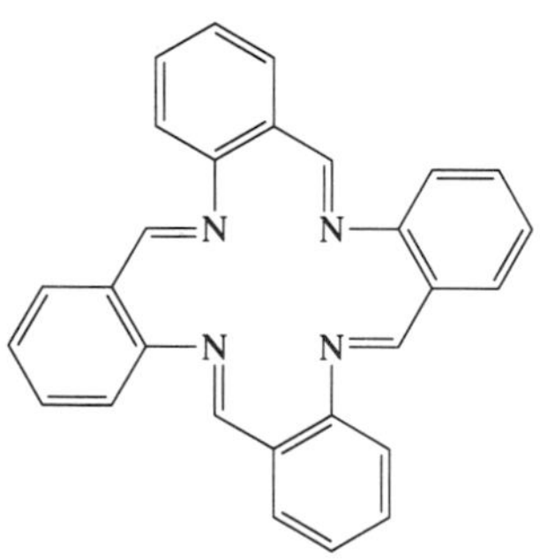 (**5**) 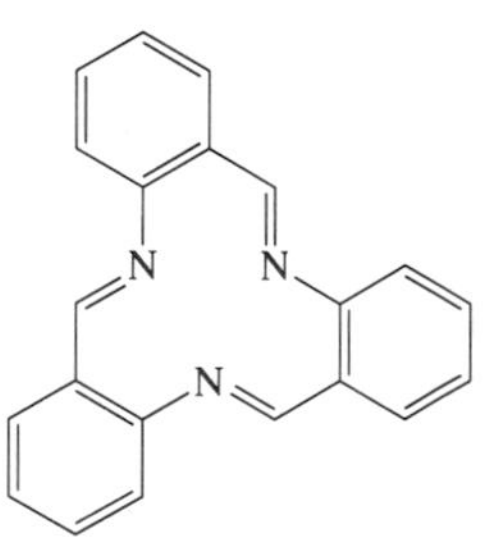 (**6**)

Illustrative of the second kind of scientific success, in the same time frame, the Curtis group serendipitously discovered and recognized a very important family of macrocyclic ligands (Structures (**7**) and (**8**)), the first nonaromatic tetraazamacrocycles.[19] It was soon concluded that the Curtis ligands may be produced by template reactions, as well as by other means.[18,20] The discovered reaction was between tris(ethylenediamine)nickel(II) salts and acetone, but various metal ions, amines, and carbonyl compounds give related products. The template effects are probably limited to the thermodynamic type, for the following reasons. (i) The reactivity depends on the lability of the metal ion, indicating that condensation of the amino groups requires their dissociation from the metal ion. Nickel(II) and copper(II) are the best, platinum(II) and palladium(II) react very slowly, and cobalt(III) does not react at all. (ii) Unsymmetrical ketones react in accord with the Claisen–Schmidt condensation, producing less-branched products under basic conditions, again indicating reaction in the absence of the metal ion. (iii) The usual procedure for preparation of the long known Curtis ligands does not use a metal ion. Template transamination processes were found for the *cis*-diene Curtis complexes of copper and nickel (Equation (1)).[21] These chelate ring replacement reactions are driven by the greater thermodynamic stability of the smaller chelate rings.

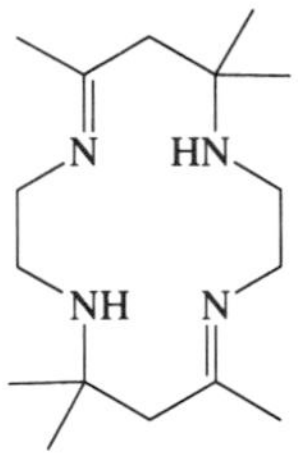

(7) 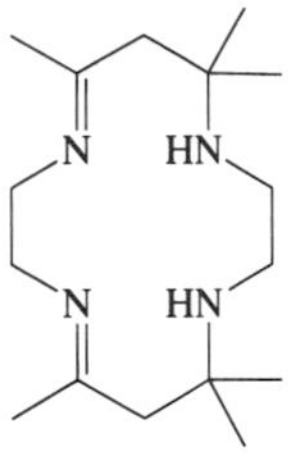 (8)

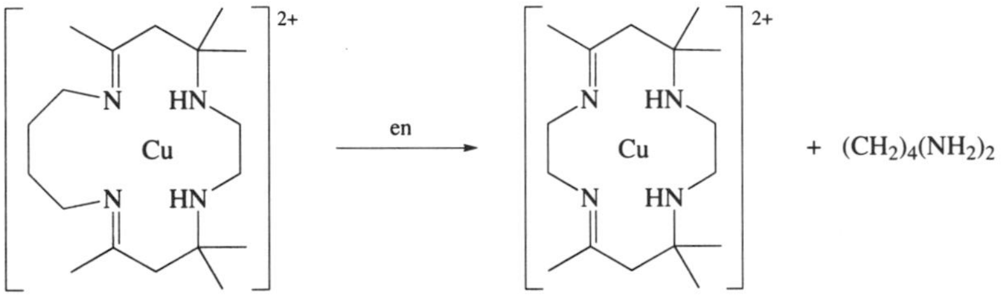

$$\xrightarrow{\text{en}} + (CH_2)_4(NH_2)_2 \qquad (1)$$

It is interesting to reflect on a number of examples where the confluence of experimental and mental processes might have uncovered template processes, but failed. Such examples are found in pioneering attempts to use metal compounds as catalysts or reagents for a variety of condensation reactions. In 1898, Posner[22a] attempted to utilize anhydrous zinc(II) chloride as a catalyst for the Henry condensation between *o*-aminobenzaldehyde and nitromethane. He obtained a crystalline product, although he also found that nitromethane did not participate in the reaction. It was suggested that the reaction involved the dehydration of *o*-aminobenzaldehyde. The empirical formula for this condensation product, $C_7H_5N{\cdot}\frac{1}{2}ZnCl_2$, was determined in 1926;[22b] however, its structure was not known until about 40 years later, following the work of Melson and Busch on the macrocyclic condensation products of *o*-aminobenzaldehyde.[16c,16d,17a] The coordination formula of the macrocyclic complex is [Zn(TAAB)][$ZnCl_4$] and the ligand has Structure (**5**).

The first synthesis of a metal phthalocyanine (Structure (**9**)) in 1927 was the adventitious result of an experiment directed at other goals. An attempt to prepare *o*-phthalonitrile from *o*-dibromobenzene and cuprous cyanide gave copper phthalocyanine (**9**) in 23% yield.[23] As the product of this condensation and other similar materials were soon recognized to be excellent pigments, they were investigated quite extensively, and their structures were established in 1933.

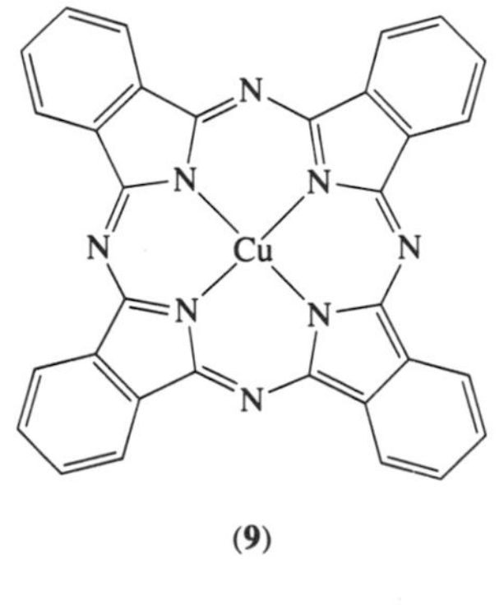

(9)

In 1928, Hein and Retter[24a] discovered an interesting example of pyridine oxidation by anhydrous ferric chloride which resulted in the formation of a 2,2′-bipyridine complex of iron(II). Four years later, this reaction was reinvestigated and, in addition to 2,2′-bipyridine, a substantial amount of 2,2′,2″-terpyridine was also detected.[24b] Eventually, about 20 products were identified. Significantly, 4,4′-bipyridine was not found in the reaction mixture. The absence of 4,4′-bipyridine in a metal-promoted reaction that yields 2,2′-bipyridine and 2,2′,2″- terpyridine suggests the binding of pyridine molecules to the iron ion during the course of the coupling reaction.

The development of a variety of analytical procedures also yielded several early examples that appear to involve template processes. One striking example of an unexpected and useful reaction which later proved to involve the template effect was found by Nilsson[25] during a study of the reaction of bis(cyclohexanoneoxalyldihydrazone) with copper(II) salts. The reagent was found to provide an even more sensitive test for copper in the presence of excess aliphatic aldehyde and oxygen. In addition to its extreme sensitivity, another advantage lay in its specificity for copper, and this reaction became a method for the determination of trace amounts of copper. The structure of the final purple product was eventually established by x-ray and spectroscopic measurements (Structure (**10**)).[26] The complex is remarkable both for the novel structure of its macrocyclic ligand and for the fact that it is water soluble and contains stable copper(III).

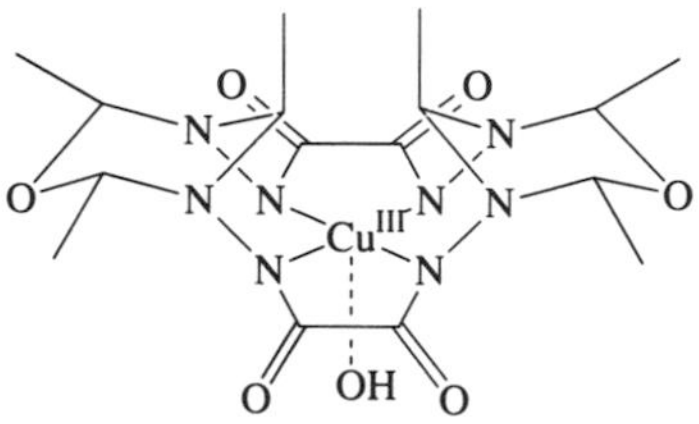

(**10**)

Cage complexes were probably first prepared by Feigl[27] during his investigation of an indirect method for the detection of tin(II). In this test, a stannous salt reduces ferric ion in the presence of dimethylglyoxime. The development of a red color during the test was attributed to the formation of an iron(II) dimethylglyoximate complex. It is now known, however, that the oxidation product of the reaction, tin(IV), reacts with iron(II) and dimethylglyoxime causing the formation of an iron(II) clathrochelate complex which also has a dark red color (Equation (2)). Complexes of this general family of clathrochelates were first synthesized and characterized by Boston and Rose[28] using boron derivatives as the bridging reagents rather than tin reagents (Structure (**11**)).

$$Fe^{2+} + 2SnCl_4 + 3H_2DMG \longrightarrow [\ldots]^{2-} \quad (2)$$

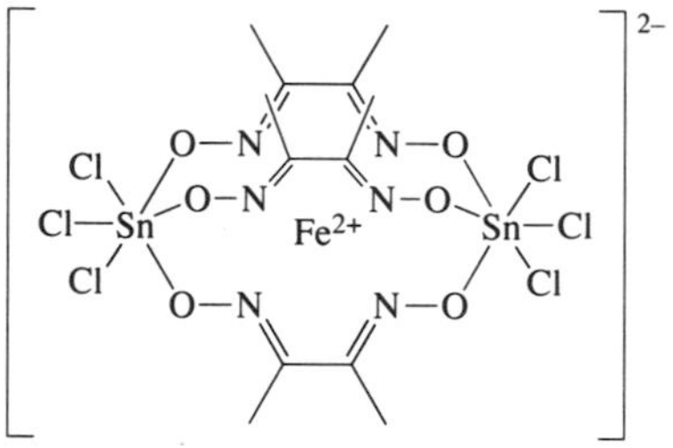

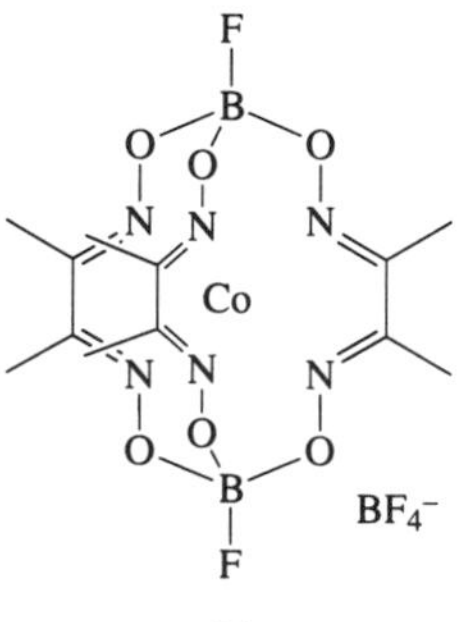

(**11**)

The investigation of the solubility of $Ni(DMG)_2$ in different media showed an anomalously high solubility in the presence of boric acid[29] and a reaction between the oxygen atoms of the dimethylglyoxime and boric acid was suggested as an explanation. Within the decade, two laboratories proved that a variety of reagents, including some based on boron, form macrocycles by replacing the bridging proton in the planar structure[30] (Structure (**12**)).

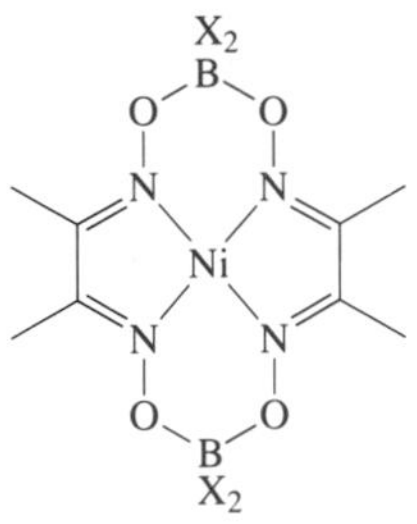

(**12**)

1.1.4 Synopsis

From these beginnings and near beginnings, molecular template processes have expanded enormously. The target motifs continue to unfold, and these are treated first in Section 1.2 where the variety of interactions generating template effects is examined. Then in Section 1.3, a summary is offered of the various topological and geometric motifs that have been synthesized successfully. Finally, the challenging potential of template control of chemical reactions is explored in Section 1.4.

1.2 KINDS OF TEMPLATING AGENTS

1.2.1 Metal Ions

Metal ions have been employed extensively as molecular-scale templating agents, and they have been used to create a variety of molecular architectures. The simplest of these is the macrocycle. In 1964, in an article describing the template hypothesis, Thompson and Busch first demonstrated the ability of a metal ion to organize reactive sites to preferentially yield a macrocyclic product (see above).[2,15–18,31] A plethora of template syntheses of macrocycles has followed, and a number of these are summarized in Section 1.3 on motifs, in order to display useful generalizations about template syntheses.

The broadest range of motifs has been achieved with metal ions as templates. Cage compounds, bicyclic ligands which encapsulate metal ions, have been synthesized by utilizing the metal ion template effect. The first such example was reported by Boston and Rose in 1968;[28] the topic was reviewed in the late 1980s.[32] The effectiveness of control by the molecular template is shown in Raymond and co-workers' synthesis of the ferric complex of bicapped TRENCAM (the macrobicyclic catechoylamide ligand derived from the reaction of three functionalized catechols with two tris(2-aminoethyl) amines is shown in Scheme 1).[32a] The cage complex was obtained in an overall yield of 50%. The catenanes and knots of Sauvage and co-workers elegantly demonstrate the ability of metal ions to control molecular architecture. Orientation of the specifically designed ligands around the tetrahedral copper(I) site enabled the Strasbourg team to produce a catenane in previously unattainable yields from complex (**13**).[33] This highly innovative work has led to the development of synthetic methodologies to tie the first molecular knot, as well as rotaxanes and other catenanes (see Section 1.3).

In the so-called "covalent template reactions," metalloid elements, such as tin, silicon, antimony, or sometimes boron, form weak covalent bonds with oxygen, nitrogen, or sulfur during the course of the reaction. The bound heteroatom remains fairly nucleophilic and reactions with electrophiles such as acyl chlorides proceed readily.[34] In the first example of such a template process (Scheme 2), the reacting functional group was bound to silicon. Initial condensation of silicon tetraisocyanate with glycol gives the spiro compound (**14**), which is subsequently caused to react with $C(=O)Im_2$ to form a macrocyclic complex. The metalloid–heteroatom bond is easily hydrolyzed and the template can easily be removed.

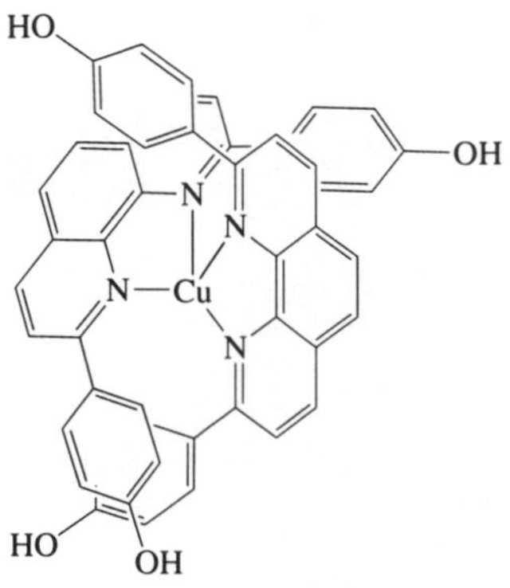

(13)

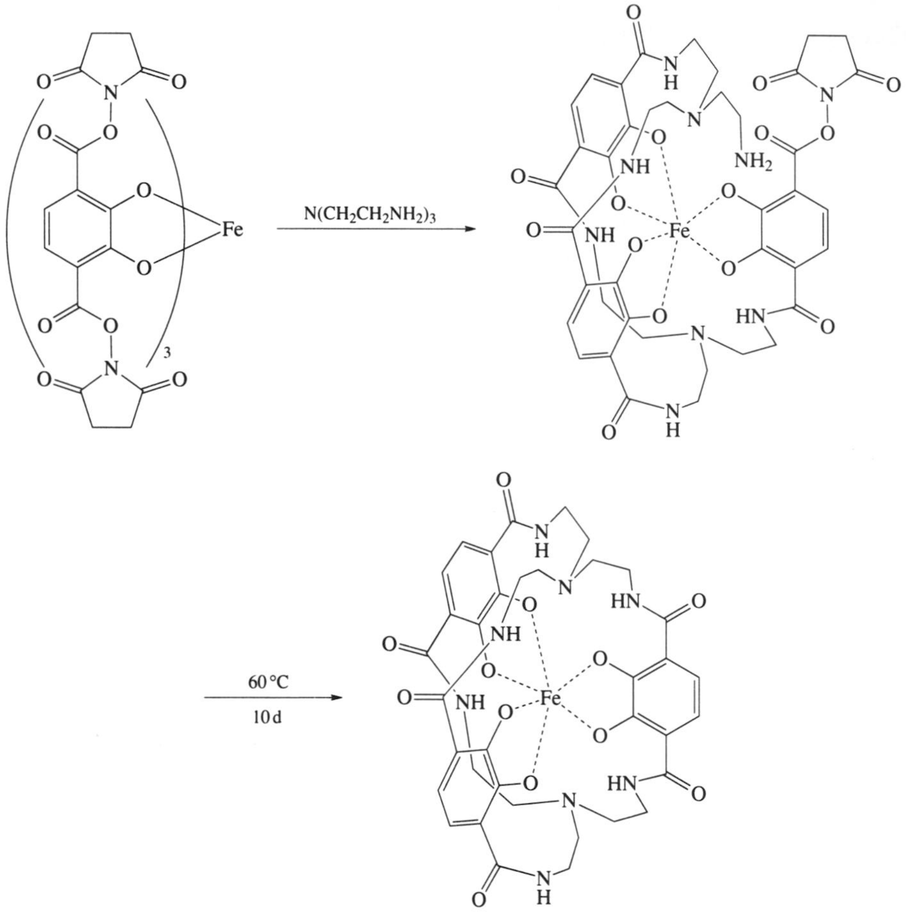

$N(CH_2CH_2NH_2)_3$

60 °C
10 d

Scheme 1

1.2.2 Electrostatic

In recent years, templates other than metal ions have been developed and put to good use in new syntheses. By taking advantage of electrostatic interactions such as π stacking between electron-rich and electron-poor aromatic systems, an organic template effect can be observed, and the template components may remain as integral parts of the resultant product.[4] The most striking example of such a synthetic approach is seen in the catenane syntheses reported by Stoddart and co-workers.[35] The template in these reactions may be viewed as the electrostatic interaction between the electropositive bipyridinium moieties and the electronegative hydroquinone units of the crown ether (Structure (**15**)).

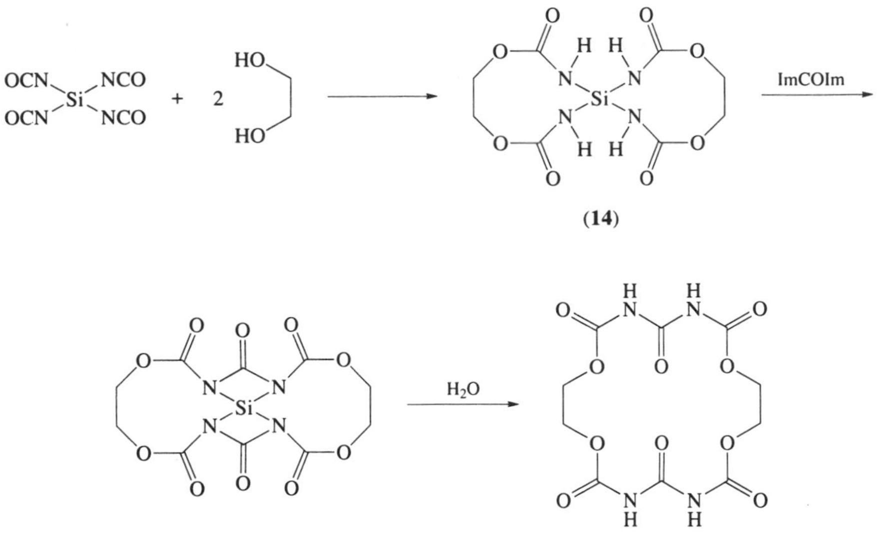

(14)

Scheme 2

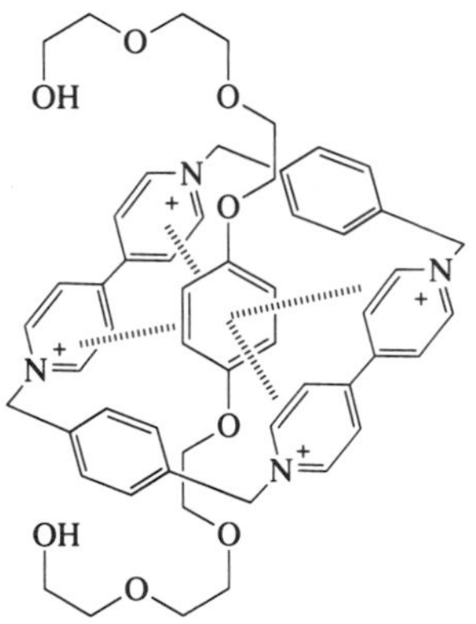

(15)

1.2.3 Hydrogen Bonding

The design of templates includes striking examples of the exploitation of hydrogen-bonding effects. The combined bond strengths of multiple hydrogen bonds lend stability to the aggregates produced in the laboratory of Whitesides and his colleagues.[36] In the fascinating self-replicating molecules reported by Rebek and co-workers, hydrogen bonds play a crucial role in the ability of the templating molecule to organize reactants for the production of its complementary molecule (see Scheme 3).[37,38] While the observed phenomenon has been confirmed, the mechanism has been questioned.[39]

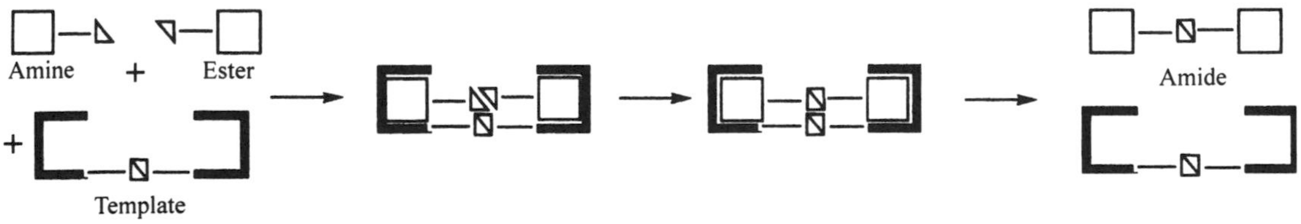

Amine + Ester

+ Template

Amide

Scheme 3

1.2.4 Other Interactions

The ability of a nonmetal ion, acting as a molecular "guest," to induce a certain arrangement of its host has further broadened the concept of a molecular template. Various anionic or neutral guests containing hydrophobic moieties were found by Fujita and co-workers to be capable of organizing the formation of the palladium(II) cage complex shown here as Structure (**16**).[40] In the absence of such a template, oligomers were formed. In addition, the cage complex, once formed, does not dissociate into the starting materials when the guest is removed. This type of complex formation may be stretching the definition of a template effect, and could be likened to the zeolite formations mentioned earlier wherein the anion guest actually serves as a mold around which the cage is formed.

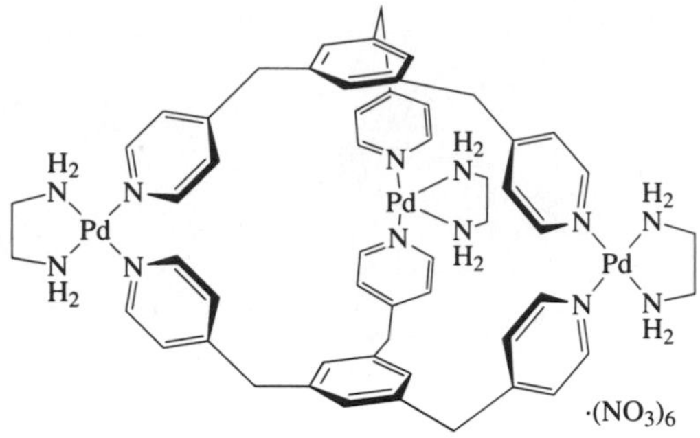

·$(NO_3)_6$

(**16**)

The coordination of pyridyl nitrogen atoms to porphyrin metal ions serves as the basis for the template syntheses of the cyclic porphyrin oligomers produced by Sanders and co-workers.[3] For example, a 4,4′-bipyridine unit has been utilized as a template in the formation of cyclic porphyrin dimers (Scheme 4). By changing the shape of the template molecule, and the length of the porphyrin oligomers, larger cyclic products were isolated. Müller has reported the synthesis of polyoxometallates in which metal oxide clusters were formed around anionic templates.[41] The size, shape, and charge of the anions (for example N_3^-, SCN^-, and ClO_4^-) were found to profoundly influence the shapes of the clusters, as shown by the cross-sections given in Figure 2. The phenomenon of encapsulating an anionic guest within an anionic host has been described as "The Taming of the Shrew," and it is believed that the repulsive forces between anions are counterbalanced by weak O=M···X interactions (X = template) which direct the cluster formation.

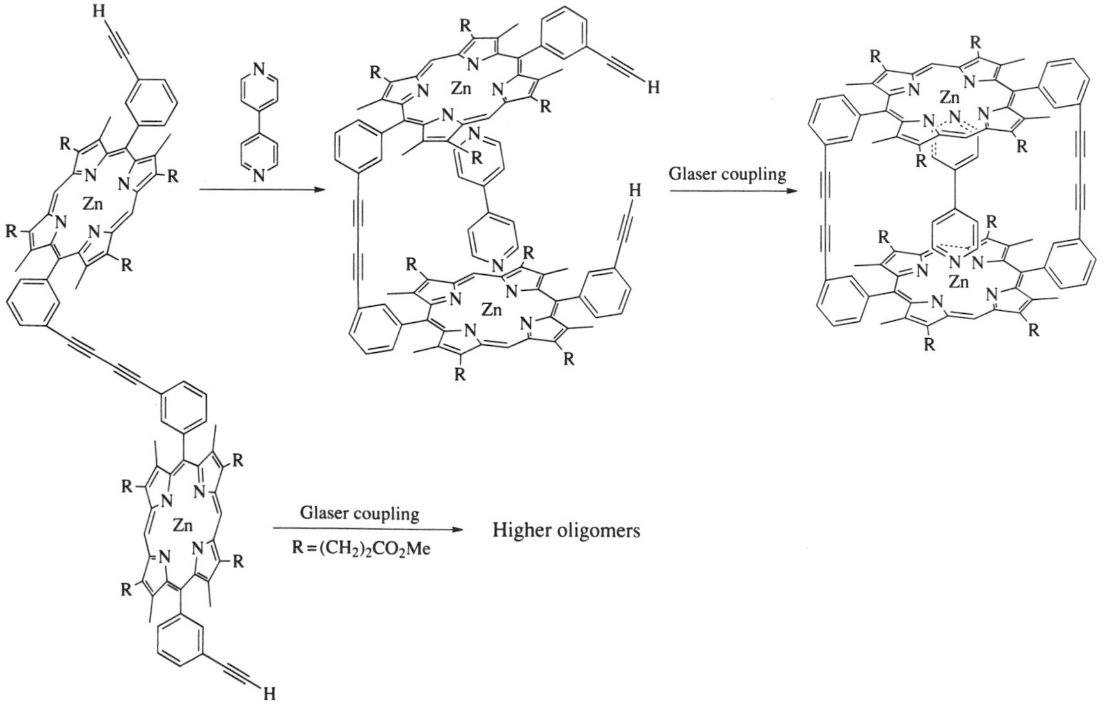

Glaser coupling

Glaser coupling

R = $(CH_2)_2CO_2Me$

Higher oligomers

Scheme 4

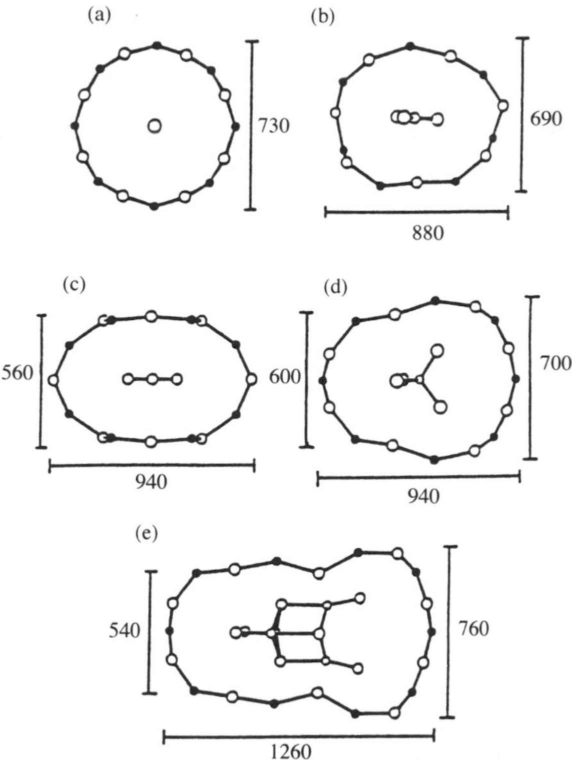

Figure 2 (a)–(e) Influence of the size, shape, and charge of anions on shapes of metal oxide clusters (lengths are in picometers) (reprinted from *J. Mol. Struct.*, **325**, A. Müller, Supramolecular inorganic species—an expedition into a fascinating, rather unknown land mesoscopia with interdisciplinary expectations and discoveries, 13–35, Copyright 1994, with kind permission of Elsevier Science-NL, Sara Burgerhartstraat 25, 1055 KV Amsterdam, The Netherlands).

In summary, essentially all those modes of binding that lead to the formation of a coordination entity can be used to generate a template effect in the synthesis of molecules. The strengths of such interactions range from that of the covalent bond to the very weak affinities attributed to hydrophobic and van der Waals interactions. The dynamics of the associated processes will vary in response to the lability of the binding process. With inert metal ions, the kinetic-template reaction product is expected to remain bound to the metal ion, while no such complexation is required for labile templating agents. This relationship is further complicated by the possibility of binding in labile systems simply because of enhanced thermodynamic stability, that is, thermodynamic templating. Furthermore, one recognizes two fundamental aspects of all template processes: (i) binding, and (ii) reaction of functional groups in the bound entity. Thus any template effect involves an anchor and a reaction. As will be seen in Section 1.3, a third most helpful feature is a molecular turn.

1.3 MOTIFS

Molecular motifs accessed through template-directed syntheses include an array of molecular architectures, some of which have not been prepared in the laboratory by other routes: acyclic molecules, simple macrocycles, macrobicycles, polycycles, catenanes, polycatenanes, doubly threaded (figure-of-eight) catenanes, a trefoil knot, and rotaxanes.

1.3.1 Selective Synthesis of Acyclic Ligands

In many cases the main evidence for the intervention of a template reaction is the formation of a product that would not have been formed in the absence of the templating reagent. This is particularly true in the case of acyclic template products, as many of these are relatively easy to prepare by means of the usual nontemplate reactions. The least complicated template processes leading to targeted noncyclic molecules is the rearrangement of single molecules. For example, condensation of 2-formylpyridine and tren in the absence of metal ions gives the urotropine-like structure (**17**), which rearranges into the tripodal ligand (**18**) in the presence of iron(II) salts.[42] In the most common category, di- or polyfunctional molecules react with other monofunctional molecules under the influence of the templating agent. For example, Carugo and Bisi-Castellani[43] have applied this approach to the preparation of complexes with the tripodal ligand (**18**) by condensation of 2-acetylpyridine and tren in the presence of lanthanoid(III) cations (Equation (3)).

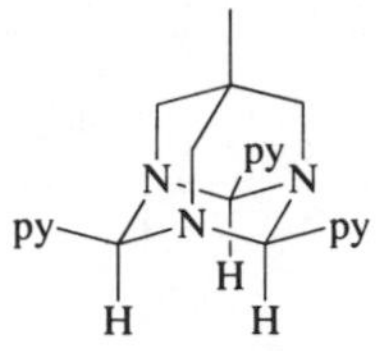

(**17**)

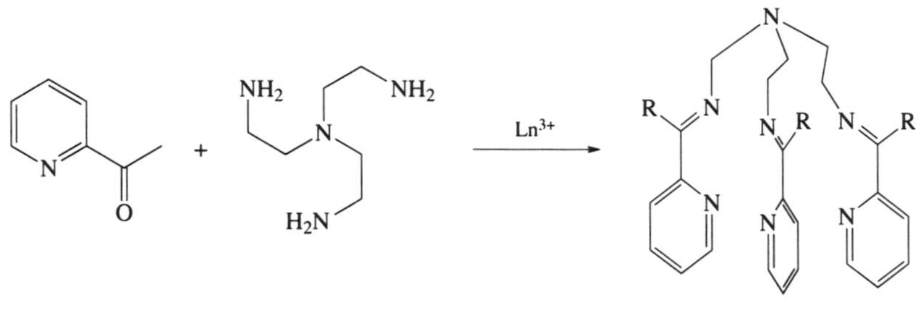

$$\xrightarrow{Ln^{3+}} \qquad (3)$$

(**18**)

More complicated systems may involve several kinds of di- or polyfunctional molecules and/or reactions, and the structures of the condensation products are neither so obvious nor predictable. Interesting template products have quite often been obtained by accident, and their structures represent discoveries rather than successful molecular design. For example, a complicated set of reactions between (*S*)-methylmethionine and ethylenediamine in the presence of cobalt salts and dioxygen leads to the elimination of dimethylsulfide and the formation of an interesting linear tetradentate ligand, in the form of its cobalt(III) complex (Scheme 5). The net result of the reaction is a transformation of the three-coordinated bidentate ligands into a substituted 2,3,2-tetramine.[44]

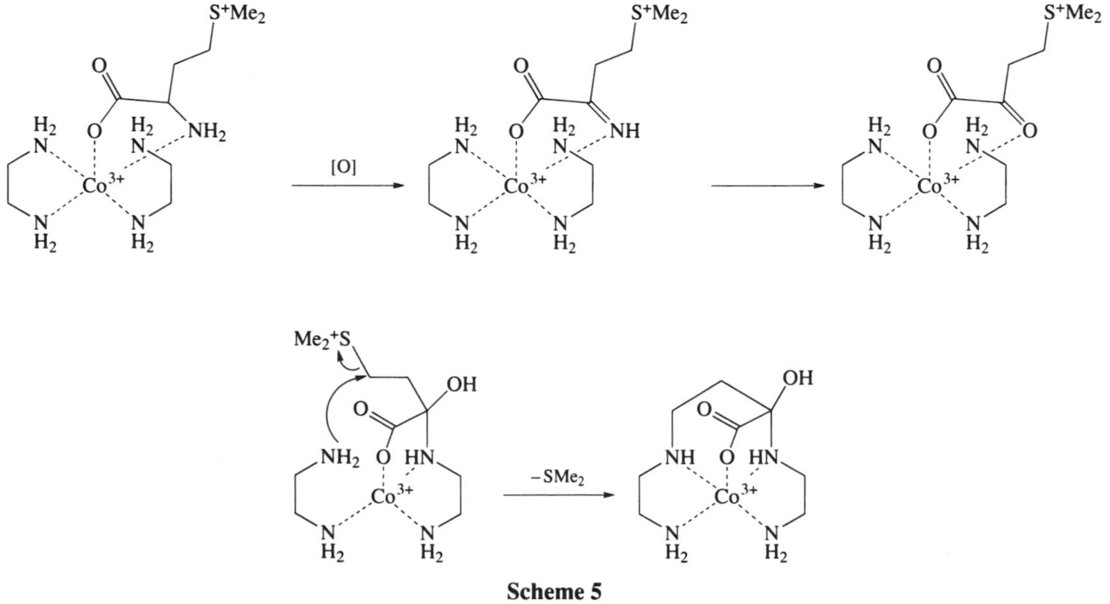

Scheme 5

Open-chain polyfunctional compounds are often the by-products (or the only product) of attempts to prepare macrocycles. Open-chain molecules representing three-quarters or one-half of the intended macrocycles can be used for further cyclization and preparation of unsymmetrical macrocycles[45] (Scheme 6). Metal ion coordination may cause ring opening with formation of complexes of polydentate chelate ligands. These processes proceed especially easily at the tetrahedral carbon atom connected to two heteroatoms[46] (Scheme 7).

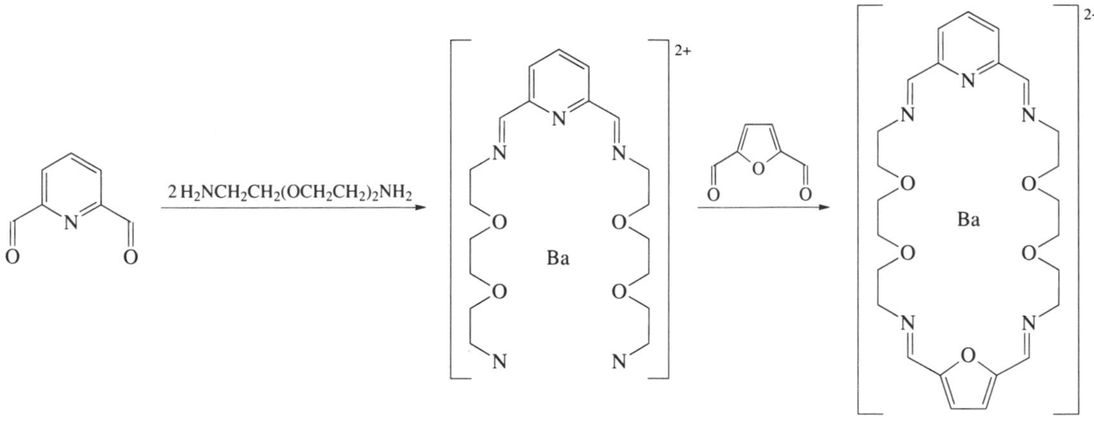

$2\,H_2NCH_2CH_2(OCH_2CH_2)_2NH_2$

Scheme 6

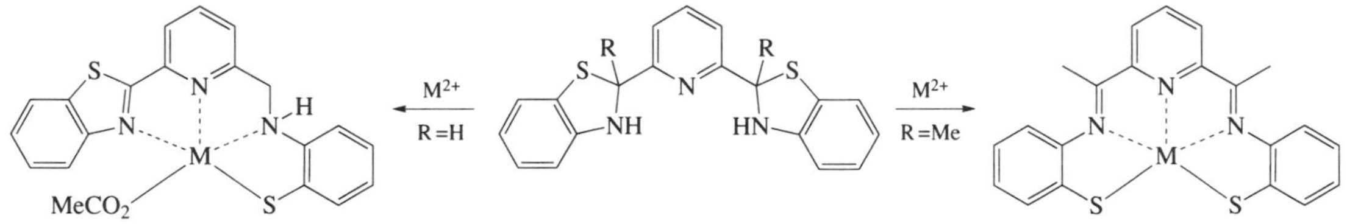

Scheme 7

1.3.2 Preparation of Macrocyclic Compounds

A minimal feature at some point in the synthesis of macrocycles is the evident or hidden bifunctionality of reactants. This bifunctionality is accompanied by the very general problem of competing reactions that produce polymers and cyclic oligomers. Many cases are further complicated by still higher functionality, greatly multiplying the number of possible products. Several approaches have been developed to overcome these problems. Both high dilution techniques and entropic control methods require 1:1 or intramolecular condensations. In addition, there are some practical restrictions, depending on the system in question, including the rates of second-order reactions in dilute solutions and the difficulty of handling very large volumes of solutions. Template syntheses do not suffer from these restrictions; however, their use is limited by other considerations. To a large extent, the templating agent dictates the nature and stoichiometry of cyclization reactions. Astute selection of the templating agent and noncyclic precursors are central factors to successful template-controlled syntheses. Although the number of successful template cyclization processes is extremely large, the main features can be displayed by using a few traditional classes of macrocyclic compounds as examples.

1.3.2.1 Monocycles — phthalocyanines

These ligands provide good examples of the thermodynamic template effect and illustrate its use in the synthesis of unsymmetrically substituted macrocycles. Historically, phthalocyanines were the first known synthetic macrocycles, and, industrially, they are among the most important macrocyclic complexes. Furthermore, they are usually synthesized with metal ion control (Scheme 8).

Despite the age of the subject, new synthetic work continues on industrially important phthalocyanines.[47] While for the overwhelming majority of transition and representative metals the reactions depicted in Scheme 8 give the familar tetradentate phthalocyanine, isomers of other dentate numbers are known. Boron[48] and uranyl[49] templates yield subphthalocyanine and superphthalocyanine, which are tridentate and pentadentate, respectively (Scheme 9). In the absence of their templating ions, both undergo rearrangement to the common tetramer. This reaction of the subphthalocyanine has been exploited to synthesize monosubstituted or unsymmetrically substituted phthalocyanines of the type $R_1{}^1R_2{}^2R_3{}^2R_4{}^2$, where the subscripts represent the number of corresponding benzene rings and the superscripts represent the type of substitution pattern. Previously known methodologies[50] provided routes to $R_1{}^1R_2{}^1R_3{}^2R_4{}^2$ or $R_1{}^1R_2{}^2R_3{}^1R_4{}^2$.

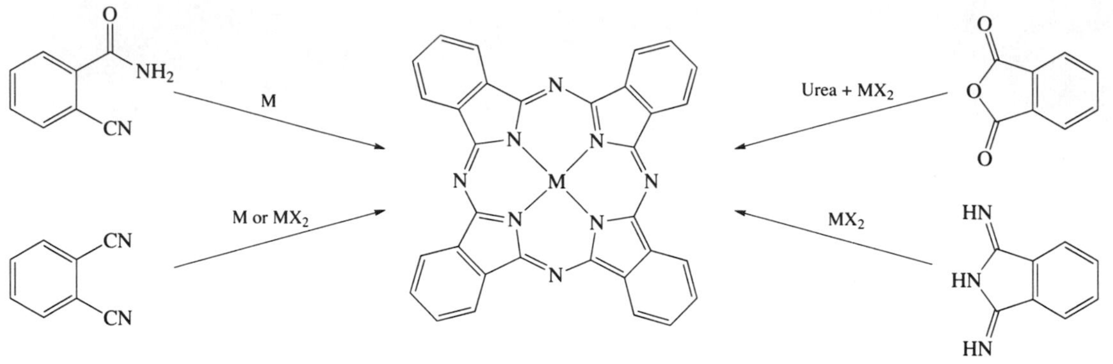

Scheme 8

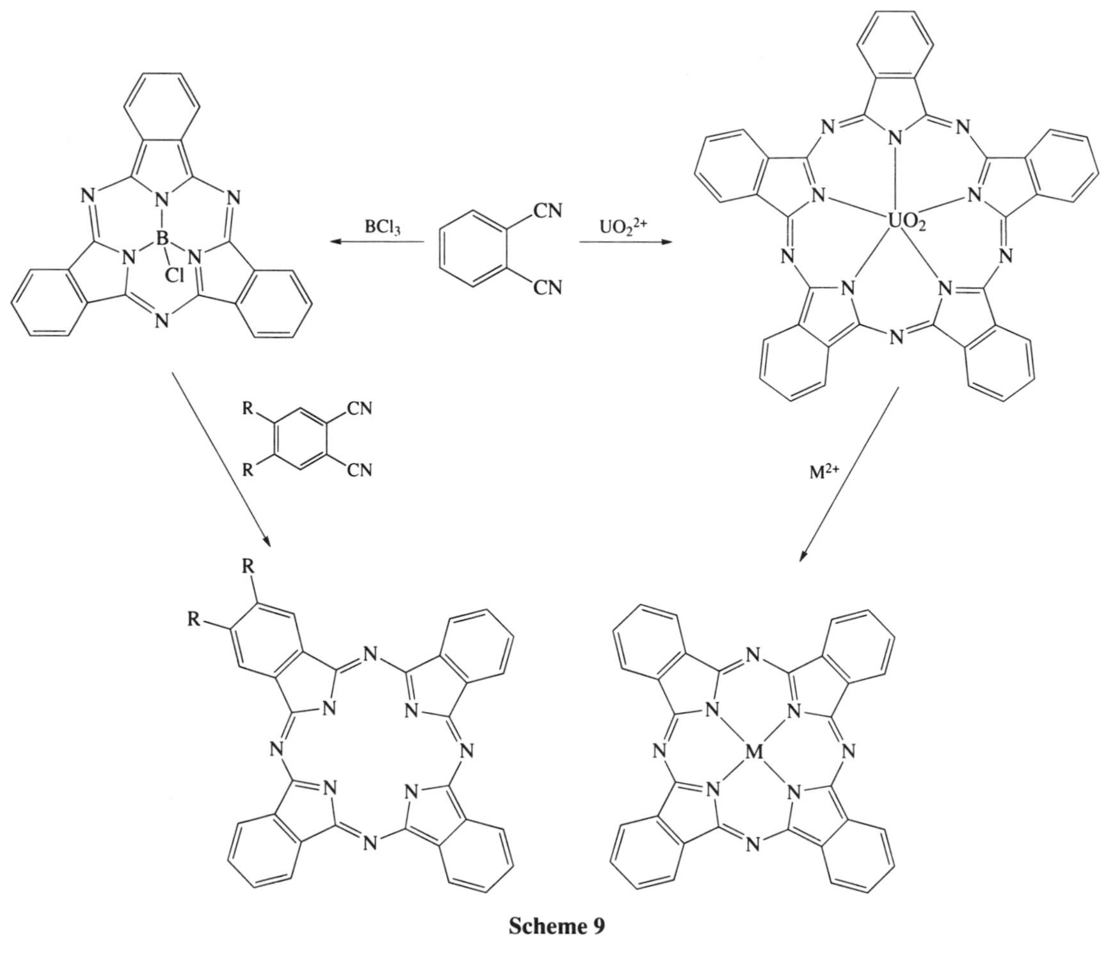

Scheme 9

1.3.2.2 Cyclic derivatives of aromatic o*-aminoaldehydes*

The formation of macrocyclic complexes from aromatic *o*-aminoaldehydes provides an unambiguous example of the thermodynamic template effect. The macrocyclic ligands TAAB (**5**) and TRI (**6**), the cyclic anhydrotetramer of *ortho*-aminobenzaldehyde, are believed to be unstable with respect to such alternative structures[51] as (**19**) and (**20**) (Scheme 10), and all attempts to isolate them have failed. In contrast, their complexes are a virtual thermodynamic sink. Almost every transformation of *o*-aminobenzaldehyde or its self-condensation products in the presence of metal ions gives these macrocyclic complexes (Scheme 10). The metal ion template determines which cyclic oligomer is formed, TAAB or TRI; iron(II), cobalt(II), zinc(II), palladium(II), and platinum(II) appear to give only TAAB complexes,[52] while vanadyl(IV) ions cause exclusive trimerization.[53] Nickel(II) templates generate complexes of both ligands,[16c,16d,54] $Ni(TRI)_2^{2+}$, $Ni(TRI)(H_2O)_3^{2+}$, and $Ni(TAAB)^{2+}$, and the reaction of cobalt(II) with subsequent oxidation of the templating ion to cobalt(III) yields[52a] a mixture of $Co(TAAB)^{3+}$ and $Co(TRI)_2^{3+}$. $Co(TRI)_2^{3+}$ and $Ni(TRI)_2^{2+}$ exist as *meso* and racemic forms (Equation (4)). This TAAB/TRI chemistry shows that the size of the template ion is not the only factor determining stoichiometry in template-directed cyclooligomerization. Some evidence suggests that the cyclooligomerization of *o*-aminobenzaldehyde proceeds in a stepwise manner rather than depending strictly on the number of aminobenzaldehyde fragments in the organic precursor. In the case of substitution-inert platinum(II), an intermediate complex with the deprotonated anhydrodimer of aminobenzaldehyde was isolated[52c] in almost equal yields from reactions with monomeric *o*-aminobenzaldehyde and with the diquinozolinium salt (Structure (**20**)). Also, in attempts to prepare the monomethylated complex, $Ni(Me\text{-}TAAB)^{2+}$, from equimolar amounts of bis(anhydrotrimer) (Structure (**19**)) and 4-methyl-2-aminobenzaldehyde, a statistical mixture of mono-, di-, tri-, and tetramethylated cyclic tetramers was obtained;[55] this stands in contrast to the formation of individual asymmetrical phthalocyanine from subphthalocyanine and substituted phthalonitrile (Scheme 9).

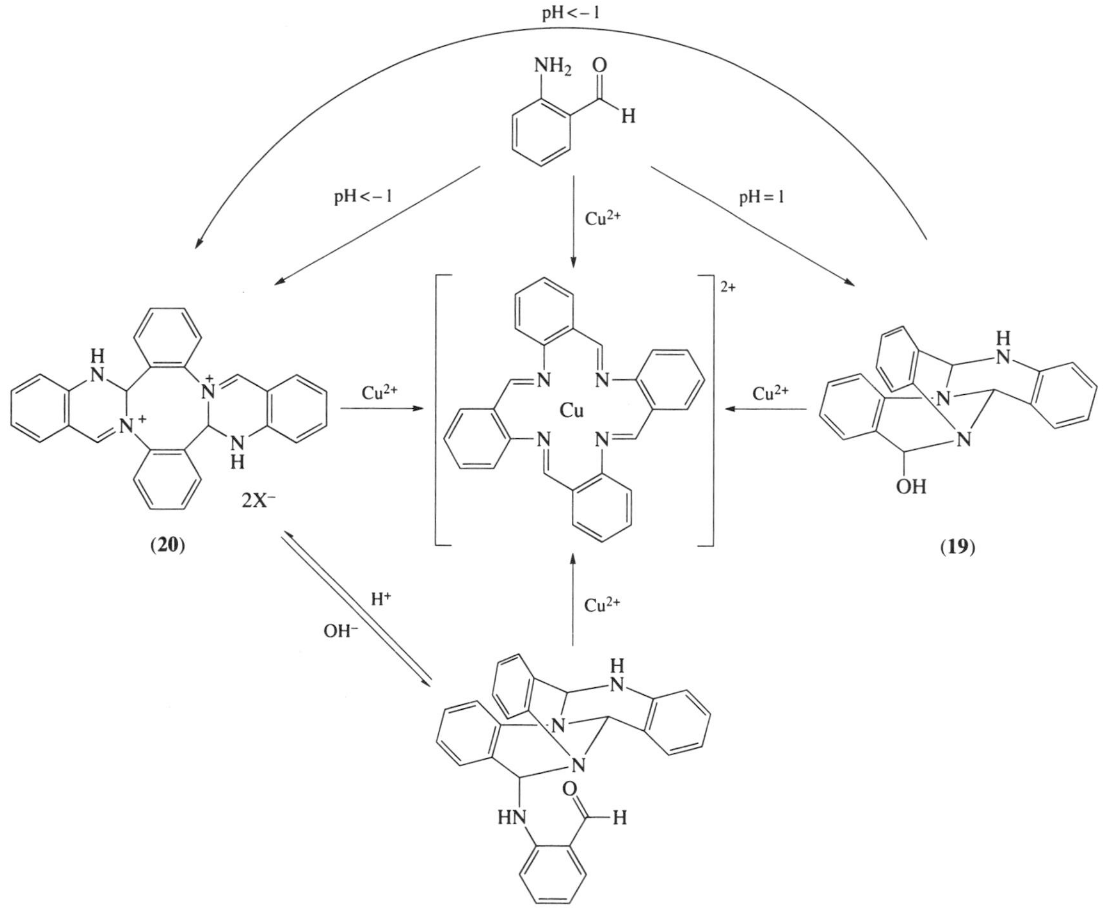

Scheme 10

1.3.2.3 Crown ethers and related compounds

Despite the labilities of the metal ions to which they are usually bound, the crown ethers somewhat paradoxically provide good examples of kinetic template effects. The first syntheses of crown ethers were performed under medium or high dilution conditions, so that the role of metal ions in the cyclization processes might not have been obvious, but Charlie Pedersen stated to the senior author his belief that the good yields were in large measure attributable to metal ion template effects. The template effect for the Williamson reaction, which is widely used for the preparation of crown ethers, was first clearly shown by Greene.[55] The condensation of triethylene glycol dialkoxide with triethylene glycol ditosylate gives high (up to 93%) yield of 18-crown-6 in the presence of potassium ion (Equation (5)). The yield was dramatically decreased, with predominant formation of polymeric products, when potassium was replaced by tetrabutylammonium. Many examples of the use of the template effect in crown ether preparation are given by Gokel and Korzeniowski.[56] A similar, but less pronounced, template effect was detected for the Richman–Atkins cyclization to form polyazamacrocycles (Equation (6)).[57] Replacement of tetramethylammonium counterion by sodium increased the yield from 50% to 80%. Later, Mandolini and co-workers[58] demonstrated the kinetic template effect in crown ether preparation.

According to Mandolini, metal ions (i) organize the open chain reactant(s) in a conformation(s) favorable for ring closure, and (ii) mask the reactivity of phenoxide or alkoxide ions by forming relatively stable ion pairs. The second function hinders both cyclization and polymerization. Kinetic studies (Figure 3) show that, when the open-chain precursor contains oxygen donor atoms, the first function predominates for all alkali ions except lithium, which tends to form very strong ion pairs. Tetraethylammonium ion is without effect and provides a baseline. The relative effectiveness of the different templating ions results from a combination of these two factors. For a reactant having no ether oxygen atoms (Figure 4), the first factor (organization) is eliminated and metal ions clearly suppress the cyclization processes. A similar effect was also detected for the alkylation of simple unsubstituted phenolate.

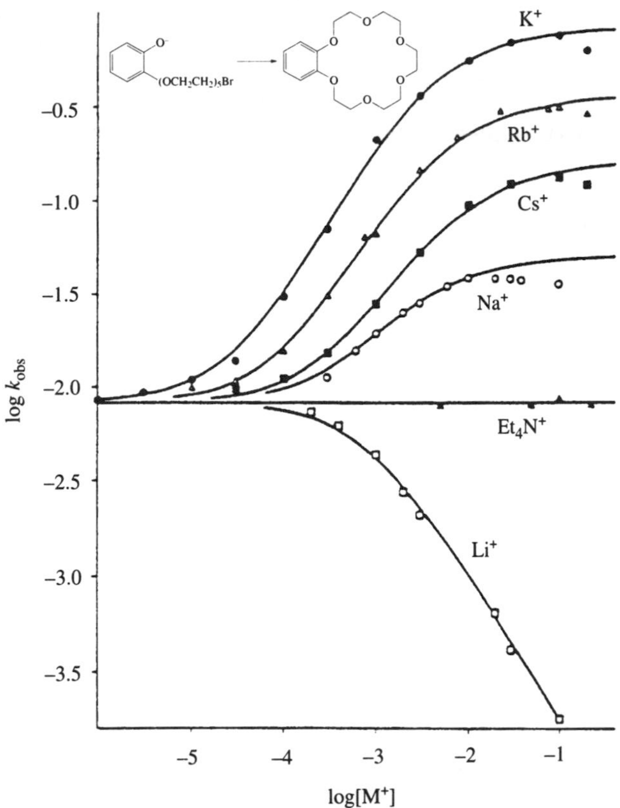

Figure 3 Kinetic studies of the effects of metal ions on cyclization (reprinted with permission from *J. Am. Chem. Soc.*, 1983, **105**, 555 Copyright 1983 American Chemical Society).

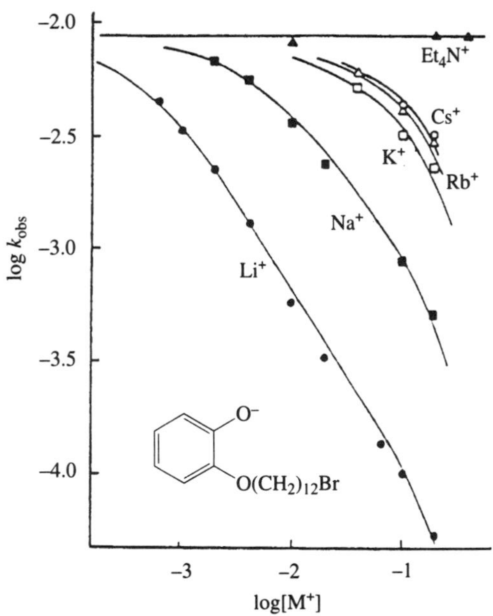

Figure 4 Effects of metal ions when no ether oxygen atoms are present (reprinted with permission from *J. Am. Chem. Soc.*, 1983, **105**, 555 Copyright 1983 American Chemical Society).

The synthetic chemistry of crown ethers reveals a connection between the size of the templating ion and macrocyclic ring size. Lithium ion[59] is the best template for the synthesis of Structure (**21**), while sodium and potassium are better for Structures (**22**) and (**23**). The large guanidinium ion was used as a template for the preparation of a 27-membered crown ether.[60] As is true of most experimental correlations, there are many exceptions to this apparent size-complementarity rule. The ability of alkali metal ions to form 2:1 complexes with crown ethers makes possible the template synthesis of large macrocycles by the cooperating action of two metal ions.[58] Another complication is anticipated because of the ability of crown ethers to form complexes in which the guest is situated above the cavity. Triphenylantimony, which is known to form complexes with crown ethers, was recently applied as a template in crown ether ester formation. In this case, the very large template facilitates 1:1 rather than 2:2 condensations.[61] The remarkable ability of caesium ions to facilitate, under certain conditions, intramolecular cyclization reactions is generally accepted. Kellogg and co-workers have been intensively developing this synthetic approach, especially for the preparation of sulfur-containing macrocycles,[62] but the mechanism is still a subject of discussion.[63]

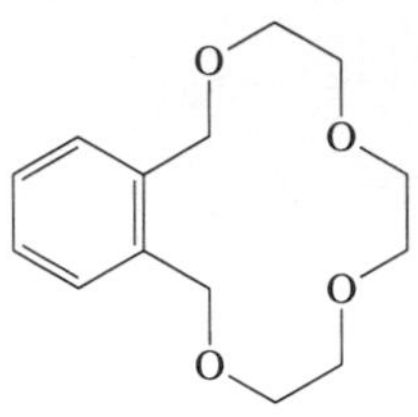

(21) 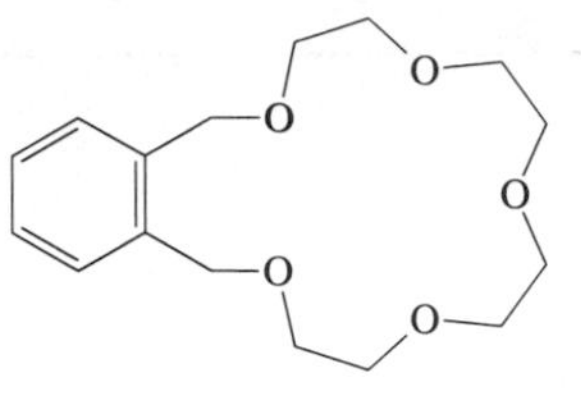 **(22)** 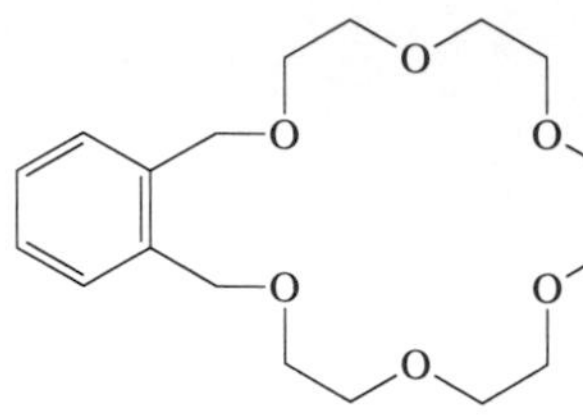 **(23)**

1.3.2.4 Macrocyclic derivatives of m*-dicarbonyl compounds*

2,6-Diacylpyridines, 2,6-diacylphenols, and related reagents are very important reactants in the design of ligands and their template-controlled synthesis.[64] Extremely broad ranges of compounds have been prepared from these precursors. These reactions illustrate [1 + 1], [2 + 2], and [3 + 3] condensations, template control of ring contraction, and the intervention of the metal ion in determining the nature of the condensation reaction.

An important issue in condensations of this type is stoichiometry. The nature of the templating agent, and the length and rigidity of the amino component can influence the stoichiometry of the condensation. When the amino component has fewer than seven atoms in the chain, a 1:1 reaction is not normally observed. Transition metals have been used most often for 1:1 condensations, while representative elements and manganese(II) have been used for 2:2 condensations. Scheme 11 illustrates the preparation of different compounds when barium(II) and lead(II) are used as templates.[65] Condensation of 2,6-bis(aminomethyl)-4-methylphenol with 2,6-diformyl-4-methylphenol in the presence of nickel(II) or copper(II) acetate gives the tetranickel complex of Structure **(24)**, containing the [2 + 2] condensation product as its ligand, or the hexacopper complex of Structure **(25)**, which contains the [3 + 3] condensation product.[66]

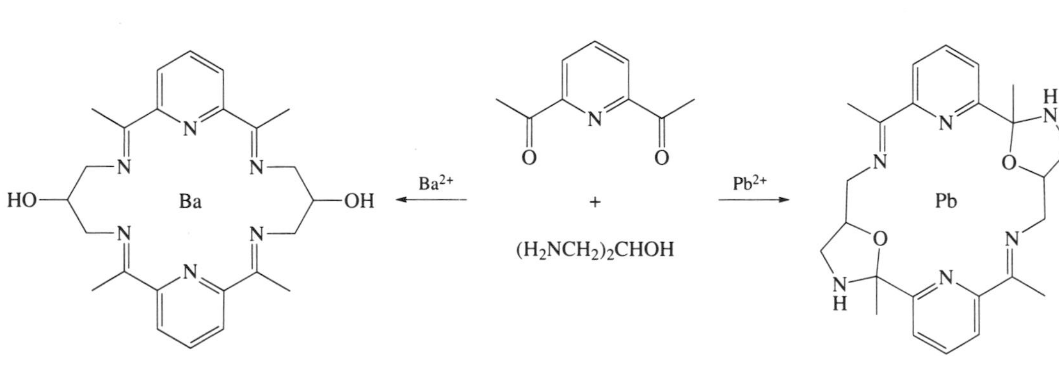

HO Ba OH $\xleftarrow{Ba^{2+}}$ + $(H_2NCH_2)_2CHOH$ $\xrightarrow{Pb^{2+}}$ Pb

Scheme 11

Addition of amino, alkoxy, or hydroxy groups to unsaturated linkages provides a route for ring contraction which may lead to a better metal–ligand size match[67] (Equation (7)). Instead of the expected tetraazomethine ligand, the template condensation of 2,6-diformyl-4-methylphenol and 1,9-diamino-3,7-diazanonane in the presence of lead(II) gives **(26)** (Equation (8)). This result was attributed to ring-size requirements.[68] Even though specifying the size and electronic structure

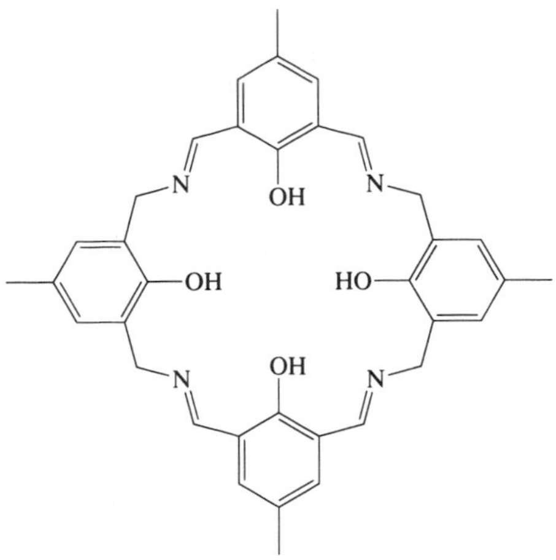

(24)

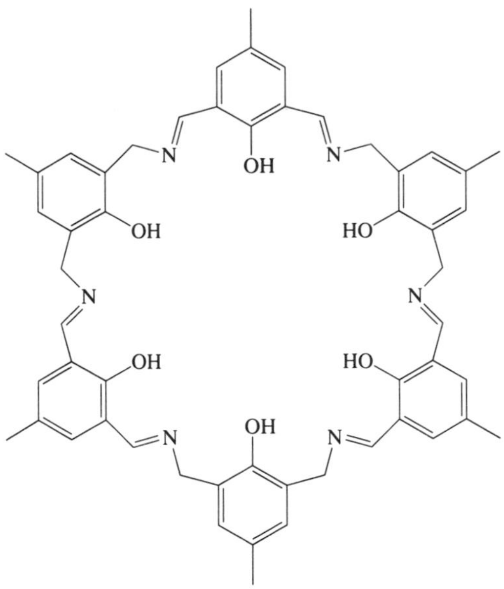

(25)

of a templating ion may improve the probability of success in predicting the structures of many condensation products, the factors regulating the stoichiometries of these reactions are not well understood.

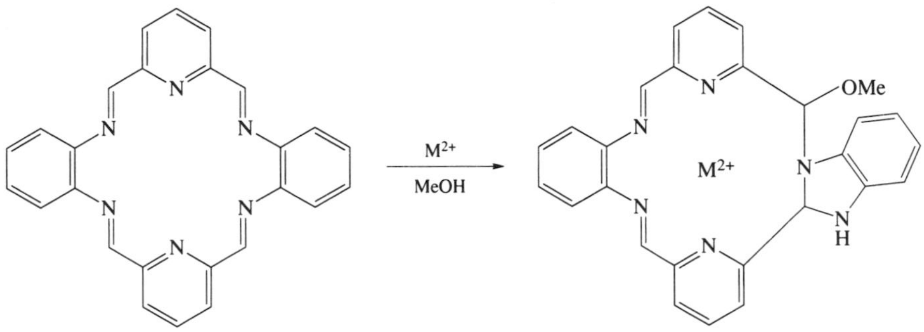

M^{2+} / MeOH (7)

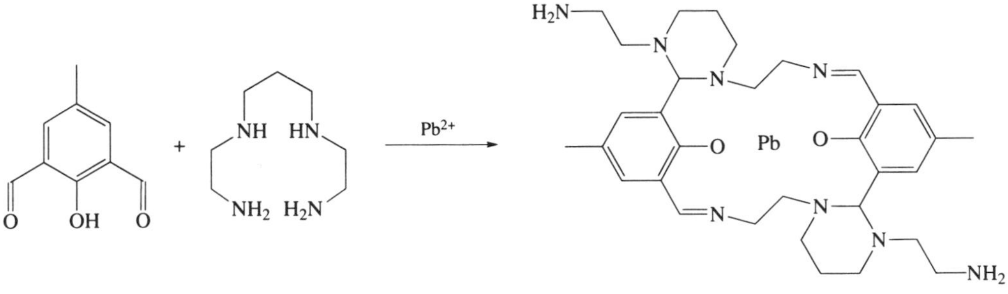

Pb^{2+} (8)

(26)

1.3.3 Bi- and Polycyclic Compounds

The overwhelming majority of bi- and polycyclic compounds prepared by template methods are cage compounds or clathro-chelates, and the typical families of compounds are polydentate oximes and amines. The modern chemistry of these compounds spans from the initial speculations[18a] and the preparation of the first representative by Boston and Rose[28] in the 1960s to the present day when reviews and special monographs describe the chemistry of these complexes.[32,69–2] In the preparation of the still more profuse families of bi- and polycyclic compounds (e.g., the cryptands, bicyclic calixarenes, and related organic hosts), the template approach plays a rather secondary role. However, the template effect sometimes provides a unique opportunity to improve yields or selectivities. Several common approaches have applied template methods to the preparation of bicyclic complexes. These are given in Scheme 12 and can be labeled as sides + caps, caps + sides, cycle + side, and cup + cap.

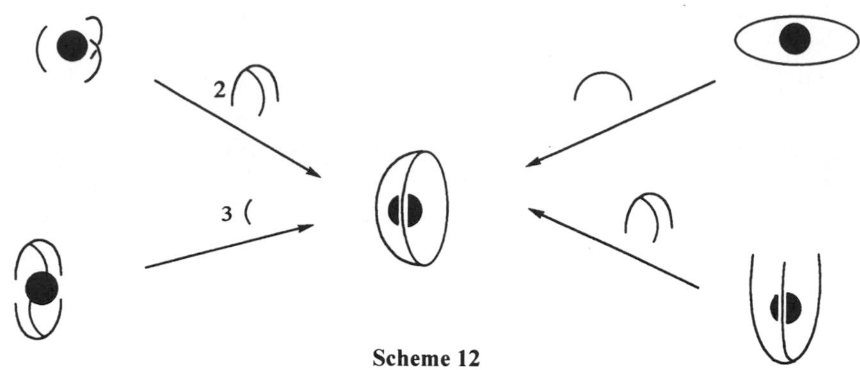

Scheme 12

1.3.3.1 *Tris(dioximate) cage compounds*

These compounds illustrate the use of Lewis acid reagents to cap the three sides of tris(didentate) complexes. They also provide examples of cage formation at the expense of disproportionation. In early studies, the tris(dioximato) complexes of the transition metals iron and cobalt, and more recently ruthenium and technicium, were found to react with Lewis acids in a capping reaction wherein the three oxime oxygen atoms became bound to the central atom of the Lewis acid[28] (Equation (9)). A wide variety of aromatic, alicyclic, and aliphatic dioximes can be used in place of dimethylglyoxime. In addition to boron trifluoride, other boron halides, boric acid, miscellaneous organoboron compounds, silicon, and tin tetrahalogenides have been used as capping reagents.[72] The reactions are probably thermodynamically driven, as relatively stable bis(dimethylglyoximato) complexes disproportionate with the formation of free metal ion and tris(glyoximato) cage complexes.[73] The thermal stability of the cage compounds complicates the preparation of bridged complexes from bis(glyoximate)s and bridging ligands.[74] Attempts to prepare unsymmetrically capped dioximate cages have failed.[75]

$$K_3CoDMG_3 + 6BF_3 \xrightarrow{\text{ether}} [\text{FB(ON=CMeCMe=NO)}_3\text{BF}]\text{Co}\ BF_4^- + 3KBF_4 \qquad (9)$$

1.3.3.2 *Cage complexes formed by Mannich chemistry*

The development of these cage complexes, under the leadership of Sargeson and his collaborators, has brought new reactions and reagents to template syntheses, but the product structures have generally been difficult to predict because of the versatility of the condensation process. Because ionization of coordinated amines is involved, the substitution–lability requirements of the template are opposite to those of, for example, the formation of Curtis' macrocycles. For the most part, complexes of this type have been prepared by condensation of various hexaaminecobalt(III) complexes with formaldehyde and weak, often C—H, acids. Scheme 13 represents the sides +

caps strategy.[32b] Condensation of tris(ethylenediamine)cobalt(III) with two common combinations of condensation reagents (formaldehyde/ammonia and formaldehyde/nitromethane) gives the cage complexes (**27**) and (**28**).

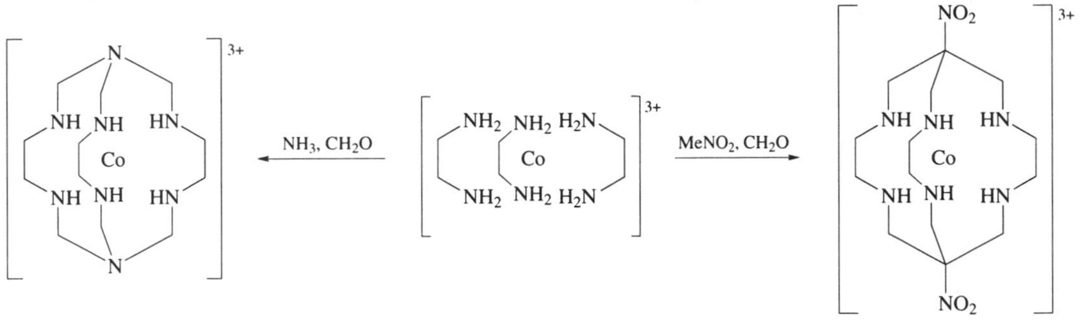

(**27**) (**28**)

Scheme 13

The cup + cap approach has been used even more extensively. The cobalt(III) complex of 4,4′,4″-ethylidynetris(3-azabutan-1-amine) (Structure (**29**)) and its several analogues were explored in miscellaneous condensations[70] (Scheme 14). In several cases, such as condensations with arsine or aldehydes, tris(formaldimine) (Structure (**30**)) was used as a starting material.[76] The condensation of similar starting materials, having trimethylenediamine "sides" instead of those based on ethylenediamine, was used[77] to cage Rh^{3+}.

The caps + sides strategy provides another method for producing cage complexes. A recent example involves the reaction of the cobalt(III) complex of ethylidynetris(methanamine) (Structure (**31**)) with propanal in a basic medium. Nucleophilic attack of the corresponding enolate on coordinated formaldimine followed by Schiff base condensation (Equation (10)) gives[78] the cage complex (**32**).

As a further example, condensations of three different isomers of bis(diethylenetriamine)cobalt(III) with formaldehyde and ammonia give different products:[79] the *s-fac* isomer gives the product with two molecules of ligand (**33**), the *u-fac* complex transforms into a 1:1 complex with ligand (**34**), and the *mer* complex gives a mixture of complexes of cobalt(III) with (**34**) and (**35**).

Platinum(IV), rhodium(III), iridium(III), chromium(III), and to some extent nickel(II), can play the role of templating ions in the preparation of Sargeson-type cage complexes.[80] The reactions are most successful with substitution-inert metal ions. For nickel(II) the yield of the cage complex is less than 1%.[81] This kinetic requirement is precisely the opposite of that found for the synthesis of Curtis complexes.

1.3.3.3 Capped bis(tren) cages—siderophore models

In their investigations of siderophore models, Raymond and co-workers prepared several cage compounds using triaminotriethylamine as a capping reagent and iron(III) and lanthanoids as templating ions.[32a] The preparation of these encapsulating siderophore models involved reactions at centers that are substantially remote from the metal ion. Activated esters were used to form amide linkages in the bicyclic products. In the presence of triethylamine, a complex was formed between 3 mol of disuccinimido-2,3-dihydroxy-terephthalate and 1 mol of $FeCl_3$. The resulting complex was then subjected to reaction with tren; an intermediate monocyclic complex was detected (Scheme 1), which was subsequently ring-closed to the desired bicycle.

1.3.3.4 Bicycyles based on Jäger macrocycles

The term "Jäger macrocycles" is used for highly unsaturated complexes (Structure (**36**); (Equation (11)).[82] Peripheral acetyl groups undergo two important reactions which can be used for the preparation of bicyclic complexes. They can be transformed into enol ethers, or the acetyl groups

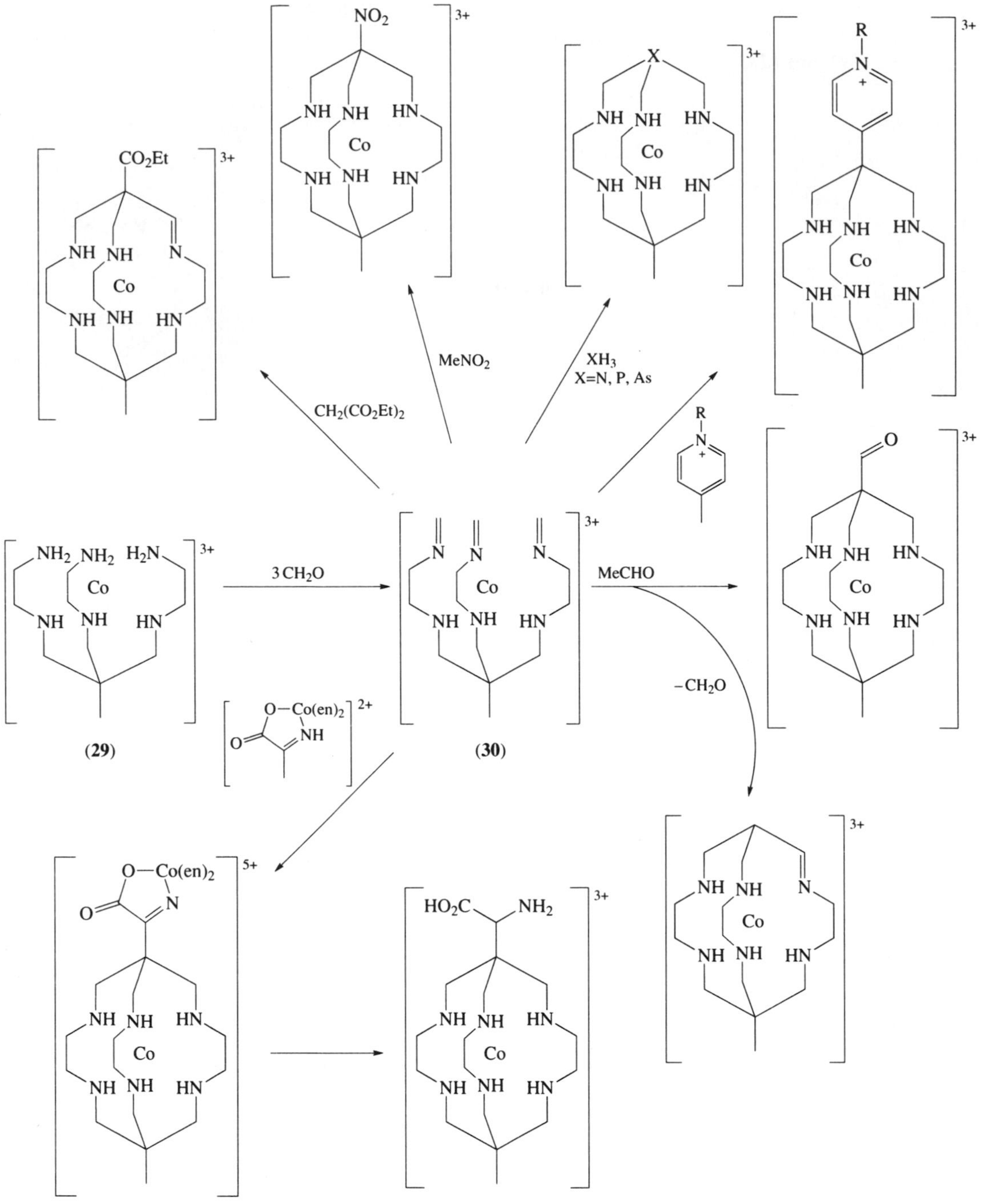

Scheme 14

can be removed under relatively mild conditions to give compounds of Structure (**37**) (Scheme 15), and replaced by other electrophilic groups. For the 15- and 16-membered parent macrocycles, these two monocyclic complexes can be easily transformed into bicyclic systems. Because of the deep saddle conformation of the 16-membered derivative (**38**), the functional groups are well oriented for ring closure.[69,83] Thus, this is a remote template system and, in this respect, is similar to the siderophore models of Raymond. The available synthetic routes for the preparation of these bicyclic complexes are summarized in Scheme 15.

Iron(II) and cobalt(II) cyclidene complexes having Structure (**39**) with hydrogens on the exocyclic nitrogen atoms can undergo an unusual rearrangement into cage complexes (Equation (12)), a process that can be viewed as resulting from a thermodynamic template effect.

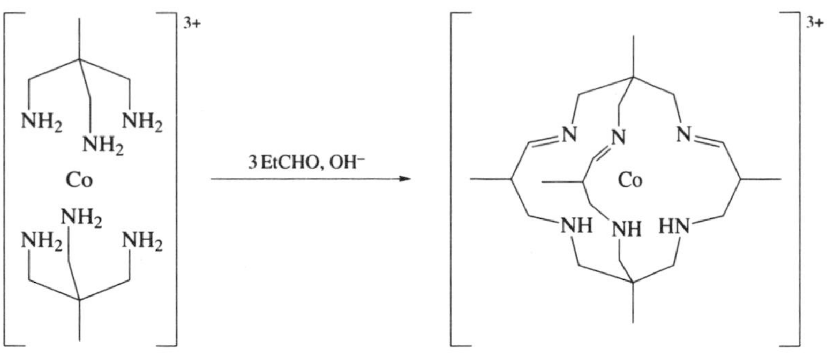

(31) $\xrightarrow{\text{3 EtCHO, OH}^-}$ (32) (10)

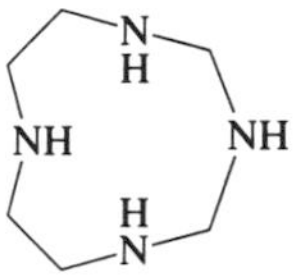

(33) 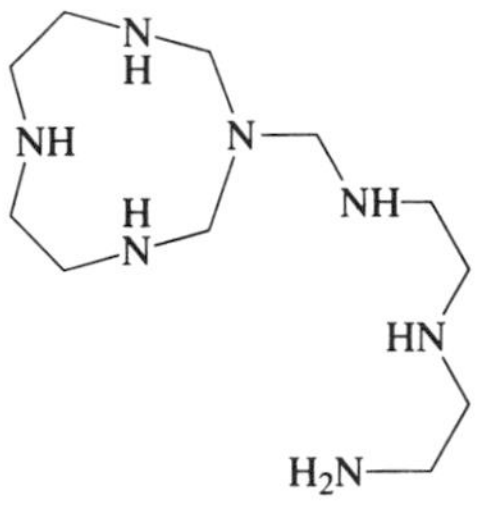 (34) 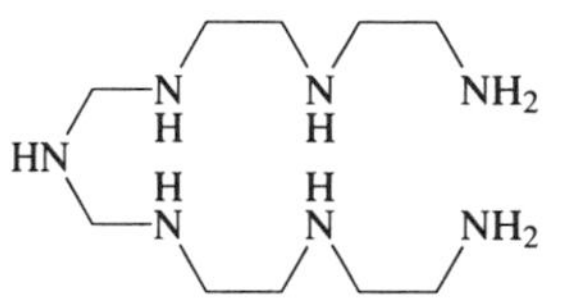 (35)

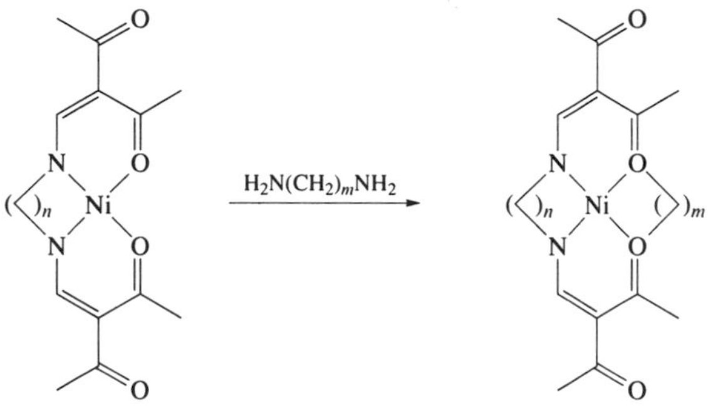

$\xrightarrow{\text{H}_2\text{N(CH}_2)_m\text{NH}_2}$ (36) (11)

1.3.3.5 Cryptands

Traditional methods of cryptand preparation do not involve a template effect. However, template reactions have been used to produce fairly complicated molecules of these kinds in one or two steps. Krakowiak, Bradshaw, and co-workers have reported the preparation of cryptands [2.2.1], [2.2.2], [3.2.2], and [3.3.3] by the reaction (Equation (13)) of 1 mol of an appropriate diamine and 2 mol of ditosylate, dimesylate, or diiodide.[84] When K^+ was used as a template ion in the preparation of the [3.3.3] species, the cryptand was the only product. Use of Cs^+ gave a mixture of [3.3.3] and the bis(macrocyclic) product (**40**).

Condensation of equimolar amounts of 4,10-diaza-[15]crown-5 and α,α′-dibromo-*m*-xylene in the presence of K^+ gives a mixture of the 1:1 condensation product (**41**) and the 2:2 condensation product (**42**). When carried out in the presence of Li^+, the only product is (**41**). These examples illustrate the high selectivity of certain templates in cryptand preparations. In comparison with monocycle preparations, the better selectivity observed for these systems might be connected to the restricted ability of cryptands to form 1:2 or out-of-cavity complexes. It is a striking observation that both the template effect and size selectivity are more generally observed with cryptands than with crown ethers.

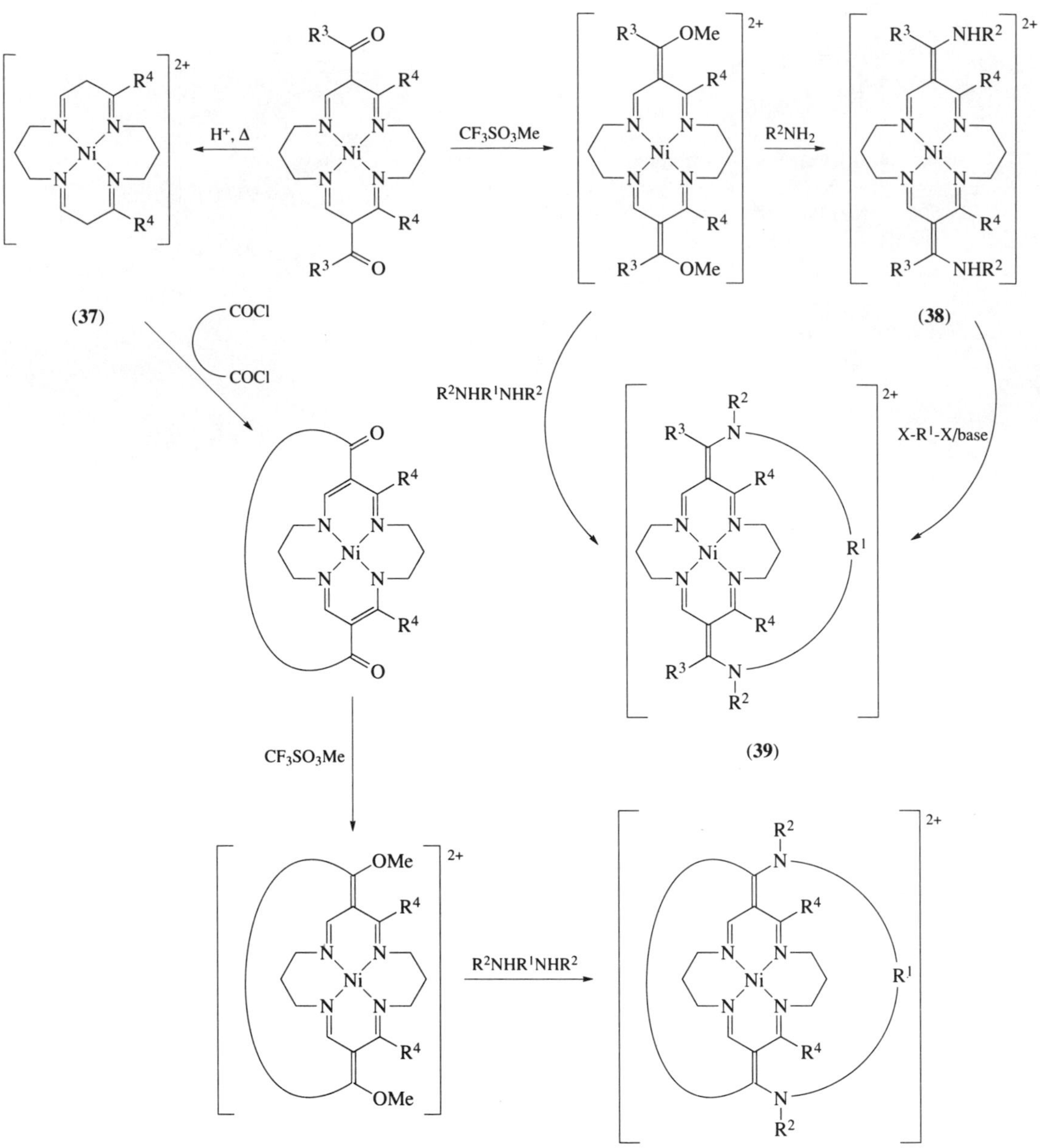

(37)

(38)

(39)

Scheme 15

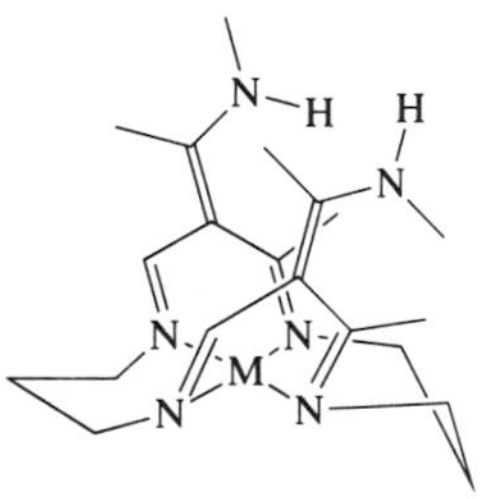

(38)

1.3.4 Catenanes and Knots

Catenanes (from *catena* meaning "chain") are interlocking macrocyclic molecules, as represented in Figure 5(a). Thus, the two rings are united into a single molecule only by mechanical means, rather than chemical bonding. The word "knot" is used in the usual sense of the *Webster's*

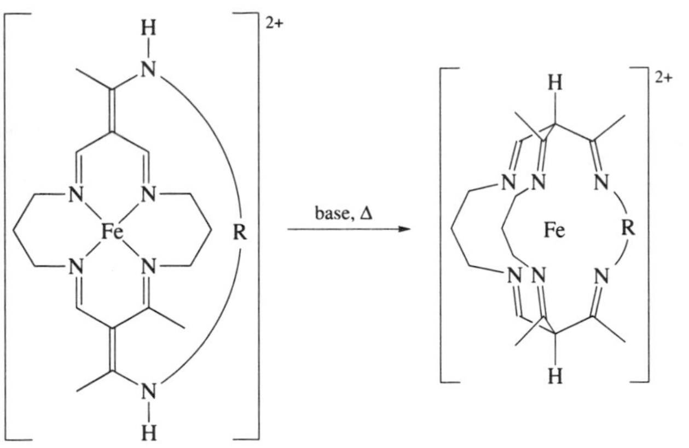

(39)

(12)

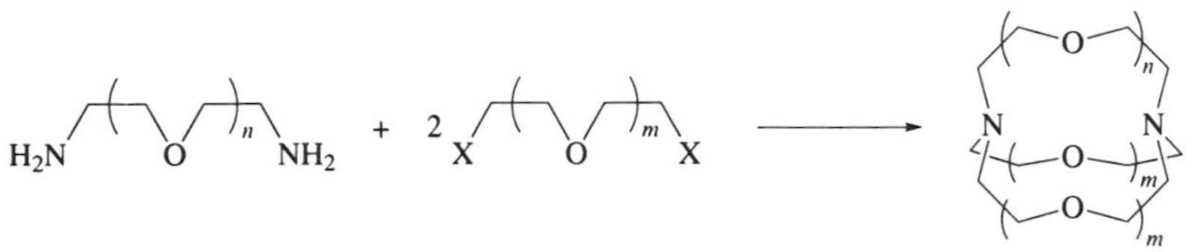

(13)

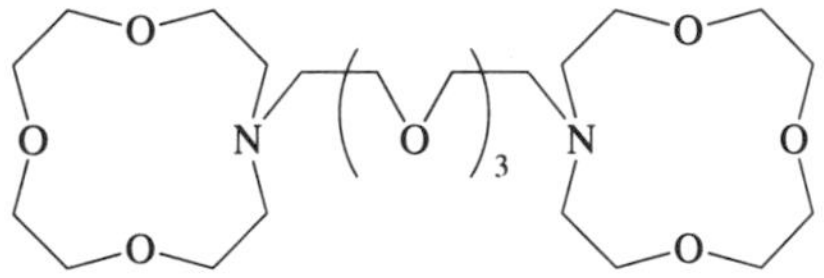

(40)

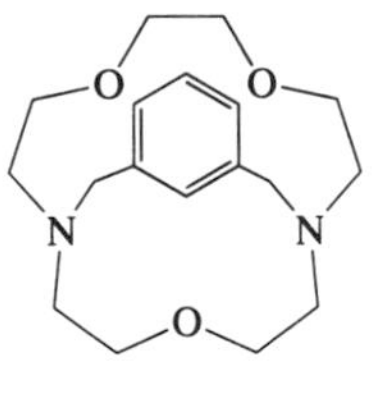

(41)

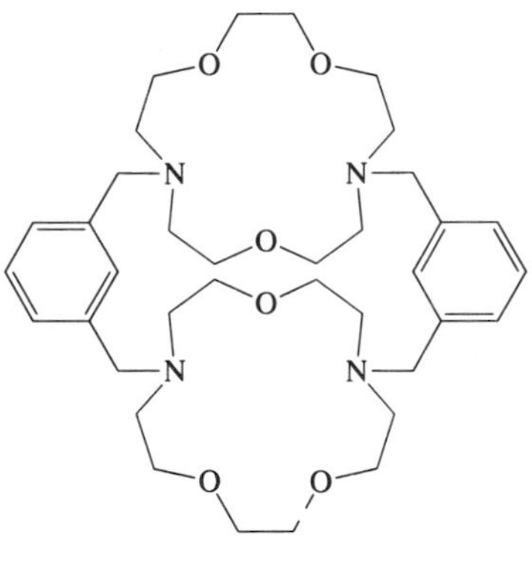

(42)

Dictionary most general definition: "an interlacement of parts of a cord, rope, or the like, forming a lump or knob." Figure 5(b) shows the motif for the trefoil knot, the first molecular knot to be tied by a chemist. Among the best displays of the utility of the template effect are the syntheses of catenanes and knots in yields that could never be approached without the level of control over the topology of the reactants which a well-designed template provides. Several reviews on the subject are available,[33b,35,85] and an issue of the *New Journal of Chemistry* has been dedicated to molecular topology.[86]

Initial attempts to produce interlocking molecular rings relied on a statistical threading approach;[87] however, early researchers realized that better yields could be obtained if the catenanes could be constructed around some sort of central core.[88] Most of these cores were envisaged as organic moieties such as benzene, but in an interesting footnote, Frisch and Wasserman wrote: "Another interesting suggestion has been that of W. Closson to utilize the geometry of the ligands about a metal atom as a core. Both approaches are being investigated in Dr. Lemal's laboratory at

Figure 5 (a) Catenane (chain) structure; (b) knot structure.

the University of Wisconsin." (D. Lemal, private communication.) The results of this work do not seem to have been published.

Interest in catenanes and knots was rekindled when reports on the copper(I) template synthesis of a [2]-catenane were presented by Sauvage and co-workers in the early 1980s.[33a] Structure (**43**) shows the metallated "catenate." The tetrahedral coordination environment of the copper(I) ion has enabled researchers in the Strasbourg group to construct a variety of topologically fascinating molecules including the trefoil knot (**44**) and doubly threaded catenane (**45**),[89a] and an octahedral Ru^{II} catenate was also prepared[89b] (see Chapter 2). [2]-Catenates have now been synthesized from the ligand shown in Equation (14).[90] The central template ion is the pseudooctahedral iron(II), and the rings are closed by coordination of the "arms" to silver(I). Two isomers were crystallized, and proton NMR studies showed that they undergo rapid exchange in solution. Octahedral metal ions should prove as useful as tetrahedral ions in the construction of interlocking rings.

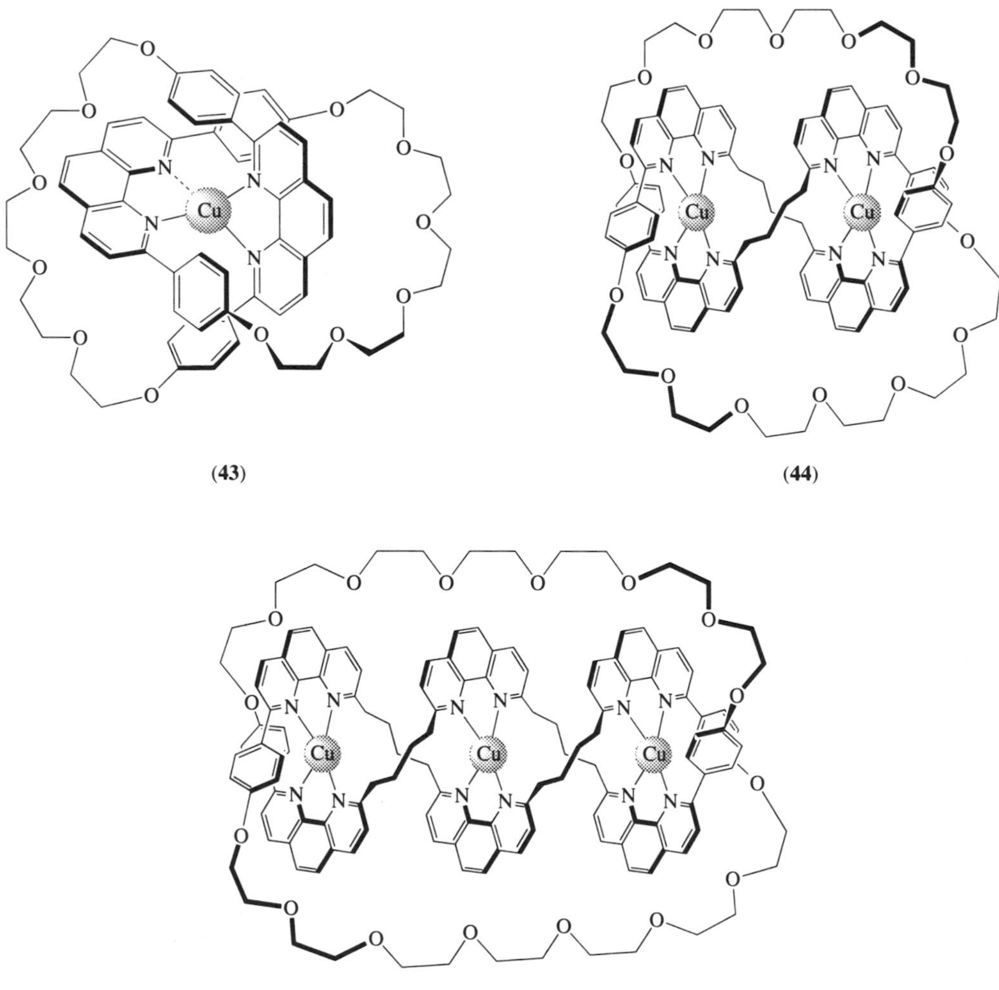

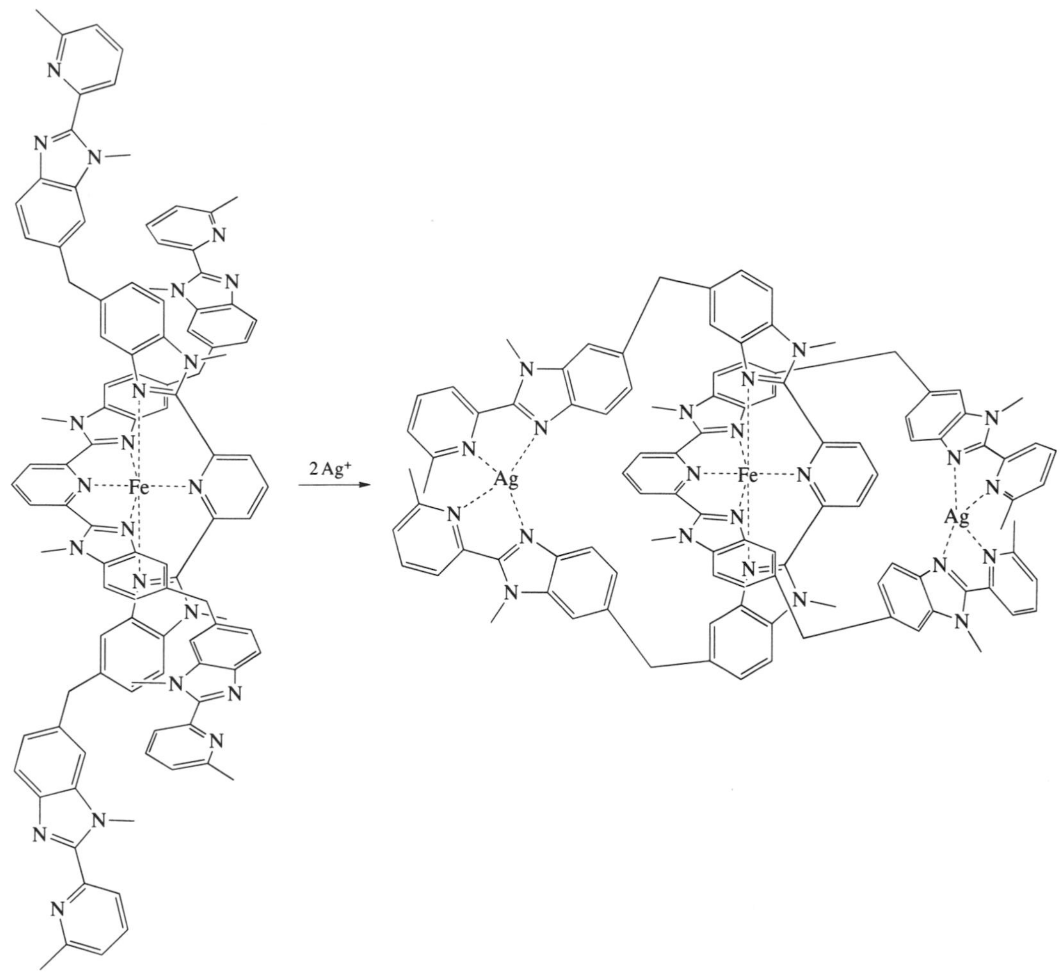

(14)

The electrostatically driven "molecular meccano" of Stoddart and co-workers has shown the astounding potential of noncovalent interactions as templates.[35] Thus far, [2]-, [3]-, [4]-catenanes, and the "olympiadane" [5]-catenane (**46**) have been reported.[91] The principles governing these syntheses have been applied to produce a variety of related species, including catenated cyclodextrins,[92] surface attached catenanes (**47**),[93] porphyrin [2]-catenanes (**48**),[94] and photoswitchable catenanes.[95]

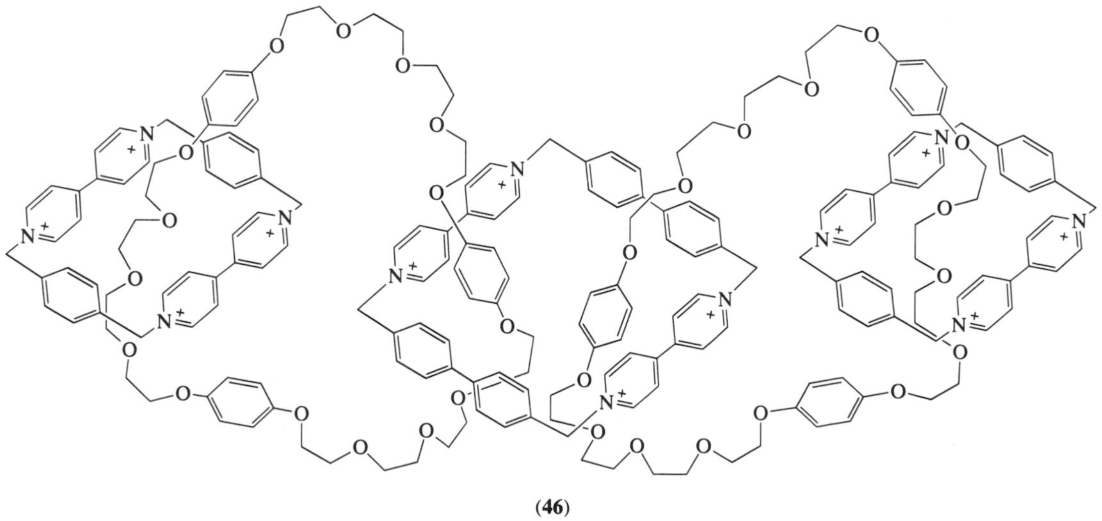

(**46**)

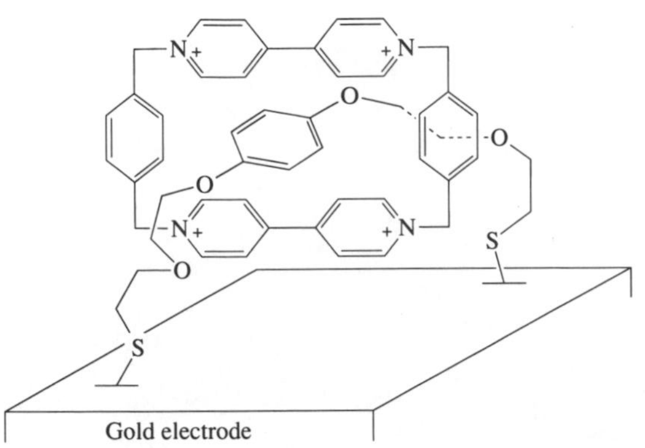

(47)

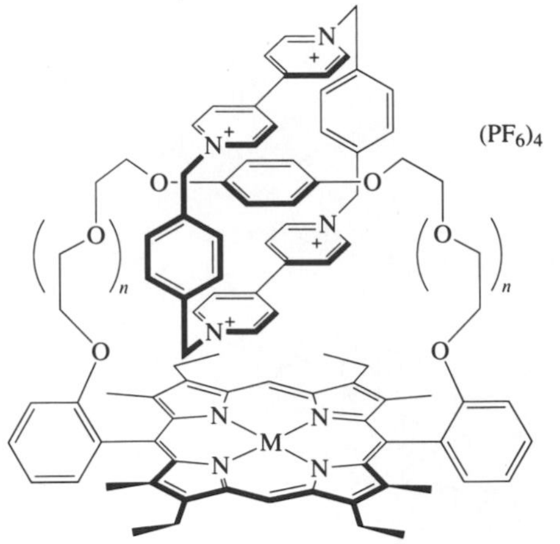

(48)

Hydrogen-bonding interactions have been attributed to the formation of catenanes fortuitously found by the independently working groups of Hunter and Vögtle.[96,97] Scheme 16 shows the catenane found by Vögtle and co-workers as well as the hydrogen-bonding interactions which serve as the template for catenane formation.

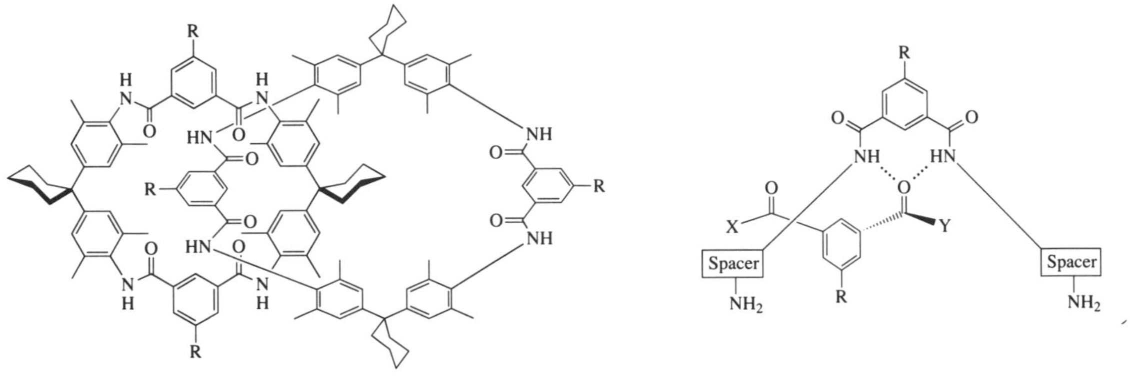

Scheme 16

In an attempt to prove a proposed mechanism for a related rotaxane formation (see below), Bickelhaupt and co-workers synthesized the first organometallic catenane.[98] As shown in Scheme 17, the diphenylmagnesium can dissociate, whereupon the cation coordinates inside the crown ether. Subsequent recombination of the fragments gives a catenane which was observed by the shift of the signal of H-2 in the proton NMR spectrum.

Another example of a catenane synthesized from preformed rings has been provided by Fujita and co-workers.[99] By a process which demonstrates a thermodynamic template effect, the platinum(II)–pyridine bond in ring 1 shown in Equation (15) is reversibly broken upon heating, and the polar medium pushes the equilibrium toward the catenane. Upon cooling, the rings are irreversibly locked.

1.3.5 Rotaxanes

Rotaxanes are composed of at least one cyclic component encircling an axle component. In common simple cases, bulky groups on the ends of the axle prevent its removal from the ring. In polymers, the extent of the polymer chain and its bends and twists may retain encircling components. Other methods of retention, such as linkages between the axle and the ring, are also possible. Several of the distinctive routes for rotaxane preparation (Scheme 18) can be template directed.

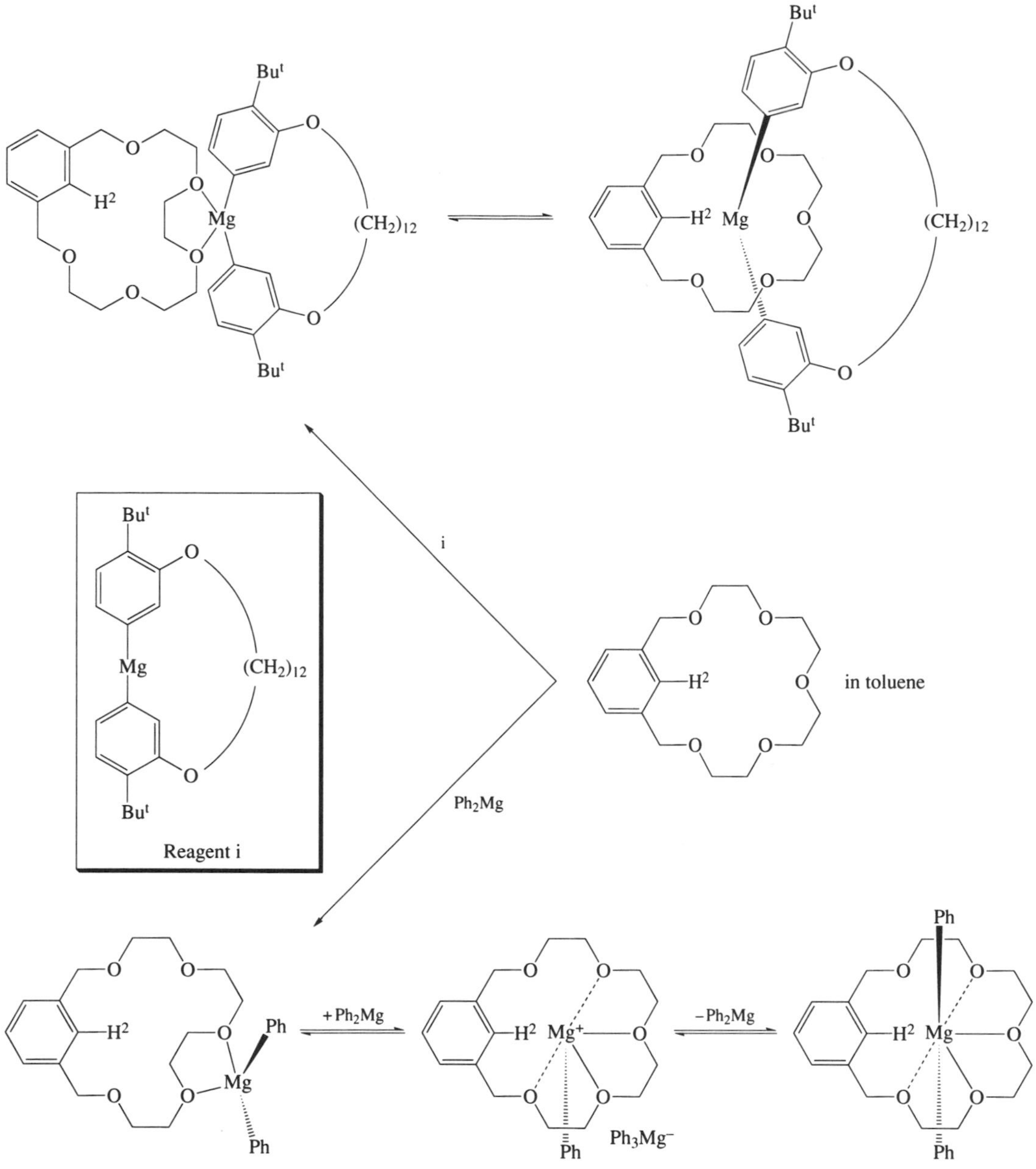

Scheme 17

In the approach labeled "clipping" in Scheme 18 (Equation (16)), the linear component is able to organize the precyclic component around the template center in a conformation favorable for axle-in-torus cyclization.[100]

An organometallic rotaxane has been reported to form by a process which could be called "snapping," in which the linear component which is capped with two bulky substituents dissociates, allowing one of the fragments to form a complex with the cyclic component (Equation (17)). When the fragments of the linear molecule subsequently recombine, a rotaxane forms. Favorable binding of the crown ether to the metal ion makes this appear to be a thermodynamic template process.[101]

The "threading" approach is clearly divided into statistical and template-directed methods. "Template threading involves an attractive interaction between the linear species and the macrocycle, such as metal chelation, charge transfer interaction, hydrogen bonding, π-stacking interactions, or the like. Hence, the equilibrium . . . is enthalpically driven . . ."[102] Metal-ion mediated template threading based on Sauvage's copper(I) bis(phenanthroline) templating system (see

2 [Ring 1] $4NO_3^-$ $\xrightarrow[(M = Pt)]{\text{irreversibly}}$ [Ring 2] $8NO_3^-$ (15)

Ring 1 Ring 2

(**a**) M = Pt
(**b**) M = Pd

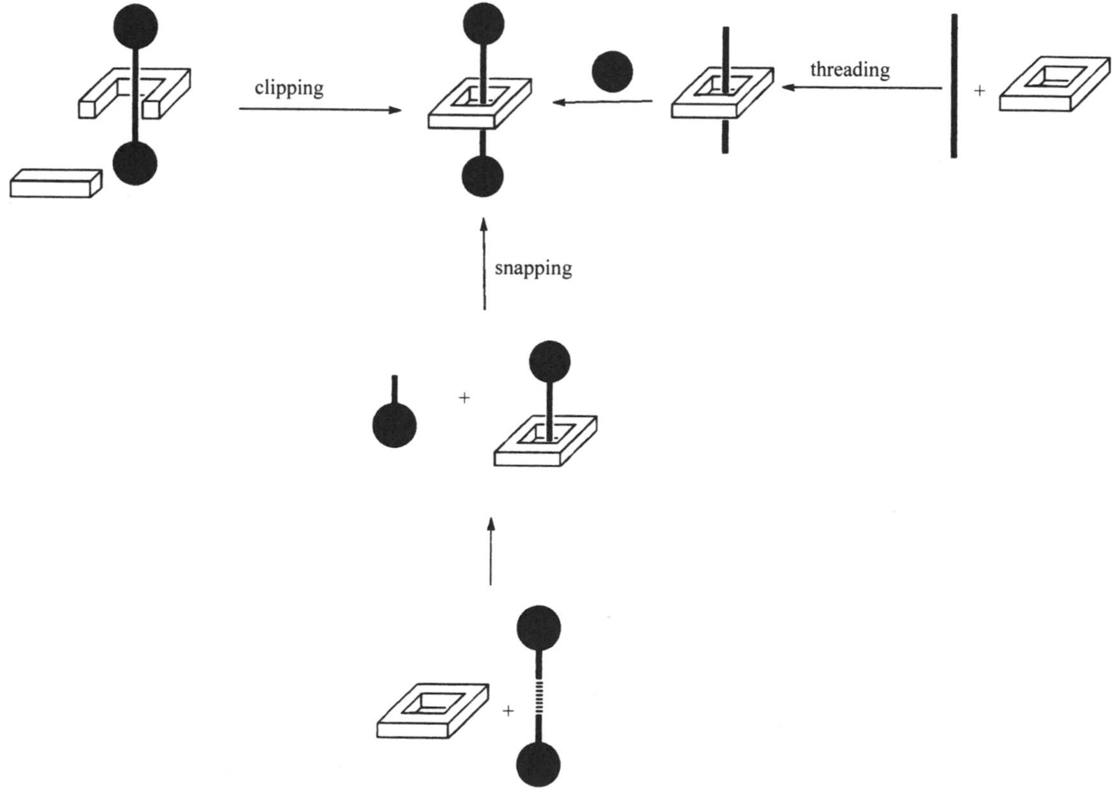

Scheme 18

Chapter 2) has been performed by Gibson and co-workers,[103] Sauvage and co-workers,[104] and J.-M. Lehn and co-workers.[105] An example of this strategy is shown in Scheme 19. Another well-developed approach for template threading in the preparation of rotaxanes is illustrated by Stoddart's electron donor–acceptor stacking template (see Chapter 3) (Scheme 20). The threading approach based on donor–acceptor stacks sometimes gives better results than the clipping method.[106]

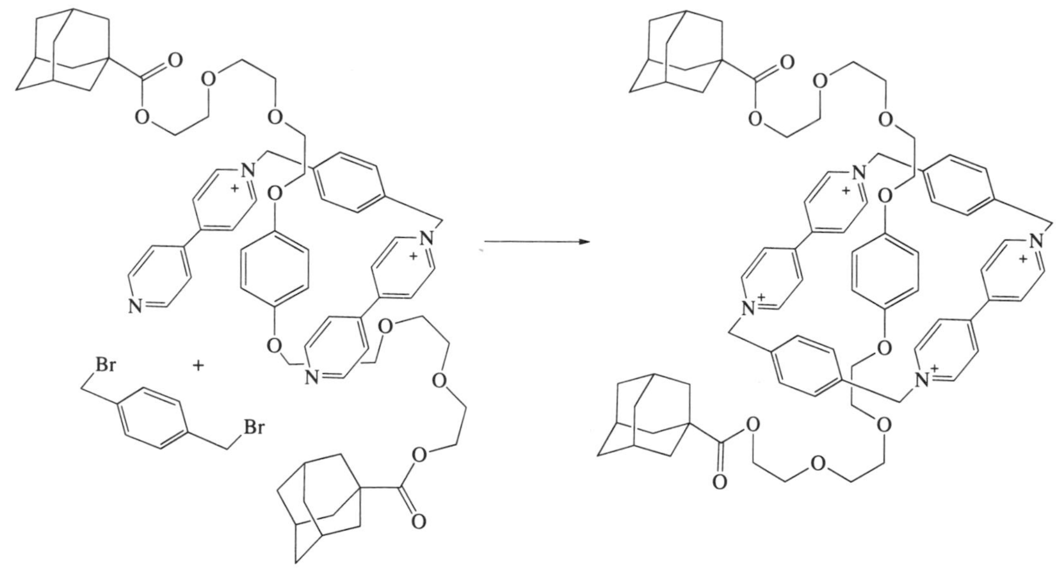

(16)

+ $MgPh^+$ + Ph^-

$MgPh_2$

(17)

Rotaxanes have been templated by the formation of stable inclusion complexes between cyclodextrins and aliphatic linear chains or bipyridinium units.[102] It has been suggested that the multiple complexation sites of polymers facilitate the threading of interacting macrocycles and their movement from the ends of the polymer chains. To test this idea of a kind of fluxional template in small molecules, the yields of rotaxane formation were compared for an axle molecule having two amino groups separated by a dimethylene chain and an axle molecule having the same composition and chain length except the secondary amine was replaced with a methylene group (Scheme 21). Remarkably good yields of rotaxane were produced for the diamino axle, but only traces were detected for the monoamine.[107]

1.4 PRINCIPLES AND PROSPECTS

1.4.1 Perspective

The preceding discussions present many of the generalizations that can be made about both kinetic and thermodynamic templates and the striking variety of molecular motifs they have helped generate. The simple macrocycles, macrobicycles, and macropolycycles reveal certain very general aspects of template reactions that have been extended to a small number of more exotic molecular motifs, rotaxanes, catenanes, figure-of-eight or doubly threaded catenanes, a trefoil knot, and multiple catenanes in which several macrocycles are successively interlocked like ordinary mechanical chains. The simple elements that are essential components of templates are examined in this section. The process of combining these intrinsic elements of templates to create the known

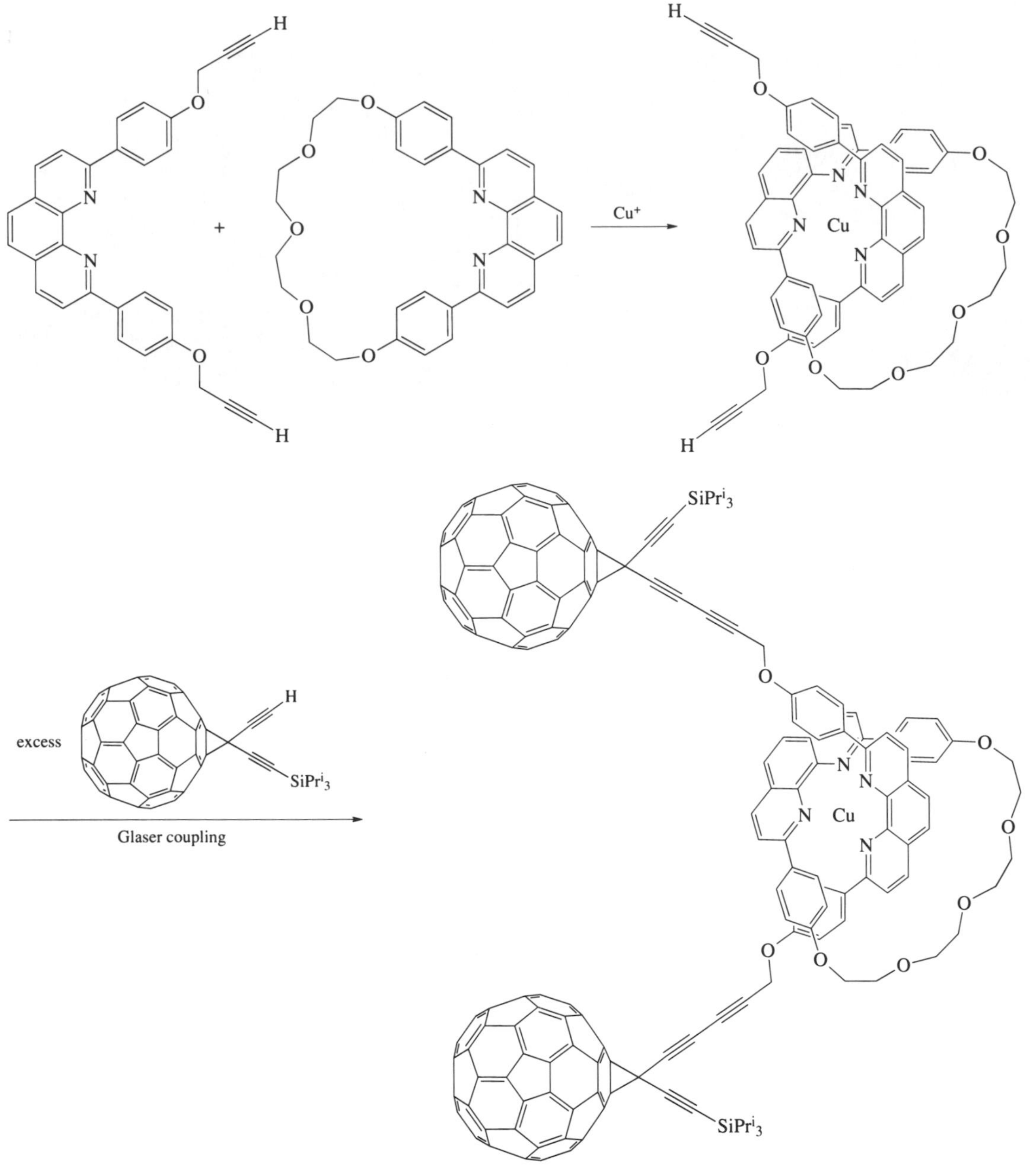

Cu⁺

excess

Glaser coupling

Scheme 19

architectures should be extendable to far more complicated molecular architectures. It has been suggested[1] that anything that can be done with a macroscopic thread should be possible in molecular structures. This notion underlies the prediction of broad families of new materials, all of which may be considered to be orderly molecular entanglements. At present, materials such as plastics and elastomers are composed of random or disorderly entanglements of molecules. Knots, braids, and knitted, crocheted, and woven cloth involve the orderly entanglement of macroscopic threads. Each should have a molecular analogue. The vision of materials containing mixtures of the many molecular motifs mentioned here has led to the notion of a "molecular macramé" as the symbol of success in this as yet poorly defined field. Success in producing such elaborate molecular arrays will depend on understanding and controlling the topological and geometric relationships between molecules to an extent competitive with nature's manipulation of biopolymers. Structural, mechanistic, topological, and stereochemical relationships will be critical, and template effects will have a central role. A few simple beginnings are discussed here.

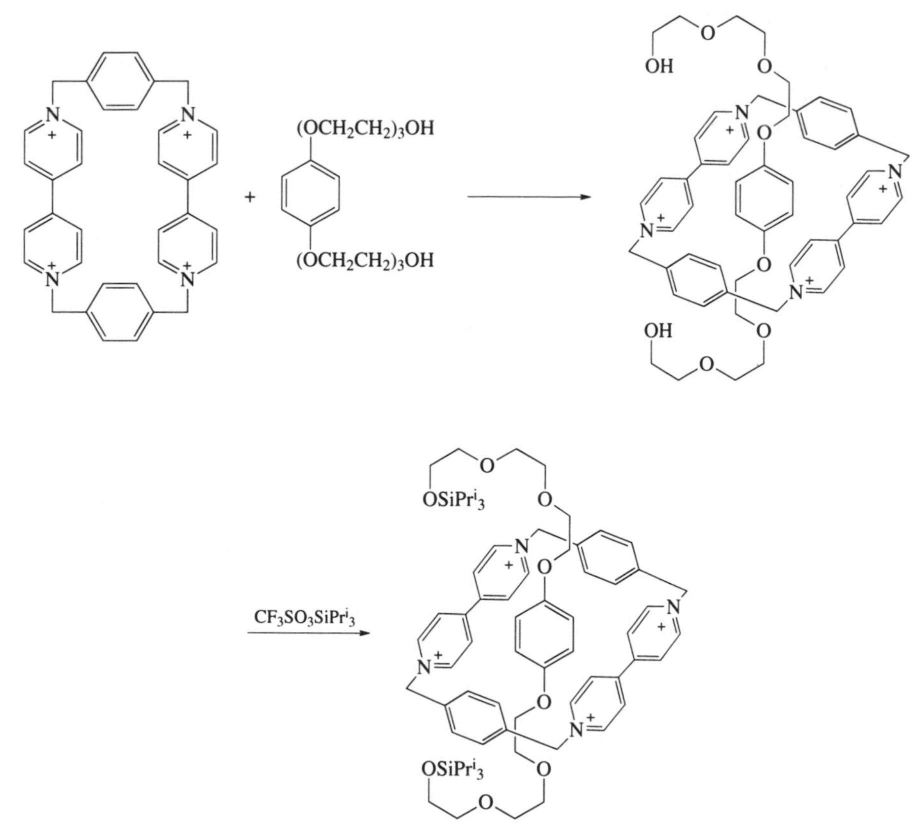

$CF_3SO_3SiPr^i_3$

Scheme 20

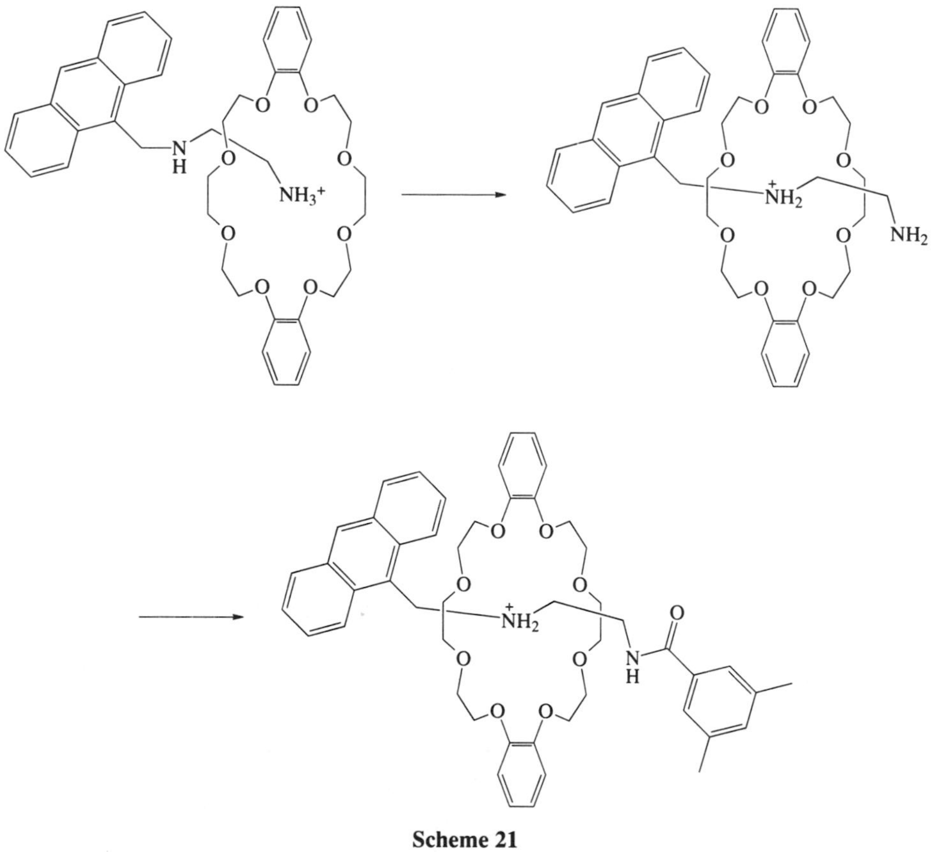

Scheme 21

1.4.2 Elements for Templates and the Stereocontrol of Reactions

In the preceding summaries, two kinds of molecular templates have been described: kinetic and thermodynamic. In the discussion to follow, kinetic templates are our main interest because they are particularly useful in designing for the purpose of controlling chemical processes. The molecular templates described here share two kinds of elements: units to be linked, and anchors to orient those units properly. The word "anchor" is a particularly convenient one to describe this function of the templating agent. Anchors are among the critical components that will be necessary during the synthesis of orderly molecular entanglements. As shown here, a variety of anchoring forces and kinds of anchors participate in template processes. Some anchors are robust and serve as constant cores during the building of a molecular architecture. Others are weak and provide a fleeting orientation during single molecular events. As more intricate and more extended molecular architectures are mastered, the appropriate kinds of anchors will be recognized and used. To repeat from the contents of Section 1.2, molecular anchors include metal and metalloid atoms and ions, hydrogen-bonded groups, electrostatic binding between molecular entities, π-stacking between aromatic rings, and unions brought about by still weaker hydrophobic and van der Waals interactions, and by combinations from this list.

The design of the units to be linked is critical to the success of template reactions, and the detail required in this design will increase in importance as the molecular architectural goals of chemists expand. The most common unit shape is that of a "turn." This is obvious in the great success of the 2,9-disubstituted-1,10-phenanthrolines reported by Sauvage and Dietrich-Buchecker (Structure (**49**)). The cartoons given by these authors, showing the actions of their templates (Figure 6), dramatize how the interlocking (at their centers) of pairs of these turns, using their only anchor (the tetrahedral copper(I) ion), greatly favors the achievement of their topological goals. The pattern is so compelling that the molecule seems destined to ring close with the formation of a catenane or a knot; however, the yields are not very good and the topologically less constrained pathways still compete. Very early in template chemistry, the advantage of using turns for both reactants was shown. In the seminal Thompson–Busch macrocyclization reaction (Structures (**1**) and (**2**)), when the organic dibromide reacting with the *cis*-bound sulfur nucleophiles was itself a turn (i.e., α,α′-dibromo-*o*-xylene), only a single product was detectable in solution,[15] while the corresponding flexible reagent (1,4-dibromobutane) gave a mixture of products.

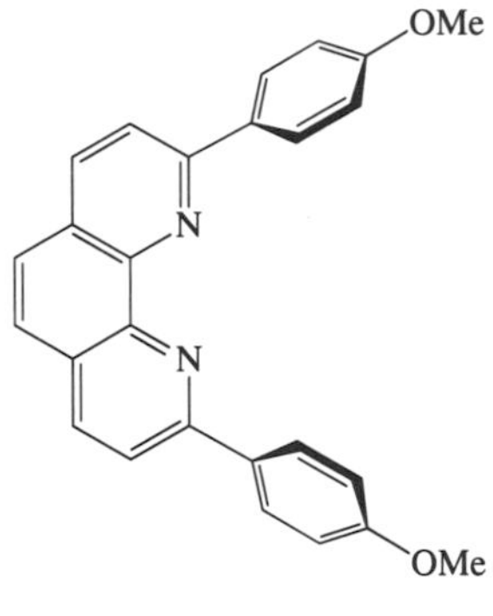

OMe

N

N

OMe

(**49**)

Examples of turns like α,α′-dibromo-*o*-xylene are common in chemistry, but those like the 2,9-disubstituted 1,10-phenanthrolines are not. Rebek has exercised elegant stereocontrol using molecular turns (Structure (**50**)) derived from Kemp's acid (Structure (**51**)).[37] The potential of molecular turns in template processes has not yet been fully exploited. Vance has designed a convenient molecular turn, specifically for use with octahedral metal ions.[108] Pairs of tridentate ligands bound to an octahedron bear the same geometrical relationships to each other and topological implications (Figure 7) as pairs of didentate ligands bound to a tetrahedral ion. The families of ligands are illustrated by Structure (**52**) and the crystal structure of the first such metal template is shown in Structure (**53**). Ease of synthesis and variety of available metal ions are advantages of these new molecular turns.

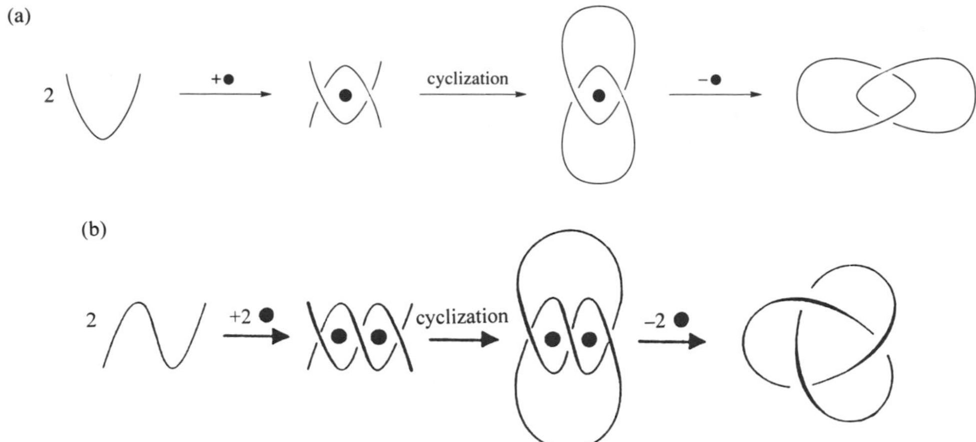

Figure 6 Formation of (a) catenane and (b) knot molecules.

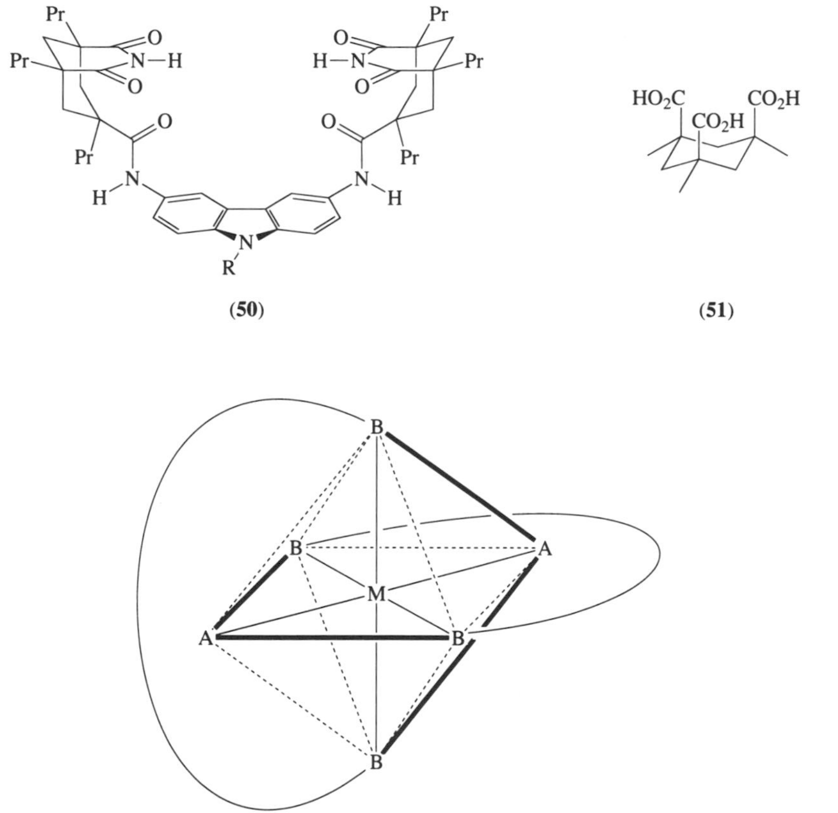

Figure 7 Topology of pairs of tridentate ligands bound to an octahedral ion.

In Scheme 12, which summarizes the combining units in the template synthesis of bicyclic molecules, it is apparent that caps are dominant reactants. Geometrically, these capping reagents produce corners, in the usual sense of that word, but their topological achievements are conveniently described by referring to them as "double turns." As with single turns, in the absence of a requirement that the molecule both binds to a metal ion and contains overextending functional groups, double turns are not extremely rare. α,α',α''-Trisubstituted trimethylmethanes (Structure **(31)**, Equation (10)) and 1,3,5-trisubstituted cyclohexanes provide examples. In fact, the cups in Scheme 12 constitute the more rare case of double turns with internal metal ion binding sites, analogously to the Sauvage single turn.

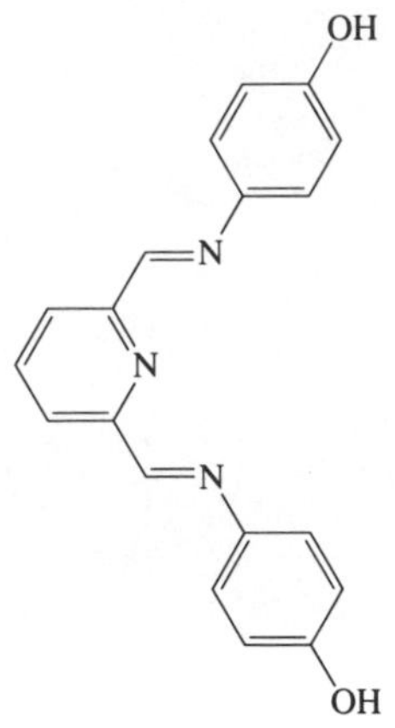

(52)

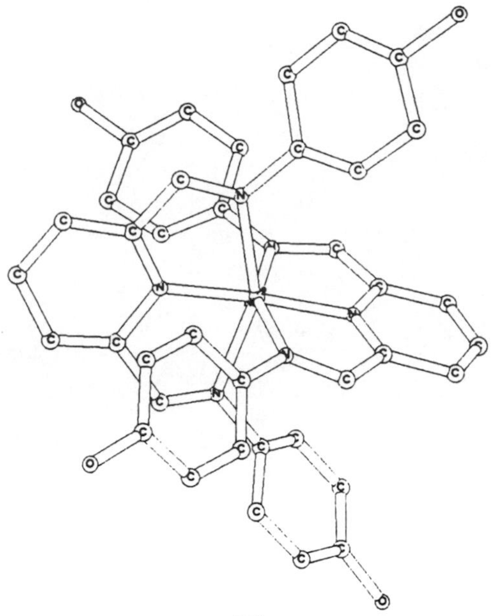

(53)

Additional elements that may prove critical in the use of templating principles in the design of orderly molecular entanglements include molecular crossovers and "divergent turns." The turns described in the previous section are convergent in the sense that the turns bend toward the anchor in the template. Divergent turns will bend away from the anchor; O'Brien has shown how divergent turns could provide an alternative route to the figure-of-eight doubly interlocked catenane (Structure (**54**)).[109]

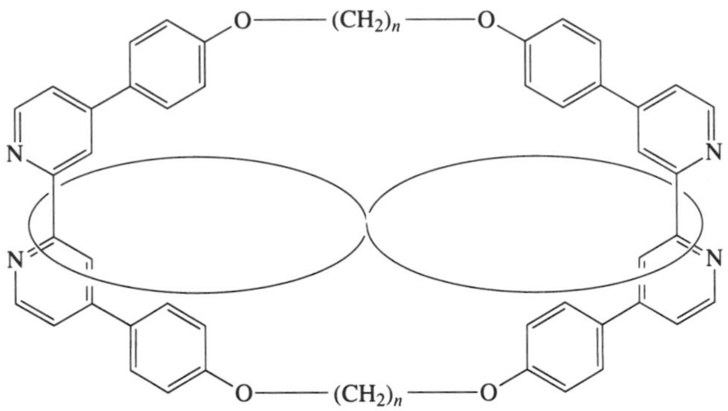

(54)

The relationships between single and double turns within a templating architecture, or multiple templating centers or processes, will be an unfolding story for some time. Protein chemists have many lessons to teach organic and inorganic chemists in this regard. As in the effects of linked turns on the helical content of peptides and proteins, propagation of the structural consequences of a turn will be very important in the development of extended orderly molecular entanglements. In the anchor–turn complex containing two turns bound to a single anchor, the angle between the centers of the turns (the turn–anchor–turn angle) and the relative torsional angles of the planes of the turns (turn plane–turn plane angle) (Figure 8) will be critical to the success of templates. In both Sauvage's and Vance's templates, the turns are arrayed at 180° with respect to each other, while the torsional angle between their planes is 90°. Consider the classic topological problem of three interlocking rings (Figure 9). Looking down the threefold axis in an octahedral template proposed to produce such a molecular motif, one sees that the turn–anchor–turn angle is 120°, but the turn plane–turn plane angle is still 90°. One anticipates increased difficulty in the use of templates where these angles are less favorable.

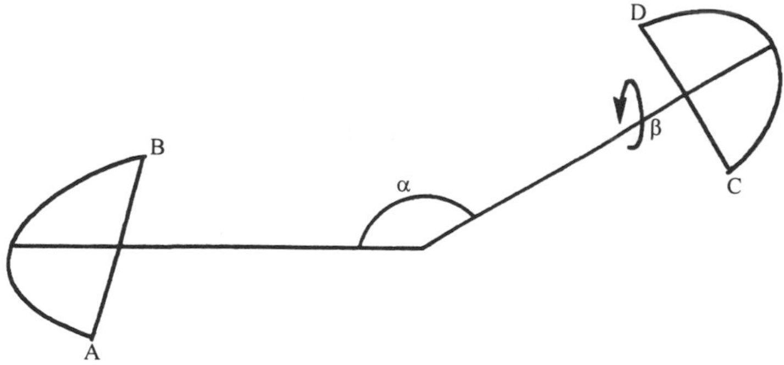

Figure 8 Illustration of the turn plane–turn plane angle.

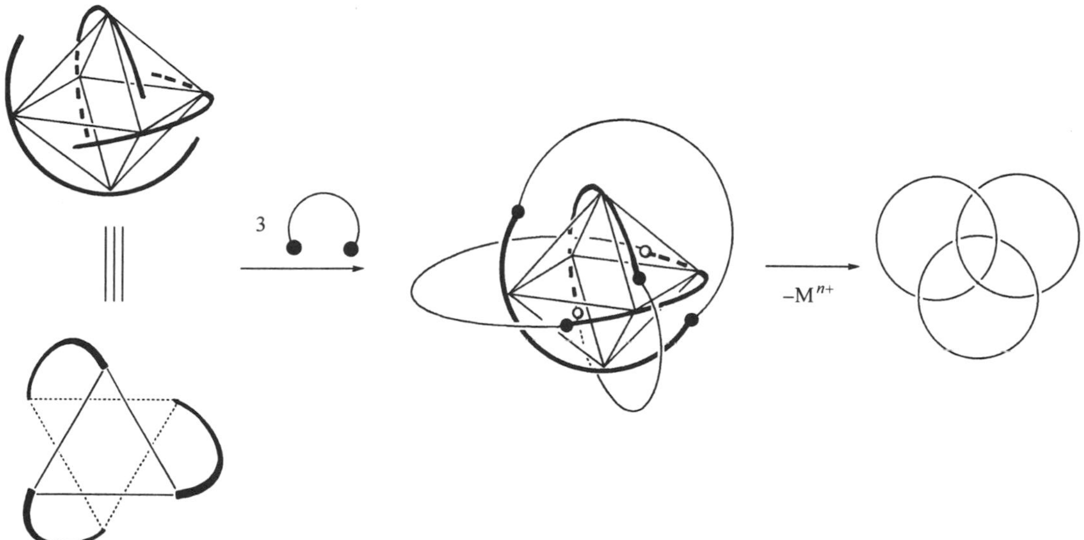

Figure 9 Topology of three interlocking rings.

Threading into a molecular void, while not an element of a template, is viewed as an elemental process critical to the ultimate routine production of materials based on orderly molecular entanglements. Catenanes and rotaxanes can, as described above, be made by threading processes, but in ultimate designs the threadings may be through loops created by crossovers within molecular strands, as in the tying of an overhand knot (Figure 10) or the formation of some cloth. Gaining template or template-like control over single and double turns, crossovers, and threading processes should be a high priority in the development of new molecular materials.

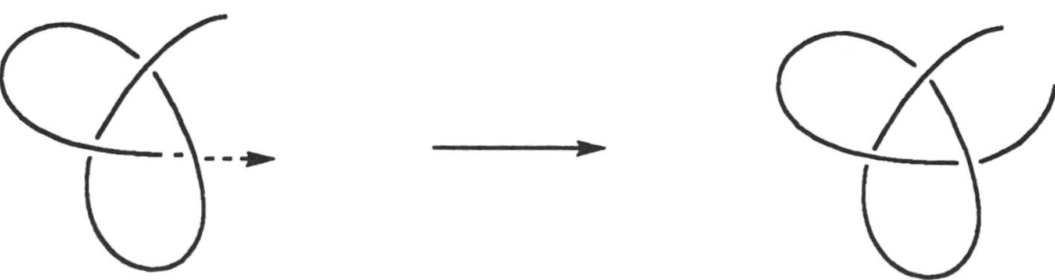

Figure 10 Tying of an overhand knot.

1.4.3 Orderly Molecular Entanglements

It may be exciting to envision, like an animated cartoon, a growing weave with so-called "living polymers" at its growing ends, and with each end moving predictably in space to produce a desired orderly entanglement, but it is not possible to do a relevant experiment today. Whereas synthesis of the ultimate orderly molecular entanglements is distant, many beginnings are possible. Most obvious among these are catenane chains, a subject already well under way, and mechanical cross-links. Molecular stitching and stapling are thinkable today as are the first class of materials that could be properly named "molecular cloth." Other investigators have also considered relatively complicated arrays.[110]

1.4.3.1 Mechanical cross-links

A knot between two molecules would be far more interesting than a knot within a molecule. The simplest image of such a process would involve a girth hitch. As shown in Figure 11, such a knot could be attained by a double threading through a single macrocycle, controlled by a templating molecular turn; a blocking reaction might be necessary to complete the rotaxane. Even simpler is the intermolecular snap, which comprises a rotaxane formed between two polymer molecules, one with a macrocycle on one end and the other acting as the axle component.

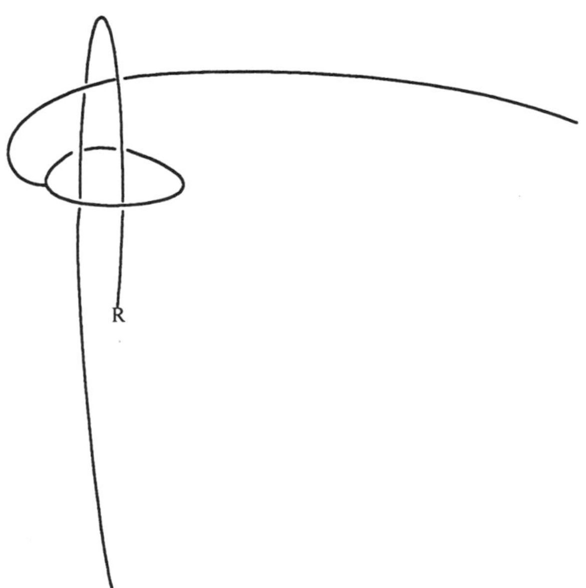

Figure 11 Girth hitch knot.

Most cross-links simply double the molecular weight of a polymer, but a molecular sheaf might multiply the molecular weight by a larger factor by enclosing a number of polymer molecules within a single loop, forming a multiply threaded rotaxane (Figure 12). Such molecular aggregates would be difficult to produce, but should be possible by combining template chemistry with mesophase chemistry.

1.4.3.2 Stitching and stapling molecules and molecular cloth

A double threading molecule opens the possibility of joining molecular species together by mechanical linkages. Such linkages would have two advantages that would greatly affect the structures of the products: the linkages would be very strong and they would be highly flexible. Figure 13(a) shows how the double threading molecule staples macrocycles together when the double rotaxane is formed. In Figure 13(b), it is suggested that the combination of threading and

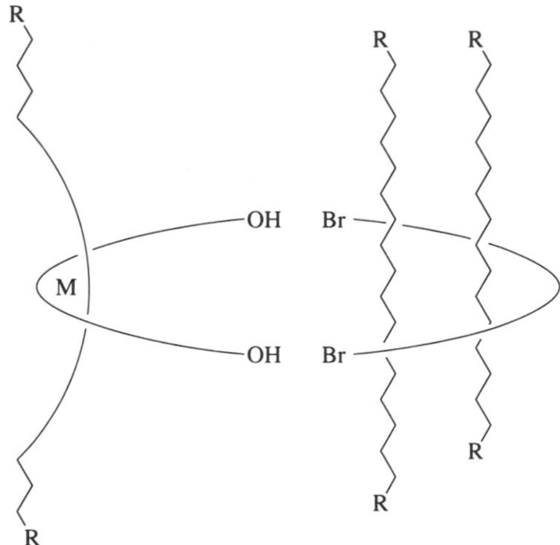

Figure 12 Multiply threaded rotaxanes.

reaction of a difunctional threading reagent with a second difunctional reagent might produce a molecular stitch, linking chains of macrocycles together. Figure 14 envisions the stitching together of chains of aligned mesophase molecules that contain macrocycles as parts of their structures. This final hypothetical product would be a legitimate example of a molecular cloth, an unknown kind of material. It might best be called a "molecular chain mail," as that early form of body armor linked its metal cloth together with rings.

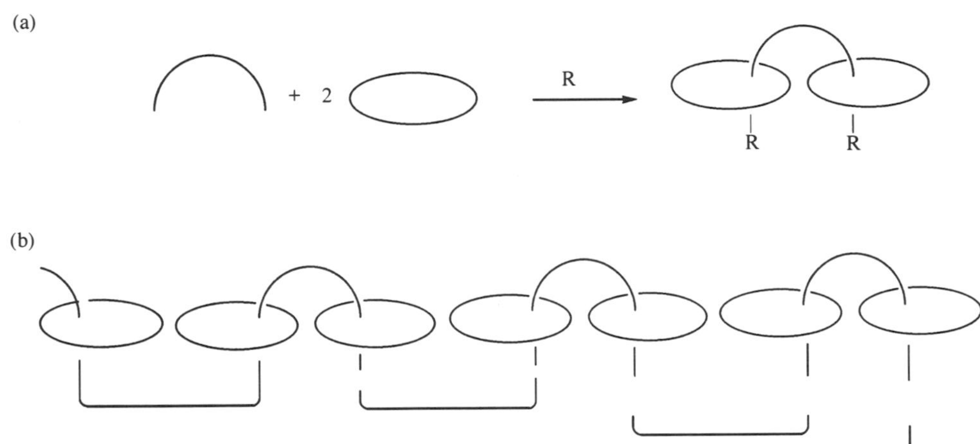

Figure 13 (a) Stapling together of macrocycles to form a double rotaxane; (b) stitching together of chains of macrocycles.

1.4.3.3 Toward learning to knit and weave

True scientific frontiers always challenge the imagination. If it is easy to design experiments, then the research is low risk and the probable results are of corresponding value. What seems impossible, even absurd, today will be routine tomorrow. A molecular cloth will have properties that are astounding. A major difficulty will be stiffness, but very loose mechanical linkages may give some such materials remarkable flexibility. A continuous macroscopic film of such a flexible molecular cloth, of monolayer thickness, may have unbelievable strength. Chemical bonds must break to rupture such a material, but a great deal of conformational rearrangement will be possible before the breakage is forced.

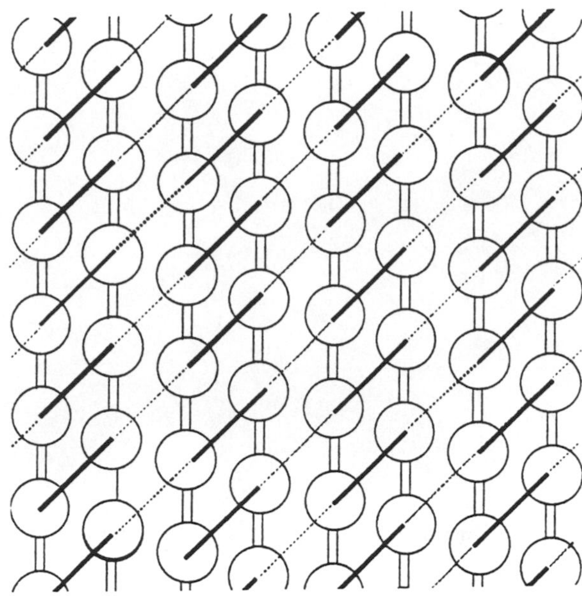

Figure 14 Stitching together of chains of aligned mesophase molecules.

1.5 REFERENCES

1. D. H. Busch, *J. Inclusion Phenom.*, 1992, **12**, 389.
2. M. C. Thompson and D. H. Busch, *J. Am. Chem. Soc.*, 1964, **86**, 213.
3. S. Anderson, H. L. Anderson, and J. K. M. Sanders, *Acc. Chem. Res.*, 1993, **26**, 469.
4. R. Hoss and F. Vögtle, *Angew. Chem., Int. Ed. Engl.*, 1994, **33**, 375.
5. (a) N. V. Gerbeleu, "Reactions on Matrices," Akademiya Nauk Moldavskoi SSR, Kishinev, 1980 (in Russian) (*Chem. Abstr.*, **93**, 237953w); (b) N. V. Gerbeleu and V. B. Arion, "Template Synthesis of Macrocyclic Compounds," Shtiintsa, Kishinev, 1990 (in Russian) (*Chem. Abstr.*, **114**, 163215k).
6. B. M. Trost, *Pure Appl. Chem.*, 1981, **53**, 2357.
7. C. R. Martin, *Acc. Chem. Res.*, 1995, **28**, 61.
8. J. S. Beck, J. C. Vartuli, W. J. Roth, M. E. Leonowicz, C. T. Kresge, K. D. Schmitt, C. T-W. Chu, D. H. Olson, E. W. Sheppard, S. B. McCullen, J. B. Higgins, and J. L. Schlenker, *J. Am. Chem. Soc.*, 1992, **114**, 10834.
9. V. A. Kabanov, in "Polymerization in Organized Media," ed. C. M. Paleos, Gordon and Breach, Philadelphia, PA, 1992, pp. 369–454.
10. J. S. Lindsey, *New. J. Chem.*, 1991, **15**, 153.
11. (a) S. Doherty, J. F. Corrigan, A. J. Carty, and E. Sappa, *Adv. Organomet. Chem.*, 1995, **37**, 39; (b) G. Deganello, "Transition Metal Complexes of Cyclic Polyolefins," Academic Press, London, 1979.
12. F. W. Lichtenthaler, *Angew. Chem., Int. Ed. Engl.*, 1994, **33**, 2364.
13. D. H. Busch, *Chem. Rev.*, 1993, **93**, 847.
14. W. P. Schammel, L. L. Zimmer, and D. H. Busch, *Inorg. Chem.*, 1980, **19**, 3159.
15. (a) C&EN, p. 57 Sept. 17, 1962; (b) M. C. Thompson and D. H. Busch, *J. Am. Chem. Soc.*, 1964, **86**, 3651; (c) E. L. Blinn and D. H. Busch, *Inorg. Chem.*, 1968, **7**, 820.
16. (a) M. C. Thompson and D. H. Busch, *J. Am. Chem. Soc.*, 1962, **84**, 1762; (b) J. D. Curry and D. H. Busch, *J. Am. Chem. Soc.*, 1964, **86**, 592; (c) G. A. Melson and D. H. Busch, *J. Am. Chem. Soc.*, 1964, **86**, 4834; (d) G. A. Melson and D. H. Busch, *J. Am. Chem. Soc.*, 1965, **87**, 1706.
17. (a) G. A. Melson and D. H. Busch, *J. Am. Chem. Soc.*, 1964, **86**, 4830; (b) G. R. Brubaker and D. H. Busch, *Inorg. Chem.*, 1966, **5**, 2114; (c) D. H. Busch, *Adv. Chem. Ser.*, 1966, **62**, 616.
18. (a) D. H. Busch, *Adv. Chem. Ser.*, 1963, **37**, 1; (b) D. H. Busch, *Rec. Chem. Prog.*, 1964, **257**, 107; (c) D. H. Busch, *Helv. Chim. Acta*, Fasc. extraord. Alfred Werner, 1967, 174.
19. (a) N. F. Curtis and D. A. House, *Chem. Ind.*, 1961, 1708; (b) N. F. Curtis, *Coord. Chem. Rev.*, 1968, **3**, 3.
20. N. F. Curtis and R. W. Hay, *J. Chem. Soc., Chem. Commun.*, 1966, 524.
21. K. B. Yatsimirskii, V. V. Pavlishchuk, A. G. Kolchinski, G. V. Filonenko, and I. V. Voloshina, *Dokl. Akad. Nauk SSSR*, 1985, **281**, 1384.
22. (a) T. Posner, *Berichte*, 1898, 656; (b) F. Seidl, *Chem. Ber.*, 1926, **59B**, 1894.
23. H. de Diesbach and E. von der Weid, *Helv. Chim. Acta*, 1927, **10**, 886.
24. (a) F. Hein and W. Retter, *Berichte*, 1928, **61**, 1790; (b) G. T. Morgan and F. H. Burstall, *J. Chem. Soc.*, 1932, 20.
25. G. Nilsson, *Acta Chem. Scand.*, 1950, **4**, 205.
26. (a) G. R. Clark, B. W. Skelton, and T. N. Waters, *J. Chem. Soc., Dalton Trans.*, 1976, 1528; (b) W. E. Keyes, J. B. R. Dunn, and T. M. Loehr, *J. Am. Chem. Soc.*, 1977, **99**, 4527.
27. F. Feigl, *Chem. Ztg.*, 1919, **163**, 30.
28. D. R. Boston and N. J. Rose, *J. Am. Chem. Soc.*, 1968, **90**, 6859.
29. H. Christopherson and E. B. Saudell, *Anal. Chim. Acta*, 1954, **10**, 1.
30. (a) G. N. Schrauzer, *Chem. Ber.*, 1962, **95**, 1438; (b) D. Thierig and F. Umland, *Angew. Chem.*, 1962, **74**, 388.

31. D. H. Busch and N. A. Stephenson, *Coord. Chem. Rev.*, 1990, **100**, 119.
32. (a) T. J. McMurray, K. N. Raymond, and P. H. Smith, *Science*, 1989, **244**, 938; (b) A. M. Sargeson, *Pure Appl. Chem.*, 1986, **58**, 1986.
33. (a) J.-P. Sauvage, *Acc. Chem. Res.*, 1990, **23**, 319; (b) J.-C. Chambron, C. O. Dietrich-Buchecker, V. Heitz, J.-F. Nierengarten, J.-P. Sauvage, C. Pascard, and J. Guilhem, *Pure Appl. Chem.*, 1995, **67**, 233.
34. A. Shanzer, J. Libman, and F. Frolow, *Acc. Chem. Res.*, 1983, **16**, 601.
35. D. B. Amabilino and J. F. Stoddart, *Chem. Rev.*, 1995, **95**, 2725.
36. G. M. Whitesides, E. E. Simanek, J. P. Mathias, C. T. Seto, D. N. Chin, M. Mammen, and D. M. Gordon, *Acc. Chem. Res.*, 1995, **28**, 37.
37. J. Pieters, I. Huc, and J. Rebek, Jr., *Tetrahedron*, 1995, **51**, 485.
38. E. A. Wintner, M. M. Conn, and J. Rebek, Jr., *Acc. Chem. Res.*, 1994, **27**, 198.
39. F. M. Menger, A. V. Eliseev, N. A. Khanjin, and M. J. Sherrod, *J. Org. Chem.*, 1995, **60**, 2870.
40. M. Fujita, S. Nagao, and K. Ogura, *J. Am. Chem. Soc.*, 1995, **117**, 1649.
41. A. Müller, *J. Mol. Struct.*, 1994, **325**, 13.
42. S. O. Wandiga, J. E. Sarueski, and F. L. Urbach, *Inorg. Chem.*, 1972, **11**, 1349.
43. O. Carugo and C. Bisi-Castellani, *Monatsh. Chem.*, 1994, **125**, 647.
44. P. M. Angus, B. T. Golding, S. S. Jurisson, A. M. Sargeson, and A. C. Willis, *Austr. J. Chem.*, 1994, **47**, 501.
45. S. M. Nelson, C. V. Knox, M. McCann, and M. G. B. Drew, *J. Chem. Soc., Dalton Trans.*, 1981, 1669.
46. (a) D. C. Liles, M. McPartlin, and P. A. Tasker, *J. Am. Chem. Soc.*, 1977, **99**, 7704; (b) L. F. Lindoy and D. H. Busch, *Inorg. Chem.*, 1974, **13**, 2494.
47. (a) S. Venkatachalam and V. N. Krishnamurthy, *Ind. J. Chem.*, 1994, **33A**, 506; (b) E. H. Mørkved, S. M. Neset, H. Kjøsen, G. Hvistendahl, and F. Mo, *Acta Chem. Scand.*, 1994, **48**, 912.
48. (a) A. Meller and A. Ossko, *Monatsh. Chem.*, 1972, **103**, 150; (b) S. Dabak, A. Gül, and Ö. Bekaroğlu, *Chem. Ber.*, 1994, **127**, 2009 and references therein; (c) M. Hanack and M. Geyer, *J. Chem. Soc., Chem. Commun.*, 1994, 2253, and references therein.
49. T. J. Marks and D. R. Stojakovic, *J. Am. Chem. Soc.*, 1978, **100**, 1695.
50. J. G. Yound and W. Onyebuagu, *J. Org. Chem.*, 1990, **55**, 2155 and references therein.
51. (a) S. G. McGeachin, *Can. J. Chem.*, 1966, 2323; (b) J. S. Skuratowicz, I. L. Madden, and D. H. Busch, *Inorg. Chem.*, 1977, **16**, 1721.
52. (a) A. M. Tait and D. H. Busch, *Inorg. Synth.*, Wiley, 1978, **18**, 30 and references therein; (b) A. Reuveni and V. Malatesta, *Can. J. Chem.*, 1977, **55**, 70; (c) M. D. Timken, R. I. Sheldon, W. G. Rohly, and K. B. Mertes, *J. Am. Chem. Soc.*, 1980, **102**, 4716.
53. G. Howley and E. L. Blinn, *Inorg. Chem.*, 1975, **14**, 2865.
54. L. T. Taylor, S. C. Vergez, and D. H. Busch, *J. Am. Chem. Soc.*, 1966, **88**, 3170.
55. R. N. Greene, *Tetrahedron Lett.*, 1972, 1793.
56. G. W. Gokel and S. H. Korzeniowski, "Macrocyclic Polyether Synthesis," Springer, Berlin, 1982.
57. J. E. Richman and T. J. Atkins, *J. Am. Chem. Soc.*, 1974, **96**, 2268.
58. R. Cacciapaglia and L. Mandolini, *Chem. Soc. Rev.*, 1993, **22**, 221.
59. D. N. Reinhoudt, R. T. Gray, C. J. Smit, and M. I. Veenstra, *Tetrahedron*, 1976, **32**, 1161.
60. K. Madan and D. J. Cram, *J. Chem. Soc., Chem. Commun.*, 1975, 427.
61. Y. Habata, F. Fujishiro, and S. Akabori, *J. Chem. Soc., Chem. Commun.*, 1994, 2217.
62. (a) J. J. H. Edema, J. Buter, F. S. Schoonbeck, R. M. Kellogg, F. van Bolhuis, and A. L. Spek, *Inorg. Chem.*, 1994, **33**, 2448; (b) J. J. H. Edema, J. Buter, and R. M. Kellogg, *Tetrahedron*, 1994, **50**, 2095 and references therein.
63. (a) A. Osbrowicki, E. Koepp, and F. Vögtle, *Top. Curr. Chem.*, 1991, **161**, 37; (b) C. Galli, *Org. Prep. Proc. Int.*, 1992, **24**, 285.
64. (a) V. Alexander, *Chem. Rev.* 1995, **95**, 273; (b) S. M. Nelson, *Pure Appl. Chem.*, 1980, **52**, 2461; (c) P. Guerriero, S. Tamburini, and P. A. Vigato, *Coord. Chem. Rev.*, 1995, **139**, 17.
65. N. A. Bailey, D. E. Fenton, I. T. Jackson, R. Moody, and C. R. de Barbarin, *J. Chem. Soc., Chem. Commun.*, 1983, 1463.
66. (a) M. Bell, A. J. Edwards, B. F. Hoskins, E. H. Kachab, and R. Robson, *J. Am. Chem. Soc.*, 1989, **111**, 3603; (b) B. F. Hoskins, R. Robson, and P. Smith, *J. Chem. Soc., Chem. Commun.*, 1990, 488.
67. S. M. Nelson, F. S. Esho, M. G. B. Drew, and P. Bird, *J. Chem. Soc., Chem. Commun.*, 1979, 1035.
68. K. Motoda, H. Sakiyama, N. Matsumoto, H. Okawa, and S. Kida, *Bull. Chem. Soc. Jpn.*, 1992, **65**(4), 1176.
69. D. H. Busch, *Pure Appl. Chem.*, 1980, **52**, 2477.
70. G. A. Lawrance, M. Maeder, and E. N. Wilkes, *Rev. Inorg. Chem.*, 1993, **13**, 199.
71. N. A. Kostromina, Y. Z. Voloshin, and A. Yu. Nazarenko, "Clathro-chelates: Synthesis, Structure, Properties," Naukova Dumka, Kiev, 1990 (in Russian).
72. (a) J. G. Muller and K. J. Takeuchi, *Polyhedron*, 1989, **8**, 1391; (b) Y. Z. Voloshin and V. V. Trachevskii, *J. Coord. Chem.*, 1994, **31**, 147 and references therein.
73. Y. Z. Voloshin, A. Yu. Nazarenko, and V. V. Trachevskii, *Ukr. Khim. Zh.*, 1985, **24**, 121.
74. K. A. Lance, K. A. Goldsby, and D. H. Busch, *Inorg. Chem.*, 1990, **29**, 4537.
75. V. E. Zavodnik, V. K. Belsky, and Y. Z. Voloshin, *J. Coord. Chem.*, 1993, **28**, 97.
76. A. Hohn, R. J. Geue, and A. M. Sargeson, *J. Chem. Soc., Chem. Commun.*, 1990, 1473 and references therein.
77. R. J. Geue, M. McDonnel, A. W. H. Mau, A. M. Sargeson, and A. C. Willis, *J. Chem. Soc., Chem. Commun.*, 1994, 667.
78. R. J. Geue, A. Hohn, S. F. Ralph, A. M. Sargeson, and A. C. Willis, *J. Chem. Soc., Chem. Commun.*, 1994, 1513.
79. S. Utsuno, R. Miyamoto, and B. E. Douglas, *Inorg. Chim. Acta*, 1987, **129**, 199.
80. (a) H. A. Boucher, G. A. Lawrance, P. A. Lay, A. M. Sargeson, A. M. Bond, D. F. Sangster, and J. C. Sullivan, *J. Am. Chem. Soc.*, 1983, **105**, 4652; (b) J. M. Harrowfield, A. J. Herlt, P. A. Lay, and A. M. Sargeson, *J. Am. Chem. Soc.*, 1983, **105**, 5503; (c) T. Ramasami, J. F. Endicott, and G. R. Brubaker, *J. Phys. Chem.*, 1983, **87**, 5057; (d) M. P. Suh, W. Shiu, D. Kim, and S. Kim, *Inorg. Chem.*, 1984, **23**, 618.
81. M. P. Suh, S. Kim, S-J. Cho, and W. Shin, *J. Chem. Soc., Dalton. Trans.*, 1994, 2765 and references therein.
82. E. Jäger, *Z. Chem.*, 1968, **8**, 392.
83. D. H. Busch and N. W. Alcock, *Chem. Rev.*, 1994, **94**, 585.

84. K. E. Krakowiak, J. S. Bradshaw, X. Kou, and N. K. Dalley, *Tetrahedron*, 1995, **51**, 1599 and references therein.
85. R. Dagani, *C&EN*, 1994, **72**(35), 28.
86. *New J. Chem.*, 1993, **17**(10/11).
87. E. Wasserman, *J. Am. Chem. Soc.*, 1960, **82**, 4433.
88. H. L. Frisch and E. Wasserman, *J. Am. Chem. Soc.*, 1961, **83**, 3789.
89. (a) J.-F. Nierengarten, C. O. Dietrich-Buchecker, and J.-P. Sauvage, *J. Am. Chem. Soc.*, 1994, **116**, 375; (b) J.-P. Sauvage and M. Ward, *Inorg. Chem.*, 1991, **30**, 3869.
90. C. Piguet, G. Bernardinelli, A. F. Williams, and B. Bocquet, *Angew. Chem., Int. Ed. Engl.*, 1995, **34**(5).
91. D. B. Amabilino, P. R. Ashton, A. S. Reder, N. Spencer, and J. F. Stoddart, *Angew. Chem., Int. Ed. Engl.*, 1994, **33**(12), 1286.
92. D. Armspach, P. R. Ashton, C. P. Moore, N. Spencer, J. F. Stoddart, T. J. Wear, and D. J. Williams, *Angew. Chem., Int. Ed. Engl.*, 1993, **32**(6), 854.
93. T. Lu, L. Zhang, G. W. Gokel, and A. E. Kaifer, *J. Am. Chem. Soc.*, 1993, **115**, 2542.
94. M. J. Gunter, D. C. R. Hockless, M. R. Johnston, B. W. Skelton, and A. H. White, *J. Am. Chem. Soc.*, 1994, **116**, 4810.
95. F. Vögtle, W. M. Müller, U. Müller, M. Bauer, and K. Rissanen, *Angew. Chem., Int. Ed. Engl.*, 1993, **32**(9), 1295.
96. F. J. Carver, C. A. Hunter, and R. J. Shannon, *J. Chem. Soc., Chem. Commun.*, 1994, 1277.
97. F. Vögtle, S. Meier, and R. Hoss, *Angew. Chem.*, 1992, **104**(12), 1628.
98. G.-J. M. Gruter, J. J. de Kanter, P. R. Markies, T. Nomoto, O. S. Akkerman, and F. Bickelhaupt, *J. Am. Chem. Soc.*, 1993, **115**, 12179.
99. M. Fujita, F. Ibukuro, K. Yamaguchi, and K. Ogura, *J. Am. Chem. Soc.*, 1995, **117**, 4175.
100. P. R. Ashton, M. Grognuz, A. M. Z. Slawin, J. F. Stoddart, and D. J. Williams, *Tetrahedron Lett.*, 1991, **32**(43), 6235.
101. P. R. Markies, T. Nomoto, O. S. Akkerman, F. Bickelhaupt, W. J. J. Smeets, and A. L. Spek, *J. Am. Chem. Soc.*, 1988, **110**, 4845.
102. H. W. Gibson, M. C. Bheda, and P. T. Engen, *Prog. Polym. Sci.*, 1994, **19**, 843.
103. C. Wu, P. R. Lecavalier, Y. X. Shen, and H. W. Gibson, *Chem. Mater.*, 1991, **3**, 569.
104. (a) J.-C. Chambron, V. Heitz, and J.-P. Sauvage, *J. Am. Chem. Soc.*, 1993, **115**, 12378, and references therein; (b) F. Diederich, C. O. Dietrich-Buchecker, J.-F. Nierengarten, and J.-P. Sauvage, *J. Chem. Soc., Chem. Commun.*, 1995, 781.
105. H. Sleiman, P. Baxter, J.-M. Lehn, and K. Rissanen, *J. Chem Soc., Chem. Commun.*, 1995, 715.
106. P. L. Anelli, P. R. Ashton, R. Ballardini, V. Balzani, M. Delgado, M. T. Gandolfi, T. T. Goodnow, A. E. Kaifer, D. Philip, M. Pietraszkiewicz, L. Prodi, M. V. Reddington, A. M. Z. Slawin, N. Spencer, J. F. Stoddart, C. Vicent, and D. J. Williams, *J. Am. Chem. Soc.*, 1992, **114**, 193 and references therein.
107. A. G. Kolchinski, D. H. Busch, and N. W. Alcock, *J. Chem. Soc., Chem. Commun.*, 1995, 1289.
108. A. L. Vance, N. W. Alcock, and D. H. Busch, work in progress, University of Kansas, 1995.
109. J. G. O'Brien, communication, University of Kansas, 1994.
110. G. Karagounis, I. Pandi-Agathokli, E. Kontaraki, and D. Nikolelis, *Prakt. Acad Athenon*, 1975, **49**, 501.

2

Transition Metals as Assembling and Templating Species: Synthesis of Catenanes and Molecular Knots

JEAN-CLAUDE CHAMBRON, CHRISTIANE DIETRICH-BUCHECKER and JEAN-PIERRE SAUVAGE
Université Louis Pasteur, Strasbourg, France

2.1 INTRODUCTION

Creating novel chemical species at the molecular level can be motivated by various reasons. For instance, the synthesis of the compound to be made is a challenging problem or the properties of the target molecule are expected to be of special interest. The synthesis of interlocking ring molecular systems and knots combines both sets of motivation but it also adds an aesthetical dimension to the chemical problem. The search for aesthetically attractive molecules has been a goal since the very origin of chemistry. What are the criteria which will lead the observer to find a molecule "beautiful" or not? Very often, symmetrical shapes analogous to the traditional geometrical objects (tetrahedron, dodecahedron, etc.) have been considered as the ultimate goal by prestigious laboratories. Dodecahedrane[1] is an example whose importance stems both from the amount of work involved and from the cultural significance of the geometrical shape reached. This attractive molecule represents the culmination of considerable effort, and it remains today as one of the most difficult patterns synthesized. Transition metals can form molecular clusters whose arrangements are particularly beautiful. In particular, platinum carbonyl clusters lead to geometrical figures whose regularity and harmony were completely unexpected during the 1970s.[2] The edifices obtained are somewhat reminiscent of modern architectural works.

However, symmetry is not a compulsory element of beauty. A remarkable example is that of the molecular arrangement found by biologists and crystallographers in the photosynthetic reaction center from the bacteria *Rhodopseudomonas viridis*.[3] Elucidation of the three-dimensional structure of this reaction center, although a long and sometimes discouraging process for the researchers involved in this adventure,[4] led to a marvellous view of the molecular edifice. It displays only limited symmetry (a very approximate C_2 axis). However, the beauty of the whole arrangement is undeniable. It rests not only on the overall morphology of the system but also on the fundamental and wonderful nature that the photosynthetic process represents for life on earth.

The aesthetic aspect of any object is usually connected to its shape in Euclidian geometry: the object is conveniently represented by points and lines, the metric properties (length of a segment, angles, etc.) being of utmost importance. In this case, the object cannot be put out of shape. However, another interesting facet of beauty rests in the topological properties of the object. Among the most fascinating objects displaying nontrivial topological properties, interlaced design and knots occupy a special position.

In pagan and early religious symbolism, interlocked and knotted rings represent continuity and eternity. They occupied a privileged position in the art of the most ancient civilizations. Developed by Egyptians, Persians, and Greeks among others, this virtually universal art reached its zenith in Celtic culture. The magnificient illuminations consisting of extremely complex interlaced designs and knots of the *Book of Kells*[5] give evidence of the fascination that braids, wreaths, and knots exert on man. In this famous manuscript, the work of Irish monks during the eighth century, inextricable geometrical figures are converted into marvellous representations. Later superb interlaced designs were created by Albrecht Dürer and Leonardo da Vinci. Interlocked rings also contain symbolic significance related to the unbreakable link between the constitutive rings. The most intriguing example is probably that of the Borromean sign, symbol of this powerful Italian family. The three interlocked rings are arranged in such a way that by any cut, the three circles will fall apart, none being interlocked to another one without the help of the third component. Some views of interlaced design and knots are presented in Figure 1. Modern art has also devoted special attention to knotted threads. The Dutch artist Cornelius Escher[6] is sometimes inspired by fantastic monsters and animals associated with impossible geometrical figures. Many of his works contain volumes and interlaces so closely related to modern molecular science that he is one of the most popular artists among the community of chemists. The simplicity of his view of the trefoil knot makes it particularly attractive (Figure 1(c)).

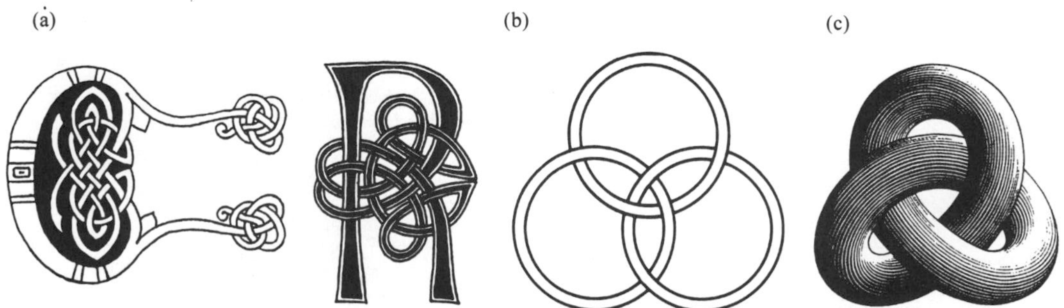

Figure 1 (a) Lettering examples taken from the *Book of Kells* (eighth century); (b) the Borromean symbol; (c) the trefoil knot, as viewed by Escher (source: Locher[6]).

In mathematics, knots and links occupy a special position. They have been the object of active thinking for more than a century. Interestingly, mathematical topology has recently undergone a revival of interest[7] and has been able to link other fields of scientific investigation such as physics, chemistry, and molecular biology. Any interested reader should look at *The Knot Book.*[8] In Figure 2 the four first prime knots are represented. These knots are single knotted loops, as opposed to links (or catenanes, in chemistry), which are sets of knotted or unknotted loops, all interlocked together.

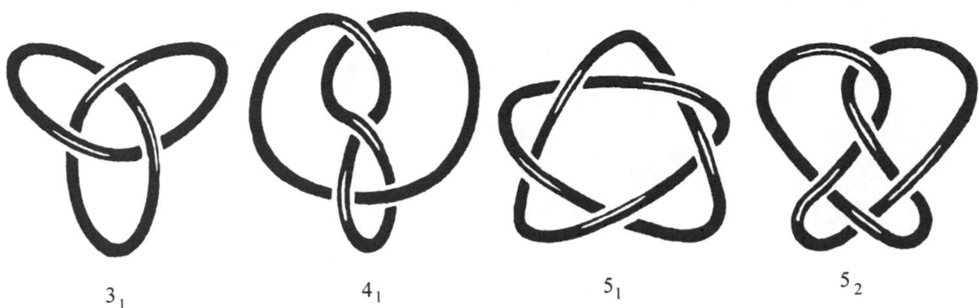

3_1 4_1 5_1 5_2

Figure 2 The first four prime knots.

The discovery that DNA forms catenanes and knots, some of them being of extreme complexity, initiated a new field of research which has been called "Biochemical Topology".[9] In 1967, Vinograd and co-workers detected in HeLa cell mitochondria "isolable DNA molecules that consist of independant, double stranded, closed circles that are topologically interlocked or catenated like the links in a chain".[10] A few years later, catenanes had been observed everywhere that circular DNA molecules were known[11] and the first knot was found by Liu *et al.* in single-stranded circular phage fd DNA treated with *Escherichia coli* ω-protein.[12] In 1980, knots could also be generated in double-stranded circular DNA.[13]

Catenanes and knots discovered in naturally occurring DNA or generated *in vitro* are molecules with a nonplanar presentation. An appropriate way to analyze such molecules is to define the number, sign, and arrangement of the crossing points or nodes. As sedimentation analyzes, agarose gel electrophoresis, and electron microscopy enabled initial recognition of the topoisomers of DNA, it is possible today to provide an unambiguous topological characterization of knots and catenanes due to improved techniques. The introduction of denaturation bubbles[14] and the thickening of strands using the RecA[15] or T4UvsX[16] protein coating method allows visualization by electronic microscopy of the orientation of each ring in a catenane, as well as identification of DNA overlay and underlay at crossing points or nodes. Combination of both latter techniques leads today to reliable determination of the absolute handedness of DNA knots and catenanes. Figure 3 shows beautiful T4UvsX protein-coated trefoil knots and the electron micrograph of a RecA protein-coated figure-8 catenane.

The ever increasing interest in DNA catenanes and knots stems not just from their widespread occurrence or their topological novelty. Determination of their structures provides precious information about the biological processes which generate them. A whole class of enzymes effect these topological transformations perfectly: they are called topoisomerases. Type I topoisomerases, such as the ω-protein from the bacterium *Escherichia coli*,[17] cut a single strand of DNA, whereas type II topoisomerases, such as the famous bacterial enzyme gyrase discovered by Gellert *et al.*[18] or the T4 DNA topoisomerase,[13] cut both strands of the double helix. Both types of enzymes work the same way: to achieve topoisomerization reactions including relaxation–supercoiling, catenation–decatenation, or knotting–unknotting, they cut the DNA strands, pass segments through the reversible transient break, and reseal the cut ends. The possible role of such enzymes in a large variety of biological functions was, and still is, intensively studied. It is today commonly assumed that topoisomerases are able to solve the topological problems arising during replication, site-specific recombination and transcription of circular DNA.[13,16,19]

Interestingly, DNA is not the only biological molecular system to have the privilege of forming catenanes and knots. Liang and Mislow examined x-ray structures of many proteins and, to the surprise of many molecular chemists and biochemists, they found catenanes and even trefoil knots.[20] This remarkable finding addresses the general question of whether the topological properties of proteins have any biological significance. It can certainly be expected that many other proteins and, in particular, metalloenzymes, will be found to display nontrivial topological properties.

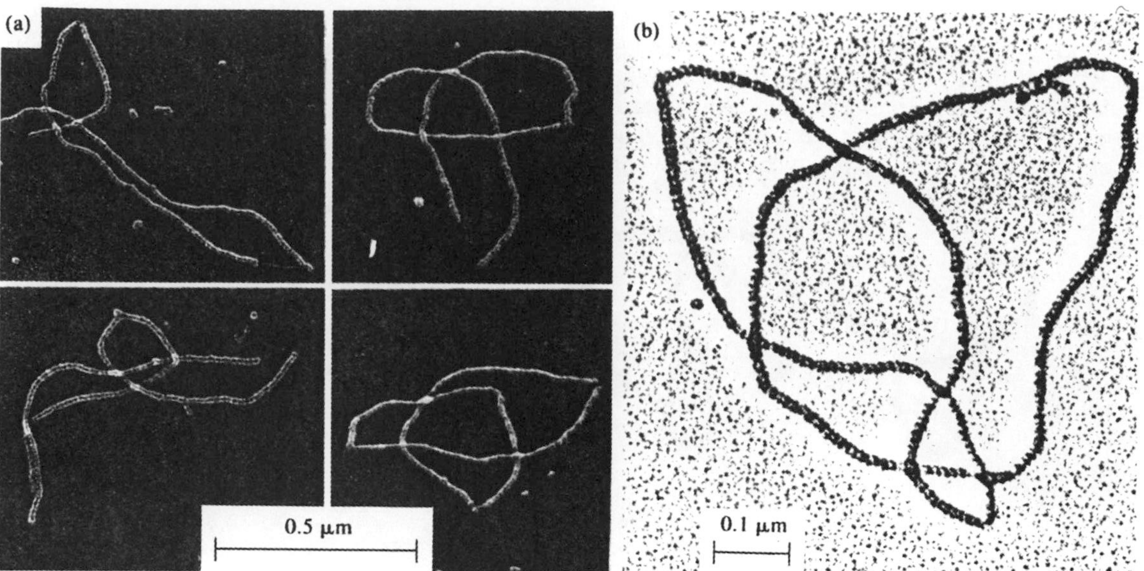

Figure 3 (a) Electron micrographs of trefoil DNA molecules (source: Griffith and Nash[16]). (b) Visualization of a figure-8 DNA catenane (reprinted by permission from *Nature*, 1983, **304**, 560 Copyright © 1983 Macmillan Magazines Limited).

Turning now to pure chemistry and synthetic molecular objects, it is clear that the concept of catenanes has fascinated chemists for several decades. Very intuitively, the topology of two separate rings is trivial, whereas that of the system consisting of the two same rings, but interlocked into one another (catenane) is of interest (Figure 4). Graph theory applied to chemistry has been called chemical topology. However, it must be stressed that it is only very recently that a rigorous mathematical treatment has been applied to molecules. Walba's use of graph theory for describing molecular systems now provides a unified view.[21]

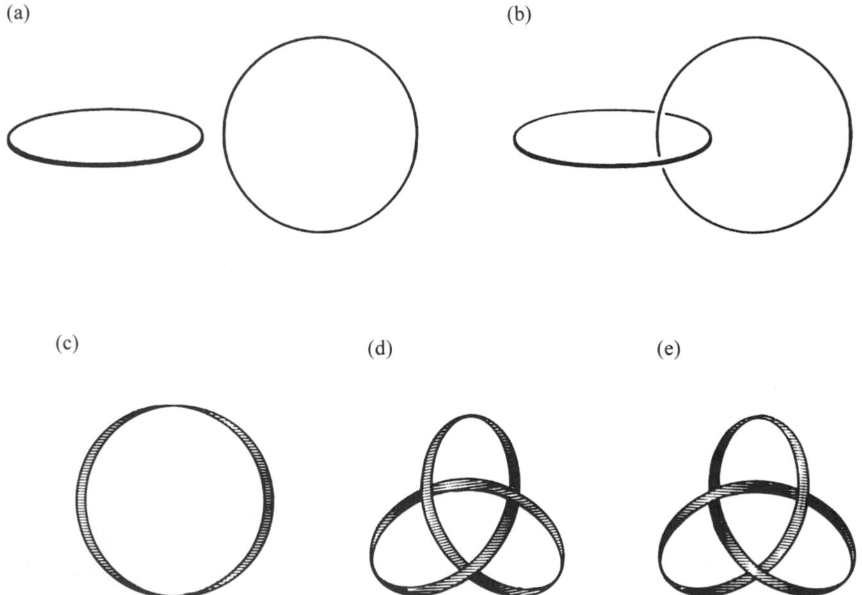

Figure 4 The set of two separate rings (a) and the catenane (b) are obviously distinct chemical species. The rings (c), (d), and (e) are topological stereoisomers; the two knots (d) and (e) are topological enantiomers.

Before being developed into rigorous theory, chemical topology received much interest from many scientists and several contributions of importance are worth mentioning.

Anyone who is interested in chemical topology must read the book written by Schill in 1971.[22] This is considered as the real Bible of the topologist. In addition to interesting theoretical considerations regarding interlocked rings (catenanes) and knots, it contains a wealth of information dealing with the numerous experimental approaches used by Lüttringhaus and, later, Schill and their co-workers in order to prepare such topologically novel systems. It is reported[23] that the possible existence of such compounds was already envisaged as early as the beginning of the twentieth century. As indicated by Prelog, Willstätter postulated in a seminar in Zürich that it should be possible one day to interlock rings. This was in fact a remarkable prediction.

The first theoretical written discussion to appear on chemical topology was a publication by Frisch and Wasserman.[24] This general article seems to be the foundation of the field and it contains, expressed in a very chemical and accessible language, most of the notions which constitute the background of chemical topology. For instance, the idea of topological isomers is introduced by comparing a [2]catenane (two interlocked rings) to the set consisting of the two separate molecules. Later, the expression topological stereoisomers was proposed.[21] It is best exemplified using a single closed curve: normal (topologically trivial) or knotted cycle (the most simple knot being the trefoil knot). The three objects of Figures 4(c)–(e) are topological stereoisomers: although they may consist of exactly the same atoms and chemical bonds connecting these atoms, they cannot be interconverted by any type of continuous deformation in three-dimensional space. In addition, the compounds of Figures 4(d) and 4(e) are topological enantiomers since they are mirror images.

Two other historically important discussions are worth mentioning. The first is a very imaginative paper written by Van Gülick at the beginning of the 1960s but, unfortunately, the manuscript was not accepted for publication at that time. It was recently published[25] in a special issue of the *New Journal of Chemistry* devoted to chemical topology and also containing many other contributions spanning from mathematical topology to polymers and DNA.[26] The second is a review by Sokolov,[27] which appeared in the Russian literature, and it is particularly relevant to the work discussed in this chapter since it contains interesting ideas regarding the possible use of transition metals to build catenanes and knots.

2.2 PREVIOUS SYNTHESES OF CATENANES: FROM PURELY STATISTICAL APPROACHES TO SCHILL'S DIRECTED SYNTHESIS

Although the existence of interlocked molecular rings was discussed, according to Schill,[23] as early as 1912 by Willstätter, it was only at the beginning of the 1960s that Frisch and Wasserman stated clearly by which routes the organic chemists could have access to such molecules:[24] "The formation of interlocking rings may be accomplished by the statistical threading of one ring by a linear molecule which is to be formed into the second ring. Such a procedure utilizes the probability that the first ring, when sufficiently large, will take on a conformation which permits the precursor of the second to pass through. Alternatively, the two rings may be constructed about a central core, a procedure which should give rise to much higher yields of interlocking rings." A third approach, which may be considered as a compromise between the pure statistical method and the directed synthesis around a central core, is the Möbius strip approach.

As will be shown below, all three approaches imply, at very different stages, the formation of macrocycles. It is interesting to note that high-yield macrocyclization methods became available a few years before Frisch and Wasserman conceived the different synthetical routes to catenanes. This coincidence is not a mere accident but confirms that the imagination of the chemist is dependent on the power of the synthetic tool.

2.2.1 Catenanes by Statistical Threading

As shown in Figure 5 this approach is based on a very simple principle. A molecular thread A–B, functionalized on both ends, may enter into a macrocycle of adequate size: subsequent cyclization of A–B leads necessarily to two interlocked rings. However, since the probability that cyclization occurs while the linear molecule A–B is threaded through the macrocycle is very small, one can only expect poor yields in this kind of synthesis. Despite this fact, the first [2]catenane ever made resulted from such a threading process. In 1960, Wasserman[28] demonstrated that the acycloin fraction obtained by acycloin condensation of diester (**1**) in the presence of partly deuteriated cyclic hydrocarbon (**2**) contained a small amount of catenane (**3**) (Scheme 1). Compound (**3**) could not be isolated, but the fact that the purified acycloin product still contained carbon–deuterium bonds (characteristic C—D stretches in IR) and that about 1% of the deuterated macrocycle was recovered in addition to diacid (**4**) after oxidative cleavage is strong evidence for its formation.

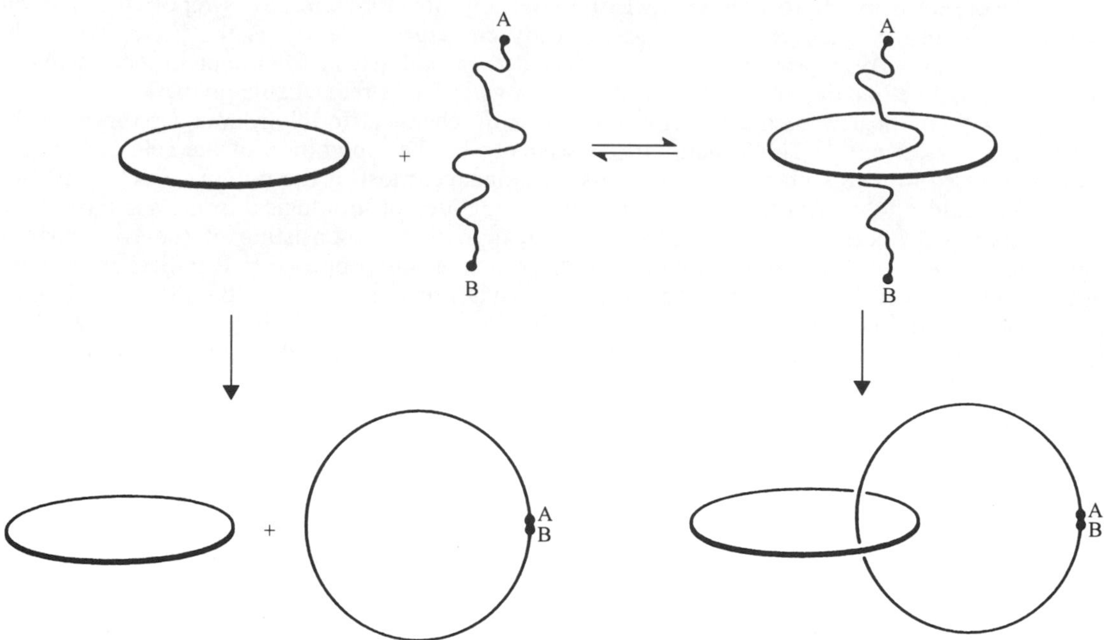

Figure 5 Catenane synthesis by statistical threading.

CO_2Et CO_2Et (**1**) + CD_2 CD_2 (**2**) $\xrightarrow[\text{Xylene}]{\text{Na}}$ (**3**) $\xrightarrow{H_2O_2/OH^-}$ (**2**) + (**4**)

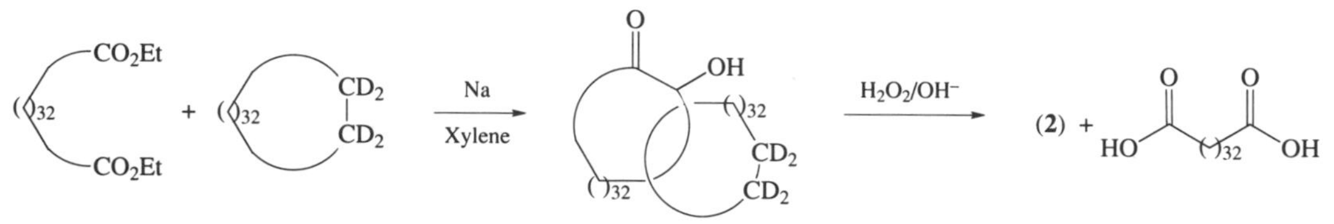

Scheme 1

Even if the previous example has no synthetic value, historically it is of great importance because it proved that Wasserman's statement was not mere speculation. From a preparative point of view, statistical threading became significant with the work of Zilkha and co-workers[29] (Scheme 2): rotaxane (**5**) obtained in 18.5% yield by statistical threading of a crown polyether by poly (ethylene glycol)400, leads to catenane (**6**) (14% yield) after cyclization in high-dilution conditions. Another double-stage catenane synthesis starting from a statistically formed rotaxane was described by Schill *et al.*[30]

Randomness, which is a highly limiting factor in the statistical syntheses of the catenanes discussed above, may become less determining if the threading process is favored by an even weaker interaction between the linear thread and the cycle. Such an attempt was described by Lüttringhaus *et al.*[31] as early as 1958. They discussed the formation of a rotaxane-like inclusion compound (**7**) between hydrophobic α- or β-cyclodextrin and an aromatic long-chain dithiol. Unfortunately, they could not achieve the expected ring closure that would have led to a catenane.

Bickelhaupt and co-workers[32] observed that the interaction between a magnesium diaryl-based macrocycle (**9**) and a crown ether (**8**) led to the equilibrium between the so-called side-on complex (**10**) and the catenane (**11**) shown in Scheme 3. The mechanism of catenane formation involves cleavage of a carbon–magnesium bond, complexation of the magnesium ion in the inner cavity of the macrocycle, and subsequent reformation of the carbon–magnesium bond.

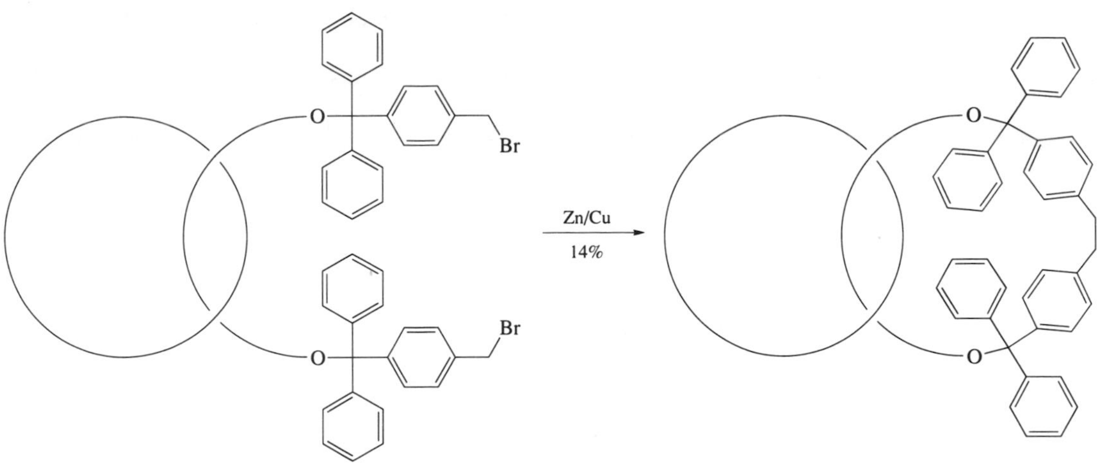

Zn/Cu

14%

(**5**) (**6**)

Scheme 2

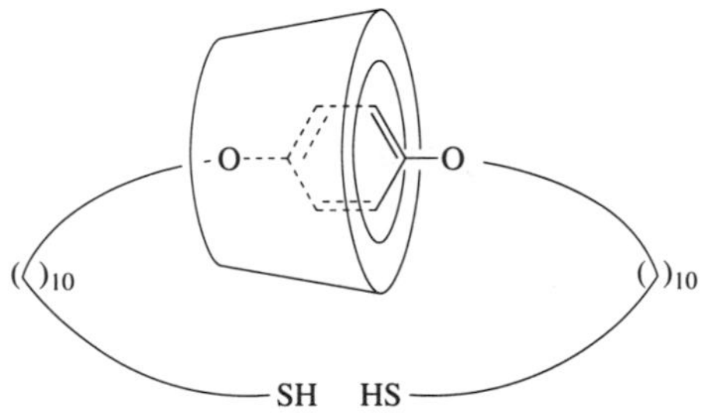

(**7**)

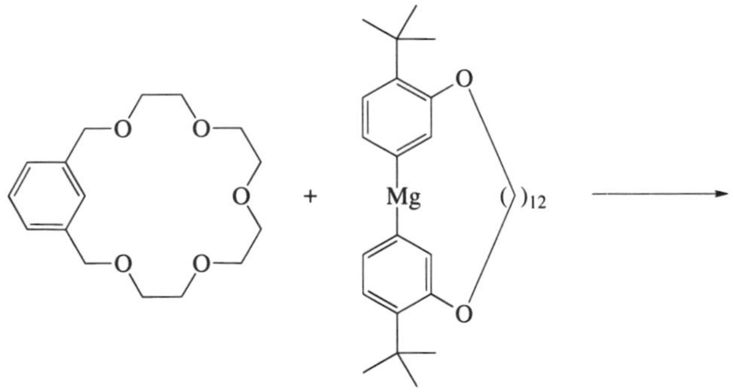

(**8**) (**9**)

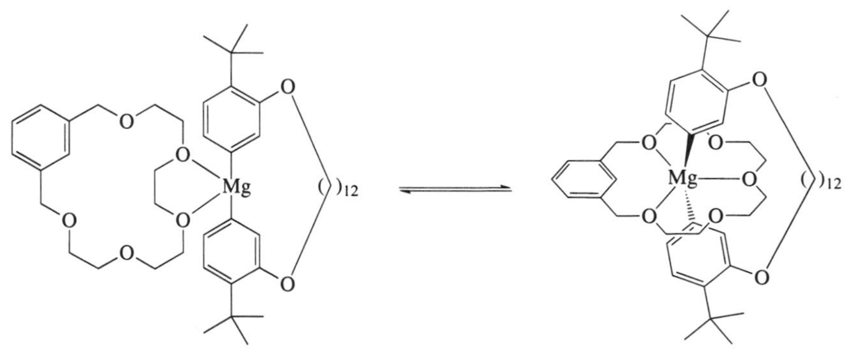

(**10**) (**11**)

Scheme 3

Attractive interactions between the linear thread and the cycle are at the basis of highly efficient synthetic strategies of rotaxanes and catenanes: the use of donor–acceptor aromatic interactions,[33] hydrogen bond formation, or hydrophobic effects have also been described.[34,35] These methods are not discussed here, since they are covered in Chapters 3 and 7.

Catenane formation, instead of occurring while a linear molecule is statistically threaded through a ring, may occur intramolecularly in a preshaped molecular edifice. Such is the conjecture expressed by the Möbius strip approach.

2.2.2 The Möbius Strip Approach

The Möbius strip approach, already considered by Frisch and Wasserman,[24] Van Gülick,[25] and Schill,[22] is based on a ladder-shaped molecule in which the ends are able to twist prior to bimacrocyclization. Scheme 4 shows that after cleavage of the vertical rungs one may have access to separate macrocycles or to a catenane, depending on the number ($n = 0, 1, 2, ...$) of half-twists occurring before double-ring closure.

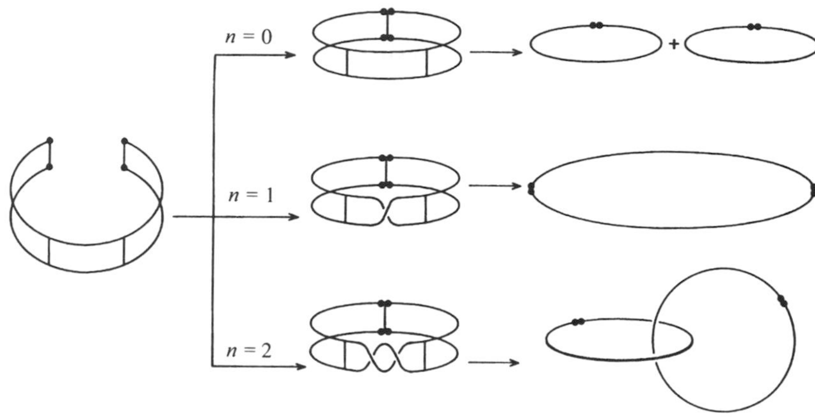

Scheme 4

Experimental work related to such an approach has been performed. Walba *et al.* achieved the synthesis of the first molecular Möbius strip in 22% yield.[36] Starting from tetrahydroxymethylethylene (THYME; (**12**)), they found that the intermediate "molecular ladder," (**13**), when submitted to cyclization, leads to an equimolecular mixture of the molecular cylinder (**14**) and its topological stereoisomer (**15**) (Scheme 5). The tris(THYME) cylinder (**14**) and the racemic Möbius strip (**15**) were readily separated by flash chromatography and their structures established unambiguously (single-crystal x-ray analysis for (**14**), ^{1}H and ^{13}C NMR spectra for (**15**)). In addition, it was possible to demonstrate that compound (**15**) was chiral: addition of the Pirkle chiral solvating reagent (+)-2,2,2-trifluoro-9-anthrylethanol allows NMR discrimination of the two enantiomers.

Although these remarkable results are good conjecture for a catenane synthesis, achievement of the latter still implies randomness: the statistical probability that the ends of a molecular ladder twist twice before cyclization, thus leading to the required single, full-twisted Möbius strip, remains quite small.

From the preceding discussion it clearly appears that both the pure statistical threading approach and the Möbius type approach are dependent on statistical probability. Randomness may be totally excluded from a synthesis in which the different subunits, precursors of the catenane, are gathered in a rigid molecular edifice so that cyclization can only occur in a predetermined way. Such a fundamentally different route to catenanes was successfully developed by Schill and Lüttringhaus.[37]

2.2.3 The Schill–Lüttringhaus Directed Catenane Synthesis

The elegant synthetic pathway conceived by Schill and Lüttringhaus is given in Scheme 6. In this multistep synthesis, which starts from 4,5-dimethoxyisophthaldehyde (**16**), compound (**18**) appears as a key intermediate: Euclidian geometry (bond angles, bond lengths, tetrahedral structure of the

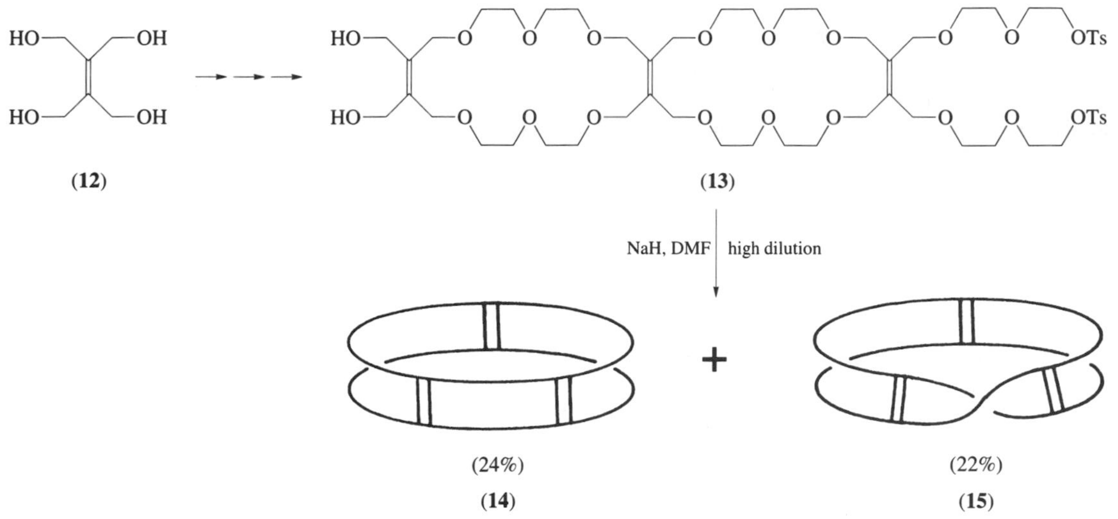

HO OH HO OH (12) → → → HO HO O O O O O O O O O O O O OTs OTs (13)

NaH, DMF high dilution

(24%) (14) + (22%) (15)

Scheme 5

ketal carbon atom, size of the polymethylene macrocycle, length of the alkyl chloride chains) imposes intramolecular macrocyclization in a unique and predetermined way. Alkylation of the amino group occurs only with the two alkyl chloride chains located one above and one below the plane of the central benzene ring. Selective cleavage of the aryl–nitrogen bond and hydrolysis of the ketal in triansa compound **(19)** leads, as expected, to [2]catenane **(20)**.

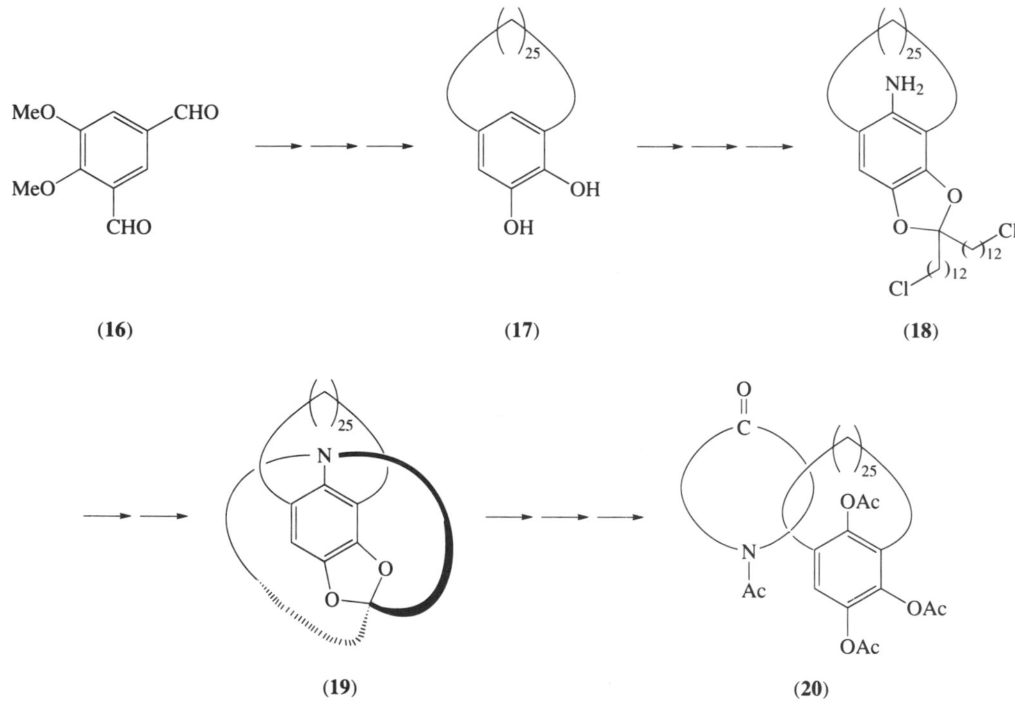

MeO MeO CHO CHO (16) → → → 25 OH OH (17) → → → 25 NH_2 O O Cl 12 12 Cl (18)

→ → 25 N O O (19) → → → O C 25 N Ac OAc OAc OAc (20)

Scheme 6

The pathway followed in the above-described syntheses ensures the formation of interlocked rings. Nevertheless, because of the numerous steps involved, the large-scale preparation of catenanes via the Schill–Lüttringhaus strategy remains a highly difficult task.

From what is shown above, Schill's directed synthesis of catenanes may be considered as relying on a central core, the benzo-ketal group, around which the different rings are built up. Another central core, in the way Frisch and Wasserman defined it,[24] could be a transition metal. Interestingly, such an approach, based on a template effect, was mentioned or evoked (the term template effect was not used in 1961 or 1972), twice in the past. In 1961, Frisch and Wasserman[24] stated "Another interesting suggestion has been that of Closson to utilize the geometry of the ligands about a metal as a core." In 1972, Sokolov reported[27] "Coordination compounds of a metal can be used as scaffolding for building up catenoid structures." Although only speculative, Sokolov's discussion is of great interest in chemical topology, presenting various routes to templated threading of molecules.

2.3 TEMPLATE SYNTHESIS OF CATENANES USING TRANSITION METALS

2.3.1 Template Synthesis: Principle and Examples

Transition metals, with their ability to gather and dispose ligands in a given predictable geometry, can induce what is generally called a template effect. This specific property of transition metals has been widely used to achieve the template syntheses of various single macrocycles,[39] one of the earliest examples probably being Reppe's cyclooctatetraene synthesis in which a nickel atom is supposed to bring together four ethyne molecules around it prior to cyclotetramerization.[40] Other well-known examples of early coordination template syntheses are given by the metallophthalocyanins preparations,[41] as well as by the copper(II) and nickel(II) aliphatic Schiff base complexes first obtained and recognized by Curtis and House.[42] Based on analogous condensation reactions between copper(II) or nickel(II) diamine complexes and aliphatic ketones or aldehydes, a great variety of tetradentate macrocyclic ligands have been subsequently synthesized.[43] Nevertheless, one may notice that the major part of the numerous template syntheses related in the literature involve transition metals bound to four coordination sites in a square-planar geometry and thus occur in two-dimensional space; in contrast, there are only a few syntheses which rely on the reaction between ligands prearranged around the metal in three-dimensional space. Such syntheses become possible, however, with transition metals which are hexacoordinated in a preferentially octahedral geometry. An early known and today almost archetypical complex suitable for a three-dimensional template synthesis appears to be the octahedral complex formed by cobalt(III) with three bidentate ethylenediamine-type ligands. A first beautiful illustration of the three-dimensional (or generalized) template effect induced by cobalt(III) is given by the results of Boston and Rose[44] who prepared the clathrochelate derived from dimethylglyoxime and boron trifluoride etherate. Soon after, analogous three-dimensional macrocyclic encapsulation reactions occurring between hexadentate tripode tris(2-aldoximo-6-pyridyl)phosphine and boron trifluoride or tetrafluoroborate were reported.[45] Another more recent example of the generalized template effect induced by cobalt(III) is given by the beautiful cobalt(III) sepulchrate synthesis performed by Sargeson and co-workers (Equation (1)).[46] The reaction between $[Co(en)_3]^{3+}$ **(21)**, HCHO, and NH_3 leads, in a single step, with 74% yield, to the macrobicyclic cobalt(III) sepulchrate complex **(22)** in which the ligand completely encapsulates the cobalt ion.

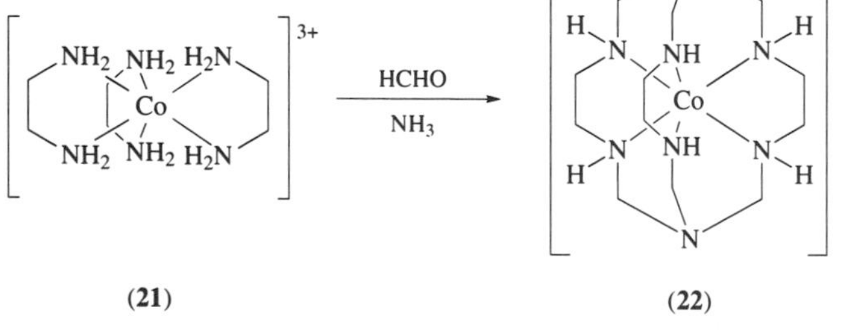

(21) $\xrightarrow[NH_3]{HCHO}$ **(22)** (1)

The various three-dimensional template syntheses mentioned above all rely on the same central core: octahedral cobalt(III) hexacoordinated to diamine chelates. This may be explained by the fact that such complexes, known for a long time, are easily accessible and very stable, the main requirements for a central core in a template synthesis. Nevertheless, recent literature reports prove that other transition metal complexes may also be perfectly suitable. Thus, iron(III) was successfully used as a template by Raymond and co-workers[47] in the high-yield synthesis (40%

overall yield) of a macrobicyclic catechoylamide ferric ion sequestering agent. The same macrobicycle synthesized by the conventional high-dilution technique is obtained in only 3.5%. Comparison between both yields highlights the essential role played by iron(III): it gathers and holds the three functionalized catechol ligands in a geometry favorable to the ultimate macrobicyclization step. Another elegant example of three-dimensional template synthesis using a metal ion other than cobalt(III) is given by the results of Von Zelewsky and co-workers,[48] who prepared the first closed cage ruthenium(II) complex from a ruthenium(II) tris(bipyridine)-like precursor.

2.3.2 Preparation of the Prototypical [2]Catenane

For our part, we envisaged the use of a generalized template effect in order to build up a catenane structure. 2,9-Dianisyl-1,10-phenanthroline forms, in the presence of copper(I), a very stable pseudotetrahedral complex in which the two ligands "fit-in" around the metallic center. Due to its very special topography, first indicated by NMR,[49] this complex appeared to be a perfect precursor or building-block for a templated catenane synthesis, as shown in Figure 6.

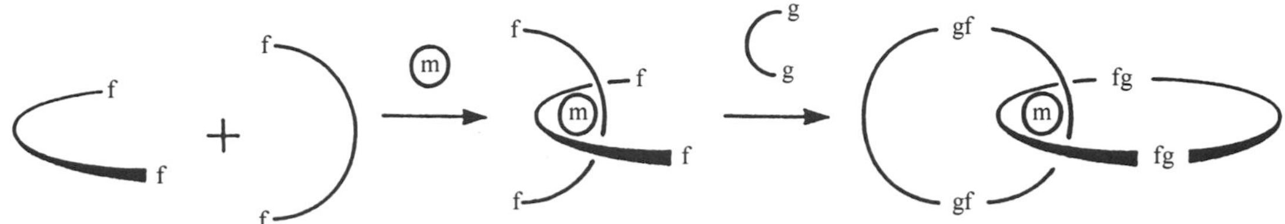

Figure 6 Three-dimensional template synthesis of interlocked ring systems induced by a transition metal. The metal center m disposes the two f–f fragments perpendicular to one another. Functions f and g react to form the links.

The functionalized ligand 2,9-bis(*p*-hydroxyphenyl)-1,10-phenanthroline (**23**), precursor of all our catenane syntheses, was prepared by the addition of lithioanisole to 1,10-phenanthroline, leading to 2,9-dianisyl-1,10-phenanthroline, which was subsequently deprotected by the pyridine hydrochloride procedure.[50] In the presence of $Cu(MeCN)_4{}^+ \cdot BF_4{}^-$, two (**23**) ligands fit together, forming the very stable copper(I) complex (**24**). The latter reacts with two g–g chains (with g–g being the diiodo derivative of pentaethylene glycol) under high-dilution conditions in the presence of a large excess of caesium carbonate in dimethylformamide (Scheme 7). By this very simple procedure, the expected copper(I) [2]catenate (**25**), was obtained in 27% yield. Gram-scale preparation of (**25**) could be performed using this very convenient one-pot synthesis. Finally, this complex could be quantitatively demetallated by treatment with potassium cyanide, affording the free ligand [2]catenand (**26**) (Equation (2)).[51]

2.3.3 Topologically Chiral Catenanes

Topological chirality of a molecule requires that its mirror image presentations be topologically distinct, that is, cannot be converted into one another by continuous deformation in three-dimensional space.[21] It must be stressed that [2]catenanes, despite the fact that they have a nonplanar molecular graph, are normally topologically achiral. However, [2]catenanes made from two directed rings become chiral, as shown in Figure 7.

In principle, orientation of a macrocycle may be achieved by several ways. Cyclic peptides[52] or macrocyclic lactones are both oriented molecular rings and thus could be suitable for the preparation of a chiral catenane. In the template synthesis of the first isolable chiral copper(I) catenate (**28**), the constituent oriented rings (**27**) were built up from an asymetrically substituted 1,10-phenanthroline.[53]

After workup, the racemic catenate was obtained in 12–15% yield. The chirality of (**28**) was demonstrated by ^1H NMR spectrocopy: in the presence of Pirkle's chiral reagent,[54] most of the signals corresponding to the aromatic protons were split. An analogous NMR resolution of topological enantiomers could not be observed after decomplexation of (**28**). However, if the chiral reagent is unable to recognize the two enantiomers of this highly flexible free ligand, topological chirality of the latter is nevertheless obvious from the nonsymmetrical nature of the starting

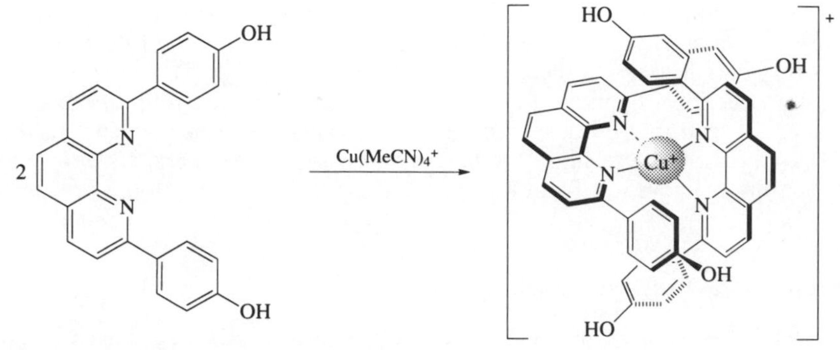

(23) (24)

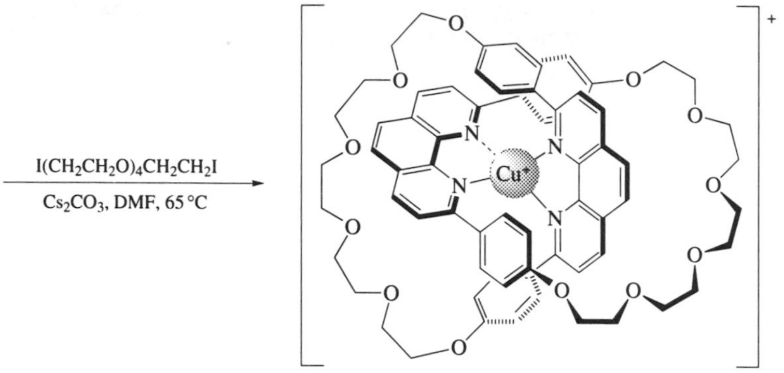

(25)

Scheme 7

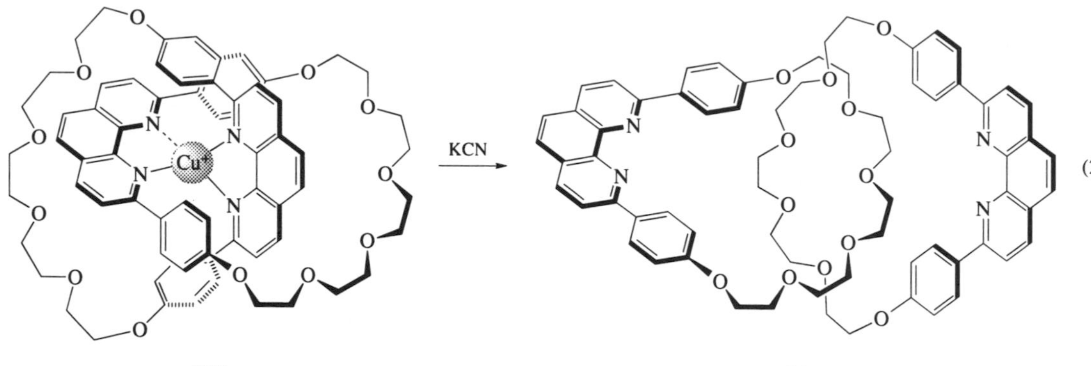

(2)

(25) (26)

compounds and from the synthetical pathway followed. Finally, enantiomerically enriched (up to 70%) optically active topologically chiral copper(I) catenate (**28**) could be obtained by HPLC of the racemate on amylose tris(3,5-dimethylphenylcarbamate) coated on silica gel.[55]

2.3.4 TTF- and Porphyrin-containing Catenanes

Catenates incorporating electro- or photoactive moieties at the ring periphery were prepared. Two examples are described in this section.

2.3.4.1 A TTF-containing catenane

The one-pot synthesis of catenates presented above is suitable for symmetrical catenates only. In the case of catenates made from two different macrocycles, a two-step strategy, requiring the preliminary and independant synthesis of one of the macrocycles, and shown schematically in Figure 8,

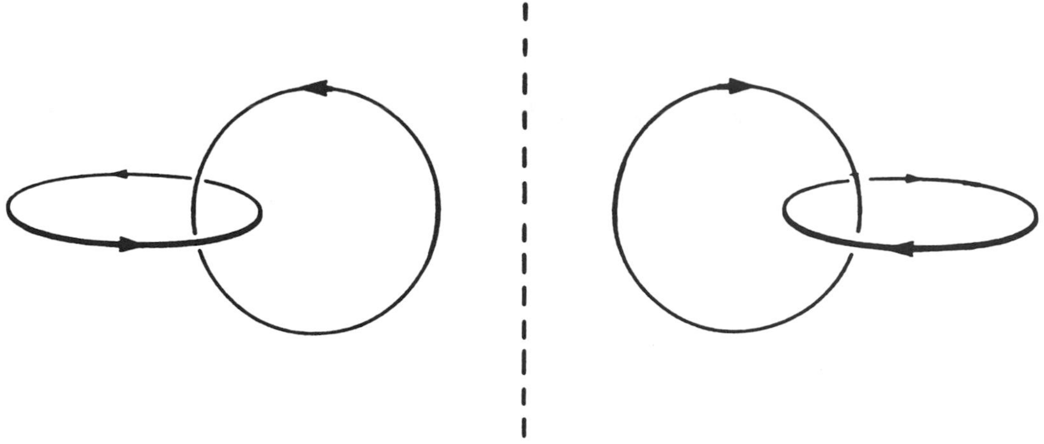

Figure 7 Topological chirality of an oriented [2]catenane.

(27)

(28)

has to be considered. In this manner the tetrathiafulvalene (TTF)-containing catenate (**30**) was prepared in order to combine the electrophoric properties of TTF and the chromophoric properties of the central copper(I) complex.[56]

This strategy has been applied to the synthesis of various nonsymmetrical [2]catenates and their corresponding [2]catenands.[57] In particular, the TTF-containing catenate (**30**) could be obtained in 14% yield, starting from a precatenate (**29**) made from the TTF-containing macrocycle (Equation (3)).

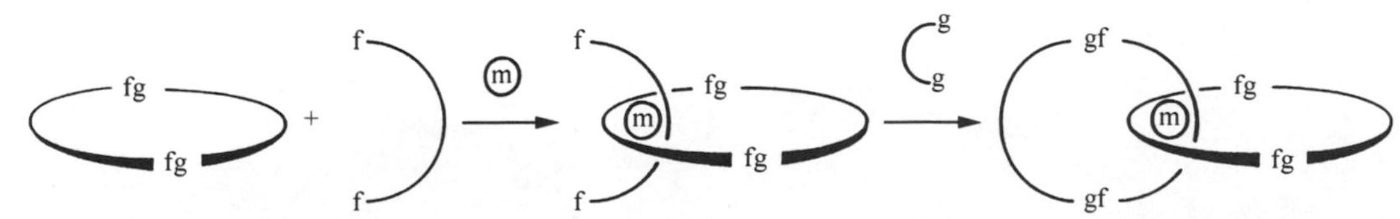

Figure 8 Preparation of a nonsymmetrical metallocatenate. Two-step synthetic strategy based on a three-dimensional template effect induced by a transition metal m. Metal m holds fragments f–f (linear or already included in a cycle) perpendicular to one another. Functions f and g react to form the links.

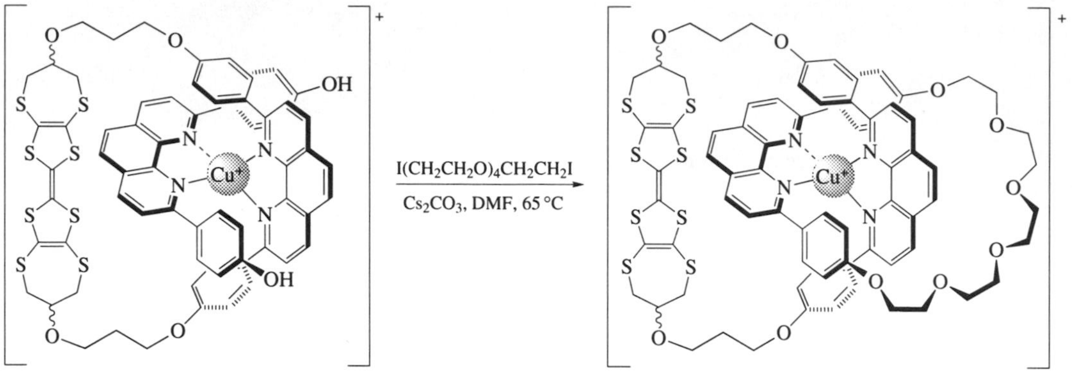

$I(CH_2CH_2O)_4CH_2CH_2I$

Cs_2CO_3, DMF, 65 °C

(3)

(29) **(30)**

2.3.4.2 A bis(porphyrin) [2]catenane

Catenane species incorporating porphyrin in one of the rings were first synthesized by Gunter and co-workers[58] by the "donor–acceptor stacks" strategy discussed in Chapter 3. The copper(I)-based template method was used by Momenteau *et al.* for making a symmetrical catenate incorporating two interlocked basket-handle porphyrins.[59] Despite its symmetrical nature, catenate **(31)** was synthesized by the two-step strategy outlined above, the last cyclization step occurring in 31% yield. It is noteworthy that the porphyrins were prepared prior to the catenation reaction. These strong metal binders were protected with zinc(II), from copper(II) possibly released by the pre-catenate species during the reaction.

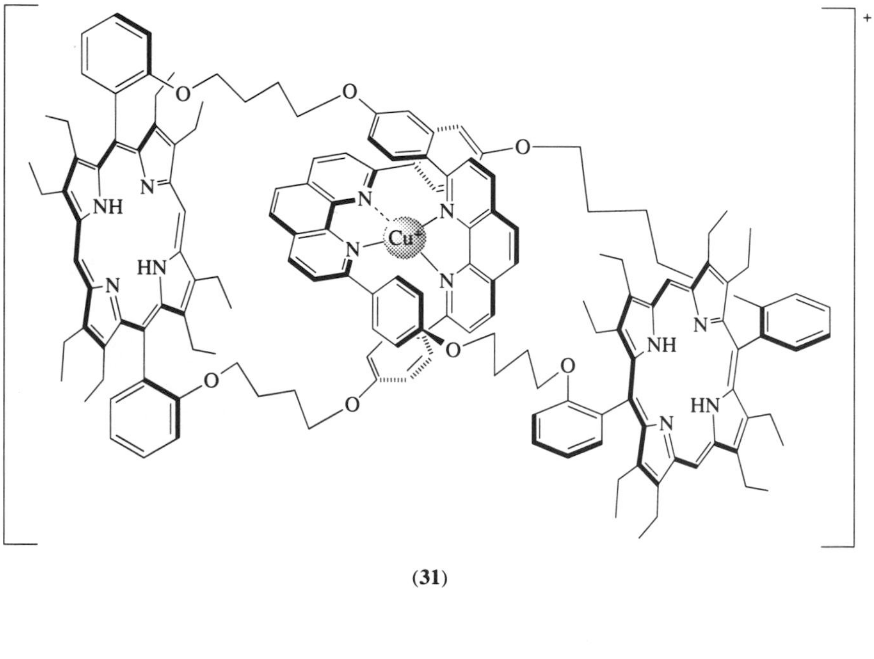

(31)

2.3.5 Multiring Interlocked Systems

A [2]catenane is the simplest element of the family of compounds formed by interlocked rings. The complexity of the systems and the number of possibilities increase rapidly with the number of rings to be interlocked. From several rings, the most straightforward combination is that of the rings arranged as the cyclic links of a chain, to form some type of unidimensional set. Clearly, many other possibilities exist, leading to very complex figures.

With three rings, there are numerous combinations. Some of the possible systems are the [3]catenane (chain-like) arrangements, which may simply be regarded as the sum of two [2]catenanes and thus be considered as a synthetically accessible system. The situation is very different for other interlocking ring systems, such as the Borromean sign. Although the synthesis of the Borromean ring system is still a challenging problem to chemists, its topological stereoisomer consisting of a chain of three interlocked rings ([3]catenane) was made by Schill and co-workers in 1970.[60]

More recently, Stoddart and co-workers made higher-order multi interlocked ring catenanes. The most impressive achievement in this field is the synthesis of a [5]catenane,[61] named olympiadane after the symbol of the Olympic games.

Since the strategy of using transition metals as templating species turned out to be efficient for preparing [2]catenanes, we tried to generalize the concept to the synthesis of [3]catenanes. We retained the copper(I) and 1,10-phenanthroline systems and simply considered a [3]catenane to be the sum of two [2]catenanes. In other words, the precursors used were virtually the same for making two- or three-interlocked ring systems. Two slightly different strategies have been developed.

2.3.5.1 The eight-reacting centers approach

The principle of the eight-reacting centers strategy is given in Figure 9. As it appears in the figure, dimerization leading to a [3]catenate may only occur if the linking fragment used in the cyclization step is too short to allow intramolecular ring formation. Such a prerequisite could be fulfilled by short chains such as the dibromo derivatives of tri- or tetraethylene glycol, (**34**) or (**35**). Reaction of precursor (**33**) (obtained from a 1:1:1 mixture of macrocycle (**32**), diphenol (**23**), and $Cu(MeCN)_4^+$) with (**34**) or (**35**) under high-dilution conditions, in the presence of Cs_2CO_3, affords poor yields of the expected dinuclear [3]catenates (**36**) (6%) or (**37**) (2%) as shown in Scheme 8.[62] Demetallation of (**36**) or (**37**) by KCN leads, respectively, to the [3]catenanes (**38**) (15%) or (**39**) (82%). Although obtained in low yield (a few per cent), the procedure is direct enough to be considered as preparative. For instance, catenand (**39**) (containing two peripheral 30-membered rings interlocked with a central 54-membered ring (Figure 10)) can be made on the 0.1 g scale, which is quite sufficient for performing various physical studies on the molecule. A detailed mass spectroscopy study was carried out on (**39**) and its homologue, which fully confirmed the molecular topology of the various catenanes prepared.[62]

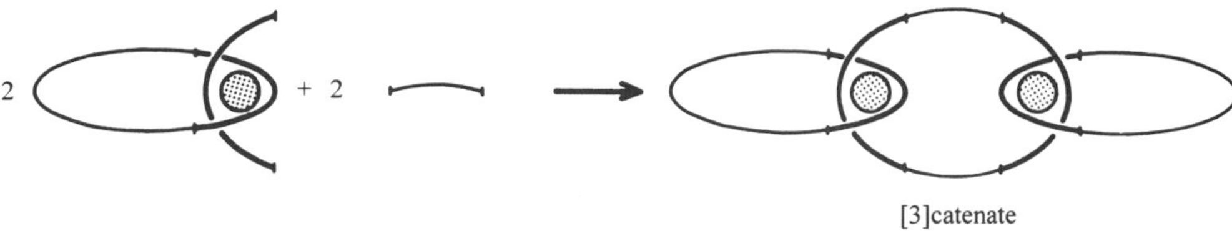

Figure 9 Principle of templated synthesis of a [3]catenate. The transition metal holds the two coordination fragments (thick line) perpendicular to one another. The cyclization step involves two additional short linear links (thin line) and requires the participation of eight reacting centers.

The conformational properties of (**39**) have been studied by ^{1}H NMR. At the same time, its crystal structure was determined by x-ray diffraction.[63] Interestingly, there is a remarkable agreement between the solution conformation of the molecule and its topography in the solid state. The intramolecular stacking interactions clearly shown by the molecular structure of Figure 10 are approximately kept in solution (CD_2Cl_2). In the crystal, the 1,10-phenanthroline subunits of the two lateral 30-membered rings are parallel and stacked with a short interplane distance (334 pm).

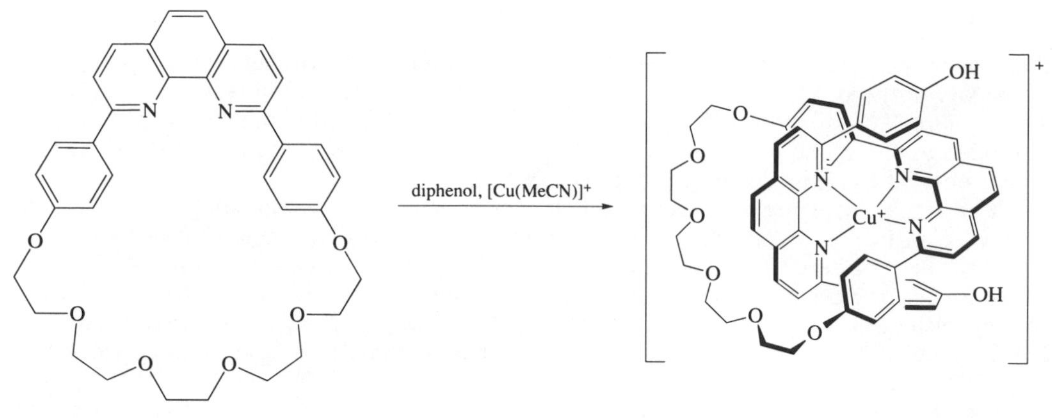

(32)

diphenol, $[Cu(MeCN)]^+$

(33)

Cs_2CO_3, DMF
$BrCH_2(CH_2OCH_2)_nCH_2Br$
(34) $n=2$
(35) $n=3$

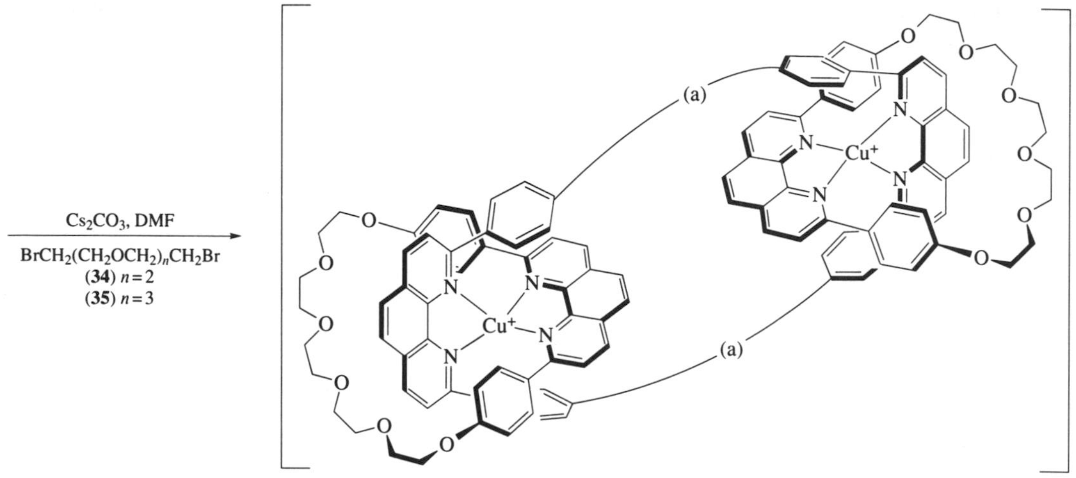

(36) (a) = $—O(CH_2CH_2O)_3—$
(37) (a) = $—O(CH_2CH_2O)_4—$

Scheme 8

The large central ring adopts a chair-like conformation and the $—(CH_2CH_2O)_3—$ unit folds, forming two molecular hollows with the 30-membered rings in which CH_2Cl_2 molecules (not represented on the figure) embed themselves.

2.3.5.2 The four-reacting centers approach

Poor yields and very tedious reaction workups for the syntheses of [3]catenates discussed above led to the development of a new and highly efficient method of preparation of [3]catenates based on alkynic oxidative coupling.[64] The strategy corresponds to a true cyclodimerization, involving only four reacting centers. The principle of this synthesis is given in Figure 11.

Oxidative coupling of terminal alkynes (Glaser reaction) has been applied to diyne systems for many years. By intramolecular coupling, macrocyclic diynes have been obtained, whereas tetra-alkynes are formed by oxidative cyclodimerization.[65,66] The latter reaction has been taken advantage of for making rigid paracyclophanes designed as molecular receptors.[67] Furthermore, cyclodimerization of a trialkynic compound has been carried out with a surprisingly high yield.[68]

The precursor used and the precise reaction scheme are represented in Scheme 9. Complex **(40)** was obtained quantitatively, as a red solid, by mixing stoichiometric amounts of the 30-membered macrocycle **(32)**, $Cu(MeCN)_4{}^+BF_4{}^-$, and a phenanthroline-based open-chain diyne in CH_2Cl_2:MeCN (2:1). The oxidative coupling leading to **(41)** was performed in DMF, with large amounts of CuCl and $CuCl_2$, in the presence of air. After workup and chromatographic separation,

(**38**) (a) = $—O(CH_2CH_2O)_3—$, (b) = $—O(CH_2CH_2O)_5—$
(**39**) (a) = $—O(CH_2CH_2O)_4—$, (b) = $—O(CH_2CH_2O)_5—$

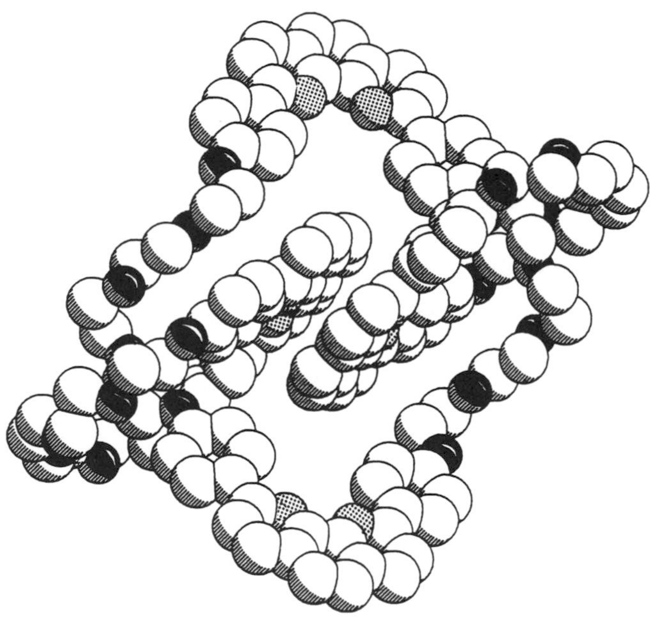

Figure 10 Molecular structure of the [3]catenand (**39**) consisting of a central 54-membered ring and two 30-membered peripheral rings stacked via their aromatic parts (reprinted with permission from *J. Am. Chem. Soc.*, 1988, **110**, 8711 Copyright 1988 American Chemical Society).

a 58% yield of (**41**) (red solid; m.p. > 300 °C, decomp.) was obtained. This strikingly high yield allows gram-scale preparation of the latter compound. In addition to (**41**), another copper(I) catenate was formed in 22% yield. Based on high-resolution ^{1}H NMR data and fine mass spectroscopy (FAB) studies, this second product was identified as the trimetallic complex formed by cyclotrimerization of (**40**) consisting of four interlocked rings: a central hexayne 66-membered ring and three peripheral rings composed of (**32**).

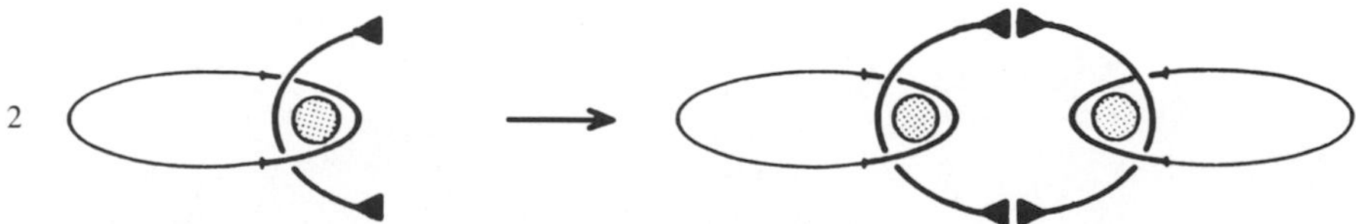

Figure 11 Principle of the four-reacting center approach to [3]catenates. A black triangle represents a chemical function able to react with itself under given experimental conditions to form a bond.

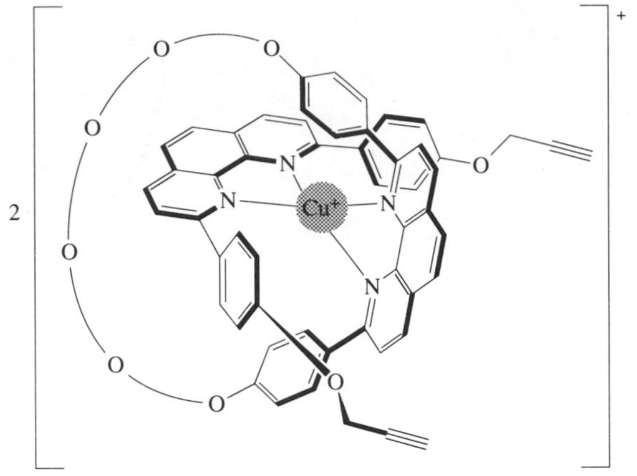

(40)

$CuCl$, $CuCl_2$ DMF, air

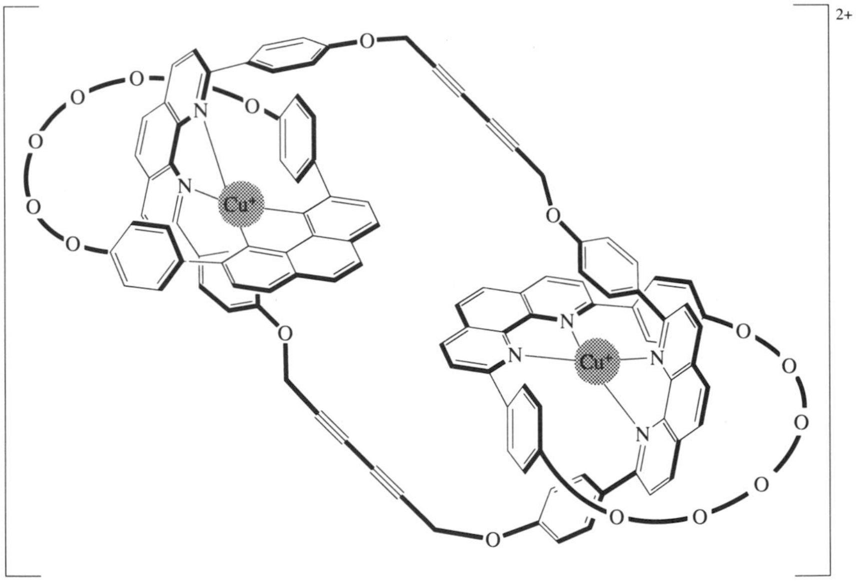

(41)

Scheme 9

Demetallation of **(41)** was performed with KCN. The corresponding free ligand was obtained in 75% yield as an insoluble white solid (m.p. 183–184 °C), its ^1H NMR spectrum (CD_3SOCD_3 at 100 °C) clearly showing the disentangling of the 2,9-diphenyl-1,10-phenanthroline fragments.

The crystallographic study of the dicopper(I) complex **(41)** and a detailed ^1H NMR solution study were carried out in parallel.[69] Both methods demonstrate that **(41)** has a dramatically different geometry from that implied in Scheme 9. The molecular structure of the [3]catenate, as determined by x-ray diffraction, is shown in Figure 12.

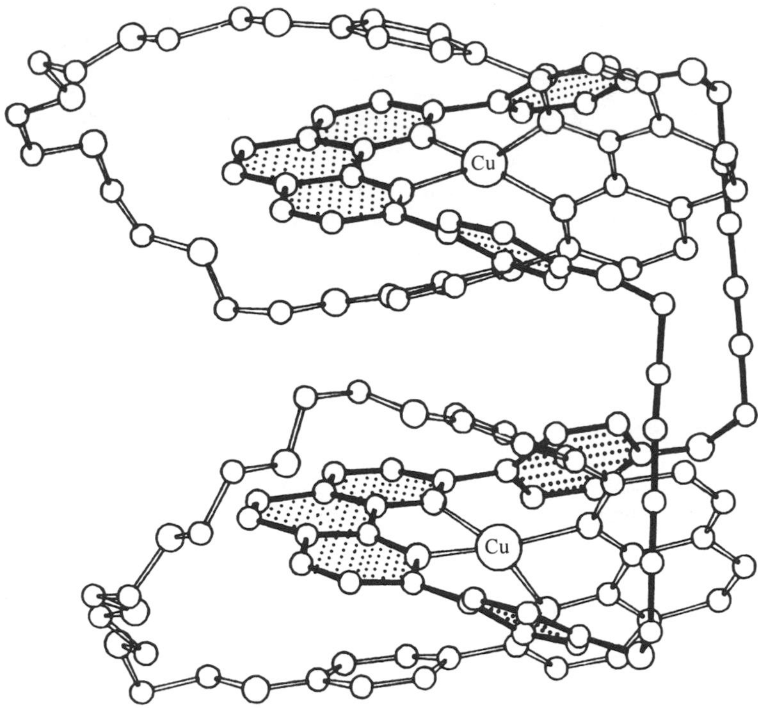

Figure 12 X-ray structure of the dicopper(I) [3]catenate (**41**) showing the curling up of the large central 44-membered cycle, thus bringing various aromatic nuclei into close contact (reproduced by permission of VCH from *Angew. Chem., Int. Ed. Engl.*, 1987, **26**, 661).

The globular shape of (**41**) is obvious from the view shown. The large central ring is folded up, bringing the 1,10-phenanthroline rings in a parallel arrangement and 770 pm apart. The dialkynic chains are also parallel to one another and located 1050 pm apart. The phenyl rings of the central cycle consist of two pairs of nearly parallel nuclei. The peripheral 30-membered rings are placed as bracelets in an approximately parallel fashion. Each aromatic ring of the molecular system is in close contact with another one (distance less than 350 pm), a rough parallelism being observed. All the phenyl rings are rotated around their link to their respective 1,10-phenanthroline nucleus, the average torsion angle being 43 °. Contrary to what could be expected from the drawing in Scheme 9, the copper(I) centers are relatively close to one another (800 pm).

The curling up of (**41**) and the proximity of the two copper(I) complex subunits could be detected by ^{1}H NMR solution measurements.[69] In particular, the upfield shift of given protons as compared to the corresponding signals of the same fragments in the [2]catenate (**25**) could only be explained by a curled-up geometry. The analogy between the molecular shapes in the solid state and in solution was also confirmed by the observation of nuclear Overhauser effects (NOE) between certain protons of the 44-membered ring and the peripheral cycles.

The folding up of the central ring and the subsequent globular shape of the molecule might be explained by stabilizing π interactions between the aromatic rings belonging to the two copper binding sites. It is noteworthy that the ternary structure of (**41**) (by analogy to proteins) is stable enough to be maintained in solution. The curled-up geometry of (**41**) and the internal stacking observed are reminiscent of systems consisting of a cationic noble metal complex linked to a large macrocyclic polyether in which a folded geometry is observed as a consequence of the formation of the transition metal complex.[70]

2.3.5.3 Molecular necklaces

The preparation of a trimetallic complex, in addition to the expected cyclodimer (**41**), when applying the strategy of Figure 11 to the diyne precursor complex (**40**), suggested the formation of even higher homologues according to the general scheme given in Figure 13.

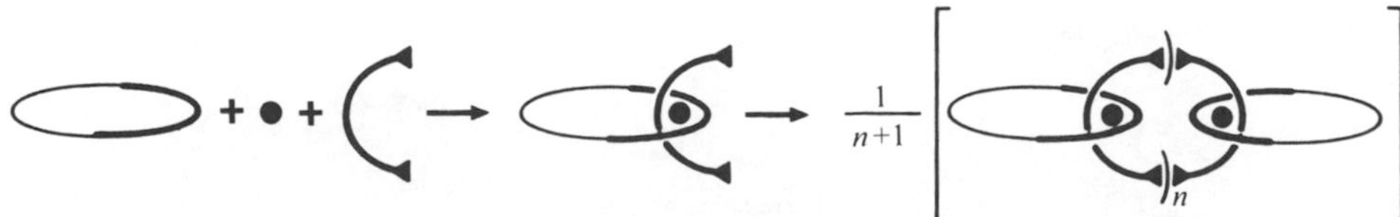

Figure 13 Template synthesis of multiring catenates. The gathering transition metal (Cu^+) is represented by a black dot, and the coordination fragments are drawn as thick lines. The triangles represent the terminal functions (—C≡CH) to be dimerized to —C≡C—C≡C—. The systems obtained consist of $n + 1$ peripheral rings separately interlocked to a central ring built during the cyclooligomerization reaction.

Multiple oxidative coupling of ($n + 1$) precursors should lead to a whole family of cyclooligomers in which ($n + 1$) peripheral rings are interlocked to a large polyalkynic central cycle. Due to an extensive electrospray mass spectrometry (ESMS) study, such multiring systems, real molecular necklaces, could indeed be identified, the small peripheral rings being either 30-membered or 27-membered.[71,72] Both series gave almost similar results so that the description of the various catenates observed (from $n = 1$–5) are restricted in Scheme 10 and Figure 14 to the family containing 27-membered peripheral rings.

A chromatographic separation on silica gel afforded (**42**)–(**44**) as pure isolated BF_4^- salts, whereas the higher homologues (**45**) and (**46**) could only be identified from a mixture. All these compounds are deep red glassy solids which display very similar ^1H NMR properties, identical elemental analyses and thus could only be distinguished from one another by mass spectrometry.

2.3.5.4 Multi-interlocking ring systems with a bicyclic core

All the systems consisting of several interlocked cycles reviewed up to now (multilink chains or olympics, molecular necklaces) are made from single monocycles, which means that their molecular graphs contain only two-order vertices.[73] An interesting extension, namely the introduction of higher-order vertices, has been achieved by using a bicyclic skeleton as the interlocking component for which the two bridgehead atoms are three-order vertices.[74] This new strategy, also based on the three-dimensional template effect of copper(I), is depicted in Figure 15.

In the first step the three arms of the bicyclic core intertwine with one, two, or three open-chain chelates around copper(I). The following cyclization step enables various ring systems comprising up to three metal centers and three rings to be obtained, each one being interlocked with an arm of the bicycle. The chemical structures of the compounds used and the reactions performed are represented in Scheme 11. The macrobicyclic endoreceptor (**47**)[75] containing three convergent 2,9-diphenyl-1,10-phenanthroline (dpp) subunits was reacted with three equivalents of $Cu(MeCN)_4^+$ BF_4^- and subsequently (**23**) (three equivalents) to yield (**48**) quantitatively. The cyclization reaction, performed in DMF in the presence of Cs_2CO_3, led, after tedious chromatographic separation, to the polycyclic complexes (**49**)–(**51**) in 8%, 4%, and 0.5% yield, respectively. A large sterical congestion inside the cage cavity (clearly evidenced by ^1H NMR studies) explains the poor yields for these three multicatenates the structures and topologies of which were determined by FAB-MS and ES-MS.

2.3.5.5 Conclusions and prospects

The recent development of synthetic strategies which rely on the template effect induced either by a transition metal or by electron donor–acceptor interactions not only gives easy access to catenanes or multiring interlocking systems but may also, in the near future, give access to polyrotaxanes and polycatenanes. Since these novel polymer architectures are expected to have unique solid-state or solution properties, much effort is being devoted to their design and synthesis. Polymeric catenanes are today still limited to the interpenetrating polymer networks (IPNs) which are obtained by cross-linking an initially linear polymer in the presence of a cross-linked network and thus rely on statistical threading. The situation appears quite different for polymeric rotaxanes:[76] numerous polyrotaxane syntheses, for which the large random statistical dependence (the weak

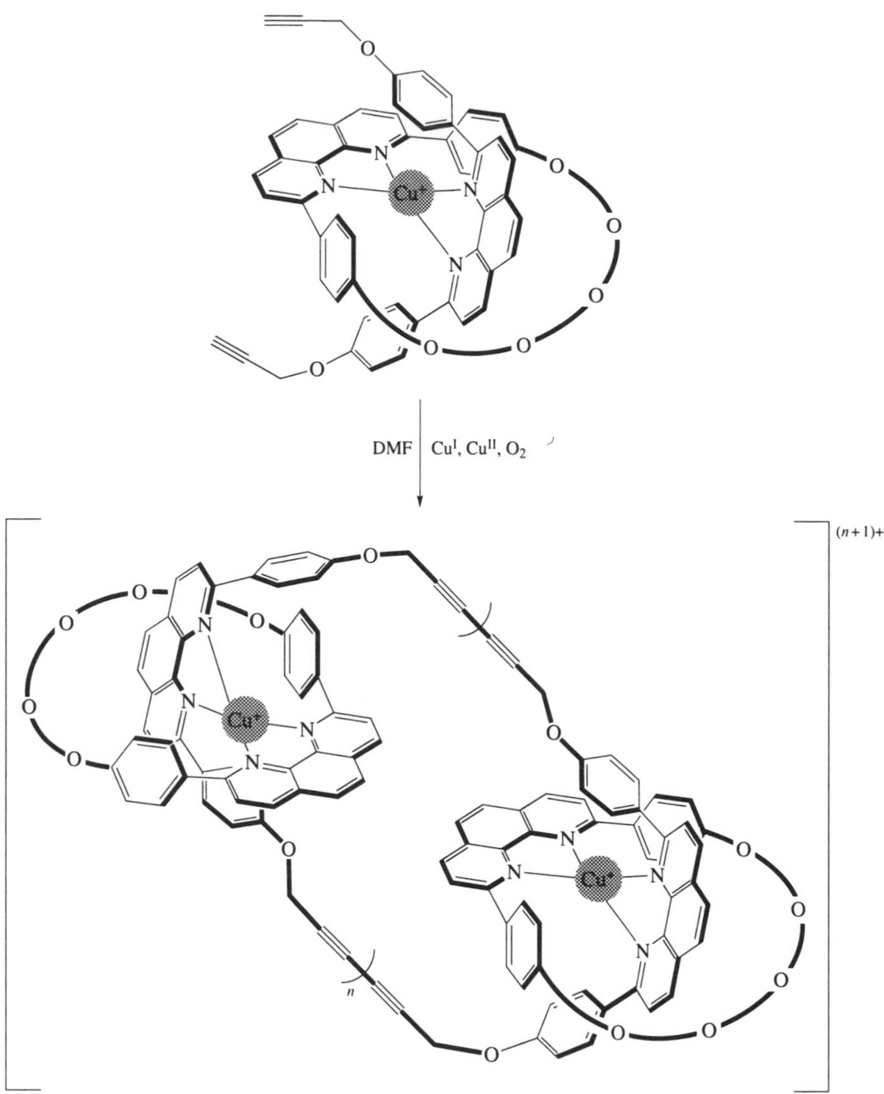

Structure	n	Product	Yield (%)	MW	Charge
(42)	1	[3]catenate	23	2222.9	2+
(43)	2	[4]catenate	23	3333.3	3+
(44)	3	[5]catenate	16	4444.7	4+
(45)	4	[6]catenate	13	5557.2	5+
(46)	5	[7]catenate	13	6668.6	6+

Scheme 10

point of the early preparations[29a,77]) could be strongly reduced, have been developed since the 1980s. The driving force for almost quantitative threading, or attractive interaction between linear components and macrocycles, is based either on metal chelation, charge-transfer complexation, or hydrophobicity. Interesting results in this rapidly developing field have been obtained by Gibson and co-workers who achieved the synthesis of polyurethane-based polyrotaxanes with a 80% threading efficiency arising from the host–guest complexation of paraquat dications,[78] whereas Shaffer and Tsay undertook the metal-coordination assisted synthesis of rotaxane polymers using a diphenyl phenanthroline as the monomeric subunit.[79] Especially noteworthy is the synthesis of a molecular necklace resulting from threading up to 40 α-cyclodextrins on a poly (iminooligomethylene) chain by Wenz and Keller.[80] Similar results were obtained simultaneously by Harada *et al.*[81]

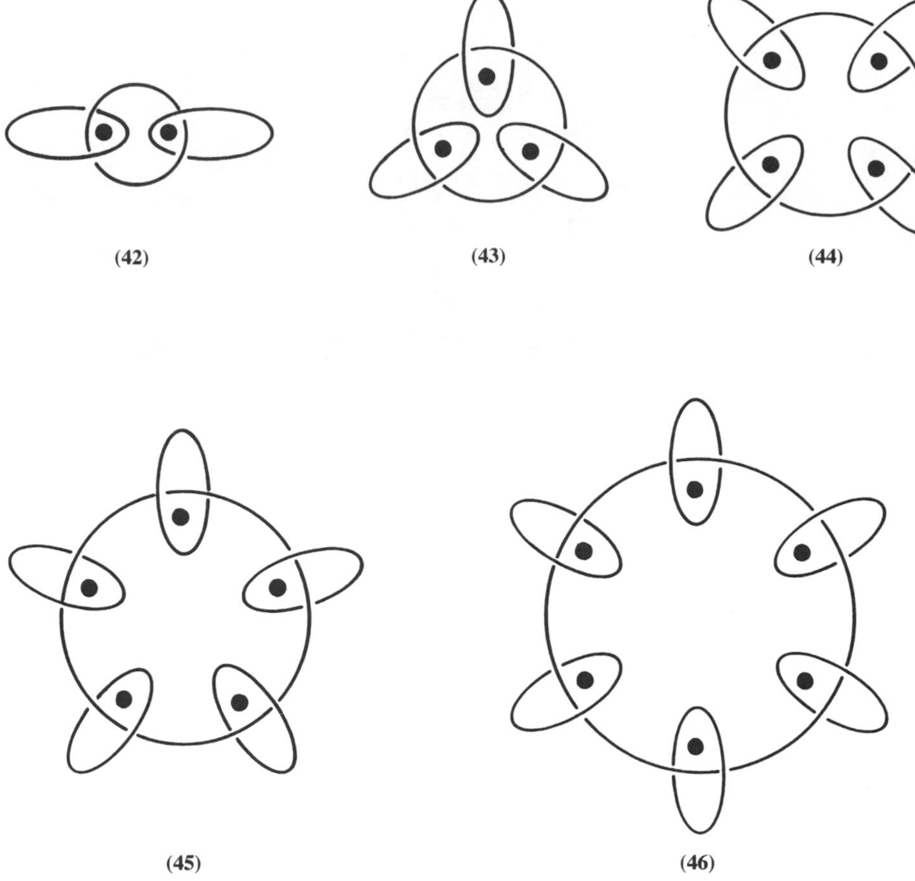

Figure 14 Schematic diagrams of multicatenates **(42)**–**(46)**.

2.3.6 Catenanes in Motion

Bistability is an essential requirement for the storage and delivery of information or images by molecular assemblies, solids, or molecules. Due to their potential future use in everyday life, several bistable systems have been designed.[82–4] Special attention is being directed towards systems in which linkage isomerism can be induced by a redox process.[85–8] The possibility of a linkage isomerism reaction with such a molecule made from two interlocked coordinating rings was envisaged. Taking advantage of the preferred coordination number (CN) for the two electrochemically easily accessible different redox states of the copper (CN = 4 for copper(I) and CN = 5 or 6 for copper(II)), it was thought that a redox process might be able to induce in the catenate a sliding motion of one ring within the other, thus leading to a complete metamorphosis of the molecule. The principle of the process, which requires a [2]catenate containing two different rings, is given in Figure 16.

The organic backbone of the nonsymmetrical catenate used consists of a dpp bidentate chelate included in one cycle and, interlocked to it, a ring containing two different subunits: a dpp moiety and a terdentate ligand, terpy. Depending on the mutual arrangement of both interlocked rings, the central copper ion can be tetrahedrally complexed (two dpp) or five-coordinate (dpp + terpy). Interconversion between these two complexing modes results from a complete pirouetting of the two-site ring. It can be electrochemically induced by taking advantage of the different geometrical requirements of the two redox states of the Cu^{II}/Cu^{I} couple. From the stable tetrahedral four-coordinate monovalent complex, oxidation leads to a four-coordinate Cu^{II} state which rearranges to the more stable five-coordinate compound. The process can be reversed by reducing the divalent state to the five-coordinate Cu^{I} complex, obtained as a transient species before a changeover process takes place to afford the starting tetrahedral monovalent state. The synthetic strategy and the structures of the compounds used are given in Figure 17.[85]

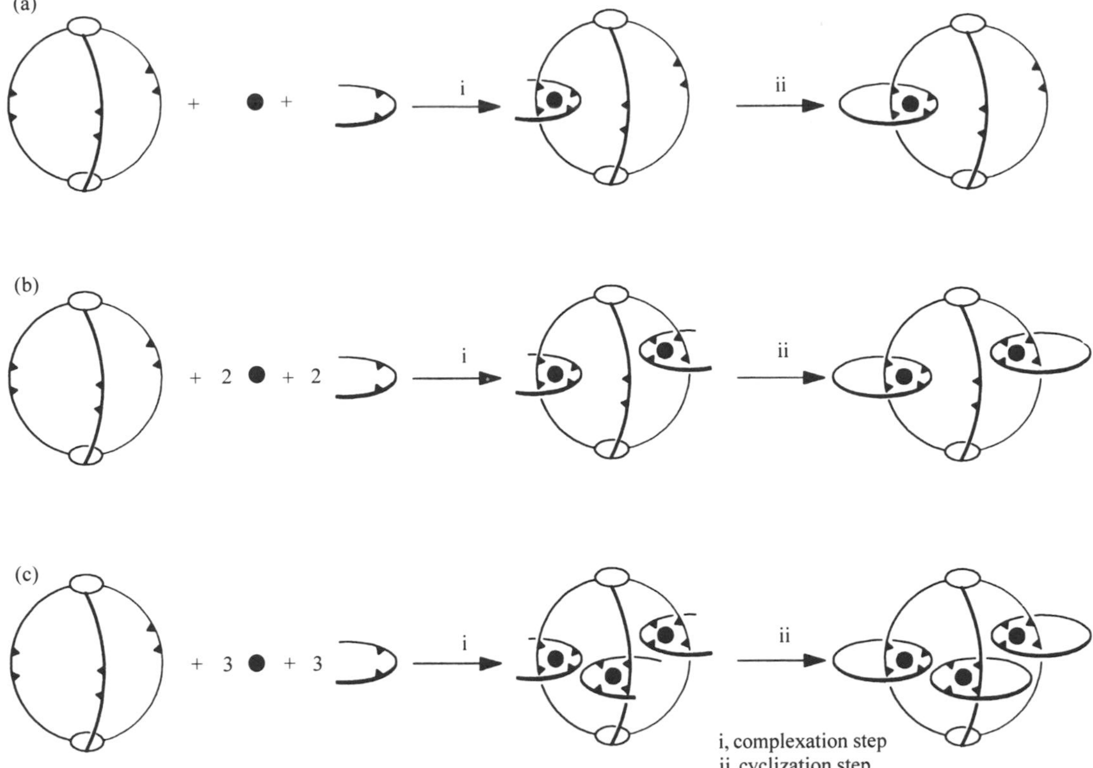

Figure 15 From a bicyclic compound with three coordinating arms (the small black triangles symbolize the 1,10-phenanthroline nitrogen atoms) to various interlocking systems. The metal center (black circle) serves to assemble and orient the various components. The three arms of the bicyclic core intertwine with (a) one, (b) two, or (c) three open chain chelates around one, two, or three metal centers, respectively.

The synthesis of the ditopic macrocycle (**52**) required the preliminary preparation (several steps starting from 2-acetyl-5-picoline) of a 5,5′-di(3-bromopropyl)-2,2′,6′,2″-terpyridine.[89] The macrocyclization, precatenate, and catenate formation steps were performed according to the previously reported procedures.[51b] After the last cyclization step and chromatographic purification, four-coordinate catenate (**53**) was obtained as its PF^-_6 salt in 10% yield as an intense red solid. According to the principle formulated in Figure 16, significantly different physical properties (in particular spectroscopic and electrochemical properties) were expected for the two possible forms of the catenate in a given oxidation state (either Cu^I or Cu^{II}).[90] This could be demonstrated by oxidizing the copper(I) catenate (**53**) and subsequently monitoring the absorption spectrum and the redox properties of the divalent complex obtained as a function of time. Catenate (**54**), the tetrahedral Cu^{II} species obtained immediately after oxidation, is a deep green complex in solution (λ_{max} = 670 nm in MeCN). It was observed that the electronic spectrum of the oxidized solution changes with time: a considerable intensity reduction occurs around 670 nm and a pale yellow-green solution results within a few days. This slow process is in agreement with the changeover reaction represented in Scheme 12 (the black circle represents Cu^{II}) leading to the five-coordinate copper(II) complex (**55**), in which a coordinated dpp chelate has been replaced by the incoming terpy unit belonging to the same cycle.[85] Interestingly, this transformation is accompanied by a change in the electrochemical properties of the complex, paralleling the spectroscopic changes. Both methods (electrochemical and spectroscopic) afford similar kinetics.

Finally, reduction of (**55**) could be carried out by electrolysis to regenerate the starting copper(I) complex.

$$(\mathbf{55}) \xrightarrow{+e} (\mathbf{55})^{-1} \rightarrow (\mathbf{53})$$

This rearrangement, which restores the initial species, is relatively fast in MeCN (seconds). The changeover process of the monovalent complex, $(\mathbf{55})^{-1}\rightarrow(\mathbf{53})$, is much faster than the reverse

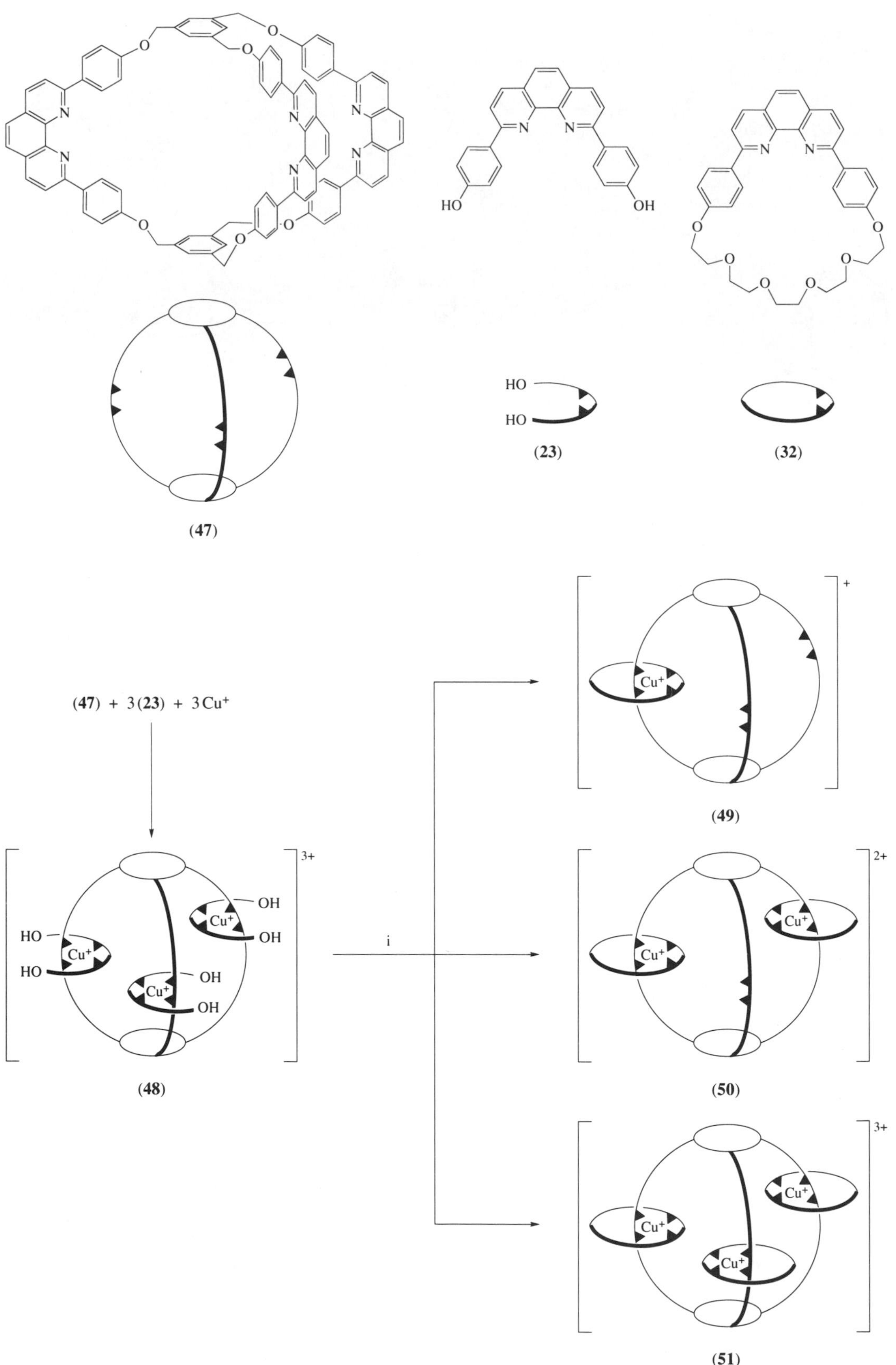

i, Cs_2CO_3, DMF, $ICH_2(CH_2OCH_2)_4CH_2I$

Scheme 11

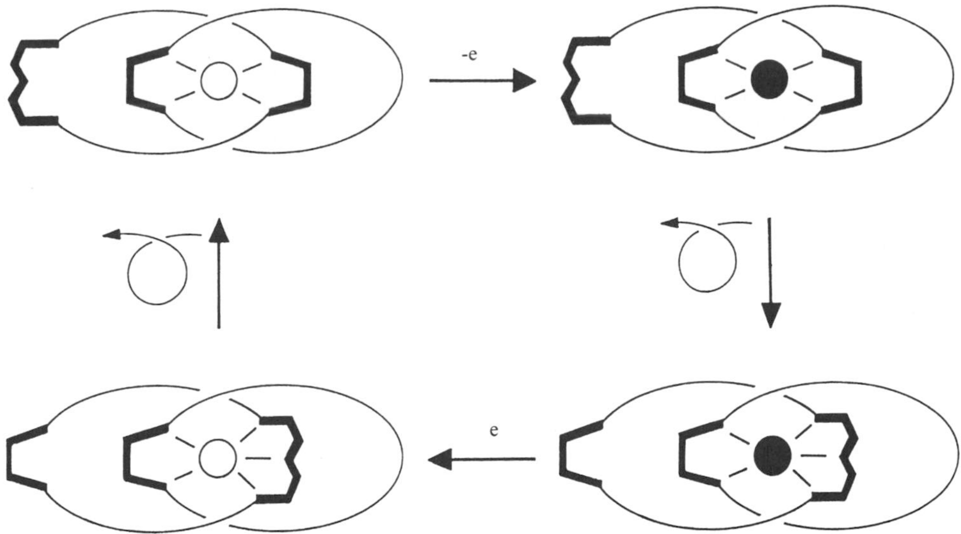

Figure 16 Principle of the electrochemically triggered rearrangement of an asymmetrical [2]catenate. The stable four-coordinate monovalent complex (top left, the white circle represents Cu^{I}) is oxidized to an intermediate tetrahedral divalent species (top right, the black circle represents Cu^{II}). This compound undergoes a complete reorganization process to afford the stable five-coordinate Cu^{II} complex (bottom right). Upon reduction, the five-coordinate monovalent state is formed as a transient (bottom left). Finally, the latter undergoes the conformational change which regenerates the starting complex.

rearrangement on the divalent copper complex. The sliding process has to involve decoordination of the metal at some stage, and, clearly, this step is expected to be much slower for Cu^{II} than for Cu^{I} due to the greater charge of the former cation. The overall process (linkage isomerism) described here can be related to some other conformational changes involving catenanes and rotaxanes, as reported by Stoddart and co-workers on purely organic compounds.[91]

2.4 MOLECULAR KNOTS AND MULTIPLY INTERLOCKED CATENANES

Made from a single knotted closed ring, the trefoil knot, which requires a minimum of three crossing points for its representation in a plane, can be considered as the prototypical example of a topologically nontrivial molecular object (Figure 18).

The building of molecules displaying novel topological properties, of a molecular trefoil knot in particular, has been a target in chemistry for a long time and several approaches have been envisaged.

2.4.1 Early Discussions and Attempts

In principle, the most obvious way to form a molecular knot relies on the direct random knotting occurring in a difunctionalized single molecular strand before its cyclization (Figure 19). Despite the simplicity of the concept, this approach has, to our knowledge, never been attempted due to the low probability a chain has to tie a knot before its ends A and B find each other and connect.[24]

On the contrary, the Möbius strip approach, depicted in Figure 20, also suggested by Frisch and Wasserman in 1961,[24] appeared to be much more realistic.

In this approach, the target molecule (**56**) should be obtained after cleavage of the vertical rungs of the intermediate compound. This originates from the ladder-shaped molecule in which the ends are able to twist prior to bimacrocyclization. In order to achieve the synthesis of a trefoil knot by this strategy, Walba *et al.*[92] prepared the four-rung diol-ditosylate molecular ladder (**57**) and submitted it to cyclization (Scheme 13). Among the different compounds formed in this synthesis, only the untwisted molecular cylinder (**58**) could be fully characterized. Twisted isomers (**59**), (**60**), and especially the three half-twist isomer (**61**) could not be identified with certainty.

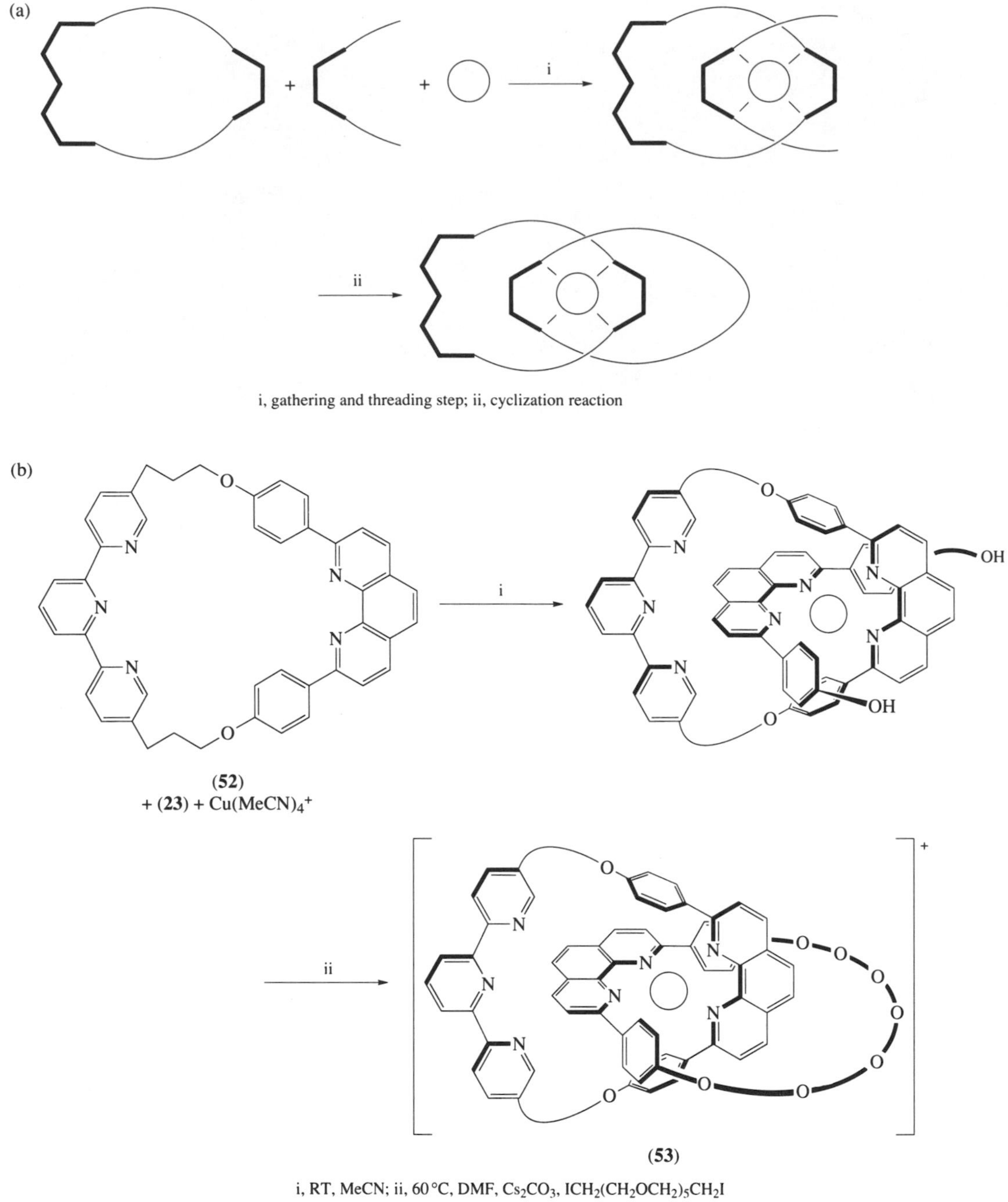

(a)

i, gathering and threading step; ii, cyclization reaction

(b)

(52)
+ (23) + $Cu(MeCN)_4^+$

(53)

i, RT, MeCN; ii, 60 °C, DMF, Cs_2CO_3, $ICH_2(CH_2OCH_2)_5CH_2I$

Figure 17 (a) Strategy used for constructing a two-binding-mode [2]catenate. The starting macrocycle incorporates two different chelates. The circle represents Cu^I. (b) The compounds made and the catenate formation reaction.

If three half-twists are very difficult to introduce in a Möbius ladder for statistical reasons, one half-twist, by contrast, occurs much more frequently as can be seen in Walba's synthesis of the molecular Möbius strip (**15**) (Scheme 5). Using this observation, these authors conceived a very ingenious strategy for the synthesis of knotted rings which requires only one half-twist of the ladder, additional torsion being already present in the starting compound. The principle of this "hook and ladder" approach is given in Scheme 14.[73]

A molecular hook can be obtained by the entwining of two phenanthroline derivatives around the copper(I) ion. Connection of the latter complex to THYME polyether ladders should lead to

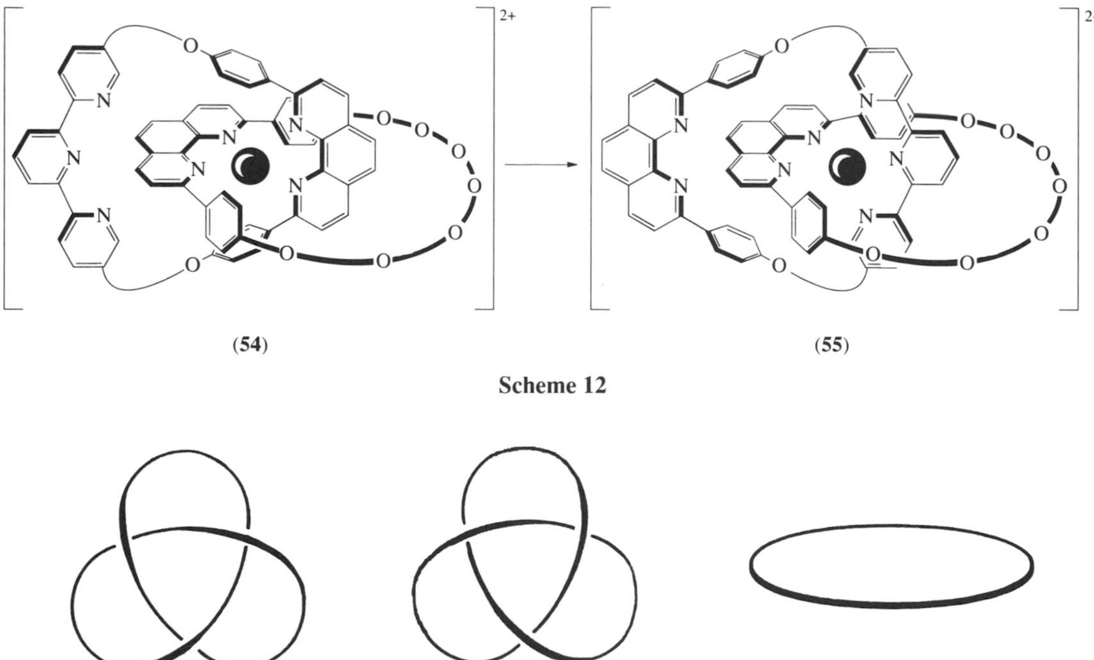

Scheme 12

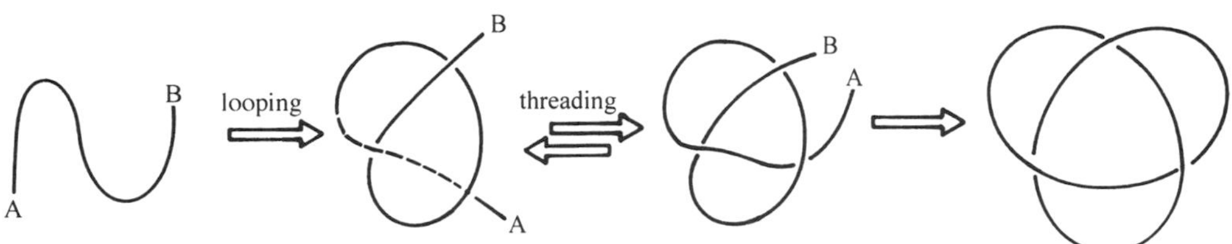

Figure 18 The trefoil knot: a closed ring with a minimum of three crossing points. Representation of its topological enantiomers.

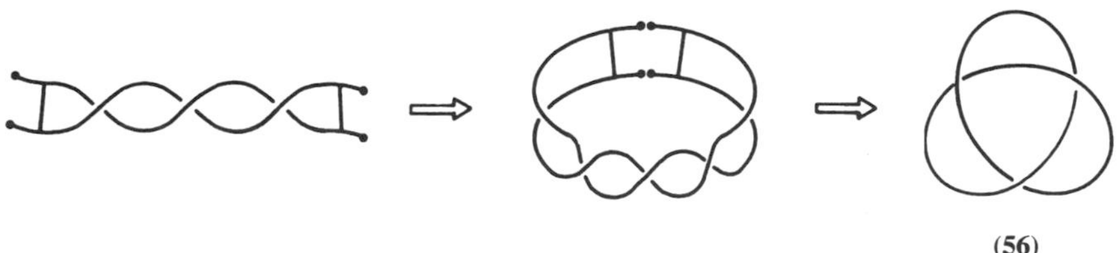

Figure 19 Tying a knot in the course of the cyclization reaction of a linear molecule A–B after statistical threading.

(56)

Figure 20 The Möbius strip approach (three half-twists) to a trefoil knot.

the desired hook and ladder precursor (**62**). Intramolecular cyclization of the twisted ends of (**62**), followed by cleavage of the vertical rungs and demetallation of the "hook" part would give the target trefoil (**64**).

Another very interesting approach towards a trefoil knot can be found in a review by Sokolov.[27] The principle of the synthesis imagined by this author as early as 1973 is given in Scheme 15. Three bidentate chelates disposed in a suitable fashion around an octahedral transition metal used as a matrix may, after connection of their ends, lead to a molecular knot. The probability that the six ends will connect (two by two) in the required fashion is quite low. Nevertheless, strict geometrical control of the coordinated fragments involved or slight changes in the above proposed scheme may one day give access to a knot using this strategy.

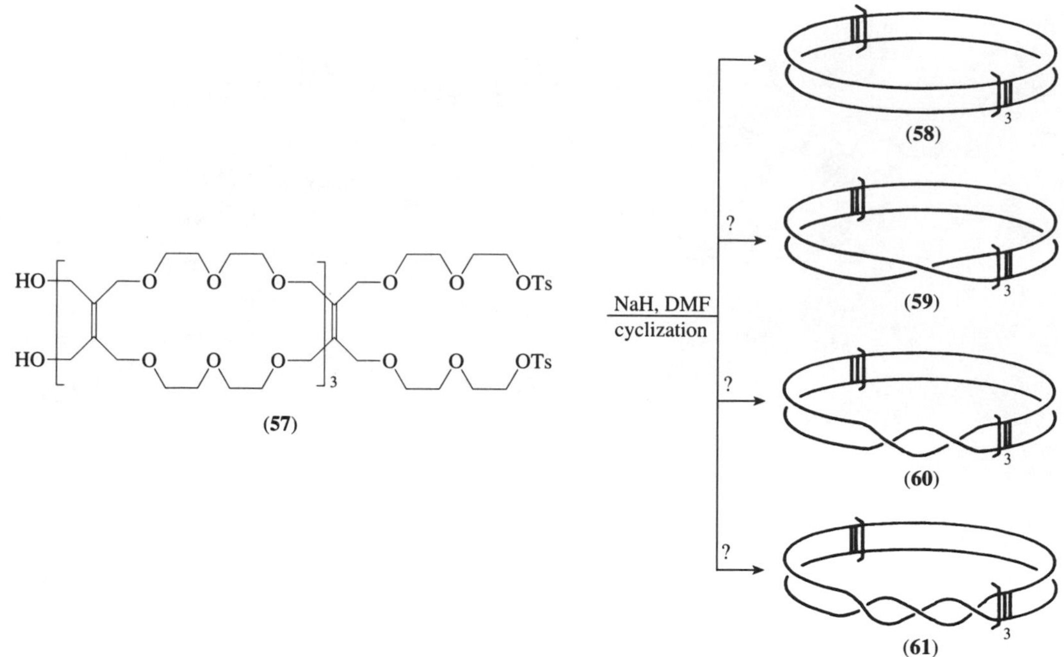

Scheme 13

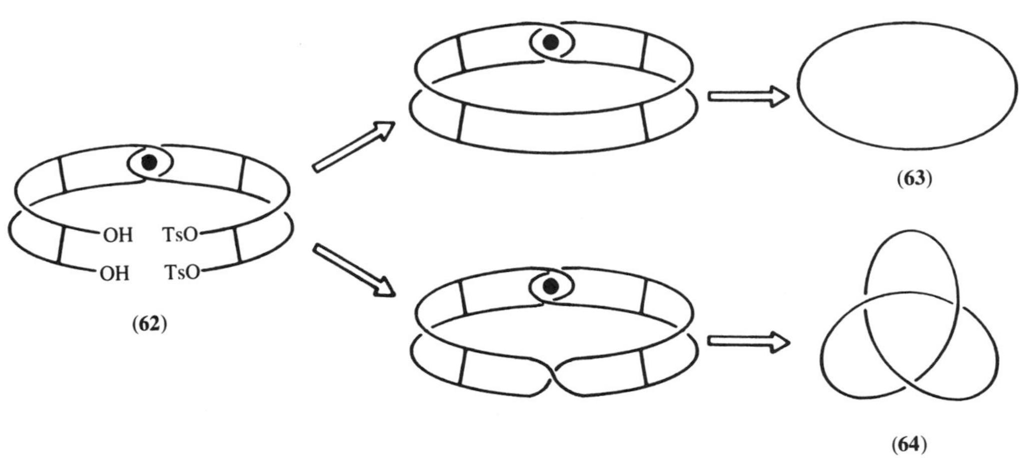

Scheme 14

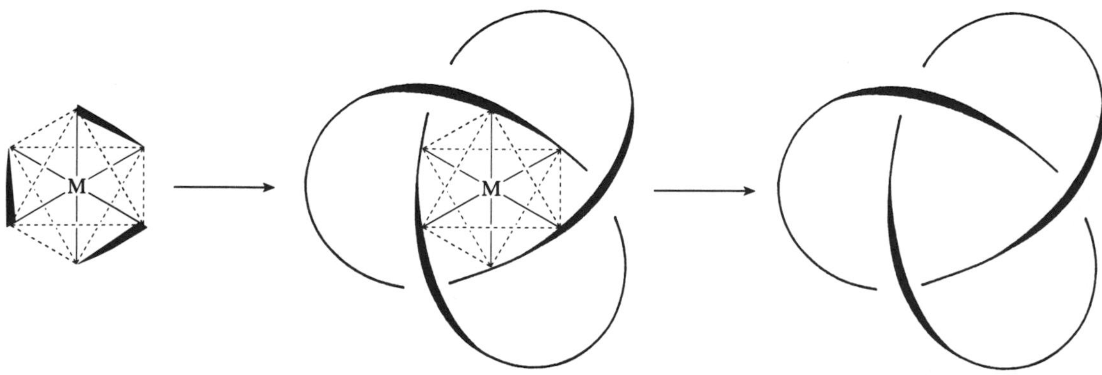

Scheme 15

For their part, Schill and co-workers conceived and attempted directed syntheses of a trefoil knot which relied on the use of a benzoacetal central core.[93] These syntheses are closely related to those used successfully to prepare various [2]- and [3]catenanes. Starting from compound **(65)** (Scheme 16), the diaminoacetal **(66)** was synthesized in a nine-step reaction sequence. Subsequent macrocyclization should yield the fourfold bridged 5,6-diamino-1,3-benzodioxole derivative **(67)**, direct precursor of a trefoil knot **(68)**, after selective cleavage of the aryl–nitrogen bonds and hydrolysis of the ketal. Unfortunately, after cyclization of diamino compound **(66)**, three isomeric products were obtained in a total yield of 1.7% and the pre-knot **(67)** could not be identified with certainty.

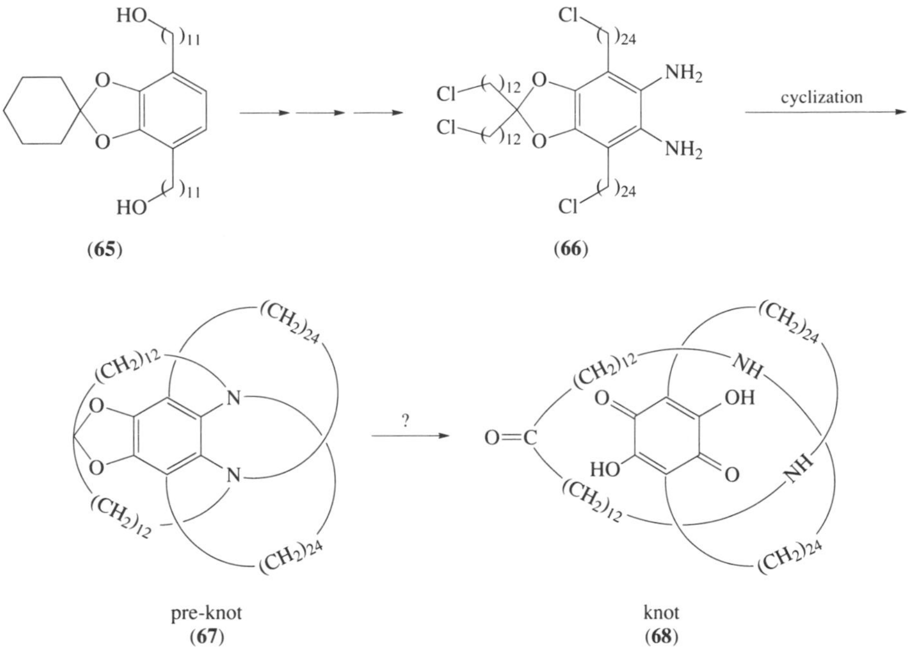

Scheme 16

As seen above, randomness in the Möbius strip approach and difficult numerous steps in Schill's directed approach appear as highly limiting factors in a trefoil knot synthesis. Both these major obstacles may be circumvented by the use of an unambiguous templated synthesis procedure.

2.4.2 Templated Syntheses of Molecular Knots

2.4.2.1 Strategy

Seeking the goal of preparing a trefoil knot, we extended the synthetic concept which had already proved successful for the synthesis of catenanes (Figure 21(a)). In this strategy towards a trefoil knot, the three-dimensional template effect induced by two transition metals was exploited. As shown in Figure 21(b), two bis-chelating molecular threads A_2 can be interlaced on two transition metal centers, leading to a double helix B_2. After cyclization to C_2 and demetallation, a knotted system D_2 should be obtained. An important prerequisite for the success of this approach is the formation of a helicoidal dinuclear complex.

Of utmost importance is the stability of the double helix B_2, which has to resist the cyclization conditions used to obtain C_2. The choice of the chemical link between the two chelating subunits of each A_2 fragment is thus crucial. Besides chemical stability, steric parameters and flexibility are factors that also have to be considered. In particular, during the reaction of A_2 with the transition metal ion, the formation of B_2 must be favored over the formation of polymeric complexes or the formation of a mononuclear complex produced from folding A_2 onto one metal center. It is

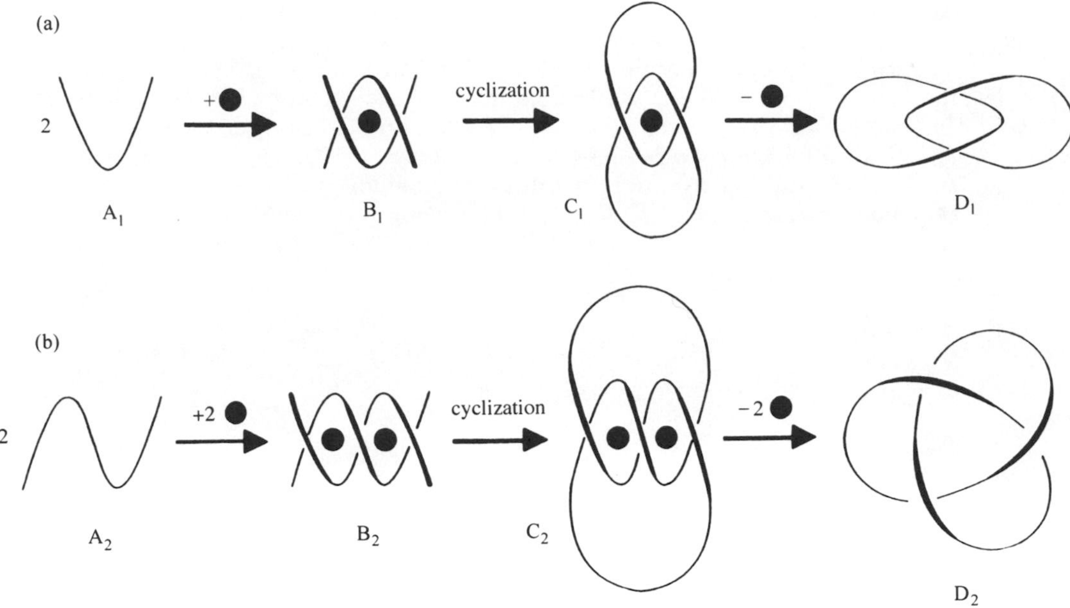

Figure 21 Template synthesis of (a) a catenane (one metal center) or (b) a trefoil knot (two metal centers).

interesting to note that only a few dinuclear helicoidal complexes have been reported in the literature up to the 1980s; since then, this domain has developed so rapidly that many numerous dinuclear double helices are now known. Although the preparation of double helices from various transition metals and bis-chelate ligands is very likely to have occurred long ago, it was only in 1976 that the first such system was recognized and characterized.[94] Moreover, the scientific interest of these arrangements was not at all obvious. One of the earliest dinuclear helical complex was identified by Fuhrhop *et al.* (Figure 22) in 1976.[94]

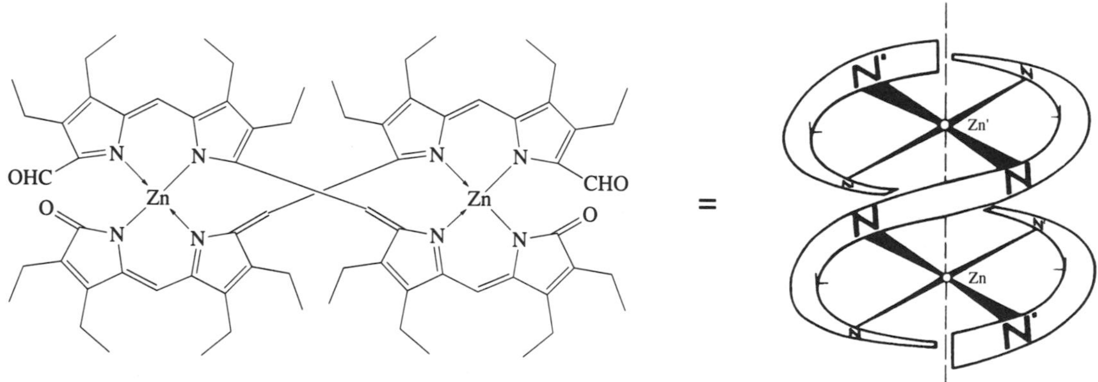

Figure 22 The first structurally characterized double helix built on two transition metal centers from two bis-chelating open-chain fragments.

During the 1980s, several laboratories prepared and investigated double-stranded helical complexes, the systems containing either pyrrolic ligands[95,96] and derivatives[97,98] (with Zn^{2+}, Ag^{+}, or Cu^{+}) or oligomers of 2,2′-bipyridine.[99,100] Helicates[99] may consist of up to five copper(I) centers and these self-assembling systems are reminiscent of the DNA double helix.

More recently, oligobidentate ligands have been used for the synthesis of various triple-stranded helical complexes around metals with a preference for octahedral coordination, such as Co^{II} [101] or Ni^{II}.[102]

2.4.2.2 The first synthetic molecular trefoil knot

After many attempts with various linkers, it was found that 1,10-phenanthroline nuclei, connected via their 2-positions by a —$(CH_2)_4$— linking unit, will indeed form a double helix when complexed to two copper(I) centers. In addition, by introducing appropriate functions at the 9-positions, the strategy of Figure 21 could be followed to achieve the synthesis of a molecular knot of the D_2 type. The precursors used and the reactions carried out are shown in Scheme 17.[103]

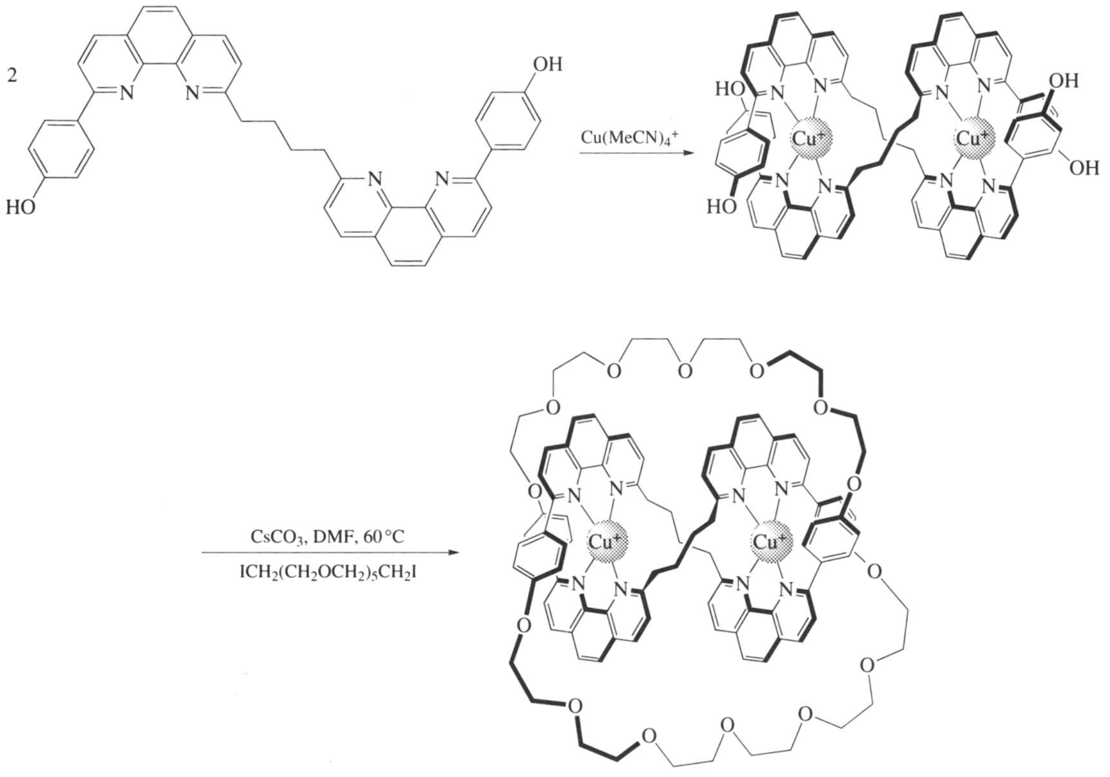

Scheme 17

The diphenolic bis-chelating molecular thread (prepared in several steps starting from 1,10-phenanthroline and Li—$(CH_2)_4$—Li[52,104,105]) was reacted with a stoichiometric amount of $Cu(MeCN)_4{\cdot}BF_4$ to afford the dinuclear precursor double helix together with an important proportion of other copper(I) complexes. The complex mixture containing the double helix was reacted under high-dilution conditions with two equivalents of the diiodo derivative of hexaethyleneglycol in the presence of Cs_2CO_3. After a tedious purification process, a small amount (3% yield) of the bis(copper) complex (**78**) could be isolated. Its knotted topology, first demonstrated by mass and NMR spectroscopy, was fully confirmed after demetallation: among the various possible dicopper complexes which can originate from a dinuclear double helix during the cyclization step, and which all display classical chirality because of the presence of a dicopper double helix in their central core, only the trefoil knot will remain chiral. Quantitative demetallation of (**78**) by KCN afforded a free ligand, whose topological chirality could be demonstrated (Figure 23).

However, the major cyclization product (24%) obtained in the latter reaction could be identified as a dicopper complex consisting of two 43-membered rings arranged around the metallic centers in an approximate face-to-face geometry.[106] This unknotted compound originates from a nonhelical precursor which stays in equilibrium with the expected double helix. Figure 24 describes in a schematic way the alternative cyclization reaction leading to the unknotted face-to-face complex and the equilibrium which interconverts the helical and the nonhelical precursors.

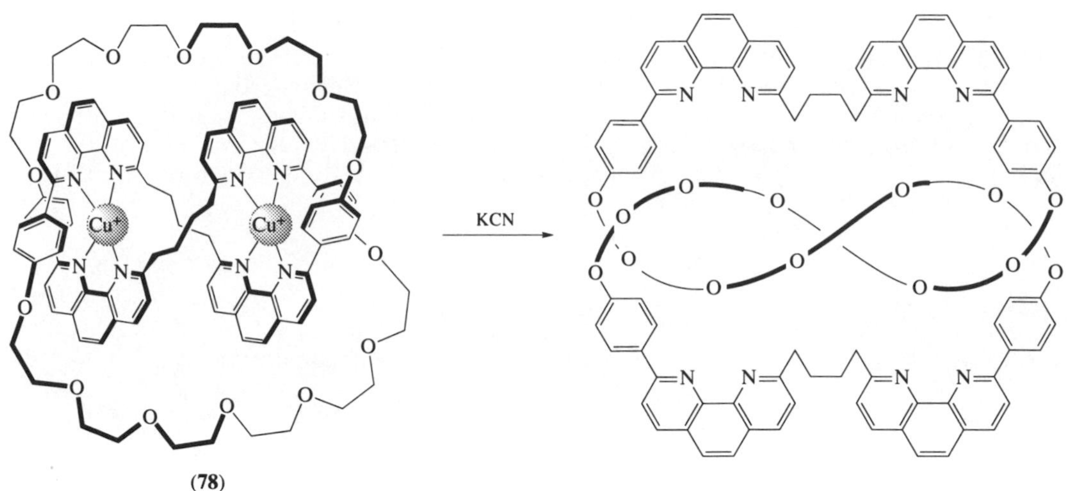

Figure 23 The free trefoil knot.

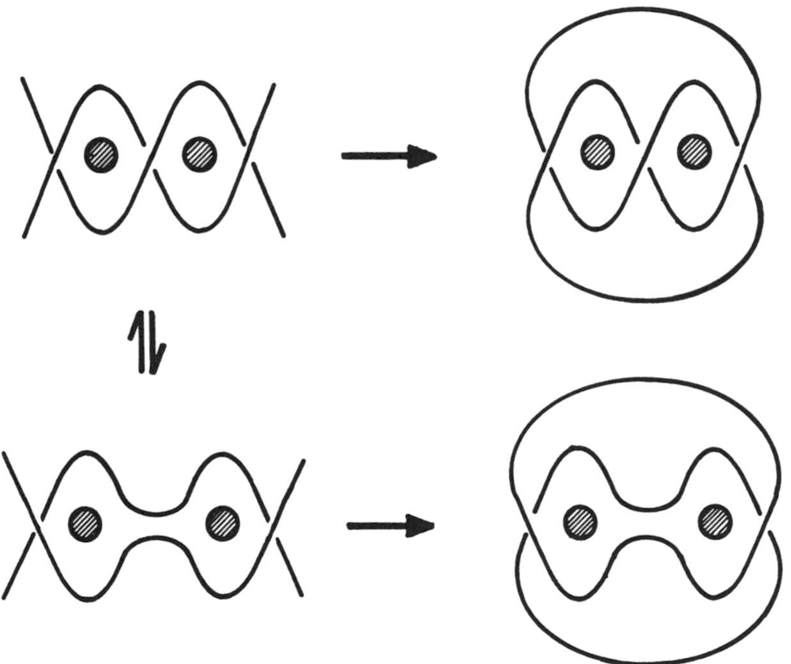

Figure 24 The equilibrium between the helicoidal interlaced system and its face-to-face analogous complex. Interconversion between the two isomeric cyclic products is not possible.

2.4.2.3 Improved syntheses using polymethylene linkers

The poor yield of only 3% prompted the search for better synthetic routes. The combination of copper(I) as templating metal and 2,9-disubstituted-1,10-phenanthroline as the chelating "brick" showed so many promising possibilities for constructing interlocking rings[107,72] that these two species were retained in the modified routes. The only modular parts were thus the linker connecting the two chelates before cyclization and the long functionalized chain used in the cyclization step.

Improvement in the yields of trefoil knots will be determined by:

(i) the proportion of double helix precursor formed vs. face-to-face open-chain complex;

(ii) the spatial arrangement of the four reacting ends of the helicoidal dinuclear complex.

This latter factor will reflect the degree of winding of the two molecular strings interlaced around the copper(I) atoms. The various complexes synthesized and studied are depicted in Scheme 18 and Figure 25.[108]

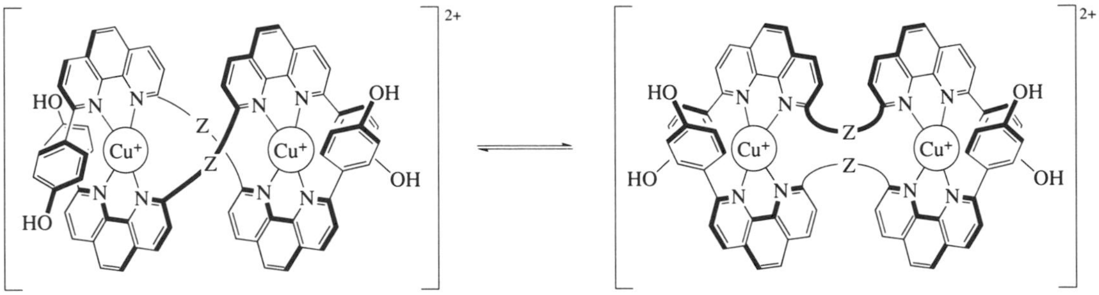

(**69**) $Z=(CH_2)_2$
(**70**) $Z=(CH_2)_4$
(**71**) $Z=(CH_2)_6$

(**72**) $Z=(CH_2)_2$
(**73**) $Z=(CH_2)_4$
(**74**) $Z=(CH_2)_6$

Scheme 18

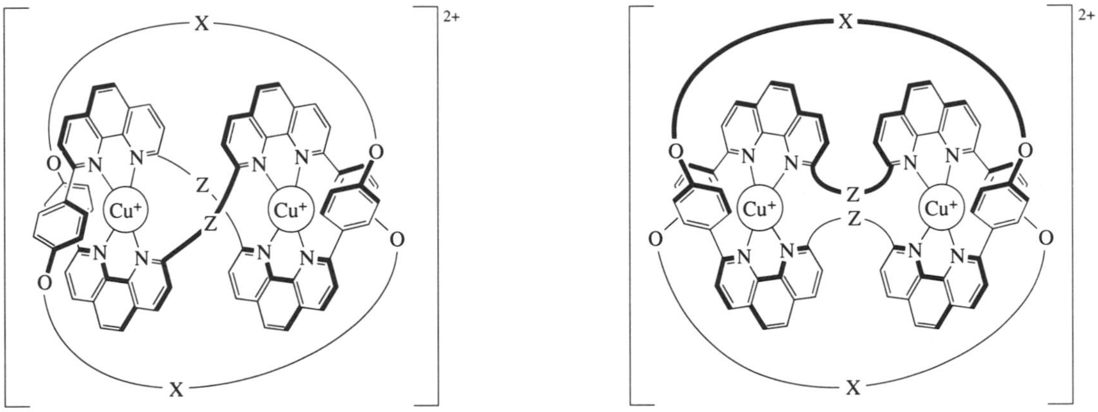

$X=CH_2(CH_2OCH_2)_nCH_2$

(**75**) $[Cu_2(k\text{-}80)]^{2+}$, $n=4$, $Z=(CH_2)_4$, 0.7%
(**76**) $[Cu_2(k\text{-}82)]^{2+}$, $n=5$, $Z=(CH_2)_2$, 0.5%
(**77**) $[Cu_2(k\text{-}84)]^{2+}$, $n=4$, $Z=(CH_2)_6$, 8.0%
(**78**) $[Cu_2(k\text{-}86)]^{2+}$, $n=5$, $Z=(CH_2)_4$, 3.0%
(**79**) $[Cu_2(k\text{-}90)]^{2+}$, $n=5$, $Z=(CH_2)_6$, 2.5%

(**80**) $[Cu_2(m\text{-}40)_2]^{2+}$, $n=4$, $Z=(CH_2)_4$,4.6%
(**81**) $[Cu_2(m\text{-}41)_2]^{2+}$, $n=5$, $Z=(CH_2)_2$, 0.5%
(**82**) $[Cu_2(m\text{-}42)_2]^{2+}$, $n=4$, $Z=(CH_2)_6$, unstable
(**83**) $[Cu_2(m\text{-}43)_2]^{2+}$, $n=5$, $Z=(CH_2)_4$, 24%
(**84**) $[Cu_2(m\text{-}45)_2]^{2+}$, $n=5$, $Z=(CH_2)_6$, 21%

Figure 25 For the cyclic compounds, the total number of atoms connecting two phenolic oxygen atoms is 16 if $n = 4$ (pentakis(ethyleneoxy) fragment) or 19 if $n = 5$ (hexakis(ethyleneoxy) linker). Each knot is represented by the letter k accompanied by the overall number of atoms included in the cycle. The face-to-face complexes contain two monocycles (letter m), the number of atoms in each ring also being indicated. It can be noted that each knot has a face-to-face counterpart. For instance $[Cu_2(k\text{-}90)]^{2+}$ and $[Cu_2(m\text{-}45)_2]^{2+}$ are constitutional isomers. They are by no means topological stereoisomers.

All the bis-chelating ligands used for the preparation of the acyclic precursor complexes (**69**)–(**74**) were prepared as previously described.[109]

The dicopper(I) complexes (**69**)–(**74**) were made by mixing $Cu(MeCN)_4^+$ with the corresponding diphenol. It must be stressed that mixtures of complexes were obtained, corresponding to the equilibrium of Scheme 18. All the cyclization reactions leading to the dicopper trefoil knots (**75**)–(**79**) and the face-to-face dinuclear complexes (**80**)–(**84**) were carried out under similar conditions, on the crude mixture of precursors (double helix and face-to-face complex). The phenolic complexes (**69**) + (**72**), (**70**) + (**73**), or (**71**) + (**74**) were reacted with the desired diiodo derivative (penta- or hexaethyleneglycol) in DMF, in the presence of Cs_2CO_3, with vigorous stirring. The knotted and the face-to-face complexes were obtained in various proportions, depending on which open-chain compounds were used: bis(phenanthroline) and diiodo derivative. It can be noted that all the preparation yields are rather low, except for some face-to-face complexes which can be obtained in more than 20% yield. As far as dicopper(I) knots are concerned, the best result was obtained for **(77)** (8% yield). A possible explanation is that the $—(CH_2)_6—$ fragment linking the two phenanthroline units is particularly favorable to either double-helix formation vs. the face-to-face precursor or to cyclization to the knot. The latter explanation seems reasonable on the basis of a recent x-ray crystallographic study on $Cu_2(k\text{-}84)^{2+}$ **(77)**.[110] The structure shows the helicoidal core to be very well adapted to the formation of a knot, particularly with a highly twisted helix, owing to the relatively long $—(CH_2)_6—$ fragment.

The x-ray structures of $Cu_2(k\text{-}84)^{2+}$ **(77)**[110] and $Cu_2(k\text{-}86)^{2+}$ **(78)**[111] are represented in Figure 26. Both molecules are very symmetrical and their overall shapes are similar. However, the central double helix of **(77)** is more wound than that of **(78)**. The Cu—Cu distances are also significantly different: 703 pm in **(77)** and 630 pm in **(78)**, in accordance with the longer methylenic fragments connecting both subcomplexes in the former system.

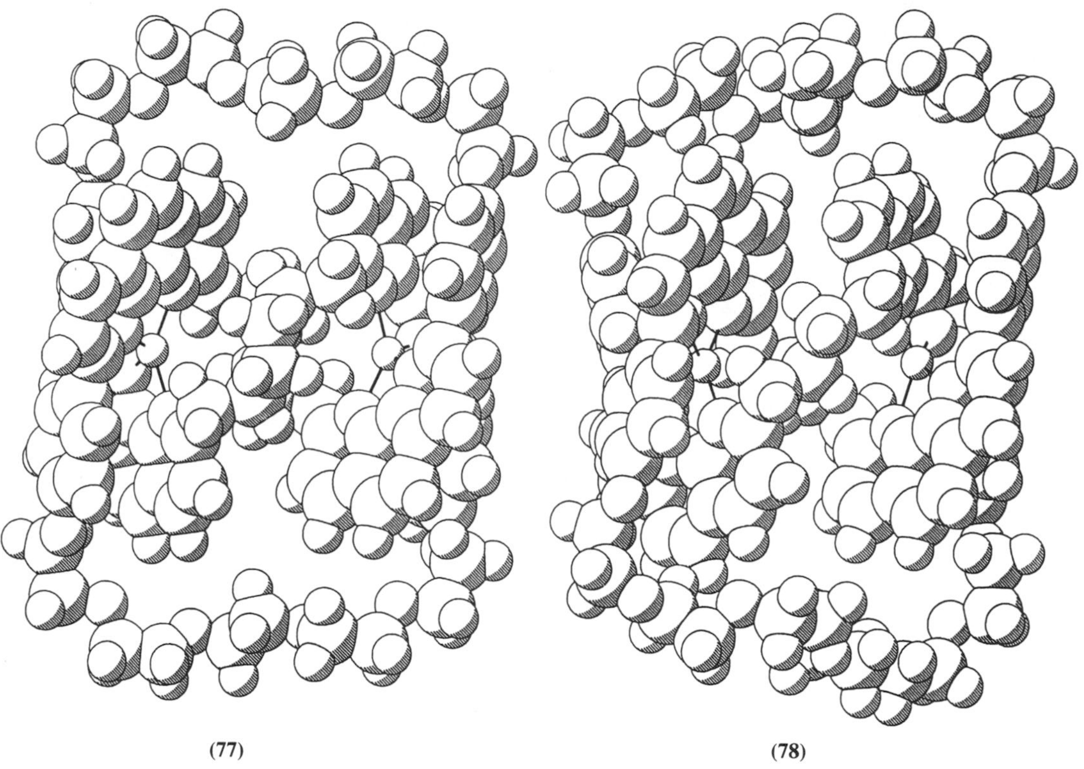

Figure 26 X-ray structures of dicopper(I) knot compounds **(77)** and **(78)** (reproduced by permission of the International Union of Pure and Applied Chemistry from *Pure Appl. Chem.*, **67**, 233).

Interestingly, the x-ray structure analysis showed that **(78)**, a *chiral* dicopper knot, crystallizes as a conglomerate of enantiomers. In other words, a given crystal contains only one enantiomer. It can thus be hoped to manually separate the right- and left-handed knots, provided sufficiently big crystals can be grown. This requirement was fulfilled by the knot **(77)**, but unfortunately it crystallizes as a racemate ($\alpha_D = 0$).

2.4.2.4 *High-yield synthesis of a dicopper(I) trefoil knot with a 1,3-phenylene spacer*

The still modest (8%) yield obtained with the hexamethylene spacer prompted the search for other possibilities. Among them a 1,3-phenylene spacer, already used for the construction of helical complexes,[112,113] was very appealing due to its expected chemical robustness (necessary in order to survive the harsh cyclization conditions) and ease of introduction to the established methodology.[107]

The organic precursors as well as the reactions leading to the dicopper(I) trefoil knot are represented in Scheme 19.[114]

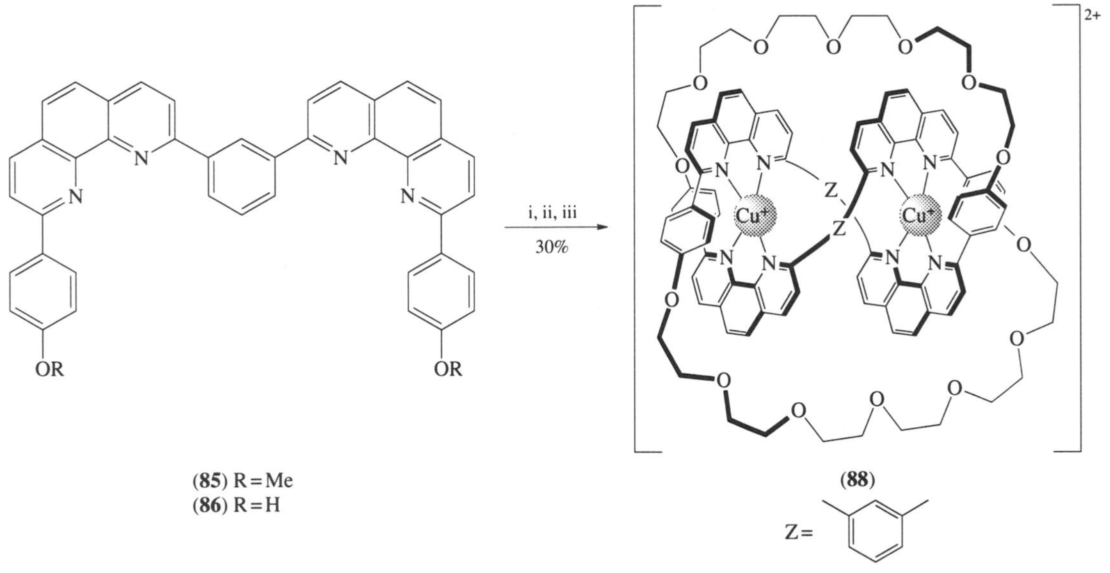

(**85**) R = Me
(**86**) R = H

(**88**)

Z =

i, demethylation; ii, $Cu(MeCN)_4{\cdot}PF_6$; iii, cyclization

Scheme 19

The bis-chelate (**85**) was obtained in 55% yield by reacting 2-(*p*-anisyl)-1,10-phenanthroline with 1,3-dilithiobenzene in THF, followed by hydrolysis and MnO_2 oxidation; diphenol (**86**) was prepared in 84% yield in usual conditions (HCl/pyridine at 210 °C for 3 hr).[109]

The use of 1,3-phenylene as the spacer group proved to be extremely beneficial. First, (**85**) was used as a model and reacted with $Cu(MeCN)_4{}^+{\cdot}PF_6{}^-$ in CH_2Cl_2/MeCN. Preparation of the double-stranded helix (**87**) turned out to be quantitative. This complex could be crystallized from MeCN/toluene and its x-ray structure was solved (Figure 27).[114]

The structure of (**87**) is nicely wound and is therefore well adapted to the formation of a knot by connecting the appropriate ends of the strands. It is noteworthy that (**87**) is a rigid and compact edifice, with a much shorter Cu···Cu distance than in previously synthesized knots (476 pm instead of 630 pm or 700 pm).[110,111] These factors all favor the successful continuation of the synthesis. Indeed, reaction of the tetraphenolic double helix (whose structure is certainly very similar to that of (**87**), as shown by ^1H NMR) with $ICH_2(CH_2OCH_2)_5CH_2I$ and Cs_2CO_3 in DMF at 60 °C afforded a single isolable copper(I) complex (Scheme 19). The dicopper(I) knot (**88**) was isolated in 29% yield after chromatography (silica gel, CH_2Cl_2/3–5% MeOH). It forms sea urchin-shaped aggregates of crystals (as the $BF_4{}^-$ salt).

Analogous to the previously synthesized dicopper(I) knots,[109] (**88**) shows characteristic ^1H NMR and FAB-MS spectroscopy data.[114] In particular, some of the aromatic protons of the bis-chelates are strongly shielded which indicates that (**88**) is compact and that its helical core is geometrically very similar to (**87**).

With the significant improvement in yield, now allowing gram-scale preparation of knots, it becomes realistic to study the specific properties related to their topology as well as the coordination chemistry of this new type of ligand.

2.4.3 Generalization: Towards Topologically More Complex Molecules Using Double-stranded Polynuclear Helical Precursors

The concept of using polynuclear double helices as precursors to more complex interlocked and knotted systems can be generalized following the principle schematically explained in Figure 28. As a general rule, if the number of metal centers is even, the number of crossing points in the

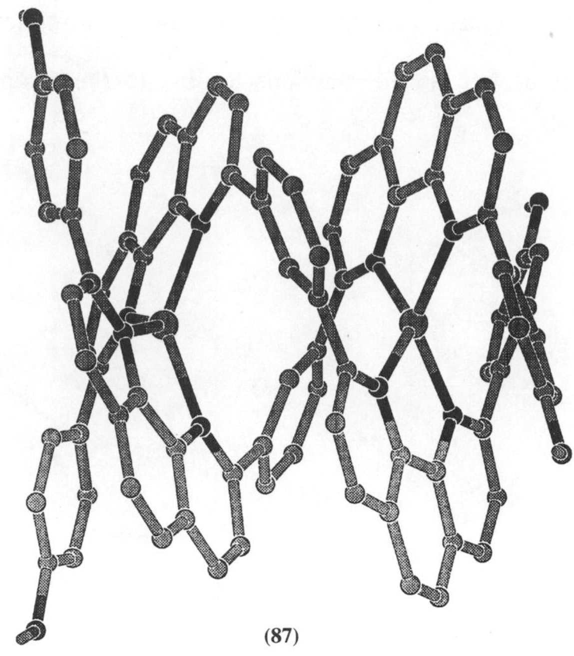

Figure 27 The crystal structure of the dicopper(I) helical precursor complex (**87**) (reproduced by permission of the Royal Society of Chemistry from *J. Chem. Soc., Chem. Commun.*, 1994, 2231).

molecular graph after cyclization is odd, and a single closed knotted curve is formed (trefoil, pentafoil, heptafoil, etc., from two, four, six, etc., metal centers, respectively). If the double helix is built around an odd number of templating metals, there will be an even number of crossing points, which will result in multiply interlocked [2]catenanes (two singly, doubly, triply, etc., interlocked rings from one, three, five, etc., metals, respectively).

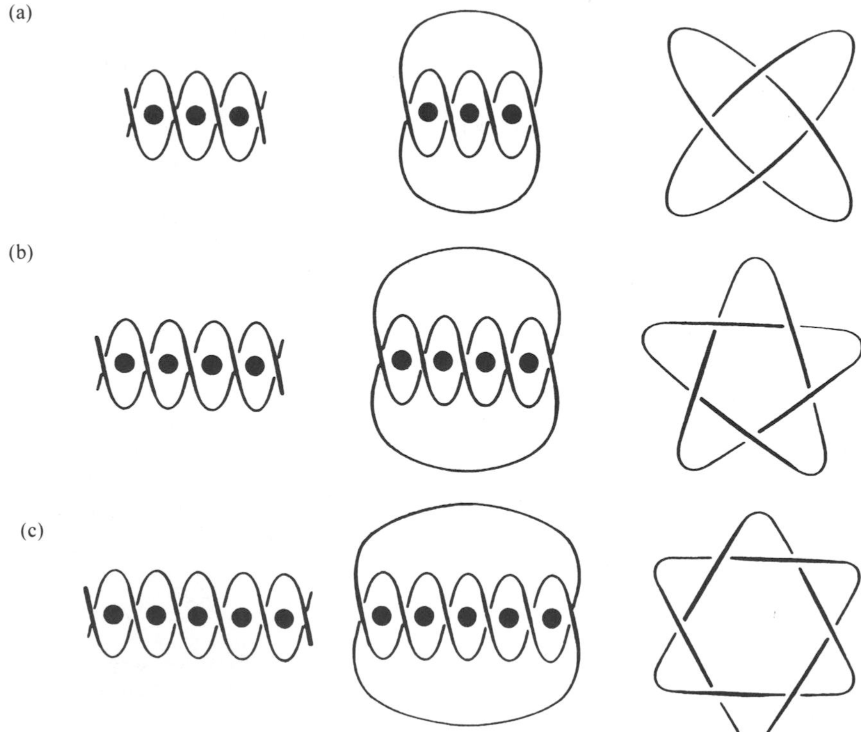

Figure 28 Concept of producing more complex interlocking and knotted systems from (a) three; (b) four; and (c) five-metal centers.

After the singly interlocked [2]catenane which requires two crossing points for its plane presentation, the second most simple [2]catenane is the four-crossing system depicted in Figure 29. It is noteworthy that a doubly interlocked [2]catenane is, like the trefoil knot, unconditionally topologically chiral.

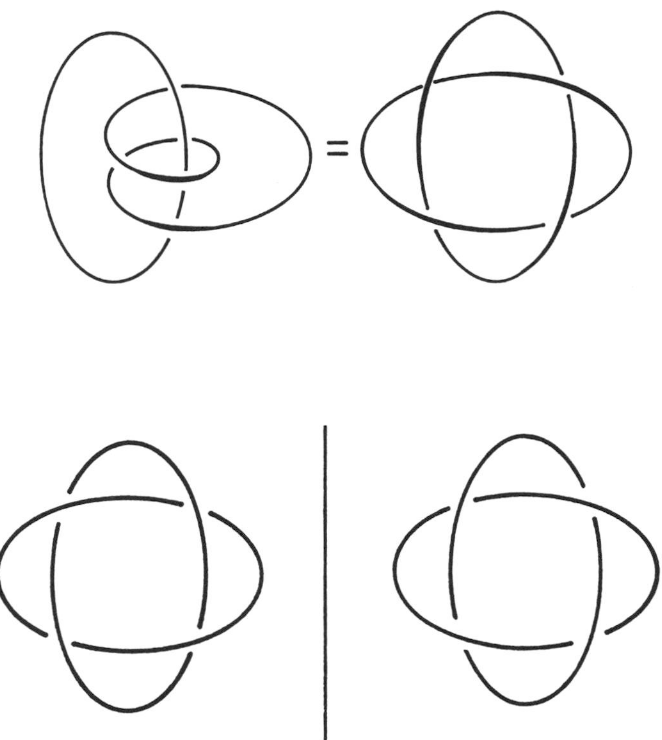

Figure 29 A doubly interlocked [2]catenane is topologically unconditionally chiral.

The strategy used to obtain the first doubly interlocked [2]catenane is depicted in Figure 30.[115] The key step here is the formation of a trinuclear copper(I) helicoidal complex C between an acyclic thread A and a macrocycle B which both contain three chelates. By cyclization of the strand originating from A, a doubly wound catenate should be obtained as a trinuclear complex. Organic precursors and reactions actually performed are represented in Scheme 20.

Compound (**89**) is the three-chelate fragment A of Figure 30 and (**90**) is the cyclic precursor B. It is assumed that by mixing (**89**), (**90**) and $Cu(MeCN)_4{}^+ \cdot PF_6{}^-$ in a 1:1:3 proportion, a significant amount of the precursor (**91**) is formed. Since several other copper(I) complexes were also formed (identified by TLC, NMR), leading to an inseparable mixture of compounds, the cyclization reaction supposed to lead to (**92**) was attempted on the crude mixture of products. The cyclization reaction was carried out in DMF with Cs_2CO_3 as base and the diiodo derivative of heptaethyleneglycol as the linker. After workup and column chromatography, a fraction containing an interesting mixture of copper(I) complexes was isolated. Due to the relatively poor stability of the complexes, the purification procedure was continued on a demetallated mixture. After several chromatographies, the doubly interlocked catenane was isolated in 2% yield and characterized as a free ligand by NMR and careful mass spectroscopy.[116]

2.5 CONCLUSION

Topologically novel molecules such as catenanes, rotaxanes, and knots have experienced a revival of interest since the mid-1980s. Following the beautiful contributions of the German[22,30,31,37,60,93] and American[23b,24,28,77] pioneers, the concept of template synthesis has indeed fully changed our view of such compounds and, for instance, led us to consider a catenane as a "normal" molecule and not as an exotic species prepared as a laboratory curiosity in exceedingly small amounts. The same is true for rotaxanes and related threaded multicomponent molecular

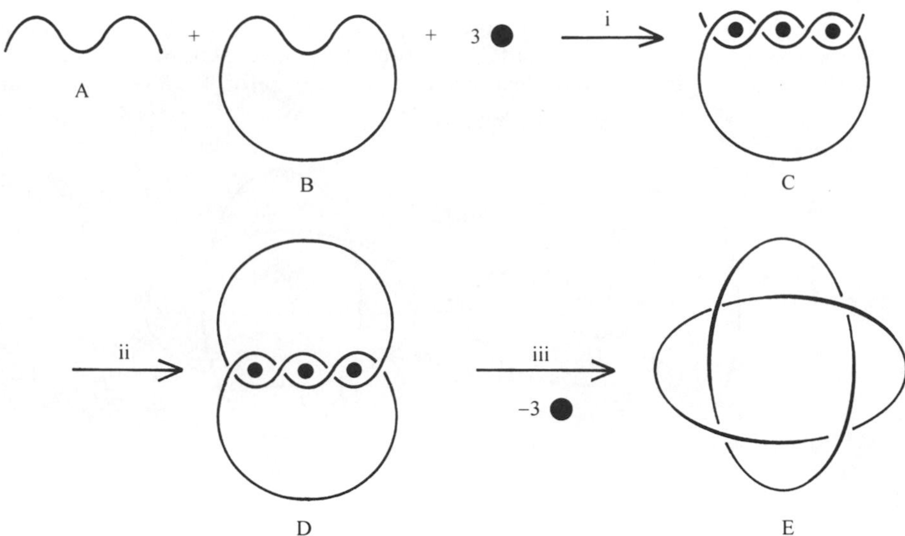

Figure 30 Compounds A and B contain three-chelate molecular fragments, with A being a simple string and B having its complexing subunit included in a cycle. (i) In presence of three metal centers (black circles) A and B can form a double-stranded double helix C. (ii) By cyclization of the strand originating from A, the doubly wound catenate is obtained as a trinuclear complex. (iii) Demetallation affords the free doubly interlocked [2]catenane E. Of course, both right- and left-handed double stranded helices should be formed, leading to D and subsequently E as racemates.

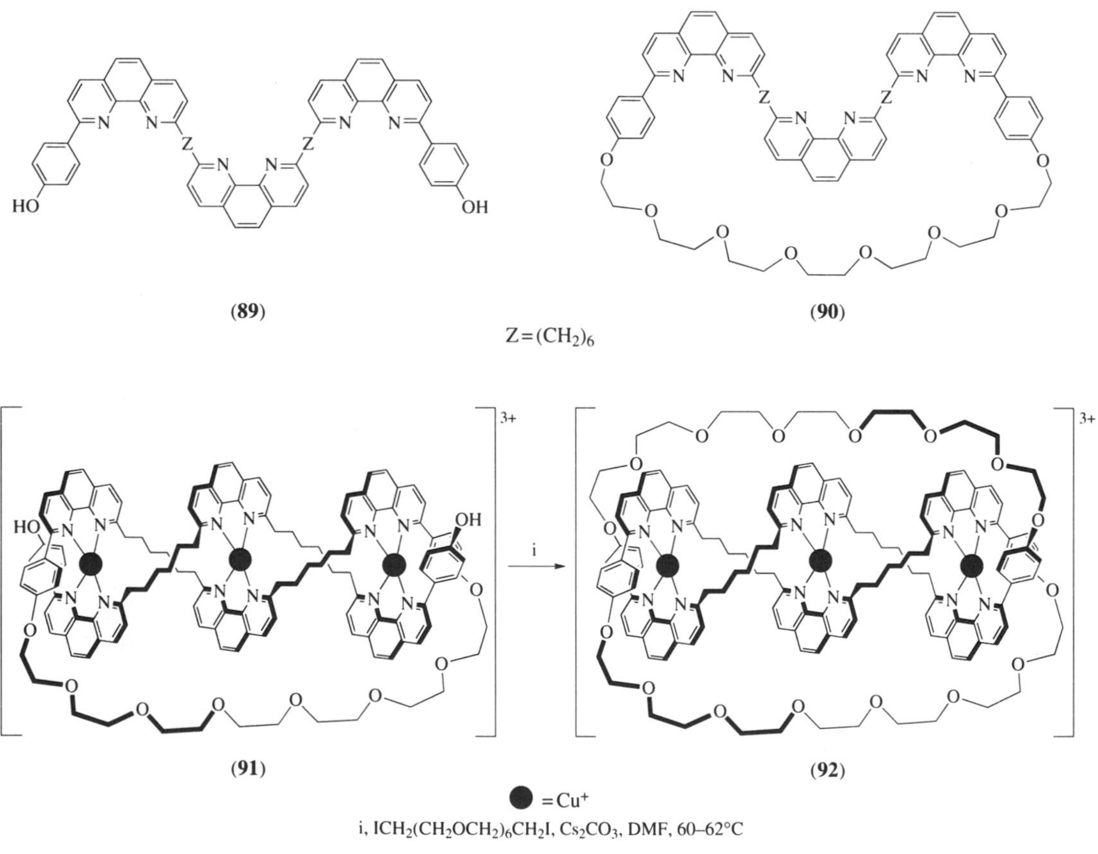

Scheme 20

systems, whose formation can be strictly controlled by using precursor fragments able to interact (acceptor–donor stacks[33,91] or transition metal complexes[85]) in a totally predictable manner.

If, until now, the main challenge has been to make these chemical objects, and to make them at a preparative scale, one can think now in terms of new properties and novel chemical, physical, or biological functions. It has already been shown that catenanes may display highly unusual coordination properties or lead to interesting copper(I) photoactive complexes. Rotaxanes have also demonstrated pronounced ability to promote ultrafast electron transfer over large distances in a way reminiscent of the natural photosynthetic system. Another potentially very important field of investigation is related to *controlled molecular motion*, under the action of light or an electrochemical signal.[33,85,91] Complex molecular systems, displaying strictly controlled photo- and electrochemical properties, and whose individual components can be reached and modified at will by an external signal, are very promising, for instance, in relation to information storage and imaging processes at the molecular level. It is hard to think of a molecular trefoil knot without immediately thinking of its chirality. Once molecular knots are resolved into their enantiomers, several lines of research can be explored, either in connection with asymmetric induction, catalysis, and enantioselective electron transfer, or in a more biologically oriented perspective. For example, positively charged multitransition metal complexes with a trefoil knot backbone could also display enantioselective interaction with DNA or other important biological molecules.

2.6 REFERENCES

1. (a) R. B. Woodward, T. Fukunaga, and R. C. Kelly, *J. Am. Chem. Soc.*, 1964, **86**, 3162; (b) P. E. Eaton, *Tetrahedron*, 1979, **35**, 2189; (c) first synthesis of dodecahedrane: R. J. Ternansky, D. W. Balogh, and L. A. Paquette, *J. Am. Chem. Soc.*, 1982, **104**, 4503; (d) crystallographic study of dodecahedrane: J. C. Gallucci, C. W. Doecke, and L. A. Paquette, *J. Am. Chem. Soc.*, 1986, **108**, 1343.
2. P. Chini, G. Longoni, and V. G. Albano, *Adv. Organomet. Chem.*, 1976, **14**, 285 and references therein.
3. (a) J. Deisenhofer, O. Epp, K. Miki, R. Huber, and H. Michel, *J. Mol. Biol.*, 1984, **180**, 385; (b) J. Deisenhofer, O. Epp, K. Miki, R. Huber, and H. Michel, *Nature*, 1985, **318**, 618.
4. (a) J. Deisenhofer and H. Michel, *EMBO J.*, 1989, **8**, 2149; (b) R. Huber, *Angew. Chem., Int. Ed. Engl.*, 1989, **28**, 848.
5. G. Bain, "The Methods of Construction of Celtic Art," Dover, New York, 1973.
6. J. C. Locher, "The world of M. C. Escher," H. N. Abram, New York, 1971.
7. (a) F. Michel and C. Weber, "Noeuds et Entrelacs," Institut de Mathématiques, University of Geneva, 1988; (b) J. H. Conway, in "Computational Problems in Abstract Algebra," Pergamon, New York, 1970, p. 329.
8. C. C. Adams, "The Knot Book," Freeman, New York, 1994.
9. S. A. Wasserman and N. R. Cozzarelli, *Science*, 1986, **232**, 951.
10. (a) B. Hudson and J. Vinograd, *Nature (London)*, 1967, **216**, 647; (b) D. A. Clayton and J. Vinograd, *Nature (London)*, 1967, **216**, 652.
11. (a) R. Radloff, W. Bauer, and J. Vinograd, *Proc. Natl. Acad. Sci. USA*, 1967, **57**, 1514; (b) W. Goebel and D. R. Helinski, *Proc. Natl. Acad. Sci. USA*, 1968, **61**, 1406; (c) G. Riou and E. Delain, *Proc. Natl. Acad. Sci. USA*, 1969, **62**, 210; (d) R. Jaenisch and A. J. Levine, *Virology*, 1971, **44**, 480; (e) H. C. MacGregor and M. Vlad, *Chromosoma*, 1972, **39**, 205; (f) K. N. Kreuzer and N. R. Cozzarelli, *Cell*, 1980, **20**, 245.
12. L. F. Liu, R. E. Depew, and J. C. Wang, *J. Mol. Biol.*, 1976, **106**, 439.
13. L. F. Liu, C.-C. Liu, and B. M. Alberts, *Cell*, 1980, **19**, 697.
14. S. A. Wasserman and N. R. Cozzarelli, *Proc. Natl. Acad. Sci. USA*, 1985, **82**, 1079.
15. M. A. Krasnow, A. Stasiak, S. J. Spengler, F. Dean, T. Koller, and N. R. Cozzarelli, *Nature*, 1983, **304**, 559.
16. J. D. Griffith and H. A. Nash, *Proc. Natl Acad. Sci. USA*, 1985, **82**, 3124.
17. J. C. Wang, *J. Mol. Biol.*, 1971, **55**, 523.
18. M. Gellert, K. Mizuuchi, M. H. O'Dea, and H. A. Nash, *Proc. Natl. Acad. Sci. USA*, 1976, **73**, 3872.
19. L. Yang, M. S. Wold, J. J. Li, T. J. Kelly, and L. F. Liu, *Proc. Natl. Acad. Sci. USA*, 1987, **84**, 1950.
20. (a) C. Liang and K. Mislow, *J. Am. Chem. Soc.*, 1994, **116**, 11189; (b) C. Liang and K. Mislow, *J. Am. Chem. Soc.*, 1995, **117**, 4201.
21. D. M. Walba, *Tetrahedron*, 1985, **41**, 3161.
22. G. Schill, "Catenanes, Rotaxanes and Knots," Academic Press, New York, 1971.
23. G. Schill, "Catenanes, Rotaxanes and Knots," Academic Press, New York, 1971, p. 1.
24. H. L. Frisch and E. Wasserman, *J. Am. Chem. Soc.*, 1961, **83**, 3789.
25. N. Van Gülick, *New. J. Chem.*, 1993, **17**, 619.
26. J.-P. Sauvage (ed.), Special issue of *New J. Chem.*, 1993, **17**.
27. V. I. Sokolov, *Russ. Chem. Rev. (Engl. Transl.)*, 1973, **42**, 452.
28. E. Wasserman, *J. Am. Chem. Soc.*, 1960, **82**, 4433.
29. (a) G. Agam, D. Graiver, and A. Zilkha, *J. Am. Chem. Soc.*, 1976, **98**, 5206; (b) G. Agam and A. Zilkha, *J. Am. Chem. Soc.*, 1976, **98**, 5214.
30. G. Schill, N. Schweickert, H. Fritz, and W. Vetter, *Angew. Chem., Int. Ed. Engl.*, 1983, **22**, 889.
31. A. Lüttringhaus, F. Cramer, H. Prinzbach, and F. M. Hengleir, *Liebigs Ann. Chem.*, 1958, **613**, 185.
32. G. J. M. Gruter, F. J. J. de Kanter, P. R. Markies, T. Nomoto, O. S. Akkerman, and F. Bickelhaupt, *J. Am. Chem. Soc.*, 1993, **115**, 12179.
33. (a) P. L. Anelli, P. R. Ashton, R. Ballardini, V. Balzani, M. Delgado, M. T. Gandolfi, T. T. Goodnow, A. E. Kaifer, D. Philp, M. Pietraszkiewicz, L. Prodi, M. V. Reddington, A. M. Z. Slawin, N. Spencer, J. F. Stoddart, C. Vicent, and D. J. Williams, *J. Am. Chem. Soc.*, 1992, **114**, 193; (b) D. B. Amabilino, P. R. Ashton, C. L. Brown, E. Cordova, L. A. Godinez, T. T. Goodnow, A. E. Kaifer, S. P. Newton, M. Pietraszkiewicz, D. Philp, F. M. Raymo, A. S. Reder, M. T. Rutland, A. M. Z. Slawin, N. Spencer, J. F. Stoddart, and D. J. Williams, *J. Am. Chem. Soc.*, 1995, **117**, 1271.

34. (a) C. A. Hunter, *J. Am. Chem. Soc.*, 1992, **114**, 5303; (b) H. Adams, F. J. Carver, and C. A. Hunter, *J. Chem. Soc., Chem. Commun.*, 1995, 809; (c) S. Ottens-Hildebrandt, M. Nieger, K. Rissaen, J. Rouvinen, S. Meier, G. Harder, and F. Vögtle, *J. Chem. Soc., Chem. Commun.*, 1995, 777 and references therein.
35. M. Fujita, F. Ibukuro, K. Yamaguchi, and K. Ogura, *J. Am. Chem. Soc.*, 1995, **117**, 4175.
36. D. M. Walba, R. M. Richards, and R. C. Haltiwanger, *J. Am. Chem. Soc.*, 1982, **104**, 3219.
37. (a) G. Schill and A. Lüttringhaus, *Angew. Chem.*, 1964, **76**, 567; (b) G. Schill, *Chem. Ber.*, 1967, **100**, 2021.
38. See footnote 7 of Ref. 24.
39. G. A. Melson (ed.) "Coordination Chemistry of Macrocyclic compounds," Plenum, New York, 1979, and references therein.
40. W. Reppe, O. Schlichting, K. Klager, and T. Toepel, *Liebigs Ann. Chem.*, 1948, **560**, 1.
41. A. B. P. Lever, *Adv Inorg. Radiochem.*, 1965, **7**, 28.
42. (a) N. F. Curtis, *J. Chem. Soc.*, 1960, 4409; (b) N. F. Curtis and D. A. House, *Chem. Ind. (London)*, 1961, 1708.
43. D. H. Busch, *Helv. Chim Acta*, 1967, 174 and references therein.
44. D. R. Boston and N. J. Rose, *J. Am. Chem. Soc.*, 1968, **90**, 6859.
45. (a) J. E. Parks, B. E. Wagner, and R. H. Holm, *J. Am. Chem. Soc.*, 1970, **92**, 3500; (b) J. E. Parks, B. E. Wagner, and R. H. Holm, *Inorg. Chem.*, 1971, **10**, 2472.
46. (a) I. I. Creaser, J. MacB. Harrowfield, A. J. Herlt, A. M. Sargeson, J. Springborg, R. J. Geue, and M. R. Snow, *J. Am. Chem. Soc.*, 1977, **99**, 3181; (b) I. I. Creaser, R. J. Geue, J. MacB. Harrowfield, A. J. Herlt, A. M. Sargeson, M. R. Snow, and J. Springborg, *J. Am. Chem. Soc.*, 1982, **104**, 6016.
47. (a) T. J. McMurry, S. J. Rodgers, and K. N. Raymond, *J. Am. Chem. Soc.*, 1987, **109**, 3451. (b) T. J. McMurry, M. W. Hosseini, T. M. Garrett, F. E. Hahn, Z. E. Reyes, and K. N. Raymond, *J. Am. Chem. Soc.*, 1987, **109**, 7196; (c) T. M. Garrett, T. J. McMurry, M. W. Hosseini, Z. E. Reyes, F. E. Hahn, and K. N. Raymond, *J. Am. Chem. Soc.*, 1991, **113**, 2965.
48. (a) P. Belser, L. De Cola, and A. Von Zelewsky, *J. Chem. Soc., Chem. Commun.*, 1988, 1057; (b) F. Barigelletti, L. De Cola, V. Balzani, P. Belser, A. Von Zelewsky, F. Vögtle, F. Ebmeyer, and S. Grammenudi, *J. Am. Chem. Soc.*, 1989, **111**, 4662.
49. C. O. Dietrich-Buchecker, P. A. Marnot, J.-P. Sauvage, J.-P. Kintzinger, and P. Maltèse, *Nouv. J. Chim.*, 1984, **8**, 573.
50. C. O. Dietrich-Buchecker and J.-P. Sauvage, *Tetrahedron Lett.*, 1983, **24**, 5091.
51. (a) C. O. Dietrich-Buchecker, J.-P. Sauvage, and J.-M. Kern, *J. Am. Chem. Soc.*, 1984, **106**, 3043; (b) C. O. Dietrich-Buchecker and J.-P. Sauvage, *Tetrahedron*, 1990, **46**, 503.
52. (a) V. Prelog and H. Gerlach, *Helv. Chim. Acta*, 1964, **47**, 2228; (b) H. Gerlach, J. A. Owtschinnikow and V. Prelog, *Helv. Chim. Acta*, 1964, **47**, 2294.
53. (a) D. K. Mitchell and J.-P. Sauvage, *Angew. Chem., Int. Ed. Engl.*, 1988, **27**, 930; (b) J.-C. Chambron, D. K. Mitchell and J.-P. Sauvage, *J. Am. Chem. Soc.*, 1992, **114**, 4625.
54. W. Pirkle and M. S. Hoekstra, *J. Am. Chem. Soc.*, 1976, **98**, 1832.
55. Y. Kaida, Y. Okamoto, J.-C. Chambron, D. K. Mitchell, and J.-P. Sauvage, *Tetrahedron Lett.*, 1993, **34**, 1019.
56. T. Jörgensen, J. Becher, J.-C. Chambron, and J.-P. Sauvage, *Tetrahedron Lett.*, 1994, **35**, 4339.
57. C. O. Dietrich-Buchecker, J.-P. Sauvage, and J. Weiss, *Tetrahedron Lett.*, 1986, **27**, 2257.
58. (a) M. J. Gunter and M. R. Johnston, *J. Chem. Soc., Chem. Commun.*, 1992, 1163; (b) M. J. Gunter, D. C. R. Hockless, M. R. Johnston, B. W. Shelton, and A. H. White, *J. Am. Chem. Soc.*, 1994, **116**, 4810 and references therein.
59. M. Momenteau, F. Le Bras, and B. Loock, *Tetrahedron Lett.*, 1994, **35**, 3289.
60. (a) G. Schill and K. Murjahn, *Liebigs Ann. Chem.*, 1970, **740**, 18; (b) G. Schill, K. Rissler, H. Fritz, and W. Vetter, *Angew. Chem., Int. Ed. Engl.*, 1981, **20**, 187.
61. D. B. Amabilino, P. R. Ashton, A. S. Reder, N. Spencer, and J. F. Stoddart, *Angew. Chem., Int. Ed. Engl.*, 1994, **33**, 1287; see also Ref. 33b.
62. J.-P. Sauvage and J. Weiss, *J. Am. Chem. Soc.*, 1985, **107**, 6108.
63. J. Guilhem, C. Pascard, J.-P. Sauvage, and J. Weiss, *J. Am. Chem. Soc.*, 1988, **110**, 8711.
64. C. O. Dietrich-Buchecker, A. Khemiss, and J.-P. Sauvage, *J. Chem. Soc., Chem. Commun.*, 1986, 1376.
65. (a) G. Eglinton and A. R. Galbraith, *Chem. Ind. (London)*, 1956, 737; (b) G. Eglinton and A. R. Galbraith, *J. Chem. Soc.*, 1959, 889.
66. (a) F. Sondheimer and Y. Amiel, *J. Am. Chem. Soc.*, 1957, **79**, 5817; (b) F. Sondheimer, Y. Amiel, and R. Wolovsky, *J. Am. Chem. Soc.*, 1957, **79**, 6263.
67. (a) E. T. Jarvi and H. W. Whitlock, Jr., *J. Am. Chem. Soc.*, 1980, **102**, 657; (b) S. P. Miller and H. W. Whitlock, Jr., *J. Am. Chem. Soc.*, 1984, **106**, 1492.
68. D. O'Krongly, S. R. Denmeade, M. Y. Chiang, and R. Breslow, *J. Am. Chem. Soc.*, 1985, **107**, 5544.
69. C. O. Dietrich-Buchecker, J. Guilhem, A. K. Khemiss, J.-P. Kintzinger, C. Pascard, and J.-P. Sauvage, *Angew. Chem. Int., Ed. Engl.*, 1987, **26**, 661.
70. H. M. Colquhoun, J. F. Stoddart, and D. J. Williams, *Angew. Chem., Int. Ed. Engl.*, 1986, **25**, 487 and references therein.
71. F. Bitsch, C. O. Dietrich-Buchecker, A. K. Khemiss, J.-P. Sauvage, and A. Van Dorsselaer, *J. Am. Chem. Soc.*, 1991, **113**, 4023.
72. F. Bitsch, G. Hegy, C. O. Dietrich-Buchecker, E. Leize, J.-P. Sauvage, and A. Van Dorsselaer, *New J. Chem.*, 1994, **18**, 801.
73. D. M. Walba, in "Graph Theory and Topology in Chemistry," eds. R. B. King and D. H. Rouvray, Elsevier, Amsterdam, 1987.
74. C. O. Dietrich-Buchecker, B. Frommberger, I. Lüer, J.-P. Sauvage, and F. Vögtle, *Angew Chem., Int. Ed. Engl.*, 1993, **32**, 1434.
75. F. Vögtle, I. Luer, V. Balzani, and N. Armaroli, *Angew. Chem., Int. Ed. Engl.*, 1991, **30**, 1333.
76. H. W. Gibson, M. K. Bheda, and P. T. Engen, *Prog. Polym. Sci.*, 1994, **19**, 843.
77. I. T. Harrison and S. Harrison, *J. Am. Chem. Soc.*, 1967, **89**, 5723.
78. Y. X. Shen, C. Lim, and H. W. Gibson, *Am. Chem. Soc. Polym. Prepr.*, 1991, **32**, 166.
79. T. D. Shaffer and L. M. Tsay, *J. Polym. Sci., Polym. Chem.*, 1991, **29**, 1213.
80. G. Wenz and B. Keller, *Angew. Chem., Int. Ed. Engl.*, 1992, **31**, 197.

81. A. Harada, J. Li, and M. Kamachi, *Nature*, 1992, **356**, 325.
82. (a) U. Kölle, *Angew. Chem., Int. Ed. Engl.*, 1991, **30**, 956 and references therein; (b) P. Gütlich and P. Poganiuch, *Angew. Chem., Int. Ed. Engl.*, 1991, **30**, 975.
83. (a) O. Kahn, J. Kröber, and C. Jay, *Adv. Mater.*, 1992, **4**, 718; (b) J. Kröber, E. Codjovi, O. Kahn, F. Grolière, and C. Jay, *J. Am. Chem. Soc.*, 1993, **115**, 9810; (c) H. Bolvin, O. Kahn, and B. Vekheter, *New. J. Chem.*, 1991, **15**, 889.
84. (a) M. Sano and H. Taube, *J. Am. Chem. Soc.*, 1991, **113**, 2327; (b) M. Sano and H. Taube, *Inorg. Chem.*, 1994, **33**, 705.
85. A. Livoreil, C. O. Dietrich-Buchecker, and J.-P. Sauvage, *J. Am. Chem. Soc.*, 1994, **116**, 9399.
86. N. E. Katz and F. Fagalde, *Inorg. Chem.*, 1993, **32**, 5391 and references therein.
87. (a) W. E. Geiger, A. Salzer, J. Edwin, W. von Philipsborn, U. Piantini, and A. L. Rheingold, *J. Am. Chem. Soc.*, 1990, **112**, 7113; (b) T. C. Richards and W. E. Geiger, *J. Am. Chem. Soc.*, 1994, **116**, 2028.
88. T. Roth and W. Kaim, *Inorg. Chem.*, 1992, **31**, 1930.
89. D. L. Jameson and L. E. Guise, *Tetrahedron Lett.*, 1991, **32**, 1999.
90. (a) J. A. Goodwin, D. M. Stanbury, L. J. Wilson, C. W. Eigenbrot, and W. R. Scheidt, *J. Am. Chem. Soc.*, 1987, **109**, 2979; (b) J. A. Goodwin, G. A. Bodager, L. J. Wilson, D. M. Stanbury, and W. R. Scheidt, *Inorg. Chem.*, 1989, **28**, 35; (c) J. A. Goodwin, L. J. Wilson, D. M. Stanbury, and R. A. Schott, *Inorg. Chem.*, 1989, **28**, 42.
91. (a) D. Philp and J. F. Stoddart, *Synlett*, 1991, 445 and references therein; (b) P. R. Ashton, C. L. Brown, E. J. T. Chrystal, K. P. Parry, M. Pietraszkiewicz, N. Spencer, and J. F. Stoddart, *Angew. Chem., Int. Ed. Engl.*, 1991, **30**, 1042; (c) R. Ballardini, V. Balzani, M. T. Gandolfi, L. Prodi, M. Venturi, D. Philp, H. G. Ricketts, and J. F. Stoddart, *Angew. Chem., Int. Ed. Engl.*, 1993, **32**, 1301.
92. D. M. Walba, J. D. Armstrong, A. E. Perry, R. M. Richards, T. C. Homan, and R. C. Haltiwanger, *Tetrahedron*, 1986, **42**, 1883.
93. (a) G. Schill, G. Doerjer, E. Logemann, and H. Fritz, *Chem. Ber.*, 1979, **112**, 3603; (b) G. Schill and J. Boeckmann, *Tetrahedron*, 1974, **30**, 1945.
94. J. H. Fuhrhop, G. Struckmeier, and U. Thewalt, *J. Am. Chem. Soc.*, 1976, **98**, 278.
95. W. S. Sheldrick and J. Engel, *J. Chem. Soc., Chem. Commun.*, 1980, 5.
96. J. V. Bonfiglio, R. Bonnet, D. G. Buckley, D. Hamzetash, M. B. Hursthouse, K. M. A. Malick, A. F. McDonald, and J. Trotter, *Tetrahedron*, 1983, **39**, 1865.
97. (a) G. C. van Stein, H. van der Poel and G. van Koten, *J. Chem. Soc., Chem. Commun.*, 1980, 1016; (b) G. C. Van Stein, G. van Koten, K. Vrieze, C. Brevard, and A. L. Spek, *J. Am. Chem. Soc.*, 1984, **106**, 4486.
98. G. C. van Stein, G. van Koten, H. Passenier, O. Steinebach, and K. Vrieze, *Inorg. Chim. Acta*, 1984, **89**, 79.
99. (a) J.-M. Lehn, A. Rigault, J. Siegel, J. Harrowfield, B. Chevrier, and D. Moras, *Proc. Natl. Acad. Sci. USA*, 1987, **84**, 2565; (b) J. M. Lehn and A. Rigault, *Angew. Chem., Int. Ed. Engl.*, 1988, **27**, 1095.
100. (a) E. C. Constable, M. G. B. Drew, and M. D. Ward, *J. Chem. Soc., Chem. Commun.*, 1987, 1600; (b) M. Barley, E. C. Constable, S. A. Corr, R. C. S. McQueen, J. C. Nutkins, M. D. Ward, and M. G. B. Drew, *J. Chem. Soc., Dalton. Trans.*, 1988, 2655; (c) E. C. Constable and M. D. Ward, *J. Am. Chem. Soc.*, 1990, **112**, 1256.
101. A. F. Williams, C. Piguet, and G. Bernardinelli, *Angew. Chem., Int. Ed. Engl.*, 1991, **30**, 1490.
102. R. Krämer, J.-M. Lehn, A. De Cian, and J. Fischer, *Angew. Chem., Int. Ed. Engl.*, 1993, **32**, 703.
103. C. O. Dietrich-Buchecker and J.-P. Sauvage, *Angew. Chem., Int. Ed. Engl.*, 1989, **28**, 189.
104. (a) R. West and E. G. Rochow, *J. Org. Chem.*, 1953, **18**, 1739; (b) J. X. McDermott, J. F. White, and G. M. Whitesides, *J. Am. Chem. Soc.*, 1976, **98**, 6529.
105. T. J. Curphey, E. J. Hoffman and C. McDonald, *Chem. Ind. (London)*, 1967, 1138.
106. C. O. Dietrich-Buchecker, J.-P. Sauvage, J. P. Kintzinger, P. Maltèse, C. Pascard, and J. Guilhem, *New. J. Chem.*, 1992, **16**, 931.
107. C. O. Dietrich-Buchecker and J.-P. Sauvage, *Bioorg. Chem. Frontiers*, 1991, **2**, 195 and references therein.
108. C. O. Dietrich-Buchecker, J.-F. Nierengarten, and J.-P. Sauvage, *Tetrahedron Lett.*, 1992, **25**, 3625.
109. C. O. Dietrich-Buchecker, J.-F. Nierengarten, J.-P. Sauvage, N. Armaroli, V. Balzani, and L. De Cola, *J. Am. Chem. Soc.*, 1993, **115**, 11 237 and references therein.
110. A.-M. Albrecht-Gary, C. O. Dietrich-Buchecker, J. Guilhem, M. Meyer, C. Pascard, and J.-P. Sauvage, *Recl. Trav. Chim. Pays-Bas*, 1993, **112**, 427.
111. C. O. Dietrich-Buchecker, J. Guilhem, C. Pascard and J.-P. Sauvage, *Angew. Chem., Int. Ed. Engl.*, 1990, **29**, 1154.
112. S. Rüttiman, C. Piguet, G. Bernardinelli, B. Bocquet, and A. F. Williams, *J. Am. Chem. Soc.*, 1992, **114**, 4230.
113. (a) E. C. Constable, M. J. Hannon, and D. A. Tocher, *Angew. Chem., Int. Ed. Engl.*, 1992, **31**, 230; (b) E. C. Constable, M. J. Hannon, and D. A. Tocher, *J. Chem. Soc., Dalton Trans.*, 1993, 1883.
114. C. O. Dietrich-Buchecker, J.-P. Sauvage, A. De Cian, and J. Fischer, *J. Chem. Soc., Chem. Commun.*, 1994, 2231.
115. J. F. Nierengarten, C. O. Dietrich-Buchecker, and J. P. Sauvage, *J. Am. Chem. Soc.*, 1994, **116**, 375.
116. C. O. Dietrich-Buchecker, E. Leize, J. F. Nierengarten, J.-P. Sauvage, and A. Van Dorsselaer, *J. Chem. Soc., Chem. Commun.*, 1994, 2257.

3

Donor–Acceptor Template-directed Synthesis of Catenanes and Rotaxanes

DAVID B. AMABILINO
Institut de Ciència de Materials de Barcelona, Spain
and
FRANÇISCO M. RAYMO and J. FRASER STODDART
University of Birmingham, UK

3.1 INTRODUCTION

3.1.1 Preamble

Supramolecular chemistry[1–7] is practised and exemplified at the ultimate level of sophistication by biological systems. Self-assembly processes[8,9] are responsible for the apparently effortless manner in which large and extremely complex natural molecular assemblies and supramolecular arrays are constructed. These processes rely upon the recognition between simple subunits, which are brought together rather precisely as a result of covalent and noncovalent bonding interactions to afford, through a series of complementary steps involving mutual templating,[10–12] a final thermodynamically stable, kinetically determined structure. The information content of the modular building blocks, expressed in terms of their stereoelectronic characteristics, imposes a high degree of control that dictates the entire self-assembly process with high overall efficiency and selectivity. The synthetic chemist has realized the potential of self-assembly,[8,9,13,14] self-organization[15–17] and self-replication[18–22] and now uses these concepts to tackle the chemical syntheses of large, well-defined molecular assemblies and supramolecular arrays. Indeed, a wide variety of abiotic self-assembling systems have been reported.[23–9] In large part, these developments have been fuelled by the desire to create synthetic chemical systems with functioning machine-like properties.

3.1.2 Catenanes, Rotaxanes and Pseudorotaxanes

The idea of imitating the features of mechanical interlocking at a molecular level was proposed for the first time as long ago as 1910 by Willstätter.[30,31] Molecules such as catenanes and rotaxanes,[32–40] composed of two or more distinct components, held together by means of a mechanical bond, have proved to be an intriguing synthetic challenge to chemists. Catenanes (Figure 1) are molecules composed of two or more macrocyclic components, mechanically interlocked as the links in a chain. Rotaxanes (Figure 1) are molecules comprising one or more macrocyclic components threaded by a rod-like component bearing large stoppers at its ends, a dumbbell in fact, in order to prevent the unthreading of the macrocyclic component(s). Pseudorotaxanes[41–52] (Figure 1) are supramolecular entities comprising two or more macrocyclic components threaded by a rod-like component. An [*n*]catenane is formed by *n* components, namely *n* macrocycles, while an [*n*]rotaxane and an [*n*]pseudorotaxane incorporate $n-1$ macrocycles and a linear component with and without stoppers, respectively. The template-directed syntheses of these compounds and complexes have become predictable and efficient. Furthermore, the investigation of their properties is attracting the interest of many researchers with the ultimate goal of generating molecular and/or supramolecular devices[1–7] at the nanoscopic[53–7] level with precise shapes and functions.

3.1.3 From Host–Guest Complexes to Catenanes, Rotaxanes and Pseudorotaxanes

The synthesis (Figure 2) of a molecular compound, such as a catenane or a rotaxane, featuring an aspect of mechanical interlocking requires the formation of a pseudorotaxane-like or precatenane-like intermediate. In other words, the threading of a macrocyclic compound over an acyclic compound is the key step for the generation, after macrocyclization, of a catenane and, after the covalent attachment of stoppers at both ends of an acyclic compound, a rotaxane.

The early syntheses of catenanes[31,58,59] and rotaxanes[60–4] relied for the most part upon statistical threading. By mixing an acyclic compound with a large excess of a macrocycle of a suitable size, it could be expected that a small portion of the acyclic compound would be inserted through the cavity of the macrocycle, leading to the required pseudorotaxane intermediate. Indeed, statistical syntheses of catenanes and rotaxanes have been achieved by several investigators.[31,58–64] However, the yields were very low. The advent of supramolecular chemistry[1–7] has led to a dramatic improvement in the efficiency of the syntheses of these intriguing compounds. In particular, cyclodextrins, since they include in their cavities linear organic molecules in a pseudorotaxane-like manner, have been widely used in the syntheses of rotaxanes,[65–70] polyrotaxanes[71–3] and catenanes.[74–6] An elegant and extremely efficient metal-templated approach to the synthesis of catenanes and rotaxanes, and even molecular knots, has been developed by Sauvage and co-workers.[35,37,39] Hydrogen-bonding interactions have also been employed[12,77–81] to generate catenated structures consisting of interlocked macrocyclic lactams. Crown ethers able to bind

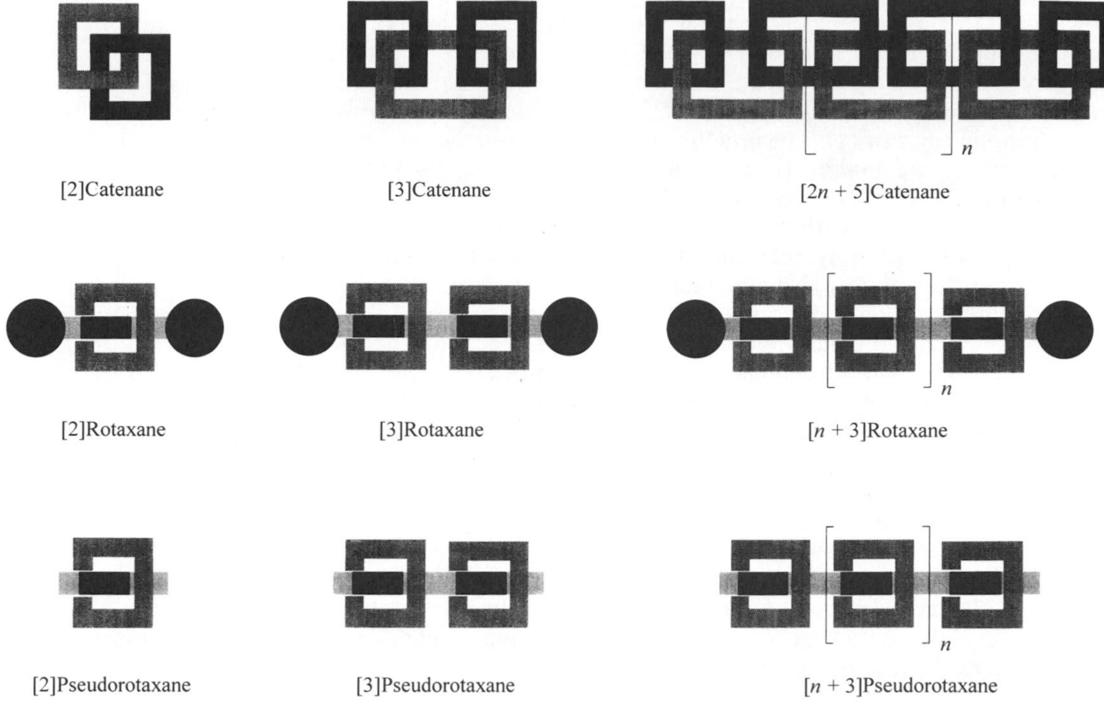

Figure 1 Schematic representation of some catenanes, rotaxanes and pseudorotaxanes.

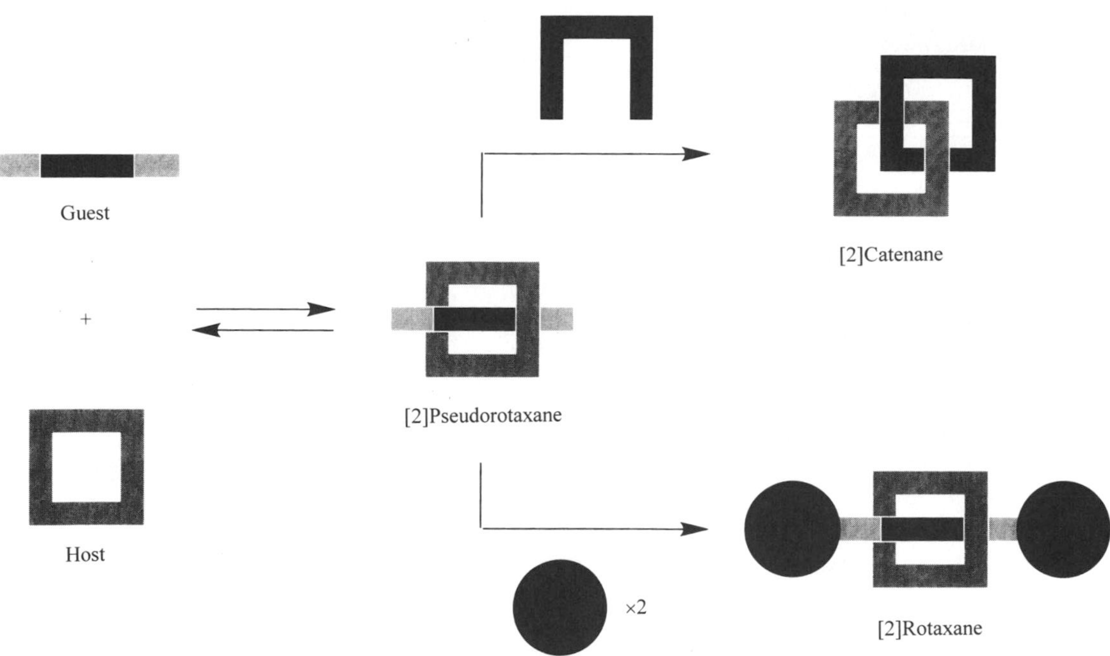

Figure 2 Host–guest approach to catenanes and rotaxanes.

organomagnesium derivatives have been employed for the syntheses[82,83] of both organometallic catenanes and rotaxanes. The self-assembly of organometallic catenanes comprising transition metal based macrocycles has been achieved by Fujita *et al.* [84,85]

In the late 1980s, we began developing a self-assembly approach[86,87] to catenanes and rotaxanes which relies upon the high complementarity that exists between π-electron-rich and π-electron-deficient aromatic building blocks. A search for a synthetic receptor for the π-electron-deficient herbicide paraquat (**2**) [88] led to the preparation[89] of the macrocyclic polyether di-*p*-phenylene-34-crown-10 (**1**). This π-electron-rich macrocyclic polyether binds[41] (Figure 3) paraquat bis(hexafluorophosphate) (**2**) in solution (K_a = 730 M^{-1} in Me_2CO), affording a [2]pseudorotaxane superstructure. The x-ray crystal structure of the complex (**1·2**) confirmed the incorporation of paraquat (**2**) inside the cavity of the macrocycle (**1**), with the bipyridinium unit sandwiched between the two hydroquinone rings of (**1**). The noncovalent bonding interactions responsible for the self-assembly of the complex are mainly (i) π–π stacking[90–4] between the bipyridinium unit and the hydroquinone rings, and (ii) hydrogen-bonding[27,28,95–7] interactions between the CH acidic hydrogen atoms in the α position with respect to the nitrogen atoms of the bipyridinium unit and the central oxygen atoms of the polyether chains in the macrocycle. In addition, the macrocycle is highly preorganized,[98] as indicated by the fact that its conformations in the free and complexed states in the solid state are remarkably similar. On reversing the nature of the recognition sites incorporated within the host and the guest, we anticipated that a π-electron-deficient cyclophane incorporating bipyridinium units would be able to bind π-electron-rich acyclic polyethers incorporating hydroquinone rings. The tetracationic cyclophane (**6**) binds[42,43] (Figure 3) various acyclic molecules incorporating π-electron-rich residues, including 1,4-dimethoxybenzene (**7**) and the acyclic polyether derivatives (**8**)–(**11**). The complexes (**6·7**)–(**6·11**) adopt pseudorotaxane geometries both in solution and in the solid state. Aromatic π–π stacking interactions between the bipyridinium units and the hydroquinone ring, as well as hydrogen-bonding interactions between some of the polyether oxygen atoms in (**8**)–(**11**) and the α-CH hydrogen atoms on the bipyridinium units, along with T-type[99–101] interactions between the hydrogen atoms of the hydroquinone ring and the *p*-xylyl spacers of the host, are responsible for the relatively high stabilities of these complexes. The rigidity of cyclobis(paraquat-*p*-phenylene) imposes upon it high preorganization, and hence gives it its ability to form strong complexes.

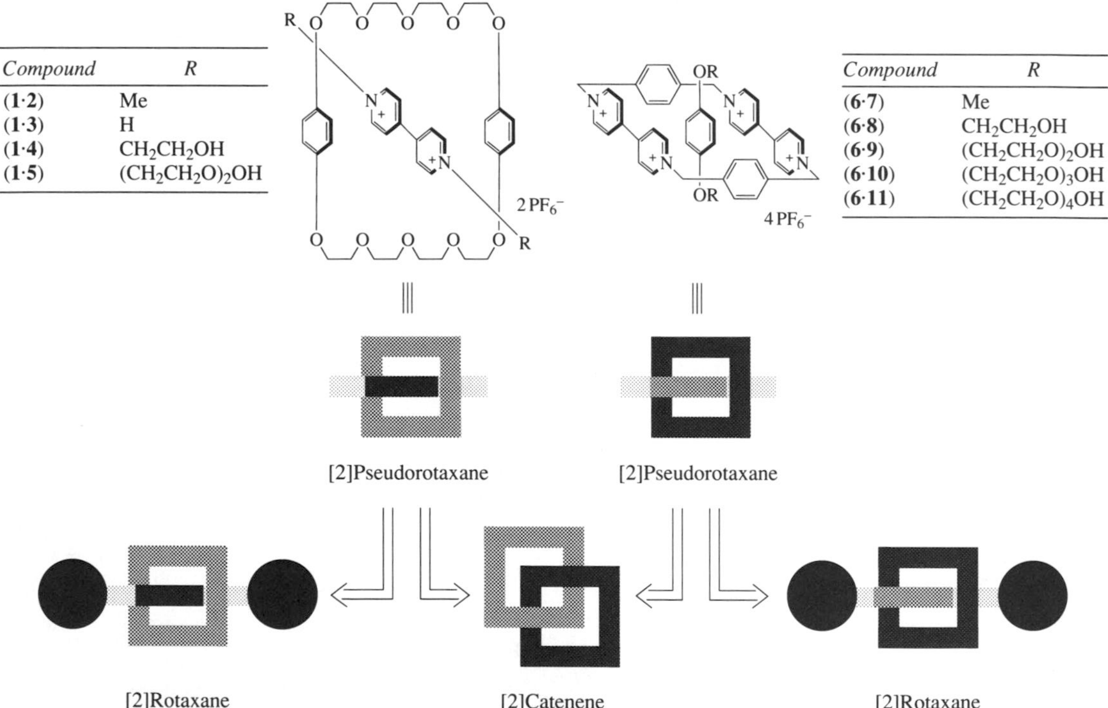

Compound	*R*
(**1·2**)	Me
(**1·3**)	H
(**1·4**)	CH_2CH_2OH
(**1·5**)	$(CH_2CH_2O)_2OH$

Compound	*R*
(**6·7**)	Me
(**6·8**)	CH_2CH_2OH
(**6·9**)	$(CH_2CH_2O)_2OH$
(**6·10**)	$(CH_2CH_2O)_3OH$
(**6·11**)	$(CH_2CH_2O)_4OH$

Figure 3 Conceptual link between supramolecular complexes such as pseudorotaxanes and the self-assembly of catenanes and rotaxanes.

The efficient formation of these 1:1 complexes in solution strongly suggests the possibility of employing them for the generation (Figure 2) of the pseudorotaxane intermediate required in the synthesis of both catenanes and rotaxanes. Indeed, the progression from these 1:1 complexes to catenanes and rotaxanes, as schematically shown in Figure 3, has been realized as a result of the appropriate combination of π-electron-rich and π-electron-deficient components.

3.2 PSEUDOROTAXANES

3.2.1 Pseudorotaxanes Incorporating π-Electron-rich Macrocycles and π-Electron-deficient Acyclic Components

In the complex (**1·2**), the π-electron-rich macrocycle (**1**) incorporates (Figure 3) within its cavity the bipyridinium-based guest (**2**), giving rise to a [2]pseudorotaxane. Inspired by this result, several 4,4′-disubstituted bipyridinium dications (**3**)–(**5**) were synthesized. Their inclusion with pseudorotaxane-like geometries inside the cavity of di-*p*-phenylene-34-crown-10 (**1**) has been observed both in solution and in the solid state.[44,45] A complex with similar geometry is formed[46] upon combination of the π-electron-deficient guest (**2**) and the macrocyclic polyether 1,5-dinaphtho-38-crown-10 (**12**). On employing the even larger macrocycle 1,5-dinaphtho-44-crown-12 (**13**), a 2:1 array (**2·13**) is formed[47] with paraquat (**2**) in the solid state. A continuous π-donor–π-acceptor stack exists (Figure 4) in the solid state for the array (**2·13**). The macrocycle (**13**) incorporates one paraquat unit (**2**) in its cavity, while a second paraquat residue (**2**) is sandwiched between adjacent 1:1 complexes. The self-organization in the solid state of this 2:1 array proved to be the inspiration for the making of the first catenane. The two different paraquat units are separated by a distance of ~700 pm, which is the length of the *p*-xylyl spacers separating the bipyridinium units in the cyclophane (**6**).

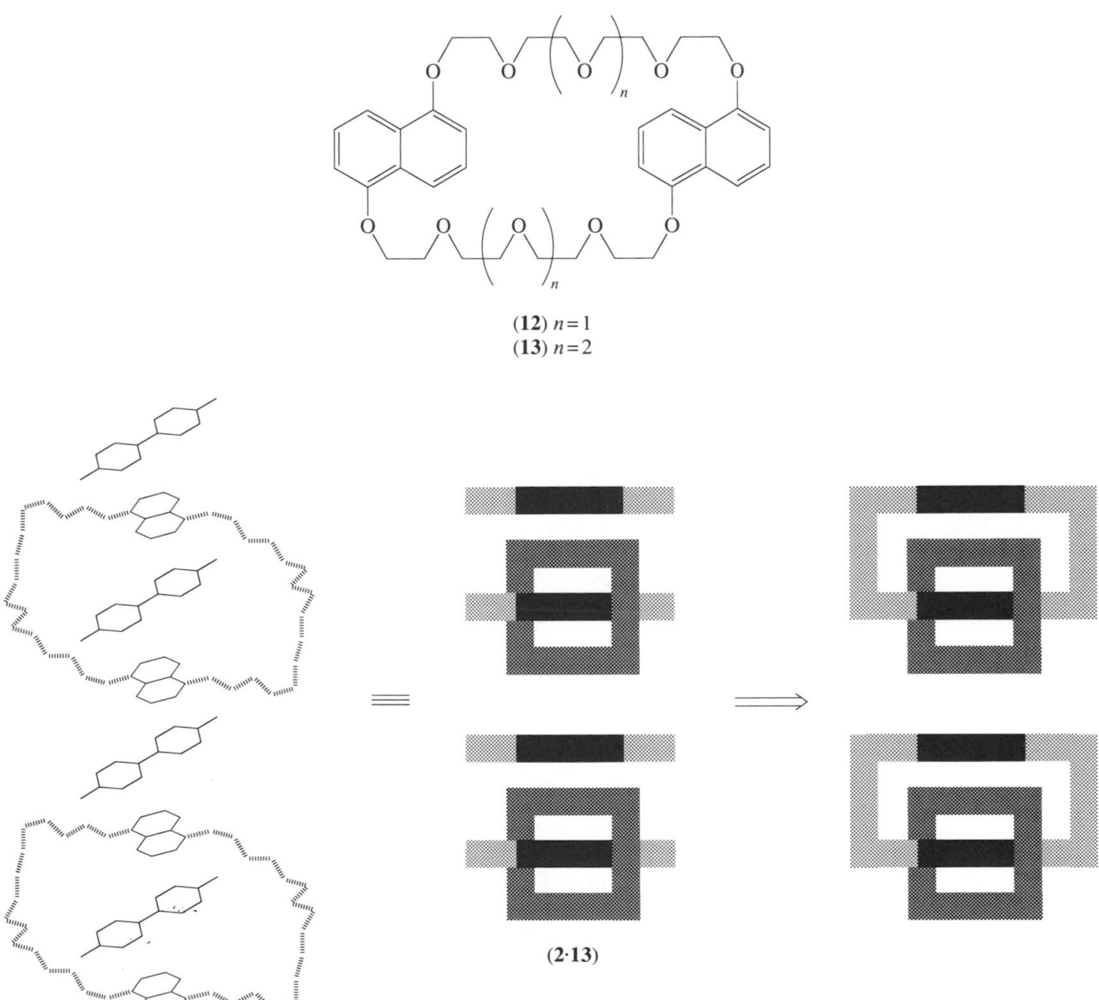

Figure 4 Self-organization in the solid state of the [2]pseudorotaxane (**2·13**) and conceptual link with the self-assembly of a [2]catenane.

3.2.2 Pseudorotaxanes Incorporating π-Electron-deficient Macrocycles and π-Electron-rich Acyclic Components

The tetracationic cyclophane (**6**) forms[42,43] (Figure 3) pseudorotaxane-like complexes both in solution and in the solid state with the hydroquinone-based guests (**7**)–(**11**). On increasing the length of the substituents attached to the 1,4-dioxybenzene by the insertion of ethylene glycol repeating units, we find that the association constants in acetonitrile for the 1:1 complexes increase

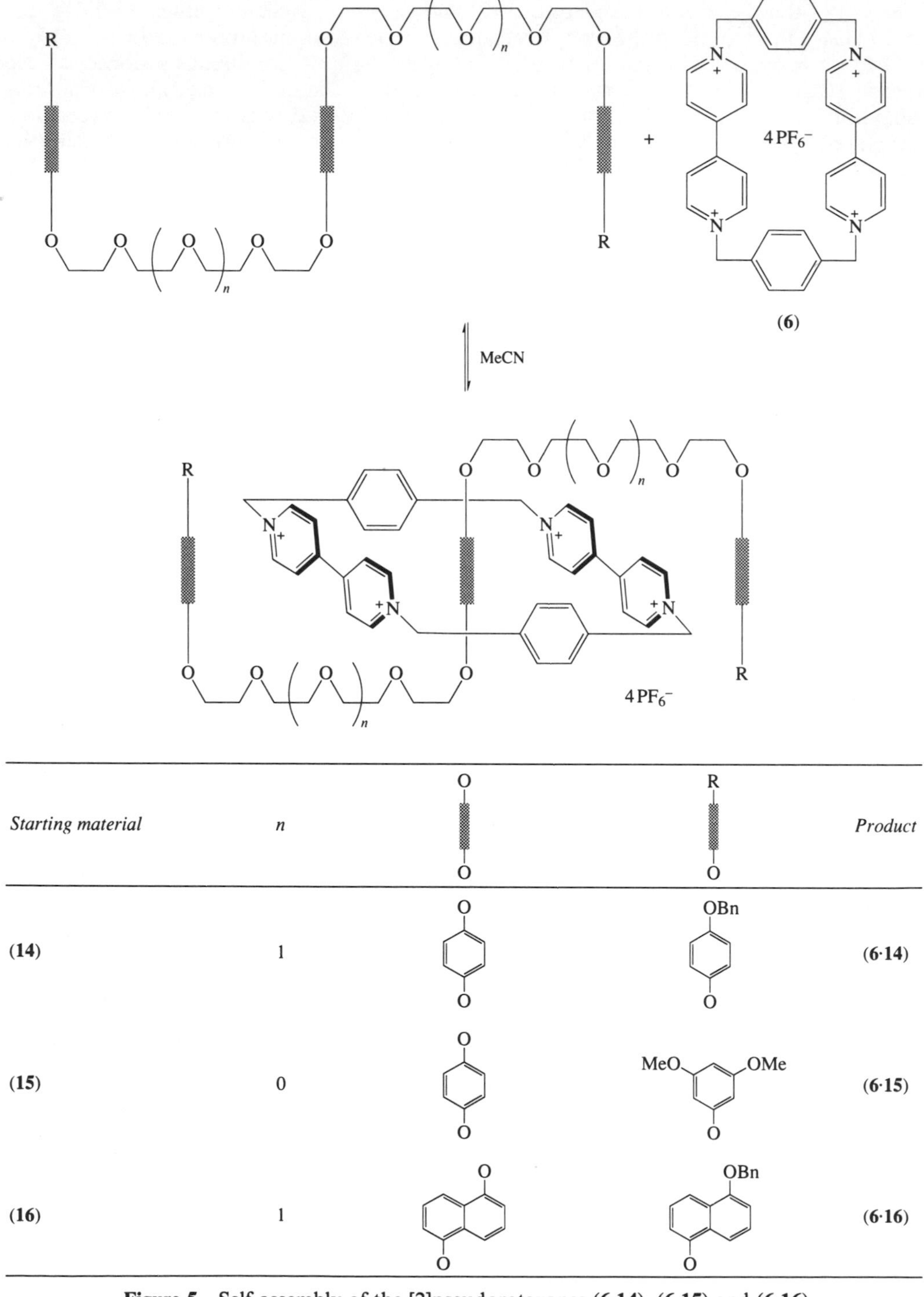

Starting material	n	O–▒–O	R–▒–O	Product
(**14**)	1	1,4-dioxybenzene	OBn	(**6·14**)
(**15**)	0	1,4-dioxybenzene	MeO, OMe	(**6·15**)
(**16**)	1	1,5-dioxynaphthalene	OBn	(**6·16**)

Figure 5 Self-assembly of the [2]pseudorotaxanes (**6·14**), (**6·15**) and (**6·16**).

from 17 M^{-1} in the case of (**6·7**) to 2520 M^{-1} for (**6·11**). Encouraged by these results, a series of molecular 'threads', incorporating from three to five π-electron-rich units, were prepared. The π-electron-rich acyclic derivatives (**14**), (**15**) and (**16**) form[48,49] (Figure 5) pseudorotaxane-like complexes with (**6**) both in solution and in the solid state. The x-ray crystal structures of the 1:1 complexes show the central π-electron-rich ring of the acyclic component in each case to be located within the cavity of the cyclophane (**6**), while the terminal π-electron-rich aromatic residues form π–π stacking interactions alongside the outer surface of the bipyridinium units in the cyclobis (paraquat-*p*-phenylene).

By increasing the number of recognition sites along the 'thread', the self-assembly of higher-order pseudorotaxanes has been realized. The acyclic derivative (**17**), incorporating five hydroquinone rings, when mixed with one molar equivalent of (**6**) forms[50] (Figure 6) the [2]pseudo-rotaxane (**6·17**). Addition of a further equivalent of the tetracationic cyclophane (**6**) affords the [3]pseudorotaxane (**6_2·17**).

The complexation (Figure 7) of 1,5-dimethoxynaphthalene (**18**) by the cyclophane (**6**) led to the design of a [3]pseudorotaxane.[51] The x-ray crystal structure of the [2]pseudorotaxane (**6·18**) revealed that the distance between the oxygen atoms of neighbouring complexes corresponds approximately to the length of a propyl chain. Indeed, the linking of two 1,5-dioxynaphthalene

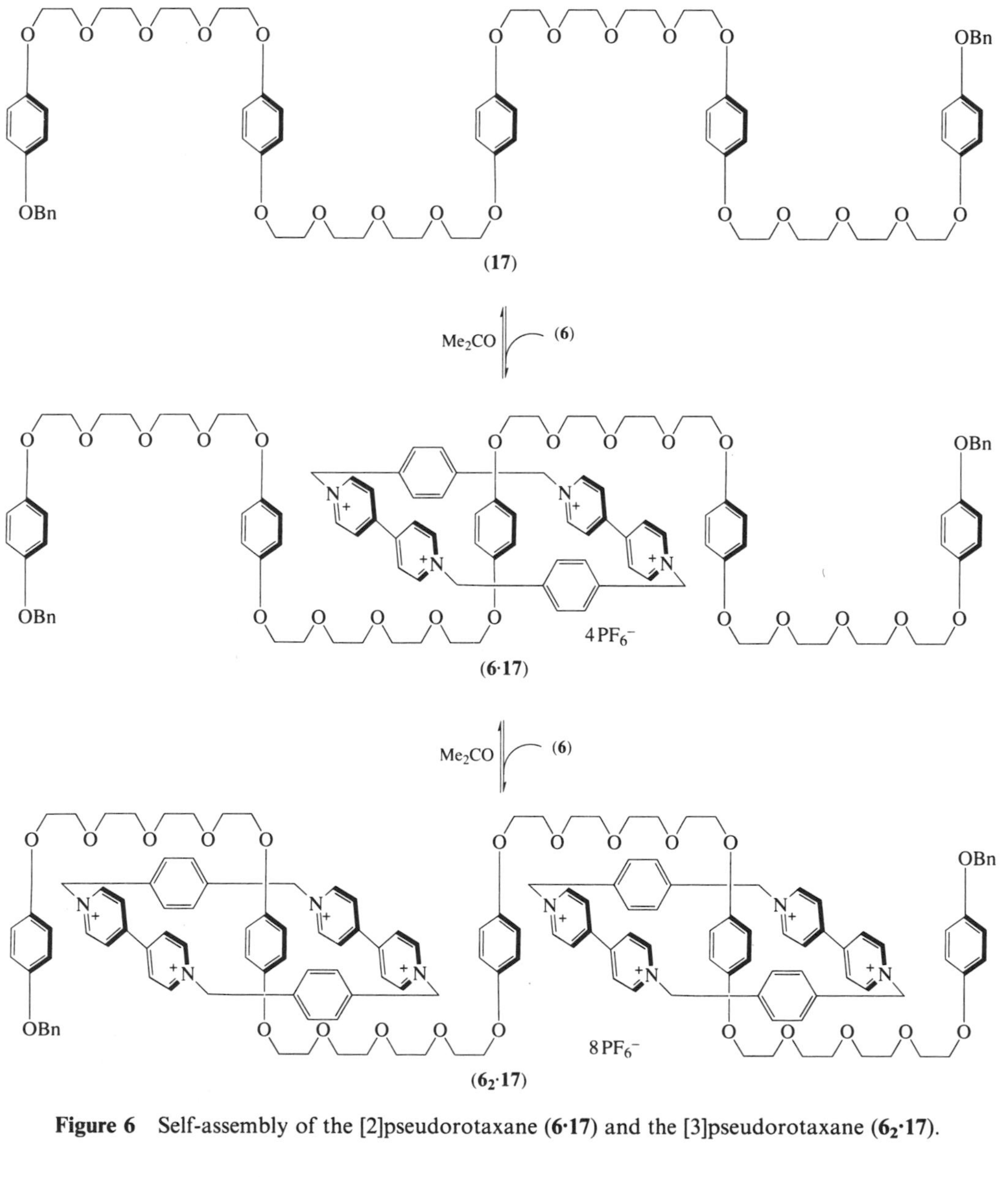

Figure 6 Self-assembly of the [2]pseudorotaxane (**6·17**) and the [3]pseudorotaxane (**6_2·17**).

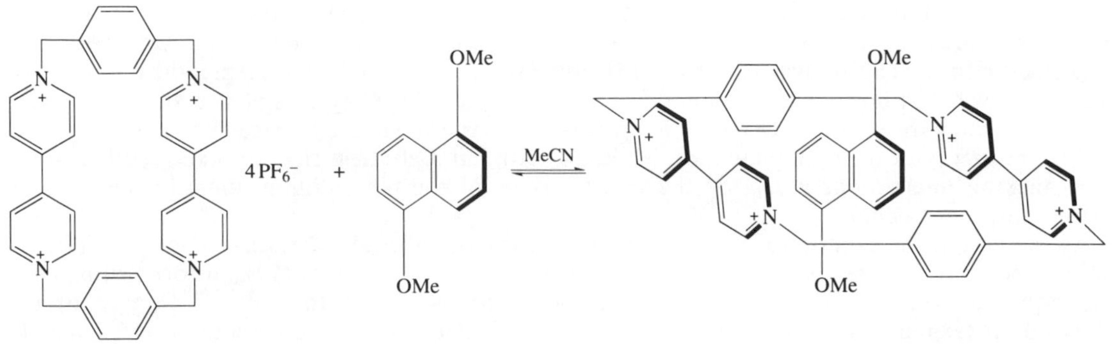

(6) (18) (6·18)

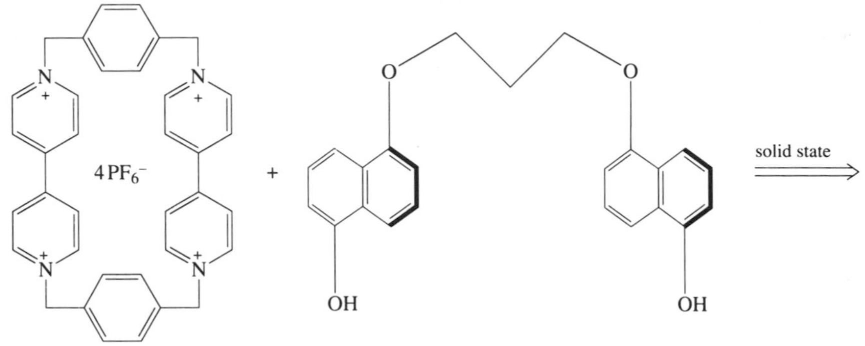

(6) (19)

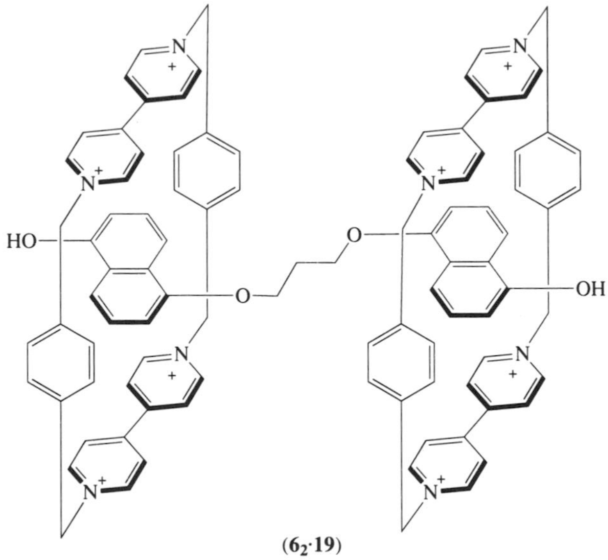

(6_2·19)

Figure 7 Self-assembly of the pseudorotaxanes (**6·18**) and (**6_2·19**).

units by an *n*-propyl chain afforded the molecular thread (**19**), which upon complexation with the tetracationic cyclophane (**6**) forms (Figure 7) the [3]pseudorotaxane (**6_2·19**).

The reversible attenuation of the conductivity of a thiophene-based polymer (**21**) has been achieved[52,102] by Swager and co-workers as a result of the formation (Figure 8) of poly-pseudorotaxane-like derivatives upon complexation of the polymer with paraquat (**2**). The highly conjugated polymer (**21**) was prepared from the macrocycle (**20**), which itself binds paraquat (**2**). Upon combination of (**21**) and (**2**), an anodic shift in the redox potential, along with the attenuation of the conductivity of the polymer, is observed as a result of polypseudorotaxane formation.

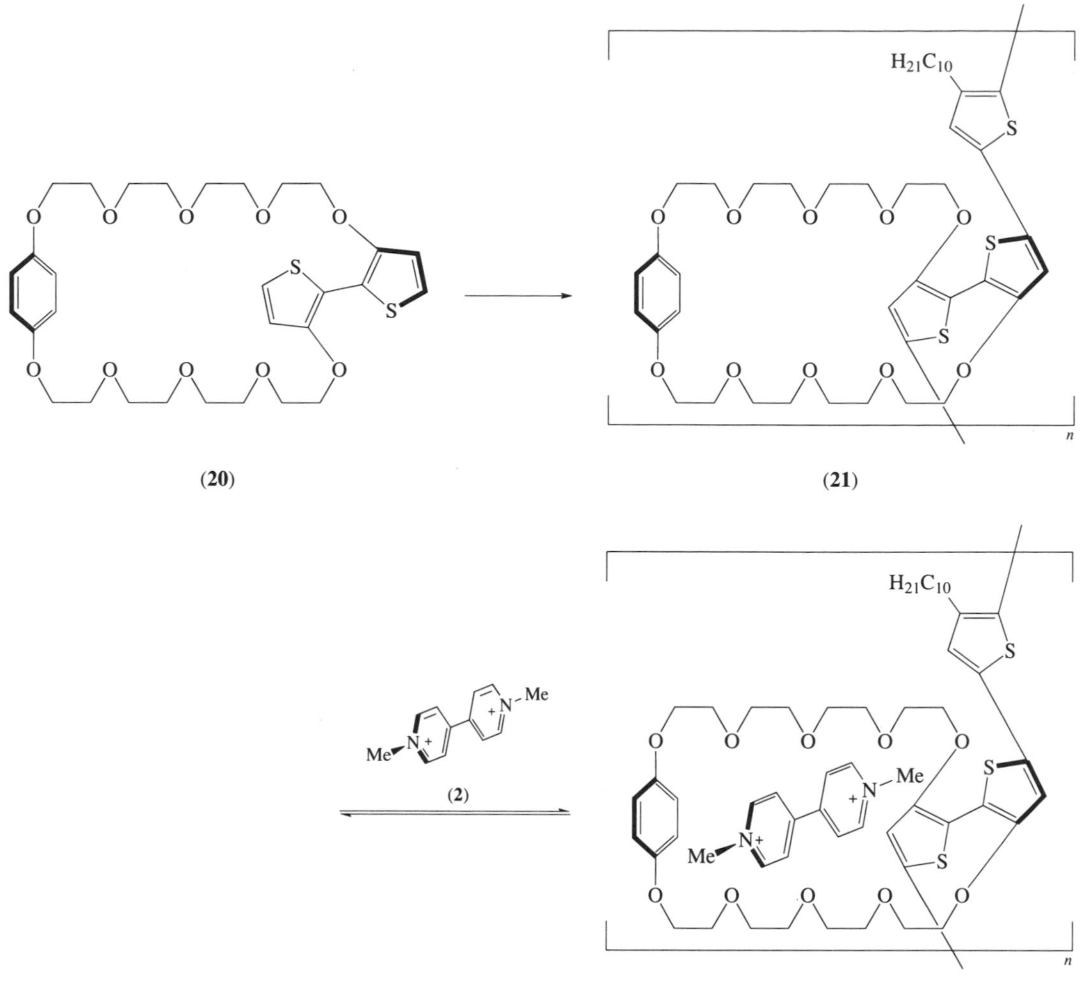

(20) **(21)**

(2)

(2_n·21)

Figure 8 Self-assembly of the polypseudorotaxanes (**2_n·21**).

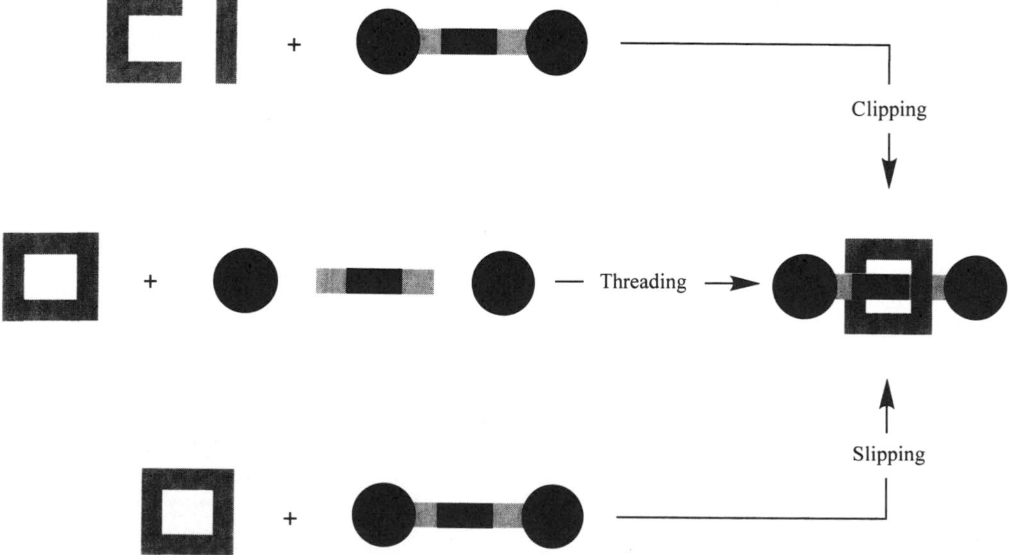

Figure 9 Synthetic approaches to self-assembling rotaxanes.

3.3 ROTAXANES

3.3.1 Self-assembling Approaches to Rotaxanes

The self-assembly of rotaxanes incorporating π-electron-rich and π-electron-deficient components can be achieved[86,87] by employing one of the three methodologies schematically represented in Figure 9. In the clipping approach, the macrocyclization of the cyclic component is performed in the presence of a preformed dumbbell-shaped component which acts as the template for ring closure. In the case of threading, the inclusion of an acyclic derivative inside the cavity of a preformed macrocycle is followed by the covalent attachment of the two stopper units which prevent the unthreading of the ring from the newly formed dumbbell-shaped component. In the slipping approach, both macrocyclic and dumbbell-shaped components are preformed separately and then they are persuaded thermodynamically into associating one with the other under the influence of just the right amount of thermal energy.

3.3.2 Rotaxanes Incorporating π-Electron-deficient Macrocycles and π-Electron-rich Acyclic Components

The self-assembly of [2]rotaxanes incorporating one hydroquinone ring within the dumbbell-shaped component has been achieved[43] (Scheme 1) by both the clipping and threading approaches. The combination of acyclic hydroquinone-containing derivatives such as (**9**) and (**10**) with the tetracationic cyclophane (**6**) affords the corresponding [2]pseudorotaxanes (**6·9**) and (**6·10**), respectively. The reaction of the terminal hydroxy groups of the acyclic components of the pseudorotaxanes (**6·9**) and (**6·10**) with triisopropylsilyl triflate in the presence of lutidine gives the corresponding [2]rotaxanes (**22**) and (**23**), both in 22% yield. The reaction of the dibromide (**25**) with the dication (**26**) in the presence of the preformed dumb-bell-shaped component incorporating one hydroquinone ring (**24**) affords the [2]rotaxane (**23**) in a yield of 14% after counterion exchange. The clipping procedure was also employed in the self-assembly[103] of a [2]rotaxane analogous to (**23**), but adamantoyl stopper groups were used to prevent the unthreading of the tetracationic macrocycle.

The [2]rotaxane (**32**), self-assembled[104] in 32% yield (Scheme 2), incorporates two stations, namely hydroquinone rings, along the dumbbell-shaped component. Therefore, the π-electron-deficient macrocyclic component moves back and forth (Figure 10) from one station to the other, giving rise to a so-called molecular shuttle. The movement of the cyclophane along the dumbbell-shaped component is clearly revealed by variable-temperature ^{1}H NMR spectroscopy. At 140 °C in CD_3SOCD_3, the process occurs rapidly on the ^{1}H NMR timescale and the protons attached to the hydroquinone rings resonate as a single AA′BB′ system centred at δ 5.16. On cooling down a CD_3COCD_3 solution of the [2]rotaxane (**33**) to −50 °C, two distinct AA′BB′ systems can be identified for the occupied station and unoccupied station at δ 3.80 and δ 6.38, respectively. By employing the coalescence method,[105,106] the energy barrier for the shuttling process was calculated to be 54.4 kJ mol^{-1}.

The [2]rotaxanes (**33**)–(**36**) have also been self-assembled[107,108] (Scheme 2) employing the clipping procedure. It is noteworthy that, as the number of π-electron-rich residues is increased, the yields of the [2]rotaxanes (**33**)–(**36**) increase also. However, no higher-order rotaxanes were formed. The low yield for the [2]rotaxane (**33**) is probably a result of the proximity of the blocking groups to the templating aromatic residue.

In order to investigate further the equilibrium dynamics of the shuttling process occurring within a [2]rotaxane, a range of [2]rotaxanes incorporating two or three different π-electron-rich stations along the dumbbell-shaped component were self-assembled. The possibility of controlling (Figure 11) by some external stimulus—chemical, electrochemical or photochemical—the equilibration between the translational isomers[109] could obviously lead to potential molecular switches in the shape of controllable molecular shuttles.

The self-assembly of the [2]rotaxane (**37**) (Figure 10) was achieved[110] by the clipping procedure in 8% yield as a result of reacting the preformed dumbbell-shaped component with the dication (**26**) and the dibromide (**25**). It was envisaged that the tetracationic cyclophane would occupy mainly the hydroquinone station, and after the oxidation of the hydroquinone unit, the macrocycle would then be obliged to move to the adjacent *p*-xylyl station. However, variable-temperature ^{1}H NMR spectroscopy revealed not only the rapid equilibration between the two translational

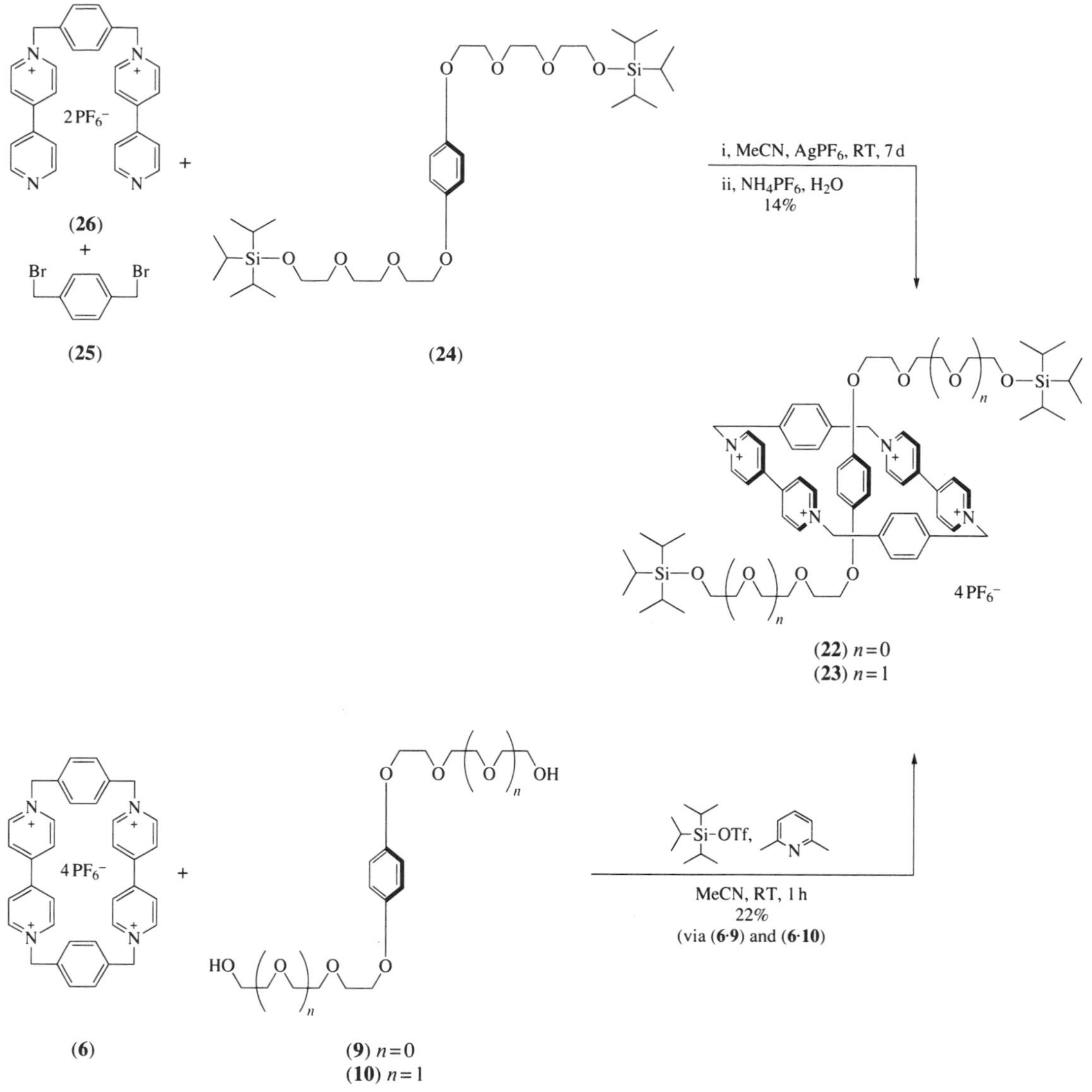

2 PF_6^-

(**26**) + (**25**) + (**24**)

i, MeCN, $AgPF_6$, RT, 7 d
ii, NH_4PF_6, H_2O
14%

4 PF_6^-

(**22**) $n=0$
(**23**) $n=1$

4 PF_6^-

(**6**) + (**9**) $n=0$, (**10**) $n=1$

Si-OTf,
MeCN, RT, 1 h
22%
(via (**6·9**) and (**6·10**))

Scheme 1

isomers at 70°C in CD_3CN, but at −40°C the ratio between the two isomers was only 7:3 in favour of the one having the more π-electron-rich station, namely the hydroquinone ring, inside the cavity of the macrocycle. By replacing the *p*-xylyl unit incorporated within the dumbbell-shaped component with an indole residue, it had been hoped to create a molecular shuttle with better equilibrium characteristics. The [2]rotaxane (**38**) (Figure 10) was self-assembled[111] via the clipping procedure in 9% yield. Surprisingly, however, it existed exclusively as the translational isomer with the hydroquinone unit located inside the cavity of the macrocycle! Presumably, this clear preference is a result of the steric size of the more π-electron-rich indole unit. The redox properties of tetrathiofulvalene and its complexation[112] by the tetracationic cyclophane (**6**) suggested the possibility of synthesizing a [2]rotaxane incorporating the tetrathiofulvalene unit. The [2]rotaxane (**40**) was self-assembled[113] (Scheme 3; 1 bar = 10^5 Pa) by reacting the preformed dumbbell-shaped component (**39**) with the dibromide (**25**) and the dication (**26**) under ultrahigh-pressure conditions. Variable-temperature ^1H NMR spectroscopic studies revealed that the equilibration between the two translational isomers is solvent dependent. In CD_3COCD_3 solution the hydroquinone station is exclusively occupied by cyclobis(paraquat-*p*-phenylene), while in CD_3SOCD_3, the tetrathiofulvalene unit is the station that is included predominantly inside the cavity of the tetracationic cyclophane (**6**). The molecular shuttling properties of this [2]rotaxane are under investigation.

Starting material	*R*	*m*	*n*	*Product*	*Yield (%)*
(27)	Si(iPr)₃	3	0	**(32)**	32
(28)	adamantyl-C(=O)–	0	0	**(33)**	1
(29)	adamantyl-C(=O)–	0	2	**(34)**	25
(30)	adamantyl-C(=O)–	0	3	**(35)**	29
(31)	adamantyl-C(=O)–	0	4	**(36)**	40

Scheme 2

Another two π-electron-rich aromatic units, benzidine and 4,4′-biphenol, have been used to synthesize[114] (Scheme 4), via the threading methodology, the [2]rotaxanes **(43)** and **(44)**. The pseudorotaxane formation resulting from the combination of the π-electron-rich acyclic derivatives **(41)** and **(42)** with the tetracationic cyclophane **(6)** afforded, after the covalent attachment of triisopropylsilyl stoppers, the [2]rotaxanes **(43)** and **(44)** in yields of 5% and 39%, respectively. The difference in these yields is believed to result from the relative binding affinities of the π-electron-rich units.

This result led to the self-assembly of a [2]rotaxane incorporating one benzidine unit and one 4,4′-biphenol unit along the dumbbell-shaped component. The [2]rotaxane **(45)** was self-assembled[115] using the threading approach in a yield of 19% (Figure 12). ^{1}H NMR spectroscopy indicates that the ratio between the two translational isomers in CD_3CN at –44 °C is 84:16 in favour of the isomer (Figure 12) having the benzidine station encircled by the cyclophane. The addition of an excess of deuterated trifluoroacetic acid to the solution results in the protonation of the two amino groups of the benzidine unit. As a result, the tetracationic cyclophane is forced to move away from the now positively charged benzidine unit and only the translational isomer with the 4,4′-biphenol unit occupied is detected by ^{1}H NMR spectroscopy. The neutralization of the solution by the addition of deuterated pyridine restores the original isomer distribution following the deprotonation of the benzidine unit. Reversible switching between the two translational isomers can also be achieved electrochemically. Oxidation of the benzidine unit drives the cyclophane exclusively to the 4,4′-biphenol station, while reduction restores the original equilibrium. Thus, the [2]rotaxane **(45)** constitutes a chemically and electrochemically controllable molecular shuttle.

The challenge of making photochemically controllable molecular shuttles led to the incorporation of porphyrins as stopper groups in [2]rotaxanes. Thus, the [2]rotaxanes **(46)** and **(47)** bearing porphyrin-based stoppers were self-assembled[116] by the clipping procedure. Variable-temperature ^{1}H NMR spectroscopy revealed the shuttling properties of these [2]rotaxanes **(46)** and **(47)** (Figure 13) to be very similar to those of the [2]rotaxane **(32)** incorporating triisopropylsilyl stoppers; that is, the degenerate shuttling process has a free energy of activation of 54.5 kJ mol^{-1} in both cases. The photochemical properties of these porphyrin-containing rotaxanes are under investigation.

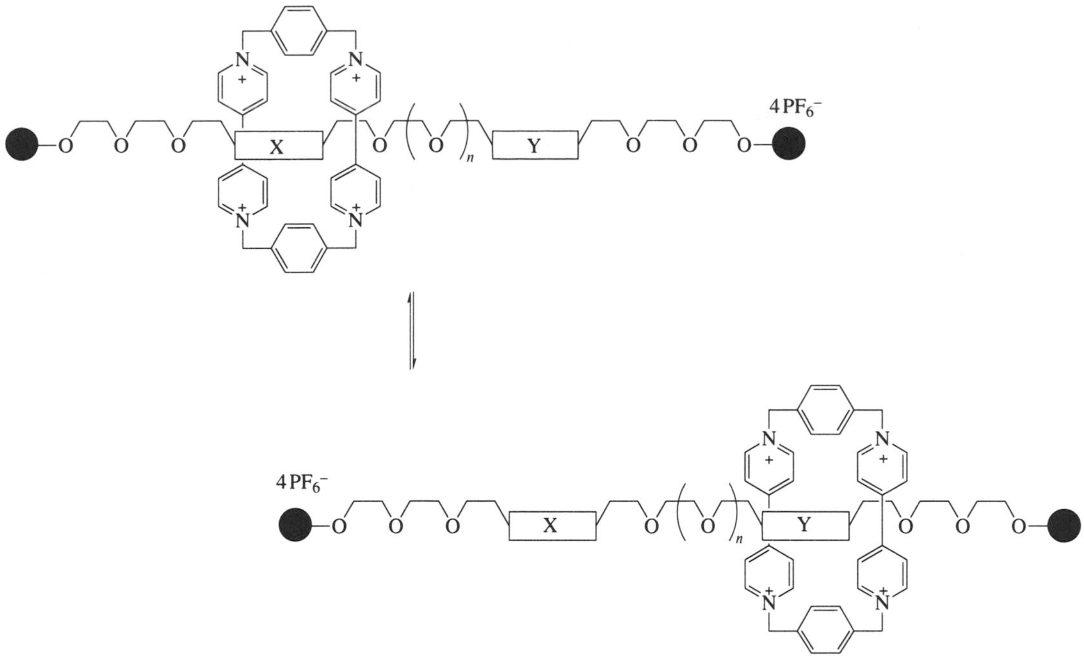

Compound	n	● / X / Y	Yield (%)
(32)	2	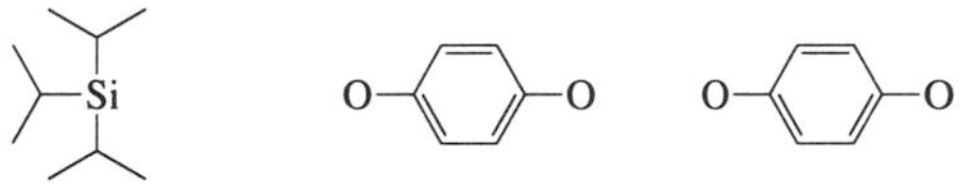	32
(37)	4	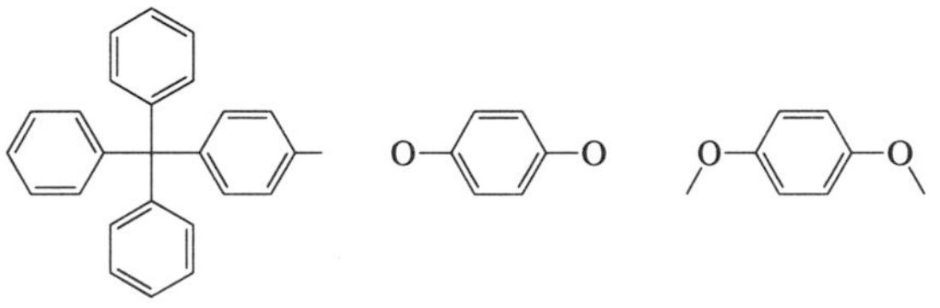	8
(38)	1	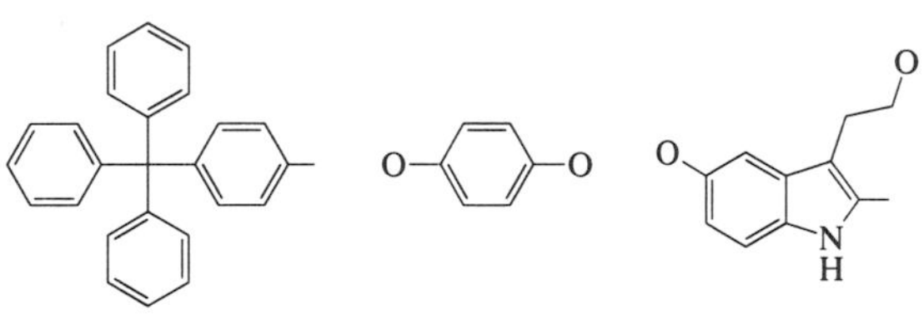	9

Figure 10 Yields for the self-assembly of the [2]rotaxanes (**32**), (**37**) and (**38**) each incorporating two π-electron-rich stations within the dumbbell-shaped component.

The self-assembly of the [2]rotaxanes (**50**) and (**51**) (Scheme 5) incorporating ferrocenyl[117] and anthracenyl[118] stoppers at both ends of the dumbbell-shaped component has been reported by Harriman and co-workers.

The fate of the radical ion pairs formed by the illumination of the charge transfer bands of the [2]rotaxanes (**50**) and (**51**) was investigated.[117,118] Irradiation of (**50**) by a laser pulse at 437 nm presumably affords a radical ion pair which decays very slowly, as schematically shown in Figure 14. The deactivation of the radical pair occurs via rapid charge recombination and via the oxidation of one of the ferrocenyl stoppers. The formation of a positively charged stopper is followed by conformational changes of the rotaxane in order to increase the spatial separation between the charged subunits—namely, cyclophane and stopper. As a result, the charge recombination involving electron transfer from the radical cation of the cyclophane to the ferrocenyl stopper is slowed down dramatically. In contrast, the anthracenyl-stoppered [2]rotaxane (**51**) exhibits immediate charge recombination of the radical ion pair upon excitation.

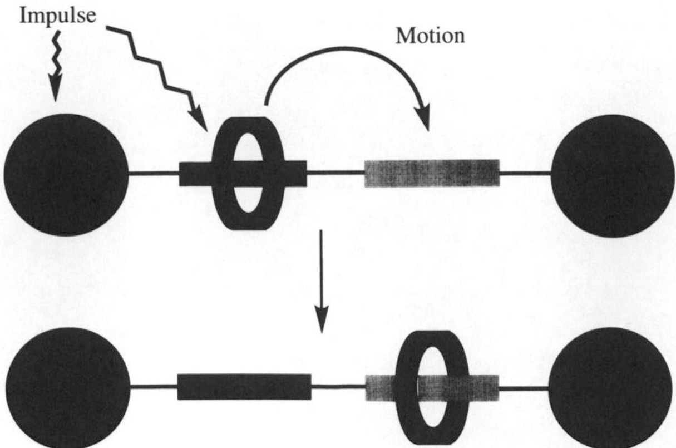

Figure 11 Control via external stimuli of the shuttling process occurring within a [2]rotaxane.

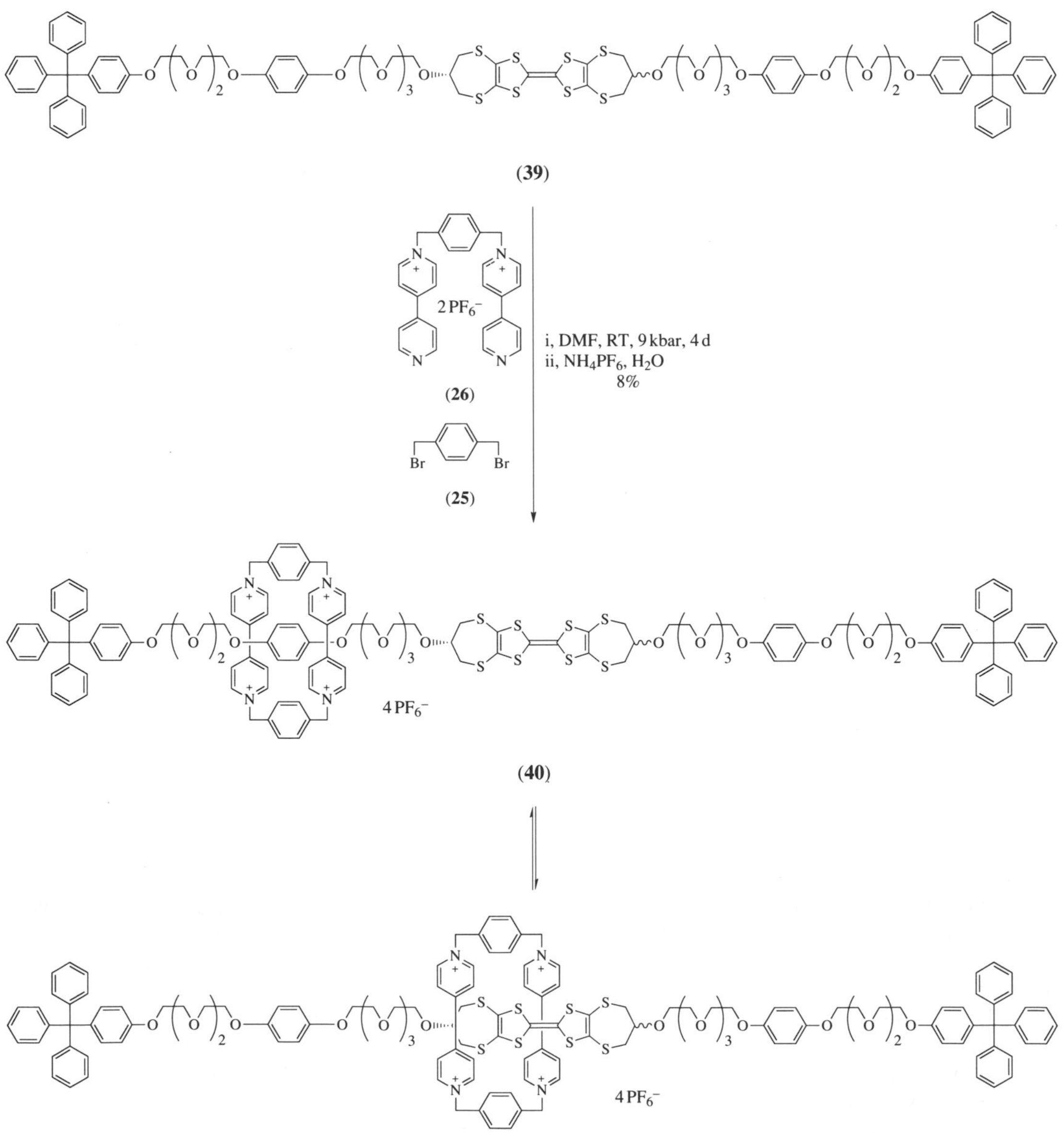

(39)

$2PF_6^-$

(26)

Br Br

(25)

i, DMF, RT, 9 kbar, 4 d
ii, NH_4PF_6, H_2O
8%

$4PF_6^-$

(40)

$4PF_6^-$

Scheme 3

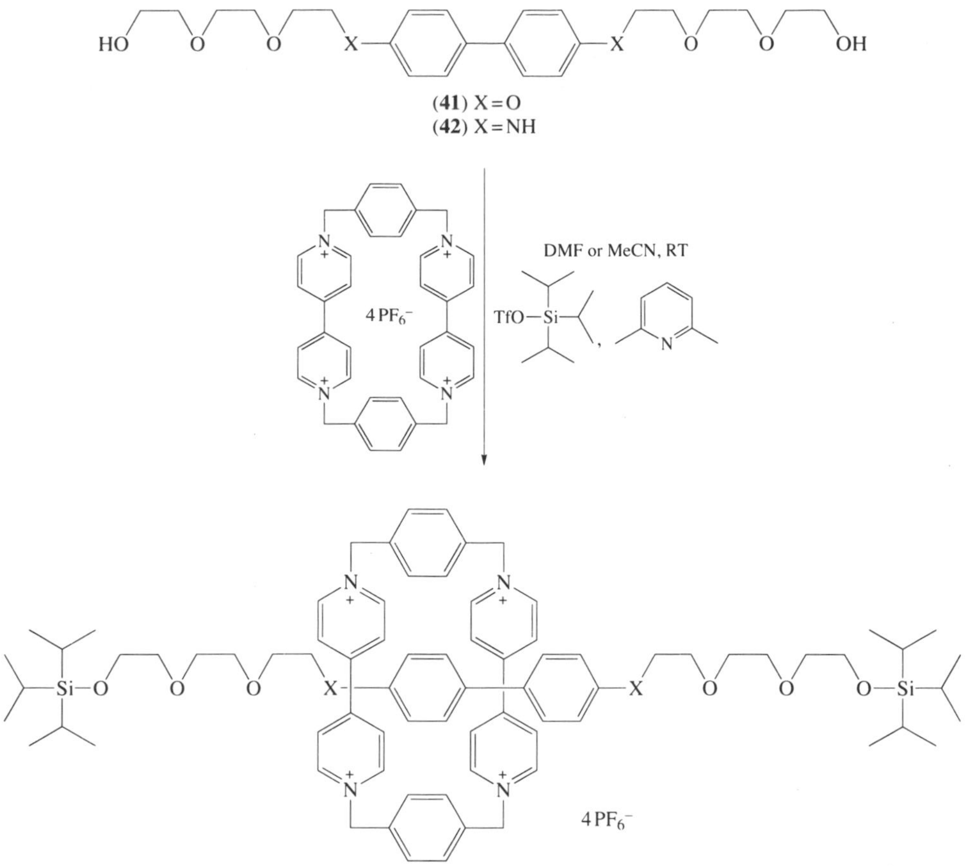

(**41**) X=O
(**42**) X=NH

DMF or MeCN, RT

$4PF_6^-$

(**43**) X=O
(**44**) X=NH

Scheme 4

The self-assembly of a bis[2]rotaxane (Scheme 6) was achieved[119] by employing 1,5-dioxynaphthalene-based dumbbell-shaped templates and the components of a bipyridinium-based biscyclophane. The reaction of the dication (**26**) with the tetrabromide (**53**) in the presence of the preformed dumbbell-shaped compound (**52**) gave the bis[2]rotaxane (**54**) in 7% yield after counterion exchange.

3.3.3 Rotaxanes Incorporating π-Electron-rich Macrocycles and π-Electron-deficient Acyclic Components

The self-assembly of rotaxanes incorporating π-electron-rich macrocycles and π-electron-deficient dumbbell-shaped components has been realized by both the threading and the slipping approaches. The reaction[120] (Scheme 7) of the dication (**26**) with the chloride (**55**) under ultrahigh-pressure conditions affords a tricationic intermediate which is bound by the π-electron-rich macrocycle (**1**). The subsequent covalent attachment of a second stopper, as a result of the nucleophilic displacement of the chloride ion by the bipyridine nitrogen atom, affords the [2]rotaxane (**56**) in 23% yield after counterion exchange.

The dynamic process occurring within the [2]rotaxane (**56**), involving the shuttling of the π-electron-rich macrocycle back and forth from one bipyridine station to the other, was investigated by variable-temperature ^1H NMR spectroscopy. The free energy of activation for the process was calculated to be $41.4\,kJ\,mol^{-1}$, corresponding to a rate of shuttling of $\sim 3 \times 10^5\,s^{-1}$ at room temperature.

The careful matching of the size of both the stopper and the macrocyclic components allows rotaxanes incorporating π-electron-deficient dumbbell-shaped derivatives and π-electron-rich macrocycles to be prepared using the slipping methodology. When the appropriate steric

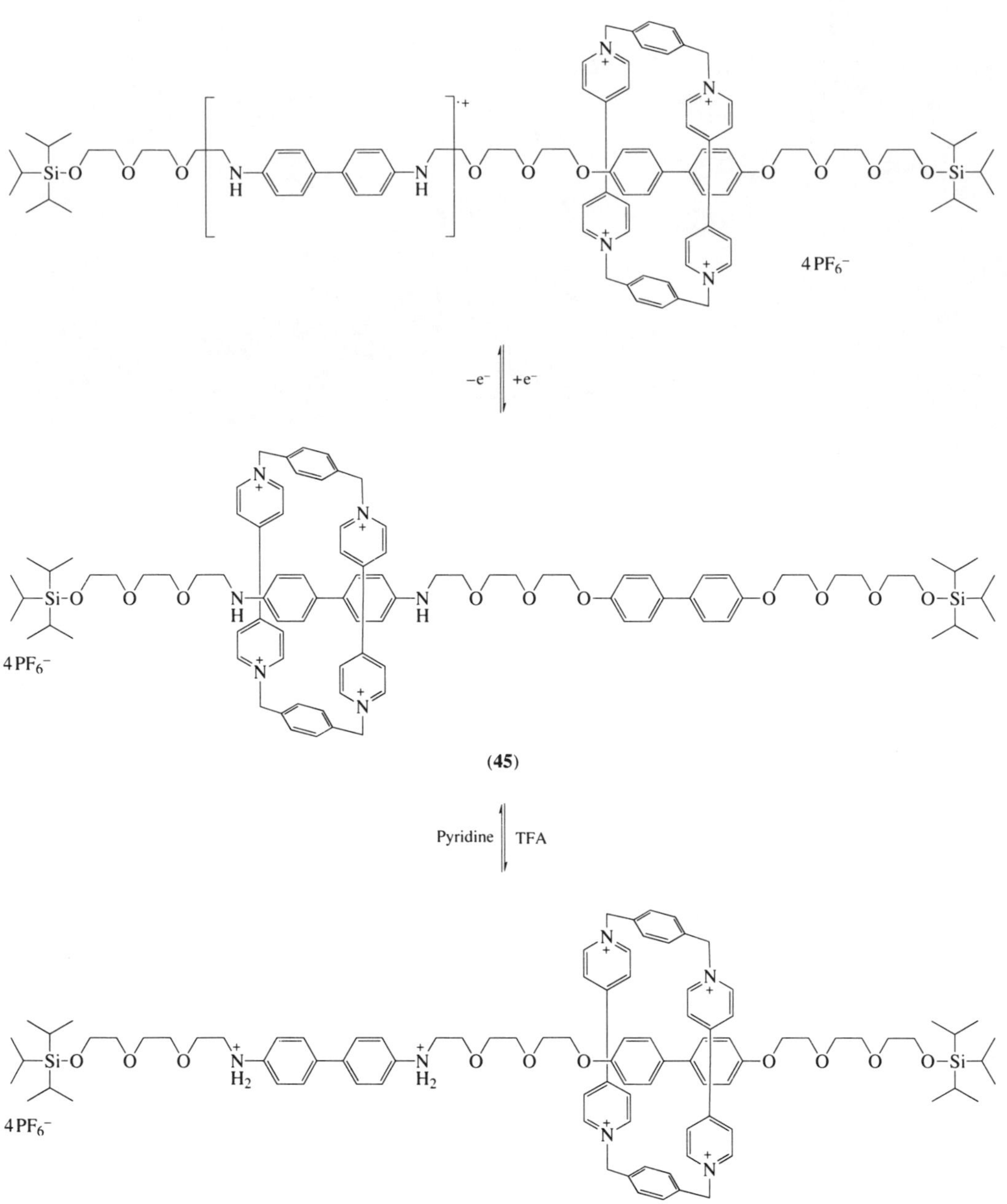

$4PF_6^-$

$-e^-$ $+e^-$

$4PF_6^-$

(**45**)

Pyridine TFA

$4PF_6^-$

Figure 12 Electrochemically and chemically reversible control upon the shuttling process occurring within the [2]rotaxane (**45**).

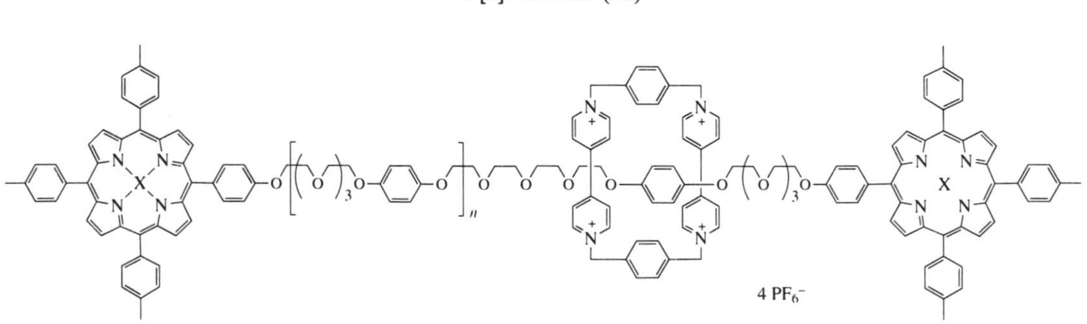

$4PF_6^-$

(**46**) $n=0$, $X=H_2$
(**47**) $n=1$, $X=Zn$

Figure 13 The [2]rotaxanes (**46**) and (**47**) incorporating porphyrins as the stopper groups.

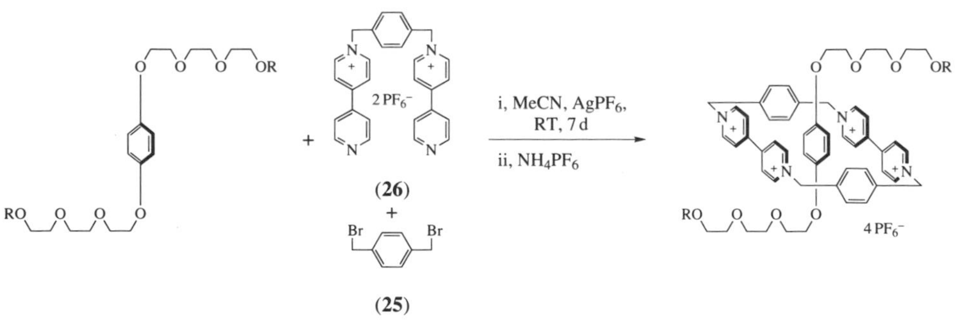

(26) + (25)

i, MeCN, $AgPF_6$, RT, 7 d

ii, NH_4PF_6

$2 PF_6^-$ $4 PF_6^-$

Starting material	*R*	*Product*	*Yield* (%)
(**48**)	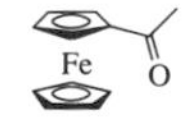	(**50**)	6
(**49**)	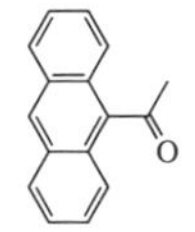	(**51**)	25

Scheme 5

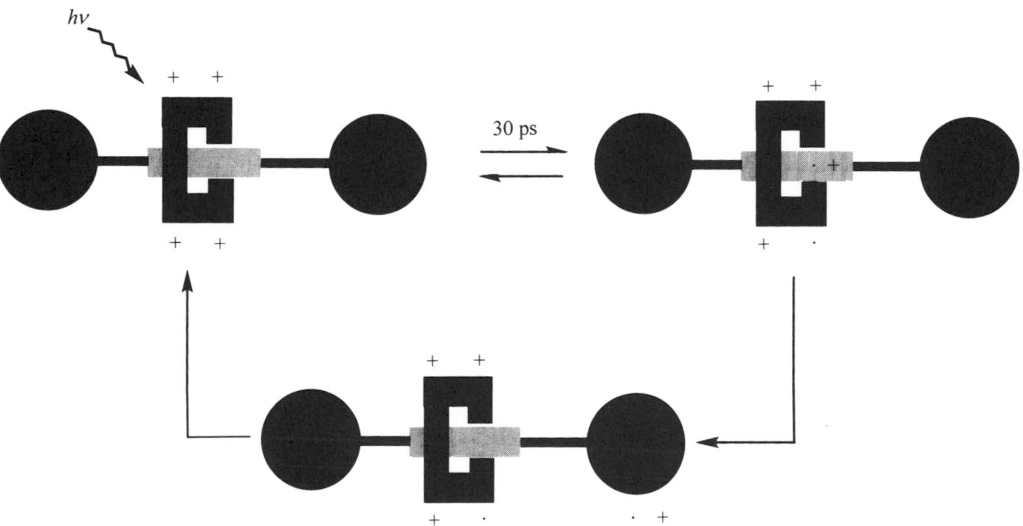

Figure 14 Schematic representation of the illumination and charge recombination of the [2]rotaxanes (**50**).

complementarity between stoppers and macrocycle is established, the use of thermal energy can be exploited to promote the slippage of the macrocycle over the stoppers. Thus, the presence of a recognition site within the dumbbell-shaped component, namely a bipyridinium unit, provides the required thermodynamic trap for the macrocycle, raising the free energy of activation for its extrusion. The synthesis[121] of the dications (**57**)–(**60**) (Scheme 8) provided a series of dumbbell-shaped compounds in which the size of the stoppers was varied systematically. On stirring acetonitrile solutions of the dications (**57**)–(**59**) and the π-electron-rich macrocycle (**1**) for 10 d at 55°C, the corresponding [2]rotaxanes (**61**)–(**63**) were self-assembled[121] (Scheme 8) in yields of 52%, 45% and 47%, respectively. In the case of the dumbbell-shaped compound (**60**) bearing an isopropyl group as part of the tetraarylmethane-substituted stoppers, no rotaxane was detected. This

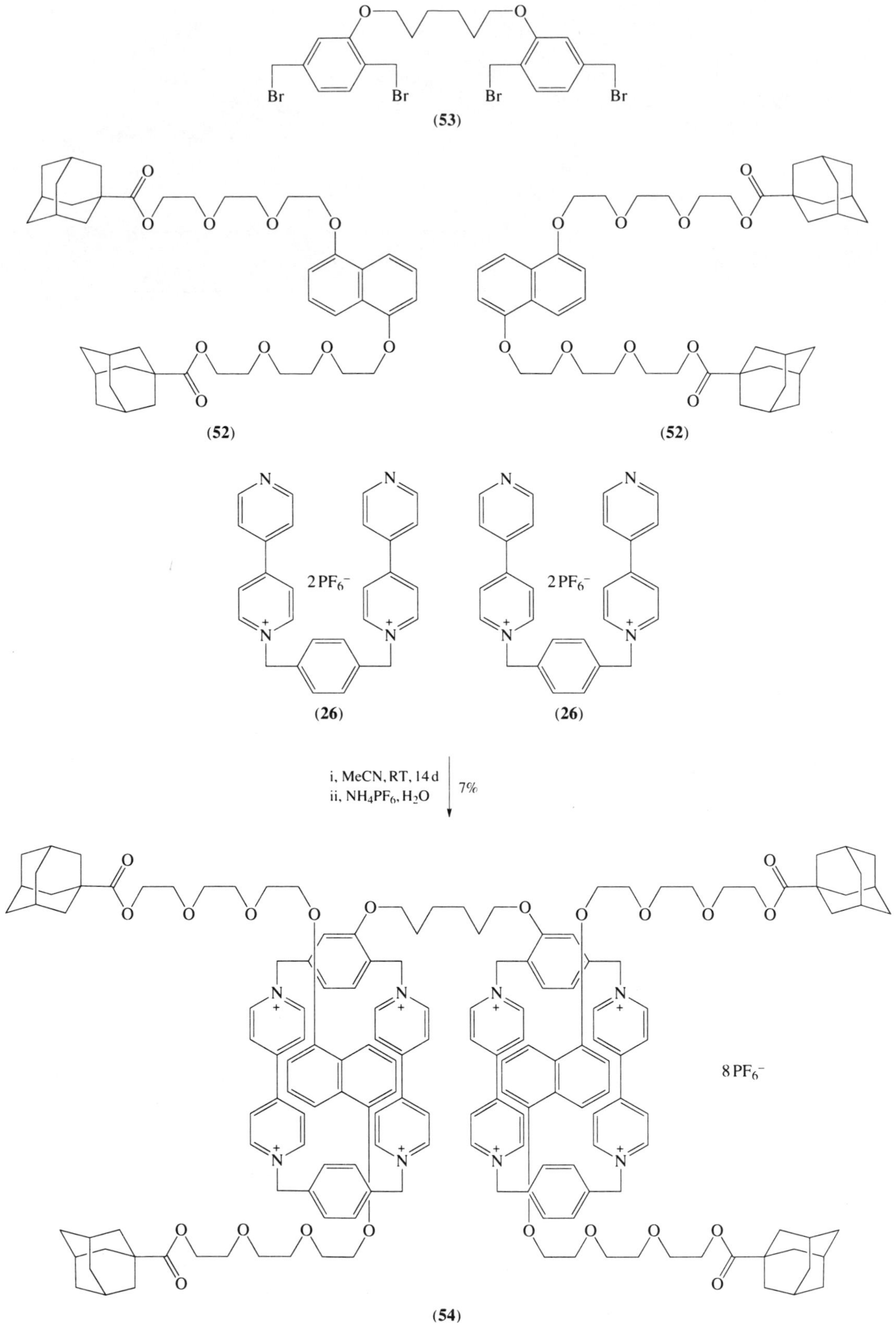

(53)

(52) (52)

$2\,PF_6^-$ $2\,PF_6^-$

(26) (26)

i, MeCN, RT, 14 d
ii, NH_4PF_6, H_2O
7%

$8\,PF_6^-$

(54)

Scheme 6

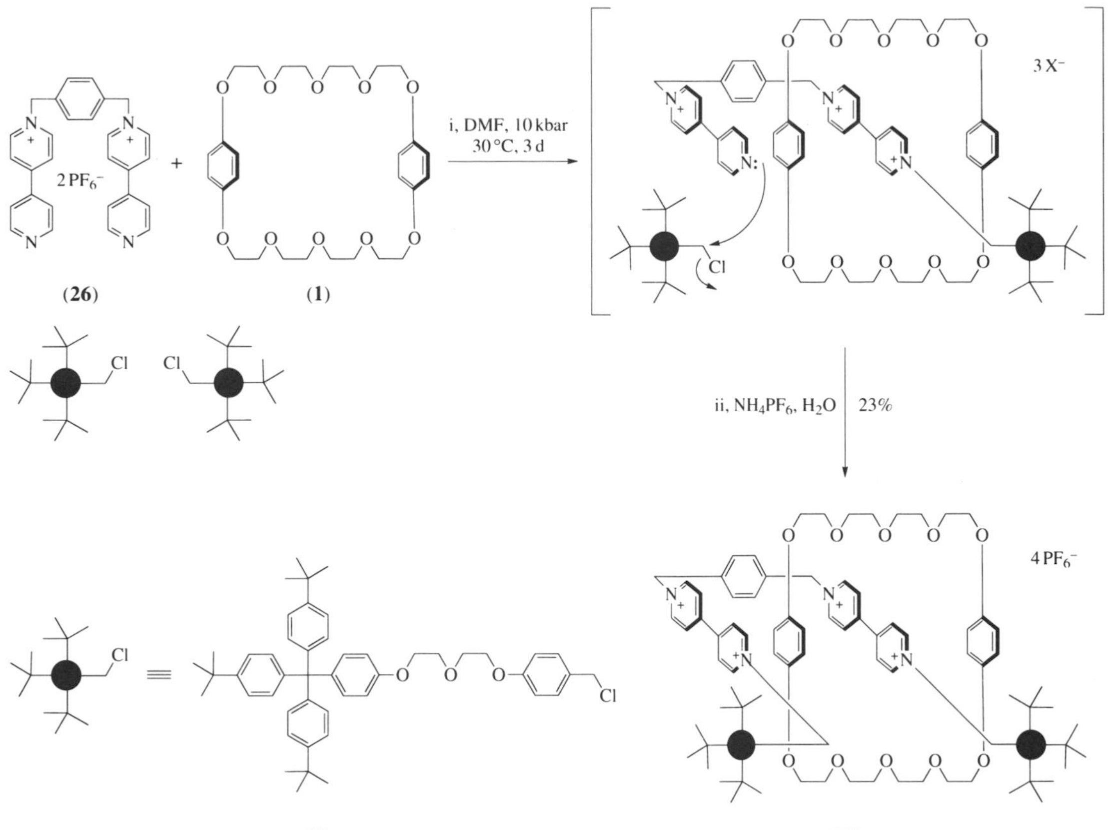

Scheme 7

observation indicates that the barrier for the slippage of the macrocycle (**1**) over the stoppers is exceeded on going from the ethyl-containing to the isopropyl-containing stoppers. However, slippage (Scheme 9) of the larger 1,5-dinaphtho-38-crown-10 (**12**) over the isopropyl-containing stoppers of the dumbbell (**60**) afforded[122] the [2]rotaxane (**65**) in a yield of 57% after only one day.

By increasing the number of recognition sites, namely bipyridinium units, incorporated within the dumbbell-shaped component, a [3]rotaxane has been self-assembled by the slipping approach. After stirring an acetonitrile solution of the tetracationic dumbbell compound (**66**) with either 4 or 10 molar equivalents of the macrocycle (**1**) at 55°C over 10 d, the [3]rotaxane (**67**) was self-assembled[123] (Scheme 10) in a yield of 8% or 51%, respectively.

The efficiency and relative simplicity of this approach suggested the possibility of synthesizing even more complex mechanically interlocked molecular compounds. To this end, the triply branched hexacationic derivative (**68**) (Scheme 11), incorporating three bipyridinium recognition sites and stoppers of suitable size for the slipping of the macrocycle (**1**), was combined in acetonitrile with either 2 or 15 molar equivalents of (**1**). Heating of the solution at 50°C for 10 d afforded[124] the [4]rotaxane (**69**)—the first step towards dendritic rotaxanes—in a yield of 6% or 22%, respectively.

The combination of threading and slipping methodologies has been exploited to self-assemble a [3]rotaxane incorporating two constitutionally different macrocyclic components. The reaction (Scheme 12) of the dication (**26**) with the chloride (**70**) under ultrahigh-pressure conditions in the presence of di-*p*-phenylene-34-crown-10 (**1**) afforded[122] the [2]rotaxane (**71**) in a yield of 19% after counterion exchange. As a result of the presence within the [2]rotaxane (**71**) of a free bipyridinium recognition site, the slippage of the 1,5-dinaphtho-38-crown-10 macrocycle (**12**) over the isopropyl-substituted stoppers can be realized. Indeed, by heating an acetonitrile solution of the [2]rotaxane (**71**) and macrocycle (**12**) at 55 °C over a period of 48 h, the [3]rotaxane (**72**) has been self-assembled[122] (Scheme 12) in 49% yield.

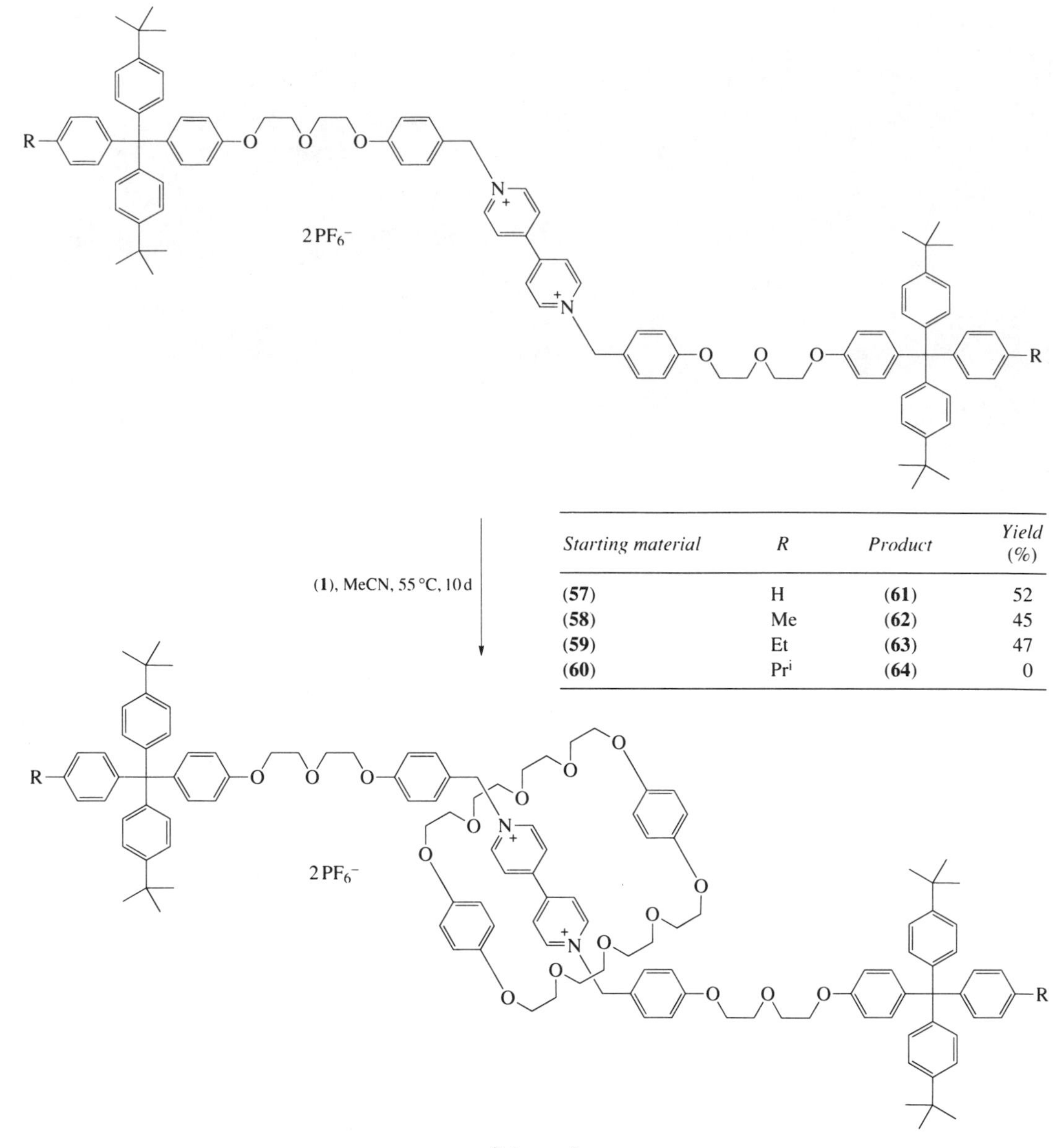

Starting material	*R*	*Product*	*Yield (%)*
(**57**)	H	(**61**)	52
(**58**)	Me	(**62**)	45
(**59**)	Et	(**63**)	47
(**60**)	Pr^i	(**64**)	0

Scheme 8

3.4 CATENANES

3.4.1 Self-assembling Approaches to Catenanes

The self-assembly of catenanes comprising π-electron-rich and π-electron-deficient components can be achieved[86,87] by following one of the approaches schematically represented in Figure 15. The macrocyclization of one of the ring components in the presence of the other preformed macrocycle affords the corresponding catenane by a clipping procedure. When the macrocyclization of both components is performed at the same time, a catenane is generated via a double clipping procedure.

3.4.2 Self-assembly and Properties of a [2]Catenane Incorporating π-Electron-rich and π-Electron-deficient Components

The synthesis of the first [2]catenane incorporating π-electron-rich and π-electron-deficient cyclophanes was realized[43,125] as illustrated in Scheme 13. Reaction of the dication (**26**) with the

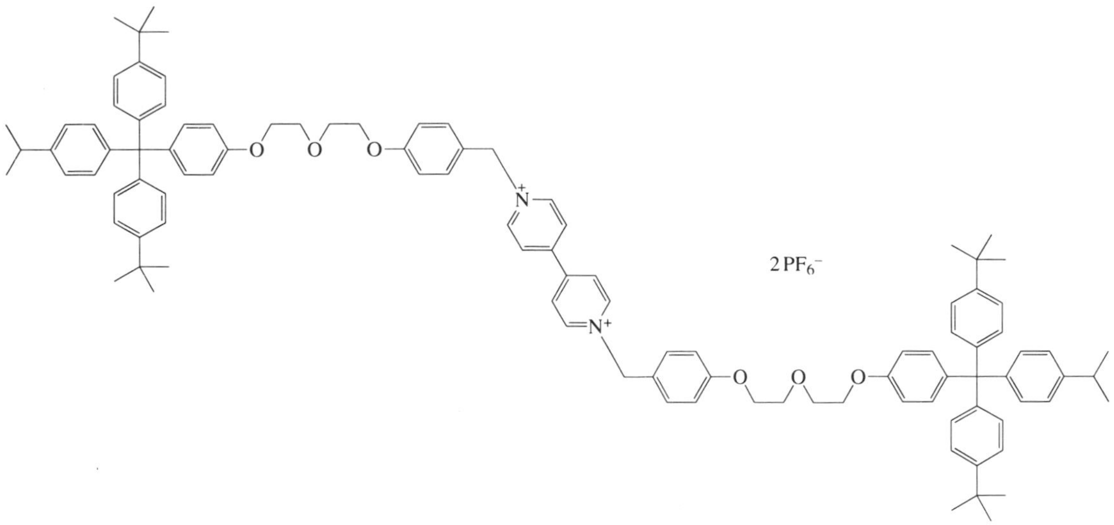

(**60**)

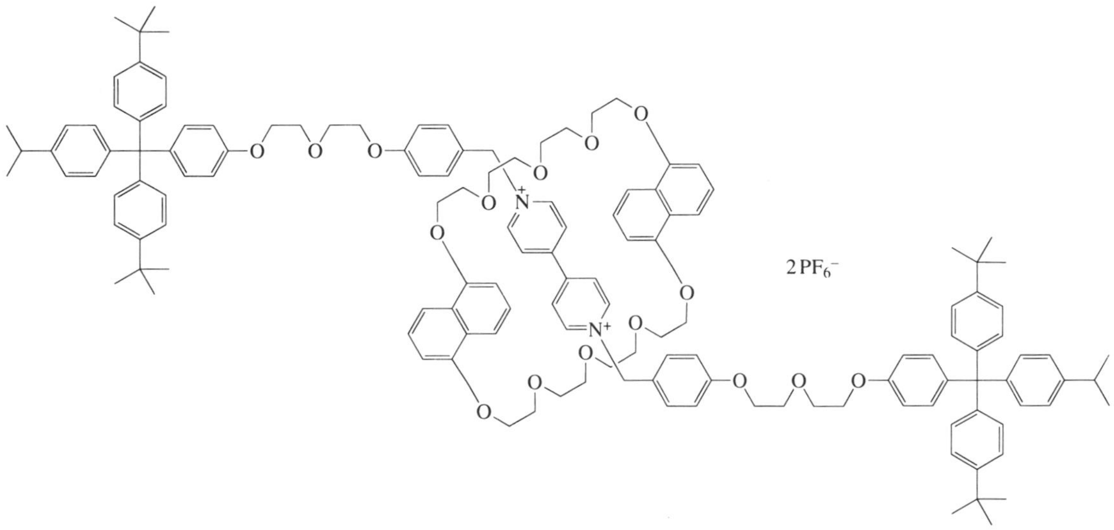

(**65**)

Scheme 9

dibromide (**25**) gives a tricationic intermediate which is bound by di-*p*-phenylene-34-crown-10 (**1**) to afford a pseudorotaxane-like, or precatenane, superstructure. The subsequent macrocyclization, as a result of nucleophilic displacement of bromide ion, affords the corresponding [2]catenane (**73**) in the amazing yield of 70%, after the exchange of counterions.

An alternative route to the [2]catenane (**73**) is the one depicted in Scheme 13. Reaction under ultrahigh pressure of the dibromide (**25**) with bipyridine (**74**) in the presence of the π-electron-rich macrocycle (**1**) affords[126] the [2]catenane (**73**) in a yield of 42%. The mutual interlocking of the two macrocyclic components in (**73**) is clearly revealed[125] by the x-ray crystal structure (Figure 16). The interplanar separations between the π-electron-deficient bipyridinium and the π-electron-rich hydroquinone aromatic units are ~350 pm. Furthermore, in the crystal, the same separation is observed between the alongside bipyridinium moiety of one catenane molecule and the alongside hydroquinone ring of an adjacent catenane molecule. Thus, the [2]catenane (**73**) forms infinite π-acceptor–π-donor stacks along one of the crystallographic directions in the solid state.

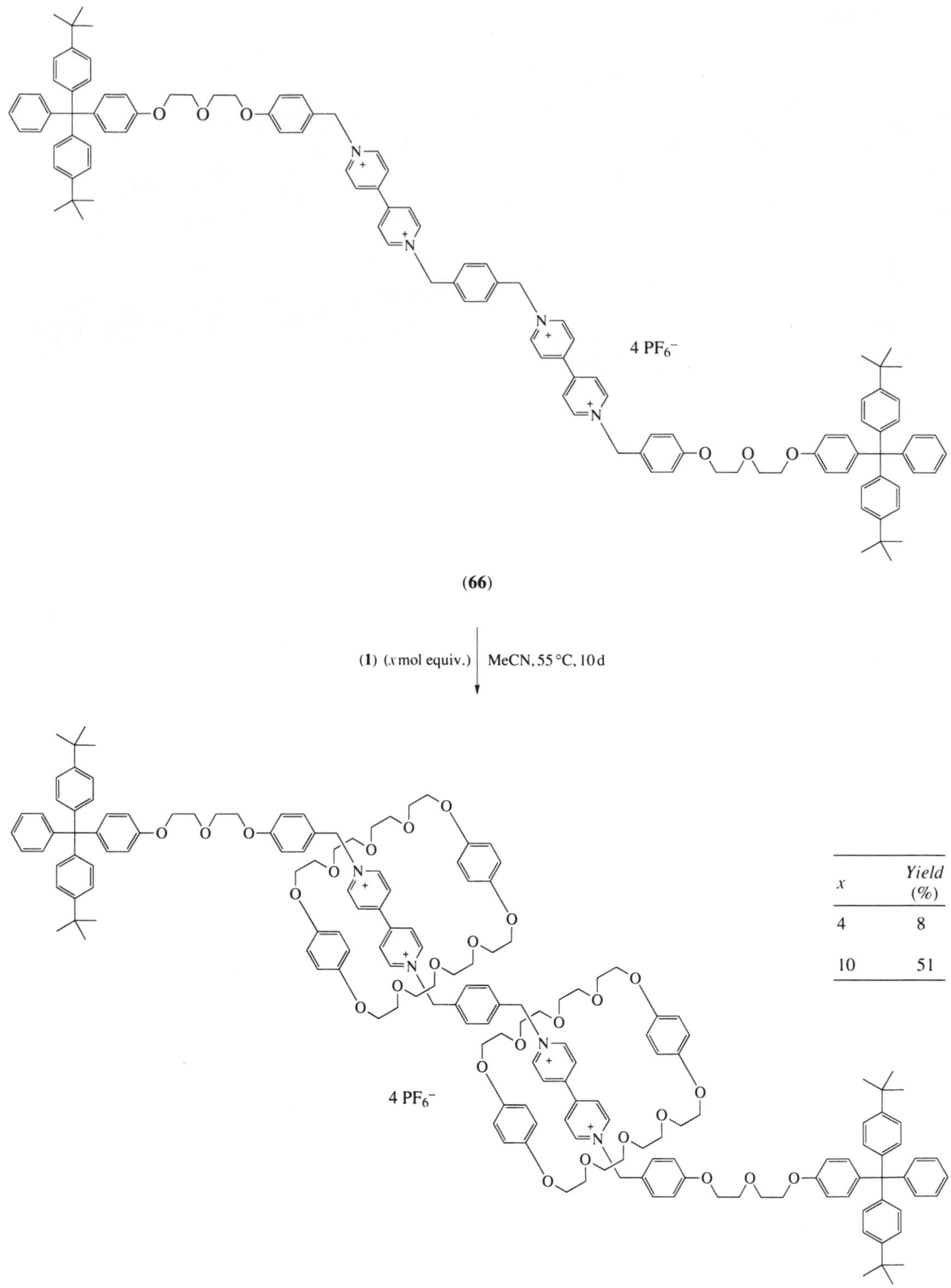

4 PF_6^-

(66)

(**1**) (*x* mol equiv.) | MeCN, 55 °C, 10 d

x	*Yield* (%)
4	8
10	51

4 PF_6^-

(67)

Scheme 10

The dynamic processes occurring in solution within the [2]catenane (**73**) are illustrated schematically in Figure 17. Although they are associated with numerous noncovalent bonding interactions, the two mechanically interlocked rings are able to circumrotate through each other. These relative motions are clearly revealed by variable-temperature ^{1}H NMR spectroscopy. The ^{1}H NMR

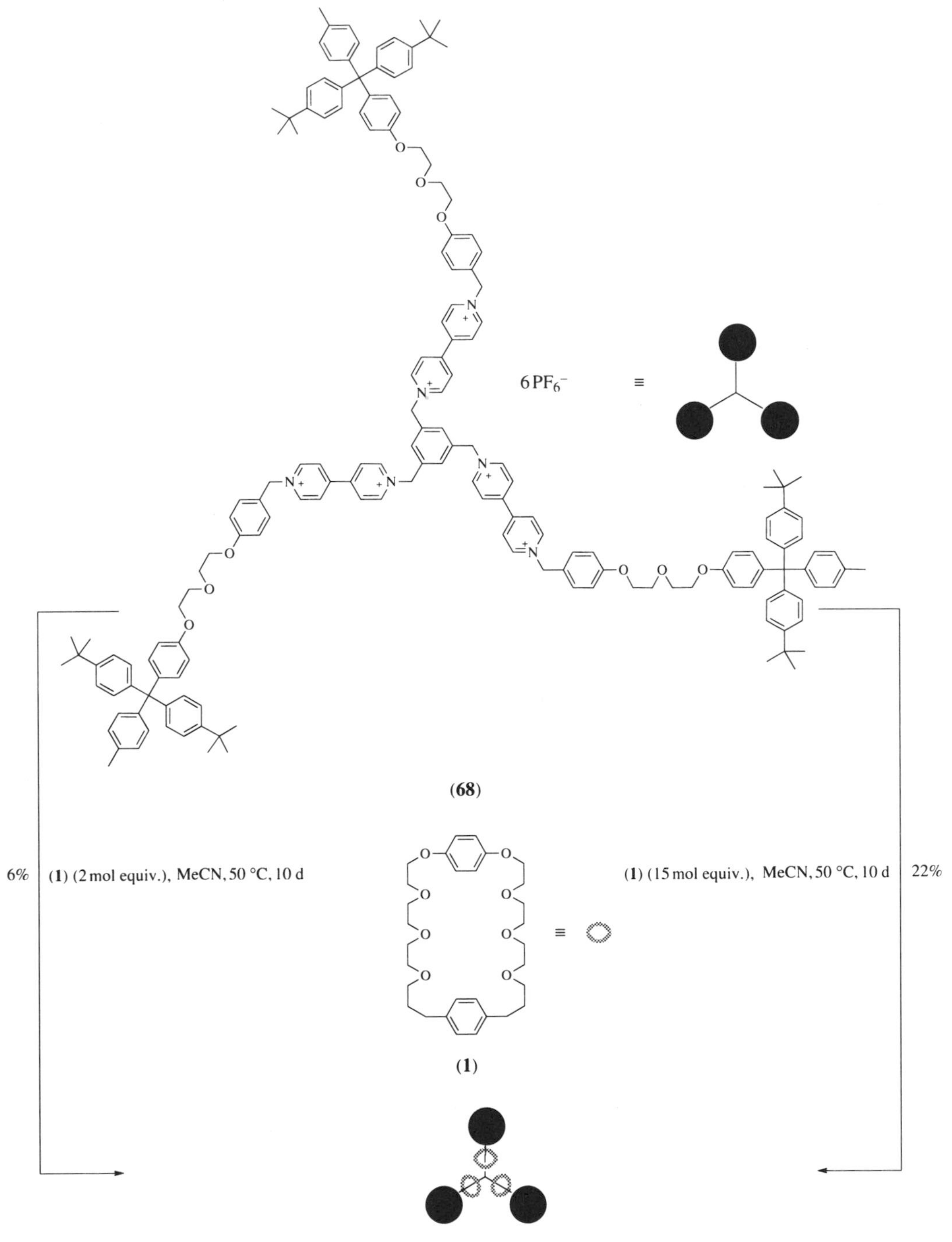

Scheme 11

spectrum of the [2]catenane (**73**) in CD_3SOCD_3 at 81 °C shows one singlet for the hydroquinone protons centred on δ 4.57. At room temperature, as a result of the slow circumrotation on the ^1H NMR timescale of the crown ether through the cyclophane (process I), two equally intense singlets resonating at δ 3.45 and δ 6.16 are observed for protons on the inside and alongside hydroquinone rings, respectively. By employing the coalescence method, the energy barrier for the process was calculated to be 65.3 kJ mol^{-1}. The circumrotation of the cyclophane through the cavity of the crown ether (process II) is fast on the ^1H NMR timescale at room temperature. As a result, the protons attached to the α positions with respect to the nitrogen atoms in the bipyridinium units of the cyclophane appear as a doublet centred on δ 9.34 at room temperature in CD_3COCD_3. On cooling the solution down to –45 °C, this resonance separates into two doublets

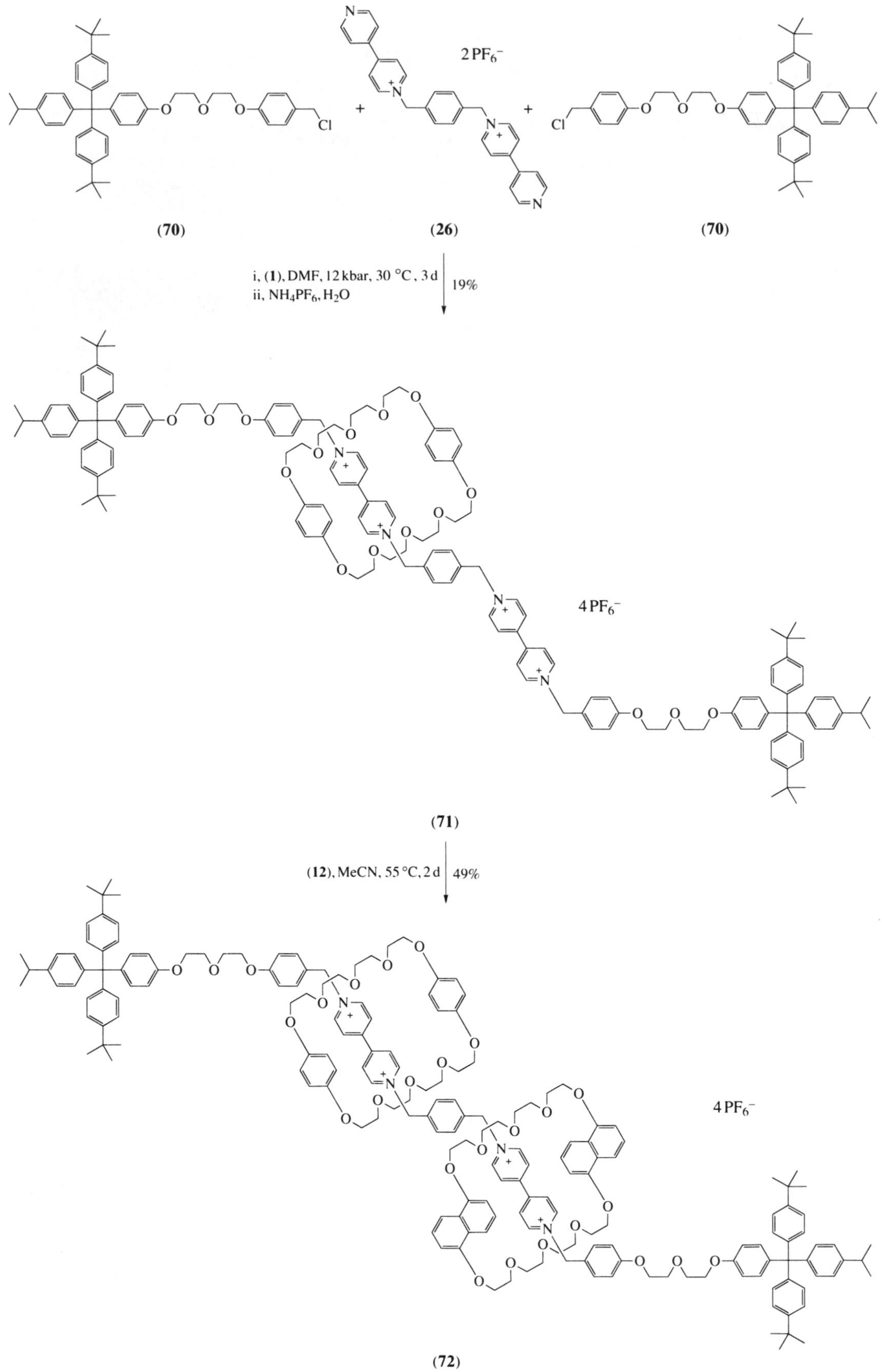

$2PF_6^-$

Cl + Cl

(**70**) (**26**) (**70**)

i, (**1**), DMF, 12 kbar, 30 °C, 3 d
ii, NH_4PF_6, H_2O
19%

$4PF_6^-$

(**71**)

(**12**), MeCN, 55 °C, 2 d 49%

$4PF_6^-$

(**72**)

Scheme 12

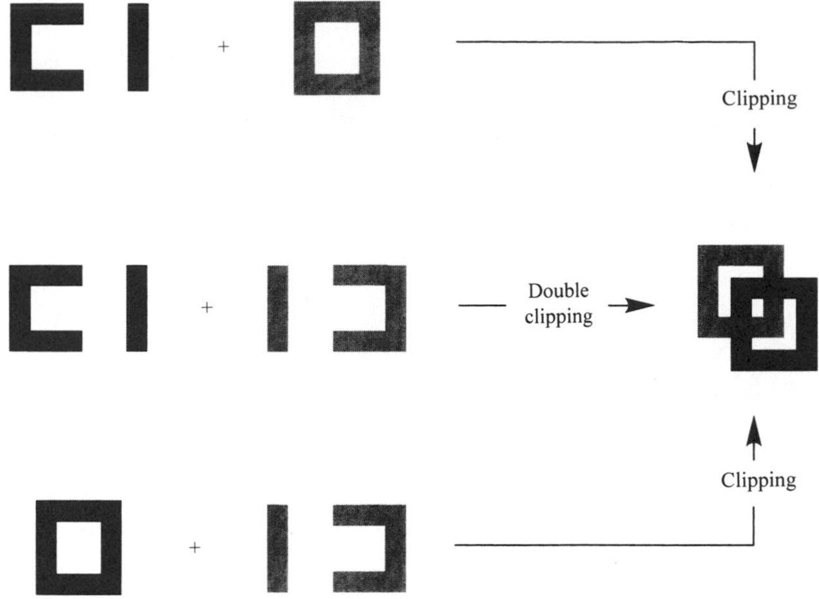

Figure 15 Synthetic approaches to self-assembling catenanes.

centred on δ 9.23 and δ 9.41, corresponding to the α-CH protons on the inside and alongside bipyridinium units, respectively. The free energy of activation associated with this process was calculated to be 51.1 kJ mol^{-1}. Besides the circumrotation of the two macrocyclic components one through the other, two more dynamic processes, namely rattling and rocking, are associated[43] in solution with the [2]catenane (**73**). The hydroquinone ring located inside the cavity of the tetracationic cyclophane can undergo translation back and forth along the long axis of the tetracationic cyclophane between two degenerate states by means of a rattling process. Furthermore, the oxygen–oxygen axis of the inside hydroquinone ring can oscillate with respect to the nitrogen–nitrogen axis of the bipyridinium units, giving rise to a degenerate process termed rocking. Thus, although in the solid state (Figure 16) the inside hydroquinone ring occupies a centrosymmetric position within the cavity of the tetracationic cyclophane, in solution it moves rapidly between degenerate states as a result of the rattling and rocking motions.

The strong noncovalent interactions that exist between the bipyridinium units and the hydroquinone rings are responsible for the very different electrochemical properties of the [2]catenane (**73**) when compared with its components. The free tetracationic cyclophane (**6**) exhibits two two-electron reduction waves for the reduction of the two identical bipyridinium units. Cyclic voltammetry of the [2]catenane (**73**) shows first a one-electron reduction wave corresponding to reduction of the alongside bipyridinium unit, then a second one-electron reduction wave for the reduction of the inside bipyridinium unit and finally a two-electron reduction wave for the complete reduction of both bipyridinium units. The splitting of the first wave, on 'going from' the cyclophane (**6**) to the catenane (**73**), is presumably a result of the 'protection' given to the inside bipyridinium unit by the two adjacent hydroquinone rings. The second reduction wave is not split, presumably because the circumrotation of the now partially reduced cyclophane through the crown ether rings is fast with respect to the timescale of the reduction process.

3.4.3 Self-assembly and Properties of [2] Catenanes Incorporating Structural Modifications in the π-Electron-rich Component

A considerable number of π-electron-rich macrocyclic polyethers have been used to template the formation of the cyclobis(paraquat-*p*-phenylene), thus generating a family of [2]catenanes. The [2]catenane incorporating the crown ether di-*m*-phenylene-32-crown-10 (**75**) (Figure 18) can be self-assembled[127] in a yield of 17%. The reduced efficiency of the self-assembly process is mainly a result of the stereoelectronic properties of the resorcinol-based macrocycle, which are also reflected in the dynamic properties of the [2]catenane (**75**). The energy barrier associated with the

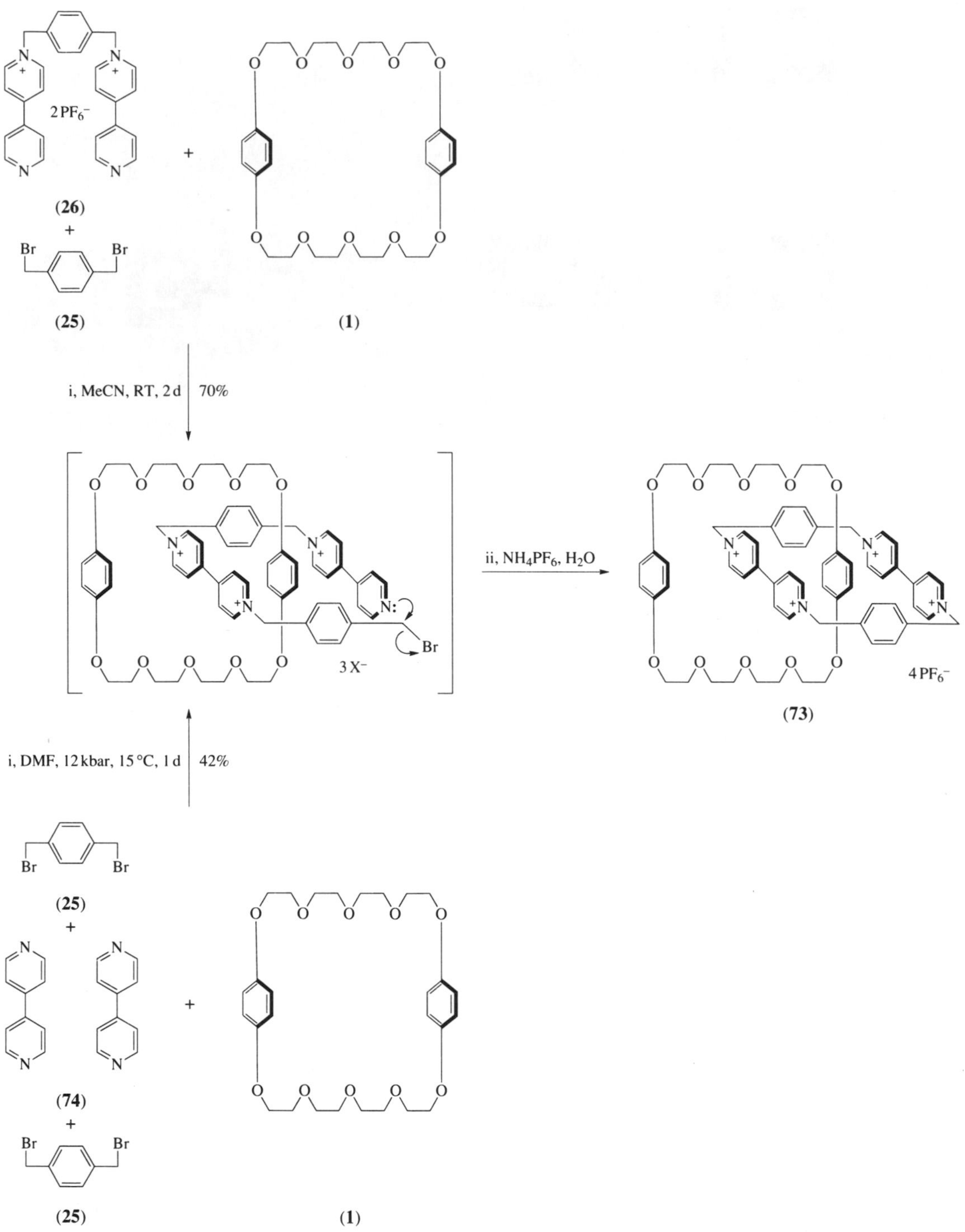

(26) + (25) + (1)

i, MeCN, RT, 2 d 70%

ii, NH_4PF_6, H_2O

(73) $4PF_6^-$

i, DMF, 12 kbar, 15 °C, 1 d 42%

(25) + (74) + (25) + (1)

Scheme 13

circumrotation of the crown ether component through the cavity of the cyclophane (cf. process I in Figure 17) is only 56.9 kJ mol^{-1}, which is ~8.4 kJ mol^{-1} lower than in the case of the hydroquinone-containing catenane **(73)**.

The [2]catenane **(76)** (Figure 18) incorporating the 1,5-dinaphtho-38-crown-10 **(12)** unit was self-assembled[128] in a yield of 51%. The energy barriers analogous to processes I and II in Figure 17 correspond to 72.0 kJ mol^{-1} and 53.2 kJ mol^{-1}, respectively. The high free energy of activation for the circumrotation of the crown ether component through the cavity of the cyclophane is presumably a result of the very strong π–π stacking interactions between the π-electron-rich 1,5-dioxynapthalene units of the crown and the bipyridinium units in the tetracationic cyclophane.

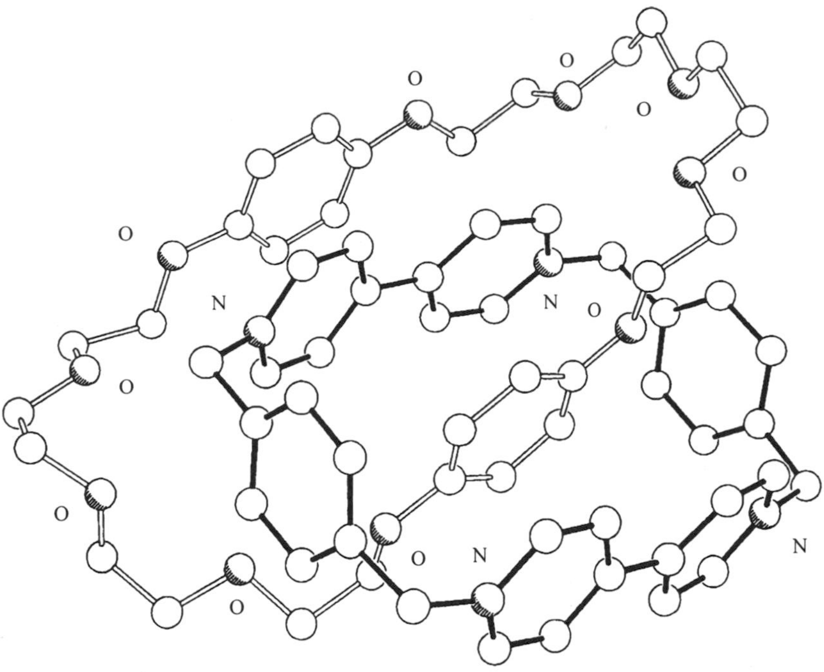

Figure 16 The x-ray crystal structure of the [2]catenane (**73**).

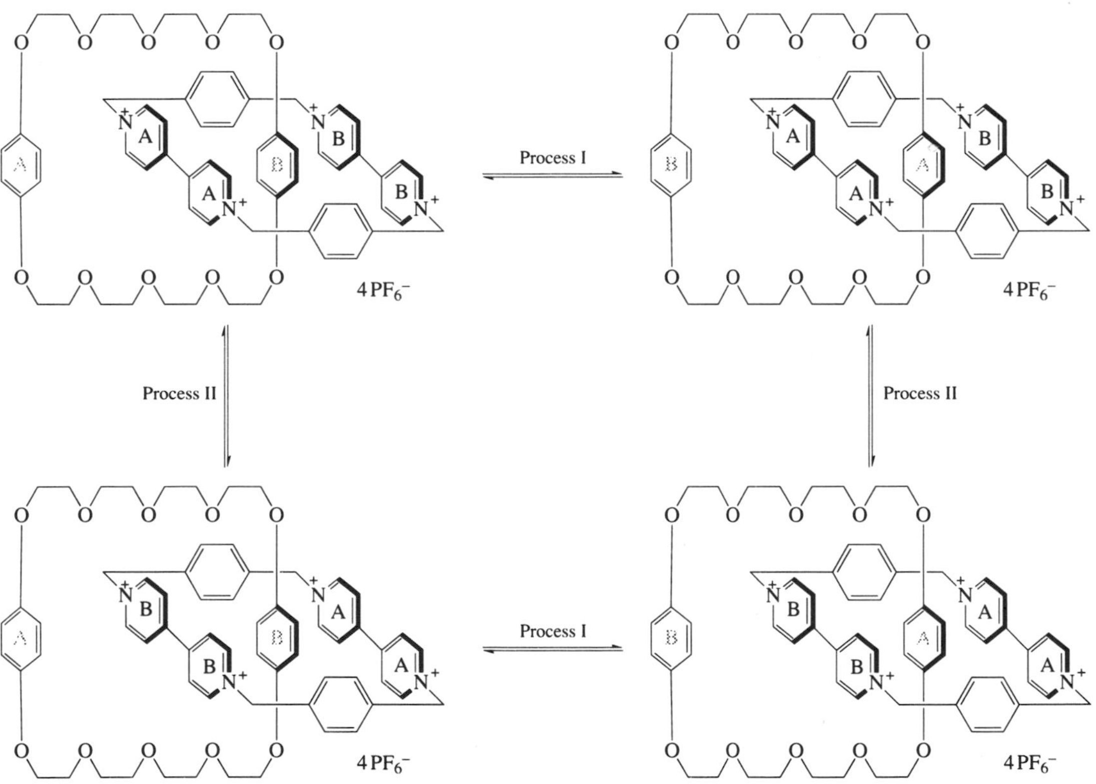

Figure 17 Dynamic processes occurring in solution within the [2]catenane (**73**).

Preparation of macrocyclic polyethers with two different π-electron-rich units allows self-assembly of [2]catenanes that exhibit translational isomerism. The [2]catenane (**77**) (Figure 18) incorporating one 1,5-dioxynaphthalene and one hydroquinone ring within the crown ether component, which

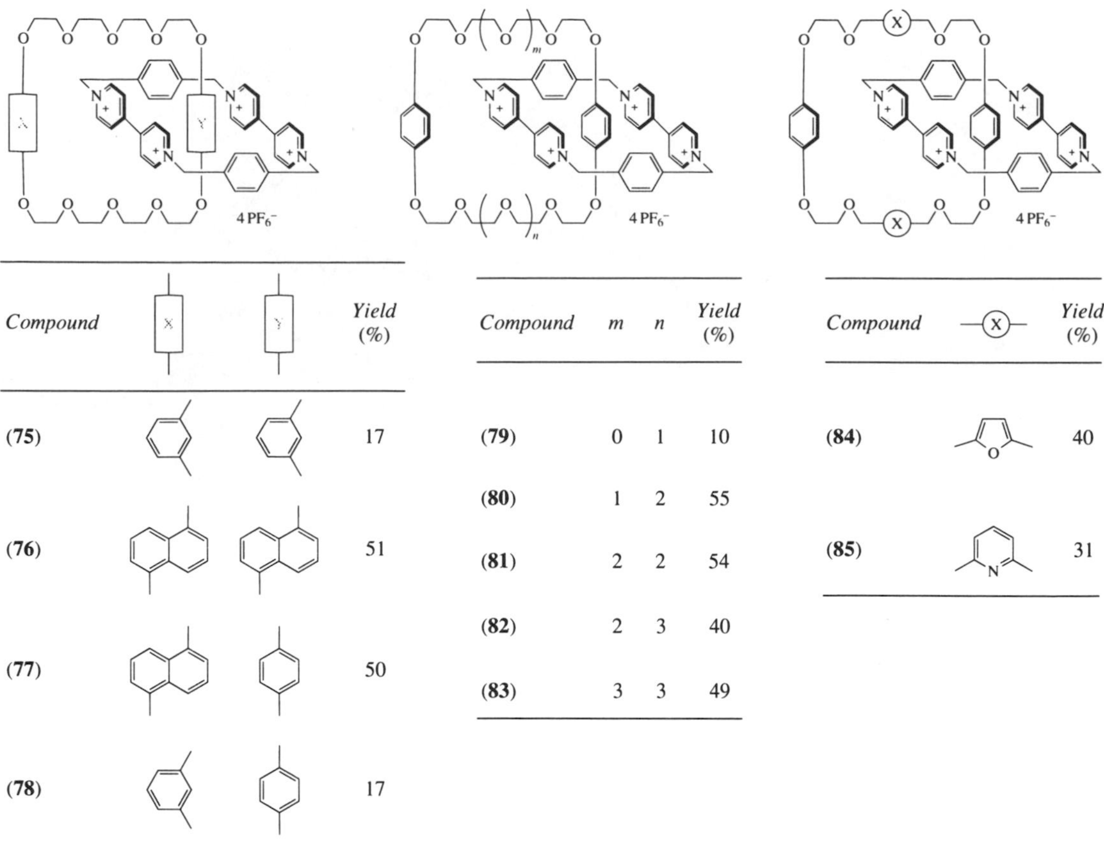

Compound	X	Y	Yield (%)
(75)			17
(76)			51
(77)			50
(78)			17

Compound	m	n	Yield (%)
(79)	0	1	10
(80)	1	2	55
(81)	2	2	54
(82)	2	3	40
(83)	3	3	49

Compound	–(X)–	Yield (%)
(84)		40
(85)		31

Figure 18 Yields for the self-assembly of the [2]catenanes (**75**)–(**85**) incorporating structural modifications within the π-electron-rich macrocyclic component.

has been self-assembled[129] in a yield of 50%, exists in solution as two translational isomers (Figure 19). In CD_3COCD_3 solution at 25 °C, the ratio between the isomers corresponds to 70:30 in favour of the one having the hydroquinone ring located inside the cavity of the cyclophane. On employing the more polar solvent CD_3SOCD_3, we find that the ratio between the isomers is reversed, the isomer having the hydroquinone ring located in the alongside position being the major one. The [2]catenane (**78**) (Figure 18) incorporating a macrocyclic polyether composed of one hydroquinone and one resorcinol ring has been self-assembled[130] in a yield of 17%. The translational isomerism exhibited (Figure 19) by the [2]catenane (**78**) is highly selective, the ratio between the two isomers being 98:2 in favour of the one positioning the hydroquinone ring inside the cavity of the cyclophane in CD_3COCD_3 at 0 °C.

The effect of modifying the construction of the spacers linking the two π-electron-rich recognition sites incorporated within the macrocyclic polyether has also been studied. A range of di-*p*-phenylene crown ethers incorporating polyether chains of different lengths have been used to self-assemble[131] the corresponding [2]catenanes (**79**)–(**83**) (Figure 18). Interestingly, the yields obtained in the self-assembly of the [2]catenanes (**79**)–(**83**) are well below the 70% yield recorded for the first [2]catenane (**73**) incorporating di-*p*-phenylene-34-crown-10 (**1**). Furthermore, the energy barriers associated with the circumrotation of the crown ether component through the cyclophane (cf. process I in Figure 17) within the [2]catenanes (**79**)–(**83**) are lower than the barrier observed in the case of (**73**). These results suggest that an optimal stereoelectronic match between tetracationic cyclophane and di-*p*-phenylene-34-crown-10 crown ethers is achieved for the [2]catenane (**73**), affording it relatively strong intercomponent interactions of a noncovalent bonding nature.

The role of the central oxygen atoms in the macrocyclic polyether, believed to be involved in hydrogen bonding to the protons attached to the bipyridinium units in the tricationic intermediate leading to the tetracationic cyclophane, has been studied. Introduction of either furan[132] or pyridine[133] units into the polyether chains of the π-electron-rich component results in the [2]catenanes (**84**) and (**85**) which have been self-assembled in yields of 40% and 31%, respectively. The energy

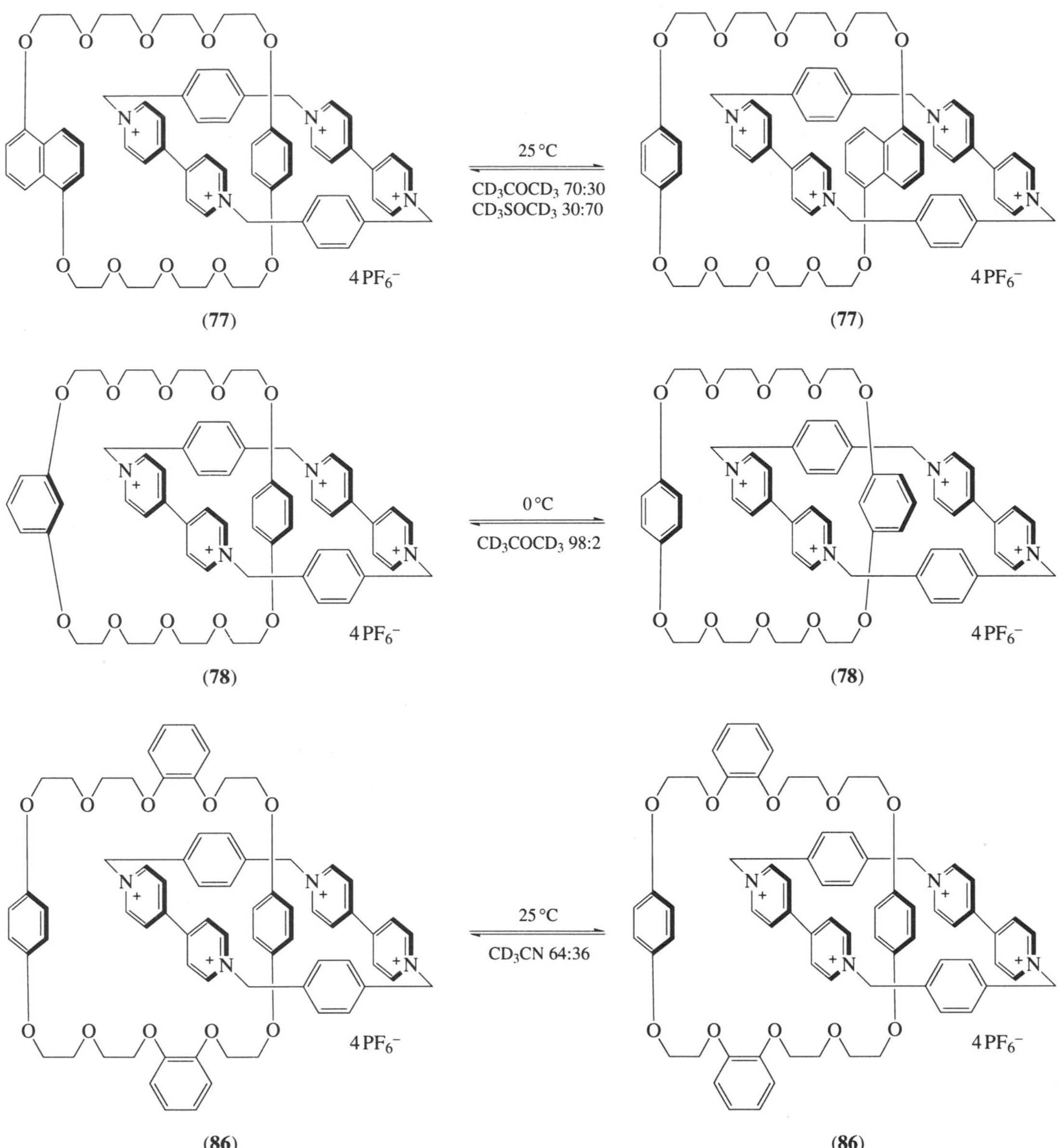

Figure 19 Equilibration between the translational isomers of the [2]catenanes (**77**), (**78**) and (**86**) incorporating dissymmetric π-electron-rich macrocyclic components.

barriers associated with the circumrotations of the crown ether components (cf. process I in Figure 17) through the cyclophane are 57.8 kJ mol^{-1} and 47.3 kJ mol^{-1} for (**84**) and (**85**), respectively. In both cases, the energy of activation is lower than that observed for the [2]catenane (**73**), presumably as a result of the weaker intercomponent interactions existing within these catenated derivatives. Translational isomerism can also be induced by introducing asymmetric spacers within the polyether chains of a crown ether component carrying identical recognition sites. The [2]catenane (**86**) incorporating catechol units within the polyether chains of the neutral component was self-assembled[134] in a yield of 16%. The ratio between the translational isomers (Figure 19) corresponds to 64:36 in favour of the isomer in which the hydroquinone ring closest to the catechol units resides inside the cavity of the cyclophane. A series of macrocyclic polyethers incorporating a strapped porphyrin and π-electron-rich hydroquinone or 1,5-dioxynaphthalene ring systems have been synthesized by Gunter and co-workers.[135–41] These macrocycles are able to bind paraquat (**2**) with a pseudorotaxane-like geometry. The [2]catenanes (**89**) and (**90**) were self-assembled[139,140] (Scheme 14) in yields of 20% and 28% starting from the π-electron-rich macrocycles (**87**) and (**88**), respectively. The circumrotation of the π-electron-rich macrocycle through the π-electron-deficient cyclophane is prevented by the size of the porphyrin unit, which cannot slide through the cavity of the tetracationic cyclophane. However, the circumrotation of the cyclophane through the polyether macrocycle is possible, and energies of activation of 60.3 kJ mol^{-1} and 54.0 kJ mol^{-1} associated with the process have been obtained for **(89)** and **(90)**, respectively. The protonation[141] of the porphyrin within the catenated environments of **(89)** and **(90)** requires much stronger acids than in the case of the corresponding macrocycles **(87)** and **(88)**. As a result of the protonation of the [2]catenane **(90)**, it changes its conformation so as to locate a *p*-xylyl spacer, rather than a bipyridinium unit, between the porphyrin and the hydroquinone ring.

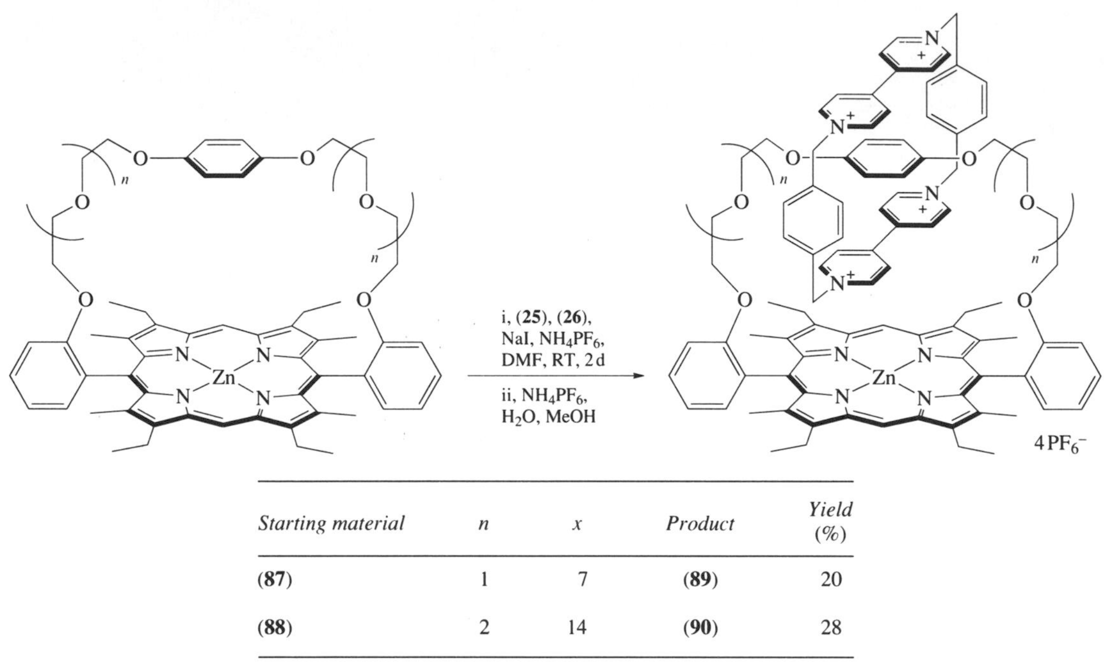

Starting material	*n*	*x*	*Product*	*Yield* (%)
(87)	1	7	**(89)**	20
(88)	2	14	**(90)**	28

Scheme 14

The self-assembly in solution of the pseudorotaxane **(6·91)** with thiol groups located at both ends of the thread-like component included in the tetracationic cyclophane **(6)** has been exploited by Gokel and co-workers[142] to generate (Scheme 15) a surface-attached [2]catenane after binding of the sulfur atoms to a gold surface. The entrapment of the cyclophane was confirmed by cyclic voltammetry, which showed reversible reduction waves for the bipyridinium units.

3.4.4 Self-assembly and Properties of [2]Catenanes Incorporating Structural Modifications in the π-Electron-deficient Component

A series of [2]catenanes incorporating di-*p*-phenylene-34-crown-10 **(1)** and a wide diversity of tetracationic cyclophanes have been synthesized in order to investigate the influence of the structural modification of the π-electron-deficient component upon the efficiency of the self-assembly process as well as upon the dynamic and other properties of these catenanes. If one of the *p*-xylyl spacers incorporated within the tetracationic cyclophane **(6)** is replaced with a *m*-xylyl unit, the [2]catenane **(94)** can be self-assembled[143] according to the two routes depicted in Scheme 16. Reaction of 1,3-bis(bromomethyl)benzene **(92)** with the dication **(26)** in the presence of the crown ether **(1)** gives, the [2]catenane **(94)** in a yield of 18%, while reaction of 1,4-bis(bromomethyl)-benzene **(25)** with the dication **(93)** in the presence of **(1)** affords the catenane **(94)** in a yield of 40%. The difference in the yields of the two self-assembly approaches to **(94)** is presumably a result of a higher efficiency characterizing the macrocyclization step involving the ring closure at the *para*-substituted end of the tricationic intermediate.

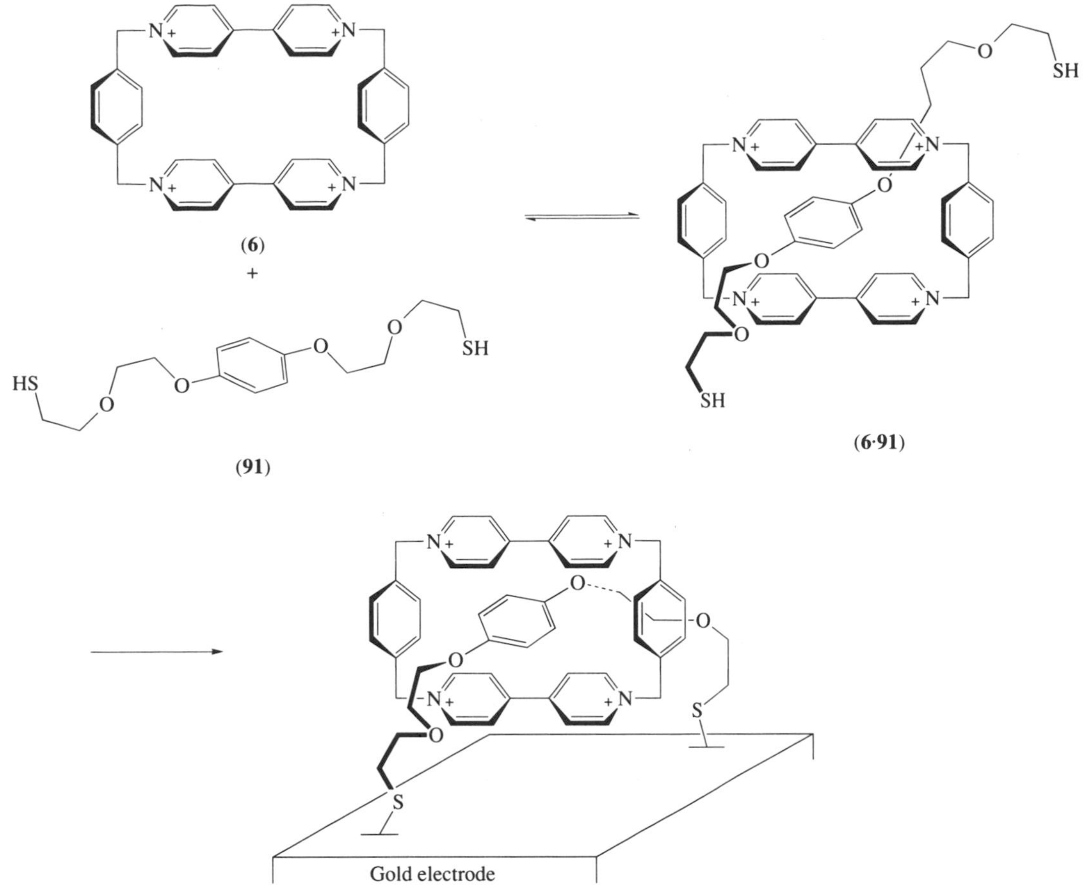

Scheme 15

A consequence of replacing both the *p*-xylyl spacers of the tetracationic component with *m*-xylyl units is to reduce the size of its cavity even further. As a result, the reaction between 1,3-bis (bromomethyl)benzene (**92**) and the dication (**93**) in the presence of the crown ether (**1**) requires ultrahigh-pressure conditions in order to afford[143] (Scheme 16) the [2]catenane (**95**) in a yield of 28%. No [2]catenane was formed at ambient pressure. The reduced distance separating the bipyridinium units incorporated within the [2]catenanes (**94**) and (**95**) is reflected also in the rates of the dynamic processes occurring within these molecules. The replacement of only one *p*-xylyl spacer with a *m*-xylyl unit results in the free energy of activation associated with the circumrotation of the neutral macrocycle through the cavity of the tetracationic component being increased from 65.3 kJ mol^{-1} to 68.2 kJ mol^{-1}. The replacement of the second *p*-xylyl spacer with a *m*-xylyl unit increases further the energy barrier to 73.7 kJ mol^{-1}, presumably as a result of steric compression in the cyclophane and the increased intercomponent interaction, which results from the proximity of the π-electron-rich and π-electron-deficient units. In a similar vein, the introduction of both one and two thiophene residues as the spacers separating the bipyridinium units of the tetracationic component has been exploited to self-assemble[144] (Scheme 16) the [2]catenanes (**97**) and (**99**) in yields of 36% and 59%, respectively. The self-assembly of an optically active [2]catenane has also been achieved by replacing one of the *p*-xylyl spacers of the tetracationic cyclophane with an (*S*,*S*)-hydrobenzoin-based diether chain. The reaction of the chiral dicationic derivative (**100**) with the dibromide (**25**) in the presence of the crown ether (**1**) afforded[145] (Scheme 16) the [2]catenane (**101**) in a yield of 8%.

The incorporation of either one or two photochemically responsive azobenzene residues in place of the *p*-xylyl spacers of the tetracationic component has been achieved by Vögtle and coworkers.[146,147] The [2]catenanes (**103**) and (**105**) have been self-assembled in yields of 34% and 16%, respectively (Scheme 17). The isomerization from the (*E*) to the (*Z*) configuration of the azobenzene spacers occurs upon irradiation of the sample at 310 nm. ^{1}H NMR spectroscopic studies showed that the half-lives for the thermal reisomerization of the [2]catenanes (**103**) and (**105**) are 20.5 h

(1)

i, see below
ii, NH_4PF_6, H_2O

Dibromide	Dication	X	Y	i	Product	Yield (%)
(92)	(26)			DMF, RT, 7 d	(94)	18
(25)	(93)			DMF, RT, 7 d	(94)	40
(92)	(93)			DMF, 10 kbar, RT, 3 d	(95)	28
(25)	(96)			MeCN, RT, 7 d	(97)	36
(98)	(96)			MeCN, RT, 7 d	(99)	59
(25)	(100)			MeCN, RT, 14 d	(101)	8

Scheme 16

and 12 d, while the half-lives for the reisomerization of the corresponding free tetracationic cyclophanes are 8.3 d and 14 d, respectively. It is evident that the presence of the hydroquinone ring inside the cavity of the π-electron-deficient macrocycle assists the reisomerization, thus reducing the time required for the process to occur. The effect is more profound in the case of the smaller tetracationic cyclophane component.

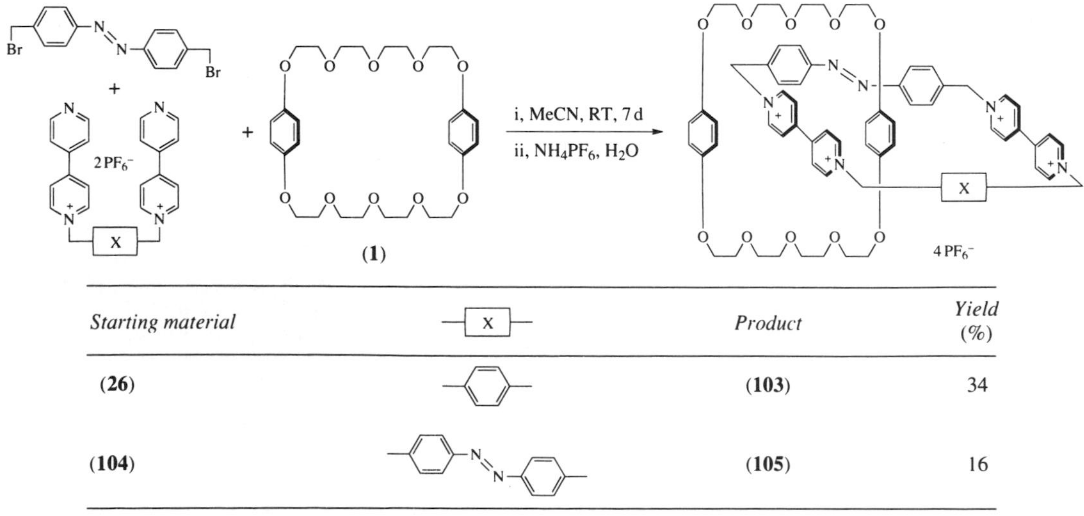

Starting material	X	Product	Yield (%)
(26)		(103)	34
(104)		(105)	16

Scheme 17

Replacement of one of the bipyridinium recognition sites of the tetracationic cyclophane of the [2]catenane (**73**) by a bis(pyridinium)ethene residue has also been investigated. The [2]catenane (**108**) was self-assembled[148] in 23% yield according to the route shown in Scheme 18. The [2]catenane (**108**) exists as two translational isomers, one with the bipyridinium unit located inside the cavity of the macrocyclic crown ether and the other in which the *trans*-bis(pyridinium)ethene residue is encircled by the crown ether. When a CD_3COCD_3 solution of (**108**) is cooled down to –60 °C, the circumrotation of the tetracationic macrocycle through the crown ether becomes slow on the ^{1}H NMR timescale and the isomer ratio at equilibrium is 92:8 in favour of the translational isomer in which the bipyridinium unit is encircled by the macrocyclic crown ether.

2 PF_6^- (**106**) (**107**)

i, MeCN, RT, 5 d
ii, NH_4PF_6, H_2O

4 PF_6^-

Starting material		*Product*	*Yield (%)*
(**1**)		(**108**)	23
(**109**)		(**110**)	58

Scheme 18

The [2]catenane (**112**) incorporating two *trans*-bis(pyridinium)ethene units in place of the bipyridinium recognition sites in cyclobis(paraquat-*p*-phenylene) was prepared[148] (Scheme 19) in a yield of 15%. The increased size of the cavity of the tetracationic cyclophane, as a result of the extension to the viologen units, is directly reflected in both the low yield of the catenane and in the rates of the dynamic processes occurring within it. The energy barrier associated with the circumrotation of the tetracationic cyclophane through the macrocyclic polyether (cf. process I in Figure 17) corresponds to 46.5 kJ mol^{-1}, while the energy of activation for the process (cf. process II in Figure 17) involving the circumrotation of the neutral macrocycle through the charged one is 52.7 kJ mol^{-1}. In both processes, the energy of activation is reduced with respect to the energies observed in the [2]catenane (**73**) incorporating cyclobis(paraquat-*p*-phenylene) and di-*p*-phenylene-34-crown-10.

Cyclic voltammetry of [2]catenane (**108**) showed four monoelectronic reduction waves. The first monoelectronic addition involves the reduction of the bipyridinium unit at a potential very similar to that of the first reduction wave of the free tetracationic cyclophane, confirming that the bipyridinium unit does not reside exclusively inside the cavity of the macrocyclic polyether. The following two monoelectronic additions involve the reduction of the bis(pyridinium)ethene unit, while the last monoelectronic wave corresponds to the second reduction of the bipyridinium unit. Interestingly, the potential of the first reduction wave of the extended π-electron-deficient residue is very similar to that observed for the [2]catenane (**112**), suggesting that after the first reduction wave, namely the one responsible for the reduction of the bipyridinium unit, the bis(pyridinium)ethene unit resides predominantly inside the cavity of the macrocyclic polyether.

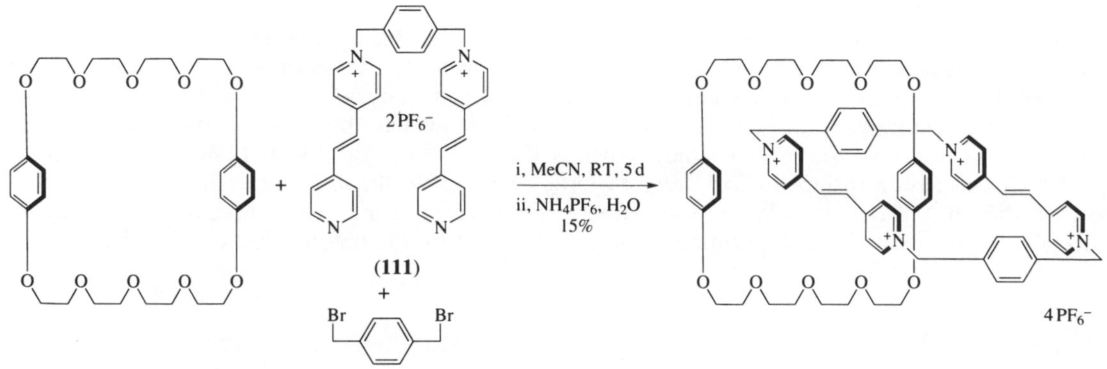

Scheme 19

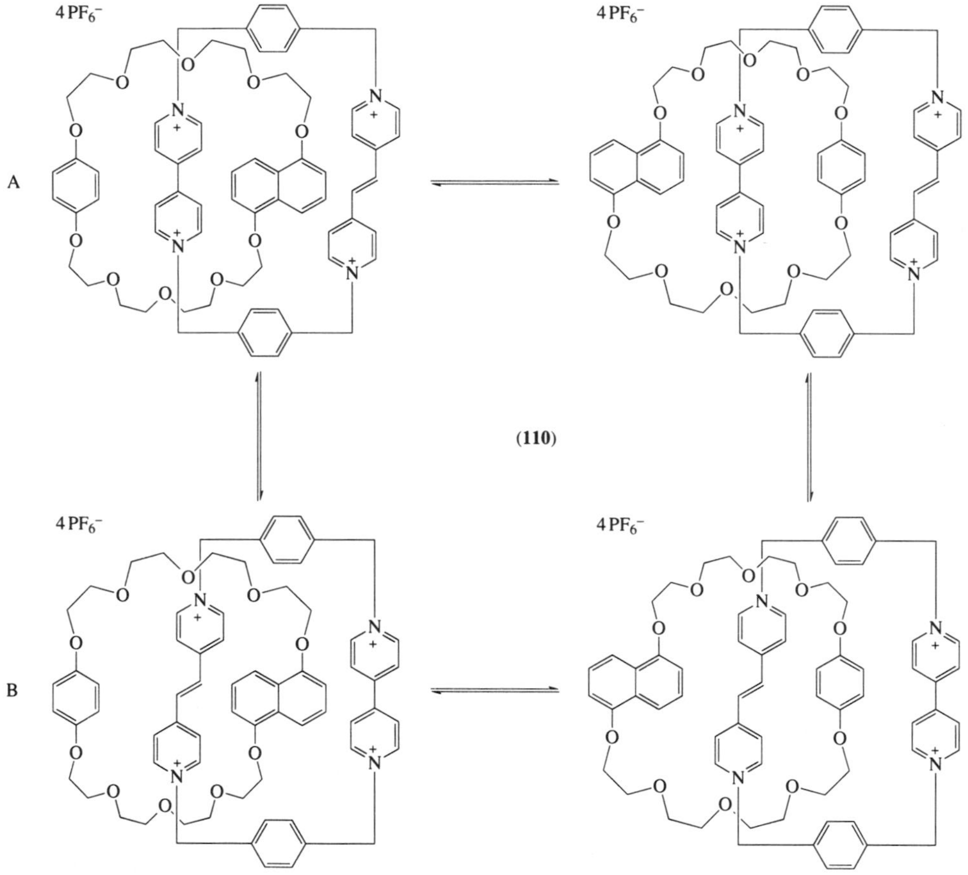

Figure 20 Equilibration between the four translational isomers of the [2]catenane (**110**) incorporating a dissymmetric π-electron-rich macrocyclic crown ether and a dissymmetric π-electron-deficient tetracationic cyclophane.

3.4.5 Self-assembly and Properties of a [2]Catenane Incorporating Structural Modifications in both the π-Electron-rich and π-Electron-deficient Components

The synthesis of a [2]catenane incorporating two dissymmetric macrocyclic components has been realized. Reaction of (**106**) with (**107**) in the presence of the macrocyclic polyether (**109**) afforded[149] (Scheme 18) the [2]catenane (**110**) in a yield of 58%. As a result of the presence of two different recognition sites in each of the two ring components, the possible existence of four translational isomers (Figure 20) can be predicted for the [2]catenane. At room temperature, the circumrotation of the crown ether through the charged macrocycle is slow on the ^{1}H NMR time-

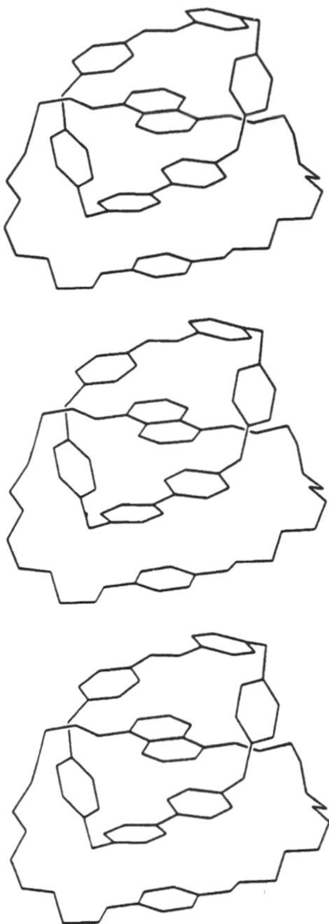

Figure 21 The x-ray crystal structure of (**110**).

scale and the ^{1}H NMR spectrum of a CD_3COCD_3 solution of (**110**) shows the preferential inclusion of the 1,5-dioxynaphthalene residue inside the cavity of the tetracationic cyclophane. If the solution is cooled down to −40 °C, the circumrotation of the charged macrocycle through the crown ether cavity is slowed down on the ^{1}H NMR timescale and only the isomers A and B are observed in a ratio of 95:5. Therefore, the most abundant isomer is the one having the bipyridinium unit inside the cavity of the crown ether with the 1,5-dioxynaphthalene residing inside the cavity of the tetracationic cyclophane.

Interestingly, the x-ray crystal structure of the [2]catenane (**110**) (Figure 21) shows only the translational isomer A in accord with the variable-temperature ^{1}H NMR spectroscopic studies. It is hoped that the rational design of such complex structures will ultimately lead to the creation of controllable catenanes.

3.4.6 Self-assembly of [3]Catenanes, [4]Catenanes and [5]Catenanes

The synthesis of [3]catenanes can be achieved by enlarging the size of either the π-electron-rich or the π-electron-deficient macrocycle. Thus, the reaction (Scheme 20; 1 bar = 10^5 Pa) under ultrahigh pressure of the dibromide (**25**) with the dication (**26**) in the presence of the macrocycle tetra-*p*-phenylene-68-crown-20 (**113**) affords[150] the [3]catenane (**114**) in a yield of 11%. High-pressure conditions are necessary for the formation of the [3]catenane, as at room pressure only the corresponding [2]catenane is formed.

Variable-temperature ^{1}H NMR spectroscopic studies showed that the two tetracationic cyclophanes incorporated within the [3]catenane (**114**) are able to move from station to station, giving rise to the degenerate equilibration process shown in Figure 22. Interestingly, only the translational isomer in which the two charged macrocycles reside on diametrically opposite stations can be detected by ^{1}H NMR spectroscopy at low temperature. Presumably, electrostatic repulsion disfavours the translational isomers in which the two tetracationic macrocycles reside on adjacent stations.

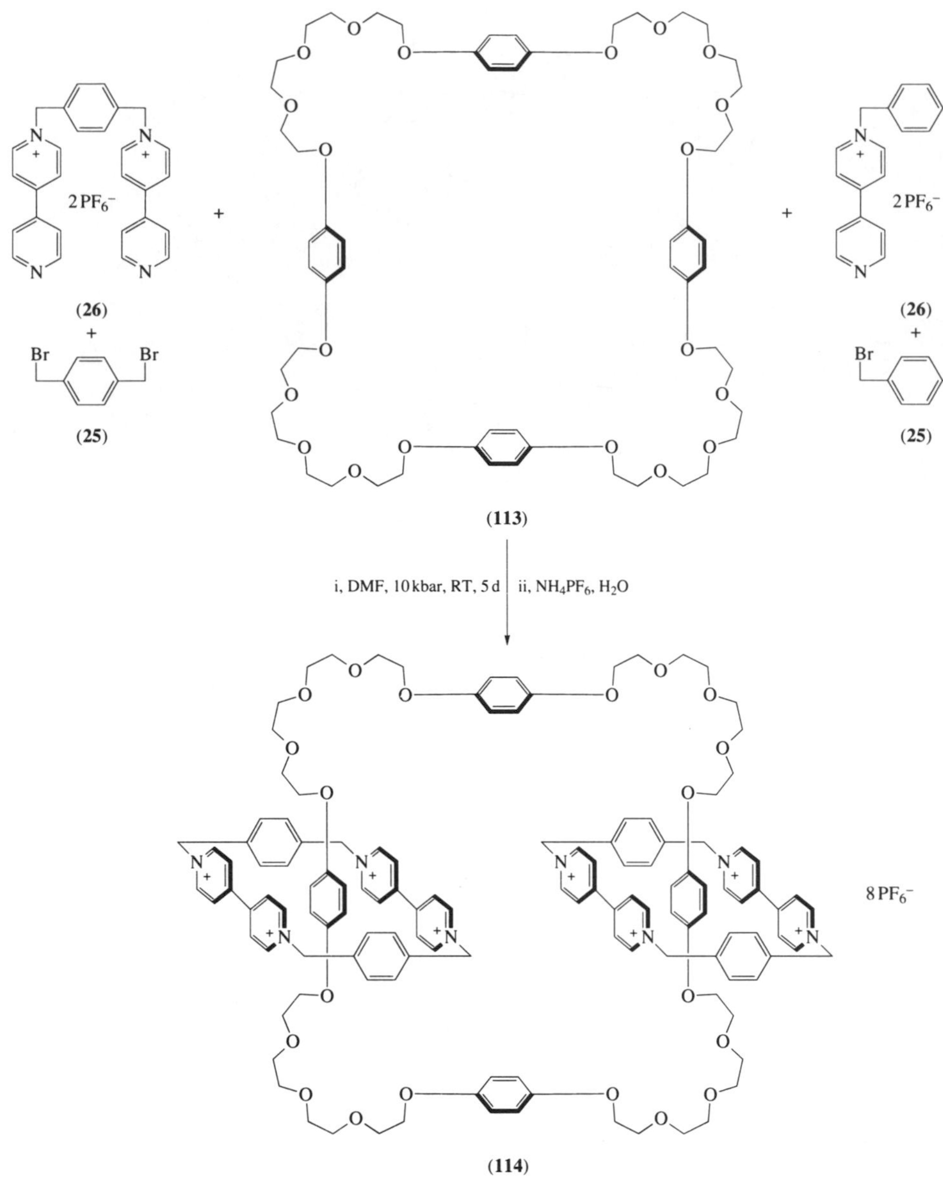

(26)

(25)

(113)

i, DMF, 10 kbar, RT, 5 d ii, NH_4PF_6, H_2O

(114)

Scheme 20

The self-assembly of a [3]catenane can also be achieved by the enlargement of the π-electron-deficient cyclophane. Replacement of the *p*-xylyl spacers separating the bipyridinium units in the tetracationic cyclophane (**6**) with 4,4′-bitolyl units gives a cyclophane large enough to accommodate simultaneously two hydroquinone rings inside its cavity. The reaction (Scheme 21) of the dication (**115**) with the dibromide (**116**) in the presence of either the hydroquinone-based (**1**) or the 1,5-dioxynaphthalene-based (**12**) affords[131,151,152] the corresponding [3]catenane (**117**) or (**118**) in yields of 25% or 31%, respectively. Interestingly, the formation of the [3]catenane is greatly favoured over that of the corresponding [2]catenane, indicating that the generation of the higher catenane involves some kind of cooperative process.

Given the success in preparing first the [3]catenane (**114**) incorporating one π-electron-rich macrocycle and two π-electron-deficient cyclophanes and second the [3]catenanes (**117**) and (**118**) incorporating two π-electron-rich macrocycles and one large π-electron-deficient cyclophane, it was reasoned that if large π-electron-rich and large π-electron-deficient macrocycles were employed, linear catenanes incorporating several linked components could be generated. The situation is shown schematically in Figure 23.

Regrettably, the reaction of the macrocycle (**113**) with the components of the large cyclophane did not yield any catenated compounds. As a result, the macrocyclic polyether was redesigned and

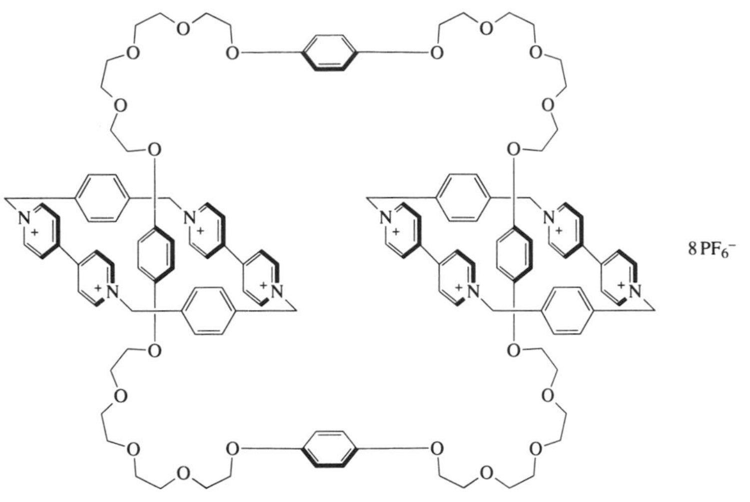

Figure 22 Equilibration between the translational isomers of the [3]catenane (**114**).

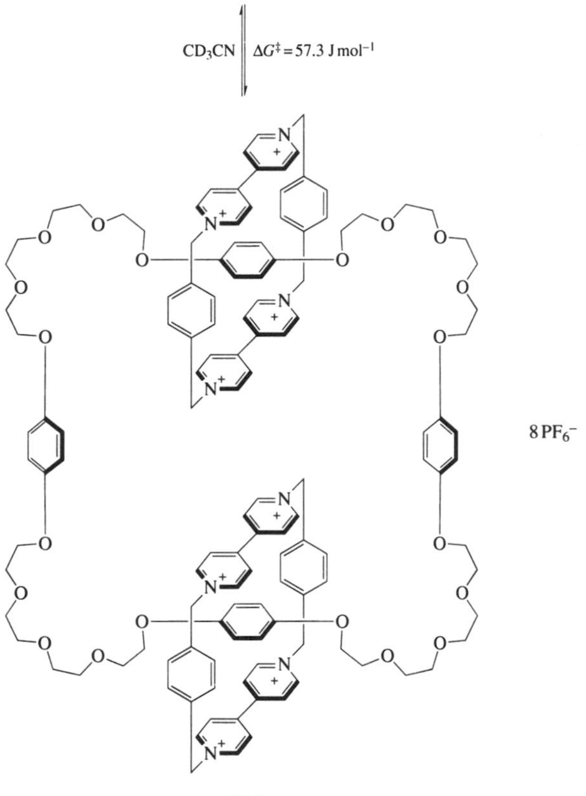

the crown ethers (**119**) and (**121**) incorporating three hydroquinone[131,153] and three 1,5-dioxynaphthalene[154] residues, respectively, were synthesized. The reaction (Scheme 22) of the dibromide (**25**) with the dication (**26**) in the presence of either (**119**) or (**121**) affords the corresponding [3]catenane (**120**)[131,153] or (**122**),[154] both in a yield of 15%. Although high-pressure conditions were necessary for the formation of (**120**), the [3]catenane (**122**) was obtained at room pressure.

The crucial step in the formation of higher catenanes was the reaction (Scheme 23) of the larger dication (**115**) incorporating a bitolyl spacer with the dibromide (**116**) in the presence of either (**119**) or (**121**). The [3]catenanes (**123**)[131,153] and (**124**)[155] were indeed formed, if only in 3% and 6% yield, respectively.

The [3]catenanes (**123**) and (**124**) each have two free recognition sites in the form of either hydroquinone rings or 1,5-dioxynaphthalene residues. Therefore, the incorporation of one or two

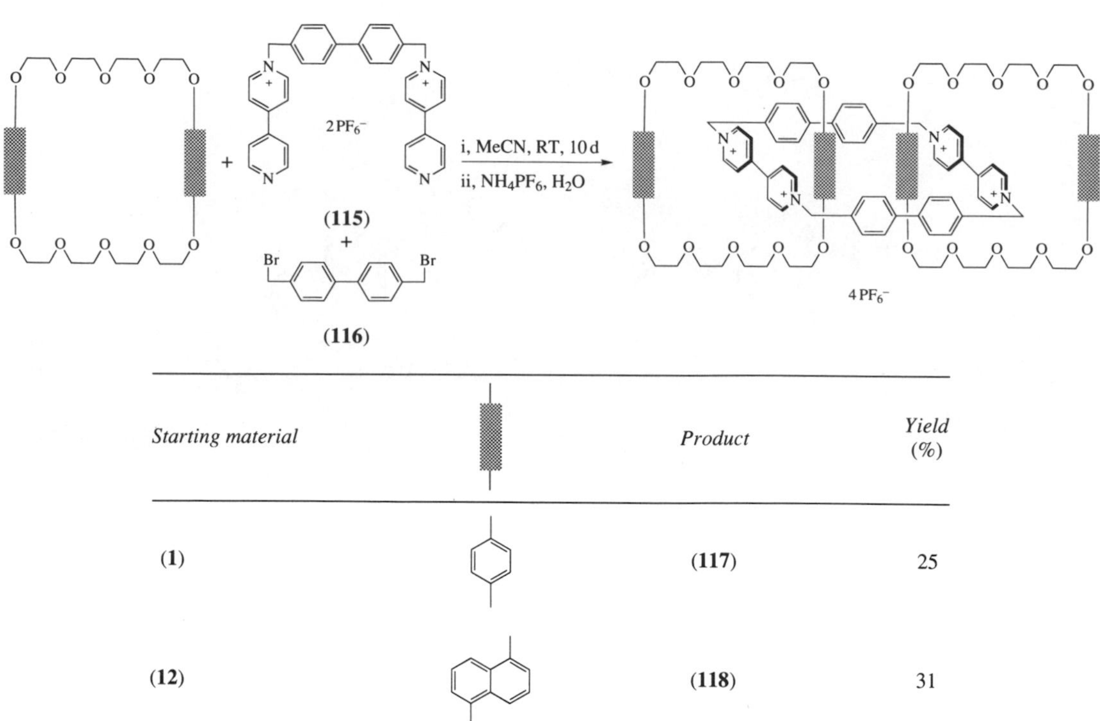

Starting material		*Product*	*Yield (%)*
(1)		(117)	25
(12)		(118)	31

Scheme 21

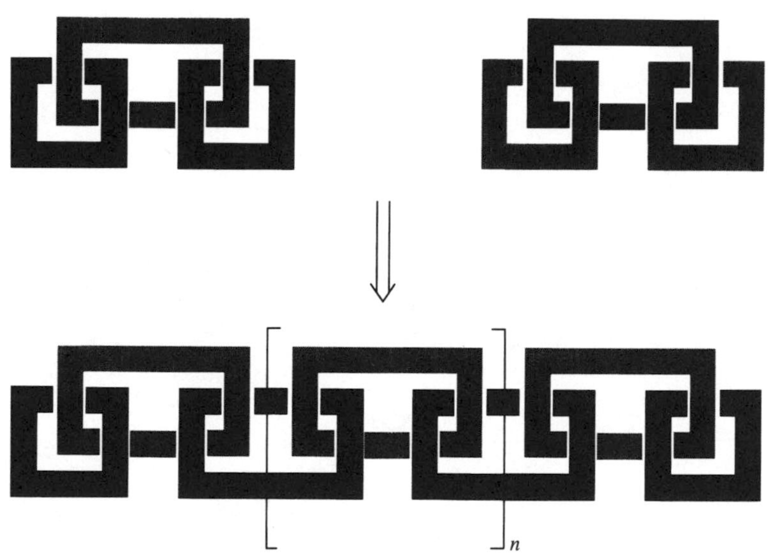

Figure 23 Progression from [3]catenanes to [*n*]catenanes following the enlargement of the macrocyclic components.

more tetracationic cyclophanes can be achieved by reacting (Scheme 24) the dibromide (**25**) with the dication (**26**) in the presence of either (**123**) or (**124**). Indeed, when the hydroquinone-containing [3]catenane (**123**) was employed, the reaction carried out under ultrahigh-pressure conditions afforded[131,153] the [4]catenane (**125**) in a yield of 22% and the [5]catenane (**127**) in a very low yield (<0.5%). In the case of the 1,5-dioxynaphthalene-containing [3]catenane (**124**), the reaction was performed at ambient pressure and temperature to afford[155] the [4]catenane (**126**) and the [5]catenane (**128**) in yields of 31% and 5%, respectively. The [5]catenane, the chemical equivalent of the symbol of the International Olympics Movement, was named Olympiadane after a suggestion made by van Gulick.[156]

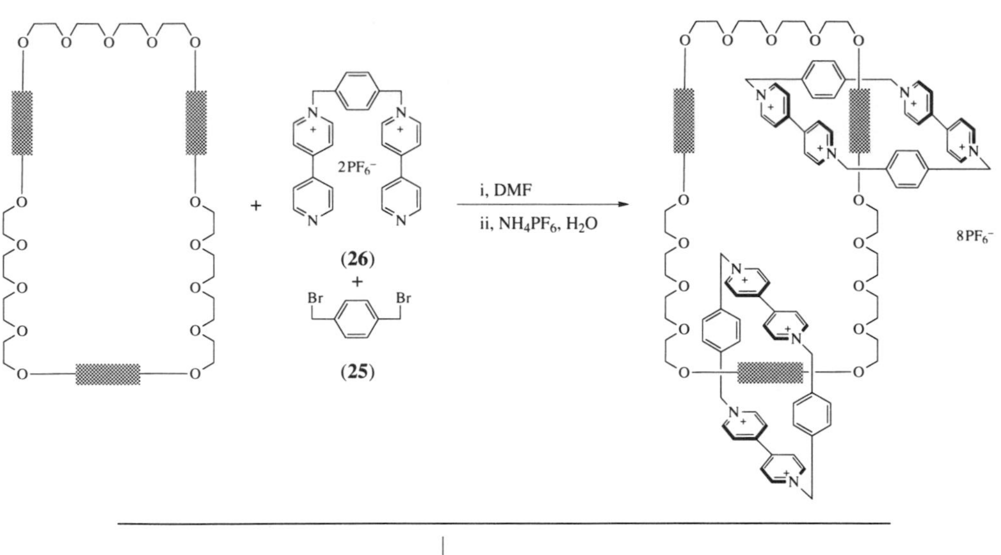

Starting material		*Product*	*Yield* (%)
(119)		**(120)**	15
(121)		**(122)**	15

Scheme 22

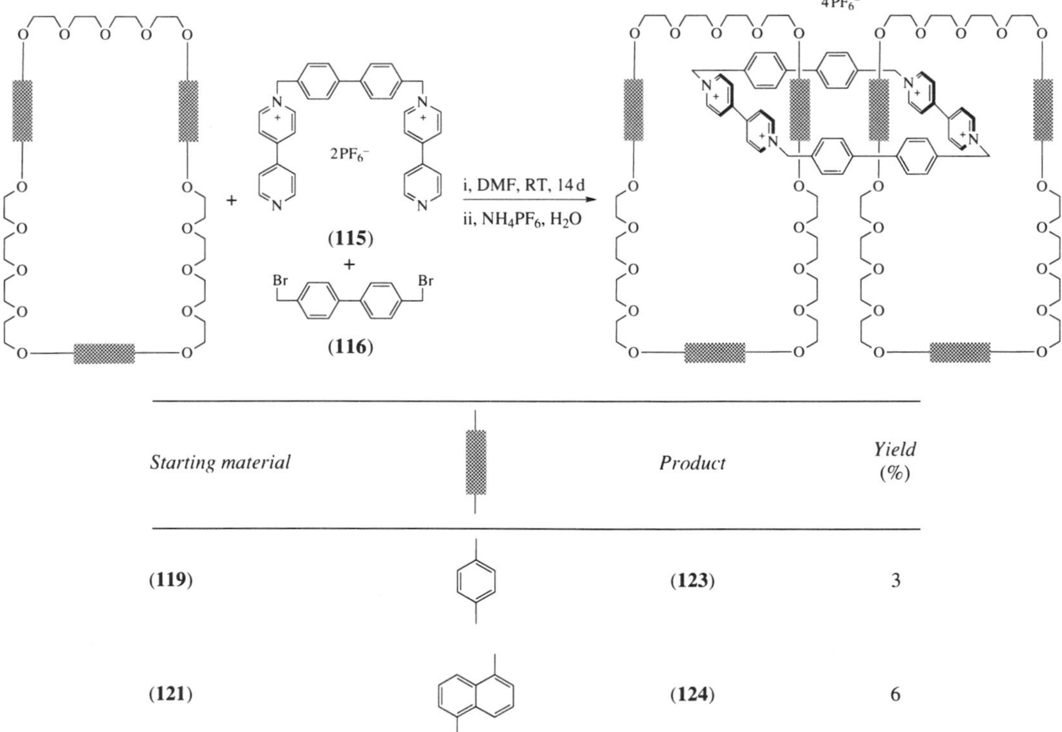

Starting material		*Product*	*Yield* (%)
(119)		**(123)**	3
(121)		**(124)**	6

Scheme 23

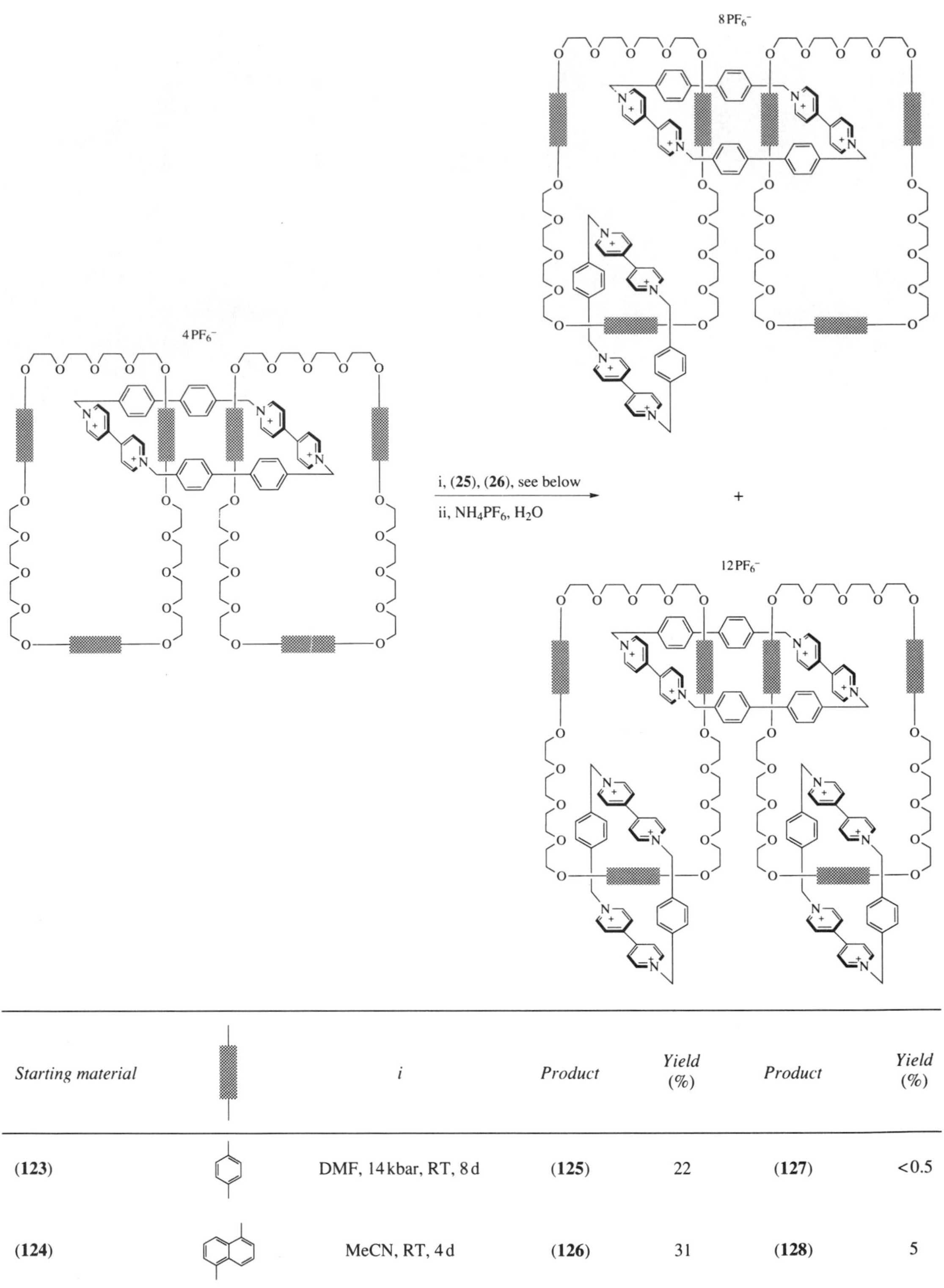

Starting material		*i*	*Product*	*Yield* (%)	*Product*	*Yield* (%)
(123)		DMF, 14 kbar, RT, 8 d	**(125)**	22	**(127)**	<0.5
(124)		MeCN, RT, 4 d	**(126)**	31	**(128)**	5

Scheme 24

3.4.7 Self-assembly of Bis[2]catenanes

An alternative approach to the construction of linear [*n*]catenanes involves the use of covalently bridged bismacrocycles, as illustrated schematically in Figure 24.

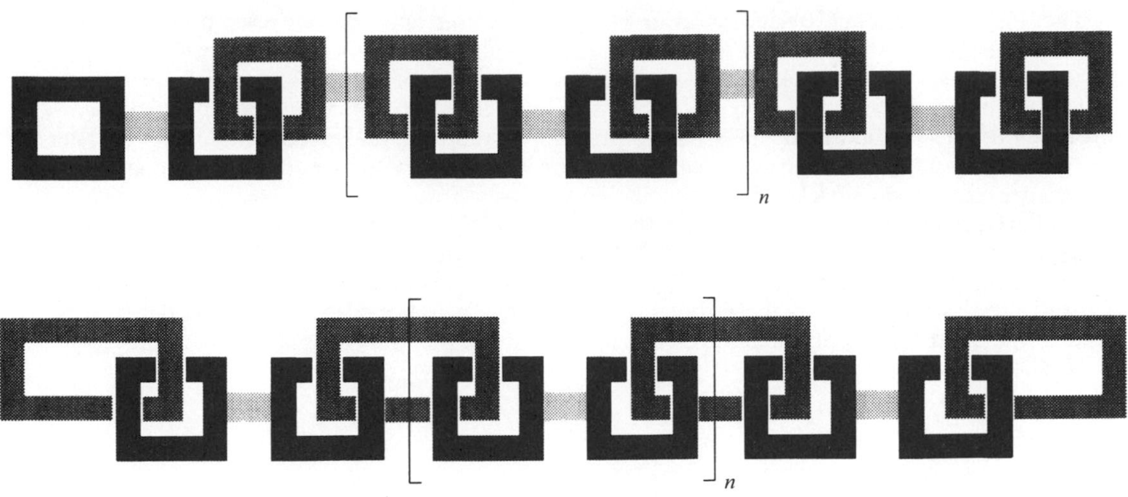

Figure 24 Schematic representation of [*n*]catenanes in which the macrocyclic components are held together via both covalent and mechanical bonds.

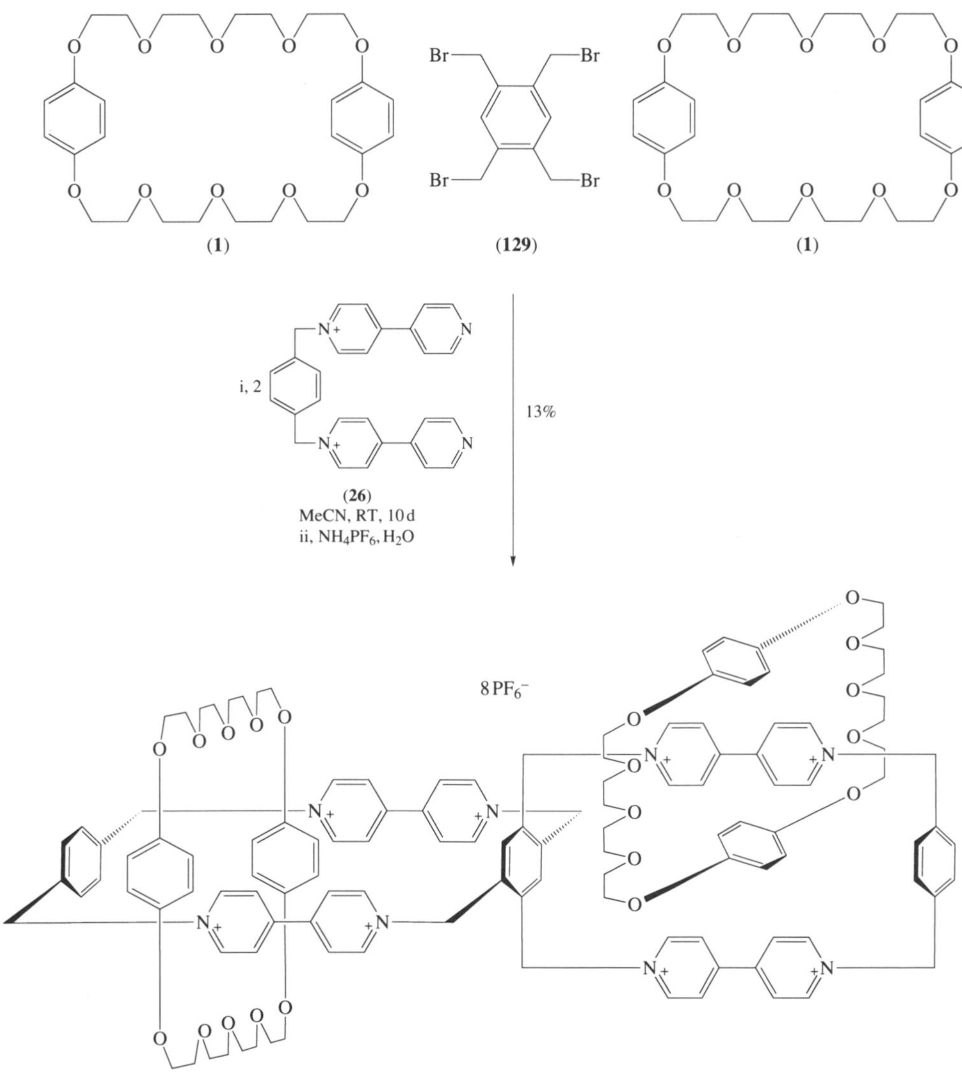

(1) (129) (1)

i, 2 (26), MeCN, RT, 10 d

ii, NH_4PF_6, H_2O

13%

$8\,PF_6^-$

(130)

Scheme 25

The bis[2]catenane (**130**) incorporating two π-electron-rich macrocycles and a π-electron-deficient biscyclophane was self-assembled[157] (Scheme 25) in a yield of 13% from five components in a single step. The self-assembly process takes advantage of the preferential formation of the *para*-substituted versus the more sterically constrained *ortho*- and *meta*-substituted bismacrocycles. Variable-temperature ^{1}H NMR spectroscopy studies revealed that the energy barrier for the circumrotation of the neutral macrocycles through the cavities of the biscyclophane component corresponds to 67.0 kJ mol^{-1}.

Similarly, reaction (Scheme 26) of the tetrabromide (**53**) with the dication (**26**) in the presence of di-*p*-phenylene-34-crown-10 (**1**) affords[158] the bis[2]catenane (**131**) in a yield of 31%. The higher yield during the self-assembly of bis[2]catenane (**131**) is presumably a result of the reduced electrostatic repulsions between the tetracationic subunits. It is hoped that these model studies will lead to the creation of polycatenanes.[159]

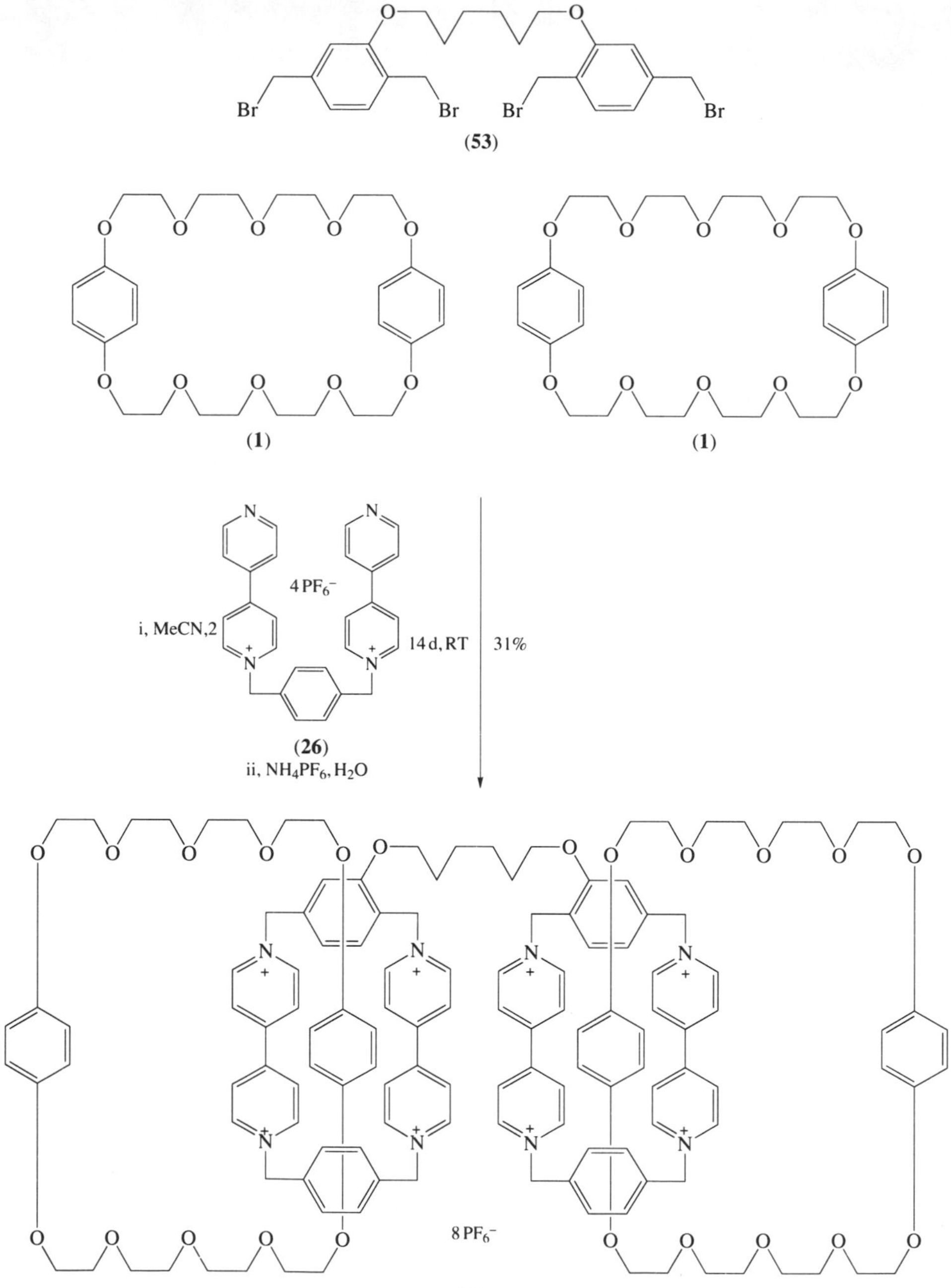

Br Br Br Br

(**53**)

(**1**) (**1**)

$4PF_6^-$

i, MeCN,2 14 d, RT 31%

(**26**)

ii, NH_4PF_6, H_2O

$8PF_6^-$

(**131**)

Scheme 26

3.5 CONCLUDING REMARKS

A synthetic approach towards the self-assembly of mechanically interlocked molecules, such as catenanes and rotaxanes, and supermolecules, such as pseudorotaxanes, has been developed[86,87] since the mid-1980s. The method relies upon the stereoelectronic complementarity between π-electron-rich and π-electron-deficient subunits. A series of noncovalent interactions—mainly (i) π–π stacking between π-electron-rich and π-electron-deficient aromatic units, (ii) hydrogen bonding between polyether oxygen atoms and CH acidic bipyridinium protons, and (iii) T-type hydrogen bonding between the π-faces of the xylyl spacers in the tetracationic cyclophane and the hydrogen atoms attached to the aromatic rings contained within the included neutral components—are responsible for the self-assembly of these structures and superstructures. These intercomponent interactions live on in the final thermodynamically stable structures and superstructures, driving both the self-organization observed in the solid state and the dynamic processes that occur in solution. The methodology has been successfully employed by the authors and many other investigators for the self-assembly of numerous [2]-, [3]- and [*n*]pseudorotaxanes, as well as [2]-, [3]- and [4]rotaxanes and a range of [2]-, [3]-, [4]- and [5]catenanes. The dynamic, electrochemical and photochemical properties of these mechanically interlocked compounds have been investigated in some detail. In particular, a chemically and electrochemically controllable molecular shuttle in the shape of a [2]rotaxane[115] and a controllable [2]catenane[148] have both been self-assembled. Thus, the relative motions within the components of a mechanically interlocked molecule can be controlled reversibly via external stimuli. These results suggest the possibility of generating electrochemically responsive[160–3] and photochemically active[163–5] molecular and supramolecular devices,[1–7] exploiting the features of mechanical constriction.[166–9] Furthermore, if the methodology described in this review is employed, the self-assembly in a precise and controlled manner of macromolecules such as polycatenanes[36,38,40,170–3] and polyrotaxanes,[36,38,40,174] can be envisaged in order to generate new materials at the nanoscopic[53–7] level with precise functions and shapes.

3.6 REFERENCES

1. J.-M. Lehn, *Angew. Chem., Int. Ed. Engl.*, 1988, **27**, 89.
2. D. J. Cram, *Angew. Chem., Int. Ed. Engl.*, 1988, **27**, 1009.
3. C. J. Pedersen, *Angew. Chem., Int. Ed. Engl.*, 1988, **27**, 1021.
4. J. Rebek, Jr., *Acc. Chem. Res.*, 1990, **23**, 399.
5. F. Vögtle, 'Supramolecular Chemistry', Wiley, New York, 1991.
6. J.-M. Lehn, *Science*, 1993, **260**, 1762.
7. J.-M. Lehn, 'Supramolecular Chemistry', VCH, Weinheim, 1994.
8. J. S. Lindsey, *New J. Chem.*, 1991, **15**, 153.
9. G. M. Whitesides, J. P. Mathias and C. T. Seto, *Science*, 1991, **254**, 1312.
10. D. H. Busch, *J. Inclusion Phenom.*, 1992, **12**, 389.
11. S. Anderson, H. L. Anderson and J. K. M. Sanders, *Acc. Chem. Res.*, 1993, **26**, 469.
12. R. Hoss and F. Vögtle, *Angew. Chem., Int. Ed. Engl.*, 1994, **33**, 375.
13. J. F. Stoddart, in 'Host–Guest Molecular Interactions: from Chemistry to Biology', Wiley, Chichester, 1991, pp. 5–22.
14. D. B. Amabilino and J. F. Stoddart, *New Sci.*, 1994, (1913), 25.
15. H. Ringsdorf, B. Schlarb and J. Venzmer, *Angew. Chem., Int. Ed. Engl.*, 1988, **27**, 113.
16. M. Ahlers, W. Müller, A. Reichert, H. Ringsdorf and J. Venzmer, *Angew. Chem., Int. Ed. Engl.*, 1990, **29**, 1269.
17. J. M. Fuhrhop and J. König, in 'Molecular Assemblies and Membranes', ed. J. F. Stoddart, Royal Society of Chemistry, Cambridge, 1995.
18. J. Rebek, Jr., *Angew. Chem., Int. Ed. Engl.*, 1990, **29**, 245.
19. M. Famulok, J. S. Nowick and J. Rebek, Jr., *Acta Chem. Scand.*, 1992, **46**, 315.
20. S. Hoffmann, *Angew. Chem., Int. Ed. Engl.*, 1992, **31**, 1013.
21. L. E. Orgel, *Nature (London)*, 1992, **358**, 203.
22. J. Rebek, Jr., *Chem. Br.*, 1994, **30**, 286.
23. J.-M. Lehn, *Angew. Chem., Int. Ed. Engl.*, 1990, **29**, 1304.
24. R. Kramer, J.-M. Lehn and A. Marquis-Rigault, *Proc. Natl. Acad. Sci. USA.*, 1993, **90**, 5394.
25. E. C. Constable, *Angew. Chem., Int. Ed. Engl.*, 1991, **30**, 1450.
26. E. C. Constable, *Tetrahedron*, 1992, **48**, 10013.
27. M. C. Etter, *Acc. Chem. Res.*, 1990, **23**, 120.
28. G. R. Desiraju, *Acc. Chem. Res.*, 1991, **24**, 290.
29. G. M. Whitesides, E. E. Simanek, J. P. Mathias, C. T. Seto, D. N. Chin, M. Mammen and D. M. Gordon, *Acc. Chem. Res.*, 1995, **28**, 37.
30. R. Willstätter, Seminar in Zürich between 1906 and 1912; see footnote 5 in ref. 31.
31. H. L. Frisch and E. Wasserman, *J. Am. Chem. Soc.*, 1961, **83**, 3789.
32. G. Schill, 'Catenanes, Rotaxanes and Knots', Academic Press, New York, 1971.
33. V. I. Sokolov, *Russ. Chem. Rev.*, 1973, **42**, 452.
34. D. M. Walba, *Tetrahedron*, 1985, **41**, 3161.

35. C. O. Dietrich-Buchecker and J.-P. Sauvage, *Chem. Rev.*, 1987, **87**, 795.
36. Y. S. Lipatov, T. E. Lipatova and L. F. Kosyanchuk, *Adv. Polym. Sci.*, 1989, **88**, 49.
37. C. O. Dietrich-Buchecker and J.-P. Sauvage, in 'Bioorganic Chemistry Frontiers', ed. H. Dugas, Springer, Weinheim, 1991, vol. 2, pp. 195–248.
38. H. W. Gibson and H. Marand, *Adv. Mater.*, 1993, **5**, 11.
39. J. C. Chambron, C. O. Dietrich-Buchecker and J.-P. Sauvage, *Top. Curr. Chem.*, 1993, **165**, 131.
40. H. W. Gibson, M. C. Bheda and P. T. Engen, *Prog. Polym. Sci.*, 1994, **19**, 843.
41. B. L. Allwood, N. Spencer, H. Shariari-Zavareh, J. F. Stoddart and D. J. Williams, *J. Chem. Soc., Chem. Commun.*, 1987, 1064.
42. P. R. Ashton, B. Odell, M. V. Reddington, A. M. Z. Slawin, J. F. Stoddart and D. J. Williams, *Angew. Chem., Int. Ed. Engl.*, 1988, **27**, 1550.
43. P. L. Anelli, P. R. Ashton, R. Ballardini, V. Balzani, M. Delgado, M. T. Gandolfi, T. T. Goodnow, A. E. Kaifer, D. Philp, M. Pietraszkiewicz, L. Prodi, M. V. Reddington, A. M. Z. Slawin, N. Spencer, J. F. Stoddart, C. Vicent and D. J. Williams, *J. Am. Chem. Soc.* 1992, **114**, 193.
44. P. R. Ashton, D. Philp, M. V. Reddington, A. M. Z. Slawin, N. Spencer, J. F. Stoddart and D. J. Williams, *J. Chem. Soc., Chem. Commun.*, 1991, 1680.
45. Y. A. Shen, P. T. Engen, M. A. G. Berg, J. S. Merola and H. W. Gibson, *Macromolecules*, 1992, **25**, 2786.
46. P. R. Ashton, E. J. T. Chrystal, J. P. Mathias, K. P. Parry, A. M. Z. Slawin, N. Spencer, J. F. Stoddart and D. J. Williams, *Tetrahedron Lett*, 1987, **28**, 6367.
47. J. Y. Ortholand, A. M. Z. Slawin, N. Spencer, J. F. Stoddart and D. J. Williams, *Angew. Chem., Int. Ed. Engl.*, 1989, **28** 1394.
48. P. L. Anelli, P. R. Ashton, N. Spencer, A. M. Z. Slawin, J. F. Stoddart and D. J. Williams, *Angew. Chem., Int. Ed. Engl.*, 1991, **30**, 1036.
49. P. R. Ashton, D. Philp, N. Spencer, J. F. Stoddart and D. J. Williams, *J. Chem. Soc., Chem. Commun.*, 1994, 181.
50. P. R. Ashton, D. Philp, N. Spencer and J. F. Stoddart, *J. Chem. Soc., Chem. Commun.*, 1991, 1677.
51. M. V. Reddington, A. M. Z. Slawin, N. Spencer, J. F. Stoddart, C. Vicent and D. J. Williams, *J. Chem. Soc., Chem. Commun.*, 1991, 630.
52. M. J. Marsella, P. J. Carroll and T. M. Swager, *J. Am. Chem. Soc.*, 1994, **116**, 9347.
53. K. E. Drexler, 'Engines of Creation', Fourth Estate, London, 1990.
54. R. C. Merkle, *Nanotechnology*, 1991, **2**, 131.
55. R. C. Merkle, *Nanotechnology*, 1993, **4**, 86.
56. D. A. Tomalia and H. D. Durst, *Top. Curr. Chem.*, 1993, **165**, 193.
57. K. E. Drexler, *Annu. Rev. Biophys. Struct.*, 1994, **23**, 377.
58. E. Wasserman, *J. Am. Chem. Soc.*, 1960, **82**, 4433.
59. E. Wasserman, *Sci. Am.*, 1962, **207**, 94.
60. I. T. Harrison and S. Harrison, *J. Am. Chem. Soc.*, 1967, **89**, 5723.
61. I. T. Harrison, *J. Chem. Soc., Chem. Commun.*, 1972, 231.
62. I. T. Harrison *J. Chem. Soc., Perkin Trans. 1*, 1974, 301.
63. G. Agam, D. Graiver and A. Zilkha, *J. Am. Chem. Soc.*, 1976, **98**, 5206.
64. G. Agam and A. Zilkha, *J. Am. Chem. Soc.*, 1976, **98**, 5214.
65. H. Ogino, *J. Am. Chem. Soc.*, 1981, **103**, 1303.
66. H. Ogino and K. Ohata, *Inorg. Chem.*, 1984, **2**, 3312.
67. H. Ogino, *New J. Chem.*, 1993, **17**, 683.
68. R. Isnin and A. E. Kaifer, *J. Am. Chem. Soc.*, 1991, **113**, 8188.
69. R. Isnin and A. E. Kaifer, *Pure Appl. Chem.*, 1993, **65**, 495.
70. G. Wenz, F. Wolf, M. Wagner and S. Kubik, *New J. Chem.*, 1993, **17**, 729.
71. G. Wenz and B. Keller, *Angew. Chem., Int. Ed. Engl.*, 1992, **31**, 197.
72. A. Harada, J. Li, T. Nakamitsu and M. Kamachi, *J. Org. Chem.*, 1993, **58**, 7524.
73. A. Harada, J. Li and M. Kamachi, *J. Am. Chem. Soc.*, 1994, **116**, 3192.
74. D. Armspach, P. R. Ashton, C. P. Moore, N. Spencer, J. F. Stoddart, T. J. Wear and D. J. Williams, *Angew. Chem., Int. Ed. Engl.*, 1993, **32**, 854.
75. D. Armspach, P. R. Ashton, R. Ballardini, V. Balzani, A. Godi, C. P. Moore, L. Prodi, N. Spencer, J. F. Stoddart, M. S. Tolley, T. J. Wear and D. J. Williams, *Chem. Eur. J.*, 1995, **1**, 33.
76. A. Harada, *Am. Chem. Soc., Div. Polym. Chem., Polym. Prepr.*, 1995, **36**, 570.
77. C. A. Hunter, *J. Am. Chem. Soc.*, 1992, **114**, 5303.
78. F. J. Carver, C. A. Hunter and R. J. Shannon, *J. Chem. Soc., Chem. Commun.*, 1994, 1277.
79. F. Vögtle, S. Meier and R. Hoss, *Angew. Chem., Int. Ed. Engl.*, 1992, **31**, 1619.
80. S. Ottens-Hildebrandt, S. Meier, W. Schmidt and F. Vögtle, *Angew. Chem., Int. Ed. Engl.*, 1994, **33**, 1767.
81. S. Ottens-Hildebrandt, M. Nieger, K. Rissanen, J. Rouvinen, S. Meier, G. Harder and F. Vögtle, *J. Chem. Soc., Chem. Commun.*, 1995, 777.
82. G. J. M. Gruter, F. J. J. de Kanter, P. R. Markies, T. Nomoto, O. S. Akkerman and F. Bicklehaupt, *J. Am. Chem. Soc.*, 1993, **115**, 12179.
83. F. Bicklehaupt, *J. Organomet. Chem.*, 1994, **475**, 1.
84. M. Fujita, F. Ibukuro, H. Hagihara and K. Ogura, *Nature (London)*, 1994, **367**, 720.
85. M. Fujita, F. Ibukuro, K. Yamaguchi and K. Ogura, *J. Am. Chem. Soc.*, 1995, **117**, 4175.
86. D. Philp and J. F. Stoddart, *Synlett*, 1991, 445.
87. D. B. Amabilino and J. F. Stoddart, *Pure Appl. Chem.*, 1993, **65**, 2351.
88. L. A. Summers, 'The Bipyridinium Herbicides', Academic Press, London, 1980.
89. R. C. Hengelson, T. L. Tarnowsky, J. M. Timko and D. J. Cram, *J. Am. Chem. Soc.*, 1977, **99**, 6411.
90. M. H. Schwartz, *J. Inclusion Phenom.*, 1990, **9**, 1.
91. C. A. Hunter and J. K. M. Sanders, *J. Am. Chem. Soc.*, 1990, **112**, 5525.
92. C. A. Hunter, *J. Mol. Biol.*, 1993, **230**, 1025.
93. C. A. Hunter, *Angew. Chem., Int. Ed. Engl.*, 1993, **32**, 1584.

94. C. A. Hunter, *Chem. Soc. Rev.*, 1994, **23**, 101.
95. A. D. Hamilton, *J. Chem. Educ.*, 1990, **67**, 821.
96. C. B. Aakeröy and K. R. Seddon, *Chem. Soc. Rev.*, 1993, **22**, 397.
97. M. J. Zawarotko, *Chem. Soc. Rev.*, 1994, **23**, 283.
98. A. M. Z. Slawin, N. Spencer, J. F. Stoddart and D. J. Williams, *J. Chem. Soc., Chem. Commun.*, 1987, 1070.
99. R. O. Gould, A. M. Gray, P. Taylor and M. D. Walkinshaw, *J. Am. Chem. Soc.*, 1985, **107**, 5921.
100. S. K. Burley and G. A. Petsko, *J. Am. Chem. Soc.*, 1986, **108**, 7995.
101. W. L. Jorgensen and D. L. Severance, *J. Am. Chem. Soc.*, 1990, **112**, 4786.
102. T. M. Swager, M. J. Marsella, R. J. Newland and Q. Zhou, *Am. Chem. Soc., Div. Polym. Chem., Polym. Prepr.*, 1995, **36**, 546.
103. P. R. Ashton, M. Grognuz, A. M. Z. Slawin, J. F. Stoddart and D. J. Williams, *Tetrahedron Lett.*, 1991, **32**, 6235.
104. P. L. Anelli, N. Spencer and J. F. Stoddart, *J. Am. Chem. Soc.*, 1991, **113**, 5131.
105. I. O. Sutherland, *Annu. Rep. NMR Spectrosc.*, 1971, **4**, 71.
106. J. Sandström, 'Dynamic NMR Spectroscopy', Academic Press, London, 1982.
107. X. Sun, D. B. Amabilino, I. W. Parsons and J. F. Stoddart, *Am. Chem. Soc., Div. Polym. Chem., Polym. Prepr.*, 1993, **32**, 104.
108. X. Sun, D. B. Amabilino, I. W. Parsons, J. F. Stoddart and M. S. Tolley, *Macromol. Symp.*, 1994, **77**, 191.
109. G. Schill, K. Rissler, H. Fritz and W. Vetter, *Angew. Chem., Int. Ed. Engl.*, 1981, **20**, 187.
110. P. R. Ashton, R. A. Bissell, N. Spencer, J. F. Stoddart and M. S. Tolley, *Synlett*, 1992, 914.
111. P. R. Ashton, R. A. Bissell, R. Górski, D. Philp, N. Spencer, J. F. Stoddart and M. S. Tolley, *Synlett*, 1992, 919.
112. D. Philp, A. M. Z. Slawin, N. Spencer, J. F. Stoddart and D. J. Williams, *J. Chem. Soc., Chem. Commun.*, 1991, 1584.
113. P. R. Ashton, R. A. Bissell, N. Spencer, J. F. Stoddart and M. S. Tolley, *Synlett*, 1992, 923.
114. E. Córdova, R. A. Bissell, N. Spencer, P. R. Ashton, J. F. Stoddart and A. E. Kaifer, *J. Org. Chem.*, 1993, **58**, 6550.
115. R. A. Bissell, E. Córdova, A. E. Kaifer and J. F. Stoddart, *Nature (London)*, 1994, **369**, 133.
116. P. R. Ashton, M. R. Johnston, J. F. Stoddart, M. S. Tolley and J. W. Wheeler, *J. Chem. Soc., Chem. Commun.*, 1992, 1128.
117. A. C. Benniston and A. Harriman, *Angew. Chem., Int. Ed. Engl.*, 1993, **32**, 1459.
118. A. C. Benniston, A. Harriman and V. M. Lynch, *Tetrahedron Lett.*, 1994, **35**, 1473.
119. P. R. Ashton, J. A. Preece, J. F. Stoddart and M. S. Tolley, *Synlett*, 1994, 789.
120. P. R. Ashton, D. Philp, N. Spencer and J. F. Stoddart, *J. Chem. Soc., Chem. Commun.*, 1992, 1124.
121. P. R. Ashton, M. Bělohradský, D. Philp and J. F. Stoddart, *J. Chem. Soc., Chem. Commun.*, 1993, 1269.
122. D. B. Amabilino, P. R. Ashton, M. Bělohradský, F. M. Raymo and J. F. Stoddart, *J. Chem. Soc., Chem. Commun.*, 1995, 747.
123. P. R. Ashton, M. Bělohradský, D. Philp and J. F. Stoddart, *J. Chem. Soc., Chem. Commun.*, 1993, 1274.
124. D. B. Amabilino, P. R. Ashton, M. Bělohradský, F. M. Raymo and J. F. Stoddart, *J. Chem. Soc., Chem. Commun.*, 1995, 751.
125. P. R. Ashton, T. T. Goodnow, A. E. Kaifer, M. V. Reddington, A. M. Z. Slawin, N. Spencer, J. F. Stoddart, C. Vicent and D. J. Williams, *Angew. Chem., Int. Ed. Engl.*, 1989, **28**, 1396.
126. C. L. Brown, D. Philp and J. F. Stoddart, *Synlett*, 1991, 459.
127. D. B. Amabilino, P. R. Ashton and J. F. Stoddart, *Supramol. Chem.*, 1995, **5**, 5.
128. P. R. Ashton, C. L. Brown, E. J. T. Chrystal, T. T. Goodnow, A. E. Kaifer, K. P. Parry, D. Philp, A. M. Z. Slawin, N. Spencer, J. F. Stoddart and D. J. Williams, *J. Chem. Soc., Chem. Commun.*, 1991, 634.
129. P. R. Ashton, M. A. Blower, D. Philp, N. Spencer, J. F. Stoddart, M. S. Tolley, R. Ballardini, M. Ciano, V. Balzani, M. T. Gandolfi, L. Prodi and C. H. McLean, *New J. Chem.*, 1993, **17**, 689.
130. D. B. Amabilino, P. R. Ashton, G. R. Brown, W. Hayes, J. F. Stoddart, M. S. Tolley and D. J. Williams, *J. Chem. Soc., Chem. Commun.*, 1994, 2475.
131. D. B. Amabilino, P. R. Ashton, C. L. Brown, E. Córdova, L. A. Godinez, T. T. Goodnow, A. E. Kaifer, S. P. Newton, M. Pietraszkiewicz, D. Philp, F. M. Raymo, A. S. Reder, M. T. Rutland, A. M. Z. Slawin, N. Spencer, J. F. Stoddart and D. J. Williams, *J. Am. Chem. Soc.*, 1995, **117**, 1271.
132. P. R. Ashton, M. A. Blower, S. Iqbal, C. H. McLean, J. F. Stoddart, M. S. Tolley and D. J. Williams, *Synlett*, 1994, 1059.
133. P. R. Ashton, M. A. Blower, C. H. McLean, J. F. Stoddart and M. S. Tolley, *Synlett*, 1994, 1063.
134. D. B. Amabilino and J. F. Stoddart, *Recl. Trav. Chim. Pays-Bas*, 1993, **112**, 429.
135. M. J. Gunter and M. R. Johnston, *Tetrahedron Lett.*, 1990, **31**, 4801.
136. M. J. Gunter and M. R. Johnston, *Tetrahedron Lett.*, 1992, **33**, 1771.
137. M. J. Gunter and M. R. Johnston, *J. Chem. Soc., Perkin Trans. 1*, 1994, 995.
138. M. J. Gunter, M. R. Johnston, B. W. Skelton and A. H. White, *J. Chem. Soc., Perkin Trans. 1*, 1994, 1009.
139. M. J. Gunter and M. R. Johnston, *J. Chem. Soc., Chem. Commun.*, 1992, 1163.
140. M. J. Gunter, D. C. R. Hockless, M. R. Johnston, B. W. Skelton and A. H. White, *J. Am. Chem. Soc.*, 1994, **116**, 4810.
141. M. J. Gunter and M. R. Johnston, *J. Chem. Soc., Chem. Commun.*, 1994, 829.
142. T. Lu, L. Zhang, G. W. Gokel and A. E. Kaifer, *J. Am. Chem. Soc.*, 1993, **115**, 2542.
143. D. B. Amabilino, P. R. Ashton, M. S. Tolley, J. F. Stoddart and D. J. Williams, *Angew. Chem., Int. Ed. Engl.*, 1993, **32**, 1297.
144. P. R. Ashton, J. A. Preece, J. F. Stoddart, M. S. Tolley, A. J. P. White and D. J. Williams, *Synthesis*, 1994, 1344.
145. P. R. Ashton, I. Iriepa, M. V. Reddington, N. Spencer, A. M. Z. Slawin, J. F. Stoddart and D. J. Williams, *Tetrahedron Lett.*, 1994, **35**, 4835.
146. F. Vögtle, W. M. Müller, U. Müller, M. Bauer and K. Rissanen, *Angew. Chem., Int. Ed. Engl.*, 1993, **32**, 1295.
147. M. Bauer, W. M. Müller, U. Müller, K. Rissanen and F. Vögtle, *Liebigs Ann. Chem.*, 1995, 649.
148. P. R. Ashton, R. Ballardini, V. Balzani, M. T. Gandolfi, D. J. F. Marquis, L. Pérez-García, L. Prodi, J. F. Stoddart and M. Venturi, *J. Chem. Soc., Chem. Commun.*, 1994, 177.
149. P. R. Ashton, L. Pérez-García, J. F. Stoddart, A. J. P. White and D. J. Williams, *Angew. Chem., Int. Ed. Engl.*, 1995, **34**, 571.

150. P. R. Ashton, C. L. Brown, E. J. T. Chrystal, K. P. Parry, M. Pietraszkiewicz, N. Spencer and J. F. Stoddart, *Angew. Chem., Int. Ed. Engl.*, 1991, **30**, 1042.
151. P. R. Ashton, C. L. Brown, E. J. T. Chrystal, T. T. Goodnow, A. E. Kaifer, K. P. Parry, A. M. Z. Slawin, N. Spencer, J. F. Stoddart and D. J. Williams, *Angew. Chem., Int. Ed. Engl.*, 1991, **30**, 1039.
152. P. R. Ashton, C. L. Brown, J. R. Chapman, R. T. Gallagher and J. F. Stoddart, *Tetrahedron Lett.*, 1992, **33**, 777.
153. D. B. Amabilino, P. R. Ashton, A. S. Reder, N. Spencer and J. F. Stoddart, *Angew. Chem., Int. Ed. Engl.*, 1994, **33**, 433.
154. D. B. Amabilino, P. R. Ashton, J. F. Stoddart, S. Menzer and D. J. Williams, *J. Chem. Soc., Chem. Commun.*, 1994, 2475.
155. D. B. Amabilino, P. R. Ashton, A. S. Reder, N. Spencer and J. F. Stoddart, *Angew. Chem., Int. Ed. Engl.*, 1994, **33**, 1286.
156. N. van Gulick, *New J. Chem.*, 1993, **17**, 619.
157. P. R. Ashton, A. S. Reder, N. Spencer and J. F. Stoddart, *J. Am. Chem. Soc.*, 1993, **115**, 5286.
158. P. R. Ashton, J. A. Preece, J. F. Stoddart and M. S. Tolley, *Synlett*, 1994, 789.
159. D. B. Amabilino, P. R. Ashton, J. A. Preece, J. F. Stoddart and M. S. Tolley, *Am. Chem. Soc., Div. Polym. Chem., Polym. Prepr.*, 1995, **36**, 570.
160. P. D. Beer, *Adv. Inorg. Chem.*, 1992, **39**, 79.
161. P. D. Beer, *Adv. Mater.*, 1994, **6**, 607.
162. J. A. Preece and J. F. Stoddart, in 'Molecular Engineering for Advanced Materials', eds. J. Becher and K. Schaumburg, Kluwer, Dordrecht, 1995, pp. 1–28.
163. R. A. Bissell, E. Córdova, J. F. Stoddart and A. E. Kaifer, in 'Molecular Engineering for Advanced Materials', eds. J. Becher and K. Schaumburg, Kluwer, Dordrecht, 1995, pp. 29–40.
164. V. Balzani, *Tetrahedron*, 1992, **48**, 10443.
165. S. Misumi, *Top. Curr. Chem.*, 1993, **165**, 163.
166. H. Hopf, *Angew. Chem., Int. Ed. Engl.*, 1991, **30**, 1117.
167. D. J. Cram, *Nature (London)*, 1992, **356**, 29.
168. H. J. Choi, D. J. Cram, C. B. Knobler and E. F. Maverick, *Pure Appl. Chem.*, 1993, **65**, 539.
169. J. C. Sherman, *Tetrahedron*, 1995, **51**, 3395.
170. J. E. Mark, *Acc. Chem. Res.*, 1985, **18**, 202.
171. J. E. Mark, *New J. Chem.*, 1993, **17**, 703.
172. S. J. Clarson, *New J. Chem.*, 1993, **17**, 711.
173. Y. Lipatov and Y. Nizel'sky, *New J. Chem.*, 1993, **17**, 715.
174. D. B. Amabilino, I. W. Parsons and J. F. Stoddart, *Trends Polym. Sci.*, 1994, **2**, 146.

4

Templated Chemistry of Porphyrin Oligomers

JEREMY K. M. SANDERS
University of Cambridge, UK

4.1 INTRODUCTION

The work described in this chapter derives from the Cambridge project to create enzyme mimics that are capable of binding two or more substrates within a cavity, catalysing reactions between them and releasing the product(s). The long-term aim is not to mimic any particular natural enzyme. However, inspired by the example of nature, the aim of the project is to gain an understanding of the principles of how to design and create homogeneous catalytic systems with subtle recognition properties.

It was clear at the outset that any host system capable of catalysing interesting reactions would have to enclose a large volume, and therefore, if it was to be synthetically readily accessible, it would have to be constructed from large active building blocks. At an early stage the porphyrin unit was chosen as one of the primary building blocks, as explained in detail in Section 4.2. In summary, porphyrins are ideal building blocks for assembly of large host architectures because of their central binding sites, nominally planar geometry, stability, ease of synthesis, and spectroscopic eloquence deriving from the delocalized π-system. Their properties can also be fine-tuned by varying the complexed metal atom. Following the pioneering work of Burrows and of Davis, cholic acid has also been found to be a versatile building block for binding and catalytic systems;[1,2] the concave shape and hydrogen-bonding features of cholic acid complement well the properties of porphyrins described here (see Chapter 7, Volume 4).

The strategy used in Cambridge was to construct cyclic porphyrin oligomers that enclose a large but open cavity and which are linked by groups of controllable length and rigidity. The ligand-binding properties of these systems were then to be investigated with two complementary short-term aims in view: (i) development of selective, ligand-templated syntheses, and (ii) understanding of what types of ligands might lead to intracavity catalysis. The complementarity or symmetry relating these two processes can be summarized as templating outwards from inside, and templating inwards from outside (Figure 1).

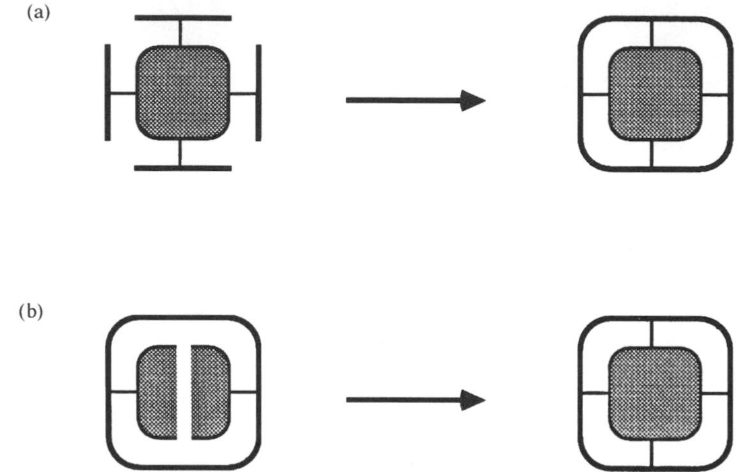

Figure 1 Schematic representation of templating: (a) inside outwards; and (b) outside inwards.

A template in a particular reaction can be classified according to many different aspects of its activity.

(i) Does it operate under thermodynamic or kinetic conditions?

(ii) Does it bring reactants together using covalent or noncovalent interactions, and does it generate covalent or noncovalent interactions in the products?

(iii) What topology of products does it induce—linear, cyclic or interweaved?

(iv) Does it act as a 'positive' template by accelerating a desired reaction or as a 'negative' template by inhibiting an undesired reaction?

These general questions were reviewed with historical examples in 1993,[3] and they are explored in several other chapters in this volume. This chapter restricts the discussion to the particular oligoporphyrin systems studied by Sanders and co-workers. The order of presentation is intended to be logical in part, rather than strictly historical—many of the binding experiments and mechanistic rationalizations actually occurred after some of the successful templated syntheses were developed. The long-term aim was, and remains, development and understanding of catalytic systems.

4.2 PORPHYRINS—A BRIEF GUIDE

4.2.1 Properties of Porphyrins

The porphyrin unit (**1**) (Scheme 1) is an attractive and versatile building block for supramolecular chemistry.[4] Indeed it has been used in nature for hundreds of millions of years precisely because it is so versatile: metalloporphyrins are used in cytochromes to transport electrons around the cell and in myoglobins and haemoglobins to transport coordinated molecular oxygen around whole organisms, while many enzymes use the coordinating properties of the metalloporphyrin to control oxygenation and other reactions.

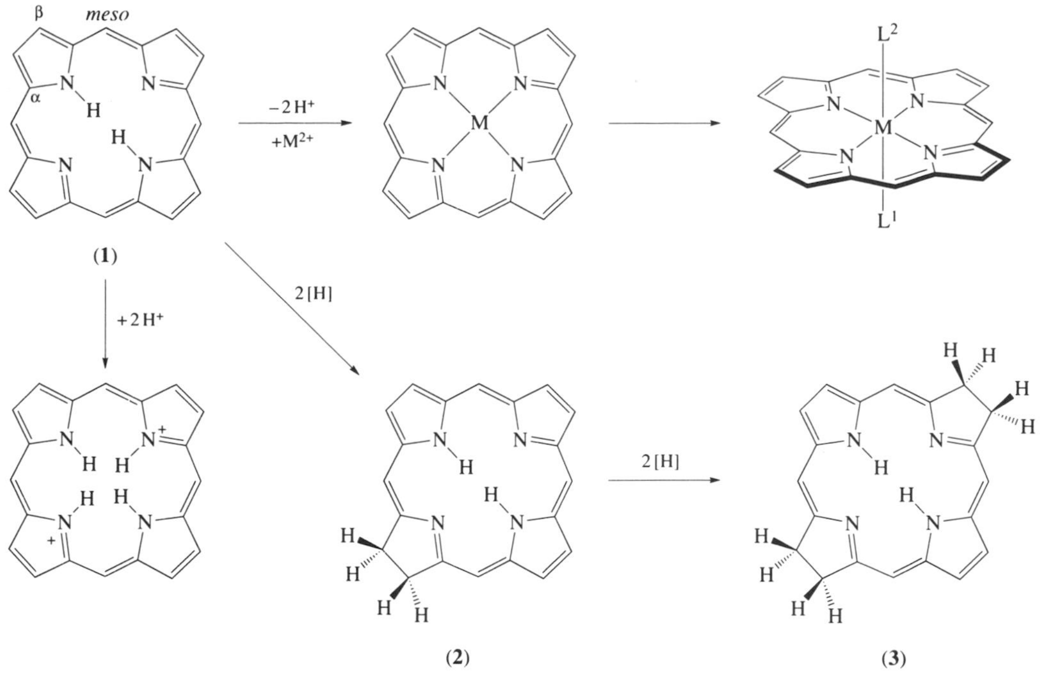

Scheme 1

Two features of metalloporphyrins make them attractive to nature and supramolecular chemists. The first is the delocalized, aromatic π-system which makes the porphyrin a good source and sink for electrons, and an efficient processor of light energy. The second is the central metal ion which is capable of binding one or two axial ligands and which may also have a rich redox chemistry of its own. Both of these properties are discussed in more detail below. Increasingly, chemists are also attracted to the additional features that make porphyrins good building blocks for the assembly of large host architectures: porphyrins have a relatively rigid and reasonably planar geometry, are quite stable, are easy to synthesize in a variety of substitution patterns and have rich colours and spectroscopy deriving from the delocalized π-system.

A porphyrin consists of four pyrrole-derived rings connected by four *meso*-carbons. The peripheral carbons are known as the β-carbons, while those adjacent to the nitrogens are the α-carbons. Two of the nitrogens are pyrrole-like and bear slightly acidic protons. These protons can be removed by base to give a dianion, which binds metal ions to give a metalloporphyrin. Divalent metal ions such as zinc, copper or nickel give neutral products, while trivalent ions such as Fe^{3+} carry counterions such as Cl^-. Small ions sit in, or close to, the plane of the porphyrin, but large ions such as the lanthanides sit on top of the porphyrin. Most metals of the periodic table have been inserted into porphyrins; some yield very stable complexes from which the metal can be removed only under forcing conditions, while other complexes are extremely labile.

Most metal ions in porphyrins are Lewis acids, binding one or two basic ligands in an axial fashion as shown in Scheme 1. Most of the work in the Sanders group has concentrated on zinc porphyrins for a variety of reasons: they are diamagnetic so are readily studied by NMR; they are easy to prepare from free-base porphyrins and are readily demetallated with dilute acid but they are stable to moisture, air and silica; they are neutral so avoid the complications associated with counteranions; they have simple coordination chemistry, forming 1:1 complexes with amines, while complexes with oxygen, sulfur and phosphorus donor ligands are much less stable; and they are kinetically labile, with very fast ligand association and dissociation.

Returning to the metal-free (free-base) porphyrin in Scheme 1, it can be seen that the two nitrogens that do not bear protons are pyridine-like and slightly basic. They can be protonated by acid as shown to yield the dication; this has some potential for recognizing anions.[5] Finally, note that the free-base porphyrin is in tautomeric equilibrium, so that each nitrogen alternates between pyrrole-like and pyridine-like, and the central protons migrate from one nitrogen to another.[6]

The aromatic, delocalized π-system extending over the whole framework gives the porphyrin much more stability than might otherwise have been expected. If nine cyclically conjugated double bonds are drawn as shown to give an 18π-electron aromatic system, then the peripheral double bonds in the 'south-west' and 'north-east' rings are not required for aromaticity. These extra double bonds can be reduced once to give a chlorin (**2**), or further to give a bacteriochlorin (**3**), (or an isomeric isobacteriochlorin with adjacent reduced rings) without destroying the aromaticity; reduced macrocycles are used in nature in chlorophylls and some enzyme cofactors, and have many similar properties to their parent porphyrins. There has been very little supramolecular exploration of synthetic chlorins because reduction of substituted porphyrins generally leads to multiple products of low symmetry.

An inevitable property of the delocalized π-system is that the HOMO is high energy, while the LUMO is low energy. Therefore it is easy to oxidize a porphyrin P by removing one electron from the HOMO to give a radical cation ($P^{+\cdot}$) or to reduce porphyrins to a radical anion ($P^{-\cdot}$) by adding one electron into the LUMO; these properties are heavily exploited in enzymes,[7] electron-transfer proteins[8] and synthetic models of the natural photosynthetic process.[9]

A further important property of the delocalized π-system is a small HOMO–LUMO energy gap, giving electronic transitions which correspond to long wavelength absorptions in the visible region. The most intense absorption of porphyrins is called the Soret band, and it occurs around 400 nm with an extinction coefficient of around 10^5, while there are also weaker but important 'Q-band' absorptions in the 500–600 nm region. All of these absorptions are exquisitely sensitive to the substitution pattern, the central metal ion, coordination state of the metal ion, and solvent polarity. Thus haem is red, zinc porphyrins are a beautiful deep pink and chlorophylls are green, the precise colours depending on circumstances; for example, the zinc diarylporphyrin system that is used as the main building block in this chapter has a Soret absorption at 409 nm, which moves to 422 nm on binding pyridine. This dependence of absorption frequency on environment coupled with very intense absorption allows one to use electronic spectroscopy to monitor ligand-binding processes in extremely dilute solutions and to measure binding constants as large as $10^9\,L\,mol^{-1}$ with reasonable precision. A further attraction of the intense colours, beyond the merely aesthetic, is that even a few micrograms can be followed on a TLC plate or down a column.

The final property of the π-system which is important in porphyrins is its diamagnetic ring current. This is similar to the familiar ring current in benzene but much larger. Protons and carbons in the porphyrin plane experience a deshielding, so that a *meso*-proton generally resonates around 10 ppm (Equation (1)), while there is a shielding zone above and below the plane. When a pyridine molecule binds to a zinc–porphyrin, it falls into the shielding zone, leading to shifts of 6.0 ppm and 1.8 ppm for its α- and β-protons respectively as shown in Equation (1). These shifts are due to through-space field effects, rather than through-bond electronic effects, and the shape of the shielding and deshielding zones is well known,[10] so the geometrical relationship can be mapped between ligands and porphyrin, or between different porphyrins in an oligomer. Much of the evidence for the host–guest geometries presented in this chapter depends on the shifts observed in such complexes. Where the binding is strong ($>10^7\,L\,mol^{-1}$) there is slow exchange between free and bound species on the chemical shift timescale[11] and the bound chemical shifts can be observed directly; in cases of weaker binding ($<10^5\,L\,mol^{-1}$) exchange is fast, averaged spectra are obtained and titration curves need to be analysed to obtain limiting bound shifts.

δ 10 ppm H (meso) — Zn porphyrin + pyridine (δ 8.5 ppm H_α, δ 7.5 ppm H_β, R) free ⇌ Zn porphyrin–pyridine (δ 2.5 ppm H_α, δ 5.3 ppm H_β, R) bound (1)

Porphin itself, the fully unsubstituted molecule shown in Scheme 1, is virtually insoluble in any solvent and lacks any attachment points, so peripherally substituted systems must be used. Historically the most common synthetic molecules were the octaalkylporphyrins (**4**), which are similar to natural haems in structure and properties but have rather flexible links to other components, and the tetraarylporphyrins (**5**) which readily can be made on a large scale.

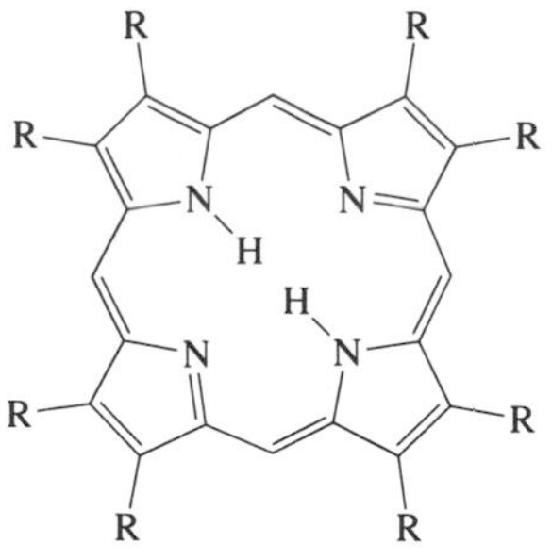

(4)

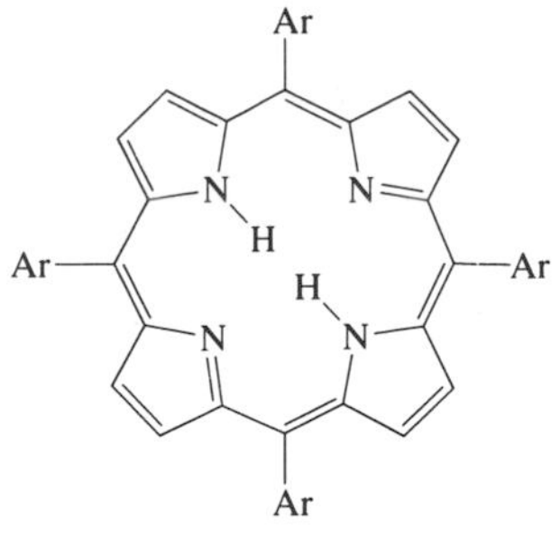

(5)

Many research groups, including that of Sanders, have recently focused on *meso*-diarylporphyrins (**6**).[12] These have several attractions including efficient synthesis, the ability to extend the architecture via the aryl substituents and the potential to control solubility independently through the peripheral alkyl groups. The need for relatively rigid linkers and an open accessible cavity led to use of the diarylporphyrin (**7**) with linkers placed at the *meta*-position of the aryl group. This substitution pattern directs the linker perpendicular to the plane of the porphyrin and generates a large cavity; the lack of *ortho*-substituents leads to relatively rapid aromatic ring rotation and avoids the need to separate atropoisomers. Furthermore, simple geometrical considerations show that the cyclic trimer should be strain-free if a linear linker is used. The attractions of a trimer include the large working volume available and the possibility of termolecular chemistry, although at the price of complex binding and kinetics. Methyl β-pyrrole substituents adjacent to the aryl groups lead to little steric compression at the porphyrin periphery while large nonpolar substituents were desirable at the other β-pyrrole positions to improve the solubility; initially porphyrins were made with R = ethyl but most of the results described below were obtained with material where R = $CH_2CH_2CO_2Me$. The ester-functionalized series is slightly more difficult to make[12] but has the great advantage that the solubility characteristics of any oligomer can be tuned simply by transesterification with appropriate alcohols. For example, the NMR competition experiments and titrations carried out on the cyclic tetramer described in Section 4.2.3.4 required the use of solubilizing *iso*-decyl ester sidechains. This change in esterifying sidechain generally has no detectable effect on the strength of binding.

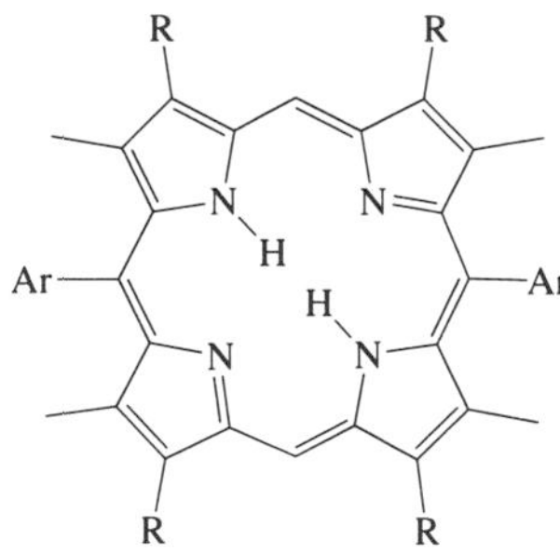

(6)

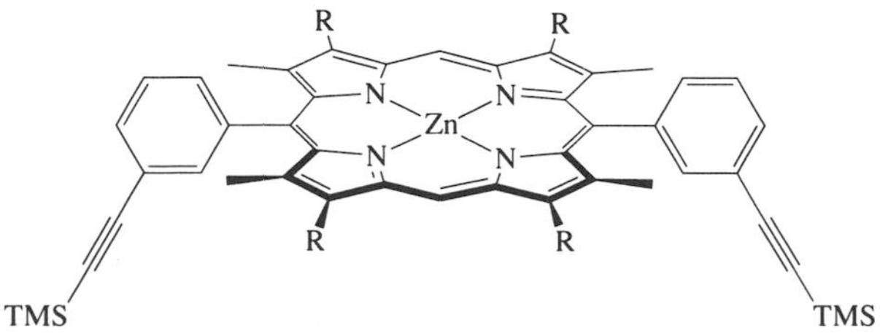

R = Et or $CH_2CH_2CO_2Me$
(7)

The monomer (**7**) binds pyridine ligands in chlorinated solvents with binding constants of $2–5 \times 10^3\,M^{-1}$ at ambient temperature. All the binding constants discussed below were measured under these conditions except where stated otherwise. The electron-withdrawing —$CH_2CH_2CO_2R$ sidechain leads to a slightly larger binding constant than the ethyl analogue by increasing the Lewis acidity of the zinc ion. Variable temperature experiments on (**7**) show that the binding is enthalpy driven as would be expected: $\Delta H = -38\,kJ\,mol^{-1}$ and $T\Delta S = -20\,kJ\,mol^{-1}$.[13] More basic ligands such as imidazole are bound more strongly, while oxygen ligands are bound only very weakly, with K generally less than 10.[1]

4.2.2 Nontemplated Syntheses of Oligomers

4.2.2.1 Statistical syntheses of symmetrical oligomers

The strategy of this work has been to incorporate the monomer (**7**) into a range of oligomers with the same fundamental recognition properties, but with a range of linkers varying in length and rigidity. Scheme 2 summarizes the symmetrical alkyne-linked oligomers prepared from (**7**) in nontemplated syntheses; for clarity, sidechains and individual pyrrole rings are omitted in this simplified cartoon representation. It was decided in the first instance to use butadiyne linkers generated by Glaser–Hay coupling of a terminal alkyne (Equation (2)), a popular reaction for constructing macrocycles.

$$Ar—C\equiv C—H + H—C\equiv C—Ar \rightarrow Ar—C\equiv C—C\equiv C—Ar \qquad (2)$$

This reaction, carried out— with a porphyrin concentration of 0.4 mM, gives a mixture of cyclic oligomers from which the 2,2-dimer (**8**) (~30% yield), 2,2,2-trimer (**9**) (~35%) and 2,2,2,2-tetramer (**10**) (~15%) can be separated by chromatography;[12] mass spectrometry shows the presence of small amounts of even higher oligomers, but they could not be separated from each other. Each '2' signifies the number of alkyne links between each porphyrin unit. This proves a useful shorthand notation as one changes the linker as described below.

The mechanism of the actual coupling reaction is not known in detail, but the oligomerization must follow the pathway shown in Scheme 3. Two monomers couple to give linear dimer, (**15**), which is then subject to two competing processes: intramolecular cyclization can occur, giving 2,2-dimer, while coupling with another porphyrin leads to chain extension. One factor that will influence this competition is the conformational distribution of (**15**): the 30% yield of 2,2-dimer compared with more than 50% yield of higher oligomers implies that open conformations are dominant. The proposed intermediate (**15**) was synthesized independently,[14,15] and indeed gave only 21% cyclic dimer and more than 70% higher oligomers when subjected to coupling conditions. The dramatic effect of templates on this reaction is described in Section 4.3.

The efficiency of cyclization is easily expressed in terms of a kinetic effective molarity, EM_{kin};[16] this is closely related to the thermodynamic effective molarity discussed below. EM_{kin} is defined as the concentration at which the rate of intramolecular coupling equals that of intermolecular coupling; it is the effective molarity of one end of the linear oligomer experienced by the other during the reaction. It can be calculated from the product distribution, as shown in Scheme 4 and Equations (3) and (4) where $[SM]_0$ is the initial concentration of starting material, which can be approximated as the average concentration of terminal alkene groups [C≡C–H] since $2[SM]_0$ is the initial [C≡C—H], and k is a second-order rate constant.

Using these equations, the EM_{kin} for cyclization of linear dimer (**15**) is determined to be 0.05 mM;[15] carrying out the coupling above this concentration gives primarily intermolecular reaction, while intramolecular cyclization should be the dominant process when the concentration is below 0.05 mM.

Symmetrical oligomers with linkers other than butadiyne have also been prepared. For example, (**7**) could be elongated via a mixed Glaser–Hay coupling to give a 'stretched' monomer (**11**) (Equation (5)), which was coupled to give 4,4-dimer (**13**) and 4,4,4-trimer (**14**) (Equation (6) and Scheme 2).[17]

A platinum-linked analogue derived from the same monomer (**7**) was designed to exhibit the same topology but with a cavity size intermediate between that provided by 2,2- and 4,4,4-linkers (Equation (7) and Scheme 2).[18] This reaction gave trimer (**12**), but no corresponding dimer or tetramer could be convincingly characterized.

Molecular mechanics calculations and model building indicated that the 2,2,2-trimer (**9**) should enclose a cavity where the metal–metal distance is around 1.5 nm, while in the 2,2-dimer (**8**) the metal–metal distance is around 1.2 nm and the 2,2,2,2-tetramer (**10**) is so flexible that no meaningful distance can be quoted. The bulk of the work described below has been in this butadiyne-linked

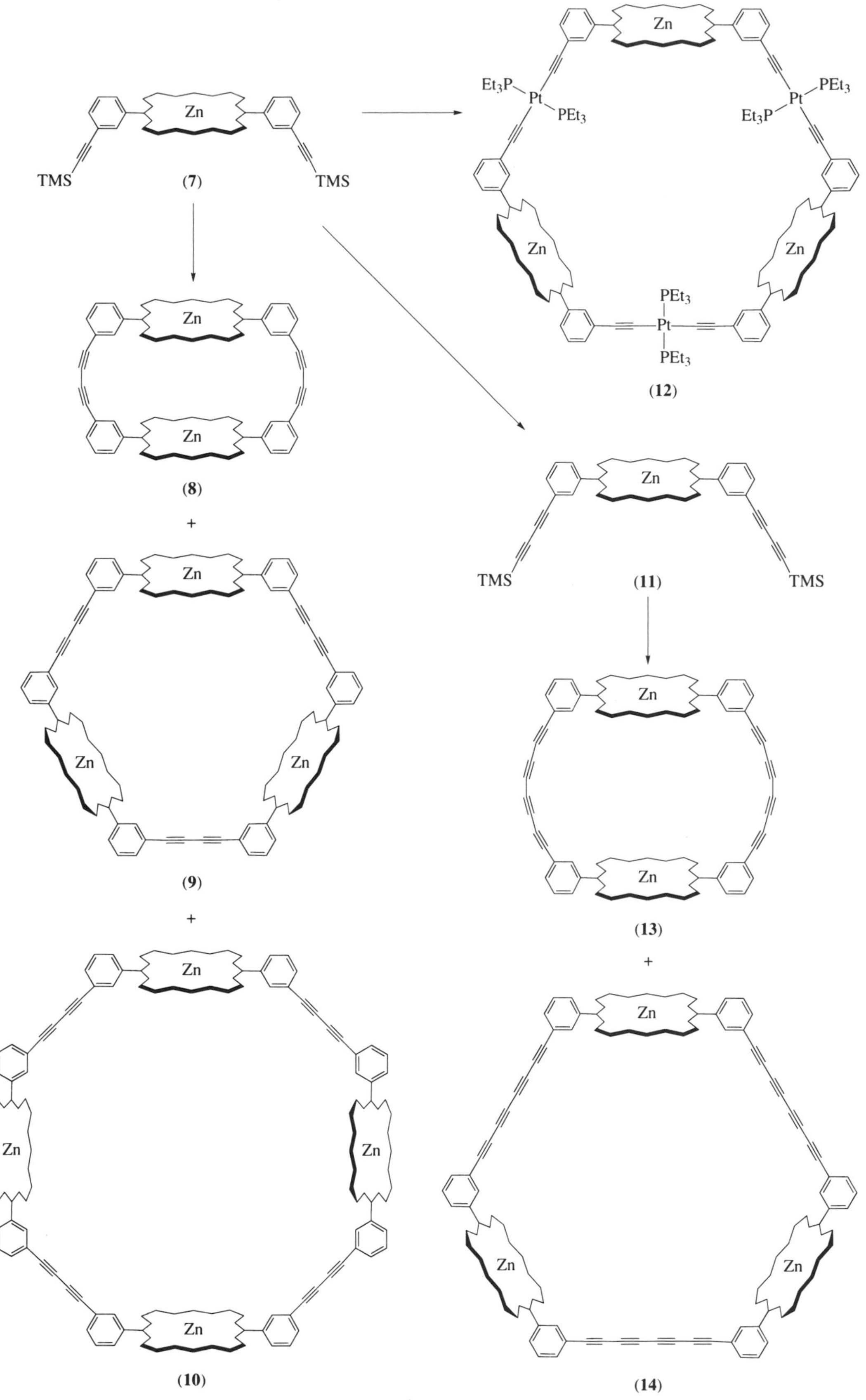

Zn
TMS
(7)
TMS
Et3P
Pt
PEt3
PEt3
Pt
Et3P
PEt3
Pt
PEt3
(12)
(8)
+
(9)
+
(10)
(11)
(13)
+
(14)

Scheme 2

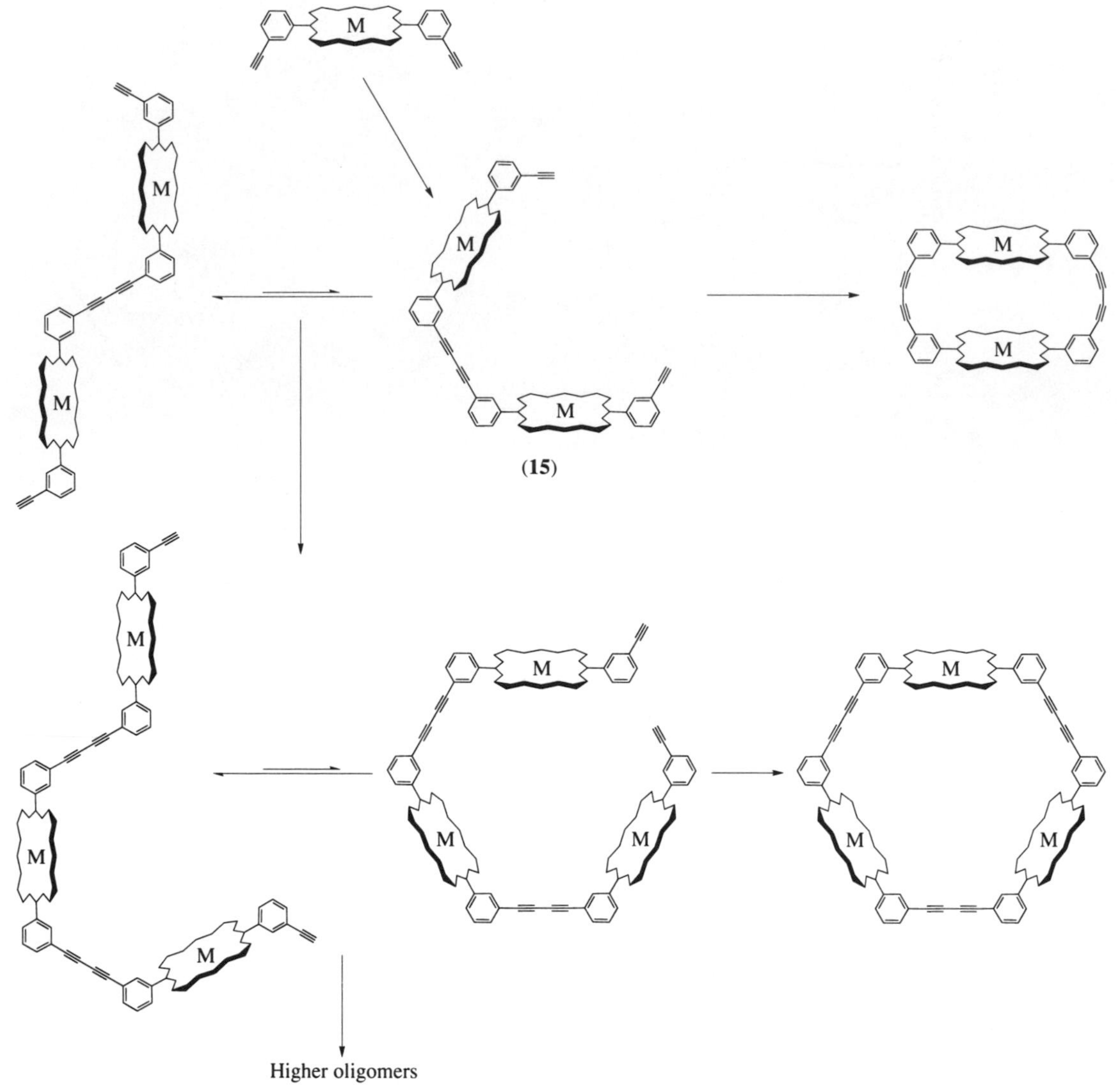

(15)

Higher oligomers

Scheme 3

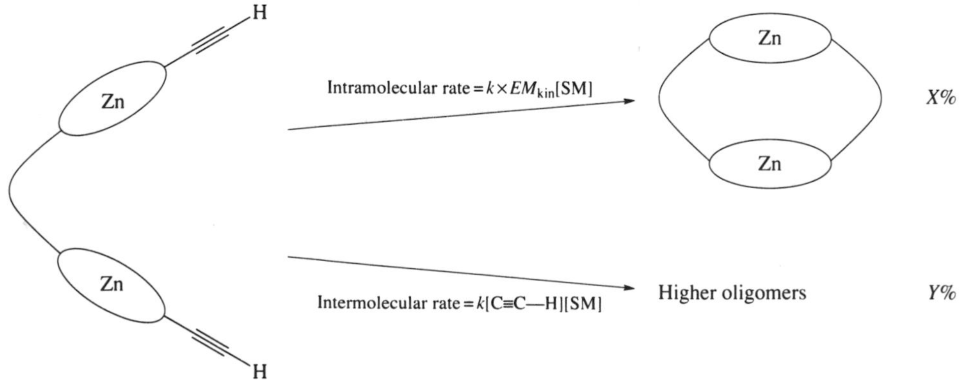

Scheme 4

$$\frac{X\%}{Y\%} = \frac{\text{Rate of intramolecular reaction}}{\text{Rate of intermolecular reaction}} = \frac{k \times EM_{\text{kin}}}{k[\text{SM}]_0} = \frac{EM_{\text{kin}}}{[\text{SM}]_0} \quad (3)$$

$$EM_{\text{kin}} = \frac{X[\text{SM}]_0}{Y} \quad (4)$$

$$\text{Ar—C}\equiv\text{C—H} + \text{H—C}\equiv\text{C—TMS} \rightarrow \text{Ar—C}\equiv\text{C—C}\equiv\text{C—TMS} \quad (5)$$

$$\text{Ar—C}\equiv\text{C—C}\equiv\text{C—H} + \text{H—C}\equiv\text{C—C}\equiv\text{C—Ar} \rightarrow \text{Ar—C}\equiv\text{C—C}\equiv\text{C—C}\equiv\text{C—C}\equiv\text{C—Ar} \quad (6)$$

$$\text{Ar—C}\equiv\text{C—H} + \text{Cl—Pt(PEt}_3)_2\text{—Cl} + \text{H—C}\equiv\text{C—Ar} \rightarrow \text{Ar—C}\equiv\text{C—Pt(PEt}_3)_2\text{—C}\equiv\text{C—Ar} \quad (7)$$

series. The Zn–Zn distance in the platinum-linked trimer (**12**) is around 1.8 nm, while in the rather flexible 4,4,4-trimer (**14**) it is around 2.0 nm. In the rigid 4,4-dimer (**13**), the Zn–Zn distance is calculated to be comparable to that in the 2,2,2-trimer.

Hydrogenation of the butadiyne linkers in the 2,2-dimer and 2,2,2-trimer gives the corresponding $(CH_2)_4$-linked molecules in good yield.[19] These are highly flexible molecules, the trimer exhibiting very different conformational, ligand-binding and catalytic properties from its 2,2,2-parent as discussed below.

All attempts to prepare the smaller 1,1-dimer (**16**) and 1,1,1-trimer (**17**) by oligomerizing suitable monomers failed, but small amounts (5–10% dimer; 1–3% trimer) can be isolated from a porphyrin synthesis using benzaldehyde dimers connected by preformed linkers as shown in Scheme 5.[20] This may be the first synthesis of a cyclic porphyrin oligomer using preformed linkers, but attempts are being made to devise more convenient routes.

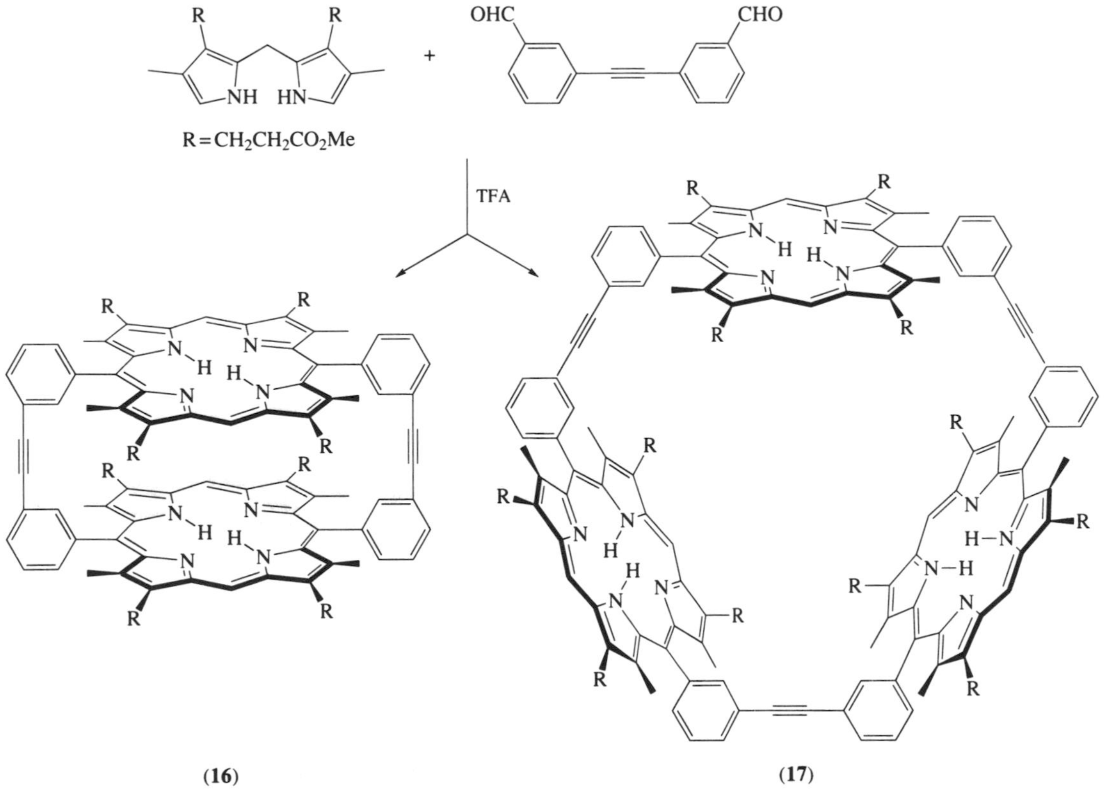

R = $CH_2CH_2CO_2Me$

TFA

(**16**) (**17**)

Scheme 5

4.2.2.2 Stepwise syntheses

Linear porphyrin dimer (**15**) was synthesized from monomer (**7**) by partial deprotection, careful chromatography and coupling.[14] The corresponding linear tetramer could not be prepared by partial deprotection of (**15**) followed by coupling because chromatographic separation of the unprotected dimer from the mixture proved impossible. Coupling of the unseparated mixture led to a complex mixture of products. There was some linear tetramer in this mixture but it was difficult to separate it cleanly from the other components. The idea of a scavenger template was devised to overcome this problem; see Section 4.3.2.2.

The statistical synthesis of oligomers from a monomer limits the range of architectures which can be readily prepared, even when templating is used to direct the reaction. One modified architecture which appeared attractive was linear dimer (**15**) that is capped by a central unit. The cap can be used either to force some change on the geometrical relationship between the two porphyrin units and so probe the sensitivity of binding and catalysis to subtle spatial effects, or to carry additional recognition and catalytic features. For example, model building suggested that a binaphthyl unit would impose a helical twist between the two porphyrin units; this might lead to enantioselectivity in binding, and should in addition give useful insight into the geometry dependence of the rates of Diels–Alder and acyl transfer reactions within the cavity.

The obvious route, adding a cap across the ends of (**15**), was not attractive because the linear dimer preparation is not readily scaled up, and because the efficiency of capping was not likely to be very high. The stepwise route summarized in Scheme 6 was therefore chosen.[21] The unsymmetrical porphyrin monomer can be made on a gram scale, the palladium-catalysed coupling of this monomer to a central core possessing two identical terminal alkynes is high-yielding, and the final step is an intramolecular Glaser–Hay coupling of the porphyrin alkyne units, so that the central unit becomes the cap. Given that the final relationship between these two porphyrins is effectively the same as in the 2,2,2-trimer (**9**), there was good reason to believe that this intramolecular coupling would be high-yielding and controllable by templating if necessary. Using this and related routes the 1,1,2- and 2,2,2-binaphthyl-capped dimers (**18**) and 1,1,2-porphyrin trimers (**19**) with a range of different metal ions (Zn, Ni, Ru) in the central porphyrin were prepared.[22] The final ring closures in these syntheses can be templated as described below.

4.2.3 Ligand-binding Properties of Oligomers

4.2.3.1 General considerations

The shape and flexibility of the porphyrin oligomers have been explored by testing their ability to bind a range of monodentate and multidentate ligands (Figure 2).[1,2,13,15,17–23] Most of the ligands were commercially available or readily synthesized, but two, $(pyacac)_3Al$ and $py_3C_{12}H_3$, were specially designed and synthesized to be complementary to their hosts, the platinum-linked trimer and 4,4,4-trimer, respectively.

Scheme 7 shows a schematic view of the binding of a bidentate ligand, L—L, to a metalloporphyrin dimer, M—M.[23] The measured binding constant for the symmetrical bifunctional ligand L—L to the dimer is $K_{LL\cdot MM}$, while the microscopic binding constant for each M—L interaction is K_1. The equilibrium constants for the two processes are $2K_1$ and $EM \times K_1/2$. The numerical factors arise from the fact that the ligand has two equivalent sites for binding. *EM* is the thermodynamic effective molarity or chelation factor for the second binding,[16] which can be defined as in Equation (8).

$$EM = K_{LL\cdot MM}/K_1^2 \tag{8}$$

Effective molarities are a useful way of quantifying host–guest complementarity, and they also give an indication of the degree of rate acceleration that might be achieved if this binding energy could be directed into transition state stabilization for a bimolecular reaction.

When considering binding of a tridentate ligand to a tritopic host the definition in Equation (9) is used.[23] Here *EM* is actually the square root of the product of two chelation factors.

$$EM = \sqrt{\left(\frac{K_{LLL\cdot MMM}}{K_1^3}\right)} \tag{9}$$

In trying to calculate an effective molarity the problem is encountered that K_1 cannot be measured directly; it is not possible to measure the affinity of just one end of L—L for M—M without doing something to prevent chelation, and any tampering with the structure is likely to affect the microscopic binding properties. It is rarely true that the microscopic binding constant of one end of a dipyridyl ligand is the same as that of pyridine, because the basicity of pyridine is affected by substituents. It is better to approximate the microscopic binding constant of one end of a zinc–porphyrin dimer to that of the analogous monomer, because the zinc sites are further apart and so likely to behave more independently, provided that the cavity is sufficiently large. The best approach

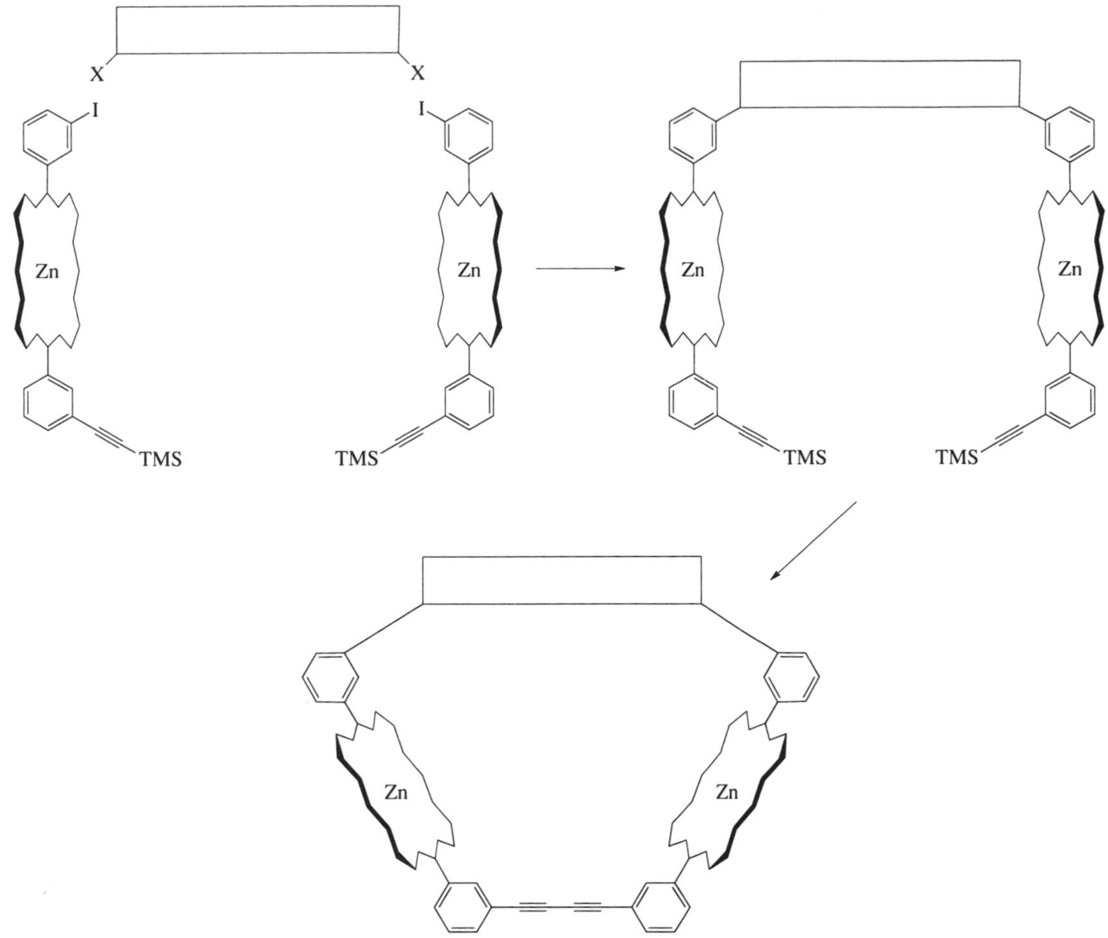

Scheme 6

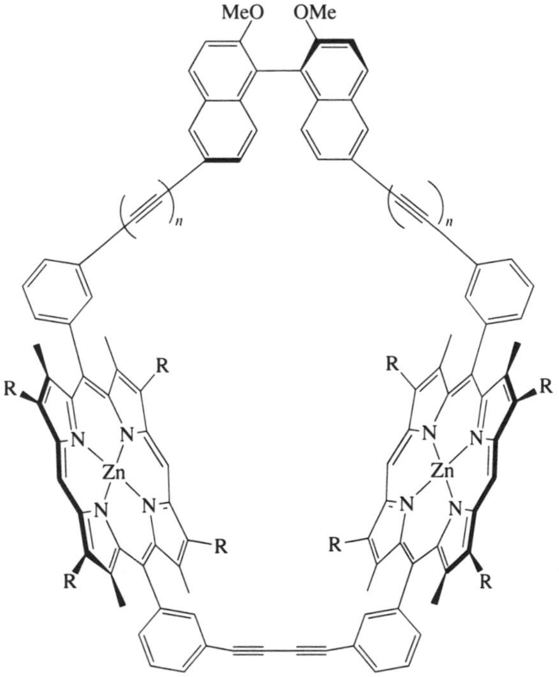

(**18**) R = $CH_2CH_2CO_2Me$, n = 1 or 2

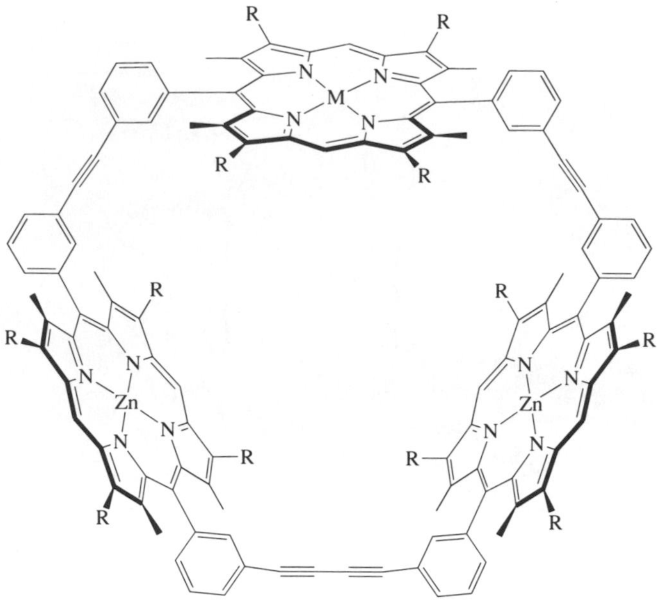

(**19**) R = $CH_2CH_2CO_2Me$, M = Ni, Zn, Ru–CO....

is to assume that one end of a zinc–porphyrin dimer binds any pyridine ligand more, or less, strongly than the corresponding monomer by some constant factor, so Equation (10) can be used to estimate the affinity of one end of the bidentate ligand, L—L, for one end of the dimer, M—M.

$$K_1 = \frac{K_{LL \cdot M}}{2} \times \frac{K_{L \cdot MM}}{K_{L \cdot M}} \tag{10}$$

Entropic considerations[23] associated with the loose Zn—N bond suggest that a strain-free chelate between a porphyrin dimer host and a perfectly complementary dipyridine should exhibit a maximum effective molarity in the region of 100–400 M, a value that is not far from some of those observed experimentally as described below.

4.2.3.2 Dimers

Table 1 summarizes the ligand-binding properties of most of the dimers discussed above. Two general points emerge. (i) For a given host–guest pair, binding is 5–10 times stronger in CH_2Cl_2 solution than in $C_2H_2Cl_4$ solution, reflecting the greater solvating ability of the latter. Binding increases even more along the series $CHCl_3 \rightarrow CCl_4 \rightarrow$ hydrocarbon.[1] (ii) For these rather rigid hosts and guests, binding strength reaches a well-defined maximum when host and guest are complementary. So bipy is the optimum ligand for 2,2-dimer, with an effective molarity of 76 for the second binding. This is close to the limit calculated above, implying that the stability of the complex is limited more by the loose zinc porphyrin–pyridine interaction than by the imperfection of the shape complementarity between host and guest. A comparable result was obtained by Sutherland's group for bipy and a similar dimer.[24] (**20**), (**21**) and (**22**) are the optimum complexes found so far for Zn_2-1,1-, 2,2- and 4,4-dimers, respectively.

Table 1 Binding constants for dimers in CH_2Cl_2 solution, 30 °C.

Host	*Pyridine* ($L\,mol^{-1}$)	*bipy* ($L\,mol^{-1}$)	*py_2C_2* ($L\,mol^{-1}$)	*py_2Et* ($L\,mol^{-1}$)	*py_2Pr* ($L\,mol^{-1}$)	*py_2C_4* ($L\,mol^{-1}$)	*py_2py* ($L\,mol^{-1}$)	*py_4P* ($L\,mol^{-1}$)
Monomer (**7**)	5×10^3	8×10^3		1×10^4	2×10^4		1×10^4	
Linear dimer (**15**)		6×10^6						3×10^6
Zn_2-2,2,-dimer (**8**)	5×10^3	1×10^9		1×10^6	1×10^5		6×10^3	
		$1 \times 10^{8\,a}$	$4 \times 10^{2\,a}$	$1 \times 10^{5\,a}$	$2 \times 10^{4\,a}$	$6 \times 10^{2\,a}$		
Zn_2-4,4-dimer (**13**)		$1 \times 10^{4\,a}$	$7 \times 10^{6\,a}$	$7 \times 10^{6\,a}$	$1 \times 10^{6\,a}$	$1 \times 10^{6\,a}$		
Zn_2-1,1,2-binap dimer (**18**)	5×10^3				2×10^8		4×10^7	
Zn_2-2,2,2-binap dimer (**18**)	5×10^3				4×10^8		6×10^8	
	$9 \times 10^{2\,a}$				$7 \times 10^{7\,a}$		$3 \times 10^{7\,a}$	

[a] $C_2H_2Cl_4$ solution, 30 °C.

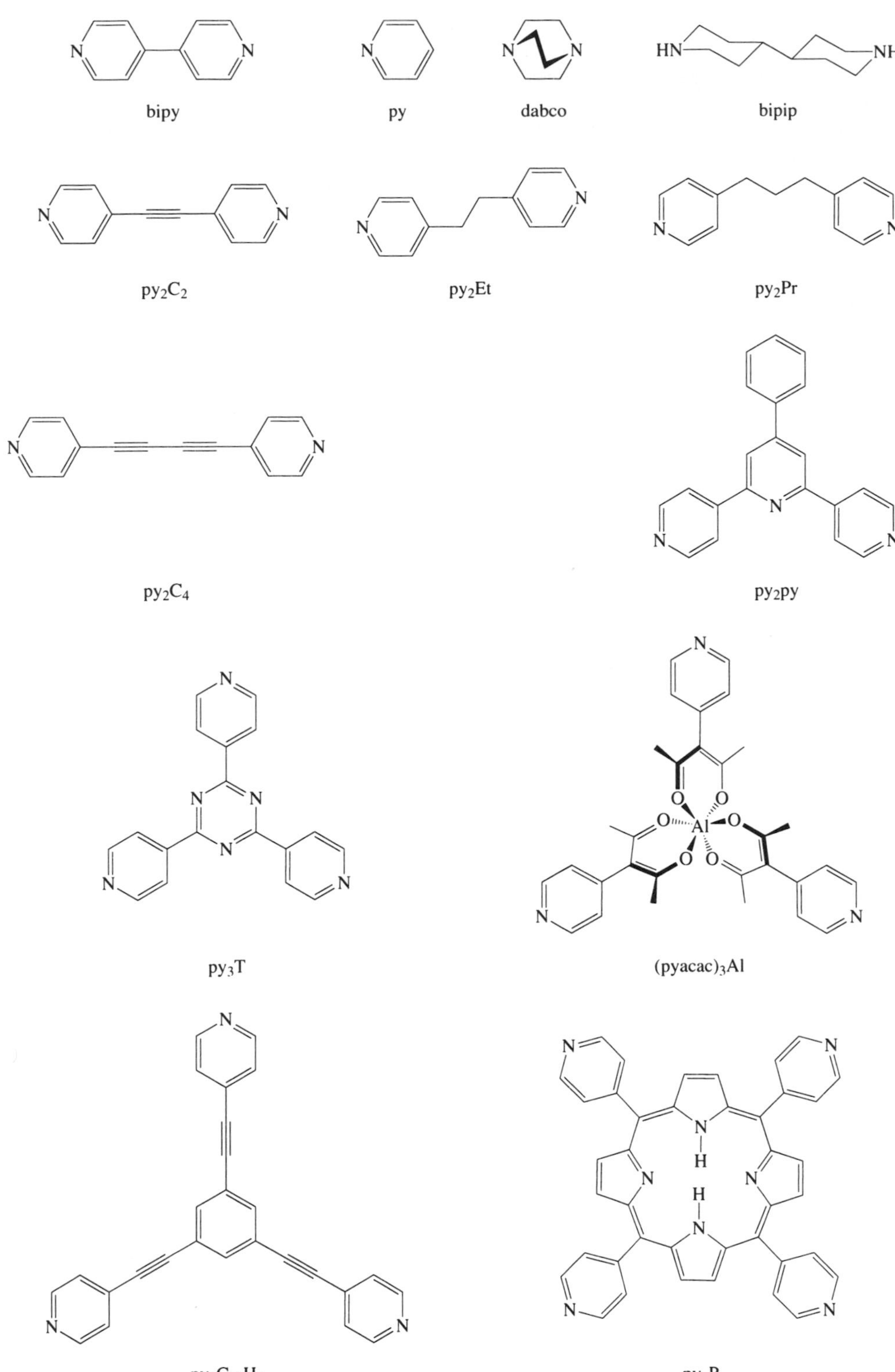

Figure 2 Ligands used in the work discussed in this chapter.

The data in Table 1 show that the long bidentate ligands py_2C_2, py_2C_4, py_2Et and py_2Pr are all more complementary to the 4,4-dimer than to the 2,2-dimer; this is hardly surprising, given the rigidity of the smaller cyclic dimer. However even the best ligands for the larger dimer have binding constants of only 10^6–10^7 rather than the 10^8–10^9 which would have been expected for a good fit. Since the larger dimer has linkages which are longer by one butadiyne (—C≡C—C≡C—) unit, it had been expected that ligands with four-carbon atom links would be optimal. In fact, the shorter py_2C_2 appears to be the best ligand complex; (**22**). This may be rationalized if the Zn—Zn distance is less than expected: there is evidence[23] that the porphyrin units in the 2,2-dimer are domed, presumably because the butadiyne units are too short to accommodate significant distortion. However the longer octatetrayne linker may be able to bend more, leading to less porphyrin doming and a smaller Zn—Zn distance. The optimum ligand for 4,4-dimer shows an effective molarity of no more than 1 M. Most of the linear and cyclic dimers prepared by other groups have sufficient flexibility or rotational freedom that effective molarities even for binding rigid guests are smaller than 1 M, while the effective molarity for the binding of flexible alkyl diamines is always much smaller.[24–6]

Solubility and aggregation problems hinder the accurate measurement of binding constants for the 1,1-cyclic dimer, but NMR competition experiments with 2,2,-dimer show, as expected, that dabco is strongly preferred by the smaller dimer complex, (**20**), while bipy is too large to fit into the smaller cavity.

An interesting observation is that py_2Pr binds with essentially the same affinity to both binaphthyl-capped dimers, presumably because it can adjust its conformation to fit the host geometry, while the more rigid py_2py has a significant preference for the less twisted 2,2,2-host.

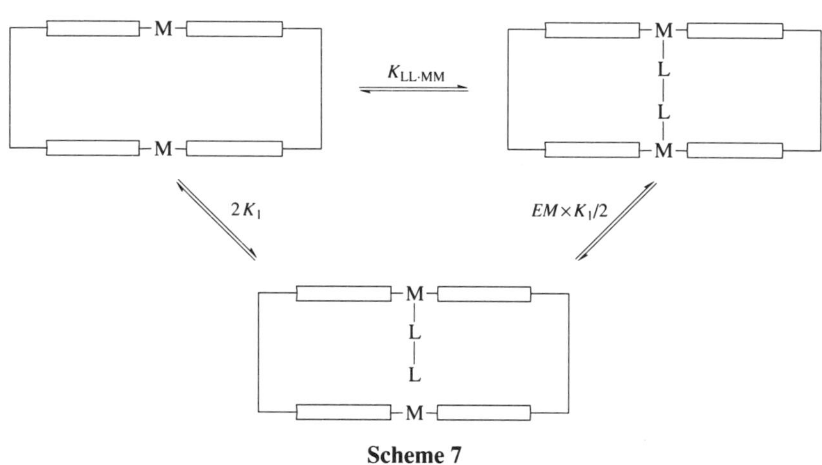

Scheme 7

Importantly for the experimental success of templating, and our understanding, linear dimer (**15**) binds bipy much more strongly than monomer but less strongly than cyclic 2,2-dimer (see Section 4.3.2.1). Intriguingly, the linear dimer derived from *meso*-dinitroporphyrin does not bind bipy well, and is not subject to bipy-templated cyclization; this is discussed in Section 4.3.3.

4.2.3.3 Trimers

Table 2 summarizes the binding properties of our larger trimers. The most thoroughly studied trimer is the 2,2,2-Zn_3-trimer (**9**). This has more than one binding site but careful analysis of the binding curves with pyridine shows that each binding site is behaving independently. The empirical binding constants are identical, within experimental error, to those of the corresponding monomer (**7**), indicating that there is no hindrance to binding inside the cavity. Statistically, a mixture of inside and outside binding would be expected, but crystal structures of the pyridine and quinuclidine complexes show all three ligands bound inside the cavity, presumably in order to improve packing.[27] Ligands capable of hydrogen bonding to each other within the cavity, such as 4-hydroxymethylpyridine, do show some cooperative binding behaviour; the scope and use of such cooperativity is being explored.[28]

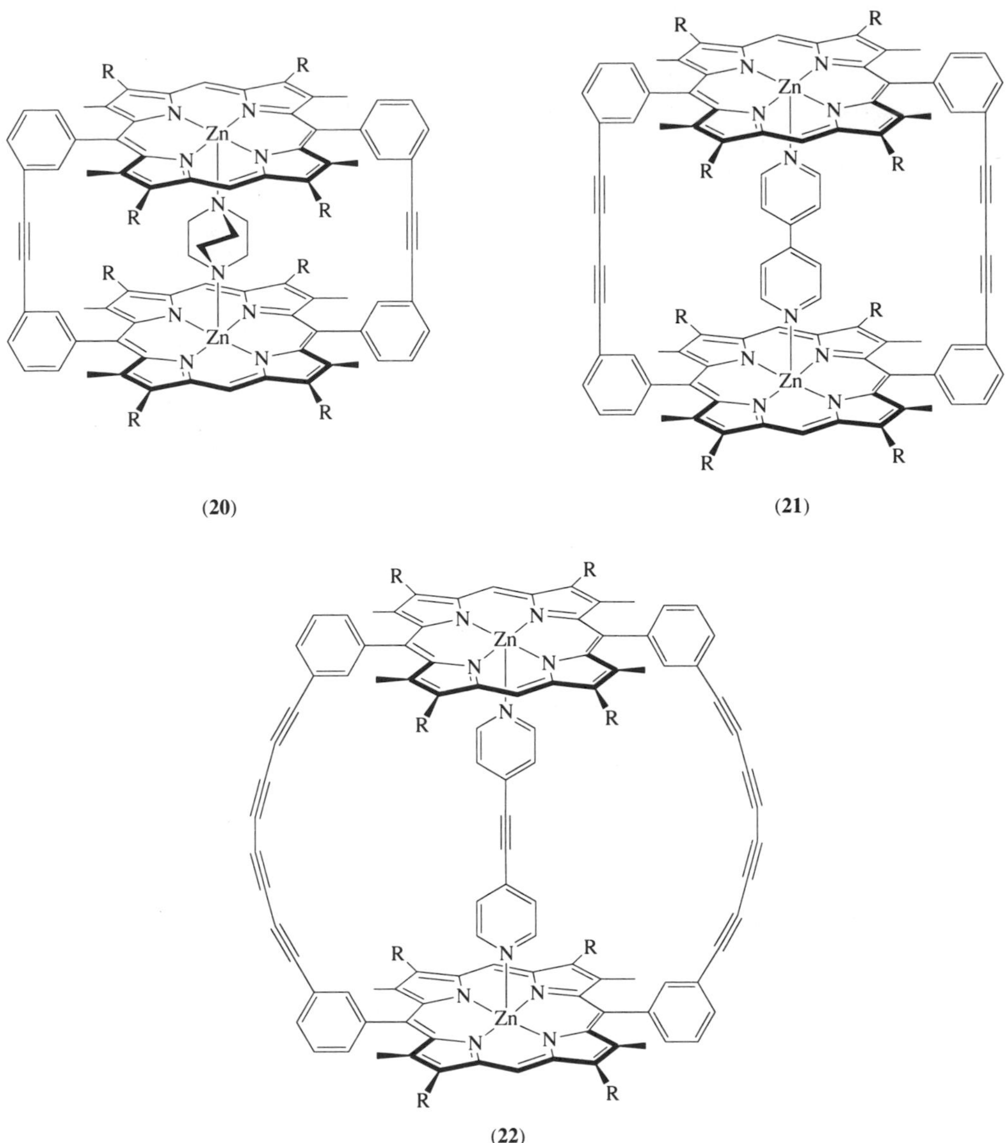

(20) **(21)**

(22)

Table 2 Binding constants, K_1, for trimers in CH_2Cl_2 solution, 30 °C.

Host	*Pyridine* (L mol^{-1})	*bipy* (L mol^{-1})	*py_2Et* (L mol^{-1})	*py_2Pr* (L mol^{-1})	*py_2Py* (L mol^{-1})	*py_3T* (L mol^{-1})	*$(pyacac)_3Al$* (L mol^{-1})	*$py_3C_{12}H_3$* (L mol^{-1})
Monomer (**7**)	5×10^3	8×10^3	1×10^4	2×10^4	1×10^4	8×10^3	1×10^4	
Zn_3-2,2,2-timer (**9**)	4.4×10^3	8×10^4	3×10^7	6×10^8	5×10^8	$\sim 10^{10}$		
						$1 \times 10^{9\,a}$	$1 \times 10^{6\,a}$	$2 \times 10^{5\,a}$
Zn_3-Pt_3-trimer (**12**)						3×10^7	2×10^{10}	
Zn_3-4,4,4-trimer (**14**)						$2 \times 10^{5\,a}$	$3 \times 10^{6\,a}$	$2 \times 10^{8\,a}$

[a]$C_2H_2Cl_4$ solution, 30 °C.

Binding curves (from UV–visible binding data) for coordination of bidentate ligands to (**9**) are biphasic: at low ligand concentrations one ligand molecule binds across two zinc atoms with binding constant K_1, and at higher ligand concentrations a second ligand molecule binds to the third zinc site with binding constant K_2. Only the first binding process is observed with Zn_2H_2-trimer. The K_1 value of almost 10^5 for bipy binding to Zn_3-trimer is much greater than the 10^3 characteristic of a single binding contact, but model building demonstrates that considerable host distortion is required to allow simultaneous two-point binding. The energetic cost of this distortion is reflected in the value of K_1, which is low for two-point binding. However, this shows that the trimers are elastic hosts, able to respond to the geometrical demands of ligand binding. The K_1 values increase in the series bipy, py_2Et, py_2Pr as the ligands become longer and more complementary to the size of the relaxed trimer cavity. The effective molarity for binding of py_2py (16 M) is only slightly greater than that for py_2Pr (9 M) despite the fact that both ligands are about the same size and py_2py must be more preorganized for binding; perhaps there is an unfavourable interaction between the py_2py phenyl group and the third, unbound porphyrin.

The py_3T ligand possesses good shape complementarity with the cavity of the trimeric hosts, and as expected it binds extremely strongly to Zn_3-2,2,2-trimer. The formation constant of the complex is so high that the measured binding constant is only approximate, probably lying in the range $1–5 \times 10^{10}\,M^{-1}$. Clearly the effective molarity for binding of the third site is no more than 1 M, probably a result of strain associated with coordination of the third site; inspection of models indicates that the N—N distance in py_3T (~960 pm) is 200–300 pm too small to fit the relaxed cavity perfectly, and it must be more difficult for the host to bind misfit-guests at three points rather than just two. NMR competition experiments with the 1,1,2- and 1,1,1-trimers show that py_3T is overwhelmingly bound within these smaller cavities, confirming that the 2,2,2-cavity is larger than the optimum size. At the time of writing there are no good estimates of these larger binding constants.

Some ring-current induced shifts for these ligand–trimer complexes are shown in Figure 3. The terminal proton of py_2py experiences an unusually large ring-current because it is forced into close proximity with the third unbound porphyrin. The Zn_3-2,2,2-trimer·py_3T complex (**24**) is in slow exchange on the NMR timescale with excess of either component at room temperature. The ring-current shifts of the α- and β-pyridine protons show that all three pyridines bind simultaneously. NMR spectra acquired in presence of excess trimer at increasingly elevated temperature until coalescence occurs indicate that association is fast, probably at a rate approaching diffusion control, while dissociation is rather slow. In the more strongly binding trimers, dissociation is even slower, occurring over minutes or even hours.[13,29]

The py_3T ligand binds with a much smaller affinity to the platinum-linked trimer (Table 2), suggesting that the ligand can only coordinate to two of the porphyrin units in this trimer, while the larger, specially designed ligand $(pyacac)_3Al$ binds extremely strongly. Remarkably, the chirality of the remote propeller-shaped, octahedral guest ligand is detectable in the ^1H-NMR spectrum of the host, the porphyrin methyl and ethyl side-chain resonances each being split into two signals of equal intensity, as the adjacent side-chains (a and a′, b and b′) are now in different environments in the Zn_3-Pt_3-trimer·$(pyacac)_3Al$ complex (**25**).

The results in Table 2 show that py_3T and even $(pyacac)_3Al$ are such small ligands relative to the cavity of 4,4,4-trimer that only bidentate binding is possible. However, $py_3C_{12}H_3$ is large enough to achieve three-point binding, giving the complex (**26**). The observed binding constant of $2 \times 10^8\,M^{-1}$ is significantly less than that for the smaller trimers and their complementary guests; while this may reflect imperfect fit, it is probably more a measure of the very large entropic costs involved in organizing such a large cavity.

The conformational and binding properties of the flexible $(CH_2)_4$-linked trimer (and its related dimer) are very complex; they are summarized in simplified form in Scheme 8. In the absence of ligand the *cis, cis, cis* (CCC) isomer constitutes around 30% of the total, while the *trans, trans, trans* isomer, which is not shown, makes up just 5%. One- and two-dimensional NMR spectra of this mixture are enormously complicated but the different isomers have been identified and assigned by a combination of chemical shift, symmetry, NOE and exchange arguments. The illustrated interconversion (which involves rotation of the aryl groups relative to their attached porphyrins) is rather slow, and can be followed by trapping the CCC isomer with py_3T or the *trans, trans, cis*, (TTC) isomer with dabco. Competition experiments show that the CCC isomer actually binds tripyridyltriazine rather strongly. As might have been expected, this trimer is completely ineffective in the Diels–Alder reaction; surprisingly, however, it is quite a good catalyst for the acyl transfer as described in Section 4.4.2.

4.2.3.4 *Tetramers*

CPK models show that the 2,2,2,2-tetramer (**10**) is the smallest of the butadiyne-linked cyclic oligomers with a strain-free transoid arrangement of the aryl linkers; the expected conformational mixture is indeed observed in the NMR spectrum of ligand-free tetramer.

The addition of bipy gives a 2:1 complex, the binding of the second ligand showing strong positive cooperativity. The value of K_1K_2 of $4 \times 10^{16}\,M^{-2}$ from the UV titration is close to the

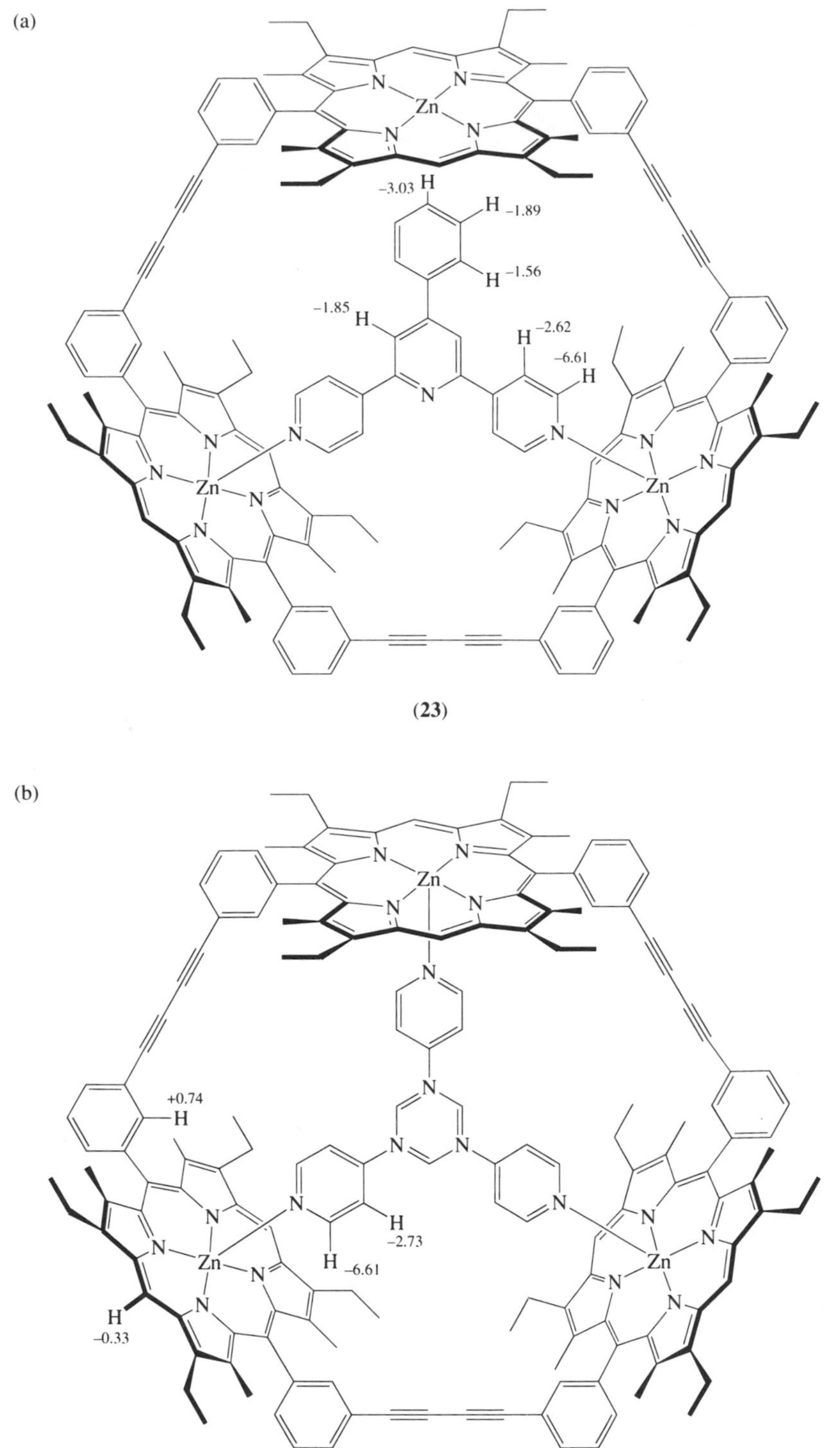

Figure 3 Ring-current shifts (ppm) observed for (a) py_2py and (b) py_3T within the cavity of Zn_3-2,2,2-trimer.

value ($3 \times 10^{15}\,M^{-2}$) determined by a ^{1}H-NMR competition experiment between cyclic dimer and cyclic tetramer. ^{1}H-NMR spectra show that the complexes formed with bipy and 4,4′-bipiperidine (bipip) possess the same symmetry and 2:1 stoichiometry. The binding of one bidentate ligand sets up the second binding site so that it is more complementary to a second bidentate ligand; the second binding constant must be greater than that for the first, so that the 1:1 complex is not

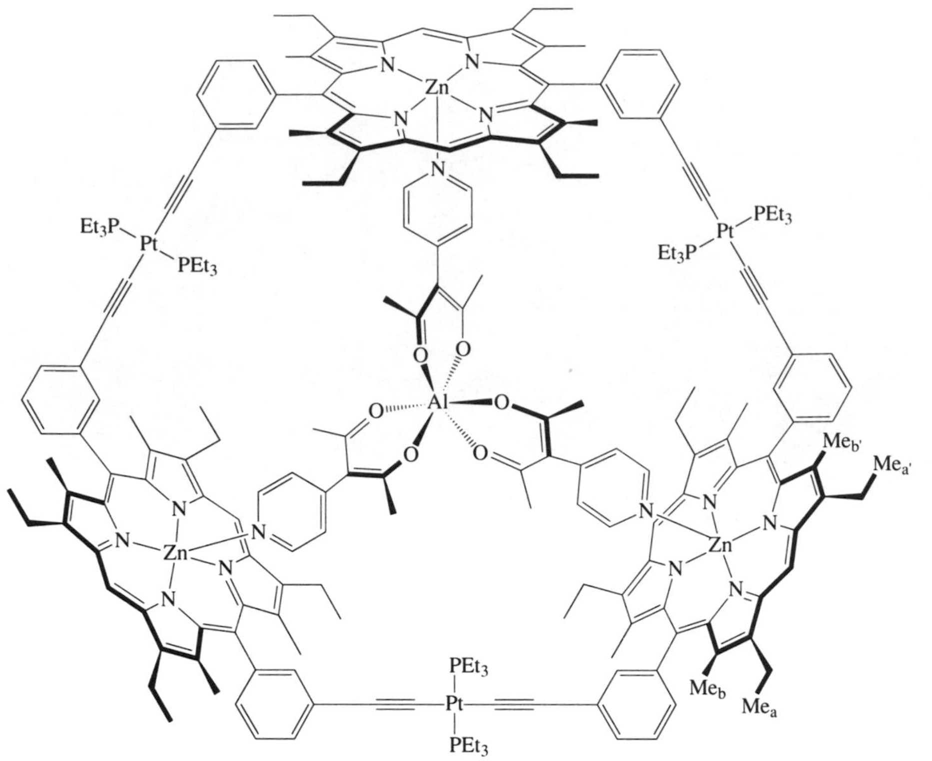

(25)

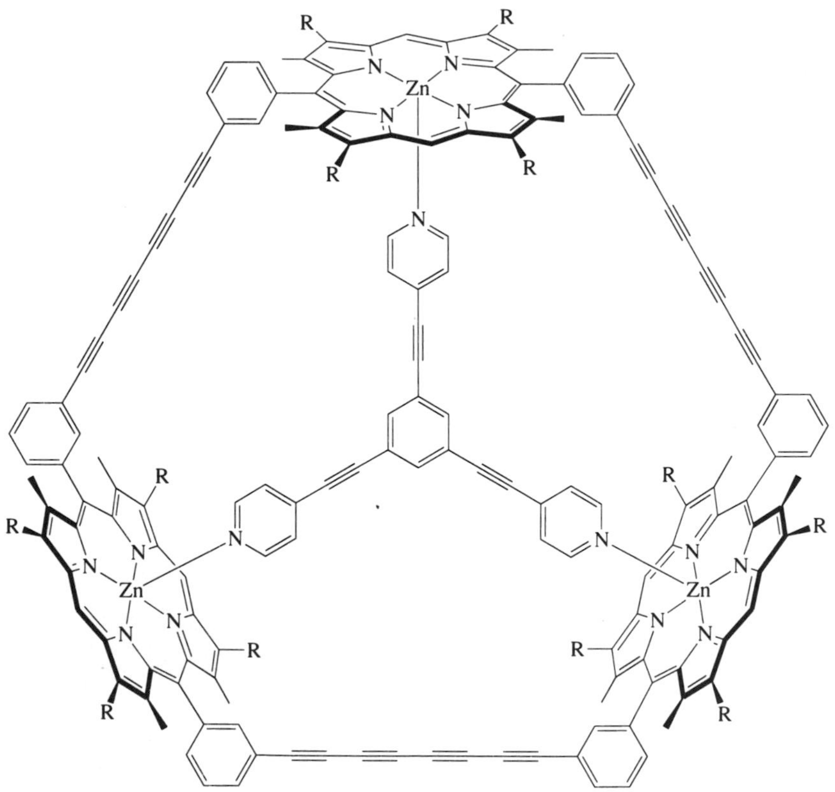

(26)

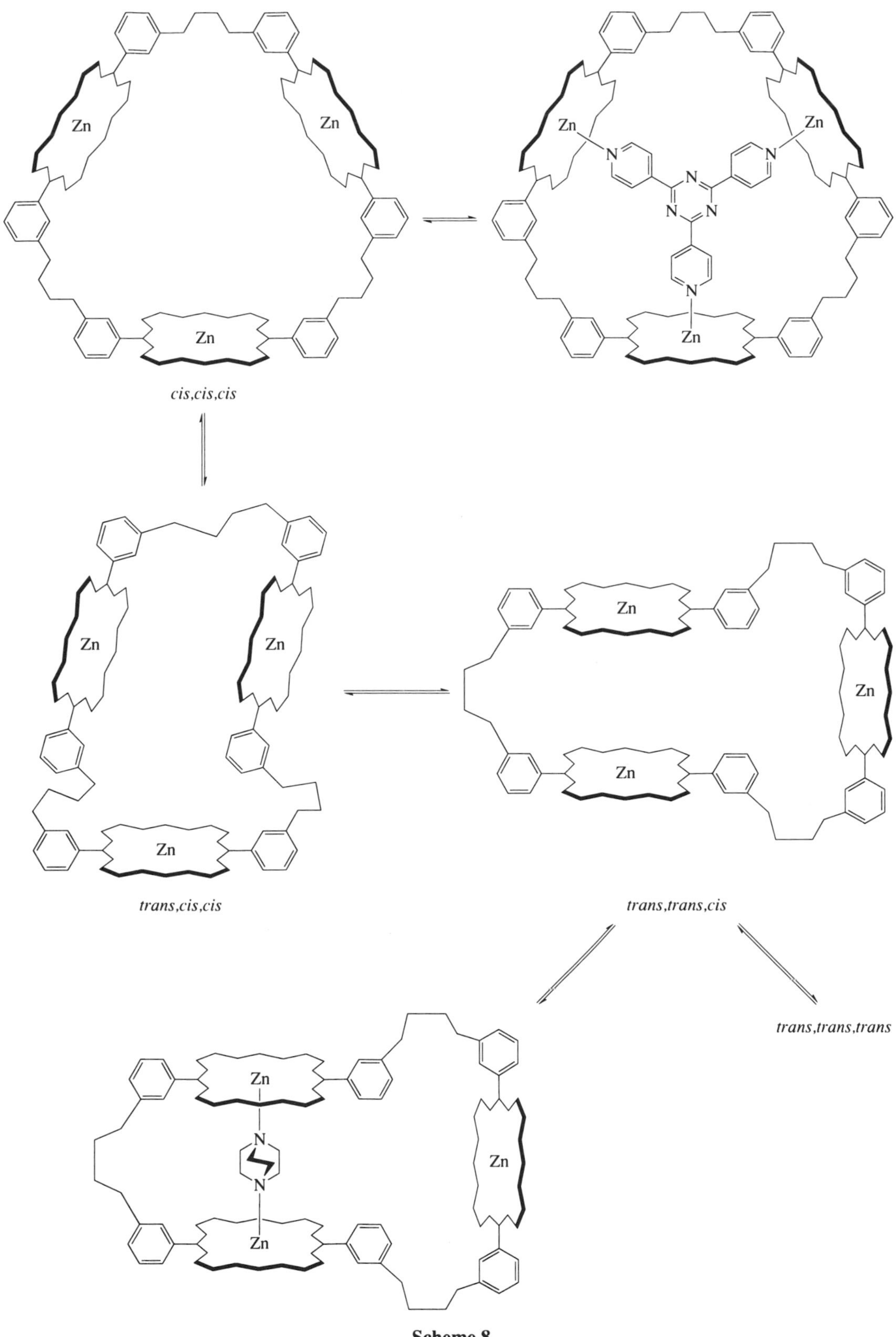

Scheme 8

formed in detectable concentrations. The numbers of signals observed, and their chemical shifts, all indicate that these complexes have C_{2h} symmetry, with two *cis*- and two *trans*-substituted porphyrins (Figure 4).

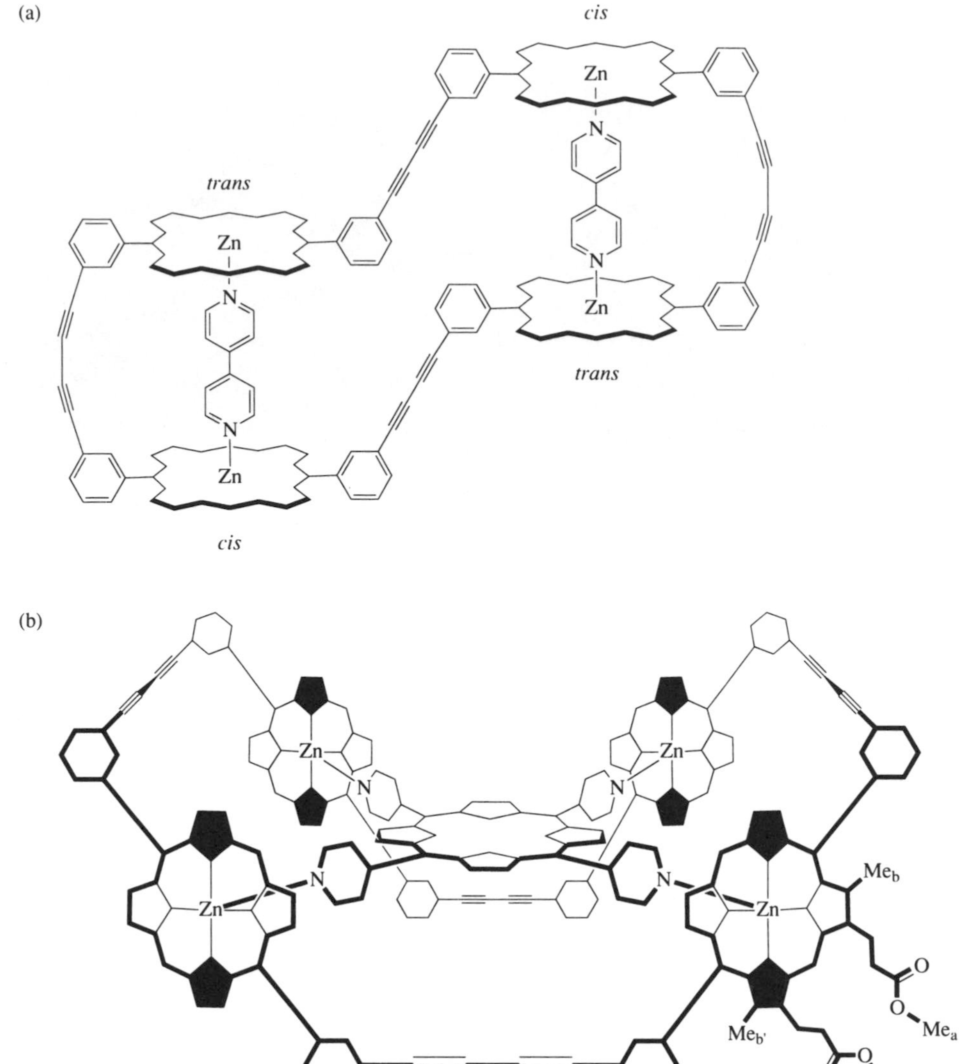

Figure 4 Complexes of Zn_4-2,2,2,2-tetramer (**10**): (a) the 2:1 complex with bipy, showing alternating *cis*- and *trans*-substituted porphyrins; and (b) the 1:1 complex with py_4P, with 'axial' pyrrole groups shaded and 'equatorial' pyrrole groups unshaded.

The same tetramer binds to tetrapyridyl porphyrin, py_4P, with a binding constant of $2 \times 10^{10}\,M^{-1}$, giving a 1:1 complex that chromatographs on silica as a single band. Measurement of such a large binding constant by UV–visible spectroscopy was facilitated by the extinction coefficient of more than $10^6\,cm^{-1}\,M^{-1}$ at λ_{max} for cyclic porphyrin tetramer. This 1:1 complex is in slow exchange with excess guest on the ^{1}H-NMR chemical shift timescale and since the ester methyl (Me_a) and the ring methyl (Me_b) signals each give rise to two equal intensity singlets separated by 0.28 ppm and 0.05 ppm, respectively, a D_{4h} conformation can be ruled out. Similar splittings are also observed in the ^{13}C-NMR spectrum. Molecular mechanics calculations and CPK models indicate the complex prefers to adopt a tub-shaped geometry reminiscent of cyclooctatetraene (Figure 4). The D_{2d} symmetry is consistent with the split signals observed in the ^{1}H and ^{13}C NMR, these splittings resulting from the presence of the two pyrrole environments (Figure 4); the axial pyrroles are shaded black. The α-pyridine protons of py_4P are shifted upfield by 6.81 ppm and the β-protons by 2.41 ppm on complexation to cyclic tetramer, showing that all four porphyrins must be bound simultaneously. The N—H signal of the py_4P is shifted by 2.9 ppm upfield to −4.8 ppm. This upfield shift is consistent with four porphyrin ring currents perpendicular to the ligand plane and in close proximity. The symmetry of the complex in the crystal structure is essentially identical to that observed in solution.[9]

(i) Binding properties of linear tetramer.

The linear tetramer whose synthesis is described in Section 4.3.2.2 binds to bipy with an affinity of $7 \times 10^6\,M^{-1}$; the UV–visible data were analysed for a simple statistical 1:1 binding process and the simulated curve was found to fit the experimental data very well, indicating that the two binding events are either completely statistical or cooperative.

The same linear tetramer may form 1:1 or 1:2 complexes with py_4P, depending on absolute and relative concentrations, a feature which is important for understanding templating efficiency; the 1:1 complex will be favoured at low concentrations. Titration gives a binding constant of $7 \times 10^9\,M^{-1}$. The binding constant is very similar to, but clearly smaller than, that for the complex formed between cyclic tetramer and py_4P. This difference was qualitatively confirmed by an NMR competition experiment for py_4P between linear and cyclic tetramer. The NMR properties can be rationalized by the formation of a 1:1 complex in which the linear tetramer wraps itself around the central py_4P, but with several TMS proton environments due to rotation about the porphyrin–aryl bond that is slow on the ^{1}H-NMR timescale. When more than 1 equiv. of py_4P was added the spectrum changed in a way that is consistent with the formation of a 1:2 complex, but it has not been possible to elucidate fully the structure of the complexes formed between linear tetramer and py_4P.

4.3 TEMPLATED SYNTHESES OF OLIGOMERS

4.3.1 First Attempts and Successes

The very different ligand-binding characteristics of the butadiyne-linked oligomers tempted Sanders and co-workers to try using these differences to influence the synthesis by templating.[14,30] Addition of py_3T or bipy as potential templates during pyridine–CuCl Glaser coupling had no effect on the product distribution: template-binding is too weak in the presence of a large excess of pyridine. New reaction conditions were therefore sought under which efficient coupling and templating could be achieved, screening many solvent/reagent combinations; many conditions were rejected because they resulted in copper metallation of the porphyrin. Pyridine, 2-picoline, 2,5-lutidine, DMF, MeCN, acetone, THF, CH_2Cl_2, $CHCl_3$ and toluene were explored as solvents, using CuCl, CuCl·TMEDA and CuCl·2,2′-bipy as reagents, in an atmosphere of dry air or oxygen, with py_3T or bipy as template. With CuCl as reagent, coupling only occurs in pyridine, 2-picoline, 2,5-lutidine and DMF. A small template effect was detected in DMF, but not in pyridine, 2-picoline or 2,5-lutidine; copper-metallation was fastest in DMF. Chelating ligands such as TMEDA and 2,2′-bipy increase the solubility of the copper(I) chloride and extend the range of solvents which can be used; with CuCl·TMEDA and CuCl·2,2′-bipy as reagents coupling worked in all the solvents, except $CHCl_3$. Template effects increase as the solvent is made less coordinating and less polar in the order THF $<$ acetone $\simeq$ MeCN $<$ CH_2Cl_2 $<$ toluene. Coupling goes more rapidly and cleanly in CH_2Cl_2 than toluene. Use of TMEDA rather than 2,2′-bipy generally gave faster coupling; the CuCl·2,2′-bipy reaction showed slightly larger template effects, but did not always go to completion. In all cases coupling occurs more cleanly under air than oxygen although no coupling occurs under argon.

Coupling with CuCl · TMEDA/CH_2Cl_2 is almost as efficient as with CuCl/pyridine and there is a dramatic template effect. The most useful cyclization template for the synthesis of cyclic porphyrin dimer was bipy. The addition of a sixfold excess of bipy relative to 2×10^{-4} M porphyrin led to an increase in the yield of cyclic dimer from 23% to 72%, whilst the yield of cyclic trimer was reduced to 4% from 34%. Similarly when py_3T was used to template the synthesis of cyclic trimer the yield of trimer increased to 55% from 34% and the yield of dimer was reduced to 6% from 23%. Using this route 2,2-dimer (**8**) and 2,2,2-trimer (**9**) can be prepared in 70% and 50% isolated yield respectively, on the 200 mg scale.

The first step in this templated reaction must be the combination of two porphyrin units to yield linear dimer (**15**). Then either an intramolecular cyclization occurs or reaction with a further monomer porphyrin yields linear porphyrin trimer. The latter process is prevented by the bipy template, which induces the reactive ends of the linear dimer intermediate to come into close proximity and so increases the rate of intramolecular cyclization. This reaction therefore proceeds in the same way as classical metal-cation templated macrocyclizations.

In the less successful py_3T-templated synthesis of cyclic trimer, the first step must again be the coupling of monomer units to yield linear dimer, but the conformation adopted when this is bound to py_3T has the reactive ends held apart; intramolecular cyclization is now disfavoured, leaving the way open for intermolecular reaction to take place. However, the proportion of monomer in the reaction mixture will decline rapidly and therefore the probability of linear dimer yielding linear trimer will also be low towards the end of the reaction. The coupling of monomer in the presence of py_2py, which should bind to the linear dimer in a similar way to py_3T but which lacks the third binding site, gives a similar product distribution. This confirms that the template in the synthesis of cyclic trimer plays a mainly preventative role; it also incidentally indicates that the 'obvious' templating mechanism—preassembly of all the monomeric components around the ligand before coupling takes place—is not operating in this series. Genuine examples of the preassembly mechanism are discussed in Section 4.3.2.5.

Analysis of the differences in the templating abilities of bipy and py_3T led (i) to efficient syntheses of porphyrin tetramers and an octamer (Section 4.3.2.2),[14] (ii) to recognition of the various template roles described in Section 4.3.2,[3] and (iii) to a series of experiments designed to confirm the operation of these possible roles of templates in these syntheses.[15]

4.3.2 Several Roles for Templates

4.3.2.1 Positive cyclization

This is the effect exhibited in the bipy-templated synthesis of cyclic porphyrin dimer from monomer. In order to understand the proposed role of bipy in more detail, the linear dimer (**15**) intermediate was synthesized and its cyclization behaviour was investigated. The bipy ligand increased the yield of cyclic dimer from 21% to 83%, with a corresponding drop in the yield of larger oligomers. So, the ratio of cyclic dimer to other products increased 20-fold (from 21:79 to 83:17), corresponding to a 20-fold rate acceleration and an EM_{kin} of 1 mM. This increase in effective molarity can be attributed to the bipy ligand's influence on the conformation of the linear dimer and its ability to bind preferentially to the transition state for the intramolecular reaction. When bound to linear dimer, bipy forces the reactive alkyne ends of the dimer to be brought into close proximity, and intramolecular coupling to the cyclic dimer occurs more quickly than intermolecular coupling. The measured binding constants show that the cyclization reaction is driven by a progressive increase in binding energy as the reaction proceeds from linear dimer ($K = 6 \times 10^6\,M^{-1}$) to cyclic dimer ($K = 1 \times 10^9\,M^{-1}$).

Similarly, py_4P enhanced the cyclization yield of linear porphyrin tetramer from 60% to 90%, corresponding to an increase in EM from 0.15 mM to 0.9 mM, or a sixfold increase in rate. Not surprisingly, given the more flexible complex that must be created, bipy was less effective, but it was able to double the cyclization rate to give 75% yield.

The effective molarities in the previous two paragraphs show that manipulation of the synthesis using templates was only possible in a narrow concentration range: below 0.05 mM, cyclization of linear dimer would be the dominant process anyway, while above 1 mM even bipy would not prevent intermolecular coupling from being the most important process. The concentration range where py_4P can enhance the cyclization yield of linear tetramer is even narrower.

Glaser–Hay cyclization of the 1,1-linked linear Zn_3- or $Ru(CO)Zn_2$-trimers to give the 1,1,2-cyclic trimers was effectively templated by py_3T in a process where the template appears to be acting positively.[21,22] As pointed out above, host–guest complementarity is better than in the 2,2,2-series so this should not be surprising; it has the happy consequence illustrated in Equation (11) that the CO attached to the central ruthenium-ion is inevitably on the outside, forcing any future ligand (after removal of the template) to bind inside the cavity.

4.3.2.2 Scavenging

Cyclization templates can be used to scavenge cyclizable material in a reaction mixture and thus facilitate the formation of linear oligomers. Using the strategy for synthesis of linear oligomers in Scheme 9, reaction i, dimerization of the partially deprotected material can only be carried out efficiently in the absence of fully deprotected molecules because the latter can couple with monoprotected material to generate a new reactive oligomer and ultimately a complex mixture. This problem can be avoided by separation of the doubly reactive material before coupling, but this becomes more difficult with increasing chain length. Scavenger templates overcome the problem by

(11)

enforcing intramolecular reaction of doubly-reactive molecules (Scheme 9, reaction ii); mono-protected molecules have no choice but to couple to each other. The separation is left until after coupling, when it has become much easier because the desired linear compound is twice as massive as either the starting material or the cyclic by-product.

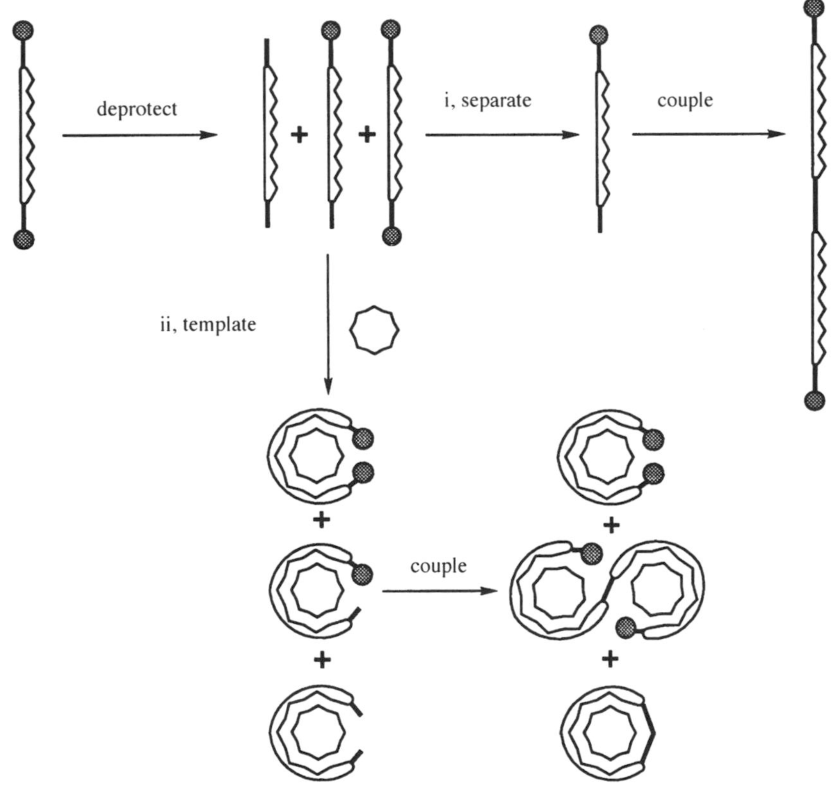

Scheme 9

Sanders and co-workers used this scavenging strategy to synthesize a linear tetramer from linear dimer (**15**) (Equation (12)) and a linear octamer from the tetramer (Equation (13)), using bipy and py_4P respectively as templates. In both cases the yields of desired product are close to the theoretical maximum of 50%; the doubly protected starting material can be recycled and the cyclic by-product is useful in its own right. In the absence of template, both reactions do give significant amounts of the desired linear coupling product, but there are so many other coupling products that separation of pure material on a preparative scale is difficult and inefficient. In the presence of scavenger template, only three products are formed, greatly easing the isolation process; this advantage is as important as the improvement in absolute yield.

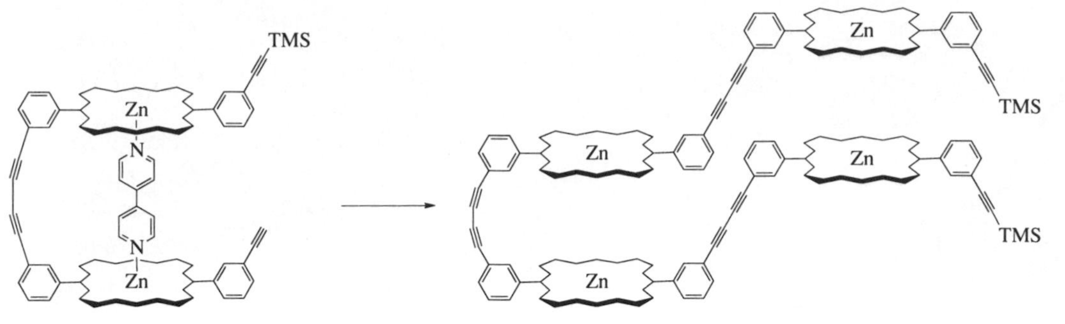

(12)

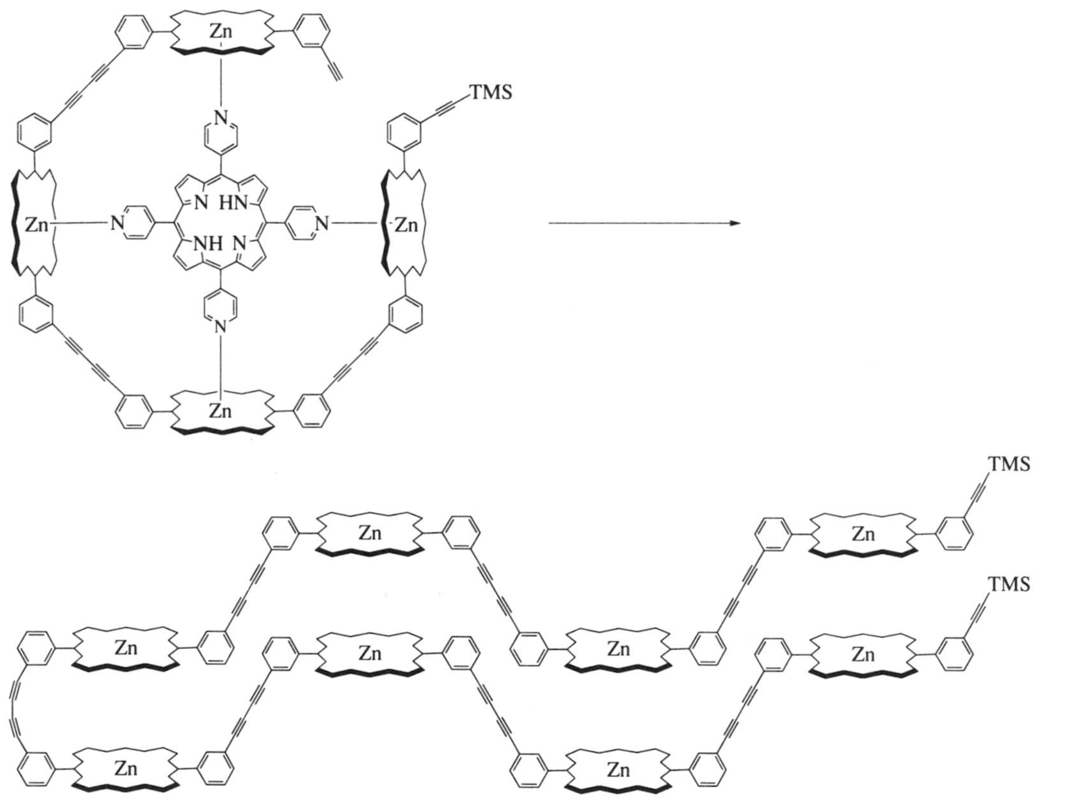

(13)

4.3.2.3 Negative cyclization

It is also possible for a template to act negatively by disfavouring a specific reaction between bound substrates; this is not the same as accelerating a competitive reaction. In some cases, as seen below, the negative template may have a second function as a positive template during a subsequent step. In principle any system which can be induced to adopt a particular molecular architecture by a positive template can be prevented from doing so by a negative template. For example, the Glaser cyclization of linear dimer (**15**) can be inhibited by the negative cyclization template py_2py, the yield of cyclic dimer being reduced to 12%, corresponding to a 40% decrease in rate. Clearly the ligand binds so that linear dimer adopts a conformation in which the reactive groups are held apart so as to discourage intramolecular cyclization to cyclic dimer. This same effect is exploited in the py_3T-templated preparation of 2,2,2-trimer from monomer (**7**).

A more satisfying example of negative templating is the efficient one-step synthesis (77% yield) of cyclic 2,2,2,2-tetramer from linear dimer (**15**) templated by py_4P (Scheme 10). Here the template modestly inhibits cyclic dimer formation, reducing the yield to 14% by acting negatively, and positively accelerates cyclization of the intermediate linear tetramer as mentioned above. Under some conditions the coupling of two dimer units to yield a linear tetramer is also accelerated by the py_4P template—this is explored in Section 4.3.2.4. Further investigation of the effect of this ligand on the linear dimer revealed that the negative templating was modest because the TMEDA in the reaction mixture competes with the template for the zinc–porphyrin; switching to 2,2′-bipy which cannot compete for the porphyrin has the disadvantage of longer reaction times and more side products (particularly due to copper-metallation of the porphyrin) but does lead to the expected suppression of cyclic dimer to just 1.5%.

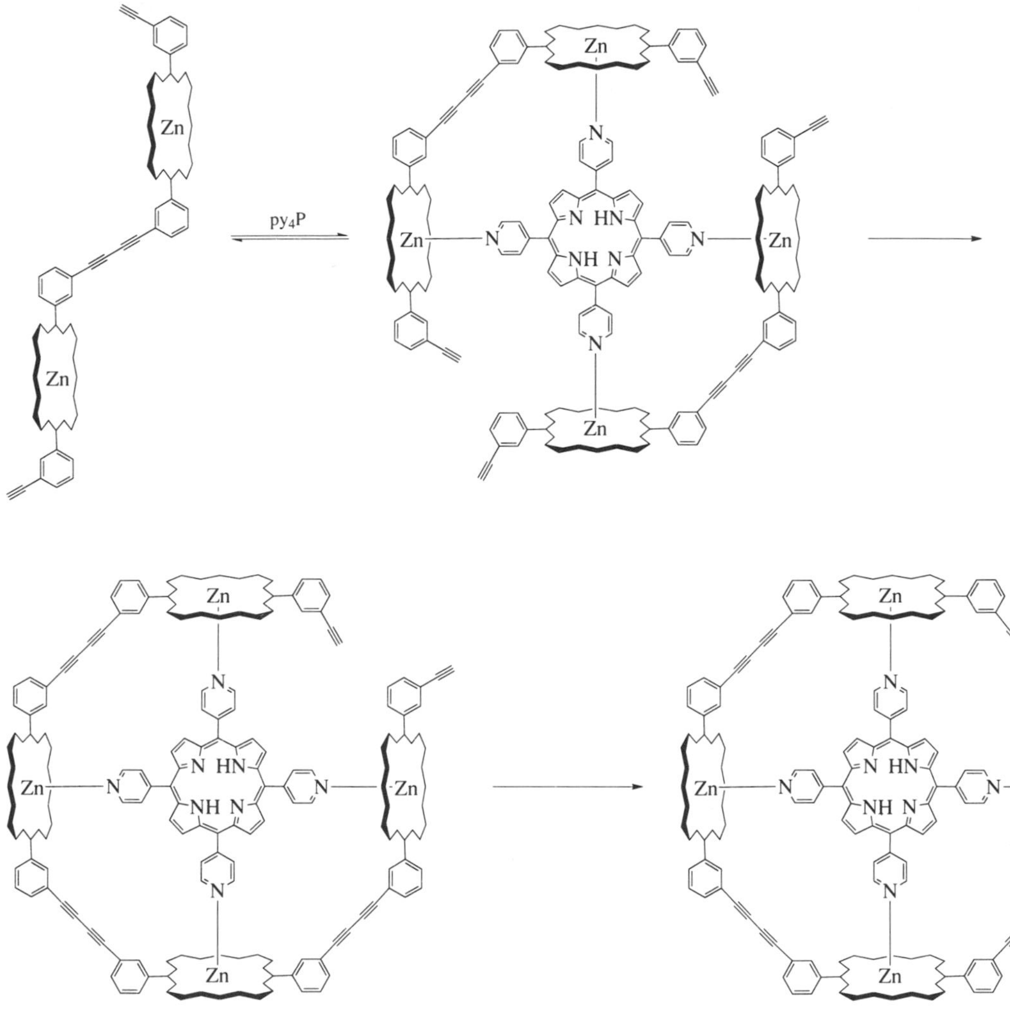

Scheme 10

4.3.2.4 *Positive linear*

Conceptually the simplest role for a template is to bring together two molecules and encourage them to combine. Sanders and co-workers have defined this as linear templating.[3] It is just the intermolecular equivalent of cyclization templating.

The one-pot synthesis of cyclic 2,2,2,2-tetramer from linear dimer described above and summarized in Scheme 10 clearly has the potential for coupling of the two linear dimers to linear tetramer to be accelerated by the template. Most of the template effects discussed so far were investigated by product distribution analysis, but this approach could not be used to probe the extent of linear templating in the reaction shown in Scheme 10 because the same linear tetramer intermediate is an inevitable first step of both the templated and untemplated reactions. There are two ways in which the extent of linear templating could be analysed. First, the reaction could be followed so that the concentration of starting materials, intermediates and products would be evident at given time intervals and therefore the rate of linear tetramer formation could be calculated. Second, a mixture of free base and metallated linear dimers which are blocked at one end could be coupled; in this experiment coupling is necessarily terminated at the linear tetramer stage, and the question becomes whether the distribution of metallated and free base sites in the coupling products is influenced by the presence of py_4P.

Sanders and co-workers pursued both approaches and have described the results in detail elsewhere.[15] In summary, it was possible to demonstrate a genuine templating effect on dimerization when 2,2′-bipy was used for copper chelation, but the effect is not significant under the more usual conditions when TMEDA is present during coupling. The roles of py_4P in such couplings are also critically dependent on relative and absolute concentrations: an excess of template will lead to most linear dimers being present as 1:1 complexes and very few template molecules being loaded with two dimer molecules, while large concentrations will tend to favour higher-order aggregates.

4.3.2.5 Preassembly of multiple components

In order to study steric and electronic effects on the Diels–Alder and acyl transfer reactions described in Section 4.4 porphyrins (**27**) and (**28**) with electron-withdrawing substituents on the *meso*-positions and (**29**) with buttressing methoxy groups were prepared.[13,29] As expected the former bind pyridine more strongly than conventional porphyrins; for example the dioxoporphyrin (**27**) was found to have a binding constant of $5.2 \pm 0.5 \times 10^5\,M^{-1}$ in dichloromethane at 25 °C.[13] A van't Hoff plot gives $\Delta H = -29\,kJ\,mol^{-1}$ and $T\Delta S = +3\,kJ\,mol^{-1}$ at 25 °C in tetrachloroethane; these values are much smaller than for (**7**), implying greater coordination of solvent in the absence of added ligand, and release of solvent on pyridine binding. The large affinity of dioxoporphyrin for pyridine ligands means that multicomponent complexes are readily assembled; thus 82% of the dioxoporphyrin in a 0.2 mM solution is calculated to be present as the 2:1 complex (**30**) with 0.1 mM bipy in dichloromethane solution, compared with only 15% for the corresponding porphyrin solution. Even more strikingly, replacement of bipy by 0.05 mM py_4P should lead to 67% of the dioxoporphyrin being present as the 4:1 complex (**32**) vs. 2% for the porphyrin analogue.

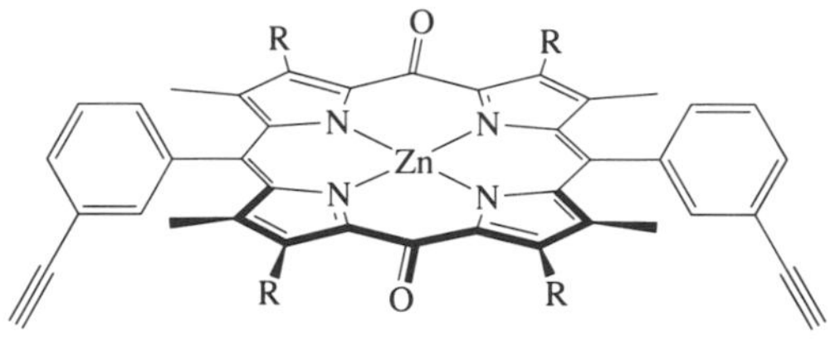

R = $CH_2CH_2CO_2Me$
(**27**)

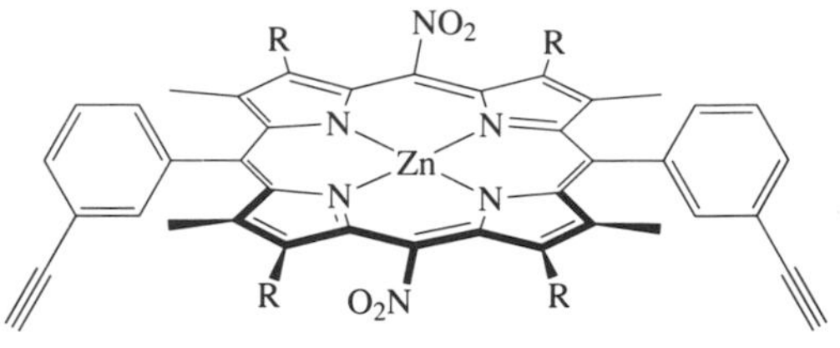

R = $CH_2CH_2CO_2Me$
(**28**)

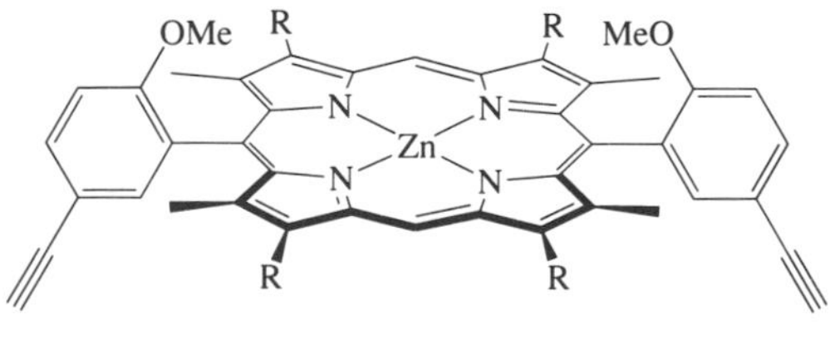

R = $CH_2CH_2CO_2Me$
(**29**)

This has major implications for the assembly of oligomers on a ligand template: Glaser–Hay coupling of 0.2 mM dioxoporphyrin in the absence of any template yields about 40% each of the

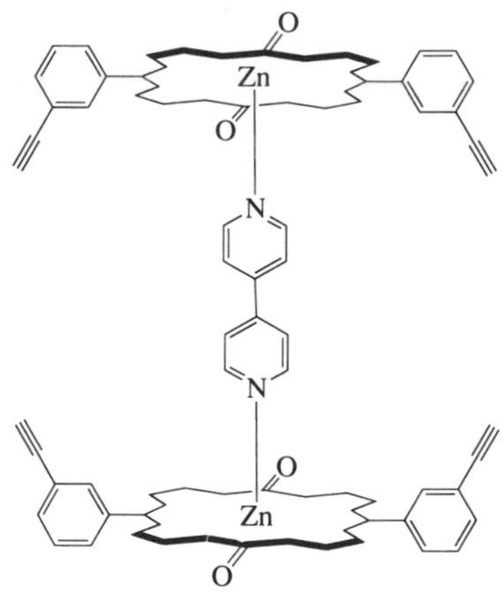

(30)

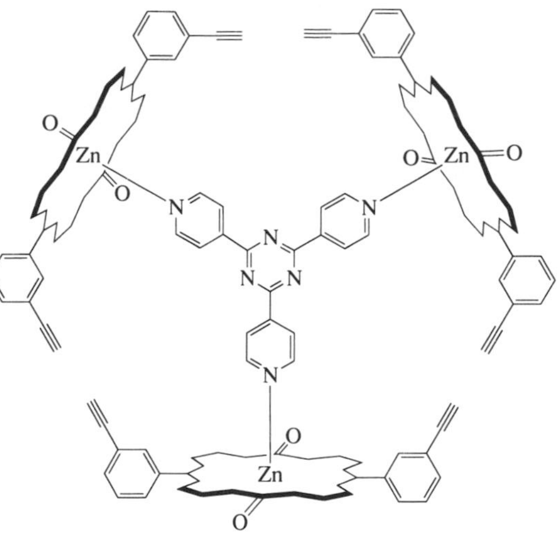

(31)

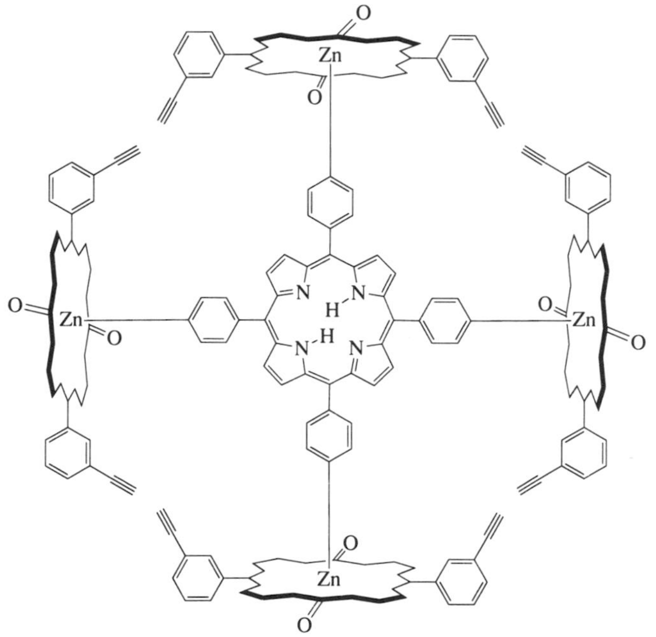

(32)

corresponding cyclic trimer and tetramer with very little dimer. However, addition of 0.5 mol. equiv. of bipy to the reaction mixture completely changes the course of the reaction: cyclic Zn_2-2,2-linked-tetraoxodimer is now produced almost quantitatively as the only detectable product. Similarly, 0.33 mol. equiv. of py_3T leads to virtually exclusive formation of cyclic-Zn_3-2,2,2-hexaoxotrimer, while 0.25 mol. equiv. of py_4P gives over 70% of cyclic Zn_4-2,2,2,2-octaoxotetramer. All of these dioxoporphyrin oligomers can readily be isolated on an arbitrarily large scale in one step and purified by recrystallization as no similarly sized molecules are present to interfere.

These synthetic results are consistent with preassembly of all the oligomer components around the ligand, (e.g., **(30)**–**(32)**), before any covalent chemistry takes place. The alkyne groups will of course be randomly oriented rather than all facing inwards, but the rate of aryl rotation ($\sim 1\ s^{-1}$) is fast compared with the reaction used to couple the alkynes together. This multiple preassembly is very unusual in synthetic supramolecular chemistry—normally no more than two reactive species are preassembled before an individual reaction step. There are no quantitative estimates of effective molarities for cyclizations in these experiments, but clearly the large binding constants greatly extend the concentration range where successful intervention with templates can be expected. Another possibility for this type of templating might be to use conventional porphyrins but with more strongly ligating central metal ions such as iron, cobalt or ruthenium. These, however, suffer several disadvantages as general building blocks: the metals are less easily removed than zinc; some of the oxidation states are paramagnetic and/or oxygen sensitive; and their coordination chemistry is more complex, with two axial ligation sites rather than one for zinc.

Dioxoporphyrins represent a simple but highly effective tuning of the porphyrin moiety, and they offer new opportunities for creating templated or self-assembled supramolecular systems. Unfortunately, the lack of overall aromaticity implied by the cross-conjugated structure leads to a complete loss of ring-current shifts: the pyrrole NH protons in freebase dioxoporphyrin resonate at +12.37 ppm compared with −2.48 ppm in the conventional porphyrin analogue. Similarly the limiting shifts for bound pyridyl ligands are scarcely different from their initial values in free solution; in all cases $\Delta\delta < 1$ ppm. The absence of dramatic NMR shifts on binding appears to be the only way in which dioxoporphyrins are less attractive as building blocks than their parent porphyrins.

A series of NMR competition experiments with a range of porphyrin trimers shows that the hexaoxo trimer binds py_3T with a binding constant at least 800 times greater than that of the corresponding porphyrin trimer, that is, more than $10^{12}\,M^{-1}$. This is consistent with the fact that the maximum possible association rate is diffusion controlled (10^9–$10^{10}\,M^{-1}\,s^{-1}$) and the dissociation rate observed in the competition experiments is slower than $10^{-3}\,s^{-1}$. This binding is so strong that molecular ions are readily obtained in positive FAB-MS (fast atom bombardment-mass spectrometry) spectra of the bipy, py_3T and py_4P complexes of the dioxoporphyrin dimer, trimer and tetramer respectively.

The corresponding *meso*-dinitroporphyrin (**28**) behaves in the same way as dioxoporphyrin with py_3T and py_4P, giving the trimer and tetramer essentially quantitatively. However, its behaviour with bipy is quite different, as described in Section 4.3.3.

4.3.3 Failures

Unfortunately, in the synthesis of the platinum- and octatetrayne-linked porphyrin oligomers, ligand binding is not strong enough to template cyclization with a useful effective molarity.[17,18] Given the marginal effective molarities operating in the smaller butadiyne systems this is, in retrospect, perhaps not surprising. It is possible that the electronically 'improved' porphyrins with better binding properties would be susceptible to templating in such large systems, but this has yet to be attempted.

More remarkably, apparently small substituent-induced changes in the geometry of individual porphyrin units or in the conformation of the linear intermediates can have a dramatic and unexpected effect on the outcome of the coupling reaction. The methoxy-substituted monomer (**29**) gives 60% yield of 2,2-dimer in the absence of template, rising to 65% in the presence of bipy, but the trimer could not be isolated, even under the best templating conditions.[31] X-ray crystal structure evidence from porphyrins with similar *ortho*-substitution patterns indicates that the porphyrin is likely[32] to be bowed in such a way as to push the alkynes together and favour dimer formation—a true 'structure-directed' synthesis that appears virtually immune to intervention by a template.

Conversely, no cyclic 2,2-dimer could be prepared from zinc–*meso*-dinitroporphyrin (**28**), even under good templating conditions with bipy.[29] Sanders and co-workers have, however, prepared the linear dimer, and shown that it binds bipy surprisingly weakly, with $K = 3 \times 10^7\,M^{-1}$; given that zinc–*meso*-dinitroporphyrin binds pyridine with $K = 1 \times 10^5\,M^{-1}$, it is clear that there is a very strong conformational bias against bringing the linear dimer into a suitable conformation for bipy binding and subsequent cyclization. Fully substituted porphyrins of this type are known from x-ray evidence to be distorted into a saddle shape,[33] and this presumably forces the alkyne linkages away from the geometry required for cyclization to dimer. This shape is no inhibition to quantitative formation of cyclic trimer and tetramer via the preassembly mechanism described in Section 4.3.2.5.

These large effects from apparently small structural changes are a powerful reminder of just how marginal are the syntheses of these giant macrocycles, delicately poised between spectacular success and dismal failure.

4.4 TEMPLATED CHEMISTRY WITHIN CAVITIES

The starting point is summarized as (**33**): when two or three pyridine ligands are bound within the cavity of Zn_3-2,2,2-trimer their effective concentration is dramatically increased while at the same time their range of relative orientations is limited by the geometry of Zn—N coordination. The trimer should, therefore, act as an 'entropic trap' and accelerate any reaction whose transition-state geometry matches the orientation of bound ligands.

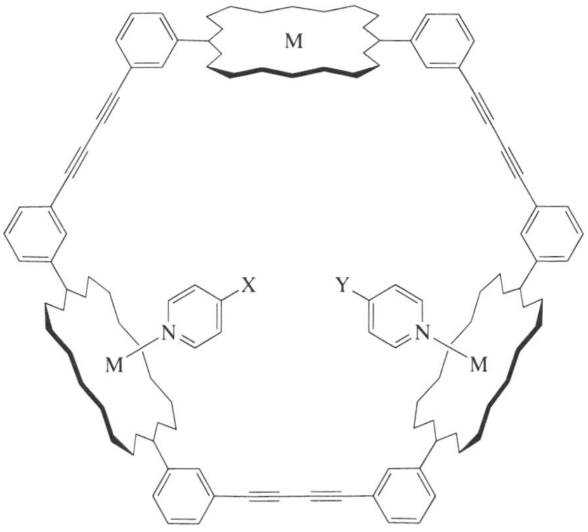

(**33**)

4.4.1 Diels–Alder Reactions

The Diels–Alder reaction proved attractive because it has rich stereo- and regiochemistry, a stringent geometrical requirement, no need for external reagents, and it has been the subject of related studies using catalytic antibodies. It also offers the possibility of altering stereo- and regiochemistry through geometry control within a cavity. A furan-derived diene and maleimide-derived dienophile were used because this combination reacts reversibly, giving the opportunity of studying the kinetics, and therefore the approach to the transition states, in both the forward and reverse directions.[34–6] The reaction in Scheme 11 leads to inhibition by strongly bound products; this could have been avoided by engineering a second step, but the apparent disadvantage of lack of turnover is outweighed by the direct access to the transition state energies available from a study of the unimolecular reverse reactions as described below.

The equilibrium constant for this reaction is quite large ($k_1(exo)/k_{-1}(exo)$ is $\sim$5000 M^{-1} in $C_2H_2Cl_4$ at 30 °C, decreasing to 400 M^{-1} at 60°C) but the equilibrium at millimolar concentrations lies on the diene–dienophile side. The relative orientations of the pyridine groups in the two adducts are very different, and model building suggested that the *endo*-adduct would fit less well into the 2,2,2-trimer cavity than the *exo*-adduct. The measured binding constants confirm this, the *exo*-adduct binding around 15 times more strongly in $C_2H_2Cl_4$ at 30 °C (9×10^6 M^{-1} vs. 6×10^5 M^{-1}) and six times more strongly at 60 °C (4.6×10^5 M^{-1} vs. 8×10^4 M^{-1}). In the absence of added porphyrin trimer, the kinetic *endo*-adduct is a significant product at low temperature while the thermodynamic *exo*-adduct is the only product detected at high temperatures; at 60 °C the *exo*/*endo* ratio is $\sim$4 initially, but the *endo*-adduct then disappears.

Addition of 1 equiv. of trimer to the two reactants (0.9 mM each in $C_2H_2Cl_4$) accelerates the forward Diels–Alder reaction around 1000-fold at 30 °C and 200-fold at 60 °C, yielding the *exo*-adduct as the only detectable product. The initial reaction rate in the presence of trimer is almost temperature independent under these conditions because as the temperature is raised the increase in the intrinsic rate of the reaction inside the porphyrin cavity is almost exactly offset by weaker substrate binding. As an aside it is worth noting that the *exo*-adduct is a good template for the synthesis of 2,2,2-trimer from monomer, while the *endo*-adduct is not; in this observation, the templating complementarity illustrated in Figure 1 receives its experimental expression.

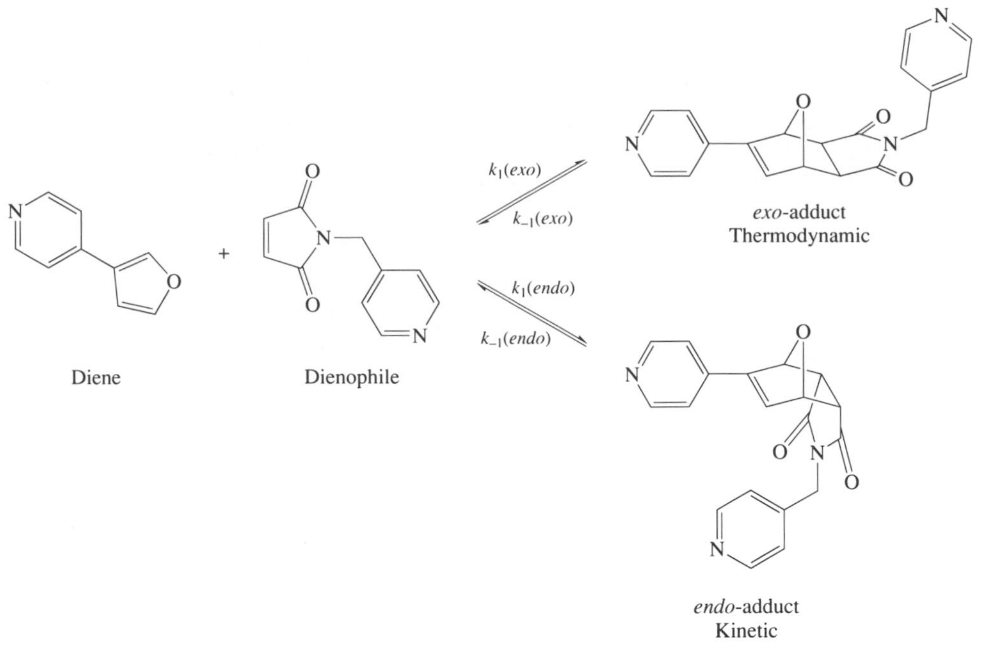

Scheme 11

Measurements of the forward and reverse reaction rates in the absence of trimer, of the binding constants for products and starting materials, and of the unimolecular reverse reaction in the presence of trimer, allow construction of the complete free energy profile for the reaction at 60 °C as shown in Figure 5. Neither the *exo*- nor *endo*- forward reaction rates within the cavity of the trimer can be measured directly (i) because the binding equilibria with mixed ligands are so complex and (ii) because the *endo*-reaction is not observed at all. However their transition state energies are accessible through observation of the reverse reaction.

The profile identifies three sources of the *exo*-selectivity in the trimer-accelerated reaction: (i) the *endo*-adduct is less stable thermodynamically by ~13 kJ mol^{-1}; (ii) the trimer binds the *endo*-adduct less strongly, so the *endo*-complex is 18 kJ mol^{-1} less favourable than the *exo*-complex; and (iii) the trimer slows the reverse reaction of the *endo*-adduct by threefold, showing that the *endo*-transition state is less well recognized by the trimer than the *endo*-adduct. By contrast, the reverse reaction of the *exo*-adduct is not slowed by trimer: the *exo*-transition state is as well recognized as the *exo*-adduct.

These three factors combine to place the *endo*-transition state within the trimer cavity some 13 kJ mol^{-1} above the *exo*-transition state, giving a predicted difference of more than 100-fold in the forward rates for formation of the two adducts inside the cavity. This prediction is consistent with an inability to detect *endo*-product. The intracavity transition state energies derived by this indirect route indicate an increase in the activation energy for the forward *endo*-reaction from 110 kJ mol^{-1} to 114 kJ mol^{-1} and a decrease in activation energy for the *exo*-reaction from 105 kJ mol^{-1} to 101 kJ mol^{-1}.

The decreased activation energy for the forward *exo*-reaction inside the cavity corresponds to an effective molarity of at least 4 M; the reaction is accelerated by a positive templating effect. It appears therefore that the macroscopic acceleration observed is partly the result of concentrating the diene and dienophile within the same cavity, and partly the result of a reduced activation energy once bound. In contrast, the larger activation energy of the forward *endo*-reaction within the cavity corresponds to an effective molarity of 0.2 M within the cavity. Since this *EM* is greater than the free substrate concentration, a macroscopic acceleration of the forward *endo*-reaction is predicted by virtue of the high concentration of substrate pairs in the cavity. However it cannot be observed experimentally because the *exo*-reaction is so dominant.

The design of host and guests in this reaction are closely matched, and the question arises as to how important that design actually is. The isomeric Diels–Alder reaction shown in Equation (14) was therefore studied. In this reaction the maleimide portion is the same as in the first reaction (Scheme 11), and the porphyrin-binding pyridine is also the same, but their geometrical relationship

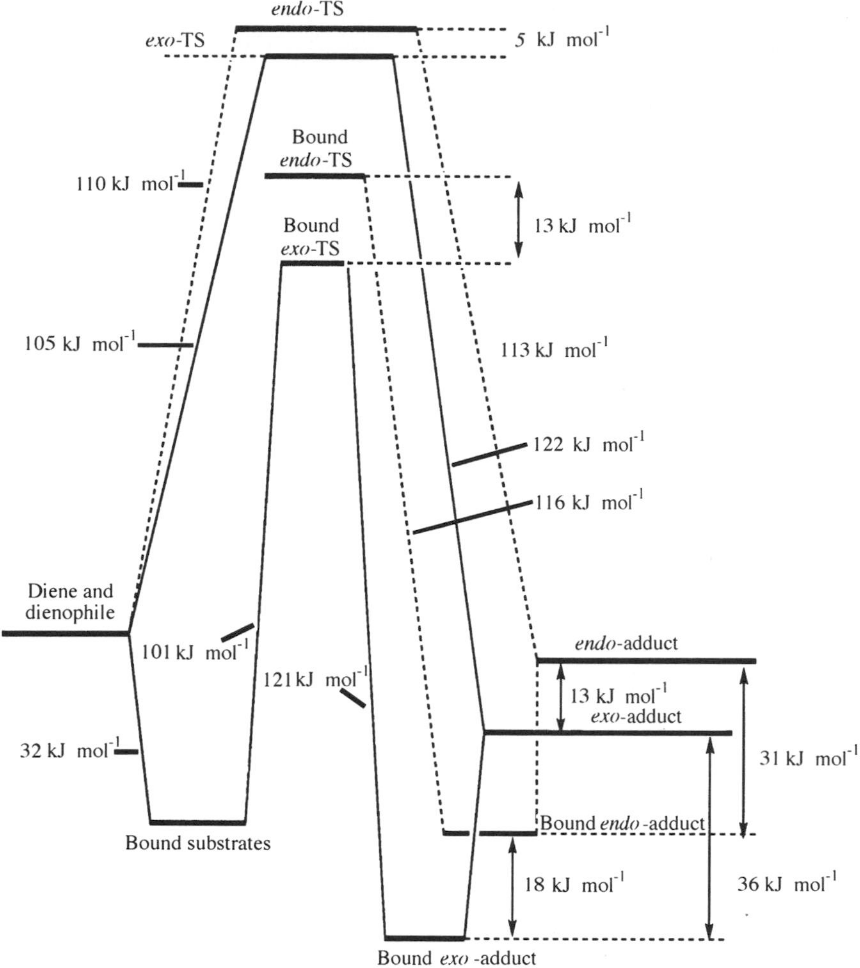

Figure 5 Free-energy profiles for the Diels–Alder reaction in Scheme 11 at 60 °C in $C_2H_2Cl_4$ solution.

is different: the maleimide is now attached to the pyridine is at the 3-position. Models suggested a worse fit inside the trimer cavity for the new Diels–Alder adducts, and indeed the binding of the *exo*-adduct is 20–30 times weaker and the acceleration is now only around 1/10 that of the first reaction.[36]

(14)

However models suggested that Zn_2-4,4-dimer (**13**) would be a better host for this isomeric reaction than for the original reaction, and that prediction too is borne out by experiment; it does not detectably accelerate the first reaction but it accelerates the second reaction around 50-fold at 30 °C. Quantitatively, this last result is disappointing, as molecular models suggest a rather good fit, but qualitatively it is pleasing that the relative order of reactivities for different geometries can be predicted.

Most of the dimers and trimers described in this chapter are being screened for Diels–Alderase activity; this is a complex and time-consuming study involving careful measurements of many forward and reverse rates and numerous binding constants. While it is too soon to draw detailed conclusions, some broad messages are becoming clear: not surprisingly, the platinum-linked trimer is completely ineffective as its cavity is much too large, while the floppy $(CH_2)_4$-linked trimer also shows no acceleration. More subtly, acceleration of *exo*- and *endo*-reactions are both very sensitively dependent on host geometry and flexibility, and in different ways. Compared with the standard 2,2,2-trimer, the nitrated trimers are less good at accelerating the Diels–Alder reaction at 30 °C, but better at 80 °C; they are also slightly less stereoselective, accelerating the *endo*-reaction quite significantly. The reasons for these and related observations on the binaphthyl-capped dimers are as yet unclear in detail. A single preliminary experiment indicates that the Zn_3-1,1,2-trimer lacks selectivity, while the Zn_3-1,1,1-trimer appears to be highly selective for the *endo*-reaction.

4.4.2 Acyl Transfers

The effect of these hosts on the Diels–Alder reaction is not catalytic because the product is strongly bound. However, a transfer reaction of the type A + BC → AB + C (Scheme 12) should be ideal for demonstrating catalysis and turnover: it should be accelerated by substrate proximity and should show efficient turnover because the products are no more strongly bound than the reactants.

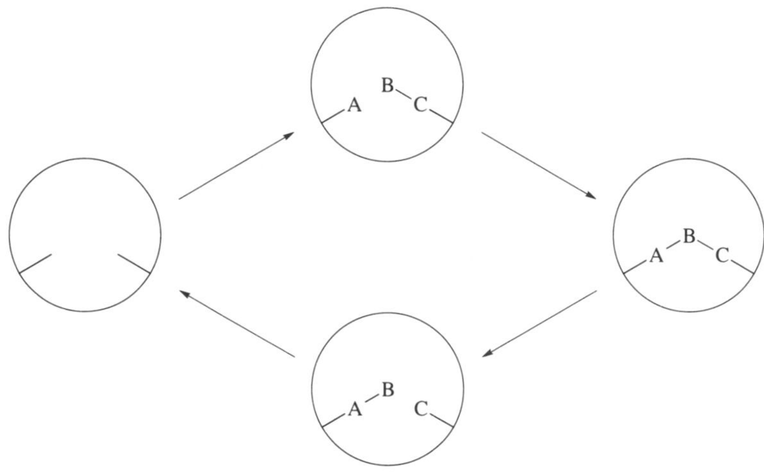

Scheme 12

Furthermore, the intermediate or transition state is stabilized because it is doubly bound to the host. The acyl-transfer reaction shown in Equation (15) is indeed catalysed by the same Zn_3-2,2,2-trimer, and control experiments indicate that the trimer cavity is a crucial component.[28,37] In accordance with expectation, the trimer is effective catalytically rather than stoichiometrically. It is tempting to believe that the tetrahedral intermediate for the reaction shown in Equation (15) is strongly bound within the trimer cavity, but there is not yet any direct evidence for this.

Sanders and co-workers have tested the same range of porphyrin hosts as for the Diels–Alder reaction, and at the gross level almost exactly the same spectrum of activity as for the first Diels–Alder reaction was found; this is hardly surprising as both reactions were designed for the same trimer. However, the flexible methylene-linked trimer is still a catalyst for this acyl-transfer reaction, and there is much less sensitivity to subtleties of geometry; for example, after correction for the smaller number of binding sites, the binaphthyl-capped dimers have similar catalytic effectiveness to the 2,2,2-trimer. The precise geometry and chain lengths of the hydroxymethylpyridine can also be varied without dramatic effects on acyl-transfer rates. To date the Zn_3-1,1,1-trimer is the best catalyst, showing an effective molarity approaching 30 M for the original acyl-transfer reaction.

It seems clear, and perhaps it is not too surprising, that simply bringing together these two reactants can give a significant acceleration because, unlike in the Diels–Alder reaction, there are many suitable pairs of substrate geometries that allow the nucleophilic attack of one on the other. It may also be that the rate-determining step in this reaction is not affected by proximity, for example, if it involves proton transfer.

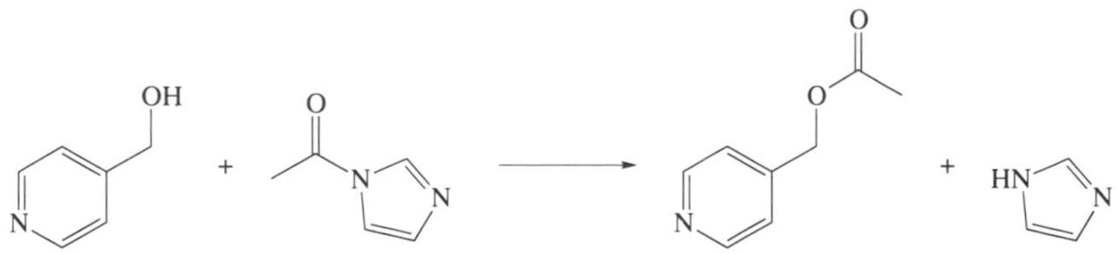

(15)

4.4.3 Prospects

The above acyl-transfer results indicate that the prospects are good for catalysing atom- or group-transfer reactions with relatively tolerant geometrical approaches to the transition state geometry. For example, oxidations that use oxygen-transfer reactions should be susceptible to catalysis and controlled regioselectivity simply by manipulation of promixity within a cavity. Reactions that require a third component, such as acid or base catalysis, will be more problematical as the geometrical requirements and binding complexities will increase rapidly with the number of components; even two ligands competing for three sites leads to extraordinary complexities if analysed carefully.[38]

If the observation that 1,1,1-trimer selectively accelerates the *endo*-Diels–Alder reaction is confirmed it will prove that suitable hosts can act as templates for thermodynamically disfavoured reactions, and support the idea that hosts can be designed to carry out new, or different, reactions; for example, changing the regio- and stereoselectivity of pericyclic reactions should be within our grasp. Unfortunately, model building gives no hint as to why the substituted-2,2,2-trimers should be less effective than their conventional porphyrin analogue. The lesson here is that, as with the templated syntheses of the oligomers themselves, there is a very fine line separating success and failure. Failure can sometimes be more instructive than success; it was the collapsed cavities exhibited by our first generation of flexible porphyrin dimers that led to a simple model of π–π interactions that has become so valuable.[39,40]

It is apparent that the subtleties of the design rules are still far from clear, a point made in the opening paragraph of this chapter. There is a long way to go before synthetic enzymes worthy of the name are produced, but it is at least clear that it is possible.

ACKNOWLEDGEMENTS

It is a pleasure to thank all my talented co-workers for their outstanding experimental and conceptual contributions to this story over the years: Harry Anderson, Richard Bonar-Law, Sally Anderson, Lindsey Mackay, Chris Walter, Anton Vidal-Ferran, Claudia Müller, Duncan McCallien, Esther Levy, Jeremy Prime, Steve Wylie, Nick Bampos, Lance Twyman, Simon Webb, Zöe Clyde-Watson. Financial support from SERC, EPSRC, BBSRC, the EU, Ethyl Corporation, Rhône-Poulenc-Rorer, and Magdalene College and Trinity College, Cambridge are also gratefully acknowledged, as is the excellent EPSRC mass spectrometry service in Swansea.

4.5 REFERENCES

1. R. P. Bonar-Law and J. K. M. Sanders, *J. Am. Chem. Soc.*, 1995, **117**, 259.
2. R. P. Bonar-Law and J. K. M. Sanders, *J. Chem. Soc., Perkin Trans 1*, 1995, 3085.
3. S. Anderson, H. L. Anderson and J. K. M. Sanders, *Acc. Chem. Res.*, 1993, **26**, 469.
4. D. Dolphin, (ed.), 'The Porphyrins', Academic Press, NY, 1978, vols. 1–7.
5. H. L. Anderson and J. K. M. Sanders, *J. Chem. Soc., Chem. Commun.*, 1992, 946.
6. M. J. Crossley, M. M. Harding and S. Sternhell, *J. Org. Chem.*, 1992, **57**, 1833.
7. M. Akhtar, *Nat. Prod. Rep.*, 1991, **8**, 527.
8. H. B. Gray and W. R. Ellis, Jr, in 'Bioinorganic Chemistry', eds. I. Bertini, H. B. Gray, S. J. Lippard and J. S. Valentine, University Science Books, Mill Valley, CA, 1994, 315.
9. S. Anderson, H. L. Anderson, A. Bashall, M. McPartlin and J. K. M. Sanders, *Angew. Chem.*, 1995, **107**, 1196; *Angew. Chem., Int. Ed. Engl.*, 1995, **34**, 1096.
10. K. J. Cross and M. J. Crossley, *Aust. J. Chem.*, 1992, **45**, 991.
11. J. K. M. Sanders and B. K. Hunter, 'Modern NMR Spectroscopy: a Guide for Chemists', 2nd edn. Oxford University Press, Oxford, 1993, chap. 7.
12. H. L. Anderson and J. K. M. Sanders, *J. Chem. Soc., Perkin Trans 1*, 1995, 2223.
13. D. W. J. McCallien and J. K. M. Sanders, *J. Am. Chem. Soc.*, 1995, **117**, 6611.
14. S. Anderson, H. L. Anderson and J. K. M. Sanders, *J. Chem. Soc., Perkin Trans 1*, 1995, 2247.
15. S. Anderson, H. L. Anderson and J. K. M. Sanders, *J. Chem. Soc., Perkin Trans 1*, 1995, 2255.

16. L. Mandolini, *Adv. Phys. Org. Chem.*, 1986, **22**, 1; A. J. Kirby, *Adv. Phys. Org. Chem.*, 1980, **17**, 183.
17. H. L. Anderson, C. J. Walter, A. Vidal-Ferran, R. A. Hay, P. A. Lowden and J. K. M. Sanders, *J. Chem. Soc., Perkin Trans 1*, 1995, 2275.
18. L. G. Mackay, H. L. Anderson and J. K. M. Sanders, *J. Chem. Soc., Perkin Trans 1*, 1995, 2269.
19. N. Bampos, V. Marvaud and J. K. M. Sanders, in preparation.
20. A. Vidal-Ferran, Z. Clyde-Watson and J. K. M. Sanders, in preparation.
21. A. Vidal-Ferran, C. M. Müller and J. K. M. Sanders, *J. Chem. Soc., Chem. Commun.*, 1994, 2657.
22. A. Vidal-Ferran, C. M. Müller, S. J. Webb, L. J. Twyman and J. K. M. Sanders, in preparation.
23. H. L. Anderson, S. Anderson and J. K. M. Sanders, *J. Chem. Soc., Perkin Trans 1*, 1995, 2231.
24. I. P. Danks, T. G. Lane, I. O. Sutherland and M. Yap, *Tetrahedron*, 1992, **48**, 7679.
25. M. J. Crossley, T. W. Hambley, L. G. Mackay, A. C. Try and R. Walton, *J. Chem. Soc., Chem. Commun.*, 1995, 1077.
26. Y. Uemori, A. Nakatsubo, H. Imai, S. Nakagawa and E. Kyuno, *Inorg. Chem.*, 1992, **31**, 5164.
27. H. L. Anderson, A. Bashall, K. Henrick, M. McPartlin and J. K. M. Sanders, *Angew. Chem.*, 1994, **106**, 445; *Angew. Chem., Int. Ed. Engl.*, 1994, **33**, 429.
28. L. G. Mackay, E. G. Levy and J. K. M. Sanders, in preparation.
29. D. W. J. McCallien and J. K. M. Sanders, in preparation.
30. H. L. Anderson and J. K. M. Sanders, *Angew. Chem.*, 1990, **102**, 1478; *Angew. Chem., Int. Ed. Engl.*, 1990, **29**, 1400.
31. J. C. Prime and J. K. M. Sanders, in preparation.
32. J. W. Sparapany, M. J. Crossley, J. E. Baldwin and J. A. Ibers, *J. Am. Chem. Soc.*, 1988, **110**, 4559.
33. M. O. Senge, *J. Chem. Soc., Dalton Trans.*, 1993, 3539.
34. R. P. Bonar-Law, L. G. Mackay, C. J. Walter, V. Marvaud and J. K. M. Sanders, *Pure Appl. Chem.*, 1994, **66**, 803.
35. C. J. Walter and J. K. M. Sanders, *Angew. Chem.*, 1995, **107**, 223; *Angew. Chem., Int. Ed. Engl.*, 1995, **34**, 217.
36. C. J. Walter, D. W. J. McCallien, A. Vidal-Ferran, C. M. Müller, L. J. Twyman, Z. Clyde-Watson and J. K. M. Sanders, in preparation.
37. L. G. Mackay, R. S. Wylie and J. K. M. Sanders, *J. Am. Chem. Soc.*, 1994, **116**, 3141.
38. R. S. Wylie and J. K. M. Sanders, *Tetrahedron*, 1995, **51**, 513.
39. C. A. Hunter and J. K. M. Sanders, *J. Am. Chem. Soc.*, 1990, **112**, 5525.
40. C. A. Hunter, *Chem. Soc. Rev.*, 1994, **23**, 101.

5

Metal Ion Directed Assembly of Complex Molecular Architectures and Nanostructures

PAUL N. W. BAXTER

Université Louis-Pasteur, Strasbourg, France

5.1 INTRODUCTION

Supramolecular chemistry, or chemistry beyond the molecule, has rapidly become a concept of major importance in 1990s chemical thought. The realization that a wide variety of complex biological and artificial structures self-assemble from a simpler set of molecular subunits in a single step under equilibrium thermodynamic conditions, layed the foundations for the controlled use of intermolecular interactions as a general building principle for the construction of supermolecules. This approach enables the rapid construction of chemical systems displaying a level of structural complexity which would be inaccessible by utilization of conventional sequential covalent synthetic methodology.

Consequently, a pathway has now been opened which allows for the unification of chemistry with biology by way of direct entry into the molecular nanostructural domain. For example, one particularly important implication of this is that molecular self-assembly must have played a crucial role in the chemical events which led to the development of the first life on earth. Many complex multifunctional biomolecules, such as microtubules, ribosomes and mitochondria, owe their very existence to self-assembly processes. In addition, higher hierarchical interactions expressed by these systems are also mediated and controlled by supramolecular forces. At some time and place on prebiotic earth, self-assembly and catalysis must have become inextricably linked, which would have made possible the creation of a pool of complex molecular structures, that is, molecular machinery. Eventually, further symbiotic cooperative association and functioning of such structures under nonequilibrium thermodynamic conditions may have led to primitive replicating life forms.

The field of supramolecular chemistry poses many intriguing futuristic possibilities, for example, in materials science, where direct access to nanostructural molecules may lead to many applications such as light harvesting and biomolecular transport/delivery systems, and also information transfer and transduction via molecular photonics, ionics, electronics and so on. Long-term goals include the possible construction of nanocomputers by way of the emergent fields of molecular electronics and ionics, or molecular mechanochemistry. The latter category may evolve from the combination of nanomolecular sized bearings, rods, gears and other molecular analogues of macroscopic mechanical components. A combination of molecular electronics and mechanochemistry may in turn lead to a type of molecular cybernetics in which electronic and mechanical subunits interact to give addressable moving mechanomolecules, both at the nanoscopic level and in the bulk phase.

Even though generally the field of supramolecular chemistry can still be considered as being in its infancy, many pioneering approaches have already been successfully developed, and a wide range of intermolecular forces investigated with a view to building complex molecular assemblies. Examples of such forces include attractive aromatic interactions,[1] hydrogen bonding,[2] interfacial phenomena in the form of liposomes, vesicles, bilayers and surface chemistry[3] and metal–ligand coordination. Of these, the last category forms the subject matter of this chapter.

The use of transition metal ions to direct the assembly of supramolecular architectures has become a popular and rapidly growing discipline. Molecules with a wide variety of topologies and shapes have been constructed in this way, with helicates and metal-containing rings (metallorings) attracting by far the greatest amount of activity. Many of these compounds have subsequently been demonstrated to have novel and unusual physicochemical properties. This chapter presents a compilation of the main structural types of such compounds reported in the literature, arranged hierarchically in order of increasing dimensionality. Thus, complexes with an overall one-dimensional shape (i.e., helicates and chains) are discussed first, followed by those consisting of two-dimensional metal–ligand arrangements (stepladders, grids and metallorings). Finally, cage complexes and a dendritic macromolecule which consist of three-dimensional metal–ligand architectures are described.

Although a fairly large number of transition metal assembled systems have been characterized, this chapter is concerned mainly with metal complexes whose overall size lies within the 1000–10 000 pm region, and the comparatively weaker coordinative metal–ligand-type interactions for their formation. They are herein referred to as nanostructures or nanoarchitectures. Several borderline cases (~700–1000 pm) have also been included in order to present the subject matter within a wider context.

Metal-containing aggregates and clusters constitute a conspicuously large family of inorganic architectures. An enormous body of work has been published since the 1960s in this area, especially concerning the preparation of, for example, chalconide, oxo, hydrido, phosphine and metal carbonyl clusters. These compounds, often of very high nuclearity, can be considered as truly nanostructural entities and as such have been the focus of intensive investigations, for example, as models for metal surface catalysis. Work in this field continues unabated, and has been reviewed elsewhere.[4] Examples from this large and mature area of chemistry have only been included in this account where particularly unusual topologies and/or supramolecular-type properties have been demonstrated.

Of the many literature examples of metallo-assembled nanostructures, it is sometimes unclear whether they have actually formed under equilibrium conditions and therefore if they can be considered as truly self-assembled. When this uncertainty exists but the resulting product has a structure and properties which are interesting and relevant to the theme of this review, then it has been included.

Finally, this review is not intended to be a complete collection of all known metallo-assembled nanoarchitectures, but rather to illustrate to the reader the range of structural motifs that have so far been synthetically accessible. The emphasis is therefore placed upon the most recent literature, that is, published since the mid-1980s.

5.2 HELICATES

One of the most rapidly expanding families of metallo-assembled architectures is that of the helicates, which consist of linear arrays of metal atoms, held together by bridging polytopic ligands, helically arranged about the central axis through the metal chain. Although evidence for the existence of such structures was reported as early as 1958, most work in this area has been accomplished since the mid-1980s. The majority of these structures consist of di- and trinuclear complexes which have already been described in several highlights and reviews.[5] For the sake of completeness, two reports of helicates of nuclearity *n*, where $n \geq 4$, and which can therefore be considered as lying well within the nanostructural domain, are described below. In both cases, the products are generated under very different physical conditions.

5.2.1 Tetranuclear and Pentanuclear Helicates

Initial pioneering investigations in this area concentrated on the preparation of di- and trinuclear complexes, structurally characterized by x-ray crystallography.[6] The work was further extended to include tetra- and pentanuclear arrays of metals which are held together by helically arranged polytopic oxomethylene-bridged bipyridines.[7] Thus, when ligands BP_4 and BP_5 (Figures 1 and 2) are treated with a slight stoichiometric excess of a solution of a Cu^I salt, the respective helicates, ds-H$[Cu_4(BP_4)_2]^{4+}$ and ds-H$[Cu_5(BP_5)_2]^{5+}$ (ds-H = double stranded helix) are formed quantitatively (Figure 3). Their total lengths are estimated to be about 2200 pm and 2700 pm, respectively, and can therefore be considered as examples of self-organized structurally well-defined nanoarchitectures. The mechanism of their formation almost certainly takes place with positive cooperativity and self-recognition.

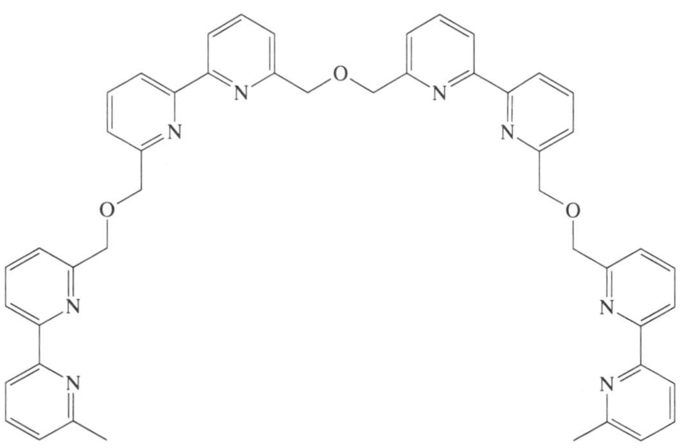

Figure 1 6,6′-Oxomethylene-bridged oligobipyridine, BP_4.

In conclusion, the metal-directed self-assembly of supramolecules such as these must involve information processing at a molecular level in which the organic ligands may be regarded as

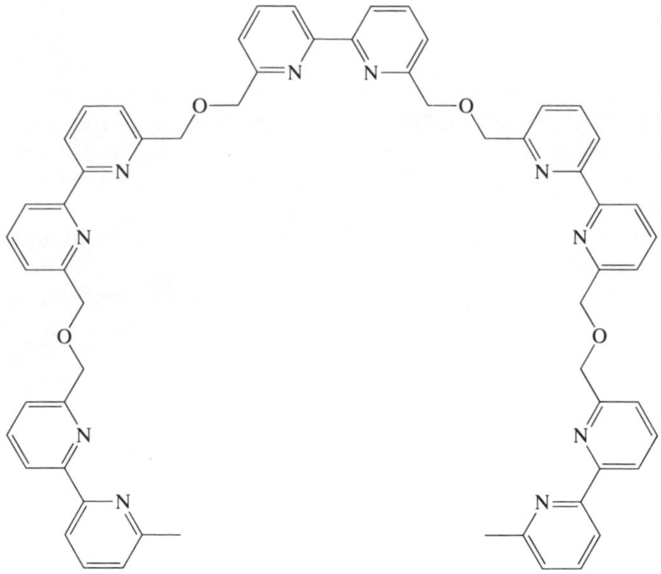

Figure 2 6,6′-Oxomethylene-bridged oligobipyridine, BP_5.

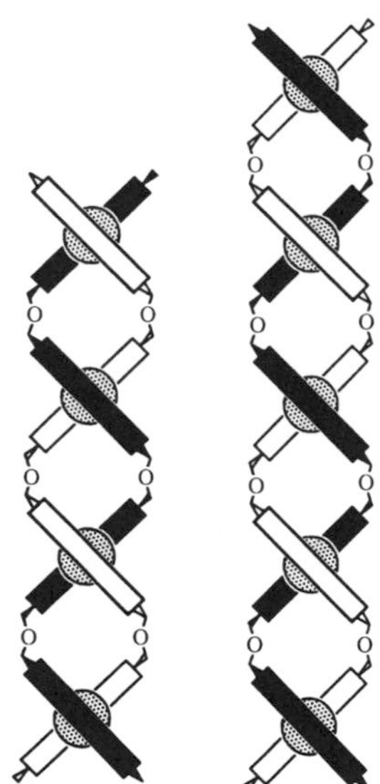

Figure 3 ds-H$[Cu_4(BP_4)_2]^{4+}$ and ds-H$[Cu_5(BP_5)_2]^{5+}$.

containing the steric program that is 'read' by the metal ions following the algorithm represented by their coordination geometry.

5.2.2 Pentadecanuclear and Higher Oligomers

The second report concerns polynuclear anions of the type $[M_x(CN)_{2x+1}]^-$ ($x \leq 27$), which are generated by laser ablation of solid $Zn(CN)_2$ and $Cd(CN)_2$.[8] The stoichiometric constitution of these species has been determined by mass spectrometry. This technique showed the most abundant ions to fall in the m/z region $15 \leq x \leq 23$ for $[Cd_x(CN)_{2x+1}]^-$ and similarly for $[Zn_x(CN)_{2x+1}]^-$. The strongest intensities occur at $x = 15$, 17 and 19 in both cases.

The structures of these species are unlikely to be fragments of the diamondoid crystal structures of $Zn(CN)_2$ and $Cd(CN)_2$, because such lattices are of low density with long linear M—CN—M connections, resulting in interpenetration and solvent encapsulation. Fragmentation patterns of this type are not consistent with the observed spectroscopic data. Density functional and force-field calculations on the lower homologues, $[Zn_2(CN)_5]^-$ and $[Zn_3(CN)_7]^-$, show that linear Zn—CN—Zn bridges, trigonal planar coordination at zinc and linear Zn—CN termini are the most stable basic structural features. Extension of these structural principles to that of the experimentally observed negative ions results in the calculation of helical structures as the most thermodynamically stable arrangements for such species. In these architectures, linear M—CN—M bridges, trigonal planar coordination around zinc and terminal Zn—CN units are preserved, and in addition, geometry optimizations reveal the presence of intrastrand secondary interactions between the zinc atoms and bridging carbon and nitrogen atoms of the contiguous strand. The length of such stabilizing intrastrand interactions was calculated as lying between 365 pm and 385 pm. Geometry optimizations also revealed the possibility of coulombic attractions between terminal $CN^{\delta-}$ ligands on the helix periphery and $Zn^{\delta+}$ atoms in adjacent strands of the helix.

The helical structures can therefore be regarded as supramolecular and resulting from non-bonded interactions. The mode of formation of the supramolecular entities discussed above is particularly unique, in that it occurs in the absence of solvent. Such a process may therefore more appropriately be described as 'supramolecular gas-phase chemistry'. The generation of these structures in solution awaits further experimentation.

5.3 CHAINS

Oligomeric, linear arrays of metals in the form of chains, held together by ligands in a non-helical arrangement, are also well represented in the literature. The nuclearity, n, of such structures most frequently appears to be $n = 3$ and 4, but a few reports of $n > 4$ chains do exist. One-dimensional compounds of this type have been most intensively studied with respect to platinum chemistry, because of the possible information such complexes may give concerning the mode of interaction between platinum anticancer therapeutics and the base pairs of DNA.

5.3.1 Tetranuclear Chains

In 1908, an important class of dark-blue platinum compounds known as 'platinum blues' were discovered. They were produced upon treatment of aqueous solutions of Pt^{II} with amides, and formulated as, for example, $Pt(MeCONH_2)_2(H_2O)$. This formulation remained unchallanged during the next 60 years, although the presence of Pt—Pt bonds was suspected.

The discovery of the anticancer properties of *cis*-$[PtCl_2(NH_3)_2]$ initiated a revival of interest in the above compounds, and a second set of platinum blue compounds containing thymine, uridine, uracil, 1-methyluracil and polyuracil were prepared in the 1970s and were also found to exhibit anticancer activity. These developments in turn led to increased efforts to obtain a more complete characterization of the platinum blues and other compounds included under this name. Although the reaction between the aquated products of *cis*-$[PtCl_2(NH_3)_2]$ and the above-named heterocycles produced oligomeric mixtures of varying degrees of hydrolytic stability, dark-blue crystals of $[Pt_4(NH_3)_8L_4](NO_3)_5$ (L = deprotonated α-pyridone) were obtained by reacting *cis*-$[Pt(NH_3)_2(H_2O)_2]^{2+}$ with α-pyridone. The tetranuclear platinum complex thus formed was characterized crystallographically and found to consist of a linear chain of platinum atoms.[9,10] The complex consists essentially of a zigzag chain of two linked Pt—Pt dimers. The central Pt—Pt bond (287.70(5) pm) is slightly longer than the outer two Pt—Pt bonds (277.45(4) pm), and is further strengthened by four hydrogen bonds between the coordinated amines and the oxygen atoms of the α-pyridonate ligands. This result suggested that a similar interaction may exist between the antitumour drug *cis*-$[PtCl_2(NH_3)_2]$ and DNA, RNA and their constituents. As the unit cell contains five nitrate ions, the tetranuclear chain must consist of a mixed-valence arrangement of three Pt^{II} and one Pt^{III}, the formal oxidation state of each platinum atom in the tetranuclear chain being 2.25. Magnetic and EPR measurements on the complex[10] showed it to be a simple Curie paramagnet with a magnetic moment of 1.8 BM and $S = 1/2$. The unpaired electron resides in an MO composed of d_{z^2} atomic orbitals directed along the platinum chain axis.

Since the pioneering structural study of the pyridone-bridged tetranuclear platinum blue complex described above, several other x-ray crystallographic investigations have subsequently been carried out on complexes of similar composition. For example, $[Pt_4(en)_4L_4](NO_3)_5{\cdot}H_2O$ has been isolated from a mixture of $[Pt_4(en)_4L_4](NO_3)_4$ and $[Pt_2(en)_2L_2(NO_2)(NO_3)](NO_3)_2$ in aqueous HNO_3,[11] and also $[Pt_4(NH_3)_8(1\text{-MeU})_4](NO_3)_5{\cdot}5H_2O$ (1-MeU = deprotonated 1-methyluracil) upon reaction of *cis*-$[Pt(NH_3)_2(1\text{-MeU})_2]{\cdot}4H_2O$ with *cis*-$[Pt(NH_3)_2(H_2O)_2]^{2+}$, also under acidic conditions.[12] Both cations are deep blue, with a formal oxidation state on each platinum of 2.25, and consist of a 'dimer of dimers' tetranuclear chain structure in which NH···O hydrogen bonds between the heterocycle oxygens and en NH_2 groups reinforce the central Pt—Pt bond (Figure 4). Spectroscopic measurements, coupled with theoretical calculations, demonstrate that the energy of the metal–metal charge-transfer (CT) transitions responsible for the intense colours described above become lower as the Pt—Pt distances increase.[13]

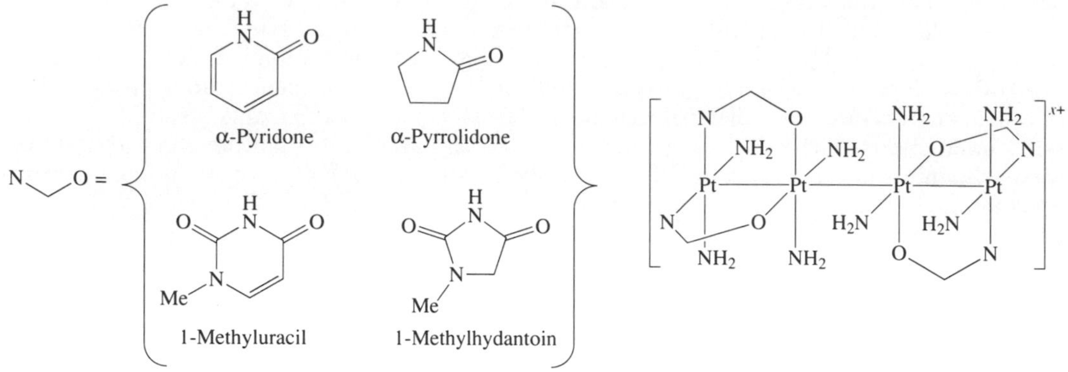

Figure 4 General structure of Pt_4 chains.

Crystal structures of tetranuclear platinum complexes in different oxidation states have also been determined. For example, $[Pt_4(NH_3)_8(hdt)_4](NO_3)_4{\cdot}H_2O$ (hdt = deprotonated 1-methylhydantoin) was formed upon successive treatment of *cis*-$[Pt(NH_3)_2Cl_2]$ with aqueous $AgNO_3$ and hdt.[14] Greenish yellow $[Pt_4(NH_3)_8L_4](NO_3)_4$ was obtained from the reaction between $[Pt_2(NH_3)_4(OH)_2](NO_3)_2$ and α-pyridone under aqueous alkaline conditions.[15]

Finally, $[Pt_4(NH_3)_8(L')_4](NO_3)_6{\cdot}2H_2O$ (L′ = deprotonated α-pyrrolidone) was isolated from an aqueous mixture of the hydrolysis product of *cis*-$[Pt(NH_3)_2Cl_2]$ and α-pyrrolidone after sequential neutralization with aqueous NaOH and acidification with aqueous HNO_3.[16,17] In the former two complexes, each platinum atom has a formal oxidation state of 2, and in the last complex, 2.5. Again, both complexes consist of a tetranuclear chain of platinum atoms, the central Pt—Pt bond of which is strengthened by additional intraligand hydrogen bonding. Interestingly, the latter higher oxidation state complex was found to oxidize water to molecular oxygen, even in the solid state (Equation (1)).[18] The above-described complexes exhibit a rich redox chemistry, the reversible oxidation-state relationships of which are summarized in Table 1 for the $[Pt_4(NH_3)_8(L')_4]^{x+}$ system.

$$2[Pt_4(NH_3)_8(L')_4]^{8+} + 2H_2O \rightleftharpoons 2[Pt_4(NH_3)_8(L')_4]^{6+} + O_2 + 4H^+ \qquad (1)$$

Other transition metals have also been demonstrated to form tetranuclear metal chains. For example, an x-ray structural characterization of a $(rhodium)_4$ chain has been described. The complex $[Rh_2(bridge)_4]X_2$ (bridge = 1,3-diisocyanopropane) has been the subject of intensive investigations, owing to the fact that irradiation of aqueous acidic solutions results in the cleavage of water to molecular hydrogen. Thus, when $[Rh_2(bridge)_4](BPh_4)_2$ is irradiated in aqueous HCl, the photoproduct is $[Rh_2(bridge)_4Cl_2]Cl_2{\cdot}8H_2O$. However, when the latter complex is dissolved in 12 M HCl, green crystals of the complex $[Rh_4(bridge)_8Cl]^{5+}$ are isolated (as the $CoCl_4^{2-}$ salt). The crystal structure shows the complex to be a linear chain of four rhodium atoms (Figure 5) with the formal oxidation states Rh^I, Rh^{II}, Rh^{II}, Rh^I, and this is believed to be the photoactive species involved in H_2 production upon irradiation.[19] The outer two Rh—Rh bonds, each connected by two bridge ligands, have lengths of 293.2(4) pm and 292.3(3) pm, respectively, whereas the inner Rh—Rh bond is much shorter, 277.5(4) pm. Terminal chloride ions connect the $[Rh_4(bridge)_8]^{6+}$ units to form an infinite chain of repeated units (Rh_4Cl) in the solid state. The complex has an intense absorption at 572 nm attributable to a $\sigma \rightarrow \sigma^*$ transition in the Rh_4^{6+} chain. The excited state actually appears to be formed by a transfer of charge from the outer to the inner Rh—Rh bonds and its electronic structure is completely delocalized.[20]

Ir_4, Cu_4 and V_4 chains have also been identified crystallographically. Treatment of $[Ir(\mu\text{-}C_7H_4NS_2)(cod)]_2$ ($C_7H_4NS_2$ = benzothiazole-2-thiolate) with carbon monoxide yields the deep-

Table 1 Redox state relationships for the $[Pt_4(NH_3)_8(L')_4]^{x+}$ system.

Complex	*Average Pt oxidation state*	*Colour (solid)*	*Magnetic state of solid*
$[Pt_4(NH_3)_8(L')_4]^{8+}$	3	Yellow	Diamagnetic
−e ⇅ +e			
$[Pt_4(NH_3)_8(L')_4]^{6+}$	2.5	Brown-red	Diamagnetic
−e ⇅ +e			
$[Pt_4(NH_3)_8(L')_4]^{5+}$	2.25	Blue	Paramagnetic
−e ⇅ +e			
$[Pt_4(NH_3)_8(L')_4]^{4+}$	2	Yellow	Not reported

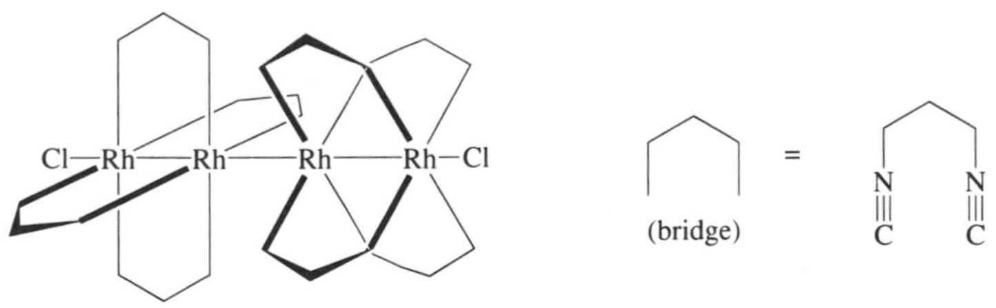

Figure 5 Pictorial representation of the Rh_4 chain unit.

purple dimeric complex $[Ir(\mu\text{-}C_7H_4NS_2)(CO)_2]_2$, which, upon addition of iodine in toluene solution, results in the rapid crystallization of dichroic, orange-brown $[Ir_4(\mu\text{-}C_7H_4NS_2)_4(CO)_8I_2]$ (Figure 6) in 85% yield. Its crystal structure showed it to be a tetranuclear Ir_4 chain composed of two binuclear halves, each bridged by two benzothiazole-2-thiolate ligands so that the outer iridium atom is N-coordinated and the inner iridium atoms are S-coordinated.[21] The outer Ir—Ir bond lengths are 273.1(2) pm and the inner Ir—Ir bond length is 282.8(2) pm and unsupported by any bridging ligand architecture. Carbonyl groups occupy positions *trans* to the $C_7H_4NS_2$ ligand, with iodine atoms capping the axial positions on the two terminal iridium atoms. Further treatment of this complex with iodine and irradiation with light causes cleavage of the central Ir—Ir bond to give the dimer $[Ir(\mu\text{-}C_7H_4NS_2)(CO)_2I]_2$. This demonstrated the previously unsuspected involvement of tetranuclear complexes in oxidative additions to binuclear complexes.

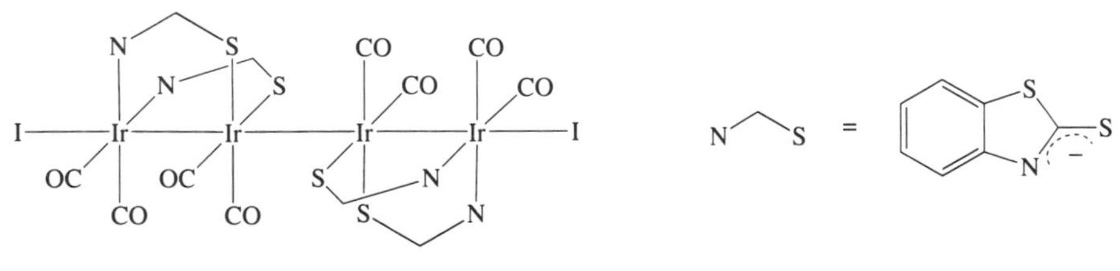

Figure 6 Pictorial representation of the tetranuclear $[Ir_4(\mu\text{-}C_7H_4NS_2)_4(CO)_8I_2]$ chain.

When the sodium salt of 6-fluoro-2-hydroxypyridine (fhp) is ground in the solid state with hydrated copper(II) nitrate in a 2:1 stoichiometric ratio and the mixture is extracted with dichloromethane, a deep-green solution is obtained, which upon addition of hexane results in the formation of yellow-brown crystals of a tetranuclear complex, $[Cu_2(fhp)_4]_2$. The crystal structure of this complex showed it to consist of a dimer of binuclear units linked by two μ-O atoms from the fhp ligands.[22] The Cu—Cu distance within each dimer unit is 254.25(14) pm and the interdimer

Cu—Cu distance is 311.6(2) pm. The fluoropyridine ligands are oriented such that all the halogens are located at the ends of the chain. Presumably, the absence of potential axial ligands results in dimerization, and thus the formation of a higher nuclearity assembly. Recrystallization of the above complex from methanol–dichloromethane causes polymerization to $[Cu_4(OMe)_4(fhp)_4]_n$, which contains the tetranuclear unit as a repeating motif.

Two unusual tetranuclear oxo-bridged $(vanadium)_4$ chains have been prepared and structurally characterized.[23] Treatment of [VO(salen)] (salen = deprotonated *N,N'*-ethylenebis(salicylideneimine)) with $HBF_4{\cdot}Et_2O$ in a 1:2 stoichiometric ratio unexpectedly yielded the tetranuclear $[V(salen)OV(salen)OV(salen)OV(salen)](BF_4)_2$. Similarly, the reaction between [VO(salen)] and hydroiodic acid resulted in the formation of $[V(salen)OV(salen)]_2(I_3)_2$, in which the central V—V distance is bridged by two salen oxygen atoms. The former complex can be considered as a $[V^{IV}{=}O{\rightarrow}V^{IV}{-}O{-}V^{IV}{\rightarrow}O{=}V^{IV}]^{2+}$ chain in which two $V^{IV}{=}O$ units donate to a central $V^{IV}{-}O{-}V^{IV}$ core.

5.3.2 Pentanuclear and Higher Nuclearity Oligomeric Chains

Metal chains with nuclearity $n > 4$ are also known, but are less common than the tetranuclear analogues. The rhodium complex $[Rh_4(bridge)_8]^{6+}$ discussed previously has been found to produce a rich variety of $Rh_m{}^{n+}$ oligomers when partially reduced by chromium(II) in dilute sulfuric acid. For example, species with the nuclearity $Rh_6{}^{8+}$, $Rh_8{}^{10+}$ and $Rh_{12}{}^{16+}$ were observed.[24] In a more in-depth study, reaction of the axially hindered $[Rh_2(TMB)_4]^{2+}$ (TMB = 2,5-dimethyl-2,5-diisocyanohexane) with $[Rh_4(bridge)_8]^{6+}$ in a 2:1 ratio resulted in the formation of the octanuclear $[Rh_2(TMB)_4Rh_4(bridge)_6Rh_2(TMB)_4]^{10+}$, in which both ends of the complex are capped by $[Rh_2(TMB)_4]^{2+}$ units, thereby disfavouring further oligomerization. When solutions of $[Rh_2(TMB)_4]^{2+}$ (10^{-6} M) were allowed to react with a large excess of $[Rh_4(bridge)_8]^{6+}$ in 1M H_2SO_4 solution, a hexanuclear species $[Rh_2(TMB)_4Rh_4(bridge)_8]^{8+}$ and its $Rh_{12}{}^{16+}$ dimer were detected. The $Rh_6{}^{8+}$, $Rh_8{}^{10+}$ and $Rh_{12}{}^{16+}$ oligomers were identified on the basis of UV and redox titration experiments, in which the energy of the intense $\sigma \rightarrow \sigma^*$ transition was found to decrease systematically with increasing rhodium chain length.[25]

An unusual example of a mixed-metal pentanuclear Pt—Pt—Ag—Pt—Pt chain has been described.[26] Thus, deep yellow crystals of $[(NH_3)_4Pt_2(1\text{-MeU})_2Ag(1\text{-MeU})_2Pt_2(NH_3)_4]$-$(NO_3)_5{\cdot}4H_2O$ formed upon slow evaporation of an aqueous solution of the dimer *cis*-$[Pt_2(NH_3)_4(1\text{-MeU})_2](NO_3)_2{\cdot}H_2O$ and $AgNO_3$. Crystallographic characterization of these crystals showed the structure of the complex to consist of two 1-MeU-bridged Pt—Pt dimers, joined together by a central silver(I) atom which is coordinated to four oxygens of the 1-MeU ligands (two from each Pt—Pt dimer) in a square-planar arrangement. Each 1-MeU ligand therefore simultaneously binds to three metal centres, two platinum and one silver (Figure 7). The Pt—Pt distances (Pt—Pt intradimer = 294.9(2) pm, interdimer = 324.6(2) pm) are longer than in the tetrameric $[Pt_4(NH_3)_8L_4](NO_3)_4$ (L = deprotonated α-pyridone; Pt—Pt intra = 288 pm and inter = 313 pm, respectively),[15] which may be an electronic consequence of additional binding of Ag^+ to the 1-MeU ligands. From the same solution a $Pt_4{}^{5+}$ complex was isolated (blue-green in aqueous solution), which was assumed to be the decomposition product of the $(Pt_4Ag)^{5+}$ complex described above.

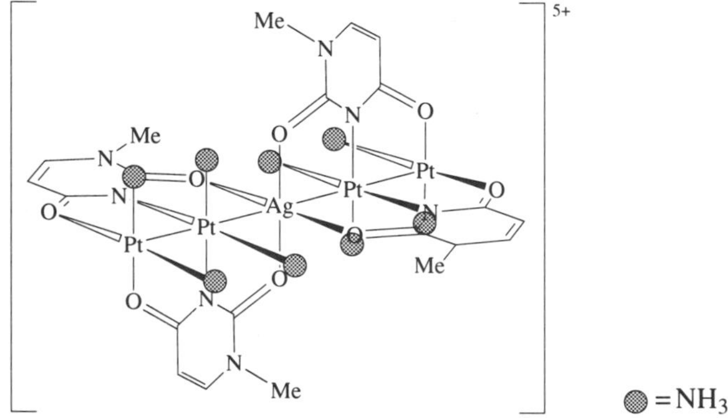

Figure 7 Pictorial representation of $[(NH_3)_4Pt_2(1\text{-MeU})_2Ag(1\text{-MeU})_2Pt_2(NH_3)_4]^{5+}$.

In the tetranuclear platinum blue complexes so far structurally characterized, all contain heterocyclic amidate bridging ligands, whose oxygens are coordinated to the platinum atoms of the central Pt—Pt bond. Other interactions between terminal platinum atoms are prevented by steric hindrances between the heterocyclic rings. However, the originally prepared platinum blues were synthesized with less sterically demanding chain amides and would therefore be expected to yield higher nuclearity chains.

One such remarkable example of a relatively high nuclearity chain has been isolated from a hot aqueous mixture of the hydrolysis product of *cis*-$[Pt(NH_3)_2Cl_2]$ and acetamide. The octanuclear mixed-valence platinum–acetamide complex $[Pt_8(NH_3)_{16}(C_2H_4NO)_8](NO_3)_{10}\cdot 4H_2O$ formed as red-purple plates upon concentration of the reaction mixture at 5 °C. The x-ray crystal structure showed the complex to be octanuclear, consisting of four dimeric units, each doubly bridged with acetamide ligands in a head-to-head manner (Figure 8).[27] The central interdimer interaction (Pt-4—Pt-4′ = 293.4(1) pm) is further strengthened by four hydrogen bonds between the acetamide oxygens and amine hydrogens in a comparable way to the central Pt—Pt interaction in the tetranuclear platinum complexes discussed above. Each interdimer interaction of Pt-2—Pt-3 = 290.0(1) pm is also reinforced by two chemically nonequivalent hydrogen bonds. The 10 nitrate anions found per cation lead to a formal oxidation state on each platinum of 2.25 (Pt^{II}_6, Pt^{III}_2). The complex is diamagnetic, with the Pt^{III} states probably delocalized over several platinum atoms as in the previously discussed platinum blues.

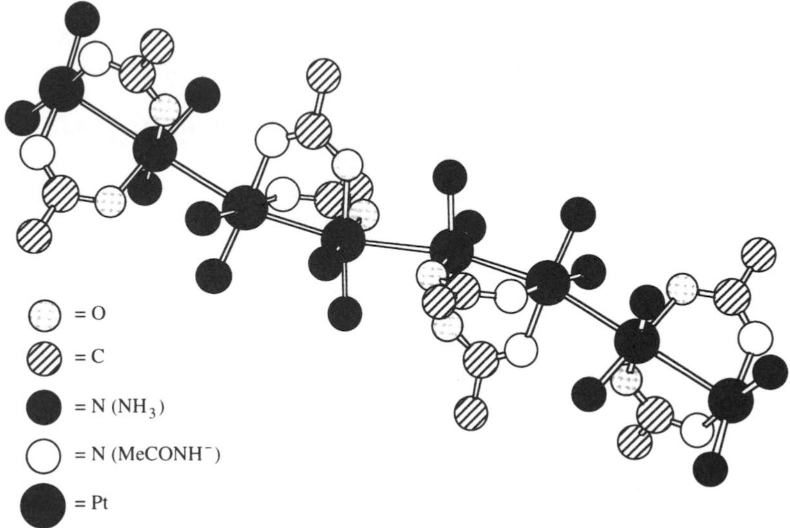

Figure 8 X-ray structure (ball-and-stick representation) of the octanuclear $[Pt_8(NH_3)_{16}(C_2H_4NO)_8]^{10+}$ chain.

Finally, a series of very unusual metal-chain complexes have been detected in the gas phase. Laser ablation of solid CuCN or AgCN yields positive and negative ions of composition $[M_n(CN)_{n+1}]^-$ (M = Cu, n = 1–5; M = Ag, n = 1–4) and $[M_n(CN)_{n-1}]^+$ (M = Cu, n = 1–6; M = Ag, n = 1–4).[28] These ions were detected by Fourier transform ion cyclotron resonance mass spectrometry (FTICR-MS). The study was undertaken in order to obtain more detailed information about the little-known chemistry of metal cyanide complexes in the gas phase. Information concerning the structure of the copper-containing ions was obtained indirectly by the use of density functional theory. Calculations showed that of the possible structural identities of these species, that is, chain, cyclic or metal-bridged arrangements, linear CN-bridged chain structures were the lowest in energy. It was concluded that the suite of copper-containing ions generated adopted a linear 'spear'-shaped geometry (Figure 9). The largest ion observed, $[Cu_5(CN)_6]^-$, would be almost 3000 pm long. Again, a future challenge is the generation and stabilization of such structures in solution.

5.4 STEPLADDERS

Compared with their polymeric analogues, two-dimensional oligonuclear metal complexes are less well represented in the literature. One fundamentally interesting topology is that of a molecular

Cu—C≡N—Cu—N≡C—Cu—C≡N—Cu $[Cu_4(CN)_3]^+$

N≡C—Cu—C≡N—Cu—C≡N—Cu—N≡C—Cu—C≡N $[Cu_4(CN)_5]^-$

Cu—C≡N—Cu—N≡C—Cu—C≡N—Cu—N≡C—Cu $[Cu_5(CN)_4]^+$

N≡C—Cu—C≡N—Cu—N≡C—Cu—C≡N—Cu—N≡C—Cu—C≡N $[Cu_5(CN)_6]^-$

Figure 9 Oligonuclear copper cyanide cationic and anionic 'spears'.

ladder, which could in principle be constructed by arranging linear chains of metals parallel to one another and connecting them together by bridging units. A possible way in which molecules of this type may be assembled is to allow two linear wrack-like ligands (ladder sides) with *n* metal binding sites to react with 2*n* metal ions in the presence of *n* bridging ligands containing 'back-to-back' coordinating sites (ladder steps). The shape of the resulting architecture could then be described as a stepladder. Molecules of this type may be expected to exhibit unusual and interesting physico-chemical properties which will be discussed in greater detail in Section 5.5 for the case of square and rectangular grids.

Thus, when nitromethane was added to a mixture of quaterpyridine (**1a**), tetraphenylbipyrimidine (**2**) and $[Cu(MeCN)_4](X)$ ($X = ClO_4^-$, PF_6^-) in a 1:1:2 stoichiometric ratio, all components dissolved to give a dark-brown solution. Proton NMR and fast atom bombardment (FAB) and electrospray ionization (ESI) mass spectrometric data of the product complex supported the formulation $[Cu_4(\mathbf{1a})_2(\mathbf{2})_2](X)_4$ (Equation (2)). Although not characterized by x-ray crystallography, the tetranuclear product (**3**) is almost certainly a metallomacrocyclic ring of four copper(I) ions and four ligands or, in other words, a 'two-rung' stepladder ((**3**) in Figure 10).[29]

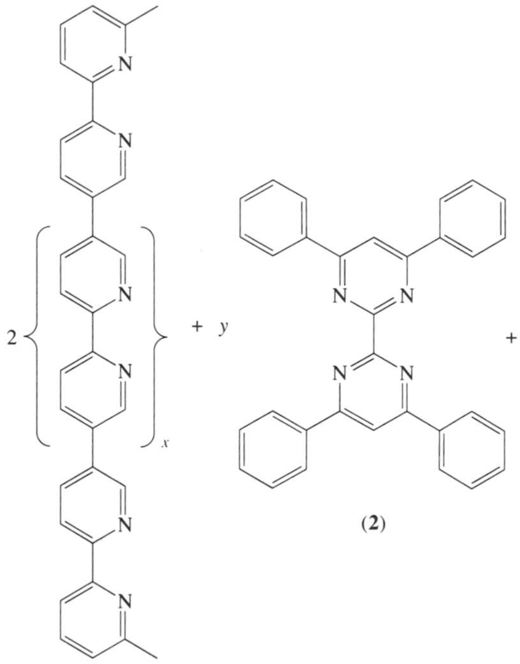

$$2\,(\mathbf{1}) + y\,(\mathbf{2}) + z\,[Cu(MeCN)_4][PF_6] \longrightarrow [Cu_z(L)_2(\mathbf{2})_y][PF_6]_z + n(\,MeCN) \quad (2)$$

(**1**)
(**a**) $x=0$
(**b**) $x=1$

(**2**)

(**3**) L=(**1a**), $y=2$, $z=4$, $n=16$
(**4**) L=(**1b**), $y=3$, $z=6$, $n=24$

Further extension of this building principle to see if true stepladders, that is, with $n > 2$ steps, could be prepared necessitated the use of the linear tritopic ligand sexipyridine (**1b**) in place of (**1a**). Thus, from a mixture of (**1b**), (**2**) and $[Cu(MeCN)_4](X)$ in a 1:1.5:3 stoichiometric ratio in

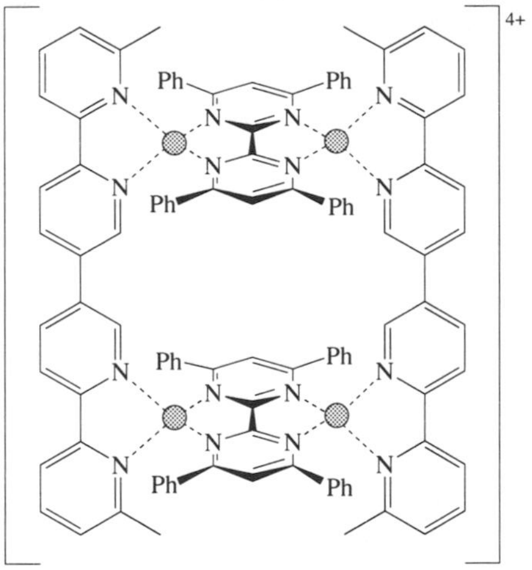

(3) 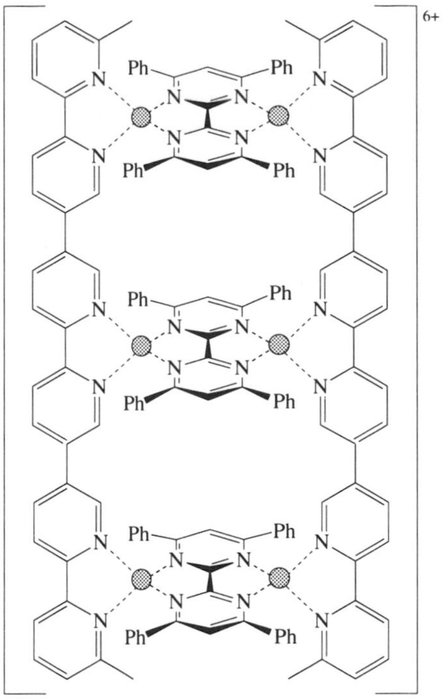 (4)

Figure 10 'Two- and three-rung' stepladders.

nitromethane as solvent was isolated a complex **(4)** formulated as $[Cu_6(\mathbf{1b})_2(\mathbf{2})_3](X)_6$ (Equation (2)). The structural formulation was based on ^1H NMR, FAB and ESI mass spectrometric data, and is consistent with a 'three-rung' stepladder-type architecture ((**4**) in Figure 10). The ^1H NMR spectrum was particularly informative in that the *ortho*- and *meta*-phenyl ring protons of the two outer and one inner ligands **(2)** (outer and inner 'rungs') were clearly distinguishable as a pair of doublets and a pair of triplets, respectively, each in the expected 2:1 ratio. This latter molecule represents an impressive example of a programmed molecular system, in which 11 'informed' components spontaneously and correctly assemble to form a single architecture of precisely defined composition.

When **(1a)**, **(2)**, **(1b)** and $[Cu(MeCN)_4](PF_6)$ were mixed in a 1:2.5:1:5 stoichiometric ratio in nitromethane, only two products were found to exist in the reaction mixture after 72 h. Proton NMR and ESI mass spectrometric measurements confirmed the identity of both products as being $[Cu_4(\mathbf{1a})_2(\mathbf{2})_2](PF_6)_4$ and $[Cu_6(\mathbf{1b})_2(\mathbf{2})_3](PF_6)_4$. In other words, a mixture of 19 particles correctly associate, without error, to give two structurally complex and precisely defined nanoarchitectures. This is an unusual example of an 'instructed mixture', in which correct recognition, growth and termination of the self-assembly process result in the formation only of products composed of two different ligands. The process may therefore be described as 'nonself-recognition' (cf., an example from helicates in which only products containing the same ligands form from the 'instructed mixture', i.e., by self-recognition).[30]

5.5 GRIDS

The almost explosive progress in the field of information technology has resulted in an ever-increasing demand for the fabrication of advanced materials which in turn display novel physicochemical properties. A direct consequence of increasing structural complexity is that of increasing functional complexity. Self-assembly processes therefore offer for the future the possibility of fabricating very complex structures which may one day function as truly intelligent materials. Highly organized architectures of this type will result from the spontaneous and correct sequential association of many different and highly chemically informed components, the process of which may more accurately be described as 'informational assembly' rather than 'self-assembly'. A process of this type is more comparable to molecular biological events.

As a first step towards this long-term goal, the controlled ordering of metal ions into two-dimensional arrays by organic components is of great interest in that it offers intriguing prospects for the future development of electroionic devices for information storage.[31] One way to achieve this would be to use rigid, linear, polytopic ligands containing bidentate coordinating sites which may be able to assemble into chessboard-type arrays upon mixing with metal ions of tetrahedral coordination geometry. A ligand with n coordinating sites would thus be able to form an $n \times n$ ion grid ($[n \times n]$G) composed of n^2 metal ions and $2n$ ligand components (Equation (3)). The structure of a prototype tetranuclear [2 × 2]G system composed of four copper(I) ions and four ditopic 3,6-bis(2-pyridyl)pyridazine (bppz) ligands had been reported[32] and served as a suitable model upon which extension to an $n > 2$ grid may be possible.

6 + 9 ⟶ (3)

5.5.1 Square Grids

Accordingly, combination of the tritopic ligand 6,6′-bis[2-(6-methylpyridyl)]-3,3′-bipyridazine (Me_2-bpbpz) and silver(I) trifluoromethanesulfonate in a 1:1.5 stoichiometric ratio in nitromethane solution resulted in the spontaneous assembly of a 3 × 3 square grid consisting of nine silver(I) ions and six ligand components (Equation (4)).

6 (Me_2-bpbpz) + 9 $AgCF_3SO_3$ ⟶ [3×3]G (4)

The [3 × 3]G structure has been identified on the basis of ^1H, ^{13}C and ^{109}Ag NMR spectroscopy and x-ray crystallography.[33] The NMR data were particularly informative in this respect. The ^1H NMR gave two sets of signals in a 2:1 ratio, corresponding to protons on the outer and inner ligands, respectively. Silver-109 NMR showed three resonances in the ratio of 4:4:1, corresponding to silver ions located at the four corners, the four edge mid-positions and the single central site in the grid. Yellow rhombohedral crystals of the Ag_9 complex could be grown from nitromethane solution. An x-ray structural analysis of these crystals showed that the structure of the cation was indeed that of a 3 × 3 grid of nine Ag^I ions (Figure 11). The grid is distorted into a diamond-type shape, with an angle of about 72(3)° between the mean planes through the ligands and an average Ag—Ag distance of approximately 372(3) pm. All silver ions are in a distorted tetrahedral environment, and collectively together sit on a slightly warped 'saddle-back'-type surface (Figure 12). This is mainly due to the curved nature of each ligand, which, in turn, arises from the fact that pyridazine rings are contracted about the N=N bond and therefore not regular hexagons. The six ligands are divided into two sets of three, one of which is above and the other below the mean plane formed by the silver ions. The average distance between the mean planes through adjacent ligands is 374(8) pm and therefore just outside that expected for van der Waals contact (~340 pm).

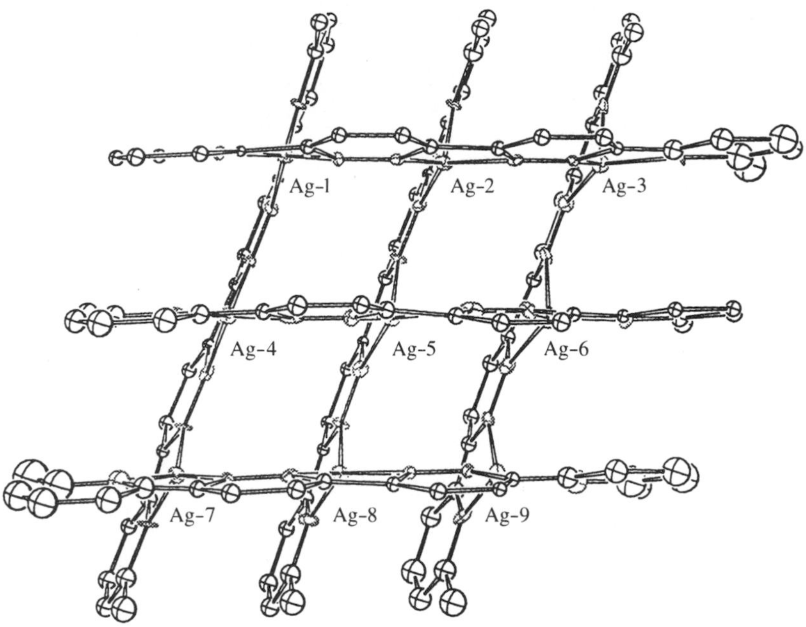

Figure 11 ORTEP view of cation, perpendicular to the mean plane through the silver atoms.

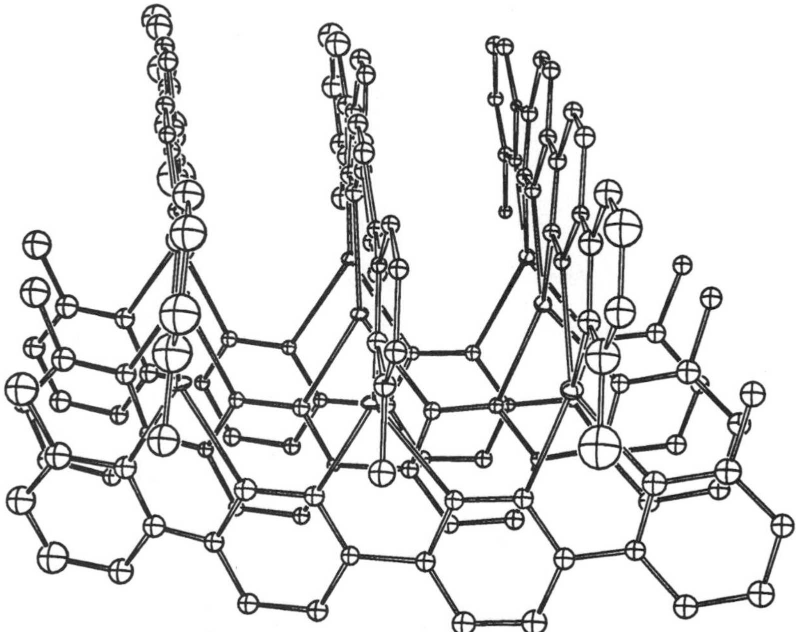

Figure 12 ORTEP side-on view of cation showing ligand curvature.

This molecule is remarkable in that its formation involves the spontaneous and correct association of 15 particles (nine metal ions and six ligand components), and as such fulfils the three basic levels of operation of a programmed supramolecular system, that is, recognition, orientation and termination.

5.5.2 Rectangular Grids

A further interesting development is the possibility of generating rectangular grids which are constructed from ligands with different numbers ($m \neq n$) of binding sites.

Thus, a mixture of 3,6-bis(6-methyl-2-pyridyl)pyridazine (Me_2-bppz), Me_2-bpbpz and silver trifluoromethanesulfonate in a 1.5:1:3 stoichiometric ratio in nitromethane solution resulted in the formation of a rectangular grid composed of three Me_2-bppz, two Me_2-bpbpz and six silver(I) ions (Equation (5)). This formulation is substantiated in particular by ^{109}Ag and ^{1}H NMR spectroscopy, the latter of which shows three sets of signals in a 2:2:1 ratio corresponding to Me_2-bppz protons in two different environments (2:1) and bpbpz ligands in a single environment. The ^{109}Ag NMR displays two resonances in a 2:1 ratio which are attributable to the four corner and two middle silver ions, respectively.[34]

$$2\,(Me_2\text{-bpbpz}) + 3\,(Me_2\text{-bppz}) + 6\,AgCF_3SO_3 \longrightarrow [2\times3]G \qquad (5)$$

The successful preparation and characterization of these [3 × 3]G and [2 × 3]G systems opens the way to a whole new class of multinuclear [$n \times n$]G and [$m \times n$]G grids. Spatially organized and structurally well-defined architectures of this type bear a close analogy to grids based on quantum dots which are of special interest in the field of microelectronics. Unlike microelectronics systems which are fabricated from larger structures, grids of the above type spontaneously assemble from a set of much smaller components. It is conceivable that information may be stored in such systems in the form of patterns generated by electron holes or occupancies either on the metal ions (as differing redox states) or on sites situated within the connecting ligands (as localized occupied or vacant orbitals). In this way, several levels of information may be simultaneously stored over the whole molecule. Thus the molecule may be addressed in the form of light or electrically and may be regarded as a nanoscale 'digital supramolecular chip'. Additionally, structures such as these may be further deposited as two-dimensional layers on surfaces or sandwiched into a surrounding matrix to form interconnected two- or three-dimensional nanopatterns.

5.6 RINGS

Cyclic architectures which are formed via the process of metal-directed assembly and which incorporate metal atoms as integral structural components represent one of the largest classes of nonpolymeric inorganically assembled topologies. In terms of reported examples, the class of metal-containing rings lies only second to that of inorganic cluster compounds and aggregates. Large rings which can be considered as nanostructural and which may incorporate four or more metals are less numerous than their smaller congeners, but still well represented in the literature. Interestingly, most reported metal-assembled rings with nuclearity $n \geq 5$ are constructed from even numbers of metal atoms, as reflected in their high symmetry. Also, although many examples of lower nuclearity rings which include guests have been reported, a large number of higher nuclearity metallocycles ($n > 4$) do not appear to include guests (as judged by x-ray structural data), and therefore presumably do not require the presence of a guest species to template their formation. Instead, many of these complexes exhibit some representative type of noncovalent weak intramolecular interaction such as attractive aromatic forces or hydrogen bonding which help to direct collectively the formation of the cyclic architecture, and prevent polymerization. In cases where both secondary intramolecular interactions and ion templating are absent, ring closure probably results from a subtle interplay between coordination geometry preferences of the metal and stereoelectronic requirements of the ligands. The fact that many of the higher nuclearity rings result from serendipitous discoveries clearly suggests that the forces which control ring formation are still poorly understood from the point of view of design.

Presented in the following account is a collection of several impressive examples of metallo-assembled rings. Their formation is mediated via a wide range of metal–ligand bonding modes, such as a metal–ligand coordination, cluster type and organometallic interactions. The metallo-rings discussed are presented in order of increasing nuclearity.

5.6.1 Tetranuclear Rings

Two particularly remarkable examples of tetranuclear metalloring complexes have been reported.

In the first example, blue crystals isolated from an aqueous solution of β-cyclodextrin, $Cu(OH)_2$ and LiOH in a 1:2:8 stoichiometric ratio were examined by x-ray crystallography.[35] They were found to be composed of a tetranuclear complex formulated as $Li_{11}[Cu_4(\beta\text{-}CDH_{11.5})_2]\cdot xH_2O$ ($\beta\text{-}CDH_{21}$ = β-cyclodextrin). The complex consisted of two cyclodextrin tori connected together at their 02–03 edges by four copper ions, each of which is oxygen bound in a distorted square-planar geometry. The resulting complex therefore appears like a sandwich-type cylinder or double torus (Figures 13 and 14). Seven of the 11 lithium ions present in the molecule are coordinated within the double torus of the complex, and in addition nine hydrogen bonds of the type $O^-\cdots HO$ further strengthen the interactions between the two cyclodextrin rings. The central hydrophobic cavity is filled with seven water molecules, arranged in the form of a heptagon, and coordinated to the lithium ions. Further, less-ordered water molecules also fill the void. Upon heating to 70 °C, two-thirds of the water content of the crystals is lost, corresponding to the water content of the cavity.

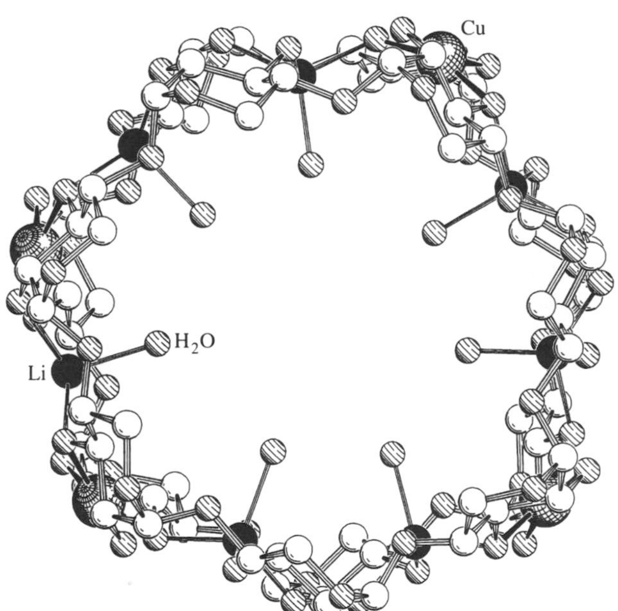

Figure 13 SCHAKAL view along central axis through double torus (all C-6 and O-6 atoms of the ligand anhydroglucose units are omitted for clarity) (reproduced by permission of VCH from *Angew. Chem.*, 1993, **105**, 895).

The second example is that of a very unusual cyclic tetrameric complex (Figure 15), composed of four divalent $1,2\text{-}C_2B_{10}H_{10}$ cages, linked by four mercury atoms. In addition, a chloride ion is located in the centre of the cavity at a distance of 294.4(2) pm from each mercury atom (i.e., significantly shorter than van der Waals contact).[36] The cyclic complex was obtained in 75% yield

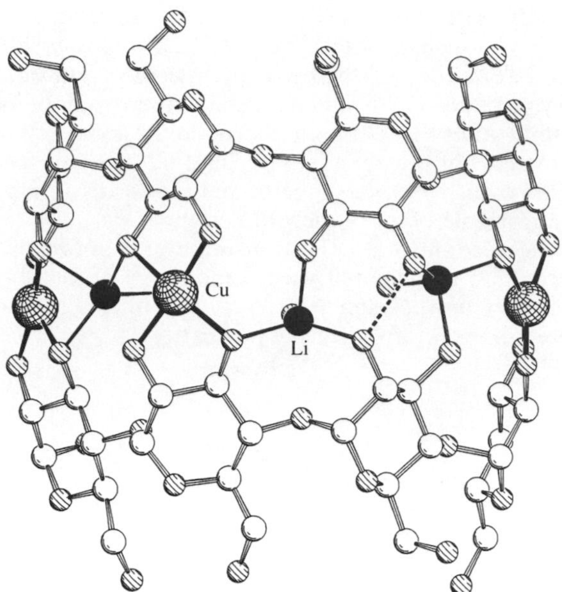

Figure 14 SCHAKAL side-view of double torus. For clarity, the back half of the structure has been omitted and Li—O bonds are drawn only for the four nearest neighbours (reproduced by permission of VCH from *Angew. Chem.*, 1993, **105**, 895).

upon initial double deprotonation of the icosahedral carborane 1,2-$C_2B_{10}H_{12}$ with 2 equiv. of *n*-butyllithium, and then reaction with mercury(II) chloride. The product was obtained in this way as the lithium chloride salt with the composition Li[$Cl\subset(HgC_2B_{10}H_{10})_4$]. The cyclization to form a tetrameric complex is presumably the result of a chloride-induced template effect and, as such, serves as an unusual example in which a guest anion, instead of a cation, acts as a template for its organic/inorganic host (see also anion-templated formation of spherical polyoxovanadates in Section 5.7.8). The preferred C—Hg—C angle in diorganomercurials is 180°, whereas the above complex exhibits C—Hg—C angles of 162.0(3)°. This therefore represents the largest deviation from 180° reported to date. Hence the formation of a tetrameric ring which is strained, and therefore energetically less favourable, further supports the intermediacy of a chloride ion template effect.

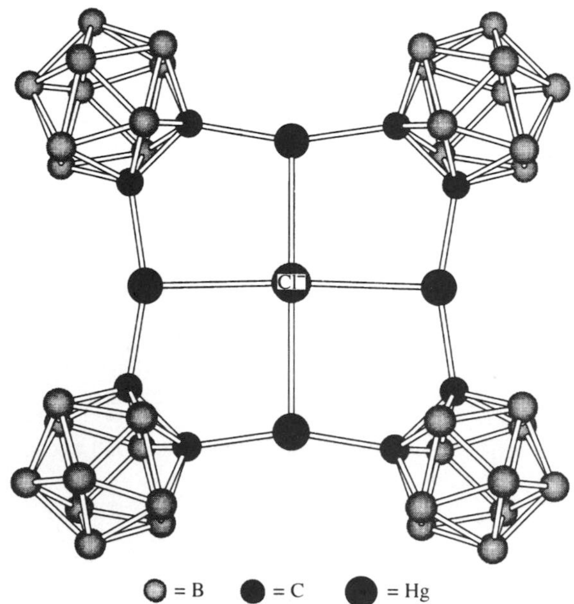

Figure 15 X-ray structure (ball-and-stick representation) of [12]mercuracarborand-4 showing the included chloride anion guest.

Further work on this system has demonstrated it to have an interesting anion-binding chemistry. Thus the coordinated Cl^- anion can be removed with silver salts without destruction of the host framework. The cyclic tetrameric ring is able to tightly bind I^- and, perhaps more interestingly, it is able to bind simultaneously two *closo*-$B_{10}H_{10}^{2-}$ anionic guests.[37] The stability of the latter supramolecular aggregate must originate from the collective, cooperative and complementary three centre–two electron B—H···Hg bonding interactions of the four mercury atoms to four equatorial BH units in each coordinated anion (Hg···H distances range between 218(12) pm and 310(1) pm). This cyclic tetrameric anion binder has been named [12]mercuracarborand-4 after the topologically similar [12]crown-4 oxoethylene macrocyclic cation binder. The preparation and characterization of this complex opens up a new and unusual approach to the field of anion binding chemistry.

5.6.2 Hexanuclear Rings

Some examples of hexanuclear metallorings have been reported. An unusual organometallic cyclic hexamer of the formula $[Al_6(\mu\text{-}Cl)_6(\mu\text{-DMB})_6]$ (DMB = 2,3-dimethylbutadiene) was prepared upon cocondensation of the reactive intermediate [AlCl] with DMB in toluene solution in a 1:1 ratio at 77 K and subsequent warming to 273 K. An x-ray structure of the crystals which formed upon standing showed the hexamer (Figure 16) to consist of a hexagonal arrangement of six aluminum atoms bridged externally by six DMB molecules and internally by six chlorine atoms.[38]

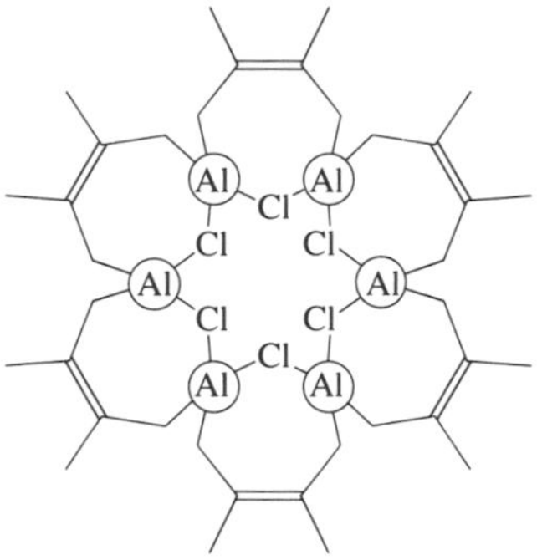

Figure 16 Pictorial representation of the $[Al_6(\mu\text{-}Cl)_6(\mu\text{-DMB})_6]$ hexameric ring.

An interesting example of a manganese(II) hexanuclear ring of formula $[Mn(hfac)_2(NITPh)]_6$ has been crystallographically characterized.[39] Dark-green crystals of this complex were isolated from the reaction between $Mn(hfac)_2{\cdot}2H_2O$ and the stable nitronyl–nitroxide radical NITPh. The cyclic hexamer (Figure 17) consists of discrete clusters of six radicals and six $Mn(hfac)_2$ units in which each metal is in a distorted octahedral environment consisting of two chelating hfac ligands and two oxygens from separate bridging NITPh groups. The phenyl rings of the NITPh bridges lie inside the 36-membered ring and within van der Waals contact with hfac ligands. The summed effect of these π-contacts may account for the formation and the stability of the ring. Magnetic and EPR measurements demonstrate that the hexamer consists of 12 spins which are antiferromagnetically coupled to give an $S = 12$ ground state. However, the molecule still behaves as a paramagnet down to 5 K.

Cyclic metal-assembled structures containing nucleobases are also known. An example was obtained from the reaction between 9-methylguanine and *cis*-$[(PMe_3)_2Pt(\mu\text{-}OH)]_2(NO_3)_2$ in water in a 2:1 ratio.[40] A crystal structure of the colourless crystals which formed from this reaction showed the product to consist of a hexameric ring of platinum(II) atoms in which six *cis*-$(PMe_3)_2Pt$ units are symmetrically bridged by the guanine ligands through their N-1 and N-7 atoms (Figure 18). The formula of the complex is *cis*-$[(PMe_3)_2Pt(9\text{-MeGu})]_6(NO_3)_6{\cdot}18H_2O$ (9-MeGu = N-1 deprotonated 9-methylguanine). An intramolecular hydrogen bond exists between the exocyclic NH_2 hydrogens and the oxygen atoms of two adjacent nucleobases (O-6···N-2 = 296 pm). Again, the cumulative effect of these intramolecular hydrogen bonding interactions may be important in stabilizing the ring architecture.

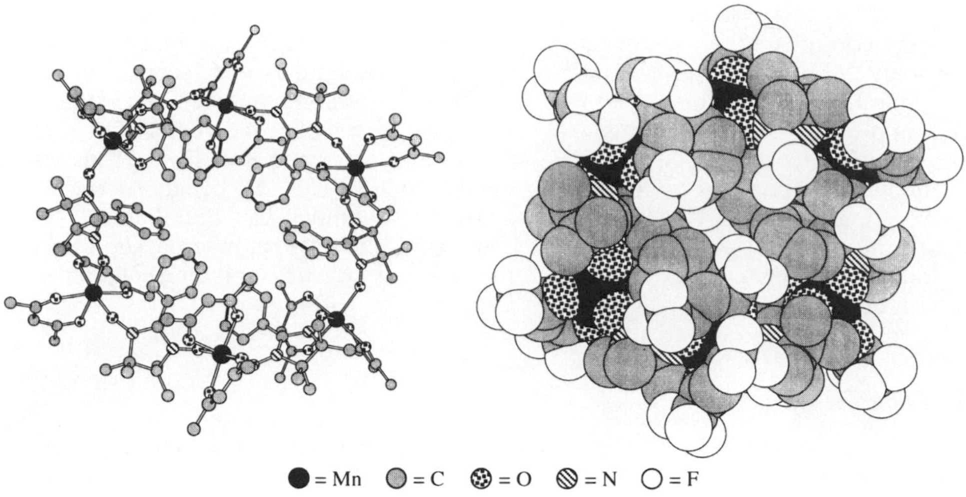

Figure 17 X-ray structure of $[Mn(hfac)_2(NITPh)]_6$ showing ball-and-stick (left) and space-filling (right) representations.

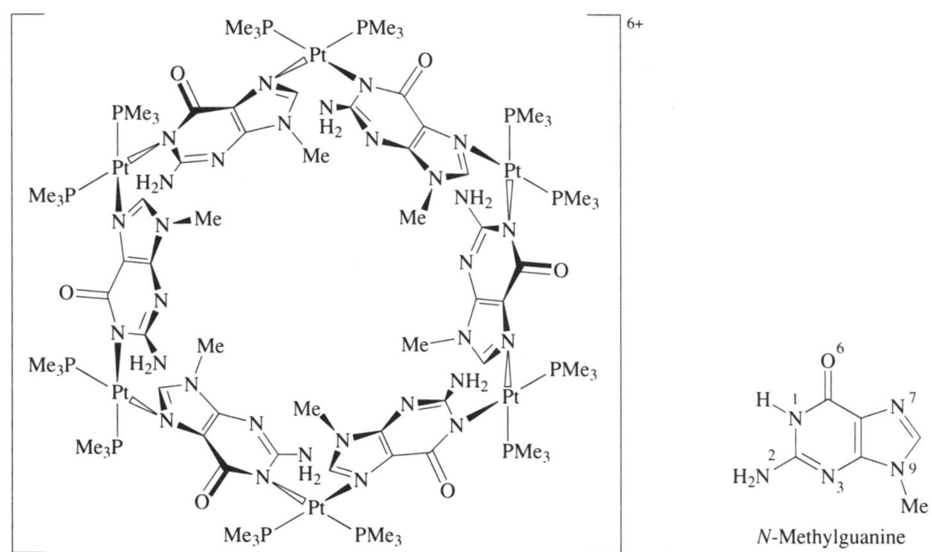

Figure 18 Pictorial representation of $[cis\text{-}[(PMe_3)_2Pt(9\text{-}MeGu)]_6]^{6+}$ cyclic hexamer.

5.6.3 Octanuclear Rings

A highly symmetrical octanuclear palladium(I) ring has been reported.[41] The metalloring of formula $[(Pd(\mu\text{-}SC_6F_5)(\mu\text{-}dppm)Pd)(\mu\text{-}SC_6F_5)]_4$ is formed in 63% yield upon reaction of $[Pd_2(dba)_3]\cdot CHCl_3$ and $[Pd(SC_6F_5)_2(dppm)]$ in a 1:2 stoichiometric ratio in diethyl ether solution. The product complex has been characterized by x-ray crystallography and is best described as being constructed from four metal–metal-bonded 'Pd(μ-SC_6F_5)(μ-dppm)Pd' units (Pd—Pd = 258.1 pm) which are bridged through four additional SC_6F_5 groups to form a 12-membered puckered cycle (Figure 19), and with four dppm ligands sitting in turn above and below the ring. Weak intramolecular Pd···F interactions (Pd—F closest distances 295–305 pm) may also possibly help to stabilize this cyclic architecture.

5.6.4 Decanuclear Rings

Perhaps one of the best known examples of large metallorings is that of the 'ferric wheel'.[42] The 'wheel' prepared from the reaction between the monochloroacetate analogue of basic iron acetate,

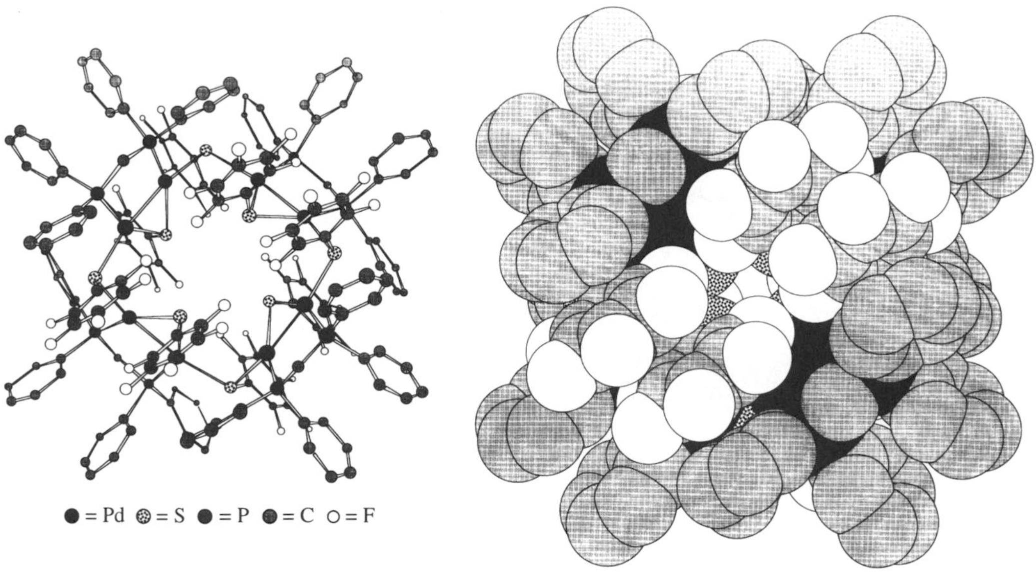

Figure 19 X-ray structure of $[(Pd(\mu\text{-}SC_6F_5)(\mu\text{-dppm})Pd)(\mu\text{-}SC_6F_5)]_4$ ring showing ball-and-stick (left) and space-filling representation (right).

$[Fe_3O(O_2CCH_2Cl)_6(H_2O)_3](NO_3)$, and $Fe(NO_3)_3{\cdot}9H_2O$ in methanol solution was isolated as yellow crystals. An x-ray structural analysis showed the complex to be a giant cyclic iron(III) decamer of the formula $[Fe(OMe)_2(O_2CCH_2Cl)]_{10}$, which contained bridging chloroacetate and methoxide ligands (Figure 20). The 'wheel' is about 120 pm in diameter with a small and unoccupied hole of 200–300 pm across. The decamers are also stacked in the crystal lattice in such a way as to produce an infinite channel which passes through the holes. Interestingly, the highest amounts of product were obtained under relatively dilute conditions in a modest yield of 17%. A detailed analysis of the crystal structure revealed that the cyclic architecture resulted from a balance between destructive (convex) curvature across each set of bridging methoxide ligands and constructive (concave) curvature across the iron centres (Figure 21). The fact that the metal atoms share octahedral faces allows the concave curvature across each iron site to be additive and dictate the formation of a ring.

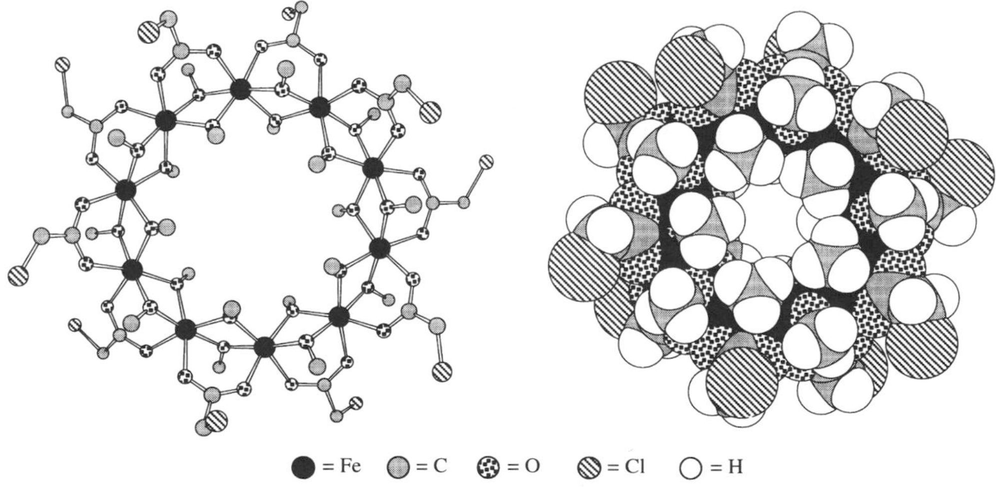

Figure 20 X-ray structure of $[Fe(OMe)_2(O_2CCH_2Cl)]_{10}$ showing ball-and-stick (left) and space-filling representation (right).

A decameric metallomacrocycle of formula $[Y(OC_2H_4OMe)_3]_{10}$ has also been reported to be formed in nearly quantitative yield, either by direct reaction between yttrium metal or alcoholysis

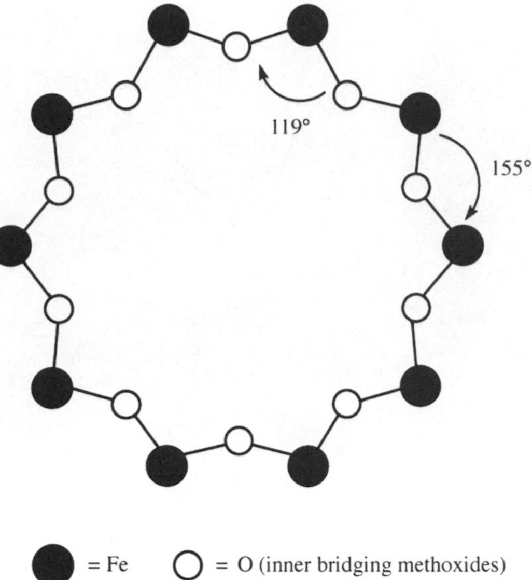

Figure 21 Pictorial representation of ring geometry perpendicular to plane through iron atoms (bridging methoxide oxygens shown are those projected on to plane of iron atoms).

of $Y_5O(OPr^i)_{13}$ and 2-methoxyethanol. The crystal structure of the complex showed the coordination polyhedron for each yttrium atom to be heptacoordinate (Figure 22). Each metal was coordinated by one terminal monodentate alkoxo, two chelating 2-methoxyethoxo and two bridging alkoxo groups from the two chelating groups on an adjacent yttrium. This decamer represents one of the highest molecular weight oligomers which has been structurally characterized for homoleptic metal alkoxides and has an internal diameter of about 700 pm, as indicated by contacts between carbon atoms.[43]

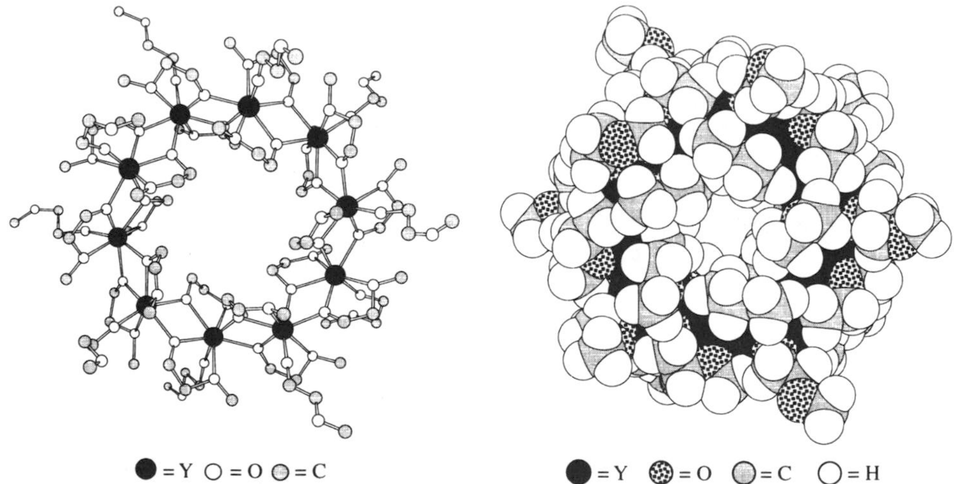

Figure 22 View of crystal structure perpendicular to plane through yttrium atoms (ball-and-stick (left) and space-filling (right) representations).

5.6.5 Dodecanuclear Rings

A particularly large, conformationally flexible 24-membered metalloring of formula $[AgSR]_{12}$ (R = cyclohexyl) was discovered upon reinvestigation of the crystal structure of (cyclohexanethiolato)silver(I). The cyclic molecules are lapped over each other in a chain-like arrangement in which some secondary intermolecular Ag···S bonding contacts occur.[44]

A beautiful example of a dodecanuclear metal-assembled iron ring from the area of chalcogen cluster chemistry has been described.[45] The dodecanuclear product formed in 90% yield as dark-green crystals when a mixture of $FeCl_3$ and PPh_3 in dichloroethane was contacted with a solution of PhSe-TMS. An x-ray crystal structure of the product $[Fe_{12}(SePh)_{24}]$ showed it to consist of a large Fe_{12} ring in which each iron atom is tetrahedrally coordinated by four bridging μ_2-SePh ligands (Figure 23). A space-filling model of the complex shows that the iron ring is deeply buried within the surrounding phenyl groups of the SePh ligands, which are arranged alternately above and below the metal ring plane. The complex is about 1200 pm in diameter. A larger iron–sulfur ring cluster, $[Na_2\subset Fe_{18}S_{30}]^{8-}$, has also been structurally characterized, and was found to include two nonsolvated Na^+ ions (Figure 24).[46]

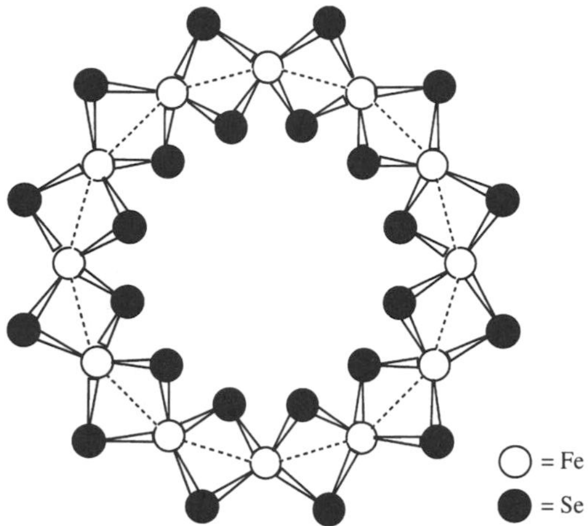

Figure 23 Pictorial representation of cyclic $Fe_{12}Se_{24}$ core.

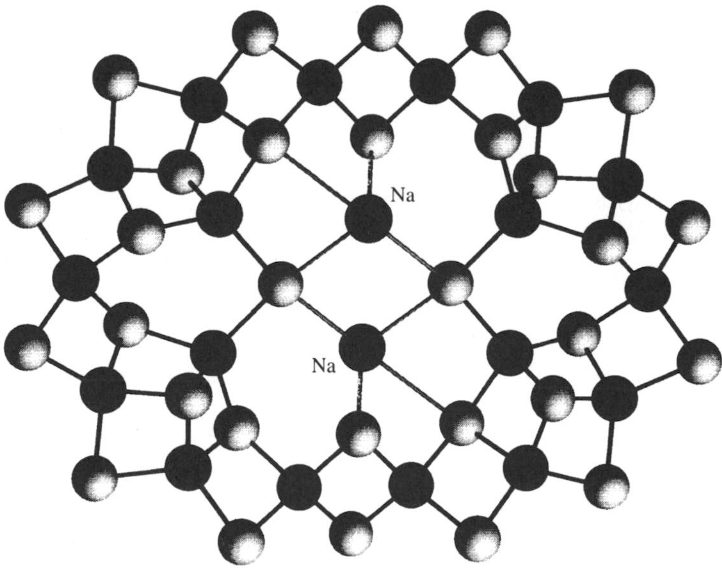

Figure 24 X-ray structure (ball-and-stick model) of cyclic $[Fe_{18}S_{30}]^{10-}$ ring showing the two included sodium cation guests. Black spheres = Fe; pale grey spheres = S.

5.6.6 Hexadecanuclear Rings

A spectacular complex was isolated from the reaction between lead(II) nitrate and γ-cyclodextrin upon addition of aqueous sodium hydroxide. X-ray structural analysis of the colourless crystals which formed upon heating the reaction mixture showed the product to be a hexadecanuclear lead complex of formula $[Pb_{16}(\gamma\text{-CDH}_{-16})_2]\cdot xH_2O$ ($x \approx 20$).[47] As in the tetranuclear copper analogue discussed above, the complex consists of two cyclodextrin tori joined edge-to-edge by a ring of metal ions (in this case, 16 lead atoms) in such a way as to produce a sandwich-type cylinder. Both cyclodextrin tori are 16-fold deprotonated with each alkoxide oxygen bridging two tetra-coordinate lead atoms. Viewed down through the centre of the cavity, the lead atoms sit in the form of an eight-point star with eight outside and eight on the inside (Figure 25). The outer lead atoms, which form part of two five-membered chelate rings, have the same coordination environment as the copper atoms of the β-cyclodextrin complex described above. The shapes of the cyclodextrin tori are that of truncated cones widened at the base by the large Pb_{16} ring (Figure 26). The cavity size of this ring is about 800 pm in diameter and intramolecular hydrogen bonds appear to be of less importance than in the β-cyclodextrin–copper complex. The lead complex crystallizes from solutions containing excess γ-cyclodextrin, which implies that its formation involves a cooperative mechanism. Also, the complex can form in solutions at pH 7, demonstrating that such ligands are extensively deprotonated at neutrality and are thus able to form stable complexes at this pH. This fact may be of particular biochemical interest. Complexes of this type, which contain large cavities, may also be expected to have an extensive host–guest inclusion chemistry.

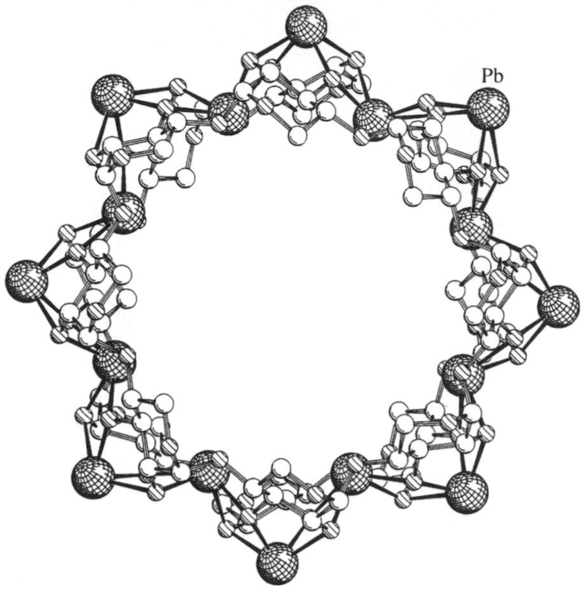

Figure 25 SCHAKAL view of double torus perpendicular to plane through lead atoms (all C-6 and O-6 atoms of the anhydroglucose units are omitted for clarity) (reproduced by permission of VCH from *Angew. Chem.*, 1994, **106**, 1925).

5.7 CAGES

Just as organic macrobicycles and macrotricycles represent a logical extension of two-dimensional macrocyclic structures into the third dimension, so does the progression from metal-assembled macrocycles to the metal-mediated formation of cage-type architectures. Three-dimensional constructs of this type which incorporate metals as architectural units may be expected to exhibit new properties such as host–guest binding interactions of much greater strength and specificity than their two-dimensional forerunners. The enhanced complexity of such structures may result in the development of unusual electrooptical, redox and catalytic properties. Possibilities also exist for controlled guest uptake/transport and altered guest physical and chemical reactivity within the cavity. An additional attraction is, of course, the fact that many of these architectures may be self-assembled in one step from a mixture of simpler 'informed' components. In this way, some very large and complex nanostructures can be produced in a single step from the correct stoichiometric mixture of ligands and metals, some of which, as will be shown, possess molecular masses similar to those of small proteins.

Again, the examples described in this part of the review are presented in order of increasing nuclearity. In Section 5.7.8 an account of the comparatively high-nuclearity polyoxovanadate cages is included. Although polyoxometallate chemistry has perhaps the longest scientific history and is therefore one of the most thoroughly investigated topics, it was only discovered in the early 1990s that spherical polyoxovanadates can exhibit a previously unsuspected host–guest chemistry.

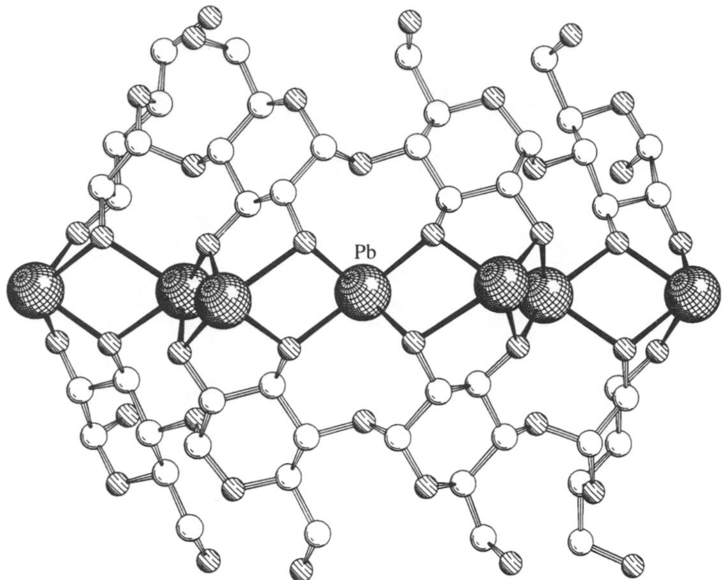

Figure 26 SCHAKAL view of front part of double torus (reproduced by permission of VCH from *Angew. Chem.*, 1994, **106**, 1925).

Compounds of this class almost certainly form under equilibrium conditions, and, in addition, the mechanism relies upon an anion template effect. Finally, a single report of the metal ion-assisted assembly of a dendritic macromolecule is described in Section 5.8.

5.7.1 Dinuclear Cages

A remarkable compound formulated as $(PPh_4)_2(NH_2Me_2)(NH_4)[Pd_2(S_7)_4]$ has been isolated from the reaction between an acetonitrile–dimethylformamide solution of palladium acetylacetonate and an ethanolic solution of polysulfide in the presence of the cations PPh_4^+ and $NH_2Me_2^+$.[48] An x-ray structure shows the anion to consist of an unusual cage architecture in which two palladium(II) ions are bridged by four terminally ligating S_7^{2-} chains such that each palladium is in a square-planar coordination environment with a Pd—Pd distance of 630.0(1) pm (Figures 27 and 28). The cavity size of the anion is therefore relatively large and encapsulates an ammonium cation guest, with a Pd—N distance of 315.0(1) pm. The product can therefore be regarded as a 'cation–anion complex' $[NH_4 \subset Pd_2(S_7)_4]^{3-}$, and its formation presumably results from a cation-templated mechanism. Investigations into varying the cage size and cation-exchange studies were not reported.

The reaction between a gold(I) solution (generated *in situ* by the reduction of $K[AuCl_4]$ in methanol) and the ligand 2,7-bis(diphenylphosphino)-1,8-naphthyridine (L) in an equimolar ratio yielded a yellow solution, from which crystals of a compound formulated as $[Au_2KL_3](ClO_4)_3$ formed upon addition of $LiClO_4$. An x-ray structural analysis of the product complex showed the cation to consist of two gold atoms bridged by three L molecules coordinated via the terminal diphenylphosphino groups, such that each gold(I) was in a trigonal planar coordination geometry.[49] A potassium cation was found to be encapsulated within the cavity via coordination to the three naphthyridine moieties. The coordination environment around the encapsulated six-coordinate K^+ ion was distorted trigonal-prismatic with K—N distances in the range 247.7(10)–254.9(10) pm. These distances, which are significantly shorter than usual, indicate tight binding of the cation, a factor which must contribute strongly to the stability of this complex. The complex can therefore be considered as a type of metallocryptand. A trinuclear silver complex of similar composition to that described above has also been prepared and structurally characterized. Both complexes exhibit room-temperature emission upon photoexcitation at 300–400 nm.

In a systematic study directed towards understanding the structural and chemical features which influence magnetic exchange coupling through multiatom bridging ligands, two series of dinuclear metallomacrobicyclic complexes were prepared, which included first-row transition metal ions as

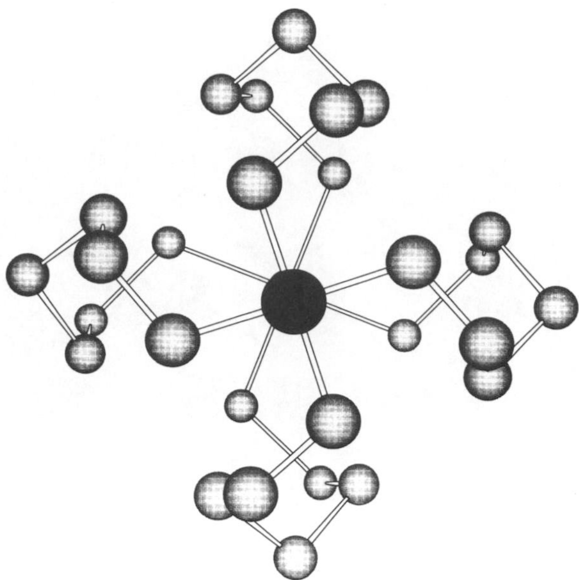

Figure 27 X-ray structure (ball-and-stick representation) of $[Pd_2(S_7)_4]^{4-}$ cage viewed along axis through palladium atoms (NH_4^+ guest omitted for clarity).

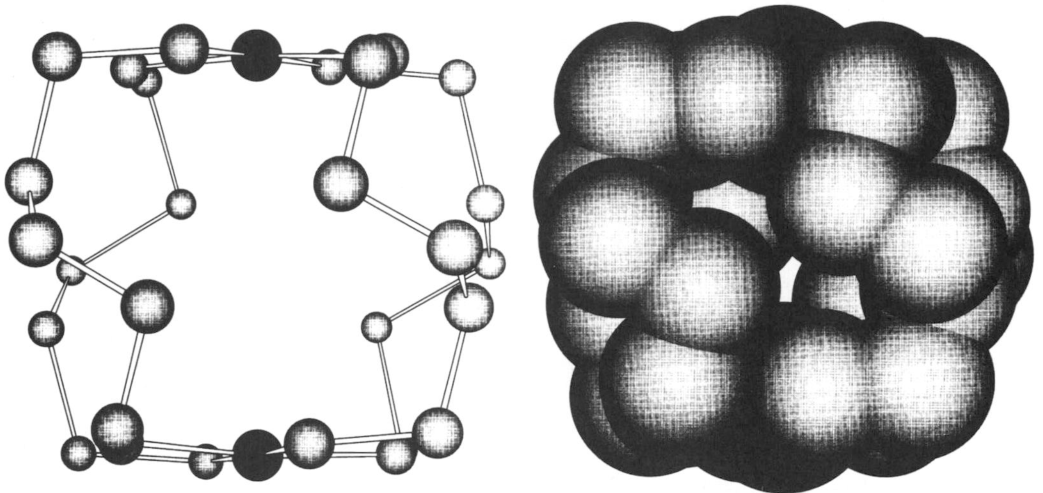

Figure 28 X-ray structure of $[Pd_2(S_7)_4]^{4-}$ cage, viewed from side (ball-and-stick (left) and space-filling (right) representations) (NH_4^+ guest omitted for clarity).

guests. The two series of complexes were found to have the compositions $[LFe^{III}(\mu\text{-}(DMG)_3M^{II})\text{-}Fe^{III}L](X)_2$ (L = 1,4,7-trimethyl-1,4,7-triazacyclononane; DMG = doubly deprotonated dimethylglyoxime; $M^{II} = Mn^{2+}$, Fe^{2+}, Co^{2+}, Ni^{2+}, Cu^{2+}, Zn^{2+}; X = ClO_4^-, PF_6^-)[50] and $[LMn^{III}(\mu\text{-}(dmg)_3M^{II})Mn^{III}L]^{2+}$ ($M^{II} = Zn^{2+}$, Cu^{2+}, Ni^{2+}, Mn^{2+}).[51] They were synthesized by treating a methanolic solution of L with sequential additions of $Fe(MeCO_2)_2$ or manganese(III) acetate, $M^{II}(MeCO_2)_2$ and dimethylglyoxime in the presence of triethylamine in a 1:1:0.5:1.5 stoichiometric ratio. Subsequent refluxing and addition of the appropriate counterion X resulted in precipitation or crystallization of the cage-like complexes. Although the iron salt was introduced as Fe^{II}, the resulting products contained exclusively Fe^{III} centres, which presumably resulted from air oxidation upon isolation. X-ray crystal structures of the complexes with Cu^{2+} and Ni^{2+} as the central metal ions were also obtained (Figure 29). Both complexes consisted of a linear arrangement of three transition metal ions, in which the outer two Fe^{III} atoms cap the central $(M^{II}(dmg)_3)^{4-}$ bridging unit, each via coordination to three deprotonated glyoxime oxygens. Both Fe^{III} centres are in trigonally distorted octahedral coordination environments. The central M^{II} ion is coordinated by six nitrogen atoms, two from each glyoxime ligand, such that its overall coordination polyhedron is that of a distorted trigonal biprism. In the copper-containing complex the Fe···Cu separation is 358 pm and in the nickel complex a slightly shorter Fe···Ni separation of 349 pm was found.

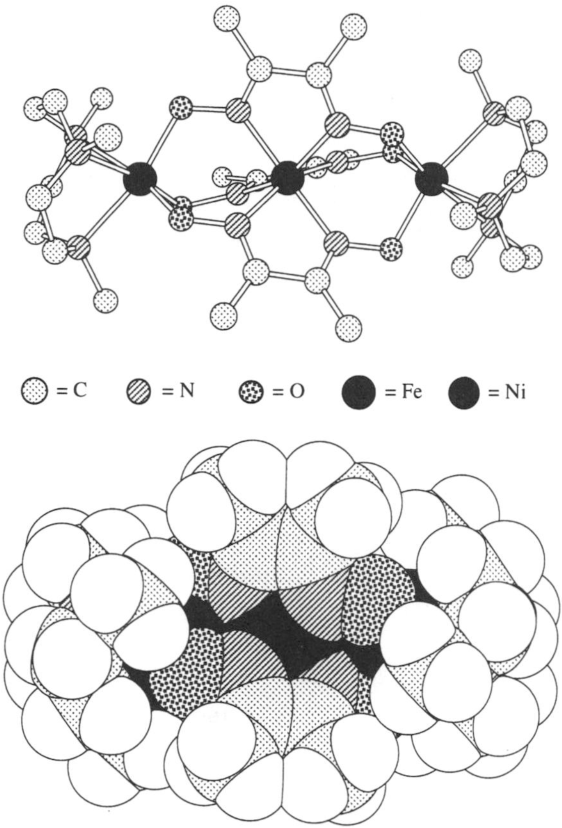

Figure 29 X-ray structure of $[LFe^{III}\{\mu\text{-}(dmg)_3Ni^{II}\}Fe^{III}L]^{2+}$ cage showing ball-and-stick (top) and space-filling (bottom) representations.

The magnetic behaviour of the Ni^{2+}-containing complex was studied in detail and it was shown that the complex contained two high-spin Fe^{III} sites, antiferromagnetically coupled to a high-spin Ni^{II}, and confirmed the essentially σ nature of this interaction.[52] In all cases examined the magnetic exchange coupling occurs between the two Fe^{III} centres, which are separated by $\sim$700 pm. This is a large, but not limiting, distance for such an interaction. Not surprisingly, some complexes also exhibit a rich electrochemistry, and may allow for the isolation of species such as $Fe^{III}Ni^{III}$-Fe^{III} and $Fe^{II}Ni^{II}Fe^{II}$.

The series of trinuclear Mn^{III} complexes described above were regarded as being of particular interest in relation to the area of photosynthetic water oxidation by redox-active manganese complexes, a field which has been the focus of intensive research. The zinc-containing complex was characterized by x-ray crystallography, which showed it to be structurally similar to the Fe^{III} analogues. Upon treatment with NO^+ in acetonitrile, the metal core of the complex is oxidized to $Mn^{IV}Zn^{II}Mn^{IV}$. Both complexes are electrochemically interconvertible without any apparent change of structure in solution.

Interestingly, a metal templating effect appears to be unnecessary for the formation of the above complexes. The successful preparation of the dichromium(III) analogue of formula $[LCr^{III}$-$(Hdmg)_2(dmg)Cr^{III}L](ClO_4)_2 \cdot MeOH$ has been described. The complex was characterized by x-ray crystallography, and shown to contain a central diprotonated cavity unoccupied by a metal ion.[53]

The synthesis and host–guest inclusion properties of organic cages composed of bowl-shaped molecules covalently linked together has been the subject of thorough and active investigations since the early 1980s. Spectacular chemical achievements in this field have been demonstrated, such as the taming of cyclobutadiene[54] and drug encapsulation[55] using a class of organic container molecules known as carceplexes.

In 1994, a self-assembly approach to molecular containers was achieved by joining together two bowl-shaped calix[4]arenes using metal coordination instead of organic spacers.[56] Thus, a dinuclear biscalix[4]arene (Figure 30) was isolated in 50% yield from the reaction between a chloroform solution of a calix[4]arene functionalized on its upper rim with two acetylacetonate ligands and an aqueous solution of $Cu(NO_3)_2$. The product complex was characterized by mass spectral and vapour-phase osmometric (VPO) measurements. It was shown to bind diamines into its cavity in much the same way as a similarly constructed macrocyclic system.[57] For example, the logarithm of the stability constant (log *K*) for the binding of ethylenediamine and dabco (1,4-diazabicyclo[2.2.2]octane) in chloroform solution was found to be 3.1 and 1.5, respectively. The weaker association between dabco and the bis(calix[4]arene) was ascribed to a greater mismatch in fit between the rigid dabco guest and the copper sites of the host. This work is being further extended to catalytic systems.

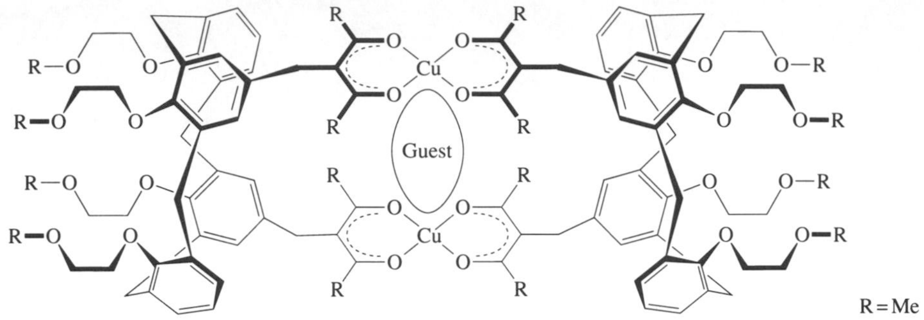

R = Me

Figure 30 Dicopper bis(calix[4]arene) complex.

5.7.2 Trinuclear Cages

A particularly interesting and unusual example of a trinuclear cage-like complex has been reported in which the cationic complex assembles in high yield only in the presence of specific guest molecules. Thus, when the tridentate ligand 1,3,5-tris(4-pyridylmethylene)benzene (L), $[Pd(en)(NO_3)_2]$ and sodium 4-methoxyphenylacetate (Nampa) were combined in water in a 2:3:5 molar ratio, the complex $[(Nampa)\subset(Pd(en))_3L_2]^{6+}$ formed, which was isolated in 94% yield as the perchlorate salt (Equation (6)).[58] The formulation of a 1:1 association between the cation and Nampa was substantiated by ESI-MS and ^1H NMR spectroscopy. In the case of the ESI-MS results, a molecular ion corresponding to a 1:1 complex (minus NO_3^-) was observed. The association constant between the cation and Nampa was found to be $\geq$104 by ^1H NMR titration experiments. Also, significant upfield shifts of the aromatic and methoxy protons of Nampa were observed, while the $\Delta\delta$ values of the $CH_2CO_2^-$ group were little affected. This was interpreted in terms of the hydrophobic region of Nampa lying within the cavity of the cage and the hydrophilic CO_2^- group remaining outside. In the absence of a guest, the host complex formed in highest yields of 60% and appeared to be in equilibrium with oligomers.

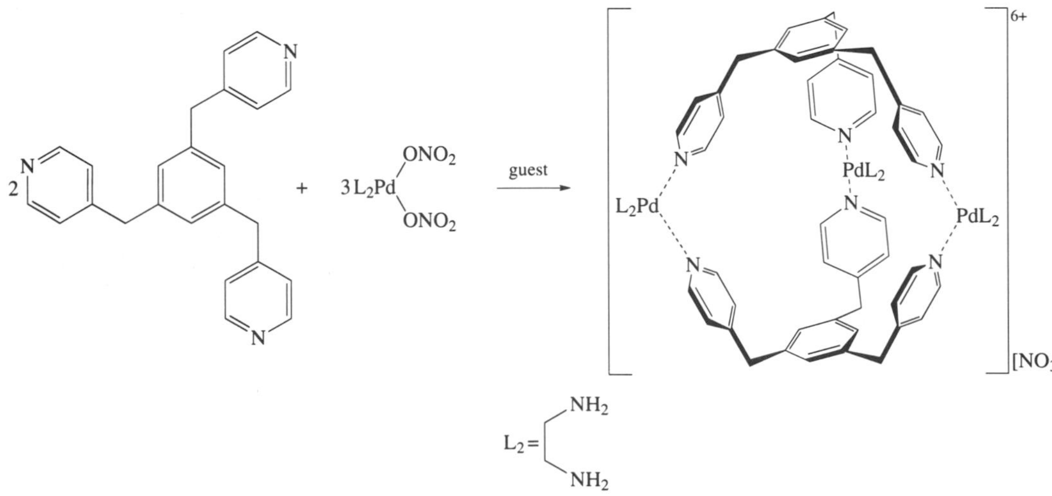

$$2\,L + 3\,L_2Pd(ONO_2)_2 \xrightarrow{\text{guest}} [(PdL_2)_3L_2]^{6+}[NO_3^-]_6 \qquad (6)$$

$L_2 = H_2NCH_2CH_2NH_2$

The effect of various anionic guests on the assembly was investigated, and it was concluded that the host forms best (> 90%) when the guests have a bulky hydrophobic group such as adamantyl or 1-phenylethyl, which are comparable in size to the host cavity. Less hydrophobic substrates such as dicarboxylates were poorly complexed. Interestingly, the neutral guest mesitylene was also found to enhance the assembly of the host, showing that stabilization of the complex by ion pairing was not essential.

In conclusion, the above work demonstrated that metal-directed assembly of a cage-like architecture which is strongly enhanced only in the presence of specific guests can serve as a model for 'induced fit' molecular recognition, the concept of which is more familiar to host–guest interactions in biological systems.

5.7.3 Tetranuclear Cages

A series of unusual 'adamantanoid' metal-assembled cage-like complexes have been prepared and their physical properties investigated. Thus, metallation of diethyl malonate with 2 equiv. of methylmagnesium iodide in tetrahydrofuran solution at –78 °C, followed by addition of 0.5 equiv. of oxalyl chloride and finally workup in the presence of aqueous ammonium chloride solution, yielded the salt $(NH_4)_4[Mg_4L_6]$ (L = (tetraethyl 2,3-dioxobutane-1,1,4,4-tetracarboxylato)$^{2-}$). This product was isolated in 40% yield as colourless crystals (but readjustment of the reactant molar ratio to 1:1:0.25 caused the yield to increase to 85%). The simplicity of the ^1H and ^{13}C NMR spectra suggested a molecule of high symmetry (T_d), but provided little further information. In order to obtain more details concerning the identity of this compound, an x-ray structural investigation was performed, which revealed that the core of the anion consisted of a tetrahedral arrangement of four magnesium(II) ions.[59]

Each cation is bridged by a tetradentate ligand (L) in such a way that each of the six bridging units rests on an edge of a regular tetrahedron. Every magnesium ion in the product is thus octahedrally coordinated by six oxygen atoms. The resulting complex can therefore be regarded as a self-assembled adamantanoid cage, with a small unoccupied central cavity (Figure 31). A zinc(II) analogue of the above complex was also prepared, but not investigated x-ray crystallographically. The above results were then extended to see if other transition metal ions could be incorporated into an identical framework. An adaptation of the above synthetic procedure which replaced MeMgI with a combination of methyllithium and $MgCl_2$ further improved the yield of the tetranuclear magnesium complex to 90%. In this way, by using the chlorides of manganese(II), cobalt(II) and nickel(II) instead of $MgCl_2$, the corresponding tetranuclear Mn^{2+}, Co^{2+} and Ni^{2+} analogues of the magnesium complex discussed above were successfully prepared. X-ray structural analyses were performed on the $[Co_4L_6]^{4-}$ and $[Ni_4L'_6]^{4-}$ complexes (L′ = tetramethyl analogue of L) and showed that they both retained the same structural composition as the corresponding magnesium complex, in which the metal ions were situated at the apices of a tetrahedron and with six ligand bridges (Figure 31). In the case of the cobalt analogue, the six atropesomeric bridging ligands are twisted in the same sense and have nearly C_2 symmetry. The resulting complex is therefore dissymmetric and chiral, the anion having nearly *T* symmetry. Only relatively complicated molecules can possess *T* symmetry, which demonstrates that the self-assembly of this entity must take place by way of a highly directed and chirally informed pathway.[60]

Further extension of the above work using MeLi–$FeCl_2$ in the reaction mixture resulted in the formation of the mixed-valence inclusion complexes $[NH_4 \subset Fe_4L_6]^-$ and $[NH_4 \subset Fe_4L' \subset Fe_4L_6']^-$, both of which have been unambiguously characterized by FAB-MS and x-ray crystallography.[61] In both cases the adamantanoid framework consists of a tetranuclear tetrahedral iron core, the corners of which are defined by one Fe^{II} and three Fe^{III} ions. In addition, a single ammonium ion guest is situated within the cavity of each complex, and probably serves as a cationic template during the assembly process (Figure 31). Mass spectrometric studies showed that alkali metal cations can also be included within the cage. The complexes crystallize as a racemic mixture of two enantiomers, which again possess approximately *T* molecular symmetry. Mössbauer spectroscopic measurements unambiguously demonstrated the complexes to be mixed valence in nature, and from a combination of EPR and electrochemical investigations it was concluded that a small but appreciable electronic interaction was occurring between the iron centres.

Finally, in an attempt to increase the cavity size of the tetranuclear iron complex in order to include larger guests, the complexation reaction was performed with $FeCl_3$ in the presence of the doubly deprotonated ligand tetramethyl 2,2′-terephthaloyldimalonate (L″). Additionally, the use of a trivalent cation would result in the formation of a neutral complex. An x-ray crystallographic

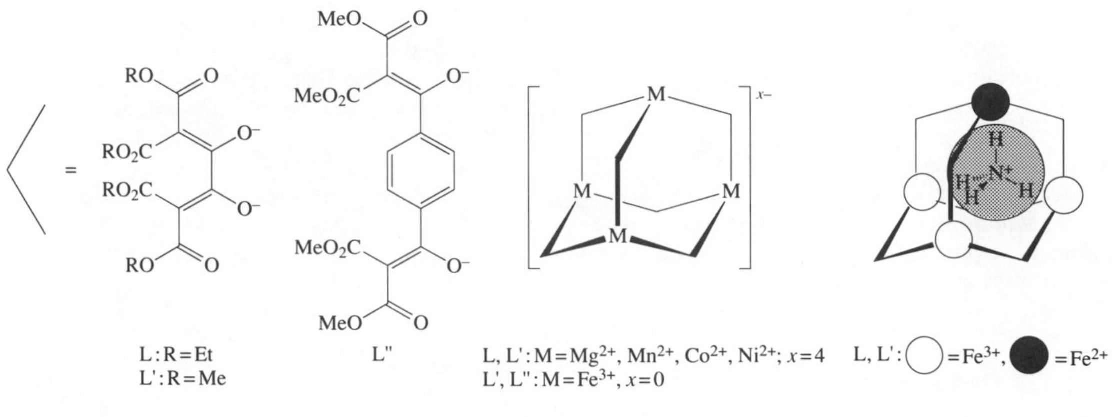

L: R = Et
L′: R = Me

L″

L, L′: M = Mg^{2+}, Mn^{2+}, Co^{2+}, Ni^{2+}; $x = 4$
L′, L″: M = Fe^{3+}, $x = 0$

L, L′: ○ = Fe^{3+}, ● = Fe^{2+}

Figure 31 Pictorial representation of self-assembled adamantanoid cages.

analysis of the ruby-red product revealed that it was also constructed of the expected adamantanoid tetranuclear framework of composition $[Fe_4L''_6]$, with each iron(III) octahedrally coordinated by oxygen atoms. The distance between facing phenyl centres was approximately 1000 pm, but the resulting void remained unoccupied by a guest. In contrast to the chiral racemic, *T*-symmetrical complexes containing tetranuclear Mg^{2+}, Mn^{2+}, Co^{2+}, Ni^{2+} and Zn^{2+} cores discussed previously, the iron complex above is achiral and in the *meso* form. Magnetic and electrochemical measurements showed that this tetranuclear complex consisted essentially of four magnetically and electronically isolated high-spin iron(III) centres.[62]

5.7.4 Penta- and Hexanuclear Cages with Sulfur Donor Ligands

Penta- and hexanuclear copper(I), silver(I) and gold(I) cage-type complexes have been prepared from the sulfur donor ligands benzenethiolate (SPh) and ethane-1,2-dithiolate, respectively.[63] The pentanuclear complexes have the composition $(Me_4N)_4[M_5(SPh)_7]$ (M = Cu^+ and Ag^+) and were prepared from mixtures of benzenethiol, trialkylamine, tetramethylammonium chloride and $Cu(NO_3)_2{\cdot}3H_2O$ or $AgNO_3$. The crystal structures of both the copper and silver complexes present a very unusual bonding arrangement of the M_5S_7 cores. In each complex the M_5S_7 core is composed of four metal atoms possessing approximate trigonal planar coordination, while the fifth has almost linear (Cu, 175.2(1)°; Ag, 176°) digonal coordination (Figure 32). Each benzenethiolate group is found to bridge two metal atoms. The M_5S_7 polyhedron may therefore be visualized as arising from a cage-like tetranuclear M_4S_6 octahedron by opening of a thiolate bridge and insertion of an M—S unit. Unfortunately, the product complexes are unstable to air and dissociate in protic solvents, and in addition present a cavity which is unoccupied.

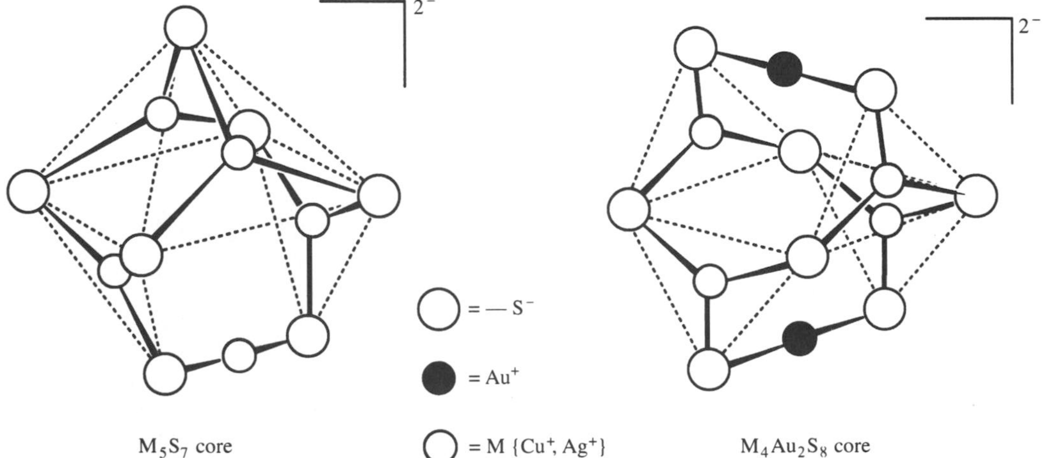

Figure 32 Structural core compositions of the M_5S_7 and $M_4Au_2S_8$ anionic polyhedra.

Hexanuclear complexes exhibiting the M_6S_8 cage skeleton were prepared upon reaction of $CuCl_2$ or $AgNO_3$ in methanol (or CuCl in acetonitrile) with ethane-1,2-dithiolate $(SCH_2CH_2S)^{2-}$, followed by addition of $(Et_4N)[AuBr_2]$. In this way, mixed metal complexes of composition $[Au_2Ag_4(SCH_2CH_2S)_4]^{2-}$ and $[(Au_2Cu_4(SCH_2CH_2S)_4)_2(Au_3Cu_3(SCH_2CH_2S)_4]^{6-}$ were isolated as their Ph_4P^+ salts. The x-ray crystal structures of all three types of anion consist of an M_6S_8 distorted polyhedral cage (Figure 32). In each case the gold(I) ions are linearly coordinated and the silver(I) ions are situated in a trigonal planar coordination environment. All ligating sulfur atoms are bridged by two metals. Again, the complexes present a central cavity, unoccupied by a guest.[64] The preparation of higher-nuclearity clusters does not appear to result in an appreciable increase in cavity size.

5.7.5 Octanuclear Cages

An unusual and particularly beautiful octanuclear cubic cage complex has been prepared and characterized by x-ray crystallography. The complex, formulated as $(Et_4N)_3[Na{\subset}Au_{12}Se_8]$, was isolated as a yellow solid from the reaction of a methanolic solution of AuCN and Na_2Se in the presence of Et_4NCl.[65] The structure of the anion consists of an almost perfect cube, with 12 linearly coordinated gold atoms situated at the midpoint of each edge and eight trigonally coordinated selenium atoms at the corners (Figure 33). A single sodium ion has been captured inside the cavity, with Au—Na distances varying between 327.2(3) pm and 346.7(3) pm. The fact that the gold atoms lie in closer contact with the included sodium ion than the selenium atoms implies that Au^+ can form attractive interactions with alkali metal cations. Another contributory factor to the inclusion of Na^+ is presumably the relatively high negative charge on the cube itself. The size of the sodium cation perfectly matches the size of the cavity. Also, the empty cube $[Au_{12}Se_8]^{4-}$ cannot be prepared in the absence of Na^+, which suggests that the pathway of formation of this complex operates via a cation template mechanism. Interestingly, attempts to prepare the potassium ion-containing analogue, $[KAu_{12}Se_8]^{3-}$ and the sulfur analogue $[NaAuS_8]^{3-}$ failed, presumably because of cation radius/cage cavity volume mismatches. Both the ^{23}Na and ^{77}Se NMR of the complex in DMF gave single resonances, showing that the cage cluster retains its integrity in solution.

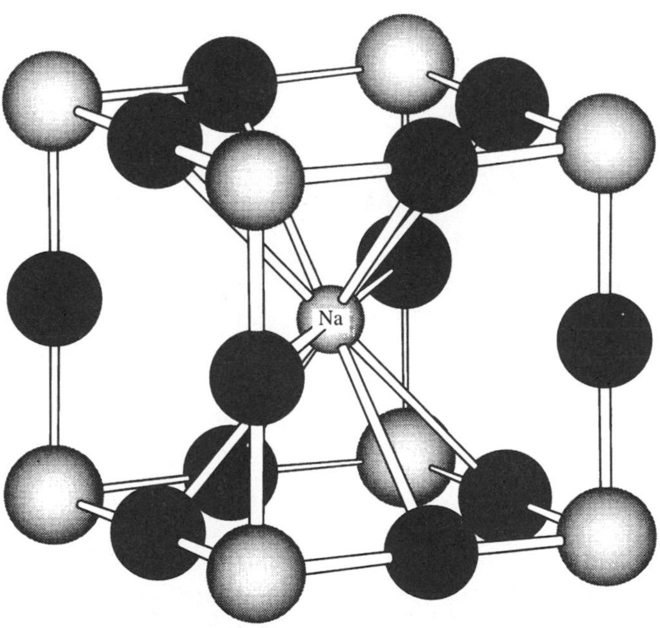

Figure 33 X-ray structure (ball-and-stick representation) of the $[Au_{12}Se_8]^{4-}$ cube showing the included sodium cation (black spheres, Au; grey spheres, Se).

Another example of an octanuclear cage-type complex is provided by the x-ray crystal structure of a compound formulated as $Na[(Na(DMF)_3)_2\{Na{\subset}(Ga(SHI))_8(\mu_2\text{-}OH)_4\}]$ (H_3SHI = salicylhydroxamic acid).[66] The complex was isolated from the reaction between H_3SHI, $Ga(NO_3)_3$ and sodium trichloroacetate in a 1:1:3 stoichiometric ratio in DMF solution. The structure, which is essentially cage-like, consists of two metallorings incorporating four gallium(III) ions, each bridged by four SHI ligands, which in addition are joined together via the gallium atoms by four bridging OH^- groups (Figure 34). The skeleton of the complex is therefore composed of an approximately cubic arrangement of Ga^{III} ions with oxygen bridges between the two tetranuclear gallium rings. The coordination polyhedron around each gallium is square pyramidal with the bridging hydroxide oxygens occupying the apical positions. The Ga—Ga separation across each OH^- bridge is 354 pm. Three sodium ions are associated with the cage complex. Two are situated in capping positions, outside the Ga_8 cube and coordinated to the SHI ligands and DMF. The third sodium is encapsulated within the cage, which has a cavity radius of 107 pm ($r_{Na^+} = 118$ pm), and is eight coordinate, with a square prismatic polyhedron. ESI-MS and Ga EXAFS studies showed that the cage-like structure of the complex remained intact in solution.

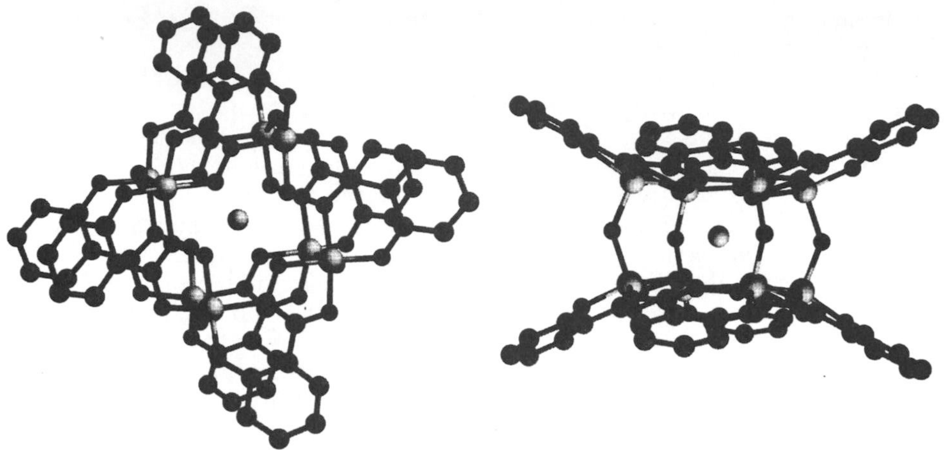

Figure 34 X-ray structure (ball-and-stick representation) of the $[(Ga(SHI))_8(\mu_2\text{-}OH)_4]^{4-}$ cage showing the included sodium cation (view perpendicular to mean plane through SHI ligands (left) and view along plane through OH^- ligands (right); H atoms are omitted for clarity).

Finally, an octanuclear cadmium complex formulated as $[I\subset Cd_8(SCH_2CH_2OH)_{12}]^{3+}3I^-\cdot H_2O$ has been crystallographically characterized.[67] The product was isolated from the reaction between an aqueous solution of cadmium acetate and thioglycol in the presence of sodium iodide. It consists of a cubic core of eight cadmium atoms, each edge of which is bridged by a sulfur atom from a thioglycolate ligand (Figure 35). Encapsulated within the metal cube is a single iodide ion. This complex exhibits the same polynuclear arrangement of one halogen, eight metal and 12 sulfur atoms as that found in the series of $[Cl\subset M\subset M^I_8M^{II}_6L_{12}]^{5-}$ complexes described in Section 5.7.7. Interestingly, it also appears to form only in the presence of halide anions, which suggests that an anion template mechanism is instrumental in the generation of this species.

5.7.6 Hexa-, Nona- and Dodecanuclear Cages with Nitrogen Donor Ligands

In almost every reported example of metal-mediated self-assembly, the product complex is composed, along with the metal ions, of only one type of coordinating organic ligand. Although in principle it should be possible to assemble fairly large architectures on combination of a single type of ligand with metal ions, this approach obviously sets limits upon the degree of structural and functional complexity of the resulting supermolecule. In living systems, many biomolecules are assembled from a cocktail of different molecular subunits through a sequence of highly specific and multiple binding events. In order to further extend the utility and functional diversity of artificial supramolecular systems, it is therefore of great importance to learn how to perform highly convergent single-step self-assembly from mixtures of structurally diverse components. As a first step towards this goal, the successful preparation of a homologous series of self-assembled cage-type structures which incorporate two different types of ligand is described below (see Sections 5.4 and 5.5.2 for related mixed-ligand complexes).

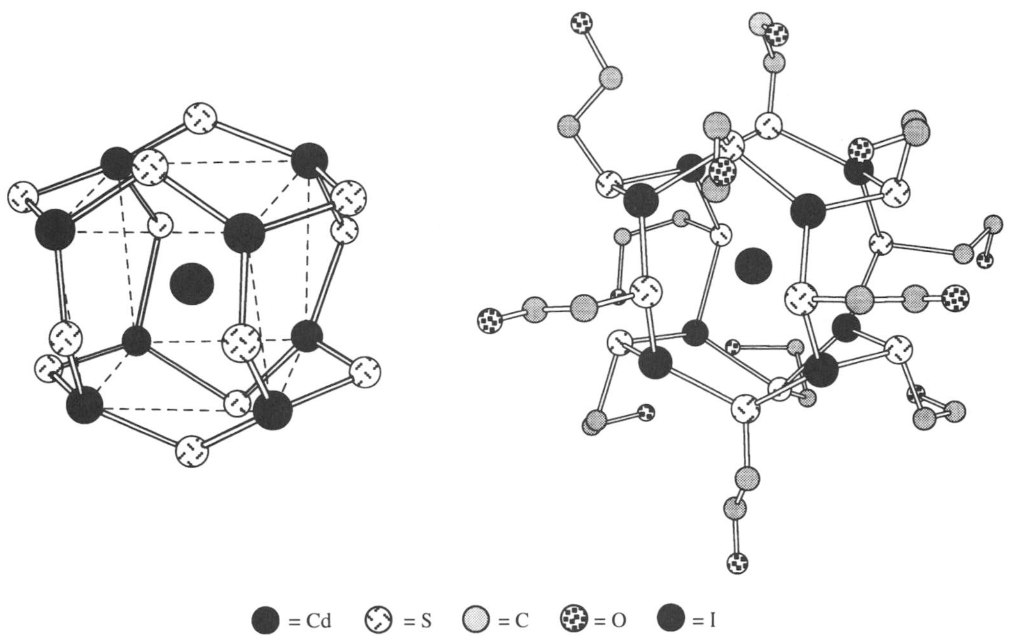

Figure 35 X-ray structure (ball-and-stick representation) of the $[Cd_8(SCH_2CH_2OH)_{12}]^{4+}$ octanuclear cage showing the included iodide anion (Cd_8S_{12} core (left) and complete ligand coordination environment (right); H atoms are omitted for clarity).

The design principle of the basic cage unit is based upon the mode of construction of the step-ladders discussed in Section 5.4. Thus, if two flat, horizontal tritopic ligands were each connected by three vertical, linear ditopic ligands using six metal ions, they would collectively constitute the top, bottom and walls of a cage or cylinder. The statistical probability of the generation of such a complex molecule as the single product from a multicomponent reaction mixture would be very small and its actual formation in quantitative yield demonstrates that the process must involve a very high degree of 'informational programming' of the component parts in terms of the specificity of their interactions.

Thus, when an acetonitrile solution of $[Cu(MeCN)_4]X$ ($X^- = BF_4^-, PF_6^-, CF_3SO_3^-$) was added to a CH_2Cl_2 solution of the ligands (**5**) and (**1a**) in a 3:1:1.5 stoichiometric ratio, a colour change from reddish brown through dark brown to finally deep purple was observed. Upon removal of solvent and workup, a deep purple solid was obtained. The simplicity of the ^1H NMR spectrum, suggestive of a single highly symmetric species in solution, coupled with FABMS results, were consistent with a product (**6a**) of formulation $[Cu_6(\mathbf{5})_2(\mathbf{1a})_3](X)_6$ (Equation (7)). In order to obtain more insight into the structure of this complex, an x-ray crystallographic investigation was carried out which showed that the product did indeed consist of the expected cage-like architecture (Figure 36). Two disc-like HAT (**5**) ligands constitute the top and bottom of the cage, while three quaterpyridine (**1a**) ligands bridge each of the HAT components in order to form the cage walls. The six copper(I) ions act as 'cement' which holds the organic scaffolding together. The complex possesses a C_2 axis which passes down the central C—C bond of one of the quaterpyridine ligands. This deviation from C_3 symmetry is presumably due to crystal-packing factors. The HAT ligands are rotated by about 27° with respect to each other, and two quaterpyridine bridges are twisted by 36° about the central C—C bond, and the third by 25°. Also, the quaterpyridine ligands are tilted by 66° with respect to the vertical axis passing through the two HAT units. Each Cu^+ ion is in a distorted tetrahedral environment. The overall effect is that the structure is twisted into a triple-helical arrangement. The complex contains a central cylindrical void of approximately 400×400 pm (taking van der Walls radii into account), which appears to be unoccupied by a guest.[68]

Similarly, replacement of the copper(I) salt by $AgCF_3SO_3$ in the above reaction mixture resulted in the formation of a silver cage (**6b**) of composition $[Ag_6(\mathbf{5})_2(\mathbf{1a})_3](CF_3SO_3)_6$. The x-ray crystal structure of this complex shows the two-component HAT ligands to be almost eclipsed (Figure 37), with the quaterpyridine ligands forming three nearly vertical bridges. The resulting shape of the cation is that of an almost perfect trigonal prism.[69] The highly symmetrical shape of the silver complex is presumably a consequence of the larger effective ionic radius of Ag^+ and also its less

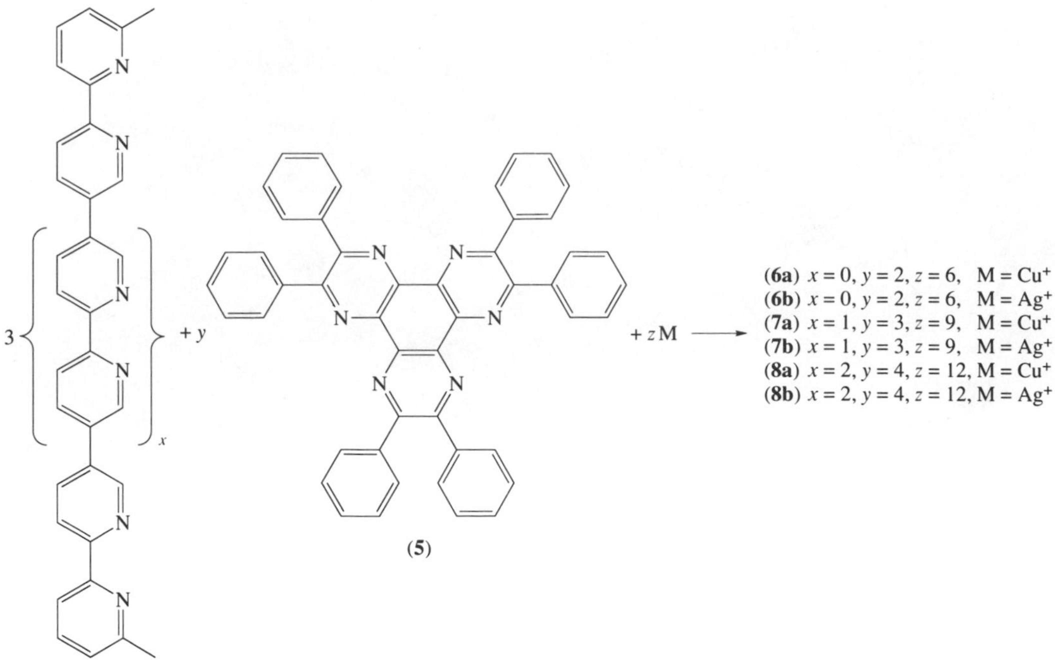

$3\,(\mathbf{1}) + y\,(\mathbf{5}) + z\mathrm{M} \longrightarrow$ (7)

(1)
(a) $x = 0$
(b) $x = 1$
(c) $x = 2$

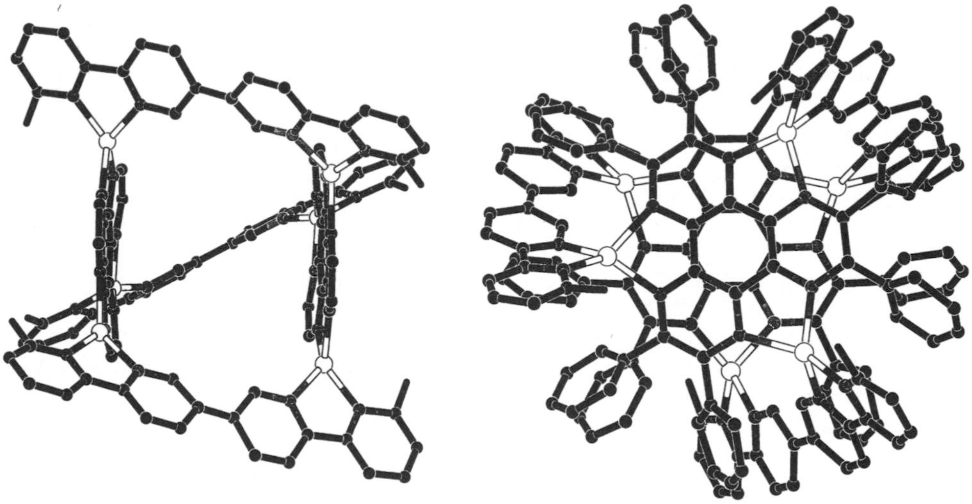

(5)

(6a) $x = 0, y = 2, z = 6$, $M = Cu^+$
(6b) $x = 0, y = 2, z = 6$, $M = Ag^+$
(7a) $x = 1, y = 3, z = 9$, $M = Cu^+$
(7b) $x = 1, y = 3, z = 9$, $M = Ag^+$
(8a) $x = 2, y = 4, z = 12$, $M = Cu^+$
(8b) $x = 2, y = 4, z = 12$, $M = Ag^+$

Figure 36 X-ray structure (bond framework representation) of the hexanuclear copper cage (**6a**) (view along planes through HAT (**5**) ligands with phenyl substituents removed (left), and view down central axis of complex perpendicular to HAT ligand planes (right); H atoms have also been removed for clarity in both pictures).

geometrically demanding coordination polyhedron. The above two complexes are formed in quantitative yield and represent a very impressive example of metal-directed self-organization which involves greater informational complexity than most previously studied systems.

More detailed investigations into the mechanism of formation of the hexanuclear copper complex by way of ESI-MS titrations allowed a possible mechanism of formation of the complex to be deduced.[70] As quaterpyridine (**1a**) is titrated into a mixture of HAT (**5**) and Cu^+ in a 1:3 stoichiometric ratio, it is found that even in the presence of relatively low ratios of quaterpyridine the cage complex has already formed. This is contrary to what might be expected when one considers that the quaterpyridine would be able to coordinate Cu^+ more strongly in the presence of HAT

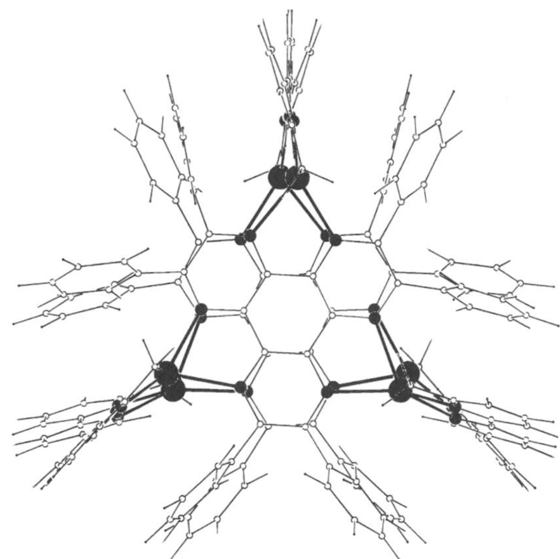

Figure 37 X-ray structure (wireframe representation) of the hexanuclear silver cage (**6b**) viewed down the central cage axis perpendicular to the HAT (**5**) ligand planes.

owing to the increased basicity of the quaterpyridine nitrogens, thereby forming species composed entirely of quaterpyridine and Cu^+. The fact that the cage complex, which involves associations between weak and strongly coordinating ligands, forms preferentially strongly implies that the assembly of the cage complex is positively cooperative.

The pathway by which the hexanuclear complex is formed appears then to begin by reaction of the initially present $[Cu(\mathbf{5})_2]^+$ complex with $[Cu(\mathbf{1a})]^+$ to give $[Cu_2(\mathbf{5})_2(\mathbf{1a})]^{2+}$ in which two HAT units have been positioned onto a single quaterpyridine ligand by two Cu^+ ions. Two further sequential insertions of $[Cu_2(\mathbf{1a})]^{2+}$ into the latter species finally results in the formation of the hexanuclear cage. Very low concentrations of acetonitrile present also presumably help to stabilize some of the species generated by weak coordination to vacant Cu^+ sites.

In order to see whether still larger systems (Figure 38) could be generated in the same way, the ligands sexipyridine (**1b**) and octapyridine (**1c**) were synthesized and reacted with $[Cu(MeCN)_4](PF_6)$ or $AgCF_3SO_3$ and HAT (**5**) in the stoichiometric ratios $3M^+$:1(**5**):1(**1b**) and $4M^+$:1.33(**5**):1(**1c**). Proton NMR and ESI-MS data for the isolated products were in agreement with the formulations $[M_9(\mathbf{5})_3(\mathbf{1b})_3](X)_9$ and $[M_{12}(\mathbf{5})_4(\mathbf{1c})_3](X)_{12}$ ($M^+ = Cu^+$, $X^- = PF_6^-$; $M^+ = Ag^+$, $X^- = CF_3SO_3^-$) (Equation (7)).

In the ^1H NMR of the nonanuclear copper (**7a**) and silver (**7b**) complexes, the resonances of the *ortho* and *meta* phenyl ring protons of the HAT ligands are clearly divided into two sets of signals in a ratio of 2:1, corresponding to the two outer and single inner HAT ligands, respectively. In the case of the dodecanuclear complexes (**8a**) and (**8b**), the peaks due to the HAT phenyl *ortho* and *meta* protons are split into two sets with an integration ratio of 2:2, corresponding to the two inner and two outer HAT ligands. An x-ray crystal structure of the nonanuclear copper complex has also been obtained,[71] and again confirms the expected cage-like arrangement in which three bridging sexipyridine ligands are found to form the walls of a cylindrical cage, closed at the top and bottom, and divided in the centre by three HAT ligands, respectively (Figures 39 and 40). The ligands are held in position by nine copper(I) ions, each in a distorted tetrahedral environment, such that the overall appearance of the complex is that of a triple helix. Remarkably, four PF_6^- anions are encapsulated within the cage architecture, two in each cavity, with the result that almost all the available space within each compartment is filled. In view of this, it is also possible that the Cu_{12} complex (**8a**) may include six PF_6^- anions and would therefore have actually been formed by the assembly of 25 particles!

The $[M_9]^{9+}$ ((**7a**) and (**7b**)) and $[M_{12}]^{12+}$ ((**8a**) and (**8b**)) complexes can be considered as multicompartmental cage-like architectures, each of which form spontaneously in one step by the correct assembly of 19 and possibly 25 particles (in the case of (**8a**)), respectively! In addition, complex (**7a**) also exhibits the relatively rare phenomenon of multiple guest inclusion. The molecular weights of the copper and silver dodecanuclear complexes are 7206 and 7787 Da, respectively, and are therefore comparable to those of small proteins. These abiological 'molecular

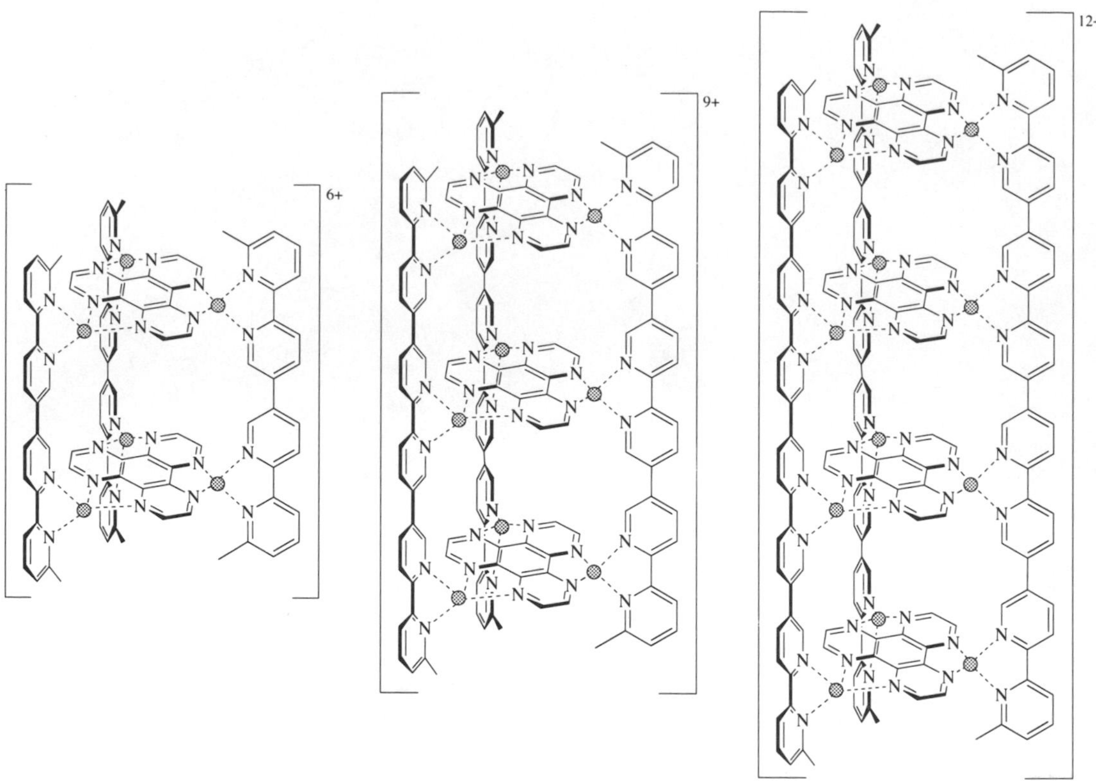

Figure 38 From left to right, the homologous series of hexa-(**6a**, **b**), nona-(**7a**, **b**) and dodecanuclear (**8a**, **b**) cages.

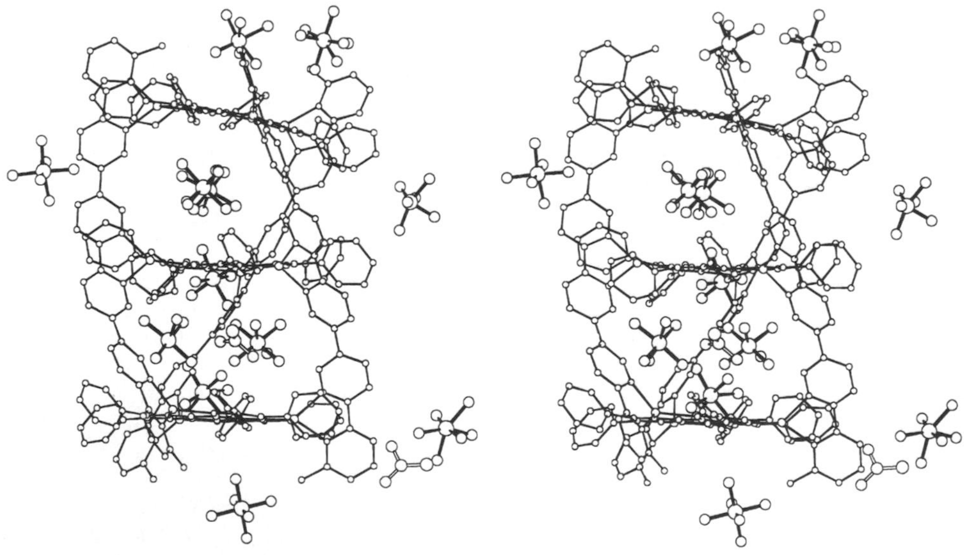

Figure 39 Stereodiagram of the $[Cu_9(\mathbf{5})_3(\mathbf{1b})_3]^{9+}$ cage (**7a**) showing four included PF_6^- anions, external PF_6^- anions and some nitromethane solvate.

skyscrapers' represent novel and relatively complex nanostructural 'unnatural products'. The fact that they can be synthesized at all amply demonstrates the truly impressive degree to which such programmed molecular systems can operate. This result also clearly shows that within the field of supramolecular coordination chemistry, it is possible to access structures of far greater complexity

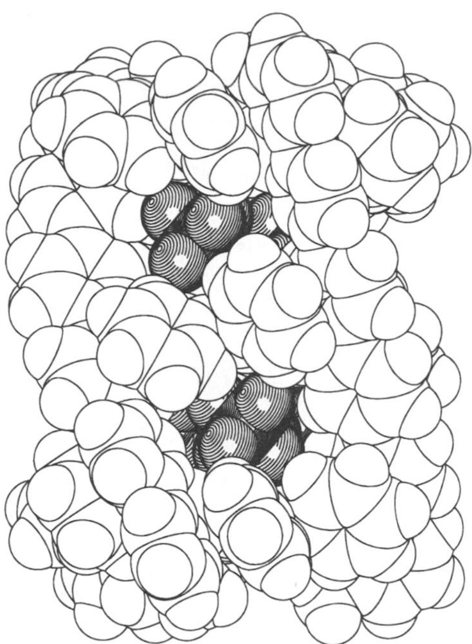

Figure 40 X-ray structure of (**7a**) in a space-filling format (included PF_6^- anions are highlighted in grey).

than was originally thought to be achievable. The one-step assembly of 'supermolecules' of this type must involve three stages: recognition between the components, correct orientation in order to allow growth and termination of the process, leading to a structurally well-defined, discrete supramolecular entity. This sequence of events can only take place if the reaction mixture initially contains a correctly chemically informed set of molecular components.

5.7.7 Tetradecanuclear and Higher Nuclearity Cages

Mention may be made of a series of particularly compact and high-nuclearity mixed-valence cage-like complexes of the general formula $[Cl{\subset}M^I{}_8M^{II}{}_6L_{12}]^{5-}$ which have been reported.[72] They were prepared by the consecutive addition of an M^I salt ($M^I = Cu^+$ or Ag^+) to an aqueous solution of the deprotonated amino acid D-penicillamine, $[SC(Me)_2CH(NH_2)CO_2]^{2-}$ (L), followed by that of an M^{II} chloride salt ($M^{II} = Cu^{2+}$, Ni^{2+}, Pd^{2+}). Four complexes with the core compositions $Cu^I{}_6Cu^{II}{}_8$, $Ag^I{}_6Ni^{II}{}_8$, $Cu^I{}_6Ni^{II}{}_8$ and $Ag^I{}_6Pd^{II}{}_8$ were prepared in this way and isolated as $[Co(NH_3)_6]^{3+}$ or Na^+ salts. The first two complexes were characterized by x-ray crystallography and showed a central cubic arrangement of eight M^I ions, each of the six faces of the cube of which is bridged via the sulfido groups of an $M^{II}L_2$ chelate unit. The resulting assembly has a small central cavity within the $M^I{}_8$ cube which is occupied by a chloride ion (Figure 41). The formation of this structure probably takes place via an anion template mechanism, as the presence of chloride was observed to be crucial for the formation of these complexes. As such, it may therefore represent an earlier example of the assembly of a complex architecture by an 'induced fit' process.

Finally, an iron–sulfur cluster of unprecedented structure was isolated in 70% yield from the reaction between Na[PhNC(O)Me], $FeCl_3$ and Li_2Se in ethanol. The x-ray crystal structure of the product, formulated as $(Bu_4N)_{4.5}Na_{4.5}[Na_9{\subset}Fe_{20}Se_{38}]\cdot 2PhNHCOMe\cdot 12EtOH$, showed the anion to consist of a macrobicyclic architecture constructed from 21 nonplanar Fe_2S_2 rhombs (Figure 42).[73a,b] Fifteen such units share vertices in order to form three curved $Fe_6(\mu_2\text{-}Se)_{10}$ chains which converge at each end via connection to an $FeSe_4$ unit. The chains sit approximately parallel to a C_3 axis which passes down through the centre of the molecule and includes two bridgehead iron and selenium atoms. The bridgehead positions of the macrobicyclic structure consist of the previously unknown tetranuclear $[Fe_4\text{–}\mu_2\text{-}Se_3\text{–}\mu_4\text{-}Se]$ fragments. The C_2 symmetric anion is prolate ellipsoidal in shape, with overall dimensions 1120×1740 pm, and is also mixed valence in nature (formally $18Fe^{III}$ and $2Fe^{II}$). Within the cavity are included nine sodium cations which are coordinated to selenium atoms and ethanol solvate molecules, and presumably help to balance the high charge on

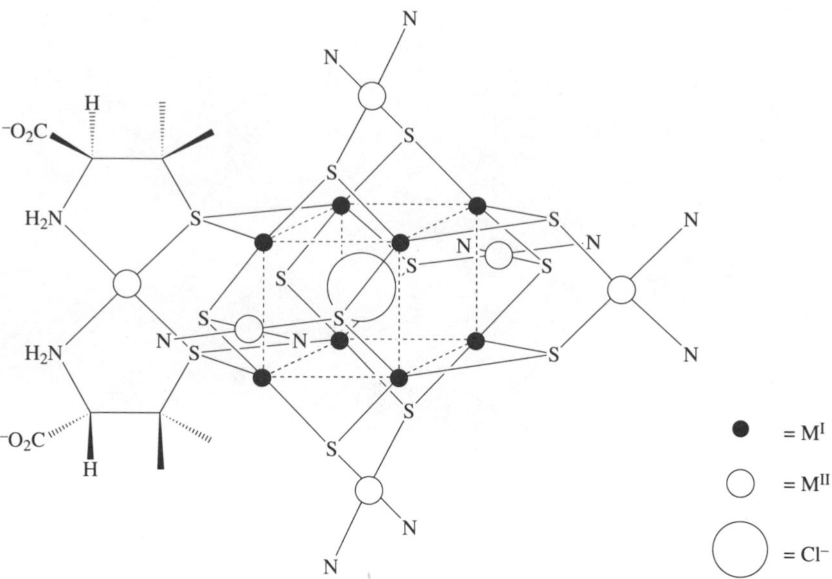

Figure 41 Schematic representation of the mixed metal D-penicillamine cage $[Cl{\subset}M^I{}_8M^{II}{}_6L_{12}]^{5-}$ core structure (for clarity, complete ligand environment has only been included on left-hand side of molecule).

the $[Fe_{20}Se_{38}]^{18-}$ cluster core. Mössbauer studies showed that the cluster was a highly electronically delocalized structure with a singlet ground state. In conclusion, the above architecture is singularly remarkable in that it represents a rare example of a totally inorganic cryptate, which in addition displays multiple guest inclusion. The isolation of this and related products, on the other hand, depends crucially upon the nature of the cations present. In the above example, Bu_4N^+ cations appear to precipitate the macrobicyclic structure from an ethanolic solution of equilibrated oligomers and polymer, formulated as $(FeSe_2)_n{}^{n-}$. Once isolated, the compound appears to be stable in solution, as evidenced by Mössbauer studies in acetonitrile.

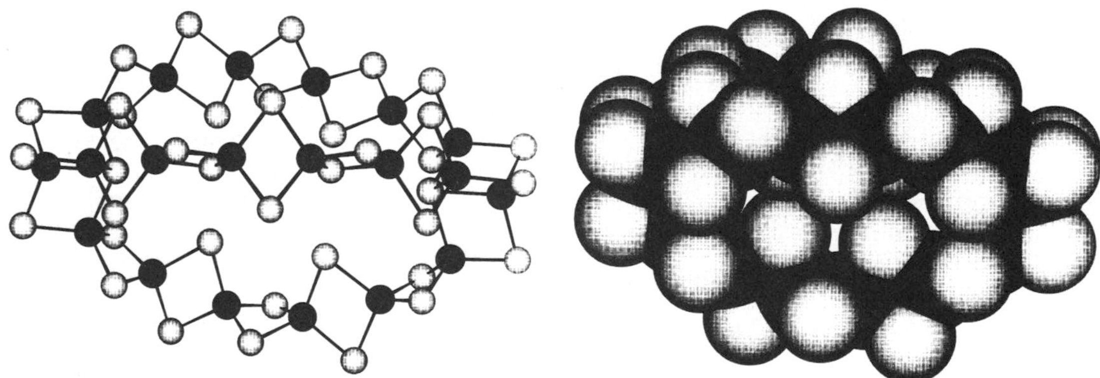

Figure 42 X-ray structure of the macrobicyclic $[Fe_{20}Se_{38}]^{18-}$ cluster core (ball-and-stick (left) and space-filling representation (right); black spheres = Fe, grey spheres = Se).

5.7.8 Polyoxometallate Cages

The chemistry of polyoxometallate clusters has had a long and distinguished scientific history, extending back to the early 1800s. The first report of this fascinating class of compounds appeared in 1826, when Berzelius described the formation of a yellow precipitate upon reaction of phosphoric acid with ammonium molybdate.[74] The reaction proceeds according to Equation (8)

$$12(NH_4)_2MoO_4 + H_3PO_4 \rightarrow (NH_4)_3[(PO_4)Mo_{12}O_{36}] + 21NH_3 + 12H_2O \tag{8}$$

and the product is now known to consist of a cage of 12 (MoO_6) octahedra connected together with a central hole that is occupied by a single phosphorus atom. In this and other related structures the central PO_4^{3-} unit is strongly covalently bonded to the surrounding oxometallate matrix such that it forms an integral part of the whole ensemble. Compounds of this type cannot therefore be considered as true supramolecular structures and have been comprehensively reviewed elsewhere.[75]

Molecular polyoxometallate clusters that bind cations are also well represented in the literature, and may be regarded as totally inorganic equivalents to the organic crowns and cryptands. Interest in these species was initiated by the discovery that the biologically active $[NaW_{21}Sb_9O_{86}]^{18-}$ anion is, in fact, a cryptate in which the sodium ion is internally encapsulated.[76] This compound, which exhibits antiviral activity, has also found applications in AIDS therapy. Many reports of related structures with cation-binding activity, sometimes exchangeable in the case of the crown analogues, have since appeared in the literature, and the interested reader is therefore referred to lead references in this area.[77a,c]

Particularly worthy of mention is a class of compounds that consists of spherical polyoxovanadate cages, which have recently been discovered to possess the hitherto unsuspected property of including anions. This is especially remarkable when one considers that the polyoxovanadate cages are themselves negatively charged, but are in addition able to encapsulate a range of anions of widely differing effective ionic radii.

A series of polyoxovanadate cages of the general composition $M_9[X\subset H_4V_{18}O_{42}]\cdot 12H_2O$ ($M^+ = Cs^+, K^+$; $X^- = Br^-, I^-$) have been prepared by treating a hot aqueous solution of MVO_3 with hydrazinium hydroxide followed by adjustment of the pH to 7.9–8.0 with aqueous HX. Alternatively, prolonged heating of an aqueous mixture of NMe_4X and $(NH_4)_3[VS_4]$ results in the crystallization of complexes of the general formula $(NMe_4)_6[X\subset V_{15}O_{36}]\cdot 4H_2O$ ($X^- = Cl^-, Br^-$). X-ray crystallographic characterization of both types of complex revealed that they consisted of a spherical oxovanadium shell in which each vanadium is coordinated by five oxygen atoms in a square-pyramidal arrangement, with the apical oxygens pointing out of the cage. A single anion X^- is included inside each cage. Of the anions Cl^-, Br^-, I^- and O^{2-}, I^- ($r = 220$ pm) gives the best fit into the cavity of the former $[X\subset H_4V_{18}O_{42}]^{9-}$ cage complex. Interestingly, the smaller Br^- ($r = 196$ pm) is also encapsulated without problem. (The distance from the centre of the cluster to the oxygen and vanadium atoms of the periphery varies between 367.5 pm and 375.0 pm.) In the case of the smaller $[X\subset V_{15}O_{36}]^{6-}$ cage (distance from cavity centre to cluster wall = 388.2–323.6 pm), Br^- has a better fit, yet Cl^- ($r = 181$ pm) is also accommodated, and in addition is not disordered within the cavity (Figure 43). Thus, even though the distances between the included halides and vanadium atoms of the cage wall are large, smaller anions are easily encapsulated and held in fixed positions. An unusual type of hitherto unrecognized weak attractive interaction must therefore exist, and is decisive in the formation of these structures.[78]

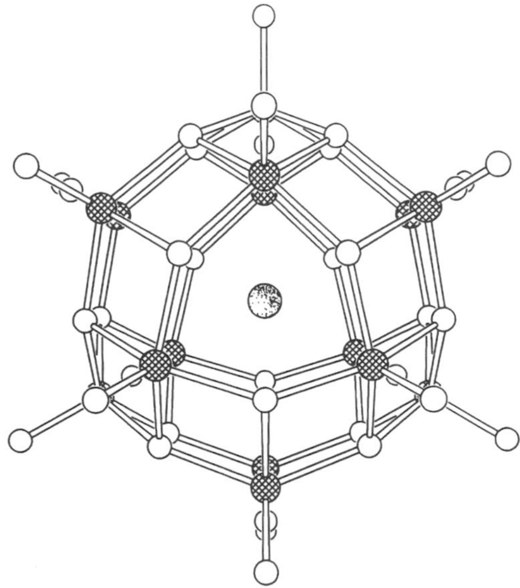

Figure 43 Structure of the polyhedral $[X\subset V_{15}O_{36}]^{6-}$ cage showing the central included halogen anion ($X = Cl^-, Br^-, I^-$) (ball-and-stick representation (V = grey, O = white)) (reproduced by permission of Elsevier Science B.V. from *J. Mol. Struct.*, 1994, **325**, 13).

Shortly after this pioneering piece of work was published, further examples of anion inclusion complexes with polyoxovanadate cages were reported. Thus, perchlorate and azide anions encapsulated inside cages with the compositions $(NEt_4)_6[ClO_4\subset HV_{22}O_{54}]$ (Figure 44) and $(NEt_4)_5[N_3\subset H_2V_{18}O_{44}]$ were characterized by x-ray crystallography.[79] The former complex was prepared by prolonged heating of an aqueous mixture of NEt_4ClO_4 and $(NH_4)_8[H_9V_{19}O_{50}]\cdot 11H_2O$, and the latter complex formed as black crystals from a hot aqueous mixture of NEt_4BF_4, $(NH_4)_8[H_9V_{19}O_{50}]\cdot 11H_2O$ and NaN_3. The cluster cages can be formally described as $[ClO_4\subset HV^{IV}{}_8V^{V}{}_{14}O_{54}]^{6-}$ and $[N_3\subset H_2V^{IV}{}_8V^{V}{}_{10}O_{44}]^{5-}$ and are of approximate D_{2d} and D_{2h} symmetry, respectively. Both complexes are constructed from edge- and corner-sharing tetragonal OVO_4 pyramids. The latter cage is elliptical in cross-section in order to accommodate the linear azide anion, with cross-sectional diameters of 560 × 940 pm. The former complex is approximately egg-shaped in cross-section, with cross-sectional diameters of 700 × 940 pm and also approximately follows the external contour of the included perchlorate anion (Figure 45).

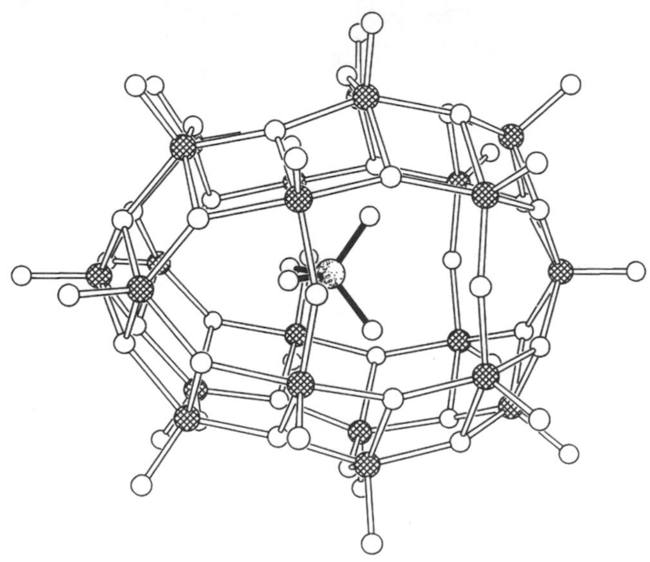

Figure 44 Structure of the $[ClO_4\subset HV_{22}O_{54}]^{6-}$ cage showing the centrally included perchlorate anion (ball-and-stick representation (V = grey, O = white)) (reproduced by permission of VCH from *Angew. Chem.*, 1991, **103**, 1720).

A polyoxovanadium cage complex similar in composition to that of the azide-containing complex discussed above has also been crystallographically characterized and shown to include a single nitrate anion.[77b,80] Again, it is elliptical in shape (Figure 46), with cross-sectional diameters of ~690 × 880 pm. The complex had the composition $K_{10}[NO_3\subset HV_{18}O_{44}]\cdot 14.5H_2O$ and was isolated as black crystals upon cooling a hot aqueous mixture of KVO_3 and hydrazinium hydroxide which had been treated with aqueous nitric acid. A polyoxovanadium cage containing a single thiocyanate anion and formulated as $(NEt_4)_6[SCN\subset HV_{22}O_{54}]$ has also been crystallographically characterized. The complex formed as black crystals upon heating an aqueous solution of NEt_4SCN and $(NH_4)_8[H_9V_{19}O_{50}]\cdot 11H_2O$ for 2–3 d. All of the above cage complexes exhibit very long distances between the walls of the cage and the included anion ($O\cdots N(N_2) > 305$ pm, $V\cdots O(ClO_3) > 2.96$ pm, $V\cdots O(NO_2) > 278$ pm, $O\cdots O(NO_2) > 2.74$ pm, $V\cdots S(CN) > 365$ pm and $O\cdots S(CN) > 353$ pm) which is indicative of only very weak interactions. However, despite this, all anions except SCN^- are ordered inside the cavity, and appear to 'hover' in space. The nature of the weak interaction presumably arises from $O{=}V^{\delta+}\cdots X^-$ attractive electrostatic forces, the summation of which must be sufficient to template the $O{=}V^{n+}O_4$ pyramidal units into a spherical shell or mould around the anion.

Anions with high charge such as $PO_4{}^{3-}$ or high basicity such as F^- are also incorporated within the polyoxovanadate cage architecture, but are strongly bonded to the cage walls and do not function as guests.

Interestingly, the preparation of the thiocyanate anion-containing cage can be repeated using acetate instead of SCN^- to give a polyoxovanadate which includes the relatively hydrophobic $MeCO_2{}^-$ anion (Figure 46). This unusual host–guest system was formulated as $(NEt_4)_5$-

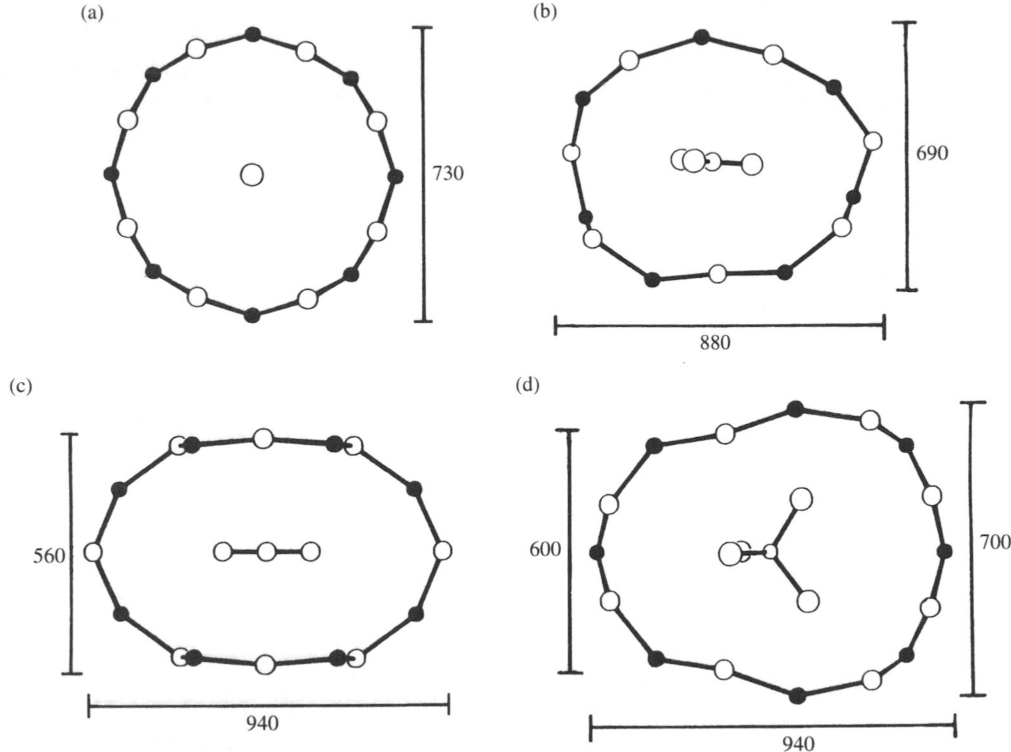

Figure 45 Illustration of the anion template effect by simplified representation of the cross-sectional shape of some V—O cluster shells (dimensions in picometres). (a) $[X{\subset}H_4V_{18}O_{42}]^{9-}$ ($X = Cl^-, Br^-, I^-$); (b) $[NO_3{\subset}HV_{18}O_{44})^{10-}$; (c) $[N_3{\subset}H_2V_{18}O_{44}]^{5-}$; (d) $[ClO_4{\subset}HV_{22}O_{54}]^{6-}$. In these 'cross-sections' the atoms in the chosen projection have the greatest distance from the 'centre' (reproduced by permission of Elsevier Science B.V. from *J. Mol. Struct.*, 1994, **325**, 13).

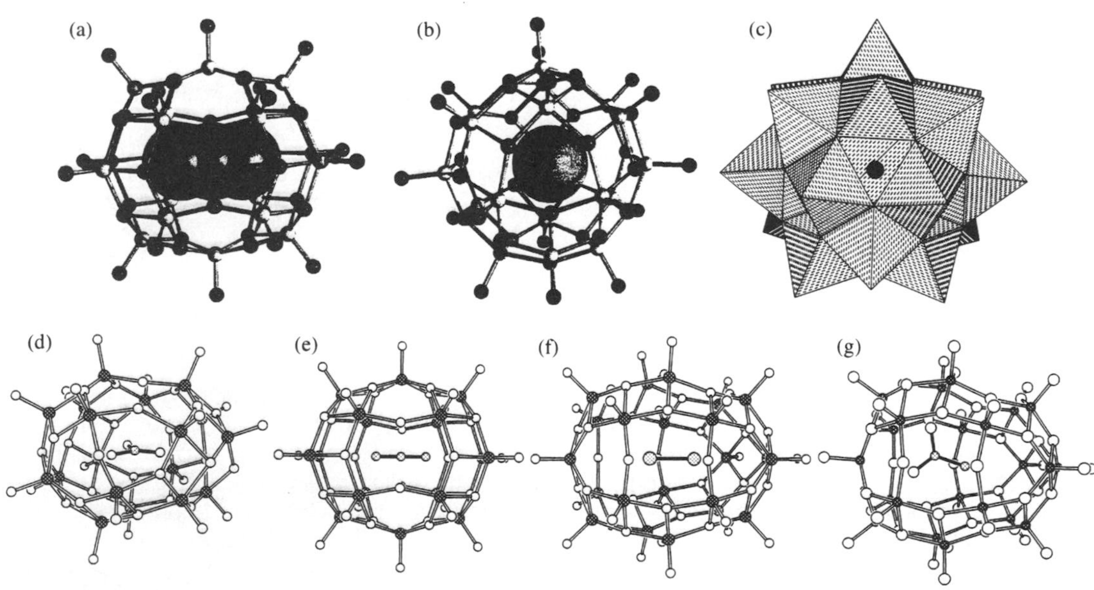

Figure 46 The host–guest relations for cluster cages. (a) $[N_3{\subset}H_2V_{18}O_{44}]^{5-}$; (b) and (c) $[X{\subset}H_4V_{18}O_{42}]^{9-}$ ($X = Cl^-, Br^-, I^-$); (d) $[NO_3{\subset}HV_{18}O_{44}]^{10-}$; (e) $[N_3{\subset}H_2V_{18}O_{44}]^{5-}$; (f) $[SCN{\subset}HV_{22}O_{54}]^{6-}$ (disordered NCS^-); (g) $[MeCO_2{\subset}H_2V_{22}O_{54}]^{7-}$. All ball-and-stick except (c) polyhedral representation (reproduced by permission of Elsevier Science B.V. from *J. Mol. Struct.*, 1994, **325**, 13).

$(NH_4)_2[MeCO_2{\subset}H_2V_{22}O_{54}]\cdot 7H_2O$.[80] An excellent review of this and related research has been published.[81]

Finally, two spectacular examples of host–guest inclusion chemistry have been reported in which multiple guest inclusion has been observed. In the first example, reaction of ammonium metavanadate with phenylphosphonic acid and hydrazine hydrate in the presence of ammonium chloride in H_2O–DMF (3:1) yielded green crystals which were characterized by x-ray crystallography. The product complex was a polyoxovanadate anion of composition $[(NH_4Cl)_2(H_2O)_2{\subset}V_{14}O_{22}(OH)_4(PhPO_3)_8]^{6-}$.[82] It consists formally of two end-capping, bowl-shaped $(V_5O_9)^{3+}$ units which are connected together via their rims by two binuclear $[V_2O_2(OH)_2(H_2O)(PhPO_3)_4]^{6-}$ aggregates (Figures 47 and 48). The resulting cylindrical shaped ensemble forms a cavity about 1200 pm long which contains two chloride ions and two ammonium ions situated in the same plane at the corners of a rhombus (Figure 49). The chloride ions are positioned at each bowl-shaped end of the cavity and in addition are bridged via Cl···HO hydrogen bonds by two encapsulated water molecules. The two chloride ions are 475 pm apart from each other. The formation of this cage is particularly noteworthy in that it presumably involves a novel synergetic template effect, in which the walls of the host are 'moulded together' by a small cation–anion aggregate.

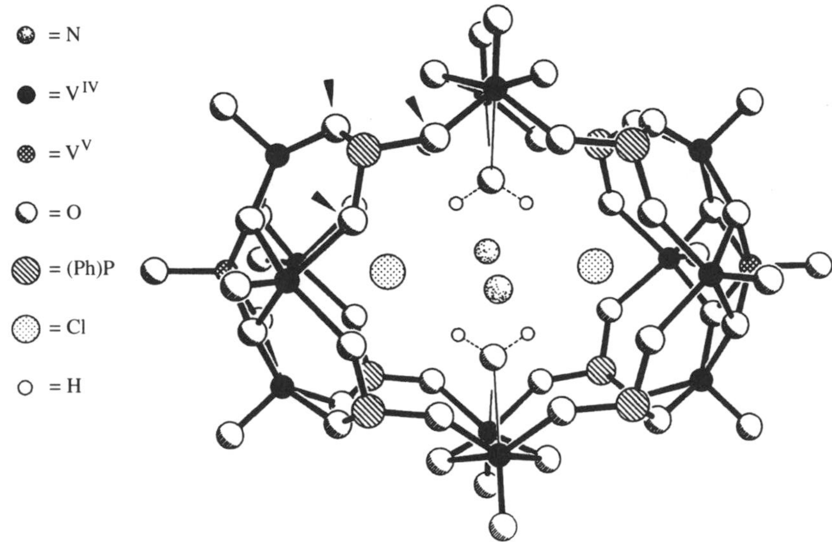

Figure 47 X-ray structure of the $[\{(NH_4{}^+Cl^-)_2(H_2O)_2\}V_{14}O_{22}(OH)_4(PhPO_3)_8]^{6-}$ anion viewed approximately along one of the idealized C_2 axes (characteristic distances (pm): V—O$_{term}$ 157.8(3)–160.0(2), V^V—O$_{br}$ 185.7(2)–188.8(2), V^{IV}—O$_{br}$ 195.1(2)–198.0(2), V—O(H) 199.9(2)–200.6(3), V—O(H)$_2$ 247.7(3); 252.0(3), P—O 151.6(3)–152.7(2), P—C 179.1(4)–179.9(3)); the oxygen atoms of a phosphonate ligand are marked by arrows (reproduced by permission of VCH from *Angew. Chem.*, 1992, **104**, 1214).

In the second example, an anionic complex formulated as $[(H_2O)_2N_3{\subset}V_{14}O_{22}(OH)_4(PhPO_3)_8]^{7-}$ was prepared and also characterized by x-ray crystallography.[83] When phenylphosphonic acid, sodium azide and hydrazine hydrate as the reducing agent are added to a hot mixture of vanadium pentoxide, *t*-butylamine and triethylamine (as deprotonating agents) in H_2O–DMF (2.5:1), dark-green crystals of the product complex formed in 79% yield. The anion is shaped approximately like a torus with a large central cavity that is occupied by two water molecules and one positionally disordered $N_3{}^-$ anion. Situated above and below the polyoxovanadate torus are two calix-like arrangements, each comprised of four phenyl rings which derive from the phenylphosphonate units in the walls of the torus and which appear as truncated cones. The upper and lower cones are separately filled by three water molecules, one DMF molecule, a sodium cation and a $Bu^tNH_3{}^+$ cation. Exposure of the complex to a 13-fold excess of rubidium bromide results in the replacement of one water and 50% of the $Bu^tNH_3{}^+$ ions by Rb^+ ions in each truncated cone. The other 50% of the $Bu^tNH_3{}^+$ ions are replaced by Na^+ ions. This, in turn, leads to a significant reorganization and migration of the $N_3{}^-$ anions and water molecules within the central cavity. Such an observation may be particularly relevant to fundamental investigations into the influence of various 'external fields' on the spatial organization of particles remotely situated inside a cavity.

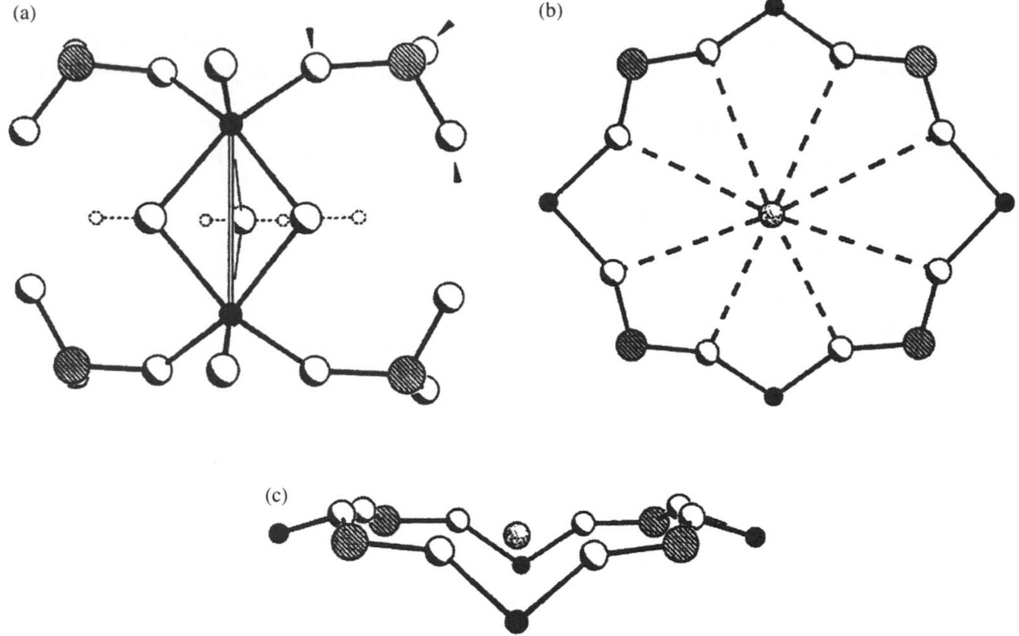

Figure 48 Component fragments of the $[\{(NH_4^+Cl^-)_2(H_2O)_2\}V_{14}O_{22}(OH)_4(PhPO_3)_8]^{6-}$ cage: (a) the $[V_2^{IV}O_2(OH)_2(H_2O)(PhPO_3)_4]^{6-}$ fragment viewed along axis passing from top to bottom of Figure 47 (oxygen atoms of one $[PhPO_3]^{2-}$ ligand marked by arrows); (b) nucleophilic 'coronand' ring viewed as in Figure 47; and (c) perpendicular to that position to illustrate the site of the NH_4^+ cation (reproduced by permission of VCH from *Angew. Chem.*, 1992, **104**, 1214).

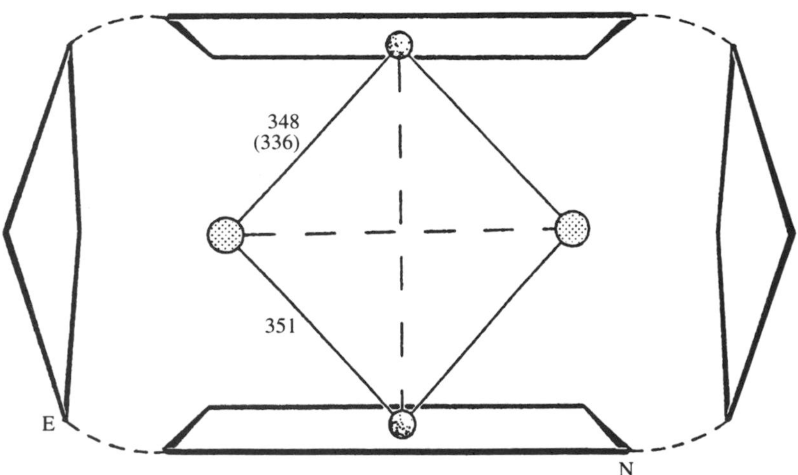

Figure 49 Host–guest relationships in Figure 47 drawn to scale for the host systems E and N which are perpendicular and electronically inverse to one another: E corresponds to the $[V_5O_9]^{3+}$ electrophilic half-shell (the illustrated plane is spanned by four V^{IV} centres), and N corresponds to the nucleophilic 'coronand' plane (the illustrated plane is spanned by four phosphorus atoms). Distances in the $[\{(NH_4^+Cl^-)_2(H_2O)_2\}$-$V_{14}O_{22}(OH)_4(PhPO_3)_8]^{6-}$ cage and in the NH_4Cl ionic lattice (in parentheses) are in picometres (reproduced by permission of VCH from *Angew. Chem.*, 1992, **104**, 1214).

Controlled interactions of the type discussed above may also be considered as models for processes occurring in ion channels, which is of considerable interest to biological systems.

5.8 DENDRITIC MACROMOLECULES

Star-burst dendrimers are a class of compounds that include some of the largest known artificially synthesized molecules of precisely defined structure and molecular weight. Constructs of this type have been prepared which incorporate a diverse range of molecular components such as redox-active metals, peptides, macrocycles and organosiloxanes. Indeed, since its conception, work in this area has progressed dendritically with time, as these substrates promise many novel and useful applications. The molecules in themselves additionally display a beauty and conceptual simplicity which presumably has also been instrumental in attracting many workers to this field. The interested reader is therefore referred to several excellent articles and reviews which deal with this rapidly developing subject.[84]

The usual approach to dendrite synthesis involves the consecutive construction of one generation at a time (divergent method), or preparing sections of the molecule and connecting together the segments in the final stages of the synthesis (convergent method). Both avenues employ sequential covalent bond-forming methodology and also require additional protection/deprotection steps at each stage of growth.

A report has appeared which described the preparation of a dendritic macromolecule in a single step using metal ion-induced self-assembly.[85] The dendrite was constructed from a precursor complex, which combined two kinetically labile MeCN ligands with one potentially kinetically inert cyanomethylene donor. Displacement of the ligated MeCN by warming a solution of the complex in nitromethane resulted in coordination of each palladium by the cyanomethylene groups. In this way, the monomer complexes were able to assemble in three dimensions to give a macromolecular aggregate (Equation (9)). The final shape and size would be determined solely by maximal steric crowding effects experienced in the last generation. Such effects would collectively prevent further growth, and together constitute the termination step. Quasi-elastic light scattering (QELS) measurements of nitromethane solutions and atomic force microscopic (AFM) examination of surface-deposited thin films revealed the presence of grains with diameters of about 200 nm. With transmission electron microscopy (TEM), globular-shaped assemblies in the 150–200 nm range were clearly visible. Application of energy-dispersive x-ray spectrometry (EDX) showed that both palladium and sulfur were present within the spherical particles, demonstrating that they were not an instrumental artifact. Interestingly, dendrimer formation and dissociation were also shown to be completely reversible and dependent upon the presence or absence of acetonitrile. The factors which control particle size are the subject of further investigations.

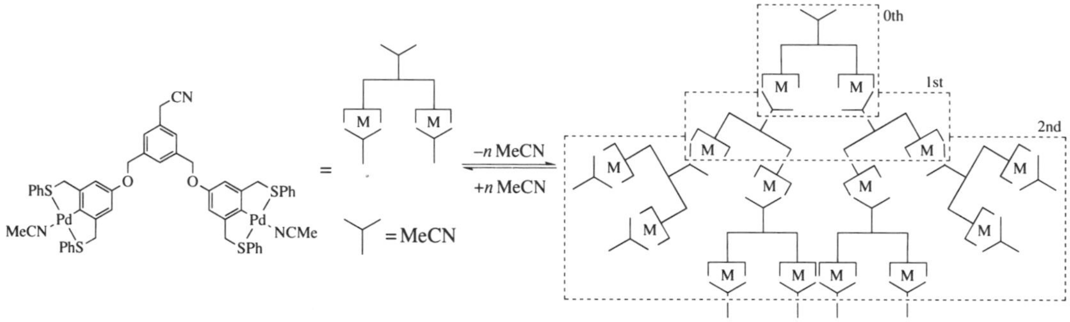

CN, PhS, SPh, Pd, MeCN, NCMe, O; = MeCN; $-n$ MeCN / $+n$ MeCN; M; 0th, 1st, 2nd (9)

5.9 CONCLUSION

5.9.1 Metal Ion-assisted Self-assembly: Summary and Present Status

The construction of molecular nanostructural materials of precisely defined chemical composition, by using metal ions as the 'cement' which collectively bonds together the final supramolecular ensemble, is an exciting and rapidly expanding field. Many preexisting organic molecules such as crowns and cryptates must be synthesized using laborious consecutive sequential steps and are obtained in poor to moderate overall yields. Structural analogues of these molecules which incorporate metal ions as integral architectural units can be prepared in a single step from a simpler set of components under equilibrium conditions. They are also usually isolated in high to quantitative yields (Sections 5.7.1 and 5.7.2). It appears that very large molecules can also be built in this way, the complex spatial organization of which would be almost impossible to achieve using conventional organic covalent synthetic methods (Sections 5.6.6 and 5.7.6–8).

The mechanism by which this takes place can be regarded as a type of molecular information transfer from a subset of chemically 'informed' components which are 'programmed' to build the desired supermolecule. The overall process is therefore under thermodynamic control, in which a combination of kinetically and thermodynamically reversible bond-forming and -breaking interactions occur between the species present within a complex mixture of molecular components. This allows for correct orientation and growth ('error checking') to take place within the mixture until, finally, the most entropically stable arrangement of molecular components is generated, which consists of a supermolecule.[86] This is the stage at which the process is terminated, and no further changes occur. In some of the examples discussed above the thermodynamic minimum appears not to consist of a single entity, but rather a mixture of interconverting oligomeric species along with the desired product. Alternatively, the product itself may be undergoing rapid or slow exchange equilibria. In the former case a single supramolecular species may still be isolated from such a mixture, for example, by selective precipitation upon addition of an appropriate counterion. This situation is often encountered in cluster chemistry, where minor adjustments to the starting conditions may direct the reaction pathway to give structurally very different products. Once isolated from the parent reaction mixture, such a complex species may be completely stable in solution. Complex molecules, once assembled, may exhibit new and unusual physicochemical properties such as electron transfer/redox, magnetic, inclusion, optical and catalytic, or combinations thereof. Functional complexity such as this arises as a direct consequence of the programmed structural complexity of the supramolecular ensemble.

Inorganically self-assembled materials with a wide variety of molecular topologies have been prepared since the early 1960s. Helicates, catenates and rings appear to have attracted the greatest attention, but chains and cages, for example, are also fairly well represented. Single examples of the metal ion-assisted assembly of a dendritic macromolecule (Section 5.8) and a $[3 \times 3]$G grid have been reported (Section 5.5). These two areas will presumably attract a great deal of attention. The latter case represents the beginning of a totally unexplored area in inorganic chemistry. Further developments in this field are likely to unearth molecules which exhibit particularly fascinating and unexpected physicochemical properties, especially if redox-active octahedral metals can be incorporated. In terms of direct applications, however, the state of the art is still in its infancy, and some of the main reasons for this are outlined below.

5.9.2 Current Problems

So far, many of the initial expectations that these systems promised, in terms of functional and practical applications, have not been realized. This is mainly due to problems inherent within the systems themselves.

For example, uses related to specific substrate recognition are limited because the high symmetry of the majority of metal-assembled hosts greatly restricts the range of guests that can be included. Also, many examples encountered in the literature consist of hosts that only assemble and remain stable in the presence of guests, and guests which cannot be removed from the host ('host–hostage complexes'). These may be of little use as molecular or ion sensors. Constitutional instability is another limitation in that many self-assembled systems undergo rapid dissociation–association equilibria under certain conditions in solution, and may be stable only within a particular concentration range. This may severely limit processibility, ease of handling and the range of external environments in which such materials may usefully operate.

In the case of very large supermolecules, poor solubility may greatly restrict potential applications. In this case, particular care will have to be devoted towards the design of ligand components or counterions, in order to increase solubility without disturbing the functional properties of the system.

Direct methods of characterization of higher molecular weight assemblies represent another obstacle. For example, the successful x-ray crystallographic solution of the structure of a large molecule is often hampered by positional disorder of solvent or counterions, and/or simply by the large number of atoms that have to be identified in the unit cell.

Finally, a potential future application of self-assembled complexes that exhibit bi- or multistability is that of information processing systems which do not suffer from the same size limitations of silicon chip devices. Unfortunately, major problems exist concerning methods of addressing and especially reading information which may be stored on such supramolecular constructs.[87]

5.9.3 Possible Solutions and Future Perspectives

It is probable, however, that many alternative applications will be found for systems which currently appear to have limited uses. For example, in the field of polyoxometallate chemistry, certain polyoxoanions are known to exhibit biological activity. Several polyoxometallate anions have been shown to have potent antiviral and anti-HIV activity, and also to bind enzymes such as dehydrogenases, phosphorylase and phosphatases.[88] This may have important implications for many of the other, large, self-assembled anionic complexes discussed above, which may, along with related systems, possess hitherto unsuspected biological properties. The mode of interaction in the case of polyoxometallates presumably results from multiple hydrogen-bonding interactions between the polyanion and the polypeptide biomolecule. In the future this may be used as a design principle for the construction of a new class of bioactive substrates which might specifically bind to and regulate protein tertiary structure. The discovery that certain mononuclear platinum complexes interact with DNA subsequently led to the development of platinum-containing anticancer therapeutics. This in turn fuelled most of the platinum chain chemistry discussed in Section 5.3. Interestingly, polynuclear double-helical copper(I) complexes have been demonstrated to bind to calf-thymus DNA, in addition to poly(dG–dC)·poly(dG–dC) and poly(dA–dT)·poly(dA–dT), probably via interaction with the major groove. The largest copper(I) helicate examined, ds-$H[Cu_5(BP_5)_2](CF_3SO_3)_5$, exhibited the strongest binding affinity.[89] In the light of this, it may be possible that many other large, polynuclear, polycationic complexes, such as those discussed in Section 5.7.6, may also show strong associations with the negatively charged surface of DNA. Further improvements in the design of these and related complexes may result in site-specific binding which, coupled with photocleaving ability, may in the future give rise to a new class of 'restriction chemzymes'. Nanosized polycationic macromolecules can be envisioned with well-defined three-dimensional shapes which may recognize and adhere to a complementary surface on RNA or DNA. Surface recognition by appropriately designed polycations may also interfere with the replicative ability of DNA with the result that these compounds may exhibit anticancer properties.

Other potential applications exist, for example, with respect to unstable host–guest complexes discussed above. It may be possible to control the uptake and release of a particular guest by a system of this type, under physiological conditions, and thereby target encapsulated biomolecules to specific biological sites. A new class of programmed self-assembling–disassembling drug delivery systems may possibly result from this approach.

Within the realm of materials science, metallo-assembled structures may find many varied applications in the future. This will result from a continued and increasing demand for compounds with novel and unusual properties. For example, the controlled assembly of large polynuclear structures which incorporate spin-coupled paramagnetic metals may give rise to new types of molecular magnetic solids. Such compounds may be reversibly dissociated and reformed at will, and incorporated into larger architectures. Other potentially useful properties include electrochemically induced positional bistability, such as that described for an iron-containing system[90] and a mononuclear copper catenate.[91] Also, an unusual example of molecular magnetic multistability has been reported[92] for an oxovanadium cluster formulated as $[V_{15}As_6O_{42}(H_2O)]^{6-}$. Molecules of this type may be considered as embryonic nanoscale switching devices, and represent the initial stages of what will be a wider search for alternative ways of inducing bi- or multistability in molecular systems. Extension of the one-dimensional metal chains such as those discussed in Section 5.3 may yield compounds that exhibit particularly interesting nonlinear optical or electrically conducting properties. Polynuclear chains of precisely defined length may serve, for example, as rigid girders in supramolecular nanoscaffolding or as electronically conducting nanowires (quantum wells).

Especially interesting is the phenomenon of multiple guest inclusion such as that exhibited by the complexes discussed at the end of Sections 5.7.6–5.7.8 and the copper–cyclodextrin complex in Section 5.6.1. Further developments in this area may yield substrates capable of growing and trapping nanocrystals. This may give some fundamental insight into the processes involved during crystal nucleation and biomineralization. Other potentialities can be envisaged such as catalysis, switching devices based upon translational multistability of encapsulated guests, and systems in which intermolecular guest interactions in the absence of solvent ('gas phase in solution') can be studied.

Finally, two areas that are of particular importance are the combination of differing types of self-assembly mechanisms within a single system and the merging of self-assembly with catalysis. In the first case, systems will have to be designed which assemble by way of a range of attractive intermolecular forces that mutually cooperate and reinforce each other. The resulting supermolecule may be constructed, for example, by the collective effects of hydrogen bonding, aromatic stacking interactions and metal coordination. Several examples have already been discussed such as platinum chains (Sections 5.3.1 and 5.3.2) and a hexanuclear platinum ring (Section 5.6.2) which involve both hydrogen bonding and metal–ligand coordination in their assembly. The successful development of this area will be of direct necessity if anything approaching the informational complexity of living systems is to be achieved.

In the second case, catalytic sites that are self-assembled,[93] and self-assembled molecules whose components require synthesis by supramolecular catalysts, are areas that also require development. Coupling of the above two situations may result in a replication cycle where the self-assembled product could catalyse the synthesis of its component parts. An exploration of this field within the framework of combinatorial chemistry and molecular templating may result in the development of a futuristic class of chemical machinery.

Whatever the nature of the technical problems encountered, concerning the development of applications for the materials discussed in this review, it is true to say that the incredible complexity of biological systems and natural phenomena provides us with a continual source of inspiration. We are reminded by this that it will be possible to overcome many of the initial obstacles confronting the development of this field and finally, in the future, to construct giant, functional supramolecular materials and machines.

ACKNOWLEDGEMENTS

I thank Bernold Hasenknopf for help with the figures, Marie-Thérèse Youinou for constructive comments, the authors who supplied pictures and especially Professor Jean-Marie Lehn for financial support.

5.10 REFERENCES

1. D. Philp and J. F. Stoddard, *Synlett*, 1991, 445.
2. G. M. Whitesides, J. P. Mathius and C. T. Sato, *Science*, 1991, **254**, 1312.
3. H. Ringsdorf, B. Schlarb and J. Venzmer, *Angew. Chem.*, 1988, **100**, 117; *Angew. Chem., Int. Ed. Engl.*, 1988, **27**, 113.
4. (a) I. G. Dance, in 'Comprehensive Coordination Chemistry', ed. G. Wilkinson, Pergamon, Oxford, 1987, vol. 1, chap. 4, pp. 135–77; (b) L. H. Gade, *Angew. Chem.*, 1993, **105**, 25; *Angew. Chem., Int. Ed. Engl.*, 1993, **32**, 24 and references therein.
5. (a) J.-M. Lehn, 'Supramolecular Chemistry: Concepts and Perspectives', VCH, Weinhein, 1995, pp. 144–54; (b) K. T. Potts, *Bull. Soc. Chim. Belg.*, 1990, **99**, 741; (c) E. C. Constable, *Nature (London)*, 1990, **346**, 314; (d) E. C. Constable, *Angew. Chem., Int. Ed. Engl.*, 1991, **30**, 1450; (e) E. C. Constable, *Tetrahedron*, 1992, **48**, 10013.
6. J.-M. Lehn, A. Rigault, J. Siegel, J. Harrowfield, B. Chevrier and D. Moras, *Proc. Natl. Acad. Sci. USA*, 1987, **84**, 2565.
7. J.-M. Lehn and A. Rigault, *Angew. Chem.*, 1988, **100**, 1121; *Angew. Chem., Int. Ed. Engl.*, 1988, **27**, 1095.
8. I. Dance, P. Dean and K. Fisher, *Angew. Chem.*, 1995, **107**, 366; *Angew. Chem., Int. Ed. Engl.*, 1995, **34**, 314.
9. J. K. Barton, H. N. Rabinowitz, D. J. Szalda and S. J. Lippard, *J. Am. Chem. Soc.*, 1977, **99**, 2827.
10. J. K. Barton, D. J. Szalda, H. N. Rabinowitz, J. V. Waszczak and S. J. Lippard, *J. Am. Chem. Soc.*, 1979, **101**, 1434.
11. T. V. O'Halloran, M. M. Roberts and S. J. Lippard, *J. Am. Chem. Soc.*, 1984, **106**, 6427.
12. P. K. Mascharak, I. D. Williams and S. J. Lippard, *J. Am. Chem. Soc.*, 1984, **106**, 6428.
13. T. V. O'Halloran, P. K. Mascharak, I. D. Williams, M. M. Roberts and S. J. Lippard, *Inorg. Chem.*, 1987, **26**, 1261.
14. J.-P. Laurent, P. Lepage and F. Dahan, *J. Am. Chem. Soc.*, 1982, **104**, 7335.
15. L. S. Hollis and S. J. Lippard, *J. Am. Chem. Soc.*, 1981, **103**, 1230.
16. K. Matsumoto and K. Fuwa, *J. Am. Chem. Soc.*, 1982, **104**, 897.
17. K. Matsumoto, H. Takahashi and K. Fuwa, *Inorg. Chem.*, 1983, **22**, 4086.
18. K. Matsumoto and T. Watanabe, *J. Am. Chem. Soc.*, 1986, **108**, 1308.
19. K. R. Mann, M. J. DiPierro and T. P. Gill, *J. Am. Chem. Soc.*, 1980, **102**, 3965.
20. V. M. Miskowski and H. B. Gray, *Inorg. Chem.*, 1987, **26**, 1108.
21. M. A. Ciriano, S. Sebastian, L. A. Oro, A. Tiripicchio, M. Tiripicchio Camellini and F. J. Lahoz, *Angew. Chem.*, 1988, **100**, 406; *Angew. Chem., Int. Ed. Engl.*, 1988, **27**, 402.
22. A. J. Blake, C. M. Grant, S. Parsons and R. E. P. Winpenny, *J. Chem. Soc., Dalton Trans.*, 1995, 1765.
23. D. L. Hughes, U. Kleinkes, G. J. Leigh, M. Maiwald, J. R. Sanders and C. Sudbrake, *J. Chem. Soc., Dalton Trans.*, 1994, 2457.
24. I. S. Sigal, K. R. Mann and H. B. Gray, *J. Am. Chem. Soc.*, 1980, **102**, 7252.
25. I. S. Sigal and H. B. Gray, *J. Am. Chem. Soc.*, 1981, **103**, 2220.
26. B. Lippert and D. Neugebauer, *Inorg. Chem.*, 1982, **21**, 451.
27. K. Sakai and K. Matsumoto, *J. Am. Chem. Soc.*, 1989, **111**, 3074.
28. I. G. Dance, P. A. W. Dean and K. J. Fisher, *Inorg. Chem.*, 1994, **33**, 6261.
29. P. N. W. Baxter, G. S. Hanan and J.-M. Lehn, manuscript in preparation.
30. R. Krämer, J.-M. Lehn and A. Marquis-Rigault, *Proc. Natl. Acad. Sci. USA*, 1993, **90**, 5394.
31. R. Fynman, *Science*, 1991, **254**, 1300.
32. M.-T. Youinou, N. Rahmouni, J. Fischer and J. A. Osborn, *Angew. Chem.*, 1992, **104**, 771; *Angew. Chem., Int. Ed. Engl.*, 1992, **31**, 733.
33. P. N. W. Baxter, J.-M. Lehn, J. Fischer and M.-T. Youinou, *Angew. Chem.*, 1994, **106**, 2432; *Angew. Chem., Int. Ed. Engl.*, 1994, **33**, 2284.

34. P. N. W. Baxter and J.-M. Lehn, to be published.
35. R. Fuchs, N. Habermann and P. Klüfers, *Angew. Chem.*, 1993, **105**, 895; *Angew. Chem., Int. Ed. Engl.*, 1993, **32**, 852.
36. X. Yang, C. B. Knobler and M. F. Hawthorne, *Angew. Chem.*, 1991, **103**, 1519; *Angew. Chem., Int. Ed. Engl.*, 1991, **30**, 1507.
37. X. Yang, C. B. Knobler and M. F. Hawthorne, *J. Am. Chem. Soc.*, 1993, **115**, 4904.
38. C. Dohmeier, R. Mattes and H. Schnöckel, *J. Chem. Soc., Chem. Commun.*, 1990, 358.
39. A. Caneschi, D. Gatteschi, J. Laugier, P. Rey, R. Sessoli and C. Zanchini, *J. Am. Chem. Soc.*, 1988, **110**, 2795.
40. B. Longato, G. Bandoli, G. Trovo, E. Marasciulo and G. Valle, *Inorg. Chem.*, 1995, **34**, 1745.
41. R. Uson, J. Fornies, L. R. Falvello, M. A. Uson, I. Uson and S. Herrero, *Inorg. Chem.*, 1993, **32**, 1066.
42. (a) K. L. Taft and S. J. Lippard, *J. Am. Chem. Soc.*, 1990, **112**, 9629; (b) K. L. Taft, C. D. Delfs, G. C. Papaefthymiou, S. Fonor, D. Gatteschi and S. J. Lippard, *J. Am. Chem. Soc.*, 1994, **116**, 283.
43. O. Poncelet, L. G. Hubert-Pfalzgraf, J.-C. Daran, R. Astier, *J. Chem. Soc., Chem. Commun.*, 1989, 1846.
44. I. G. Dance, *Inorg. Chim. Acta*, 1977, **25**, L17.
45. D. Fenske and A. Fischer, *Angew. Chem.*, 1995, **107**, 340; *Angew. Chem., Int. Ed. Engl.*, 1995, **34**, 307.
46. J.-F. You, B. S. Snyder and R. H. Holm, *J. Am. Chem. Soc.*, 1988, **110**, 6589.
47. P. Klüfers and J. Schuhmacher, *Angew. Chem.*, 1994, **106**, 1925; *Angew. Chem., Int. Ed. Engl.*, 1994, **33**, 1863.
48. A. Müller, K. Schmitz, E. Krickemeyer, M. Penk and H. Bögge, *Angew. Chem.*, 1986, **98**, 470; *Angew. Chem., Int. Ed. Engl.*, 1986, **25**, 453.
49. R. H. Uang, C. K. Chan, S. M. Peng and C. M. Che, *J. Chem. Soc., Chem. Commun.*, 1994, 2561.
50. P. Chaudhuri, M. Winter, P. Fleischhauer, W. Haase, U. Flörke and H.-J. Haupt, *J. Chem. Soc., Chem. Commun.*, 1990, 1728.
51. P. Chaudhuri, M. Winter, F. Birkelbach, P. Fleischhauer, W. Haase, U. Flörke and H.-J. Haupt, *Inorg. Chem.*, 1991, **30**, 4291.
52. P. Chaudhuri, M. Winter, B. P. C. Della Védova, P. Fleischhauer, W. Haase, U. Flörke and H.-J. Haupt, *Inorg. Chem.*, 1991, **30**, 4777.
53. D. Burdinski, F. Birkelbach, M. Gerdan, A. X. Trautwein, K. Wieghardt and P. Chaudhuri, *J. Chem. Soc., Chem. Commun.*, 1995, 963.
54. D. J. Cram, M. E. Tanner and R. Thomas, *Angew. Chem.*, 1991, **103**, 1048; *Angew. Chem., Int. Ed. Engl.*, 1991, **30**, 1024.
55. M. L. C. Quan and D. J. Cram, *J. Am. Chem. Soc.*, 1991, **113**, 2754.
56. K. Fujimoto and S. Shinkai, *Tetrahedron Lett.*, 1994, **35**, 2915.
57. A. W. Maverick, S. C. Buckingham, Q. Yao, J. R. Bradbury and G. G. Stanley, *J. Am. Chem. Soc.*, 1986, **108**, 7430.
58. M. Fujita, S. Nagao and K. Ogura, *J. Am. Chem. Soc.*, 1995, **117**, 1649.
59. R. W. Saalfrank, A. Stark, K. Peters and H.-G. von Schnering, *Angew. Chem.*, 1988, **100**, 878; *Angew. Chem., Int. Ed. Engl.*, 1988, **27**, 851.
60. R. W. Saalfrank, A. Stark, M. Bremer and H.-U. Hummel, *Angew. Chem.*, 1990, **102**, 292; *Angew. Chem., Int. Ed. Engl.*, 1990, **29**, 311.
61. R. W. Saalfrank, R. Burak, A. Breit, D. Stalke, R. Herbst-Irmer, J. Daub, M. Porsch, E. Bill, M. Müther and A. X. Trautwein, *Angew. Chem.*, 1994, **106**, 1697; *Angew. Chem., Int. Ed. Engl.*, 1994, **33**, 1621.
62. R. W. Saalfrank, B. Hörner, D. Stalke and J. Salbeck, *Angew. Chem.*, 1993, **105**, 1223; *Angew. Chem., Int. Ed. Engl.*, 1993, **32**, 1179.
63. I. G. Dance, *Aust. J. Chem.*, 1978, **31**, 2195.
64. G. Henkel, B. Krebs, P. Betz, H. Fietz and K. Saatkamp, *Angew. Chem.*, 1988, **100**, 1375; *Angew. Chem., Int. Ed. Engl.*, 1988, **27**, 1326.
65. S.-P. Huang and M. G. Kanatzidis, *Angew. Chem.*, 1992, **104**, 799; *Angew. Chem., Int. Ed. Engl.*, 1992, **31**, 787.
66. M. S. Lah, B. R. Gibney, D. L. Tierney, J. E. Penner-Hahn and V. L. Pecoraro, *J. Am. Chem. Soc.*, 1993, **115**, 5857.
67. H. B. Burgi, H. Gehrer, P. Strickler and F. K. Winkler, *Helv. Chim. Acta*, 1976, **59**, 2558.
68. P. N. W. Baxter, J.-M. Lehn, A. De Cian and J. Fischer, *Angew. Chem.*, 1993, **105**, 92; *Angew. Chem., Int. Ed. Engl.*, 1993, **32**, 69.
69. P. N. W. Baxter, J.-M. Lehn and D. Fenske, unpublished results.
70. A. Dupont, A. Marquis-Rigault, P. N. W. Baxter, J.-M. Lehn and A. Van Dorsselaer, *J. Inorg. Chem.*, 1996, accepted.
71. P. N. W. Baxter, J.-M. Lehn, G. Baum and D. Fenske, manuscript in preparation.
72. P. J. M. W. L. Birker, *J. Chem. Soc., Chem. Commun.*, 1980, 946.
73. (a) J.-F. You and R. H. Holm, *Inorg. Chem.*, 1991, **30**, 1431; (b) J.-F. You, G. C. Papaefthymiou and R. H. Holm, *J. Am. Chem. Soc.*, 1992, **114**, 2697.
74. J. Berzelius, *Ann. Phys. (Leipzig)*, 1826, **6**, 369, 380.
75. (a) M. T. Pope, 'Heteropoly and Isopoly Oxometallates', Springer, Berlin, 1983; (b) M. T. Pope, *Nature (London)*, 1992, **355**, 27.
76. J. Fischer, L. Ricard and R. Weiss, *J. Am. Chem. Soc.*, 1976, **98**, 3050.
77. (a) V. W. Day, W. G. Klemperer and O. M. Yaghi, *Nature (London)*, 1991, **352**, 115; (b) A. Müller, *Nature (London)*, 1991, **352** , 115; (c) I. Creaser, M. C. Heckel, R. J. Neitz and M. T. Pope, *Inorg. Chem.*, 1993, **32**, 1573.
78. A. Müller, M. Penk, R. Rohlfing, E. Krickemeyer and J. Döring, *Angew. Chem.*, 1990, **102**, 927; *Angew. Chem., Int. Ed. Engl.*, 1990, **29**, 926.
79. A. Müller, E. Krickemeyer, M. Penk, R. Rohlfing, A. Armatage and H. Bögge, *Angew. Chem.*, 1991, **103**, 1720; *Angew. Chem., Int. Ed. Engl.*, 1991, **30**, 1674.
80. A. Müller, R. Rohlfing, E. Krickemeyer and H. Bögge, *Angew. Chem.*, 1993, **105**, 916; *Angew. Chem., Int. Ed. Engl.*, 1993, **32**, 909.
81. A. Müller, *J. Mol. Struct.*, 1994, **325**, 13.
82. A. Müller, K. Hovemeier and R. Rohlfing, *Angew. Chem.*, 1992, **104**, 1214; *Angew. Chem., Int. Ed. Engl.*, 1992, **31**, 1192.
83. A. Müller, K. Hovemeier, E. Krickemeyer and H. Bögge, *Angew. Chem.*, 1995, **107**, 857; *Angew. Chem., Int. Ed. Engl.*, 1995, **34**, 779.

84. See, for example, (a) D. A. Tomalia, A. M. Naylor and W. A. Goddard, III, *Angew. Chem.*, 1990, **102**, 119; *Angew. Chem., Int. Ed. Engl.*, 1990, **29**, 138; (b) G. R. Newcome, C. N. Moorefield and G. R. Baker, *Aldrichim. Acta*, 1992, **25**, 31; (c) J. Issberner, R. Moors and F. Vögtle, *Angew. Chem.*, 1994, **106**, 2507; *Angew. Chem., Int. Ed. Engl.*, 1994, **33**, 2413.
85. W. T. S. Huck, F. C. J. M. van Veggel, B. L. Kropman, D. H. A. Blank, E. G. Keim, M. M. A. Smithers and D. N. Reinhoudt, *J. Am. Chem. Soc.*, 1995, **117**, 8293.
86. J. S. Lindsey, *New J. Chem.*, 1991, **15**, 153.
87. J. S. Miller, *Adv. Mater.*, 1990, **2**, 495.
88. D. C. Crans, *Comments Inorg. Chem.*, 1994, **16**, 35.
89. B. Schoentjes and J.-M. Lehn, *Helv. Chim. Acta*, 1995, **78**, 1.
90. L. Zelicovich, J. Libman and A. Shanzer, *Nature (London)*, 1995, **374**, 790.
91. A. Livoreil, C. O. Dietrich-Buchecker and J.-P. Sauvage, *J. Am. Chem. Soc.*, 1994, **116**, 9399.
92. D. Gatteschi, L. Pardi, A. L. Barra, A. Müller and J. Döring, *Nature (London)*, 1991, **354**, 463.
93. C. L. Hill and X. Zhang, *Nature (London)*, 1995, **373**, 324.

6

Polynuclear Transition Metal Helicates

EDWIN C. CONSTABLE

Institut für Anorganische Chemie der Universität Basel, Switzerland

6.1 INTRODUCTION

This chapter is concerned with the use of transition metal ions as an assembly motif for the twisting of organic ligands into multiple-helical structures. The approach described is predicated upon the concept of molecular threads, the conformations of which may be controlled by interaction with appropriate metal ions. Associated with this approach is the concept of coding within the metal ion and the molecular thread which predetermine the consequences of coordination. This coding is associated with the preferred coordination numbers and co-ordination geometries of the metal ions and the number and type of donor atoms presented by the ligand, together with the possible geometrical arrangement of these ligands. These principles of coding are the basic tenets of metallosupramolecular chemistry.[1–3] A number of reviews of the use of oligopyridine and other ligands in the formation of helicates have appeared.[2,4,5]

6.1.1 Terminology

We stated above that molecular threads play an important role in understanding the formation of multiple-helical complexes. In the context of transition metal-directed assembly, a molecular thread is an extended molecule which contains at least two separate domains which can coordinate to different metal ions—a multidentate bridging ligand.[6] A helicand is a molecular thread which gives rise to helical complexes; the helical complexes formed by helicands are termed helicates.[7] In this chapter double helicates and triple helicates are discussed.

6.1.2 Helical and Nonhelical Structures

Although this is not the place to enter into a detailed discussion, it is necessary to understand what we mean by helical and nonhelical structures. A number of excellent texts and review articles dealing with various aspects of chemical topology have been published, and the reader is referred to these for a more detailed discussion.[8–14] The essential property of a helix which distinguishes it from a nonhelical arrangement is the possession of chirality associated with a screw sense along a defined axis.[15] If we consider the arrangement of two molecular threads, the limiting nonhelical structure formed upon interaction with two metal ions is presented in Figure 1(a). The alternative, double-helical, arrangement of these same threads is presented in Figure 1(b). It should be clear that these two representations are the extremes, and the emergence of helical character occurs immediately there is any screwing of the structure in Figure 1(a). An alternative representation is presented in Figure 1(c). The double helicate can exist in left-handed (*M*) or right-handed (*P*) enantiomeric forms. Very often, the perception of a helical structure is a somewhat subjective process! The question of helical and nonhelical complexes has been discussed in some detail by Williams and co-workers.[16]

The second characteristic of a multiple-helical structure is its pitch—this is defined as the distance, measured along the principal axis of the helix, between repeat units. This is also indicated in Figure 1(b). When we consider molecular helices, the pitch will be determined by the chemical constitution of the molecular thread(s).

6.2 THE LIGAND DOMAIN MODEL AND LIGAND TYPES

6.2.1 The Ligand Domain Model in Metallosupramolecular Chemistry

The basic model for helication involves a molecular thread which may be intertwisted with one or more additional threads to give a helix. The twisting will be controlled by the coordination of the molecular thread to a metal ion. It is convenient to think of the molecular threads containing a number of discrete metal-binding domains separated by spacer groups. The choice of spacer group and the choice of metal-binding domain will then allow the specific formation of the desired helical structures with defined metal centres. These features are also illustrated in Figure 1. We will now consider the types of metal-binding domains that have proved to be popular in the design of molecular threads, before illustrating their use in the formation of helical complexes.

6.2.2 Ligand Domains for Helication

Although almost any multidentate ligand could, conceivably, be partitioned into a number of discrete metal-binding domains suitable for the formation of helical complexes, there are a number of recurrent structural themes to be found in the molecular threads which have found widespread application. This represents a number of different factors. The first, and undoubtedly the most important, is related to the accessibility and ease of synthesis of the desired ligands. As a

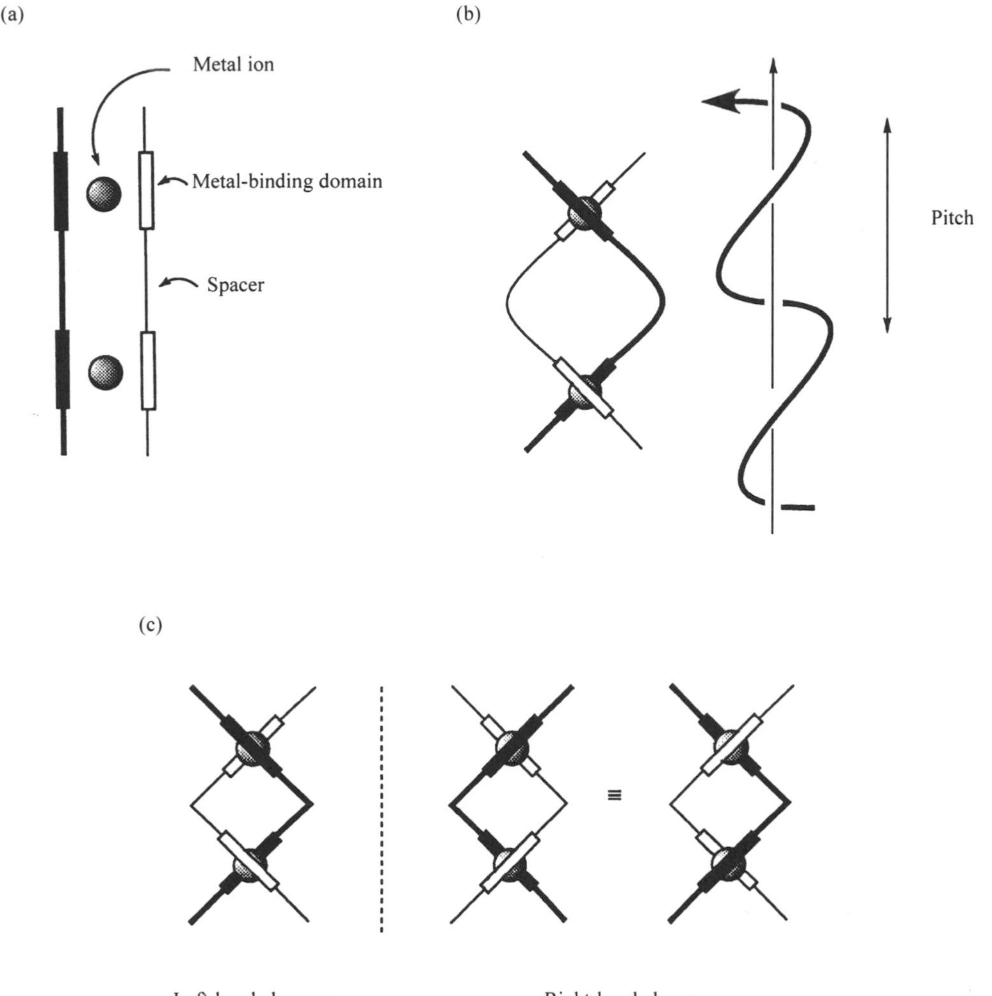

Figure 1 (a) A nonhelical or 'side-by-side' structure which results from the interaction of two molecular threads, each containing two *n*-dentate metal-binding domains, with two metal ions with a preferred coordination number of $2n$. (b) The alternative double-helical arrangement which can result from the same two molecular threads and metal ions seen in (a). An essential feature of a helix is its pitch. (c) A double helix is chiral and can exist in right-handed (*M*) and left-handed (*P*) forms.

consequence, many ligands contain metal-binding domains which are instantly familiar and which have a long and honourable history in coordination chemistry. A second factor is the predictability of the behaviour of the molecular threads upon coordination. In particular, the question of the splitting of the donor groups into the desired metal-binding domains, and the disposition of these domains in space becomes critical. The differences between the helical and nonhelical structures are often subtle and in many cases ligands which are preorganized or predisposed towards the formation of the twisted structures are adopted. In the final instance, this leads to the design of chiral molecular threads to give diastereomerically pure helicates with a single sense of direction associated with the screw thread.

At present, probably the majority of molecular threads giving rise to helicates incorporate at least one pyridine-type nitrogen donor. This represents the familiarity of the coordination chemist with such ligands and the widespread use of ligands consisting of chelating metal-binding domains containing linked pyridine rings over the past 50 years. This, combined with the facile synthetic routes available for the preparation of a wide range of derivatives of such compounds, has led to the design of numerous ligands incorporating oligopyridine and related groups such as 2,2′-bipyridyl (bipy; (**1**)), 1,10-phenanthroline (phen; (**2**)) or 2,2′:6′,2″-terpyridyl (terpy; (**3**)) metal-binding domains.

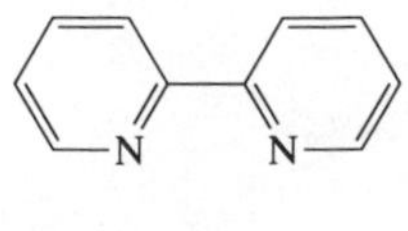

(1) 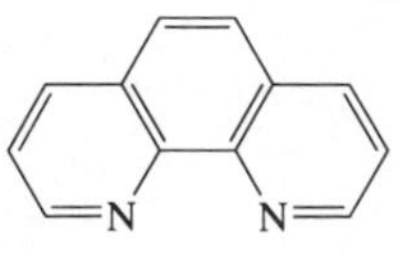 (2) 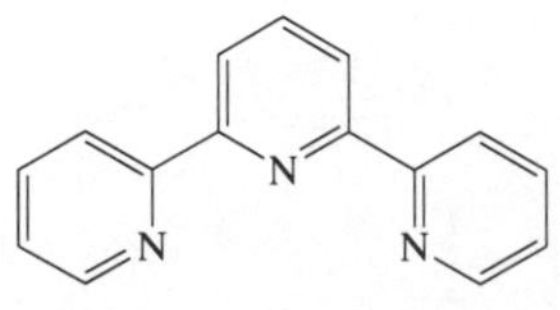 (3)

Although pyridine-type donors are extremely important, a significant number of helicates has been described in which some or all of the metal-binding domains consist of nitrogen donors in five membered rings. These may be pyrrole-type donors, such as in (**4**) or (**5**), which undergo deprotonation upon coordination to the metal centre, or neutral imidazole-type nitrogen atoms, as in (**6**). As the structural features leading to helication become better understood, an ever wider variety of metal-binding domains is to be expected.

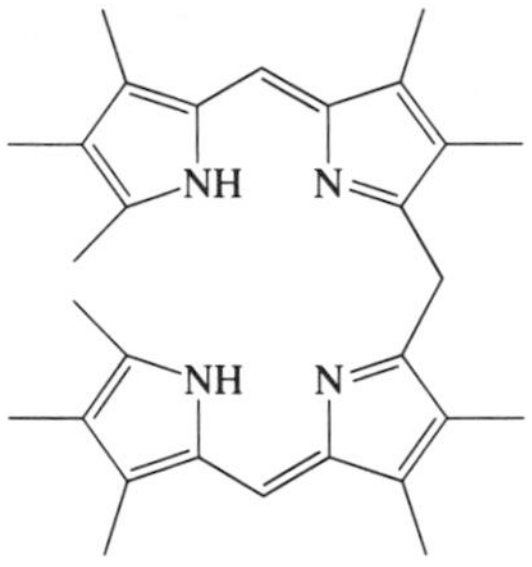

(4) 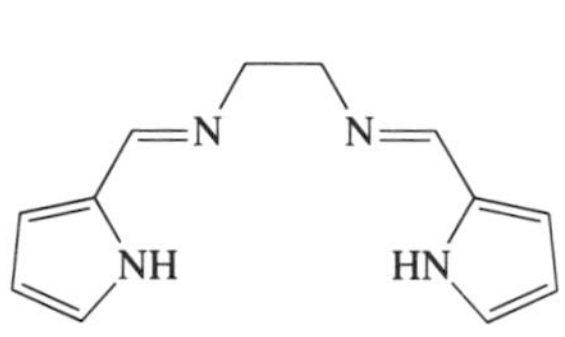 (5) 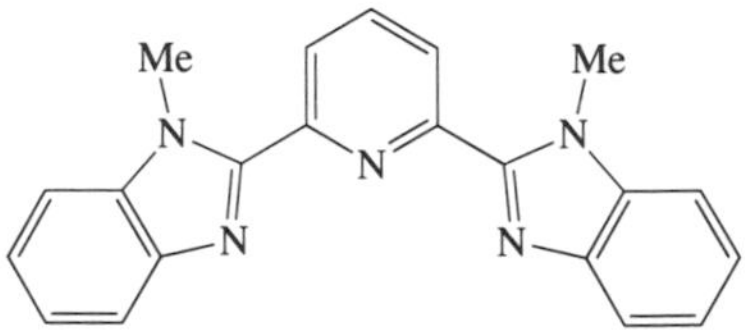 (6)

The majority of helicands contain nitrogen donor atoms, often in heterocyclic rings, but also in acyclic imines (**5**) or hydrazones (**7**). Helicands containing other donor atoms are, as yet, relatively uncommon, but isolated examples, such as (**8**) or (*S*,*S*)-(+)-$Ph_2PCH_2CH_2PPhCH_2CH_2PPhCH_2CH_2PPh_2$ have been described.

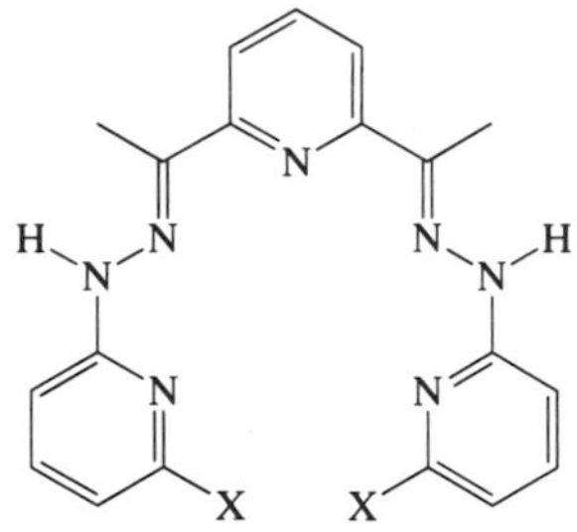

(7) X=H 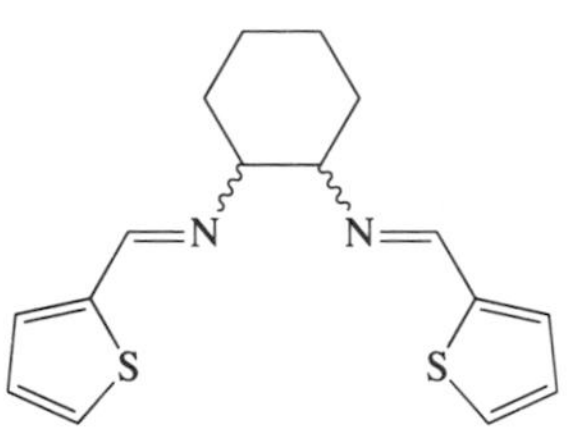 (8)

6.3 SYNTHETIC CONSIDERATIONS

6.3.1 Self-assembly

The methodology widely adopted for the synthesis of helicates relies upon the spontaneous self-assembly of the desired structure upon the interaction of the helicands and the metal centres. Self-assembly has been defined elsewhere in this volume, and it is clear that the successful self-assembly of helicates relies upon the correct matching of the molecular coding within the helicand and the metal centres. At its simplest this means that the number of donor sites available at each metal centre should be a simple multiple of the number of donor atoms within each metal-binding domain. For example, a double-helical complex could result from the interaction of two molecular threads containing two *n*-dentate metal-binding domains with two metal ions with a favoured coordination number of 2*n*. Similarly, a triple-helical complex could result from the interaction of three molecular threads containing two *m*-dentate metal-binding domains with two metal ions with a favoured coordination number of 3*m*. In practice, this very often means that four-coordinate metal centres are involved in the formation of double-helical complexes from molecular threads containing bidentate metal-binding domains. Similarly, six-coordinate metal centres are used for the assembly of double-helical complexes from molecular threads containing tridentate metal-binding domains and triple-helical complexes when the molecular thread contains bidentate metal-binding regions. Implicit in this analysis is the assumption that all of the coordination sites of the metal ion are occupied by donors from the molecular thread. If this is not the case, then the coding is not so rigidly enforced and other possibilities arise. The most common exceptions to the coding above arise when double-helical complexes are formed from the interaction of molecular threads containing bidentate domains with six-coordinate metal ions. For steric, or other reasons, it is not always possible to wrap three molecular threads around the metal centres, and the double-helical complexes arise, with the remaining donor sites of each metal centre occupied by two monodentate or one bidentate ligand. These various possibilities are illustrated in Figure 2. The above discussion assumes that helical, rather than nonhelical, structures are formed. This is by no means always the case!

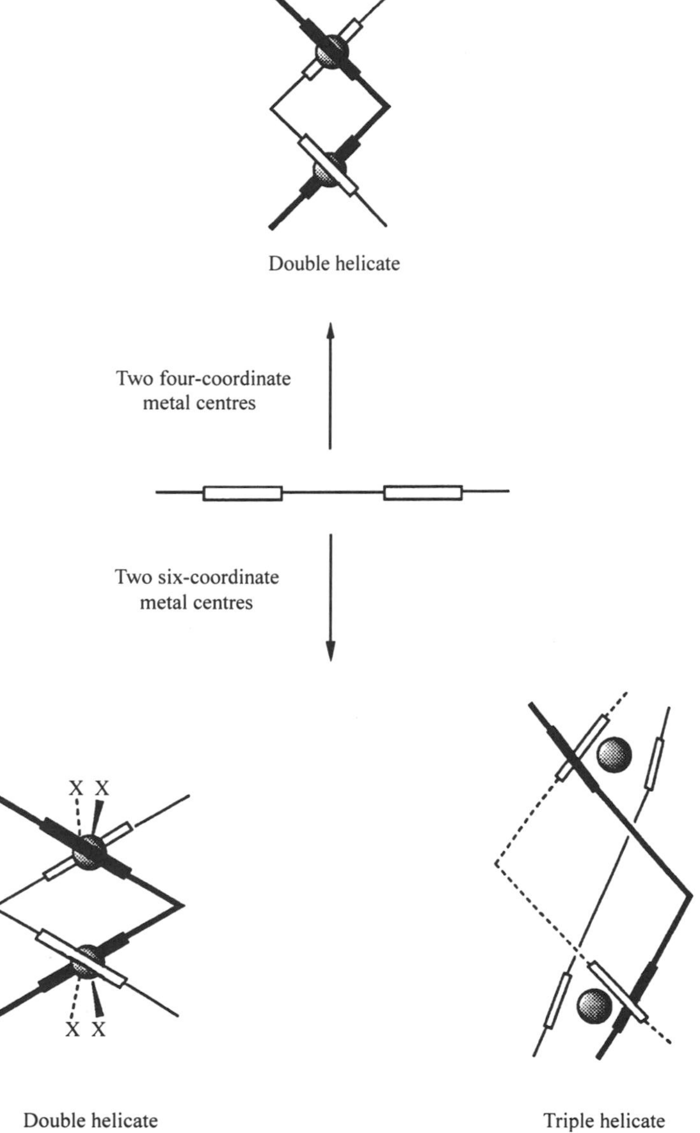

Figure 2 A molecular thread containing two bidentate metal-binding domains represents an ambiguous piece of coding. Interaction with two four-coordinate metal centres could give rise to a dinuclear double helicate, while interaction with two six-coordinate metal centres could give either a dinuclear double helicate or a dinuclear triple helicate. The course of the reaction with the six-coordinate metal ion will depend upon the nature of the spacer groups between the metal-binding domains and the presence of other potential donors in the solution.

6.3.2 Ligand-imposed Constraints—Preorganization and Predisposition

Even when the correct combination of metal-binding domains in the ligand thread and coordination requirements at the metal centres are present, it is by no means certain that their reaction will lead to the desired multiple-helical complex. One of the ways to ensure that the twisted topology is obtained is to use a preorganized molecular thread—a ligand which is already in a helical or chiral conformation. An example of such a preorganized ligand is seen in the helically chiral macrocycle (**9**), which incorporates two didentate 1,10-phenanthroline metal-binding domains. The strands containing these metal-binding domains are already in a double-helical arrangement, and coordination to a copper(I) centre gives the expected double-helical complex containing the $[Cu(\mathbf{9})]^+$ cation.[17] In a similar way, the ligand (**10**) is constrained to a helicene-like conformation, and binding to sodium ions gives mixtures of various helical complexes.[18]

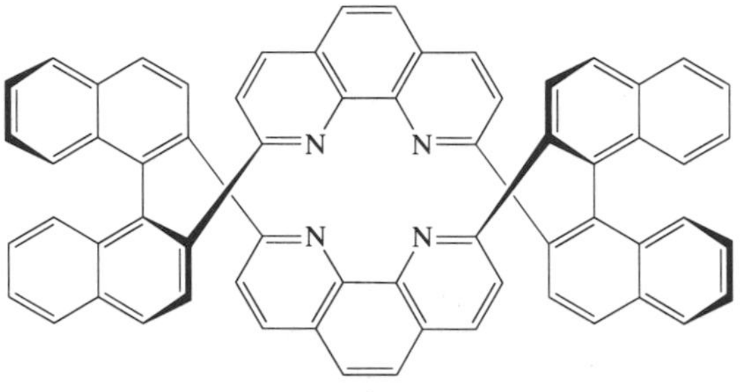

(**9**)

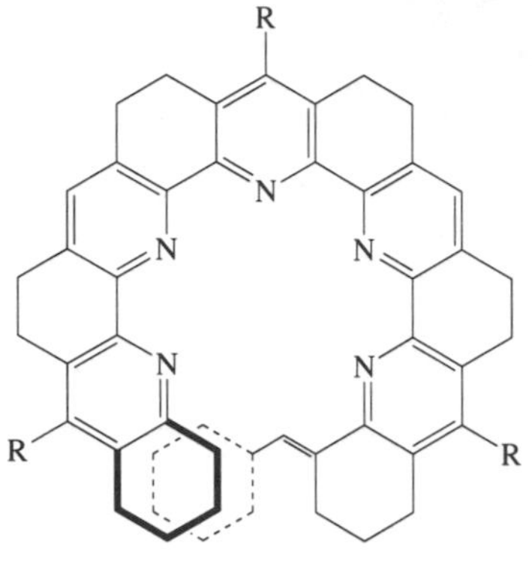

(**10**)

Ligands which are predisposed to the formation of helical complexes do not necessarily adopt a helical conformation in the free ligand, but there are structural features present such that after coordination the helical geometry is favoured over other alternatives. One of the first examples of ligand designed to be predisposed for the formation of helicates was (**11**) a substituted 2,2′:6′,2″:6″,2‴-quaterpyridine. The steric interactions between the methyl groups in the 3″- and 5′-positions prevent the ligand generating a planar N_4 donor set of the type which would favour coordination to a square-planar or octahedral metal ion.[19] The reaction of (**11**) with copper(I) salts gave the anticipated double-helical complex containing the $[Cu_2(\mathbf{11})_2]^{2+}$ cation.

6.3.3 Metal-imposed Constraints

We discussed above how the total coordination number of a metal centre could be matched to the number of donor atoms within the metal-binding domains of the molecular threads. However, we said nothing about the preferred spatial arrangement of the ligands about the metal. This is best illustrated by considering a four-coordinate metal centre. The two commonest coordination geometries associated with four coordination are square planar and tetrahedral. Obviously, the binding of two bidentate domains to a four-coordinate metal centre will have different consequences for the molecular threads, depending upon whether the favoured geometry at the metal is square planar or tetrahedral. This is well illustrated by the ligand 2,2′:6′,2″:6″,2‴-quaterpyridine (**12**), which has a total of four nitrogen donor atoms. With square planar centres, such as palladium(II), mononuclear complexes are formed, in which the four nitrogen donors all lie in a plane and are coordinated to the single metal ion. In contrast, with ions such as copper(I), which has a preference for a tetrahedral geometry, dinuclear double-helical complexes are formed. The coordination of bidentate domains to a tetrahedral metal centre has proved to be a versatile motif in metallosupramolecular chemistry and has been used for the assembly of a whole variety of novel structures in addition to the helicates discussed in this chapter.

6.4 HOMODINUCLEAR DOUBLE HELICATES

6.4.1 Oligopyridines and related ligand domains

Reference to the primary literature has only been made where it is not discussed in detail in review articles.[2,4]

6.4.1.1 Two tetrahedral metal centres—[4 + 4] helicates

One of the simplest ways to assemble a double-helicate according to Figure 2 would be the interaction of a ligand thread containing two bidentate metal-binding domains with two tetrahedral metal centres. The generic structural features for a ligand incorporating a bidentate bipy or phen domain are shown in Figure 3, and the simplest such structure would be a 2,2′:6′,2″:6″,2‴-quaterpyridine with a direct link between 6-positions of the two domains.

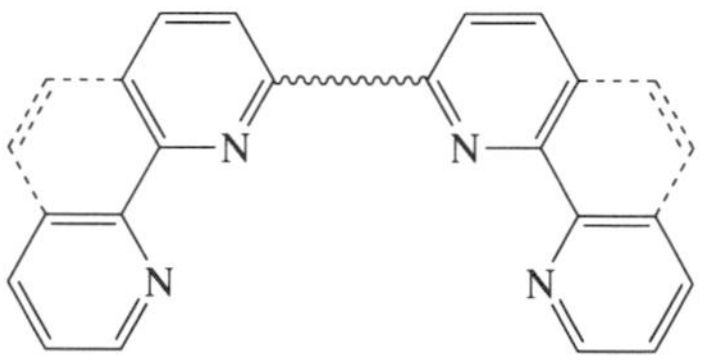

Figure 3 The generic structure of a helicand based upon two bipy or phen metal-binding domains. Such a ligand is expected to give dinuclear double helicates with four-coordinate metal centres.

As mentioned above, perhaps the first example of a designed helicate was reported by Lehn and co-workers, who used the interaction of the preorganized ligand (**11**) with copper(I) to generate the desired helical structures. The reaction with copper(I) salts resulted in the spontaneous self-assembly of the double-helical cation $[Cu_2(\mathbf{11})_2]^{2+}$. As expected, complexes with metal ions possessing a preference for six-coordinate geometries did not give double-helical complexes. The subtlety of the coding is well illustrated by the electrochemical behaviour of the double-helical complex $[Cu_2(\mathbf{11})_2]^{2+}$. Upon one-electron oxidation a mixed oxidation state complex, $[Cu_2(\mathbf{11})_2]^{3+}$, containing a copper(II) and a copper(I) centre is formed; however, upon further oxidation, or upon standing, a mononuclear copper(II) complex is formed. Detailed electrochemical studies established a reaction scheme of the type indicated in Figure 4. The implication is that the most stable copper(II) complex with (**11**) is mononuclear, and this is borne out by the structural characterization of the complex $[Cu(\mathbf{11})(H_2O)][ClO_4]_2$ which contains a five-coordinate copper(II) centre in a distorted coordination geometry.

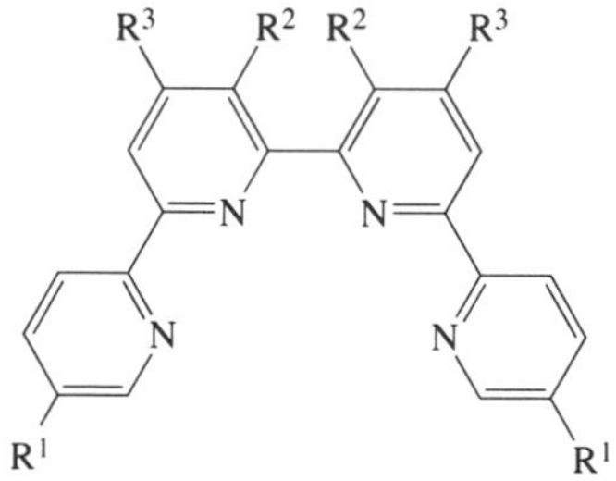

(**11**) $R^1 = R^2 = Me$, $R^3 = H$
(**12**) $R^1 = R^2 = R^3 = H$
(**13**) $R^1 = R^2 = H$, $R^3 = SMe$

It turns out that it is not necessary to use the preorganized ligand (**11**), and it has been demonstrated that (**12**) itself, and a variety of derivatives with nonsterically demanding substituents, also generate double-helical complexes upon reaction with copper(I).[2,4,20] The structure of the

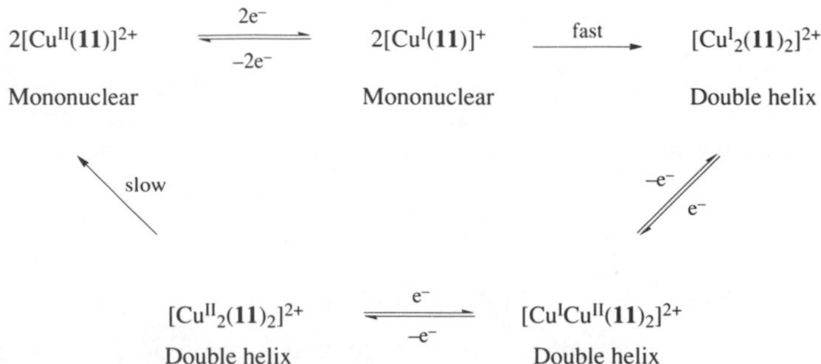

Figure 4 The electrochemical behaviour of copper complexes of 2,2′:6′,2″:6″,2‴-quaterpyridine ligands involves nuclearity changes associated with the preferred coordination geometries of the copper(II) and copper(I) oxidation states.

double-helical cation $[Cu_2(\mathbf{12})_2]^{2+}$ is presented in Figure 5. The role of the substituents is not to preorganize the ligand, but to control the pitch of the helix and the metal–metal distance. For example, in the complex $[Cu_2(\mathbf{12})_2]^{2+}$ the Cu···Cu distance is 317 pm whereas in $[Cu_2(\mathbf{11})_2]^{2+}$ it is 390 pm and in $[Cu_2(\mathbf{13})_2]^{2+}$ it is 332 pm. Electrochemical studies of these double-helical complexes parallel those with (**11**), although in most cases the mixed-oxidation state species are not observable. As expected, the precise potentials for the various processes are dependent upon the nature of the substituents present. Other metal ions with a preference for tetrahedral geometries are also expected to give double-helical complexes with these ligands, and the silver(I) cation has been shown to give a dinuclear double-helical cation $[Ag_2(\mathbf{12})_2]^{2+}$ with an Ag···Ag distance of 310 pm; this illustrates a subtle metal ion control over the pitch of the helicate as this Ag···Ag distance is shorter than the Cu···Cu distance with the same ligand. Another feature which is observed in each of these complexes is a stacking of approximately coplanar pyridine rings from the different molecular threads. This is thought to be a direct consequence of the adoption of the helical structure, rather than a necessary feature for its assembly, although it may make an additional small, but favourable, energetic contribution to the overall stability of the final structure. Although the analogous ligand containing two phen metal-binding domains, 2,2′-bi(1,10-phenanthroline), has been prepared, little is known about its coordination chemistry.

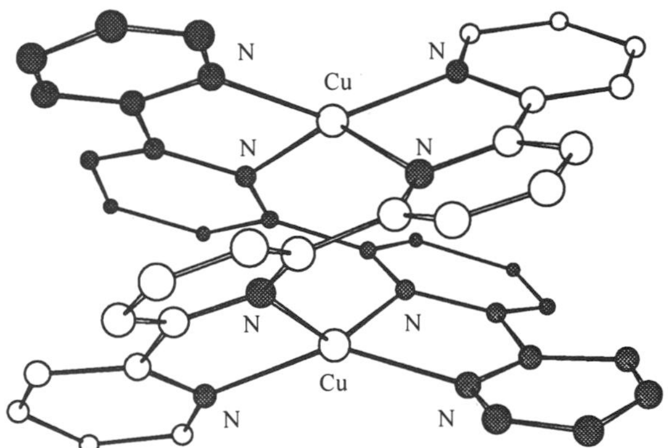

Figure 5 The interaction of 2,2′:6′,2″:6″,2‴-quaterpyridine ligands with copper(I) or silver(I) leads to the formation of dinuclear double helicates.[2,4] One of the molecular threads has been shaded for clarity.

The structural development of the helicand involves the introduction of spacer groups between the two metal-binding domains. Clearly, the position through which the rings are linked, and the chemical nature and length of the spacer group will be of paramount importance. Helicates in which the two oligopyridine domains are linked by a single atom spacer (O, S, NR or CR_2) have not yet been described, but a variety of ligands containing two bipy or phen domains linked by CH_2CH_2 spacers have been reported. The flexible CH_2CH_2 spacer linking the 6,6′-positions in (**14**) allows the formation of double-helical dinuclear complexes upon reaction with copper(I), whereas the constrained ligand (**15**) cannot be twisted into the correct conformation to give such structures.[21–3] A crystal structure determination confirmed the double-helical character of the $[Cu_2(\mathbf{14})_2]^{2+}$ cation in the solid state, and the observation in the ^1H NMR spectrum of an *ABCD* multiplet for the bridging group confirms the maintenance of this structure in solution. Double-helical structures are also formed when the CH_2CH_2 spacer is connected to the 5-position of the bipy domain, as in (**16**). Dinuclear double helicates in which the metal-binding domains are linked through the 4-positions have not yet been unambiguously established. This general pattern of reactivity is conserved in a variety of substituted derivatives of (**14**), together with analogues incorporating 1,10-phen metal-binding domains.[24] In the case of phen metal-binding domains, double-helical complexes are obtained when the CH_2CH_2 spacer links the 4-positions of the heterocycles.

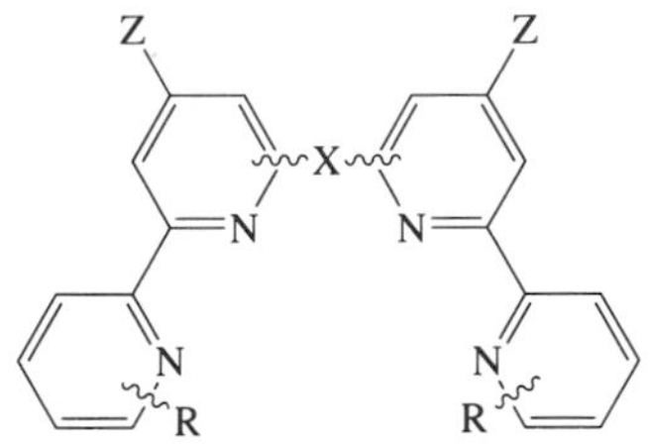

	X	R	Z
(**14**)	6,6'-CH_2CH_2	6-Me	H
(**15**)	6,6'-(*E*)-CH=CH	6-Me	H
(**16**)	5,5'-CH_2CH_2	5-Me	H
(**18**)	6,6'-CH_2OCH_2	6-Me	H
(**19**)	6,6'-(1,3-C_6H_4)	H	H or SMe
(**21**)	6,6'-(3,3'-C_6H_3-C_6H_3)	H	H or SMe
(**23**)	6,6'-S	H	H

There is a significant difference between the behaviour of ligands based upon (**12**) and those which incorporate a flexible spacer between the didentate domains. This is well illustrated by the series of ligands with two phen metal-binding domains which has been extended to the compounds (**17**) linked by a variety of α,ω-alkanediyl spacers. All of these ligands give dinuclear complexes upon reaction with copper(I). However, in many cases a mixture of the double-helical and non-helical dinuclear complexes is obtained. These complexes have been used elegantly by Sauvage as the required double-helical precursors for the synthesis of molecular knots (see Chapter 2). The yields of the desired knotted products were dependent upon the length of the spacer group and of the linker used to connect the 4-hydroxyphenyl substituents. Furthermore, the yield of the knotted product reflected the amount of the double-helical open-chain precursor present in the equilibrium mixture obtained from the reaction of the ligands with copper(I) salts (Figure 6).[8–10,25] Similar mixtures of double-helical and nonhelical complexes are expected with all ligands containing flexible, but not preorganized or predisposed linkers between the metal-binding domains.

	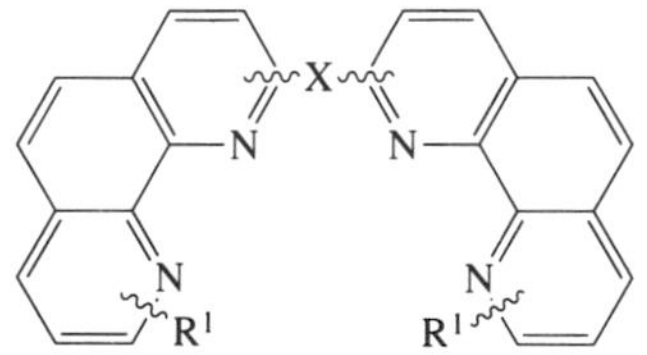 X	R^1
(**17**)	2,2'-$(CH_2)_n$ $n = 2,4,6$	4-$R^2OC_6H_4$ R^2 = Me, H
(**20**)	2,2'-(1,3-C_6H_4)	4-$R^2OC_6H_4$ R^2 = Me, H

Lehn introduced the CH_2OCH_2 group as a versatile spacer in helicands, and the prototype ligand (**18**) reacts with copper(I) to give a dinuclear double-helical complex $[Cu_2(\mathbf{18})_2]^{2+}$.[26] The ^1H NMR spectra of $[Cu_2(\mathbf{18})_2]^{2+}$ is characteristic of the double-helical species, and the methylene groups appear as an *AB* multiplet as a consequence of the nonequivalence of the two diastereotopic

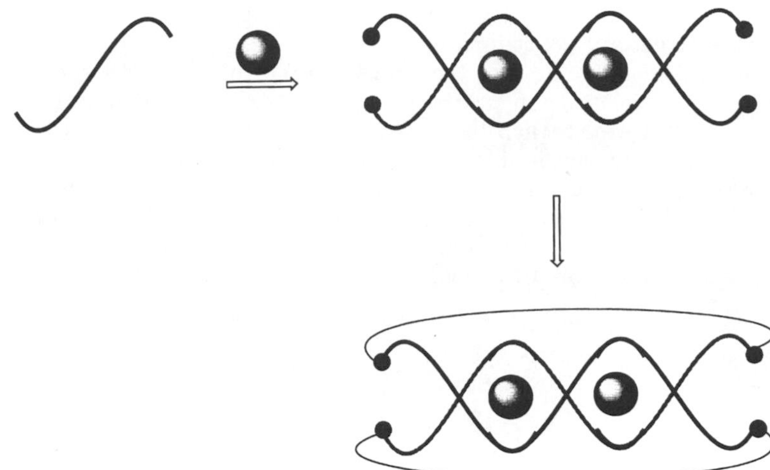

Figure 6 A knotted molecule may be prepared by the connecting together of the ends of the molecular threads in a double helix in an appropriate manner.

protons. Further evidence for the helical character of the complex was afforded from the observation that the *AB* pattern of the methylene group was further split upon the addition of (*S*)-2-anthryl-1,1,1-trifluoromethylethanol (Pirkle's reagent); this is only consistent with the presence of a racemic mixture of chiral complexes. A similar double-helical complex is also obtained with silver(I).

It should be clear from the above discussion that, while dicopper(I) and disilver(I) complexes obtained from (**12**) and its derivatives are necessarily double helical, those in which the bidentate domains are linked by more flexible spacers are not necessarily so. This problem has been addressed by a number of research groups, and the introduction of aromatic spacers between the metal-binding domains has been shown to be a valuable method for restricting the number of possibilities in conformational space and the predisposition of the ligand towards the formation of double-helical complexes.[25,27,28] In particular, 1,3-phenylene groups have proved to be useful, and the ligand (**19**) cleanly forms double-helical $[M_2(\mathbf{19})_2]^{2+}$ complexes with silver(I) and copper(I). Similarly, excellent yields of molecular trefoil knots are obtained from the double-helical complexes resulting from the coordination of (**20**) to copper(I). The process may be extended to the introduction of 3,3′-biphenylene spacer groups in (**21**), which also gives the expected double-helical complexes with copper(I) or silver(I) (Figure 7). However, to introduce a word of caution, the isomeric ligand with a 2,2′-biphenylene spacer only gives a mononuclear complex with copper(I).

The various spacers which have been introduced control the pitch of the helix, which is most easily measured in terms of the internuclear distances within the dinuclear units. For example, the internuclear Cu···Cu distance in $[Cu_2(\mathbf{21})_2][BF_4]_2$ is 626 pm. A final example of an unusual spacer group is seen in compound (**22**) in which a macrocyclic ligand is used to separate the two bidentate metal-binding domains. The reaction of (**22**) with copper(I) or silver(I) gives tetranuclear $[M_4(\mathbf{22})_2]$ complexes which possess a double-helical structure. Two of the four metal ions occupy the macrocyclic N_4 coordination sites, while the other two are each coordinated to one phen metal-binding domain from each ligand strand.[29] In one respect this ligand may be regarded as predisposed towards helication; the tetrahedral geometry about the metal ions in the macrocyclic sites orientates the ligand correctly for the required intertwisting to give the double helix. It is worth comment that the linking together of two bipy domains is not a sufficient structural requirement for helication. For example, (**23**) does not give a dinuclear double helicate upon reaction with copper(I).

6.4.1.2 Two octahedral metal centres—[6 + 6] helicates

Although the majority of the dinuclear double-helicates which have been described involve four-coordinate metal centres, a range of complexes incorporating six-coordinate metal ions are known. The molecular thread should contain two tridentate metal-binding domains, and coordination to two six-coordinate metal centres is expected to lead to the desired double-helical structures. Within the oligopyridine series, the tridentate domain will be provided by a 2,2′:6′,2″-terpyridine, and the prototype ligand is (**24**) in which the two metal-binding domains are directly linked through the

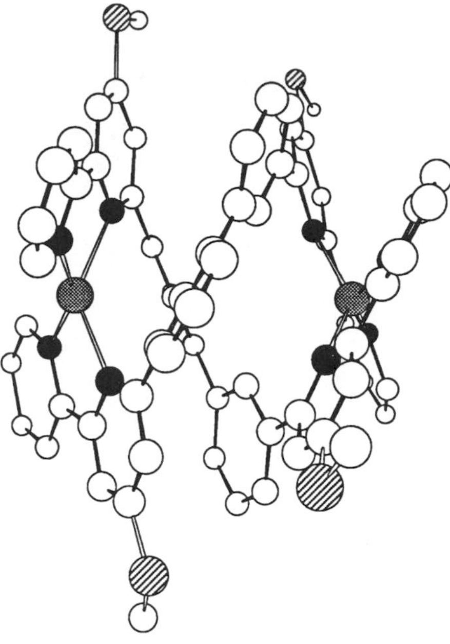

Figure 7 The double-helical cation $[Cu_2(\mathbf{21})_2]^{2+}$ is formed from the reaction of (**21**) with copper(I) salts.

(**22**)

6-positions. The ligand (**24**) partitions into two tridentate domains upon coordination to metal ions with a preference for octahedral geometries. The presence of substituents has no influence upon this behaviour, and the ligands (**25**) form exactly analogous complexes.[30] Dinuclear double helicates are formed with iron(II), cobalt(II), nickel(II), copper(II) and cadmium(II) and solid-state structural characterizations have confirmed the topology of the dicadmium(II) (Figure 8) and dicopper(II) complexes.

(**24**) X = H
(**25**) X = SR

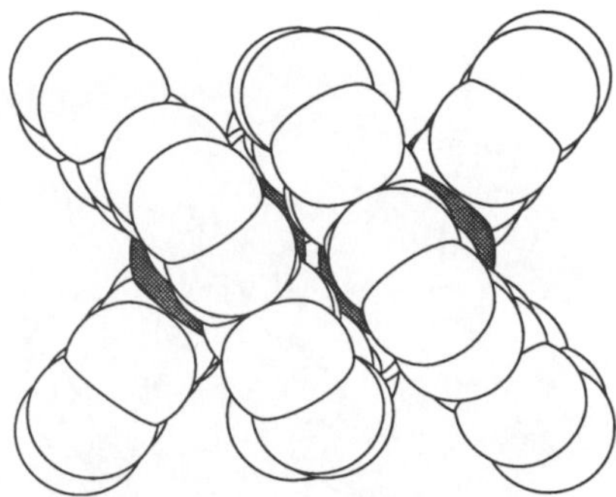

Figure 8 The double-helical cation $[Cd_2(\mathbf{24})_2]^{4+}$ obtained from the reaction of (**24**) with cadmium(II). The ligand partitions into two tridentate domains and the two molecular threads create a double-helical ligand array which can bind two octahedral cations.

In both cases, there are coplanar interactions between approximately parallel rings of the two ligands in the double-helical complex. The electrochemical behaviour of these complexes is of interest and is somewhat affected by the presence and nature of substituents on the ligand. There is a considerable degree of communication between the metal centres in these double-helical complexes, and both the dicobalt(II) and diiron(II) compounds exhibit two reversible metal(II)/metal(III) processes.

In principle, the terpy domains could be linked by a range of spacer groups to give a series of ligands analogous to those discussed in Section 6.4.1.1, designed to give dinuclear double helicates containing two octahedral centres. To date, relatively few such ligands have been developed. Compound (**26**), containing a thioether linker between the two domains does not give double-helical complexes, but (**27**) with a CH_2CH_2 bridge gives the expected double-helical $[M_2(\mathbf{27})_2]^{4+}$ complex cations (M = Fe or Ru).[31] An incomplete crystal structure determination of the complex $[Ru_2(\mathbf{27})_2][PF_6]_4$ confirmed the formation of the double helix and the presence of two $\{Ru(terpy)_2\}$ centres, although it was not possible to fully refine the structure. The two ruthenium centres are not strongly coupled and a single ruthenium(II)/ruthenium(III) process is observed at + 1.40 V (vs. SCE).

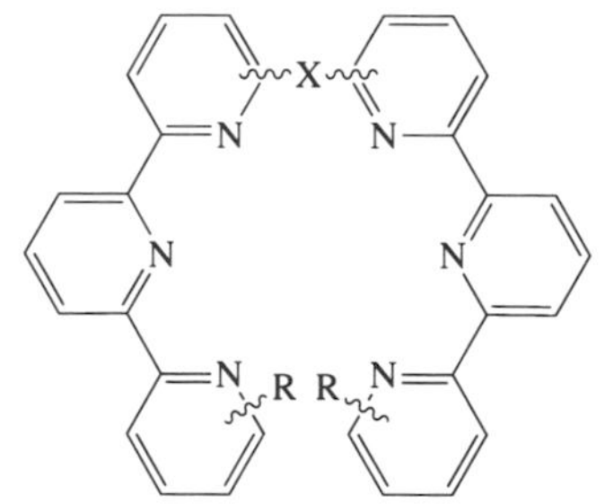

	X	R
(**26**)	6,6'-S	H
(**27**)	6,6'-CH_2CH_2	6,6'-(4-$MeOC_6H_4$)

It is also possible to incorporate a variety of aromatic spacers between the terpy domains, and this is typified by the coordination behaviour of the septipyridine ligand (**28**). A double-helical array of two (**28**) ligands will present a total of 14 donor atoms. If these are partitioned between two octahedral metal centres, two pyridine rings are left noncoordinated and could occupy terminal positions or a bridging site between the two terpy domains; that is, the ligands could be regarded as terpy–terpy–py or terpy–py–terpy. A structural analysis of the double-helical complex $[Co_2(\mathbf{28})][PF_6]_4$ reveals that this analysis of the bonding characteristics of (**28**) is correct, and one ligand strand is of the type terpy–terpy–py and the other terpy–py–terpy (Figure 9).[32] Although the two cobalt(II) centres are in different environments, only a single cobalt(II)/cobalt(III) process is observed. A dicopper(II) complex of (**28**) is thought to have an analogous structure, although the location of the noncoordinated pyridine rings in the double-helical array has not been unambiguously established.

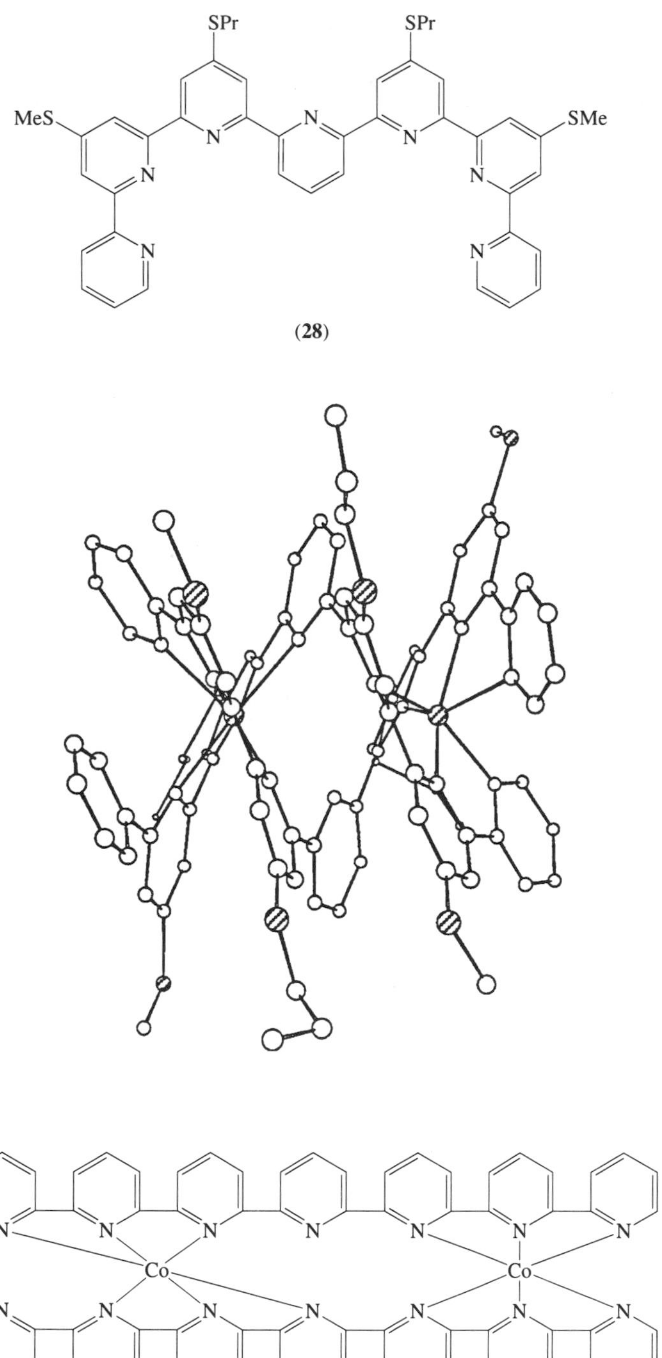

Figure 9 The coding in (**28**) is ambiguous, and a double-helical array of two (**28**) ligands contains a total of 14 donor atoms. When two six-coordinate metal ions are bound, as in the cation $[Co_2(\mathbf{28})_2]^{4+}$ two donor atoms are noncoordinated. The bonding scheme is shown in the cartoon representation underneath the crystal structural analysis. The double helicate is represented in a planar form to emphasize the coordination pattern.

6.4.1.3 Other coordination geometries

We will conclude our survey of helicands based upon oligopyridines by considering some unusual coordination modes which result in helical structures and some structures in which two or more different coordination geometries are present.

Even 2,2′:6′,2″-terpyridines can give dinuclear double-helical complexes with copper(I). For example, both (**29**) and (**30**) give rise to double-helical $[Cu_2(\mathbf{29})_2]^{2+}$ or $[Cu_2(\mathbf{30})_2]^{2+}$ species which have been structurally characterized (Figure 10).[33] The presence of the phenyl groups in the 6-position of the ligand is important in stabilizing the structure, but there is good evidence that related, air-sensitive, species are obtained with terpy itself. The coordination mode of the copper(I) ions is of some interest. A double-helical array of two terpy ligands presents a total of six donor atoms, and a [4 + 2] helicate is formed. The two copper(I) centres are in different environments with one in a distorted tetrahedral environment and the other in an approximately linear two-coordinate environment. However, the formally two-coordinate centre also shows two longer range interactions with the 'bridging' pyridine rings. Solution NMR spectroscopic studies of these and related complexes suggest that, in solution, the two metal centres are in equivalent environments, or a rapid dynamic process interconverts the [4 + 2] and [2 + 4] structures.

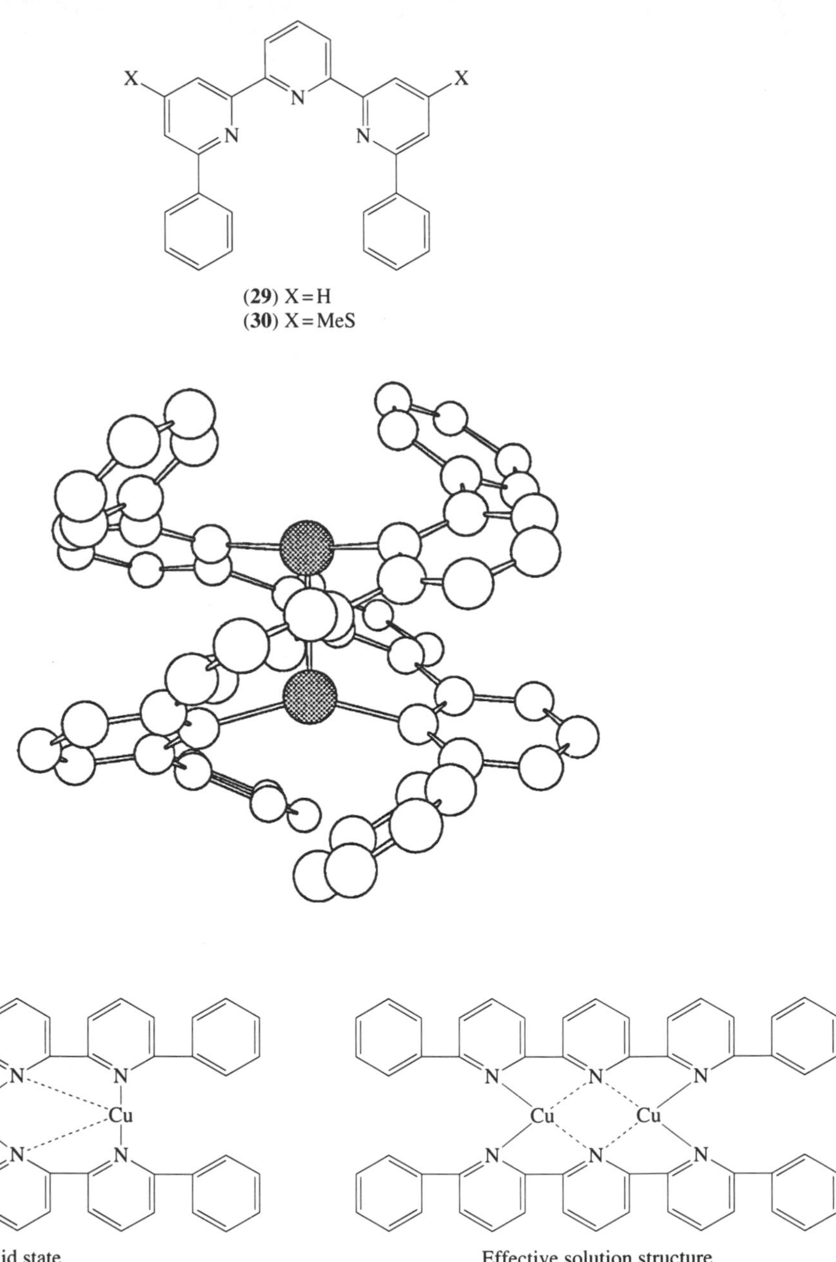

Figure 10 In the solid state $[Cu_2(\mathbf{29})_2]^{2+}$ contains one four-coordinate and one two-coordinate copper(I) centre. In solution the two centres appear to be equivalent. The double helicate is represented in a planar form to emphasize the coordination pattern.

A double-helical array formed from two 2,2′:6′,2″:6″,2‴:6‴,2⁗-quinquepyridine (qpy) (**31**) ligands will present a total of 10 donor atoms.[2,4] The most usual partitioning of this donor set is to give [6 + 4] helicates, and this general scheme is presented in Figure 11. The [6 + 4] partitioning is observed in the mixed oxidation state complexes $[Cu_2L_2]^{3+}$ obtained from the reaction of qpy or a 4′,4‴-disubstituted qpy with copper(II) acetate in methanol, or with copper(I) salts in the presence of air. The copper(II) centre is in a six-coordinate N_6 environment while the copper(I) is in a four-coordinate N_4 site. The metal–metal distance and the pitch of the double helix are critically dependent upon the presence and nature of any substituents, as is the detailed electrochemical behaviour of the complexes.

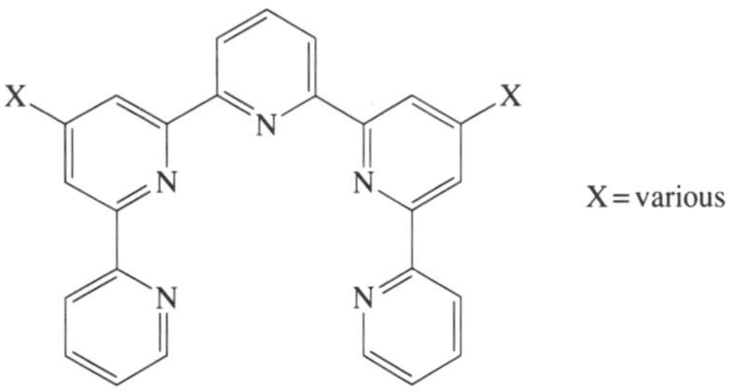

(**31**)

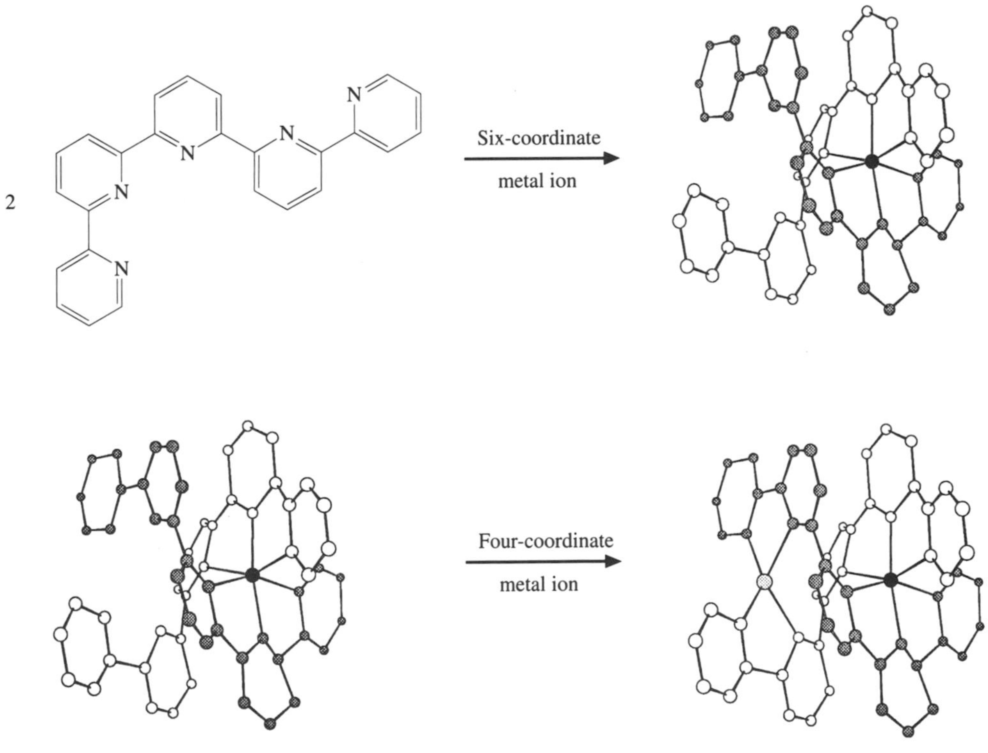

Figure 11 A cartoon representation of a [6 + 4] double helicate from two qpy helicands. The structure arises from the separate interaction with four- and six-coordinate metal centres.

Oxidation of the above complexes leads to new double-helical dicopper(II) complexes, in which each copper(II) centre is six (or five)-coordinate. One centre remains in the N_6 environment created by the ligand array, while the other is coordinated to four nitrogen donors from the double helix and two other ancillary donor atoms. This may be described as a [6 + 4 + (2)] helicate, and is the common structural feature for complexes of qpy and its derivatives with six-coordinate first row transition metals. Dinuclear 2:2 complexes are formed with manganese(II), iron(II), cobalt(II), nickel(II), copper(II) and zinc(II) and those with cobalt(II) and nickel(II) have been structurally characterized. In many cases the two ancillary donor atoms are provided by an acetate counter ion.

Some of the complexes with labile metal ions undergo a number of dynamic processes, and this will be exemplified by a brief consideration of the behaviour of cobalt(II) complexes of qpy ligands. Cobalt(II) can form double-helical dinuclear or mononuclear seven-coordinate complexes with qpy and its derivatives. In weakly donor solvents, the double-helical species are formed, but when the reaction is performed in a good donor solvent or when the double-helical complex is reacted with a good donor solvent or a donating anion such as chloride conversion to the mononuclear species occurs. Traces of water are sufficient to convert the double-helical species to the corresponding mononuclear $[CoL(H_2O)_2]^{2+}$ species. An overview of the use of qpy ligands as helicands is presented in Figure 12.

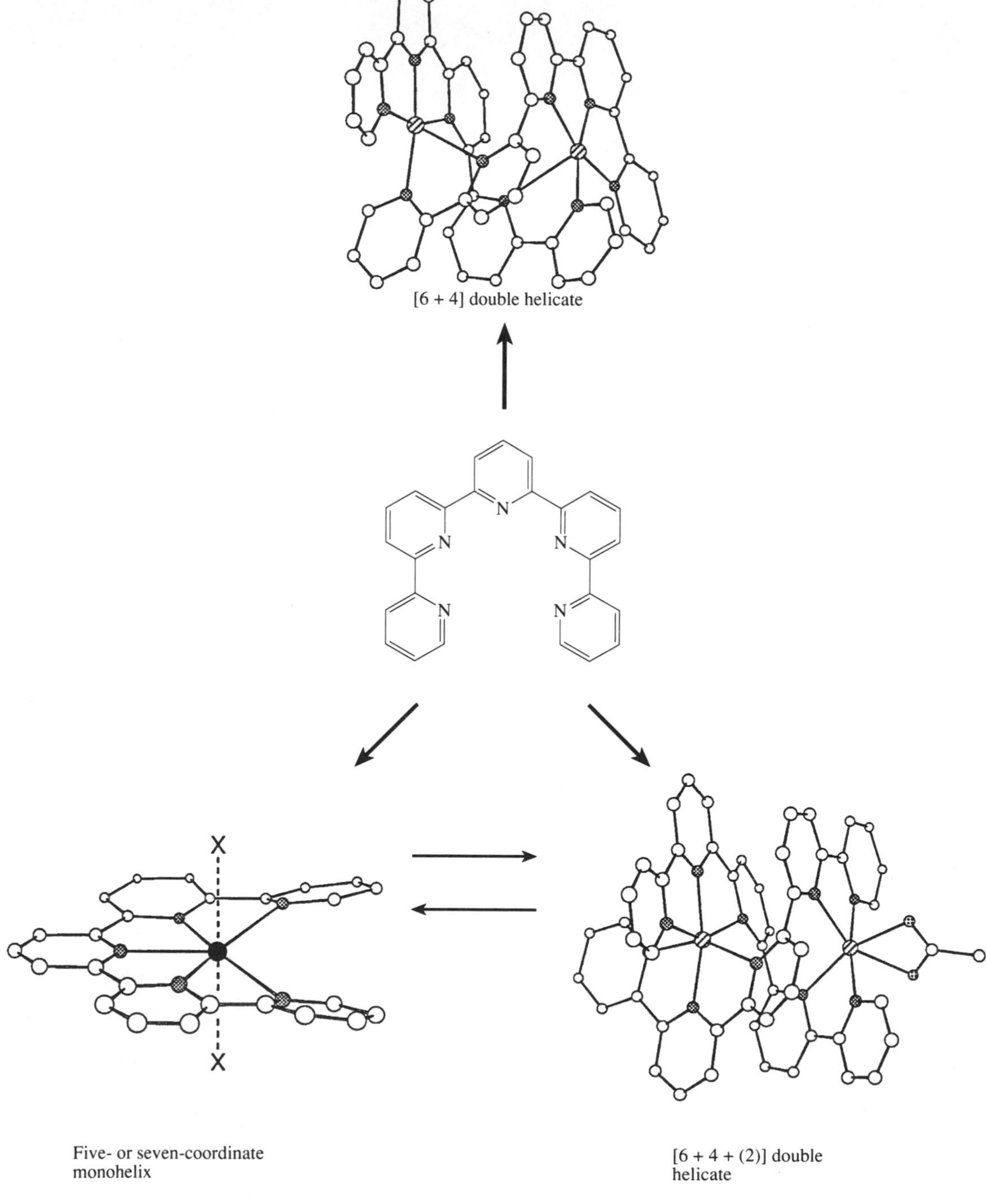

Figure 12 The coding in qpy is ambiguous, and interaction with metal centres can lead to a variety of helical structures.

The interactions of qpy ligands with heavier transition metal ions can also give rise to interesting helical structures. Palladium(II) is normally found in a four-coordinate square-planar environment and complexes with qpy are anticipated in which one of the pyridine rings is noncoordinated. In practice, the double-helical complex cation $[Pd_2(qpy)_2]^{4+}$ is obtained, and a solid-state structural determination of the hexafluorophosphate salt confirms the overall structure. Each palladium shows three short bonds to a tridentate terpy domain from one ligand strand and also a short bond (~207 pm) to a terminal pyridine of the other strand. The arrangement of these two domains about each palladium is distorted square planar. In addition, each palladium shows a longer axial interaction with the second pyridine of the monodentate strand. The structure of the cation is presented in Figure 13 and the compound may be described variously as a [5 + 5], [4 + 4] or [4 + 1 + 4 + 1] helicate.

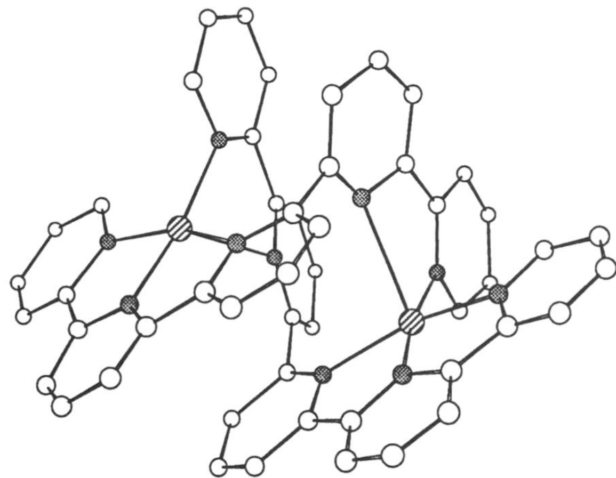

Figure 13 The interaction of qpy with palladium(II) leads to a [5 + 5] double helicate.

6.4.2 Other Heterocyclic Domains

Ligands containing five-membered heterocyclic rings have proved to be of considerable importance in developing an understanding of the features which lead to helication. Some of the very earliest examples of double-helical structures were obtained with tetrapyrrole and related ligands. The 2,2′-bipyrrole derivative (**32**) has been shown to form both 1:1 and 2:2 complexes with copper(II). The 2:2 complex arises from deprotonation of the pyrrole rings of the 1:1 complex, and is thought to be double helical, with the helication resulting from the development of a significant twist about the interannular bond between the two central pyrrole moieties.[34] The related ligand (**33**) also forms 1:1 and 2:2 complexes with zinc(II), but in this case the ligand is doubly deprotonated in each case.[35] The 1:1 complex contains a five-coordinate zinc(II) centre in which the fifth coordination site is occupied by an axial water ligand. Upon dehydration, the zinc is expected to fall back into an approximately square-planar array of the four nitrogen donor atoms of the ligand, but steric interactions between the terminal carbonyl groups result in a deviation from planarity and the formation of a double-helical $[Zn_2(\mathbf{33})_2]$ complex. In the structurally characterized double-helical 2:2 complex each zinc is coordinated to a dipyrromethene functionality from each of the two ligands so as to give a distorted tetrahedral geometry at each metal and a Zn···Zn distance of 337 pm. The steric interactions between substituents on the terminal rings are responsible for the helication in these ligands and biladienes with bulky substituents on the terminal rings also form 2:2 complexes with metal ions which can adopt tetrahedral geometries. An example of such behaviour is seen with (**4**) which doubly deprotonates upon coordination to cobalt(II), nickel(II) or zinc(II) to give dinuclear double-helical complexes $[M_2((\mathbf{4})\text{–}2H)_2]$ (M = Co, Ni or Zn) (Figure 14).[36] It should be noted that the adoption of a tetrahedral coordination geometry at the metal is of importance with this class of ligands in the same way that it is a prerequisite for helication with derivatives of (**12**).

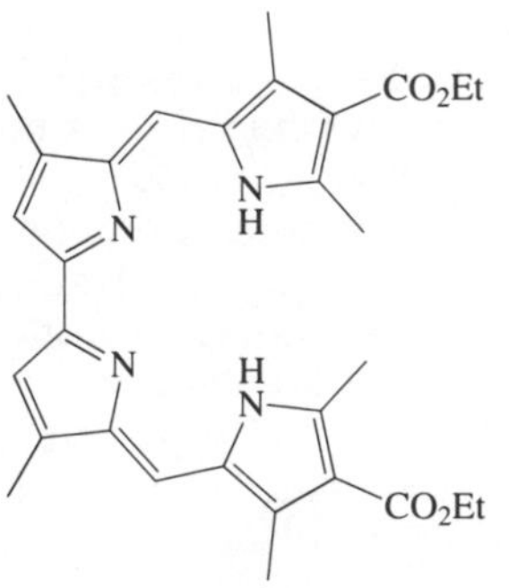

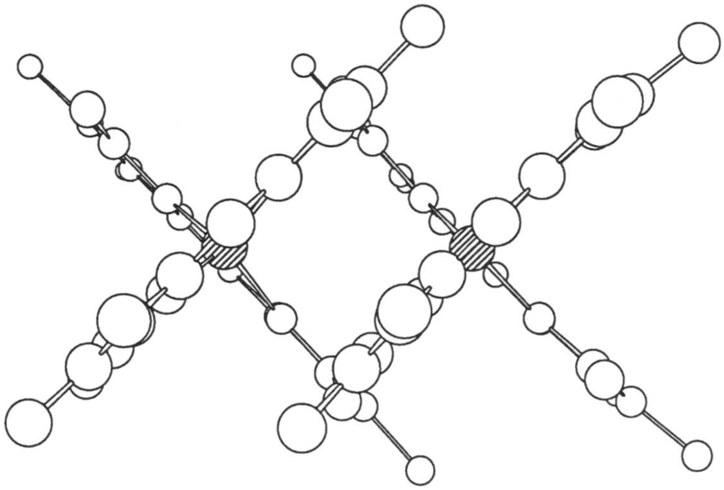

Figure 14 The tetrapyrrole compound (**4**) forms helical complexes upon deprotonation. The structure shows the compound $[Zn_2((\mathbf{4})-2H)_2]$.

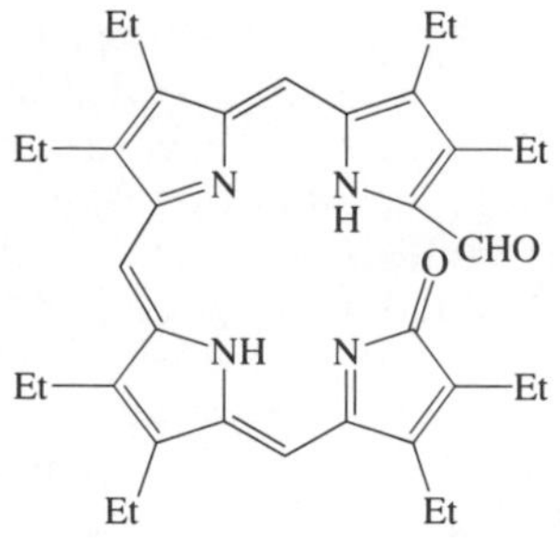

An impressive body of data has been obtained by Van Koten and Van Stein and their co-workers, who have investigated the coordination behaviour of a series of diimine ligands. These ligands were derived from the reaction of 1,2-diamino compounds with heteroaromatic aldehydes, and satisfy the gross requirements for a dinucleating helicand for two tetrahedral metals in that they possess two metal-binding domains, each composed of one of the heteroatoms and the adjacent imine nitrogen. In the case of the mixed donor ligand (**34**), 2:2 or 2:1 complexes with copper(I) or silver(I) may be obtained depending upon the stoichiometry of the reaction. A solid-state structural determination of $[Ag_2(\mathbf{34})_2][OTf]_2$ reveals it to have a double-helical structure with an intramolecular Ag···Ag distance of 291 pm. Each silver(I) is approximately linearly ($\angle$N—Ag—N $\approx$ 180°) coordinated to one nitrogen atom from each ligand strand (Ag—N $\approx$ 215 pm) with a longer contact to two thiophene sulfur atoms (Ag—S $\approx$ 290 pm) (Figure 15). A consequence of the π-interactions in the double-helical structure is that the two thienyl groups of each ligand are independent. Detailed NMR spectroscopic studies have established that the gross features of the double-helical structure persist in solution. Although the interactions with the sulfur atoms are weak, they appear to be important, since only 2:1 complexes are obtained with (**35**).[37] The related ligands (**36**) and (**37**) could also give dinuclear double-helical complexes with tetrahedral metals after deprotonation of the two pyrrole N–H groups. Both (**36**) and (**37**) react with zinc salts in the presence of base to give the double-helical species $[Zn_2L_2]$ (L = (**36**)–2H or (**37**)–2H).[38]

We now move to ligands that incorporate combinations of five- and six-membered heteroaromatic rings. Ligands of this type have also proved to be valuable helicands, and are typified by the species studied by Williams and co-workers. Compound (**38**) is expected to act as an analogue of terpy, although the bite angle is expected to vary as a result of replacing the terminal pyridine rings by the benzimidazoles. In practice, (**38**) behaves in exactly the same way as terpy and reaction with

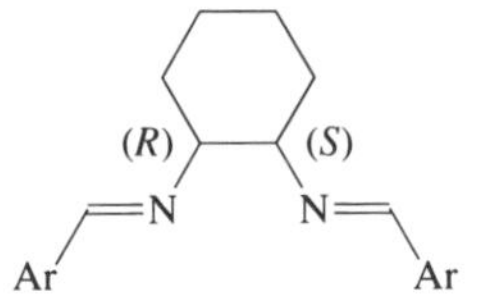

(R) (S)

(34) Ar = 5-methyl-2-thienyl or 2-thienyl
(35) Ar = 3-methylphenyl
(36) Ar = 2-pyrrolyl

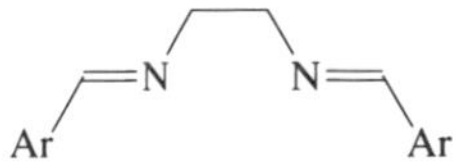

(37) Ar = 2-pyrrolyl

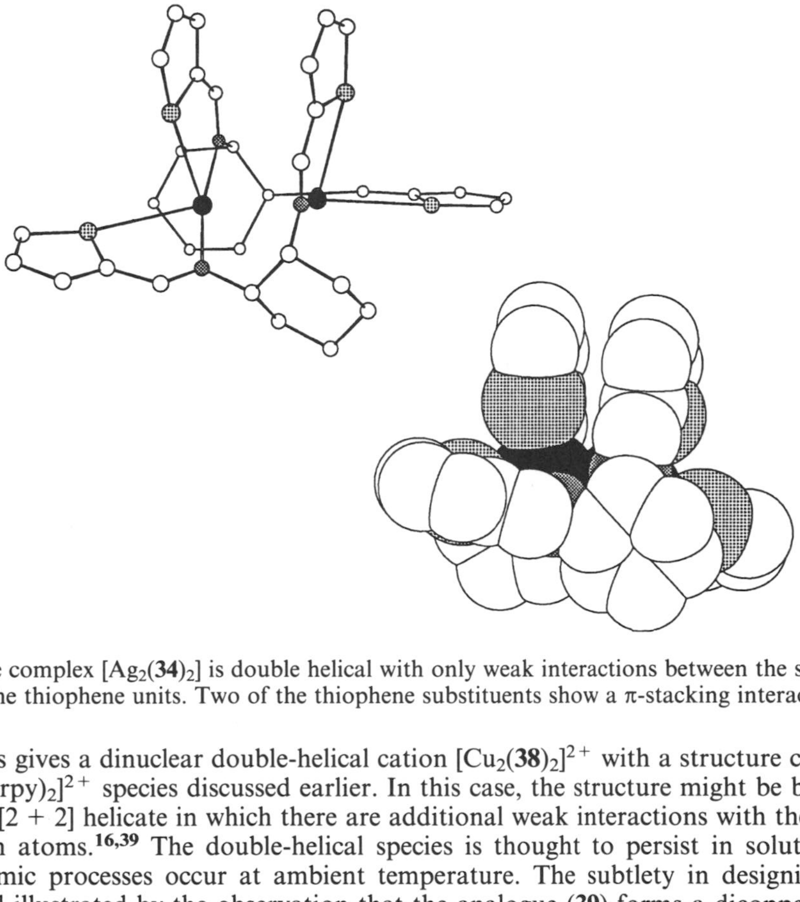

Figure 15 The complex $[Ag_2(\mathbf{34})_2]$ is double helical with only weak interactions between the silver(I) centres and the thiophene units. Two of the thiophene substituents show a π-stacking interaction.

copper(I) salts gives a dinuclear double-helical cation $[Cu_2(\mathbf{38})_2]^{2+}$ with a structure closely related to the $[Cu_2(terpy)_2]^{2+}$ species discussed earlier. In this case, the structure might be best described in terms of a [2 + 2] helicate in which there are additional weak interactions with the central pyridine nitrogen atoms.[16,39] The double-helical species is thought to persist in solution although various dynamic processes occur at ambient temperature. The subtlety in designing helicating ligands is well-illustrated by the observation that the analogue **(39)** forms a dicopper(I) 2:2 complex which is not double helical, but which possesses an open structure (Figure 16). The parent ligand **(40)** gives a double-helical cation $[Cu_2(\mathbf{40})_2]^{2+}$ which is strictly analogous in structure to that formed with **(38)**.[40] However, whereas the complex with **(38)** is stabilized by intramolecular π-stacking interactions between the benzo rings of the benzimidazoles, such interactions are not possible with the complex of **(40)**. As a result, intermolecular π-stacking interactions are optimized within the complex of **(40)** and an exceptionally attractive linear structure in which the complexes are stacked along the *b* axis is adopted. Clearly the interactions with the central nitrogen donor in **(38)** are of critical significance! The limits of helication behaviour with these ligands have been probed in some detail, and related ligands which incorporate more flexible spacer groups have been shown not to give double-helical complexes with copper(I).

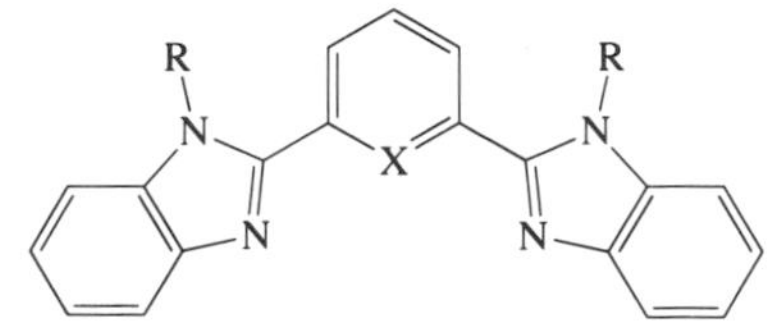

(38) X = N
(39) X = CH

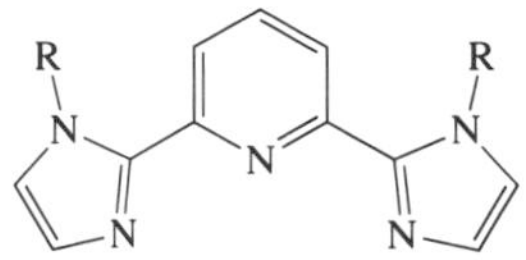

(40)

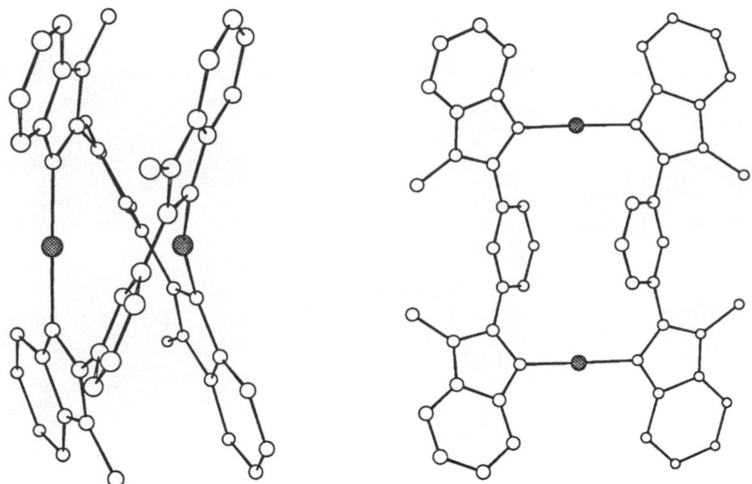

Figure 16 The coding in the molecular threads can be very subtle. Both (**38**) and (**39**) give dinuclear dicopper(I) complexes $[Cu_2L_2]^{2+}$, but only that with (**38**) is double helical.

In the same way that (**38**) is a terpy analogue, (**41**) is an analogue of (**12**) and is expected to give rise to unambiguous double-helical species upon interaction with copper(I) or silver(I). The complex $[Cu_2(\mathbf{41})_2]^{2+}$ is readily formed and detailed NMR spectroscopic studies have established that the solution structure is chiral, and hence double helical.[41]

(**41**)

Another example of a double-helical dicopper(I) complex involving pyridine donors in a bridging position between copper(I) centres is found with ligand (**42**). The dinuclear double-helical complex $[Cu_2(\mathbf{42})_2]^{2+}$ is best regarded as strictly analogous to the $[Cu_2(terpy)_2]^{2+}$ species. There is a short Cu—Cu interaction of 262.6 pm in the asymmetrical cation, and one copper(I) is essentially two-coordinate and bonded to one imine nitrogen from each ligand strand ($\angle$ N—Cu—N $\approx 176°$), although longer contacts to the central pyridine nitrogen of about 270 pm are also observed. The other copper(I) centre is best thought of as being in a tetrahedral four-coordinate environment composed of the central pyridine nitrogen and the remaining imine nitrogen from each strand, although longer contacts to the ether oxygen atoms are also present in the solid state.[42]

(**42**)

A number of other examples of helicands based upon pyridine donors have also been described. For example, the N_4 donor ligand **(43)** is clearly related to the helicands **(34)**, **(35)** and **(36)** discussed earlier. The reaction of **(43)** with silver(I) or copper(I) trifluoromethanesulfonates gives the expected double helicates $[M_2(\mathbf{43})_2]^{2+}$, and the complex $[Ag_2(\mathbf{43})_2][OTf]_2$ has been structurally characterized. Each silver(I) centre is in a distorted four-coordinate geometry and the Ag···Ag distance is 325.4 pm. NMR spectroscopic studies indicate that the dinuclear complex persists in solution and that the copper(I) complexes possess a strictly analogous structure, as do complexes of **(44)**, although a range of dynamic processes may occur, depending upon the precise structure of the ligand.[43]

(43)

(44)

Finally, we will consider some of the facets of the chemistry of some ligands derived from 2,6-diacetylpyridine. Ligand (7) can present an approximately planar N_5 donor set to a metal ion, and this is exactly the bonding mode found in the cation $[Co(7)(H_2O)_2]^{2+}$ which contains a pentagonal bipyramidal cobalt(II) centre.[44] Most other complexes of (7) with transition metal ions are thought to possess analogous pentagonal bipyramidal structures. The corresponding zinc complex may also be prepared and has been shown to possess an essentially identical pentagonal bipyramidal structure. However, the NH groups of the coordinated hydrazine ligand are acidic and treatment of the zinc complex with base generates a black species with the formula $[Zn_2((7)\text{–}2H)_2]$. The solid-state structure of this complex reveals it to be a double helix containing two approximately octahedral zinc centres (Figure 17). Each zinc forms two short Zn—N bonds to the terminal pyridine and imine nitrogen of each of the two molecular threads together with longer contacts to the central pyridine rings of each strand. The overall structure is clearly related to the dicopper(I) structures discussed earlier, in which a pyridine ligand in the centre of a molecular thread adopted a bridging role.

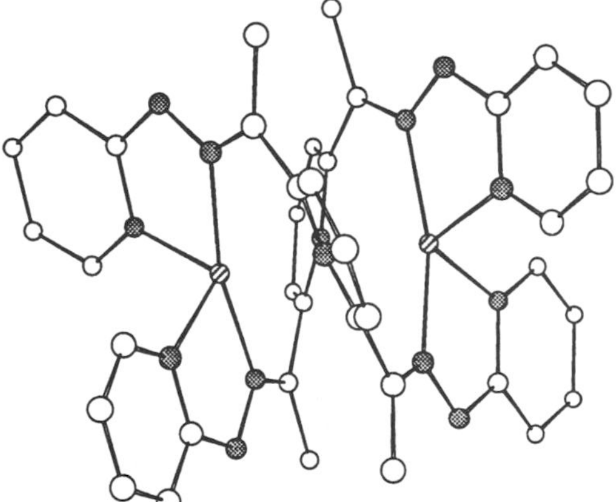

Figure 17 After deprotonation, (7) gives a double-helical complex $[Zn_2((7)\text{–}2H)_2]$. In contrast, the neutral ligand gives pentagonal bipyramidal complexes in which (7) acts as an approximately planar pentadentate donor.

A whole series of ligands based upon 2,6-diacetylpyridines are known, and a number of helicates have been characterized. The thiosemicarbazone **(45)** is expected to behave in a similar manner to (7), and present a pentagonal planar N_3S_2 donor set in pentagonal bipyramidal complexes. While this is the case with most first-row transition metal ions, the zinc complexes exhibit anomalous properties. The chemistry closely resembles that of (7); complexes with (**45**) are indeed pentagonal bipyramidal, but deprotonation of the ligand to ((**45**)–2H) gives rise to double-helical dizinc complexes. Interestingly, the precise structure of the double helix is controlled by other species present in the crystal lattice. Thus, in $[Zn_2((\mathbf{45})-2H)_2]\cdot MeOH\cdot 2DMF$, each zinc is coordinated to one sulfur and one nitrogen donor from each ligand thread, with a pseudooctahedral coordination geometry being completed by two bridging pyridine donors, of the type that are now becoming familiar. The Zn···Zn distance is 386.6 pm. In contrast, the complex $[Zn_2((\mathbf{45})-2H)_2]\cdot MeOH\cdot H_2O$ exhibits a different but related structure. The overall dinuclear double-helical structure persists, but one of the zinc atoms now remains octahedral and is coordinated to a sulfur and imine nitrogen and the pyridine nitrogen from each chain, whereas the other is now four-coordinate in an N_2S_2 environment. The difference is associated with the change in bonding mode of the central pyridine nitrogen donors from bridging to mononucleating. The structures of the two isomeric double helicates are presented in Figure 18.[45]

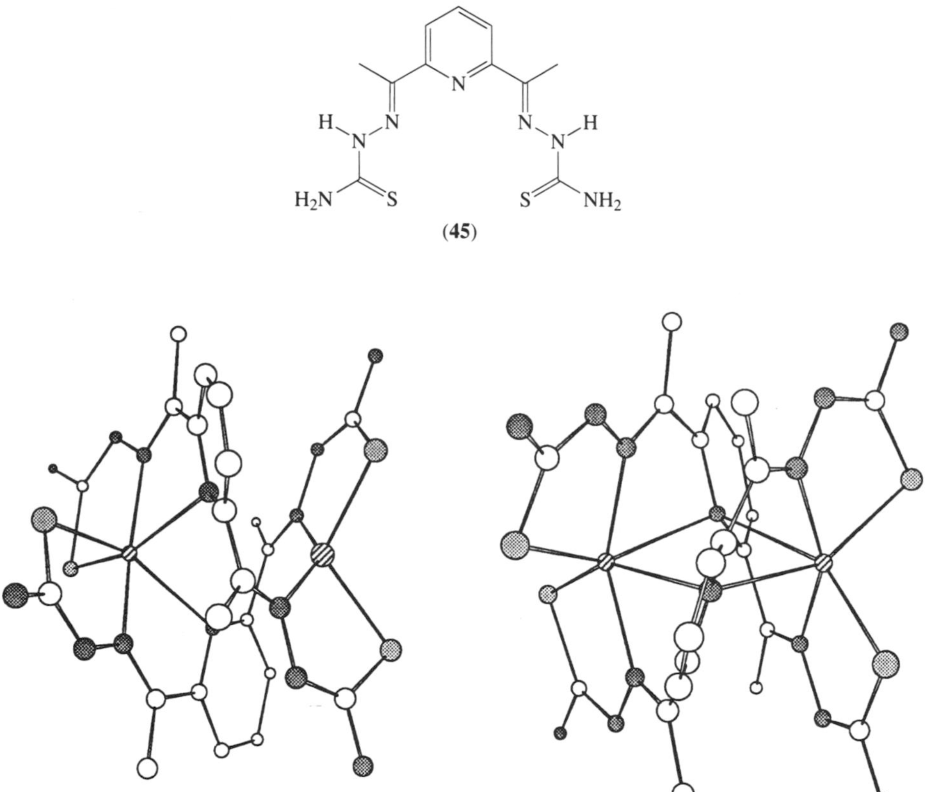

Figure 18 The double-helical array formed from two deprotonated (**45**) ligands contains a total of 10 donor atoms. Two forms of the complex $[Zn_2((\mathbf{45})-2H)_2]$ are known; in the first one zinc is six-coordinate and the other four-coordinate, while in the other each zinc is six-coordinate with the central pyridine acting as a bridging donor.

The coding in these ligand threads is clearly far from unambiguous—apart from the possibilities of the double-helical or pentagonal bipyramidal structures, there is also an ambiguity in the choice of donor atoms. For example, ligand (**46**) gives a double-helical dizinc complex $[Zn_2((\mathbf{46})-2H)_2]$ after deprotonation, but in this case each zinc is four-coordinate and bonded to one oxygen and one imine nitrogen donor from each molecular thread. The sulfur atoms of the thiophene units are not coordinated. The structure of this complex is presented in Figure 19.[46] Part of the ambiguity in the coding comes from the ability of zinc to adopt a variety of coordination geometries. If we select a metal ion with a strong ligand field-imposed preference for an octahedral geometry, we expect to see double helicates formed if the central pyridine unit acts as a bridge and there are two other didentate domains in each thread. The difficulty is in preventing the adoption of a pentagonal bipyramidal complex as opposed to the desired double helix. The bulky ligand (**47**) satisfies these criteria and forms a double helix $[Ni_2((\mathbf{47})-2H)_2]$ containing two six-coordinate nickel centres (Figure 19).[47]

R =

(**7**) (**46**) (**47**) (**48**)

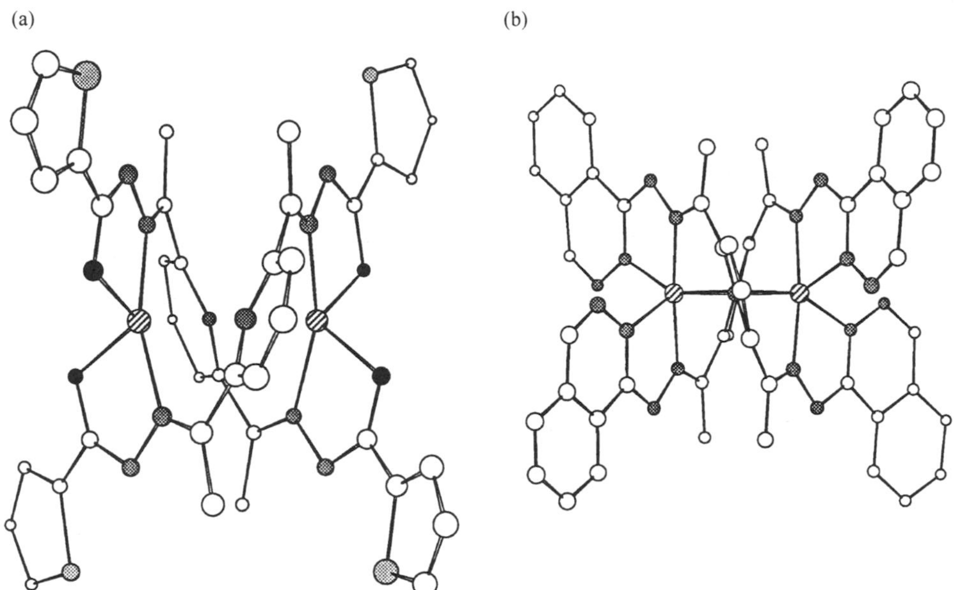

Figure 19 (a) The structure of the double-helical cation $[Zn_2((\mathbf{46})-2H)_2]$ in which each zinc is four-coordinate and there are no zinc–sulfur interactions; (b) the lower structure of $[Ni_2((\mathbf{47})-2H)_2]$ which contains two six-coordinate metal centres.

To conclude this section we return to (**7**). The five nitrogen donor atoms of (**7**) are equivalent in character and spatial arrangement to the donors of qpy (**31**). We have emphasized the coding within the ligand strands in discussing the features which are required in helicands, but we may use these ligands to probe the subtlety of the metal coding also. To date we have stressed only the preferred coordination mode of the metal ion, but have not discussed the effect of the size of metal ion in any detail. Consider (**31**)–this ligand forms a mononuclear complex with silver(I). The silver(I) has no electronic preference for any specific coordination geometry, and the five nitrogen donor atoms are distributed about the silver(I) centre in an approximately pentagonal planar manner to give optimal silver–nitrogen contacts. It is apparent that the metal and the ligand are not perfectly matched, and there is already a conflict between the steric requirements of the ligand and the optimization of the metal–ligand bonds. The silver(I) cation is a little too small for the cavity of the qpy ligand, and the two terminal pyridine rings are 'pulled' together. The consequence is a steric interaction between H-6 and H-6'''' and the development of a helical twist within the ligand (Figure 20).[48] One could visualize a process by which the metal ion is gradually made smaller. As the terminal pyridine rings are pulled in more and more to optimize the metal–ligand bonding, so a greater and greater helical twisting will develop. Eventually the twisting will be such that all of the donor atoms cannot interact with a single metal and we have arrived at the conceptual origin of a dinuclear double helix as seen in Figure 11. Although we cannot gradually change the size of a metal ion, we can keep the metal ion constant and change the structure of the ligand. We have already noted that **(7)** and **(31)** may be regarded as equivalent N_5 donors, and also that a steric interaction between H-6 and H-6'''' leads to the development of a helical twist within the ligand. If bulkier substituents are placed in these two positions, we might anticipate the formation of, eventually, a double-helical structure. This is exactly what happens in the silver(I) complex of **(48)**, which is a dinuclear double-helical species. Each silver(I) is four-coordinate and bonded to the central pyridine and an imine of one thread, and the imine and a terminal pyridine of the other thread. This arrangement minimizes any steric interactions with the 6-chloro substituents and results in two of the four 6-chloropyridyl groups being noncoordinated; these noncoordinated residues are strongly π-stacked with the central pyridine rings of the other thread. This structure is also presented in Figure 20.[48]

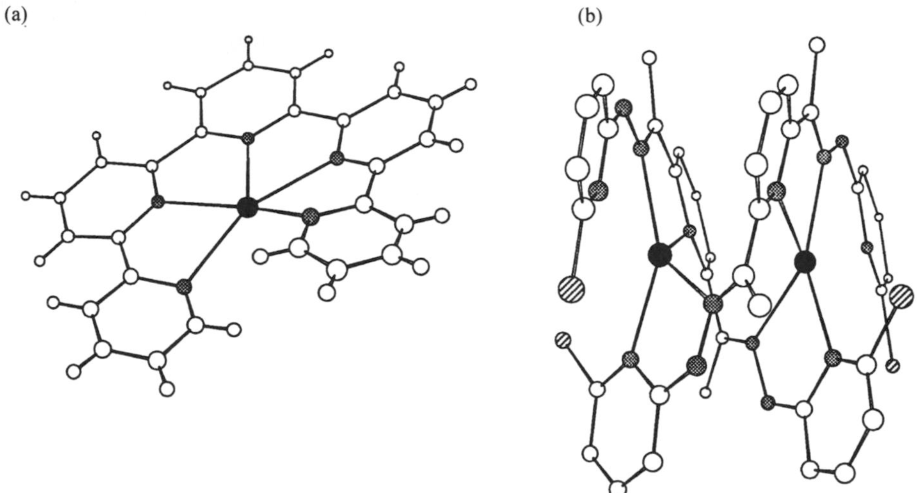

Figure 20 Two contrasting structures of silver(I) complexes of pentadentate N_5 donors. (a) The complex with **(31)** is mononuclear but with a small helical twist within the ligand, whereas (b) that with **(48)** is double helical. The helication is a result of the steric interactions that would occur between the chlorine substituents in **(48)** if a structure analogous to that with **(31)** were adopted.

6.5 HOMOPOLYNUCLEAR DOUBLE HELICATES

In the preceding section we have shown that dinuclear double helicates may be formed from the interaction of two molecular threads each containing two *n*-dentate metal-binding domains with two metal ions with a favoured coordination number of 2*n*. A simple extension of this coding allows us to design molecular threads which will give double-helical complexes containing more than two metal centres. At the simplest, we require two molecular threads each containing *x n*-dentate metal-binding domains which can react with *x* metal ions with a favoured coordination number of 2*n*. This strategy has been adopted and shown to be successful in a number of systems, all of which are based upon oligopyridine metal-binding domains.

6.5.1 Oligopyridines as Ligand Domains

The generic requirements for a molecular thread based upon bidentate oligopyridine domains were presented in Figure 3, and a simple extension leads us to the generic ligand shown in Figure 21 as the basis for the assembly of polynuclear double helicates. The interaction of these ligands with metal ions having a tetrahedral coordination preference is expected to lead to the formation of *n*-nuclear double helicates.

The simplest such system to consider is one in which there are no spacer groups and the bidentate domains are directly linked in ligands such as **(24)** or **(25)**. The reaction of these ligands with copper(I) or silver(I) salts results in the formation of the expected trinuclear double-helical

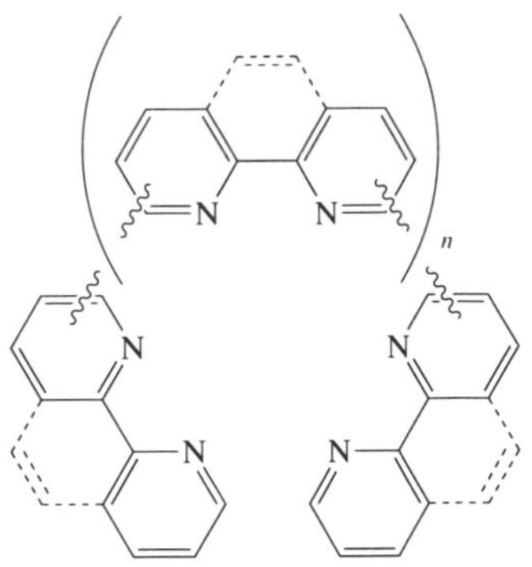

Figure 21 A generic structure of an oligopyridine-based molecular thread which might give rise to double helicates containing $n + 2$ four-coordinate metal ions.

complexes $[M_3L_2]^{3+}$ containing three four-coordinate metal centres each binding to two nitrogen donors from each thread.[30] The coding in a spy (spy = (**24**)) molecular thread is of course not unique to the binding of three tetrahedral metal ions to give a trinuclear [4 + 4 + 4] double helicate as we have already seen that they can also bind two octahedral metal ions to give a dinuclear [6 + 6] double helicate (Figure 22).

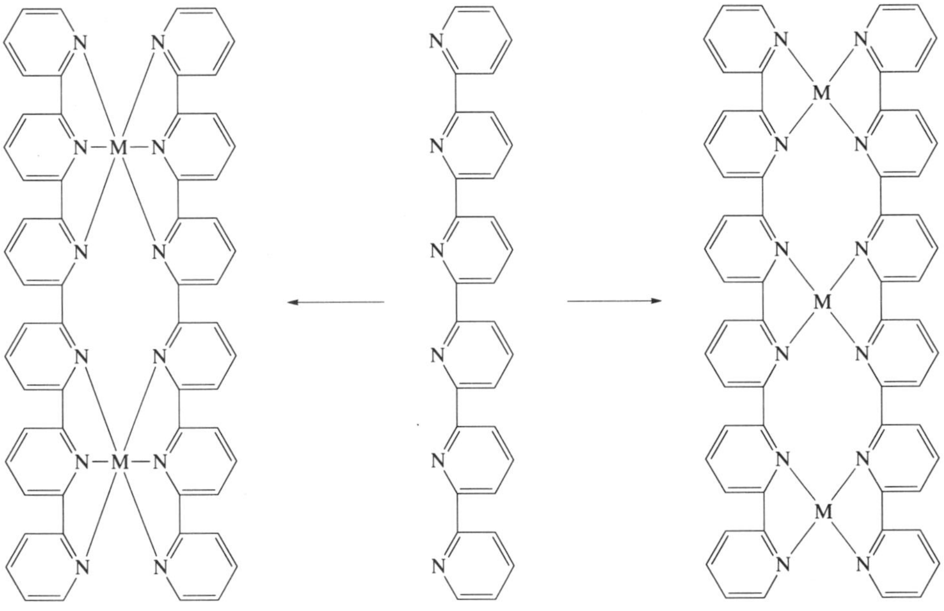

Figure 22 The ambiguity in coding in (**24**). The 12 nitrogen donors in a double-helical array of two (**24**) threads could be partitioned into two six-coordinate domains or three four-coordinate domains. The double helicate is represented in a planar form to emphasize the coordination pattern.

This coding is of particular interest when copper complexes are considered. Copper(I) is expected to form four-coordinate complexes and give the trinuclear [4 + 4 + 4] complex, whereas copper(II) gives the [6 + 6] species. An obvious question is to ask what happens when the tricopper(I) [4 + 4 + 4] double helicate is oxidized or the dicopper(II) [6 + 6] double helicate is reduced. Detailed electrochemical studies have established an extremely complex reaction scheme of the type shown in Figure 23. The key features are the formation of transient species which undergo rapid nuclearity changes in accord with the preferred coordination geometries of the new oxidation states.

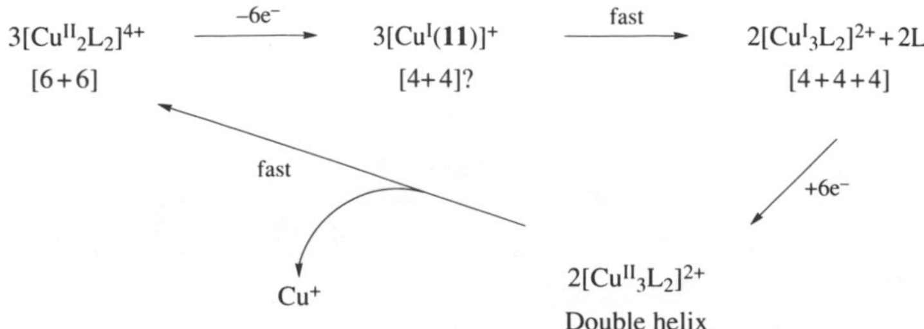

Figure 23 The electrochemical behaviour of complexes of **(24)** and **(25)** can be complex. The preference of copper(II) for six-coordinate sites and copper(I) for four-coordinate sites means that redox changes trigger interconversions between the bonding modes in Figure 22.

Unfortunately, the ambiguity in coding with the directly linked oligopyridines extends rather beyond this. We saw earlier that it was possible to form a [4 + 2] double helicate from the interaction of copper(I) with tpy ligands, and that in solution these species behaved as an equilibrating [4+2] and [2+4] pair or as a symmetrical species with the central pyridine ring of each ligand thread bridging the two metal centres. We have also seen that the incorporation of a bridging pyridine between copper(I) centres in double helicates is a relatively common structural feature. Now consider **(12)** which can give a double-helical array with a total of eight donor atoms. In Section 6.4.1.1 we partitioned these into two sets of N_4 donors and showed that two copper(I) or two silver(I) ions could be bound in such a ligand array. However, we could now predict the formation of a trinuclear double helicate of the [2 + 4 + 2] type (Figure 24). To date, no such species has been found, although intensive mass spectrometric studies have been undertaken. With the higher oligopyridines, the situation is even more complex. A double-helical array of two qpy **(31)** ligands presents a total of 10 donor atoms. In Section 6.4.1.3 we saw that these could be partitioned to give [6 + 4] or [5 + 5] dinuclear double-helicates. We can now see that with copper(I) centres it should be possible to form trinuclear [4 + 4 + 2] and tetranuclear [2 + 2 + 4 + 2] species (Figure 24).[49] In practice, the trinuclear species have been isolated[49] and there is good mass spectrometric evidence for the formation of the tetranuclear species. In general we may state that an oligopyridine with *n* donor atoms will generate a double helicate containing 2*n* donors. The maximum number of copper(I) centres which may be accommodated within this array is *n*–1, and so we can anticipate pentanuclear double helicates from derivatives of **(24)** and hexanuclear complexes from **(28)** and related compounds. Time of flight mass spectrometry of the products obtained from the reaction of ligands of this type confirms the presence of a range of species of various nuclearities, and a tetranuclear [4 + 4 + 4 + 2] complex with a derivative of **(28)** has been described.[32]

However, even this complication does not complete the story of copper complexes of the higher oligopyridines. We have already noted that copper(II) centres are usually found in six-coordinate sites while copper(I) is usually four-coordinate. Within the constraints of the 2*n* donors atoms from a double-helical array of two oligopyridines with *n* donor atoms, there is often a way, or a number of ways, in which a mixture of copper(II) and copper(I) centres may be used to utilize the 2*n* donors. We have already seen that the 10 donors from a double-helical array of two **(31)** ligands can give a [6 + 4] mixed oxidation state copper(II)–copper(I) complex. This is well illustrated by **(28)**, which will give a double-helical array consisting of 14 donor atoms. This could be occupied by a variable number of copper(I) centres in a variety of four- and two-coordinate sites, or by one copper(II) and two copper(I) centres to give a trinuclear mixed oxidation state [6 + 4 + 4] or [4 + 6 + 4] double helicate. Although such complexes have been reported, it is not known for certain whether a single isomer or mixture of the two isomers is formed.[32] By analogy with the mixed oxidation state complexes of **(31)**, these species are expected to be valence localized.

In a series of extremely elegant papers, Lehn has described the extension of the strategy used in **(18)** to the design of helicands containing three **(49)**, four **(50)** and five **(51)** metal-binding domains. These ligands behave very much as expected on the basis of the number of bidentate domains. Thus, **(49)** gives trinuclear [4 + 4 + 4] double helicates with copper(I) and silver(I), whilst **(50)** gives tetranuclear [4 + 4 + 4 + 4] species and **(51)** gives pentanuclear [4 + 4 + 4 + 4 + 4] complexes. These complexes are formed simply upon mixing the appropriate ligand with a solution of the metal salt, and represent a genuinely spontaneous self-assembly process. The tricopper(I) and trisilver(I) complexes of **(49)** have been structurally characterized, and the structure of the

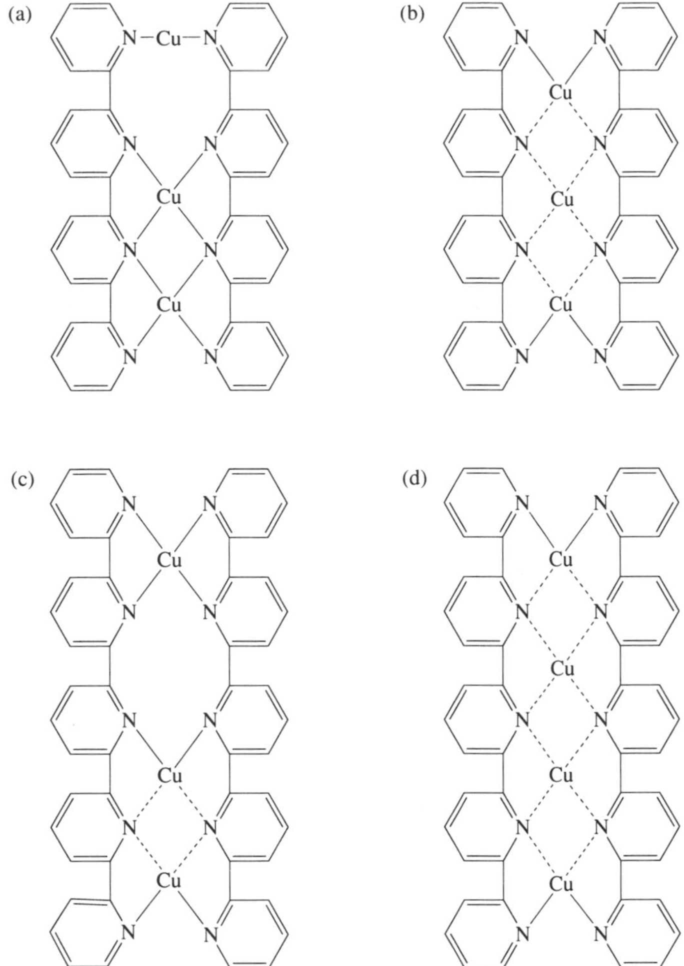

Figure 24 Additional bonding modes involving formally two-coordinate copper(I) centres are possible with oligopyridines: (a) illustrates the [4 + 4 + 2] double helicate from (**12**) while (b) gives an alternative representation of the effective solution structure; (c) and (d) illustrate that (**31**) could give trinuclear and tetranuclear complexes with copper(I). In general, an oligopyridine with n donor atoms will give a double-helical array capable of binding $n-1$ copper(I) centres. The double helicate is represented in a planar form to emphasize the coordination pattern.

trisilver(I) complex is presented in Figure 25. The tricopper(I) complex is about 1700 pm in length, which represents about one and half turns of the helix, corresponding to a pitch of about 1200 pm. The stacking of the bipy units in these complexes is an immediately noticeable feature, although the precise contribution to the stability of the double helicate is unknown.[7,50]

The assembly of these double helicates has been studied in some detail. Proton NMR spectroscopic studies have established that the double-helical structure persists in solution for tri-, tetra- and pentanuclear silver(I) and copper(I) complexes. Particularly important pieces of evidence are the observation of the diastereotopic methylene groups in the complexes but not in the ligands as AB multiplets and also that the addition of the chiral shift reagent $[Eu(tfc)_3]$ (tfc = 3-(trifluoromethylhydroxymethylene)-(+)-camphorato) results in the doubling of some of the signals. The latter observation is consistent with the presence of the two enantiomeric, but non interconverting left-handed (M) and right-handed (P) helicates in the solution. A detailed study of the self-assembly of the tricopper(I) complex of the solubilized ligand (**52**) has revealed a number of features of interest. Titration of (**52**) with copper(I) gave sharp isosbestic points in the electronic spectra, consistent with the presence of only two solution species. The data are consistent with a positive cooperation in the formation of the $[Cu_3(\mathbf{52})_2]^{3+}$ complex, a conclusion which is strictly supported by Scatchard and Hill plots. The stability constants for the various species present were

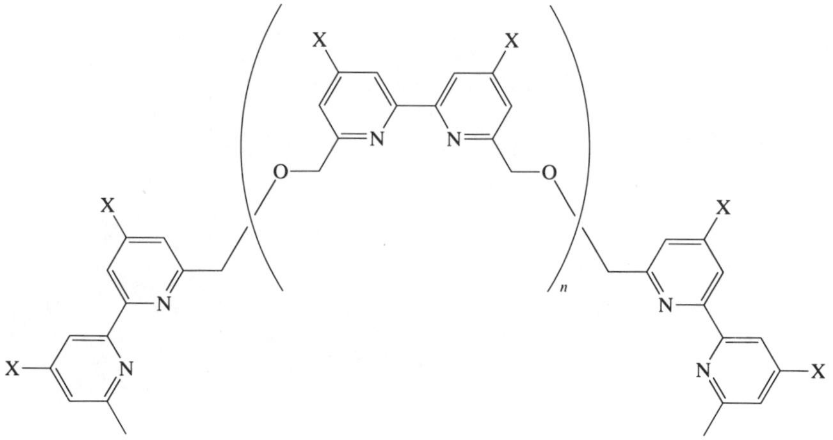

(49) $n=1$, X=H
(50) $n=2$, X=H
(51) $n=3$, X=H
(52) $n=1$, X=CO_2Et

(53) $n=1$, X=CONH—

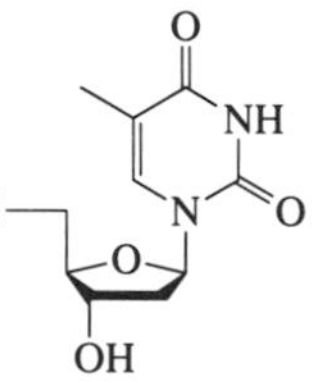

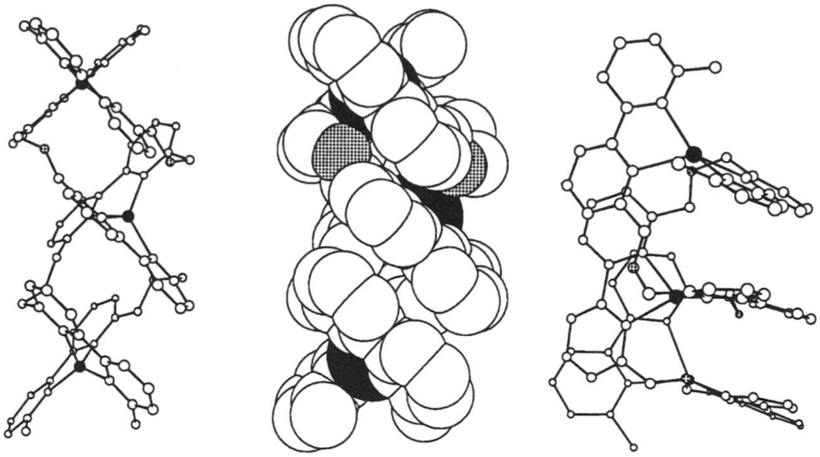

Figure 25 Three views of the trisilver(I) complex of (**49**) emphasizing the extensive stacking interactions which are present.

estimated to be log β_{11} 4.6, log β_{12} 8.2, log β_{22} 13.5 and log β_{32} 18.6 where the various log β values refer to the processes

Cu + (**52**) = Cu(**52**)	log β_{11}	(1)
Cu(**52**) + (**52**) = Cu(**52**)$_2$	log β_{12}	(2)
Cu(**52**)$_2$ + Cu(**52**) = Cu$_2$(**52**)$_2$ + (**52**)	log β_{22}	(3)
Cu$_2$(**52**)$_2$ + Cu(**52**) = Cu$_3$(**52**)$_2$ + (**52**)	log β_{32}	(4)

Similar cooperativity effects have also been reported for the silver(I) complexes with this series of ligands.

Helical structures abound in nature, and it is natural to ask whether there is any specific helix–helix interaction which might be useful in the targeting of helical biomolecules. The double helix of DNA is a particularly inviting target, and the surface phosphate groups mean that any binding will be strengthened by the electrostatic interaction of the surface anionic groups with cationic double-helical complexes. Furthermore, modelling reveals that the complex cations $[Cu_2(\mathbf{18})_2]^{2+}$, $[Cu_3(\mathbf{49})_2]^{3+}$, $[Cu_4(\mathbf{50})_2]^{4+}$ and $[Cu_5(\mathbf{51})_2]^{5+}$ have a good complementarity with the major groove of B-form DNA, and Lehn has demonstrated significant binding of these species to nucleic acid.[51] The binding was established by a variety of methods, and the site of binding (major groove) established by the inhibition of cleavage by specific restriction enzymes. Preliminary studies indicate that these complexes might be of interest as photocleavage agents for nucleic acids.

While the above results are of considerable interest and potential, it would also be desirable to be able to target a specific sequence of nucleic acid, and Lehn has also made a preliminary approach to this question. The Watson–Crick base pairing of nucleosides within DNA suggests that it might be possible to place complementary hydrogen bond donors and acceptors on the periphery of the double-helical complex. A first generation ligand (**53**) was found to give mixtures of complexes upon coordination to copper(I), presumably as a result of steric interactions between the pendant nucleoside groups.[52] Accordingly, a longer spacer group was introduced between the nucleoside and the metal-binding domain, together with a strategy in which the nucleosides were only placed on alternate metal-binding domains to give the ligands (**54**) and (**55**). With these ligands the assembly of the double-helicates was unaffected by the presence of the substituents, and the complex cations $[Cu_3(\mathbf{54})_2]^{3+}$ and $[Cu_5(\mathbf{55})_2]^{5+}$ were formed in the usual manner. Lehn describes these complexes as deoxyribonucleohelicates, and points out that they provide an interesting contrast to the double-helical structure of DNA itself. In DNA the surface is negatively charged and the bases on the inside of the helix, whereas these complexes are positively charged with the bases extrahelical. To date, no studies of the interaction of these complexes with DNA have been reported.

In a final development of these versatile ligand strands, Lehn has addressed the question of helicate chirality. The chiral ligand threads (**56**) and (**57**) have been prepared and the coordination behaviour of (**56**) studied. The reaction of (**56**) with silver(I) or copper(I) salts gave the expected trinuclear [4 + 4 + 4] double-helicates $[M_3(\mathbf{56})_2]^{3+}$.[53] The use of the (*S*,*S*)-chiral ligand thread means that instead of two enantiomeric left-handed (*M*) or right-handed (*P*) double helicates, diastereomeric (*P*,*S*,*S*) or (*M*,*S*,*S*) helicates will be formed. NMR spectroscopic studies of the complexes reveal that with both the silver(I) and the copper(I) complexes, only one of the diastereomers is formed. The similarity of the data obtained for the complex $[Cu_3(\mathbf{56})_2]^{3+}$ to those for $[Cu_3(\mathbf{49})_2]^{2+}$ suggest strongly that the double-helical complex as opposed to an alternative chiral nonhelical structure has been formed (Figure 26). It was suggested that the diastereomer formed was the right-handed *P* isomer, as a consequence of destabilizing steric interactions between the methyl groups at the chiral centres in the alternative *M* form (Figure 27).

To conclude this section, we should point out that the synthetic problems associated with these multinucleating molecular threads should not be underestimated, but that new synthetic procedures might make such ligands more accessible in the future.[54]

6.6 HETERONUCLEAR DOUBLE HELICATES

In previous sections we have shown how the metal-binding domains of molecular threads may be matched to the acceptor properties of metal ions to build double-helical structures. We also showed that in some cases the coding within the ligand thread was ambiguous (e.g. (**24**) giving [4 + 4 + 4] or [6 + 6] double helicates) or that changes in the oxidation state of a metal could have profound effects upon the preferred coordination geometry and stoichiometry of the helicate (e.g., the change from four coordination to six-coordination upon oxidation of copper(I) to copper(II)). Furthermore, we showed that in suitable cases it was possible to stabilize mixed oxidation state species when the total number of donor atoms available from the double-helical array match the total requirement of the metal ions in their various oxidation states. The obvious extension of this approach is to try to prepare heteronuclear double helicates in a specific manner coding for the particular requirements of the various metal ions.

Attempts to prepare heteronuclear complexes in self-assembly processes involving the interaction of oligopyridine-based molecular threads with mixtures of metal ions do not yet appear to have been successful. However, the first example of a heteronuclear complex to be described illustrates beautifully the principles of metallosupramolecular chemistry which are involved, and also the subtlety of control over the assembly process which may be achieved.

Consider the [6 + 4] double helicate obtained from a double-helical array of two qpy (**31**) ligands. We might expect a first row transition metal ion to favour the octahedral site and a metal

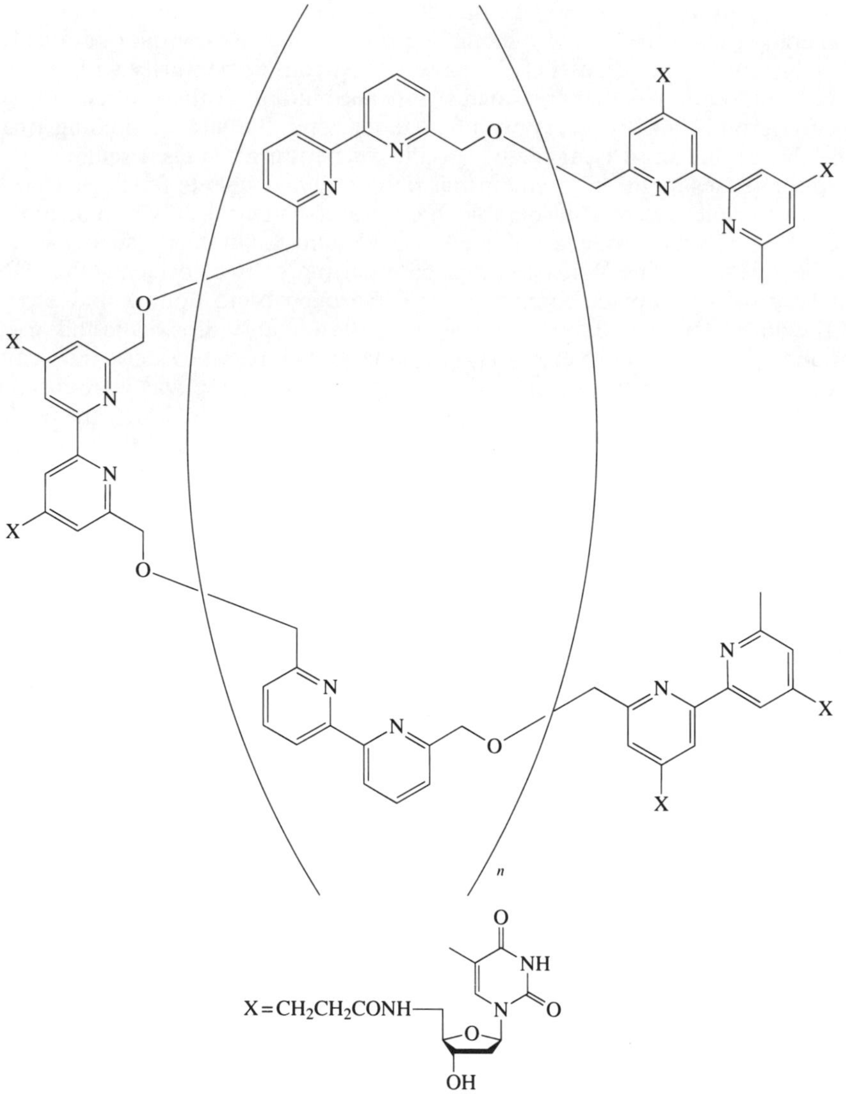

$X = CH_2CH_2CONH$—

(54) $n = 1$
(55) $n = 2$

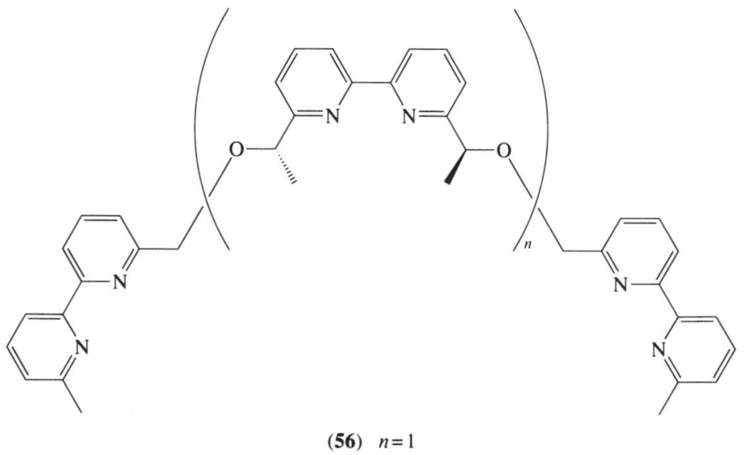

(56) $n = 1$
(57) $n = 2$

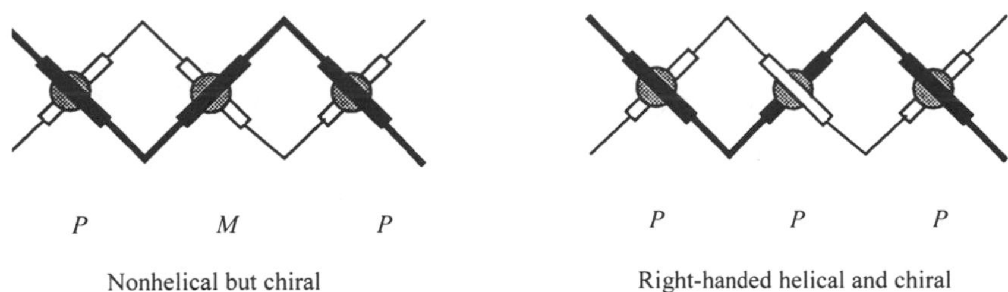

Figure 26 With trinuclear double helicates based upon four-coordinate metal centres, both the helical and the nonhelical forms may be chiral. The descriptors *P* and *M* refer to the right- or left-handedness of the helix at the individual metal centres.

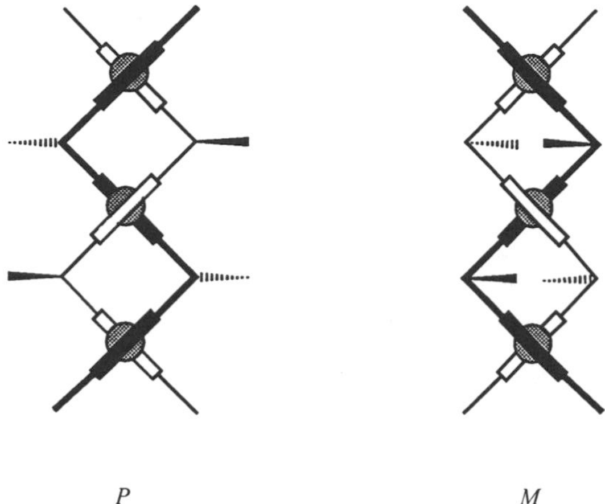

Figure 27 The trinuclear double helicates formed with the chiral molecular thread (**56**). Steric interactions between the methyl groups at the chiral centres result in differences in stability between the *P* and *M* diastereomers. With the (*S*,*S*)-ligand, the *P* form appears to be favoured.

ion such as copper(I) or silver(I) to favour the four-coordinate site. How might this partitioning be achieved? We know that solutions of cobalt(II) complexes of (**31**) in good donor solvents are mononuclear and seven-coordinate. On ligand field grounds, the cobalt(II) might be expected to prefer a six-coordinate octahedral environment. Similarly, in the complex cation $[Ag(\mathbf{31})]^+$ the silver(I) is five-coordinate, and might be expected to be four-coordinate. Now consider the reaction of $[Ag(\mathbf{31})]^+$ with $[Co(\mathbf{31})S_2]^{2+}$; the silver(I) is five-coordinate but would ideally be four-coordinate pseudotetrahedral and the cobalt(II) is seven-coordinate but would ideally be six-coordinate. Can the metals partition themselves according to Figure 28?

The reaction of one equivalent of $[Ag(\mathbf{31})]^+$ with $[Co(\mathbf{31})(MeOH)_2]^{2+}$ gives the anticipated heterodinuclear double helicate, $[CoAg(\mathbf{31})_2]^{3+}$.[55] In solution, any equilibrium between the mononuclear silver(I) and cobalt(II) complexes is displaced almost entirely towards the heterodinuclear double-helical complex. It would be nice if the same methodology could be used for the preparation of heteronuclear heteroleptic double helicates in which the two ligand threads were different. This could be achieved by the reaction of the silver(I) complex of a qpy ligand with the cobalt(II) complex of a different qpy ligand. Unfortunately the reaction of $[Ag(L^1)]^+$ and $[Co(L^2)]^{2+}$ gave a statistical mixture of the three possible heterodinuclear products:

$$[Co(L^1)S_2]^{2+} + [Ag(L^2)]^+ \rightarrow [CoAg(L^1)(L^2)]^{3+} + [CoAg(L^1)_2]^{3+} + [CoAg(L^2)_2]^{3+} \tag{5}$$

This methodology has been adopted by Williams and extended to the formation of other heterotrinuclear systems. The helicand (**58**) contains a bidentate and a tridentate domain which are spatially separated by the methylene spacer; titration with cobalt(II) or zinc(II) shows that the formation of the double-helical $[M_2(\mathbf{58})_2]^{4+}$ complexes becomes dominant at metal:ligand ratios of greater than 1.5. A detailed study of the NOE effects observed in the zinc complex unambiguously established the formation of the double-helical rather than a side-to-side nonhelical complex.[56]

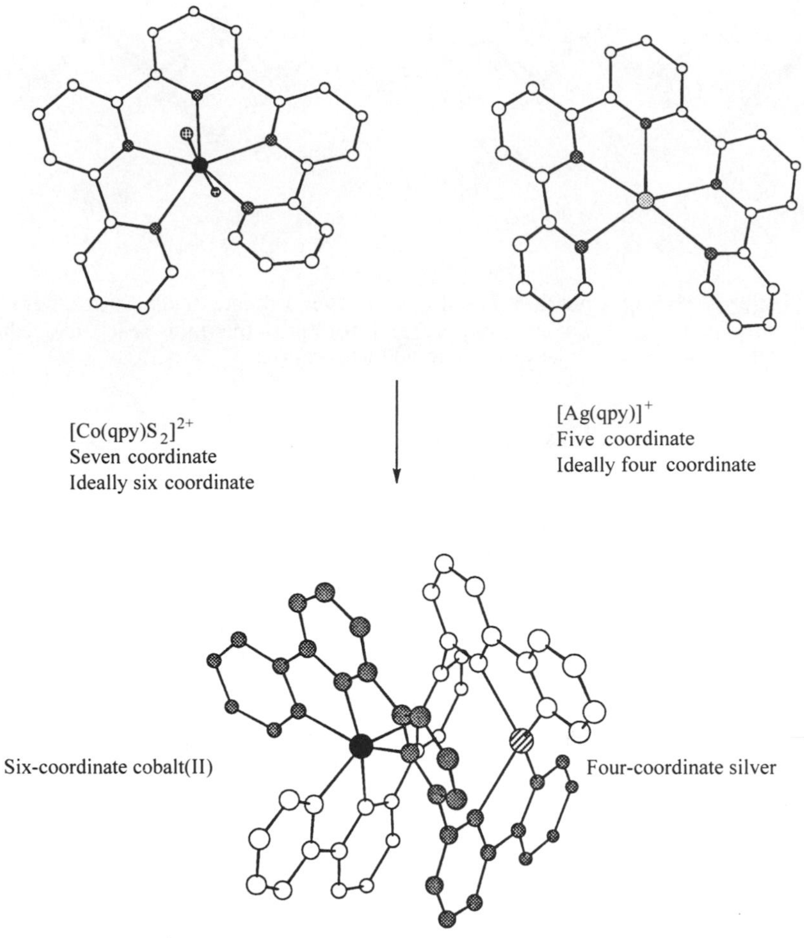

$[Co(qpy)S_2]^{2+}$
Seven coordinate
Ideally six coordinate

$[Ag(qpy)]^+$
Five coordinate
Ideally four coordinate

Six-coordinate cobalt(II)

Four-coordinate silver

Figure 28 A scheme for the formation of a heterodinuclear double helicate and the molecular structure of the complex that is formed from the reaction of mononuclear silver(I) and cobalt(II) complexes of (**31**).

Several iron(II) complexes were identified, including $[Fe(\mathbf{58})_2]^{2+}$ in which the metal is coordinated to the tridentate domain of the two ligands. The reaction of $[Fe(\mathbf{58})_2]^{2+}$ with Ag^+ salts yields an equilibrium mixture containing the heterodinuclear double-helical complex $[FeAg(\mathbf{58})_2]^{3+}$ in which the iron(II) is coordinated to the tridentate and the silver(I) to the didentate domains of the two ligand threads:

$$[Fe(\mathbf{58})_2]^{2+} + Ag^+ = [FeAg(\mathbf{58})_2]^{3+} \qquad \log \beta = 4.0(3) \qquad (6)$$

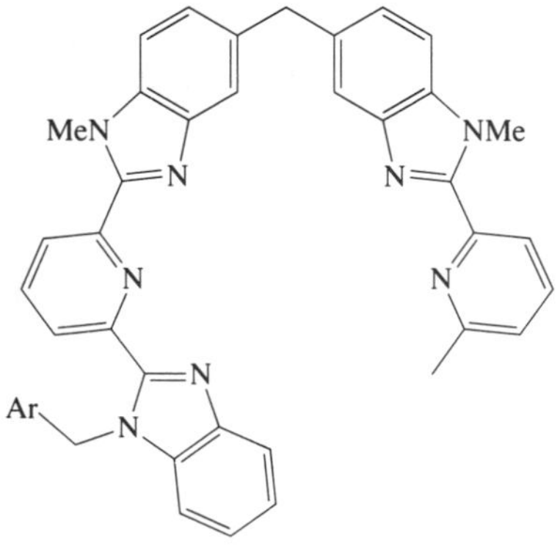

MeN NMe N N N N N Ar N

(**58**)

The next generation of ligands is represented by (**59**) which contains a central tridentate metal-binding domain which is flanked by two bidentate regions. The major species formed upon titration of (**59**) with iron(II) is the complex $[Fe(\mathbf{59})_2]^{2+}$ in which the metal is coordinated to the central tridentate domain of the two ligands. The solution equilibria are complex, but may be interpreted in terms of the following processes:

$$Fe^{2+} + 2(\mathbf{59}) = [Fe(\mathbf{59})_2]^{2+} \qquad \log \beta_{12} = 10.9(3) \qquad (7)$$

$$2Fe^{2+} + 2(\mathbf{59}) = [Fe_2(\mathbf{59})_2]^{4+} \qquad \log \beta_{22} = 14.6(3) \qquad (8)$$

$$[Fe(\mathbf{59})_2]^{2+} + Ag^{+} = [FeAg(\mathbf{59})_2]^{3+} \qquad \log \beta = 5.5(4) \qquad (9)$$

$$[Fe(\mathbf{59})_2]^{2+} + 2Ag^{+} = [FeAg_2(\mathbf{59})_2]^{4+} \qquad \log \beta = 9.9(4)(10)$$

(**59**)

The ^1H NMR spectroscopic data for the heterotrinuclear species $[FeAg_2(\mathbf{59})_2]^{4+}$ are compatible with a double-helical, a side-to-side or a catenated structure (Figure 29), although the catenated structure has now been confirmed.

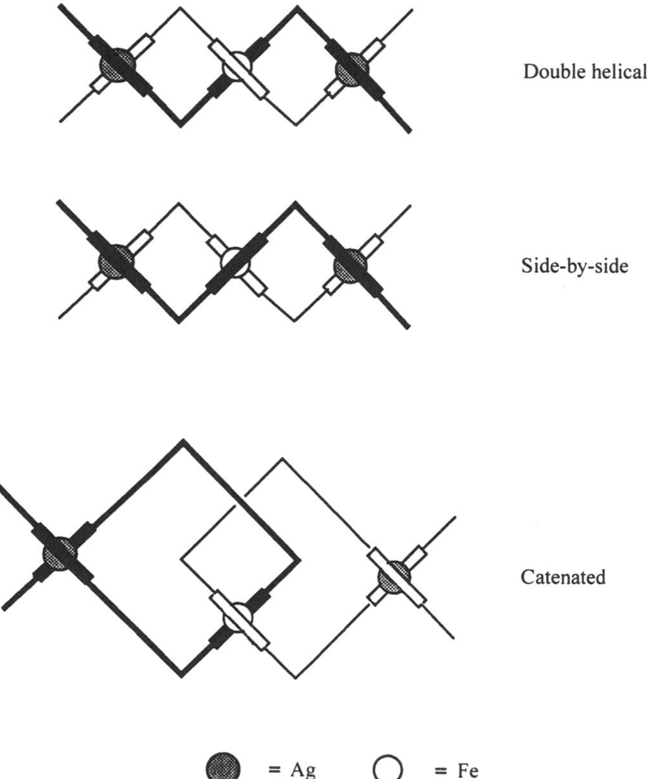

Figure 29 A further complication is observed with the heterotopic ligand (**59**). The heterotrinuclear cation $[FeAg_2(\mathbf{59})_2]^{4+}$ could be double helical, nonhelical or catenated.

6.7 TRIPLE HELICATES

The basic requirements for the formation of a triple-helical structure are relatively simply formulated: it is necessary to react a ligand thread containing *m* *n*-dentate metal-binding domains with *m* metal ions with a preference for a coordination number of 3*n* (Figure 2). As ever, the design of the ligands is critical. The principal problem is preventing the formation of a double helix, in which the remaining *n* coordination sites at each metal are occupied by ancillary ligands, or a nonhelical complex of the desired $\{M_2L_3\}$ stoichiometry.

Probably the first triple-helical dinuclear complex was prepared by the use of predisposed amines **(60)** and **(61)** with three chiral chains incorporating metal-binding hydroxamate domains. Triple-helical complexes were obtained upon reaction with iron(III). The use of the chiral strands means that diastereomerically pure triple helices are obtained, as was established from studies of the circular dichroism of the complexes. The complexes with **(60)** and **(61)** have different internal hydrogen bonding patterns, and the former gives a (Λ *cis*, Λ *cis*) left-handed helix while the latter gives a (Δ *cis*, Δ *cis*) right-handed helix.[57]

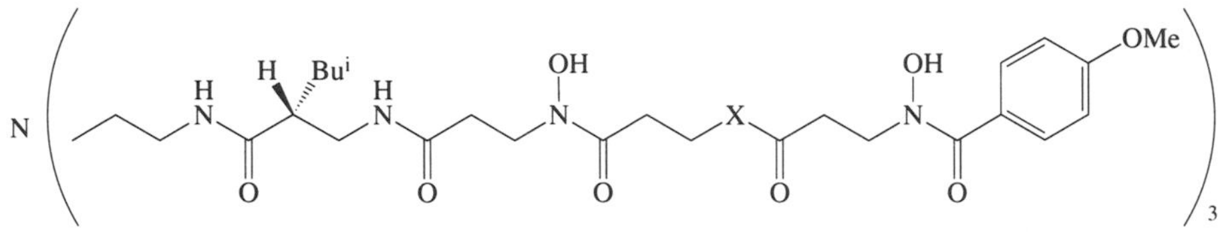

(60) X=NH
(61) X=O

The first approach to a triple-helical species containing oligopyridine metal-binding domains was probably seen with the ligand **(62)**. This ligand does react with nickel(II), cobalt(II), manganese(II) or iron(II) salts to give complexes of stoichiometry $[M_2L_3]^{4+}$, but although they were triply bridged, they were not triple helical. The structure of the iron(II) complex is presented in Figure 30; the two iron(II) centres are held in a pseudo-octahedral geometry at a distance of 765 pm.[58] The observed geometry and the triple helix are related by the mutual twisting of one six-coordinate metal ion with respect to the other; this, in turn, is controlled by the linker between the bidentate regions. In the case of **(62)** the short CH_2CH_2 linker attached in the 4-position does not allow this flexibility.

(62)

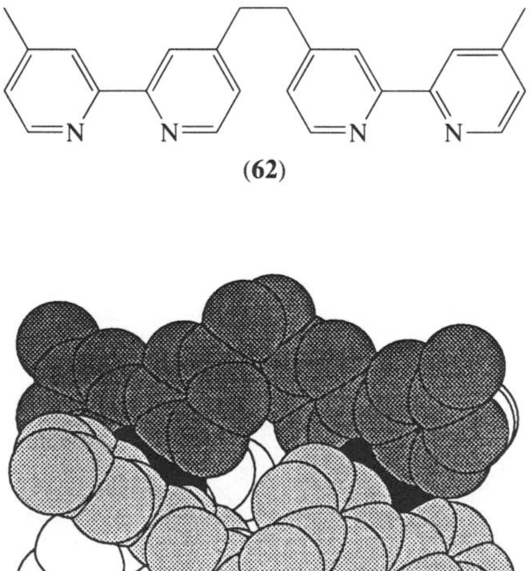

Figure 30 The dinuclear cation $[Fe_2(\mathbf{62})_3]^{4+}$ is not a triple helix. The three ligand strands have been shaded to clarify the structure.

In contrast, the related ligand (**18**) with the 1,2-ethanediyl spacer attached in the 5-position forms a 2:3 cation $[Fe_2(\mathbf{18})_3]^{4+}$ which is almost certainly triple helical, although a structural analysis has not been reported.[22]

The first example of a structurally characterized triple-helical complex was the species formed from the interaction of cobalt(II) salts with (**41**) (R = Me).[41,59] The relatively inflexible diphenylmethane spacer appears to be critical in determining the formation of the triple-helical structure. The x-ray structure of the cation $[Co_2(\mathbf{41})_3]^{4+}$ is shown in Figure 31; each cobalt(II) centre is in an approximately D_3 environment, but their mutual orientation results in the cation as a whole possessing approximate C_3 symmetry. In this triple helical complex, π-stacking does not appear to play an important role. NMR spectroscopic studies have established that in nonpolar solvents the triple-helical structure persists in solution and, most importantly, the two enantiomeric triple helicates do not interconvert in solution on the NMR timescale.[41] Although the triple helicate is the major solution species from the reaction of (**41**) with cobalt(II) (log β = 21.4(6)), the case is not so simple when complexes with other metals are considered. For example, with zinc a series of solution equilibria involving double- and triple-helical complexes occur:

$$2Zn^{2+} + 3(\mathbf{41}) = [Zn_2(\mathbf{41})_3]^{4+} \qquad \log \beta = 24.2(1.6) \tag{11}$$

$$2Zn^{2+} + 2(\mathbf{41}) = [Zn_2(\mathbf{41})_2]^{4+} \qquad \log \beta = 19.4(1.4) \tag{12}$$

$$3Zn^{2+} + 2(\mathbf{41}) = [Zn_3(\mathbf{41})_2]^{6+} \qquad \log \beta = 24.3(1.3) \tag{13}$$

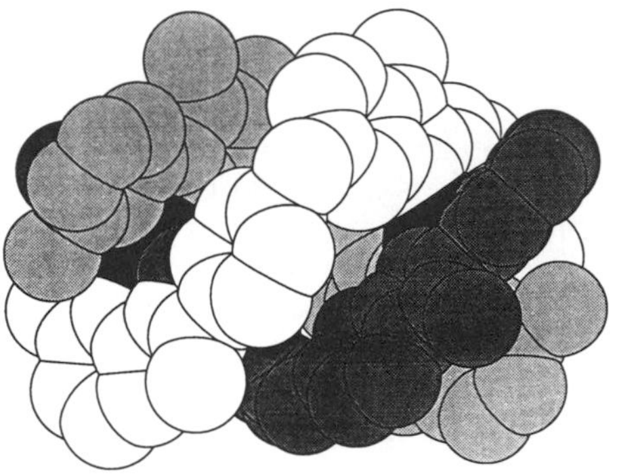

Figure 31 The triple-helical cation $[Co_2(\mathbf{41})_3]^{4+}$. The three ligand strands have been shaded and some substituents omitted in order to clarify the structure.

Although the $[Co_2(\mathbf{41})_3]^{4+}$ triple helix is stable on the NMR timescale it is still labile. However, oxidation leads to the kinetically inert dicobalt(III) complex $[Co_2(\mathbf{41})_3]^{6+}$ which has been resolved into the two enantiomers.

Remarkably, it is also possible to assemble dinuclear triple helices about lanthanoid metal cations. For example, (**63**) is a structural extension of (**41**) to a system which contains two tridentate metal-binding domains. This is the type of ligand which is expected to give dinuclear double helicates with six-coordinate metal ions. However, many lanthanoid cations exhibit higher coordination numbers than first-row transition metal ions, and nine coordination is particularly common. With two nine-coordinate metal ions, (**63**) could give a dinuclear triple helicate, and this is exactly what is observed with lanthanum(III), gadolinium(III), lutetium(III), europium(II) or terbium(III).[60] The crystal structure of the cation $[Eu_2(\mathbf{63})_3]^{6+}$ confirms the triple-helical character (Figure 32). In contrast to the dicobalt(II) triple helix discussed above, there is extensive π stacking in this cation. Detailed photophysical studies of these triple-helical lanthanoid complexes have been reported.[60]

An elegant extension of this methodology allows the formation of heteronuclear triple helicates. Helicand (**64**) is structurally related to (**58**) and contains bidentate and tridentate metal-binding domains. As expected, titration with zinc(II) reveals the formation of the complex $[Zn(\mathbf{64})_2]^{2+}$ in which the metal is coordinated to two tridentate domains from the two ligand threads, and the double-helical species $[Zn_2(\mathbf{64})_2]^{4+}$, in which one zinc is in a six-coordinate environment from the two tridentate domains and the other in a four-coordinate environment from the two bidentate domains. Similarly, a triple-helical complex $[La_2(\mathbf{64})_3]^{6+}$ is formed upon reaction with lanthanum

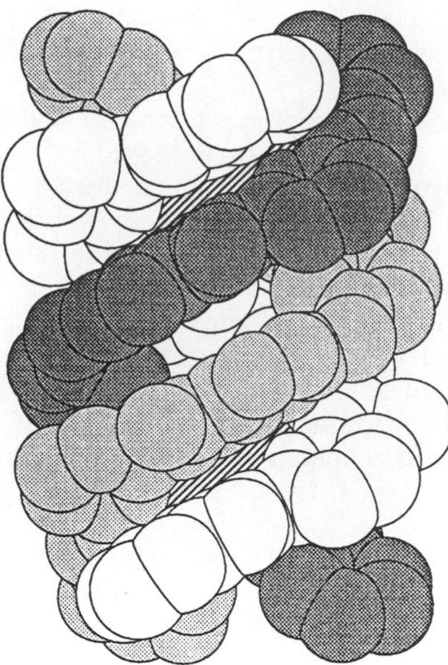

Figure 32 The triple-helical cation $[Eu_2(\mathbf{63})_3]^{4+}$ containing two nine-coordinate europium(III) centres. The three ligand strands have been shaded and some substituents omitted in order to clarify the structure.

salts. In the latter complex, one lanthanum is in an eight-coordinate environment (tridentate + tridentate + didentate) while the other is seven-coordinate (tridentate + didentate + didentate). More interestingly, the interaction of (**64**) with mixtures of zinc(II) and lanthanum(III) salts leads to the formation of the heterodinuclear triple-helical cation $[LaZn(\mathbf{64})_3]^{5+}$ with log $\beta_{113} = 26.2(3)$. The lanthanum is coordinated to the three tridentate domains, while the zinc interacts with the three bidentate domains of the three ligand strands. An analogous europium complex has also been described.[61]

(**63**)

(**64**)

Although (**62**) did not give a triple-helical structure with first-row transition metal elements, the related ligand (**65**) does form such complexes. The cation $[Fe_2(\mathbf{65})_2]^{4+}$ is triple helical and NMR data suggest that a homochiral ($\Delta\Delta$ or $\Lambda\Lambda$) system is formed.[62]

Naturally, linking together more than two bidentate regions with appropriate spacers could lead to the formation of extended polynuclear triple helicates with pseudooctahedral centres, and this has recently been demonstrated by Lehn.[63] The ligand adopted was (**66**) which reacts with nickel(II) salts to give the pink triple-helical trinuclear cation $[Ni_3(\mathbf{66})_3]^{6+}$ (Figure 33). An

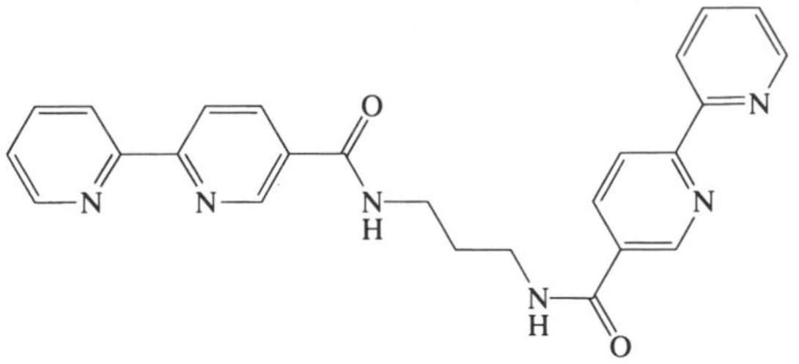

(65)

interesting feature was the observation that the salt $[Ni_3(\mathbf{66})_3][ClO_4]_6$ partially resolved upon crystallization to yield crystals which contained an excess of one of the two enantiomers. It was also demonstrated that the rate of racemization in solution was rather slow.

(66)

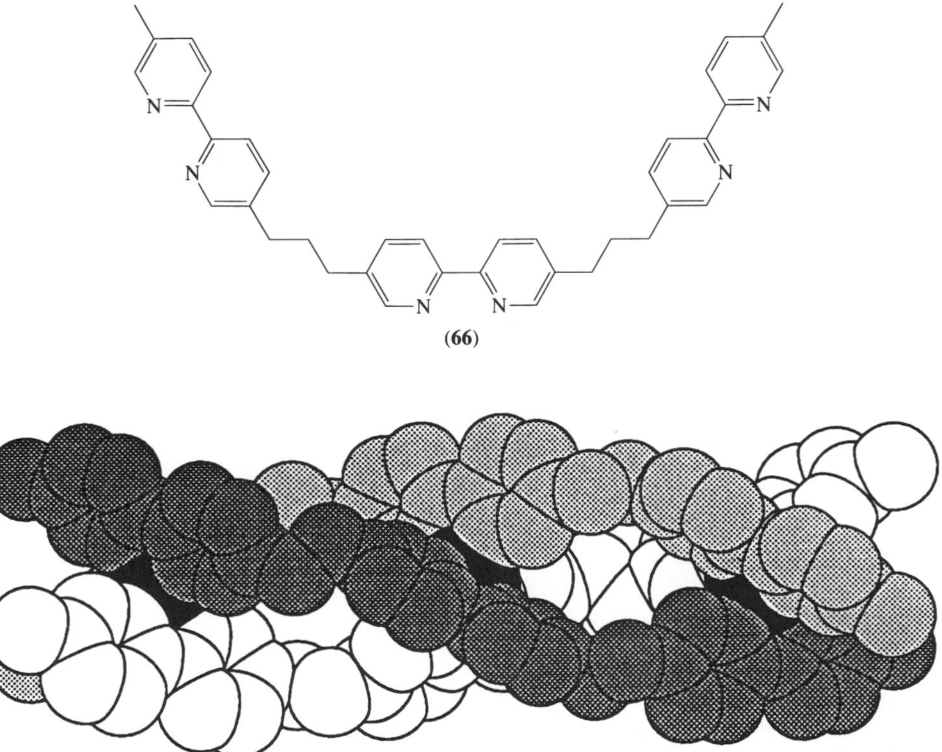

Figure 33 The trinuclear triple helicate formed from the reaction of **(66)** with nickel(II) salts. The three ligand strands have been shaded to clarify the structure.

Finally, we note that it is possible to use copper(I) centres for the assembly of triple helicates, and illustrate this by the interaction of 2,6-bis(2-pyridylethynyl)pyridine (**67**) with copper(I).[64] The resultant triple helix contains formally three-coordinate copper(I) centres (Figure 34).

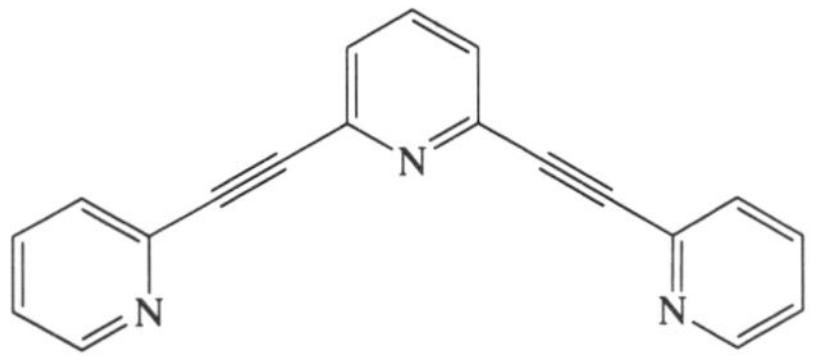

(67)

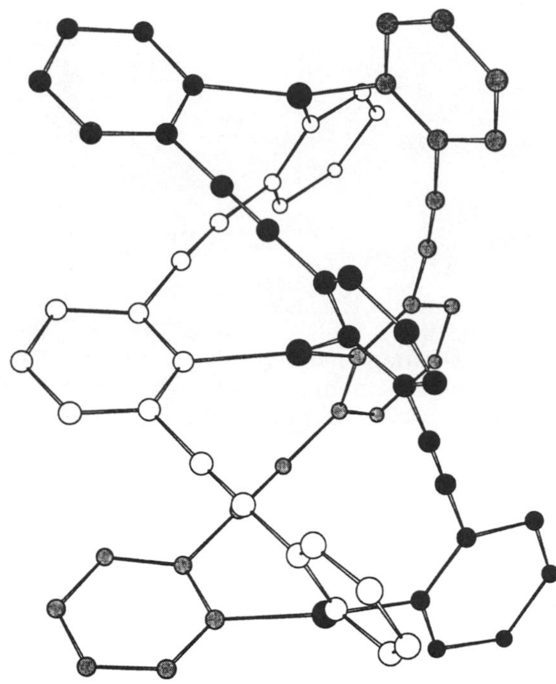

Figure 34 The triple helicate containing three three-coordinate copper(I) centres formed from (**67**). The three ligand strands have been shaded to clarify the structure.

6.8 OTHER RELATED HELICAL STRUCTURES

To conclude this chapter, it is necessary to point out that a range of other ligand types have been shown to give rise to helical, multiple-helical or recognizably helical structures as a result of interaction with metal ions. Of particular interest is the interaction of polypeptides modified with classical metal-binding domains,[65] a recent report by Shanzer describing a preorganized helical system which could act as a redox switch,[66] a range of variously organized and preorganized macrocycles,[67] and finally a series of catenated and helicated complexes based upon pyridone or pyridine *N*-oxide binding motifs.[68]

6.9 REFERENCES

1. E. C. Constable, *Chem. Ind.*, 1994, 56.
2. E. C. Constable, *Prog. Inorg. Chem.*, 1994, **42**, 67.
3. E. C. Constable and D. Smith, *Chem. Br.*, 1995, **31**, 33.
4. E. C. Constable, *Tetrahedron*, 1992, **48**, 10 013.
5. E. C. Constable and D. Smith, 'The Polymeric Materials Encyclopedia', CRC Press, Boca Raton, FL, 1995.
6. C. O. Dietrich-Buchecker and J.-P. Sauvage, *Chem. Rev.*, 1987, **87**, 795.
7. J.-M. Lehn, A. Rigault, J. Siegel, J. Harrowfield, B. Chevrier and D. Moras, *Proc. Natl. Acad. Sci.*, USA, 1987, **84**, 2565.
8. C. O. Dietrich-Buchecker and J.-P. Sauvage, *New J. Chem.*, 1992, **16**, 277.
9. C. O. Dietrich-Buchecker and J.-P. Sauvage, *Bioorg. Chem. Front.*, 1991, **2**, 195.
10. J.-P. Sauvage, *Acc. Chem. Res.*, 1990, **23**, 319.
11. G. Schill, 'Catenanes, Rotaxanes and Knots', Academic Press, New York, 1971.
12. V. I. Sokolov, *Russ. Chem. Rev. (Engl. Transl.)*, 1973, **42**, 452; M. Suffczynski, *Pol. J. Chem.*, 1995, **69**, 157.
13. J.-C. Chambron, C. Dietrich-Buchecker and J.-P. Sauvage, *Top. Curr. Chem.*, 1993, **165**, 131.
14. R. B. King, 'Chemical Applications of Topology and Graph Theory', Elsevier, New York, 1983.
15. R. S. Cahn, C. Ingold and V. Prelog, *Angew. Chem., Int. Ed. Engl.*, 1966, **5**, 385.
16. C. Piguet, G. Hopfgartner, B. Bocquet, O. Schaad and A. F. Williams, *J. Am. Chem. Soc.*, 1994, **116**, 9092; S. Ruttimann, C. Piguet, G. Bernardinelli, B. Bocquet and A. F. Williams, *J. Am. Chem. Soc.*, 1992, **114**, 4230.
17. J. K. Judice, S. J. Keipert and D. J. Cram, *J. Chem. Soc., Chem. Commun.*, 1993, 1323.
18. T. W. Bell and H. Jousselin, *Nature*, 1994, **367**, 441.
19. J.-M. Lehn, J.-P. Sauvage, J. Simon, R. Ziessel, C. Piccinni-Leopardi, G. Germain, J.-P. Declercq and M. Van Meerssche, *Nouv. J. Chim.*, 1983, **7**, 413; J.-P. Gisselbrecht, M. Gross, J.-M. Lehn, J.-P. Sauvage, R. Ziessel, C. Piccinni-Leopardi, J. M. Arrieta, G. Germain and M. Van Meerssche, *Nouv. J. Chim.*, 1984, **8**, 661; F. Arnaud-Neu, M. Sanchez and M.-J. Schwing-Weill, *New J. Chem.*, 1986, **10**, 165.

20. E. C. Constable, P. Harverson, D. R. Smith and L. A. Whall, *Tetrahedron*, 1994, **50**, 7799; E. C. Constable, M. J. Hannon and D. R. Smith, *Tetrahedron Lett.*, 1994, **35**, 6657.
21. Y. He and J.-M. Lehn, *Chem. J. Chinese Univ.*, 1990, **6**, 183.
22. M.-T. Youinou, R. Ziessel and J.-M. Lehn, *Inorg. Chem.*, 1991, **30**, 2144; J.-M. Lehn and R. Ziessel, *Helv. Chim. Acta*, 1988, **71**, 1511.
23. A. Juris and R. Ziessel, *Inorg. Chim. Acta*, 1994, **225**, 251.
24. Y. Yao, M. W. Perkovic, D. P. Rillema and C. Woods, *Inorg. Chem.*, 1992, **31**, 3956.
25. J. C. Chambron, C. O. Dietrich-Buchecker, V. Heitz, J. F. Nierengarten, J.-P. Sauvage, C. Pascard and J. Guilhem, *Pure Appl. Chem.*, 1995, **67**, 233; J. C. Chambron, C. O. Dietrich-Buchecker, J. F. Nierengarten and J.-P. Sauvage, *Pure Appl. Chem.*, 1994, **66**, 1543.
26. J.-M. Lehn, A. Rigault, J. Siegel, J. Harrowfield, B. Chevrier and D. Moras, *Proc. Natl. Acad. Sci. USA*, 1987, **84**, 2565; T. M. Garrett, U. Koert, J.-M. Lehn, A. Rigault, D. Meyer and J. Fischer, *J. Chem. Soc., Chem. Conmmun.*, 1990, 557.
27. E. C. Constable, M. J. Hannon and D. A. Tocher, *Angew. Chem., Int. Ed. Engl.*, 1992, **104**, 218; *J. Chem. Soc., Dalton Trans.*, 1993, 1883.
28. E. C. Constable, M. J. Hannon, A. J. Edwards and P. R. Raithby, *J. Chem. Soc., Dalton Trans.*, 1994, 2669.
29. R. Ziessel and M.-T. Youinou, *Angew. Chem., Int. Ed. Engl.*, 1993, **32**, 877.
30. E. C. Constable, M. D. Ward and D. A. Tocher, *J. Am. Chem. Soc.*, 1990, **112**, 1256; *J. Chem. Soc., Dalton Trans.*, 1991, 1675; E. C. Constable and R. Chotalia, *J. Chem. Soc., Chem. Commun.*, 1992, 64; K. T. Potts, M. Keshavarz-K, F. S. Tham, H. D. Abruña and C. Arana, *Inorg. Chem.*, 1993, **32**, 4436.
31. J. D. Crane and J.-P. Sauvage, *New J. Chem.*, 1992, **16**, 649.
32. K. T. Potts, M. Keshavarz-K, F. S. Tham, K. A. G. Raiford, C. Arana and H. D. Abruña, *Inorg. Chem.*, 1993, **32**, 5477.
33. E. C. Constable, A. J. Edwards, M. J. Hannon and P. R. Raithby, *J. Chem. Soc., Chem. Commun.*, 1994, 1991; K. T. Potts, M. Keshavarz-K, F. S. Tham, H. D. Abruna and C. Arana, *Inorg. Chem.*, 1993, **32**, 4450.
34. D. Dolphin, R. L. N. Harris, J. L. Huppatz, A. W. Johnson, I. T. Kay and J. Leng, *J. Chem. Soc. (C)* 1966, 98.
35. G. Struckmeier, U. Thewalt and J.-H. Fuhrhop, *J. Am. Chem. Soc.*, 1976, **98**, 278.
36. W. S. Sheldrick and J. Engel, *Acta Crystallogr., Sect. B*, 1981, **37**, 250; W. S. Sheldrick and J. Engel, *J. Chem. Soc., Chem. Commun.*, 1980, 5; D. Dolphin, R. L. N. Harris, J. L. Huppatz, A. W. Johnson and I. T. Kay, *J. Chem. Soc. (C)*, 1966, 30.
37. G. C. Van Stein, G. Van Koten, F. Blank, L. C. Taylor, K. Vrieze, A. L. Spek, A. J. M. Duisenberg, A. M. M. Schreurs, B. Kojic-Prodic and C. Brevard, *Inorg. Chim. Acta*, 1985, **98**, 107; G. C. Van Stein, G. Van Koten, A. L. Spek, A. J. M. Duisenberg and E. A. Klop, *Inorg. Chim. Acta*, 1983, **78**, L61.
38. G. C. Van Stein, G. Van Koten, H. Passenier, O. Steinebach and K. Vrieze, *Inorg. Chim. Acta*, 1984, **89**, 79.
39. C. Piquet, G. Bernardinelli and A. F. Williams, *Inorg. Chem.*, 1989, **28**, 2920.
40. R. F. Carina, G. Bernardinelli and A. F. Williams, *Angew. Chem., Int. Ed. Engl.*, 1993, **32**, 1463.
41. C. Piquet, G. Bernardinelli, B. Bocquet, A. Quattropani and A. F. Williams, *J. Am. Chem. Soc.*, 1992, **114**, 7440.
42. M. G. B. Drew, A. Lavery, V. McKee and S. M. Nelson, *J. Chem. Soc., Dalton Trans.*, 1985, 1771.
43. G. C. Van Stein, H. Van der Poel, G. Van Koten, A. L. Spek, A. J. M. Duisenberg and P. S. Pregosin, *J. Chem. Soc., Chem. Commun.*, 1980, 1016; G. C. Van Stein, G. Van Koten, K. Vrieze, C. Brevard and A. L. Spek, *J. Am. Chem. Soc.*, 1984, **106**, 4486; G. C. Van Stein, G. Van Koten, K. Vrieze and C. Brevard, *Inorg. Chem.*, 1984, **23**, 1016.
44. D. Wester and G. J. Palenik, *Inorg. Chem.*, 1976, **15**, 755; D. Wester and G. J. Palenik, *J. Chem. Soc., Chem. Commun.*, 1975, **74**.
45. A. Bino and N. Cohen, *Inorg. Chim. Acta*, 1993, **210**, 11.
46. C. Lorenzini, C. Pelizzi, G. Pelizzi and G. Predieri, *J. Chem. Soc., Dalton Trans.*, 1983, 2155.
47. G. Paolucci, S. Stelluto, S. Sitran, D. Ajo, F. Benetollo, A. Polo and G. Bombieri, *Inorg. Chim. Acta*, 1992, **193**, 57.
48. E. C. Constable, M. G. B. Drew, G. Forsyth and M. D. Ward, *J. Chem. Soc., Chem. Commun.*, 1988, 1450; E. C. Constable and J. M. Holmes, *Inorg. Chim. Acta*, 1987, **126**, 187; E. C. Constable, J. M. Holmes and P. R. Raithby, *Polyhedron*, 1991, **10**, 127.
49. K. T. Potts, M. Keshavarzk, F. S. Tham, H. D. Abruna and C. R. Arana, *Inorg. Chem.*, 1993, **32**, 4422.
50. T. M. Garrett, U. Koert, J.-M. Lehn, A. Rigault, D. Meyer and J. Fischer, *J. Chem. Soc., Chem. Commun.*, 1990, 557; J.-M. Lehn and A. Rigault, *Angew. Chem., Int. Ed. Engl.*, 1988, **27**, 1095; T. M. Garrett, U. Koert and J.-M. Lehn, *J. Phys. Org. Chem.*, 1992, **5**, 529; A. Pfeil and J.-M. Lehn, *J. Chem. Soc., Chem. Commun.*, 1992, 838; T. M. Garrett, U. Koert, J.-M. Lehn, A. Rigault, D. Meyer and J. Fischer, *J. Chem. Soc., Chem. Commun.*, 1990, 557.
51. B. Schoentjes and J. M. Lehn, *Helv. Chim. Acta*, 1995, **78**, 1.
52. U. Koert, M. M. Harding and J.-M. Lehn, *Nature (London)*, 1990, **346**, 339.
53. W. Zarges, J. Hall and J. M. Lehn, *Helv. Chim. Acta*, 1991, **74**, 1843.
54. M. M. Harding, U. Koert, J. M. Lehn, A. Marquisrigault, C. Piguet and J. Siegel, *Helv. Chim. Acta*, 1991, **74**, 594; C. D. Eisenbach, U. S. Schubert, G. R. Baker and G. R. Newkome, *J. Chem. Soc., Chem. Commun.*, 1995, 69.
55. E. C. Constable, A. J. Edwards, P. R. Raithby and J. V. Walker, *Angew. Chem., Int. Ed. Engl.*, 1993, **32**, 1465; E. C. Constable and J. V. Walker, *J. Chem. Soc., Chem. Commun.*, 1992, 884.
56. C. Piguet, G. Hopfgartner, B. Bocquet, O. Schaad and A. F. Williams, *J. Am. Chem. Soc.*, 1994, **116**, 9092; C. Piguet, G. Bernardinelli, A. F. Williams and B. Bocquet, *Angew. Chem., Int. Ed. Engl.*, 1995, **34**, 582.
57. J. Libman, Y. Tor and A. Shanzer, *J. Am. Chem. Soc.*, 1987, **109**, 5880.
58. B. R. Serr, K. A. Anderson, C. M. Elliott and O. P. Anderson, *Inorg. Chem.*, 1988, **27**, 4499.
59. A. F. Williams, C. Piguet and G. Bernardinelli, *Angew. Chem., Int. Ed. Engl.*, 1991, **30**, 1490; L. J. Charbonniere, G. Bernardinelli, C. Piguet, A. M. Sargeson and A. F. Williams, *J. Chem. Soc., Chem. Commun.*, 1994, 1419.
60. G. Bernardinelli, C. Piguet and A. F. Williams, *Angew. Chem., Int. Ed. Engl.*, 1992, **31**, 1622; G. Hopfgartner, C. Piguet, J. D. Henion and A. F. Williams, *Helv. Chim. Acta*, 1993, **76**, 1759; C. Piguet, J. C. G. Bunzli, G. Bernardinelli, G. Hopfgartner and A. F. Williams, *J. Am. Chem. Soc.*, 1993, **115**, 8197.
61. C. Piguet, G. Hopfgartner, A. F. Williams and J. C. Bunzli, *J. Chem. Soc., Chem. Commun.*, 1995, 491.
62. D. Zurita, P. Baret and J. L. Pierre, *New J. Chem.*, 1994, **18**, 1143.
63. R. Krämer, J.-M. Lehn, A. DeCian and J. Fischer, *Angew. Chem., Int Ed. Engl.*, 1993, **32**, 703.

64. K. T. Potts, C. P. Horwitz, A. Fessak, M. Keshavarz-K, E. Nash and P. J. Toscano, *J. Am. Chem. Soc.*, 1993, **115**, 10 444.
65. M. R. Ghadiri and M. A. Case, *Angew. Chem., Int. Ed. Engl.*, 1993, **32**, 1594; M. R. Ghadiri, C. Soares and C. Choi, *J. Am. Chem. Soc.*, 1992, **114**, 825; M. R. Ghadiri, C. Soares and C. Choi, *J. Am. Chem. Soc.*, 1992, **114**, 4000; M. R. Ghadiri and A. K. Fernholz, *J. Am. Chem. Soc.*, 1990, **112**, 9633; F. Q. Ruan, Y. Q. Chen and P. B. Hopkins, *J. Am. Chem. Soc.*, 1990, **112**, 9403; F. Q. Ruan, Y. Q. Chen, K. Itoh, T. Sasaki and P. B. Hopkins, *J. Org. Chem.*, 1991, **56**, 4347; M. Lieberman and T. Sasaki, *J. Am. Chem. Soc.*, 1991, **113**, 1470; H. Mihara, N. Nishino, R. Hasegawa, T. Fujimoto, S. Usui, H. Ishida and K. Ohkubo, *Chem. Lett.*, 1992, 1813.
66. L. Zelikovich, J. Libman and A. Shanzer, *Nature*, 1995, **374**, 790.
67. A. Bilyk and M. M. Harding, *J. Chem. Soc., Dalton Trans.*, 1994, 77; A. Bilyk, M. M. Harding, P. Turner and T. W. Hambley, *J. Chem. Soc., Dalton Trans.*, 1994, 2783; M. M. Harding, B. Sargeant, A. Bilyk and S. Augoloupis, *Aust. J. Chem.*, 1994, **47**, 1133; D. E. Fenton, R. W. Matthews, M. McPartlin, B. P. Murphy, I. J. Scowen and P. A. Tasker, *J. Chem. Soc., Chem. Commun.*, 1994, 1391; S. W. A. Bligh, N. Choi, E. G. Evagorou, W. S. Li and M. McPartlin, *J. Chem. Soc., Chem. Commun.*, 1994, 2399; J. K. Judice, S. J. Keipert and D. J. Cram, *J. Chem. Soc., Chem. Commun.*, 1993, 1323.
68. D. M. L. Goodgame, S. P. W. Hill and D. J. Williams, *J. Chem. Soc., Chem. Commun.*, 1993, 1019; D. M. L. Goodgame, S. Menzer, A. M. Smith and D. J. Williams, *Angew. Chem., Int. Ed. Engl.*, 1995, **34**, 574; G. A. Doyle, D. M. L. Goodgame, S. P. W. Hill and D. J. Williams, *J. Chem. Soc., Chem. Commun.*, 1993, 207.

7

Self-assembled Macrocycles, Cages, and Catenanes Containing Transition Metals in Their Backbones

MAKOTO FUJITA

Chiba University, Japan

7.1 INTRODUCTION

7.1.1 Supramolecular Self-assembly through Coordination

Weak interaction has become a key phrase in a wide area of current chemistry because it has been recognized that weak bonds have never been fully understood, despite their great significance in biological systems. The important role of weak bonds in biology is to induce the organization of biological structures, as in the case of hydrogen bonds between nucleic acids inducing organization of the double helices of DNA. The application of these phenomena to artificial systems led to the concept of supramolecular self-assembly.[1–3] In early studies of supramolecular self-assembly, hydrogen bonds were most frequently employed as they are in all biological systems. Incorporation of coordinate bonds into supramolecular self-assembly, however, triggered the development of an entirely new area of study in the late 1980s. A striking example is the double helical supramolecular copper(I) complex, Structure (**1**), offered by Lehn and co-workers in 1987.[4] In this complex, oligo(2,2′-bipyridine) ligands are spontaneously assembled around the tetrahedral coordination geometry of copper(I), resulting in the double helix due to thermodynamic equilibration. This and related double helical complexes[5–7] emphasized that coordinate bonds are the most useful weak bonds in supramolecular self-assembly due to their versatile geometrical modes (e.g., linear, trigonal, square planar, tetrahedral, etc.) in bond formations.

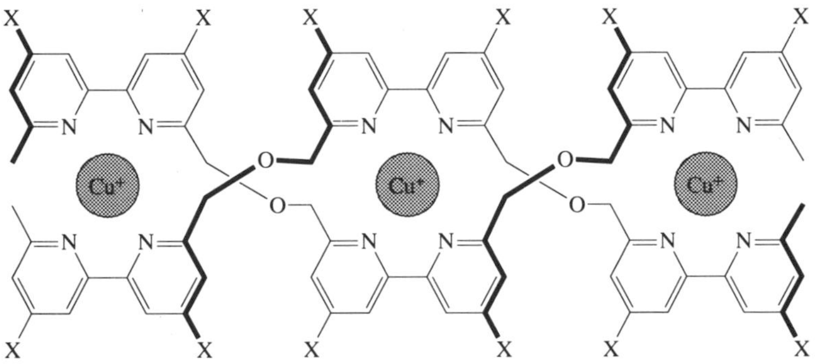

X X X X X X N N N N N N O O Cu+ Cu+ Cu+ N N N N N N O O X X X X X X

(**1**)

Following the double helical complexes (see Chapter 6),[8] self-assembly phenomena through metal coordination have shown remarkable potential in the construction of molecular architectures (see Chapter 5) such as triple helices, macrocycles, cages, tubes, grids, rods, and interlocked systems. Of these architectures, this chapter deals with macrocyclic frameworks involving transition metals.[9] Macrocycles have obviously taken an important role in molecular recognition chemistry.[10–16] Here the authors clarify how the chemistry of macrocycles has been changed by the introduction of supramolecular self-assembly. In particular, the following topics will be discussed: (i) self-assembling macrocycles and their ability as artificial receptors, (ii) self-assembling cages (three-dimensionally expanded macrocycles), (iii) self-assembling catenanes (interlocked macrocyclic systems), and (iv) infinite metal complexes containing macrocyclic, cagelike, or catenated frameworks.

7.2 SELF-ASSEMBLED MACROCYCLIC COMPLEXES

7.2.1 Background

Despite their remarkable potential utility, the practical uses of macrocycles in current chemistry have been extremely limited, mainly because of difficulties encountered in the synthesis of macrocycles. In fact, macrocyclization is an undoubtedly unfavorable process under kinetic conditions unless special conditions or techniques are employed. By contrast, the high-yield formation of calixarenes[17] that takes place under equilibrium between arenes and formaldehyde implies the facilitation of macrocyclization under thermodynamic conditions.

Since self-assembly phenomena take place under thermodynamic equilibration, macrocyclic structures fit the self-assembly system. There were even a few reports on the spontaneous formation of macrocyclic complexes in the 1970s.[18–23] For example, Shaw and others reported the formation of macrocyclic mononuclear iridium(I), platinum(II), and palladium(II) complexes (Structures (**2**),[19] (**3**),[19] and (**4**),[22] respectively) and dinuclear rhodium(I) complexes (Structures (**5**)[20] and (**6**)[19]). However, these complexes were prepared during systematic study of the complexation of transition metals with $Ph_2P(CH_2)_nPPh_2$ ($n > 4$) to obtain large ring coordination complexes with a *trans*-bonding phosphine ligand, and were not designed as host compounds for molecular recognition. Thus, no binding properties were reported for these molecules.

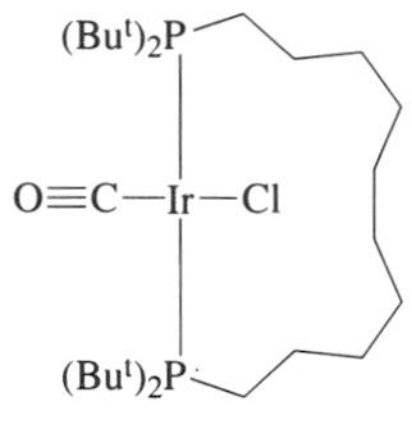

(**2**) 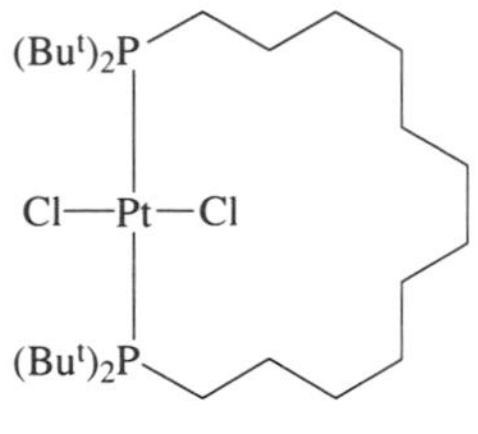 (**3**) 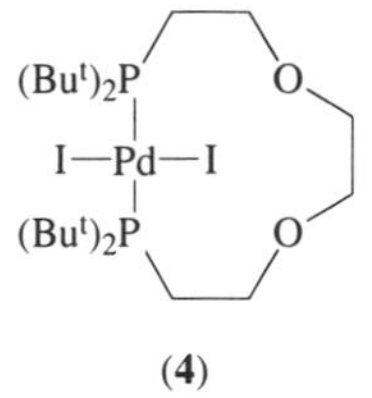 (**4**) 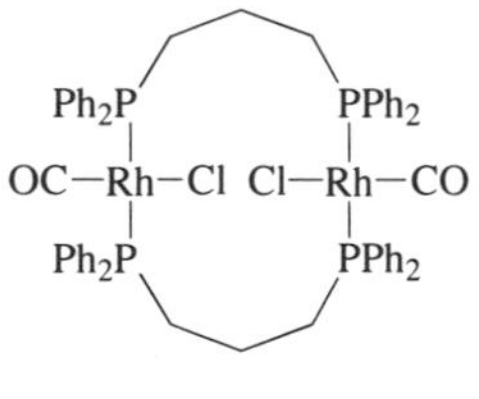 (**5**)

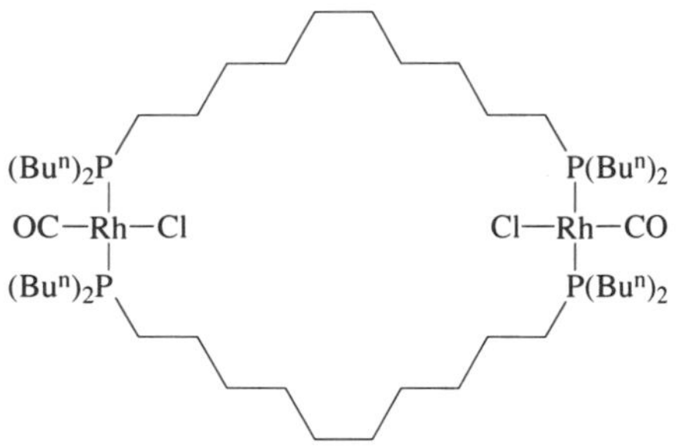

(**6**)

7.2.2 Macrocycles Containing Naked Transition Metals

The first example of inorganic macrocycles designed as synthetic receptors was most likely a dinuclear Cu^{II} complex, (**7**), reported by Maverick and Klavetter in 1984.[24] In this complex, an aromatic spacer exists between two bis(β-diketone) coordination sites so that a cavity is made through cyclization. Employment of a 2,7-naphthyl spacer made the cavity large enough to recognize small organic molecules. Structure (**8**) selectively recognized dabco, while monoamines (pyridine or quinuclidine) were hardly bound by (**8**) (Table 1). The selective recognition of dabco is consistent with internal coordination at two Cu^{II} binding sites which was confirmed by x-ray crystallography (see Figure 1).[25–7]

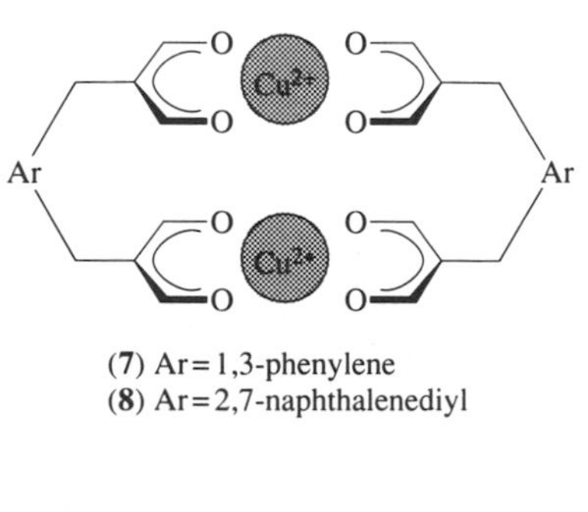

(**7**) Ar = 1,3-phenylene
(**8**) Ar = 2,7-naphthalenediyl

Various self-assembled inorganic macrocycles were reported in the 1990s. Some of them showed host characters. Self-assembled macrocycle (**9**) (Equation (1)), with a hydrophobic binding site, promoted pyrene transport through a liquid membrane.[28,29] Of the many metal salts examined in

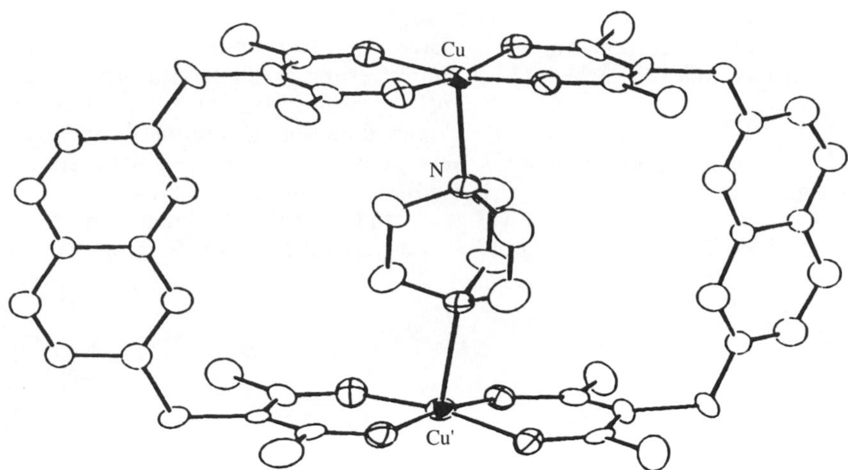

Figure 1 The crystal structure of (**8**)·dabco.

Table 1 Association constants between self-assembled macrocycles and various guests.

Host	*Guest*	*Association constant* (L mol^{-1})	*Ref.*
(**8**)	Pyridine	0.5	25
	Pyradine	5	25
	Quinuclidine	7	25
	dabco	220	25
(**9**)(M = Co)	Pyrene	14500	28
	1-(Naphthyloxy)acetate	6060	29
	Indole	848	29
	Tryptamene	60	29
	D or L-*N*-Acetyltryptophan	6	29
(**10**)	1,8-Anilinonaphthalenesulfonate	2300	30
(**14**)	*N,N'*-Dihexylterephthalamide	>1400	31
	Dimethyl terephthalate	40	31
	Dimethyl isophthalate	<1	31
(**24**)	*N*-(2-Naphthyl)acetamide	1800	32
	1,3,5-Tri(methoxy)benzene	750	33
	p-Dimethoxybenzene	330	34
	m-Dimethoxybenzene	580	34
	o-Dimethoxybenzene	30	34
	p-Bis(methoxymethyl)benzene	10	34
	1,4-(Dimethoxy)cyclohexane	not complexed	34
(**27a**)	1,3,5-Tri(methoxy)benzene	420	35
	p-Dimethoxybenzene	130	35
(**30a**)	1,3,5-Tri(methoxy)benzene	2500	32
	p-Dimethoxybenzene	2580	32
	p-Bis(methoxymethyl)benzene	560	32
	p-Dicyanobenzene	80	32
	p-Dinitrobenzene	30	32
(**30b**)	*p*-Dimethoxybenzene	200	36
	p-Dicyanobenzene	201	36

Equation (1), CoX_2 worked most effectively. M^{2+} (transport rate relative to background rate) are as follows: Mn^{2+} (1.15), Fe^{2+} (1.12), Co^{2+} (9.96), Ni^{2+} (9.92), Cu^{2+} (1.07), Zn^{2+} (1.43), Cd^{2+} (1.33). Macrocyclic copper(I) complex, (**10**), selectively catalyzed the hydrolysis of β-amino acids by cooperative effects of two Cu^{+} ions, but inhibited that of α-amino acid.[30] Crown-type compounds involving a transition metal, such as (**11**), have also been reported, and were shown to extract the second metal ion with high size selectivity.[37,38] Having a hydrophobic cavity, self-assembled, macrocyclic dinuclear Pd^{II}–phosphine, complex, (**12**), might show the abilities for both molecular recognition and catalysis.[39] Equation (2) shows a dimeric porphyrin host, (**14**), self-assembled from two porphyrin components, (**13**), was found to be a specific receptor for diamide derivatives of terephthalic acid (Table 1).[31]

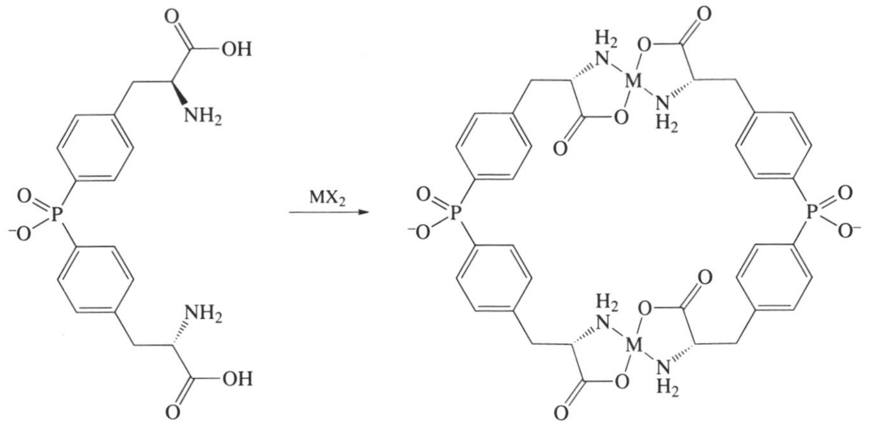

MX_2 (1)

(9)

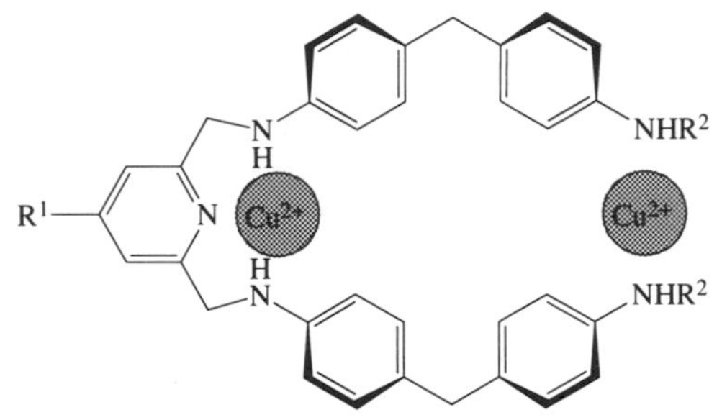

$R^1 = OCH_2NMe_3Br$
$R^2 = CH_2CH_2NMe_3Br$
(10)·$(Cu^{2+})_2$

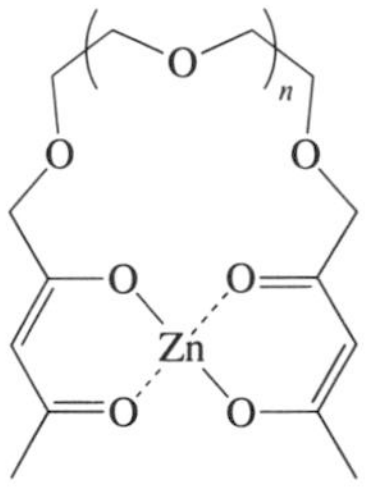

(11)

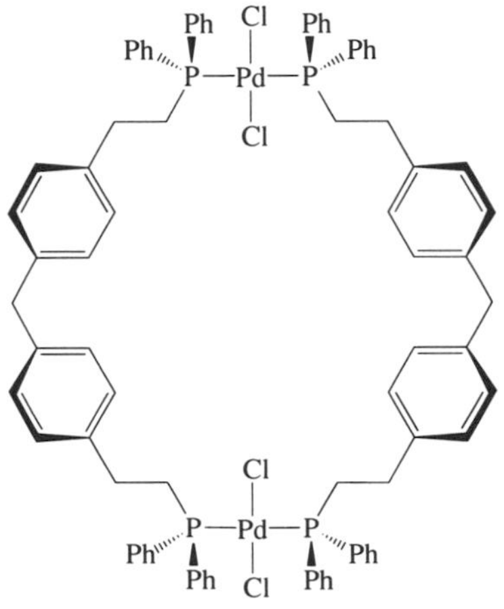

(12)

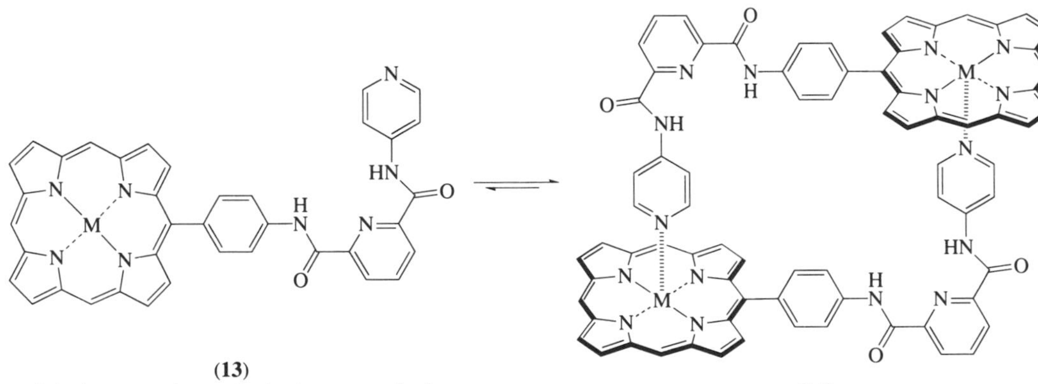

(2)

(13)
Substituents on the porphyrin rings are omitted

(14)

Macrocyclic frameworks were also constructed with more than two transition metal nuclei. Trinuclear Pd^{II} complex, **(15)**, is the first self-assembled macrocycle that involves carbon–metal bonds,[40] which have remarkably high stability in contrast to usual organometallics. Since a hydrophobic cavity exists in this complex, it may have the ability for molecular recognition; in fact, crystallography showed that an acetonitrile molecule was clathrated in the cavity. A similar "tricone" involving three Pd^{II} nuclei was also prepared from $PdCl_2$ and 1,2-bis(3,5-dimethyl-pyrazol-1-yl)ethane.[41] A few tetranuclear complexes (e.g., **(16)** and **(17)**) with a square shape have also been reported.[42,43]

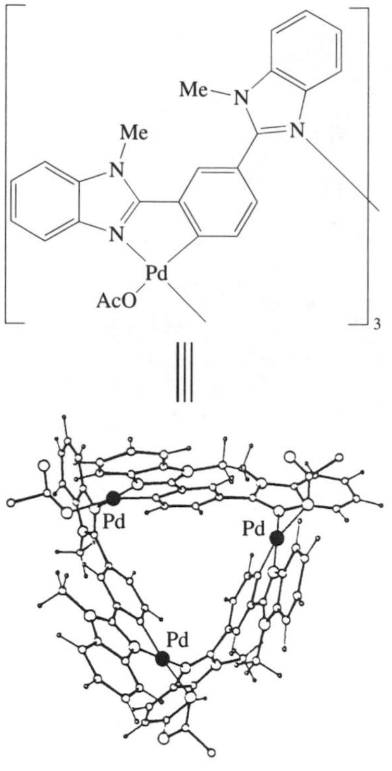

(15)

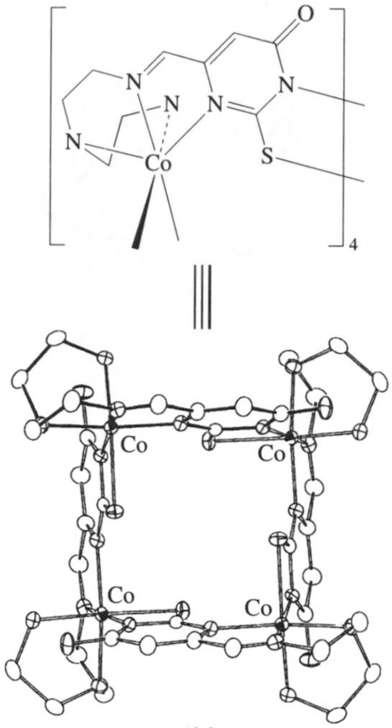

(16)

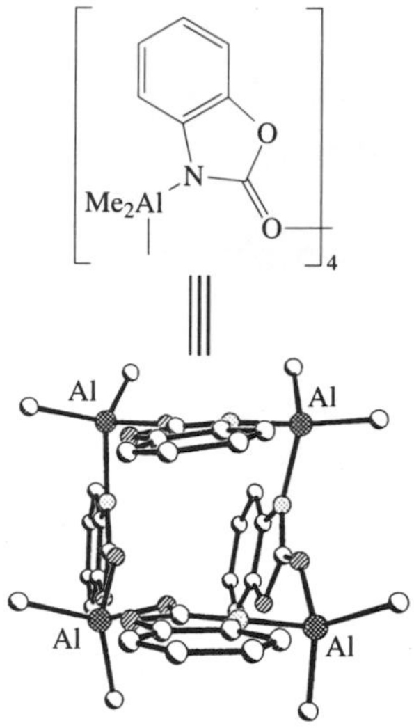

(17)

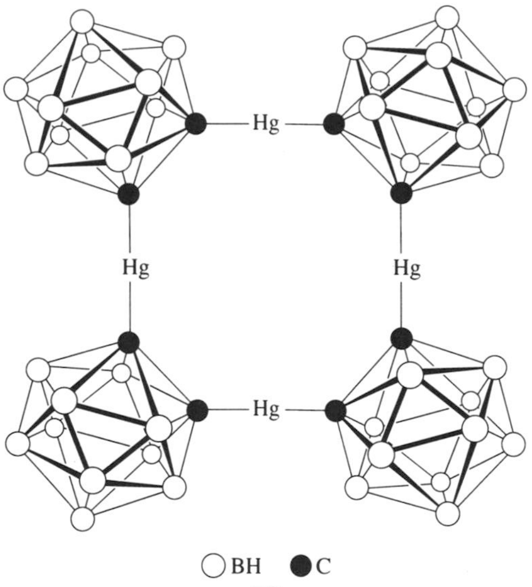

○BH ●C

(18)

Unique macrocycles involving Hg^{II} atoms and carborane units, (**18**), were reported by Hawthorne and co-workers.[44–7] Their novel properties were their properties as anion receptors[44–6] and their formation into supramolecular aggregates with *closo*-$B_{10}H_{10}^{2-}$, (**19**).[47]

In the synthesis of organic macrocycles, it should be noted that the precursors are often assembled through metal coordination (templated synthesis). Recent topical examples of the template synthesis of macrocycles are porphyrin-containing macrocycles reported by Sanders and co-workers. Scheme 1 shows that macrocyclic dimer (**20**) or trimer (**21**) are dominantly formed by employing 4,4′-bipyridine or 1,3,5-tris(4-pyridyl)triazine, respectively, as the template.[48] Enzyme-like catalysis of these macrocycles has been reported.[49–51]

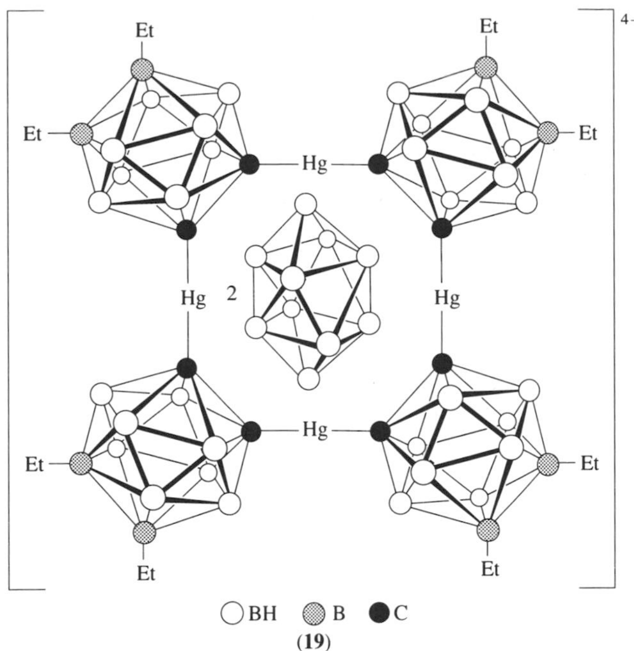

(19)

(20) **(21)**

i, CuCl, TMEDA, CH_2Cl_2, air, 4,4′-bipyridine; ii, CuCl, TMEDA, CH_2Cl_2, air, 1,3,5-tris(4-pyridyl)triazine

Scheme 1

7.2.3 Macrocycles Containing Protected Transition Metals

7.2.3.1 Tetranuclear macrocyclic complexes

As long as naked metals are employed in the self-assembly of macrocycles, the bridging ligands must have convergent donative sites at their termini such as 2,2′-bipyridyl or acetylacetate units (see Scheme 2(a)). On the other hand, if only a *cis* coordination site of the metals is utilized by the appropriate protection of the remaining coordination sites, the divergent nature of the metal components becomes rather convergent. The great advantage of this modification is that the bridging ligands can be simpler divergent structures (Scheme 2(b)).

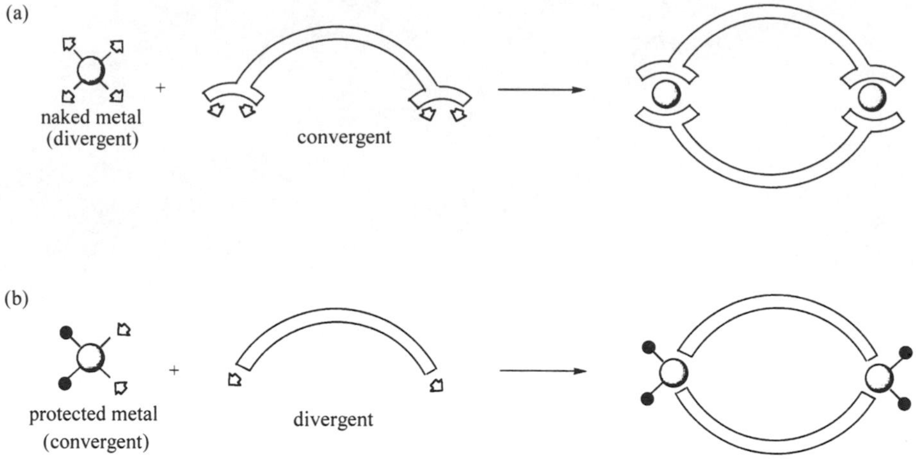

(a)

naked metal (divergent) + convergent →

(b)

protected metal (convergent) + divergent →

Scheme 2

Twenty-membered tetranuclear metal complexes, (**22**), were prepared from $(OC)_4M(nbd)$ and $P(OCH_2)_3P$ in 1983.[52] In 1990, the strategy shown in Scheme 2(b) was first applied to the self-assembly of a synthetic receptor; that is, macrocyclic tetranuclear Pd^{II} complex, (**24**), which was quantitatively assembled from protected Pd^{II}, (**23**), and bipy in water, as shown in Scheme 3.[33] The structure was mainly assigned by NMR, ESI-MS, and x-ray analysis.[53] In aqueous media, this macrocycle recognizes neutral or anionic guests (Table 1).[34] The self-assembly of a Pt^{II} analogue of (**24**) was inordinately slow due to the inactivity of Pt^{II}–pyridine bond. Consequently, its assembly from kinetically distributed oligomers was monitored by time-dependent NMR measurement at an elevated temperature.[54,55] Following these works, related square complexes (**25**)[56] and (**26**)[57–9] were reported. Incorporation of a spacer into the bipy skeleton gave tetranuclear complexes, (**27**), with expanded square cavities.[35] The largest cavities were obtained through square multiporphyrin arrays in (**28**) and (**29**), although their structures have not been determined by a convincing method.[60]

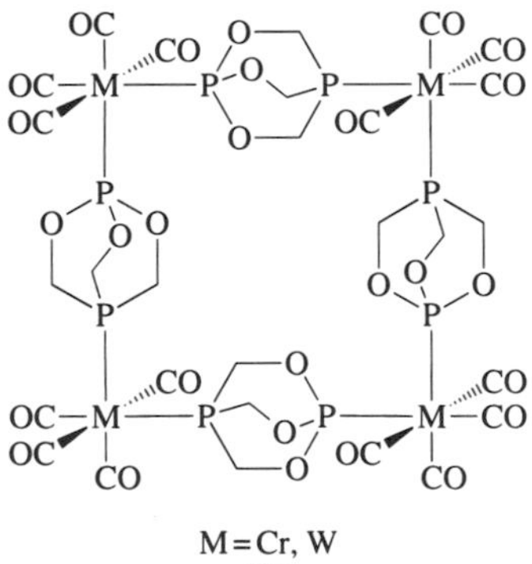

M = Cr, W
(**22**)

7.2.3.2 *Dinuclear macrocyclic complexes*

While linear rigid ligands give macrocyclic tetranuclear complexes on complexation with a *cis*-protected transition metal, flexible bridging ligands favor the self-assembly of dinuclear macrocycles. Complex (**30**) was the first dinuclear macrocycle assembled according to Scheme 2(b).[32] The macrocycle (**30a**), involving tetrafluorophenylene units, exhibited remarkable ability for the molecular recognition of electron-rich neutral aromatics. The association constants between (**30a**) and various aromatic guests increased with increasing electron density of the aromatic ring (Table 1). The specific recognition of electron-rich aromatics by (**30a**) can be most likely attributed to the

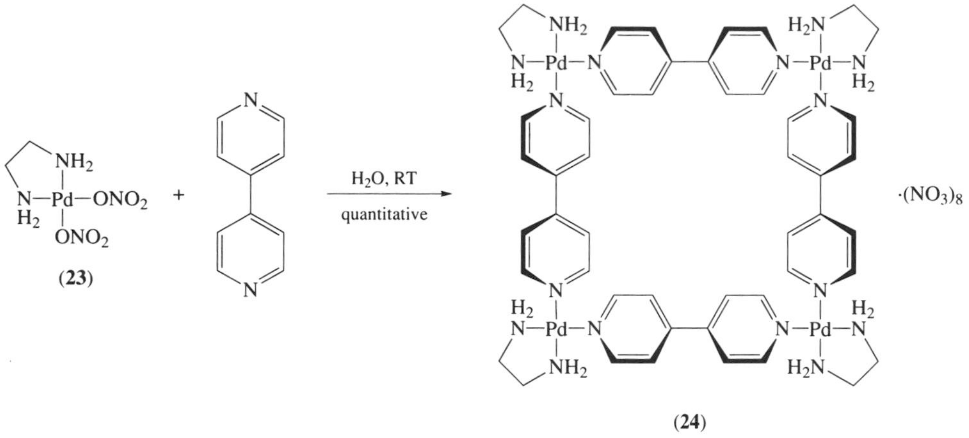

H_2O, RT
quantitative

(23)

(24)

$\cdot(\text{NO}_3)_8$

Scheme 3

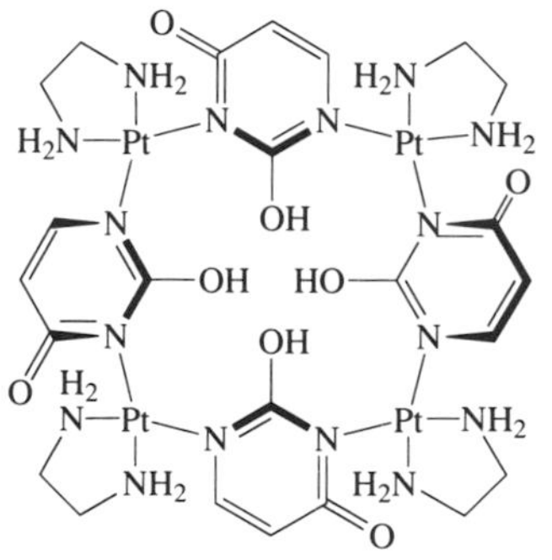

(25)

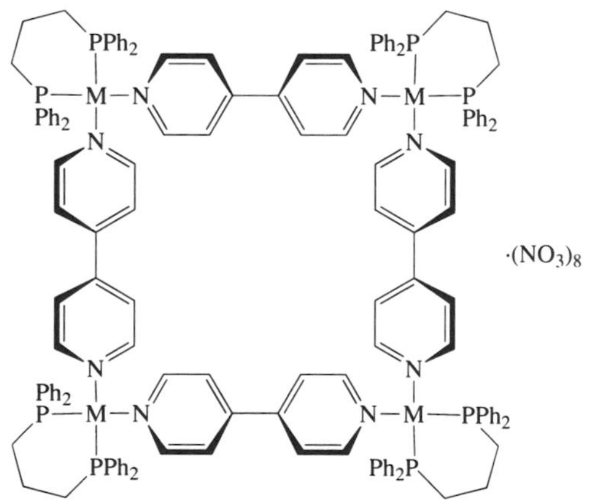

(26) M = Pd, Pt

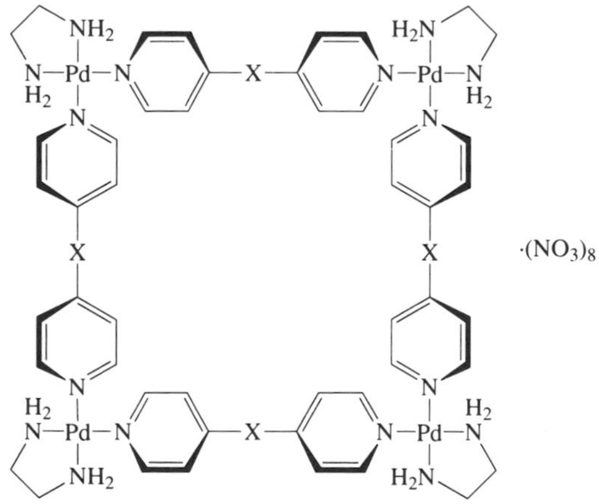

$\cdot(\text{NO}_3)_8$

(27)
(a) X = CH=CH
(b) X = C≡C
(c) X = *p*-phenylene

$\cdot(\text{NO}_3)_8$

electronic effect of the fluorine substituent because no specificity was observed in the binding results with the nonfluorinated counterpart **(30b)** (Table 1).[35] Related macrocycles **(31)**,[61] **(32)**, and **(33)**[62] were also quantitatively assembled and some of them were crystallographically characterized. The structure of **(31b)** is interesting because the carbonyl group of the bridging ligand is hydrated through cyclization. Unfavorable hydration implies that the reduction of an sp^2 angle to an sp^3 angle at the carbonyl carbon is essential to remove the ring strain from the macrocyclic structure.

Even if the ligands are rigid, some of them also afford dinuclear macrocycles, provided the lone pairs do not have a linear array. A rigid ligand, 1,4-bis(3-pyridyl)benzene, gave the dinuclear complex (**34**) upon complexation with (**23**); crystallography showed that (**34**) possesses a molecular trench. The dinuclear Au^I complex (**35**) is a unique example of organometallic macrocycles.[63] A hybrid square complex involving both Pd^{II} and hypervalent iodine, (**36**), has been reported.[64]

7.2.4 Fully Inorganic Macrocycles

Self-assembled macrocyclic metal complexes are made from both organic and inorganic components. On the other hand, the supramolecular assembly of fully inorganic macrocycles has been reported (Figure 2). $[Fe(OMe)_2(OCOCH_2Cl)]_{10}$ (**37**) (ferric wheel),[65] $[NaFe_6(OCH_3)_{12}(dbm)_6]Cl$ (**38**) (Hdbm = dibenzoylmethane),[66] $Ni_{12}(OCOMe)_{12}(chp)_{12}(H_2O)_6(THF)_6$ (**39**) (Hchp = 6-chloro-2-pyridone),[67] and $Fe_{12}(SePh)_{24}$ (**40**)[68] are recent topical compounds with fully inorganic macrocyclic skeletons.

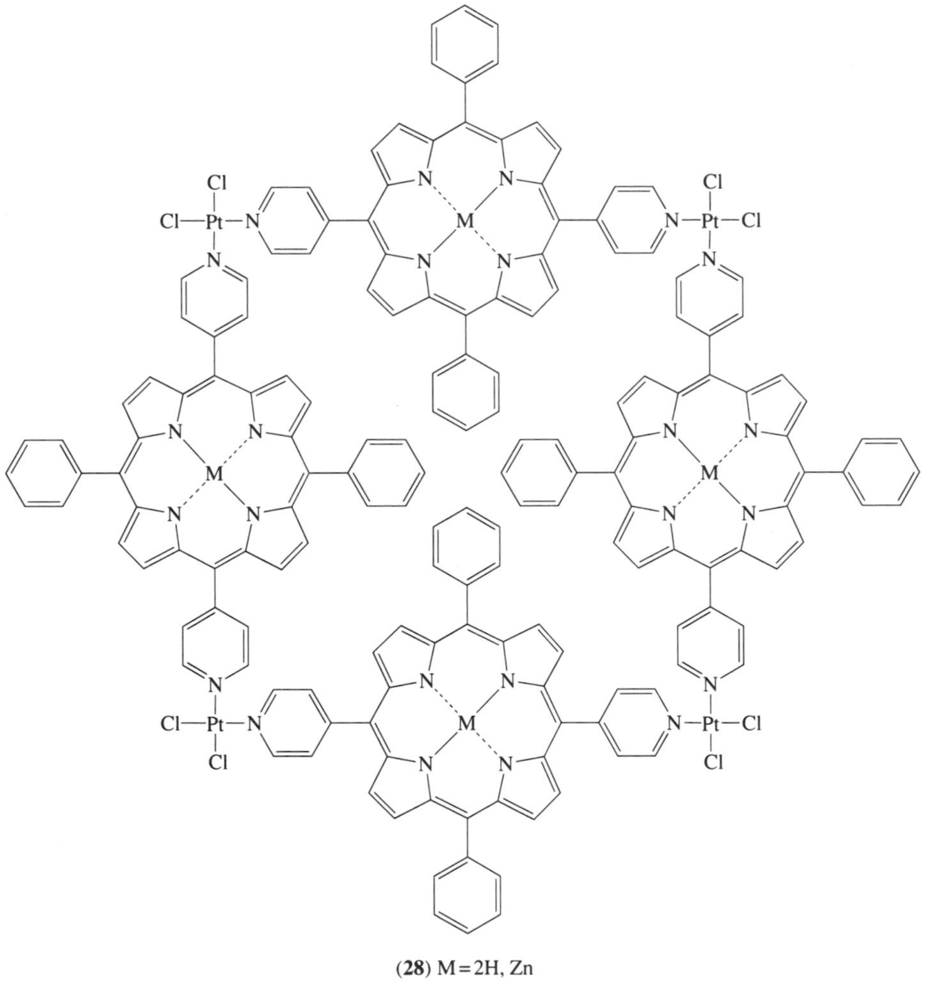

(**28**) M = 2H, Zn

7.2.5 Binding Properties of Self-assembled Macrocycles

As already discussed, self-assembled macrocycles have hydrophobic or Lewis acidic cavities and small molecules are often bound in the cavities. Table 1 summarizes the binding properties of the self-assembled macrocyclic complexes reported to date.

7.3 SELF-ASSEMBLED CAGELIKE COMPLEXES

7.3.1 Background

Since the development of cryptands opened the door to the chemistry of three-dimensional receptors,[69] various cage- and bowl-like complexes with a three-dimensional void have been prepared.[70,71] The molecular recognition ability was significantly improved by the three-dimensional receptor in comparison to the two-dimensional macrocyclic systems. For example, highly efficient asymmetric recognition of amino acids was achieved by the use of a chiral cage, (**41**),[72,73] and bowls, (**42**).[74–6] One of the most striking studies that definitely differentiates the three-dimensional cavity from the two-dimensional system is the chemistry of carcerands as studied by Cram. A guest molecule encapsulated in the carcerand was completely isolated from the outside and an extremely unstable cyclobutadiene derivative, for instance, was tamed in the three-dimensional cavity at room temperature.[77] Stereoisomerism has been reported in the encapsulation of a small guest by the carcerand.[78] These results show that three-dimensional receptors provide chemically created endohedral fields in which chemists can expect entirely new chemistries never before experienced.

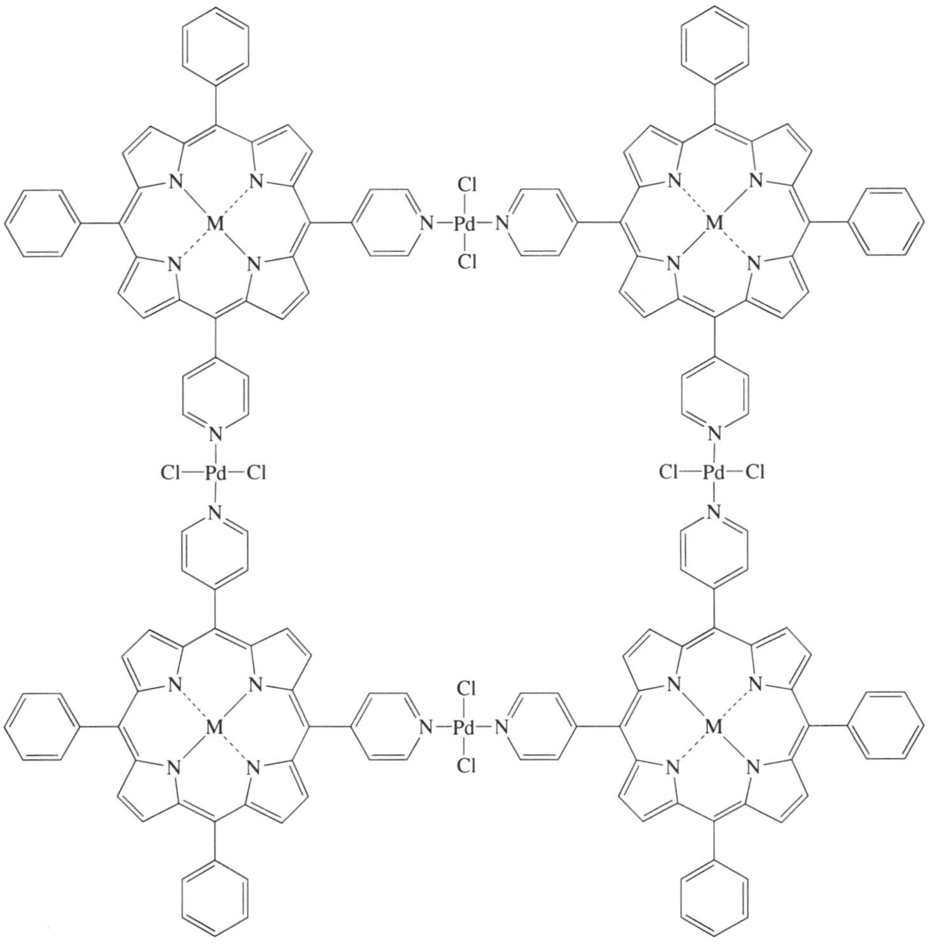

(**29**) M = 2H, Zn

More significantly, incorporation of the concept of supramolecular self-assembly into these chemistries enables the spontaneous generation of three-dimensional systems. Scheme 4 shows that a very unique example of the self-assembling three-dimensional receptor is a "tennis ball" (**44**) dimer of cyclic urea (**43**).[79,80] This sphere-like molecule was assembled from two monomers through complemental hydrogen bond formation. The following sections emphasize that coordination chemistry has provided remarkable three-dimensional systems.

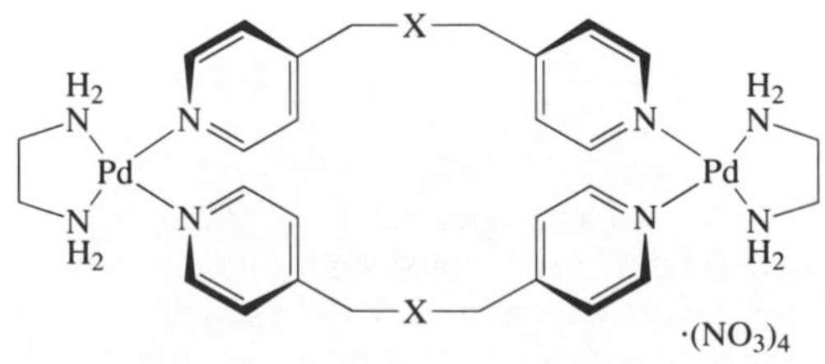

·$(NO_3)_4$

(30)

(a) X = (b) X = (c) X = (d) X = (e) X = none

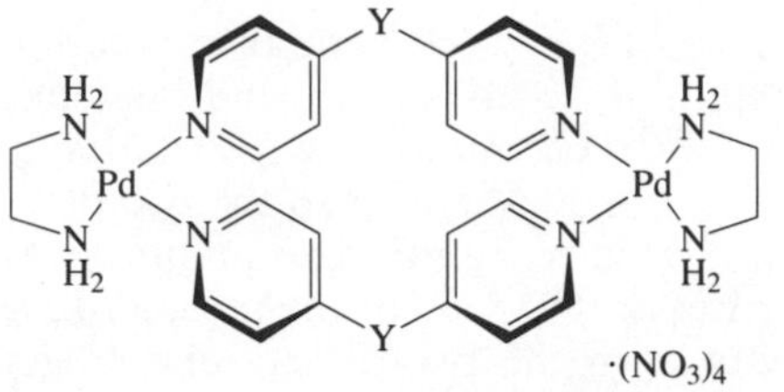

·$(NO_3)_4$

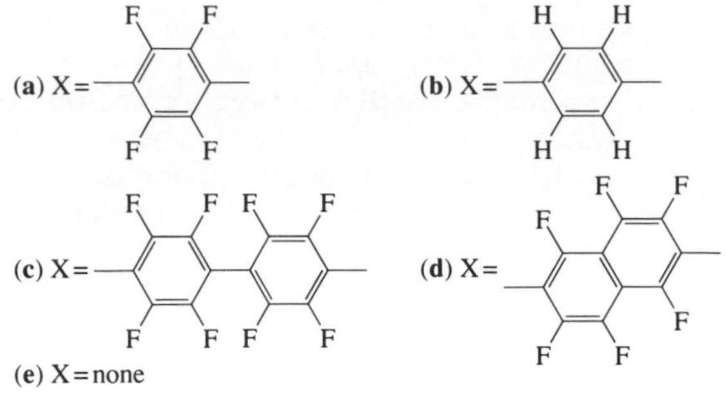

(31a) $Y=CH_2$, (b) $Y=C(OH)_2$, (c) $Y=C(C=CH_2)$

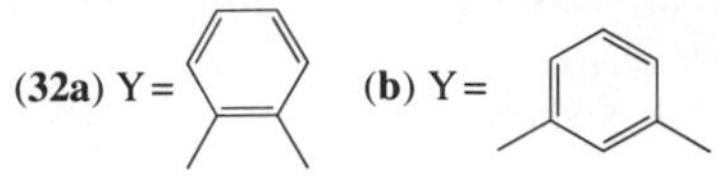

(32a) Y = (b) Y =

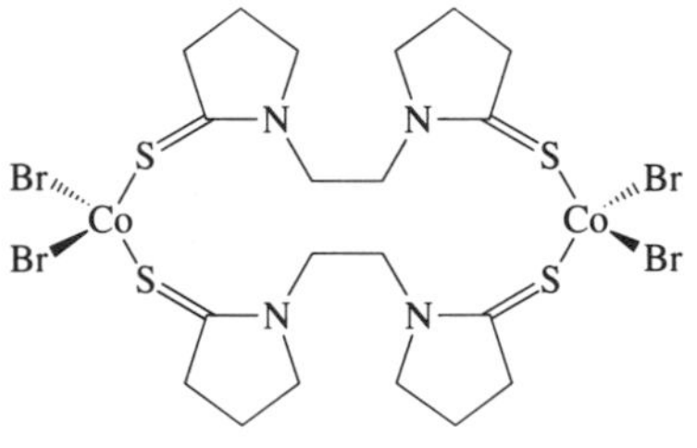

(33)

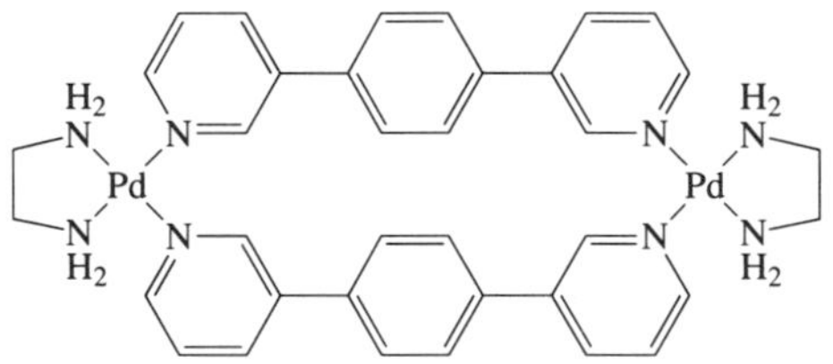

(34)

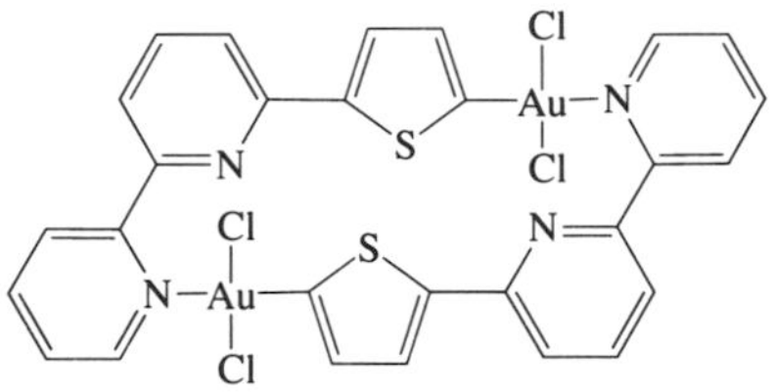

(35)

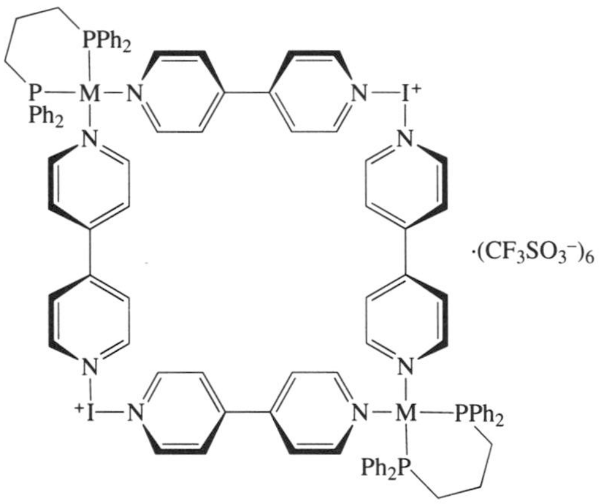

·$(CF_3SO_3^-)_6$

(36) $M=Pd^{2+}$, Pt^{2+}

7.3.2 Self-assembled Adamantanoid Complexes

In 1988, Saalfrank reported a novel self-assembled supramolecular three-dimensional complex whose framework has essentially the same symmetry as adamantane. Metallation of diethyl malonate in the presence of MgI_2 in THF followed by addition of oxalyl chloride gave an adamantane-like tetraanionic aggregate **(45)**[81] (85% yield) consisting of four Mg^{2+} ions and six chelating ligands. The structure of **(45)** was fully characterized by x-ray crystallography. The metal center can be replaced by transition metals such as manganese, cobalt, or nickel.[82] A neutral adamantanoid Fe^{III} complex was also obtained.[83] The three-dimensional void can be expanded with a phenyl as the spacer of the ligand. Thus, complex **(46)** was prepared in 57% yield and fully characterized by crystallography. Monoanionic, mixed-valence complex **(47)** involving three Fe^{III} and one Fe^{II} ions encapsulates ammonium ion for charge compensation.[84]

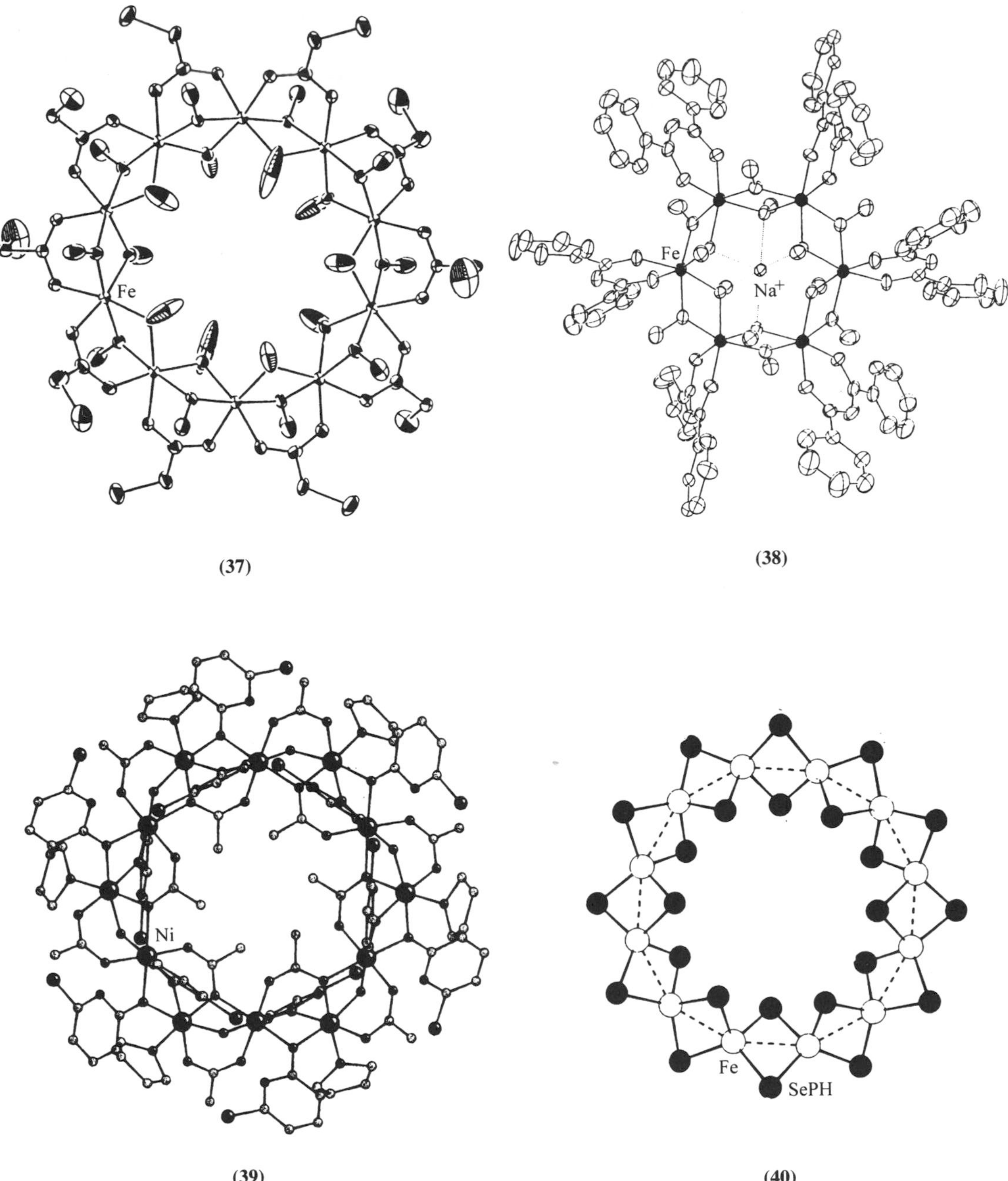

Figure 2 Inorganic macrocycles.

7.3.3 Guest-induced Assembly of a Trinuclear Palladium(II) Cage Complex

The use of a protected transition metal simplifies the design and construction of three-dimensional cages as well as two-dimensional macrocycles discussed in Section 7.2.3. This is demonstrated in Scheme 5 by the cage complex **(49)** assembled from three molecules of **(23)** and two simple, pyridine-based ligands, **(48)**.[85] The behavior of this complex is quite different from that of other self-assembling receptors because **(49)** is assembled only in the presence of an appropriate guest, providing an entire model for "induced-fit" molecular recognition. If **(23)** and **(48)** are simply combined, an intractable mixture of oligomeric products is obtained. However, the addition of an appropriate guest induces the assembly of cage like complex **(49)**. This guest-induced assembly process was monitored by a time-dependent ^{1}H NMR measurement. NMR titration experiments showed 1:1 host–guest complexation between **(49)** and various organic carboxylates. In fact,

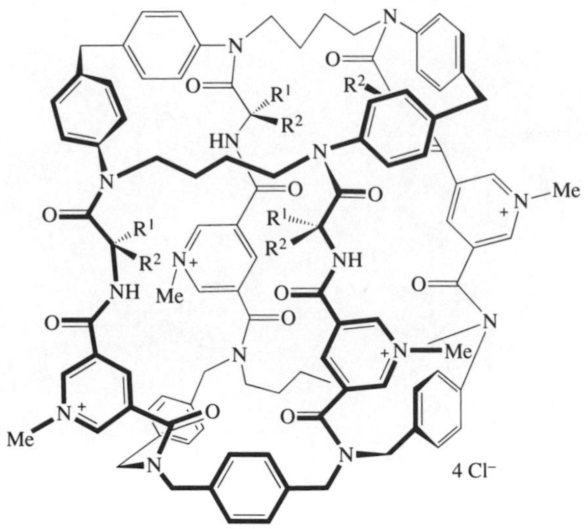

(+)-(**41**) $R^1 = Pr^i$, $R^2 = H$

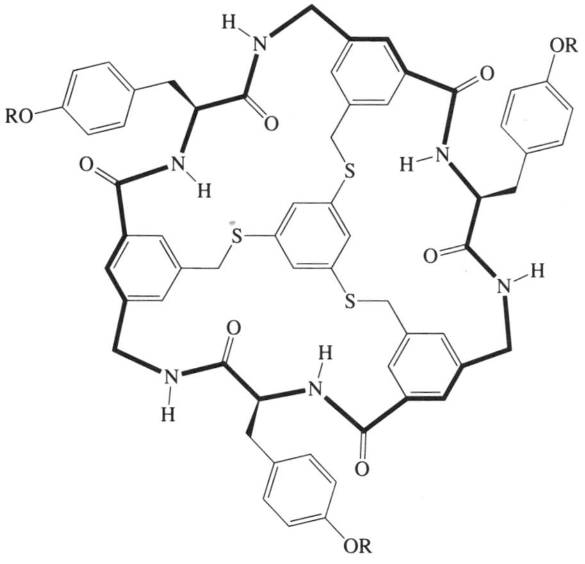

(**42**)

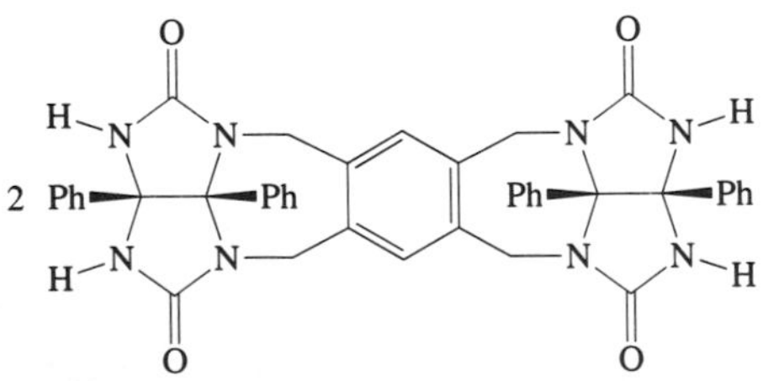

(**43**)

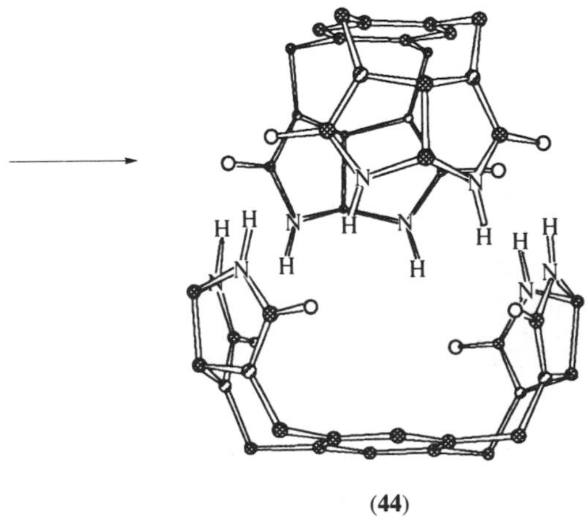

(**44**)

Scheme 4

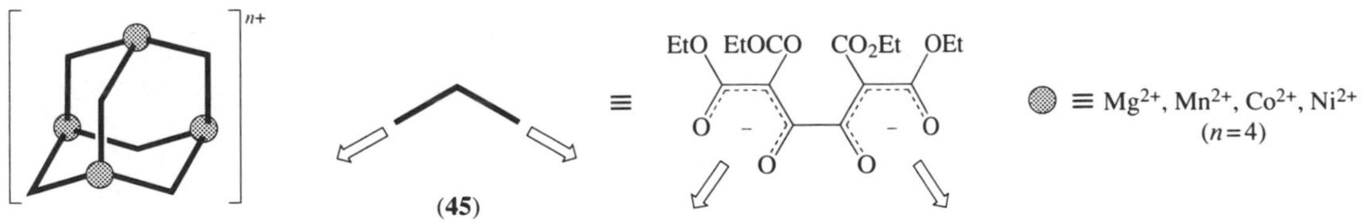

(45)

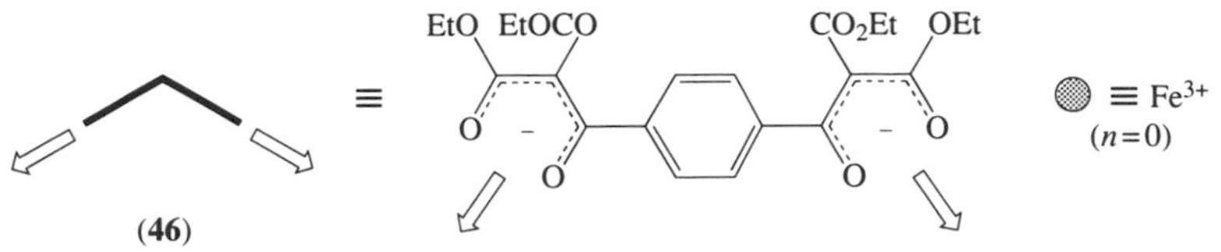

(46)

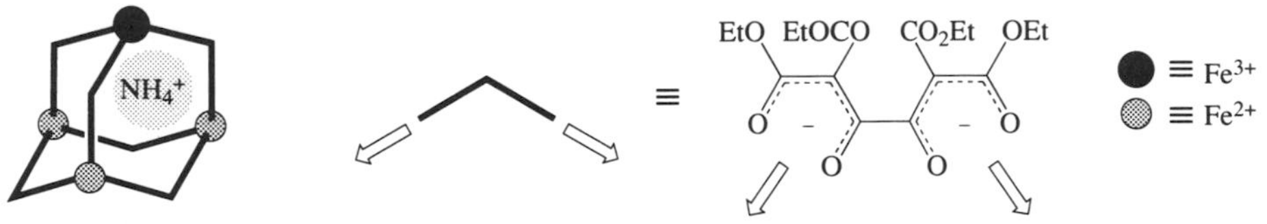

(47)

the 1:1 complex with 4-(methoxy)phenylacetate ion was isolated in 94% yield. Besides anionic hosts, a neutral hydrophobic guest such as *p*-xylene was also effective. Significant upfield shifts of guest signals up to ~3 ppm were observed when the guests effectively induced the organization of **(49)**.

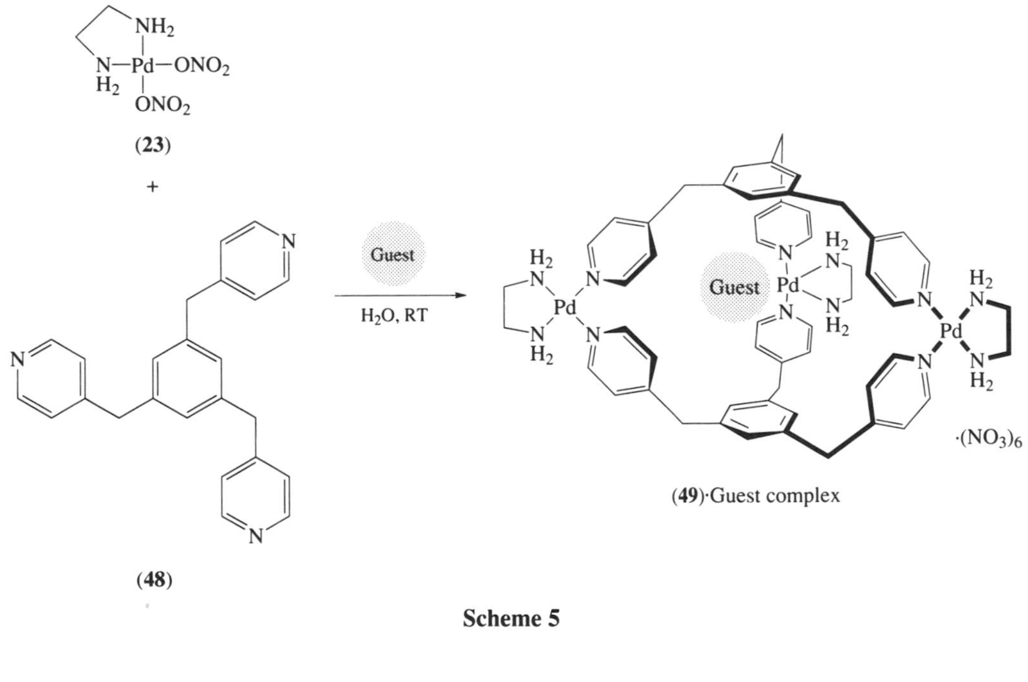

(23)

+

(48)

(49)·Guest complex

Scheme 5

7.3.4 Other Self-assembled Complexes with a Three-dimensional Cavity

Other fascinating three-dimensional structures self-assembling from transition metals and tailored ligands have been reported. A cylindrical complex, **(52)**, was assembled from six Cu^I ions, two central ligands, **(50)**, and three outer ligands, **(51)**, (see Equation (3)).[86,87] Two molecules of β-cyclodextrin were successfully coupled at their lower rims via assembly around Cu^{II} or Pb^{II}, giving large three-dimensional cavity compounds, **(53)**.[88,89]

Like the cavity of carcerands, innershell spaces completely isolated by the shell are often found in quasispherical metal–oxygen clusters.[90] They frequently encapsulate ions or small molecules inside the shell. Müller *et al.* reported the template-controlled formation of vanadium–oxygen cluster shells, in which the template ion is encapsulated. That is, the oxidation of $[H_9V_{19}O_{50}]^{8-}$ in the presence of N_3^- or ClO_4^- gave a closed shell of structure $[H_2V_{18}O_{44}(N_3)]^{5-}$ **(54)** or $[HV_{22}O_{54}(ClO_4)]^{6-}$ **(55)**, respectively. Compounds **(54)** and **(55)** are built from edge- and corner-sharing tetragonal OVO_4 pyramids and encapsulate N_3^- and ClO_4^- ions, respectively.[91] The template synthesis of three-dimensional shells was also achieved using NO_3^-, SCN^-, H_2O, or $MeCO_2$.[92] Similarly, the self-assembled cluster $[V_{14}O_{22}(OH)_4(H_2O)_2(PhPO_3)_8]^{6-}$ **(56)** has a nanometer-scale cavity that encloses two NH_4^+ and two Cl^- ions.[93] Cation inclusion in a spherical Mo—O compound has also been reported.[94] Self-assembled vanadium bowls, **(57)**, although three-dimensionally not closed, have effectively captured organic nitriles in their cavity.[95–7]

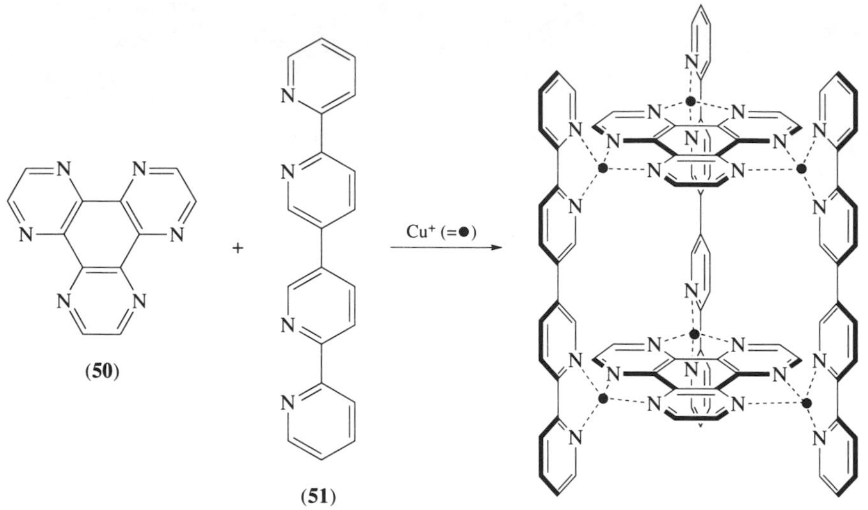

(50) **(51)** **(52)**

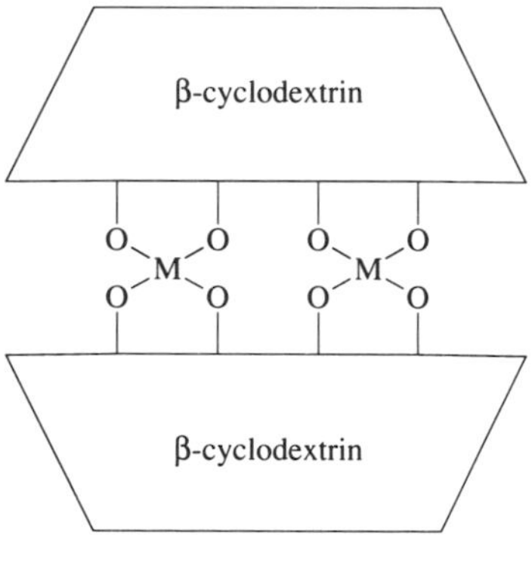

(53)

7.4 SELF-ASSEMBLED CATENANES

7.4.1 Background

The threading of a string on a ring is an extremely difficult manipulation in a molecular-scale small world because there are no instruments that can fix the molecular rings or tweeze the molecular strings. The development of such a molecular manipulation technique, however, breeds "topological chemistry"[98] that deals with topological isomerism of molecules (e.g., separate molecular rings vs. interlocking molecular rings). Although an interlocking two-ring compound ([2]catenane) was first synthesized in 1960 by Wasserman,[99] his synthetic strategy was not manipulative: he expected the statistical formation of a catenane during acyloin condensation of a long-chain diester in the presence of a cyclic hydrocarbon.

The first molecular manipulation leading to the inevitable formation of a catenane was realized in 1983 by the transition metal-template strategy of Sauvage and co-workers (see also Chapter 2).[100–3] The assembly of two phenanthroline-based ligands around the tetrahedral geometry of Cu^I gave

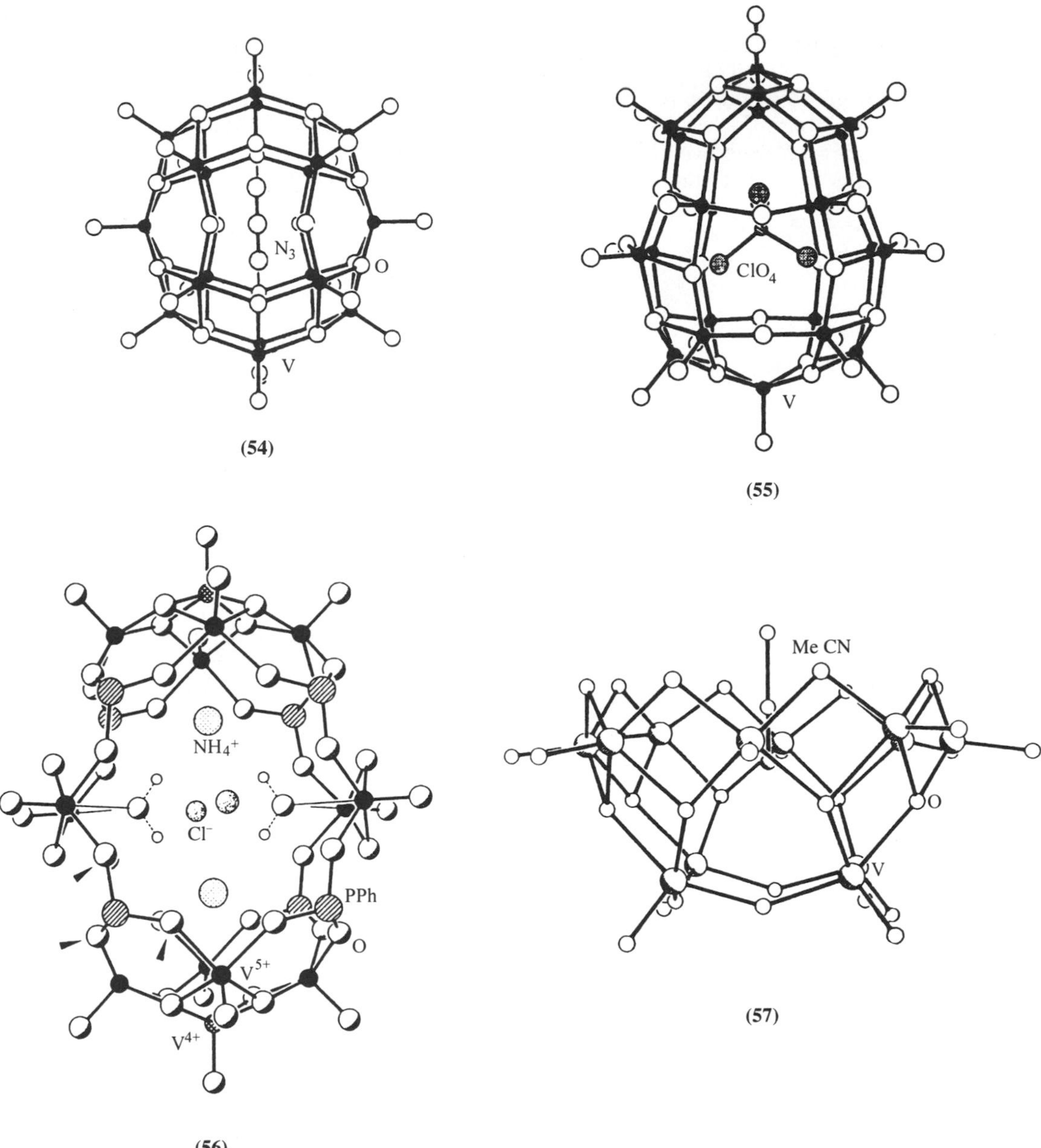

(54)

(55)

(56)

(57)

the catenane precursor, (**58**), which was transformed into a catenane, (**59**), after cyclization with polyether threads followed by demetallation (see Equation (4)).[100] Based on this strategy, related topologically interesting molecules have been developed (e.g., chiral catenanes, polyring catenanes, rotaxanes, knots, doubly-interlocked catenanes) (see Chapter 2).

Another highly efficient approach to catenanes was developed by Stoddart. The self-assembly of organic aromatic systems through charge transfer interaction led to the threading of an electron-rich molecular thread on an electron-deficient rectangular receptor. Subsequent cyclization gave [2]catenane, (**60**), in 70% yield (see Equation (5)).[103] This strategy has shown its remarkable applicability to the construction of a variety of catenane and rotaxane systems (see Chapter 3).[104–6] Hunter and co-workers[107,108] and Vögtle and co-workers[109,110] have reported the efficient formation of [2]catenanes (**61**) and (**62**), respectively.

7.4.2 Grignard Catenane

The strategies of Sauvage and Stoddart in catenane synthesis involve the self-assembly of the catenane precursors. On the other hand, the supramolecular assembly of a catenane framework itself was first realized in an organometallic system in 1993. Bickelhaupt and co-workers reported that a side-on complex of a macrocycle containing a diarylmagnesium moiety, **(63)**, with a crown ether was in equilibrium with Grignard catenane **(64)** (see Equation (6)).[111] Prior to this work, they also prepared Grignard rotaxane **(65)** (Equation (7)), whose structure was confirmed by an x-ray crystallographic analysis.[112] The structure of **(64)** is mainly based on NMR data and by analogy to **(65)**. It is interesting that the two rings would have no way to interlock unless the carbon–magnesium bond of the Grignard ring dissociates momentarily.

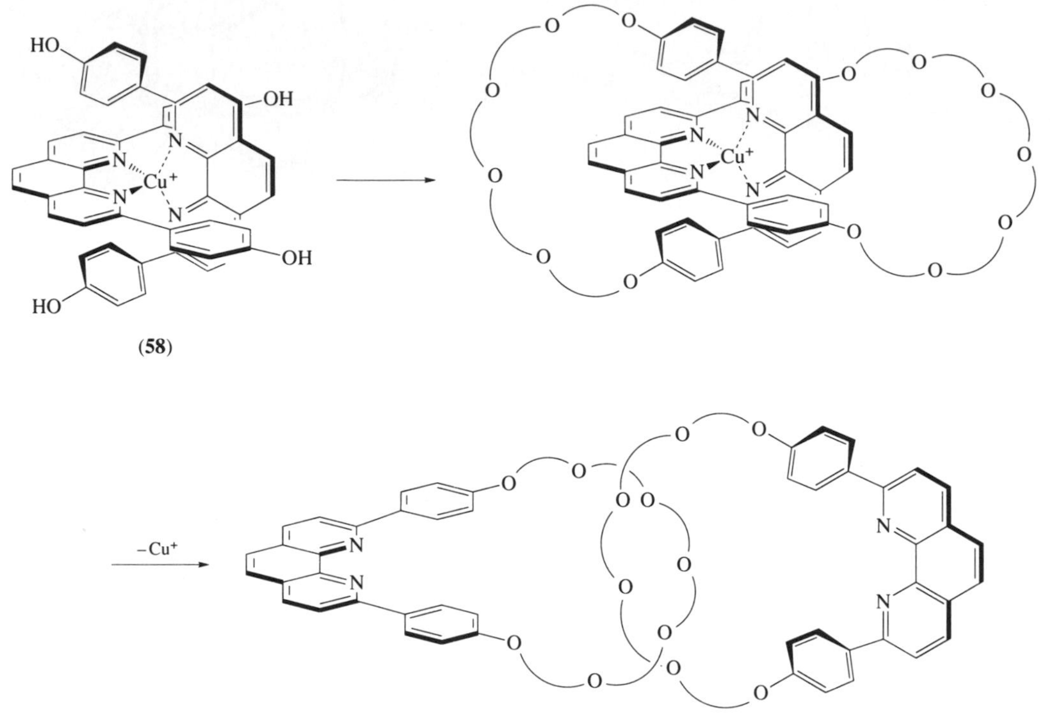

(58)

$-Cu^+$

(59) (4)

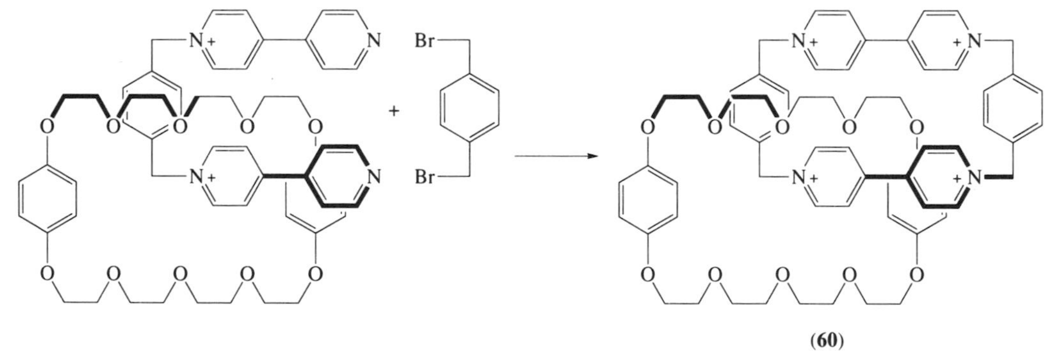

(60) (5)

7.4.3 Self-assembled Catenanes Containing Palladium(II) or Platinum(II)

7.4.3.1 Quantitative self-assembly of a [2]catenane containing palladium(II)

A self-assembled catenane involving transition metals was first reported in 1994. [2]Catenane **(66)** was found to self-assemble along with monomer ring **(67)** from **(23)** and ligand **(68)** (see Scheme 6).[36,113,114] Although the structure of **(66)** was first deduced by spectroscopy, the crystallographic study of Pt^{II} complex, **(69)**, which is analogous to **(66)**, confirmed the catenated structure.[55] Since Pd—N bonds are reversible, a rapid equilibrium was observed between **(66)** and **(67)**. While the equilibrium lay toward **(67)** at a lower concentration (<2 mM), catenane **(66)** became the overwhelmingly dominant component at higher concentration (>50 mM).

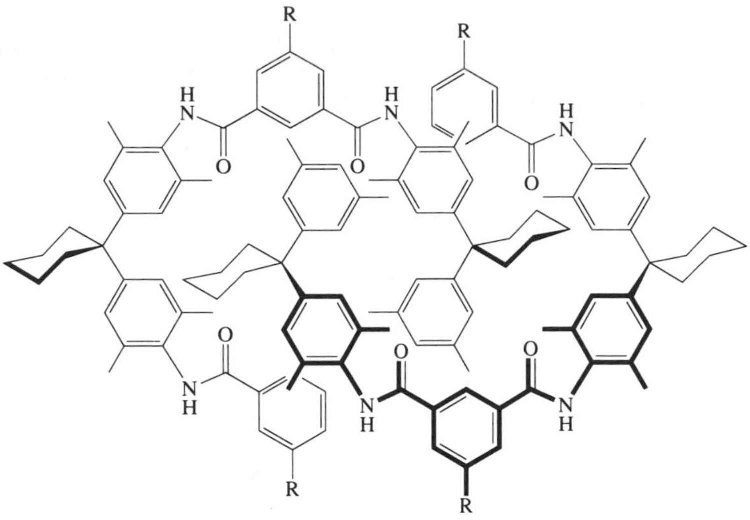

(**61**) R = H
(**62**) R = OMe

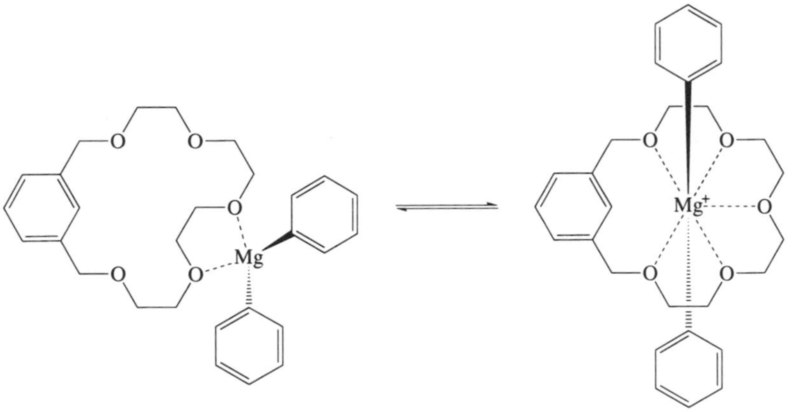

(**63**) (**64**) (6)

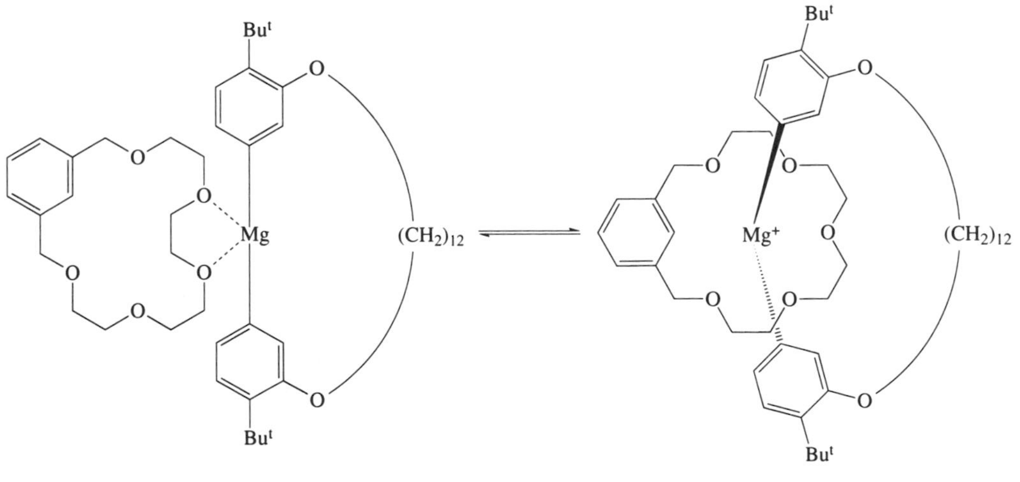

(**65**) (7)

The assembly of catenane (**66**) is promoted by "double molecular recognition," in which two molecules of (**67**) bind each other in their cavities; approximately double the magnitude of ΔG is produced making the structure of (**66**) stable enough to assemble quantitatively. Such an interpretation is consistent with a remarkable medium effect that enables modulation of the equilibrium ratio (**66**):(**67**). Thus, the employment of a more polar medium (D_2O solution of $NaNO_3$) increased the proportion of (**66**) up to >99% even at a low concentration, most probably due to enhanced hydrophobic interaction in the catenane formation. In contrast, the proportion of (**66**) diminished with a less polar medium (CD_3OD–D_2O). Selective stabilization of (**67**) by adding sodium (*p*-methoxy)phenylacetate, a specific guest for (**67**), also reduced the (**66**):(**67**) ratio.

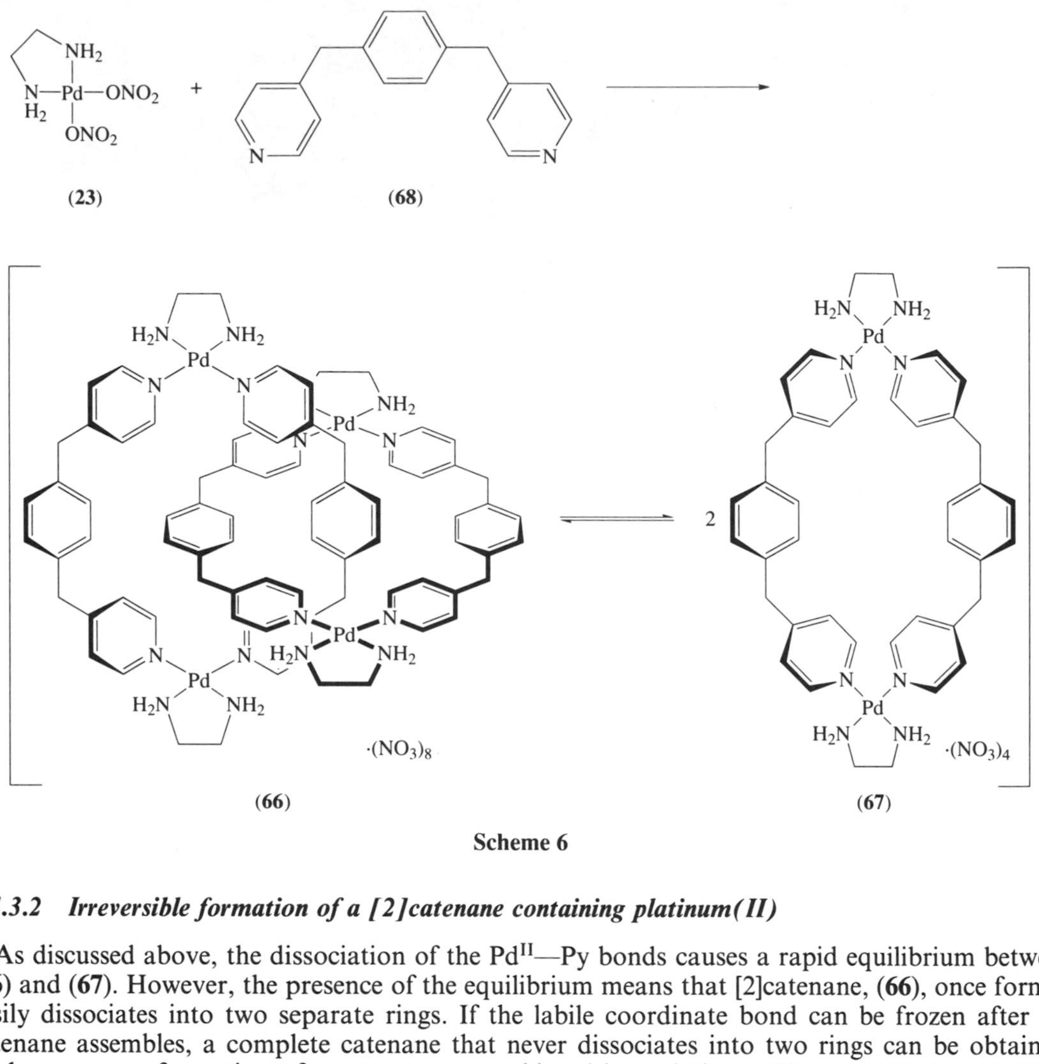

(**23**) (**68**)

(**66**) (**67**)

Scheme 6

7.4.3.2 Irreversible formation of a [2]catenane containing platinum(II)

As discussed above, the dissociation of the Pd^{II}—Py bonds causes a rapid equilibrium between (**66**) and (**67**). However, the presence of the equilibrium means that [2]catenane, (**66**), once formed easily dissociates into two separate rings. If the labile coordinate bond can be frozen after the catenane assembles, a complete catenane that never dissociates into two rings can be obtained. Such a one-way formation of a catenane was achieved in a platinum(II) counterpart system (see Equation (8)) using the concept of "molecular lock."[55]

The concept of "molecular lock" stems from exploitation of the ephemeral nature of the Pt^{II}—Py bond. This bond can be likened to a lock because it is irreversible ("locked") under ordinary conditions but becomes reversible ("released") in a highly polar medium at high concentration. Thus, Scheme 7 illustrates the overall one-way transformation of (**70**) into (**69**) using the molecular lock. Initially, a molecular ring is on the lock (A). The lock is then released by adding a salt ($NaNO_3$) and heating to 100 °C (B), allowing the self-assembly of a catenated framework within 24 h (C). Finally, this framework is locked by removing the salt and cooling (D). This chemical manipulation of molecular rings was monitored by ^1H NMR.

7.4.4 Self-assembled Catenanes Containing Iron(II) and Silver(I)

More recently, the self-assembly of [2]catenanes involving two different metal ions was reported by Piguet *et al.*[115] The ligand (**71**) has a tridentate coordination site at the center and two bidentate coordination sites at both termini. By treating it with Fe^{II} (1 equiv.) and Ag^I (2 equiv.), the Fe^{II}-center is octahedrally coordinated by the two tridentate units and each Ag^I is tetrahedrally bound

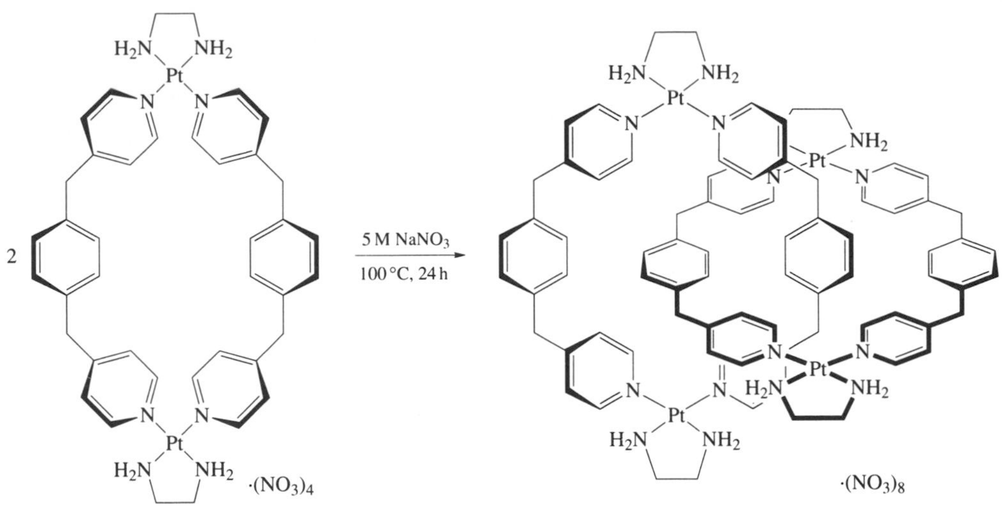

(8)

(70)

(69)

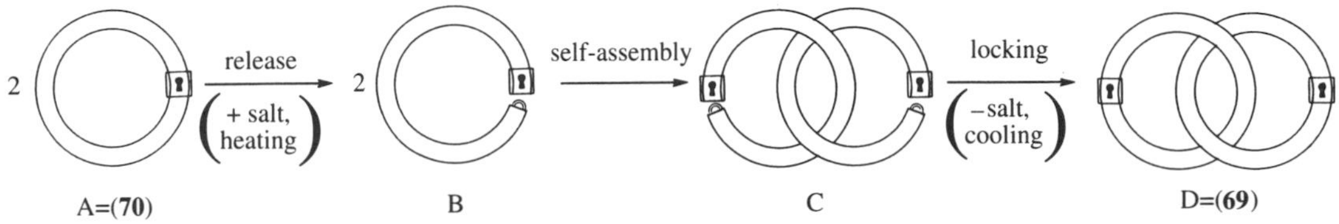

Scheme 7

to two bidentate sites, giving isomeric [2]catenanes (Scheme 8). The isomerism is due to *P* or *M* helicity around Ag^{I}: for example, (**72**) and (**73**) have *P*,*M* (*meso*) and *M*,*M* (or *P*,*P* for its enantiomer) structures, respectively. Both (**72**) and (**73**) were obtained as single crystals and their structures were fully solved by x-ray crystallography. In solution, rapid equilibration on the NMR timescale exists between (**72**) and (**73**). However, the observation of interstrand NOEs still supports the catenated structure in solution.

7.4.5 Self-assembled Rotaxanes Containing Transition Metals

There are a few examples of transition metal-based self-assembly of rotaxanes. In 1981, Ogino reported rotaxanes (**74**) consisting of β-cyclodextrin threaded by α,ω-diaminoalkanes bearing cobalt(III) complexes as stoppers, (see Equation (9)).[116,117] The quantitative self-assembly of rotaxane, (**75**), from α-cyclodextrin, $[Fe(CN)_5(H_2O)]^{3-}$, and bridging ligand (**76**) was reported by Macartney and Wylie.[118] Rotaxane (**75**) also forms spontaneously upon addition of α-cyclodextrin to a solution of the bridged dimer, $[(CN)_5Fe(\text{bipy-}(CH_2)_n\text{-bipy})Fe(CN)_5]^{4-}$, and this slippage process can be monitored by ^1H NMR spectroscopy, (see Equation (10)).

7.5 RELATED NETWORK COMPLEXES

7.5.1 Background

As discussed earlier, closed structures (macrocycles, cages, catenanes) self-assemble from divergent metals and convergent ligands, or convergent metals and divergent ligands. If both metals and ligands have divergent character, the assemblies of infinite structures that contain macrocyclic, cagelike, or catenated frameworks would be expected.[119–21] For example, the coordination of linear rigid ligands to square-planar transition metals gives two-dimensional molecular grids as illustrated in Scheme 9.[122,123] Chemistries of finite and infinite frameworks complement each other (e.g., infinite vs. finite or solid vs. solution chemistry), and thus, information from one system is predictive for the phenomena in another. Therefore, Section 7.5.2 will discuss infinite complexes that contain macrocyclic, cagelike, or catenated frameworks, putting emphasis on solid-state molecular recognition or clathrate formation.

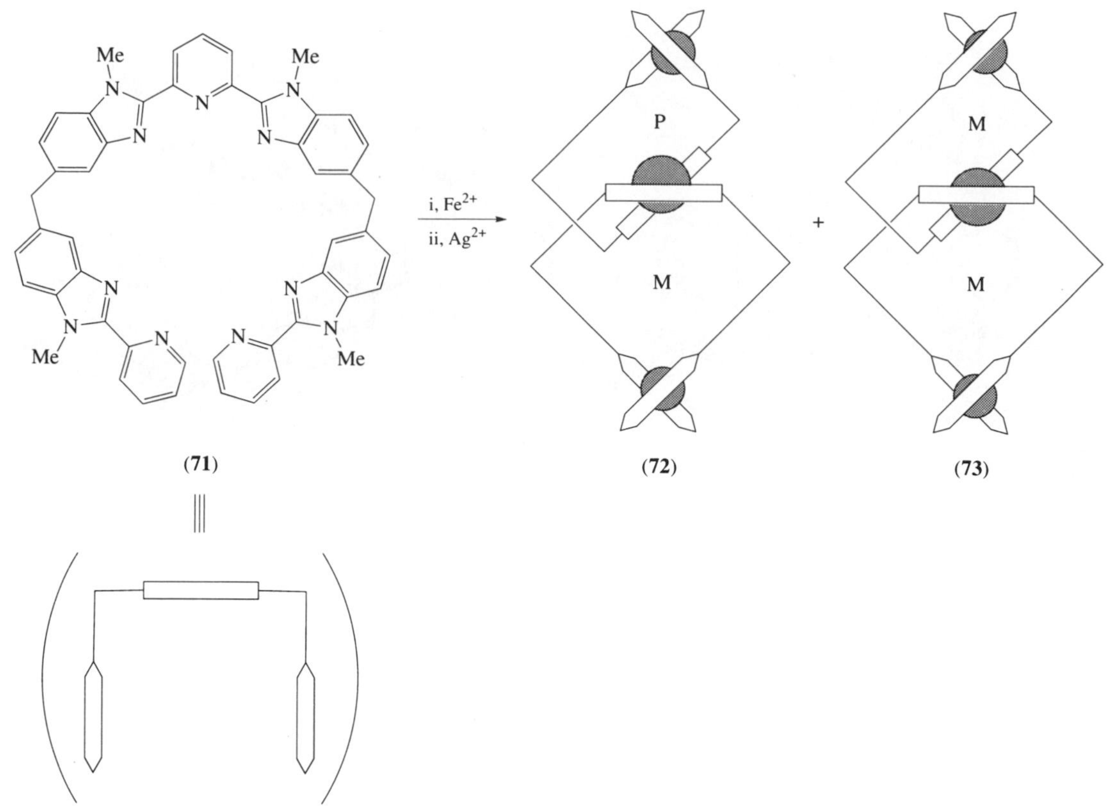

(71) (72) (73)

i, Fe^{2+}
ii, Ag^{2+}

Scheme 8

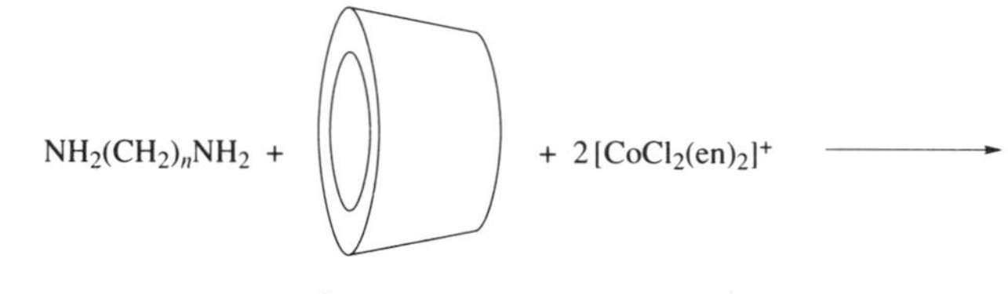

$NH_2(CH_2)_nNH_2$ + (β-cyclodextrin) + $2[CoCl_2(en)_2]^+$ ⟶

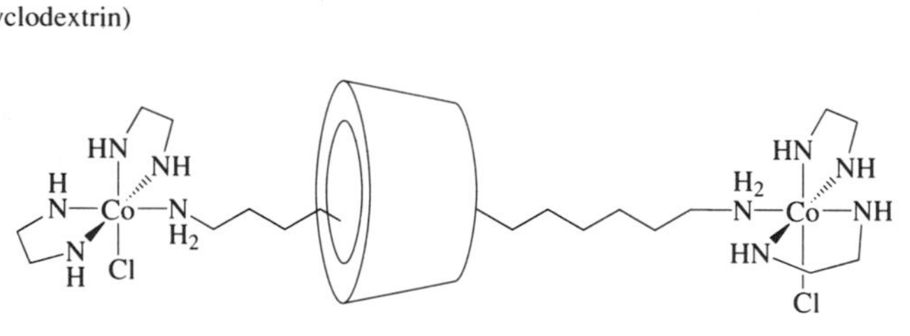

(74) (9)

7.5.2 Network Complexes Involving Macrocyclic Frameworks

One of the simplest networks involving macrocyclic frameworks is formed by square-planar coordination of linear bridging ligands around transition metals. The resulting grid networks (as shown above) have often been seen in the infinite metal cyanide complexes known as the Hoffmann complex $(NH_3)_2Ni(CN)_2{\cdot}C_6H_6$, **(77)**[122] and its derivatives.[120,121] By employing 4,4-bipy-ridine as an organic bridging ligand, a large square grid is obtained upon complexation with Zn^{2+}, Cu^{2+}, or Cd^{2+}, **(78)**. The whole crystal structure consists of two perpendicular and equivalent stacks of square grid two-dimensional sheets which interpenetrate each other (see Section 7.5.3).[119,123]

The molecular recognition ability of these grids in solid or clathrate formation characterizes them as macrocycles. In the presence of appropriate aromatic guests, the grids enclathratearomatic compounds with high shape specificity.[124] Square grid sheet $[Cd(bipy)_2(H_2O)_2](NO_3)_2$ **(78c)** enclathrates mono- and *o*-dihalobenzenes (halo = chloro or bromo), while *m*- and *p*-dihalobenzenes were not included. This high specificity was applied to isomer separation: from a mixture

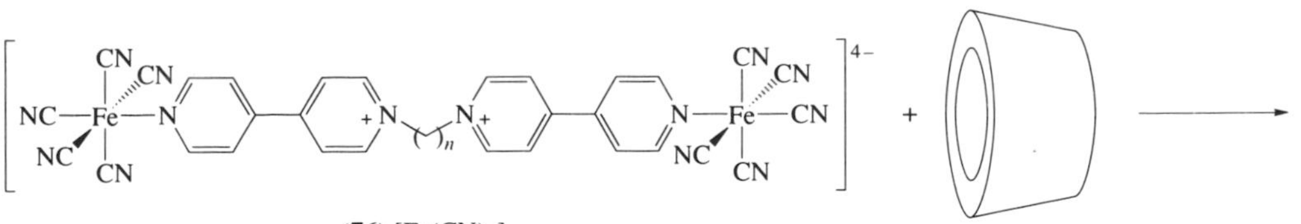

(76)·[Fe(CN)5]2

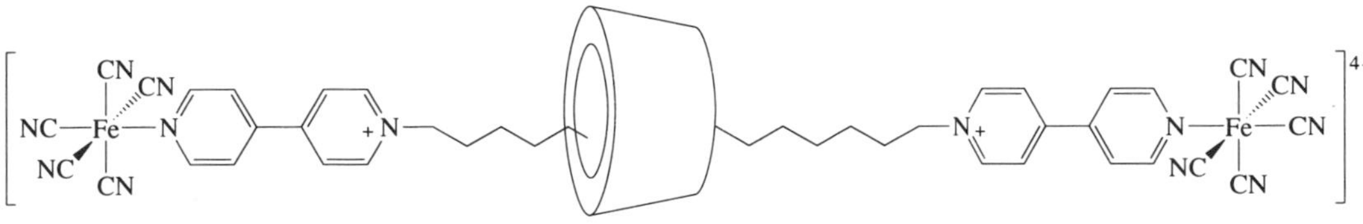

(75)

(10)

a closed structure

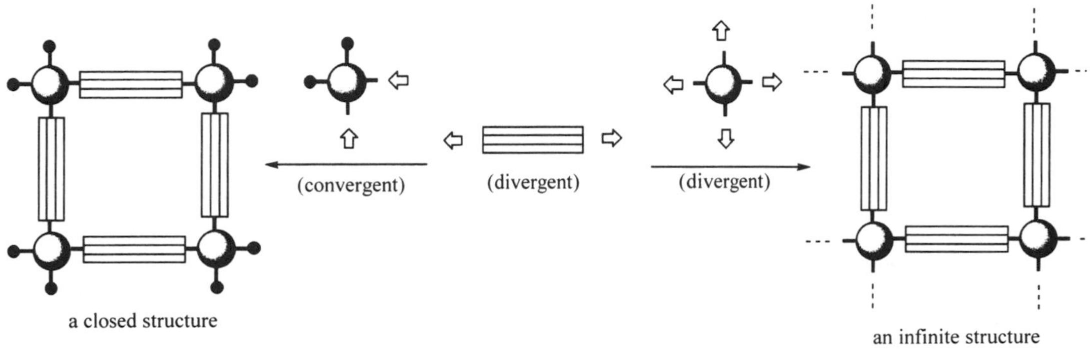

(convergent) (divergent) (divergent)

an infinite structure

Scheme 9

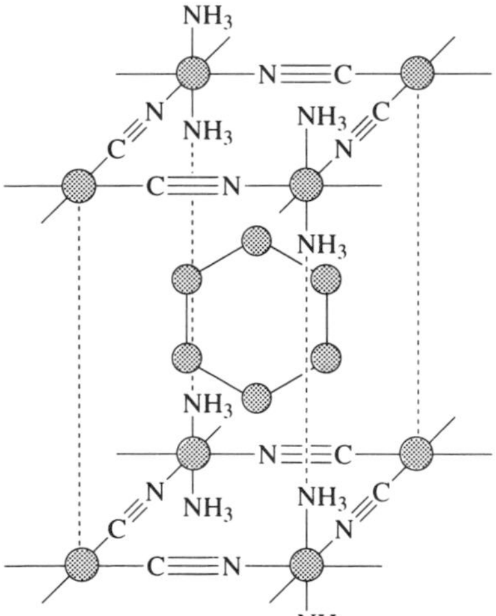

(77)

$\cdot(NO_3)_8$

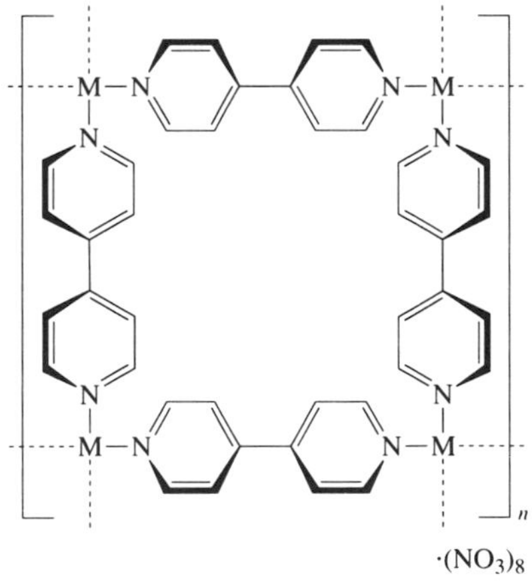

(**78a**) M = Zn^{2+}
(**78b**) M = Cu^{2+}
(**78c**) M = Cd^{2+}
(**78d**) M = Fe^{2+}

of *o*- and *m*- (or *p*-)dibromobenzene, only the clathrate (**78c**)$_2$·(*o*-$C_6H_4Br_2$)$_2$ was obtained, from which the *o*-isomer was recovered in a pure form. Clathrate complex (**78d**)$_2$·(bipy)$_2$·(H_2O)$_4$, in which bipy itself is enclathrated, is also reported.[119] Zeolite-like catalysis was first provided by (**78c**)-catalyzed cyanosilylation of aldehydes (see Scheme 10).[124] Shape specificity was observed in this reaction, implying promotion of the reaction in the network cavity. Besides the grid framework, a two-dimensional complex with a honeycomb structure is also reported.[125] Two-dimensional complexes showing magnetic properties have also been reported.[126–8]

i, TMS–C≡N (5 equiv.), (**78c**) (20 mol.%), 40 °C, 15 h, CH_2Cl_2

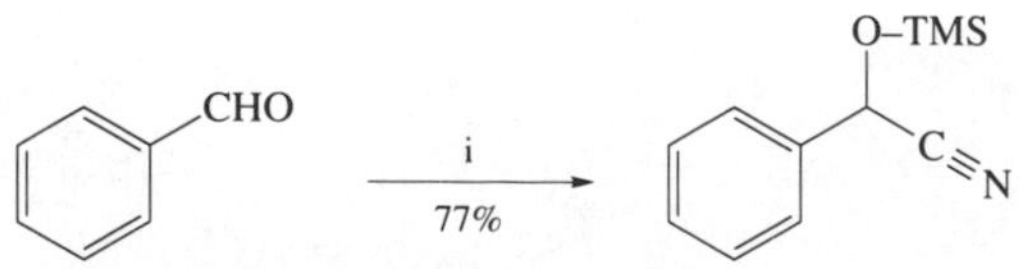

Scheme 10

Two-dimensional networks capable of clathration are also obtained from flexible ligands and transition metals. From ligand (**79**), $Cd(NO_3)_2$, and an appropriate guest, a two-dimensional network, (**80**), appeared, each cavity of which could enclathrate various guest molecules. In the absence of guest, the network enclathrated ligand (**79**) itself. In contrast to the two-dimensional grid sheets, the network of (**80**) is "swellable": the cavity inflates depending on the guest size (Figure 3).

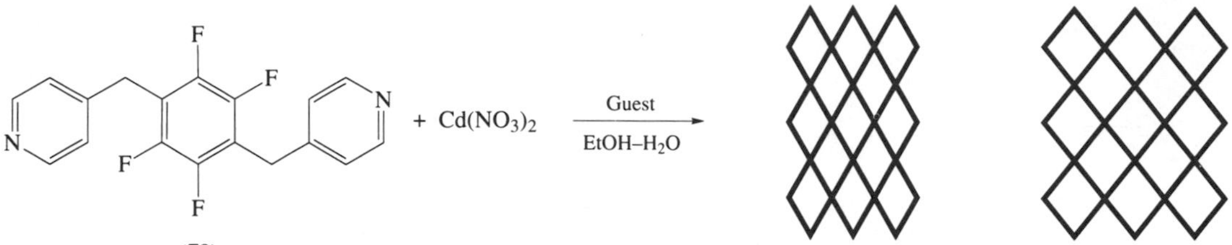

(**79**)

(**80a**) for a small guest (**80b**) for a large guest

Figure 3 Schematic illustration of "swellable" two-dimensional networks $\{[Cd(\mathbf{79})_2](NO_3)_2{\cdot}G\}_n$ (**80**, G = guest). In network patterns (**80a**) and (**80b**) Cd^{2+} is present at every intersection. Heavy connections between adjacent intersections indicate ligand (**79**). A guest is enclathrated in each cavity.

Besides these complexes, infinite complexes involving macrocyclic frameworks are often assembled through metal coordination of pyridone-based bridging ligands (e.g., $\{[Cu(ebpt)_2]PF_6\}_n$ (ebpt = *N*,*N*′-ethylenebis(pyrrolidine-2-thione)), $[\{Hg_3Co(C_4H_6NO)_6\}(NO_3)_2]_n$, $[NiNd(4\text{-picolylpyrrolidine-2-one})_4(NCS)_2(NO_3)_3]_n)$[129–33] (networks (**81**), (**82**), and (**83**), respectively), or pyridine-based bridging ligands (e.g., $[Cd(bpe)_{2.5}](NO_3)_2$ (**84**, bpe = 1,2-bis(4-pyridyl)ethane).[134]

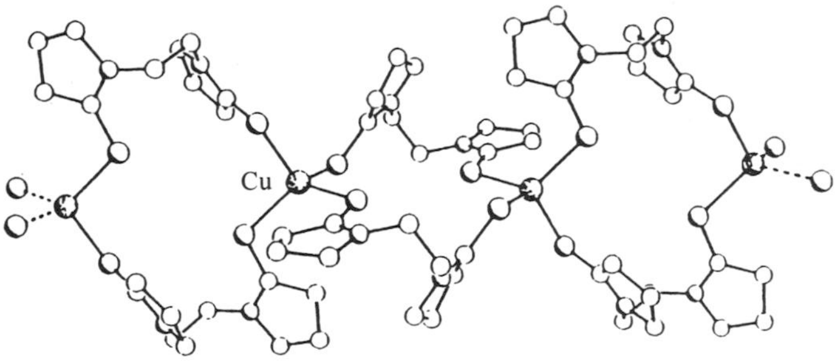

(**81**)

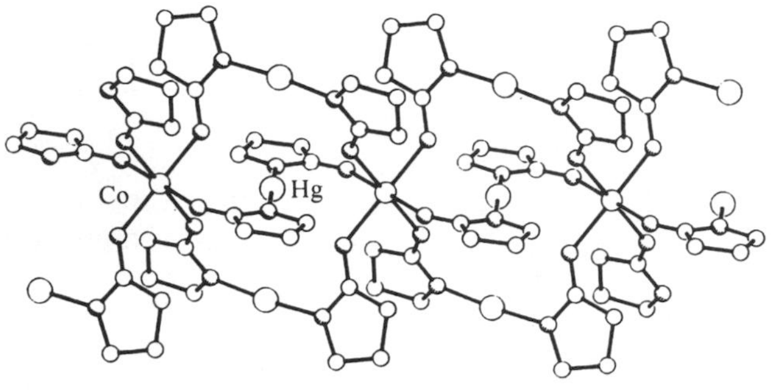

(**82**)

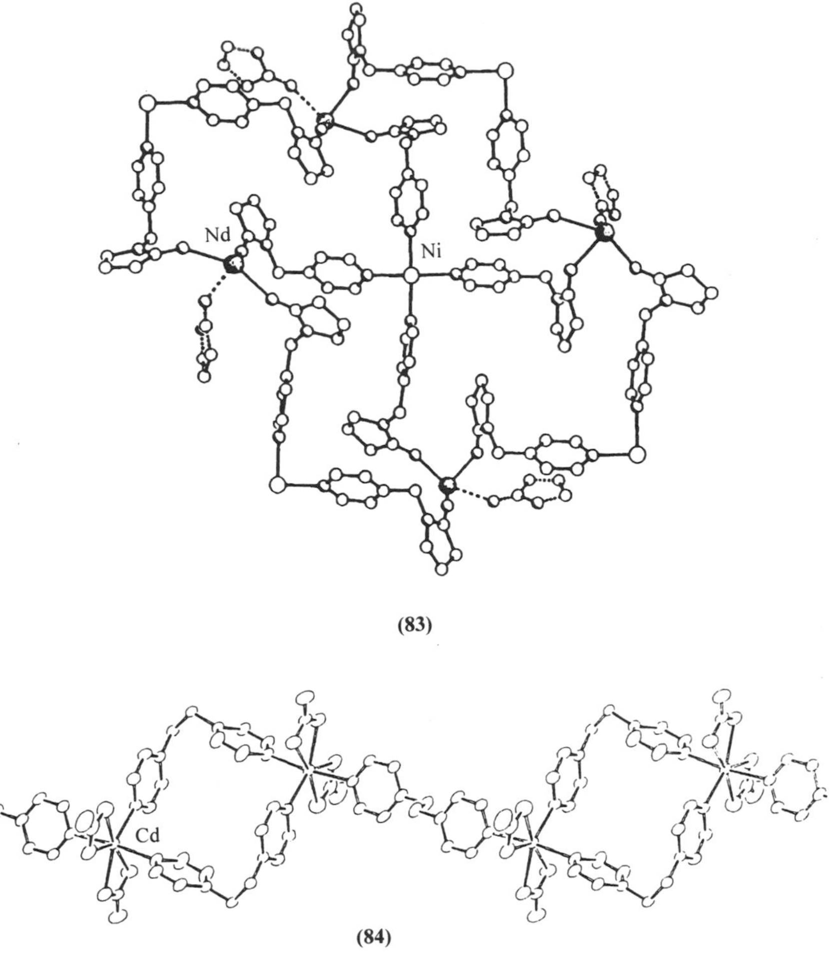

(83)

(84)

7.5.3 Network Complexes Involving Polycatenane Frameworks

Instead of simple [2]catenanes, multiring catenanes (or polycatenanes, Figure 4) have become challenging targets,[135–7] not only because of their elegance and complexity but also because of the interesting properties they might have, notably the preparation of a linear chain of five interlocking rings, Olympiadane, (**85**), by Stoddart and co-workers.[135] To apply their method to polycatenane synthesis, however, improvement in the efficiency of each step is still awaited. An attractive approach to the polycatenane might involve the supramolecular self-assembly of infinite, catenated complexes. Actually, in the interpenetrated two-dimensional network structure, catenated topology can be found. Here, interpenetrating infinite complexes involving polycatenated frameworks are discussed.

Figure 4 Schematic representation of polycatenanes.

Since the first report on the interpenetrated structure in 1922,[138] most known cases of interpenetration have involved inorganic or small organic ligands.[139–47] However, the past several years have seen many examples of interpenetrated complexes with large and tailored organic

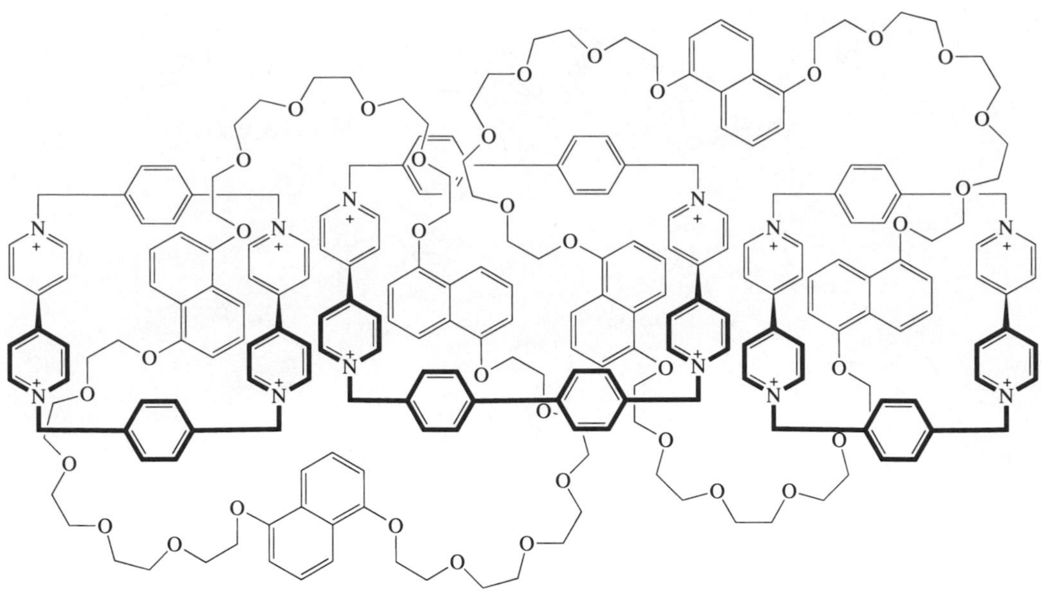

(85)

ligands.[147–54] While interpenetration with a diamond-related framework is relatively common,[142,143,146–50] there are fewer reports on the interpenetrated two-dimensional network. Rigid ligands (**86**) and (**87**) gave interpenetrated two-dimensional networks essentially the same as or similar to those of (**78a**) and (**78b**) (Figure 5) upon complexation with the metals indicated below their structures.[149–51] Among them, interpenetrated hexagonal two-dimensional network sheets containing organic radical cation material showed interesting magnetic properties.[150] Very large organic ligands, tetrakis(4-pyridyl)porphyrin (**88**) and tris[(4-cyanophenyl)ethynyl]benzene (**89**), afforded impressively large interpenetrated structures.

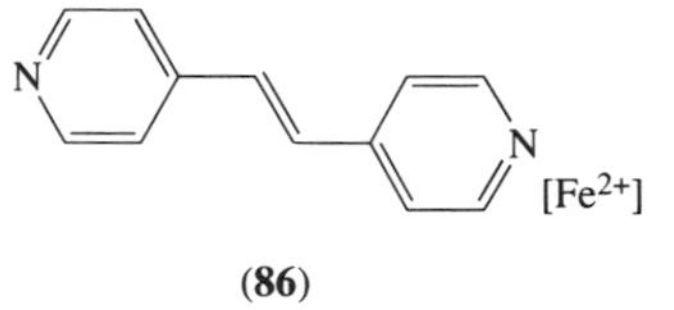

$[Fe^{2+}]$

(**86**)

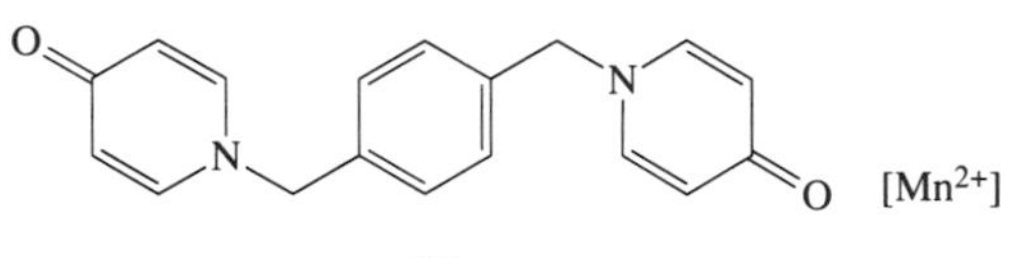

$[Mn^{2+}]$

(**87**)

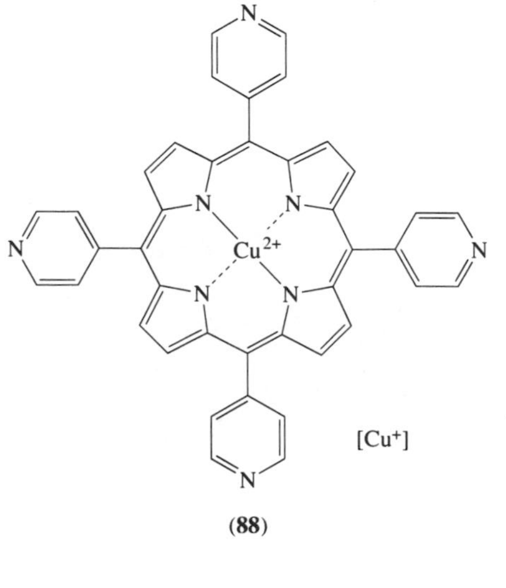

$[Cu^{+}]$

(**88**)

$[Ag^{+}]$

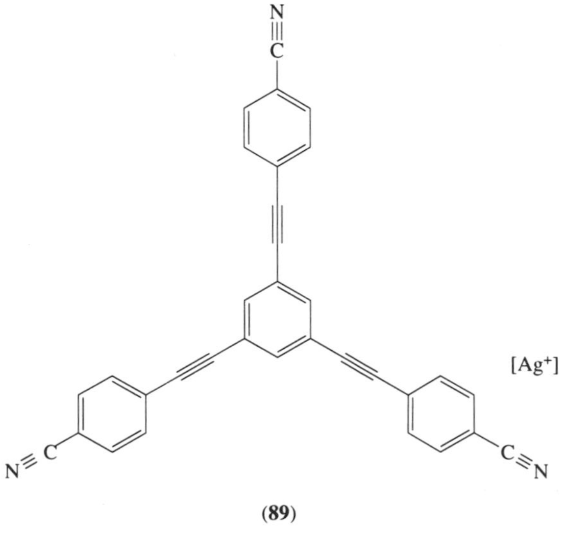

(**89**)

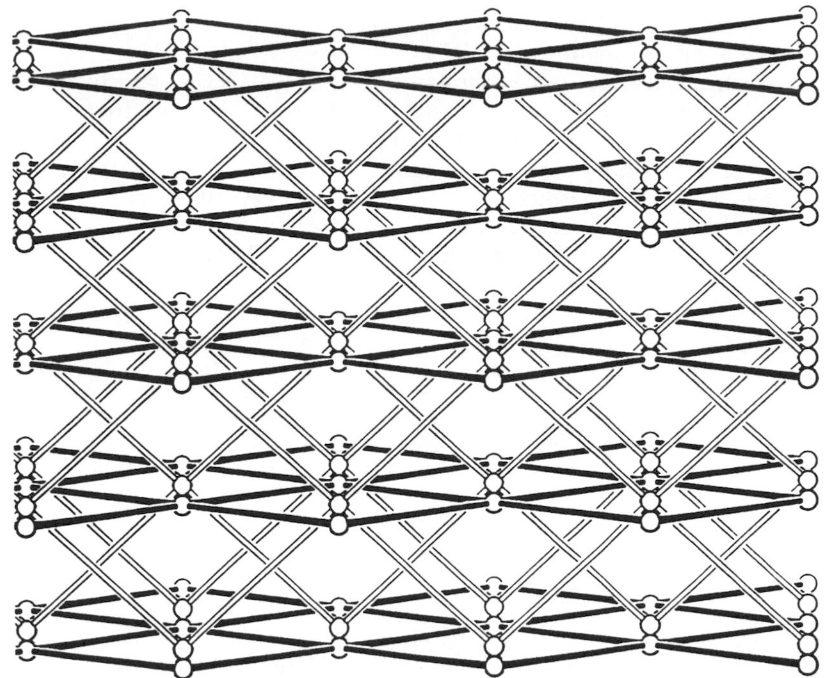

Figure 5 Interpenetrating square grid sheets common to $[M(H_2O)_2(bipy)_2](SiF_6)$ (M = Zn, Cu, Cd) (**78**). Only metal centers and bipy connections are shown (reproduced by permission of the Royal Society of Chemistry from *J. Chem. Soc., Chem. Commun.*, 1990, 1677).

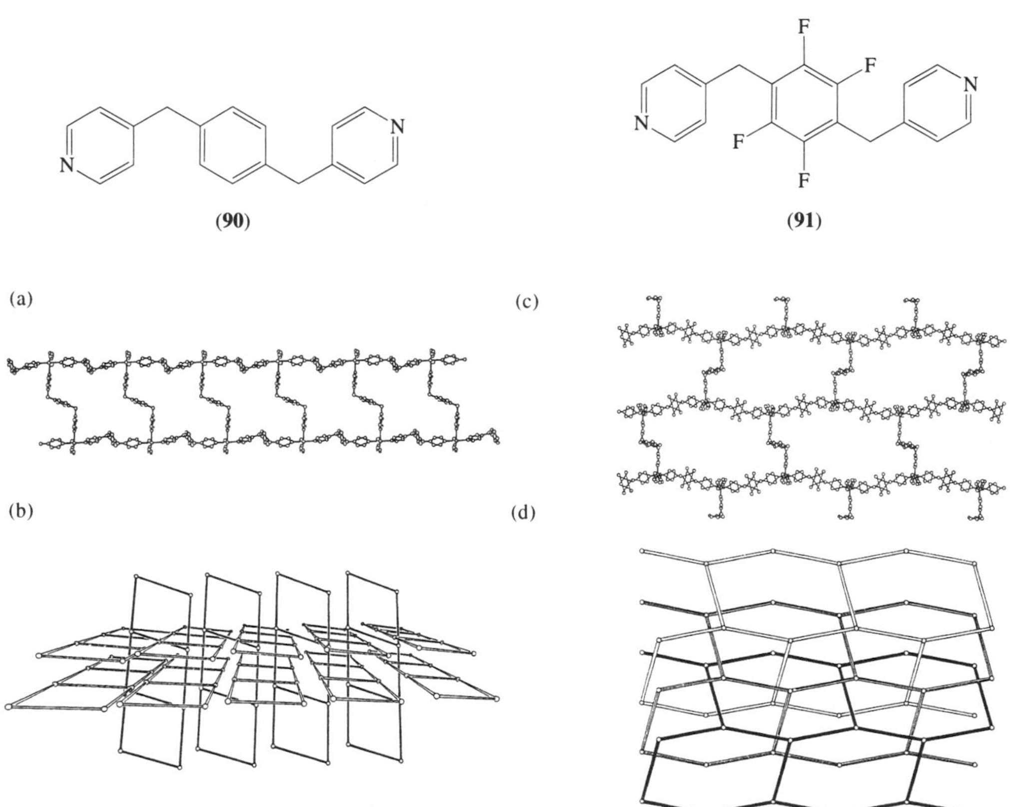

Figure 6 (a) and (b) The crystal structure of infinite molecular ladder $[Cd(\mathbf{90})_{1.5}](NO_3)_2$ (**91**). (c) and (d) Crystal structures of infinite molecular ladder $[Cd(\mathbf{79})_{1.5}](NO_3)_2$ (**92**). In (b) and (d) only cadmium atoms are shown and the heavy connections indicate the ligands (reprinted with permission from *J. Am. Chem. Soc.*, 1995, **117**, 7287. Copyright 1995 American Chemical Society).

The complexes discussed above do not include polycatenane topology (Figure 5) in which a ring is interlocked by two other *separate* rings. Two-dimensional networks involving polycatenane topologies have been reported.[154] Upon complexation with $Cd(NO_3)_2$, pyridine-based ligand (**90**) gave an infinite molecular ladder (**91**) (Figure 6(a)). In the solid structure, independent molecular ladders interpenetrate each other, in which four interlocks are found at each 60-membered ring framework (Figure 6(b)). On the other hand, ligand (**79**), a fluorinated analogue of (**90**), afforded an infinite molecular brick, (**92**), despite the very similar unit crystal structure to that of (**91**) (i.e., both complexes involve T-shaped connection of three pyridyl groups at each heptacoordinated cadmium atom) (Figure 6(c)). Again polycatenane topology is found in the triple interpenetration of these bricks (Figure 6(d)).

7.6 REFERENCES

1. J.-M. Lehn, *Angew. Chem., Int. Ed. Engl.*, 1988, **27**, 89.
2. J.-M. Lehn, *Angew. Chem., Int. Ed. Engl.*, 1990, **29**, 1304.
3. V. Balzani and L. DeCola (eds.), "Supramolecular Chemistry," Kluwer, Dordrecht, 1992.
4. J.-M. Lehn, A. M. Rigault, J. Siegel, J. Harrowfield, and B. Chevrier, *Proc. Natl. Acad Sci. USA*, 1987, **84**, 2565.
5. J.-M. Lehn and A. M. Rigault, *Angew. Chem., Int. Ed. Engl.*, 1988, **27**, 1095.
6. R. Kramer, J.-M. Lehn, and A. M. Rigault, *Proc. Natl. Acad. Sci. USA*, 1993, **90**, 5394.
7. U. Koert, M. M. Harding, and J.-M. Lehn, *Nature*, 1990, **346**, 339.
8. E. C. Constable, *Tetrahedron*, 1992, **48**, 10013.
9. U. Belluco (ed.), *Inorg Chim. Acta.*, 1996, in press (special issue on macrocyclic metal complexes).
10. M. Fujita and K. Ogura, *Bull. Chem. Soc. Jpn.*, 1996, in press.
11. F. Vögtle and E. Weber (eds.), "Host–Guest Complex Chemistry: Synthesis, Structure, Applications," Springer, Berlin, 1985.
12. M. Hirooka (ed.), "Crown Ethers and Analogous Compounds," Elsevier, New York, 1992, vol. 45.
13. D. J. Cram, *Angew. Chem., Int. Ed. Engl.*, 1988, **27**, 1009.
14. C. J. Pedersen, *Angew. Chem., Int. Ed. Engl.*, 1988, **27**, 1021.
15. F. Diederich, *Angew. Chem., Int. Ed. Engl.*, 1988, **27**, 362.
16. H.-J. Schneider, *Angew. Chem., Int. Ed. Engl.*, 1991, **30**, 1417–36.
17. V. Böhmer, *Angew. Chem., Int. Ed. Engl.*, 1995, **34**, 713.
18. A. J. Pryde, B. L. Shaw, and B. Weeks, *J. Chem. Soc., Chem. Commun.*, 1973, 947.
19. F. C. March, R. Mason, K. M. Thomas, and B. L. Shaw, *J. Chem. Soc., Chem. Commun.*, 1975, 584.
20. A. R. Sanger, *J. Chem. Soc., Chem. Commun.*, 1975, 893.
21. N. A. Al-Salem, H. D. Empsall, R. Markham, B. L. Shaw, and B. Weeks, *J. Chem. Soc., Dalton Trans.*, 1979, 1972.
22. W. E. Hill, J. G. Taylor, C. P. Falshaw, T. J. King, B. Beagley, D. M. Tonge, R. G. Pritchard, and C. A. McAuliffe, *J. Chem. Soc., Dalton Trans.*, 1986, 2289.
23. R. J. Puddephatt, *Chem. Soc. Rev.*, 1983, **12**, 99.
24. A. W. Maverick and F. E. Klavetter, *Inorg. Chem.*, 1984, **23**, 4129.
25. A. W. Maverick, S. C. Buckingham, Q. Yao, J. R. Bradbury, and G. G. Stanley, *J. Am. Chem. Soc.*, 1986, **108**, 7430.
26. J. R. Bradbury, J. L. Hampton, D. P. Martone, and A. W. Maverick, *Inorg. Chem.*, 1989, **28**, 2392.
27. A. W. Maverick, M. L. Ivie, J. H. Waggenspack, and F. R. Fronczek, *Inorg Chem.*, 1990, **29**, 2403.
28. A. W. Schwabacher, J. Lee, and H. Lei, *J. Am. Chem. Soc.*, 1992, **114**, 7597.
29. J. Lee and A. W. Schwabacher, *J. Am. Chem. Soc.*, 1994, **116**, 8382.
30. P. Scrimin, P. Tecilla, U. Tonellato, and M. Vignana, *J. Chem. Soc., Chem. Commun.*, 1991, 449.
31. C. A. Hunter and L. D. Sarson, *Angew. Chem., Int. Ed. Engl.*, 1994, **33**, 2313.
32. M. Fujita, S. Nagao, M. Iida, K. Ogata, and K. Ogura, *J. Am. Chem. Soc.*, 1993, **115**, 1574.
33. M. Fujita, J. Yazaki, and K. Ogura, *J. Am. Chem. Soc.*, 1990, **112**, 5645.
34. M. Fujita, J. Yazaki, and K. Ogura, *Tetrahedron Lett.*, 1991, **32**, 5589.
35. M. Fujita and K. Ogura, *Coord. Chem. Rev.*, 1996, in press.
36. M. Fujita, F. Ibukuro, H. Hagihara, and K. Ogura, *Nature*, 1994, **367**, 720.
37. Y. Kobuke and Y. Satoh, *J. Am. Chem. Soc.*, 1992, **114**, 789.
38. Y. Kobuke, Y. Sumida, M. Hayashi, and H. Ogoshi, *Angew. Chem., Int. Ed. Engl.*, 1991, **230**, 1496.
39. M. Fujita, J. Yazaki, T. Kuramochi, and K. Ogura, *Bull. Chem. Soc. Jpn.*, 1993, **66**, 1837.
40. S. Rüttimann, G. Bernardinelli, and A. F. Williams, *Angew. Chem., Int. Ed. Engl.*, 1993, **32**, 392.
41. A. T. Baker, J. K. Crass, M. Maniska, and D. C. Craig, *Inorg Chim. Acta*, 1995, **230**, 225.
42. T. Kajiwara and T. Ito, *J. Chem. Soc., Chem. Commun.*, 1994, 1773.
43. C. M. Bird, C. Brehene, M. G. Davidson, A. J. Edwards, S. C. Llewellyn, P. R. Raithby, and R. Snaith, *Angew. Chem., Int. Ed. Engl.*, 1993, **32**, 1483.
44. X. Yang, C. B. Knobler, and M. F. Hawthorne, *Angew. Chem., Int. Ed. Engl.*, 1991, **30**, 1507.
45. X. Yang, C. B. Knobler, and M. F. Hawthorne, *J. Am. Chem. Soc.*, 1992, **114**, 380.
46. Z. Zheng, X. Yang, C. B. Knobler, and M. F. Hawthorne, *J. Am. Chem. Soc.*, 1993, **115**, 5320.
47. X. Yang, C. B. Knobler, and M. F. Hawthorne, *J. Am. Chem. Soc.*, 1993, **115**, 4904.
48. H. L. Anderson and J. K. M. Sanders, *Angew. Chem., Int. Ed. Engl.*, 1990, **29**, 1400.
49. S. Anderson, H. L. Anderson, and J. K. M. Sanders, *Acc. Chem. Res.*, 1993, **26**, 469.
50. H. L. Anderson, A. Bashall, K. Henrick, M. McPartlin, and J. K. M. Sanders, *Angew. Chem., Int. Ed. Engl.*, 1994, **33**, 429.
51. L. G. Mackay, R. S. Wylie, and J. K. M. Sanders, *J. Am. Chem. Soc.*, 1994, **116**, 3141.
52. P. M. Stricklen, E. J. Volcko, and J. G. Verkade, *J. Am. Chem. Soc.*, 1983, **105**, 2494.

53. M. Fujita, O. Sasaki, T. Mitsuhashi, T. Fujita, J. Yazaki, K. Yamaguchi, and K. Ogura, *J. Chem. Soc., Chem. Commun.*, in press.
54. M. Fujita, J. Yazaki, and K. Ogura, *Chem. Lett.*, 1991, 1031.
55. M. Fujita, F. Ibukuro, K. Yamaguchi, and K. Ogura, *J. Am. Chem. Soc.*, 1995, **117**, 4175.
56. H. Rauter, E. C. Hillgeris, A. Erxleben, and B. Lippert, *J. Am. Chem. Soc.*, 1994, **116**, 616.
57. P. J. Stang and D. H. Cao, *J. Am. Chem. Soc.*, 1994, **116**, 4981.
58. P. J. Stang and J. A. Whiteford, *Organometallics*, 1994, **13**, 3776.
59. P. J. Stang and V. V. Zhdankin, *J. Am. Chem. Soc.*, 1993, **115**, 9808.
60. C. M. Drain and J.-M. Lehn, *J. Chem. Soc., Chem. Commun.*, 1994, 2313.
61. M. Fujita, M. Aoyagi, and K. Ogura, *Inorg Chim. Acta*, 1996, in press.
62. Z. Atherton, M. L. Goodgame, D. A. Katahira, S. Manzer, and D. J. Williams, *J. Chem. Soc., Chem. Commun.*, 1994, 1423.
63. E. C. Constable, R. P. G. Henney, P. R. Raithby, and L. R. Sousa, *Angew. Chem., Int. Ed. Engl.*, 1991, **30**, 1363.
64. P. J. Stang and K. Chen, *J. Am. Chem. Soc.*, 1995, **116**, 1667.
65. K. L. Taft, C. D. Delfs, G. C. Papaefthymiou, S. Foner, D. Gatteschi, and S. J. Lippard, *J. Am. Chem. Soc.*, 1994, **116**, 823.
66. A. Caneschi, A. Cornia, and S. J. Lippard, *Angew. Chem., Int. Ed. Engl.*, 1995, **34**, 467.
67. A. J. Blake, C. M. Grant, S. Parsons, J. M. Rawson, and R. E. P. Winpenny, *J. Chem. Soc., Chem. Commun.*, 1994, 2363.
68. D. Fenske and A. Fisher, *Angew. Chem., Int. Ed. Engl.*, 1995, **34**, 307.
69. R. M. Izatt and J. J. Christensen (eds.), "Synthetic Multidentate Macrocyclic Compounds," Academic Press, New York, 1978.
70. F. Evmeyer and F. Vögtle, in "Inclusion Compounds," eds. J. L. Atwood, J. E. D. Davies and D. D. MacNicol, Academic Press, London, 1991, vol 4, chap. 6, pp. 263–82.
71. C. Seel and F. Vögtle, *Angew. Chem., Int. Ed. Engl.*, 1992, **31**, 528.
72. Y. Murakami and O. Hayashida, *Proc. Natl. Acad. Sci. USA*, 1993, **90**, 1140.
73. Y. Murakami, O. Hayashida, and Y. Nagai, *J. Am. Chem. Soc.*, 1994, **116**, 2611.
74. A. Borchardt and W. C. Still, *J. Am. Chem. Soc.*, 1994, **116**, 373.
75. S. S. Yoon and W. C. Still, *J. Am. Chem. Soc.*, 1993, **115**, 823.
76. J.-I. Hong, S. K. Namgoong, A. Bernardi, and W. C. Still, *J. Am. Chem. Soc.*, 1991, **113**, 5111.
77. D. J. Cram, M. E. Tanner, and R. Thomas, *Angew. Chem., Int. Ed. Engl.*, 1991, **30**, 1024.
78. P. Timmerman, W. Verboom, F. C. J. M. Veggel, J. P. M. van Duynhoven, and D. N. Reinhoudt, *Angew. Chem., Int. Ed. Engl.*, 1994, **33**, 2345.
79. R. Wyler, J. de Mendoza, and J. Rebek, Jr., *Angew. Chem., Int. Ed. Engl.*, 1993, **32**, 1699.
80. N. Branda, R. Wyler, and J. Rebek, Jr., *Science*, 1994, **263**, 1267.
81. R. W. Saalfrank, A. Stark, K. Peters, and H. G. von Schnering, *Angew. Chem., Int. Ed. Engl.*, 1988, **27**, 851.
82. R. W. Saalfrank, A. Stark, M. Bremer, and H.-U. Hummel, *Angew. Chem., Int. Ed. Engl.*, 1990, **29**, 311.
83. R. W. Saalfrank, B. Hörner, D. Stalke, and J. Salbeck, *Angew. Chem., Int. Ed. Engl.*, 1993, **32**, 1179.
84. R. W. Saalfrank, R. Burak, A. Breit, D. Stalke, R. Herbst-Irmer, J. Daub, M. Porsch, E. Bill, M. Muther, and A. X. Trautwein, *Angew. Chem., Int. Ed. Engl.*, 1994, **33**, 1621.
85. M. Fujita, S. Nagao, and K. Ogura, *J. Am. Chem. Soc.*, 1995, **117**, 1649.
86. P. Baxter, J.-M. Lehn, and A. DeCian, *Angew. Chem., Int. Ed. Engl.*, 1993, **32**, 69.
87. E. Leize, A. V. Dorsselaer, R. Kramer, and J.-M. Lehn, *J. Chem. Soc., Chem. Commun.*, 1993, 990.
88. R. Fuchs, N. Habermann, and P. Klüfers, *Angew. Chem., Int. Ed. Engl.*, 1993, **32**, 852–4.
89. P. Klüfers and J. Schuhmacher, *Angew. Chem., Int. Ed. Engl.*, 1994, **33**, 1863.
90. M. T. Pope and A. Müller, *Angew. Chem., Int. Ed. Engl.*, 1991, **30**, 34.
91. A. Müller, R. Rohlfing, E. Krickemeyer, and H. Bögge, *Angew. Chem., Int. Ed. Engl.*, 1993, **32**, 909.
92. M. I. Khan, A. Müller, S. Dillinger, H. Bögge, Q. Chen, and J. Zubieta, *Angew. Chem., Int. Ed. Engl.*, 1993, **32**, 1780.
93. A. Müller, K. Hovemeier, and R. Rohlfing, *Angew. Chem., Int. Ed. Engl.*, 1992, **31**, 1192.
94. A. Müller, E. Krickemeyer, M. Penk, R. Rohlfing, A. Armatage, and H. Bögge, *Angew. Chem., Int. Ed. Engl.*, 1991, **30**, 1674.
95. V. W. Day, W. G. Klemperer, and O. M. Yaghi, *J. Am. Chem. Soc.*, 1989, **111**, 5959.
96. W. G. Klemperer, T. A. Marquart, and O. M. Yaghi, *Angew. Chem., Int. Ed. Engl.*, 1992, **31**, 49.
97. M.-M. Rhomer and M. Benard, *J. Am. Chem. Soc.*, 1994, **116**, 6959.
98. J.-P. Sauvage (ed.), *New. J. Chem.*, Special Issue, 1993, **17**(10), 11.
99. E. Wasserman, *J. Am. Chem. Soc.*, 1960, **82**, 4433.
100. C. O. D. Buchecker, J.-P. Sauvage, and J. P. Kintzinger, *Tetrahedron Lett.*, 1983, **24**, 5095.
101. C. O. D. Buchecker and J.-P. Sauvage, *Chem. Rev.*, 1987, **87**, 795.
102. J.-P. Sauvage, *Acc. Chem. Res.*, 1990, **23**, 319.
103. P. R. Ashton, T. T. Goodnow, A. E. Kaifer, M. V. Reddington, A. M. Z. Slawin, N. Spencer, J. F. Stoddart, C. Vicent, and D. J. Williams, *Angew. Chem., Int. Ed. Engl.*, 1989, **28**, 1396.
104. P. L. Anelli, P. R. Ashton, R. Ballardini, V. Balzani, M. Delgado, T. Fandolfi, T. T. Goodnow, A. E. Kaifer, D. Philp, M. Pietraszkiewicz, L. Prodi, M. V. Reddington, A. M. Z. Slawin, N. Spencer, J. F. Stoddart, C. Vicent, and D. J. Williams, *J. Am. Chem. Soc.*, 1992, **114**, 193.
105. D. B. Amabilino, P. R. Ashton, C. L. Brown, E. Cordova, L. A. Godinez, T. T. Goodnow, A. E. Kaifer, S. P. Newton, M. Pietraszkiewicz, D. Philp, F. M. Raymo, A. S. Reder, M. T. Rutland, A. M. Z. Slawin, N. Spencer, J. F. Stoddart, and D. J. Williams, *J. Am. Chem. Soc.*, 1995, **117**, 1271.
106. D. Armspach, P. R. Ashton, R. Ballardini, V. Balzani, A. Godi, C. P. Moore, L. Prodi, N. Spencer, J. F. Stoddart, M. S. Tolley, T. J. Wear, and D. J. Williams, *Chem. Eur. J.*, 1995, **1**, 33.
107. C. A. Hunter, *J. Am. Chem. Soc.*, 1992, **114**, 5303.
108. H. Adams, F. J. Carver, and C. A. Hunter, *J. Chem. Soc., Chem. Commun.*, 1995, 809.
109. F. Vögtle, S. Meier, and R. Hoss, *Angew. Chem., Int. Ed. Engl.*, 1992, **31**, 1619.

110. S. Ottens-Hildebrandt, M. Nieger, K. Rissanen, J. Rouvinen, S. Meier, G. Harder, and F. Vögtle, *J. Chem. Soc., Chem. Commun.*, 1995, 809.
111. G.-J. Guter, F. J. J. de Kanter, P. R. Markies, T. Nomoto, O. S. Akkerman, and F. Bickelhaupt, *J. Am. Chem. Soc.*, 1993, **115**, 12 179.
112. P. R. Markies, T. Nomoto, O. S. Akkerman, F. Bickelhaupt, W. J. J. Smeets, and A. L. Spek, *Angew. Chem., Int. Ed. Engl.*, 1988, **27**, 1084.
113. M. Fujita, F. Ibukura, H. Seke, O. Kamo, M. Imanari, and K. Ogura, *J. Am. Chem. Soc.*, 1996, **118**, 899.
114. M. Fujita and K. Ogura, *Supramol. Sci.*, 1996, in press.
115. C. Piguet, G. Bernardinelli, A. F. Williams, and B. Bocquet, *Angew. Chem., Int. Ed. Engl.*, 1995, **34**, 582.
116. H. Ogino, *J. Am. Chem. Soc.*, 1981, **103**, 1303.
117. H. Ogino, *New. J. Chem.*, 1993, **17**, 683.
118. R. S. Wylie and D. H. Macartney, *J. Am. Chem. Soc.*, 1992, **114**, 3136.
119. R. Robson, B. F. Abrahams, S. R. Batten, R. W. Gable, B. F. Hoskins, and J. Liu, in "Supramolecular Architecture," ed. T. Bein, ACS Symposium Series 499, American Chemical Society, Washington DC, 1992, chap. 19.
120. T. Iwamoto, in "Inclusion Compounds", eds. J. L. Atwood, J. E. D. Davies, and D. D. MacNicol, Academic Press, London, 1984, vol. 1, chap. 2; J. Liplowski, in "Inclusion Compounds," eds. J. L. Atwood, J. E. D. Davies, and D. D. MacNicol, Academic Press, London, 1984, vol. 1, chap. 3; J. Hanotier and P. de Radzitzky, in "Inclusion Compounds," eds. J. L. Atwood, J. E. D. Davies, and D. D. MacNicol, Academic Press, London, 1984, vol. 1, chap. 4.
121. T. Iwamoto, in "Inclusion Compounds," eds. J. L. Atwood, J. E. D. Davies and D. D. MacNicol, Academic Press: London, 1991, vol. 5, chap. 6, pp. 177–212.
122. J. H. Rayner and H. M. Powell, *J. Chem. Soc.*, 1952, 319.
123. R. W. Gable, B. F. Hoskins, and R. Robson, *J. Chem. Soc., Chem. Commun.*, 1990, 1677.
124. M. Fujita, Y. J. Kwon, S. Washizu, and K. Ogura, *J. Am. Chem. Soc.*, 1994, **116**, 1151.
125. M. Munakata, T. Kuroda-Sowa, M. Maekawa, M. Nakamura, S. Akiyama, and S. Kitagawa, *Inorg Chem.*, 1994, **33**, 1284.
126. S. Kawata, S. Kitagawa, M. Kondo, I. Furuchi, and M. Munakata, *Angew. Chem., Int. Ed. Engl.*, 1759.
127. K. Inoue and H. Iwamura, *J. Am. Chem. Soc.*, 1994, **116**, 3173.
128. G. D. Munno, M. Julve, F. Nicolo, F. Lloret, J. Faus, R. Ruiz, and E. Sinn, *Angew. Chem., Int. Ed. Engl.*, 1993, **32**, 613.
129. D. M. L. Goodgame, D. J. Williams, and R. E. P. Winpenny, *Angew. Chem., Int. Ed. Engl.*, 1988, **27**, 261.
130. D. M. L. Goodgame, S. P. W. Hill, and D. J. Williams, *J. Chem. Soc., Chem. Commun.*, 1993, 1019.
131. Z. Atherton, D. M. L. Goodgame, D. A. Katahira, S. Menzer, and D. J. Williams, *J. Chem. Soc., Chem. Commun.*, 1994, 1423.
132. D. M. L. Goodgame, S. Menzer, A. M. Smith, and D. J. Williams, *J. Chem. Soc., Chem. Commun.*, 1994, 1825.
133. D. M. L. Goodgame, S. Menzer, A. T. Ross, and D. J. Williams, *J. Chem. Soc., Chem. Commun.*, 1994, 2605.
134. M. Fujita, Y. J. Kwon, M. Miyazawa, and K. Ogura, *J. Chem. Soc., Chem. Commun.*, 1994, 1977.
135. D. B. Amabilino, P. R. Ashton, A. S. Reder, N. Spencer, and J. F. Stoddart, *Angew. Chem., Int. Ed. Engl.*, 1994, **33**, 433.
136. F. Bitsch, C. O. Dietrich-Buchecker, A.-K. Khémiss, J.-P. Sauvage, and A. V. Dorsselaer, *J. Am. Chem. Soc.*, 1991, **113**, 4023.
137. C. O. Dietrich-Buchecker, B. Frommberger, I. Luer, J.-P. Sauvage, and F. Vögtle, *Angew. Chem., Int. Ed. Engl.*, 1993, **32**, 1434.
138. P. Niggli, *Z. Kristallogr.*, 1922, **57**, 253.
139. J. Konnert and D. Britton, *Inorg. Chem.*, 1966, **5**, 1193.
140. T. Soma, H. Yuge, and T. Iwamoto, *Angew. Chem., Int. Ed. Engl.*, 1994, **33**, 1665.
141. A. Michaelides, V. Kiritsis, S. Skoulika, and A. Aubry, *Angew. Chem., Int. Ed. Engl.*, 1993, **32**, 1495.
142. K.-W. Kim and M. G. Kanatzidis, *J. Am. Chem. Soc.*, 1992, **114**, 4878.
143. B. F. Hoskins, R. Robson, and N. V. Y. Scarlett, *J. Chem. Soc., Chem. Commun.* 1994, 2025.
144. S. R. Batten, B. F. Hoskins, and R. Robson, *J. Chem. Soc., Chem. Commun.*, 1991, 445.
145. D. M. Proserpio, R. Hoffmann, and P. Preuss, *J. Am. Chem. Soc.*, 1994, **116**, 9634.
146. L. Carlucci, G. Ciani, D. M. Proserpio, and A. Sironi, *J. Chem. Soc., Chem. Commun.*, 1994, 2755.
147. S. B. Copp, S. Subramanian, and M. J. Zaworotko, *J. Am. Chem. Soc.*, 1992, **114**, 8719.
148. K. Sinzger, S. Hunig, M. Jopp, D. Bauer, W. Bietsch, J. U. von Schutz, H. C. Wulf, R. K. Kremer, T. Metzenthin, R. Bau, S. I. Khan, K. A. Lindbaum, C. L. Lengauer, and E. Tillmans, *J. Am. Chem. Soc.*, 1993, **115**, 7696.
149. J. A. Real, E. Andrés, M. C. Muñoz, M. Julve, T. Granier, T. A. Bousseksou, and F. Varret, *Science*, 1995, **268**, 265.
150. H. O. Stampf, L. Ouahab, Y. Pei, D. Grandjean, and O. Kahn, *Science*, 1993, **261**, 447.
151. D. M. L. Goodgame, S. Menzer, A. M. Smith, and D. J. Williams, *Angew. Chem., Int. Ed. Engl.*, 1995, **34**, 574.
152. B. F. Abrahams, B. F. Hoskins, D. M. Michail, and R. Robson, *Nature*, 1994, **369**, 727.
153. G. B. Gardner, D. Venkataraman, J. S. Moore, and S. Lee, *Nature*, 1995, **374**, 792.
154. M. Fujita, Y. J. Kwon, O. Sasaki, K. Yamaguchi, and K. Ogura, *J. Am. Chem. Soc.*, 1995, **117**, in press.

8

Towards Semiconducting Polynuclear Bridged Complexes Incorporating Phthalocyanines

ULF DRECHSLER and MICHAEL HANACK
Universität Tübingen, Germany

8.1 INTRODUCTION

Phthalocyanine and its metal complexes have been investigated in great detail. Their main application was their use as dyes or catalysts.[1] Currently, phthalocyanine and many of its derivatives find applications in a widespread field. Due to their special properties, phthalocyanines and structurally related compounds are of great interest in materials science.[2] They are in use in nonlinear optics,[3] Langmuir–Blodgett (LB) films,[4,5] optical data storage,[6] rectifying devices,[7] as liquid crystals,[8] electrochromic substances,[9] low-dimensional metals [5,10–12] or gas sensors.[13] Peripherally substituted derivatives of phthalocyanines find applications in various processes involving visible light: photoredox or photooxidation reactions in solution,[1,14,15] as sensitizers in the photodynamic therapy of cancer,[1,16] in photoelectrochemical[17,18] or photovoltaic cells.[19,20]

Phthalocyanines in the solid state show semiconducting behaviour.[19] The main difference between these molecular organic, van der Waals bonded and inorganic, covalently bonded semiconductors is the extent of orbital overlap along the conducting pathway. In molecular organic semiconductors the binding interactions by orbital overlap are about 0.1 eV, in inorganic semiconductors they are several electron volts.[2]

Figure 1 shows the chemical structures of the metal-free phthalocyanine (PcH_2), its metal complex (PcM), and some of its related macrocycles. The phthalocyanine itself is a planar, 18 π-electron macrocycle, which is built up from four isoindoline units, linked together in the 1,3-position by aza-bridges, to form a cyclic tetramer. Annulation of additional benzene rings leads to 2,3-naphthalocyanines (2,3-Nc), in the case of linear annulation, or to 1,2-naphthalocyanines (1,2-Nc) by angular annulation. A similar molecule is the tetrabenzoporphyrin (TBP) in which the four isoindolines are linked over carbon bridges. An example for a related macrocycle is the hemiporphyrazine (Hp), which is, contrary to the phthalocyanines, not planar and rather alkenic in character.

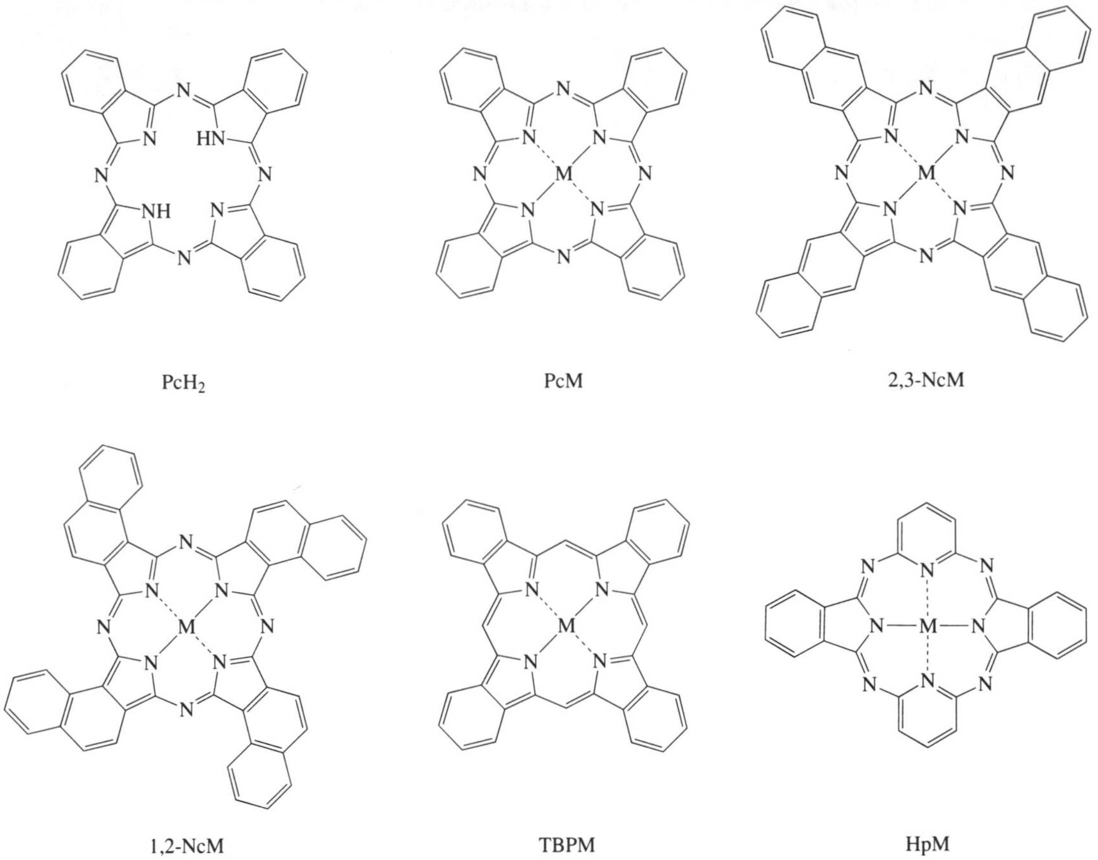

PcH_2 PcM 2,3-NcM

1,2-NcM TBPM HpM

Figure 1 Phthalocyanine and its derivatives. PcH_2, phthalocyanine; PcM, metallophthalocyanine; 2,3-NcM, 2,3-metallonaphthalocyanine; 1,2-NcM, 1,2-metallonaphthalocyanine; TBPM, metallotetrabenzoporphyrin; HpM, metallohemiporphyrazine.

In the following sections we will concentrate on the conductive properties of phthalocyanine and its metal complexes. These compounds possess electronic and morphological characteristics which are very favourable for semiconductive properties. Phthalocyaninatometal complexes show great thermal and chemical stability and many of them can be easily synthesized in high yields and purity, starting from inexpensive precursors.

8.1.1 Synthesis of Phthalocyanines

Phthalocyanines have been synthesized up to the mid-1990s using about 70 elements as the central atom: besides main group elements, transition metals, lanthanoids and actinoids as well as semimetals such as boron or silicon, also nonmetals, for example, phosphorus, have been employed. The choice of the central atom enables control of the oxidation potential of the phthalocyanine–element compounds, which leads to very different electrical properties.

A selection of basic methods to synthesize phthalocyanines is shown in Scheme 1.[1,2,5,21] Many starting materials can be used to prepare phthalocyaninatometal compounds. Most of these complexes can be synthesized from phthalonitrile or its derivatives and the corresponding metals or metal salts in high boiling solvents, such as quinoline or 1-chloronaphthalene. They are also obtained by metal insertion into the metal-free phthalocyanine. A mild method for preparing phthalocyaninatometal complexes in good yields is the direct condensation of phthalonitrile derivatives with metal salts by heating in higher boiling alcohols, for example, 1-pentanol, in the presence of 1,8-diazabicyclo[5.4.0]undec-1-ene (dbu).[22]

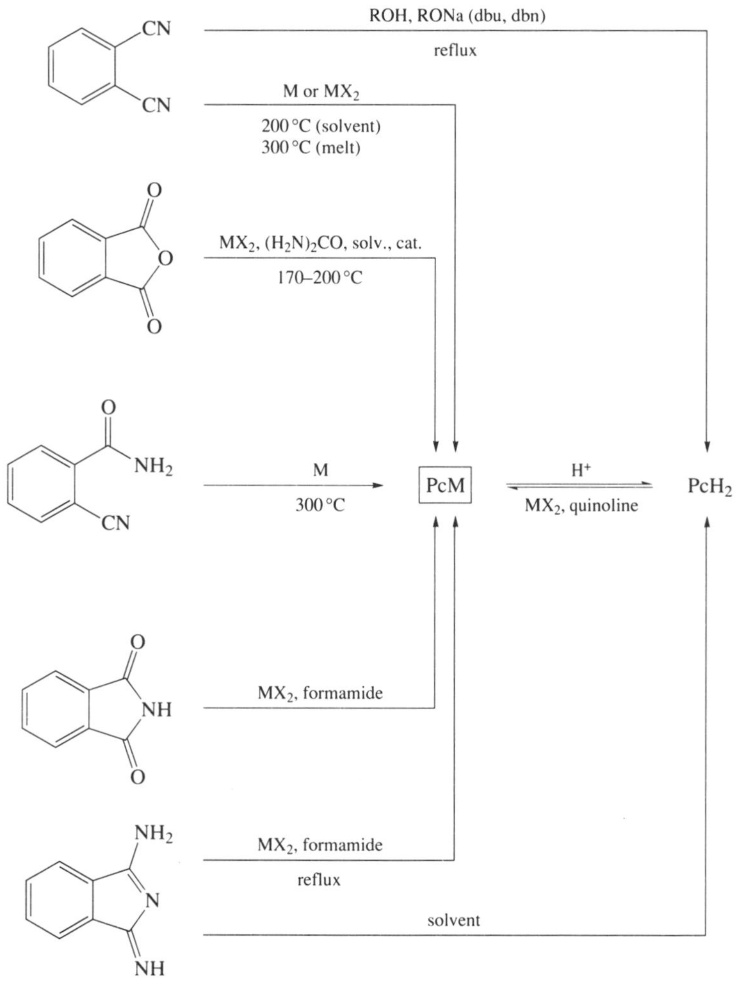

Scheme 1

The synthesis of 2,3-naphthalocyaninato- and 1,2-naphthalocyaninatometal complexes follow similar routes starting from 2,3-dicyanonaphthalene or 1,2-dicyanonaphthalene, respectively.[23–5] Synthetic routes for the preparation of tetrabenzoporphyrins and hemiporphyrazines are given in the literature.[5]

Some phthalocyaninatometal compounds cannot be obtained as easily as shown in Scheme 1. For example, the phthalocyaninatoruthenium (PcRu) or -osmium (PcOs)[26] complexes are more stable against oxidation of the central metal atom ($M^{2+} \rightarrow M^{3+}$) compared to the corresponding iron complexes and show stronger complex stabilities due to the larger radius of the metal ion. Analytically pure PcRu can be prepared by the reaction of *o*-cyanobenzamide with ruthenium trichloride in a melt of naphthalene.[27] The crude PcRu obtained is converted into the soluble complex $PcRu(DMSO)_2 \cdot 2DMSO$ with DMSO, which can be thermally decomposed after chromatographic purification to yield pure PcRu. Another simple route to obtain pure PcRu is by using quinoline (qnl) instead of naphthalene. It was suggested that the bis(axially) coordinated $PcRu(qnl)_2$ complex was formed.[28] Some results, however, show a different situation. The quinoline used as solvent in this reaction always contains a percentage of isoquinoline (iqnl) as impurity. It is shown by ^{1}H NMR spectroscopic investigations that the bis(axially) coordinated species carries two isoquinoline molecules as axial ligands, its composition is therefore $PcRu(iqnl)_2$.[29]

This soluble bis(axially) coordinated complex can be purified by column chromatography, before being thermally decomposed (250 °C) *in vacuo* to give pure phthalocyaninatoruthenium(II) (see Scheme 2).

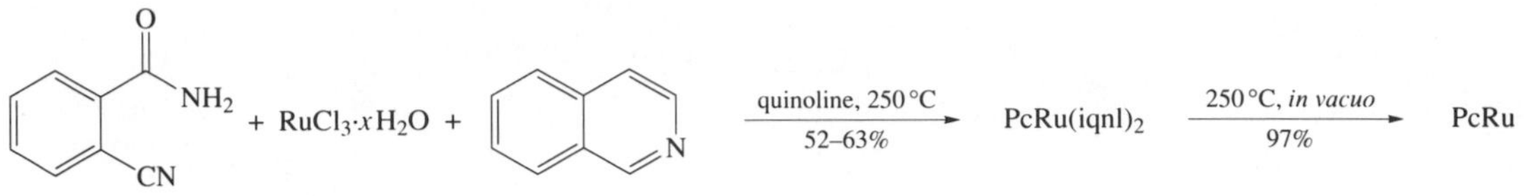

$$+ RuCl_3 \cdot xH_2O + \xrightarrow[52\text{–}63\%]{\text{quinoline, } 250\,^\circ C} PcRu(iqnl)_2 \xrightarrow[97\%]{250\,^\circ C,\ in\ vacuo} PcRu$$

Scheme 2

8.2 SOLUBLE PHTHALOCYANINES

Due to intermolecular interactions between the macrocycles, peripherally unsubstituted phthalocyanines show aggregation effects, and are therefore practically insoluble in common organic solvents. To improve their solubility it is possible to introduce substituents in the periphery of the macrocycles, which lead to a larger distance between the macrocycles, thus enabling their solvation. Especially bulky substituents, for example, Bu^t,[30] or electrically charged substituents, like carboxylato[31] or quarternary ammonium groups,[32] lead to phthalocyanines with relatively high solubility in organic or even aqueous solvents.

The most common and best investigated soluble phthalocyanines are the tetra- and octasubstituted phthalocyanines. Figures 2 and 3 show the structures of tetra- and octasubstituted phthalocyaninatometal complexes, respectively. As central metals mainly transition metals but also main group metals, for example, lead, have been used. As peripheral substituents, alkyl,[33–6] alkoxy,[27,34,37–40] sulfonyl,[41] carboxyl[31] groups or others, even crown ethers,[42] have been employed. A more detailed description of soluble metallophthalocyanines is given in the literature.[1,5,43]

Due to the location of the substituents either in the 2, 3, 9, 10, 16, 17, 23, 24 or in the 1, 4, 8, 11, 15, 18, 22, 25 positions, the tetra- and octasubstituted complexes shown in Figures 2 and 3 are called 2,3- and 1,4-substituted phthalocyanines, respectively.

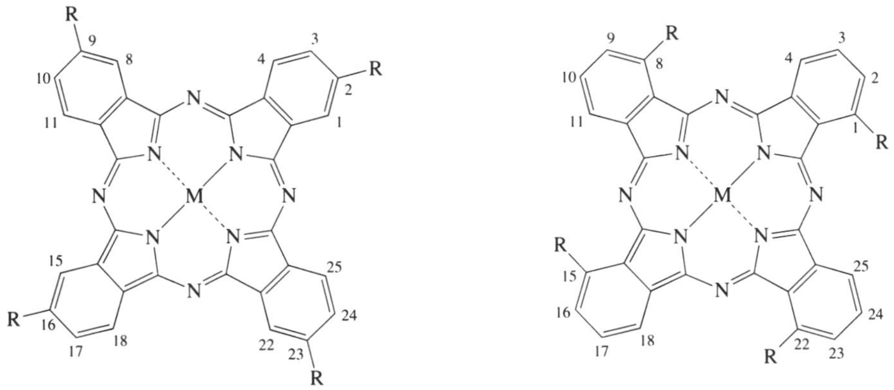

Figure 2 Tetrasubstituted phthalocyanines (only the C_{4h} isomer is shown).

As shown in Scheme 3, the 2,3-tetrasubstituted phthalocyanines are obtained from 4-substituted phthalonitriles, and the 1,4-tetrasubstituted phthalocyanines are synthesized from 3-substituted phthalonitriles, respectively.

The condensation of monosubstituted phthalonitriles always leads to a mixture of four constitutional isomers of tetrasubstituted phthalocyanines. These four isomers are depicted in Figure 4 for the 2,3-substituted systems; for the 1,4-substituted complexes, similar isomers are obtained. It is possible to separate these isomers by HPLC methods.[37,44]

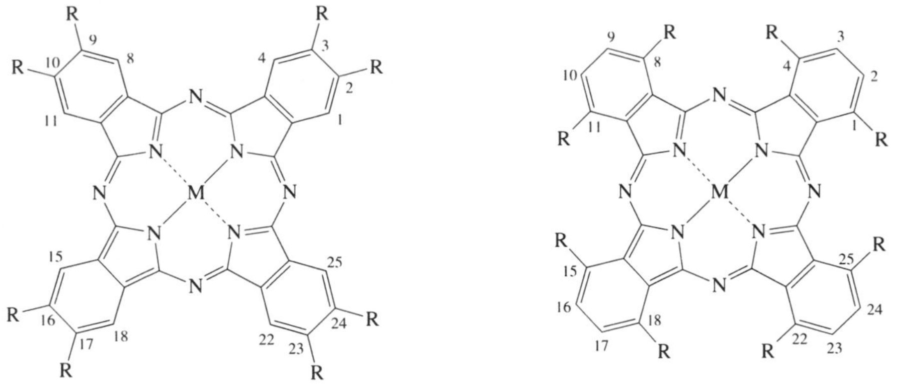

Figure 3 Octasubstituted phthalocyanines.

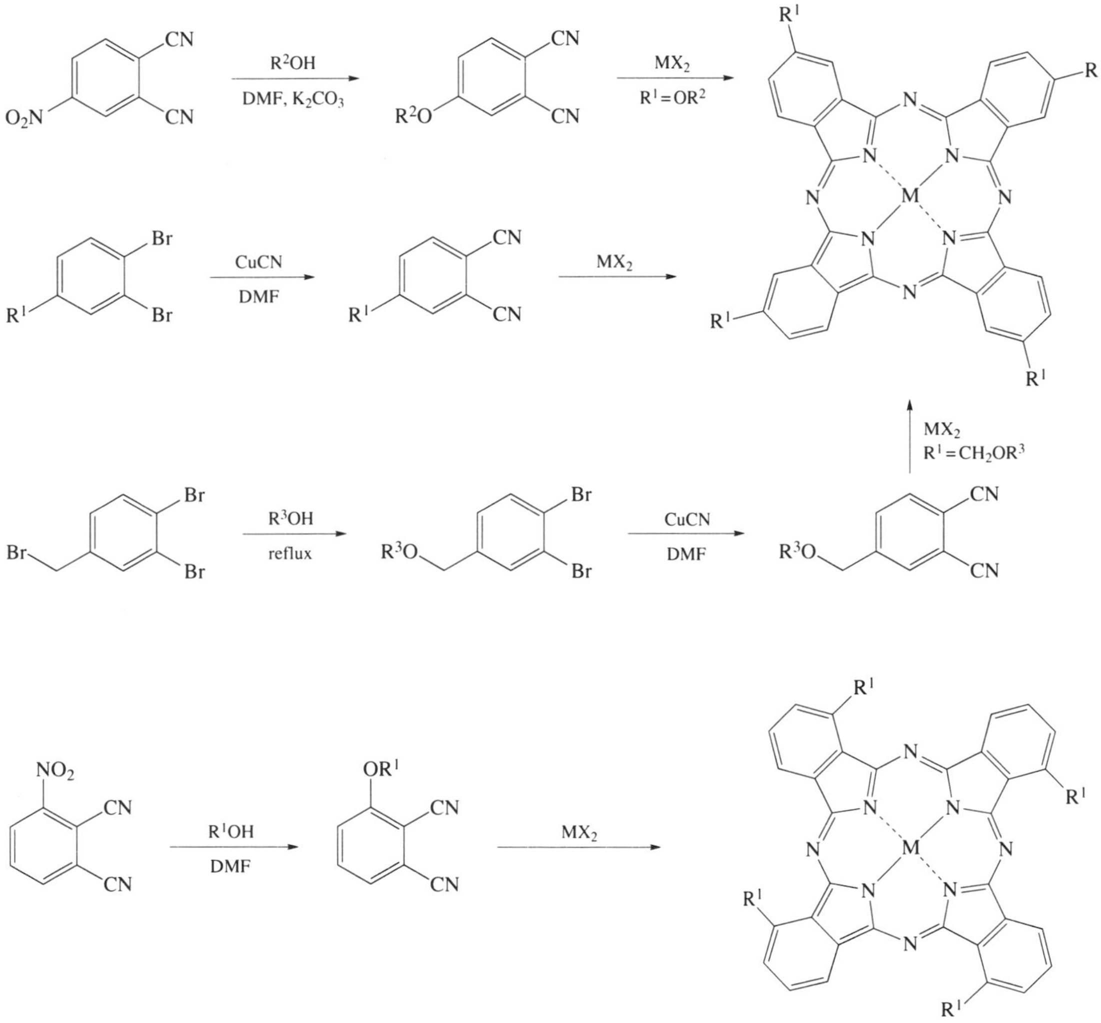

Scheme 3

2,3,9,10,16,17,23,24-Octaalkyl-[45] and 1,4,8,11,15,18,22,25-octaalkyl-substituted phthalocyanines are synthesized as shown in Schemes 4 and 5, respectively, starting from 4,5- or 3,6-substituted phthalonitriles. The synthetic pathway to these dinitriles is also shown in Schemes 4 and 5.

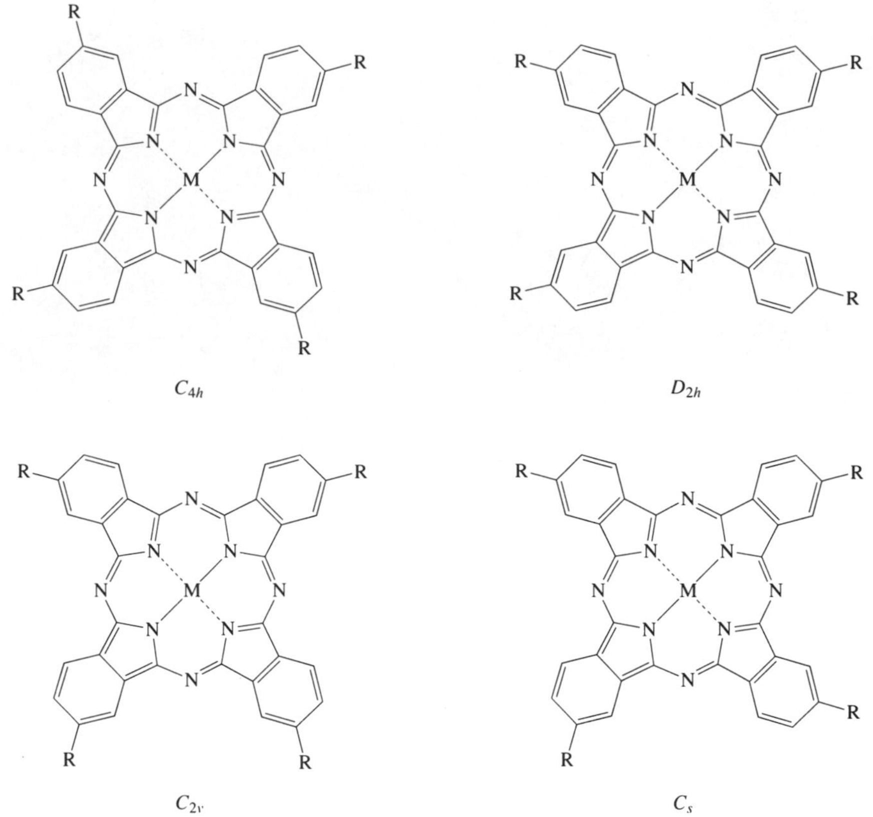

Figure 4 Constitutional isomers of 2,3-tetrasubstituted phthalocyanines.

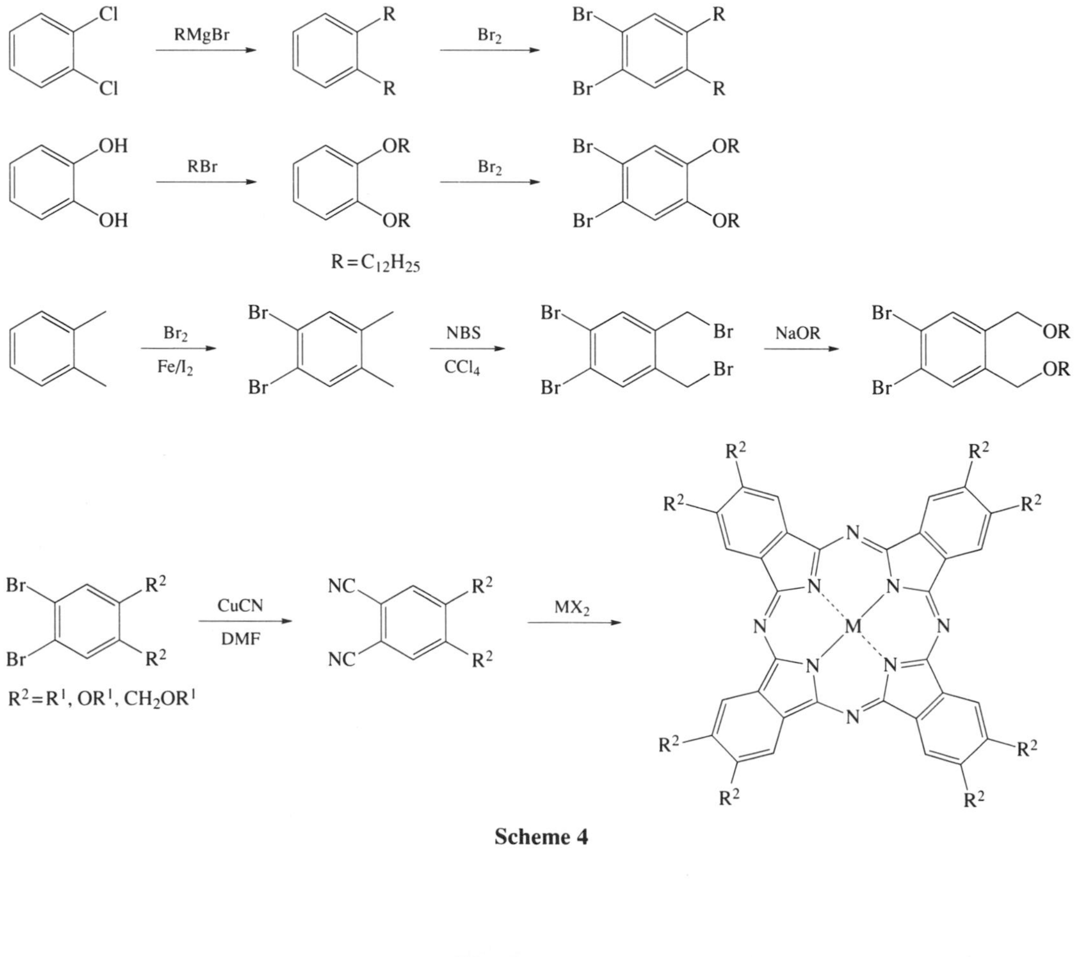

Scheme 4

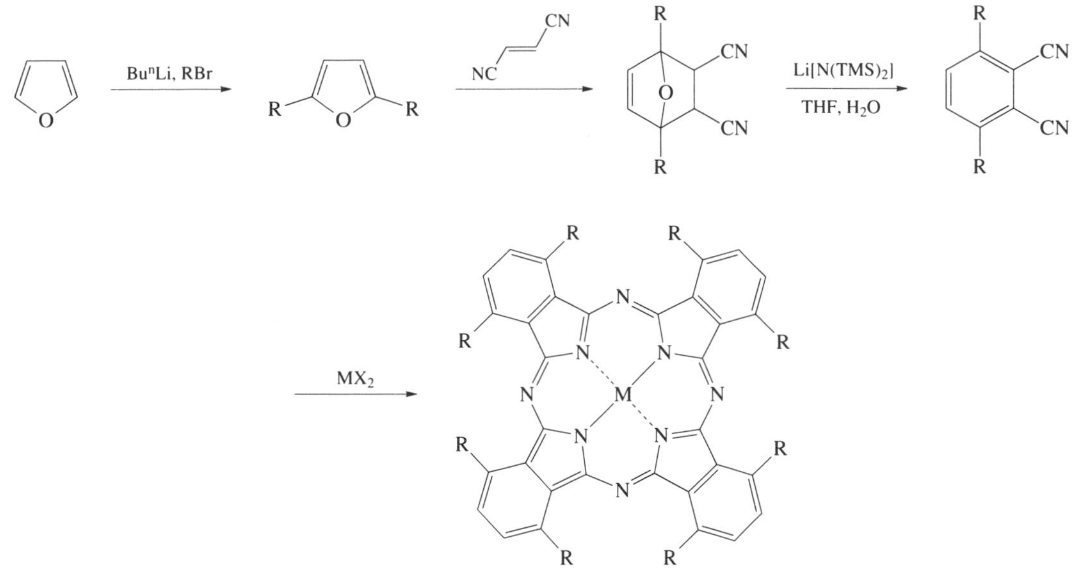

Scheme 5

Tetrasubstituted phthalocyanines usually exhibit higher solubilities in comparison with the octasubstituted phthalocyanines. Most of the substituted phthalocyanines exhibit insulating behaviour, but they are important precursors for the synthesis of bridged phthalocyaninato–transition metal complexes (see below). Phthalocyanines carrying long alkyl or alkoxy chains as peripheral substituents form liquid discotic or columnar mesophases.[46–8] Therefore these complexes are of interest as liquid crystals in materials science.

In some cases it might be useful that the phthalocyanine ring system carries nonidentical peripheral substituents. These unsymmetrically substituted phthalocyanines can be prepared by several synthetic routes. The most common pathway to obtain these compounds is to use two different substituted phthalonitriles A and B as starting material. In a statistical condensation of these two different phthalonitriles six kinds of phthalocyanines with different substitution patterns are expected: phthalocyanines containing four A or B units (AAAA or BBBB), three A units and one B unit (AAAB), three B units and one A unit (BBBA) and two A units and two B units as constitutional isomers (AABB and ABAB). The compound with the desired substitution pattern must be separated from the mixture by column chromatographic or preparative HPLC methods.[1,49,50]

A procedure which reduces the number of possible phthalocyanines is by using a phthalonitrile carrying bulky substituents such as phenyl groups in the 3,6-positions.[51] Due to steric hindrance, two units of such phthalonitriles cannot be ordered adjacently in a coplanar arrangement. By treating such a phthalonitrile (A) with phthalonitriles having no bulky substituents (B), only three different phthalocyanines should be obtained (see Scheme 6 which shows the synthesis of nonidentically substituted phthalocyanines—only the main products are shown): four B units (BBBB), three B units and one A unit (BBBA) and two B and A units with only the D_{2h} isomer (BABA). The statistical condensation of tetraphenylphthalonitrile with various alkyl- and alkoxy-substituted phthalonitriles followed by subsequent column chromatographic separation of the obtained phthalocyanines was successful.[50]

For phthalocyanines containing three identical and one different substituent other synthetic pathways may be used. One possibility to obtain these systems is by solid phase synthesis.[1,52] One phthalonitrile component is linked to a polymeric solid phase carrier and reacted with a large excess of a second substituted phthalonitrile derivative in solution. The nonidentically substituted phthalocyanine can be formed in excess on the carrier and subsequently liberated from the polymer.[1]

Another preparative method for these systems is to react a suitably substituted subphthalocyanine (SubPc)[53,54] which is composed of three isoindoline units with 1,3-diiminoisoindoline carrying the other substituent, thereby yielding the desired phthalocyanine by a ring expansion reaction.[55]

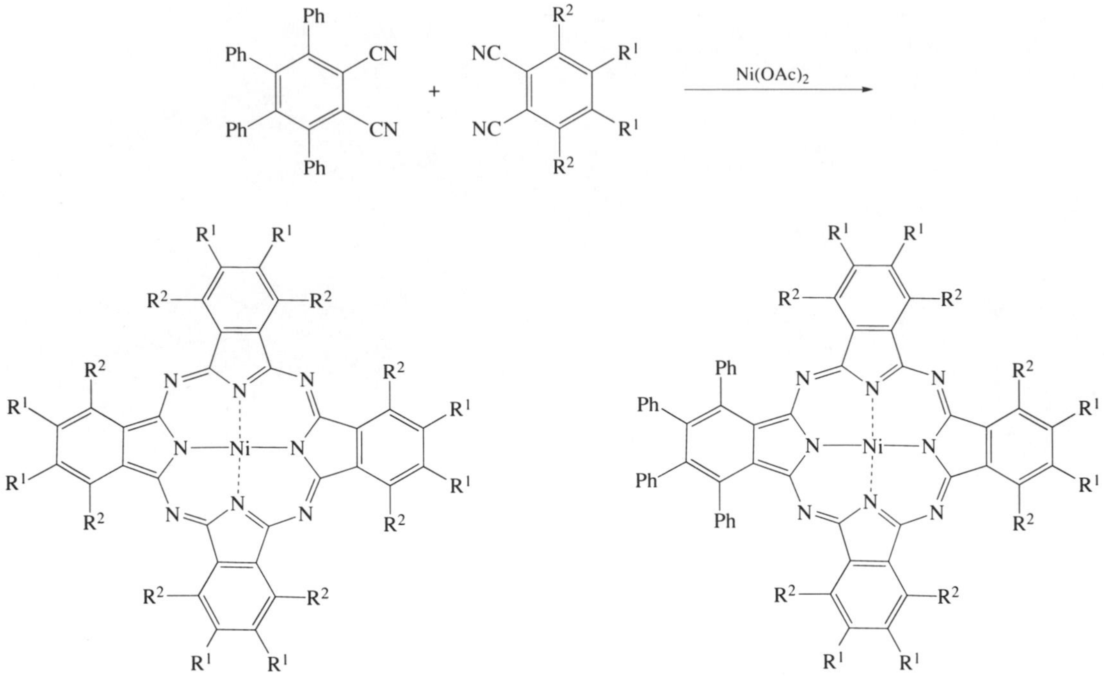

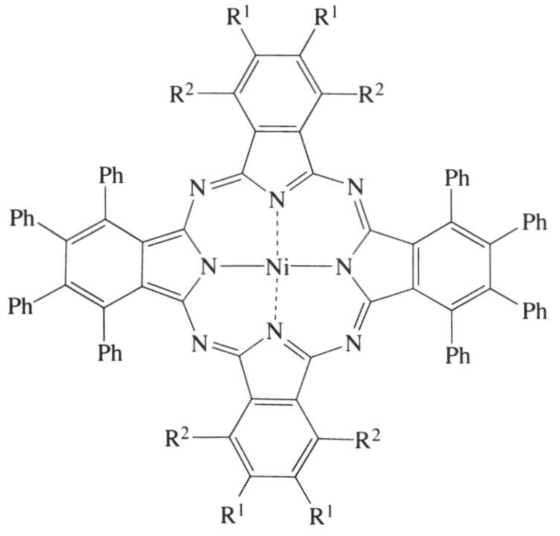

R^1	R^2
OC_5H_{11}	H
C_7H_{15}	H
H	OC_6H_{13}
H	C_7H_{15}

Scheme 6

8.3 STACKED PHTHALOCYANINES IN THE SOLID STATE AND THE 'SHISH-KEBAB' ARRANGEMENT

8.3.1 General Remarks About Organic Semiconductors

To achieve electrical conductivity in organic or organometallic materials, two major features are required.[11]

First, the molecules must be grouped in a close spatial arrangement and in similar crystallographic and electronic environments to achieve an extended pathway for electronic charge movement. That means that a band structure with suitable bandwidth (> 0.5 eV) must be built up. One possibility to realize such a situation is to order planar macrocyclic molecules in a stacked arrangement, leading to an extended π–π overlap between the cofacially arranged planar molecules, thereby building up a band structure from the HOMOs to form the valence band (VB) and from the LUMOs of the monomolecular components to form the conduction band (CB).

The second requisite is that free charge carriers must be generated, which means that the molecules must formally have partially occupied valence shells. This situation is achieved by partial oxidation or, less frequently, reduction. In accordance with inorganic semiconductors these are called doping processes.

The charge transport, finally, can be visualized as 'hopping' of charge carriers (electrons and holes) from one macrocycle to its direct neighbour (see Figure 5) in the direction perpendicular to the plane of the macrocycles (the stacking axis). This mechanism is therefore called the hopping mechanism.

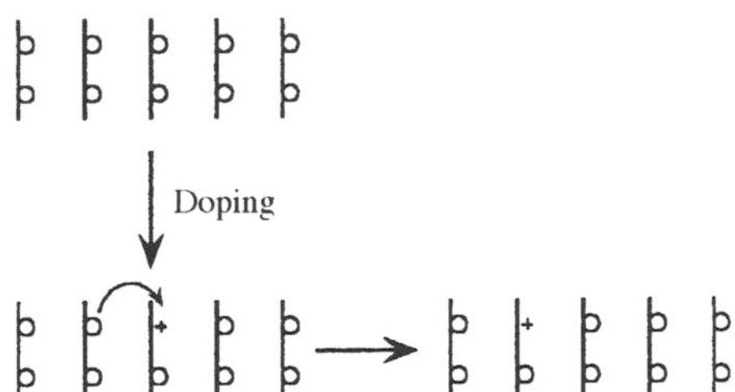

Figure 5 Schematic drawing of the charge transport mechanism in partially oxidized stacked macrocyclic complexes (hopping mechanism).

8.3.2 Stacking of Phthalocyanines in the Solid State

Phthalocyanines and related complexes, however, normally do not crystallize in the stacked arrangement described above. In general these complexes crystallize in an inclined stacked arrangement, the α- or the thermodynamically more stable β-modification, as shown in Figures 6(a) and (b). This inclined stacking order does not lead to a sufficent orbital overlap between cofacially arranged molecules. With phthalocyaninatolead(II) (PcPb), only one exception for a monoclinic modification is known.[56]

In order to achieve the desirable perpendicular stacking, several attempts have been successful. The most simple is the cocrystallization of a macrocyclic donor, for example, phthalocyanine with iodine, leading to stacks of cofacially arranged macrocycles, surrounded by parallel channels containing linear chains of the disordered counterions, for example, I_3^- or I_5^- (see Figure 6(c)). A particularly impressive example of this approach is the PcH_2I system. This system consists of stacks of $PcH_2^{0.33+}$ cations with an inter-ring distance of 325.1 pm and parallel chains of disordered off-axis I_3^- counterions.[57–9] This material illustrates conclusively that a central metal ion is not required for electrical conductivity, contrary to other systems, like the Krogmann salts, $K_2[Pt(CN)_4Cl_y]$,[60] where the band structure is composed from the platinum d_{z^2} orbitals. A single crystal sample of the PcH_2I system shows a room temperature conductivity along the stacking axis (the crystallographic *c*-axis) of approximately 700 S cm^{-1}.

Cocrystallization of metallophthalocyanines (PcM) with iodine leads to similar colinearly stacked structures.[61] Occupation of the central position of the macrocycle by a metal atom, however, can lead to a substantially changed band structure. Due to the existence of metal *d*-orbitals in a similar energy level to the macrocycle π-orbitals, mixing of these orbitals is possible. This effect has only a minor influence on phthalocyaninatometal complexes containing late transition metals, for example, nickel or copper. Due to orbital contraction, the *d*-orbitals of these metals are lower in energy, which prevents a suitable mixing with the macrocycle orbitals. Therefore, the conduction in the doped metallophthalocyanines (PcNiI, PcCuI) must be mainly centred to the macrocycle π-orbitals, similar to the metal-free PcH_2I. This situation is altered by changing the central metal ion. By moving to the left in the first transition period of the periodic table, the energy of the metal *d*-orbitals increases due to the decreasing orbital contraction. Therefore metal–macrocycle orbital mixing becomes possible or, in the case of earlier transition metals, for example, cobalt, the valence band is formed by the metal d_{z^2} orbitals.[62]

This trend suggests that metal centred conduction is most probable for doped phthalocyaninatometal complexes of the early transition elements. Electrochemical studies combined with ESR spectra[63] of PcMI indicate a metal centred oxidation for M = Cr, Mn, Fe and Co, mixed macrocycle–metal oxidation for M = Ni, and macrocycle oxidation for M = Cu, Zn or H_2. Also theoretical calculations of the band structures confirm these effects.[64]

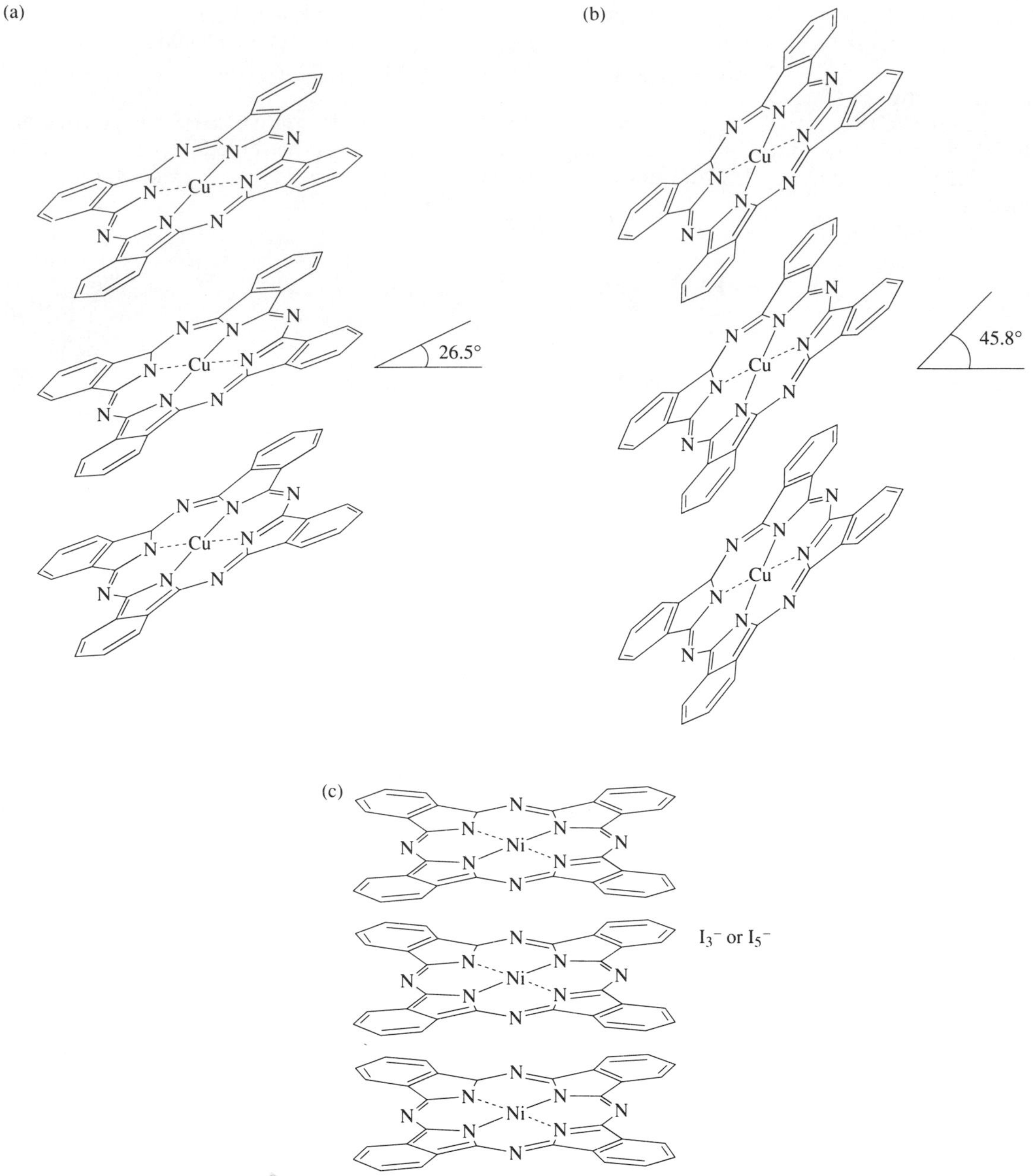

Figure 6 Different stacking arrangements for phthalocyanines in the solid state: (a) α-PcCu, (b) β-PcCu and (c) PcNiI.

8.3.3 Enforced Stacking

Another attractive approach to enforce perpendicular stacking is by covalently linking the macrocyclic subunits in a cofacial orientation. The bridged stacking is achieved by using an element of the fourth (silicon, germanium or tin)[65] or third (aluminum, gallium)[66] main group, as the central atom, and oxygen or fluorine as bridging atoms, respectively. The structure of this class of compound is given in Figure 7. These polymers are extremely robust; $[PcSiO]_n$, for example, is prepared at 400 °C and can be recovered unchanged after dissolution in strong acids.

The interplanar distances increase with increasing ionic radius, as expected. It was shown by theoretical calculations and ESR studies that the valence band is composed of the HOMO of each macrocycle, which is made up of only the carbon *p*-orbitals. The resulting band possesses a_{1u} symmetry and shows, by symmetry, no mixing with the orbitals of the central atom. To generate

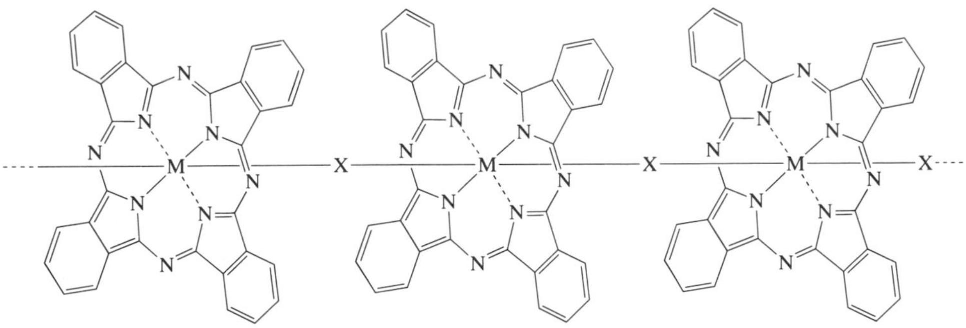

M = Si, Ge, Sn; X = O, S
M = Al, Ga, Cr; X = F

Figure 7 Bridged main group phthalocyaninatometal complexes.

charge carriers, these materials must also be doped by chemical or electrochemical methods. Consequently, the width of the valence band is determined by the inter-ring separation and the inter-ring rotation (the staggering angle) of the neighbouring macrocycles. With halogen doping the valence band becomes partially empty, thereby leading to a large increase in the electrical conductivity.

The σ_{M-X} bond orbitals lead to a band which lies far beyond the HOMO level of the macrocycle and which is largely responsible for holding the chain together.[67]

8.3.4 'Shish-kebab' Polymers

A different approach for producing bridged stacked polymers is by using bidentate linear π-electron containing molecules, for example, pyrazine, *p*-diisocyanobenzene, tetrazine, and also electrically charged groups such as CN^-, SCN^- or others (see below) as bridging ligands and a bi- or trivalent transition metal as the central atom.[68] The resulting bridged $[MacM(L)]_n$ compounds are called 'shish-kebab' polymers.

This class of compounds possesses a completely different band structure compared to the phthalocyaninatometalloxanes. The band structure of these $[MacM(L)]_n$ compounds is altered by the *d*-orbitals of the central metal. Theoretical calculations of the band structure of $[PcFe(pyz)]_n$ (pyz = pyrazine), which was calculated by means of the tight binding (LCAO) method, using a simplified structure of the macrocycle, lead to the following result.[64]

The highest occupied d_{xy} band is very narrow, and is largely composed of the iron d_{xy} orbitals. The band composed of the iron d_{xz} orbitals is also narrow because there are no appropriate orbitals of the bridging ligand for interaction. The broad d_{yz} band is the result of a strong mixing between the bridging ligand π- and the iron d_{yz} orbitals according to their coordination axes.[69] The lowest unoccupied (conduction) band is composed of the bridging ligand LUMO π*-orbitals. A bandgap of $\Delta E_{gap} < 1$ eV results, and semiconducting behaviour should therefore be possible. Due to a charge transfer from the metal d_{yz} orbital to the π*-orbital of the bridging ligand, the d_{yz} band becomes partially empty. The resulting electrical conductivities may be interpreted by two mechanisms.

In the first mechanism the partial electron loss in the d_{yz} band is immediately spread over the π*-orbital of the macrocycles. Therefore, the factor responsible for the electrical conductivity appears to be the overlap of the π*-orbitals with macrocycles in the neighbouring polymer chains by intercalation. In the second mechanism the partial electron loss in the d_{yz} band is only spread over the bridging ligand π-orbital. The electrical conductivity mainly arises from the overlap between the M d_{yz} and the bridging ligand π- or π*-orbitals in one polymer chain without the overlap of the π*-orbitals between the macrocycles in the neighbouring polymer chains.[70]

Consideration of the complete phthalocyanine system as a macrocycle in extended Hückel calculations, however, leads to a different result.[71] Figure 8 shows a schematic drawing of the band structures of $[PcFe(pyz)]_n$ and $[PcFe(tz)]_n$ (tz = tetrazine).

The highest occupied (valence) band is composed of the macrocycle π- and the metal *d*-orbitals. The composition of the lowest unoccupied (conduction) band is strongly dependent on the choice of the bridging ligand. For the pyrazine bridged system the conduction band is formed from the

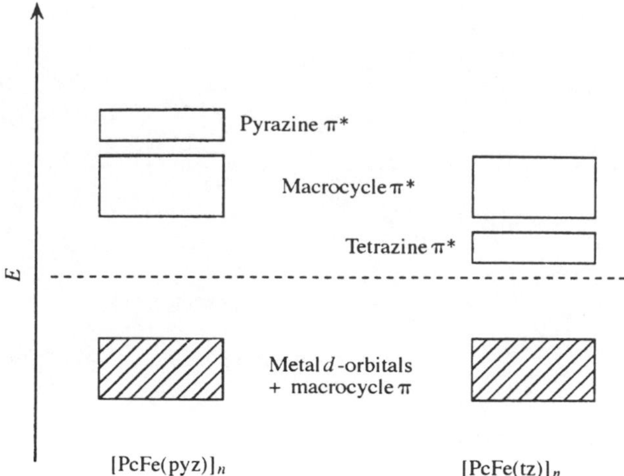

Figure 8 Schematic drawing of the band structures of $[PcFe(pyz)]_n$ and $[PcFe(tz)]_n$ in the region of the highest occupied band.

phthalocyanine π*-orbitals. Due to the lower lying LUMO in the tetrazine molecule, the conduction band in the tetrazine bridged oligomers is built up from the tetrazine π*-orbitals.

8.4 BRIDGED MAIN GROUP PHTHALOCYANINATOMETAL COMPLEXES

Some of the best investigated types of cofacially linked stacked macrocycles are the polyphthalocyaninatometalloxanes $[PcMO]_n$ (M = Si, Ge, Sn) shown in Figure 7.[65,72–5] The synthesis of $[PcMO]_n$ (M = Si, Ge, Sn) is carried out starting with $PcMCl_2$ (M = Si, Ge, Sn) which is prepared by standard procedures.[74,75] Hydrolysis of $PcMCl_2$ leads to $PcM(OH)_2$ (M = Si, Ge, Sn). The polycondensation is achieved by dehydration, either by heating $PcM(OH)_2$ *in vacuo* at 325–440 °C or by heating it in refluxing 1-chloronaphthalene or quinoline (see Scheme 7). A topotactic polymerization mechanism has been proposed.[11] Random reaction of all end groups with each other has also been discussed.[75]

Dehydration *in vacuo* provides the highest degree of polymerization. Both $[PcSiO]_n$ and $[PcGeO]_n$ have high thermal and chemical stabilities. $[PcSiO]_n$ can be recovered without any change after dissolving it in strong concentrated acids like H_2SO_4 or CF_3CO_2H. Estimation of the average molecular weights for $[PcMO]_n$ (M = Si, Ge, Sn) by IR end group analysis, tritium labelling and laser light scattering experiments yields a degree of polymerization of 70–140 subunits for $[PcSiO]_n$ and slightly lower values for $[PcGeO]_n$ and $[PcSnO]_n$. Longer polymerization reaction times and higher temperatures seem to increase the molecular weights.[76]

Based on a model structure for $[PcSiO]_n$ a one-dimensional stacking of the PcM^{2+} subunits linked by O^{2-} bridges is derived from powder diffraction data by computer simulation techniques. The M–O–M distances vary from 333 pm for M = Si to 353 pm for M = Ge, and to 382 pm for M = Sn. The best fit of the data indicates a 39° macrocycle staggering angle for $[PcSiO]_n$, and 0° for $[PcGeO]_n$ and $[PcSnO]_n$. These data are confirmed by ^{13}C CP-MAS NMR spectroscopy,[77] electron diffraction[78] and electron transmission spectroscopy.[79]

To achieve conductivity, charge carriers must be generated, either by oxidation (*p*-doping) or reduction (*n*-doping), as shown in Section 8.3. For the *n*-doped phthalocyanines only very little experimental material is available.[80] The most frequently used doping method is oxidation with iodine, which can be carried out heterogeneously with iodine vapour, by treatment with iodine solutions or by grinding of both components. Partial oxidation is also achieved with other electron acceptors, for example, chlorine, bromine, quinones and nitrosyl compounds[81] or by electrochemical methods.[75] Oxidation leads to well-defined, air stable, conducting polymers of relatively high thermal stability up to 120 °C. The best investigated iodine doped compounds lead to stoichiometries $[(PcMO)I_y]_n$ (M = Si, Ge) with a y_{max} of about 1.1, which represents a degree of maximum oxidation of 1/3. For peripherally substituted polymers, for example, $[\{(Bu^t)_4PcSiO\}I_y]_n$, a y_{max} of up to 2.0 is possible. As in the case of iodine doping of PcNi, iodine is thereby reduced to I_3^- or I_5^-, which was proved by resonance Raman spectroscopy and ^{129}I Mössbauer studies.

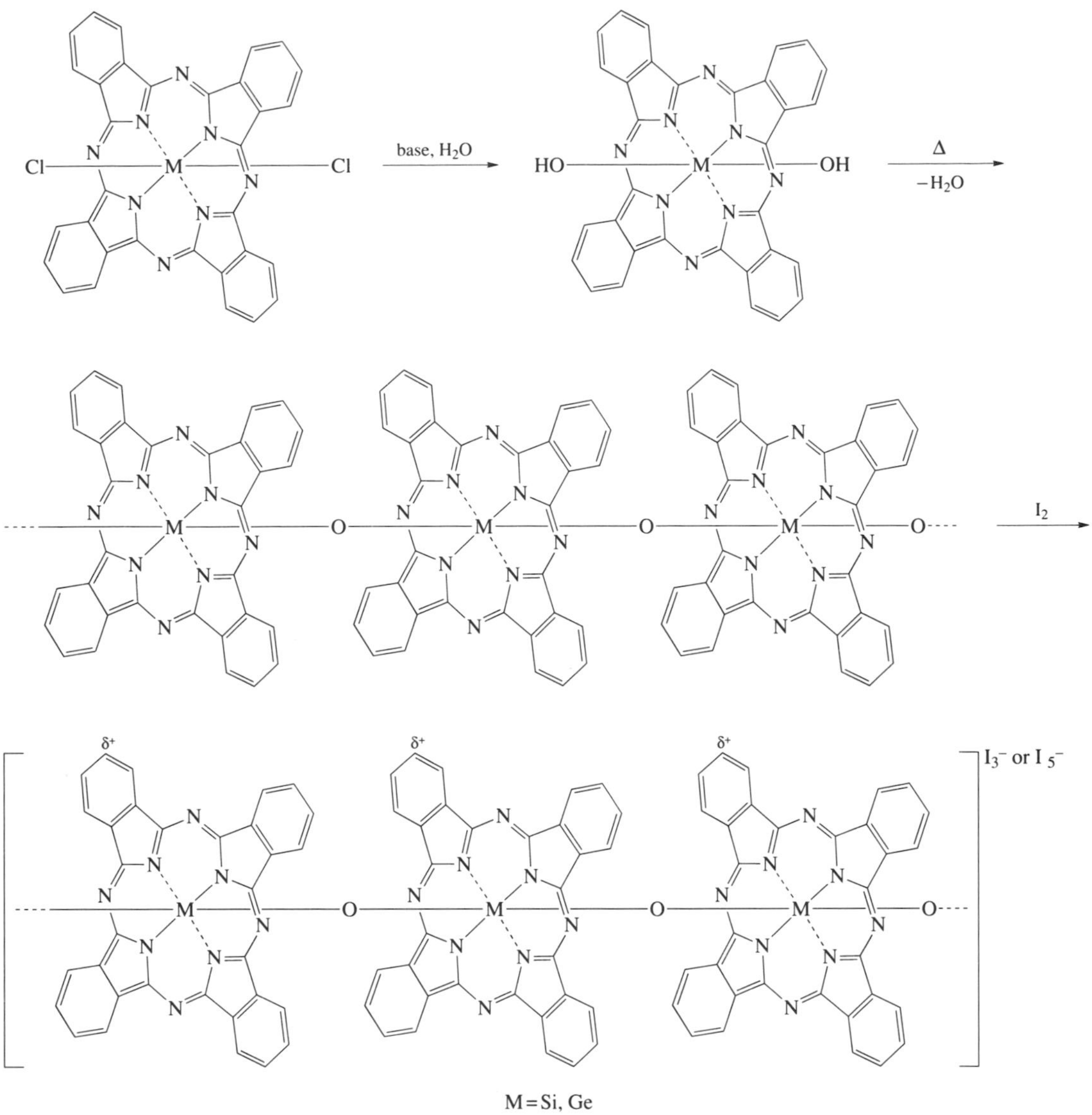

M = Si, Ge

Scheme 7

Chains of these counterions are disordered in channels parallel to the crystallographic *c*-axis. Increasing the inter-ring distance leads to the formation of I_5^- counterions.[72] In the case of $[PcSnO]_n$, doping leads to destruction of the polymeric structure.[82] It is also possible to use sulfur as a bridging ligand, for instance, in $[PcGeS]_n$, but the Ge–S bond is cleaved by doping.[83] However, the undoped polymer shows photoconductivity.[84]

Electrochemical oxidation using counterions such as BF_4^- or ClO_4^- has not led to higher conductivities thus far.[75] Analysis of powder data of $[(PcMO)I_y]_n$ (M = Si, Ge), which was compared to the model compound PcNiI, shows similarities to the model compound. In all cases a staggered arrangement of the macrocycles with a staggering angle of 39–40° and the above described parallel chains of disordered I_3^- counterions are evident. Doping also decreases the inter-ring distances in the range 3–5 pm.[72] Single crystal data are not available due to the lack of crystals of suitable size.

The room temperature conductivities of polycrystalline samples of $[(PcMO)I_y]_n$ (M = Si, Ge) for various stoichiometries are given in Table 1. The nature of dopant, for instance, iodine, bromine, quinones, and so on, has no significant effect on the conductivities.[72,81] In addition to the charge transport mechanism described in Section 8.3, other mechanisms, for example, percolation theory and fluctuation-induced carrier tunnelling through potential barriers separating metal-like regions, have also been discussed.[85]

Table 1 Room temperature conductivity data of doped and nondoped polycrystalline polyphthalocyaninatometalloxanes $[(PcMO)I]_n$ and $[(R_4PcMO)I]_n$.

Compound	*y*	*Conductivity* $(S\,cm^{-1})$
$[(PcSiO)I_y]_n$	0	5.5×10^{-6}
	1.1	6.7×10^{-1}
$[(PcGeO)I_y]_n$	0	2.2×10^{-10}
	1.1	1.1×10^{-1}
$[(PcSnO)I_y]_n$	0	1.2×10^{-9}
	1.1	2.2×10^{-6}
$[(Bu^t_4PcSiO)I_y]_n$	0	8.0×10^{-9}
	2.0	2.0×10^{-3}
$[(Bu^t_4PcGeO)I_y]_n$	0	6.0×10^{-11}
	1.9	1.0×10^{-3}
$[(Bu^t_4PcSnO)I_y]_n$	0	4.0×10^{-12}
	2.0	1.0×10^{-6}

The presence of bulky substituents in the periphery of the phthalocyanine ring system raises their solubility in common organic solvents. The synthesis of the peripherally alkylated μ-oxo-polymers $[R_4PcMo]_n$ (R = Bu^t, TMS; M = Si, Ge, Sn) was described in 1983,[86] the formation of alkoxymethylene- and alkoxy-substituted polymeric phthalocyaninatometalloxanes $[R_8PcSiO]_n$ ($R = CH_2OC_{12}H_{25}$, $OC_{12}H_{25}$) has also been reported.[87–9]

The synthesis of these complexes $[R_xPcSiO]_n$ ($x = 4, 8$) is carried out from the corresponding monomeric phthalocyaninatodihydroxysilanes $R_xPcSi(OH)_2$ with trifluoroacetic anhydride followed by thermal polymerization at 200 °C.[88] The dihydroxides $R_xPcSi(OH)_2$ can be obtained following routes similar to the unsubstituted derivatives.[5] The undoped materials show electrical conductivities which are similar or somewhat lower than those of the peripherally unsubstituted $[PcMO]_n$ (M = Si, Ge, Sn). The doped polymer $[(R_4PcSIO)I_y]_n$ (R = Bu^t, TMS) is thermally stable up to 140 °C. Above this temperature a loss of the doping agent occurs, which is finished at about 380 °C. The residue is pure $[R_4PcSiO]_n$.

Independent of the doping procedure all silicon and germanium $[R_4PcMO]_n$ samples exhibit the characteristic features reported for the conducting $[(PcSiO)I_y]_n$ materials. In the case of $[R_4PcSiO]_n$, the nature of the partially oxidized state was investigated by ESR spectroscopy, which supported a macrocycle centred oxidation process.[86]

Other examples of stacked phthalocyanines are the bridged fluorophthalocyaninatometal complexes $[PcMF]_n$ (M = Al, Ga, Cr) (see Figure 7).[66,90–2] In their preparation, PcAlCl or PcGaCl, for example, is converted into PcAl(OH) or PcGa(OH), respectively. These PcM(OH) (M = Al, Ga) complexes react with concentrated hydrofluoric acid to form polymeric $[PcMF]_n$ (M = Al, Ga), which can be purified by sublimation *in vacuo*.[90,91,93] As shown by single crystal x-ray analysis, $[PcGaF]_n$ crystallizes in perpendicular stacks of nearly eclipsed macrocycles connected by linear Ga–F–Ga bridges, with an inter-ring distance of 387 pm.[90,94] All known polymeric fluorophthalocyaninatometal compounds $[PcMF]_n$ (M = Al, Ga, Cr) can be doped with iodine to yield the partially oxidized $[(PcMF)I_y]_n$, where y is between 0.012 and 3.3, depending on the central metal. These doped systems contain I_3^- or I_5^- as counterions.[91]

The conductivities of the $[(PcMF)I_y]_n$ compounds (M = Al, Ga, Cr) increase with increasing iodine content. The highest conductivity is observed for $[(PcAlF)I_{3.3}]_n$ with a room temperature conductivity of $5\,S\,cm^{-1}$ and an activation energy of 0.017 eV. This complex is prepared from sublimed $[PcAlF]_n$. Due to the increased inter-ring distances when using gallium instead of aluminum, the π-orbital overlap between the cofacially arranged macrocycles decreases and therefore the conductivity becomes lower.[90]

In general, the fluoro-bridged $[(PcMF)I_y]_n$ complexes (M = Al, Ga, Cr) show lower thermal stability with regard to loss of iodine, compared to $[(PcSiO)I_y]_n$. $[PcMF]_n$ (M = Al, Ga,) can also be oxidized with nitrosyl salts, for instance, $NO^+BF_4^-$, to give $[(PcMF)(BF_4)_y]_n$. The conductivities are approximately $0.3\,S\,cm^{-1}$.[92]

Besides the phthalocyaninatometal complexes $[PcMO]_n$, $[PcMS]_n$ and $[PcMF]_n$, complexes with covalently linked alkynyl ligands have also been prepared, for example, μ-ethynylphthalocyaninatosilicon $[PcSi(C{\equiv}C)]_n$, which is obtained by treating $PcSiCl_2$ with bis(bromomagnesium acetylide) as the Grignard reagent. $[PcSi(C{\equiv}C)]_n$ is an insulator with a powder conductivity of less than $10^{-12}\,S\,cm^{-1}$. By doping under the conditions described above, the polymer decomposes.[95]

8.5 BRIDGED TRANSITION METAL COMPLEXES WITH MACROCYCLIC LIGANDS

As already mentioned, perpendicular stacking can also be enforced by bis(axially) connecting the central transition metal atoms of the macrocycles with bidentate π-electron containing bridging ligands (L), which allow electron migration along the central axis.[5,68] This bridging ligand must be linear, or at least form a sufficiently small M–L–M angle, thereby retaining the quasi one-dimensional arrangement of the bridged system. A schematic drawing of this class of complex is shown in Figure 9.

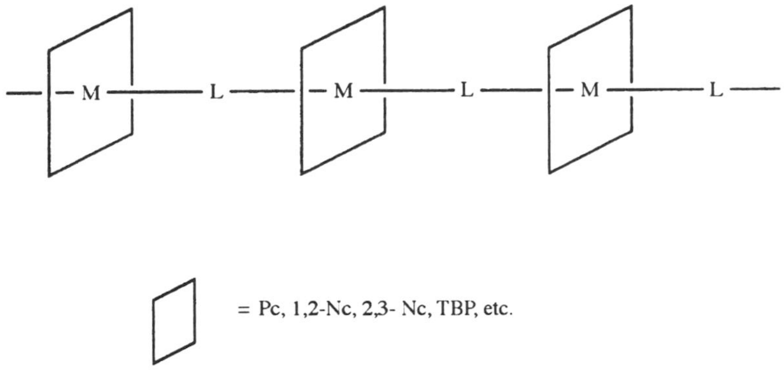

— M —— L —— M —— L —— M —— L ——

= Pc, 1,2-Nc, 2,3- Nc, TBP, etc.

M = transition metal (+2, Fe, Ru, Os; +3, Co, Rh)

L = pyz, tz, bipy, dib, Me_2dib, CN, SCN

Figure 9 Macrocyclic metal complexes bridged via ligand L ('shish-kebab' polymers).

The bridging ligand may be a bifunctional organic donor molecule such as pyz, bipy, *p*-diisocyanobenzene (dib) and substituted diisocyanobenzenes, tz, *p*-phenylenediamine (ppd) and others (see Figure 10). In the case of trivalent central metal atoms, for instance, Fe^{3+} and Co^{3+}, charged bridging ligands such as cyanide or thiocyanate can be used. In all cases the interplanar distances between the macrocycles are larger in comparison with the above described oxo- or fluoro-bridged compounds, thereby excluding π–π-overlap of the macrocycles. The central metal atom of the macrocycle must be a transition metal which prefers an octahedral coordination, for example, iron, ruthenium, cobalt, rhodium, manganese or chromium, and which is able to coordinate two bridging ligands in the axial positions. As macrocycles (Mac), phthalocyanines, substituted phthalocyanines, naphthalocyanines and others have been used.[10]

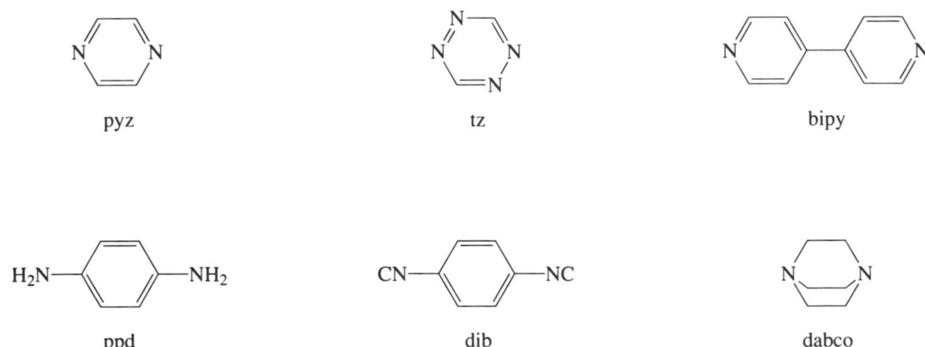

Figure 10 Selection of bidentate bridging ligands.

Due to their great structural variety these complexes provide an interesting possibility for systematically studying their physical properties and for developing new semiconducting materials.

For this purpose a large number of structurally similar, but chemically different, compounds have been synthesized.[10]

8.5.1 Pyrazine and Isonitrile Bridged Polymers

In order to prepare bridged macrocycle oligomers ('shish-kebab' polymers), a detailed knowledge of the coordination behaviour of the employed ligands is necessary. Therefore, mononuclear complexes of composition $PcML_2$ (M = Fe; L = pyz, 2-methylpyrazine (Mepyz), etc.) have been synthesized by stirring PcM complexes in the pure ligand at elevated temperatures.[10,96,97]

The prepared materials were examined by combined TGA–DTA measurements, ^1H NMR and ^{57}Fe Mössbauer spectroscopy. The results indicate that all adducts contain two axial ligands. The ^1H NMR spectroscopic data show that the phthalocyanine ring current causes shifts of the proton signals of the axial ligand to higher fields,[97] compared to the noncoordinated ligand. Iron-57 Mössbauer spectroscopy confirms the hexacoordination of the iron atom. Both isomer shifts and quadrupole splittings are lowered in comparison with PcFe, they confirm a low spin octahedral Fe^{2+} environment.[98–100]

Many of the 'shish-kebab' polymers $[MacM(L)]_n$ can be prepared in high yields and purity by treating the monomeric metallomacrocycle MacM with the pure ligand or with the ligand in an appropriate solvent such as acetone, chlorobenzene and so on. For example, the preparation of $[PcFe(pyz)]_n$ is carried out by reacting PcFe in a melt of pyrazine or in a benzene or chlorobenzene solution of PcFe and pyrazine. $[PcFe(pyz)]_n$ can also be obtained by cleaving pyrazine from the mononuclear $PcFe(pyz)_2$ in solvents like chloroform, benzene or chlorobenzene (see Scheme 8 for the synthesis of oligomeric pyrazine bridged phthalocyaninatoiron $[PcFe(pyz)]_n$).[97,101] $[PcFe(dib)]_n$ is formed quantitatively by the reaction of PcFe and 1,4-diisocyanobenzene in acetone solution under reflux. Pure $[PcRu(pyz)]_n$ is accessible by heating monomeric $PcRu(pyz)_2$ to 300 °C. Under these conditions one pyrazine molecule is eliminated and the coordinatively unsaturated intermediate polymerizes to form $[PcRu(pyz)]_n$.[27] Powder diffraction data indicate a high crystallinity for many of these coordination polymers.[97,101]

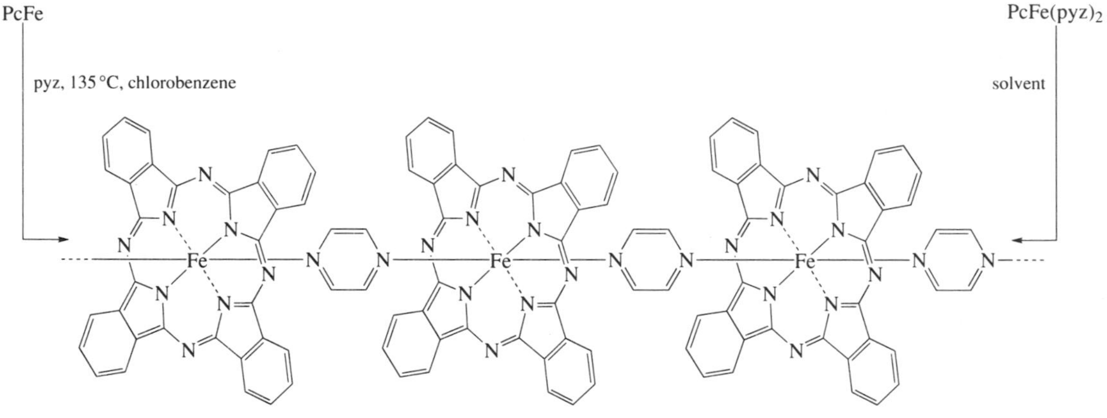

PcFe — pyz, 135 °C, chlorobenzene →

$PcFe(pyz)_2$ — solvent →

Scheme 8

The bridged structure shown in Figure 9 was confirmed by a variety of physical methods, such as thermogravimetry, detailed IR investigations,[102] Mössbauer spectroscopy,[103] ^1H and ^{13}C NMR spectroscopy[104] for soluble species (see below) and scanning tunnelling microscopy (STM).[105]

For the similar compound $[DMG\text{-}Co(pyz)]_n$ (DMG = dimethylglyoximato), a single crystal x-ray diffraction analysis shows that the pyrazine molecules within the chain are all arranged within a plane perpendicular to the plane of the DMG-Co units.[106]

As described in Section 8.3, the bandgap is determined by the difference in energies between the LUMO of the bridging ligand and the transition metal *d*-orbitals. Therefore, the conductivities can be 'adjusted' by varying the LUMO level. The higher conductivities of the tetrazine bridged complexes $[PcM(tz)]_n$ in comparison with the pyrazine bridged complexes $[PcM(pyz)]_n$ is explained by a lower lying LUMO in the tetrazine system (see below).[64,71]

The conductivities of many of the bridged phthalocyaninato transition metal complexes $[PcML]_n$, however, are relatively low, in the range 10^{-6}–10^{-7} S cm^{-1}. Many of these compounds can be doped with iodine.[99] Electrochemical doping is also possible using BF_4^-, PF_6^-, HSO_4^- or ClO_4^- as a counterion.[107] The iodine doped systems $[MacM(L)I_y]_n$ always possess a drastically increased conductivity relative to the undoped species. The room temperature conductivities are in the range 10^{-5}–10^{-1} S cm^{-1}. The doped polymers show thermal stabilities up to 120–130 °C. For example, doping $[PcFe(pyz)]_n$ with iodine yields stable compounds of stoichiometry $[PcFe(pyz)I_y]_n$, where $y = 0$–2.6.[99] Iodine doping is carried out by heterogeneous oxidation either in benzene or in chloroform.

The composition of the doped polymers was established by elemental analysis and TGA–DTA measurements. It has been proved by ^{57}Fe Mössbauer spectroscopy that doping does not destroy the bridged structure of the polymers; isomer shifts and quadrupole splitting of the chemically or electrochemically doped compounds show nearly identical values as found for the undoped species $[PcFe(pyz)]_n$. These results indicate that oxidation does not take place at metal centred orbitals. The conductivities of the polymers can be increased to 10^{-3}–10^{-1} S cm^{-1} by doping, depending on the iodine content (see Table 2).

Table 2 Room temperature powder conductivities of undoped and doped $[MacM(L)]_n$ compounds.

Compound	*Conductivity* (S cm^{-1})
Chemically doped compounds	
$[PcFe(pyz)]_n$	1×10^{-6}
$[PcFe(pyz)I_{2.5}]_n$	2×10^{-1}
$[PcFe(dib)]_n$	2×10^{-5}
$[PcFe(dib)I_{1.4}]_n$	7×10^{-3}
$[PcFe(dib)I_{3.0}]_n$	3×10^{-2}
$[PcFe(Me_4dib)]_n$	1×10^{-7}
$[PcFe(Me_4dib)I_{1.5}]_n$	1×10^{-3}
$[PcFe(Me_4dib)I_{3.0}]_n$	2×10^{-2}
$[PcRu(pyz)]_n$	1×10^{-7}
$[PcRu(pyz)I_{2.0}]_n$	2×10^{-2}
$[PcRu(dib)]_n$	2×10^{-6}
$[PcRu(dib)I_{1.0}]_n$	1×10^{-3}
$[PcRu(dib)I_{2.0}]_n$	7×10^{-3}
Electrochemically doped compounds	
$[PcFe(pyz)(BF_4)_{0.45}]_n$	5×10^{-2}
$[PcFe(pyz)(PF_6)_{0.5}]_n$	4×10^{-2}
$[PcFe(pyz)(HSO_4)_{0.4}]_n$	1×10^{-5}
$[PcFe(pyz)(ClO_4)_{0.3}]_n$	3×10^{-3}

Other types of bridging ligand which have been used to prepare bridging complexes $[MacM(L)]_n$ are dib, its substituted derivatives tetramethyl- and tetrachlorodiisocyanobenzene (Me_4dib, Cl_4dib) and 1,3-diisocyanobenzene (*m*-dib).[96] Diisocyanobenzene as the bridging ligand in $[MacM(L)]_n$ complexes leads to a larger inter-ring distance of about 1190 pm compared with the pyrazine bridged polymer, of which the distance is about 680 pm.

The conductivities of some bridged $[MacM(dib)]_n$ complexes are given in Table 2. Despite the fact that the inter-ring distance in the dib oligomers is larger, their conductivities are in the same range ($\sim 10^{-5}$ S cm^{-1}) as those of the pyz oligomers. Peripheral substitution of the phthalocyanine ring by electron donating substituents has only a small effect on the conductivity. Only substitution of the phthalocyanine with chlorine leads to a decreased conductivity of 3×10^{-11} S cm^{-1} for $[Cl_{16}PcFe(dib)]_n$.[108]

Theoretical calculations of the electronic properties of annulated phthalocyanines show that linear annulation of benzene rings in phthalocyanines produces a continuous destabilization of the HOMO level, thereby narrowing the HOMO–LUMO energy gap.[109] Calculations for one-dimensional stacks of linear annulated phthalocyanines show that these oligomers have lower oxidation potentials and also a lower gap between the VB and the CB than the nonannulated systems. Angular annulation, however, has no significant effect on the oxidation potentials.[110] The oxidation potentials of, for example, 1,2-NcFe are in the same range as those of PcFe. The lower oxidation potentials of the annulated systems are demonstrated by cyclic voltammetry investigations.[5] Due to the lower oxidation potential, these systems can be oxidized by relatively mild oxidants such as air oxygen, thereby leading to increased conductivity. These results are confirmed by studies of the redox potentials of PcFe, 1,2-NcFe, 2,3-NcFe, TBP-Fe and 2,3-tetranaphthoporphyrinatoiron (2,3-TNP-Fe) as well as the powder conductivities of their corresponding 1,4-diisocyanobenzene bridged polymers (Table 3). The oxidation potentials can be directly related to the powder conductivities of the bridged polymers.[2,5]

Table 3 Redox potentials of some iron macrocycles in pyridine–Bu_4ClO_4 vs. SCE and room temperature powder conductivities of their corresponding $[MacFe(dib)]_n$ complexes.

Assignment	Mac^{2-}/Mac^{1-} $E^1_{1/2}$	Fe^{II}/Fe^{III} $E^2_{1/2}$	Fe^{I}/Fe^{II} $E^3_{1/2}$	Mac^{3-}/Mac^{2-} $E^4_{1/2}$	Mac^{4-}/Mac^{3-} $E^5_{1/2}$	$\sigma_{RT}[MacFe(dib)]_n$ ($S\,cm^{-1}$)
PcFe	1.10	0.69	−1.085	−1.39	−1.93	2×10^{-5}
1,2-NcFe	1.01	0.68	−0.95	−1.21	−1.80	6×10^{-10}
2,3-NcFe	0.81	0.43	−1.09	−1.32	−1.86	2×10^{-3}
TBP-Fe	0.82	0.34	−0.90	−1.59	−1.87	2×10^{-6}
2,3-TNP-Fe	0.61	0.15	−0.09	−1.49	−1.87	5×10^{-3}

In order to improve the poor solubility of the $[MacM(L)]_n$ complexes, peripherally substituted oligomers $[R_4PcM(L)]_n$ and $[R_8PcM(L)]_n$ have been synthesized, which are soluble in common organic solvents. The synthesis is carried out by converting the substituted monomeric R_4PcM (R = Bu^t, Et, OC_5H_{11}—$OC_{12}H_{25}$) and R_8PcM (R = C_5H_{11}—$C_{12}H_{25}$, OC_5H_{11}—$OC_{12}H_{25}$) units with M = Fe, Ru[28,35,45,48,107,111] into the bridged oligomers (see Scheme 9 for synthesis of $[(RO)_8PcFe(dib)]_n$).[2] Due to their solubility in most common organic solvents, determination of the chain length is possible by the usual methods, ^1H NMR spectroscopy,[35] for instance. Depending on the method of preparation, oligomers which contain 20–50 MacM subunits can be obtained. The soluble substituted monomeric phthalocyaninatometal compounds are obtained as described above.

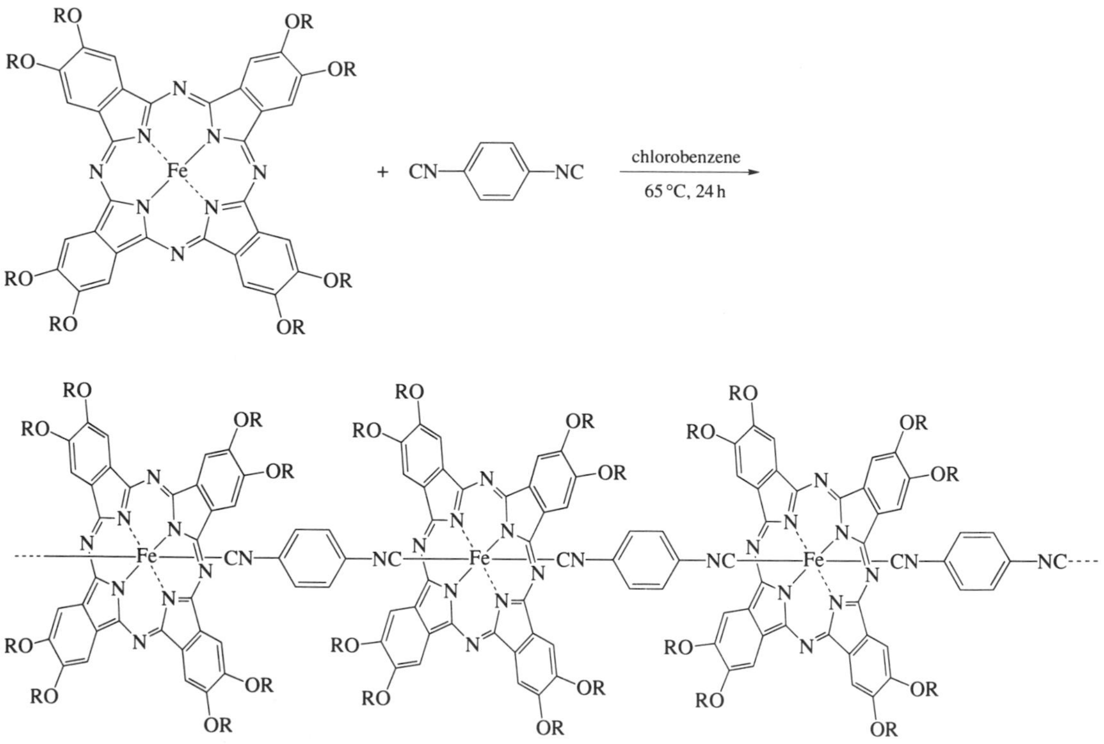

chlorobenzene, 65 °C, 24 h

Scheme 9

Doping of the soluble oligomers with iodine also leads to semiconducting systems. As for the peripherally unsubstituted polymers, spectroscopic investigations show that the bridged structure is not destroyed by the doping process.

The conductivities of the doped polymers are in the range 10^{-7}–10^{-4} S cm^{-1} (see Table 4). The conductivities of the peripherally substituted doped oligomers are somewhat lower than those of the unsubstituted systems. Furthermore, the size and location of the substituents on the macrocycles seem to play a role in determining the conductivity.

Table 4 Room temperature powder conductivities of undoped and doped soluble substituted $[R_4MacM(L)]_n$ and $[R_8MacM(L)]_n$ compounds.

Compound	*Conductivity* (S cm^{-1})
$[Bu^t_4PcRu(dib)]_n$	2×10^{-7}
$[Bu^t_4PcRu(dib)I_{1.2}]_n$	1×10^{-4}
$[Bu^t_4PcRu(Me_4dib)]_n$	1×10^{-11}
$[Bu^t_4PcRu(Me_4dib)I_{1.3}]_n$	1×10^{-4}
$[(EHO)_4PcRu(pyz)]_n$[a]	$< 10^{-12}$
$[(EHO)_4PcRu(pyz)I_{1.3}]_n$	6×10^{-5}
$[(n\text{-}C_5H_{11})_8PcFe(Me_4dib)]_n$	$< 10^{-12}$
$[(n\text{-}C_5H_{11})_8PcFe(Me_4dib)I_{1.8}]_n$	9×10^{-7}
$[(EHO)_4PcFe(Me_4dib)]_n$[a]	$< 10^{-12}$
$[(EHO)_4PcFe(Me_4dib)I_{2.5}]_n$	2×10^{-10}

[a] EHO = ethylhexyloxy.

8.5.2 Doping Processes

The doping process has been examined very carefully using different macrocycles and metals, as well as different bridging ligands. These investigations have shown that the oxidation, especially by using bridging ligands with a low lying oxidation potential, is mainly ligand centred. If, for example, in the polymeric $[PcFe(L)]_n$ system the bridging ligand L is 1,4-diisocyanoanthracene (dia), a ligand with a comparatively low oxidation potential, doping with iodine leads to an oxidation of the bridging ligand, thereby increasing the powder conductivity from 3×10^{-7} S cm^{-1} in $[PcFe(dia)]_n$ to 8×10^{-3} S cm^{-1} in $[PcFe(dia)I_{2.6}]_n$.[112] Another doping process involving a bridging ligand is shown for *p*-phenylenediamine (ppd) bridged oligomers of PcRu in Scheme 10. As in the case of polyaniline,[113] quinoid structures are formed in the bridging ligand and an increase of conductivity of five orders of magnitude in comparison with undoped $[PcRu(ppd)]_n$ is observed.[114]

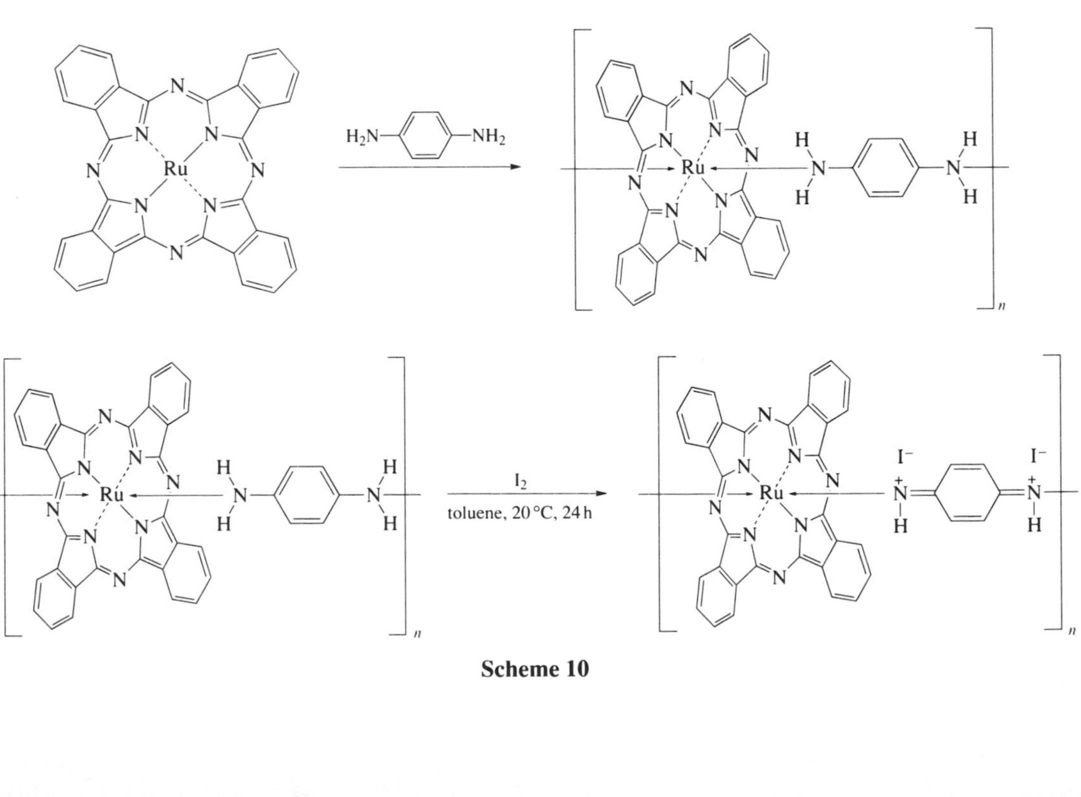

H_2N–C_6H_4–NH_2

I_2, toluene, 20 °C, 24 h

Scheme 10

8.5.3 Bridged Polymers Containing Tetrazine, its Derivatives and other Bridging Ligands as Intrinsic Semiconductors

Transition metallomacrocycle complexes, for example, PcFe, PcRu, PcOs or 2,3-NcFe, react easily with tetrazine or its derivatives to form the corresponding monomeric bis(axially) coordinated $MacM(L)_2$ (L = tz, Me_2tz) subunits, which can be converted into the bridged oligomers $[MacM(L)]_n$ (L = tz, Me_2tz). As mentioned above, the tz bridged polymers exhibit good semiconducting properties in contrast to other bridged compounds $[MacM(L)]_n$ (M = Fe, Ru, Os; L = pyz, dib), because of the lower lying LUMO of the tetrazine molecule.[64,71] The conductivities for the tetrazine bridged complexes are in the range of 5×10^{-2}–3×10^{-1} S cm^{-1}, without external oxidative doping. The powder conductivities of a selection of monomeric and bridged macrocycles in the nondoped state are given in Table 5.

Table 5 Room temperature powder conductivities of monomeric and bridged transition metallomacrocycle complexes.

Compound	*Conductivity* (S cm^{-1})
$PcFe(dabco)_2$	1×10^{-10}
$[PcFe(dabco)\cdot 1.4CHCl_3]_n$	1×10^{-9}
$PcRu(dabco)_2$	8×10^{-12}
$[PcRu(dabco)]_n$	1×10^{-9}
$PcFe(bipy)_2$	5×10^{-13}
$[PcFe(bipy)]_n$	2×10^{-8}
$[PcRu(bipy)]_n$	2×10^{-8}
$PcFe(pyz)_2$	3×10^{-12}
$[PcFe(pyz)]_n$	1×10^{-6}
$PcRu(pyz)_2$	2×10^{-11}
$[PcRu(pyz)]_n$	1×10^{-7}
$[2,3\text{-}NcFe(pyz)]_n$	5×10^{-5}
$Me_8PcFe(pyz)_2$	3×10^{-9}
$[Me_8PcFe(pyz)]_n$	9×10^{-6}
$[(CN)_4PcFe(pyz)]_n$	5×10^{-9}
$[Bu^t_4PcFe(pyz)]_n$	5×10^{-11}
$[Et_4PcRu(pyz)]_n$	5×10^{-10}
$[Bu^t_4PcRu(pyz)]_n$	7×10^{-8}
$PcFe(tz)_2$	$< 10^{-9}$
$[PcFe(tz)]_n$	2×10^{-2}
$[PcFe(Me_2(tz)]_n$	4×10^{-3}
$PcRu(tz)_2$	$< 10^{-11}$
$[PcRu(tz)]_n$	1×10^{-2}
$[PcRu\{p\text{-}(NH_2)_2tz\}]_n$	4×10^{-3}
$[PcRu(ppd)]_n$	5×10^{-9}
$[PcRu(Cl_2tz)]_n$	3×10^{-3}
$[PcRu(Me_2tz)]_n$	4×10^{-3}
$PcOS(tz)_2$	4×10^{-8}
$[PcOs(tz)]_n$	1×10^{-2}
$[2,3\text{-}NcFe(tz)]_n$	3×10^{-1}
$[Me_8PcFe(tz)]_n$	1×10^{-2}
$[(CN)_4PcFe(tz)]_n$	1×10^{-6}
$[Bu^t_4PcFe(tz)]_n$	9×10^{-9}
$[Bu^t_4PcRu(tz)]_n$	1×10^{-6}
$[Et_4PcFe(tz)]_n$	2×10^{-4}
$[Et_4PcRu(tz)]_n$	5×10^{-9}
$[PcRu(tri)]_n$[a]	2×10^{-4}
$[PcRu(p\text{-}(CN)_2C_6F_4)]_n$	1×10^{-3}
$[PcRu((CN)_2C_2]_n$	3×10^{-2}

[a]Here, tri = triazine.

1,4-Diazabicyclo[2.2.2]octane (dabco)[96,115,116] was also used as a bridging ligand. The resulting bridged polymers are obtained by reacting PcM, for example, PcFe, with free dabco in $CHCl_3$. In contrast to pyz or tz, the dabco system does not contain any π electrons. The room temperature conductivity therefore is very low (about 10^{-9} S cm^{-1}) in comparison with the pyz or tz bridged systems.

While pyz, tz and dabco cause nearly the same inter-ring distance of about 680 pm between two macrocycles in a polymer chain, other ligands such as bipy can be used to enlarge the inter-ring distances up to 1110–1320 pm. The conductivities of the $[PcM(bipy)]_n$ (M = Fe, Ru) are somewhat lower than those of the pyrazine bridged species ($\sim10^{-8}$ S cm^{-1}).[117]

While the monomeric complexes $PcM(L)_2$ (L = pyz, tz, dabco, bipy; M = Fe, Ru, Os) exhibit insulating behaviour, it is observed that the bridging ligand has a significant effect on the conductivity of the bridged polymers $[MacM(L)]_n$ (see Table 5). In addition to tz and its derivatives, some other interesting attempts to achieve intrinsic conductivity have been made in the bridged systems. The *p*-diaminotetrazine bridged polymer $[PcRu\{p\text{-}(NH_2)_2tz\}]_n$ exhibits a powder conductivity which is about six orders of magnitude higher than that of $[PcRu(ppd)]_n$, due to the fact that *p*-diaminotetrazine contains heteroatoms in the aromatic ring and therefore possesses a lower lying LUMO in comparison with *p*-phenylenediamine.[2]

Also peripherally substituted, soluble tz bridged phthalocyaninato transition metal complexes have been synthesized using a variety of tetraalkyl and tetraalkoxy substituted PcFe and PcRu complexes. These tz bridged complexes again show higher electrical conductivities than the pyrazine bridged analogues, but to a lower extent than the peripherally unsubstituted ones. Triazine is also being used as a bridging ligand in bridged phthalocyaninatoruthenium complexes. These bridged oligomers also show good semiconducting properties.[118]

Besides the above described nitrogen containing heterocycles, other bridging ligands can be used to synthesize intrinsically semiconducting bridged phthalocyaninatometal polymers, such as tetrafluoroterephthalic acid dinitrile (1,4-$(CN)_2C_6F_4$) in $[PcRu\{1,4\text{-}(CN)_2C_6F_4\}]_n$ which exhibits a powder conductivity of 10^{-3} S cm^{-1} without external oxidative doping. Also fumaronitrile (NC—CH=CH—CN), dicyanoacetylene (NC—C≡C—CN) and even dicyan (NC—CN) have been used as bridging ligands in the corresponding phthalocyaninatoruthenium polymers. The powder conductivities of all those compounds are approximately 10^{-3} S cm^{-1} (see Table 5).[2]

The low bandgaps of all tetrazine bridged complexes with group 8 transition metals, $[MacM(L)]_n$ (M = Fe, Ru, Os; Mac = Pc, 2,3-Nc; L = tz, Me_2tz, *p*-$(NH_2)_2tz$, etc.), lead to special physical properties, which are not present in the corresponding systems $[MacM(L)]_n$ (L = pyz, dib, ppd, etc.). All bridged $[MacM(L)]_n$ containing tz or its derivatives as bridging ligands show broad absorption bands in the UV–visible–NIR spectra with different maxima: $[PcFe(tz)]_n$ at 1650 nm (0.75 eV) and $[PcRu(tz)]_n$ at $\sim$1300 nm (0.95 eV). This absorption band is assigned to a charge-transfer transition from the metal d_{yz} to the tetrazine π^*-orbital.[70] The corresponding pyrazine bridged systems $[PcM(pyz)]_n$ exhibit typical phthalocyanine UV–visible spectra with Soret and Q bands at 245 nm and 700 nm, respectively.

The $[PcRu\{p\text{-}(NH_2)_2tz\}]_n$ system shows a broad absorption band in the near IR region between 1000 nm and 2000 nm with two maxima at 1525 nm (0.81 eV) and 1340 nm (0.93 eV). The corresponding *p*-phenylenediamine bridged system $[PcRu(ppd)]_n$ exhibits a simple phthalocyanine UV–visible spectrum. The absorption band in the near IR correlates well with the electrochemically estimated bandgap between the metallomacrocycle HOMO and the bridging ligand LUMO in all the tetrazine bridged systems mentioned above.

Iron-57 Mössbauer data of tetrazine coordinated phthalocyaninatoiron complexes also indicate a low bandgap in the bridged polymers. These data are compared with the data for the monomeric complexes and the corresponding pyrazine complexes. For the tetrazine bridged polymers $[PcFe(tz)]_n$, the isomer shift δ is similar to the monomeric $PcFe(tz)_2$, whereas an increase of the quadrupole splitting Δ_{EQ} even compared to $[PcFe(pyz)]_n$ is observed. This effect can be explained by the thermally activated electron transition from the highest occupied band, which leads to a diminished occupation of the metal centred *d*-orbitals and therefore to the increase in Δ_{EQ}.[2]

The increased quadrupole splitting in the Mössbauer spectrum, the charge-transfer transition in the near IR and also the electrochemical investigations correlate well to a low bandgap (ΔE_{gap} < 1 eV) quasi one-dimensional chain structure in the tz bridged complexes.

8.6 BRIDGED MACROCYCLIC COMPLEXES WITH TRIVALENT TRANSITION METALS

As described above, bridged systems may also contain the central metal atom in oxidation state +3. Appropriate bridging ligands are negatively charged molecules, such as cyanide, thiocyanate or azide. A direct route to coordination oligomers $[PcM(L)]_n$ is to displace the axial anion X^- in a coordinatively unsaturated compound PcMX (X = Cl, O_2CMe, O_2CCCl_3) by the bidentate ligands CN^-, SCN^- or N_3^-. This route has been utilized for the synthesis of cyano complexes $[PcM(CN)]_n$ (M = Fe, Mn),[119,120] thiocyanato complexes $[PcM(SCN)]_n$ (M = Fe, Mn, Co) and azido complexes $[PcM(N_3)]_n$ (M = Cr, Mn).[121]

The starting materials PcFeCl and PcMnCl were prepared from the phthalocyaninatometal complexes PcFe and PcMn,[122,123] respectively, by treatment with chlorinating agents such as hydrochloric acid or thionyl chloride and simultaneous oxidation with air oxygen or nitrobenzene. $PcM(O_2CMe)$ (M = Cr, Mn) is obtained by air oxidation of PcM (M = Cr, Mn) in acetic acid.[124] The chlorides and acetates PcMX (X = Cl, O_2CMe, O_2CCCl_3) are converted into bridged oligomers $[PcM(CN)]_n$ in aqueous or ethanolic alkali metal–cyanide, –thiocyanate or –azide solutions.[119–21]

Suitable halo precursors for the cobalt and chromium compounds $[PcCo(CN)]_n$ and $[PcCr(CN)]_n$ are not known.[125] A convenient method for the introduction of a $Pc^{2-}M^{3+}$ unit into the reaction path is by using the known dichloro derivatives $PcMCl_2$ (M = Co, Cr).[123] Again the precursors are converted into the oligomers in an aqueous alkali metal–cyanide, –thiocyanate or –azide solution. The $Pc^{-}M^{3+}$ subunit is simultaneously reduced and oligomerized under the reaction conditions.

A relatively easy pathway to synthesize $[PcM(CN)]_n$ complexes is by suspending, for example, PcCo in ethanol, which is oxidized by a stream of air to form Co^{3+} and subsequent reaction with an excess of alkali metal cyanide to form soluble $M'[PcCo(CN)_2]$ (M′ = Na, K; see Scheme 11).[125]

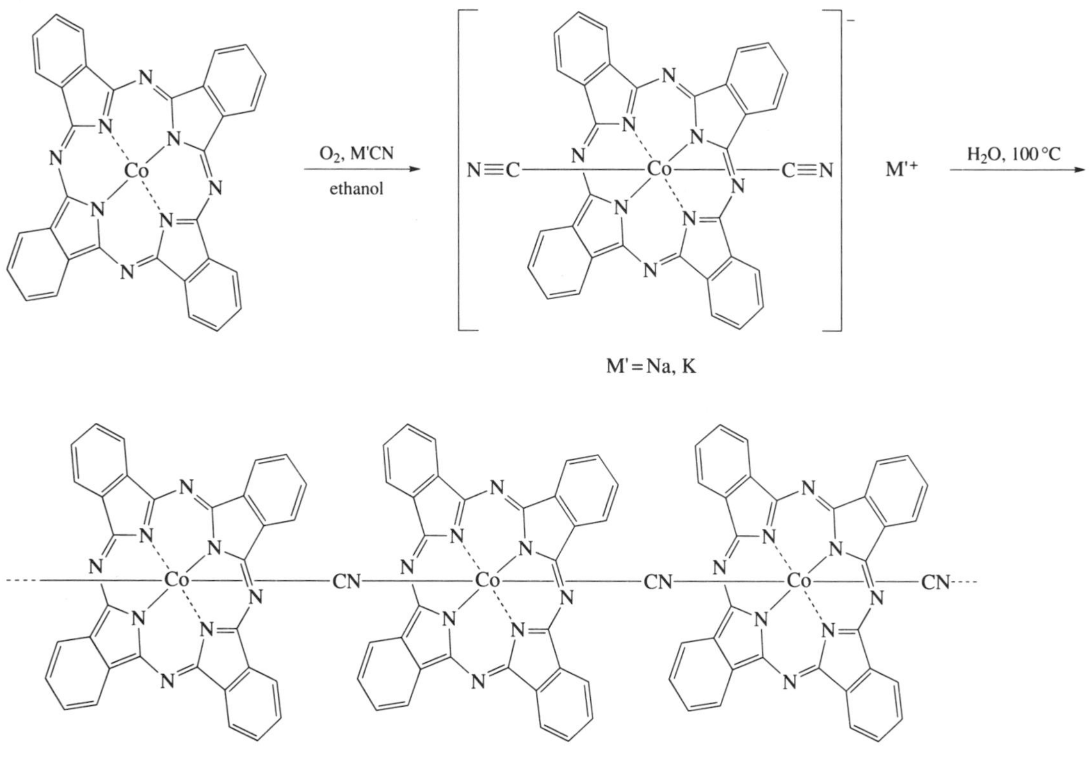

Scheme 11

Formation of the oligomer $[PcCo(CN)]_n$ is achieved by cleaving CN^- from the monomer by different methods, for example, treatment with water at 100 °C. $[PcCo(CN)]_n$ is directly obtained in almost quantitative yield.

The crystal structure of $K[PcCo(CN)_2]_2$ obtained by oxidative electrocrystallization of $K[PcCo(CN)_2]$ is known and confirms the bis(axial) coordination of the cyano groups.[126] The monomeric complexes $M'[PcM(CN)_2]$ (M′ = Na, K; M = Cr, Mn, Fe, Co, Rh) were also characterized by IR, FIR, UV–visible and in some cases ^1H NMR spectroscopy. The IR data exhibit C≡N valence frequencies which are in the anticipated region (around $2130\,cm^{-1}$) for terminal Co^{3+}, Cr^{3+} and Rh^{3+} cyano groups.[127]

The C≡N valence frequencies of the bridged oligomers shift to higher energy compared with the mononuclear complexes. This increase is a good indication of the presence of a cyano bridge in $[PcCo(CN)]_n$. Powder x-ray diffraction analysis gives evidence for an isostructural lattice in the polymeric $[PcM(CN)]_n$ complexes.

The oligomers show thermal stabilities up to 200–250 °C and are almost insoluble in noncoordinating solvents. By treating these materials with coordinating bases (L) such as pyridine, 2-methylpyrazine, piperidine and *n*-butylamine, the bridged structure is destroyed, thereby yielding mononuclear complexes with the composition PcM(L)CN.[125,128]

The $[PcM(CN)]_n$ polymers exhibit comparatively high electrical conductivities of 10^{-6}–10^{-2} S cm^{-1} without external oxidative doping (see Table 6). When the bridged structure is destroyed by treatment with coordinating ligands to form PcM(L)CN, the conductivities drop six to 10 orders of magnitude down to 10^{-12} S cm^{-1}. The $[PcCo(CN)]_n$ system also shows good photoconductivity.[129]

Table 6 Room temperature powder conductivities of mono- and polynuclear cyano–phthalocyaninato transition metal complexes.

Compound	*Conductivity* ($S\,cm^{-1}$)
$M'[PcCo(CN)_2]^a$	–
$M'[PcRh(CN)_2]$	3×10^{-9}
$K[PcCr(CN)_2]$	6×10^{-9}
$K[PcMn(CN)_2]$	5×10^{-6}
$K_2[PcFe(CN)_2]$	–
$H[TBP\text{-}Co(CN)_2]$	–
$[PcCo(CN)]_n$	2×10^{-2}
$[PcFe(CN)]_n$	6×10^{-3}
$[2,3\text{-}NcFe(CN)]_n$	1×10^{-3}
$[PcCr(CN)]_n$	3×10^{-6}
$[PcMn(CN)]_n$	1×10^{-5}
$[TBP\text{-}Co(CN)]_n$	4×10^{-2}
$[2,3\text{-}TNP\text{-}Co(CN)]_n{}^b$	2×10^{-3}

[a] M′ = Na, K, [b] TNP = tetranaphthoporphyrin.

The corresponding thiocyanato and azido bridged complexes $[PcM(SCN)]_n$ (M = Mn, Fe, Co) and $[PcM(N_3)]_n$ (M = Cr, Mn), as well as their mononuclear analogues $K[PcM(NCS)_2]$ and $K[PcM(N_3)_2]$, were characterized by IR and FIR spectroscopy, magnetic measurements, thermogravimetry, microanalysis and powder diffraction analysis.[120] All thiocyanato bridged systems show comparable diffraction patterns which give evidence for an isostructural lattice.[121]

With the exception of $[PcMn(SCN)]_n$ which loses the thiocyanate unit between 350 °C and 485 °C, all $[PcM(SCN)]_n$ (M = Fe, Co) and $[PcM(N_3)]_n$ (M = Cr, Mn) compounds are less stable than their corresponding cyano bridged systems.

As with the cyano bridged oligomers, the thiocyanato and azido bridged complexes are almost insoluble in all noncoordinating solvents: by treatment with nitrogen-donor bases (L) they are also destroyed, either by reaction to give mononuclear complexes PcM(L)SCN or by reduction to the M^{2+} species $PcML_2$. An exception is the very stable $[PcMn(N_3)]_n$ which does not react with pyridine even under forced conditions. The electrical conductivities of the thiocyanato bridged polymers are similar to or somewhat lower than those of the cyano bridged systems (10^{-7}–10^{-3} S cm^{-1}, see Table 7).[120,121]

Table 7 Room temperature powder conductivities of mono- and polynuclear thiocyanato–phthalocyaninato transition metal complexes.

Compound	*Conductivity* ($S\,cm^{-1}$)
$K[PcCo(NCS)_2]\cdot EtOH$	–
$[PcCo(SCN)]_n$	6×10^{-3}
PcCo(py)SCN	1×10^{-7}
$[PcFe(SCN)]_n$	9×10^{-5}

The cyano and thiocyanato bridged polymers, as well as the above described bridged systems containing tz, its derivatives or the ligands listed in Table 5, are of special interest because they are among the first stable organic systems which exhibit intrinsically semiconducting behaviour without requiring doping. These polymeric bridged transition metal complexes are of interest for technical applications due to their high thermal stabilities and their good semiconducting properties. They can be used for the preparation of, for example, antistatic foils or fibres. Some of these polymers also stand out for their photoconducting properties.[130]

8.7 LADDER POLYMERS FROM PHTHALOCYANINES AND THEIR RELATED COMPOUNDS

A completely different approach to achieve electrically conductive polymers containing macrocycles is to order the mononuclear subunits in a planar arrangement. This arrangement must contain a widespread π-electron system, thereby enabling electron delocalization over large regions. One possibility for realizing such a situation is to combine phthalocyaninatometal subunits in a planar arrangement. The planar polymerized phthalocyaninatocopper $[PcCu]_n$ is prepared from 1,2,4,5-tetracyanobenzene;[1,131] however, the purification and identification of this class of compounds is quite difficult. These systems are practically insoluble in organic solvents. $[PcCu]_n$, for instance, is only soluble in concentrated sulfuric acid. Also the degree of polymerization is comparatively low, at about nine phthalocyanine subunits in one oligomer.[131]

Theoretical calculations[132] indicate that planar phthalocyaninatometal polymers should exhibit good electrical conductivities; values of 10^{-11}–10^{-1} S cm^{-1} have been measured, meaning that they could be classified as semiconductors.[133–6] The conductivities are influenced by adsorbed gases, their methods of preparation and the metals used. The conductivities can be increased by up to two or three orders of magnitude by doping.[133,137] Thin films of phthalocyanines with thicknesses of 45–1200 nm, prepared from 1,2,4,5-tetracyanobenzene and copper films, exhibit conductivities up to 10^{-2} S cm^{-1}. Due to their high thermal stability, these phthalocyanines find applications in the preparation of materials exhibiting endurance at high temperatures. They are also used as electrodes, catalysts or electrocatalysts.[138,139]

In order to simplify the preparation and purification of planar conducting systems, several attempts have been proposed. One approach to achieve planar macrocyclic polymers is to order the subunits in a so-called ladder arrangement, as shown in Figure 11. It was shown by SCF calculations that the hypothetical polyphthalocyaninatonickel(II) exhibits the highest occupied band which is similar to that of a one-dimensional columnar structure.[140] One method for synthesizing phthalocyaninato ladder polymers is by linking soluble nonidentically substituted phthalocyanines over suitably functionalized bridging units, for example, by reacting a dienophilic phthalocyanine with a bifunctional diene in a Diels–Alder reaction (Equation (1)). Ladder polymers can also be obtained by Diels–Alder reaction of dienophilic and enophilic phthalocyanines (Equation (2)). The resulting oligomers must then be converted into a system containing conjugated π-electrons.[49,50]

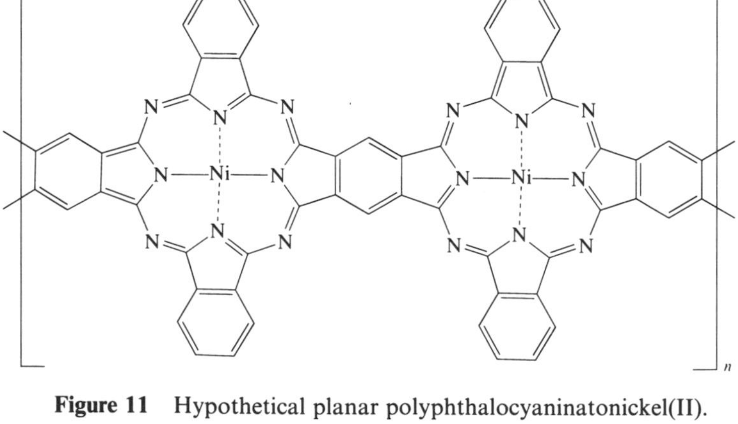

Figure 11 Hypothetical planar polyphthalocyaninatonickel(II).

Beside the preparation of ladder polymers, unsymmetrically substituted phthalocyaninatometal complexes with an enophilic substituent can be added to other dienophiles, for example, [60]fullerene by Diels–Alder reaction.[141] Dienes always add to the 6–6 double bond of the fullerene

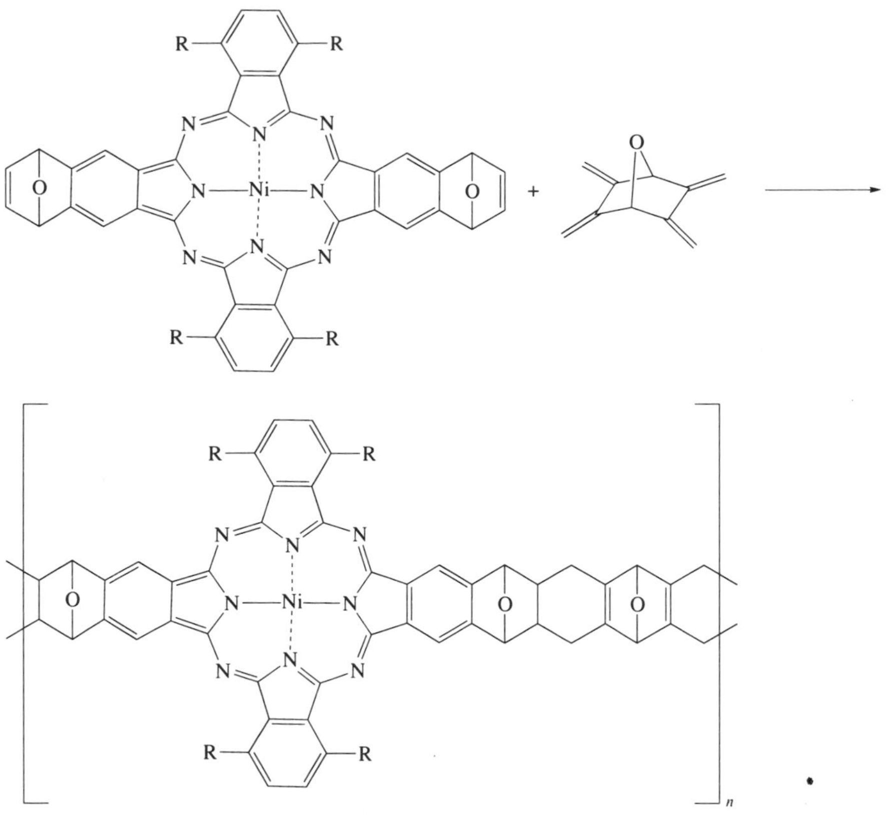

(1)

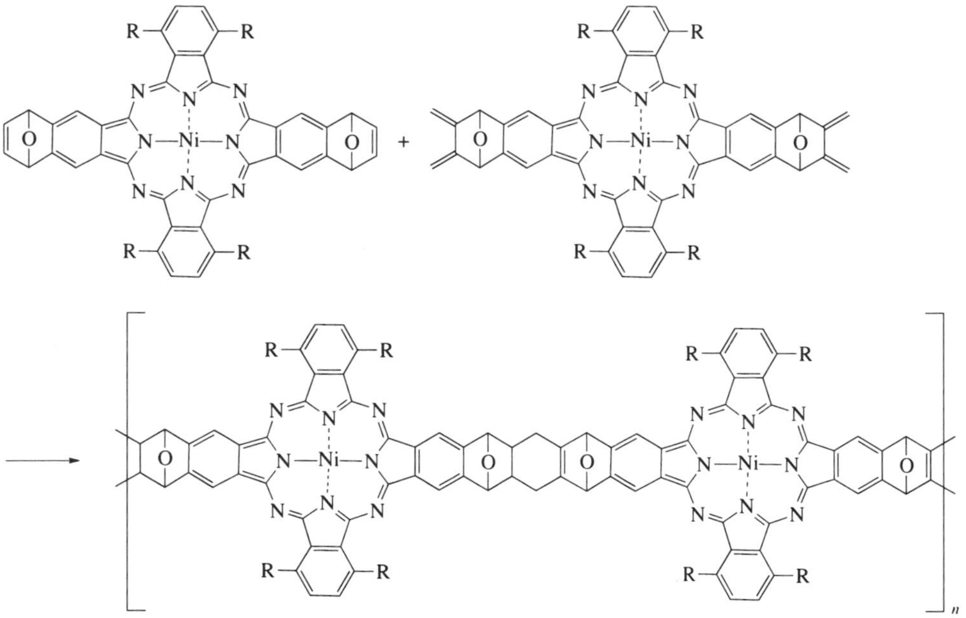

(2)

score.[142] Scheme 12 shows the synthetic route to a phthalocyaninatonickel–[60]fullerene adduct. The enophilic phthalocyaninatonickel complex is prepared by statistical condensation of three equivalents of 3,6-diheptylphthalonitrile with one equivalent of 1,2,3,4-tetrahydro-2,3-dimethylene-1,4-epoxynaphthalene-6,7-dicarbonitrile in the presence of a nickel salt, with subsequent column chromatographic separation of the diene-substituted phthalocyanine from the other phthalocyanines obtained.

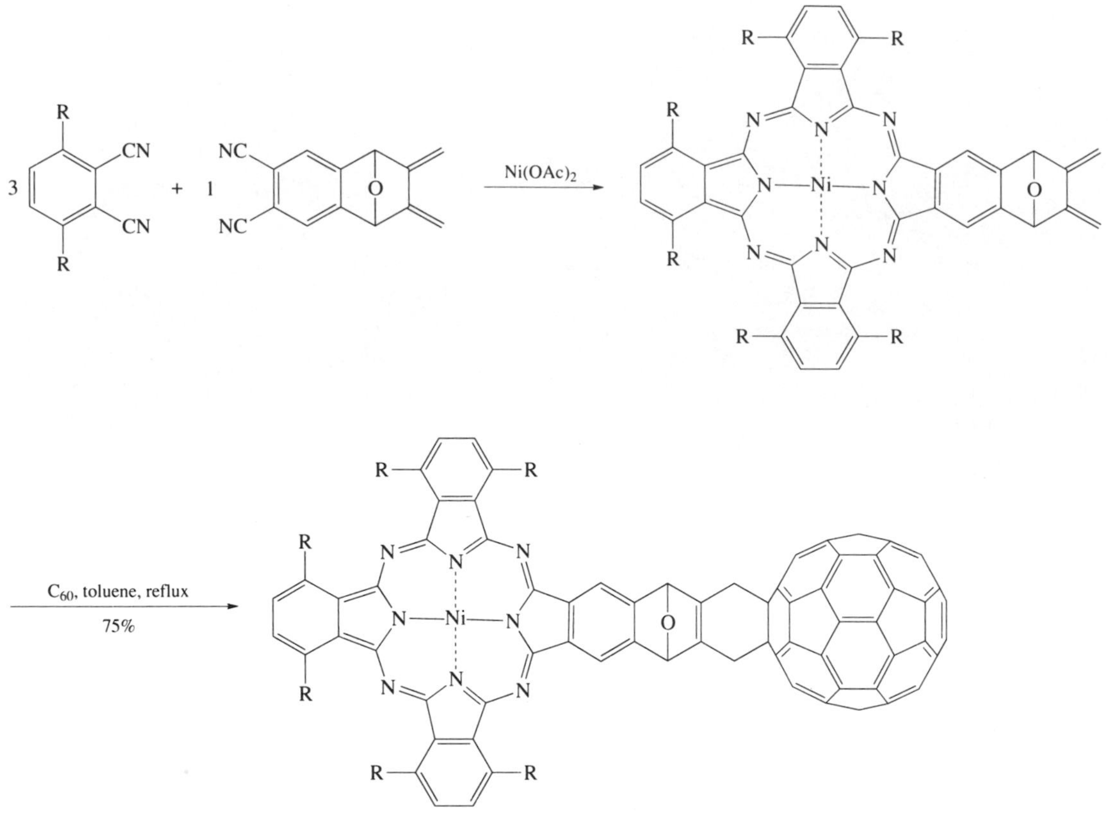

3 + 1 → Ni(OAc)$_2$

→ C_{60}, toluene, reflux 75%

$R = C_7H_{15}$

Scheme 12

As described in Section 8.2, the synthesis of phthalocyanines containing different peripheral substituents is rather difficult. In order to avoid the more or less difficult synthesis of the above described phthalocyanines, other macrocyclic systems have been used. A particularly suitable system is the hemiporphyrazine or its metal chelates (HpM), which exhibit the desired D_{2h} point group. As with the above described phthalocyanines, the hemiporphyrazinatometal complexes can be functionalized with dienophilic or enophilic peripheral substituents, thereby enabling oligomerization by repeated Diels–Alder reactions [143,144]

In order to increase the solubility of the resulting ladder polymers, additional peripheral substituents can be introduced into the monomeric subunits. The hemiporphyrazinatometal ladder polymers obtained by Diels–Alder couplings must also be converted into a π-electron containing system to achieve electrical conductivity.

8.8 REFERENCES

1. A. B. P. Lever and C. C. Leznoff (eds.), 'Phthalocyanines, Properties and Applications', VCH, New York, 1989–1993, vols. 1–3.
2. M. Hanack and M. Lang, *Adv. Mater.*, 1994, **6**, 819.
3. M. Casstevens, M. Samok, J. Pfleger and P. N. Prasad, *J. Chem. Phys.*, 1990, **92**, 2019; J. Simon, P. Bassoul and S. Norvez, *New J. Chem.*, 1989, **13**, 13.
4. G. G. Roberts, M. C. Petty, S. Baker, M. T. Fowler and N. J. Thomas, *Thin Solid Films*, 1985, **132**, 113; M. J. Cook, A. J. Dunn, M. F. Daniel, R. C. Hart, R. M. Richardson and S. J. Roser, *Thin Solid Films*, 1988, **159**, 395; S. Palacin, P. Lesieur, I. Stefanelli and A. Barraud, *Thin Solid Films*, 1988, **159**, 83; M. A. Mohammad, P. Ottenbreit, W. Prass, G. Schnurpfeil and D. Wöhrle, *Thin Solid Films*, 1992, **213**, 285.
5. H. Schultz, H. Lehmann, M. Rein and M. Hanack, *Struct. Bonding (Berlin)*, 1991, **74**, 41.
6. J. E. Kuder, *J. Imaging Sci.*, 1988, **32**, 51.
7. K. Abe, H. Saito, T. Kimura, Y. Ohkatsu and T. Kusano, *Makromol. Chem.*, 1989, **190**, 2693.
8. J. F. Van der Pol, E. Neelemann, J. W. Zwikker, R. J. M. Nolte, W. Drenth, J. Aerts, R. Visser and S. J. Picken, *Liq. Cryst.*, 1989, **6**, 577; J. Simon and C. Sirlin, *Pure Appl. Chem.*, 1989, **61**, 1625.

9. M.-T. Riou and C. Clarisse, *J. Electroanal. Chem.*, 1988, **249**, 181; D. Schlettwein, D. Wöhrle and N. I. Jaeger, *J. Electrochem. Soc.*, 1989, **136**, 2882.
10. M. Hanack, A. Datz, R. Fay, K. Fischer, U. Keppeler, J. Koch, J. Metz, M. Mezger, O. Schneider and H.-J. Schulze, in 'Handbook of Conducting Polymers', ed. T. A. Skotheim, Dekker, New York, 1986, vol. 1; M. Hanack, S. Deger and A. Lange, *Coord. Chem. Rev.*, 1988, **83**, 115.
11. T. J. Marks, *Science*, 1985, **227**, 881.
12. T. J. Marks, *Angew. Chem., Int. Ed. Engl.*, 1990, **29**, 857; B. M. Hoffman and J. A. Ibers, *Acc. Chem. Res.*, 1983, **16**, 15.
13. R. A. Collins and K. A. Mohamed, *J. Phys. D*, 1988, **21**, 154; T. A. Temofonte and K. F. Schoch, *J. Appl. Phys.*, 1989, **65**, 1350; Y. Sadaoka, T. A. Jones and W. Göpel, *Sens. Actuat. B*, 1990, **1**, 148; K.-D. Schierbaum, R. Zhou, S. Knecht, R. Dieing, M. Hanack and W. Göpel, *Sens. Actuat. B*, 1995, **24–25**, 69.
14. J. R. Darwent, P. Douglas, A. Harriman, G. Porter and M. C. Richoux, *Coord. Chem. Rev.*, 1982, **44**, 83.
15. D. Wöhrle, J. Gitzel, G. Krawczyk, E. Tsuchida, H. Ohno and T. Nishisaka, *J. Macromol. Sci. A*, 1988, **25**, 1227.
16. B. A. Hendersson and T. J. Dougherty, *Photochem. Photobiol.*, 1992, **55**, 145; G. Jori, *Photochem. Photobiol.*, 1990, **52**, 439; S. Brown and G. Truscott, *Chem. Br.*, 1993, **11**, 955.
17. T. J. Klofta, J. Danzinger, P. Lee, J. Pankow, K. W. Nebesny and N. R. Armstrong, *J. Phys. Chem.*, 1987, **91**, 5646.
18. D. Schlettwein, M. Kaneko, A. Yamada, D. Wöhrle and N. I. Jaeger, *J. Phys. Chem.*, 1991, **95**, 1748; D. Schlettwein, N. I. Jaeger and D. Wöhrle, *Ber. Bunsenges. Phys. Chem.*, 1991, **95**, 1526; D. Schlettwein, D. Wöhrle, E. Karmann and U. Melville, *Chem. Mater.*, 1994, **6**, 3.
19. J. Simon and H. J. André, 'Molecular Semiconductors', Springer, Berlin, 1985.
20. D. Wöhrle and D. Meissner, *Adv. Mater.*, 1991, **3**, 129.
21. D. Wöhrle and G. Meyer, *Kontakte*, 1985, **3**, 38.
22. H. Tomoda, S. Saito and S. Shiraishi, *Chem. Lett.*, 1983, 313; H. Tomoda, S. Saito, S. Ogawa and S. Shiraishi, *Chem. Lett.*, 1980, 1277; D. Wöhrle, G. Schnurpfeil, and G. Knothe, *Dyes Pigm.*, 1992, **18**, 91.
23. S. Deger and M. Hanack, *Synth. Met.*, 1986, **30**, 319; M. Hanack, S. Deger, U. Keppeler, A. Lange and A. Leverenz, *Synth. Met.*, 1987, **19**, 739; S. Deger and M. Hanack, *Isr. J. Chem.*, 1986, **2**, 347.
24. M. Hanack, G. Renz, J. Strähle and S. Schmid, *Chem. Ber.*, 1988, **121**, 1479.
25. M. Hanack, A. Hirsch, A. Lange, M. Rein, G. Renz and P. Vermehren. *J. Mater. Res.*, 1991, **6**, 385; M. Hanack, G. Renz, J. Strähle and S. Schmid, *J. Org. Chem.*, 1991, **56**, 3501.
26. M. Hanack and P. Vermehren, *Inorg. Chem.*, 1990, **29**, 134.
27. W. Kobel and M. Hanack, *Inorg. Chem.*, 1986, **25**, 103.
28. M. Hanack, J. Osio-Barcina, E. Witke and J. Pohmer, *Synthesis*, 1992, 211.
29. M. Hanack and R. Polley, submitted for publication.
30. S. A. Mikhalenko, S. U. Barkanova, O. L. Lebedev and E. A. Luk'yanets, *J. Gen. Chem. USSR*, 1971, **41**, 2770.
31. H. Shirai, A. Maruyama, K. Kobayashi, N. Hojo and K. Urushido, *Makromol. Chem.*, 1980, **181**, 575.
32. C. C. Leznoff, S. Vigh, P. I. Svirskaya, S. Greenberg, D. M. Drew, E. Ben-Hur and I. Rosenthal, *Photochem. Photobiol.*, 1989, **49**, 279.
33. M. Hanack and P. Vermehren, *Synth. Met.*, 1989, **32**, 257; M. Hanack and P. Vermehren, *Chem. Ber.*, 1991, **124**, 1733; A. Beck, K.-M. Mangold and M. Hanack, *Chem. Ber.*, 1991, **124**, 2315.
34. J. Metz, O. Schneider and M. Hanack, *Inorg. Chem.*, 1984, **23**, 1065.
35. M. Hanack, A. Hirsch and H. Lehmann, *Angew. Chem., Int. Ed. Engl.*, 1990, **29**, 1467.
36. M. J. Cook, S. J. Cracknell and K. J. Harrison, *J. Mater. Chem.*, 1991, **1**, 703.
37. M. Hanack, G. Schmid and M. Sommerauer, *Angew. Chem., Int. Ed. Engl.*, 1993, **32**, 1422.
38. V. M. Derkacheva, O. L. Kaliya and E. A. Luk'yanets, *J. Gen. Chem. USSR*, 1983, **53**, 163.
39. M. Hanack and R. Fay, *Recl. Trav. Chim. Pays-Bas*, 1986, **105**, 427.
40. L. R. Subramanian, A. Gül, M. Hanack, B. K. Mandal and E. Witke, *Synth. Met.*, 1991, **41–43**, 2669; M. Hanack, A. Gül, A. Hirsch, B. K. Mandal, L. R. Subramanian and E. Witke, *Mol. Cryst. Liq. Cryst.*, 1990, **187**, 365; M. Hanack, P. Haisch, H. Lehmann and L. R. Subramanian, *Synthesis*, 1993, 387; M. J. Cook, A. J. Dunn, S. D. Howe, A. J. Thomson and K. J. Harrison, *J. Chem. Soc., Perkin Trans. 1*, 1988, 2453.
41. R. P. Linstead and F. T. Weiss, *J. Chem. Soc.*, 1950, 2975; J. H. Weber and D. J. Busch, *Inorg. Chem.*, 1965, **4**, 469.
42. V. Ahsen, A. Gürek, E. Musluoglu and Ö. Bekaroglu, *Chem. Rev.*, 1993, **122**, 1073.
43. D. Wöhrle, M. Eskes, K. Shigehara and A. Yamada, *Synthesis*, 1993, 194.
44. M. Hanack, D. Meng, A. Beck, M. Sommerauer and L. R. Subramanian, *J. Chem. Soc., Chem. Commun.*, 1993, 58.
45. M. Kumada, K. Tamao and K. Sunitani, *Org. Synth.*, 1978, **58**, 127; K. Otha, L. Jaquemin, C. Sirlin, L. Bosio and J. Simon, *New J. Chem.*, 1988, **12**, 751.
46. A.-M. Giroud-Godquin and P. M. Maitlis, *Angew. Chem., Int. Ed. Engl.*, 1991, **30**, 375.
47. M. K. Engel, P. Bassoul, L. Bosio, H. Lehmann, M. Hanack and J. Simon, *Liq. Cryst.*, 1993, **15**, 709.
48. M. J. Cook, M. F. Daniel, K. J. Harrison, N. B. McKnoewn and A. J. Thomson, *J. Chem. Soc., Chem. Commun.*, 1987, 1086.
49. C. Feucht, T. Linßen and M. Hanack, *Chem. Ber.*, 1994, **12**, 113.
50. T. Linßen and M. Hanack, *Chem. Ber.*, 1994, **127**, 2051.
51. N. Kobayashi, T. Ashida and T. Osa, *Chem. Lett.*, 1992, 2031.
52. C. C. Leznoff, P. I. Svirskaya, B. Khouw, R. L. Cerny, P. Seymour and A. B. P. Lever, *J. Org. Chem.*, 1991, **56**, 82.
53. A. Meller and A. Ossko, *Monatsh. Chem.*, 1972, **103**, 150.
54. M. Hanack and M. Geyer, *J. Chem. Soc., Chem. Commun.*, 1994, 2253; J. Rauschnabel and M. Hanack, *Tetrahedron Lett.*, 1995, **36**, 1629.
55. N. Kobayashi, R. Kondo, S. Nakajima and T. Osa, *J. Am. Chem. Soc.*, 1990, **112**, 9640; A. Sastre, T. Torres and M. Hanack, *Tetrahedron Lett.*, 1995, **36**, 8501.
56. K. Ukei, *Acta Crystallogr., Sect. B*, 1973, **29**, 2290.
57. J. A. Thompson, K. Murata, D. C. Miller, J. L. Stanton, W. E. Broderick, B. M. Hoffman and J. A. Ibers, *Inorg. Chem.*, 1993, **32**, 3546.
58. T. J. Marks and D. W. Kalina, in 'Extended Linear Chain Compounds', ed. J. S. Miller, Plenum, New York, 1982, vol. 1, p. 197.

59. M. Hanack, *Chimia*, 1983, **37**, 238.
60. K. Krogmann, *Angew. Chem., Int. Ed. Engl.*, 1969, **8**, 35.
61. J. L. Petersen, C. S. Schramm, D. R. Stojakovic, B. M. Hoffman and T. J. Marks, *J. Am. Chem. Soc.*, 1977, **99**, 286.
62. J. Martinsen, J. L. Stanton, R. L. Greene, J. Tanaka, B. M. Hoffman and J. A. Ibers, *J. Am. Chem. Soc.*, 1985, **107**, 6915.
63. A. B. P. Lever, P. C. Minor and J. P. Wilshire, *Inorg. Chem.*, 1981, **20**, 2550.
64. E. Canadell and S. Alvarez, *Inorg. Chem.*, 1984, **23**, 573.
65. R. D. Joyner and M. E. Kenney, *J. Am. Chem. Soc.*, 1960, **82**, 5790; R. D. Joyner and M. E. Kenney, *Inorg. Chem.*, 1962, **1**, 236; R. D. Joyner and M. E. Kenney, *Inorg. Chem.*, 1962, **1**, 717.
66. J. P. Linsky, T. R. Paul, R. S. Nohr and M. E. Kenney, *Inorg. Chem.*, 1980, **19**, 3131.
67. M.-H. Whangbo and K. R. Stewart, *Isr. J. Chem.*, 1983, **23**, 133.
68. M. Hanack, F. F. Seelig and J. Strähle, *Z. Naturforsch., Teil A*, 1979, **34**, 983.
69. A. B. Anderson, T. L. Gordon and M. E. Kenney, *J. Am. Chem. Soc.*, 1985, **107**, 192.
70. S. Hayashida and M. Hanack, *Synth. Met.*, 1992, **52**, 241.
71. W. Koch, Ph.D. Thesis, University of Tübingen, 1986.
72. C. A. Dirk, T. Inabe, K. F. Schoch, Jr. and T. J. Marks, *J. Am. Chem. Soc.*, 1983, **105**, 1539.
73. D. W. DeWulf, J. K. Leland, B. L. Wheeler, A. J. Bard, D. A. Batzel, D. R. Dininny and M. E. Kenney, *Inorg. Chem.*, 1987, **26**, 266.
74. X. Zhou, T. J. Marks and S. H. Carr, *Polym. Mater. Sci. Eng.*, 1984, **51**, 651.
75. E. Orthmann, V. Enkelmann and G. Wegner, *Makromol. Chem., Rapid Commun.*, 1983, **4**, 687.
76. T. J. Marks, K. F. Schoch and B. R. Kundalkar, *Synth. Met.*, 1979/80, **1**, 337.
77. P. J. Toscano and T. J. Marks, *Mol. Cryst. Liq. Cryst.*, 1985, **118**, 337; P. J. Toscano and T. J. Marks, *J. Am. Chem. Soc.*, 1986, **108**, 437; B. Wehrle, H.-H. Limbach, T. Zipplies and M. Hanack, *Angew. Chem. Adv. Mater.*, 1989, **101**, 1783.
78. X. Zhou, T. J. Marks and S. H. Carr, *J. Polym. Sci.*, 1985, **23**, 305.
79. X. Zhou, T. J. Marks and S. H. Carr, *Mol. Cryst. Liq. Cryst.*, 1985, **118**, 337.
80. J. A. Ibers, L. J. Pace, J. Martinsen and B. M. Hoffman, *Struct. Bond.*, 1982, **50**, 47.
81. J. Martinsen, J. L. Stanton, R. L. Greene, J. Tanaka, B. M. Hoffman and J. A. Ibers, *J. Am. Chem. Soc.*, 1985, **107**, 6915; C. W. Dirk, E. A. Mintz, K. F. Schoch and T. J. Marks, *J. Macromol. Sci. A*, 1981, **16**, 275; T. Inabe, M. K. Moguel, T. J. Marks, R. Burton, J. W. Lyding and C. R. Kannewurf, *Mol. Cryst. Liq. Cryst.*, 1985, **118**, 349.
82. O. Schneider, J. Metz and M. Hanack, *Mol. Cryst. Liq. Cryst.*, 1982, **81**, 273.
83. M. Hanack and K. Fischer, *Chem. Ber.*, 1983, **116**, 1860.
84. M. Meier, W. Albrecht, E. Zimmerhackl, M. Hanack and K. Fischer, *J. Mol. Electron.*, 1985, **1**, 47.
85. P. Sheng, *Phys. Rev. B*, 1980, **21**, 2180.
86. J. Metz, G. Pawlowski and M. Hanack, *Z. Naturforsch., Teil B*, 1983, **38**, 378.
87. C. Sirlin, L. Bosio and J. Simon, *Mol. Cryst. Liq. Cryst.*, 1988, **155**, 231.
88. W. Caseri, T. Sauer and G. Wegner, *Makromol. Chem., Rapid Commun.*, 1988, **9**, 651.
89. E. Orthmann and G. Wegner, *Angew. Chem., Int. Ed. Engl.*, 1986, **25**, 1105.
90. P. M. Kuznesof, R. S. Nohr, K. J. Wynne and M. E. Kenney, *J. Macromol. Sci. Chem. A*, 1981, **16**, 299; R. S. Nohr and K. J. Wynne, *J. Chem. Soc., Chem. Commun.*, 1981, 1210.
91. K. J. Wynne and R. S. Nohr, *Mol. Cryst. Liq. Cryst.*, 1983, **81**, 243.
92. P. Brant, R. S. Nohr, K. J. Wynne and D. C. Weber, *Mol. Cryst. Liq. Cryst.*, 1982, **81**, 255.
93. R. S. Nohr, P. M. Kuznesof, K. J. Wynne, M. E. Kenney and P. G. Siebermann, *J. Am. Chem. Soc.*, 1981, **103**, 4371.
94. K. J. Wynne, *Inorg. Chem.*, 1985, **24**, 1339.
95. M. Hanack, K. Mitulla, G. Pawlowski and L. R. Subramanian, *Angew. Chem., Int. Ed. Engl.*, 1979, **18**, 322; K. Mitulla and M. Hanack, *Z. Naturforsch., Teil B*, 1980, **35**, 1111; M. Hanack, K. Mitulla, G. Pawlowski and L. R. Subramanian, *J. Organomet. Chem.*, 1981, **204**, 315; M. Hanack, W. Kobel, J. Metz, M. Mezger, G. Pawlowski, O. Schneider, H.-J. Schulze and L. R. Subramanian, *Mater. Sci.*, 1981, 185; M. Hanack, K. Mitulla and O. Schneider, *Chem. Scr.*, 1981, 17.
96. O. Schneider and M. Hanack, *Angew. Chem.*, 1980, **92**, 391; O. Schneider and M. Hanack, *Angew. Chem., Int. Ed. Engl.*, 1982, **19**, 2392; M. Hanack, A. Datz, W. Kobel, J. Koch, J. Metz, M. Mezger, O. Schneider and H.-J. Schulze, *J. Phys. C*, 1983, **3**, 633.
97. O. Schneider and M. Hanack, *Chem. Ber.*, 1983, **116**, 2088.
98. U. Keppeler, S. Deger, A. Lange and M. Hanack, *Angew. Chem., Int. Ed. Engl.*, 1987, **26**, 344.
99. B. N. Diel, T. Inabe, N. K. Jaggi, J. W. Lyding, O. Schneider, M. Hanack, C. R. Kannewurf, T. J. Marks and L. H. Schwartz, *Am. Chem. Soc.*, 1984, **106**, 3207.
100. H.-H. Wei and H.-L. Shyu, *Polyhedron*, 1985, **4**, 979.
101. O. Schneider and M. Hanack, *Angew. Chem., Int. Ed. Engl.*, 1983, **22**, 784.
102. J. Metz, O. Schneider and M. Hanack, *Spectrochim. Acta, Part A*, 1982, **38**, 1265.
103. M. Hanack, U. Keppeler, A. Lange, A. Hirsch and R. Dieing, in 'Phthalocyanines, Properties and Applications', eds. A. B. P. Lever and C. C. Leznoff, VCH, New York, 1993, vol. 2.
104. U. Keppeler, W. Kobel, H.-U. Siehl and M. Hanack, *Chem. Ber.*, 1985, **118**, 2095.
105. R. Aldinger, M. Hanack, K.-H. Herrmann, A. Hirsch and K. Kasper, *Synth. Met.*, 1993, **60**, 265.
106. F. Kubel and J. Strähle, *Z. Naturforsch., Teil B.*, 1981, **36**, 441.
107. M. Hanack and A. Leverenz, *Synth. Met.*, 1987, **22**, 9; M. Hanack and A. Leverenz, in 'Solid State Science 76', eds. H. Kuzmany, M. Mehring and S. Roth, Springer, Berlin, 1987.
108. O. Schneider, Ph.D. Thesis, University of Tübingen, 1983.
109. E. Orti, J. L. Brédas, M. C. Piqueras and R. Crespo, *Chem. Mater.*, 1990, **2**, 110.
110. E. Orti, J. L. Brédas, M. C. Piqueras and R. Crespo, *Synth. Met.*, 1991, **41–43**, 2647.
111. M. Hanack, A. Beck and H. Lehmann, *Synthesis*, 1987, 703.
112. M. Hanack and H. Ryu, *Synth. Met.*, 1992, **46**, 113.
113. A. G. MacDiarmid and A. J. Epstein, in 'Science and Application of Conducting Polymers', eds. W. R. Salaneck, D. T. Clark and E. J. Samuelsen, Hilger, Bristol, 1991, p. 117.

114. M. Hanack and Y.-G. Kang, *Synth. Met.*, 1992, **48**, 79.
115. M. Hanack and J. Metz, *Chem. Ber.*, 1987, **120**, 231.
116. J. P. Collman, J. T. McDevitt, C. R. Leidner, G. T. Yee, J. B. Torrance and W. A. Little, *J. Am. Chem. Soc.*, 1987, **109**, 4606.
117. W. Kobel and M. Hanack, *Inorg. Chem.*, 1986, **25**, 103.
118. J. Pohmer and M. Hanack, *J. Mater. Chem.*, in press.
119. O. Schneider and M. Hanack, *Z. Naturforsch., Teil B*, 1984, **39**, 265.
120. A. Datz, J. Metz, O. Schneider and M. Hanack, *Synth. Met.*, 1984, **9**, 31.
121. C. Hedtmann-Rein, M. Hanack, K. Peters, E.-M. Peters and H. G. Schnering, *Inorg. Chem.*, 1987, **26**, 2647.
122. P. A. Barrett, D. A. Frye and R. P. Linstead, *J. Chem. Soc.*, 1938, 1157.
123. J. F. Myers, G. W. R. Canham and A. B. P. Lever, *Inorg. Chem.*, 1975, **14**, 461.
124. M. L. Khidekel and E. L. Zhilyaeva, *Synth. Met.*, 1981, **4**, 1.
125. J. Metz and M. Hanack, *J. Am. Chem. Soc.*, 1983, **105**, 828.
126. T. Inabe and Y. Maruyama, *Chem. Lett.*, 1989, 55.
127. K. Nakamoto, in 'Infrared Spectra of Inorganic Coordination Compounds', Wiley, New York, 1983.
128. M. Hanack and X. Münz, *Synth. Met.*, 1985, **10**, 357.
129. H. Meier, W. Albrecht, E. Zimmerhackl, M. Hanack and J. Metz, *Synth. Met.*, 1985, **11**, 333.
130. H. Meier, W. Albrecht, E. Zimmerhackl, M. Hanack and K. Fischer, *J. Mol. Electron.*, 1985, **1**, 47.
131. A. J. Epstein and B. S. Wildi, *J. Chem. Phys.*, 1960, **32**, 324; D. Wöhrle, U. Marose and R. Knoop, *Makromol. Chem.*, 1985, **186**, 2209; D. Wöhrle and B. Schulze, *Makromol. Chem.*, 1988, **189**, 1229.
132. P. Gomez-Romero, Y.-S. Lee and M. Kertesz, *Inorg. Chem.*, 1988, **121**, 3672.
133. C. S. Marvel and J. H. Rassweiler, *J. Am. Chem. Soc.*, 1958, **80**, 1196.
134. G. Manecke and D. Wöhrle, *Makromol. Chem.*, 1967, **102**, 1.
135. H. Shirai, I. Takemae, K. Kobayashi", Y. Kondo, O. Hirabaru and N. Hoyo, *Makromol. Chem.*, 1984, **185**, 1395.
136. D. Wöhrle, *Adv. Polym. Sci.*, 1983, **50**, 46.
137. D. Wöhrle, *Adv. Polym. Sci.*, 1972, **10**, 35.
138. D. Wöhrle, V. Schmidt, B. Schumann, A. Yamada and K. Shigehara, *Ber. Bunsenges. Phys. Chem.*, 1987, **91**, 975; D. Wöhrle, R. Bannehr, B. Schumann and G. Meyer, *J. Mol. Catal.*, 1983, **21**, 255; D. Wöhrle, B. Schumann and N. I. Jaeger, *Makromol. Chem., Macromol. Symp.*, 1987, **8**, 195.
139. D. Wöhrle and M. Kirschenmann, *J. Electrochem. Soc.*, 1985, **132**, 1150.
140. P. Ramirez and M. C. Böhm, *Int. J. Quantum Chem.*, 1988, **33**, 73.
141. T. Linßen, K. Dürr, M. Hanack and A. Hirsch, *J. Chem. Soc., Chem. Commun.*, 1995, 103.
142. A. Hirsch, 'The Chemistry of the Fullerenes', Thieme, Stuttgart, 1994.
143. M. Hanack, K. Haberroth and M. Rack, *Chem. Ber.*, 1993, **126**, 1201.
144. M. Rack and M. Hanack, *Angew. Chem.*, 1994, **106**, 1712.

9

Molecular Architecture and Function Based on Molecular Recognition and Self-organization

ANKE REICHERT, HELMUT RINGSDORF and PETER SCHUHMACHER
Johannes-Gutenberg-Universität Mainz, Germany

WOLFGANG BAUMEISTER and TSCHANGIZ SCHEYBANI
Max-Planck-Institut für Biochemie, Martinsried, Germany

9.1 INTRODUCTION: MOLECULAR ARCHITECTURE OF SELF-ORGANIZED SYSTEMS

9.1.1 Functional Supramolecular Systems

In recent years 'supramolecular science' has become the cumulative title to describe the rapidly emerging achievements at the interfaces between chemistry, physics and biology.[1] This area is intensively studied on an interdisciplinary basis in many laboratories. Since classical organic chemistry is now able to construct highly complex molecules and provide biologists and biochemists with detailed insights into biological processes, chemists are becoming increasingly interested in investigating organic chemistry beyond the covalent bond, as coined by J.-M. Lehn.[1–3] The function of supramolecular systems is achieved by the interplay between different functional units (Figure 1). In nature, a unit that demonstrates this perfectly is the cell membrane,[4] where the interplay between molecular self-organization and molecular recognition of the individual constituents (phospholipids, glycolipids, glycoproteins, membrane spanning peptide helices, the cytoskeleton, etc.) leads to the construction of this natural supramolecular system. It combines order and mobility, and its function is based on its self-organization. The scientific challenge to understand, construct and mimic natural molecular assemblies can no longer only focus on single molecular performances, but has to be based on self-organized molecular aggregates: the whole being more than the sum of its parts. This principle of nature was already adopted as a basic idea of the ancient philosophies of Asia and in Europe: only the mutuality of the parts creates the whole and its ability to function.

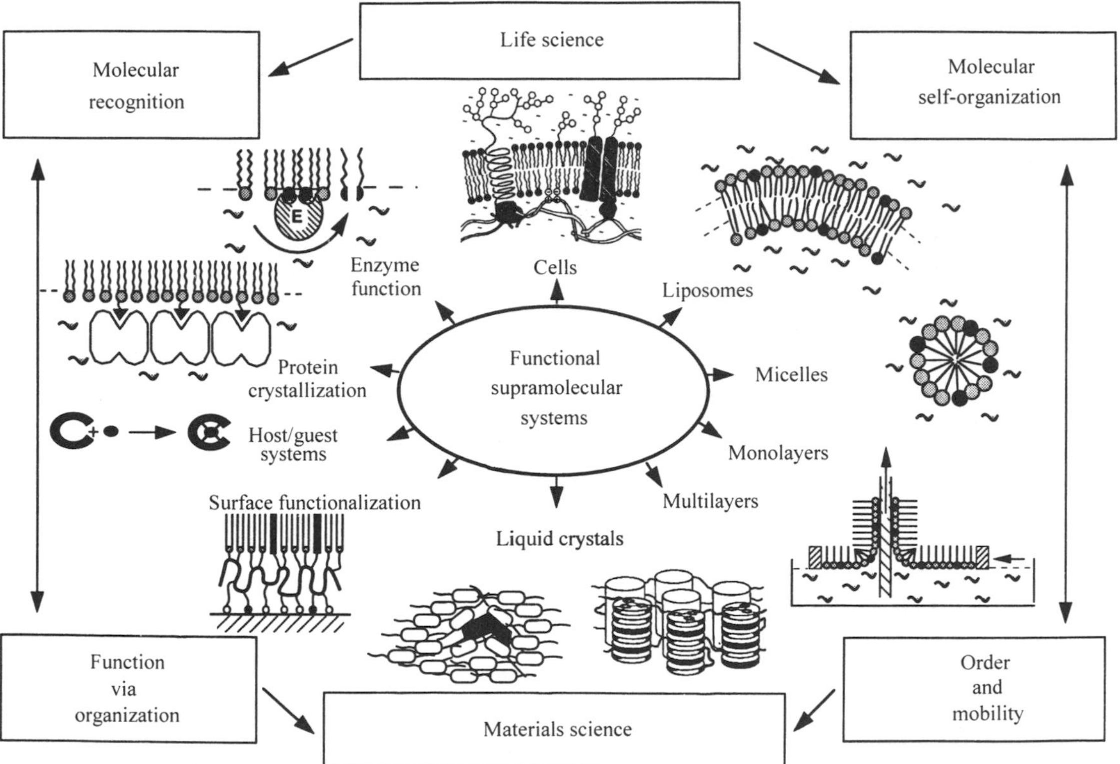

Figure 1 Functional supramolecular systems—connecting link between life science and materials science.

Figure 1 illustrates common perspectives of life science and materials science by showing examples of functional supramolecular systems, in which many of their properties, structures and functions are determined by their supramolecular order. The cell membrane with its carbohydrate recognition structures on the surface is structured by different phospholipids and stabilized by functional proteins and lipids, which extend through the membrane, and by polymer networks, for example, the cytoskeleton. Descending in complexity from this perfect natural system one finds liposomes[5] with their relatively simple spherical bilayer structure and, finally, micelles[6] with their various lyotropic structures.[7] These superstructures have long been important in studies of synthetic and natural systems and, indeed, colloid science determined the course of physical chemistry in the 1920s. Recent developments in monolayers and multilayers (Langmuir and Langmuir–Blodgett systems)[8] have resulted in ultrathin films in which order and mobility are combined. Of more interest to materials science are liquid crystals[9] which are already of industrial importance, both in low molar mass and macromolecular forms. Their order is based on their form-anisotropy (molecular shape), and it is possible for rods,[10] discs[11] and boards[12] to self-organize. Too often, it is forgotten that nature abounds with liquid crystalline materials. Many of the natural polymers, and cell membranes themselves, should be regarded as lyotropic liquid crystalline systems.[13] Besides molecular self-organization, another crucial property of supramolecular systems is their capability of specific molecular interactions. Many processes occurring at natural or synthetic membrane surfaces start with a specific molecular recognition event leading on to enzymatic reactions such as the cleavage of phospholipids by various phospholipases[3,14] or to protein crystallization, like in the interaction of biotinylated membranes with streptavidin.[3,15] The study of molecular recognition processes has for some time been an important part of organic chemistry as seen by the many studies on synthetic host–guest systems.[1b,16] These studies also include the alteration of surfaces and the structuring of ultrathin layers, which are of particular interest in catalysis research, sensor technology and tribology. Surface functionalization can be achieved by incorporation of functional groups, represented by the black units in Figure 1, into the tethered monolayers. As Figure 1 schematically shows, it is also possible to functionalize liquid crystals, again indicated by the black parts in the liquid crystalline systems. The function of such addressable liquid crystals is, in principle, based on their self-organization, and it is introduced by reactive groups, for example, photosensitive groups.[17] Intensive studies of these functionalized liquid crystals are now in progress in the areas of nonlinear optics[18] and optical storage.[19]

In this contribution, it is not intended to review the field of supramolecular science.[1] Only two supramolecular systems will be described, related to enzyme function and protein crystallization on the one hand, and discotic liquid crystals on the other. The common denominator of both systems is that they consist of similar subunits, which self-organize into complex cylindrical structures that are responsible for their function.

9.1.2 From Spheres and Discs to Cylinders and Columns

Spheres, discs, cylinders and columns are the characteristic structural features of both the natural and the synthetic supramolecular system, described here. Spheres form rings, discs form columns and rings and discs organize themselves into cylinders and hexagonal superstructures, as schematically shown in Figure 2. The molecular architecture of these two supramolecular systems leads to structures in which their function is based on their organization. The two examples, as schematically shown in Figure 2, are chosen—according to Figure 1—from life science (Figure 2(a)) and materials science (Figure 2(b)). The first example deals with proteasomes, multienzyme complexes, showing significant enzymatic activity only if all protein subunits are correctly assembled to form a cylindrical or barrel-shaped multienzyme complex (see Section 9.2). The second example, taken from the field of materials science, demonstrates that the high photoconductivity of discotic triphenylene derivatives is only found in the liquid crystalline D_h mesophase of these column-forming molecules (see Section 9.3).

9.2 INTERACTION OF PROTEASOMES WITH FUNCTIONALIZED MODEL MEMBRANE SYSTEMS

9.2.1 Properties and Structure of Proteasomes

When two or more enzymes catalyse two or more steps in a metabolic cascade, they often form noncovalently associated multienzyme complexes.[20] These protein assemblies have molar masses ranging from several hundred to several thousand kDa. The formation of these multienzyme complexes results in a greater catalytic efficiency. The organized complexes allow a much more efficient catalytic turnover relative to the nonassociated enzymes. In addition it cannot be ruled out that multicatalytic enzyme complexes may play a more important role than is known so far.

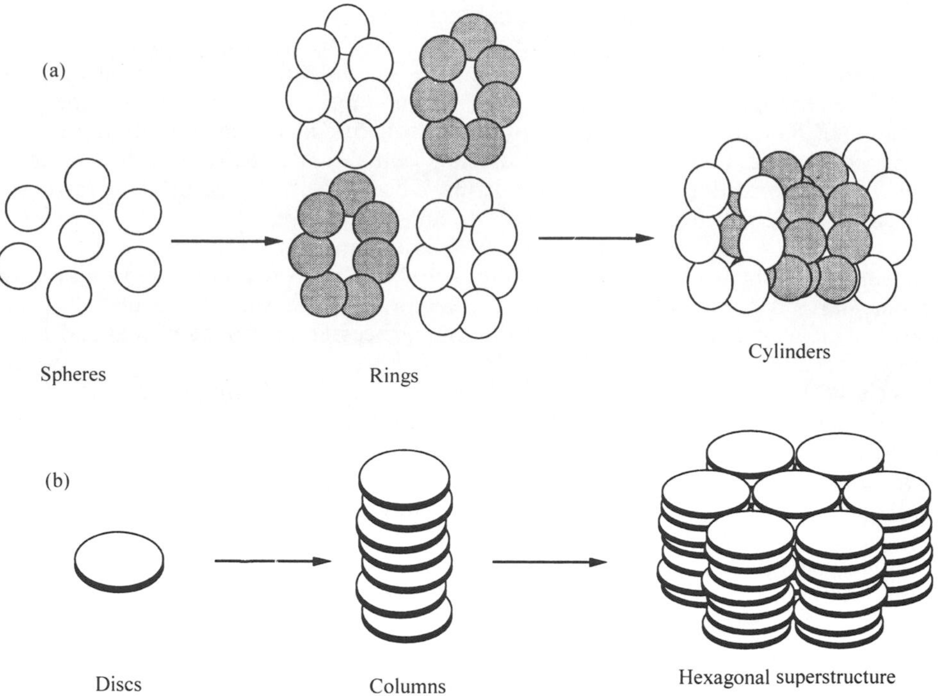

Figure 2 Representation of the self-organization of two supramolecular systems in which function is based on self-organization. (a) The barrel- or cylinder-shaped proteasomes, self-organized from four rings, which in turn have been self-organized from seven spherical-shaped protein subunits, show their highest catalytic activity only in the completely assembled form. (b) The hexagonal columnar superstructure in the D_h phase of discotic liquid crystals, which are self-organized from discs that form columns, and columns form the hexagonal superstructure, is a prerequisite for the observed high charge carrier mobilities of these compounds.

Many proteins without known functions may in fact be "building blocks" for assembling these protein complexes, which then disassemble when the specific function is finished. An analogy to this process may be the immune cascade,[21] which can be envisaged as a temporary multisubunit complex.

One of the best-characterized multienzyme complexes is the pyruvate dehydrogenase,[20] but the existence of several others with unknown structure and function is already known. Examples that have found increasing interest are the so-called proteasomes.[22] Proteasomes are high-molecular-weight multisubunit enzyme complexes (approximately 700 kDa) with at least three distinct proteolytic activities (trypsinlike, chymotrypsin-like and peptidyl-glutamyl-peptide hydrolysing).[22,23] They are highly conserved and ubiquitous in eukaryotic cells from yeast to man.[22] More recently, proteasomes were also found in the archaebacterium *Thermoplasma acidophilum*.[24] Eukariotic proteasomes have a rather complex subunit composition; they typically contain 14 different though related subunits, apparently encoded by one gene family.[25] Proteasomes from the archaebacterium *Thermoplasma acidophilum* have the same quarternary structure, in which seven of the protein subunits assemble into one ring and four of these rings collectively form the cylindrical or barrel-shaped complex (Plate 1).[26] However, in contrast to the eukariotic proteasomes, the *Thermoplasma acidophilum* proteasome consists of only two different protein subunits, α and β.[24,27] Therefore, each proteasome particle contains 14 α-subunits and 14 β-subunits. Recently it could be shown by immunoelectron microscopy that the 14 α-subunits are located in the two outer rings (7 subunits per ring) of the proteasome, while the 14 β-subunits constitute the two juxtaposed inner rings (see Plate 1).[28]

An issue of utmost importance, which was not clear at the beginning of this study, is the identification of the functional roles of the α- and β-subunits in the *Thermoplasma acidophilum* proteasome. Although an unambiguous assignment of functions was not possible at the time, it was proposed that the α-subunits serve a regulatory and targeting function, while the β-subunits alone or in combination with the α-subunits carry the active site responsible for the catalytic function.[27] Dissociation and reassociation experiments have shown that neither single α-subunits nor single β-subunits are able to catalyse significantly the hydrolysis of a polypeptide.[29] Even alone α-subunits self-organized into double rings do not show any catalytic activity. These experiments show clearly that only the completely self-organized cylindrical proteasome particle has significant catalytic activity. Therefore, it can be speculated that the active site is located between one α- and one β-subunit or between two β-subunits and that correct subunit assembly is necessary to form the binding pocket for the enzyme substrate.

To further locate the catalytic site in the proteasome particle, the interaction of proteasomes with specifically functionalized model membranes was studied.[30] Amphiphilic molecules with substrate as well as inhibitor headgroups were incorporated in different model membrane systems. Interaction of the proteasomes with the substrate-derivatized membranes should give information about their ability to specifically recognize and hydrolyse the substrate lipid headgroup at the membrane surfaces, which is dependent on the physical state of the membrane, determining the accessibility of the substrate headgroup for the enzymes.[3] Interaction of the proteasomes with the membrane-incorporated inhibitor lipid headgroups may result in binding and orientation of the enzyme complexes relative to the membrane.

Proteasomes from *Thermoplasma acidophilum* show a chymotrypsin-like activity.[31] Therefore, a substrate and an inhibitor of chymotrypsin have been used as headgroups in the synthesis of functionalized ligand lipids. With these kind of lipids, optimal conditions for the specific interaction of the proteasomes with the membrane incorporated ligand lipids could be tested, by using chymotrypsin as the receptor protein.[32] These conditions could then be used for the experiments with the proteasomes. This procedure was favourable because in contrast to the amount of chymotrypsin, the amount of proteasomes was limited.

9.2.2 Functionalized Ligand Lipids with Substrate and Inhibitor Headgroups

Different substrate and inhibitor functionalized lipids have been synthesized to study their interaction with chymotrypsin and proteasomes in model membranes.[3c] The substrate (**1**) (succinyl-alanine-alanine-phenylalanine-7-amido-4-methyl-coumarin; Suc-Ala-Ala-Phe-AMC) is an oligopeptide linked to a fluorescence dye. Chymotrypsin hydrolyses the amide bond on the carboxyl side of aromatic amino acids.[20,32,33] This hydrolysis means that the enzyme cleaves the amide bond between the phenylalanine and the fluorescence dye to release the chromophore (**3**) 7-amino-4-methyl-coumarin (Figure 3(a)). Hydrolysis of the substrate can be easily detected by fluorescence spectroscopy, because the amide linked chromophore shows a maximum fluorescence at 395 nm, whereas the free chromophore has its maximum fluorescence at 435 nm. The inhibitor (**4**) contains the same oligopeptide sequence as the substrate but has the reactive α-chloromethylketone unit as the endgroup instead of the fluorescence dye. Inhibition of chymotrypsin is based on the irreversible alkylation of histidine 57,[34] which is part of the active site of the enzyme (Figure 3(b)).

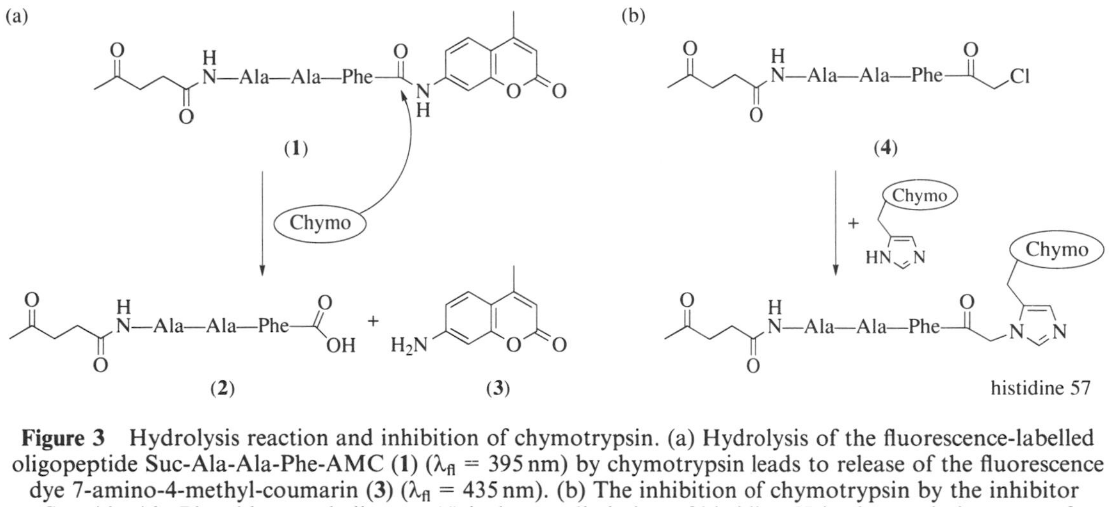

(a)

O H N—Ala—Ala—Phe— O N H O O O

(**1**)

Chymo

O H N—Ala—Ala—Phe— O OH + H_2N O O O

(**2**) (**3**)

(b)

O H N—Ala—Ala—Phe— O Cl O

(**4**)

+ Chymo HN N

Chymo

O H N—Ala—Ala—Phe— O N N O

histidine 57

Figure 3 Hydrolysis reaction and inhibition of chymotrypsin. (a) Hydrolysis of the fluorescence-labelled oligopeptide Suc-Ala-Ala-Phe-AMC (**1**) (λ_{fl} = 395 nm) by chymotrypsin leads to release of the fluorescence dye 7-amino-4-methyl-coumarin (**3**) (λ_{fl} = 435 nm). (b) The inhibition of chymotrypsin by the inhibitor Suc-Ala-Ala-Phe-chloromethylketone (**4**) is due to alkylation of histidine 57 in the catalytic centre of chymotrypsin.

Two substrate lipids (**5**), (**7**) and two inhibitor lipids (**6**), (**8**) have been synthesized. Lipids (**5**) (dioctadecylamine-substrate, DODA-S) and (**6**) (dioctadecylamine-inhibitor, DODA-I) have a very short spacer between the membrane-forming alkyl chains and the lipid headgroup. It is expected that investigations with these lipids will show that the specific interaction of receptor proteins with the membrane-incorporated ligands is strongly dependent on the physical state of the membrane. In lipids (**7**) (dioctadecylamine-diethylenoxide-substrate, DODA-EO_2-S) and (**8**) (dioctadecylamine-diethylenoxide-inhibitor, DODA-EO_2-I), the lipid headgroup is decoupled from the membrane forming alkyl chains by a long, flexible, hydrophilic spacer. In this case, the accessibility of the headgroups for proteins is greatly enhanced because of the greater distance between the membrane surface and the ligand.

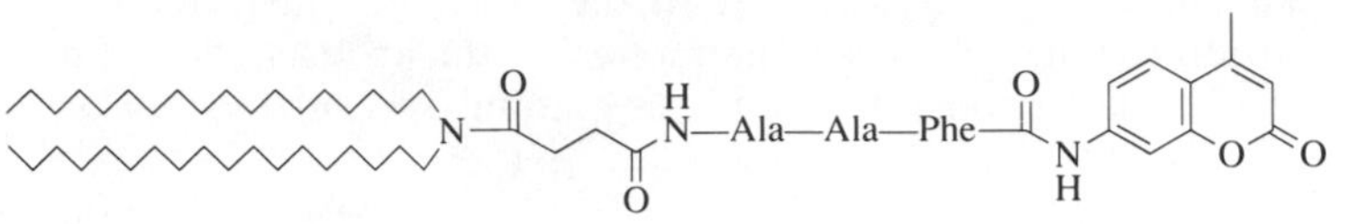

DODA-S (**5**)

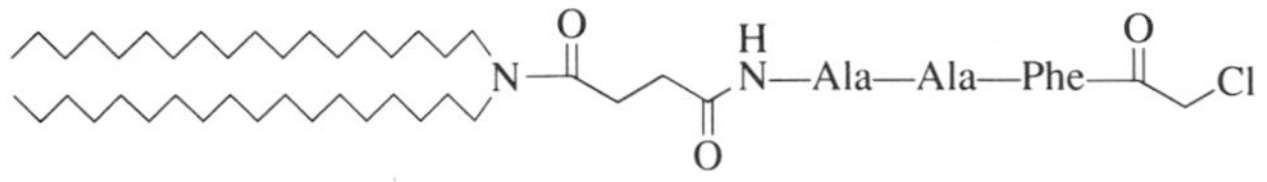

DODA-I (**6**)

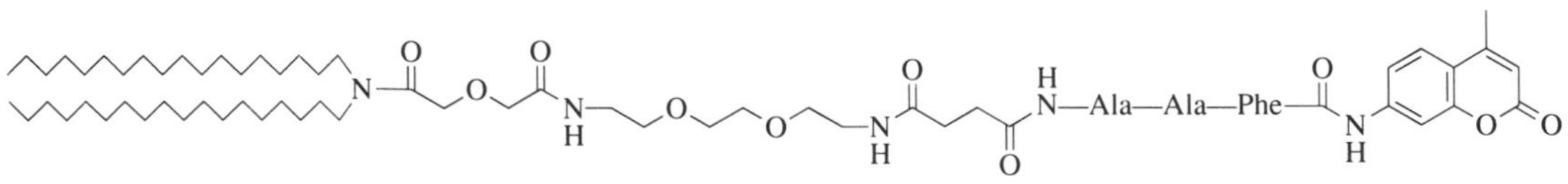

DODA-EO_2-S (**7**)

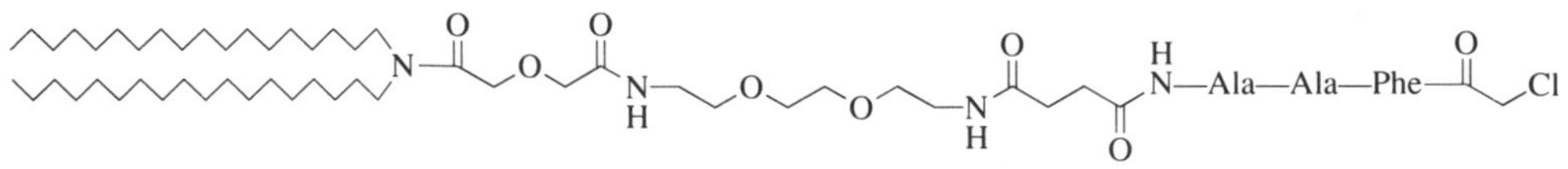

DODA-EO_2-I (**8**)

The lipids (**5**)–(**8**) have been characterized by pressure/area isotherms.[35] As one example, the pressure/area isotherms of (**8**) at different temperatures are shown in Figure 4. At 20 °C, the lipid exists only in a liquid analogous phase, which collapses at approximately 40 mN m^{-1}. At 5 °C, 10 °C and 15 °C, the isotherms show the existence of a liquid-expanded phase, a phase transition region and a solid analogous phase. The pressure/area isotherms for the other three lipids (**5**)–(**7**) look similar.[3c] These experiments demonstrate that all four ligand lipids form monomolecular films at the air–water interface and thus can be incorporated in model membrane systems.

Interaction of the receptor proteins with the functionalized ligand lipids (**5**)–(**8**) was studied in different model biomembranes.[3c] The results obtained with lipids (**5**) and (**6**) with the short spacer show clearly that the interaction of proteins with these membrane-incorporated ligand lipids is strongly dependent on the physical state of the membrane, because the headgroup is directly linked to the membrane-forming alkyl chains. For the results described in this contribution, we consider only lipids (**7**) and (**8**) with the longer spacer, because binding of the barrel-shaped proteasomes to monolayers and liposomes requires a long spacer to allow interaction of the headgroups with the, presumably, deep binding pocket of the multienzyme complexes.[36]

9.2.3 Interaction of Chymotrypsin and Proteasomes with Functionalized Model Membranes

9.2.3.1 Hydrolysis of substrate lipids with chymotrypsin and proteasomes

The accessibility of the membrane-linked substrate lipid headgroups for chymotrypsin and proteasomes was tested by measuring hydrolysis of the substrate headgroups by the proteins. Hydrolysis of the substrate was monitored in homogenous solution,[3c,20] in micelles,[3c] in

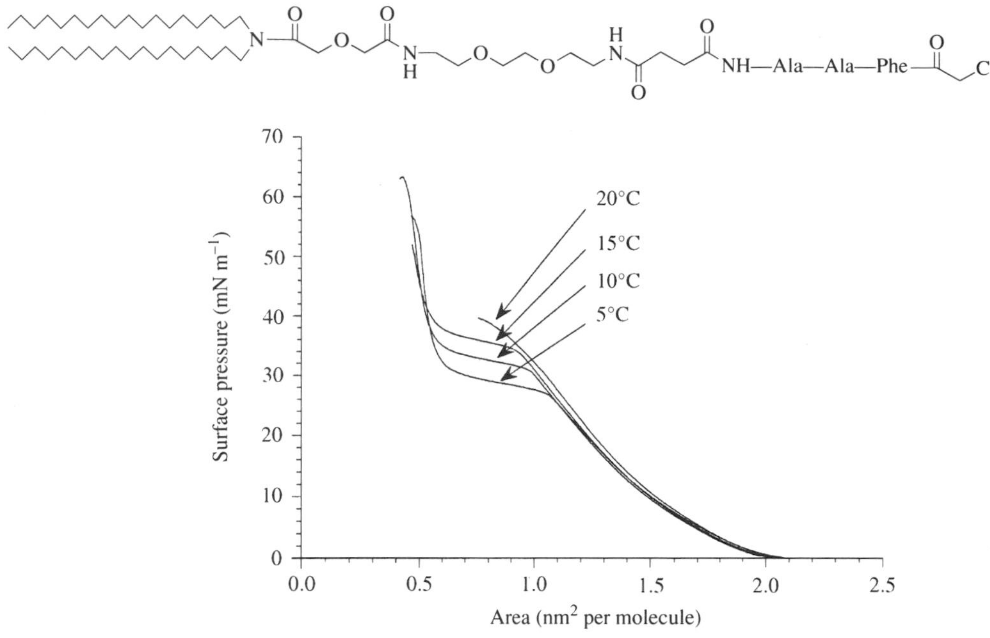

Figure 4 Surface pressure/area isotherms of the inhibitor lipid DODA-EO_2-I (**8**) at different temperatures.

liposomes[3c] and in monolayers[3c] (schematically shown in Figure 5). In this contribution, we will focus on the results obtained in solution, in liposomes and in monolayers, because these are the relevant experiments for the inhibition studies, discussed in Section 9.2.3.2 and Section 9.2.3.3.

(i) Hydrolysis in homogenous solution

Hydrolysis of the amide bond in the water-soluble oligopeptide Suc-Ala-Ala-Phe-AMC (**1**) (reaction shown in Figure 3(a)) by chymotrypsin was monitored in aqueous phosphate buffer.[20] The fluorescence spectrum of the covalently bound chromophore in (**1**) shows its fluorescence maximum at 395 nm (Figure 6(a)). After addition of the enzyme, the fluorescence intensity at 395 nm decreases and a new fluorescence band with a maximum at 435 nm appears (Figure 6(b)–6(j)), demonstrating the release of the chromophore (**3**) (7-amino-4-methyl-coumarin). The fluorescence spectra show that the hydrolysis reaction in isotropic solution is fast and the presence of an isosbestic point indicates a homogenous process.[37]

Hydrolysis of the water-soluble substrate (**1**) by proteasomes can be monitored in the same way.[24,38] The resulting spectra show similar characteristics—the only difference is the timescale. The hydrolytic activity of the proteasomes is much lower compared to chymotrypsin. Therefore, the time necessary for total conversion of the substrate is longer.

(ii) Hydrolysis in mixed liposomes

Hydrolysis of the substrate lipid DODA-EO_2-S (**7**) by chymotrypsin and proteasomes was monitored in mixed liposomes consisting of 1–2% of the ligand lipid (**7**), mixed with 98–99% of dimyristoyl-phosphatidyl-choline (DMPC). These functionalized mixed liposomes were prepared by sonication in phosphate or Mops buffer.[5,39] Enzymatic hydrolysis of the substrate lipid (**7**) was initiated by injection of the proteins into the liposome solutions.

Figure 7 shows the fluorescence spectra obtained during hydrolysis of (**7**) incorporated into liposomes, by chymotrypsin. Again, the fluorescence spectrum of the bound chromophore shows its fluorescence maximum at 395 nm (Figure 7(a)). During hydrolysis of the substrate headgroup of lipid (**7**), the intensity of this band decreases, commensurate with an intensity increase in the fluorescence band for the released chromophore at 435 nm (Figure 7(b)–7(j)). In comparison to the hydrolysis of the water-soluble substrate (**1**) in solution, the time necessary for the hydrolysis of the substrate lipid (**7**) in liposomes is much longer. Presumably, the slower hydrolysis is a result of

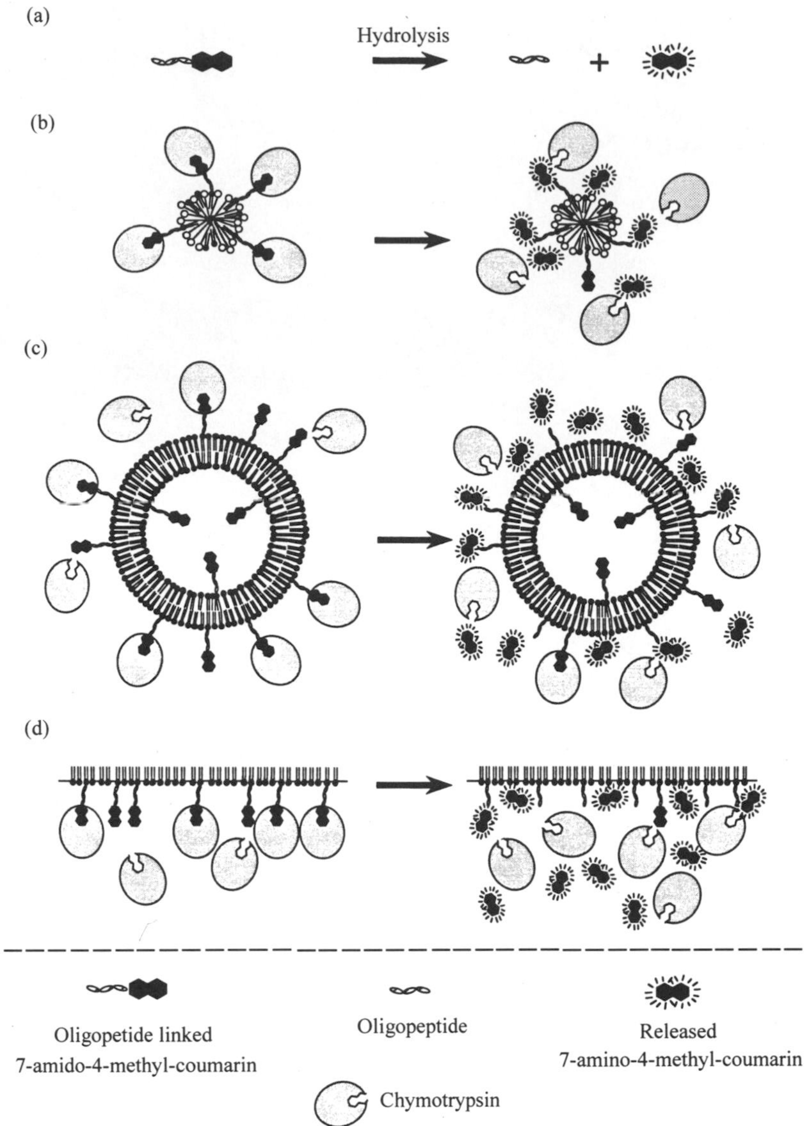

Figure 5 Representation of the hydrolysis of the soluble and lipid-linked substrate Suc-Ala-Ala-Phe-AMC by chymotrypsin and proteasomes (only drawn with chymotrypsin) in four different model systems: (a) water-soluble substrate (**1**); (b) DODA-EO_2-S (**7**) in micelles; (c) DODA-EO_2-S (**7**) in liposomes; (d) DODA-EO_2-S (**7**) in monolayers.

the hindered accessibility of the substrate headgroup being at a membrane. This slow hydrolysis leads to a slow photobleaching of the chromophore. Therefore, no isosbestic point is visible in Figure 7. In addition, it is noteworthy that full hydrolysis of all liposome-linked substrate headgroups is not possible, as shown by the remaining fluorescence intensity at 395 nm. This partial hydrolysis is a result of the preparation method for the liposomes. Some of the substrate headgroups are trapped inside the liposomes (approximately 40%) and, therefore, are not accessible to the enzyme. However, all the peptide headgroups at the outer surface of the liposomes are recognized and cleaved by chymotrypsin, which proves that the enzyme can specifically interact with its substrate in organized systems.

The hydrolysis experiments with both enzymes in homogenous solution showed that the chymotryptic activity of proteasomes is much lower compared to the activity of chymotrypsin itself.[24,27,31] Increasing the temperature can significantly increase the hydrolytic activity of the multienzyme complexes, which show their highest catalytic turnover at approximately 90 °C. Therefore hydrolysis of (7) in liposomes by proteasomes was performed at 70 °C, where the liposomes are stable but, also, the hydrolysis occurs at a rate which can be monitored. Figure 8(a)

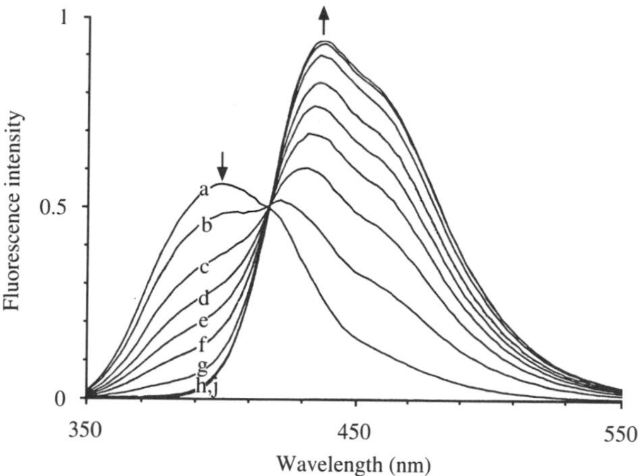

Figure 6 The fluorescence spectra as a function of time during hydrolysis of the water-soluble substrate Suc-Ala-Ala-Phe-AMC (**1**) by chymotrypsin: a, t = 0 s; b, t = 10 s; c, t = 90 s; d, t = 170 s; e, t = 250 s; f, t = 360 s; g, t = 600 s; h, t = 1200 s; j, t = 2400 s; T = 23 °C; phosphate buffer, pH = 7.5; enzyme concentration = 5 μg mL^{-1}.

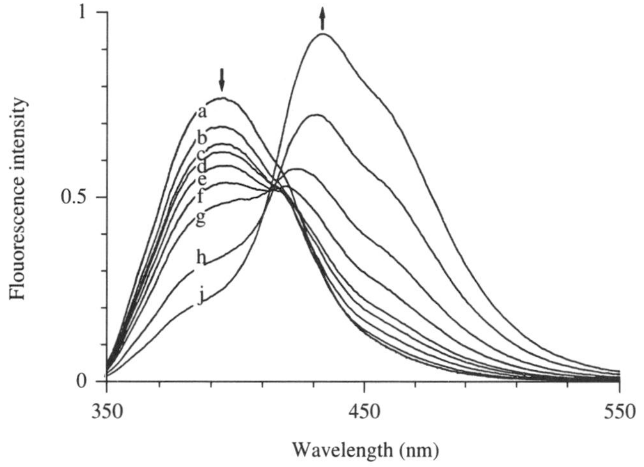

Figure 7 The fluorescence spectra as a function of time during hydrolysis of substrate lipid (**7**) by chymotrypsin in mixed liposomes (DMPC: substrate lipid (**7**) = 98:2): a, t = 0 min; b, t = 10 min; c, t = 1 h; d, t = 2 h; *e*, t = 3 h; f, t = 6 h; g, t = 10 h; h, t = 23 h; j, t = 47 h; T = 23 °C; phosphate buffer, pH = 7. 5; enzyme concentration = 100 μg mL^{-1}.

shows the fluorescence spectrum of the liposomes before addition of the proteasomes (covalently bound chromophore). After incubation of these liposomes with proteasomes for 3 d, about 50% of the substrate headgroups were cleaved by the enzyme (Figure 8(b)). This result clearly demonstrates that the highly organized cylindrical multienzyme complexes can also specifically recognize and hydrolyse membrane-bound substrates linked to the membrane with a long flexible spacer.

(iii) Hydrolysis in monolayers

After demonstrating that chymotrypsin as well as proteasomes can hydrolyse the substrate headgroup of (**7**) in curved liposome membranes, the hydrolysis reaction will also be investigated in monolayers. This model membrane system has already been used successfully for the preparation of two-dimensional protein crystals[30] and, thus, it will also be used to induce binding and orientation of proteasomes at inhibitor lipid (**8**) containing monolayers. Therefore, it is relevant to prove the accessibility of the substrate lipid headgroup for the proteins after incorporation of

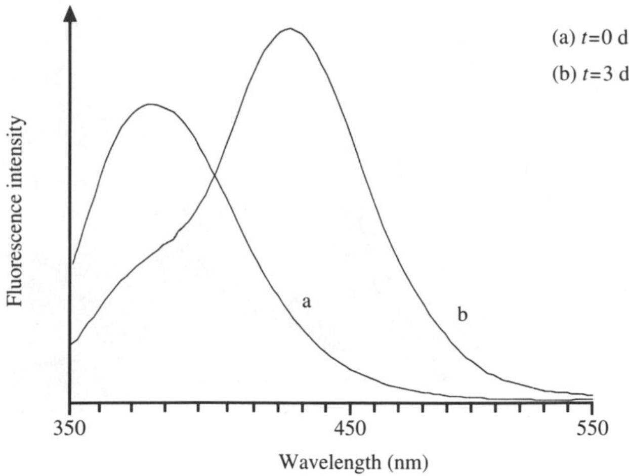

Figure 8 The fluorescence spectra as a function of time during hydrolysis of the substrate lipid (**7**) by proteasomes in mixed liposomes (DMPC: substrate lipid (**7**) = 99:1); Mops (morpholinopropansulton) buffer, pH = 7.5; proteasome concentration = 300 μg mL^{-1}.

substrate lipid (**7**) into monolayers. The hydrolysis experiments were only performed with chymotrypsin as the receptor protein, because large amounts of protein were available and the hydrolysis rate monitored at room temperature was high enough. As already shown for the liposomes, it can be assumed that proteasomes will give similar results.

Hydrolysis of substrate lipid (**7**) containing monolayers by chymotrypsin was monitored with fluorescence microscopy[3,40] and fluorescence spectroscopy at the air–water interface.[3c] With both methods, it could be shown that substrate lipid (**7**), pure and in a mixture with DMPC or palmitoyl-oleyl-phosphatidyl-choline (POPC), can be hydrolysed by chymotrypsin in monolayers. The fluorescence spectra show the same characteristic features as those obtained with liposomes (see Figures 7 and 8) and with the water-soluble substrate (**1**) in solution (see Figure 6). Upon addition of the enzyme, the intensity of the fluorescence band corresponding to the covalently linked substrate decreases, whereas the intensity of the fluorescence corresponding to the released chromophore (**3**) increases simultaneously. Therefore, these experiments show that the substrate headgroup of (**7**) is also accessible for protein binding in monolayers up to surface pressures of 30 mN m^{-1}.

All hydrolysis experiments have clearly demonstrated that the substrate headgroup of (**7**) can be recognized and hydrolysed, by chymotrypsin as well as proteasomes, after incorporation into different model biomembranes. Therefore, it can be concluded that binding of these proteins to functionalized model membrane systems containing the inhibitor lipid (**8**) should also be possible, because of their structural similarity. These experiments are described in the following sections.

9.2.3.2 Fixation of chymotrypsin at inhibitor lipid containing liposomes

The interaction of chymotrypsin with the inhibitor lipid headgroup of (**8**) was investigated in monolayers and giant vesicles,[3c] as schematically shown in Figure 9. The chymotrypsin was labelled with fluorescein-isothiocyanate (FITC)[41] in order that it could be detected upon binding to the inhibitor lipid (**8**) containing model membranes. In this contribution only the results obtained with the liposomes will be described, but all experiments performed with monolayers as model membrane systems support these results.[3c]

The specific binding of FITC-labelled chymotrypsin to inhibitor lipid containing liposomes was performed with giant vesicles.[3,42] These liposomes are visible with a light microscope (diameter = 10–40 μm). Therefore, interaction of the labelled protein with the liposomes could be easily detected with a normal fluorescence microscope.

Giant vesicles made of 10% DODA-EO_2-I (**8**) and 90% DMPC were prepared in a sealed Teflon chamber, schematically shown in Figure 10.[3,42] After 30 min incubation of these liposomes with FITC-labelled chymotrypsin, the excess protein was removed, and the liposomes were imaged

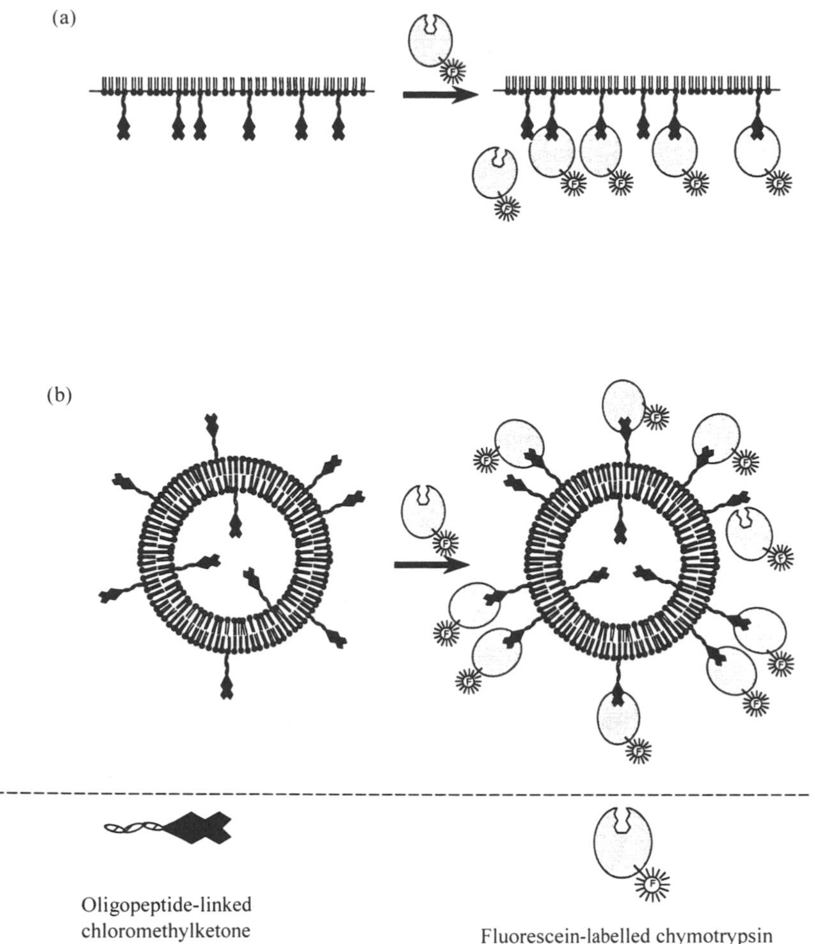

Figure 9 Representation of the interaction of fluorescein-labelled chymotrypsin with inhibitor lipid DODA-EO_2-I (**8**) containing (a) monolayers and (b) giant liposomes.

with the fluorescence microscope. A schematic representation of the molecular recognition process between the fluorescein-labelled chymotrypsin and the inhibitor lipid containing liposomes is shown in Figure 9(b). Plate 2 shows the images of the giant vesicles as observed in the light microscope in the phase contrast and the fluorescein filter mode. The outer membrane of the liposomes, visible in the phase contrast mode (Plate 2(a)), shows a very intensive fluorescein fluorescence in the fluorescein filter (Plate 2(b)). This demonstrates that a large amount of FITC-labelled chymotrypsin is linked to the membrane surface. In contrast to this, no vesicles with fluorescent membranes could be detected after incubation of pure DMPC liposomes with FITC-labelled chymotrypsin or incubation of DODA-EO_2-I (**8**) containing liposomes with inhibited FITC-labelled chymotrypsin (presaturated with the soluble inhibitor alanine-alanine-phenylalanine-chloromethylketone). As expected, the enzyme cannot bind to the surface of the vesicles under these conditions. These control experiments demonstrate that the intensive membrane fluorescence, visible in Plate 2(b), is only due to the specific interaction of chymotrypsin with the inhibitor headgroups of DODA-EO_2-I exposed at the surface of the liposome membranes.

Similar experiments were conducted with FITC-labelled proteasomes. After labelling of these multienzyme complexes with about 100 FITC molecules per enzyme complex, the proteasomes were fluorescent, but electron microscopy studies revealed that the particle structure was not intact any more. They did not retain their cylindrical structure (see Plate 1[43]), upon which their interaction with the giant liposomes depended. Proteasomes with a smaller number of FITC molecules per particle ($\sim$10) retained their structural integrity, but they did not show enough fluorescence to detect their surface binding with the fluorescence microscope.

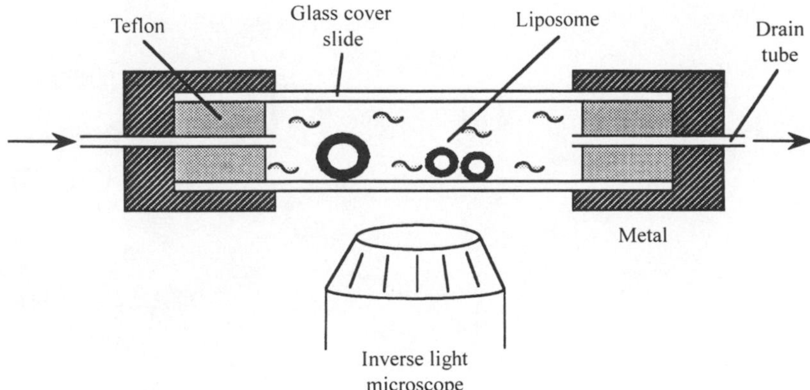

Figure 10 Schematic representation of a chamber for the preparation of giant liposomes (sectional drawing). The medium in the chamber is exchanged through the drain tubes and the liposomes are observed with an inverse light microscope (phase contrast or fluorescence mode); volume of the chamber = 200 μL.

9.2.3.3 Fixation and orientation of proteasomes from **Thermoplasma acidophilum** *at inhibitor lipid containing monolayers*

In the previous sections, it was demonstrated that proteasomes and chymotrypsin can specifically interact and hydrolyse substrate functionalized liposomes. To investigate the ability of the proteasomes to interact specifically with inhibitor lipid containing monolayers, electron microscopy was used.[3c] Binding of these multienzyme complexes to the functionalized model membranes, containing inhibitor lipid DODA-EO_2-I (**8**), may induce a preferred orientation of the proteasomes relative to the membrane. This orientation may depend on the position of the active site within the proteasome particles (see Figure 12).[30]

Several methods have already been used to obtain structural information about the proteasomes.[26,44] A resolution of 2 nm could be obtained by image analysis of single proteasome particles on carbon-coated grids. Electron micrographs of these negatively stained particles are shown in Figure 11 in different magnifications.[24] The proteins are shown either side-on (rectangular shape; indicated by an arrow) or end-on (ring shape; indicated by a circle). The four different rings, which form the proteasome particle (compare Plate 1), can be clearly identified in the side-on view.

Several investigations,[27] demonstrating that proteasomes show significant enzymatic activity only in the completely assembled state, have indicated that the active site of the enzyme is located either between one α-ring and one β-ring or at the interface of β-subunits within or between β-rings. In the case that this assumption is correct, there are two possible locations for the catalytic center. First, the active site could be inside the particle, only accessible through the top of the proteasome, or second at the outside surface of the protein between the different rings. Depending on the location of the binding site, specific interaction of the proteasomes with inhibitor lipid containing model membranes could induce a preferred orientation of the particles relative to the membrane. Two orientations are possible: location of the active site inside the particle might induce orientation of the proteasomes perpendicular to the membrane surface (Figure 12(a)); parallel orientation of the protein relative to the membrane would suggest binding of the inhibitor headgroups to the outside of the proteasomes (Figure 12(b)). Therefore, experiments demonstrating that the interaction of proteasomes with inhibitor lipids incorporated into monolayers lead to binding and a preferred orientation of the protein complexes relative to the membrane, were devised in order to elucidate the position of the active site in the enzyme particle.

Monolayer–proteasome complexes for the electron microscopy investigations were prepared using a Langmuir film balance with a subphase volume of 5 mL.[45] After spreading of the appropriate amount of the inhibitor lipid (**8**) pure or in a mixture with stearyl-oleyl-phosphatidyl-choline (SOPC) on a buffer subphase, the monolayer was compressed to a surface pressure of 20 mN m^{-1}. The proteasomes were injected into the subphase underneath the inhibitor-functionalized monolayer and incubated for approximately 24 h. Interaction of the proteasomes with the inhibitor lipid headgroups in the membrane should induce binding and orientation of the protein particles relative to the lipid film (see Figure 12). The resulting monolayer–proteasome complexes were lifted onto carbon-coated, hydrophobic electron microscopy grids using the Langmuir–

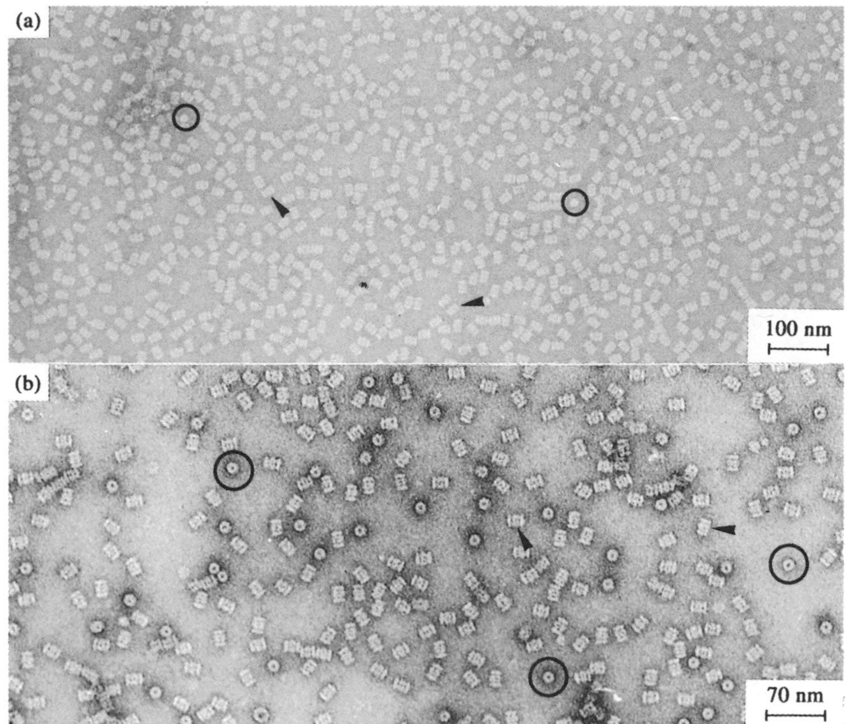

Figure 11 Two electron micrographs with different magnifications of negatively stained (uranyl acetate) proteasome particles on carbon-coated grids, seen side-on (rectangular-shape; arrows) and end-on (ring-shape; circles).

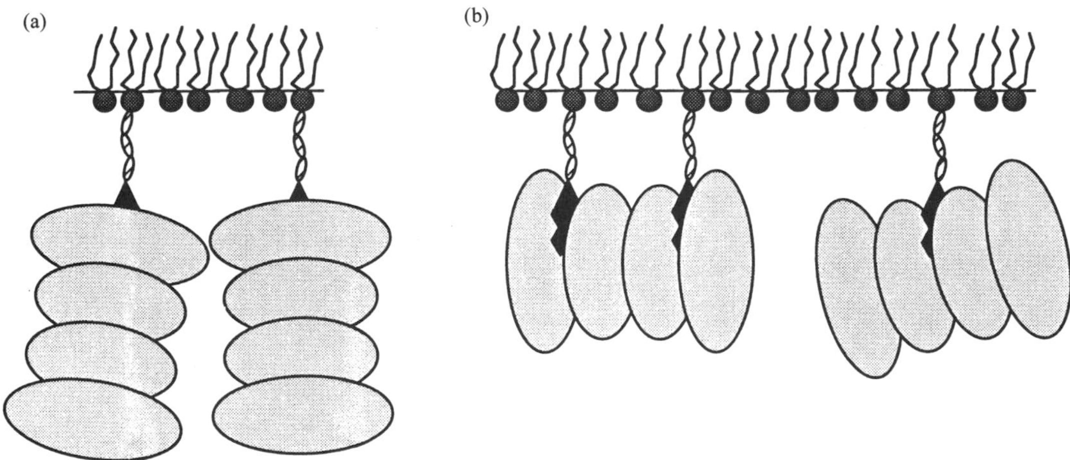

Figure 12 Representations of the possible orientation of proteasomes relative to the membrane surface after specific interaction of the membrane-incorporated inhibitor lipid headgroups with the active site of the proteasomes. (a) Location of the active site inside the particles will induce a preferred orientation of the proteasomes perpendicular to the membrane. (b) Binding of the membrane-incorporated inhibitor headgroups to an active site on the outside of the particles will induce a preferred orientation of the proteasomes parallel to the membrane.

Schäfer technique (Figure 13).[46] The samples were negatively stained with uranylacetate, air-dried and subsequently imaged with an electron microscope.

Figure 14 shows electron micrographs of four different areas of transferred lipid–proteasome layers (90% SOPC and 10% DODA-EO_2-I (**8**)). Densely packed proteasomes, which are all oriented with the cylinder axis perpendicular to the membrane, as schematically represented in Figure 12(a), are visible in the electron micrographs in Figure 14(a) and (b). In these two images,

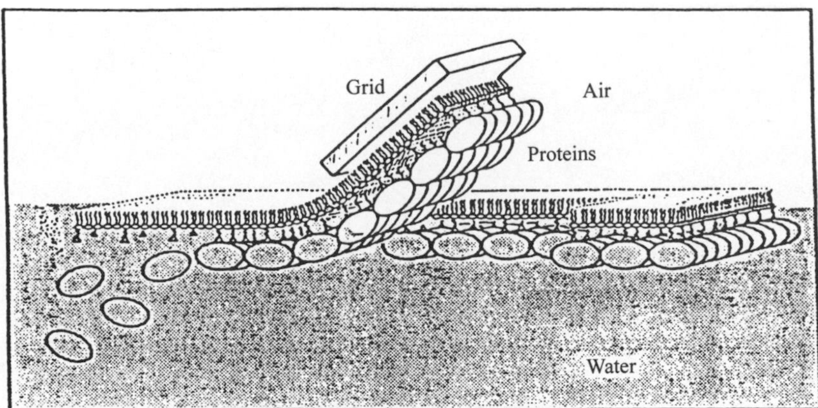

Figure 13 Schematic representation of a Langmuir–Schäfer transfer of monolayer—protein complexes onto electron microscopy grids: After placing a hydrophobized electron microscopy grid onto the lipid layer, the grid will be incubated for about 5 min, to allow interaction of the alkyl chains with the hydrophobic grid surface, and subsequently lifted. Then the transferred protein layer can be negatively stained with uranyl acetate.

there are only a few proteasomes, which are seen side-on, as schematically represented in Figure 12(b), and therefore, are oriented parallel to the film surface (circle and arrows). Just the opposite is shown in the two electron micrographs shown in Figure 14(c) and (d). Within the proteasome aggregates, visible in these two micrographs, all proteasome particles are oriented parallel to the membrane. In addition, the proteasomes within these aggregates are regularly ordered on a two-dimensional lattice.

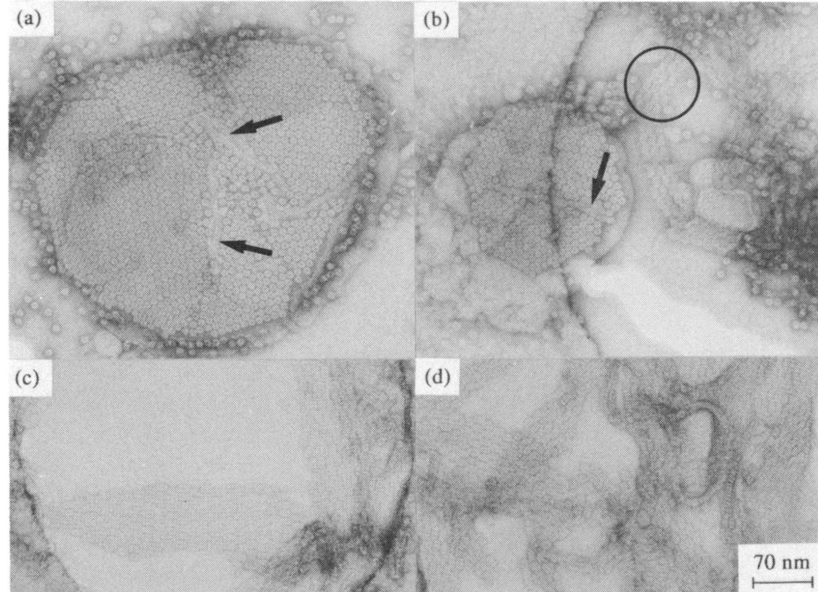

Figure 14 Electron micrographs of proteasome domains oriented at a monolayer of SOPC/DODA-EO_2-I (**9**) (9:1). (a), (b) The hexagonal packed proteasomes within the domains are all oriented perpendicular to the membrane; only a few particles in the domains (arrows) as well as some outside the domains (circle) are oriented parallel to the membrane; (c), (d) All aggregated proteasomes are oriented parallel to the membrane. The regular order within these protein aggregates already indicates that these aggregates are two-dimensional protein crystals.

This regular order within the proteasome domains, shown in Figure 14(c) and (d), indicates that the enzyme particles within these domains are highly organized. With image analysis, it could be demonstrated that these proteasome arrays are indeed two-dimensional protein crystals (Figure

15). All four rings within one particle can be clearly identified, but the positions of the single subunits in one ring (the mass distribution) are not exactly defined. The resolution that could be obtained with these images was limited to 1.6 nm, because the number of particles within one domain was too small to attain a higher resolution.

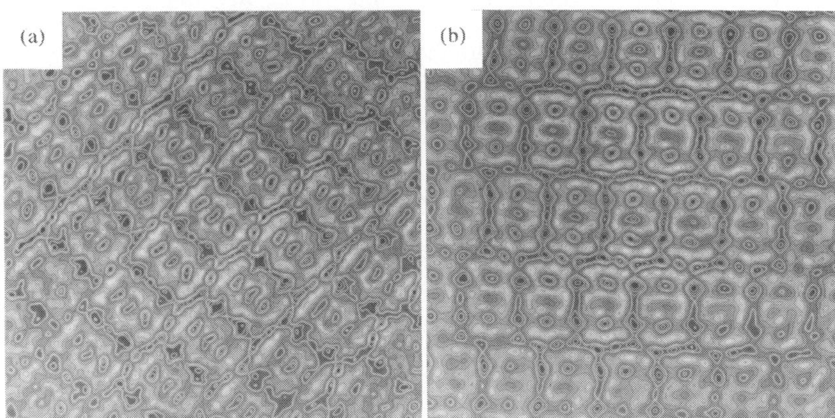

Figure 15 The side-view of two proteasome domains after averaging and image analysis: (a) average based on 22 particles; (b) average based on 67 particles.

The results described above show that proteasome aggregates with only one orientation of the enzyme particles relative to the membrane could not be obtained. Both types of orientations could be observed in one preparation. Therefore, it was not possible to get a clear indication regarding the position of the active site within the proteasomes. The reason for this may be that the spacer between the membrane-forming alkyl chains and the inhibitor headgroup was too long, allowing an orientation of the proteasomes both parallel and perpendicular to the membrane, as seen in the aggregates (as discussed in Figure 14), such that the mode of binding (Figure 12) had no effect on the orientation of the proteasomes at the membrane surface.

In the meantime, highly diffracting three-dimensional crystals of proteasome–inhibitor complexes from *Thermoplasma acidophilum* could be prepared. With these crystals, the structure of these multienzyme complexes was determined with a resolution of 0.34 nm.[47] Furthermore, the position of the active sites could be exactly defined. The catalytic centres, which are located in the interior of the particle, can be unequivocally assigned to the 14 β-subunits. Therefore, the active sites are only accessible from the cylinder ends, formed by the α-rings. In conclusion, these results support the hypothesis about the length of the spacer, allowing all orientations of the proteasomes relative to the membrane surface after binding.

In this section, one example, taken from the field of life sciences, where function is based on self-organization, was discussed. The described proteasomes are multienzyme complexes, which are only catalytically active if all 28 protein subunits are assembled to form the characteristically shaped cylindrical supramolecular complex (see Plate 1)—the catalytic function is dependent on the correct self-organization. Another example from the field of materials science, where function is directly correlated to supramolecular order, is described in the next section.

9.3 PHOTOCONDUCTIVE DISCOTIC LIQUID CRYSTALS

The second example, in which the correct self-organization determines the function of a system, is photoconductive discotic liquid crystals. It will be pointed out that the high charge carrier mobility of these compounds is only found in the liquid crystalline state and only if the compound is able to form a highly ordered liquid crystalline mesophase,[48–53] in which order and mobility are combined.

As schematically shown in Figure 16, it is interesting to follow the influence of increasing order and decreasing mobility on the charge carrier transport of photoconducting systems. While polyvinylcarbazol, which is widely used as a polymeric photoconductor,[54] is poorly ordered and has a low charge carrier mobility, organic single crystals such as anthracen, which are highly organized, show the highest photoconductivity in the field of organics.[54c,55] The major problem in using organic single crystals as photoconductors for technical applications is that they are not processible.[56] In addition, the molecules within one crystal lack any mobility, meaning that any mismatch in the crystal structure cannot be corrected and, thus, suppresses photoconduction. In this connection—as indicated in Figure 16—it was interesting to investigate the photoconductive properties of liquid crystals.[57] In the liquid crystalline phase, the molecules are ordered but still have a certain mobility. Compounds that are able to form differently organized liquid crystalline phases with varying mobilities should show different charge carrier mobilities depending on the type of mesophase. Before the results of our investigations will be discussed in detail, a short introduction to liquid crystals and a description of the basic principles of photoconductivity will be presented.

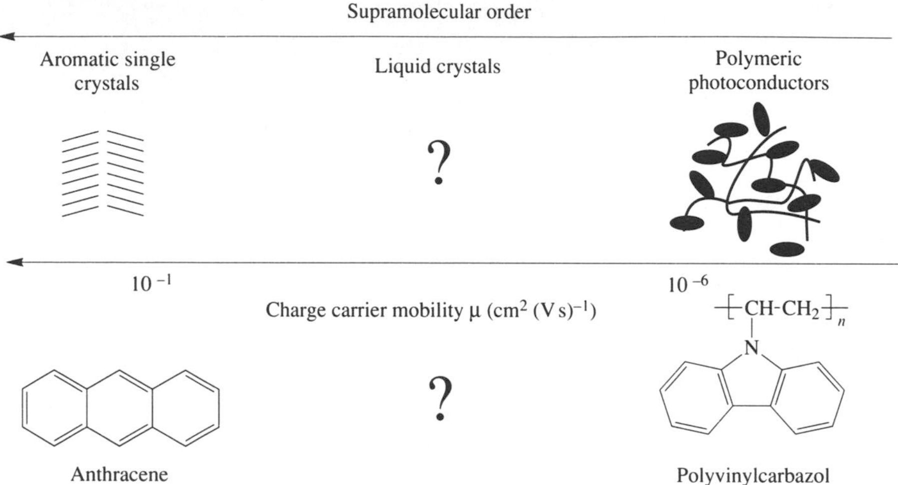

Figure 16 Influence of the supramolecular order on the charge carrier mobility of organic photoconductors.

9.3.1 Properties of Liquid Crystals

9.3.1.1 Liquid crystals—a fourth state of matter

Liquid crystals constitute one of the additional states to which the self-organization of matter may lead.[9] Although there are many possibilities for self-organization of matter, the molecular basis for the process is usually simple: formanisotropic molecules play the role of the simplest building blocks. For example, many organic compounds with rigid, rodlike molecular geometry, known as mesogens, fail to pass directly from the crystalline into the isotropic state upon heating (Figure 17).[9,58] Instead, they form intermediate liquid crystalline (LC) mesophases that are called thermotropic because mesophase changes are temperature dependent.[9,58,59] The relative order of the molecules in the liquid crystalline state determines the mesophase characteristics that are termed smectic and nematic phases (see Figure 17), involving different arrangements of essentially parallel rods.[58] Liquid crystals combine the properties of crystals and liquids—most characteristically, those of order and mobility. Within short distances, they show a similar orientational and positional order to that of a crystal but because of the remaining mobility they have only a certain degree of long-range positional order.[60]

Calamitic mesophases formed from rodlike mesogens are well known and have been realized in many monomeric and polymeric systems (Figure 18(a)).[9,58,59] Newer developments have led to the discovery that disclike molecules are also able to form different discotic liquid crystalline mesophases in monomers as well as in polymers (Figure 18(b)).[11,61–3] A third type of fundamentally different molecular structure is a boardlike shape (Figure 18(c)).[12,64] Boards do not show rotational symmetry about any axis at all, neither about the long axis as for rods nor about the short axis as for discs. Molecular engineering has succeeded in making boardlike structures by linking disclike mesogens rigidly in a main-chain polymer.

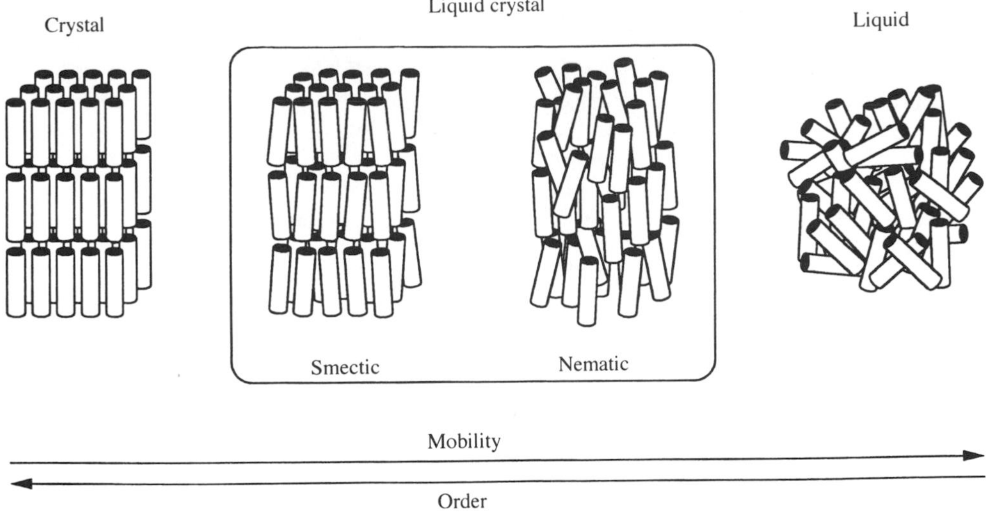

Figure 17 Liquid crystals combine order and mobility: typical phase behaviour of calamitic thermotropic liquid crystals.

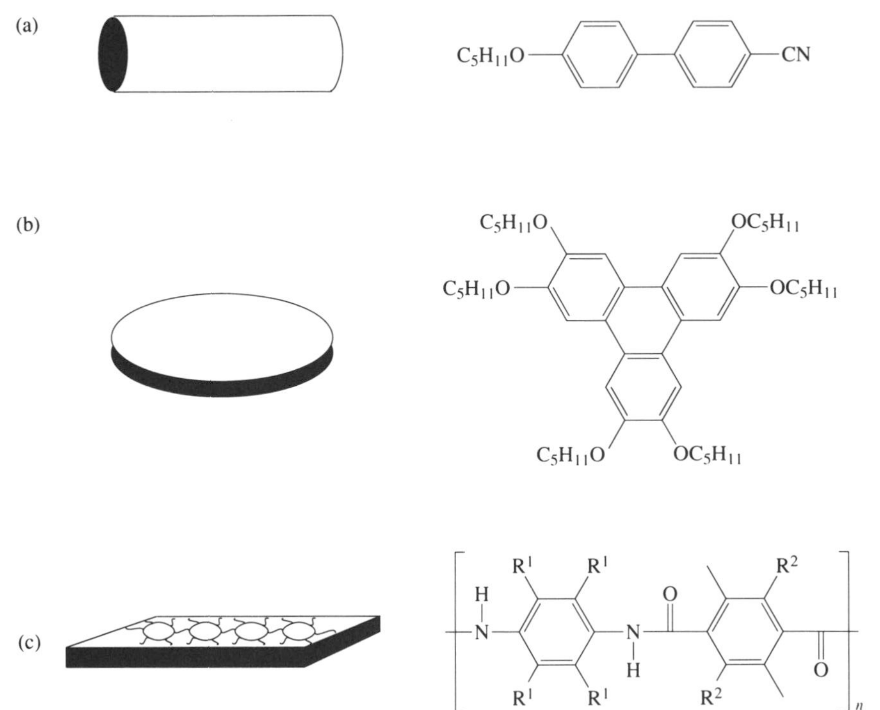

Figure 18 Formanisotropic molecules like (a) rods, (b) discs and (c) boards can form liquid crystalline phases.

9.3.1.2 Molecular engineering of liquid crystalline polymers

Low molar mass liquid crystals are already widely used for technical applications.[9c,9g,65] For better processibility of these materials, LCs have also been incorporated into polymers.[66] The development of polymeric liquid crystals began with stiff main-chain polymers, in which liquid crystallinity was observed.[67] Subsequently, besides the synthesis of new mesogens like discotics[11] and sanidics,[12] molecular engineering of liquid crystalline polymers[61] with respect to molecular architecture and functionalization has become increasingly important (Figure 19). Molecular architecture of LC polymers, for example, the variation of the arrangement of mesogens, the variation of their shapes[10–12] (rod, disc and board) and the variation of the polymer backbone, leads to polymers with new LC phases and new properties (Figure 19).[61] Parallel to this variation in molecular architecture, LC polymers have been functionalized in different ways. For this purpose, dye-containing groups,[68] groups undergoing chemical[69] or photochemical[70] reactions, and groups carrying chiral centers and strong dipole moments[71] have been either incorporated into the mesogens or were added as comonomers to form functionalized copolymers. In addition, functionalization of LC polymers was also achieved by incorporation of charge-transfer complexes.[61,62]

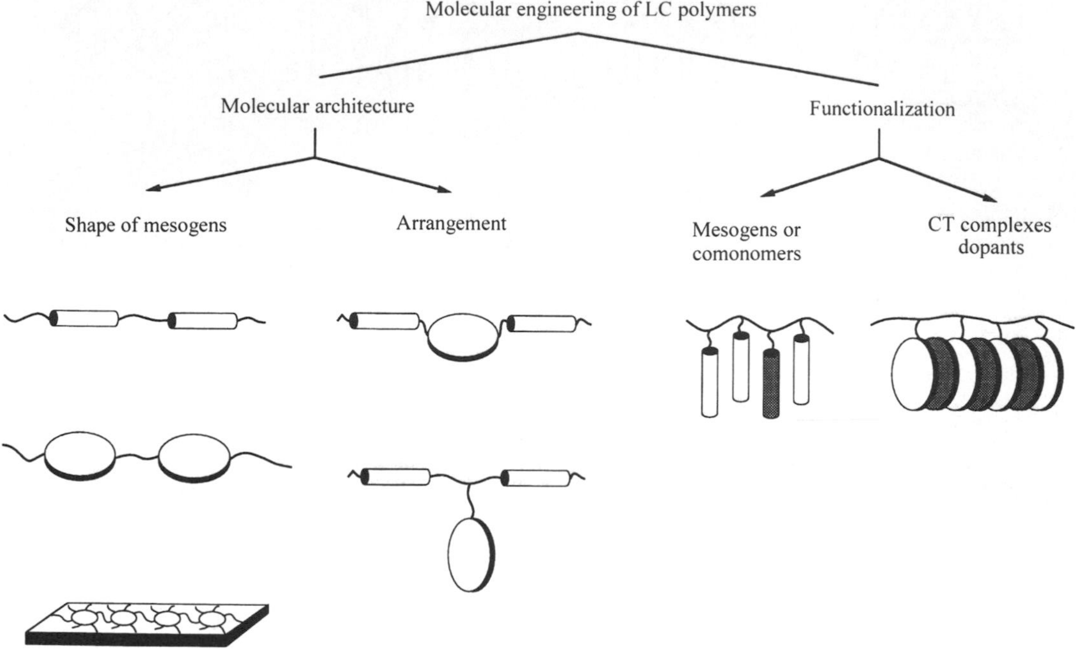

Figure 19 Molecular architecture and functionalization are two important aspects in molecular engineering.

9.3.1.3 *Properties and structures of discotic liquid crystals*

This section is concerned with the photoconductive properties of discotic liquid crystals. Therefore, only this class of liquid crystalline compounds will be described in more detail.

Analogously to calamitic phases, which are divided into nematics and smectics,[58,59] discotic liquid crystals can also form different mesophases.[63] They have either orientational order only, as in the nematic-discotic phase (N_D), or they show a columnar order with orientational as well as short-range positional order of the mesogens.[72] The columnar phases can be further divided into one mesophase in which the columns themselves have no positional order—the nematic columnar phase (N_C)—and mesophases in which the columns are ordered in a two-dimensional lattice. The name of the mesophase corresponds to the lattice type.[73] In the case of a hexagonal lattice, the mesophase is called discotic-hexagonal (D_h). Three of the possible mesophases for discotic liquid crystals are shown in Figure 20.

Figure 20 Schematic representation of discotic mesophases: N_D = nematic discotic; N_C = nematic columnar; D_h = discotic hexagonal.

Molecules that are able to form discotic mesophases are normally flat, rigid aromatic molecules surrounded by aliphatic side chains of varying lengths. Core–core interactions as well as van der Waals interactions between the aliphatic side chains are responsible for the formation of discotic mesophases. A number of different molecules that form discotic liquid crystalline mesophases are shown in Figure 21. They demonstrate the high structural variety for discotic mesogens.[11,74]

In this section, the focus will be on discotic liquid crystals that are derivatives of triphenylene.[75] Compounds like hexapentyloxytriphenylene belong to the low molar mass liquid crystals. They have also been successfully incorporated into side-chain and main-chain polymers to form polymeric discotic liquid crystals, like the triphenylene main-chain polymer (**9**) shown in Figure 22(a).[61–3] The low molar mass as well as the polymeric discotic liquid crystals build discotic-hexagonal (D_h) mesophases. Proof for the formation of the D_h phases was obtained by x-ray diffraction, electron diffraction (figure 22(b)) and high-resolution electron microscopy (Figure 22(c)).[76] In the x-ray spectrum, the intercolumnar distance as well as the distance between the discs in each column can be identified. In the electron diffraction pattern, obtained with a sample of the polymeric triphenylene (**9**) in the D_h phase (structure is shown in Figure 22(a)), the hexagonal order of the discotic mesogens is already indicated (Figure 22(b)). The high-resolution electron micrograph, obtained with a sample of compound (**9**), clearly shows that the columns formed by the triphenylene units are arranged on a hexagonal lattice (Figure 22(c)).

After demonstrating that monomeric and polymeric liquid crystals are indeed able to form highly organized liquid crystalline mesophases, in which order and mobility are combined, the basic principles of photoconductivity will be presented.

9.3.2 Photoconductivity: Principles and Characterization Methods

9.3.2.1 Principles of photoconductivity

Irradiation of photoconducting materials with light increases their conductivity by several orders of magnitude.[54c] Application of such materials for technologies such as xerography and laser-printing is limited to compounds that have a low conductivity in the dark (10^{-12}–10^{-14} S cm^{-1}).[77] The principal mechanism of charge transport in photoconductors is schematically illustrated in Figure 23. In the first step, the surface of the photoconducting material is electrostatically charged by an external electrical field. Irradiation of the charged surface with light leads to the formation of electron–hole pairs.[78,56] Depending upon the polarity of the external field, holes (Figure 23(a)) or electrons (Figure 23(b)) will migrate across the sample. The charge transport itself can be described as a thermally activated hopping of electrons between localized transport molecules.[79]

9.3.2.2 Characterization of photoconducting materials

Two methods have been employed for the characterization of our photoconductive materials.[54a,54c,80,81]

In the first method, the stationary photocurrent is measured upon illumination of the sample. A schematic representation of the experimental setup is shown in Figure 24(a). With this method, it is possible to measure the induced photocurrent of a sample as a function of the temperature. Therefore, temperature-induced changes in the photoconductive properties of the sample (i.e., in different mesophases of a liquid crystal) can be directly monitored with this procedure.

A more detailed understanding of the charge transport in photoconductors can be obtained with time-resolved measurements of the photoconductivity, using the so-called time-of-flight (TOF) method. The experimental setup for such experiments is shown in Figure 24(b). After an electric field is applied, the sample is irradiated with a laser pulse, which is shorter than the transit time for the electrons through the sample. Subsequently the time-dependent change in the induced photocurrent is recorded. More details about the experimental setup have been reported elsewhere.[81,54a] By analysing the measured transient photocurrents, two main aspects of photoconductive materials can be investigated: (i) conclusions can be drawn about the underlying charge carrier transport mechanism; (ii) the effective carrier mobilities μ can be calculated, using the well-known relation

$$\mu = l^2/t_T U \qquad (1)$$

where t_T is the carrier transit time as derived from the transient current, l is the sample thickness and U is the applied voltage across the sample.

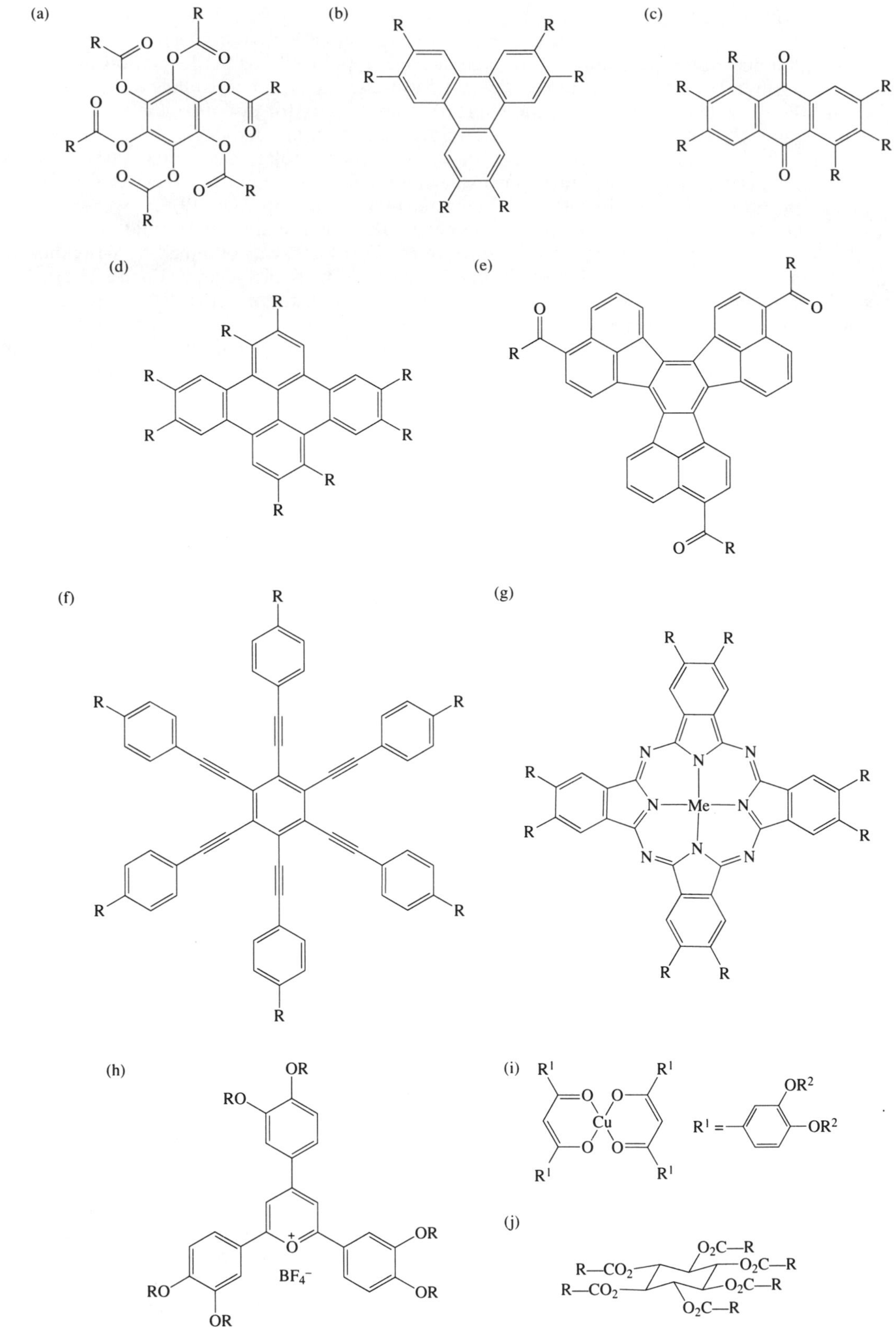

Figure 21 Structures of different discotic mesogens: (a) hexakisacyloxybenzene;[74a,74b] (b) triphenylene;[74c,74d] (c) rufigallol;[74e] (d) dibenzopyrene;[74f] (e) decacyclene;[74g] (f) hexaalkinylbenzene;[74h,74i] (g) phthalocyanine;[74j,74k] (h) 2,4,6-triarylpyrilium salts[74l,74m] (i) metal–bis(β-diketonate) complex;[74n,74o] (j) inositol.[74p]

(a)

$C_5H_{11}O$ O O O O OC_5H_{11} $C_5H_{11}O$ OC_5H_{11} $(\ )_{10}$ m

(9)

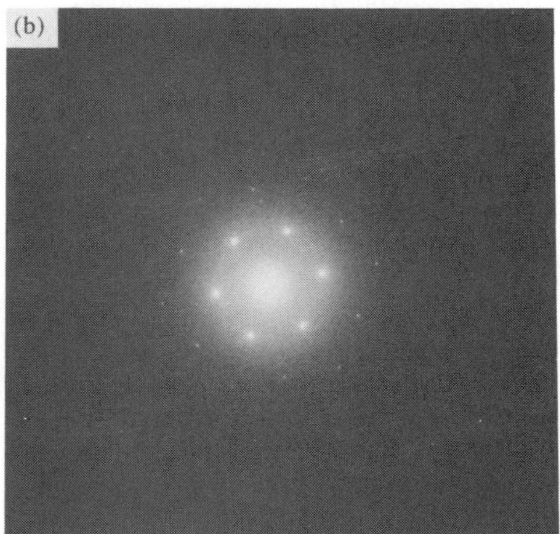

Figure 22 Experimental proof for the formation of a discotic hexagonal mesophase. (a) Structure of the discotic main-chain polymer (**9**) based on triphenylenes. (b) Electron diffraction pattern of the discotic polymer (**9**) in the D_h phase, viewed along the columnar axis; the hexagonal order of the diffraction pattern indicates a hexagonal order. (c) High-resolution electron micrograph of the discotic polymer (**9**) in the D_h phase, viewed along the columnar axis; the image shows clearly the hexagonal order in the discotic mesophase.

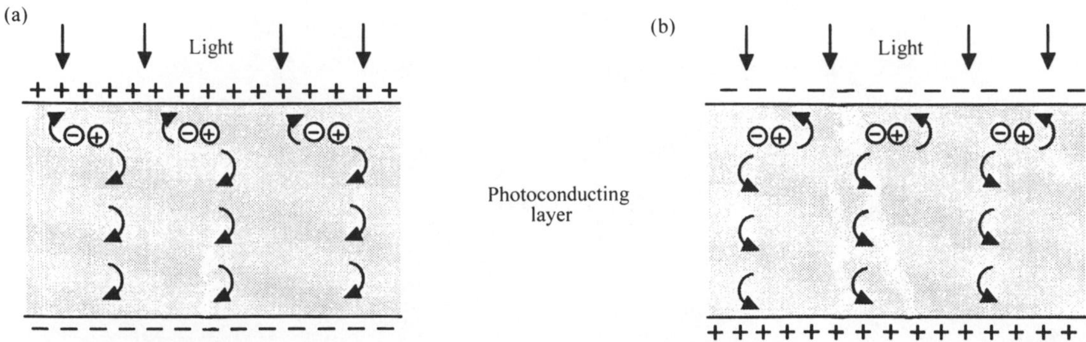

Figure 23 Principle of photoconductivity: after charging of the photoconducting material by an external electric field, irradiation with light leads to the formation of electron–hole pairs. Subsequently, depending on the polarity of the external field, (a) holes or (b) electrons will migrate across the sample to the counter electrode, causing displacement currents that can be recorded in an external circuit.

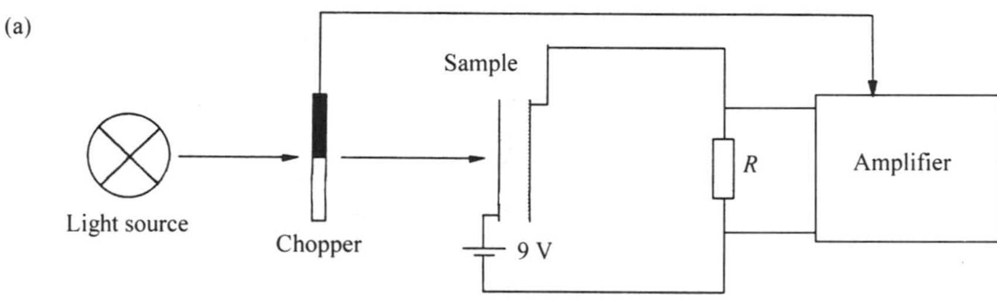

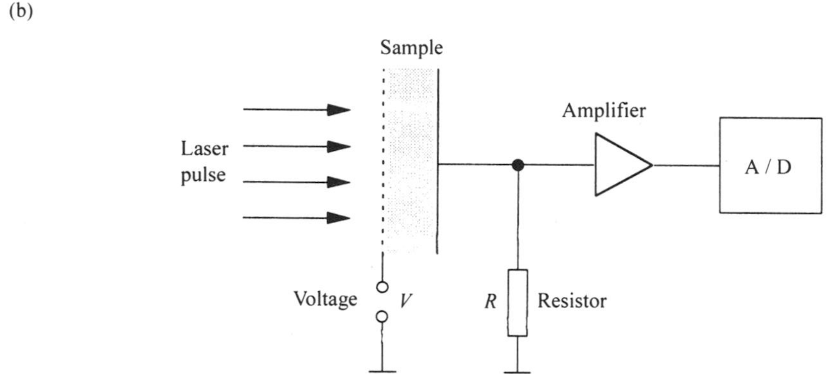

Figure 24 Schematic representation of the experimental setup (a) for the temperature-dependent determination of the stationary photocurrent and (b) for the measurement of transient photocurrents in a typical time-of-flight (TOF) experiment.

9.3.2.3 Tailoring of new photoconducting materials

Organic photoconductors such as the polymer polyvinylcarbazol (PVK) belong to the few examples where organic materials have found technological application, for instance, in xerography and laser-printing techniques. However, one main disadvantage of the conventional polymeric materials is their low charge carrier mobility, which is approximately $10^{-6}\,cm^2\,(Vs)^{-1}$.[81,54a] In general, the low charge carrier mobilities of organic photoconductors are due to a trap-dominated hopping transport between the photoconductive groups, giving rise to a certain localization of charge carriers.[77b] The only organic systems known so far to show electronic charge carrier mobilities comparable to the amorphous inorganic semiconductors that are the mainstay of the microelectronics industry are zone-refined organic single crystals.[82,55] However, single crystals are difficult and costly to process and, therefore, are not suitable for device applications.

For the development of new photoconducting materials, it is necessary to combine the high charge carrier mobility of aromatic single crystals with the processibility of polymeric photoconductors (schematically shown in Figure 25) in one material. One possible way to overcome this dichotomy, between the high mobility demand on the one hand and good mechanical properties on the other, could be the use of low molar mass and polymeric discotic liquid crystals.[48–53,57] The columnar structure of these materials is similar to that of many aromatic columnar single crystals, which have been extensively studied, and show high charge carrier mobilities.[82,55] Thus, such discotic systems and especially those exhibiting the highly ordered columnar phases (D_h) seem to be well suited for an electronic transport parallel to the columnar axis because the conjugated aromatic cores are packed in a favourable face-to-face orientation with good overlap of molecular wavefunctions over several repeat units along the columnar stacks. In addition, the liquid crystalline character of these compounds guarantees easy processibility. This all indicates that the LC concept of achieving high carrier mobilities in processible materials offers a new avenue for establishing organic high charge carrier mobility systems.

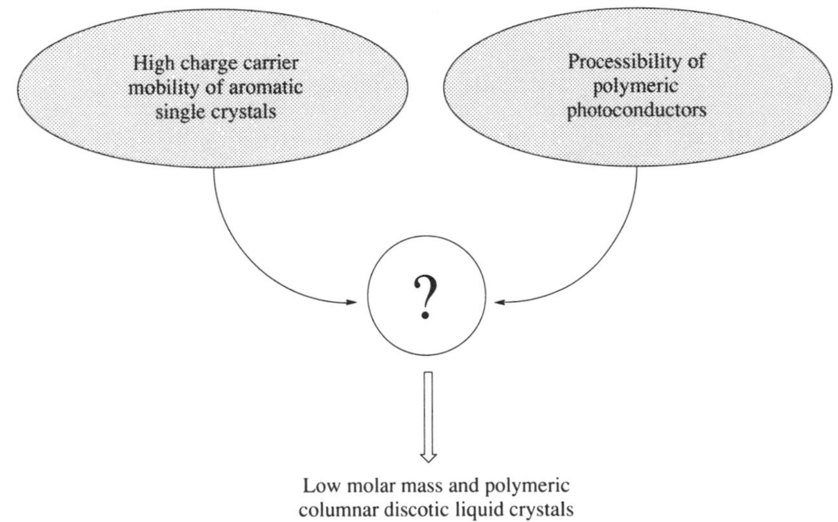

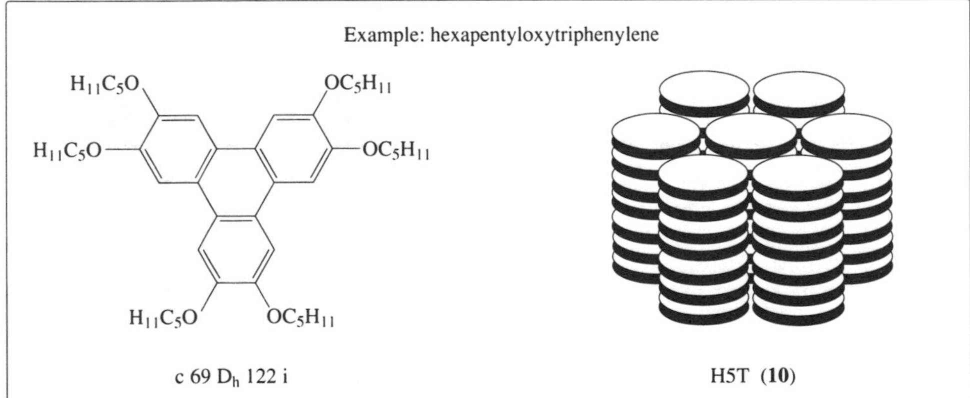

Figure 25 Tailoring of new photoconducting materials: low molar mass and polymeric discotic liquid crystals as new organic photoconductors combine the high charge carrier mobility of aromatic single crystals and the processibility of polymeric photoconductors. As one example, the structure of hexapentyloxytriphenylene (**10**) and a schematic representation of its liquid crystalline mesophase is shown.

The first substance under investigation was hexapentyloxytriphenylene (H5T) (**10**).[48–50,52,75b,83] These molecules form a columnar discotic hexagonal ordered phase (D_h) between 69 °C and 122 °C.

9.3.3 Photoconductivity of Hexapentyloxytriphenylene (10)[48–50]

The photoconductive properties of discotic liquid crystals were first tested using hexapentyloxytriphenylene (**10**) (H5T)[83,75b] as the photoconducting material. Its polycyclic aromatic structure with six donor substituents stabilizes the formation of a radical cation.[84] Such radical cations occur very often in transport processes of photoconducting materials. In the columnar discotic hexagonal mesophase (D_h) between 69 °C and 122 °C, the disclike H5T (**10**) molecules build up stacked arrays with an intramolecular distance between the mesogens of only 350 pm, allowing good overlap of the π-electron systems. Therefore, analogous to the aromatic single crystals, H5T (**10**) should show an anisotropic charge transport along the columnar axis.[85] The intercolumnar distance of 2020 pm[48] is too large to allow electronic π–π interactions between the triphenylene cores in different columns. A schematic representation of the expected charge transport along the columnar axis of the triphenylene units in the discotic hexagonal mesophase upon irradiation is shown in Figure 26.

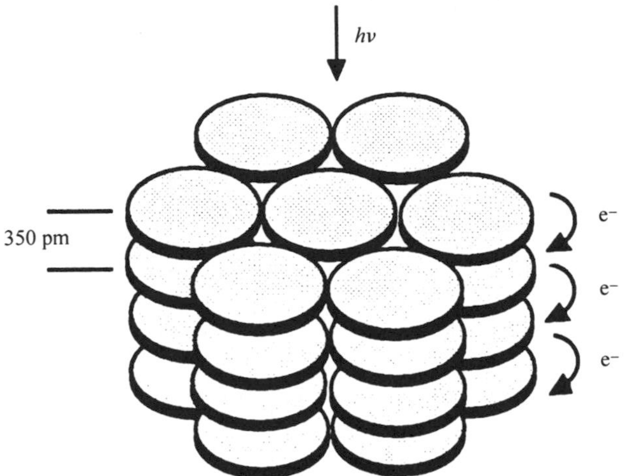

Figure 26 Schematic representation of the proposed charge transport along the columnar axis in hexagonal-ordered discotic liquid crystals upon irradiation.

9.3.3.1 Stationary photocurrent of hexapentyloxytriphenylene (10)[48]

The experimental setup used for the determination of stationary photocurrents in photoconductive materials is shown in Figure 24(a). The photocurrent of H5T (**10**) was measured in dependence of the temperature (Figure 27). The temperature interval was chosen to include all three different phases—crystalline, liquid crystalline and isotropic—in which this compound exists. Starting in the crystalline phase, there was no photocurrent measurable until approximately 65 °C, where 15 nA were recorded –4 °C below the formation of the D_h mesophase. When the mesophase was reached, the photocurrent increased continuously, nonlinearly with increasing temperature, until a maximum photocurrent of 60 nA was measured at 110 °C, 12 °C below the clearing point. At about 115 °C, the photocurrent dropped to about 10 nA and stayed constant upon further heating into the isotropic melt. Subsequent cooling of the sample leads to a qualitatively similar trend in the photocurrent; only the measured currents are much higher and the transition from the D_h phase into the crystalline phase is shifted to lower temperatures because of the supercooling effect. Upon crystallization of the material the photocurrent dropped to zero again.

The higher photocurrents observed during the cooling cycle can be explained by the fact that, while cooling the sample from the isotropic melt, an orientation effect takes place.[48] Microscopic observations reveal that the homotropic alignment[73] of the triphenylene columns perpendicular to the electrode surface becomes much better on cooling. Therefore, the higher photocurrent during the cooling cycle originates in the better orientation of the liquid crystalline phase. This better orientation may result either in a reduction in the number of defects in the mesophase which otherwise would act as traps for the charge carriers or a larger number of charge transporting channels of H5T (**10**) oriented properly or both. A more detailed characterization of the photoconductivity of H5T (**10**), using the TOF-method, is described in the next section.

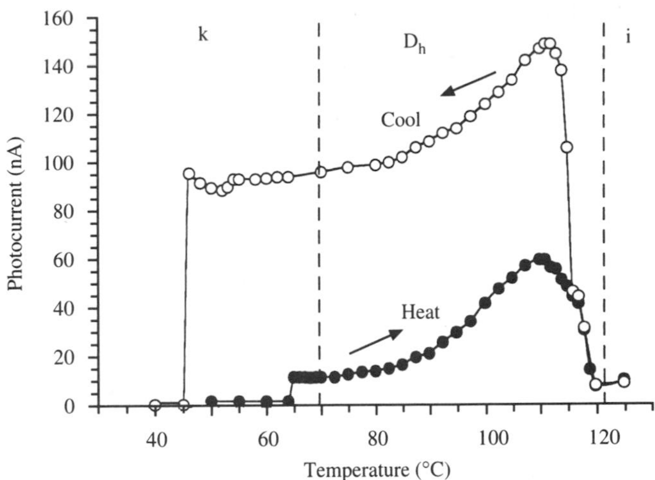

Figure 27 Temperature-dependent changes in the photocurrent of hexapentyloxytriphenylene (**10**) (filled circles: heating curve; open circles: cooling curve).

9.3.3.2 Time-resolved photoconductivity measurements with hexapentyloxytriphenylene[49,50]

For determination of the charge carrier mobility in H5T (**10**), time-resolved photoconductivity measurements were performed. The compound was filled into quartz cells of typically 10 μm thickness, which had been coated with semitransparent aluminum electrodes. The transient photocurrents generated by strongly absorbed short laser pulses were recorded using the typical TOF setup shown in Figure 24(b).

In Figure 28(a), the transient photocurrents for the crystalline and the D_h phase are shown in a linear plot. The irradiated electrode was positively charged, implying the detection of hole-transport behaviour (see Figure 23(a)). H5T (**10**) is displaying—as do a few photoconducting systems—the whole spectrum of photoconductive transient features. The hole current in the D_h mesophase is characteristic of a trap-free transport mechanism that is also referred to as nondispersive transport in the literature (Figure 28(a), right-hand side). The current is constant, even at shortest times, until it drops to zero rather abruptly at the transit time t_T. This behaviour is normally typical for many inorganic semiconductors but very exceptional for organic systems and resembles, with minor variations, data published, for example, for thin films of amorphous selenium.[86] In contrast, the photocurrent in the crystalline phase shows a totally featureless decay (Figure 28(a), left-hand side). This suggests a transport mechanism dominated by deep trapping; the latter is also called totally dispersive transport in the literature.[87] Responsible for this phenomenon, which is described in more detail below, are the grain boundaries within the microcrystalline structure, which are formed upon crystallization.[88]

To better understand the transport mechanisms in the discotic-hexagonal mesophase and in the crystalline phase of H5T (**10**), Figure 28(b) shows the time-dependent movement of the charge carriers through the sample. In the hexagonal mesophase, the charge carriers migrate through the sample in a 'packet' until they reach the counter electrode at the transit time t_T. In the crystalline phase only a few charge carriers reach the electrode—most of them stay near the irradiated surface and get trapped at grain boundaries.

The charge carrier mobilities for H5T (**10**) in the D_h mesophase were calculated using Equation (1). The calculation gives carrier mobilities for holes in the D_h mesophase of $10^{-3}\,cm^2\,(Vs)^{-1}$, which are comparable to those obtained for conjugated and quasiconjugated polymers.[89] These values are two orders of magnitude lower than aromatic single crystals and exceed those obtained for conventional amorphous systems by several orders of magnitude (Figure 29).[49]

In summary, H5T (**10**) is the first photoconductive system that exhibits both nondispersive transport—similar as for semiconductors—and deep trapping transport mechanisms—typical for amorphous polymers. One of the most interesting aspects of the liquid crystalline H5T (**10**) system is the temperature dependence of the transport phenomena. The high mobility transport in the mesophase—as compared to the inefficient, trap-dominated transport in the crystalline phase—suggests that certain fluctuations of the disclike molecules within the columnar stacks may lead to a transport mechanism, in which 'static' trapping phenomena are suppressed by the dynamic

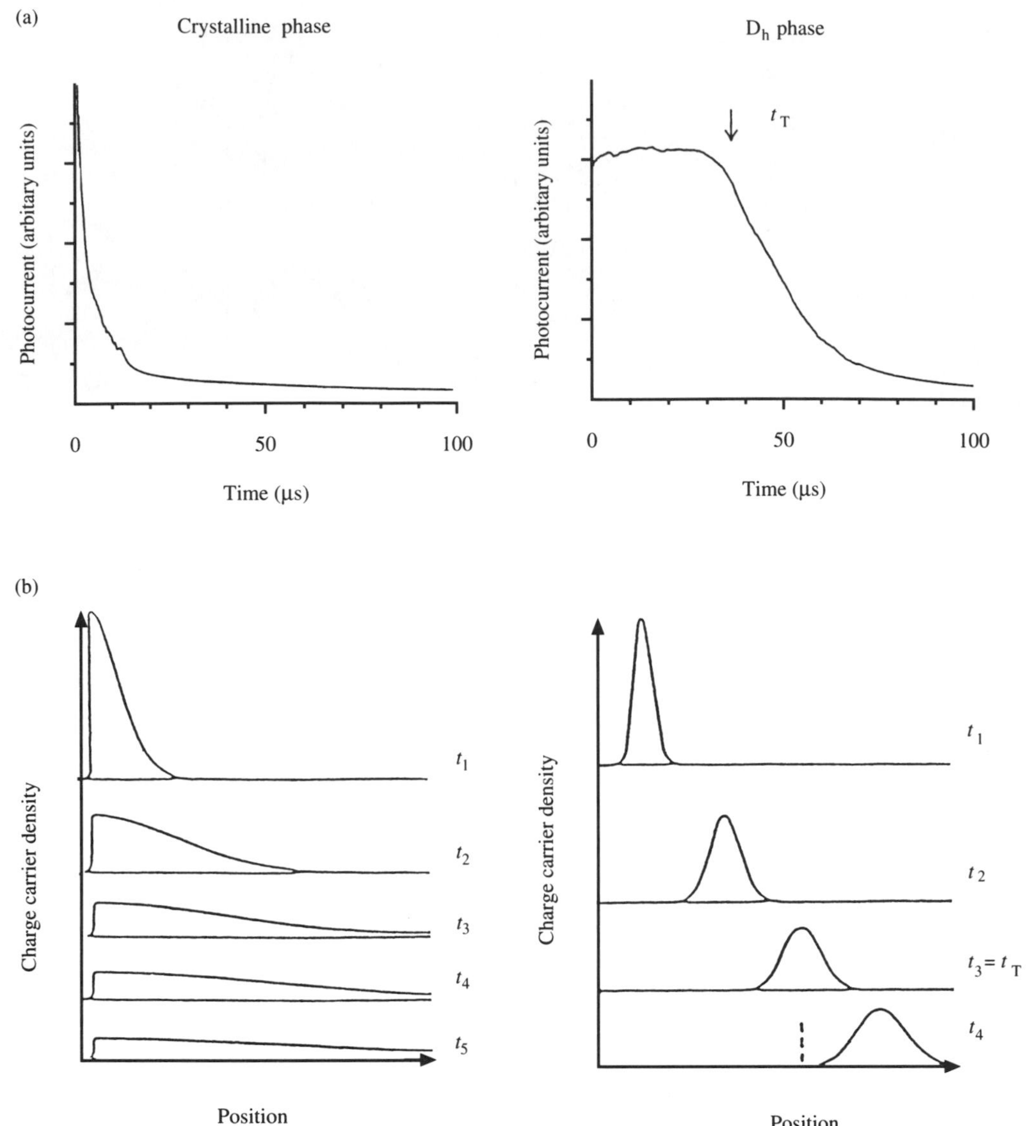

Figure 28 (a) Linear plots of the transient photocurrent I of hexapentyloxytriphenylene (**10**) as a function of time t at an electric field of $0.4 \times 10^5\,V\,cm^{-1}$ at $T = 26\,°C$ (crystalline phase) and at $T = 77\,°C$ (discotic hexagonal phase), and (b) a schematic representation of the transport mechanism of the charge carriers in the different phases—transport of holes.

properties of the LC system. In addition, these results indicate that there are no grain boundaries in the D_h phase.[88] The question arises as to whether it is possible to further increase the charge carrier mobility in a LC system to form a mesophase with the long-range positional and orientational order of the mesogens in crystals, but still retaining a certain degree of mobility to avoid the formation of grain boundaries.[88]

9.3.4 Photoconductivity of Hexahexylthiotriphenylene (11)[51]

As mentioned in the last section, the short-range intracolumnar order and considerable molecular fluctuations within the liquid crystalline phase are probably factors that limit the charge carrier mobility of H5T (**10**) in the discotic hexagonal mesophase compared to perfectly ordered single crystals. The compound 2,3,6,7,10,11-hexahexylthiotriphenylene (**11**) (H6ST) overcomes the above limitations by forming a self-organized, helical columnar (H) phase with nearly crystalline order[90] in addition to the normal D_h phase (Figure 30). The H phase can be obtained, without

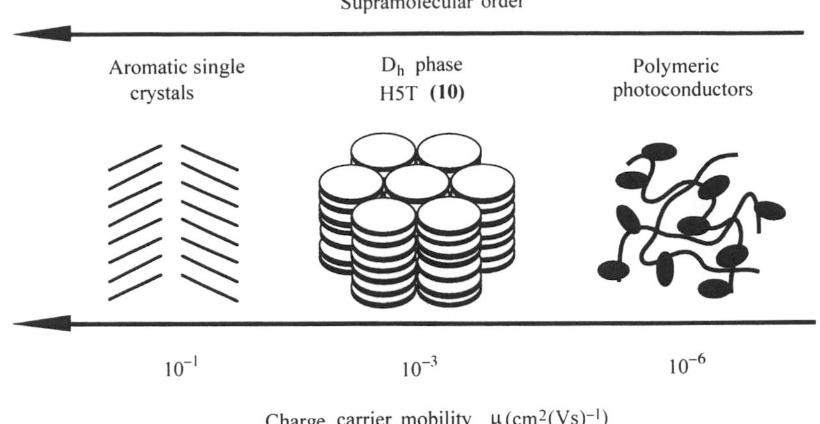

Figure 29 Charge carrier mobility of hexapentyloxytriphenylene (**10**).

major pertubations by grain- and domain-boundaries (as revealed by the determination of the photoconductive properties), by simple cooling of the sample from the liquid melt. It forms, in passing, a D_h liquid crystalline phase. The $D_h \rightarrow$ H transition is not accompanied by any major symmetry changes of the two-dimensional hexagonal lattice.

In the H phase, the columns of the discotic mesogens are ordered on a hexagonal lattice. The position of the single mesogenes within one column as well as the position of the columns itself are well defined. In addition, the positions of the mesogenes within different columns are correlated to each other, which is not the case for the mesogens in the normal D_h phase. This long-range positional order within the H phase,[90] which is comparable to the perfect order in single crystals, should lead to an increase of the charge carrier mobility in the H phase compared to the D_h phase. This hypothesis will be investigated by temperature-dependent measurements of the photocurrent transients in the different phases in a normal TOF experiment (described in Section 9.3.2.2) and subsequent calculation of the corresponding charge carrier mobilities.

The shape of the photocurrents in the H and the D_h phase of H6ST (**11**) (Figure 31) are similar to those obtained for H5T (**10**) in the D_h phase (see Figure 26) and can be explained by assuming nondispersive transport. This involves a narrow charge carrier package drifting across the sample at a constant velocity, broadened only by normal thermal diffusion leading to a moderate broadening of the photocurrent decay when the package reaches the counter electrode. This is in contrast to a dispersive behaviour as found in the polycrystalline (k) phase (Figure 31). Here, a charge carrier package is heavily smeared out in time and space because of trapping events, resulting in a featureless current decay. Finally, the current shape in the isotropic (i) phase shows a cross-over from Gaussian to dispersive behaviour, the transport now being governed by shallow trapping events (Figure 31).[81,54a]

These results are qualitatively the same as obtained with H5T (**10**): nondispersive transport in the ordered phases and dispersive behaviour in the crystalline phase. Differences in the charge carrier transport within the two mesophases become more obvious after calculation of the corresponding charge carrier mobilities. The hole mobilities as a function of temperature and for various phase regimes are shown in Figure 32. Starting in the liquid melt, where short-range order is still preserved,[90c] carrier mobilities are in an intermediate range of the order of $10^{-4}\,cm^2\,(Vs)^{-1}$. At the transition into the liquid crystalline D_h phase, a hexagonal columnar arrangement of molecules is established. This is reflected in the charge carrier mobilities as a steplike increase of one order of magnitude at the phase transition. A further increase of the charge carrier mobilities with decreasing temperature within the D_h phase is correlated with a further increase in the columnar correlation—which is on a short-range scale—as concluded from x-ray data.[60,90a] The mobility values in the D_h phase of H6ST (**11**) are comparable to those obtained in the D_h phase of H5T (**10**).[49,50] This is a clear indication that the charge carrier mobility is mainly determined by the supramolecular order of a system and not by the molecular structure of the compound. Just below the transition to the H phase, the charge carrier mobilities again increase by almost two orders of magnitude and reach values of the order of $0.1\,cm^2\,(Vs)^{-1}$. With the exception of organic single crystals,[82,55] these are the highest electronic mobility values for photoinduced charge carriers so far reported for organic systems. These mobility data in the H phase, together with the time dependence

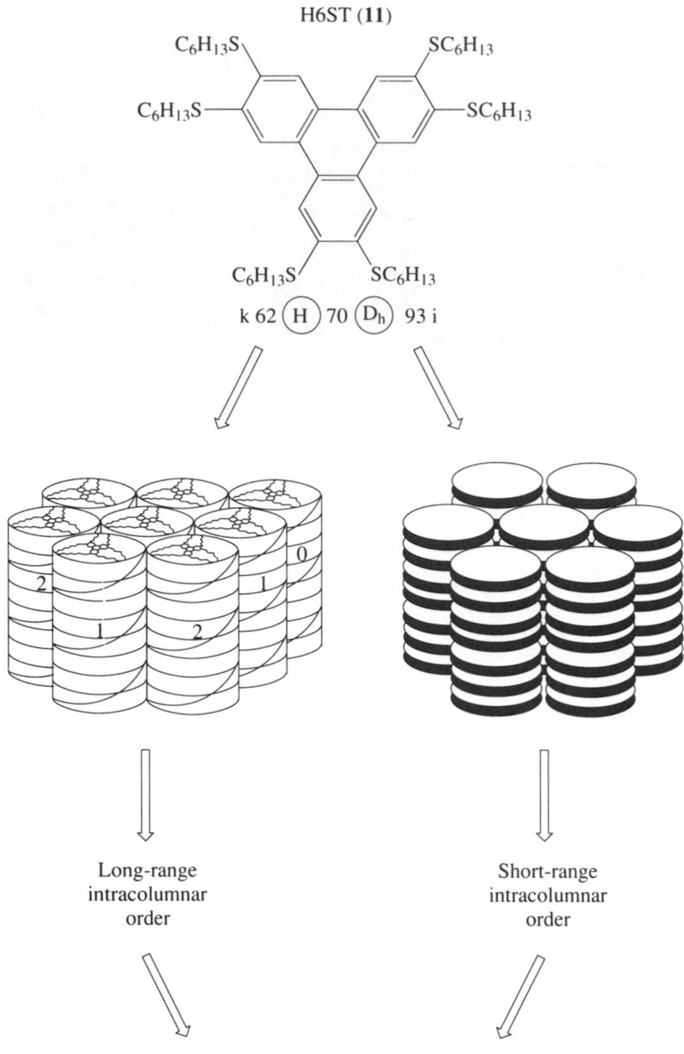

Figure 30 Hexahexylthiotriphenylene (**11**)—discotic mesogenic system with both a short-range intracolumnar stacking in a conventional discotic hexagonal liquid crystalline phase (D_h) and a long-range intracolumnar order in a helical columnar phase (H). In addition, H6ST (**11**) also exhibits a normal polycrystalline (k) phase below 62 °C and an isotropic (i) phase above 93 °C.

of the photocurrents, suggest an efficient charge transport mechanism with no major perturbation by disorder effects. Presumably, the high mobility results from the fact that in the H phase the triphenylene units lie face-to-face with good π-orbital overlap along the columns. A comparison with the dispersive transport in the polycrystalline k phase shows that charge transport in the highly ordered H phase is the result of a very delicate balance. The phase transition from the H to the k phase—which can be supercooled by 20 °C—is accompanied by somewhat reduced molecular dynamics and by minute rearrangement of the respective molecular positions. Yet these small changes in structure inhibit efficient charge transport, and result in severe charge carrier trapping. It can be speculated that the columnar transport channels that traverse the sample in the H phase (and also in the D_h phase) are somehow disarranged in the crystallization process, possibly by the formation of grain boundaries. The latter suggestion is supported by charge carrier mobility measurements with increasing temperature, starting in the polycrystalline k phase. Here the photocurrent transients in the H phase still remain dispersive—as in the k phase. Intracolumnar order on a macroscopic scale can obviously not be achieved when entering the H phase via the k phase. Further heating into the D_h phase then gives essentially the same charge carrier mobilities as found in the cooling cycle.

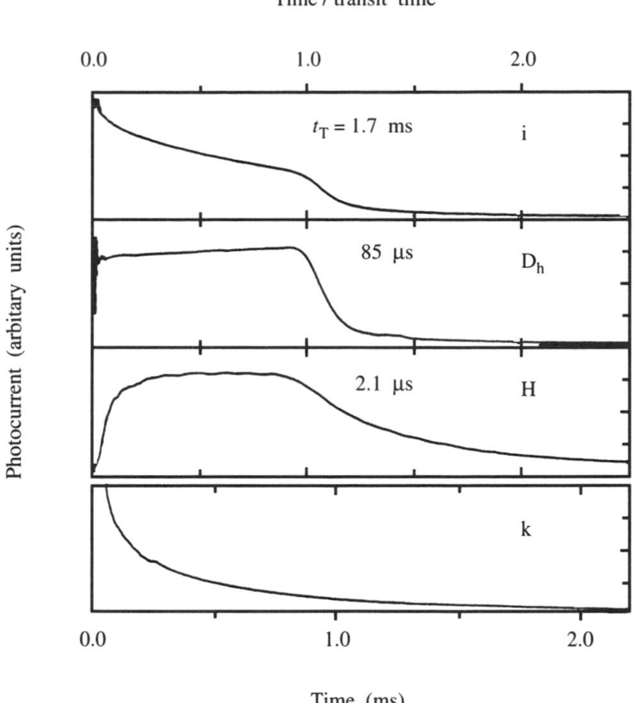

Figure 31 Transient photocurrents I for holes as a function of time t measured in the different phase regions of H6ST (**11**) by the time-of-flight method. The stepped current decays in the i, D_h and H phases mark the respective transit time t_T of the charge-carrier package. In the k phase, a transit time cannot be determined because of a featureless current decay. Temperatures: (i) 103 °C, (D_h) 79 °C, (H) 65 °C, (k) 28 °C.

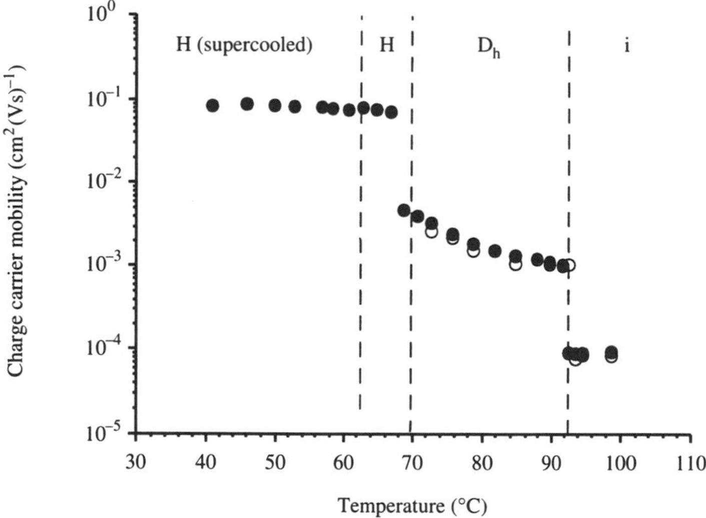

Figure 32 Arrhenius plot of H6ST (**11**) hole mobility μ vs. temperature at an electric field of 2.0×10^4 V cm^{-1}. Filled circles correspond to a cooling cycle starting in the isotropic phase. At 40 °C in the supercooled H phase, the H → k phase transition took place resulting in the dispersive current decay as shown in Figure 31. The open circles show measurements obtained for a heating cycle starting in the k phase; in this case, carrier mobilities cannot be obtained both for the k and the H phase because of featureless dispersive photocurrents.

These results have clearly demonstrated that the enhanced long-range positional and orientational order of H6ST (**11**) mesogens in the H phase,[90a] in comparison to their short-range positional order in the D_h phase, induce an increase in the charge carrier mobilities of almost two orders of magnitude. Thus, this system has a charge carrier mobility that is higher than that of H5T (**10**) and nearly as good as for aromatic single crystals (Figure 33).

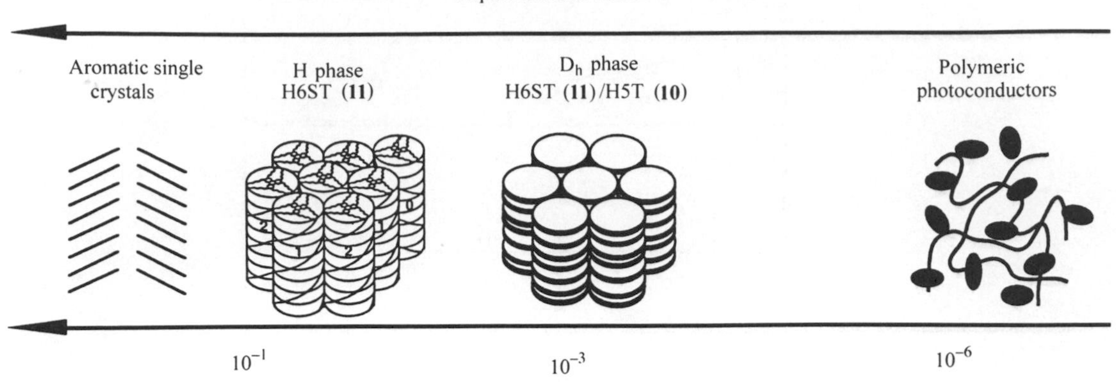

Figure 33 Comparison of the charge carrier mobility of low molar mass discotic liquid crystals H5T (**10**) and H6ST (**11**) with amorphous polymers and aromatic single crystals.

9.3.5 Morphology-controlled Charge Transport

In the last two sections, it was clearly demonstrated that the photoconductive properties of liquid crystalline materials are mainly determined by the supramolecular order and packing of the molecular units involved. The overlap of molecular wavefunctions is often more important than the molecular structure (Figure 34). The charge carrier transport in the crystalline phase, which has a lot of grain boundaries within the sample, is characterized by deep trapping (dispersive transport). In the very mobile liquid phase, charge carrier mobilities are influenced by short-range orientational order ($\mu = 10^{-4} cm^2 (Vs)^{-1}$) or by no order as in the isotropic melt. The regular orientation of the mesogens in the liquid crystalline mesophases allows a nondispersive charge transport, because the conjugated triphenylene cores provide an efficient transport path along the columns due to their quasi-one-dimensional order. The charge carrier mobility in the D_h phase is only limited by the remaining mobility in this LC phase and the structural defects resulting from these dynamic properties. On the one hand, there are no grain boundaries[88] and some defects can probably be corrected because of the possible dynamical fluctuations, for example, molecular rotations, bending or density fluctuations.[91] On the other hand, these phenomena decrease the supramolecular order and subsequently reduce the charge carrier mobility. The remaining mobility in the D_h phase is nearly completely suppressed in the H phase of H6ST (**11**). Thus, the charge carrier mobilities in this phase are in the same range as organic single crystals.

In contrast to aromatic single crystals, which show their high charge carrier mobilities only in zone-refined substances, H6ST (**11**) can be easily purified by chromatography and recrystallization. In addition, it is basically possible to prepare thin films or layers by simple cooling of the compound from the liquid melt. For application in photocopiers, laser printers, transistors or other electronic devices, it will be necessary to preserve and stabilize the columnar order. One possible way to obtain, for example, stable H or D_h phases, is the freezing-in of the mesophase structure in the glassy state of an oligomeric or polymeric liquid crystalline system. Different possibilities for the incorporation of discotic liquid crystals into oligomers and polymers are shown in Figure 35.[61,92] Some of these possibilities will be discussed in the next section.

9.3.6 Photoconductivity of Glassy Discotics: Oligomers and Polymers

In order to utilize the findings described in Section 9.3.3. and Section 9.3.4. for technical applications, the columnar order of the discotic mesophases would have to be transferred into a stable glassy state of an oligomer or a polymer or into a polymer network, while preserving the high charge carrier mobilities. Two possibilities for such discotic solids are schematically shown in Figure 36.

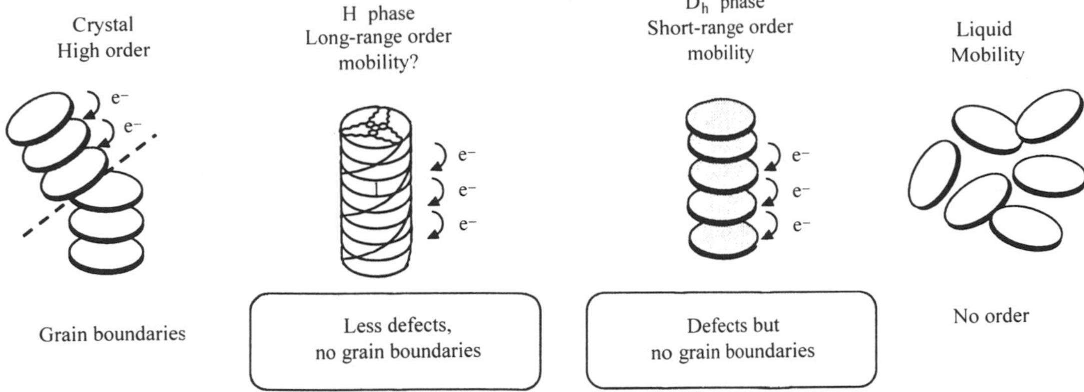

Figure 34 Morphology controls charge transport in discotic liquid crystals: the charge carrier transport in the crystalline phase is determined by the existence of grain boundaries—charge carriers get trapped and the transport is dispersive. The missing order determines the charge carrier transport in the liquid phase. The regular order in the liquid crystalline mesophases allows a nondispersive charge transport along the columns. Increasing order within the liquid crystalline phase leads to increasing charge carrier mobilities.

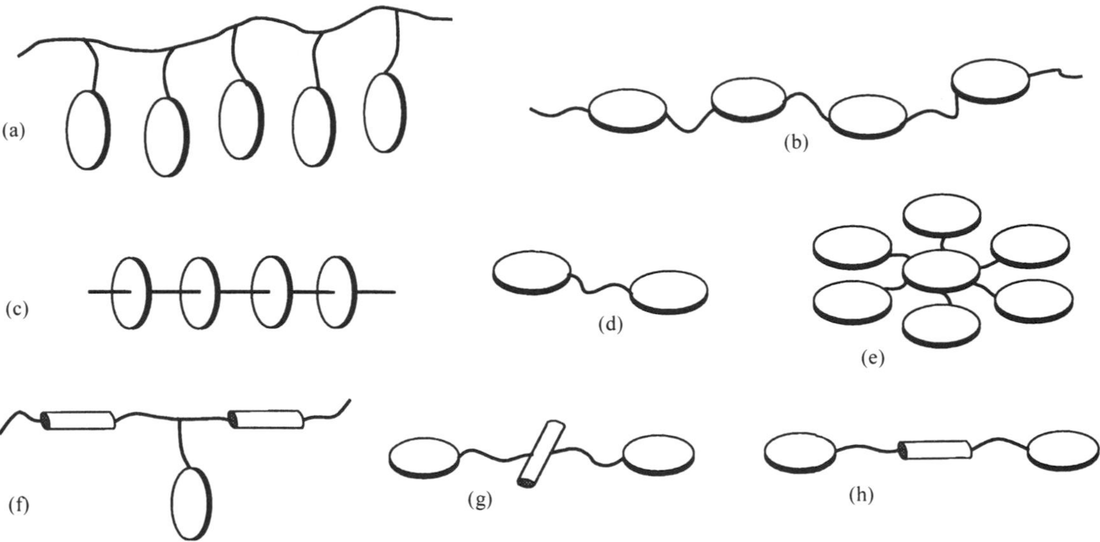

Figure 35 Structural variations of oligomeric and polymeric discotic liquid crystals: (a) discotic side-chain polymer;[61a,92a–g] (b) discotic main-chain polymer;[61a,92a–g] (c) hairy rod polymer;[92g–k] (d) discotic twin;[92l–o] (e) discotic heptamer;[61a,92p] (f) combined main-chain and side-chain polymer with discotic and calamitic mesogens;[92q] (g) two discotic mesogens laterally linked via a calamitic mesogen (wheel of Mainz);[92r] (h) two discotic mesogens terminally linked via a calamitic mesogen.[92r]

9.3.6.1 *Twins and tetramers*[53]

The first step towards the goal described above is the synthesis and characterization of a discotic twin based on the hexapentyloxytriphenylene.[53] Such dimers in particular, or discotic oligomers in general,[61,92p,93] represent ideal model compounds for discotic polymers and networks due to the ease of their purification, due to the simple tailoring of the range of the mesophase by just varying the spacer length, and due to the possibility to freeze-in their discotic phase in a glassy state. It is therefore possible to obtain charge carrier mobility data over a temperature range of about 250 °C. In the case of a discotic twin of triphenylene units, linked by a decameric spacer (shown in Figure 37), investigations were carried out from −100 °C to 165 °C, beginning with a glassy state, followed in sequence by a columnar mesophase and an isotropic phase.

To characterize the electronic charge carrier transport properties of the dimer Di 10 (**12**), transient photocurrents were measured using the TOF method. Furthermore, the corresponding values for the charge carrier mobility μ were calculated for the different phases.

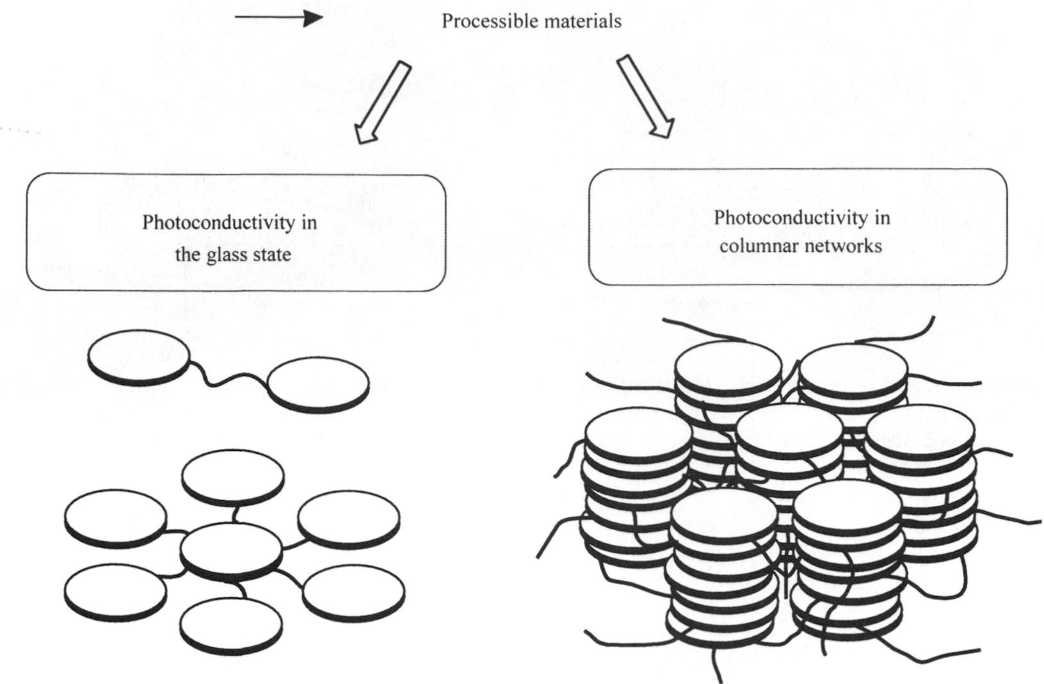

Figure 36 Mechanically and thermally stable columnar assemblies based on discotic oligomers or cross-linked discotic liquid crystals.

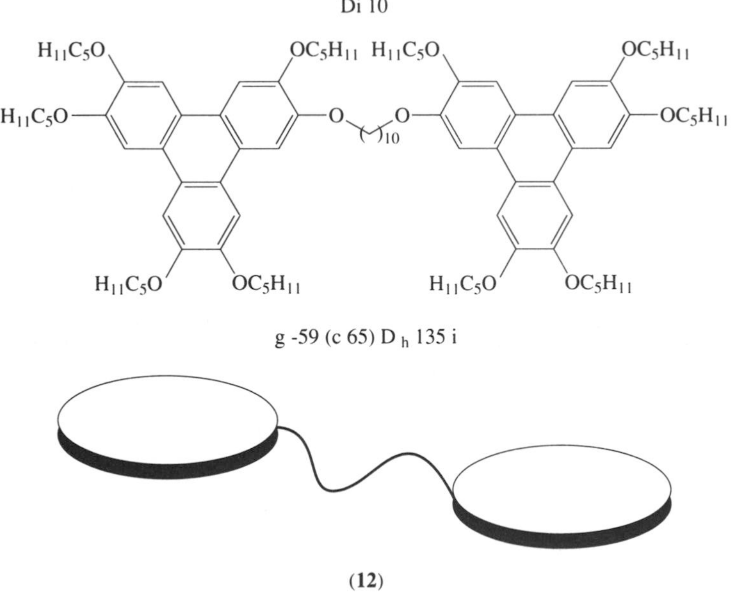

Figure 37 Structure and schematic representation of a discotic twin molecule (**12**).

The charge carrier mobilities for various temperatures are shown in Figure 38. They reach their highest values in the temperature region between 60 °C and 90 °C. It can be speculated that this is due to a dynamical healing of defects, which is caused by the moderate molecular dynamics[94] in this temperature range. By lowering the temperature into the supercooled D_h phase region, defects develop in the crystalline phase, due to a decrease in the molecular dynamics;[95] the latter is reflected in lower values for the charge carrier mobilities. In the glassy (g) phase region, one can assume a steady defect structure and thus one finds a normal thermally activated mobility with a comparatively low activation energy, indicating that the charge transport is still governed by the favourable columnar arrangement of the discs. Increasing the temperature above 90 °C also results in decreasing mobilities, since the healing of defects is probably overcompensated by an additional induced disorder due to enhanced molecular dynamics.[94] At the transition into the isotropic phase, charge carrier mobilities decrease again by two orders of magnitude as compared to the values in the mesophase. Here, the columnar order is broken up and, although the system is in its isotropic phase, there still prevails a short-range order of the dimers.[90c]

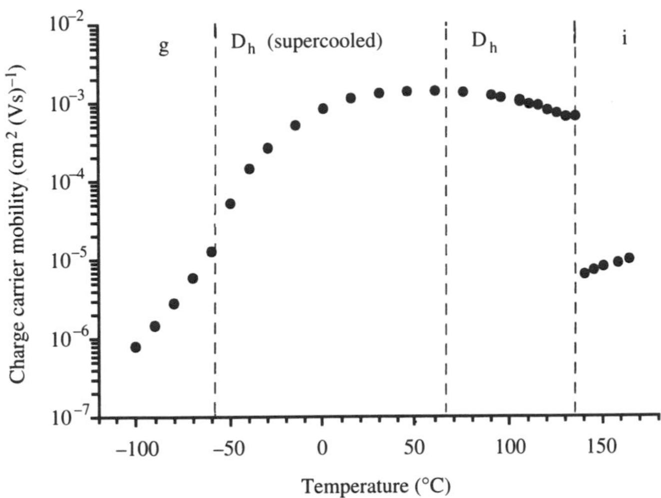

Figure 38 Arrhenius plot of Di 10 (**12**) hole mobility μ vs. temperature T at an electric field of 4.0×10^4 V cm^{-1}.

It is remarkable that the charge carrier mobilities in the D_h phase, both for the monomeric model compound H5T (**10**)[48,49] and for the discotic twin Di 10 (**12**),[53] are essentially of the same order (10^{-3} cm^2 (Vs)$^{-1}$). These similar values imply that the free rotation of the discotic mesogens about the columnar axis, which is possible in the H5T molecule (**10**) but which is impossible for the Di 10 twin (**12**), neither hinders nor promotes the observed charge carrier transport.

The influence of the molecular architecture on the charge transport properties of discotic liquid crystals was further characterized using a tetrameric triphenylene derivative (**13**), shown below. The transient photocurrents are similar to those of H5T (**10**) and Di 10 (**12**) and the calculated charge carrier mobilities are in the same range as for H5T (**10**) and Di 10 (**12**) in the D_h phase.

In conclusion, a further step towards high mobilities on the one hand and processible organic materials on the other has been described, by developing the chemistry of the liquid crystalline photoconductors.[53,94] By synthesizing a discotic twin and a discotic tetramer of the triphenylene model compound, it has been possible to obtain charge carrier mobility data in glassy columnar phases, in which the charge transport properties of the mesophases are basically maintained. The highest charge carrier mobilities, in the order of 10^{-3} cm^2 (Vs)$^{-1}$, have been obtained in the D_h mesophase.

9.3.6.2 Polymer networks

The next step in the preparation of mechanically and thermally stable columnar assemblies of discotic liquid crystals is the incorporation of the triphenylene mesogens into polymers or polymer networks.[96] For technical applications, a mechanically stable arrangement of the columnar assembly is necessary, preferably involving a macroscopically uniform alignment of the columns in one direction. Such polymeric columnar structures can be obtained using two different strategies, which are schematically shown in Figure 39. In the first possibility, mono- and difunctionalized polymerizable triphenylene monomers will be oriented in the columnar D_h phase and subsequently polymerized. In the second possibility, monofunctionalized triphenylene, difunctionalized triphenylene and poly[oxy(methylsilylene)] will be mixed and cross-linked as a first step. After

(13)

mechanical orientation of the sample, the polymer is cross-linked in a second step to give the oriented polymer network. Preliminary studies have shown that these two strategies are indeed suitable for the preparation of macroscopically ordered columnar networks.[96]

Photoconductivity in columnar networks

Enhanced mechanical stability by freezing of columnar order

In situ polymerization of oriented functionalized discotics

Highly oriented discotic elastomers

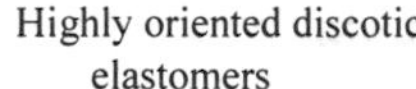

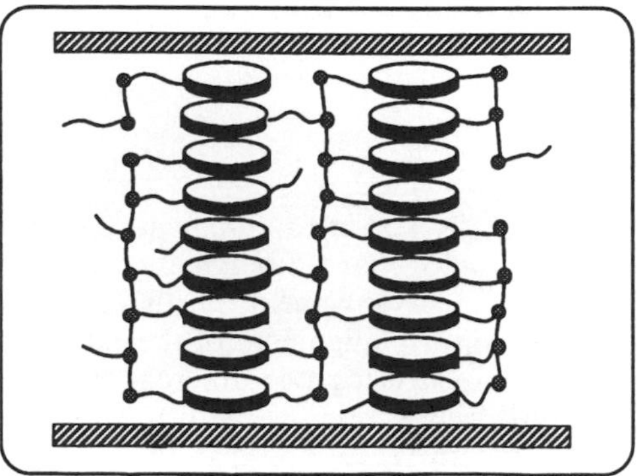

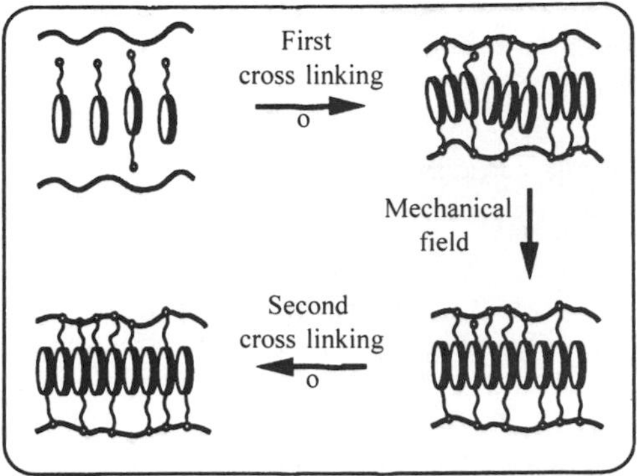

Figure 39 Concept for the preparation of new, highly ordered, processible, stable photoconductors.

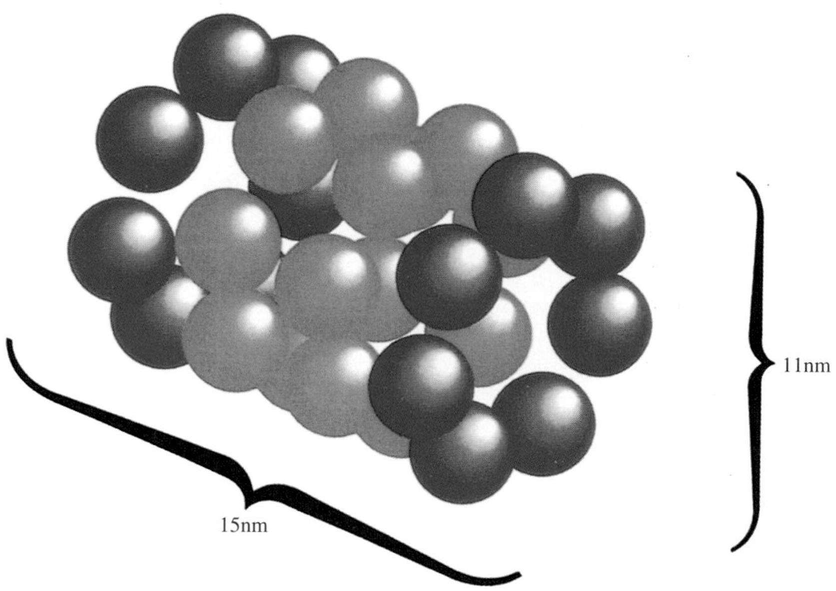

Plate 1 Model of the *Thermoplasma acidophilum* proteasome, showing the arrangement and stoichiometry of the α-subunits (two outer rings) and the β-subunits (two inner rings) in one proteasome particle. The real shape of the subunits will deviate significantly from the spheres used to construct this model resulting in a more closed barrel-shaped structure as revealed by the three-dimensional reconstruction of individual proteasomes (reproduced by permission of Oxford University Press from *EMBO J.*, 1992, **11**, 1607).

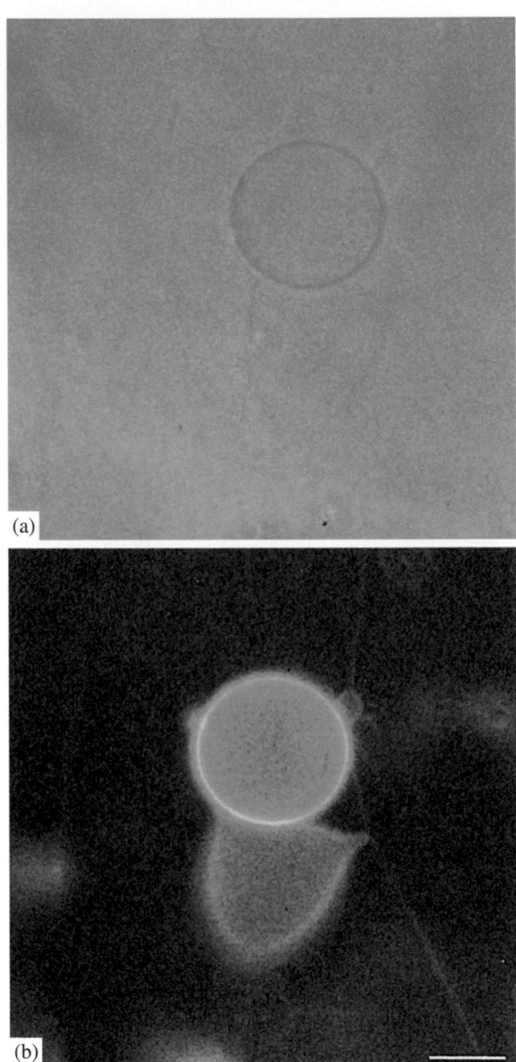

Plate 2 Interaction of giant liposomes (90% DMPC/10% DODA-EO_2-I) with fluorescein-labeled chymotrypsin: (a) the image of a giant liposome in the phase contrast modus; (b) the image of a giant liposome with covalently linked fluorescein-labeled chymotrypsin as seen with the fluorescein filter after 30 min incubation with the fluorescence-labeled enzyme.

9.4 REFERENCES

1. (a) J.-M. Lehn, 'Supramolecular Chemistry, Concepts and Perspectives', Verlag Chemie, Weinheim, 1995; (b) F. Vögtle, 'Supramolekulare Chemie', eds. C. Elschenbroich, F. Hensel and H. Hopf, Teubner, Stuttgart, 1989.
2. (a) J.-M. Lehn, *Science*, 1985, **227**, 849; (b) J.-M. Lehn, *Angew. Chem., Int. Ed. Engl.*, 1988, **27**, 89.
3. (a) H. Ringsdorf, B. Schlarb and J. Venzmer, *Angew. Chem., Int. Ed. Engl.*, 1988, **27**, 113; (b) M. Ahlers, W. Müller, A. Reichert, H. Ringsdorf and J. Venzmer, *Angew. Chem., Int. Ed. Engl.*, 1990, **29**, 1269; (c) A. Reichert, PhD Thesis, Mainz, 1993.
4. B. Alberts, D. Bray, J. Lewis, M. Raff, K. Roberts and J. D. Watson, 'Molecular Biology of the Cell', Garland, New York, 1983.
5. (a) G. Gregoriadis and A. C. Allison (eds.), 'Liposomes in Biological Systems', Wiley, Chichester, 1980; (b) C. G. Knight (ed.) 'Liposomes: From Physical Structure to Therapeutic Applications', Elsevier, Amsterdam, 1981; (c) G. Gregoriadis (ed.), 'Liposome Technology', CRC Press, Boca Raton, FL, 1984 vols. I–III; (d) M. J. Ostro (ed.), 'Liposomes', Dekker, New York, 1983; (e) M. J. Ostro (ed.), 'Liposomes: From Biophysics to Therapeutics', Dekker, New York, 1987.
6. J. N. Israelachvili, in 'Physics of Amphiphiles: Micelles, Vesicles and Micro-emulsions', eds. V. Degiorgio and M. Corti, North-Holland, Amsterdam, 1985, p. 24.
7. (a) G. W. Gray and P. A. Winsor (eds.), 'Liquid Crystals and Plastic Crystals', Horwood, Chichester, 1974 vols. I, II; (b) G. J. T. Tiddy, *Phys. Rev.*, 1980, **57**, 1; (c) B. J. Forrest and L. W. Reeves, *Chem. Rev.*, 1981, **81**, 1.
8. (a) A. Ullman, 'Langmuir-Blodgett Films', ed. G. Roberts, Plenum Press, New York, 1990; (b) *Thin Solid Films*, 1980, **68**; *Thin Solid Films*, 1983, **99**; *Thin Solid Films*, 1985, **132–134**; *Thin Solid Films*, 1988, **159–160**; *Thin Solid Films*, 1989, **178–180**.
9. (a) F. D. Saeva (ed.), 'Liquid Crystals: The Fourth State of Matter', Dekker, New York, 1979; (b) H. Kelker and R. Hatz (eds.), 'Handbook of Liquid Crystals', Verlag Chemie, Weinheim, 1980; (c) C. B. McArdle (ed.), 'Side Chain Liquid Crystal Polymers', Blackie, Glasgow, 1989; (d) P. G. de Gennes (ed.), 'The Physics of Liquid Crystals', Clarendon Press, Oxford, 1974; (e) H.-D. Koswig (ed.), 'Selected Topics in Liquid Crystal Research', Akademie Verlag, Berlin, 1990; (f) W. Helfrich and G. Heppke (eds.), 'Liquid Crystals of One- and Two-Dimensional Order', Springer-Verlag, Berlin, 1980; (g) C. Hilsum and E. P. Raynes (eds.), 'Liquid Crystals: Their Physics, Chemistry and Applications', The Royal Society, London, 1983.
10. V. Percec and C. Pugh in 'Side Chain Liquid Crystal Polymers', ed. C. B. McArdle, Blackie, Glasgow, 1989, p. 30.
11. (a) S. Chandrasekhar, B. K. Sadashiva and K. A. Suresh, *Pramana*, 1977, **9**, 471; (b) C. Destrade, P. Foucher, H. Gasparoux, N. H. Tinh, A.-M. Levelut and J. Malthête, *Mol. Cryst. Liq. Cryst.*, 1984, **106**, 121; (c) W. Kreuder and H. Ringsdorf, *Makromol. Chem., Rapid Commun.*, 1983, **4**, 807; (d) W. Kreuder, H. Ringsdorf and P. Tschirner, *Makromol. Chem., Rapid Commun.*, 1985, **6**, 367; (e) G. Wenz, *Makromol. Chem., Rapid Commun.*, 1985, **6**, 577; (f) S. Chandrasekhar, *Liq. Cryst.*, 1993, **14**, 3; (g) S. Chandrasekhar and G. S. Ranganath, *Rep. Prog. Phys.*, 1990, **53**, 57.
12. (a) H. Ringsdorf, P. Tschirner, O. H. Schönherr and J. H. Wendorff, *Makromol. Chem.*, 1987, **188**, 1431; (b) M. Ebert, O. H. Schönherr, J. H. Wendorff, H. Ringsdorf and P. Tschirner, *Liq. Cryst.*, 1990, **7**, 63; (c) M. Ballauff and G. F. Schmidt, *Makromol. Chem., Rapid Commun.*, 1987, **8**, 93; (d) I. G. Voigt-Martin, P. Simon, S. Bauer and H. Ringsdorf, *Macromolecules*, 1995, **28**, 236; (e) I. G. Voigt-Martin, P. Simon, D. Yan, A. Yakimansky, S. Bauer and H. Ringsdorf, *Macromolecules*, 1995, **28**, 243.
13. (a) J. D. Barnal, *Trans. Faraday Soc.*, 1933, **29**, 1022; (b) G. H. Brown and J. J. Wolken (eds.), 'Liquid Crystals in Biological Systems', Academic Press, New York, 1979.
14. (a) M. Waite, 'The Phospholipases', Plenum, New York, 1987; (b) E. A. Dennis, 'Phospholipases', *Methods Enzymol.*, 1991, **197**; (c) D. W. Grainger, A. Reichert, H. Ringsdorf and C. Salesse, *FEBS Lett.*, 1989, **252**, 74; (d) D. W. Grainger, A. Reichert, H. Ringsdorf and C. Salesse, *Biochim. Biophys. Acta*, 1990, **1023**, 365; (e) A. Reichert, H. Ringsdorf and A. Wagenknecht, *Biochim. Biophys. Acta*, 1992, **1106**, 178.
15. (a) R. Blankenburg, P. Meller, H. Ringsdorf and C. Salesse, *Biochemistry*, 1989, **28**, 8214; (b) M. Ahlers, R. Blankenburg, D. W. Grainger, P. Meller, H. Ringsdorf and C. Salesse, *Thin Solid Films*, 1989, **180**, 93; (c) M. Ahlers, R. Blankenburg, H. Haas, D. Möbius, H. Möhwald, W. Müller, H. Ringsdorf and H.-U. Siegmund, *Adv. Mater.*, 1991, **3**, 39; (d) M. Ahlers, M. Hoffmann, H. Ringsdorf, A. M. Rourke and E. Rump, *Makromol. Chem., Macromol. Symp.*, 1991, **46**, 307; (e) S. A. Darst, M. Ahlers, P. H. Meller, E. W. Kubalek, R. Blankenburg, H. O. Ribi, H. Ringsdorf and R. D. Kornberg, *Biophys. J.*, 1991, **59**, 387.
16. (a) F. Vögtle, H.-G. Löhr, J. Franke and D. Worsch, *Angew. Chem., Int. Ed. Engl.*, 1985, **24**, 727; (b) F. Vögtle and E. Weber, *Top. Curr. Chem.*, 1981, **98**; 1982, **101**; 1984, **121**; (c) *Top. Curr. Chem.*, 1985, **128**; 1986, **132**; 1986, **136**; (d) D. J. Cram, *Angew. Chem., Int. Ed. Engl.*, 1986, **25**, 1039; (e) F. H. Kohnke, J. P. Mathias and J. F. Stoddart, *Adv. Mater.*, 1989, **101**, 1124.
17. (a) I. Cabrera, M. Engel and H. Ringsdorf in 'Photoconversion Processes for Energy and Chemicals, Energy from Biomass 5', eds. D. O. Hall and G. Grassi, Elsevier, London 1989, pp. 68–78; (b) I. Cabrera, M. Engel, L. Häußling, C. Mertesdorf and H. Ringsdorf, in 'Frontiers in Supramolecular Organic Chemistry and Photo-chemistry', eds. H.-J. Schneider and H. Dürr, Verlag Chemie, Weinheim, 1991, pp. 311–36; (c) I. Cabrera, A. Dittrich and H. Ringsdorf, *Angew. Chem., Int. Ed. Engl.*, 1991, **30**, 76.
18. (a) D. Lupo, W. Prass, U. Scheunemann, A. Laschewsky, H. Ringsdorf and I. Ledoux, *J. Opt. Soc. Am.*, 1988, **B5**, 300; (b) W. Groh, D. Lupo and H. Sixl, *Adv. Mater.*, 1989, **101**, 366; (c) E. F. Aust, W. Hickel, W. H. Meyer, H. Ringsdorf, M. Sawodny, C. Urban, G. Wegner and W. Knoll, *Mater. Res. Soc. Symp. Proc.*, 1992, **227**, 173; (d) C. Bubeck, A. Laschewsky, D. Lupo, D. Neher, P. Ottenbreit, W. Paulus, W. Prass, H. Ringsdorf and G. Wegner, *Adv. Mater.*, 1991, **3**, 54.
19. (a) H. J. Coles and R. Simon, *Polymer*, 1985, **26**, 1801; (b) M. Eich, J. H. Wendorff, B. Reck and H. Ringsdorf, *Makromol. Chem., Rapid Commun.*, 1987, **8**, 59.
20. (a) A. Maelicke and W. Müller-Esterl (eds.), Biochemie, Verlag Chemie, Weinheim, 1992; (b) L. Stryer 'Biochemie', Spektrum der Wissenschaft, Heidelberg, 1990; (c) G. Zubay (ed.), 'Biochemistry', MacMillan, London, 1988.
21. (a) I. M. Roitt, J. Brostoff and D. K. Male (eds.), 'Kurzes Lehrbuch der Immunologie', Thieme Verlag, New York, 1991; (b) J. G. van den Tweel (ed.), 'Immunologie', Spektrum der Wissenschaft, Heidelberg, 1991.

22. (a) A.-P. Arrigo, K. Tanaka, A. L. Goldberg and W. J. Welch, *Nature*, 1988, **331**, 192; (b) B. Dahlmann, M. Rutschmann and H. Reinauer, *Biochem. J.*, 1985, **228**, 171; (c) P.-E. Falkenburg, C. Haass, P.-M. Kloetzel, B. Niedel, F. Kopp, L. Kuehn and B. Dahlmann, *Nature*, 1988, **331**, 190; (d) J. R. Harris, *Biochim. Biophys. Acta*, 1968, **150**, 534; (e) J. R. Harris, *Micron Microsc. Acta*, 1983, **14**, 193; (f) J. A. Rivett, *J. Biol. Chem.*, 1985, **260**, 12600; (g) G. Spohr, N. Granboulan, C. Morel and K. Scherrer, *Eur. J. Biochem.*, 1970, **17**, 296; (h) S. Wilk and M. Orlowski, *J. Neurochem.*, 1980, **35**, 1172.
23. (a) M. Orlowski, *Biochemistry*, 1990, **29**, 10290; (b) J. A. Rivett, *Arch. Biochem. Biophys.*, 1989, **268**, 1; (c) K. Tanaka and A. Ichihara, *Cell Struct. Funct.*, 1990, **15**, 127.
24. B. Dahlmann, F. Kopp, L. Kühn, B. Niedel, G. Pfeifer, R. Hegerl and W. Baumeister, *FEBS Lett.*, 1989, **251**, 125.
25. (a) T. Fujiwara, K. Tanaka, E. Orino, T. Yoshimura, A. Kumatori, T. Tamura, C. H. Chung, T. Nakai, K. Yamaguchi, S. Shin, A. Kakizuka, S. Nakanishi and A. Ichihara, *J. Biol. Chem.*, 1990, **265**, 16604; (b) C. Haass, B. Pesold-Hurt, G. Multhaupt, K. Beyreuther and P. M. Kloetzel, *Gene*, 1990, **90**, 235.
26. R. Hegerl, G. Pfeifer, G. Pühler, B. Dahlmann and W. Baumeister, *FEBS Lett.*, 1991, **283**, 117.
27. P. Zwickl, A. Grziwa, G. Pühler, B. Dahlmann, F. Lottspeich and W. Baumeister, *Biochemistry*, 1992, **31**, 964.
28. (a) A. Grizwa, W. Baumeister, B. Dahlmann and F. Kopp, *FEBS Lett.*, 1991, **290**, 186; (b) A. Grizwa, B. Dahlmann, Z. Cejka, U. Santarius and W. Baumeister, *J. Struct. Biol.*, 1992, **109**, 168.
29. P. Zwickl, J. Kleinz and W. Baumeister, *Nature Struct. Biol.*, 1994, **1**, 765.
30. B. K. Jap, M. Zulauf, T. Scheybani, A. Hefti, W. Baumeister, U. Aebi and A. Engel, *Ultramicroscopy*, 1992, **46**, 45.
31. B. Dahlmann, L. Kuehn, A. Grizwa, P. Zwickl and W. Baumeister, *Eur. J. Biochem.*, 1992, **208**, 789.
32. D. M. Blow, *Acc. Chem. Res.*, 1976, **9**, 145.
33. (a) R. A. Mumford, A. W. Strauss, J. C. Powers, P. A. Pierzchala, N. Nishino and M. Zimmermann, *J. Biol. Chem.*, 1980, **255**, 2227; (b) R. A. Mumford, P. A. Pierzchala, A. W. Strauss and M. Zimmermann, *Proc. Natl. Acad. Sci. USA*, 1981, **78**, 6623; (c) J. O'Donnell-Torney and J. P. Quigley, *Proc. Natl. Acad. Sci. USA*, 1983, **80**, 344.
34. D. M. Segal, J. C. Powers, G. H. Cohen, D. R. Davies and P. E. Wilcox, *Biochemistry*, 1971, **10**, 3728.
35. (a) G. L. Gaines, 'Insoluble Monolayers at Liquid–Gas Interfaces', Interscience, New York, 1966; (b) E. D. Goddard and R. F. Gould (eds.), Monolayers, American Chemical Society, Washington, DC, 1975; (c) A. W. Adamson, 'Physical Chemistry of Surfaces', Wiley, New York, 1982; (d) A. Pockels, *Nature*, 1891, **43**, 437; (e) A. Pockels, *Nature*, 1892, **46**, 418; (f) A. Pockels, *Nature*, 1893, **48**, 152; (g) I. Langmuir, *J. Am. Chem. Soc.*, 1917, **39**, 1848; (h) I. Langmuir, *Trans. Faraday Soc.*, 1920, **15**, 62; (i) I. Langmuir, *Proc. R. Soc. London, Ser. A*, 1939, **170**, 1.
36. (a) M. Ahlers, D. W. Grainger, J. N. Herron, K. Lim, H. Ringsdorf and C. Salesse, *Biophys. J.*, 1992, **63**, 823; (b) K. Kimura, Y. Arata, T. Yasuda, K. Kinosita and M. Nakanishi, *Biochim. Biophys. Acta*, 1992, **1104**, 9.
37. (a) M. Hesse, H. Maier and B. Zeeh, 'Spektroskopische Methoden in der Organischen Chemie', Thieme Verlag, Stuttgart, 1979; (b) R. C. Ahuja, P.-L. Caruso, D. Möbius, W. Paulus, H. Ringsdorf and G. Wildburg, *Angew. Chem., Int. Ed. Engl.*, 1993, **32**, 1033.
38. B. Dahlmann, L. Kuehn, M. Rutschmann and H. Reinauer, *Biochem. J.*, 1985, **228**, 161.
39. A. Reichert, W. Spevak, J. O. Nagy and D. H. Charych, *J. Am. Chem. Soc.*, 1995, **117**, 829.
40. (a) V. von Tscharner and H. M. McConnell, *Biophys. J.*, 1981, **36**, 409; (b) M. Lösche and H. Möhwald, *Rev. Sci. Instrum.*, 1984, **55**, 1986; (c) P. Meller, *Rev. Sci. Instrum.*, 1988, **59**, 2225.
41. R. D. Nargessi and D. S. Smith, *Methods Enzymol.*, 1986, **104**, 67.
42. (a) G. Decher, E. Kuchinka, H. Ringsdorf, J. Venzmer, D. Bitter-Suermann and C. Weisgerber, *Angew. Makromol. Chem.*, 1989, **166/167**, 71; (b) G. Decher, H. Ringsdorf, J. Venzmer, D. Bitter-Suermann and C. Weisgerber, *Biochim. Biophys. Acta*, 1990, **1023**, 357.
43. G. Pühler, S. Weinkauf, L. Bachmann, S. Müller, A. Engel, R. Hegerl and W. Baumeister, *EMBO J.*, 1992, **11**, 1607.
44. D. Typke, R. Hegerl and J. Kleinz, *Ultramicroscopy*, 1992, **46**, 157.
45. P. Meller, *J. Microsc.*, 1989, **156**, 241.
46. I. Langmuir, V. J. Schäfer and D. M. Wrinch, *Science*, 1937, **85**, 76.
47. J. Löwe, D. Stock, B. Jap, P. Zwickl, W. Baumeister and R. Huber, *Science*, 1995, **268**, 533.
48. H. Bengs, F. Closs, T. Frey, D. Funhoff, H. Ringsdorf and K. Siemensmeyer, *Liq. Cryst.*, 1993, **15**, 565.
49. D. Adam, F. Closs, T. Frey, D. Funhoff, D. Haarer, H. Ringsdorf, P. Schuhmacher and K. Siemensmeyer, *Phys. Rev. Lett.*, 1993, **70**, 457.
50. D. Adam, D. Haarer, F. Closs, D. Funhoff, L. Häußling, K. Siemensmeyer, H. Ringsdorf and P. Schuhmacher, *Mol. Cryst. Liq. Cryst.*, 1994, **252**, 155.
51. D. Adam, P. Schuhmacher, J. Simmerer, L. Häußling, K. Siemensmeyer, K. H. Etzbach, H. Ringsdorf and D. Haarer, *Nature*, 1994, **371**, 141.
52. D. Adam, D. Haarer, F. Closs, T. Grey, D. Funhoff, K. Siemensmeyer, P. Schuhmacher and H. Ringsdorf, *Ber. Bunsen-Ges. Phys. Chem.*, 1993, **97**, 1366.
53. D. Adam, P. Schuhmacher, J. Simmerer, L. Häußling, W. Paulus, K. Siemensmeyer, K. H. Etzbach, H. Ringsdorf and D. Haarer, *Adv. Mat.*, 1995, **7**, 276.
54. (a) D. Haarer, *Angew. Makromol. Chem.*, 1990, **183**, 197; (b) M. Stolka, J. F. Yanus and D. M. Pai, *J. Phys. Chem.*, 1984, **88**, 4707; (c) W. Wiedmann, *Chemiker Ztg.*, 1982, **106**, 275; (d) H. Hoegl, *J. Phys. Chem.*, 1965, **69**, 755.
55. (a) D. Möhwald, D. Haarer and G. Castro, *Chem. Phys. Lett.*, 1975, **32**, 433; (b) O. H. LeBlanc, Jr., *J. Chem. Phys.*, 1960, **33**, 626; (c) R. G. Kepler, *Phys. Rev.*, 1960, **119**, 1226; (d) N. Karl and J. Ziegler, *Chem. Phys. Lett.*, 1975, **32**, 438; W. Warta, R. Stehle, and N. Karl, *Appl. Phys. A*, 1985, **36**, 163.
56. D. Haarer, 'Photoconductive Polymers; A Comparison with Amorphous Inorganic Materials', Vieweg-Verlag, Braunschweig, 1990.
57. (a) L. L. Chapoy, D. K. Munck, K. H. Rasmussen, E. J. Diekmann, R. K. Sethi and D. Biddle, *Mol. Cryst. Liq. Cryst.*, 1984, **105**, 353; (b) L. L. Chapoy, D. Biddle, D. K. Munck, J. Halstrom, K. Kovacs, K. Brunfeldt, M. A. Qasim and T. Christensen, *Macromolecules*, 1983, **16**, 181; (c) M. Lux, P. Strohriegel and H. Höcker, *Makromol. Chem.*, 1987, **188**, 811; (d) Y. Shimizu, A. Ishikawa and S. Kusabayashi, *Chem. Lett.*, 1986, 1041.
58. (a) G. W. Gray, 'Critical Reports on Applied Chemistry 22', Wiley, Chichester, 1987; (b) G. W. Gray and J. W. G. Goodby (eds.), 'Smectic Liquid Crystals—Textures and Structures', Leonhard Hill, Glasgow, 1984.
59. G. Vertogen, W. H. de Jeu, 'Thermotropic Liquid Crystals, Fundamentals', Springer, Berlin, 1988.
60. E. Fontes, P. A. Heiney and W. H. de Jeu, *Phys. Rev. Lett.*, 1988, **61**, 1202.

61. (a) S. Bauer, T. Plesnivy, H. Ringsdorf and P. Schuhmacher, *Makromol. Chem., Macromol. Symp.*, 1992, **64**, 19; (b) H. Ringsdorf and R. Wüstefeld, *Philos. Trans. R. Soc. London.*, 1990, **330**, 95.
62. (a) H. Ringsdorf, R. Wüstefeld, E. Zerta, M. Ebert and J. H. Wendorff, *Angew. Chem., Int. Ed. Engl.*, 1989, **28**, 914; (b) H. Bengs, R. Renkel and H. Ringsdorf, *Makromol. Chem., Rapid Commun.*, 1991, **12**, 439.
63. (a) J. Billard, 'Discotic Mesophases: A Review', in: 'Liquid Crystals of One- and Two-Dimensional Order', eds. W. Helfrich and G. Heppke, Springer, Berlin, 1980, pp. 383–95; (b) S. Chandrasekhar, *Liq. Cryst.*, 1993, **14**, 3.
64. (a) M. Ebert, O. Herrmann-Schönherr, J. H. Wendorff, H. Ringsdorf and P. Tschirner, *Liq. Cryst.*, 1990, **7**, 63; (b) I. G. Voigt-Martin, P. Simon, S. Bauer and H. Ringsdorf, *Macromolecules*, 1995, **28**, 236; (c) I. G. Voigt-Martin, P. Simon, D. Yan, A. Yakimansky, S. Bauer and H. Ringsdorf, *Macromolecules*, 1995, **28**, 243.
65. (a) G. Elliot, *Chem. Ber.*, 1973, **9**, 213; (b) R. Williams, *J. Chem. Phys.*, 1963, **39**, 384.
66. (a) J. L. White, *J. Appl. Polym. Sci. Appl. Polym. Symp.*, 1985, **41**, 3; (b) E. T. Samulski, *Faraday Discuss., Chem. Soc.*, 1985, **79**, 7; (c) R. Zentel, in Topics in Physical Chemistry, ed. H. Stegemeyer, Springer, New York, 1994, vol. 3, pp. 103–41; (d) L. L. Chapoy, 'Recent Advances in Liquid Crystalline Polymers', Elsevier Applied Science Publishers, London, 1985.
67. (a) P. W. Morgan, *Macromolecules*, 1977, **10**, 1381; (b) S. L. Kwolek, P. W. Morgan, J. R. Schaefgen and L. W. Gulrich, *Macromolecules*, 1977, **10**, 1390; (c) T. I. Bair, P. W. Morgan and F. L. Killian, *Macromolecules*, 1977, **10**, 1396; (d) M. Panar and L. F. Beste, *Macromolecules*, 1977, **10**, 1401; (e) W. J. Jackson and H. F. Kuhfuss, *J. Polym. Sci., Polym. Chem. Ed.*, 1976, **14**, 2043; (f) W. J. Jackson, *Br. Polym. J.*, 1980, **12**, 154; (g) S. L. Kwolek, P. W. Morgan and J. R. Schaefgen, *Encyl. Polym. Sci.*, 1987, **9**, 1.
68. (a) H. Ringsdorf, H. W. Schmitt, H. Eilingsfeld and K.-H. Etzbach, *Makromol. Chem.*, 1987, **188**, 1355; (b) H. Ringsdorf, H. W. Schmitt, G. Baur, R. Kiefer and F. Windscheid, *Liq. Cryst.*, 1986, **1**, 319.
69. R. Zentel and G. Reckert, *Makromol. Chem.*, 1986, **187**, 1915.
70. I. Cabrera, V. Krongauz and H. Ringsdorf, *Angew. Chem., Int. Ed. Engl.*, 1987, **26**, 1178.
71. (a) S. Bualek and R. Zentel, *Makromol. Chem.*, 1988, **189**, 979; (b) H. Kapitza and R. Zentel, *Makromol. Chem.*, 1988, **189**, 1793.
72. (a) H. Bengs, O. Karthaus, H. Ringsdorf, C. Baehr, M. Ebert and J. H. Wendorff, *Liq. Cryst.*, 1991, **10**, 161; (b) K. Praefcke, D. Singer, B. Kohne, M. Ebert, A. Liebmann and J. H. Wendorff, *Liq. Cryst.*, 1991, **10**, 147; (c) K. Praefcke, D. Singer, M. Langner, B. Kohne, M. Ebert, A. Liebmann and J. H. Wendorff, *Mol. Cryst. Liq. Cryst.*, 1992, **215**, 121.
73. C. Destrade, P. Foucher, H. Gasparoux, N. H. Tinh, A. M. Levelut and J. Malthête, *Mol. Cryst. Liq. Cryst.*, 1984, **106**, 121.
74. (a) I. Tabushi, K. Yamamura and Y. Okada, *J. Org. Chem.*, 1987, **52**, 2502; (b) D. M. Collard and C. P. Lillya, *J. Am. Chem. Soc.*, 1989, **111**, 1829; (c) N. H. Tinh, H. Gasparoux and C. Destrade, *Mol. Cryst. Liq. Cryst.*, 1981, **68**, 101; (d) N. H. Tinh, M. C. Bernaud, G. Sigaud and C. Destrade, *Mol. Cryst. Liq. Cryst.*, 1981, **65**, 307; (e) C. Carfagna, P. Iannelli, A. Roviello and A. Sirigu, *Liq. Cryst.*, 1987, **2**, 611; (f) H. Bock and W. Helfrich, *Liq. Cryst.*, 1992, **12**, 697; (g) E. Keinan, S. Kumar, R. Moshenberg, R. Ghirlando and E. J. Wachtel, *Adv. Mater.*, 1991, **3**, 251; (h) B. Kohne and K. Praefcke, *Chimia*, 1987, **41**, 196; (i) M. Ebert, D. A. Jungbauer, R. Kleppinger and J. H. Wendorff, *Liq. Cryst.*, 1989, **4**, 53; (j) I. Cho and Y. Lim, *Mol. Cryst. Liq. Cryst.*, 1988, **154**, 9; (k) W. T. Ford, L. Sumner, W. Zhu, Y. H. Chang, P.-J. Um, K. H. Choi, P. A. Heiney and N. C. Maliszewskyj, *New. J. Chem.*, 1994, **18**, 495; (l) P. Davidson, C. Jallabert, A. M. Levelut, H. Strzelecka and M. Veber, *Liq. Cryst.*, 1988, **3**, 133; (m) M. Veber, P. Sotta, P. Davidson, A. M. Levelut, C. Jallabert and H. Strzelecka, *J. Phys. (Paris)*, 1990, **51**, 1283; (n) A. M. Giroud-Godquin, M. M. Gauthier, G. Sigaud, F. Hardouin and M. F. Achard, *Mol. Cryst. Liq. Cryst.*, 1986, **132**; (o) S. N. Poelsma, A. H. Servante, F. P. Fanizzi and P. M. Maitlis, *Liq. Cryst.*, 1994, **16**, 675; (p) B. Kohne and K. Praefcke, *Chem. Ztg.*, 1985, **109**, 121.
75. (a) J. Billard, J. C. Dubois, N. H. Tinh and A. Zann, *Nouv. J. Chim.*, 1978, **2**, 535; (b) C. Destrade, M. C. Mondon and J. Malthête, *J. Phys. (Paris)*, 1979, **40**, C3.
76. I. G. Voigt-Martin, H. Durst, V. Brzezinski, H. Krug, W. Kreuder and H. Ringsdorf, *Angew. Chem., Int. Ed. Engl.*, 1989, **28**, 323.
77. (a) M. von Ardenne, G. Musiol and S. Reball, 'Effekte der Physik und ihre Anwendungen', Verlag Harri Deutsch, Frankfurt/Main, 1990; (b) D. M. Burland and L. B. Schein, *Phys. Today*, **46**, 1986.
78. M. Stolka, 'Photoconductive Polymers', Wiley, New York, 1988, vol. 11.
79. H. Bässler, *Phys. Stat. Solidi B*, 1993, **175**, 15.
80. F. Closs, K. Siemensmeyer, T. Frey and D. Funhoff, *Liq. Cryst.*, 1993, **14**, 629.
81. E. Müller-Horsche, D. Haarer and H. Scher, *Phys. Rev. B*, 1987, **35**, 1273.
82. D. Haarer and H. Möhwald, *Phys. Rev. Lett.*, 1975, **34**, 1447.
83. (a) C. Destrade, N. H. Tinh, H. Gasparoux, J. Malthête and A. M. Levelut, *Mol. Cryst. Liq. Cryst.*, 1981, **71**, 111; (b) A. M. Levelut, *J. Phys. (Paris)*, 1979, **40**, L81; (c) S. Chandrasekhar and G. S. Ranganath, *Rep. Prog. Phys.*, 1990, **53**, 57; (d) H. Bengs, M. Ebert, O. Karthaus, B. Kohne, K. Praefcke, H. Ringsdorf, J. H. Wendorff and R. Wüstefeld, *Adv. Mat.*, 1990, **2**, 141; (e) W. Kreuder and H. Ringsdorf, *Makromol. Chem.*, 1983, **4**, 807.
84. K. Bechgaard and V. D. Parker, *J. Am. Chem. Soc.*, 1972, **94**, 4749.
85. (a) J. Simon, F. Tournilhac and J. J. Andre, *New J. Chem.*, 1987, **11**, 383; (b) J. Simon, in 'Nanostructures Based on Molecular Materials', Verlag Chemie, Weinheim 1992, pp.267–284.
86. (a) J. Noolandi, *Solid State Commun.*, 1977, **24**, 477; (b) J. Noolandi, *Phys. Rev. B*, 1977, **16**, 4466; (c) M. Lutz, *J. Imaging Technol.*, 1985, **11**, 254.
87. (a) J. C. Scott, B. A. Jones and L. T. Pautmeier, *Mol. Cryst. Liq. Cryst.*, 1994, **253**, 183; (b) H. Scher and E. W. Montroll, *Phys. Rev. B*, 1975, **12**, 2455.
88. (a) I. G. Voigt-Martin, R. W. Garbella and M. Schumacher, *Macromolecules*, 1992, **25**, 961; (b) I. G. Voigt-Martin, R. W. Garbella and M. Schumacher, *Liq. Cryst.*, 1994, **17**, 775.
89. R. R. Chance, J. L. Bredas and R. Silbey, *Phys. Rev. B*, 1984, **29**, 4491.
90. (a) P. A. Heiney, E. Fontes, W. H. de Jeu, A. Riera, P. Carrol and A. B. Smith, III, *J. Phys. (Paris)*, 1989, **50**, 461; (b) H. J. Gramsbergen, H. J. Hoving, W. H. de Jeu, K. Praefcke and B. Kohne, *Liq. Cryst.*, 1986, **1**, 397; (c) W. K. Lee, P. A. Heiney, J. P. McCauley and A. B. Smith, III, *Mol. Cryst. Liq. Cryst.*, 1991, **198**, 273; (d) G. B. M. Vaughan, P. A. Heiney, J. P. McCauley, Jr. and A. B. Smith, III, *Phys. Rev. B*, 1992, **46**, 2787.
91. (a) P. G. de Gennes, *J. Phys. Lett.*, 1983, **44**, L-657; (b) M. Werth, J. Leisen, C. Boeffel, R. Y. Dong and H. W. Spiess, *J. Phys.*, 1993, **3**, 53; (c) J. Leisen, M. Werth, C. Boeffel and H. W. Spiess, *J. Chem. Phys.*, 1992, **97**, 3749.

92. (a) W. Kreuder and H. Ringsdorf, *Makromol. Chem., Rapid Commun.*, 1983, **4**, 807; (b) W. Kreuder, H. Ringsdorf and P. Tschirner, *Makromol. Chem., Rapid Commun.*, 1985, **6**, 367; (c) O. Herrmann-Schönherr, J. H. Wendorff, W. Kreuder and H. Ringsdorf, *Makromol. Chem., Rapid Commun.*, 1986, **7**, 97; (d) G. Wenz, *Makromol. Chem., Rapid Commun.*, 1985, **6**, 577; (e) B. Kohne, K. Praefcke, H. Ringsdorf and P. Tschirner, *Liq. Cryst.*, 1989, **4**, 165; (f) W. Kranig, B. Hüser, H. W. Spiess, W. Kreuder, H. Ringsdorf and H. Zimmermann, *Adv. Mat.*, 1990, **1**, 36; (g) J. F. van der Pol, E. Neelemann, R. J. M. Nolte, J. W. Zwikker and W. Drenth, *Makromol. Chem.*, 1989, **190**, 2727; (h) C. Sirlin, L. Bosio and J. Simon, *Mol. Cryst. Liq. Cryst.*, 1988, **155**, 231; (i) W. Caseri, T. Sauer and G. Wegner, *Makromol. Chem., Rapid Commun.*, 1988, **9**, 651; (j) M. Hanack, A. Beck and H. Lehmann, *Synthesis*, 1987, **8**, 703; (k) C. F. van Nostrum and R. J. M. Nolte, *Macromol. Chem. Macromol. Symp.*, 1994, **77**, 267; (l) C. P. Lillya and Y. L. N. Murthy, *Mol. Cryst., Liq. Cryst. Lett.*, 1985, **2**, 121; (m) K. Praefcke, B. Kohne, D. Singer, D. Demus, G. Pelzl and S. Diele, *Liq. Cryst.*, 1990, **7**, 589; (n) S. Zamir, R. Poupko, Z. Luz, B. Hüser, C. Boeffel and H. Zimmermann, *J. Am. Chem. Soc.*, 1994, **116**, 1973; (o) G. C. Bryant, M. C. Cook, S. D. Haslam, R. M. Richardson, T. G. Ryan and A. J. Thorne, *J. Mater. Chem.*, 1994, **4**, 209; (p) T. Plesnivy, H. Ringsdorf, P. Schuhmacher, U. Nütz and S. Diele, *Liq. Cryst.*, 1995, **18**, 185; (q) O. Karthaus, H. Ringsdorf, M. Ebert and J. H. Wendorff, *Makromol. Chem.*, 1992, **193**, 507; (r) W. Kreuder, H. Ringsdorf, O. Herrmann-Schönherr and J. H. Wendorff, *Angew. Chem., Int. Ed. Engl.*, 1987, **26**, 1249.
93. (a) P. Henderson, H. Ringsdorf and P. Schuhmacher, *Liq. Cryst.*, 1995, **18**, 191; (b) F. Closs, L. Häußling, P. Henderson, H. Ringsdorf and P. Schuhmacher, *J. Chem. Soc. Perkin Trans.*, 1995, **1**, 829.
94. S. Zamir, R. Poupko, Z. Luz, B. Hüser, C. Boeffel and H. Zimmermann, *J. Am. Chem. Soc.*, 1994, **116**, 1973.
95. W. Kranig, B. Hüser, H. W. Spiess, W. Kreuder, H. Ringsdorf and H. Zimmermann, *Adv. Mater.*, 1990, **2**, 36.
96. S. Disch, H. Finkelmann, H. Ringsdorf and P. Schuhmacher, *Macromolecules*, 1995, **28**, 2424.

10
Synthetic Bilayer Membranes: Molecular Design and Molecular Organization

TOYOKI KUNITAKE
Kyushu University, Fukuoka, Japan

10.1 HISTORICAL BACKGROUND AS INTRODUCTION

This chapter discusses self-organization of synthetic bilayer membranes in various forms and their unique physicochemical properties.

In 1946, Bangham and Horne[1] observed aqueous suspensions of egg-yolk phosphatidylcholine by electron and optical microscopies, and found that the shape of the suspension was changed by osmotic pressure. This implied that the suspension was made of closed vesicles that acted as barriers against permeation of ions. Their subsequent experiments established that, upon dispersion in water, phospholipids produced closed vesicles (liposomes) with an inner water core. At that time, biomembranes were not necessarily considered to be composed of lipid bilayer structures, and therefore the liposome research was not directly related to model biomembranes. However, the fluid mosaic model that accounts for many properties of biomembranes (Figure 1) was proposed by Singer and Nicolson in 1972[2] and became generally accepted. Since then, liposomes have been extensively studied as models of biomembranes and for practical purposes.

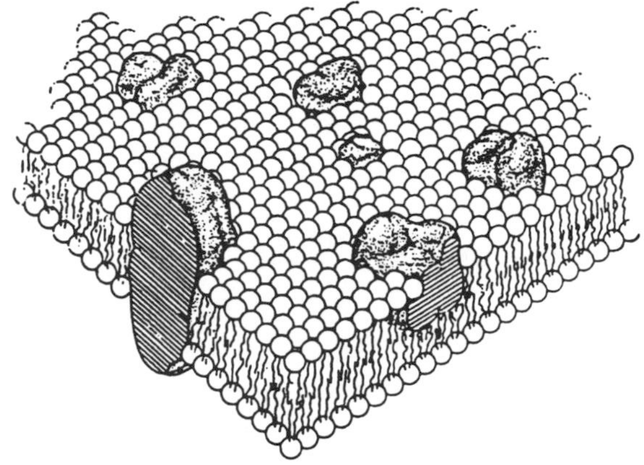

Figure 1 Fluid mosaic model of a biological membrane (after Singer and Nicolson[2]).

Attempts to prepare a bilayer assembly from amphiphilic compounds different from the natural lipid were conducted concurrently. For example, Gebicki and Hicks observed formation of globular aggregates ("Ufasome") when they dispersed thin films made of oleic acid and linoleic acid in water.[3] Hargreaves and Deamer[4] obtained vesicles by codispersion of fatty acids and higher alcohols. Vesicular aggregates have also been described for 1:1 mixtures of saturated fatty acids (C_{12}–C_{18}) and lysolecithin. However, these aggregates were not sufficiently stable and provided no definite evidence for the isolated bilayer structure.

10.2 SOME THEORETICAL CONSIDERATIONS

The self-assembly of lipid molecules into well-defined bilayer structures has attracted intensive studies by physical chemists. Tanford[5] was probably the first investigator to systematically discuss bilayer formation based on geometric considerations of aqueous aggregates. The size and shape of micelles are determined by the volume of the hydrophobic core that consists entirely of portions of hydrocarbon chains (m') and the surface area per amphiphile head group (S/m') that is a measure of the separation between adjacent head groups. Figure 2 represents the surface area per hydrocarbon chain as a function of aggregate size and shape for hydrocarbon chains with 12 carbon atoms. As the number of hydrocarbon chains per aggregate increases, the aggregate shape changes from globular to cylindrical to bilayer. Amphiphiles with two sufficiently long hydrocarbon chains tend to form bilayers.

A more detailed argument on the relationship between shapes of component amphiphiles and resulting aggregates was presented by Israelachvili *et al.* in their widely quoted 1976 paper.[6]

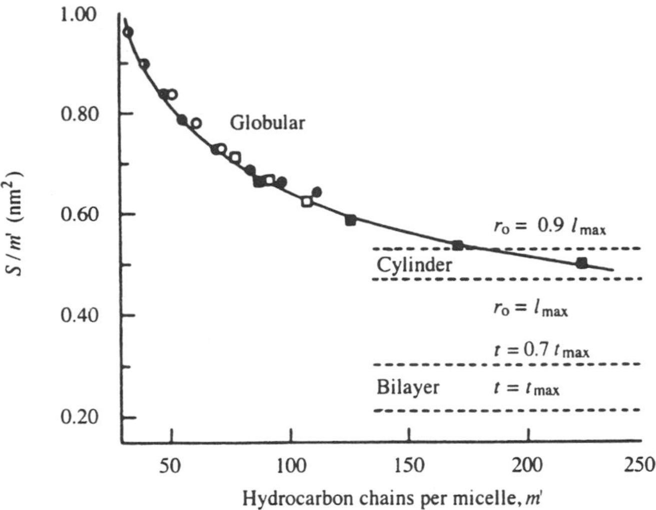

Figure 2 Surface area per hydrocarbon chain as a function of micelle size and shape. Surface areas for cylindrical micelles and bilayers are independent of aggregation number, apart from end effects, which have not been taken into consideration. In using these calculations it is important to distinguish between micelles formed by amphiphiles with a single hydrocarbon chain, for which the number of head groups is equal to the number of hydrocarbon chains, and amphiphiles with two hydrocarbon chains, for which the number of head groups is only $m'/2$, so that the area per head group is twice the area per hydrocarbon chain (after Tanford[5]).

Israelachvili subsequently reviewed thermodynamic forces that govern the self-assembly of fluid lipids into well-defined structures such as micelles and bilayers. Figure 3 illustrates a schematic representation of the balance of forces in a phospholipid bilayer. The repulsive forces arising from head-group and chain interactions balance the attractive hydrophobic surface tension. The two repulsive force contributions are equivalent to a net force of ~50 mN effectively centered near the hydrocarbon–water interface, although it should be noted that the attractive and repulsive forces do not act in the same plane. Brockerhoff[7] additionally emphasized the role of hydrogen bonding among component phospholipids. He interpreted the structural features of membrane lipids in terms of a tripartite structure consisting of a hydrophobic aliphatic double chain, a hydrophilic head group, and the hydrogen belt region linking these two moieties.

10.3 SYNTHETIC BILAYER MEMBRANES

The preceding theories certainly explain the formation of bilayer membranes from biological lipids and their close derivatives, but they do not provide guiding rules that enable novel bilayer-forming compounds to be designed. Complex intermolecular interactions cannot be

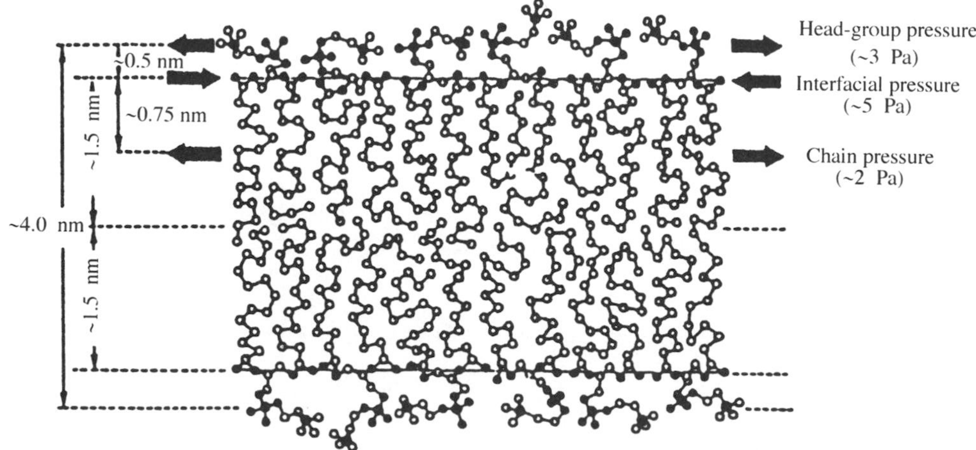

Figure 3 Schematic representation of balance of forces in a phospholipid bilayer (reproduced by permission of the Royal Society of Chemistry from *J. Chem. Soc., Faraday Trans.*, 1976, 1525).

covered by simplified thermodynamic considerations alone. Highly sophisticated molecular mechanics calculations may not yet reflect the complexity of component amphiphilic molecules.

When Kunitake and Okahata set out to develop new bilayer-forming compounds in 1976, it was assumed that the unique structure of the polar head group of biolipid molecules was determined by the biosynthetic and physiological requirements rather than by the physical chemistry of membrane formation. The first candidates were, therefore, simple double-chain ammonium salts. Bis(dodecyl)dimethylammonium bromide (**1**) (n = 12) gives a clear aqueous dispersion by sonication. When this dispersion was negatively stained by uranyl acetate and observed by electron microscopy, the formation of single- and multiwalled vesicles with a layer thickness of 3–5 nm was recognized (Figure 4). Their aggregation characteristics such as critical aggregate concentration and molecular weight were consistent with a bilayer vesicle. This result was the first example of a totally synthetic bilayer membrane.[8] The exceptional behavior of double-chain ammonium salts as aqueous dispersions had been noted many years ago. Ralston *et al.*, studied the aggregation behavior of these salts by conductometry and found an unusual dependence of the equivalent conductivity at low concentrations.[9] Kunieda and Shinoda prepared the phase diagram of these ammonium salts and indicated the transitions between molecular dispersion, liquid crystalline dispersion, and liquid crystal.[10] These dispersions were, however, never referred to as bilayer membranes.

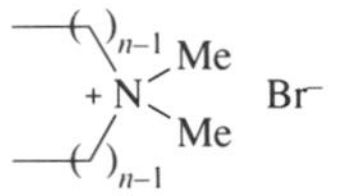

$\mathrm{(CH_2)_{n-1}}$ Me, +N, Me, Br⁻, $\mathrm{(CH_2)_{n-1}}$

n = 12, 14, 16, 18, 22

(**1**)

It is important to make a distinction between liquid crystalline dispersion and bilayer membrane. As pointed out by Gray and Winsor,[11] it is the intermicellar forces—that is, the lattice forces—rather than forces from jointing or close packing which determine the viscosities and stabilities of the mesophases. In contrast, a bilayer membrane should be able to exist as an isolated entity without relying on the lattice forces for maintaining its structural integrity. The presence of lamellar multibilayer structures does not warrant the formation of bilayer membranes. It is necessary to show the existence of isolated bilayer structures. The formation of bilayer membranes requires a self-assembling property greater than that of liquid crystalline dispersions.

In the following, the molecular design of bilayer-forming molecules, their aggregate morphologies, and their basic physicochemical properties are reviewed. Instead of attempting to prepare an exhaustive catalog of those compounds, the design principle of component amphiphiles and the representative molecular organizations they produce are presented. The molecular design will be discussed mainly in terms of the module concept given in Figure 5. This concept can cover most, if not all, of the types of bilayer-forming compounds.

Excellent reviews on various aspects of synthetic bilayer membranes have been published: Fendler prepared an extensive list of amphiphiles that are capable of bilayer formation (up to 1982);[12] routes to functional vesicle membranes have been discussed by Fuhrhop and Mathieu;[13] and an extensive review on polymeric bilayer membranes was published by Ringsdorf *et al.*[14] Fuhrhop and Köning[15] reviewed membranes and molecular assemblies according to the synkinetic approach.

The functional aspects will not be discussed in this chapter. They are most interesting parts of the bilayer membrane research, since the unique molecular assembly produces a large variety of supramolecular functions that cannot be replaced by other molecular systems. These unique features of synthetic bilayer membranes include, for example, substrate trapping in the inner water core of vesicles, slow release and permeation control of the trapped molecules, and controlled electron/energy transfer on and across the bilayer structure, reactions characteristic of organized molecules. Vigorous research activities are being carried out in these fields. However, this chapter will concentrate on the organizational aspect, because of limited time and space.

10.4 BILAYER MEMBRANES OF DOUBLE-CHAIN AMPHIPHILES

10.4.1 Molecular Design and Aggregation Behavior

Many double-chain amphiphiles have been synthesized as analogues of natural lecithin molecules, and their bilayer formation examined. The required molecular modules are tail, connector, spacer, and head groups.

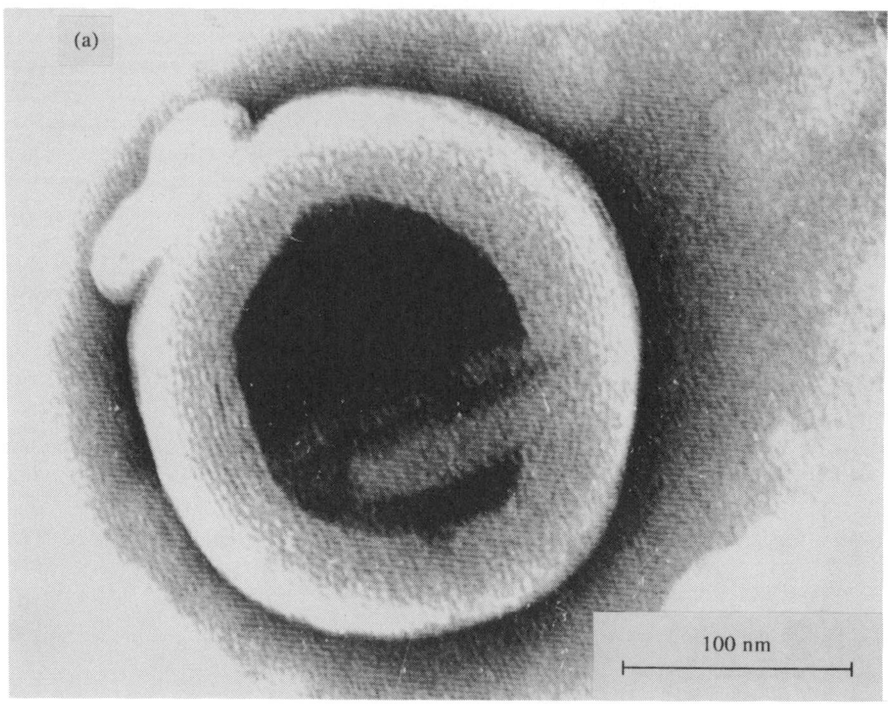

(b)

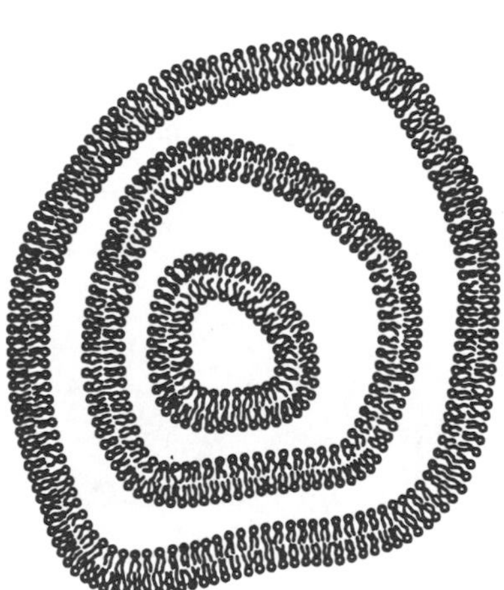

Figure 4 (a) Electron micrograph (negatively stained with UO_2^{2+}) and (b) schematic illustration of the multiwalled vesicle of (**1**) (n = 12).

A systematic survey was conducted on the aggregation behavior of dialkylammonium salts (**1**) with different tail lengths (Table 1).[16] All these double-chain compounds give colloidal solutions by sonication, and the molecular (aggregate) weights, determined by light scattering, lie between 10^6 and 10^7. Electron microscopy indicates the presence of vesicles (Figure 4) and lamellae (Figure 6), when both alkyl chains are at least 10 carbon atoms long. Their layer thicknesses are 3–4 nm, in agreement with the bilayer assemblage. Detection of the gel-to-liquid crystal phase transition by differential scanning calorimetry (DSC) strongly suggests the presence of aligned alkyl chains in the aggregate. These ammonium bilayers were extensively characterized.[17–20] The aggregate morphology is naturally affected by the method of dispersion.[19] Carmona-Ribeiro and Chaimovich reported that large unilamellar vesicles (diameter, 0.5 μm) were obtainable by the solvent vaporization method.[21]

The dimethylammonium head can be replaced by other substituents.[22] The connector portion that links the hydrophobic alkyl chains and the hydrophilic head group can be varied as shown in Figure 7.[23–7] The major function of this unit is to help promote alignment of the alkyl chains. A

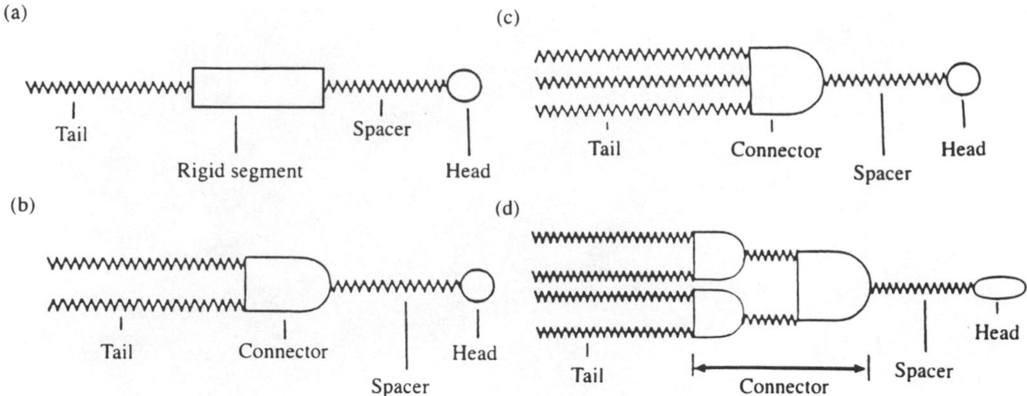

Figure 5 Structural elements (modules) of bilayer-forming amphiphiles: (a) single-chain amphiphile; (b) double-chain amphiphile; (c) triple-chain amphiphile; and (d) quadruple-chain amphiphile.

Table 1 Aggregation behavior of dialkyldimethylammonium bromides $R^1R^2Me_2N^+Br^-$ (**1**).[a]

R^1	R^2	*Solution*[b]	*Electron micrograph*	M^c	T_c^d (°C)
$C_{22}H_{45}$	$C_{22}H_{45}$	colloidal	lamella		70
$C_{18}H_{37}$	$C_{18}H_{37}$	colloidal	lamella	10^7	45
$C_{16}H_{33}$	$C_{16}H_{33}$	colloidal	lamella & vesicle		
$C_{14}H_{29}$	$C_{14}H_{29}$	colloidal	vesicle & lamella	5×10^6	16
$C_{12}H_{25}$	$C_{12}H_{25}$	clear	vesicle	$7 \times 10^5–1 \times 10^6$	n.d.[e]
$C_{18}H_{37}$	$C_{16}H_{33}$	colloidal	lamella	2×10^8	31
$C_{18}H_{37}$	$C_{14}H_{29}$	colloidal	vesicle & lamella		
$C_{18}H_{37}$	$C_{12}H_{25}$	clear	vesicle	8×10^6	n.d.[e]
$C_{18}H_{37}$	$C_{10}H_{21}$	clear	vesicle	3×10^6	
$C_{18}H_{37}$	C_8H_{17}	clear	unstructured		
$C_{16}H_{33}$	Me	clear	unstructured		

[a]The data are mostly taken from Kunitake *et al.*[16] and appended by later experiments. [b]Appearance of the 10 mM solution after sonication. [c]Molecular weight by light scattering. [d]Phase transition temperature. [e]Not determined.

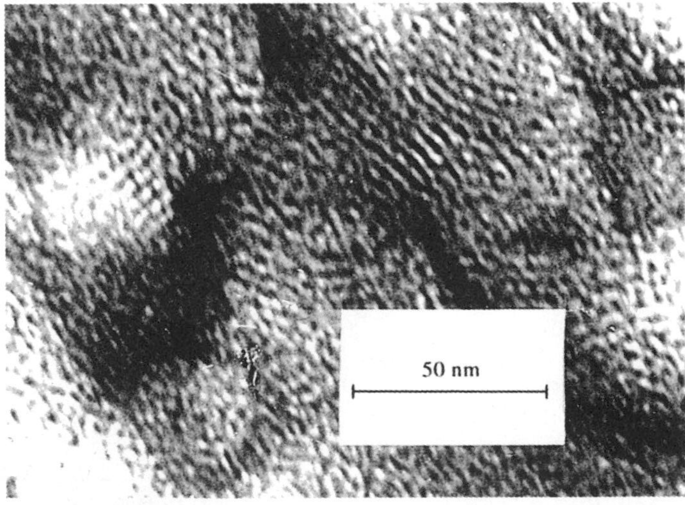

Figure 6 Electron micrograph (negatively stained with UO_2^{2+}) of lamellar bilayers of (**1**) (n = 14).

spacer unit that intervenes between the head group and the connector unit exerts significant influence on the molecular orientation in the bilayer. Double-chain amphiphiles with anionic head groups form bilayer membranes. The bilayer characteristics of amphiphiles with sulfonate (**2**), phosphate (**3**), and carboxylate (**4**) head groups are given in Table 2.[28] Stable bilayer membranes are formed again when the alkyl chain contains more than 10 carbon atoms. The presence of vesicles and lamellae is confirmed by electron microscopy, and their molecular weights are 10^6–10^7. Mortara *et al.*[29] similarly noted vesicle formation from dihexadecyl phosphate at an early stage in this research.

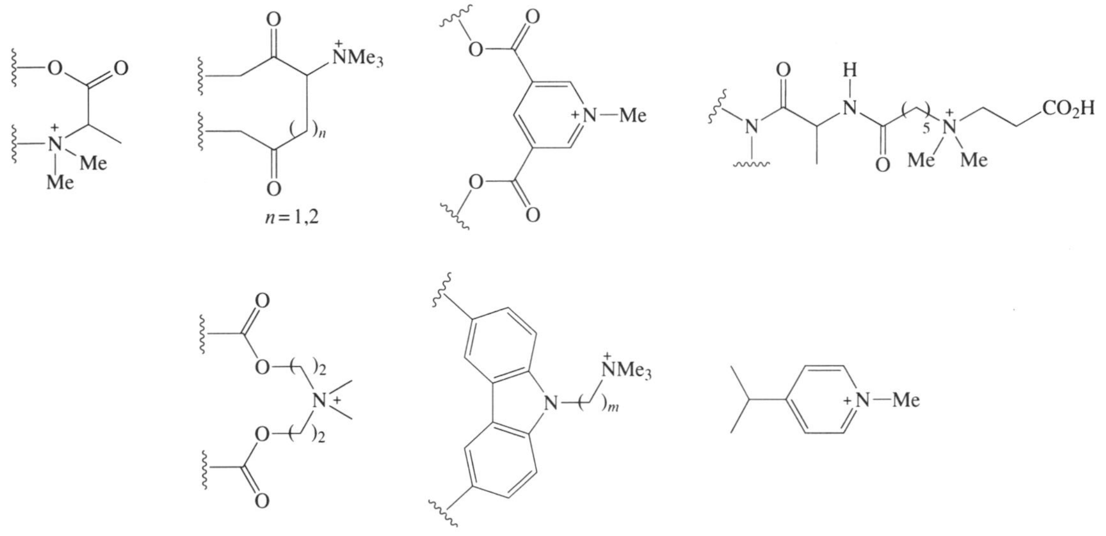

$n = 1,2$

Figure 7 Connectors of double-chain ammonium amphiphiles.

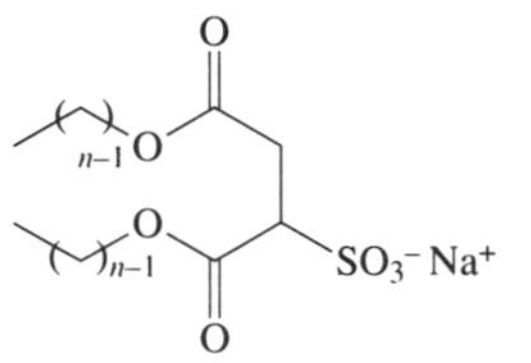

$n = 10, 12, 16$
(2)

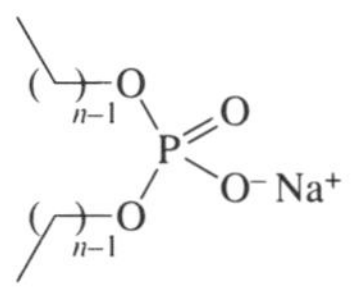

$n = 10, 12, 16$
(3)

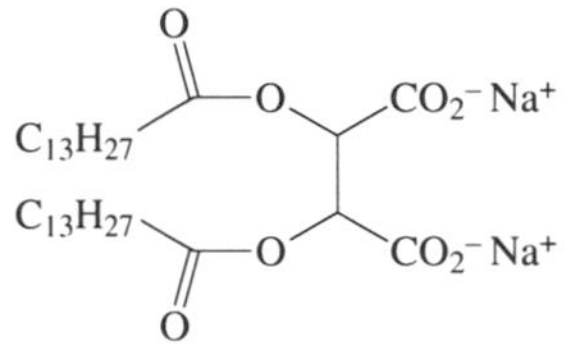

(4)

Table 2 Aggregation behavior of anionic, dialkyl amphiphiles.[a]

Compound	*Solution*	*Electron micrograph*	*M*	T_c (°C)
(2) (n = 16)	colloidal	lamella		45
(2) (n = 12)	colloidal	large vesicle	2.2×10^7	28
(2) (n = 10)	clear	large vesicle	1.9×10^6	
(3) (n = 16)	colloidal	lamella	1.2×10^7	67
(3) (n = 12)	colloidal	vesicle	2×10^7	33
(3) (n = 10)	clear	vesicle	4.4×10^6	
(4)	slightly turbid	vesicle		

Source: Kunitake and Okahata.[28]
[a]See footnotes to Table 1.

The oxyethene oligomer is conveniently used for preparing nonionic, bilayer-forming compounds.[30] The aggregation behavior of these compounds is summarized in Table 3 for **(5)** and **(6)**. In the case of **(5)** ($n = 8$), bilayer formation is not detected and the molecular weight of the aggregates ($\sim 10^5$) is smaller than those for typical bilayer aggregates ($> 10^6$). The longer chain analogues **(5)** ($n = 12$) and **(5)** ($n = 16$) produce bilayer aggregates which exhibit different morphologies. For instance, **(5)** ($n = 12$, $x = 15$) gives single-compartment vesicles, whereas **(5)** ($n = 16$, $x = 15$) produces paired bilayers (Figure 8). Zwitterionic groups such as **(7)**–**(10)** were also used as the head group of double-chain amphiphiles.[25,30]

Table 3 Aggregation behavior of nonionic amphiphiles.[a]

Compound (*n,x*)	*cmc*[b] (μM)	*Electron micrograph*	*M*	T_c (°C)
(**5**) (8, 6)		unstructured	5×10^5	
(**5**) (8, 10)	50	unstructured	4×10^5	
(**5**) (8, 12)	60	unstructured	6×10^5	
(**5**) (12, 10)	2	irregular vesicles	turbid	
(**5**) (12, 15)	2	vesicles and lamellae	1.2×10^7	n.d.
(**5**) (12, 28)	1	lamellae	1.2×10^7	
(**5**) (14, 15)		vesicles		22[c], 30
(**5**) (16, 15)			turbid	40–44
(**5**) (16, 12)	1	paired bilayers	2.7×10^7	41
(**5**) (16, 30)	1	lamellae	4.6×10^7	n.d.
(**5**) (18, 15)		disks		51,55[c]
(**6**)		fragments of lamellae	1.5×10^7	

Source: Okahata *et al.*[30]
[a]See also footnotes to Table 1. [b]Critical aggregate concentration. [c]Major DSC peak.

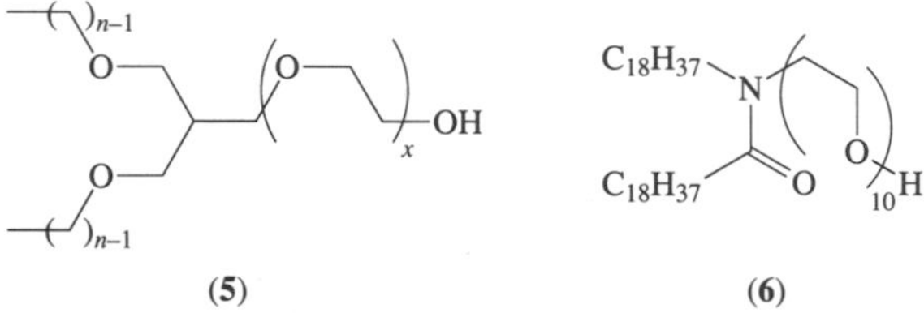

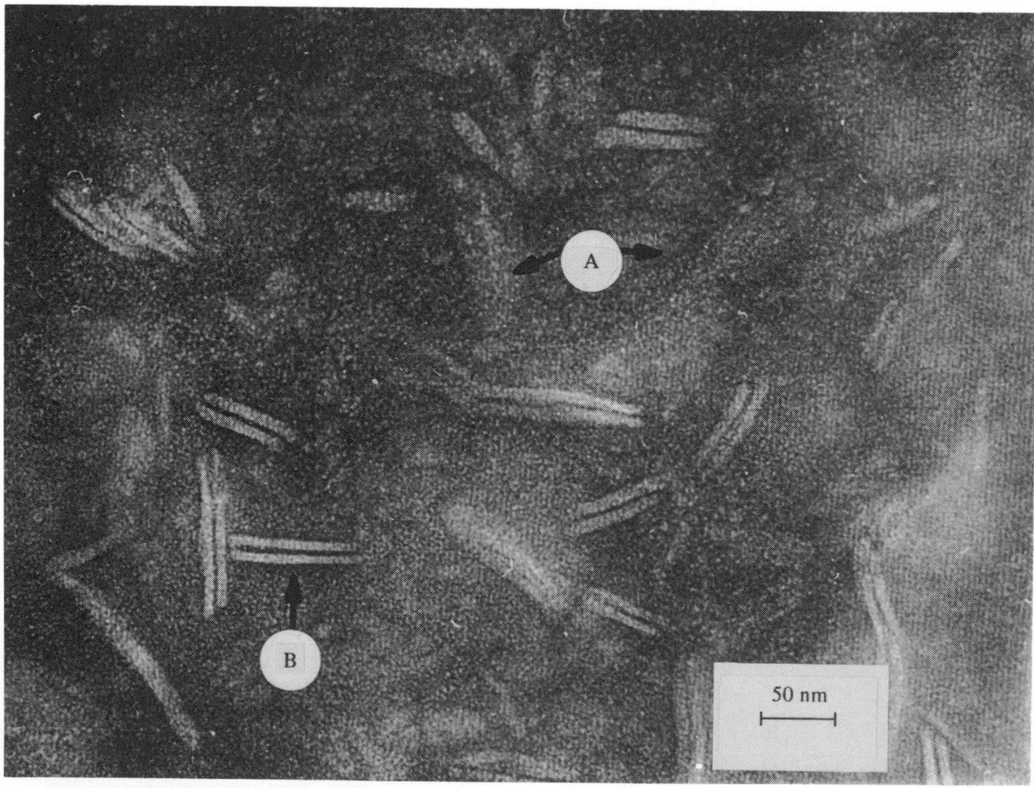

Figure 8 Electron micrograph of an aqueous dispersion of (**5**) (n = 16, x = 15): a top view (arrow A), and a side view (arrow B) of paired bilayers.

(**7**) (**8**) (**9**) (**10**)

10.4.2 Molecular Alignment

The detailed molecular arrangement in some of the synthetic bilayers has been elucidated by x-ray crystallography. Kajiyama *et al.*[31] estimated the bilayer thickness of a series of simple double-chain ammonium amphiphiles (**1**) from widths between dark striations in electron micrographs of negatively stained samples. They also determined the layer thickness of powder samples. Comparisons of these data with the molecular lengths suggested that the component molecules were tilted with respect to the bilayer surface by 32–47°, depending on the alkyl chain length.

Okuyama *et al.* succeeded for the first time in preparing single crystals of a synthetic bilayer compound that are amenable to x-ray structural analysis.[32] The structure determination has been conducted for (**1**) (n = 14, 16, 18). They all give intrinsically identical structures,[33,34] apart from the chain length. As an example, the packing structure of the monohydrate of (**1**) (n = 18) is given in Figure 9. The most important aspects in this structure are as follows: (i) Br^- and N^+ centers form a hydrophilic plane parallel to the bilayer surface; (ii) the component molecules are tilted by 45° relative to the bilayer surface; and (iii) one of the octadecyl chains assumes the *trans* zig-zag conformation (*ggtgg*) near the hydrophilic group and the *trans* zig-zag conformation towards the alkyl chain end. The chain tilting is caused by the balance between the molecular cross-section of the hydrophobic portion (0.40 nm^2 for two methylene chains) and the hydrophilic portion (0.57 nm^2 for the *bc* plane of the unit lattice). This packing structure is very similar to that found for phosphatidylcholine.[12]

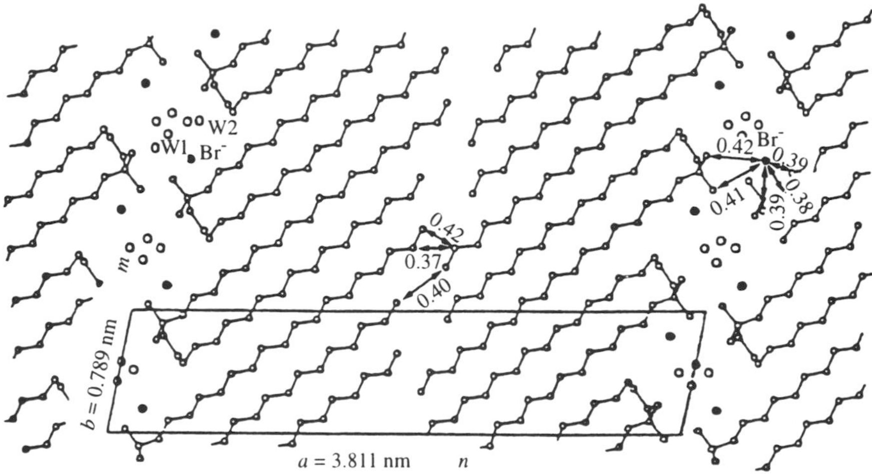

Figure 9 The packing structure of a single crystal of the monohydrate of (**1**) (n = 18) (after Okuyama *et al.*[33,34]).

When single crystals are not available, the structural simulation can be conducted by electron density matching of x-ray diffraction data.[35,36] In the case of a double-chain ammonium amphiphile with a glutamate connector,[37] the most plausible structures were selected as shown in Figure 10 for the smallest R factors of 0.11. In this model, the alkyl tail is tilted by 55° against the layer plane and the spacer methylene is tilted by an additional 25°. This difference in the tilt angles is required for packing adjustment of the double-chain tail and the single-chain spacer.

It is noteworthy that ammonium amphiphiles with two identical chains assume unsymmetrical chain packing. This seems to be a common requirement of chain packing among double-chain amphiphiles in order to attain the two-dimensionality.

10.4.3 Phase Transition, Phase Separation, and Flip-Flop

10.4.3.1 Phase transition

Phase transition and phase separation are the two most fundamental properties that describe the physical state of bilayer membranes. Flip-flop of component amphiphiles is also an important bilayer property. These three phenomena are schematically illustrated in Figure 11.

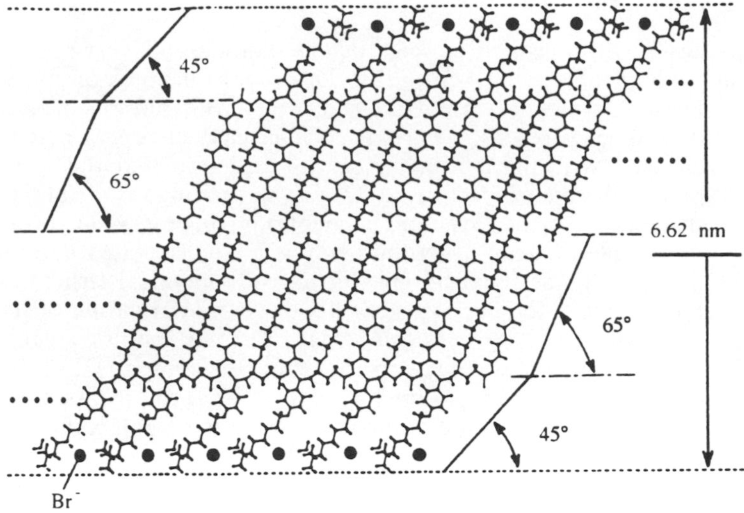

Figure 10 Plausible model of the molecular packing of the bilayer of (**68**), as produced by electron-density matching.

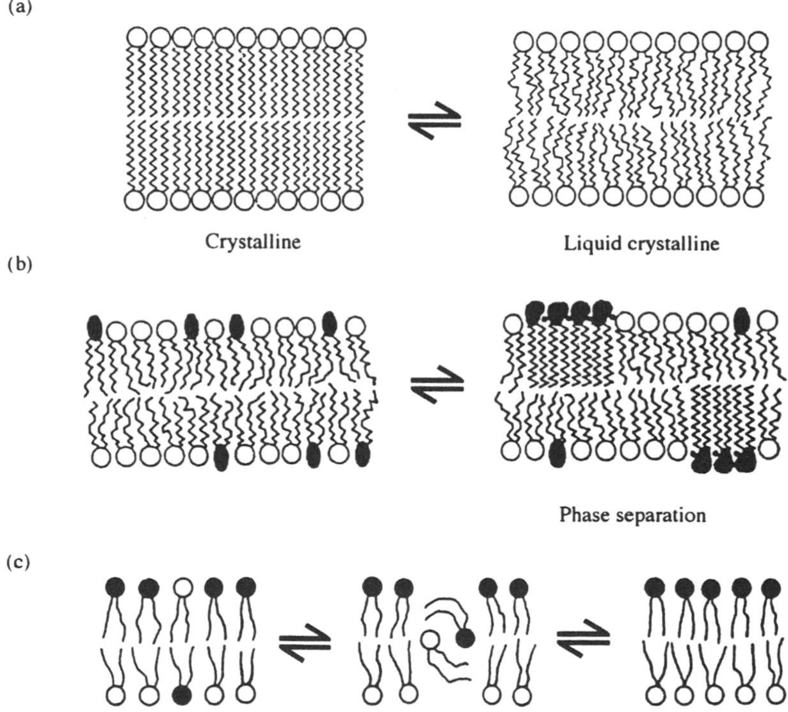

Figure 11 Schematic illustrations of (a) crystal-to-liquid crystal phase transition, (b) phase separation, and (c) flip-flop.

One of the simplest bilayer-forming amphiphiles, (**1**) (n = 18), shows a remarkably complex phase behavior at a wide concentration range.[10,38] This chapter is interested only in dilute dispersions, in which bilayer membranes are present. The gel-to-liquid crystal phase transition is commonly observed for dilute aqueous dispersions of double-chain amphiphiles. Figure 12 depicts DSC thermograms for aqueous dispersions of (**1**) (n = 12, 14, 16, 18, 22). The phase transition temperature, T_c, increases with increasing lengths of the alkyl tail. Many of the bilayer dispersions prepared by sonication give broader DSC peaks. The molecular organization in these well-dispersed samples may not be uniform. In contrast, DSC samples frozen to −50 °C give sharper single peaks, probably because more regular, multilamellar phases are formed.

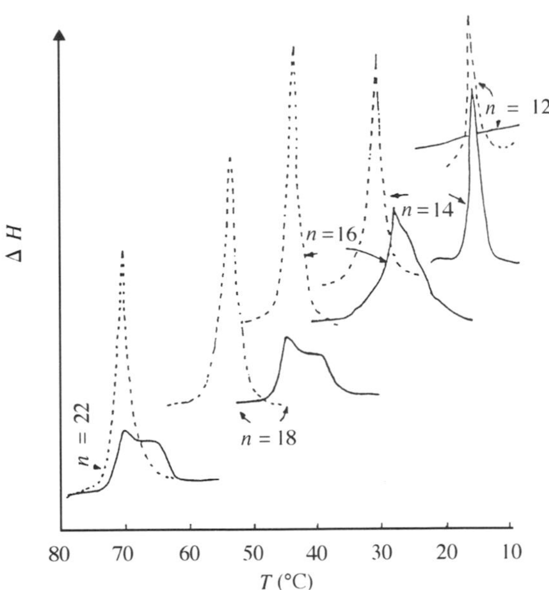

Figure 12 DSC curves of bilayers of 1–2 wt.% of (**1**) in water: (——) samples prepared by the sonication method; (- - - - -) samples prepared by the dispersion method or the freezing method.

These data are consistent with the phase transition ranges estimated by other methods: reaction rates in the bilayer matrix, NMR line broadening of the crystalline bilayer, fluorescence spectral changes of probe molecules, turbidity changes, and the extent of positron annihilation.[39] The difference in T_c between the sonicated and frozen samples depends on the molecular structure of the component amphiphile, as shown in Table 4. In the case of the simple dimethylbis(octadecyl)-ammonium amphiphile (**1**) ($n = 18$), the sonicated sample has a T_c that is 9 °C lower than that of the frozen sample. In other bis(octadecyl)ammonium amphiphiles, which possess more complex head and/or connector groups, the differences are much smaller. Uniform bilayer organizations appear to be maintained when hydrogen bonding and/or dipolar interactions are expected between component molecules.

The T_c dependence on alkyl chain lengths is affected by the spacer structure. A series of double-chain ammonium amphiphiles (**11**)–(**13**) that possess the alanine residue as the connector gives DSC figures that are enhanced with increasing tail lengths, as in the other cases.[40] It is strange, however, that the spacer length does not affect these values. The stability of the bilayer packing appears to be determined solely by the tail and the Ala connector. Analogous results have been found for phospholipids. Bach *et al.*[41] synthesized dipalmitoylphosphatidylcholines (DPPC) (**14**), in which the methylene chain between the phosphonate and ammonium groups is extended. Their T_c values are 42–45 °C in all cases, with enthalpy changes of $40 \pm 10\,\mathrm{kJ\,mol^{-1}}$. The unique effects of spacers for these series of amphiphiles are compared in Figure 13. These data give the following trends: (i) T_c, ΔH, and ΔS values are enhanced in proportion to the spacer lengths for Glu connectors; (ii) these values decrease with increasing spacer lengths for a Dea connector; and (iii) these values are virtually unaffected by the spacer length for an Ala connector.

The above data indicate that the connector structure is directly related to the tail packing as well as the spacer packing. Longer spacers in (**12**) apparently promote regular packing in the crystalline state and give rise to larger structural changes during the phase transition, as reflected in higher values of T_c, ΔH, and ΔS. This supposition is consistent with the fact that formation of helical superstructures from (**12**) is facilitated by longer spacers. However, longer spacers appear to increase the extent of disorder in the crystalline state of bilayers (**13**), as shown by smaller ΔH and ΔS values.

DSC data (T_c, enthalpy, and entropy changes) for the whole family of bilayer-forming, double-chain amphiphiles have been compiled.[42,43] The design of a novel bilayer with the desired phase transition behavior is now possible. The correlations between this behavior and molecular structure are summarized as follows. First, the entropy change of the phase transition falls within the range of $60\text{–}220\,\mathrm{J\,K^{-1}\,mol^{-1}}$ for stable bilayers of all double-chain amphiphiles including natural lipids. This indicates that all phase transition (chain melting) processes are closely related. Second,

Table 4 T_c difference between sonicated and frozen samples.

Amphiphile	ΔT_c (K)
$(C_{18}H_{37})_2N^+Me_2$	−9
$(C_{18}H_{37})_2N^+(Me)CH_2(CHOH)_4CH_2OH$	−1
$(C_{18}H_{37})_2N^+(Me)CH_2C(O)NHCH(CH_3)CO_2H$	0
$C_{18}H_{37}OC(O)CH(N^+Me_3)CH_2CH_2C(O)OC_{18}H_{37}$	0

(11) (12) (13)

the variation of the DSC data can be discussed in terms of the structural elements of component amphiphiles: (i) T_c, ΔH, and ΔS values increase with increasing lengths of tails and spacers, except when diethanolamine is included as the connector; (ii) enhanced interactions among connectors raise T_c values; however, the corresponding changes in ΔS are not straightforward; (iii) head-group interactions (electrostatic and/or hydrogen bonding) raise T_c; when the interaction is not very strong, the DSC data for bilayers of different surface charges (cationic, anionic, and neutral) do not vary; bulky substituents at the head group cause T_c lowering; and (iv) the presence of flexible units (ether and ester) either in the tail or in the spacer leads to T_c lowering; a *cis* double bond (as in the oleoyl group) has the same effect.

10.4.3.2 Phase separation

Phase separation of synthetic bilayer membranes has been detected by the presence of separate component peaks in DSC, by λ_{max} shifts in absorption spectroscopy, and by excimer emission in fluorescence spectroscopy.[44] An example of the absorption spectral method is shown in Figure 14.[45] Azobenzene-containing bilayers give large spectral shifts caused by relative orientation and stacking of the azobenzene units (see Figure 37). A 1:10 mixed dispersion of azobenzene-containing amphiphile (**15**) and dialkyldimethylammonium amphiphile (**1**) (n = 18) shows λ_{max} at 355 nm

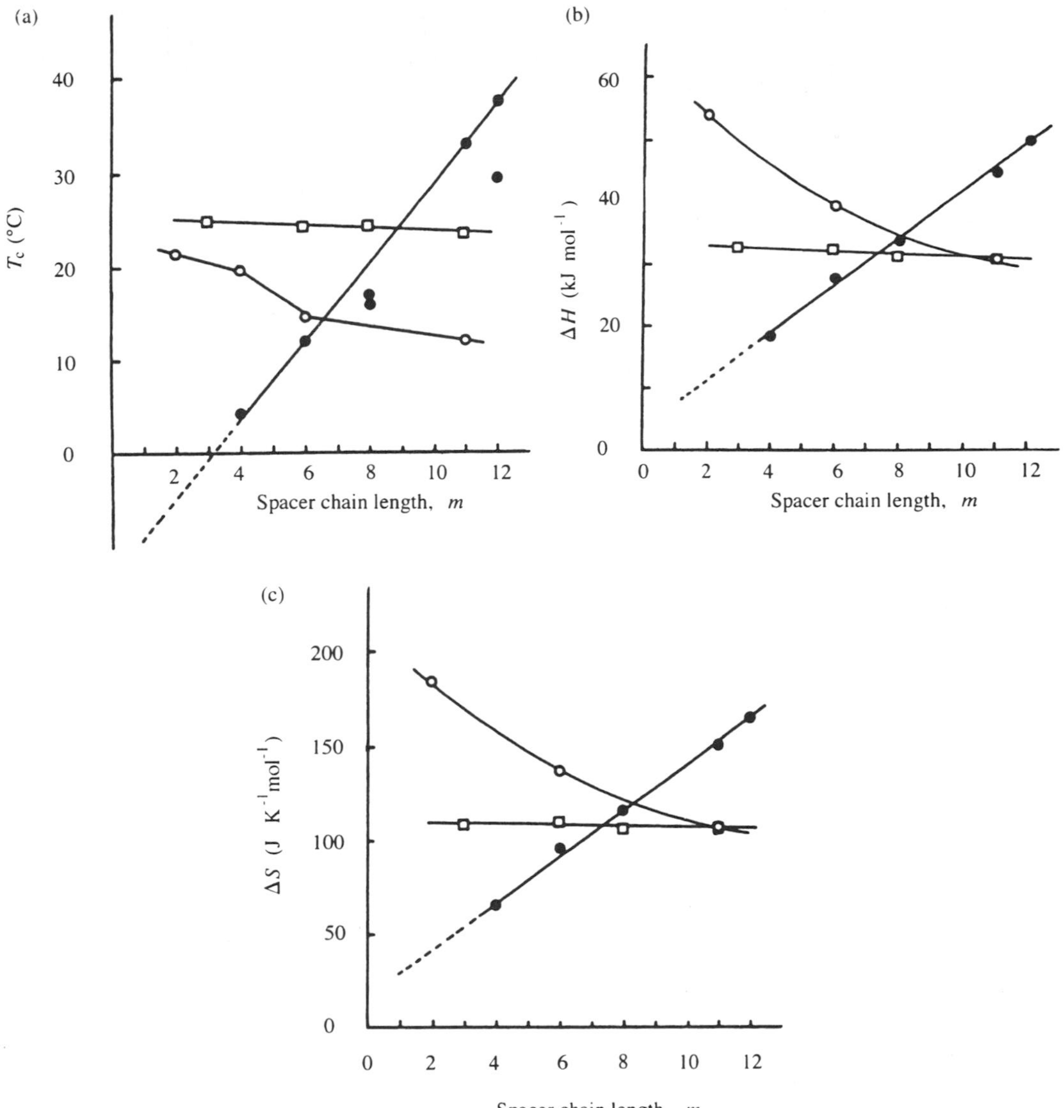

Figure 13 Influence of spacer methylene lengths on DSC behavior: variation of (a) T_c, (b) ΔH, and (c) ΔS with spacer chain length, *m*. □ = $2C_{16}AlaC_mN^+3C_1$ (**11**); ● = $2C_{12}GluC_mN^+3C_1$ (**12**); ○ = $2C_{12}DeaC_mN^+3C_1$ (**13**).

$C_{15}H_{31}$, $C_{15}H_{31}$, $\overset{+}{N}Me_3\ Br^-$

(**14**)

in the high-temperature region, indicating the presence of the isolated azobenzene species. By lowering the temperature, a new peak due to formation of the clustered azobenzene component appears at 315 nm. The spectral change corresponds to the phase transition of the matrix bilayer. This property has been exploited in novel applications of bilayers such as detection of chemical signals[46] and control of chemical reactions.[47]

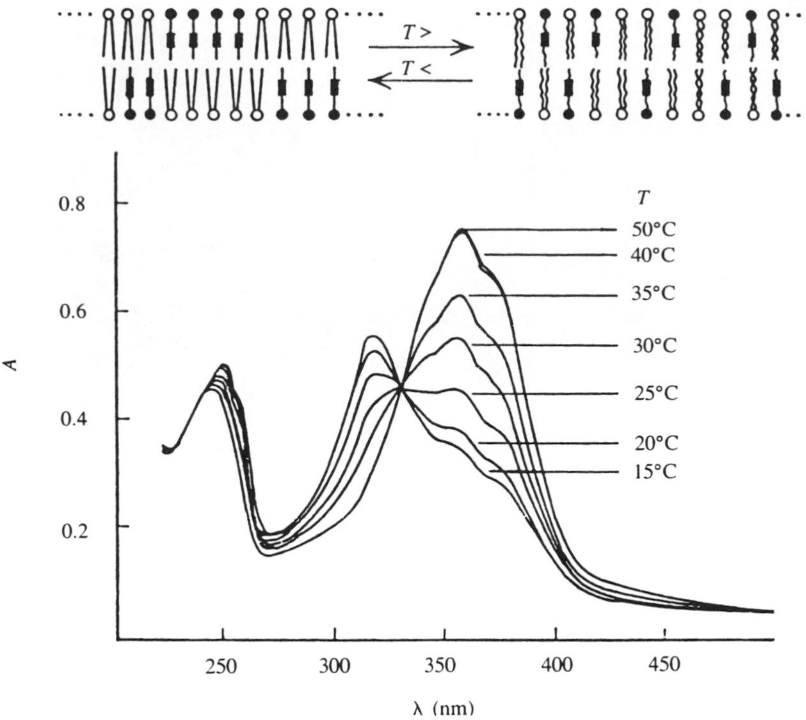

Figure 14 Absorption spectral changes as a result of phase separation in a 1:10 mixed bilayer of **(15)** and **(1)** ($n = 18$).

$C_{12}H_{25}O$–C_6H_4–N=N–C_6H_4–O–$(CH_2)_{10}$–$\overset{+}{N}Me_3\ Br^-$

(15)

The limited miscibility of hydrocarbons and fluorocarbons is used for controlled formation of separate domains, as will be discussed later.

10.4.3.3 Flip-flop

In bilayer membranes, component molecules diffuse in the lateral direction. This is largely determined by the fluidity of the bilayer. The movement of component molecules in the vertical direction is represented by flip-flop phenomena illustrated in Figure 10. Its rate constant for phosphatidylcholine vesicles was found to be $0.1\ h^{-1}$ at 20 °C by following the disappearance of a spin probe.[48] In the case of synthetic systems, Moss and co-workers estimated the flip-flop rate by discriminating *exo*-vesicular (outside) and *endo*-vesicular (inside) reactions. The hydrolysis (nitrophenol release) at the polar head of a phospholipid analogue **(16)** in mixed vesicles with a dialkylammonium salt or a phospholipid obeys a two-step process. The first step is over within a few minutes, but the second, slower process takes place in a few hours. The rate of the second step is related to the flip-flop of the *endo*-vesicular site.[49]

The flip-flop process is accelerated by transition to the liquid crystalline phase and by addition of monoalkyl amphiphiles. Polyacrylate with hydrophobic side chains embedded in bilayer vesicles enhances the flip-flop rate remarkably by disturbing the membrane surface.[50]

10.4.4 Bilayer Morphology—Static and Dynamic

Bilayer membranes dispersed in water give vesicular and lamellar aggregates in many cases. Unusual morphologies like the paired leaflet of Figure 7 are also noted. Nonbilayer aggregates

(16)

have been known for phospholipids.[51] Similarly, Murakami *et al.* reported that there were interconversions between bilayer and nonbilayer morphologies.[52] In contrast to the charged lipids **(17)** and **(18)**, which form typical vesicles and lamellae, their nonionic counterpart **(19)** is dispersed as nonlamellar network aggregates in the neutral pH range (Figure 15), but is transformed into multilamellar vesicles in the deprotonated form. Addition of cationic **(17)** also transforms the nonlamellar aggregate into vesicles. An equimolar mixture of the cationic and anionic amphiphiles leads to nonlamellar networks with repeating distances of 7.0 nm and 13.0 nm in the liquid crystalline temperature region. Small-angle x-ray diffraction suggested the formation of a face-centered cubic lattice.[53]

$n = 12, 14, 16$
(17)

$n = 14, 16$
(18)

(19)

Figure 15 Nonlamellar network aggregates of the neutral form of **(19)** (after Murakami *et al.*[52]).

Dynamic morphological changes are intrinsic to many physiological functions of the biological cell. The corresponding study on synthetic bilayer membranes is limited. Rupert *et al.* examined the ability of vesicles of (**1**) ($n = 12$) to undergo fusion.[54,55] They found that dianions such as dipicolinate promoted fusion of vesicles with sizes of at least 300 nm diameter, whereas smaller vesicles did not readily fuse despite their aggregation. The fusion was facile when the vesicles were in an overall fluid state. This research group also showed that the lamellar-to-hexagonal II phase transition proceeded by Ca^{2+}-induced fusion of didodecyl phosphate vesicles.[56] Salt-induced aggregation of dimethylbis(octadecyl)ammonium chloride and sodium bis(hexadecyl) phosphate vesicles has been discussed by Carmona-Ribeiro and Chaimovich.[57]

Optical microscopy is a convenient tool for real-time observation of the morphology change. Evans *et al.* used video-enhanced contrast–differential interference contrast microscopy to observe dynamic changes of aggregate morphologies of simple double-chain ammonium amphiphiles (**1**) ($n = 12$).[58] Hotani showed in 1984 that dark-field optical microscopy was particularly advantageous for studying the finer details of biological structures.[59] These investigations confirmed that several transformation pathways existed for various morphologies of liposomes: biconcave, elliptical, regular polygonal, and so on.

This technique was applied to examine the morphology of (**1**) ($n = 12$).[60] The dynamic changes observed are remarkably similar to those of the phospholipids mentioned above. For example, a tubule attached to a vesicle is elongated with time while the vesicle shrinks (Figure 16(a)). A vesicle with a protuberance is converted into a spherical vesicle via an irregular intermediate (Figure 16(b)), and a tadpole and a tripod interconvert (Figure 16(c)). Electron microscopy of freeze-fracture replicas of this dispersion shows that elongated tubules branch into smaller tubules and fibers, which, in turn, are transformed into trains of very small vesicles (diameter, 30–40 nm). Dark-field optical microscopy was further employed for observation of myelin figures and double-helical fibers of bilayers of double-chain ammonium amphiphiles (**20**) (Figure 17).[61] The latter morphology was observed for egg-yolk lecithin.

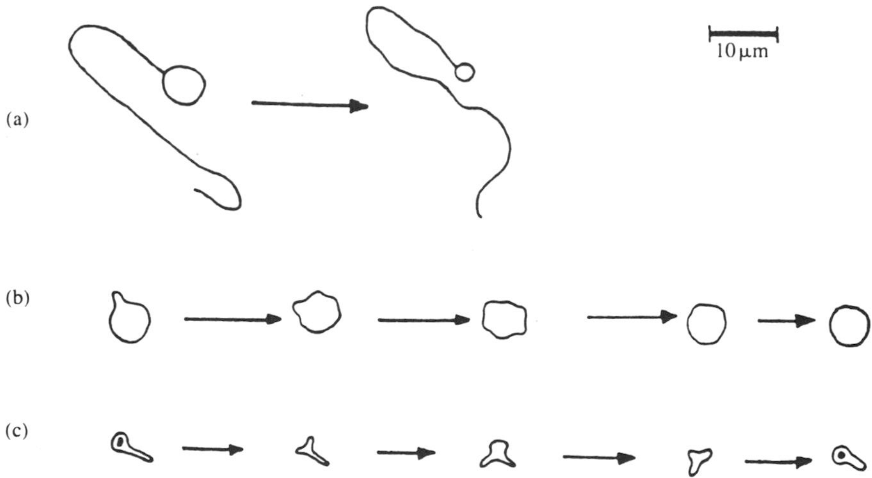

Figure 16 Schematic illustration of the morphological transformations of an aqueous dispersion of (**1**) ($n = 12$): (a) a tubule attached to a vesicle is elongated with time which the vesicle shrinks; (b) a vesicle with protuberance is converted into a spherical vesicle; and (c) a tadpole and a tripod interconvert.

10.5 BILAYER MEMBRANES OF SINGLE-CHAIN AMPHIPHILES

10.5.1 Molecular Design

Attachment of two alkyl chains, instead of one, to a single head group should enhance the molecular orientation in aggregates because their conformational mobility is restricted. The larger liquid crystalline region in the phase diagram of double-chain amphiphiles compared with that of single-chain counterparts is consistent with this view. If this were the case, stable bilayer membranes would be produced even from single-chain amphiphiles whose conformations are restricted by incorporation of rigid segments or by intermolecular interactions.

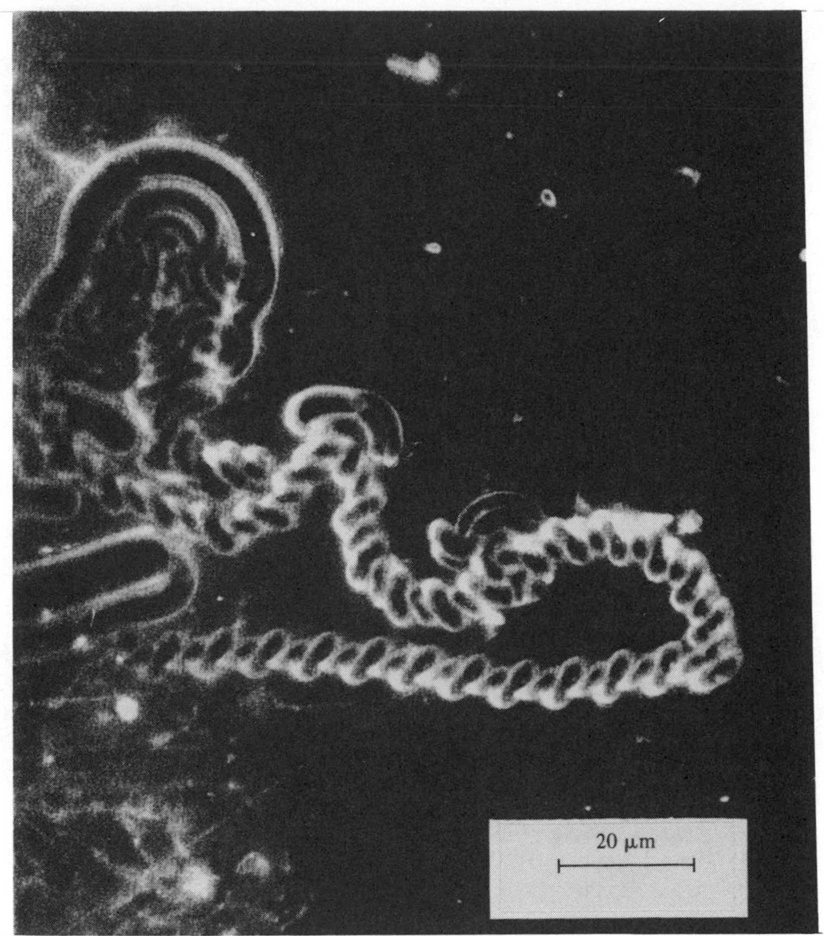

Figure 17 Dark-field optical micrograph of an aqueous dispersion of (**20**) (n = 2).

(**a**) n=2, X=Cl; (**b**) n=11, X=Br

(**20**)

Scheme 1 illustrates a design principle of a bilayer-forming, single-chain amphiphile. A conventional liquid crystalline material, MBBA, contains a rigid Schiff base unit connected to a flexible alkyl chain. A micelle-forming surfactant, CTAB, contains a trimethylammonium head group and a hexadecyl chain. The combination of the structural features of these two compounds results in a novel amphiphile that should form aqueous aggregates with a molecular orientation better than that of the conventional micelle. It contains a hydrophilic head, a spacer, a rigid segment, and a flexible tail as structural modules.

Within this structural framework (as in (**21**) and (**22**)), the lengths of the methylene chain of the spacer and the hydrocarbon tails were varied, and their aggregation behavior examined (Table 5).[62] Molecular aggregation detected by surface tension measurements occurs at concentrations of

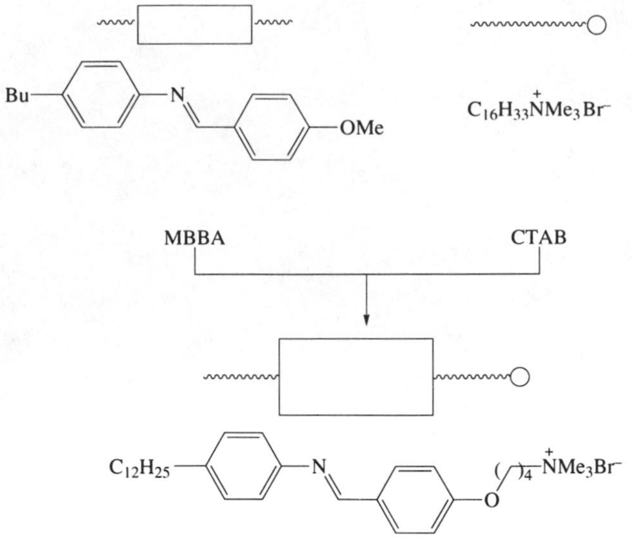

MBBA CTAB

Scheme 1

10^{-4}–10^{-5} M except for **(21)** (n = 0) and **(22)** (n = 0, x = 10), which did not show significant lowering of surface tension. The critical aggregate concentration is lowered by lengthening of the hydrocarbon tail (C_n portion), but is affected little by the length of the intervening methylene group (C_m portion). The critical aggregate concentrations for **(21)** (n = 4) and **(22)** (n = 4, m = 10) lie in the common range of (2–3) × 10^{-4} M. The critical aggregate concentration for **(21)** (n = 12) contrasts sharply with that of **(22)** (n = 0, m = 10) or **(22)** (n = 4, m = 10). In terms of the hydrophile–lipophile balance, these compounds should show similar results. Since this is not so, the location of the rigid segment must determine, to a large extent, the tendency of these amphiphiles to align at the air–water interface.

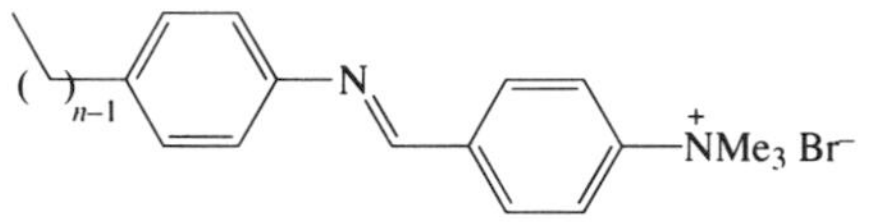

n=0, 4, 7, 12
(21)

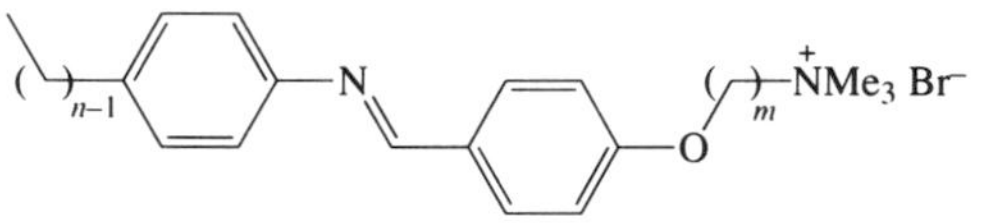

n=0, 4, 7, 12; m=4, 10
(22)

Table 5 Aggregate behavior of single-chain amphiphilies.

Amphiphile	*n*	*m*	*cmc*[a] (mM)	*M*[b]	*Electron micrograph*
(21)	0		> 10	< 10^4	unstructured
	4		0.22	1 × 10^5	unstructured
	7		0.06	8 × 10^6	lamella
	12		0.01	4 × 10^6	vesicle and lamella
(22)	4	4	0.16	8 × 10^6	unstructured
	7	4	0.01	2 × 10^7	lamella
	12	4	0.01	4 × 10^6	vesicle
	0	10	> 10	4 × 10^6	unstructured
	4	10	0.33	6 × 10^6	unstructured
	7	10	0.02	2 × 10^7	vesicle
	12	10	0.01	1 × 10^7	lamella

[a]Critical aggregate concentration. [b]Molecular weight determined by light scattering.

The average molecular weight of most of the aggregates is 10^6–10^7. Therefore, the average aggregation numbers are in the range of 10^3–10^5. The electron microscopic observation establishes the structural requirement for the formation of a stable bilayer. The development of the bilayer structure is improved with increasing chain lengths of the flexible tail (C_n portion), the minimal chain length probably being between $n = 7$ and $n = 4$. It is worthy of emphasis that **(21)** ($n = 4$) and **(22)** ($n = 4$, $m = 10$) fail to form well-organized assemblies in spite of their formation of very large aggregates. A certain tail length is required for an amphiphile to align in aqueous aggregates as well as at the air–water interface.

A systematic variation of lengths of flexible tail and spacer was also conducted for azobenzene-containing, single-chain amphiphiles **(23a)**.[63] Among the total of 21 compounds, the formation of stable bilayer membranes as judged by three criteria was concluded for the following combinations of tail length n and spacer length m: $n = 12$ and $m = 2$–12; $n = 10$ and $m = 5$–10; and $n = 8$ and $m = 10$. Clearly, the tail length is more critical than the spacer length in bilayer formation.

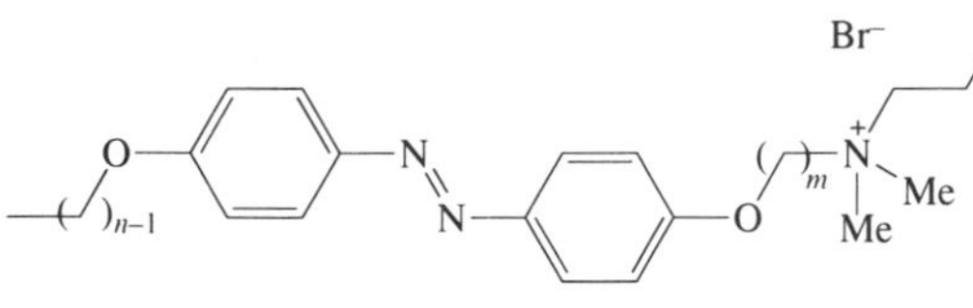

(23a) 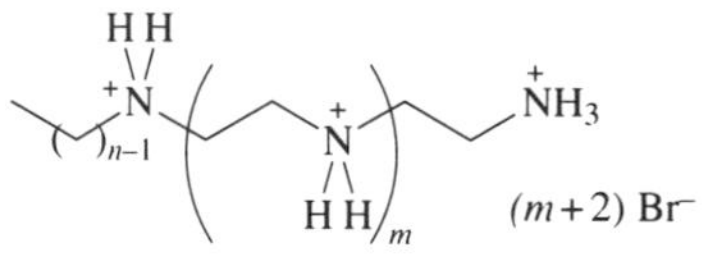 **(23b)**

The subsequent studies on molecular design of the single-chain amphiphile established that the kind of head group could be varied as much as those of the double-chain counterpart (cationic, anionic, nonionic, and zwitterionic)[64,65] and that the rigid segment is extensively variable, as shown in Figure 18.[66,67] An extensive list of this class of single-chain compounds and their phase transition behavior has been compiled.[68]

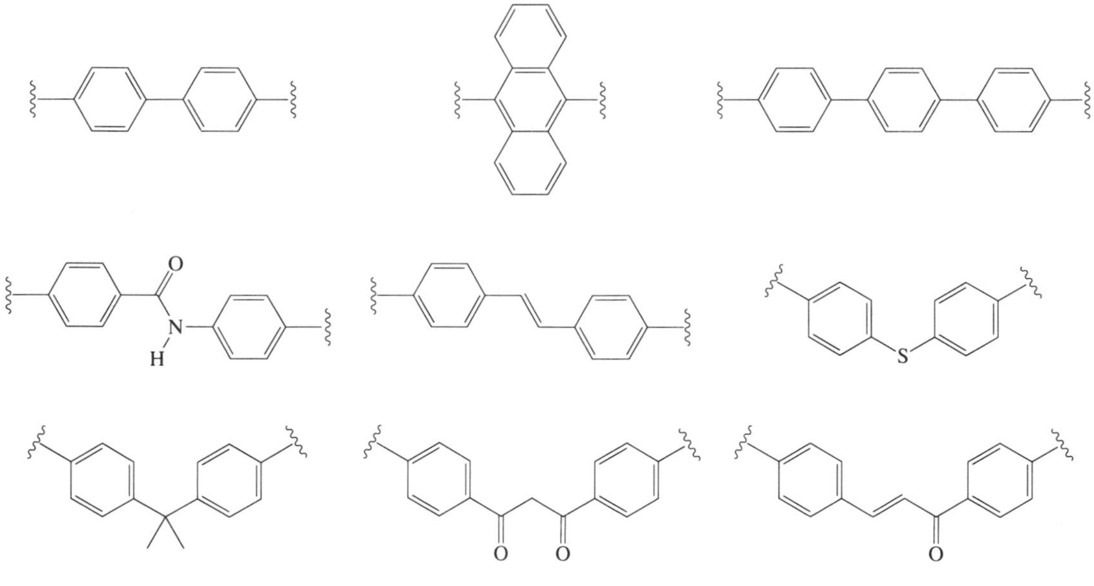

Figure 18 Examples of rigid segments for single-chain bilayer-forming amphiphiles.

Single-chain amphiphiles that are not included in these categories have been found to form bilayer membranes. Cho *et al.* reported that an ammonium derivative of cholesterol formed bilayer vesicles.[69] The concept of the "rigid segment" may be extended to the structural unit that promotes molecular alignment by intermolecular stacking and other intermolecular interactions. The aromatic "rigid segment" of Figure 18 promotes intermolecular stacking. The molecular alignment can be improved in single-chain ammonium amphiphiles by a strong hydrogen-bonding capability.[70]

Menger and Yamasaki[71] even claimed that hyperextended single-alkyl ammonium amphiphiles formed stable bilayer assemblies in water. According to their results, multicationic compounds **(23b)** (where $n = 24$, 28, 35 and $m = 3$, 5) formed large spherical aggregates, on the basis of dynamic light scattering and scanning electron microscopy (SEM), with diameters in the tens to hundreds of namometers range. However, the SEM picture does not indicate the presence of an independent bilayer structure, though the spherical aggregate is wrinkled, as is often observed for typical bilayer vesicles. The aggregate may as well be described as a liquid crystalline dispersion.

10.5.2 Morphology

Single-chain amphiphiles form bilayer vesicles and lamellae, similar to those of double-chain amphiphiles. The rigid segment in these amphiphiles promotes regular molecular orientation. The effect of the chemical structure of the rigid segment on aggregate morphologies has been studied systematically for a series of single-chain ammonium amphiphiles.[66] Electron microscopy indicates the formation of aqueous aggregates with various morphologies: the representative cases are globule, vesicle, lamella, rod, tube, and disk (Figure 19). These morphologies can be explained as arising from different characteristics of the rigid segment, as summarized in Figure 20. When the rigid segment is biphenyl or azobenzene, globular aggregates (probably collections of small bilayer fragments) are formed. The diphenylazomethine group is structurally similar to the azobenzene group, but the behavior as a rigid segment is quite different. The amphiphile with a diphenylazomethine rigid segment produces much better developed bilayer membranes as a result of the dipolar interaction between the diphenylazomethine groups. In the case of the symmetrical rigid segments (biphenyl and azobenzene), extensive organization appears less favored.

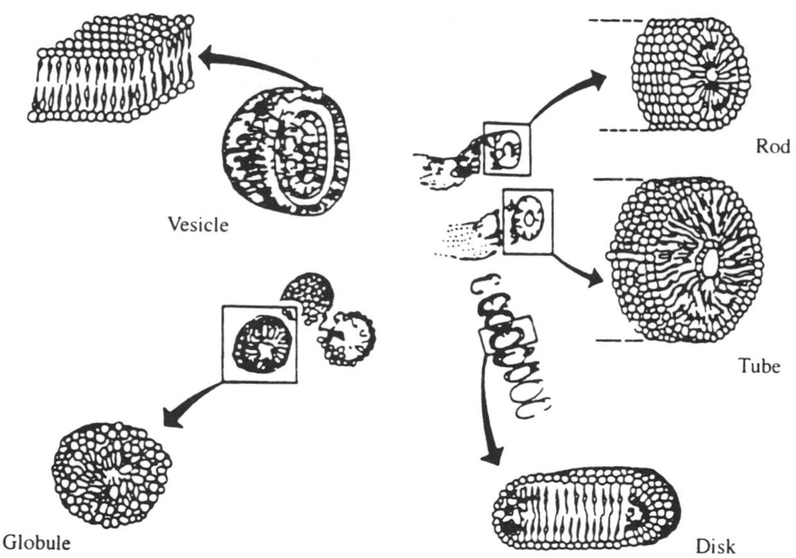

Figure 19 Aggregate morphologies of single-chain amphiphiles.

A structural feature common to the third group of single-chain ammonium amphiphiles is that they contain bent rigid segments. The bending is either due to single-atom connection of two benzene rings, due to rotation of the single bond, or due to *meta* substitution. The contrasting behavior of the straight and bent ammonium amphiphiles is interesting. The straight molecules tend to assemble in a neat, two-dimensional arrangement (membrane), but the bent molecules give rise to curvatures which result in a radial molecular arrangement (rod-like and tubular structures). Double *meta* substitution at the rigid segment produces less organized aggregates.

The disk structure contains two kinds of molecular arrangement. One is the two-dimensional arrangement (flat portion) as in the typical bilayer membrane, and the other is a highly curved arrangement (side of the disk) as in the rod-like structure. The fourth group of amphiphiles can assume both the extended and bent conformations, which may be used advantageously to produce the two kinds of molecular packing involved in a disk.

These morphological correlations suggest that the surface curvature of an aggregate can be controlled by appropriate mixing of different classes of amphiphiles. Amphiphile (**22**) ($n = 12$, $m = 4$) gives well-developed, multilamellar vesicles (Figure 21(a)). Replacement of this rigid segment with the diphenyl ether unit results in a tubular morphology (Figure 21(b)). A 1:1 mixture of

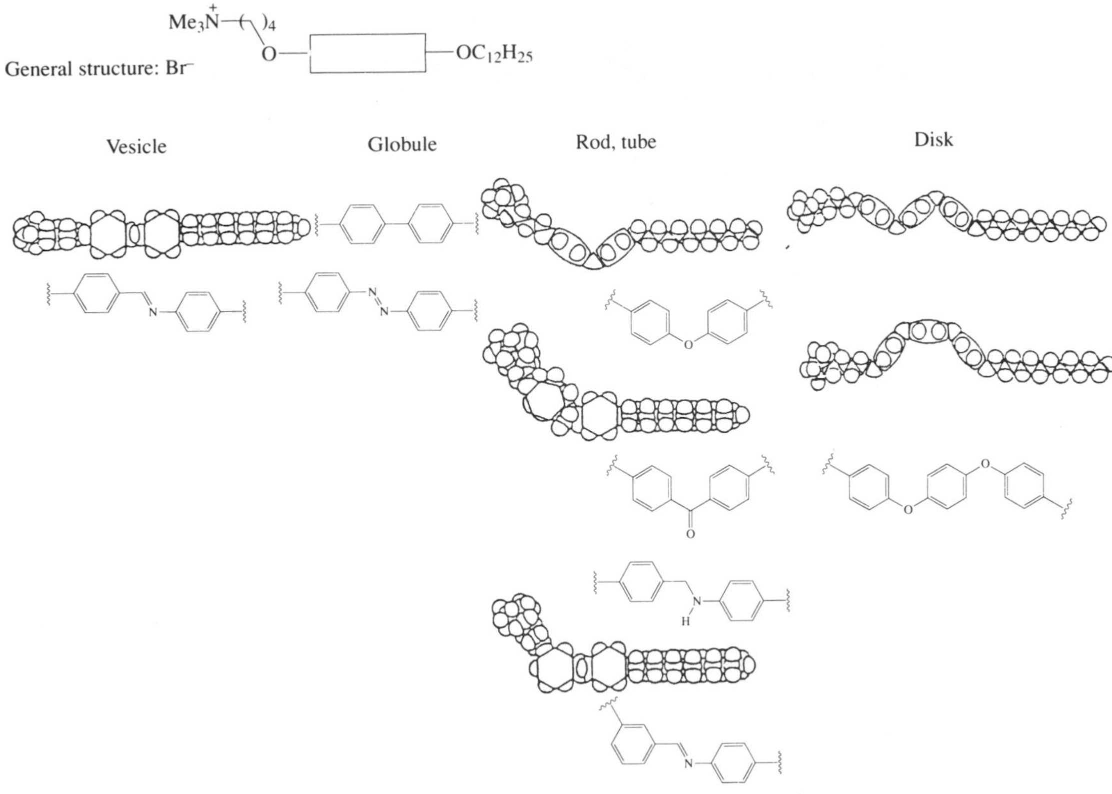

Figure 20 Relationship between molecular geometry and aggregation morphology.

these amphiphiles gives tubular aggregates very similar to Figure 21(b). When the molar ratio is 7:3, an obscure mixture of tubes and lamellae is found. A 9:1 mixture produces stacked disks (or stacked patches of bilayers) with a layer thickness of 4–5 nm (Figure 21(c)). It is conceivable that the flat portion of the disk is mainly composed of (**22**) (n = 12, m = 4), and that (**24**) is concentrated along the side of the disk where large curvature is required.

A related morphology control is found for a mixture of (**22**) (n = 12, m = 4), and (**25**). Cosonication of multiwalled vesicles of (**22**) (n = 12, m = 4) and globular (**25**) gives single-walled vesicles.

10.6 MOLECULAR MEMBRANES OF MULTICHAIN AND TWO-HEADED AMPHIPHILES

10.6.1 Triple-chain Amphiphiles

As discussed at the beginning of this chapter, the hydrophile–lipophile balance is not the most crucial factor that determines bilayer formation. Certain degrees of imbalance will be accommodated by the stabilization gained by molecular alignment. Compounds (**26**)–(**28**) are examples of triple-chain ammonium amphiphiles that were tested for bilayer formation.[72] Simple triple-chain compounds (**26**) do not seem to give well-developed bilayer structures. Amphiphiles (**27**) and (**28**) form stable bilayer membranes in water, as inferred from their huge molecular weights, the presence of phase transitions, and electron microscopy observations. The CPK molecular models of these amphiphiles with different connector structures (Figure 22) endorse these experimental observations. In the case of (**26**) (n = 12), the long alkyl chains cannot be aligned at locations close to the ammonium group because of its tetrahedral configuration. In contrast, the alkyl chains of (**27**) (n = 12) can be well aligned. The compact chain packing is produced by the presence of the ester unit, and the ammonium head can protrude without conformational constraint. The same situation is found for (**28**) (n = 12, m = 2), again due to the presence of the ester linkage.

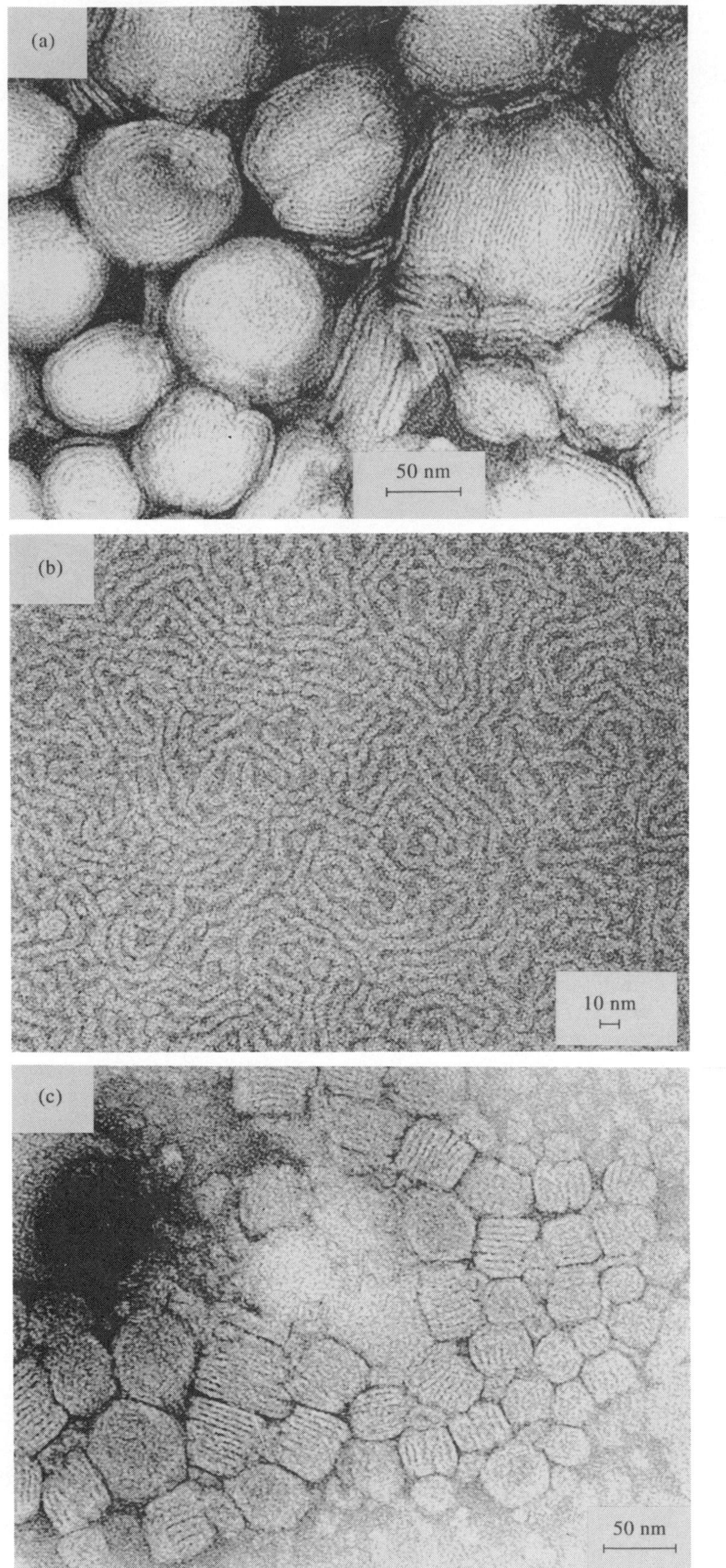

Figure 21 Electron micrographs of aqueous bilayer dispersions of single-chain ammonium amphiphiles (negatively stained by UO_2^{2+}): (a) **(22)** (n = 12, m = 4); (b) **(24)**; and (c) a 9:1 mixture of **(22)** (n = 12, m = 4) and **(24)**.

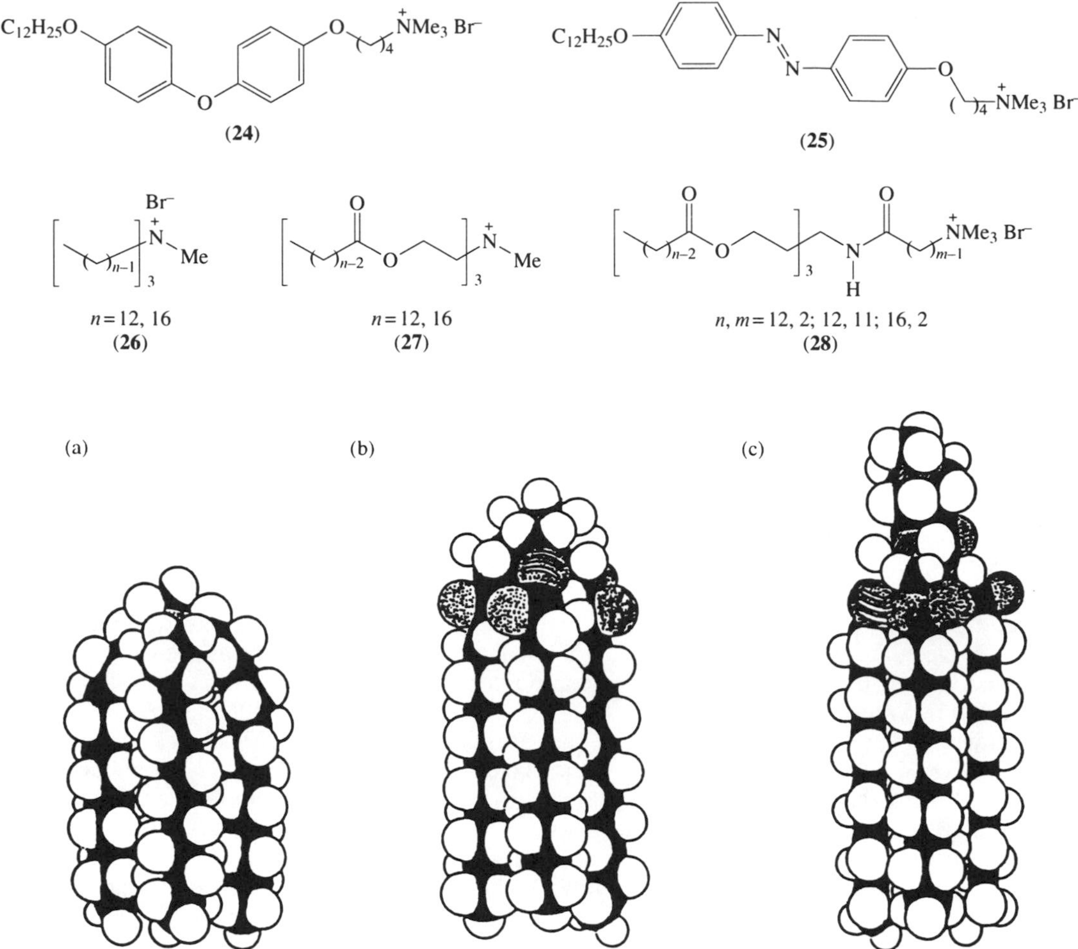

Figure 22 CPK model of triple-chain amphiphiles with different connectors. The parallel alignment of the alkyl chains in (a) (**26**) is inferior to that in (b) (**27**) and (c) (**28**).

10.6.2 Quadruple-chain Amphiphiles

Even quadruple-chain ammonium amphiphiles such as (**29**) are suitable for two-dimensional molecular alignment.[73] These compounds give transparent aqueous dispersions at 20 mM, and display typical bilayer characteristics. Electron microscopy indicates the formation of vesicles and tape-like aggregates. A DSC experiment showed the presence of a phase transition peak. The transition behavior of (**29**) is not very different from that of the corresponding double-chain amphiphile (**30**). The chain alignment is strongly dependent on the type of connector. The chiral unit in (**29**) and (**30**) should produce different chiral binding sites at the membrane surface.

(**29**)

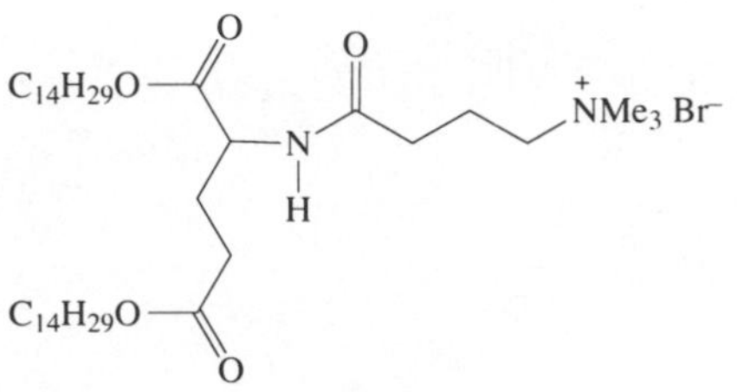

(30)

The native counterpart of these amphiphiles is four-chained cardiolipin. It is interesting to note, however, that cardiolipin cannot form a stable bilayer assembly by itself.

10.6.3 Two-headed Ammonium Amphiphiles

Some archeobacteria contain another class of membrane lipids. One of those lipid components is a macrocyclic amphiphile that spans the bilayer thickness.[74] Covalent bonding of alkyl chain ends of synthetic bilayer components produces two-headed amphiphiles which may form monolayer membranes in water.

Structural extention of single-chain, bilayer-forming amphiphiles to two-headed amphiphiles results in molecular structures such as **(31)**–**(33)**.[75] Two ammonium groups at the ends of a molecule are connected by a single chain that is composed of methylene and aromatic units. They are all dispersed in water as stable aggregates with molecular weights of 10^6–10^7. Electron microscopy shows that lamellae and rod-like structures are formed. As shown in Figure 23, structure **(31)** (10 mM) gives highly developed lamellae with a layer thickness of 3–4 nm. This thickness corresponds to the extended molecular length; therefore, the bis(ammonium) molecule is aligned normal to the layer plane. Molecular elongation and increase in the flexible hydrocarbon chain as in **(32)** (molecular length, 6.7 nm) leads to the formation of tubular structures (diameter 7–10 nm), with dark striations (deposits of uranyl acetate) in the middle. The flexible methylene chain in the center of the molecule would allow the bent molecular packing required for the tubular morphology (Figure 23(b)).

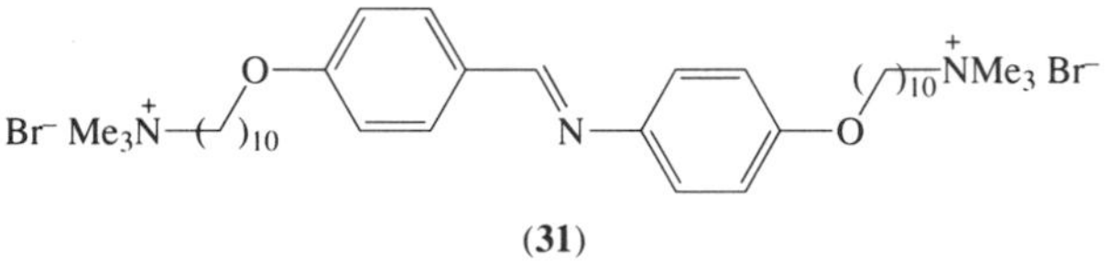

(31)

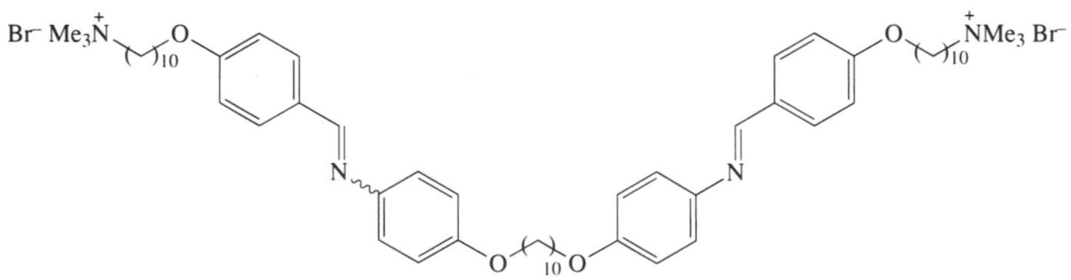

(32)

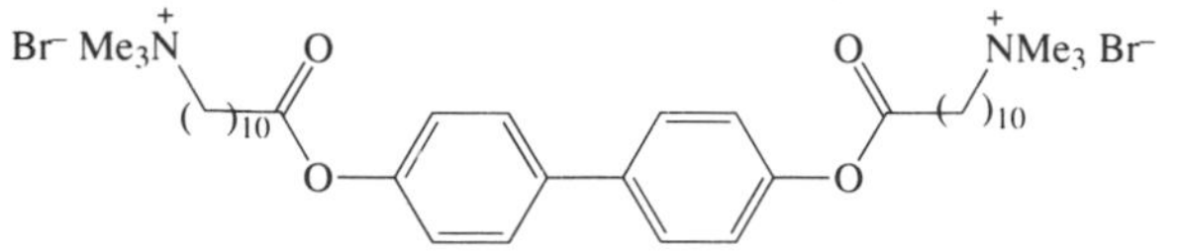

(33)

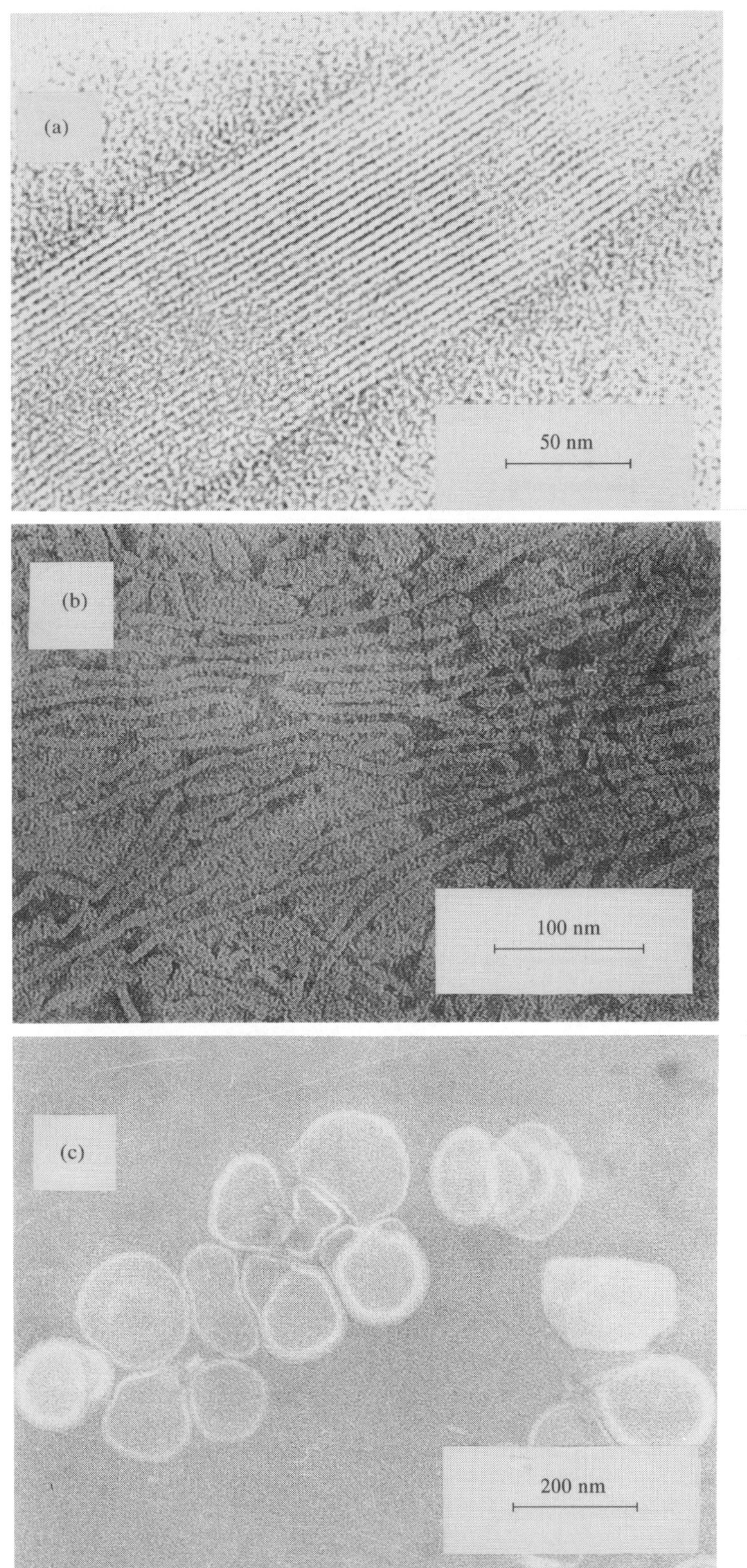

Figure 23 Electron micrographs of aqueous dispersions of two-headed ammonium amphiphiles (stained by UO_2^{2+}): (a) (**31**); (b) (**32**); and (c) (**31**) + 1/3 cholesterol.

The rigid lamella of Figure 23(a) can be transformed into single-walled vesicles as in Figure 23(c) by cosonication with cholesterol in a 3:1 ratio. The vesicles are 100–300 nm in diameter and 6–7 nm in layer thickness. It is probable that cholesterol molecules are preferentially located in the outer half of the membrane, thus creating a surface curvature suitable for vesicle formation (see Figure 24(a)). Remarkable transformations of morphology of lecithin vesicles by addition of lysolecithin and cholesterol were reported in a landmark paper of Bangham and Horne in 1964.[1]

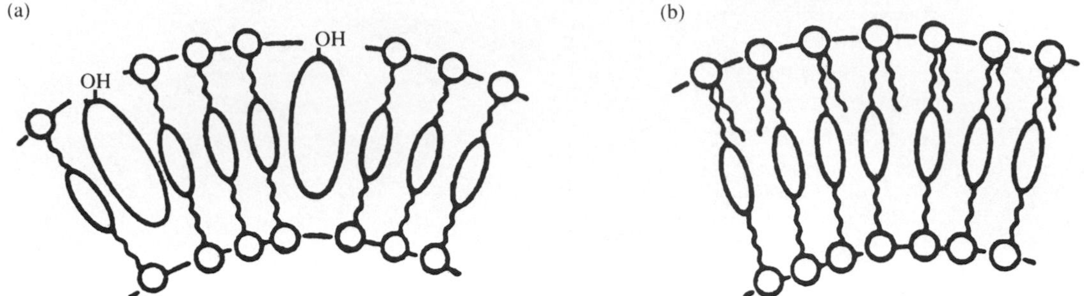

Figure 24 Creation of surface curvatures in monolayer membranes: (a) (**31**) + 1/3 cholesterol; (b) (**34**).

Another way of creating curvature is the use of unsymmetrical bis(ammonium) compounds. Attachment of a decyl chain to one of the ammonium heads of (**31**) changes the aggregate morphology from lamella to multiwalled vesicles.[76] The decyl-substituted ammonium head in (**34**) seems to be more populous in the outer surface (Figure 24(b)).

(**34**)

Amphiphiles related to (**31**)–(**34**) can be designed by combining only one end of the alkyl chain in double-chain amphiphiles.[77] Bis(ammonium) compound (**35**) and related ionene oligomers (**36**) can form monolayer membranes.

(**35**) (**36**)

Fuhrhop and co-workers have carried out extensive investigations on the formation of stable vesicles from two-headed amphiphiles.[13,15,78] They showed that macrocyclic compounds (**37**) with two hydrophilic units and single-chain amphiphiles (**38**) with two cationic terminals formed monolayer vesicles. The former compounds were the shortest and most easily accessible bolaamphiphiles (bola = two-headed), which produce vesicles with an aggregation number of about 10^4 upon sonication.[79,80] It was found that the monolayer vesicles are highly efficient barriers for ions and electrons. Small regular spacings which are filled presumably by a tilting of the hydrocarbon chains at the outer ends can explain the good barrier property.

Asymmetry between the outer and inner surface of vesicles is produced by asymmetric incorporation of a second component or by asymmetric modification of polar head groups at the two sites (Figure 24). One-sided precipitation of the bolaamphiphile is also one of the simplest procedures It has been reported that α, ω-dicarboxylic acids are often soluble at pH > 8 and spontaneously form vesicles upon acidification at pH 5. Similarly, the α,ω-bis(dipyridinium) tetrachloride salt

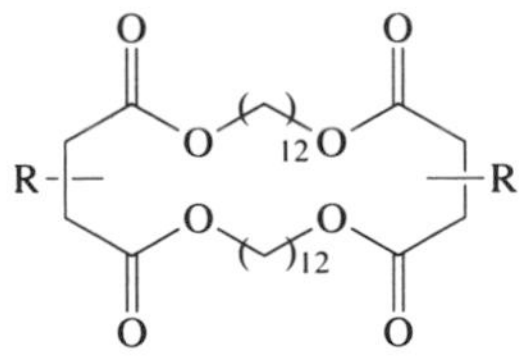

R = CH_2CO_2H, CH_2OH, SO_3H
(37)

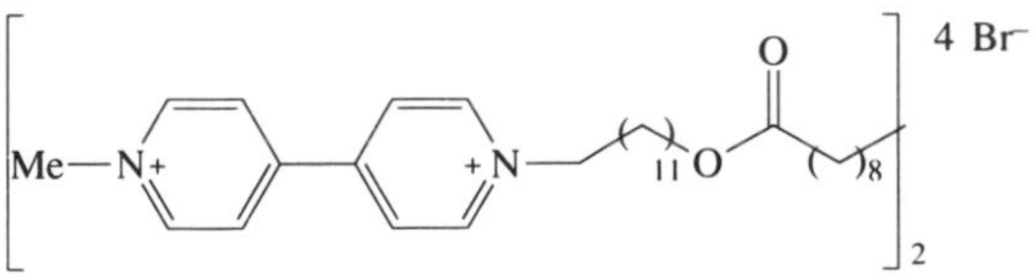

(38)

(38) is easily soluble in water, but upon addition of 1–2 moles of perchlorate or iodide, a one-sided precipitation occurs.[81]

Asymmetrical monolayer vesicles are also formed from bolaamphiphiles such as **(39)** with one large head group and one small head group. More than 99% of the small sulfonate groups are located on the smaller inner surface, with the large succinate head group on the large outer surface. Proof of this asymmetric arrangement comes from quantitative comparisons of the "metachromatic effect" of surface-adsorbed acridine orange.[82]

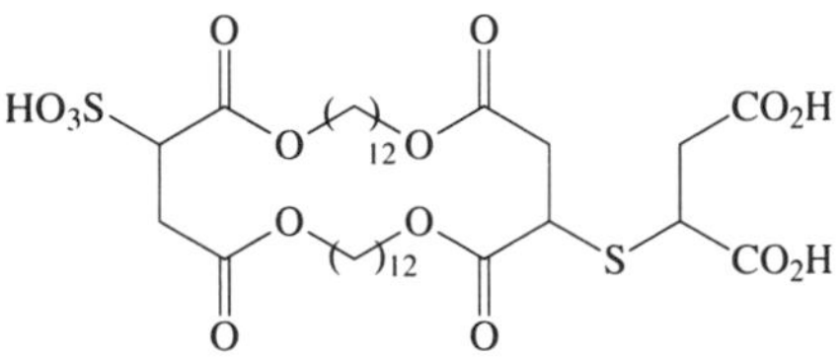

(39)

10.7 BILAYER MEMBRANES OF FLUOROCARBON AMPHIPHILES

A new class of bilayer membranes are derived by replacement of hydrocarbon chains with fluorocarbon chains. They are single-, double-, and triple-chain compounds; for example, **(40)**–**(44)**.[44,83–5]

Sonication of these amphiphiles in water produces clear to translucent dispersions. Electron microscopy observations after the negative staining reveals the existence of the bilayer structure (vesicles and/or lamellae). For instance, **(41)** gives well-dispersed, single-walled vesicles. A related hydrocarbon (dodecanoate) amphiphile produces multiwalled vesicles. Double-chain **(45)**, which possesses both fluorocarbon and hydrocarbon tails, gives even better developed single-walled vesicles. Those amphiphiles, which differ only in the tail portion (fluorocarbon vs. hydrocarbon), show aggregation morphologies that are fairly similar. The single-chain amphiphile **(40)** and triple-chain amphiphiles like **(44)** also produce bilayer aggregates.

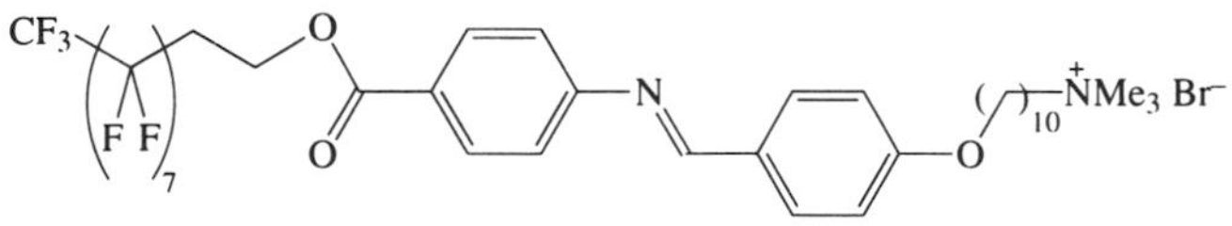

(40)

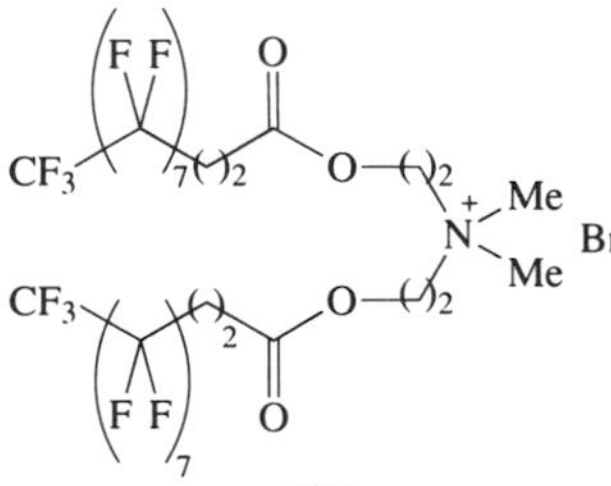

(41)

Fluorocarbon bilayers exhibit small endothermic peaks in DSC thermograms. The phase transition temperature (T_c) of the fluorocarbon bilayers is detected at temperatures 20–30 °C higher with smaller enthalpy changes than those of hydrocarbon bilayers. The smaller enthalpy changes appear to be an inherent property of bilayers made from fluorocarbon amphiphiles.

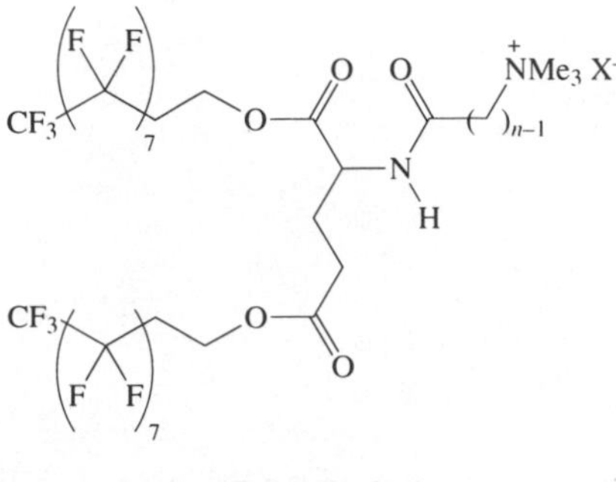

$n=2$–11, X=Cl, Br

(42)

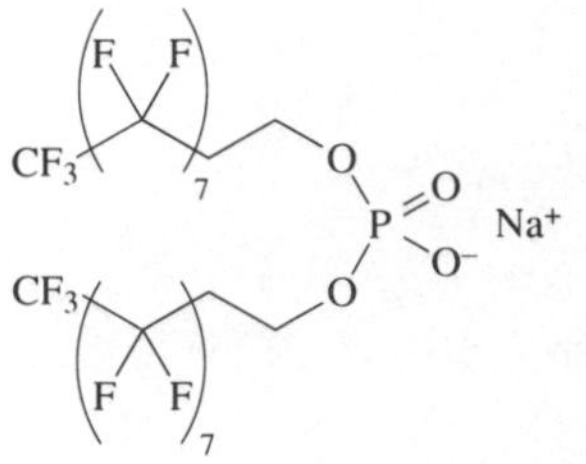

(43)

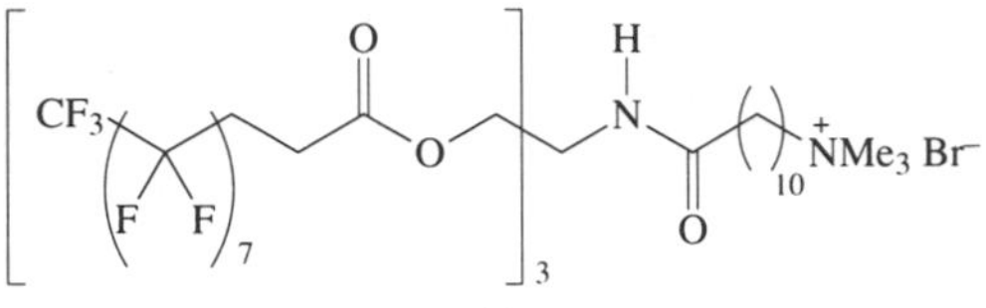

(44)

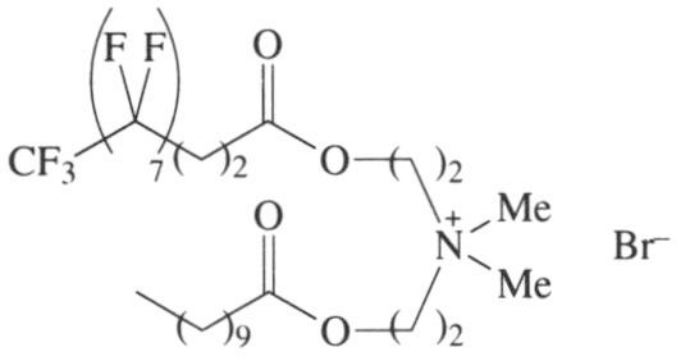

(45)

Figure 25 illustrates, as an example, the variation of fluorescence depolarization P of diphenylhexatriene embedded in bilayer membranes of triple-chain amphiphiles.[85] At low temperatures, the P values are large irrespective of the number of fluorocarbon chains, but they decrease sharply near T_c, leading to large P differences. These data indicate that: (i) the membranes are equally rigid at low temperatures; (ii) the fluidity increases (decrease in P) at the temperature regions which correspond to the respective DSC peaks; and (iii) the membrane fluidity of the liquid crystalline bilayer (at $T > T_c$) decreases with increasing numbers of fluorocarbon chains.

Phospholipid analogues (**46**) with fluorocarbon chains were synthesized by Riess and co-workers.[86] These amphiphiles form highly stable vesicles (shelf-life much longer and thermal stability much higher than those of the hydrocarbon counterpart). The vesicles have average sizes of 60–200 nm. Liang and Hui, in contrast, prepared a monolayer vesicle from a bolaamphiphile (**47**) with a semifluorocarbon chain.[87] Apparently, all the fluorosulfonate halves are on the outer side of the vesicle membrane, because of the immiscibility of hydrocarbon and fluorocarbon and the larger cross-section of the latter. This asymmetry was probed by the methachromasic effect with the sulfonate group and by spin-labeling experiments.

In the same vein, fluorocarbon and hydrocarbon bilayers show limited miscibilities. However, this is improved by adding unsymmetrical components such as (**45**).[44] This situation is more clearly demonstrated in the case of triple-chain bilayer components. As shown schematically in Figure 26, the hydrocarbon component (3H) and the fluorocarbon component (3F) are hardly miscible. Upon addition of a partially fluorocarbon-tailed amphiphile 1F2H, the original two components become more miscible.

General aggregation behavior of the fluorocarbon amphiphiles is closely related to that of the corresponding hydrocarbon compounds. However, the bilayer properties are very different for the two systems. First, fluorocarbon bilayers are much less permeable than hydrocarbon bilayers toward ions and small molecules.[83,86] Second, these two membrane components tend to form separate domains, and the phase separation has been effectively used for controlled permeation and catalysis.[84]

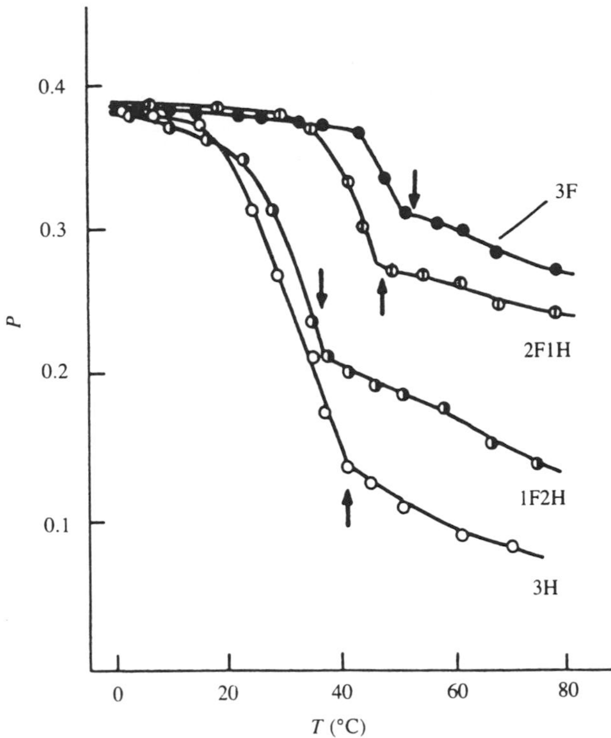

Figure 25 Fluidity of fluorocarbon bilayers measured by fluorescence depolarization P of a diphenylhexatriene probe. The arrows indicate the T_c values determined by DSC. 3F = (**44**); 3H = (**28**) (n = 12, m = 11).

(**a**) $n=4$, $m=10$, $X=CO_2$
(**b**) $n=8$, $m=10$, $X=CO_2$
(**c**) $n=8$, $m=4$, $X=O$
(**46**)

(**47**)

10.8 CHIRAL BILAYER MEMBRANES

As discussed above, the amino acid unit acts as a useful module in the design of bilayer-forming compounds. Some of the resulting chiral bilayers of single-chain and double-chain amphiphiles display much enhanced circular dichroism (c.d.).[88,89] These data suggested that the chiral components are arranged asymmetrically in fixed spatial dispositions as a result of the formation of higher order structures.

A fairly simple, double-chain amphiphile, L-(**48**), forms single- and double-walled vesicles with diameters of 30–100 nm when dispersed in water by sonication (Figure 27(a)).[90] These vesicles are slowly transformed into the helical morphology which is made of twisted tapes, as shown in Figure 27(b). Further aging gives tubes with helical streaks. These helices and tubes are better observed by dark-field optical microscopy (Figure 27(c))[91] and by high-resolution scanning electron microscopy (Figure 27(d)).[92] Direct images are obtainable by the latter two microscopic methods. The growth and degradation processes of the helices in water is observable by optical

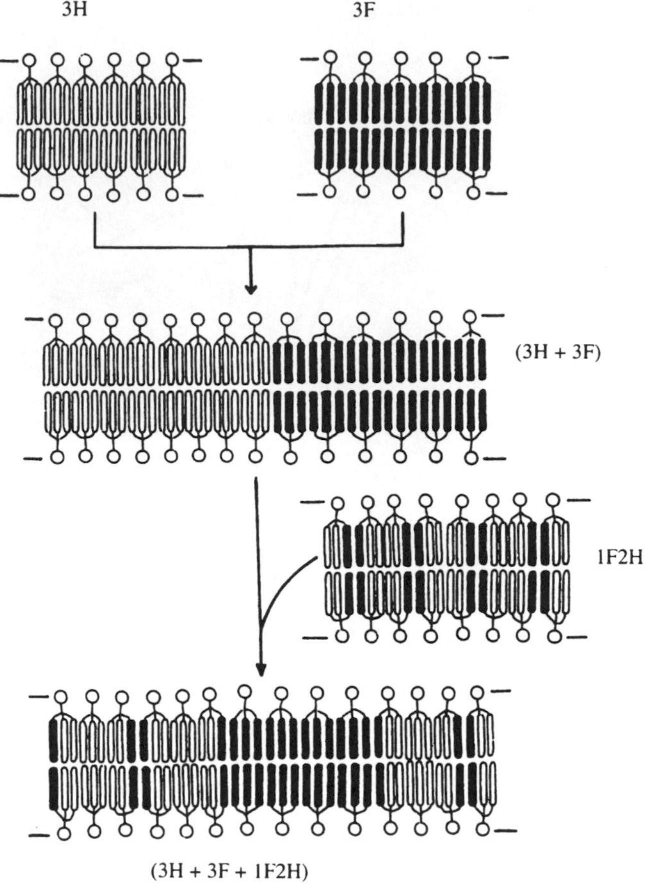

Figure 26 Miscibility of bilayer membranes of triple-chain amphiphiles containing fluorocarbon and hydrocarbon chains, before and after addition of a mixed-chain amphiphile.

microscopy at a resolution of 1 μm, and the three-dimensional structure of the molecular aggregate can be determined by the scanning electron microscope at a theoretical resolution of less than 1 nm.

(48)

The helical superstructure is maintained only in the crystalline state. When the aqueous dispersion of L-(**48**) (T_c of the bilayer, 33 °C) is warmed to 35 °C, the helices and tubes are all converted to large flexible vesicles. The melting process can be observed directly by optical microscopy.

The enantiomeric amphiphile D-(**48**) similarly showed a slow growth of helices. These are left-handed, in contrast to the right-handed helices of the L-enantiomer. In the case of its racemic bilayer, initially formed vesicles are converted into rod-like aggregates without formation of helices.

Helix formation is sensitive to the molecular structure. It is not observed for a closely related amphiphile with a shorter spacer length: CH_2 instead of $(CH_2)_{10}$. The shorter spacer may interfere

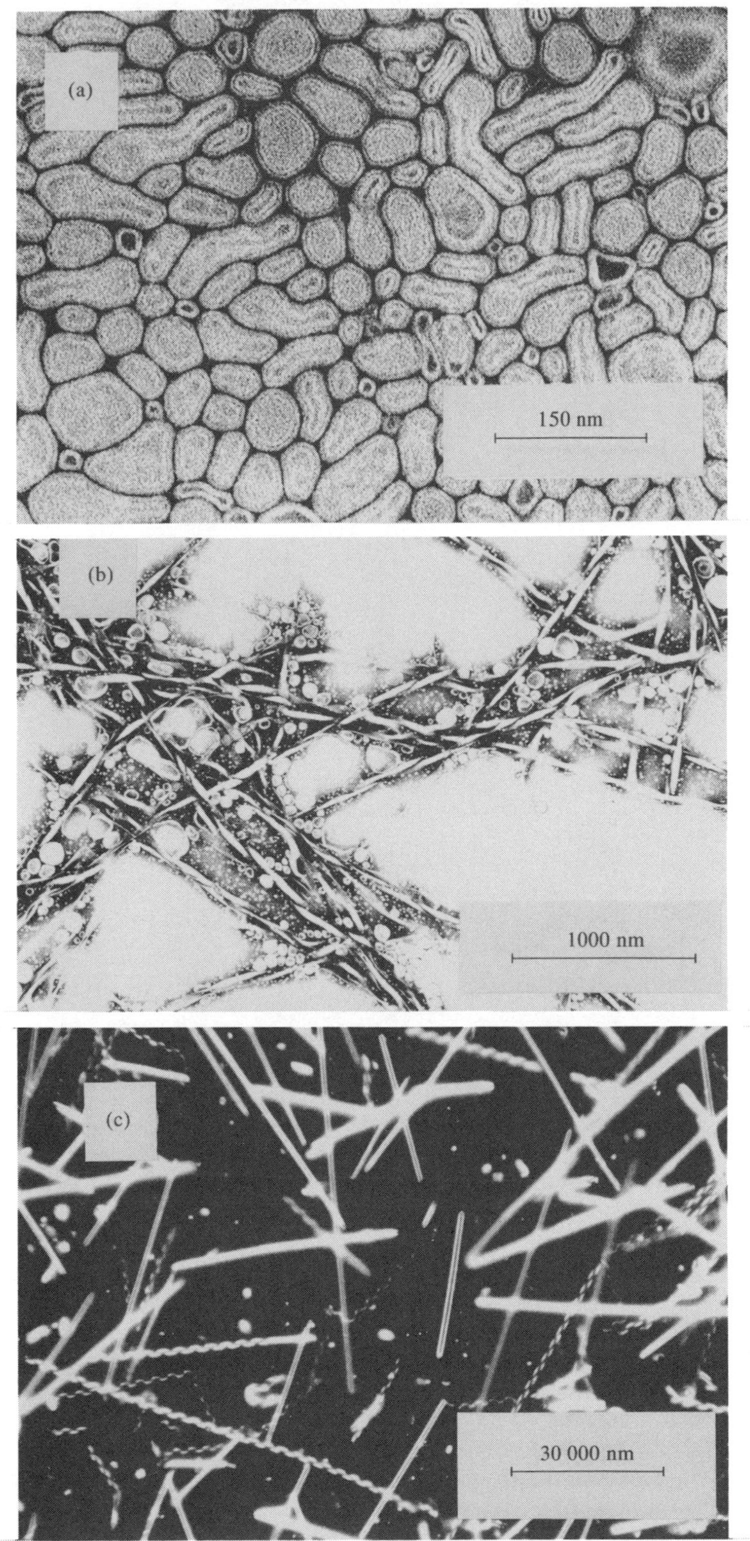

Figure 27 Formation of helical superstructures from bilayer membranes of chiral ammonium amphiphiles, L-(**48**). (a) Electron micrograph of aqueous vesicles: no aging, stained with UO_2^{2+}. (b) Electron micrograph of the helical superstructure: stained by UO_2^{2+}. (c) Dark-field optical micrograph of the helical superstructure. (d) Electron micrograph (high-resolution scanning electron microscopy) of the helical superstructure. (e) A schematic illustration of helix formation.

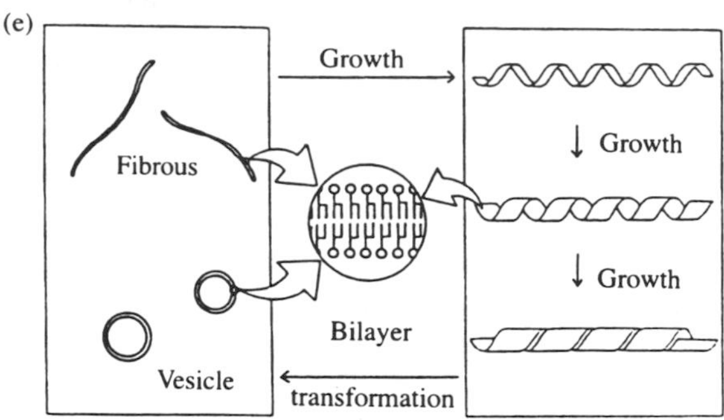

Figure 27 (continued).

with the regular molecular arrangement, since this amphiphile instead forms smooth single-walled vesicles.

These chiral molecular assemblies give rise to unique chiroptical properties. Circular dichroism of chiral synthetic bilayers which contain aromatic chromophores such as *p*-phenylene, biphenyl, naphthalene, and azobenzene moieties is remarkably enhanced due to the dipole–dipole coupling of the organized chromophores.[88,89] Huge c.d. intensities are induced when dye molecules such as cyanines and methyl orange were bound to chiral bilayers,[93] but is lost when the chiral bilayer undergoes gel-to-liquid crystal phase transition. For example, 9-anthracenecarboxylate bound to chiral **(48)** (with helical superstructure) gives a molecular ellipticity [θ] of almost 10^7.[94] The extent of orientational fixation of the chromophore is one of the highest values ever observed for guest molecules embedded in chiral microenvironments. The importance of helical growth in the induction of large c.d. values is demonstrated by Figure 28, in which the huge [θ] value is found for the helical bilayer with long aging but not for the rapidly cooled bilayer (no helical growth).

The helical superstructure has been obtained for other bilayer-forming amphiphiles. The single-chain ammonium amphiphile **(49)** gives helical superstructures accompanied by c.d. enhancement.[95]

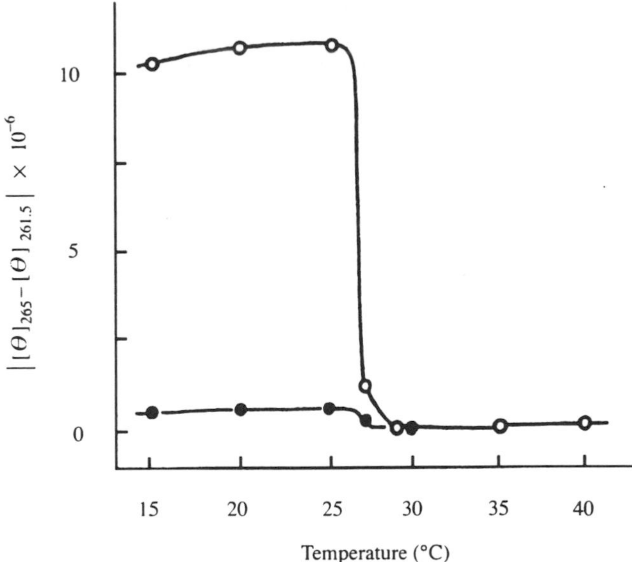

Figure 28 Temperature dependence of the induced circular dichroism of 9-anthracenecarboxylate bound to the aqueous chiral bilayer of L-(**48**) prepared by long aging (○) and rapid cooling (●).

Cho and Park reported helical superstructures formed by cationic cholesterol-containing polymers.[69] Yamada *et al.*,[96] concurrently with the initial finding of Kunitake and co-workers, observed that dialkyl amphiphiles with the oligoglutamate head group (**50**) produced helical superstructures (twisted ribbons with a layer thickness of 3–4 nm). Single- and double-chain phosphate amphiphiles can form helical superstructures as well.[97] Tubular aggregates, which are products of helical growth, are found in a dispersion of a diacetylenic phosphocholine lipid.[98] Related helical fibers of lipid bilayers were obtained from *N*-octyladonamides, and have been extensively examined in subsequent years.[15]

(**49**)

$m = 12, 16$
(**50**)

10.9 LESS CONVENTIONAL, AQUEOUS BILAYER MEMBRANES

10.9.1 Bilayer Formation Assisted by Complementary Hydrogen Bonds

In the preceding examples, bilayer components are independent molecules in which necessary modules are connected by covalent bonding. Construction of bilayer membranes via complementary hydrogen bonding would provide a new class of self-assembly in which novel molecular control is achieved. In general, hydrogen bonding in artificial molecular systems is most effective in the solid state or in noncompetitive (aprotic) organic media. However, it can be used efficiently even in water when combined with hydrophobic association.

Substituted melamines (**51**) and (**52**) and isocyanuric acid derivatives (**53**) and (**54**) are suitable for this purpose because they form extended arrays of complementary hydrogen bonds.[99] Among the equimolar complexes of **(51)**·**(53)**, **(51)**·**(54)**, **(52)**·**(53)**, and **(52)**·**(54)** that are prepared in ethanol and dried, only **(51)**·**(53)** gives a transparent dispersion in water (~30 mM) upon sonication. This dispersion is stable over a period of 1 month. In contrast, the other complexes and single components **(51)**, **(52)**, and **(53)** display poor solubilities in water even at a lower concentration of 5 mM. From comparisons of the component structures, it is suggested that improved molecular orientation by the phenyl group in **(53)** and facilitated alkyl chain alignment by the ether linkage in **(51)** are crucial to give an ordered, stable assembly. Similar structural effects have been noticed in conventional bilayer assemblies.[100,101] The equimolar composition is essential for producing a stable dispersion, since nonequimolar mixtures do not give homogeneous dispersions. A transmission electron micrograph of complex **(51)**·**(53)** dispersed in water shows disk-like aggregates with diameters of several tens of nanometers and thicknesses of ~10 nm.

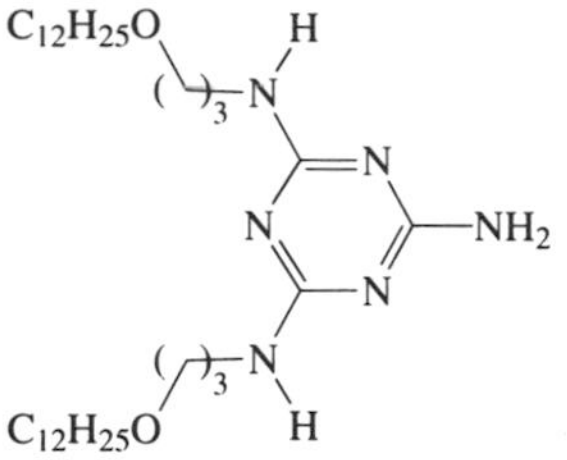

$C_{12}H_{25}O$ H $(\ )_3$ N N N NH$_2$ N $(\ )_3$ N $C_{12}H_{25}O$ H

(51)

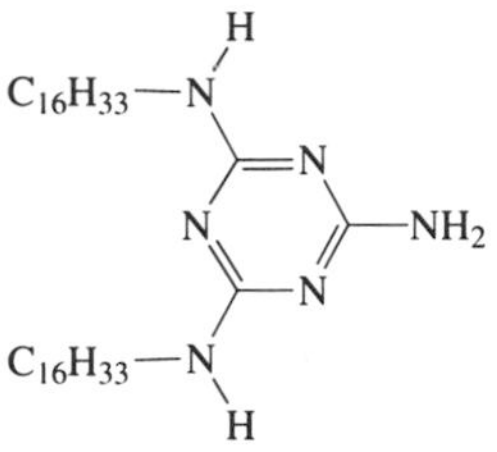

H $C_{16}H_{33}$—N N N NH$_2$ N $C_{16}H_{33}$—N H

(52)

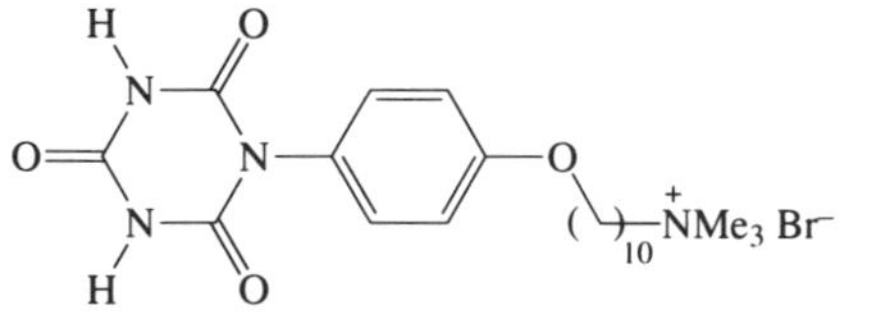

H O N O N O N H O $(\ )_{10}$ $\overset{+}{N}Me_3$ Br$^-$

(53)

H O N O N $(\ )_{10}$ N $\overset{+}{N}Me_3$ Br$^-$ H O

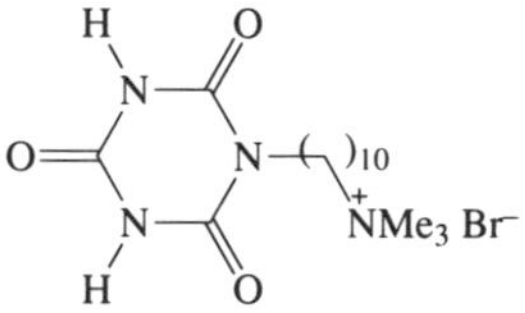

(54)

A self-supporting film is obtainable by casting the aqueous dispersion on a glass plate. Its IR spectrum suggests that complementary hydrogen bonds are formed between the two components. In the transmission x-ray diffraction pattern, equatorial diffractions corresponding to the bilayer thickness are observed up to 10th order with a long period of 9 nm. It is clear that regular multibilayers of hydrogen-bonded molecular arrays exist parallel to the film plane. The identical network structure is probably present in the aqueous dispersion, since the single-walled bilayer (9–10 nm thickness) is suggested by electron microscopy. Remarkable stabilization of the complementary hydrogen bonds in the aqueous environment can be ascribed to the ordered stacking of the extended array of hydrogen-bonding units in the bilayer.

The physical state and surface structure characteristics of conventional bilayer membranes are found in this hydrogen bond mediated bilayer.[102] Differential scanning calorimetry of aqueous **(51)**·**(53)** shows concentration-dependent endothermic peaks: a broad endothermic peak between 70 °C and 100 °C (ΔH, ~50 kJ mol^{-1}) at a concentration of 20 mM. When the sample is heated beyond 80 °C, there appears a new endothermic peak at 53 °C ascribable to the uncomplexed aggregate of **(53)**. Thus, it is concluded that the **(51)**·**(53)** composite undergoes a gel-to-liquid crystal transition at 50–80 °C and partial dissociation to individual components at 80–100 °C.

Figure 29 displays the effects of temperature on the fluorescence intensity of cyanine dye NK2012 and fluorescence depolarization of 1,6-diphenylhexatriene (DPH) bound to aqueous **(51)**·**(53)**. The fluorescence intensity due to the J-aggregate formation decreases at 40–50 °C, reflecting the onset of the gel-to-liquid crystal phase transition of the bilayer. Fluorescence depolarization proceeds in the same temperature region from 0.35 to 0.12. These spectroscopic changes are comparable to those reported for the phase transition of conventional aqueous bilayer aggregates. NK2012 is conceivably bound electrostatically to the bilayer surface, while DPH is solubilized in the hydrophobic interior. A schematic illustration of the physical state of bilayer **(51)**·**(53)** and membrane-bound dyes is given in Figure 30.

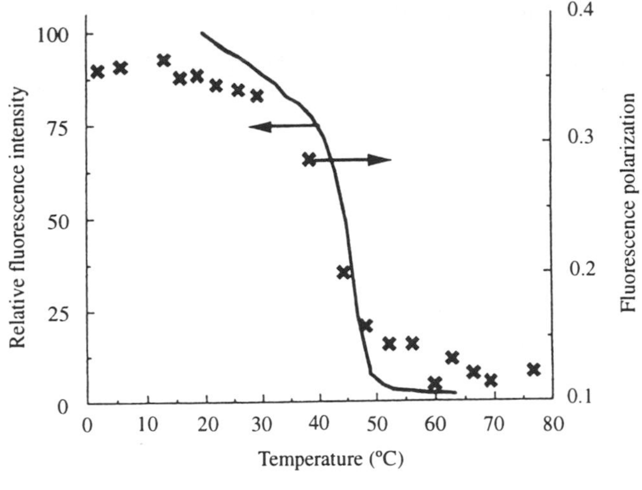

NK 2012

DPH

Figure 29 Temperature dependence of the fluorescence intensity of NK2012 (solid line) and fluorescence depolarization of DPH (X) embedded in hydrogen-bond mediated bilayer membrane (**51**)·(**53**).

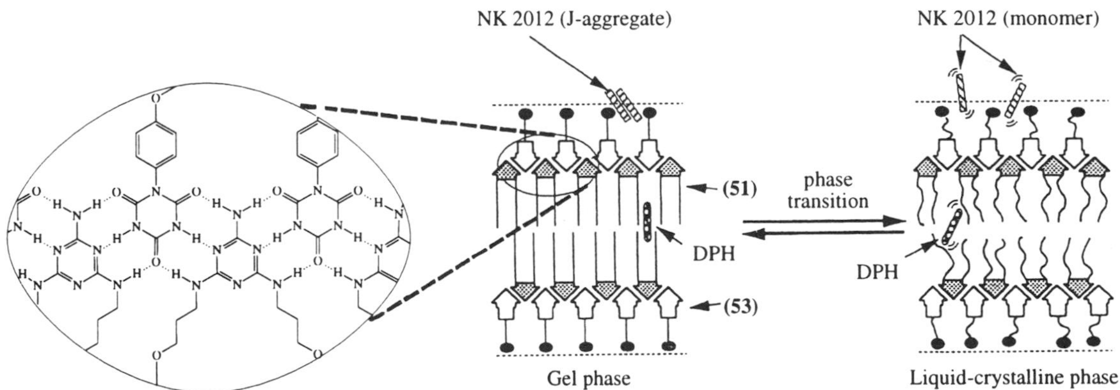

Figure 30 Schematic illustration of the physical states of bilayer (**51**)·(**53**) and membrane-bound dyes.

In another combination of (**55**) and (**56**), circular dichroism is observed in the absorption region of the azobenzene unit. This implies that chirality is transmitted across the hydrogen bond network from the hydrophobic unit to the hydrophilic unit.[103]

(**55**)

(**56**)

10.9.2 Bilayer Membranes of Ion-paired Amphiphiles

Hydrogen bonding is not the only secondary valence force that leads to bilayer assemblage from otherwise less organizing molecules. Ion pairing has been used for improving aggregate stability. Earlier work of Jokura *et al.*[104] showed that the combination of oppositely charged surfactants gave rise to lamellar phases. CTAB, a common micelle-forming surfactant, was converted to a bilayer-forming polyelectrolyte when the bromide ion is replaced with polyacrylate.[105] Rapid injection of its ethanolic solution into water produces a translucent solution that contains particles with diameters ranging between 80 nm and 350 nm, as inferred from dynamic light scattering and electron microscopy. It is stable for more than 48 h at ambient temperature. The thickness of the unit layer was estimated to be 3.35 nm.

Shortly afterwards, Kaler and others claimed spontaneous formation of vesicles by the mixing of cationic and anionic, single-tail surfactants.[106] Aqueous mixtures of cetyltrimethylammonium tosylate and sodium dodecylbenzenesulfonate apparently contain vesicles of 30–80 nm in radius, as estimated from light scattering, freeze fracture transmission electron microscopy, and glucose entrapment experiments. It was also shown by Fukuda *et al.* that amphiphilic ion pairs derived from a series of trimethyl-*n*-alkylammonium bromides and saturated fatty acids (chain length, 14, 16, 18) formed vesicles that were stable for more than 48 h (diameter, 20–80 nm) upon ultrasonic dispersal in water.[107] The ability of these ion-paired amphiphiles to form a lamellar phase is critically dependent upon the absence of salts. DSC measurements indicate that their phase transition temperatures are close to those of double-chain phosphatidylcholines of the same chain length. In both of their gel (23 °C) and fluid (50 °C) states, DPPC membranes are significantly less permeable toward sucrose than the ion-paired (C_{16} + C_{16}) membranes.[108]

10.9.3 Bilayer Assembly Assisted by Metal Complexation

Dithiooxamide derivatives readily form stable coordination polymers in combination with divalent metal ions: Cu^{2+}, Ni^{2+}, and so on. Two ammonium terminal groups in (**57**) are connected with hydrocarbon chains and the dithiooxamide ligand.[109] This amphiphile, by itself, cannot form stable monolayer membranes in water. However, its 1:1 complex with Cu^{2+} gives a coordination polymer with a degree of polymerization of 30 (GPC measurement in $CHCl_3$). Addition of excess Cu^{2+} does not lower the molecular weight. Electron microscopic observation of the aqueous dispersion indicated the existence of vesicle-like monolayer structures, as illustrated schematically in Figure 31(a).

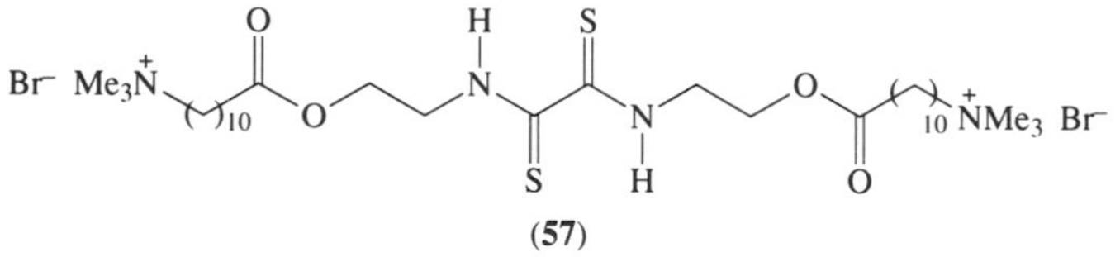

Br⁻ Me₃N⁺–(CH₂)₁₀–C(=O)–O–CH₂CH₂–NH–C(=S)–C(=S)–NH–CH₂CH₂–O–C(=O)–(CH₂)₁₀–N⁺Me₃ Br⁻

(**57**)

Suh *et al.* prepared a sulfonate amphiphile containing two liganding sites.[110] This amphiphile apparently forms bilayer membranes upon sonication in water, as shown by electron microscopy. Complexation of Fe^{2+}, Fe^{3+}, Co^{2+}, or Co^{3+} ions leads to the formation of coordinatively polymerized bilayer membranes (Figure 31(b)). Helical superstructures are obtained with Co^{2+} ions, although the ligand membrane does not contain any chiral center, whereas rod-like structures are formed with the other ions. Stabilization of the bilayer by metal coordination was demonstrated by enhanced resistance to disruption in aqueous ethanol.

10.10 BILAYER ASSEMBLY IN WATER/ALCOHOL MIXTURES AND IN NONPOLAR ORGANIC MEDIA

10.10.1 Bilayer Assembly in Water/Ethanol Mixtures

In the self-organization of bilayer assembly in aqueous media, the amphiphilic nature of component molecules is essential for the hydrophobic assembling phenomenon. It is possible to design "amphiphilic compounds" that would form stable bilayer organizations in other media without

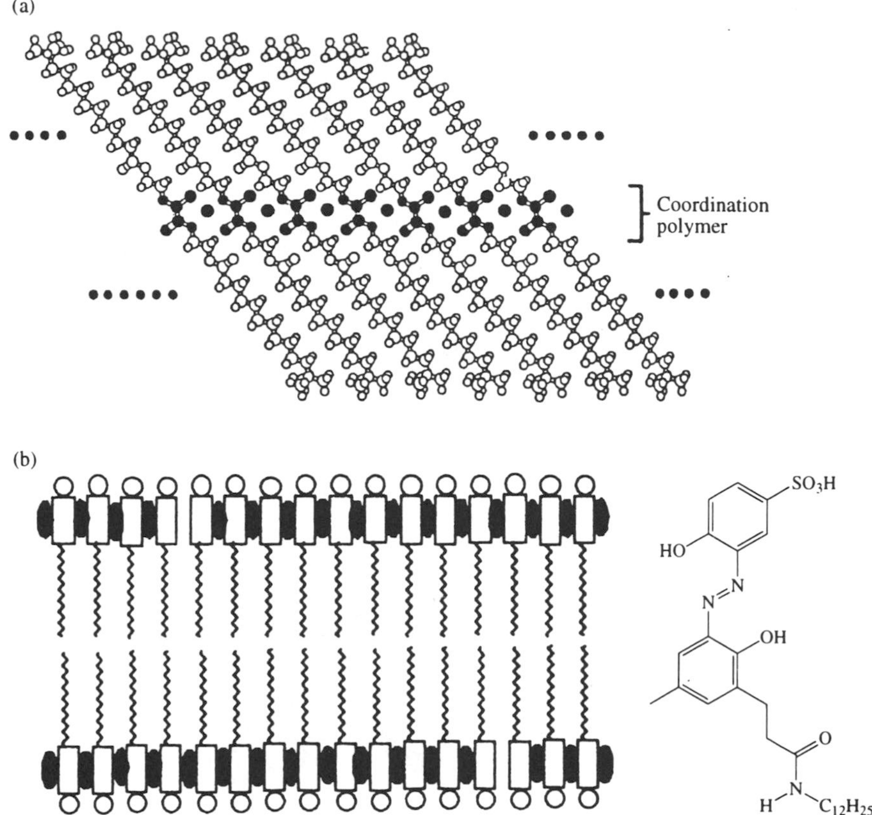

Figure 31 Schematic illustrations of the monolayer membrane of: (a) $Cu^{2+}\cdot$(**57**) and (b) a coordinately polymerized bilayer membrane.

overriding hydrophobic forces. Conventional surfactants such as alkylammonium salts are known to form micelles even in nonaqueous, protic environments such as hydrazine, formamides, glycols, glycerols, and ethylammonium nitrate (a low-melting fused salt),[111] although their critical micelle concentrations (CMCs) in these media are markedly higher and the corresponding aggregate numbers are much smaller than those of aqueous micelles. It is clear that sufficiently stable aggregates are not obtainable in nonaqueous media from conventional amphiphilic compounds.

Ammonium amphiphile (**58**) that is protonated by HBr, HCl, or *p*-toluenesulfonic acid (PTS) can be dispersed in water or in ethanol/water mixtures.[112] The (**58**)·HBr complex dispersed in water gives electron micrographs that indicate formation of globular aggregates with diameters of 30–200 nm. In 40 vol.% ethanol/water, however, helical tape-like structures that are typical of bilayers are found. Absorption spectra of these dispersions show a remarkable medium dependency. The absorption maxima of (**58**) in ethanol and in water are essentially identical to those of nonassociating (**59**), because the molecularly dispersed species and/or noninteracting chromophores are present. In contrast, a remarkably red-shifted peak is found for (**58**) at 462 nm in 40 vol.% $EtOH/H_2O$. The corresponding exciton-coupled c.d. spectrum was found only in this particular medium, but not in water or in ethanol. These spectral data strongly support the formation of the J-aggregated chromophore.

Endothermic peaks are observed at 51.7 °C (ΔH, 31 kJ mol^{-1}) and at 44.6 °C (ΔH, 54 kJ mol^{-1}) for its dispersion in pure water and in 40 vol.% $EtOH/H_2O$, respectively. The UV–visible absorption of the J-like aggregate in 40% $EtOH/H_2O$ decreases with increasing temperature and gives a broad peak at 58 °C. It appears that the bilayer structure is maintained even at high temperatures, and the endothermic peaks are attributed to the gel-to-liquid crystal phase transition.

Figure 32 contains probable structures of the bilayer organization based on electron microscopic and spectral evidence. In water, the polar head groups are in a less-ordered alignment, while the tilted head group stacking is induced in appropriate ethanol/water mixtures. It has been reported that ethanol/water and acetonitrile/water mixtures contain specific molecular clusters. The content of the hydrated ethanol cluster is greatest at 43 vol.% $EtOH/H_2O$.[113] This unique feature may be related to the particular bilayer organization observed.

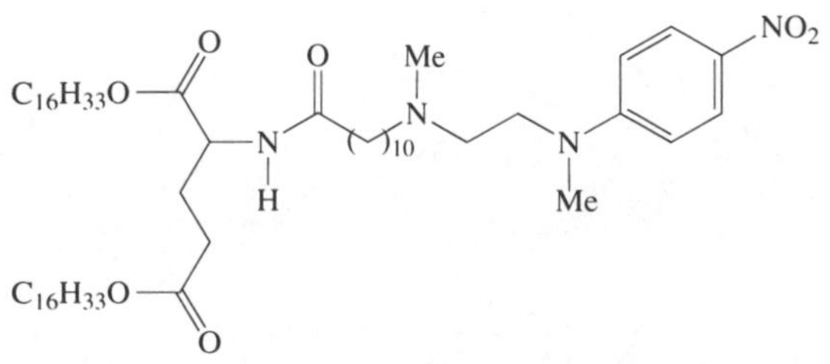

(58)

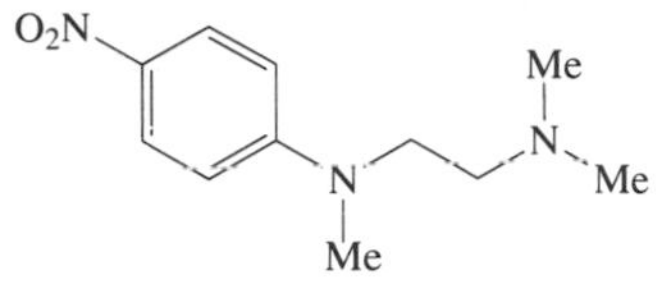

(59a)

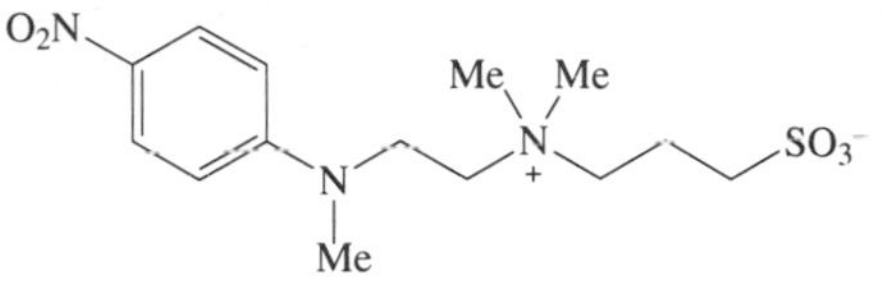

(59b)

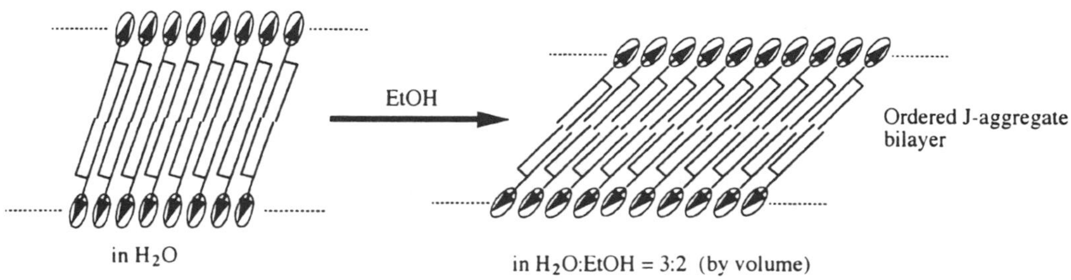

Figure 32 Medium dependency of the bilayer HBr·(**58**). Transition dipoles are indicated by arrows.

10.10.2 Bilayer Assembly in Aprotic Organic Solvents

Bilayer vesicles may be formed in organic media either as reversed vesicles or as fluorocarbon bilayers. In the former case, Kunieda *et al.*[114] demonstrated that reversed vesicles as illustrated in Figure 33(a) were formed in hydrocarbon solvents by shaking a lamellar phase made of tetraethylene glycol dodecyl ether in dodecane and a small amount of water. Video-enhanced microscopy and fluorescence imaging show the existence of layer-inverted vesicles (diameter 1–10 μm).

Kim and Kunitake prepared a stoichiometric complex of Ca^{2+} and bilayer-forming (in water) phosphate (**60**).[115] The equimolar complex prepared at pH 7.5 possessed a molecular weight of over 10^6 in $CHCl_3$, spectroscopic data suggest considerable alignment of alkyl chains, and the electron microscopic observation of a layer structure (diameter, ~10 nm) is consistent with 2–4 molecular layers. It is probable that the reversed bilayer assembly given in Figure 33(b) is present in $CHCl_3$ dispersion.

Long perfluoroalkyl compounds generally display low solubilities in hydrocarbon media due to their small cohesive forces. Thus, perfluoroalkyl chains in the liquid state are forced to remain assembled by hydrocarbon surroundings. Turberg and Brady in fact reported that semifluorinated alkanes, $C_8F_{17}—C_nH_{2n+1}$ ($n = 12, 16$), assembled into micellar aggregates in toluene and perfluorooctane.[116]

It is possible to design stable bilayer assemblies to be formed in aprotic, organic media such as cyclohexane, benzene, and 2-butanone.[117] Compounds (**61**)–(**63**) are composed of the following four molecular modules: (i) double fluorocarbon chains as the solvophobic portion; (ii) the glutamate connector, which is chiral and capable of hydrogen bonding; (iii) an aromatic unit next to the connector module as chromophoric reporter and a supplier of the van der Waals force; and (iv) a flexible hydrocarbon chain as the solvophilic module. Dark-field optical microscopy of dispersions of (**61**), (**62**), and (**63**) (5 mM) in benzene and chlorocyclohexane displays the presence of fibrous aggregates (length, 10 μm; diameter, 0.1 μm) that are stacked to yield huge rods or thicker fibers without regard to the molecular structures of these compounds.

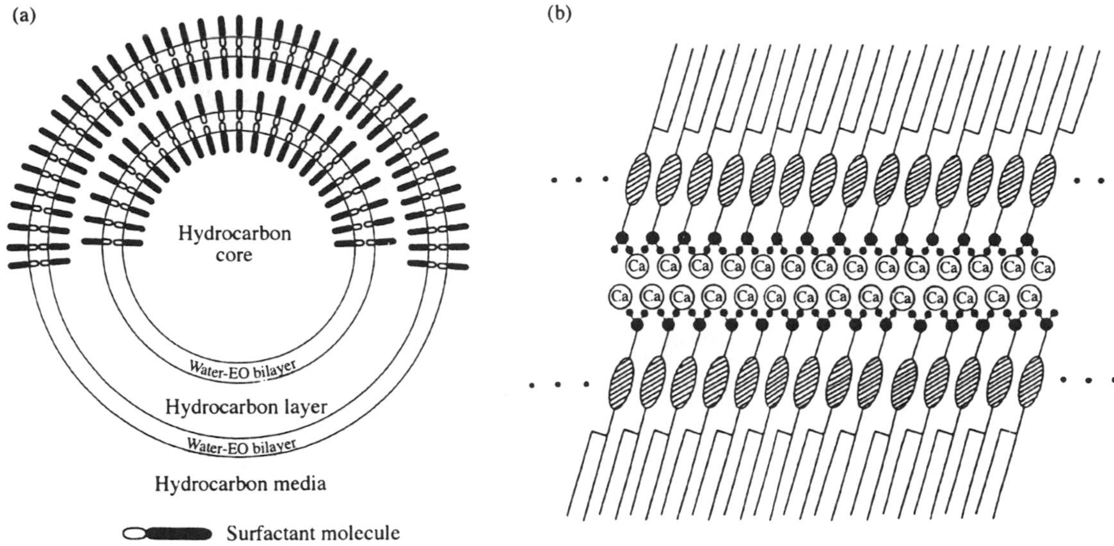

Figure 33 (a) Schematic structure of a reversed vesicle. (b) Schematic illustration of the Ca^{2+}·**(60)** bilayer in organic media.

(60)

(61)

(62)

(63)

Figure 34 shows an example of transmission electron microscopy (TEM) photographs of organic aggregates. The staining agent gives dark shadows. A chlorocyclohexane dispersion of (**61**) (1 mM) contains oval structures of ~10 nm thickness and 20–200 nm diameter. By comparison, its benzene dispersion shows rods with 12 nm thickness aligned with each other. Compound (**62**) in chlorocyclohexane (1 mM) formed clearly defined tube-like aggregates of 8 nm thickness. The narrowest part of these aggregates is about twice as thick as the molecular length (4–6 nm) of the component molecule, and the width of the other part of the assembly is roughly comparable to the several molecular layers.

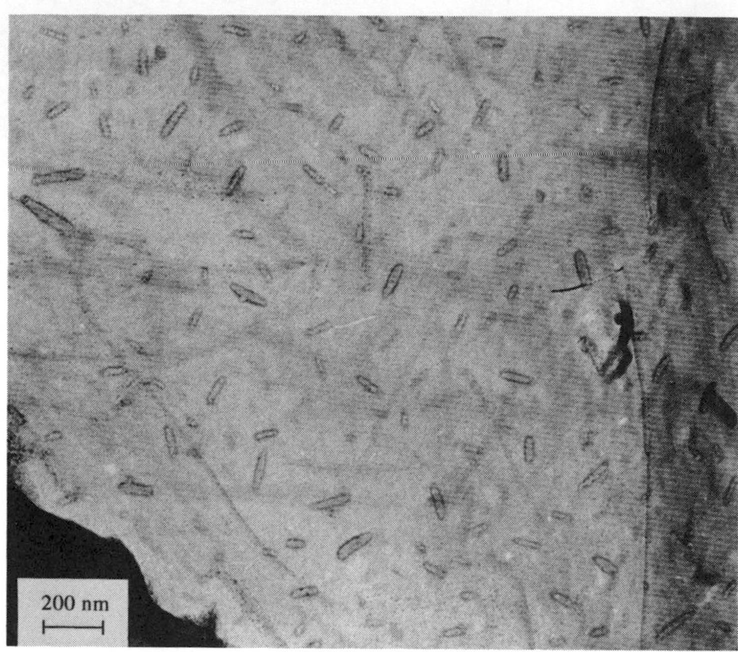

Figure 34 Transmission electron micrograph of (**61**) dispersed in chlorocyclohexane (1 mM).

The molecular alignment in these bilayer dispersions was examined with the help of DSC, c.d., and IR spectroscopy. DSC thermograms show rather broad endothermic peaks (in the heating scan) at 17–42 °C for (**61**) with an enthalpy change of 39 kJ mol^{-1}. A typical fluorocarbon bilayer in water gives a much smallar enthalpy change. Molecular weight measurement by vapor pressure osmometry indicates that (**61**) is present as a monomeric species at temperatures above the endothermic peak. Spectroscopic data are consistent with the bilayer–monomer transition.

The thermodynamic relations among monomer, surface monolayer, and dispersed bilayer can be summarized as illustrated in Figure 35 in the case of (**63**) on the basis of these data and thermodynamic analysis of the surface tension. The ΔH^{B}_{M} and ΔS^{B}_{M} terms for the molecular association in bulk solvent assume large, negative values. This indicates that the enthalpic force governs the formation of the nonaqueous bilayer, and it is a thermodynamic situation similar to the hydrophobic force in water and the solvophobic force in water-like organic media. There are no detectable changes in the thermodynamic values for the conversion between surface monolayer and dispersed bilayer. The two organized states, therefore, are thermodynamically identical, and they must possess analogous structural characteristics. The entropy change in the monomer-to-monolayer conversion is negatively much greater than that of alkylammonium adsorption at the air–water interface. Therefore, the fluorocarbon monolayer is much more ordered.

The molecular assembly in these cases is commonly derived from immiscibilities of solutes against solvents (i.e., cohesive energy differences), regardless of the hydrogen-bonding ability of solvents. The cohesive energy, intermolecular force, of a liquid can be considered in terms of surface tension. The surface tension of liquids decreases in the following order: water (72 mN m^{-1}) > water-like media (60–70 mN m^{-1}) > > hydrocarbons (around 30 mNm^{-1}) > perfluorocarbons (10–15 mN m^{-1}). The assembling systems in water, in water-like media, and in aprotic hydrocarbons possess the common feature that the cohesive force of the medium molecules is stronger than that of the solute molecules (fluorocarbon or hydrocarbon chains). The solute

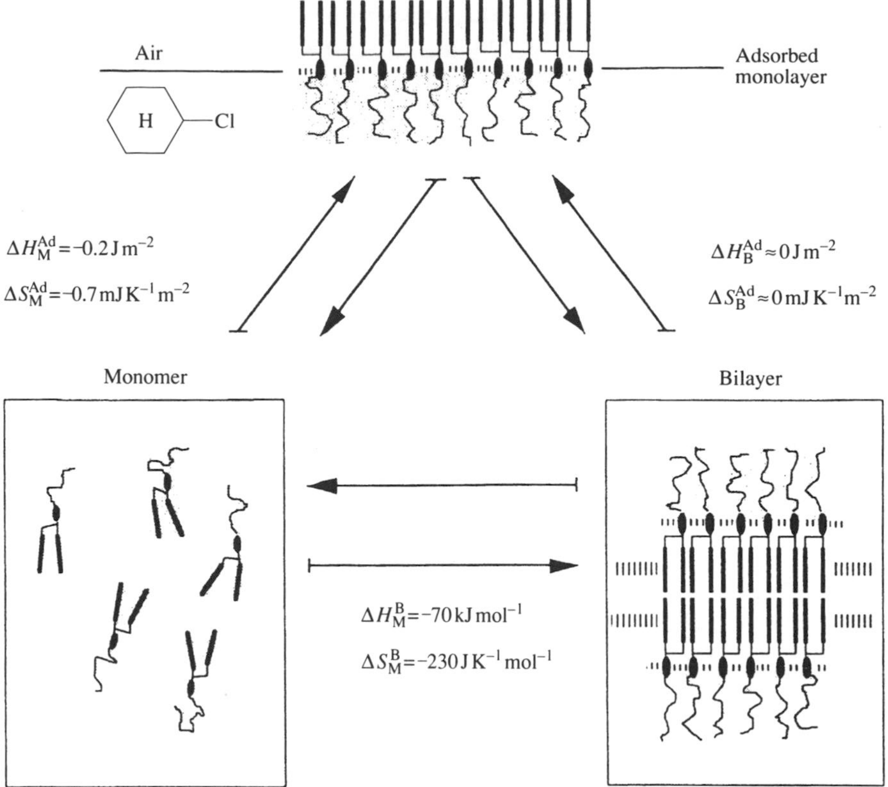

Figure 35 Thermodynamic relations of surface adsorption and bilayer formation of fluorocarbon–hydrocarbon amphiphile (**63**).

chains are forced away from tighter aggregates of the medium molecules, resulting in molecular assemblies. The active role in the assembling process lies in the medium rather than in the alkyl chains. In this sense, the present fluorocarbon bilayer can be perceived as showing "passive" solvophobicity because of the weaker cohesive forces of solutes.

This solvophilic–solvophobic relationship may be reversed. Formation of bilayer assemblies in an aprotic fluorocarbon medium was reported.[118] In this case, amphiphilic molecules, such as (**64**), are composed of double hydrocarbon chains as the solvophobic moiety and flexible perfluoroalkyl chains as the solvophilic moiety. These particular compounds possess "reversed" amphiphilicity relative to the preceding example.

(**64**)

These results establish that assembly of molecular bilayers is a widely observable phenomenon. Organized molecular assemblies are produced in any medium if the "amphiphilic–amphiphobic" nature of the solute molecules is properly designed. Figure 36 represents the generalized bilayer concept.[117c] Self assembling bilayers can be formed in protic media. The common driving forces of the molecular assembly are the solute–solvent immiscibilities (enthalpic forces) that arise from differences in cohesive energy between solute (amphiphile) and solvent. The cohesive energy of the solvent is greater than that of the solute in the cases of Figures 36(a) and (b), but this relation is reversed in Figure 36(c). Therefore, the magnitude of the cohesive energy of the solute *per se* is not relevant for promoting effective molecular assembly.

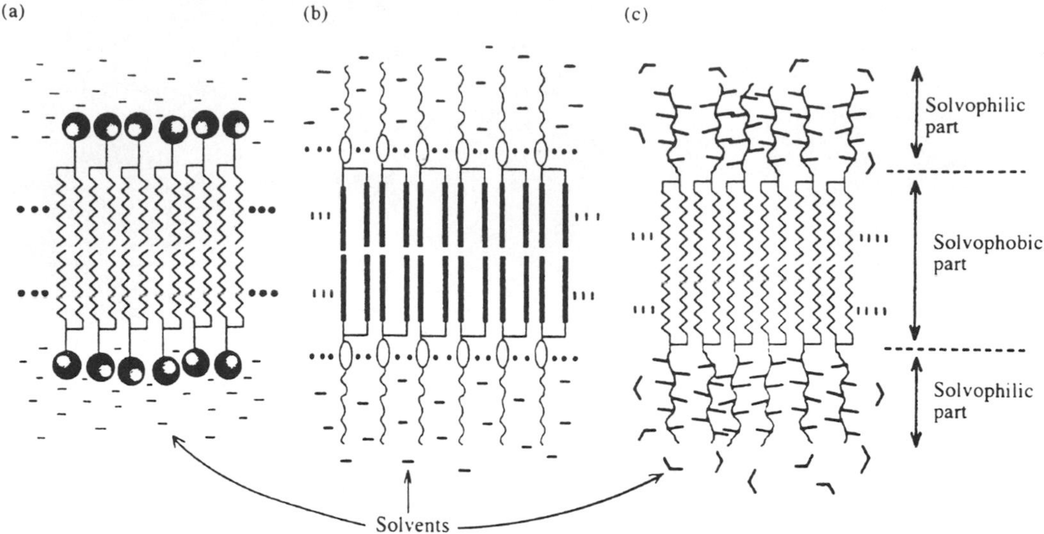

Figure 36 Generalized bilayer concept. The cohesive force of medium molecules is stronger than that of solute alkyl chains in (a) protic media and (b) aprotic hydrocarbon media. The force of medium molecules is weaker than that of solute alkyl chains in (c) aprotic fluorocarbon media.

10.11 CHROMOPHORE ORGANIZATION BASED ON BILAYER ASSEMBLY

The regular two-dimensional organization of synthetic bilayer membranes has been advantageously used for the ordering of covalently or noncovalently bound chromophores on bilayer templates. The orientation of covalently bound chromophores in a bilayer assembly was systematically investigated for the first time in the case of single-chain azobenzene amphiphiles (**23a**). Figure 37 shows representative spectral variations of aqueous dispersions with the component molecular structure. All the bilayer components contain the identical azobenzene chromophore, yet their spectral patterns are quite different. λ_{max} covers a wide range from 300 nm for (**23a**) ($n = 8$, $m = 10$) to 390 nm for (**23a**) ($n = 12$, $m = 5$). The observed λ_{max} variation needs to be explained in terms of the chromophore interaction (orientation and aggregation) within the aggregate. According to the molecular exciton model of Kasha,[119] the blue shift for the bilayer of (**23a**) ($n = 8$, $m = 10$) relative to the absorption of the isolated chromophore ($\lambda_{max} = 355$ nm) results from a parallel orientation of the neighboring transition dipoles (the H-aggregate, Figure 37(a)), and a red shift for the bilayer of (**23a**) ($n = 12$, $m = 5$) is attributable to the tilted chromophore orientation (the J-like aggregate, Figure 37(d)). The other bilayers of (**23a**) ($n = 10$, m = 4), (**23a**) ($n = 12$, $m = 10$), and (**23a**) ($n = 12$, $m = 6$) apparently have intermediate orientations (Figure 37(b)).

The interdigitated bilayer structure was proposed for the bilayer of (**23a**) ($n = 8$, $m = 10$) first on inspection of the x-ray diffraction of its cast multilayer film.[120] Subsequently, Okuyama and coworkers determined its structure from a single crystal (Figure 38).[121] The parallel azobenzene packing is appropriate for its particular molecular structure, since the cross-section of the ammonium head group is about twice the cross-section of the hydrophobic portion and the lengths of the spacer and tail are very similar.

The crystal structure of the bilayer of (**23a**) ($n = 12$, $m = 5$) is shown in Figure 39.[122] A typical bilayer structure is formed. The tilt angle of the molecular axis from the layer normal amounts to 62°. This inclination is required for the matching of the cross-sections of the head group and the hydrophobic chain. The J-aggregate formation inferred from the spectroscopic data is consistent with the crystal structure. The spacer length of $m = 5$ appears essential for the tilted packing, because analogous tilted arrangements are observed for the bilayers with the common $m = 5$ spacer and different tail lengths of $n = 6$, 8, and 10.[123,124]

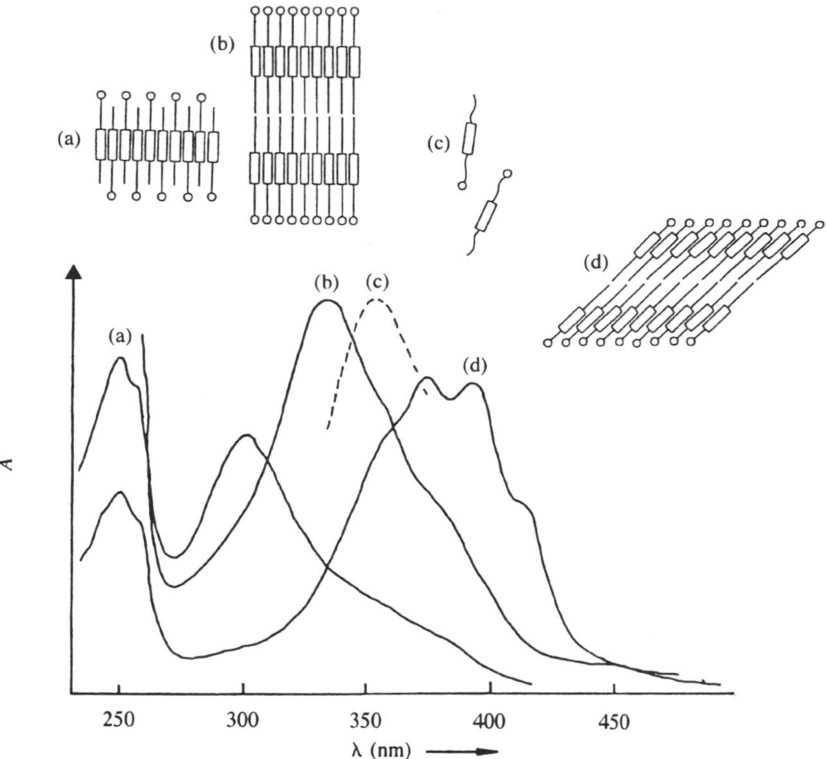

Figure 37 Absorption spectra and packing modes of bilayers of **(23a)**: (a) $n = 8$, $m = 10$, H-aggregate; (b) $n = 12$, $m = 10$; (c) absorption of monomeric azobenzene derivatives; and (d) $n = 12$, $m = 5$, J-aggregate.

It was inferred by Okuyama that the relative length of methylene chains in the spacer and tail portions affects the molecular arrangement. The interdigitated arrangement is favored when $m-n > 2$.

X-ray photoelectron spectroscopy (XPS) is also useful for estimation of the molecular packing in single crystals. Nakayama *et al.*[125] examined the N_{1s} spectrum of a single crystal of **(23a)** ($n = 12$, $m = 5$). The calculated intensity ratio of the azobenzene nitrogen to the ammonium nitrogen agreed with the experimental pattern if the molecular axis was assumed to incline by ~60° relative to the bilayer normal at the outermost surface of the crystal. This inclination is in close agreement with that determined by x-ray diffraction.

The variation of the molecular alignment with methylene chain lengths is not restricted to bilayers of azobenzene derivatives. Partially interdigitated structures were proposed for **(21)** ($n = 10$, 12, 14).[35] A series of single-chain amphiphiles **(65)** which contain the *o*-hydroxydiphenylazomethine unit as the rigid segment were prepared.[126] The spectral shifts in the absorption spectrum and the layer thickness determined by x-ray diffraction of cast films indicated that the effect of alkyl chain lengths on the molecular packing is virtually identical with those of the azobenzene-containing bilayer.[127]

These changes in the chromophore arrangement can lead to a remarkable variation in their physical properties and chemical reactivities.[128] For example, some of the azobenzene bilayers were found to be fluorescent, and the fluorescence intensity decreased as the chromophore orientation changed from the tilted head-to-tail type to the parallel type. Emission quenching was observed in the presence of extremely small amounts of a bound cyanine dye. In the *trans–cis* photoisomerization of the bilayers, the rate in the gel state decreased with changing chromophore orientation from the head-to-tail type to the parallel type. The rate was much larger and unaffected by the molecular structure in the case of the liquid crystalline bilayers and of the azobenzene amphiphiles isolated in inert bilayer matrices. The emission was quickly lost by the formation of the *cis* isomer. The photoisomerization was suppressed in the presence of the cyanine, probably due to energy transfer to the cyanine and sensitization of the reverse photoisomerization by the cyanine (Figure 40).

The decreased photoreactivity was also found for stilbene-containing bilayer dispersions, due to parallel chromophore stacking.[129]

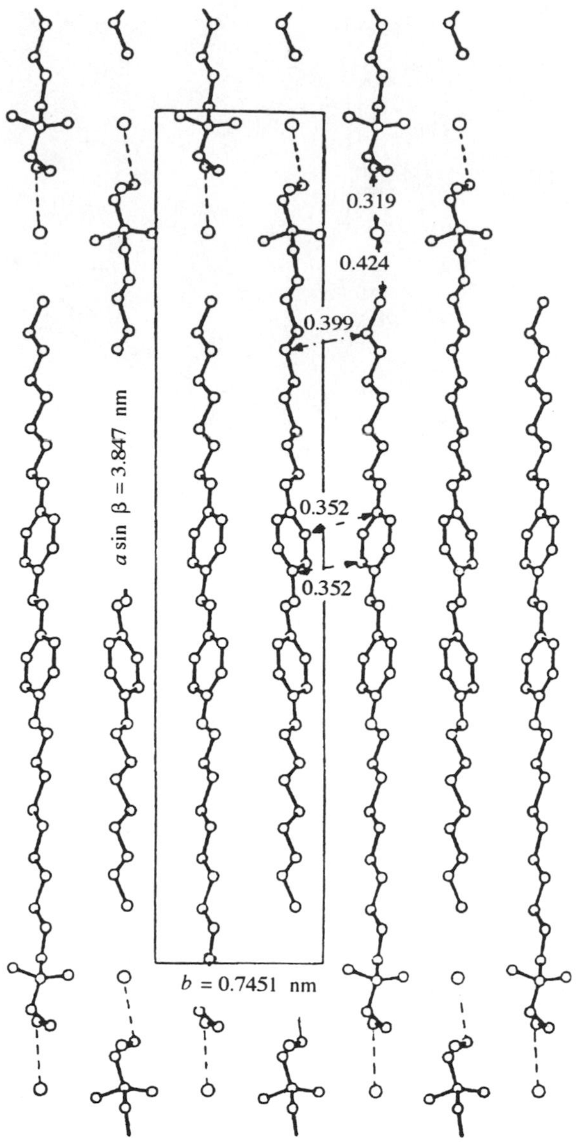

Figure 38 The packing structure of a single crystal of (**23a**) (n = 8, m = 10).

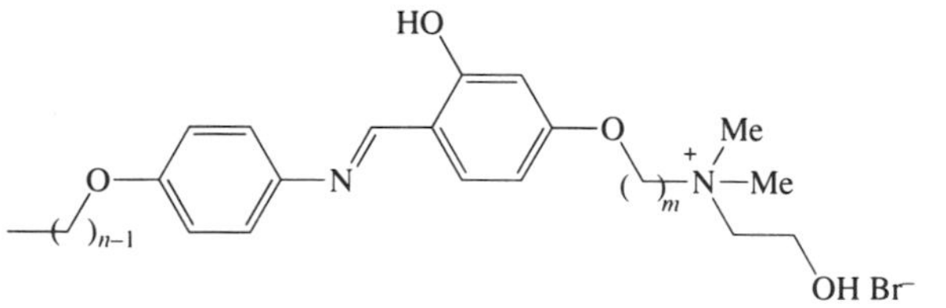

n, m = 8, 10; 12, 5; 12, 10
(**65**)

10.12 CAST FILMS OF SYNTHETIC BILAYER MEMBRANES AND THEIR USE AS MOLECULAR TEMPLATES

10.12.1 Cast Films of Synthetic Bilayer Membranes

The two-dimensional molecular (microscopic) ordering of synthetic bilayer membranes is readily transformed to macroscopic ordering in the form of cast films, because of their structural stability as independent entities (Figure 41). Although cast films of phospholipids have been prepared

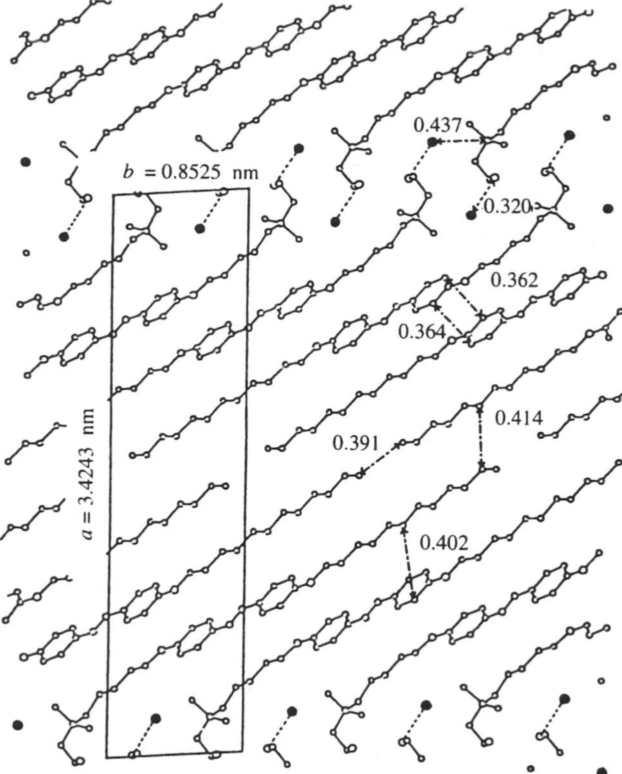

Figure 39 The packing structure of a single crystal of (**23a**) (n = 12, m = 5).

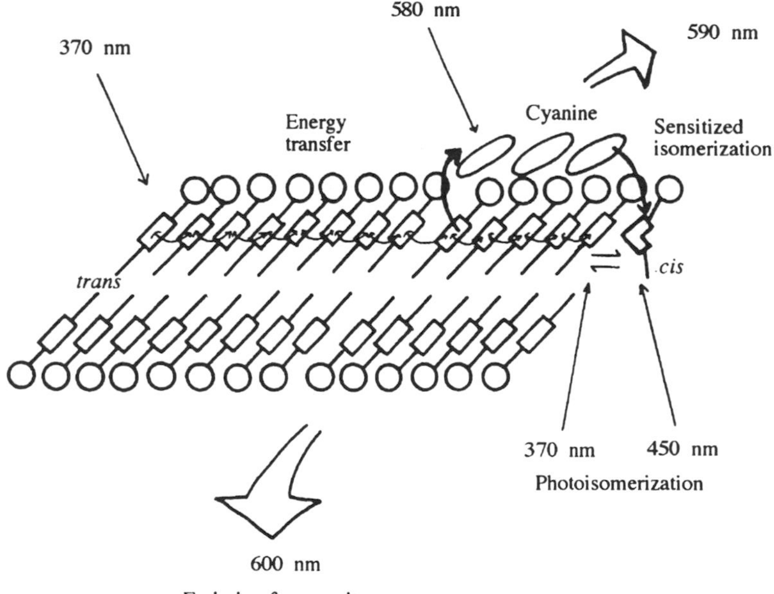

Figure 40 Photophysical and photochemical processes of an azobenzene-containing bilayer membrane of (**23a**) (n = 12, m = 5).

and used for determination of the anisotropic orientation of metalloporphyrins,[130] their structural regularity is not sufficient and the molecular design of film materials is quite limited.

Nakashima *et al.*[131] conducted for the first time a systematic study of solvent casting of aqueous bilayer dispersions. For example, a few drops of clear dispersions of (**66**) (n = 2–6, 10) and (**1**) were spread on glass or quartz plates, and water was removed by one of three methods: (i) keeping

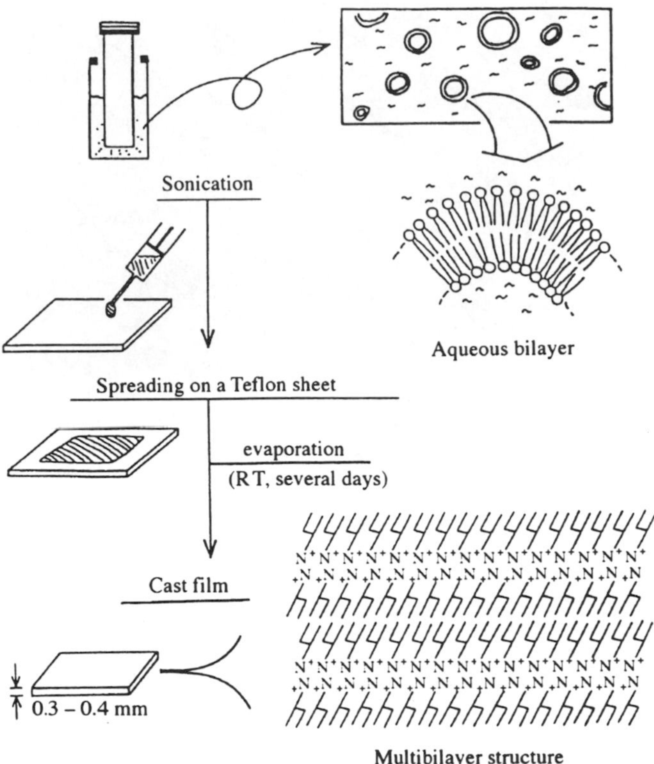

Figure 41 Casting process of aqueous bilayer dispersion.

the plate *in vacuo* (~20 mmHg) at room temperature; (ii) heating at 50–60 °C; and (iii) allowing to stand at room temperature (~20 °C). Transparent films were obtainable from **(66)** by either of these methods, but only method (ii) was effective for **(1)** (n = 14, 16). Maintenance of the bilayer characteristics is confirmed by the presence of the crystal-to-liquid crystal phase transition similar to that of aqueous dispersions. It is known that the spectral property of cyanine and merocyanine dyes bound to aqueous bilayers are highly sensitive to the chemical structure and physical state of the bilayers.[132] These dye molecules bound to cast films display very similar spectral patterns and temperature dependence to those of aqueous dispersions. It can be concluded from these data that the basic physicochemical characteristics of bilayers are maintained in cast films.

$C_{14}H_{29}O$ — $C_{14}H_{29}O$ — $\overset{+}{N}Me_3\ Br^-$

n=2–6, 10
(66)

This casting technique was subsequently extended to other bilayer membranes. In the case of ammonium azobenzene amphiphiles,[63] a series of these amphiphiles with different alkyl chain (spacer and methylene) lengths were successfully cast to form self-supporting films. A scanning electron micrograph of a cast film of **(23)** (n = 8, m = 10) shows highly developed layer structures in the cleaved film edge. X-ray diffraction of the cross-section gives a long spacing of 3.9 nm, which agrees with the interdigited structure. An absorption maximum is observed at 300 nm, consistent with this structure.

Kuo and O'Brien prepared free-standing cast films from double-chain ammonium amphiphiles with a dialkyne polymerizable group.[133] They noticed that the cast multilayer film had sufficient molecular order to allow the topotatic polymerization to proceed. The molecular order was retained best when the casting was conducted below its T_c.

In 1993, transmission electron microscopy was used to observe individual bilayers in cast films directly.[134] Ammonium amphiphiles (**67**) and (**68**) dispersed in water were cast on Teflon sheets and treated with aqueous potassium sodium tartrate and OsO_4 vapor. Whereas cast film (**67**) gives well-developed, flexible layers, cast film (**68**) possesses a much more regular, rigid-looking multilayer (Figure 42(a)). The layers run parallel to the film plane. The presence of the benzene ring in the spacer portion has been shown to improve the regularity of bilayer organization.[100] The improved molecular orientation appears to be related to the better multibilayer regularity. The multilayer structure contains defects and dislocations. Figure 42(b) gives a schematic illustration of typical dislocations. The individual layer is highly regular, and the dislocations appear to be produced by discontinuous layers and are localized, unlike gradual translocations of molecules in more flexible layers.

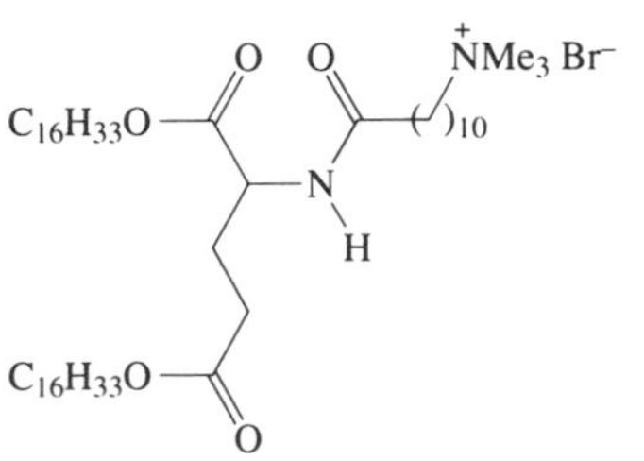

$C_{16}H_{33}O$ O O $\overset{+}{N}Me_3$ Br^- $)_{10}$ N H $C_{16}H_{33}O$ O

(67)

$C_{14}H_{29}O$ O O O $)_6$ $\overset{+}{N}Me_3$ Br^- N H $C_{14}H_{29}O$ O

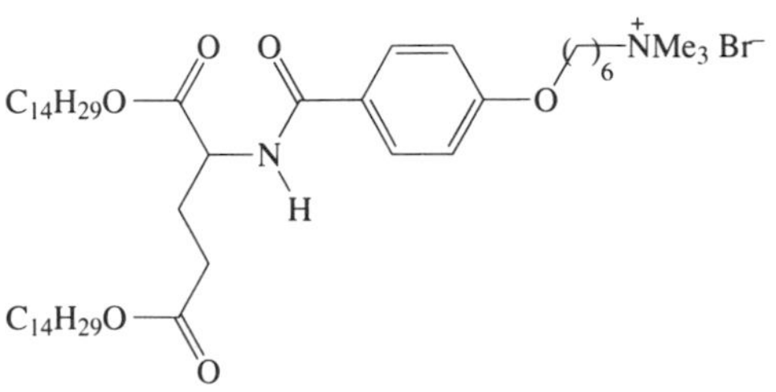

(68)

10.12.2 Stabilization of Cast Films

Conventional bilayer membranes are water soluble, and their cast films may disintegrate eventually in the presence of water. Since the cast films are expected to be useful as novel molecular-precision materials, improvement of their self-supporting properties and stability against water is desirable. Major stabilization techniques are as follows.

10.12.2.1 Polymer–bilayer composite film

Kajiyama and co-workers[135,136] prepared water-insoluble cast films from tetrahydrofuran solutions of dialkylammonium salts, (**1**), and poly(vinyl chloride). The two components are not miscible and phase-separated domains of bilayer and polymer are produced in the film, as examined by x-ray diffraction, thermal measurement, and electron microscopy. By comparison, an aqueous mixture of poly(vinyl alcohol) and dialkylammonium salts gives stable cast films, which, upon treatment with formaldehyde gas or by γ-irradiation, become water insoluble without destruction of the bilayer structure.[137]

10.12.2.2 Polyion complex

When aqueous solutions of charged bilayer membranes are mixed with oppositely charged polymers in water, precipitates of polyion complexes are formed instantaneously. Aging above T_c of the bilayer is effective for complete complexation. Casting of the precipitate from $CHCl_3$ and washing with water produces water-insoluble films that contain multilayer structures composed of bilayer domains.[138] Their physicochemical characterization was further conducted together with their use as permselective membranes and ion sensors.[139]

10.12.2.3 Immobilization in porous polymer supports

Porous polymer films and capsules are convenient supports for bilayer membranes. For example, Okahata *et al.* applied solutions of dialkylammonium salts in dodecane to porous microcapsules and measured ion permeation in the aqueous system.[140]

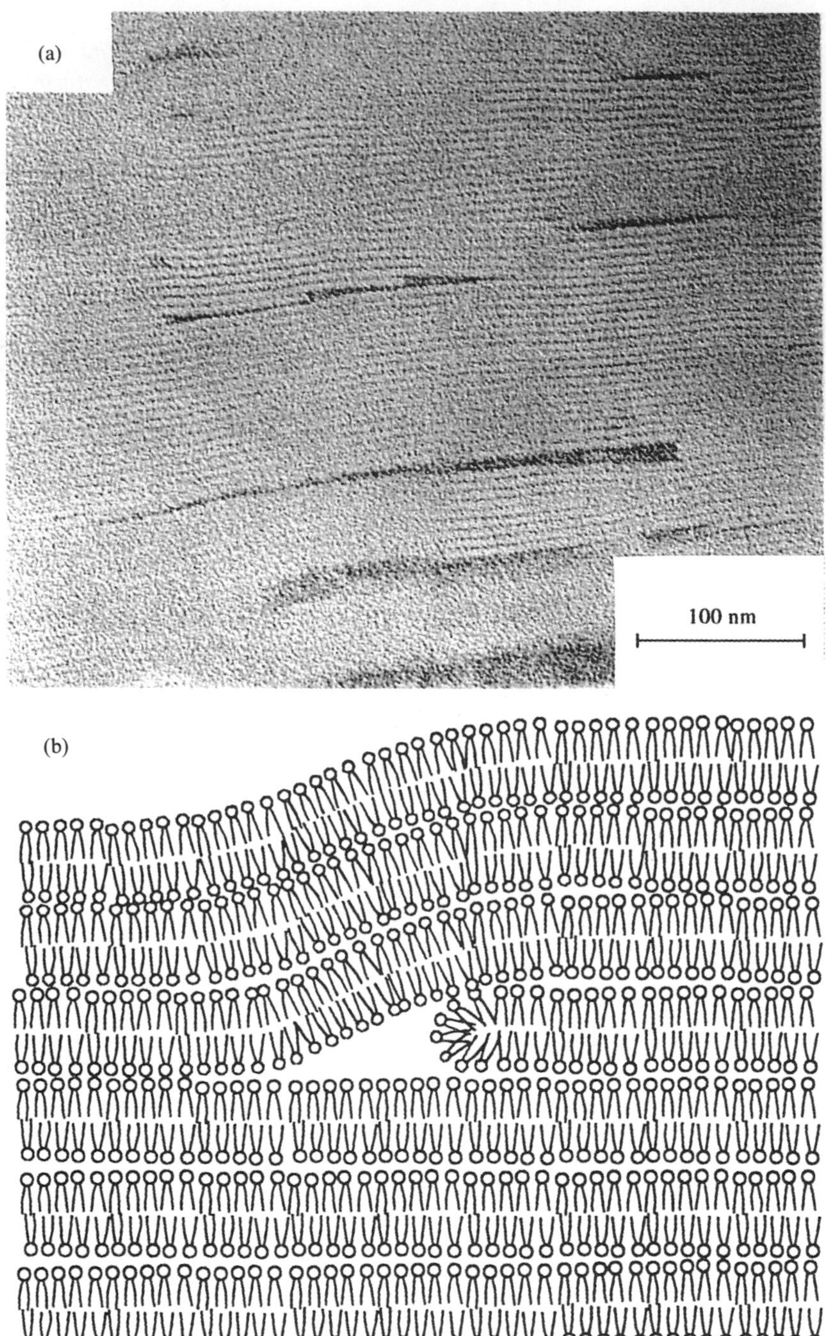

Figure 42 (a) Cross-section TEM image of a cast film of amphiphile (**68**). (b) Schematic illustration of the microscopic morphology.

10.12.2.4 Polymer–bilayer tridecker film

A mixture of dialkylammonium salts and poly(vinyl alcohol) is cast on a thin film of acetylcellulose and dried. Acetylcellulose in acetone is spread on top and dried slowly. The obtained film possesses a tridecker cross-section. Dissolution of the bilayer component is less than 5% when the film is kept in water at 60 °C for 1 h. The bilayer characteristics such as phase transition and phase separation are maintained.[141]

10.12.3 Anisotropic Orientation of Functional Units

Macroscopic ordering in the cast film can be used for producing anisotropic orientation of functional materials. For example, square-planar complexes of [Cu(zincon)] (**69**) or [Cu(tiron)$_2$] (**70**) are dispersed in water together with (**23**) (n = 8, m = 10).[142] When the ESR spectrum was measured for a rapidly quenched sample of the aqueous dispersion, it was typical of a randomly oriented square-planar copper(II) complex. In contrast, when prepared slowly, the cast film showed peculiar ESR patterns that depend only on the angle between the film z-axis and the magnetic field. Apparently, the planar copper(II) complex (**70**) is placed flat on the membrane surface but the in-plane arrangement is random. The angle dependence of the spectral pattern of [Cu(zincon)] in a cast film is totally different. The [Cu(zincon)] molecule is inserted vertically into the bilayer. There is only one negative charge on [Cu(zincon)], in contrast to the six negative charges on [Cu(tiron)$_2$]. This structural difference must be the cause of the different arrangement of these copper(II) chelates in the bilayer membrane.

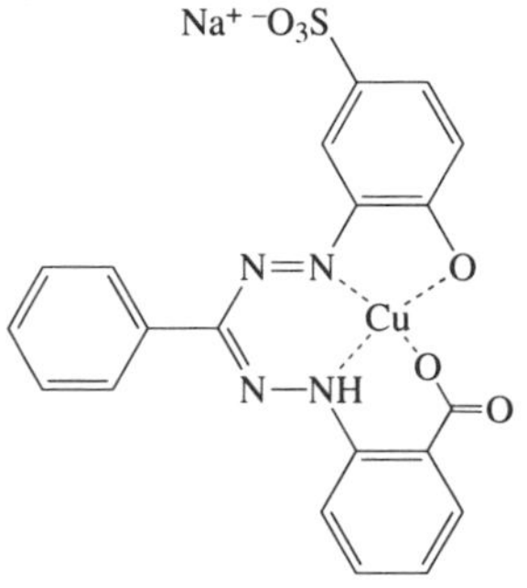

(**69**) 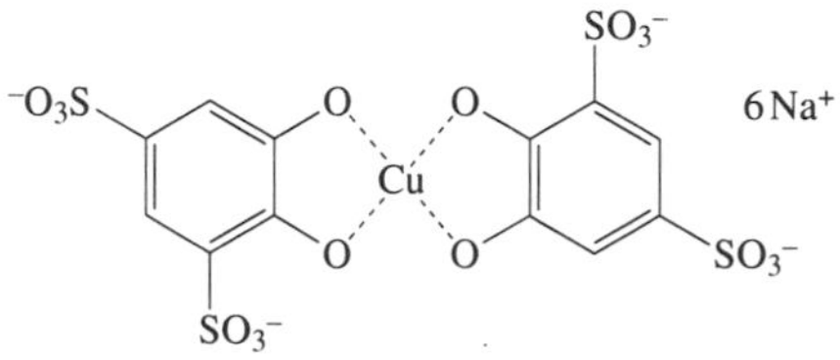 (**70**)

Subsequently, more systematic study led to a methodology to place charged porphyrin derivatives in specific spatial arrangements in matrices of ammonium bilayer membranes.[143] Aqueous mixtures of anionic copper(II) porphyrins and bilayer dispersions of single- and double-chain ammonium amphiphiles were cast on Teflon sheets to produce regular multilayer films. The orientation of doped copper(II) porphyrins is determined by anisotropies of the ESR spectral patterns. Quantitative estimates of the porphyrin orientation are possible by elaborate computer simulation of the observed spectra. The mode of porphyrin orientation is determined again by the distribution of anionic substituents on guest porphyrins and the supramolecular structure of host bilayers. As shown in Figure 43, type III porphyrins, which possess evenly distributed anionic substituents, are incorporated horizontally on the ammonium bilayer surface. Type I porphyrins, in which anionic substituents are localized on one side of the porphyrin ring, are incorporated into the spacer portion of the bilayer parallel to the molecular axis in the case of the double-chain ammonium amphiphile. In contrast, these porphyrins cannot penetrate into the bilayer interior of the single-chain amphiphile and show random incorporation. A type II porphyrin with three sulfonate substituents gives a horizontal orientation on the nonpenetrable bilayer of the single-chain amphiphile and a random orientation in the bilayer of the double-chain amphiphile. The microscopic orientation is highly specific.

This approach was further extended to anisotropic arrangements of globular proteins.[144] The heme group in myoglobin (Mb), an O_2-binding, water-soluble protein, is a convenient probe for absorption spectral detection of denaturation and ESR spectral detection of protein orientation. The cast film containing met-Mb is self-supporting and shows a gel-to-liquid crystal phase transition behavior very similar to that of the bilayer alone. This confirms that the bilayer structure is maintained even in the presence of immobilized met-Mb. ESR spectra of the cast film show strong anisotropy, depending on the angle (θ) between the normal of the film plane (z-axis) and the applied magnetic field. This angular dependence clearly indicates that the heme plane of met-Mb is oriented at an angle of 15–20° against the bilayer surface, as in Figure 44.

This specific orientation is maintained when the molar ratio of met-Mb over amphiphile is varied from 1:40 to 1:330. The overall dimension of myoglobin is about $4.5 \times 3.5 \times 2.5$ nm, and 30 basic amino acid residues and 20 acidic amino acid residues are distributed on the protein surface, forming a few charged domains. The molar ratio of 1:40 corresponds to a situation where met-Mb molecules almost completely cover the polar bilayer surface.

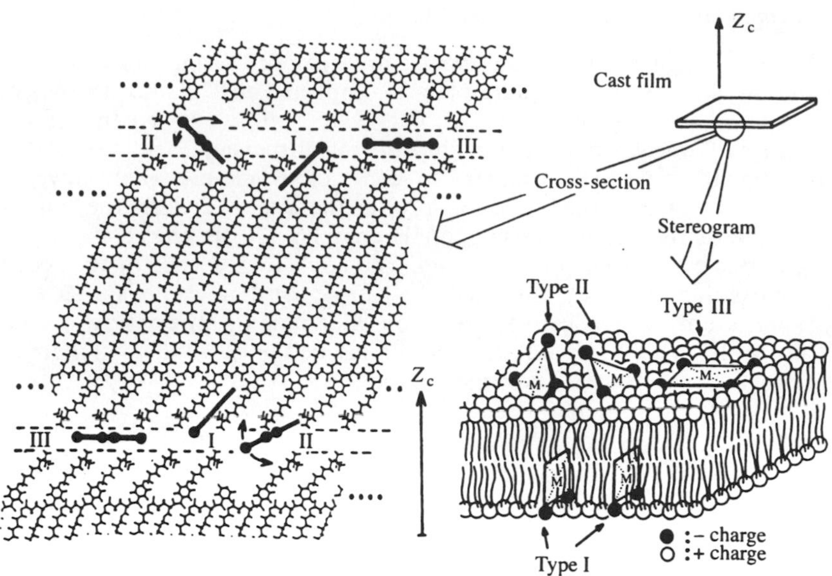

Figure 43 Schematic representation of three types of anionic porphyrins in a cast multibilayer film of (**68**). The overall bilayer organization is assumed to be the same as that of Figure 10. The spacer portion is not shown in the "Stereogram." Type I porphyrins are inserted into the bilayer along the molecular axis of the spacer chain. Type II porphyrins are randomly placed on the bilayer surface. Type III porphyrins lie flat on the bilayer.

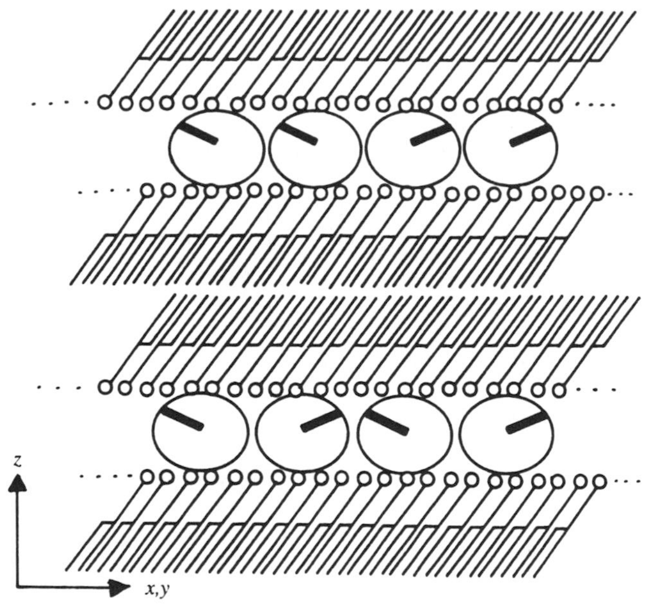

Figure 44 Schematic illustration of met-Mb molecules immobilized in cast multibilayer films. The protein molecules show one-dimensional ordering along the *z*-axis. The heme plane is expressed by filled rectangles.

The ability to synthesize mesoscopic inorganic materials is of crucial importance not only in the development of novel materials for electronics, optics, magnetism, and catalysis but also for understanding their size-dependent properties. Self-organization of inorganic precursors in predetermined dimensional patterns is applicable to fabrication and organization of varied inorganic nanostructures at an atomic/molecular level. The precursor may be single ions, molecular complexes, or clusters that are amenable to further derivatization. Biological systems are replete with examples of inorganic materials in which the assembly of the mineral constituents are precisely controlled at the organic–inorganic interfaces.[145] These biomineralization processes inspired considerable efforts in the construction of organic–inorganic superlattices under mild conditions.

Lead halide clusters of $[PbBr_4]^{2-}$ are formed by ion exchange of counterions of an ammonium cast film.[146] The ion exchange is quantitative (bilayer:$[PbBr_4]$ = 2:1), and the cast film maintains its regular multibilayer structure. This film showed a remarkable blue shift (λ_{max} = 324 nm) relative to those of the two- or three-dimensionally extended perovskites of $PbBr_2$. This implies that a less than two-dimensional cluster or a monomeric complex of $[PbBr_4]^{2-}$ is formed and electrostatically stabilized.

An alternate method to control the mesoscopic structure is to couple preorganization of inorganic precursors with their covalent linking by metal cyanide bonds.[147] Stepwise synthesis of the cyano-bridged bimetallic complex $[Ni(CN)_4{-}Cu^{II}]$ is achieved by successively dipping an ammonium cast film first in aqueous $[Ni(CN)_4]^{2-}$ and then in aqueous copper(II) nitrate. ESR spectra of copper(II) in the resulting film show macroscopic anisotropy, and the bimetallic coordination compound has a highly oriented two-dimensional structure with long-range order in the direction of *z*-axis of the film, as schematically shown in Figure 45. The two-dimensional cyanide network is sufficiently large such that orientational fluctuation does not arise in spite of probable undulations of the bilayer surface.

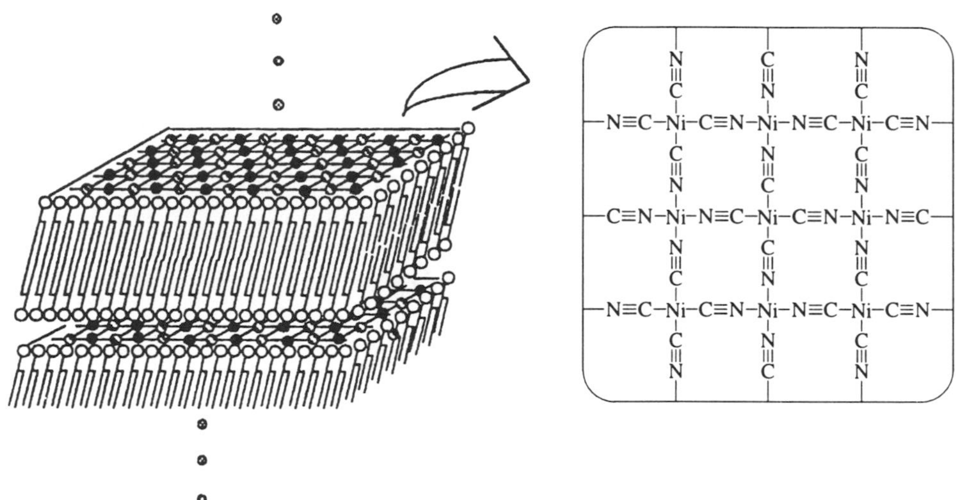

Figure 45 Bimetallic (Cu–Ni) cyanide network in cast multibilayer film.

10.12.4 Molecular Templates

As mentioned above, the cast multibilayer films are stable enough to accommodate a variety of hydrophilic guest molecules (small molecules, proteins, nanoparticles) at the interbilayer space without significant disruption of the regular structure.

Bis(acrylate) monomer (**71**) is readily accommodated in an equimolar cast film of an ammonium multilayer matrix. The regular multilayer structure is not disturbed upon cross-linking of this monomer by photopolymerization, as confirmed by x-ray diffraction. The bilayer matrix is removed by immersing in methanol. The remaining film is flexible and self-supporting. Raman scattering spectra indicate almost complete consumption of the vinyl group in the two-dimensional film. The greatest structural difference of the extracted cast film relative to the three-dimensional polymer gel of (**71**) is seen in an SEM micrograph of its cross-section, showing that wavy layer structures are formed parallel to the film surface. The anisotropic structural characteristics are evident in anisotropic swelling and mechanical properties. It is known that formation of microgels leads to nonuniform cross-linking in three-dimensional gels, which result in limited strength and elongation. In contrast, the monomer molecules in an equimolar cast film are uniformly distributed with a thickness of several nanometers and cross-linking is uniformly extended in two-dimensional space (see Figure 46).[37]

Similar two-dimensional network films are obtainable by cross-linking of linear polymers which are incorporated in the interlayer space. For example, poly(allylamine) and chitosan in a cast film of an ammonium amphiphile can be cross-linked by treatment with glutaraldehyde.[148] Scanning electron microscopy after washing out the matrix clearly shows that the resulting film is composed

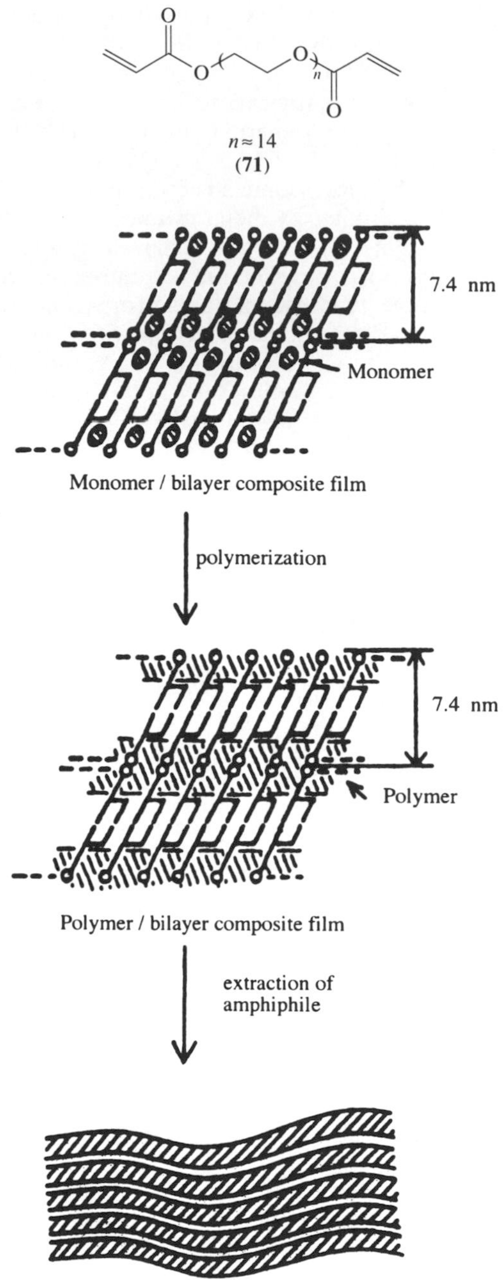

Figure 46 Template synthesis of a multilayered two-dimensional polymer network.

of a multilayer with less than 10 nm unit thickness. A multilayered film of two-dimensionally cross-linked poly(tetraallylammonium bromide) has been similarly prepared.[149]

The preceding approach is applicable to the synthesis of molecularly controlled siloxane networks. When mixed dispersions of alkoxysilanes (e.g., $(MeO)_3SiMe$) and bilayer components are cast and subjected to ammonia treatment, hydrolysis and condensation of the alkoxysilane unit proceed, and stable films of a polysiloxane two-dimensional network are formed upon extraction of matrix bilayers. The SEM observation of the cross-section of the film demonstrates that ultrathin layers with 2 nm thickness are produced parallel to the film plane under proper conditions.[150] The morphology of the cross-section is highly variable (thin layer, plate, pore, etc.), depending on the casting conditions. The pore distribution of the multilayer film as measured by N_2 adsorption changes correspondingly.

In cast films of ammonium bilayers, bromide counterions are readily replaced with silicate anions. Subsequent treatment with hydrobromic acid generates silicate and bromide counteranions. This procedure can be repeated at least four times and the increment of the anionic sillicate unit is stoichiometric in each cycle. Removal of the matrix bilayer by $CHCl_3$ extraction gives a self-supporting silicate multilayer.

Ultrafine particles of metal oxides can be used as starting materials of inorganic multilayers. A homogeneous mixture of a double-chain ammonium amphiphile and an aqueous solution of amorphous Al_2O_3 particles (diameter, 10–100 nm) produces a composite cast film, which is subsequently converted to a multilayered Al_2O_3 film upon calcination at temperatures above 300 °C.[151] Crystalline anisotropy is found for the calcined film by x-ray diffraction and SEM observation, as opposed to an Al_2O_3 film obtained without the bilayer template. A large surface area is retained in the template-synthesized film even after calcination at 1500 °C.

Other metal oxide solutions similarly give rise to self-supporting multilayer films.

10.13 SUMMARY AND PERSPECTIVES

In this chapter, molecular design and unique features of synthetic bilayer membranes have been discussed. The concept of the synthetic bilayer was created through an attempt to replace the structure of biological lipids by simplified synthetic analogues. It is now established that bilayer formation is observable for a variety of synthetic amphiphiles. Their molecular structures are derived from combinations of several molecular modules. For example, double-chain bilayer compounds are composed of hydrophobic tails, connectors, spacers, and hydrophilic heads. Since varied structures are conceivable for each module, the modular combination produces a large number of candidate molecules. In addition to these conventional bilayer compounds where structural modules are connected by covalent bonding, noncovalent forces such as hydrogen bonding, ion pairing, and metal coordination can be used to combine structural modules. In particular, connection of hydrophilic and hydrophobic units by complementary hydrogen bonding provides a very interesting bilayer system.

Further extension of this concept was achieved by bilayer formation in water/alcohol mixtures and in aprotic organic solvents. In the latter case, novel "amphiphilic" compounds composed of solvophobic fluorocarbon chains and solvophilic flexible hydrocarbon chains spontaneously form bilayer structures in cyclohexane solvents. Thus, it is concluded that bilayer formation is a widely observable physicochemical phenomenon that is not limited to particular media like water.

Finally, preparation of cast films and their use as molecular templates has been discussed. Owing to their strong self-organization capability, highly regular multilayers are formed by simple solvent casting and the microscopic (molecular) anisotropy of the bilayer is transformed to macroscopic anisotropy. Functional units incorporated in cast films are therefore anisotropically oriented. The multilayer structure is also effective as two-dimensional templates, and two dimensional organic and inorganic polymer networks and clusters can be obtained.

Biological membranes are basic building blocks of highly organized biomolecular systems. Similarly, synthetic bilayer membranes are a basis for highly organized, artificial supramolecular systems. The expanded bilayer concept is particularly useful in this respect. The biological membranes can maintain their unique organizational features only in the aqueous system. In contrast, properly designed synthetic bilayers retain their unique organizations in aqueous and nonaqueous media. Unfortunately, the synthetic molecular organization remains rudimentary at the current level, and further elaboration of structural control is indispensable. If a methodology to control relative dispositions of molecules within the two-dimensional assembly can be found, much more elaborate molecular systems will be designed.

ACKNOWLEDGMENTS

The author is grateful to Dr. I. Ichinose, Ms. R. Ando, and Ms. C. Ito for their devoted help in the manuscript preparation.

10.14 REFERENCES

1. A. D. Bangham and R. W. Horne, *J. Mol. Biol.*, 164, **8**, 660.
2. S. J. Singer and G. L. Nicolson, *Science*, 1972, **175**, 720.

3. J. M. Gebicki and M. Hicks, *Nature*, 1973, **3243**, 232.
4. W. R. Hargreaves and D. W. Deamer, *Biochemistry*, 1978, **17**, 3759.
5. C. Tanford, in "The Hydrophobic Effect, Formation of Micelles, Polymers and Biological Membranes," Wiley-Interscience, New York, 1973, chap. 9.
6. J. N. Israelachvili, D. J. Mitchell, and B. W. Ninham, *J. Chem. Soc., Faraday Trans. 2*, 1976, 1525.
7. H. Brockerhoff, in "Bioorganic Chemistry," ed. E. E. von Tamelen, Academic Press, New York, 1977, vol. 3, p.1.
8. T. Kunitake and Y. Okahata, *J. Am. Chem. Soc.*, 1977, **99**, 3860.
9. A. W. Ralston, D. N. Eggenberger, and P. L. DuBrow, *J. Am. Chem. Soc.*, 1948, **70**, 977.
10. H. Kunieda and K. Shinoda, *J. Phys. Chem.*, 1978, **82**, 1710.
11. W. G. Gray and P. A. Winsor, in "Lyotropic Liquid Crystals," ed. S. Friberg, American Chemical Society, Washington, 1976, chap. 1, p. 10.
12. J. H. Fendler, in "Membrane Mimetic Chemistry," Wiley-Interscience, New York, 1982, chap. 6.
13. J.-H. Fuhrhop and J. Mathieu, *Angew. Chem.*, 1984, **96**, 124; *Angew. Chem., Int. Ed. Engl.*, 1984, **23**, 100.
14. H. Ringsdorf, B. Schlaub, and J. Venzmer, *Angew. Chem.*, 1988, **100**, 117; *Angew. Chem., Int. Ed. Engl.*, 1988, **27**, 113.
15. J.-H. Fuhrhop and J. Köning, in "Membranes and Molecular Assemblies, The Synthetic Approach," Royal Society of Chemistry, Cambridge, 1994.
16. T. Kunitake, Y. Okahata, K. Tamaki, F. Kumamaru, and M. Takayanagi, *Chem. Lett.*, 1977, 378.
17. K. Deguchi and J. Mino, *J. Colloid Interface Sci.*, 1978, **65**, 155.
18. U. Herrmann and J. H. Fendler, *Chem. Phys. Lett.*, 1979, **64**, 270.
19. Y. Y. Lim and J. H. Fendler, *J. Am. Chem. Soc.*, 1979, **101**, 4023.
20. K. Kano, A. Romero, B. Djermoune, H. H. Ache, and J. H. Fendler, *J. Am. Chem. Soc.*, 1979, **101**, 4030.
21. A. M. Carmona-Ribeiro and H. Chaimovitch, *Biochim. Biophys. Acta*, 1983, **733**, 132.
22. T. Kunitake and Y. Okahata, *Chem. Lett.*, 1977, 1337.
23. T. Kunitake, N. Nakashima, S. Hayashida, and K. Yonemori, *Chem. Lett.*, 1979, 1413.
24. E. J. R. Sudholter, J. B. F. N. Engberts, and D. Hoekistra, *J. Am. Chem. Soc.*, 1980, **102**, 2467.
25. Y. Murakami, A. Nakano, and K. Fukuya, *J. Am. Chem. Soc.*, 1980, **102**, 4235.
26. J.-H. Fuhrhop, H. Bartsch, and D. Fritzsch, *Angew. Chem.*, 1981, **93**, 797; *Angew. Chem., Int. Ed. Engl.*, 1981, **20**, 804.
27. T. Kunitake, M. Shimomura, Y. Hashiguchi, and T. Kawanaka, *J. Chem. Soc., Chem. Commun.*, 1985, 833.
28. T. Kunitake and Y. Okahata, *Bull. Chem. Soc. Jpn.*, 1978, **51**, 1877.
29. R. A. Mortara, R. H. Quina, and H. Chaimovich, *Biochem. Biophys. Res. Commun.*, 1978, **81**, 1080.
30. Y. Okahata, S. Tanamachi, M. Nagai, and T. Kunitake, *J. Colloid Interface Sci.*, 1981, **82**, 401.
31. T. Kajiyama, A. Kumano, M. Takayanagi, Y. Okahata, and T. Kunitake, *Contemp. Top. Polym. Sci.*, 1984, **4**, 829.
32. K. Okuyama, Y. Soboi, K. Hirabayashi, A. Harada, A. Kumano, T. Kajiyama, M. Takayanagi, and T. Kunitake, *Chem. Lett.*, 1984, 2117.
33. K. Okuyama, Y. Soboi, N. Niijima, K. Hirabayashi, T. Kunitake, and T. Kajiyama, *Bull. Chem. Soc. Jpn.*, 1988, **61**, 1485.
34. K. Okuyama, N. Niijima, K. Hirabayashi, T. Kunitake, and M. Kusunoki, *Bull. Chem. Soc. Jpn.*, 1988, **61**, 2337.
35. H. Okada, Molecular Architecture Project, Japan, unpublished results.
36. K. Okuyama, Y. Ogawa, and T. Kajiyama, *Nippon Kagaku Kaishi*, 1987, 2199.
37. A. Asakuma, H. Okada, and T. Kunitake, *J. Am. Chem. Soc.*, 1991, **113**, 1749.
38. M. Kodama, T. Kunitake, and S. Seki, *J. Phys. Chem.*, 1991, **94**, 1550.
39. Y. Okahata, R. Ando, and T. Kunitake, *Ber. Bunsenges. Phys. Chem.*, 1981, **85**, 789.
40. Y. Murakami, A. Nakano, A. Yoshimatsu, K. Uchitomi, and Y. Matsuda, *J. Am. Chem. Soc.*, 1985, **106**, 3613.
41. D. Bach, I. Bursuker, H. Eible, and I. R. Miller, *Biochim. Biophys. Acta*, 1978, **514**, 310.
42. T. Kunitake, R. Ando, and Y. Ishikawa, *Mem. Fac. Eng. Kyushu Univ.*, 1986, **46**, 221.
43. J.-M. Kim and T. Kunitake, *Mem. Fac. Eng. Kyushu Univ.*, 1989, **49**, 93.
44. T. Kunitake, S. Tawaki, and N. Nakashima, *Bull. Chem. Soc. Jpn.*, 1983, **56**, 3235.
45. M. Shimomura and T. Kunitake, *Chem. Lett.*, 1981, 1001.
46. M. Shimomura and T. Kunitake, *J. Am. Chem. Soc.*, 1982, **104**, 1757.
47. T. Kunitake, H. Ihara, and Y. Okahata, *J. Am. Chem. Soc.*, 1983, **105**, 6070.
48. R. D. Kornberg and H. M. MacConnelle, *Biochemistry*, 1971, **10**, 1111.
49. R. A. Moss, S. Bhattacharya, and S. Chatterjee, *J. Am. Chem. Soc.*, 1989, **111**, 3680; R. A. Moss, T. Fujita, and S. Gangli, *Langmuir*, 1990, **6**, 1197; R. A. Moss and Y. Okumura, *J. Am. Chem. Soc.*, 1992, **114**, 1750.
50. S. Bhattacharya, R. A. Moss, H. Ringsdorf, and J. Simon, *J. Am. Chem. Soc.*, 1993, **115**, 3812.
51. For example, V. Luzzati, in "Biological Membranes," ed. D. Chapman, Academic Press, New York, 1968, p. 71.
52. Y. Murakami, A. Nakano, J. Kikuchi, T. Takaki, and K. Uchimura, *Chem. Lett.*, 1983, 1891; Y. Murakami, J. Kikuchi, T. Takaki, K. Uchimura, and A. Nakano, *J. Am. Chem. Soc.*, 1985, **107**, 2161.
53. Y. Murakami, J. Kikuchi, T. Takaki, and K. Uchimura, *Bull. Chem. Soc. Jpn.*, 1986, **59**, 515.
54. L. A. M. Rupert, D. Hoekistra, and J. B. F. N. Engberts, *J. Am. Chem. Soc.*, 1985, **107**, 2628.
55. L. A. M. Rupert, J. B. F. N. Engberts, and D. Hoekistra, *J. Am. Chem. Soc.*, 1986, **108**, 3920.
56. L. A. M. Rupert, J. F. L. van Breemen, E. F. J. van Bruggen, J. B. F. N. Engberts, and D. Hoekistra, *J. Membr. Biol.*, 1987, **95**, 255.
57. A. M. Carmona-Ribeiro and H. Chaimovich, *Biophys. J.*, 1986, **50**, 621.
58. For example, D. F. Evans and B. W. Ninham, *J. Phys. Chem.*, 1986, **90**, 226.
59. H. Hotani, *J. Mol. Biol.*, 1984, **178**, 113.
60. N. Nakashima, S. Asakuma, T. Kunitake, and H. Hotani, *Chem. Lett.*, 1984, 227.
61. N. Kimizuka, T. Takasaki, and T. Kunitake, *Chem. Lett.*, 1988, 1911.
62. T. Kunitake and Y. Okahata, *J. Am. Chem. Soc.*, 1980, **102**, 549.
63. M. Shimomura, R. Ando, and T. Kunitake, *Ber. Bunsenges. Phys. Chem.*, 1983, **87**, 1134.
64. Y. Okahata and T. Kunitake, *Ber. Bunsenges. Phys. Chem.*, 1980, **84**, 550.
65. Y. Okahata, H. Ihara, M. Shimomura, S. Tawaki, and T. Kunitake, *Chem. Lett.*, 1980, 1169.
66. T. Kunitake, Y. Okahata, M. Shimomura, S. Yasunami, and K. Takarabe, *J. Am. Chem. Soc.*, 1981, **103**, 5401.
67. M. Shimomura, H. Hashimoto, and T. Kunitake, *Chem. Lett.*, 1982, 1285.

68. T. Kunitake, R. Ando, and Y. Ishikawa, *Mem. Fac. Eng. Kyushu Univ.*, 1986, **46**, 245.
69. I. Cho and J. G. Park, *Chem. Lett.*, 1987, 977.
70. T. Kunitake, Y. Yamada, and N. Fukunaga, *Chem. Lett.*, 1984, 1089.
71. F. M. Menger and Y. Yamasaki, *J. Am. Chem. Soc.*, 1993, **115**, 3840.
72. T. Kunitake, N. Kimizuka, N. Higashi, and N. Nakashima, *J. Am. Chem. Soc.*, 1984, **106**, 1978.
73. N. Kimizuka, H. Ohira, M. Tanaka, and T. Kunitake, *Chem. Lett.*, 1990, 29.
74. For example, S. C. Kushuwaha, M. Kates, G. D. Sprott, and I. C. P. Smith, *Biochim. Biophys. Acta*, 1981, **664**, 156.
75. Y. Okahata and T. Kunitake, *J. Am. Chem. Soc.*, 1979, **101**, 5231.
76. S. Yasunami, Masters Thesis, Kyushu University, 1980.
77. T. Kunitake, A. Tsuge, and K. Takarabe, *Polym. J.*, 1985, **17**, 633.
78. J.-H. Fuhrhop and D. Fritzsch, *Acc. Chem. Res.*, 1986, **19**, 130.
79. J.-H. Fuhrhop, K. Ellermann, H.-H. David, and J. Mathieu, *Angew. Chem., Int. Ed. Engl.*, 1982, **21**, 440.
80. J.-H. Fuhrhop, H.-H. David, J. Mathieu, V. Liman, H.-J. Winter, and E. Boekema, *J. Am. Chem. Soc.*, 1986, **108**, 1785.
81. (a) J.-H. Fuhrhop, D. Fritzsch, B. Tesche, and H. Schmiady, *J. Am. Chem. Soc.*, 1984, **106**, 1998; (b) J.-H. Fuhrhop and D. Fritzsch, *J. Am. Chem. Soc.*, 1984, **106**, 4287.
82. J.-H. Fuhrhop and J. Mathieu, *J. Chem. Soc., Chem. Commun.*, 1983, 144.
83. T. Kunitake, Y. Okahata, and S. Yasunami, *J. Am. Chem. Soc.*, 1982, **104**, 5547.
84. T. Kunitake and N. Higashi, *Makromol. Chem. Suppl.*, 1985, **14**, 81.
85. T. Kunitake and N. Higashi, *J. Am. Chem. Soc.*, 1985, **107**, 692.
86. C. Santaella, P. Vierling, and J. G. Riess, *Angew. Chem., Int. Ed. Engl.*, 1991, **30**, 567.
87. K. Liang and Y. Hui, *J. Am. Chem. Soc.*, 1992, **114**, 6588.
88. T. Kunitake, N. Nakashima, M. Shimomura, Y. Okahata, K. Kano, and T. Ogawa, *J. Am. Chem. Soc.*, 1980, **102**, 6642.
89. T. Kunitake, N. Nakashima, and K. Morimitsu, *Chem. Lett.*, 1980, 1347.
90. N. Nakashima, S. Asakuma, J.-M. Kim, and T. Kunitake, *Chem. Lett.*, 1984, 1709.
91. N. Nakashima, S. Asakuma, and T. Kunitake, *J. Am. Chem. Soc.*, 1985, **107**, 509.
92. Y. Ishikawa, T. Nishimi, and T. Kunitake, *Chem. Lett.*, 1990, 25.
93. N. Nakashima, H. Fukushima, and T. Kunitake, *Chem. Lett.*, 1981, 1207.
94. N. Nakashima, R. Ando, T. Muramatsu, and T. Kunitake, *Langmuir*, 1994, **10**, 232.
95. T. Kunitake and Y. Yamada, *J. Chem. Soc., Chem. Commun.*, 1986, 655.
96. K. Yamada, H. Ihara, T. Ide, and T. Fukumoto, *Chem. Lett.*, 1984, 1713.
97. J.-M. Kim, Y. Ishikawa, and T. Kunitake, *J. Chem. Soc., Perkin Trans. 2*, 1991, 885.
98. P. Yager and P. E. Shoen, *Mol. Cryst. Liq. Cryst.*, 1964, **106**, 371, and subsequent publications by these authors.
99. N. Kimizuka, T. Kawasaki, and T. Kunitake, *J. Am. Chem. Soc.*, 1993, **105**, 4387.
100. N. Nakashima and T. Kunitake, *J. Am. Chem. Soc.*, 1982, **104**, 4261.
101. S. Kato and T. Kunitake, *Polym. J.*, 1991, **23**, 135.
102. N. Kimizuka, T. Kawasaki, and T. Kunitake, *Chem. Lett.*, 1994, 33.
103. N. Kimizuka, T. Kawasaki, and T. Kunitake, *Chem. Lett.*, 1994, 1399.
104. P. Jokura, B Jonsson, and A. Khan, *J. Phys. Chem.*, 1987, **91**, 3291.
105. M. Wakita, K. A. Edwards, and S. L. Regen, *J. Am. Chem. Soc.*, 1982, **110**, 5221.
106. E. W. Kaler, A. K. Murthy, B. E. Rodringuez, and J. A. N. Zasadzinski, *Science*, 1989, **245**, 1371.
107. H. Fukuda, K. Kawata, H. Okuda, and S. L. Regen, *J. Am. Chem. Soc.*, 1990, **112**, 1635.
108. Y.-C. Chung, S. L. Regen, H. Fukuda, and K. Hirano, *Langmuir*, 1992, **8**, 2843.
109. I. Ichinose, Y. Ishikawa, and T. Kunitake, *Complex Salt Chem. Prepr. Jpn.*, 1989, **43**, 293.
110. J. Suh, K.-J. Lee, G. Bae, O.-B. Kwon, and S. Oh, *Langmuir*, 1995, **11**, 2626.
111. For example, D. F. Evans, A. Yamauchi, G. Wel, and V. A. Bloomfield, *J. Phys. Chem.*, 1983, **87**, 3537 and papers cited therein.
112. N. Kimizuka, T. Wakiyama, K. Yoshimi, H. Miyauchi, and T. Kunitake, *Polym. Prepr. Jpn.*, 1994, **43**, 3752.
113. N. Nishi, K. Koga, C. Ohshima, K. Yamamoto, U. Nagashima, and K. Nagami, *J. Am. Chem. Soc.*, 1988, **110**, 5246.
114. H. Kunieda, K. Nakamura, and D. F. Evans, *J. Am. Chem. Soc.*, 1991, **113**, 1051.
115. J.-M. Kim and T. Kunitake, *Chem Lett.*, 1989, 959.
116. M. P. Turberg and J.-E. Brady, *J. Am. Chem. Soc.*, 1988, **110**, 7794.
117. (a) Y. Ishikawa, H. Kuwahara, and T. Kunitake, *J. Am. Chem. Soc.*, 1991, **111**, 8350; (b) Y. Ishikawa, H. Kuwahara, and T. Kunitake, *Chem. Lett.*, 1989, 1737.
118. H. Kuwahara, M. Hamada, Y. Ishikawa, and T. Kunitake, *J. Am. Chem. Soc.*, 1993, **115**, 3002.
119. M. Kasha, in "Spectroscopy of the Excited State," ed. B. D. Bartolo, Plenum, New York, 1976, p. 337.
120. T. Kunitake, M. Shimomura, T. Kajiyama, A. Harada, K. Okuyama, and M. Takayanagi, *Thin Solid Films*, 1984, **121**, L89.
121. G. Xu, K. Okuyama, and M. Shimomura, *Polym. Prepr. Jpn.*, 1989, **38**, 2407.
122. K. Okuyama, H. Watanabe, M. Shimomura, K. Hirabayashi, T. Kunitake, T. Kajiyama, and N. Yasuoka, *Bull. Chem. Soc. Jpn.*, 1986, **59**, 3351.
123. H. Watanabe, K. Okuyama, H. Ozawa, K. Hirabayashi, M. Shimomura, and N. Yasuoka, *Nippon Kagaku Kaishi*, 1988, 55.
124. K. Okuyama, C. Mizuguchi, G. Xu, and M. Shimomura, *Bull. Chem. Soc. Jpn.*, 1989, **62**, 3211.
125. Y. Nakayama, T. Takahagi, F. Soeda, A. Ishitani, M. Shimomura, K. Okuyama, and T. Kunitake, *Appl. Surf. Sci.*, 1988, **33/34**, 1307.
126. Y. Ishikawa, T. Nishimi, and T. Kunitake, *Chem. Lett.*, 1990, 165.
127. T. Nishimi, Y. Ishikawa, R. Ando, and T. Kunitake, *Recl. Trav. Chim. Pays-Bas*, 1994, **113**, 201.
128. M. Shimomura and T. Kunitake, *J. Am. Chem. Soc.*, 1987, **109**, 5175.
129. M. Shimomura, H. Hashimoto, and T. Kunitake, *Langmuir*, 1989, **5**, 174.
130. R. K. Poole, H. Blum, R. I. Scott, A. Collinge, and T. Ohnishi, *J. Gen. Microbiol.*, 1980, **119**, 145.
131. N. Nakashima, R. Ando, and T. Kunitake, *Chem. Lett.*, 1983, 1577.

132. N. Nakashima, R. Ando, H. Fukushima, and T. Kunitake, *J. Chem. Soc., Chem. Commun.*, 1982, 707.
133. T. Kuo and D. F. O'Brien, *J. Am. Chem. Soc.*, 1988, **110**, 7570; *Macromolecules*, 1990, **23**, 3225; *Langmuir*, 1991, **7**, 584.
134. Y. Wakayama and T. Kunitake, *Chem. Lett.*, 1993, 1425.
135. T. Kajiyama, A. Kumano, M. Takayanagi, Y. Okahata, and T. Kunitake, *Chem. Lett.*, 1979, 645.
136. A. Kumano, O. Niwa, T. Kajiyama, M. Takayanagi, T. Kunitake, and K. Kano, *Polym. J.*, 1984, **16**, 461.
137. N. Higashi and T. Kunitake, *Polym. J.*, 1984, **16**, 538.
138. T. Kunitake, A. Tsuge, and N. Nakashima, *Chem. Lett.*, 1984, 1783.
139. For example, (a) Y. Okahata and G. Enna, *J. Phys. Chem.*, 1988, **92**, 4546; (b) Y. Okahata and O. Shimizu, *Langmuir*, 1978, **3**, 1171.
140. Y. Okahata, *Acc. Chem. Res.*, 1986, **19**, 57.
141. M. Shimomura and T. Kunitake, *Polym. J.*, 1984, **16**, 187.
142. Y. Ishikawa and T. Kunitake, *J. Am. Chem. Soc.*, 1986, **108**, 8300.
143. Y. Ishikawa and T. Kunitake, *J. Am. Chem. Soc.*, 1991, **113**, 621.
144. I. Hamachi, S. Noda, and T. Kunitake, *J. Am. Chem. Soc.*, 1990, **112**, 6744.
145. L. Addadi and S. Weins, *Angew. Chem., Int. Ed. Engl.*, 1992, **31**, 153.
146. N. Kimizuka, T. Maeda, I. Ichinose, and T. Kunitake, *Chem. Lett.*, 1993, 941.
147. N. Kimizuka, T. Handa, I. Ichinose, and T. Kunitake, *Angew. Chem., Int. Ed. Engl.*, 1994, **33**, 2483.
148. K. Kohyama, K. Hirayama, and T. Kunitake, *Polym. Prepr. Jpn.*, 1990, **39**, 653.
149. M. Marek, Jr., K. Fukuta, and T. Kunitake, *Chem. Lett.*, 1993, 291.
150. K. Sakata and T. Kunitake, *J. Chem. Soc., Chem. Commun.*, 1990, 504.
151. N. Tsutsumi, K. Sakata, and T. Kunitake, *Chem. Lett.*, 1992, 1465.

11
Synkinetic Micellar and Vesicular Fibers Made of Amphiphilic Lipids and Porphyrins

JÜRGEN-HINRICH FUHRHOP
Freie Universität Berlin, Germany

11.1 INTRODUCTION

The main components of biological organisms are water, protein and/or carbohydrate fibers, and bilayer lipid membranes. Fibers provide, for example, the rigidity of tree trunks and finger nails; they allow the propagation of electric currents in nerves and the contraction and gliding against each other of muscles. The lipid membranes, on the other hand, dissolve membrane proteins and various cofactors to allow vectorial ion flow, light-induced charge separation, and various enzymatic reactions. Furthermore, some extremely sensitive organelles such as mitochondria and chloroplasts as well as the genetic apparatus are separated by lipid membranes from the rest of the cell lumen and are thus protected.[1]

All of these fibrous and membraneous structures have in common that they are closely linked to each other by noncovalent binding forces. Organization does not occur by stepwise addition of small synthons to a growing molecule, but by a self-assembly process of complete molecules within ultrathin lipid monolayers or bilayers.

This assembly stops at the gel-like state of membrane structures and does not proceed to three-dimensional crystals, because most constituents are curved: neither the spherical and undulating lipid membranes, nor the helical protein, polysaccharide, or nucleic acid fibers have a tendency to crystallize from water. Living systems are predominantly neither thermodynamic stable solutions nor crystals, but metastable, perfectly organized gels. A particular illustrative example for a biological gel structure is the eye-lens (Figure 1),[2] which does not crystallize within a century. As an example of a composite fiber we name the 30 nm queue of the tobacco mosaic virus (virion), which consists of a protein coat and a nucleic acid core.

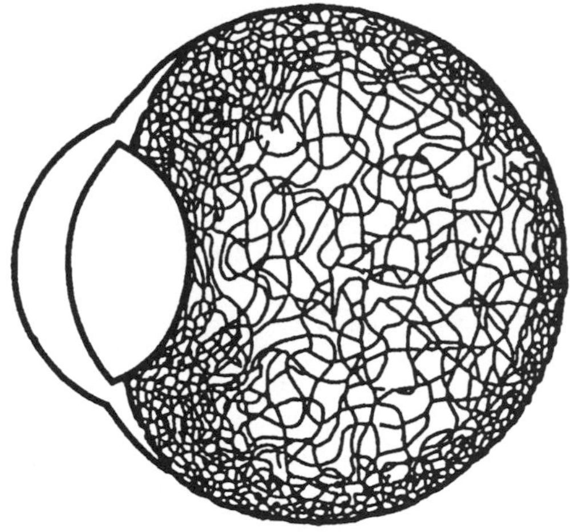

Figure 1 Model of the human eye lens as an example of long-lived aqueous gels made of chiral organic fibers (here, anionic polysaccharides and collagen). Such gels may contain >99% (w/w) water and be (meta)stable for many years.

In this chapter, we shall discuss artificial supramolecular assemblies made of amphiphilic lipids and porphyrins, which combine the properties of polymeric fibers and membraneous assemblies. We shall call the assembly of molecules to noncovalent polymer fibers "synkinesis,"[2] because the standard expression "self-organization" should be reserved to micelles and vesicles in which no binding interactions except for solvophobic effects occur. At the moment one starts to apply regio- and stereoselective binding interactions between molecules, in addition to "self-organization", we speak of "synkinesis," a term which is closely related to covalent "synthesis." The comparison of a micelle, a hydrated bilayer, and a hydrogen-bonded structure in Figure 2[3] shows the difference between self-organized head group regions with no binding head group interactions and a typical synkinetic structure, which is organized by various hydrogen bonds.

11.2 PRINCIPLES OF SYNKINESIS

Synkinesis is defined as the synthesis of noncovalent compounds.[2] In this chapter noncovalent polymers are discussed. The monomers needed for synkinesis are called "synkinons," replacing the

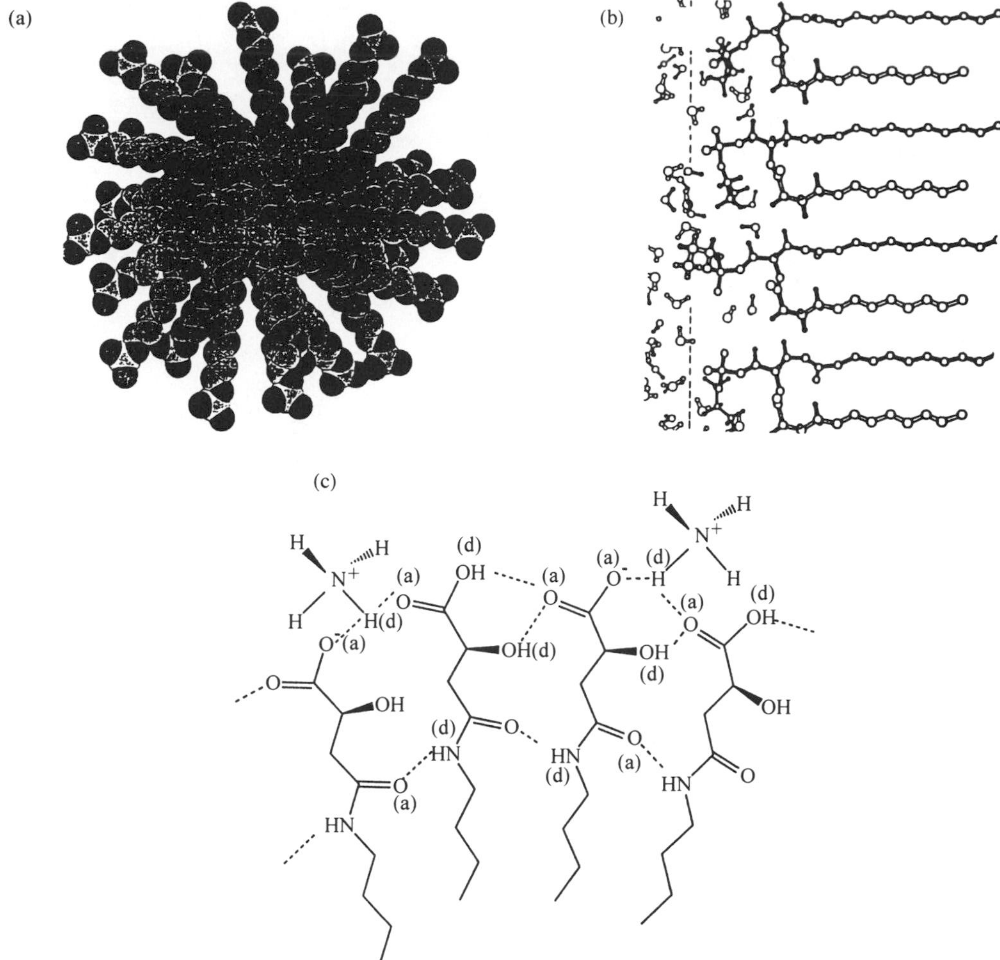

Figure 2 (a) Models of a spherical micelle and (b) part of a vesicle bilayer membrane. Nondirected repulsive interactions between the head groups and the hydrophobic effect lead to fluid molecular assemblies of spherical or planar shape. (c) Strong linear hydrogen bond chains ((d) = proton donor, (a) = acceptor) and similar directed bonding interactions convert spherical liquid assemblies first to disks and then to solid fibers (see Figure 20). Part of a circular disk is shown.

conventional term "building blocks." Synkinesis in dilute solutions does not lead to only spherical molecular assemblies, in which the surface area towards the bulk solvent is minimized, but defined pores within vesicle membranes, asymmetric vesicle membranes with different head groups on the inner and outer surfaces and, as discussed in this chapter, well-defined mono- and bilayer fibers (Figure 3) can also be obtained by optimization of synkinons and synkinesis conditions.

In order to describe the simple synkinetic methods which are at our disposal today, we first sketch the typical course and results of statistical aggregation in bulk water and then introduce the most important ordering forces which lead to well-organized molecular assemblies and, at the same time, prevent three-dimensional crystallization.

11.2.1 Nonordered Aggregates

The primary assembly of amphiphiles in water is presumably always a micelle as shown in Figure 2. These micelles may, however, rapidly rearrange to form regular vesicles or even fibers if the critical micellar concentration of the monomeric amphiphiles[4] is below 10^{-5} M. If, however, the insoluble amphiphiles occur as mixtures with different head groups or different lengths of the hydrophobic chains, then the formation of vesicles or fibers from micelles may become very slow. Heterogenous lumps of micelles, which are then often observed (Figure 4), produce clear gels. Examples are mixtures of *N*-octyl and *N*-dodecyl-gluconamides[5] and of 1-palmitoyl-*sn*-glycero-3-phosphocholine[6] in water.

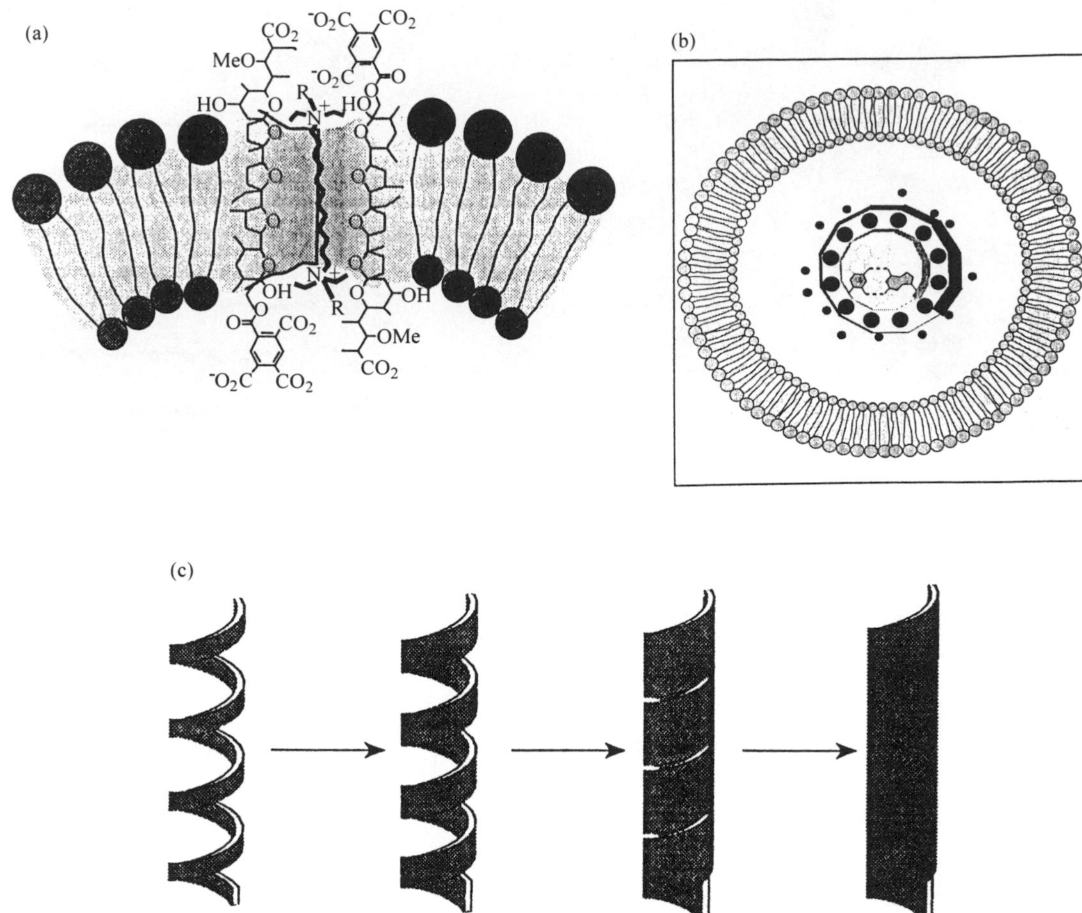

Figure 3 Typical synkinetic target-assemblies. (a) Fluid vesicle membranes with ordered domains acting as an ion pore[2] which can be sealed and opened. (b) Fluid, asymmetric tubular membrane entrapping a polymeric or colloidal counterion,[2] for example, DNA. (c) Solid tubule made of chiral amphiphiles.[2] Twisted ribbons with a uniform pitch may be formed at first, which then are closed to vesicular tubules.

Inorganic colloidal particles, which are coated by lipid membranes, on the other hand, do not form such compact lumps. They rather stick together as dendrites or fractal assemblies (Figure 5).[7] This is the same behavior which is sometimes observed for noncoated gold particles[8] and similar inorganic colloids.[9] These hard particles assemble by nonreversible, diffusion-controlled aggregation and tend to form much more extended structures.

11.2.2 The Hydrophobic Effect and Curvature

As shown in the micelle structure in Figure 2, the hydrophobic effect forces the amphiphilic molecules together in spherical clusters or planar bilayers in order to minimize the interface between water and the hydrophobic chains. The latter occur mainly in the all-*trans* configuration,[10] but they cannot crystallize because hydration and steric forces keep the head groups of the amphiphiles apart.[11] The environment of each molecule is different and rapidly changing.[3] Nevertheless, the aggregation number and overall spherical shape are relatively well defined in short time intervals and a statistical curvature $1/r$ can be defined, which is close to the reciprocal lengths of the amphiphilic molecules.[12] The micelle has the highest possible surface area for a bilayer assembly and remains in this state of low aggregation number because the micelle disintegrates

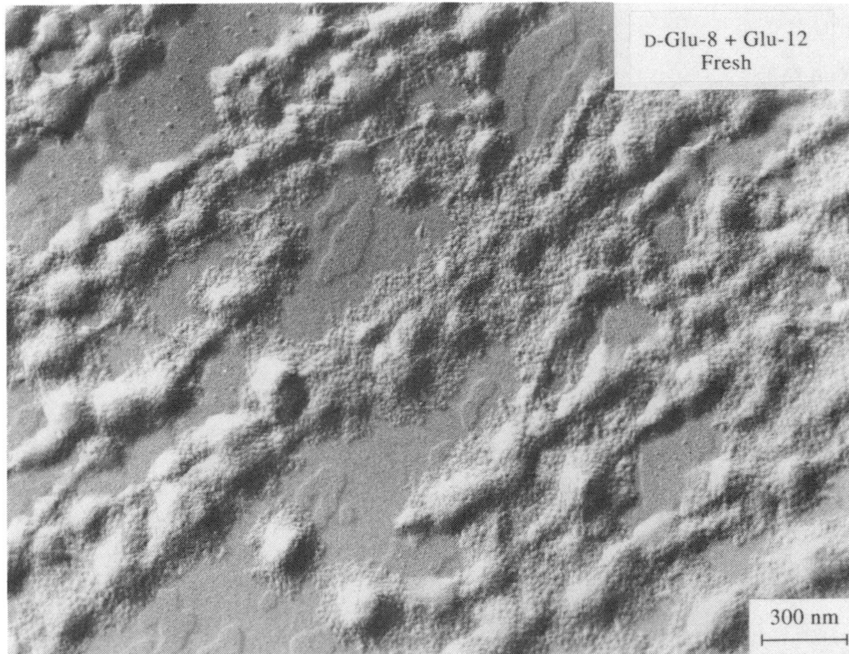

Figure 4 Electron micrograph of a clear aqueous gel containing disordered micellar lumps made of a 1:1 mixture of *N*-octyl- and *N*-dodecyl-D-gluconamides,[5] which form solid helices as pure compounds (see Figure 19).

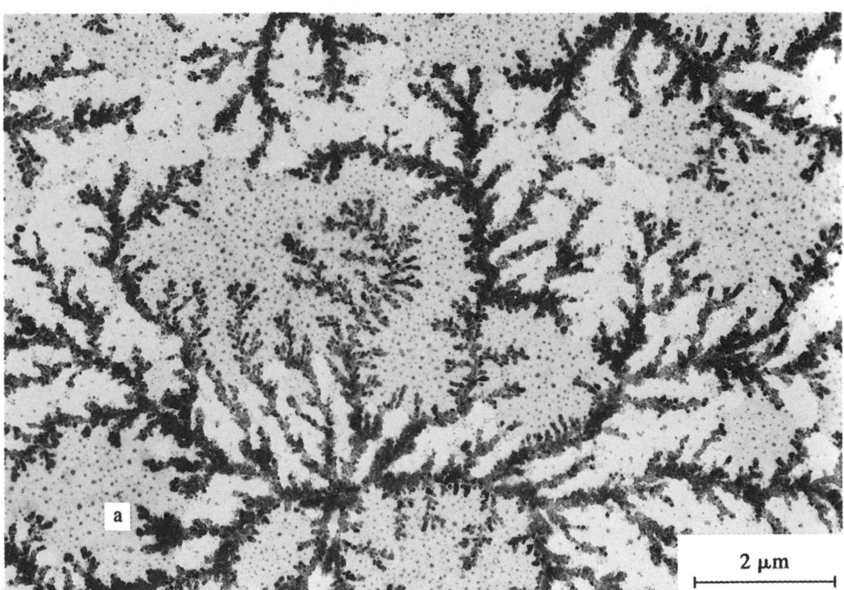

Figure 5 Electron micrograph of a membrane-coated manganite colloid.[7] The fractal pattern only appears upon evaporation of water on the electron microscope grid, but not in cryomicroscopic preparations. Note the lack of small particles at the central parts of the aggregate. This is typical for diffusion-controlled aggregation which is characterized by irreversible adsorption of small particles to the growing branches.

into monomers and smaller aggregates which then rapidly reassemble. Typical half-life times of micelles are between milliseconds for charged amphiphiles and seconds for electroneutral amphiphiles.[13a]

Less soluble amphiphiles ($< 10^{-5}$ M) tend to form planar molecular bilayers in water. They can be transformed into spherical vesicles with minimal radii of about 10–20 nm.[13] Both the planar and spherical bilayer membranes may be indefinitely stable, because undulation disturbs membrane–membrane interactions and fusion.[14] Hydration forces[11] prevent their precipitation as three-dimensional crystals. In the case of bolaamphiphile assemblies,[15,16] which are amphiphiles with a head group on each end of a hydrophobic skeleton, one may not only find uniform curvature, an inherent consequence of the tendency to form the smallest possible membrane surface, but a totally asymmetric distribution of different head groups can also be enforced.[17–19] Large head groups tend to be on the larger outer surface, the smaller head groups are then located at the inner surface (Figure 6(a)).[17,18] One can also play with counterions. The less soluble head group–counterion pair will often occur preferably at the inner surface (Figure 6(b)).[20]

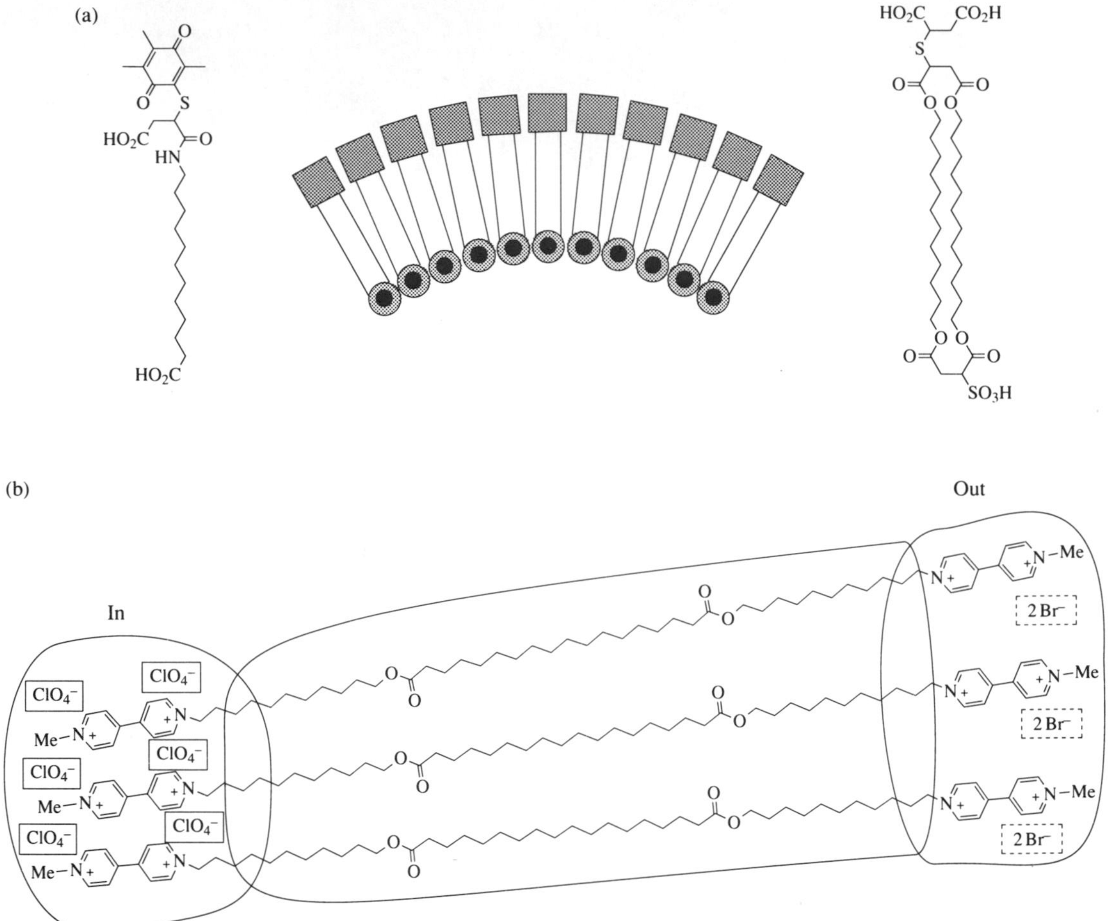

Figure 6 In monolayered vesicular assemblies, large head groups can be localized[17] exclusively on the outer (a), and less soluble ammonium salts[20] on the inner (b) membrane surface. In tubular vesicles, such an asymmetric membrane has not been characterized to date.

The hydrophobic effect can thus not only be used to form curved molecular membrane structures, to produce an organic medium in water, and to separate inner and outer water volumes, but it also allows the separation of large and small head groups within one molecular monolayer structure. This is a very simple, membrane specific stereoselective separation effect. Spherical micelles and vesicles are, however, only formed in defined concentration ranges and under low salt conditions. If one strongly reduces the hydration forces acting on the head groups, the spheres assemble to form extended threads or tubules (see Section 11.4).

11.2.3 Amide Hydrogen Bond Chains and Fiber Formation

The intermolecular interaction between the relatively acidic hydrogen atom and the strongly basic oxygen atom of a secondary amide group leads to the strongest hydrogen bond in supramolecular assemblies. Its main effect is to convert spherical structures to fibrous structures; the linearity of hydrogen bond chains overcome repulsive hydration energies and lead to linearly fixated arrays. The effect has been thoroughly studied in detail by x-ray crystallography in protein main chains.[21] The main results are summarized here and will be applied in Sections 11.3–11.5.

Main chain–main chain hydrogen bonding, NH···O=C, in α-helices of proteins are uniform and well-defined. The in-plane angle component γ (Figure 7) is usually close to 18°, the C=O···H out-of-plane angle 147° instead of the 60° and 120° angles expected for an sp^2 geometry around carbonyl oxygen atoms. The distribution of hydrogen bond lengths is much broader in proteins than that of bond angles and lies in the range 170–250 pm (see Figures 2(c) and 15). The hydrogen bond is thus distorted in the α-helices from equilibrium values observed in crystal structures close to 206 pm in small molecules. This is consistent with the assumption of softness of the hydrogen bonds. Irregular main chain–side chain and main chain–water hydrogen bonds do not usually change the geometry of the main chain–main chain interactions very much, although side chains tend to fully satisfy their hydrogen-bonding acceptor and donor potentials by extensive side chain–side chain interactions. In these interactions, one finds considerable angular spread, whereas the main chain hydrogen bonds remain uncharged. One may therefore classify the main-chain interactions as "supramolecular," the side-chain interactions as "flexible" and "solution-like."

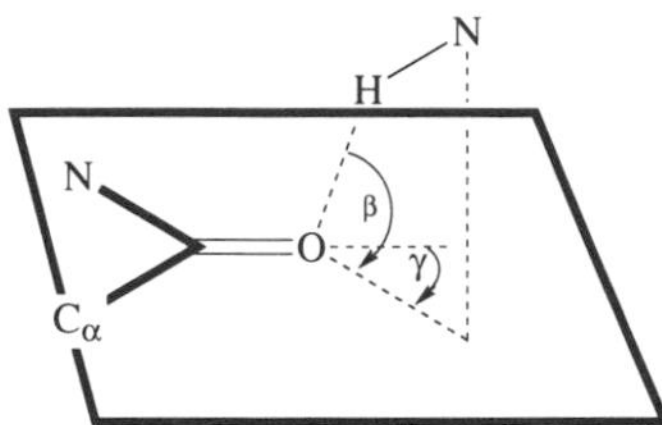

Figure 7 Out-of-plane (β) and in-plane (γ) amide hydrogen bonds[21] may stabilize supramolecular helical structures in the plane of micellar disks and at an angle β in a micellar helix (see Figure 19 for an example).

β-Pleated sheet structures are even less flexible than α-helices and much more regular. The C=O···H angle is close to 120° (±10°) and the NH···O bond length 210 pm (±10 pm).

Thermodynamics indicate that the formation of amide–amide hydrogen bonds in water is largely entropy driven.[22] Three to four water molecules are released from solvated amides. In hydrophobic environments, however, the driving force comes directly from the negative free energy change of amide–amide hydrogen bond formation. It is therefore equally favorable in both hydrophilic and hydrophobic environments ($\Delta G = -28 \pm 4\,\text{kJ mol}^{-1}$) and can occur on membrane surfaces as well as in its interior.

Bending of protein fibers, whose appearances are fixed by hydrogen bond chains, occurs if hydrophobic amino acids can be buried inside the helix and hydrophilic amino acids dominate on the outside. Hydrogen bonds in the hydrophobic face are 10–20 pm shorter and more linear than those in the hydrophilic face. This difference causes "superhelical" bending of the α-helices.[23]

In amphiphiles with carbohydrate head groups, the stereoselective formation of hydrogen bonds between hydroxyl groups is also of importance for the determination of molecular conformations and supramolecular structures (see Section 11.4.4). If only one chiral center containing a hydroxyl group is present (Figure 2(c)), only dimers can be formed within a molecular assembly. With at least two hydroxyl groups (see tartaric acid model in Figure 15), one can form chiral chain structures such as helical rods or twisted ribbons with continuous hydrogen bonded chains in aqueous media.

11.2.4 Acid–Base and Counterion Effects

Similar hydrogen donor–acceptor chains are probably involved in half protonated carboxylate or phosphate assemblies ($CO_2H\cdots{}^{-}O_2C$ and $OPO_3H\cdots{}^{-}O_3PO$-), but are less well defined by x-ray crystal structures than the amide hydrogen bond in proteins. The two hydrogen bond chains, however, can be manipulated extensively by a change of pH. The amide hydrogen bonds discussed above only respond to temperature changes and, often difficult to control, addition of organic solvents.

The classical acid–base equilibrium involves the conversion of micelles made of fatty acid salts at pH ≥ 7 to vesicles made of protonated fatty acids at pH ≤ 4.[24] In the case of α,ω-dicarboxylate (**1**) the salt is water soluble in monomeric form at pH > 8, and spontaneously forms vesicles at pH 4. At lower pH, the vesicle's surfaces become electroneutral and precipitation occurs.[18] One-sided salt formation with bivalent metal ions does not lead to spontaneous vesiculation but to ill-defined precipitates. The same experiment with the bis(bipyridinium bolaamphiphile) (**2**) and iodide or perchlorate ions on one side and bromide on the other is, however, successful. Quantitative vesiculation is observed upon addition of 1–2 equimol of these ions, which enforce close contact of half of the bipyridinium head groups and thus enforce curvature (see Figure 6).[20]

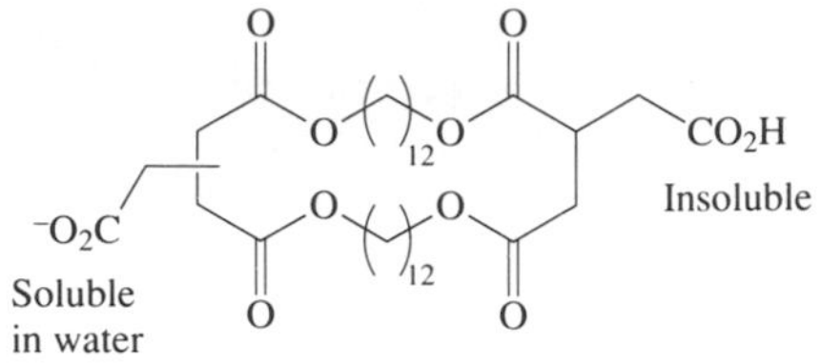

(**1**)

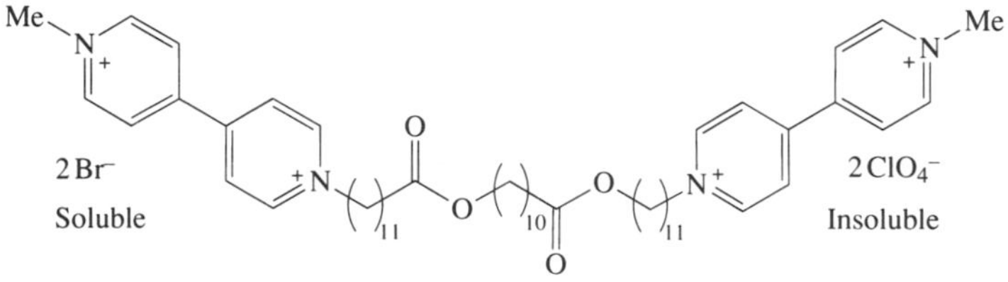

(**2**)

One may also use polymeric counterions in order to stabilize vesicles either on the inside ("polymer skeleton") or the outer surface ("vesicle in a net"). The "vesicles in a net"[25] have surface properties which largely correspond to the properties of the polyelectrolyte, for example, they become less vulnerable to precipitation by salts.

11.2.5 Typical Synkinons for Noncovalent Fibers

Synkinons for the formation of fibers may have a charged head group (carboxylate, phosphate, sulfonate, ammonium), but they must be half-neutralized or be connected to tightly bound counterions. They may be linearly connected in chains by hydrogen bonds at pH values close to the pK_a value or by appropriate organic counterions. Formulas (**1**) and (**2**) as well as Table 1 show a few synkinon ion pairs which have been used successfully.

Even more successful are synkinons with strongly hydrogen bonding nucleotide head groups or amino acid head groups (Table 2). One may also add small chromophores into the hydrophobic skeleton. Phenols and azobenzene derivatives are popular examples.[26,27] Successful fiber formation is, however, almost guaranteed if accessible secondary amide groups are introduced into carbohydrate-based amphiphiles.[28] This helps in the case of single-headed amphiphiles as well as in bolaamphiphiles with either one or two hydrocarbon chains.

Fluorinated chains, which are very helpful in the formation of domains in vesicle membranes[19] as well as in the synthesis of unsymmetrical membranes have, to the best of our knowledge, not been applied in fiber synthesis.

The best way to convert fibrous aggregates to polymers of very high molecular weight ($>10^6$) is to use dialkynic units at appropriate positions. Table 3 shows an example of such an amphiphile which has been successfully polymerized.[29–31] Similar dialkynic amphiphiles, which form very similar supramolecular assemblies, however, may not react at all under UV light.[30,31] The stereochemical differences between reactive and inert fibers are not yet known, but have been clarified in three-dimensional crystals.[32] Hydrosulfides can be reversibly polymerized by oxidation with peroxide,[33] but have so far not been used in fibers. Vinyl esters, on the other hand, produce small polymers in vesicles, but very high molecular weight polymers in fibers.[34]

Table 1 Charged amphiphiles which form rods and tubules.

NMe_3^+ CO_2^- OH **(3)** Cetyltrimethylammonium salicylate produces fluid, viscoelastic fibers in dilute aqueous solutions
CO_2^- OH $^+MeH_2N$ **(4)** Ephedrinium myristate produces fluid fibers in bulk water, which dissolve metalloporphyrins
O O^- N H O Na^+ or K^+ or NH_4^+ **(5)** The secondary amide group in tartaric acid dodecylamides enforces the formation of solid fibers. Depending on the counterion rods or tubules are formed

Table 2 Chiral amphilphiles which form rods and tubules.

O O O $)_{12}$ $)_{11}$ O NH_2 N O N O^- O P O O Na^+ O HO H **(6)** Glyceronucleotides form solid twisted ribbons
O N H $)_{11}$ O $)_{10}$ $N(Me)_3^+$ $)_2$ O O $)_{11}$ O Br^- **(7)** Glutamic diester monoamide amphiphiles form solid helices of light microscopic dimensions and twisted ribbons of molecular dimensions
OH OH H N HO OH OH O **(8)** Gluconamides produce solid quadruple helices of bimolecular thickness

Table 3 Polymerizable amphilphiles which form rods and tubules.

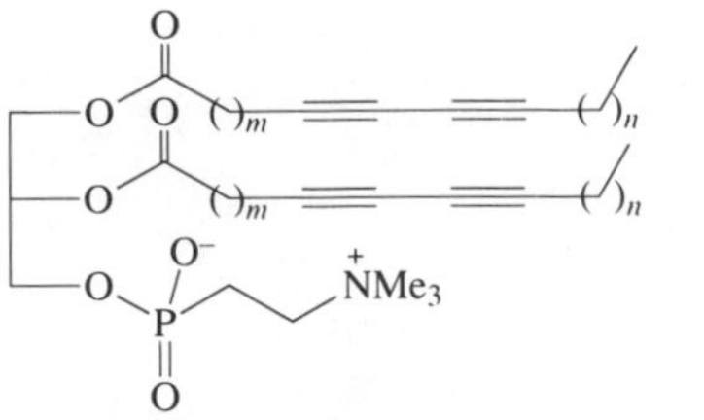

(**9**)
Dialkynic phospholipids polymerize to give stable tubules.

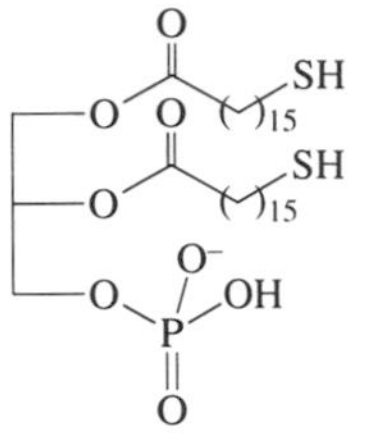

(**10**)
Dihydrosulfides polymerize reversibly on oxidation. So far only demonstrated in vesicle membranes.

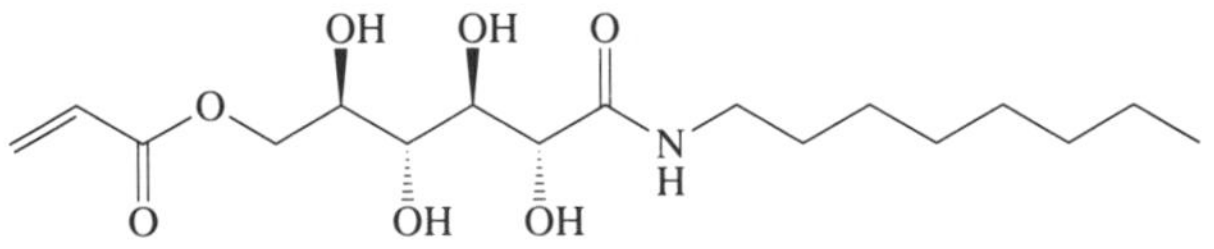

(**11**)
Terminal vinyl group polymerization leads to high molecular weight fibers. Linear chains parallel to the axis are probably formed.

11.3 FLUID MICELLAR THREADS AND VESICULAR TUBULES

Fluid fibers are formed from amphiphiles with weak interactions between the head groups. They cannot be considered as well-organized, supramolecular assemblies. However, they provide the unique possibility for dissolving reactive dyes in long-lived molecular bilayers and combining them with counterions that are tightly bound on the bilayer surface. Furthermore, liquid bilayer fibers often form transparent gels, which help in spectroscopic investigations and which are indefinitely stable. Last but not least, wax tubules on hydrophobic surfaces play an important role in the protection of tiny pore structures on the hydrophobic surfaces of plant tissues.

11.3.1 Viscoelastic Micellar Threads

The longest-lived fibers in aqueous media are made from cetyltrimethylammonium bromide (CTAB) and equimolar amounts of salicylic acid ((**3**), see Table 1). These fibers are formed on simple mixing of the components at room temperature and have a diameter of 5 nm in electron micrographs. They readily aggregate to form fibers with a uniform diameter of 12 nm and a length:diameter ratio of $>10^4$ (Figure 8). It might well be that the latter fiber is made of three or four bilayer threads.[35,36] Salicylic acid can be replaced by *o*- and *p*-iodophenol or *p*-toluic acid in a 1:1 molar ratio with respect to CTAB and at concentrations between 10^{-3} M and 10^{-2} M. Other hydroxybenzoic acids and diphenols do not induce fiber formation, but the salting out effect of 1 M sodium chloride has the same effect.[37,38]

The slightly viscous solutions show the amusing effect of viscoelasticity: air bubbles in a rotating solution change their sense of rotation upon a stop. They seem to bounce against a rubber-like wall of fibers.[37a]

The liquid CTAB fibers are very good solvents for small molecules, for example, cyclohexane,[37] but they do not dissolve metalloporphyrins.[38] Here, a similar viscoelastic system made from

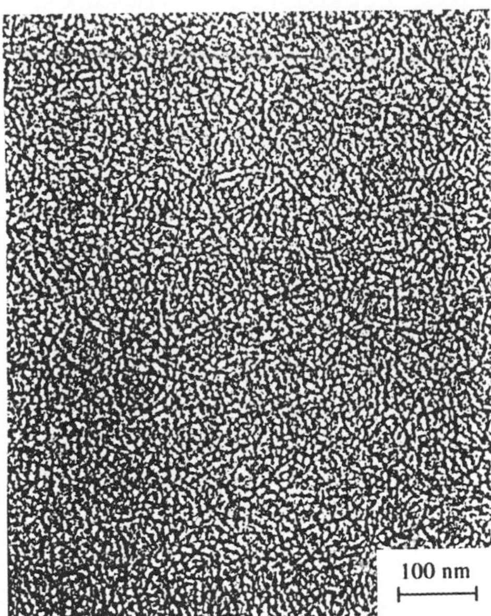

Figure 8 Cryoelectron micrograph of typical micellar threads made from ionic single-chain amphiphiles and organic counterions. Such threads appear at relatively low amphiphile concentrations (0.5–1% w/w), produce slightly viscous, clear solutions which may or may not be viscoelastic.[36]

myristic acid and ephedrine counterions (see Table 1) is much more potent.[38] This mixture dissolves up to 5% of hydrophobic magnesium octaethylporphyrin and this solution produces a circular-dichroic effect of medium intensity. The magnesium porphyrinate is dissolved in the form of single molecules or small aggregates and may carry the ephedrine base as axial ligand. Zinc octaethylporphyrin does not dissolve. The arrangement of amphiphiles and counterions is not known. In the case of ephedrine and myristic acid mixtures, one sees an optimal resolution of the aromatic proton signals in the NMR spectra at a molar ratio of 1:1, much less resolution at 2:1, and a broad line at 1:4. Similar sharpening effects are observed for the methyl protons at the chiral center. CTAB–salicylic acid mixtures also show the aromatic proton signals to shift to higher field ($\approx$0.2 ppm) if one goes from an aqueous solution of the acid to a micellar solution in CTAB and broaden when the latter assemble to fibers.[39]

Additions of small quantities of poly(vinyl methylether) or poly(propylene oxide) completely eliminate the viscoelasticity and, presumably, the fibers.[40] More polar polymers such as poly(ethylene oxide) or poly(vinylpyrrolidone) have no effect. The hydrophobic polymers may wrap around the fibers, reduce the unfavorable core water contact in reef-like micelles and therefore stabilize the smaller spherical micelles. The hydrophilic polymers, on the other hand, remain dissolved in water and are not enriched on the surface or within the fibers.

11.3.2 Lecithin Tubules

Fusion of unilamellar phosphatidylserine vesicles in the presence of calcium ions creates lamellae which roll up spontaneously to form multilayered short cylinders. Calcium ions act in two ways: they crystallize the individual lamellae and induce tight folding. The edges of the cylinders may then provide nucleation sites for further growth of the cylinders.[41] The addition of edta to the cylinders restores the negative charge of the bilayers and the fluidity. Large, unilamellar vesicles are formed (Figure 9).

The same calcium-ion induced precipitation of acidic phospholipids, for example, **(12)** may also lead to double helices of light microscopic dimensions (Figure 10). This was achieved with a 1:2 mixture of cardiolipin and dimyristoyl phosphatidylcholine. Again the driving force for helix formation is the calcium-mediated membrane–membrane binding and on higher calcium ion concentrations the helices rapidly collapse on themselves.[42]

Similar changes in the morphology of a dimyristoyl phosphatidylcholine (DMPC **(13)**) or lecithin tubule are observed in the swelling processes of lecithin crystals in water. In the first step, rods of 20–40 μm diameter ($\sim$250) molecular bilayers grow into the medium and fold together (Figure

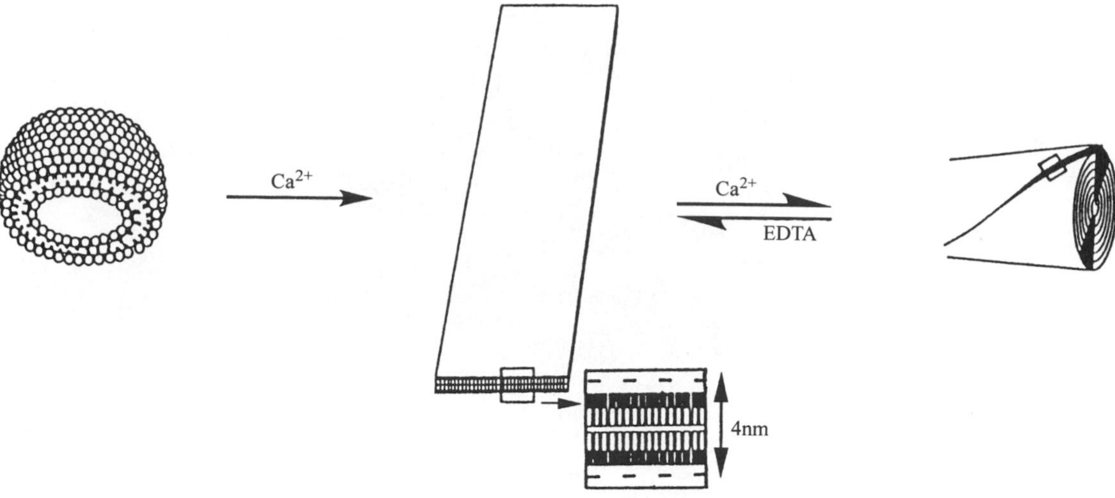

Figure 9 Negatively charged vesicle surfaces fuse to form large bilayer sheets upon addition of calcium ions. These neutralized sheets roll up to form scrolls. The latter process is reversible upon addition of edta, which removes the calcium ions and charges up the membrane surface.[40]

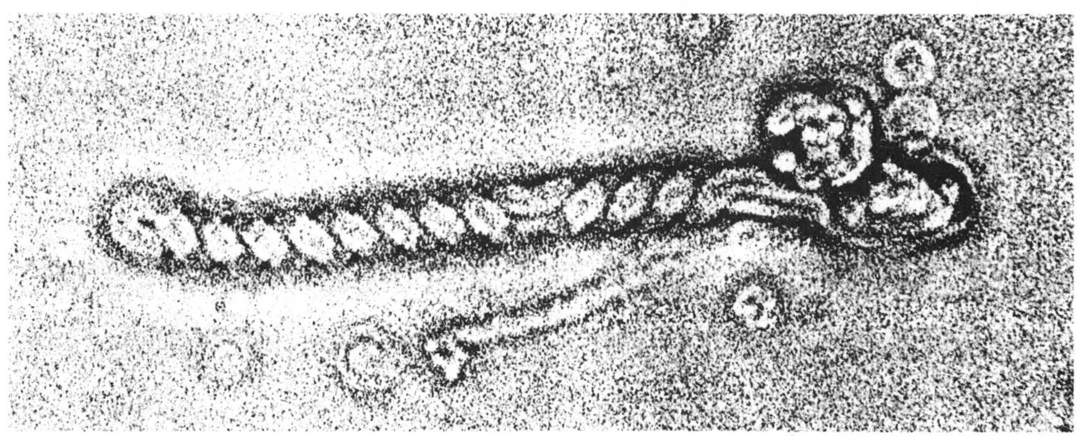

Figure 10 Helices are formed from linear bilayer tubes upon addition of calcium ions. The thermodynamic driving force for helix formation is the calcium ion mediated membrane–membrane binding. Left- (M) and right- (P) handed helices are formed in a statistical distribution.[41] If an M- and P-helix meet, they may annihilate each other, which is exemplified in the center of the micrograph.

11(a)). In the second step, several helical and coiling forms appear (Figure 11(b)). It is typical for these fluid assemblies that they continuously change their shape and that the surface tends to become smaller and smaller.[43] Therefore, such assemblies are difficult to control and should not be applied in the synkineses of complex structures.

However, it is also possible to split the growing lecithin tubules into small vesicle components by the addition of amphiphilic polymers, for example, dextran with hydrocarbon side chains (Figure 12). These polymers permanently dissolve only in the outer part of hydrophobic bilayer and thus introduce curvature.[44] This behavior can be regarded as a counterpart for the action of calcium ions described above and is analogous to the "budding of vesicles" in the presence of sodium acetate, where acetate only binds to the outer layers.[45] The techniques to use spacers and counter-ions in order to change curvature are very important tools in the synkineses of curved supramolecular assemblies on molecular bilayers.

Longer-lived vesicular tubules have been obtained from synthetic diacetylenic lecithins such as 1,2-bis(10,12-tricosadiyoyl)-*sn*-glycero-3-phosphocholine[46,47] (**9**) (see Table 3) in methanol–water. Electron micrographs show short tubules and twisted ribbons. The helical structures produced by precipitation at high solvent concentrations could be rendered permanent by polymerization under UV light. Water-free tubules have also been obtained in acetonitrile. These tubules are typically

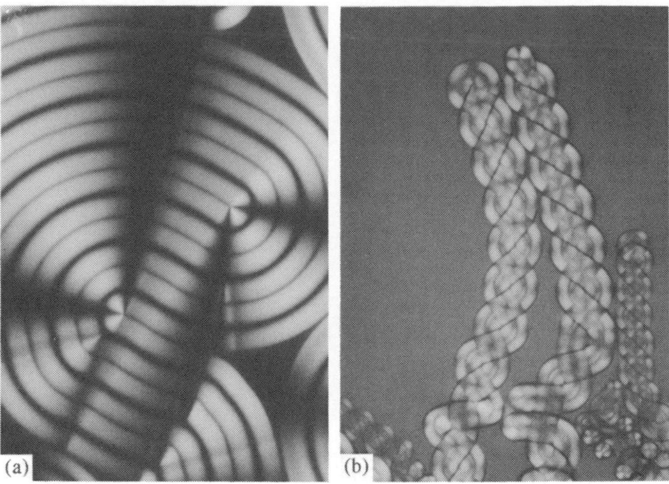

Figure 11 Swelling of crystals of insoluble amphiphiles in water produces myelin figures of multiple bilayers (a), which often separate into single, short-lived helices (b).[42] Such swelling or hydration processes do not lead to stable supramolecular structures. The helices eventually formed are even shorter-lived than those of the calcium-bridged system in Figure 10 (courtesy of Professor T. Sakurai).

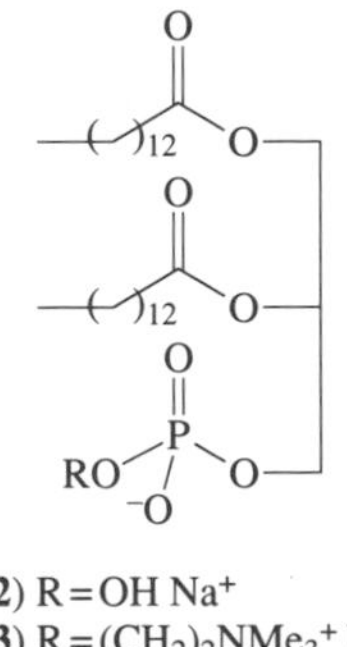

(**12**) $R = OH\ Na^+$
(**13**) $R = (CH_2)_2NMe_3^+\ Br^-$

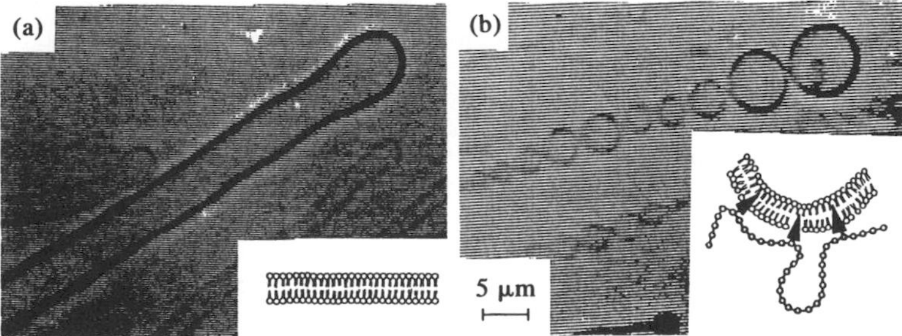

Figure 12 Polymers with hydrophobic side chains separate liquid bilayer fibers into vesicles. Monomeric hydrocarbons do not have such an effect because they distribute rapidly in both the inner and outer layer of the fibers (courtesy of Professor H. Ringsdorf).

5–150 μm in length, with a 0.5 μm core surrounded by two to seven bilayers.[48] An interesting example of an organic–inorganic combination of ultrathin layers on a molecular scale has been prepared by the electroless plating of polymerized multilayer tubules with nickel metal.[49]

The polymerizable *L*-configured (**9**) with a chiral center at C-2 of glycerol, formed right-handed helices and tubules,[48–50] whereas the racemate gave tubules as well as right- and left-handed helices in equal abundance. In no case, however, was a regular pitch observed. It was the alkynic groups which introduced some order into the bilayer and which even induced a partial splitting of the racemate. The lack of uniform helices, however, showed that there was certainly no defined binding interaction between the monomers in the sense of one- or two-dimensional crystal formation.

It was also shown that racemic and pure enantiomeric phospholipids were functionally indistinguishable.[51]

11.3.3 Wax Tubules

Fluid lipid tubules are also observed on the surface of natural wax deposits, for example, on pine needles or barley leaves. From transmission electron microscope (TEM) examination of the replica, the exact dimensions of such tubules could be determined in the case of *Pseudotsuga menziesii* as 2.3 µm in length, 136 nm outer diameter, and 40 nm wall thickness. The surface of the tubules had a striated appearance with the striae oriented helically.[52] Structures with similar habits could be produced *in vitro* when chloroform extracts of needles were subjected to slow evaporation.[53] As to the chemical composition of the wax tubules, the most uniform materials were found on the surface of *Picea abies* and other conifer needles (Figure 13(a)). Here they consisted exclusively of the chiral alcohol 10-nonacosanol **(14)**[54] which also formed tubules upon crystallization.[55,56] 10-Nonacosanol was chiral, but the enantiomeric purity in natural materials could not be proven by c.d.[57] or by ^{1}H NMR[56] spectroscopies, because the C_{18} and C_9 alkyl chains were too similar. A pure (*R*) enantiomer **(14)** (*ee* > 90%) and the racemate were therefore synthesized and it was found that only the pure enantiomer formed tubules from chloroform solutions (Figure 13(b)) (see also Section 11.6) and the racemate precipitated as crystals.[56] All of these tubules were formed from semisolid plant waxes or apolar organic solvents. The hydrogen bridges between the hydroxyl groups should be responsible for curvature, because there was no solvophobic effect involved in the alkyl chains.

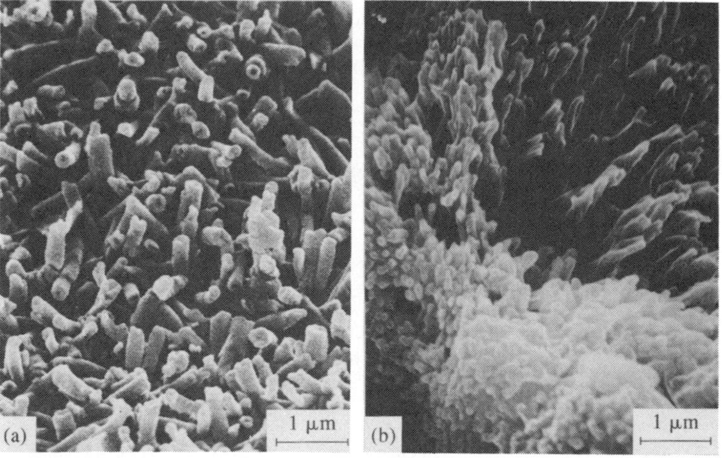

Figure 13 The best-known examples of liquid tubules in nature occur in pine needle pore surfaces. Such tubules (a) are made of 10-eicosanol and are also formed from chloroform solution (b), if the compound is a pure enantiomer, for example, the natural product or synthetic (*R*)-eicosanol. The racemate only gives platelets upon crystallization. This is the most simple example of the chiral bilayer effect (see Section 11.6) (the scanning electromicrograph of the natural tubules was kindly provided by Professor Riederer).

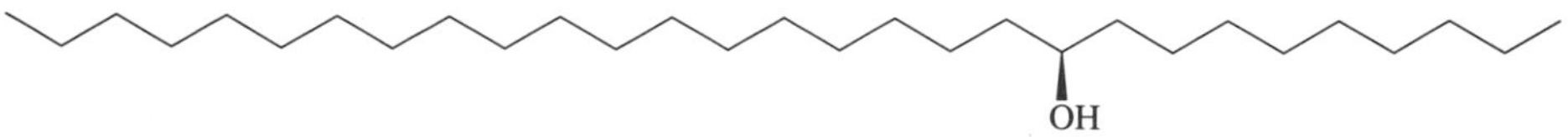

(14)(*R*)

A chiral hydrophobic polypeptide, namely poly(γ-benzyl-L(and-D)-glutamate), has also been studied extensively in organic solvents. Helical fibers have been obtained from DMF by precipitation with propionic acid or butoxyethanol.[58] From bad solvents, rings were predominantly precipitated, better solvents gave helical strands. Playing with solvent mixtures is another important tool in synkinesis. Small amphiphiles usually produce smaller assemblies in good solvents, for example, micellar instead of vesicular fibers, whereas with polymers often the opposite is observed, for example, long fibers instead of rings.

11.3.4 Scheibe (Jelley) Aggregates

Heterocyclic or polyene dyes containing electron-poor and electron-rich regions often self-organize to long fibers in organic solvents or water. The most simple case is 1,1′-diethyl-2,2′-pyranine or pseudoisocyanine (**15**) which is monomeric and nonfluorescent in ethanol but in water above 10^{-3} M forms a fluorescent polymer. The dye molecules are presumably ordered laterally (*J*-aggregate) with a slip angle $\alpha \ll 54°$ and produce in addition to fluorescence a sharp, long-wavelength band.[59,60] The fluorescence of a whole fiber with typical aggregation numbers of 10^4–10^5 and molecular masses $>10^6$ Da could be drastically reduced with a single quencher molecule.[62] The Scheibe aggregates could also be combined with helical polymers, for example, DNA,[61] or chiral counterions, for example, tartrate[63] to form CD-active fibers. Regular stirring without any chiral additives was also reported to produce CD effects which depended on the sense of stirring.[64] Impressive electron micrographs of Scheibe aggregates have been obtained in the case of 3,3′-bis(β-carboxyethyl)-5,5′-dichloro-9-methyl-thiacarbocyanine (**16**). Fibers of 2.8 nm diameter aligned to form rods which were several micrometers long.[65] The driving force for the formation of such lateral dye assemblies is very likely an electron donor–acceptor interaction in the ground state (Figure 14).[66] The origin of the sharp, long-wavelength band and fluorescence is a coupling of the transition dipole moments (exciton model) at a fixed angle $\ll 54°$.[67] Planar Scheibe aggregates have also been realized on a silver surface and were characterized by scanning tunneling electron microscopy[68] and are extensively used as spectral sensitizers in silver halide photoimaging systems.[69]

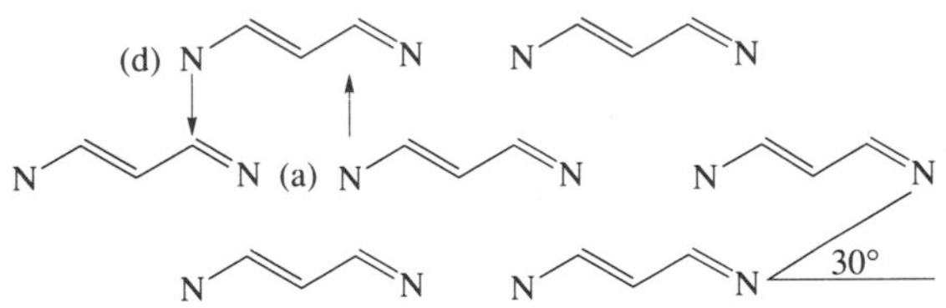

Figure 14 Some heterocyclic dyes with an electron-donating enamine nitrogen atom at one end and an electron-accepting immonium nitrogen atom at the other end of a short polyene chain form very long fibers (Scheibe aggregates) in water or polar organic solvents. Exciton theory requires a tilted stacking where $\alpha \ll 54°$.

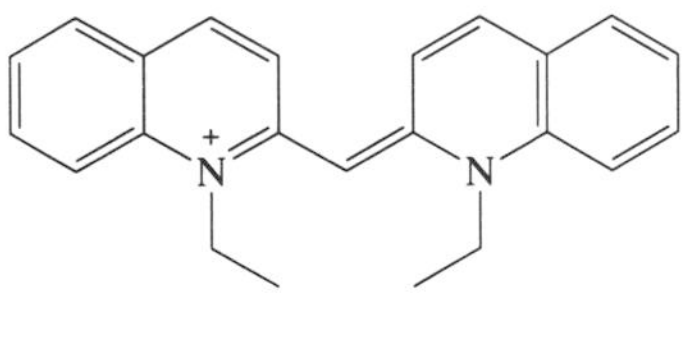

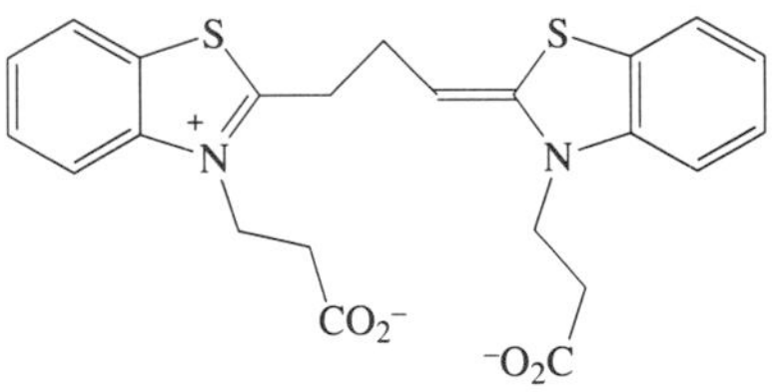

11.4 SOLID MICELLAR RODS AND RIBBONS

In the following sections, micellar rods and ribbons, which are supposed to consist of molecular mono- or bilayers without an entrapped water volume, are distinguished from vesicular tubules (Section 11.5), in which a mono- or bilayer contains an internal water cylinder. This distinction usually comes from the inspection of electron micrographs and should not be taken too strictly. In some of the micellar fibers, a void with diameters of about 1 nm or less is probably also present, although it has not been detected on electron micrographs because of low resolution or inaccessibility to staining.

All of the membrane structures discussed in the following sections depend on a combination of the hydrophobic effect, which induces curvature, directed binding interactions between the head groups, which rearrange micellar and vesicular spheres to linear fibers, and hydration forces, which prevent the formation of three-dimensional crystals. The hydrogen bond patterns within a hydrated fiber may not be as regular and fixed as in three-dimensional crystals, but head group conformations are usually uniform (Figure 15).

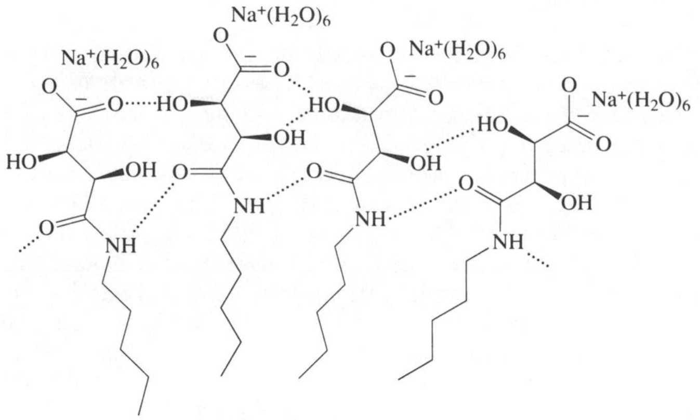

Figure 15 Solid micellar threads and vesicular tubules appear as stiff rods and are characterized by strong binding interactions between the head groups and linear conformations of the hydrophobic chains in the core. The hypothetical model given corresponds to micellar or vesicular fibers (see Figures 28 and 38). The model indicates various possible hydrogen bonds within the hydrated fiber surface. They may be broken and rearrange, but the overall molecular and supramolecular structures remain stable or "solid."

The binding interactions between the head groups not only change the morphology of lipid membranes, but they also lead to a completely ordered arrangement of the alkyl chains. It is not possible to use these lipid membranes as solvents for dyes and edge-amphiphilic, pore-forming domains any more. Solutes have to be co-crystallized with the membrane components.

11.4.1 Chiral Fatty Acid Fibers

Section 11.3 introduced two examples of organogels: the fibers and curvature were not produced by the hydrophobic effect, but by weak interactions between functional groups at a chiral carbon center. These tubules are characterized as "fluid," because no well-defined repetitive units were detectable on electron micrographs.[63] The first micellar fibers which showed such well-defined patterns were described by Tachibana and co-workers[70–2] many years ago and come from chiral hydroxy fatty acid metal salts in organic solvents. These "solid" fibers were produced by dissolving the soap in a solvent at high temperature and then by cooling the hot solution to room temperature.[70] Helical ribbons of 20–30 nm width were the smallest fiber diameters obtained. The usual solvents were ethanol, long-chain alcohols, nujol, and toluene. When (*R*)12-hydroxystearic acid (**17**) was used, potassium salts from hexanol with a right-handed twist were obtained, whereas the rubidium and caesium salt fibers were left-handed.[70] The racemic soaps always form non-twisted fibers, which is not in accord with an explanation of helicities by simple dislocations in whisker growth.[73] They should rather be determined by the chiral centers, although there is obviously a delicate interplay between the metallated carboxylate groups and the more or less solvated hydroxyl groups. There is, however, no chiral hydrogen bond chain possible in such monoalcohol fiber (compare Figures 2(c) and 15). Tachibana also showed that the fiber occurred on water surfaces in collapsed monolayers when the (*R*)enantiomer was applied, whereas only platelets were formed from the racemate.[72]

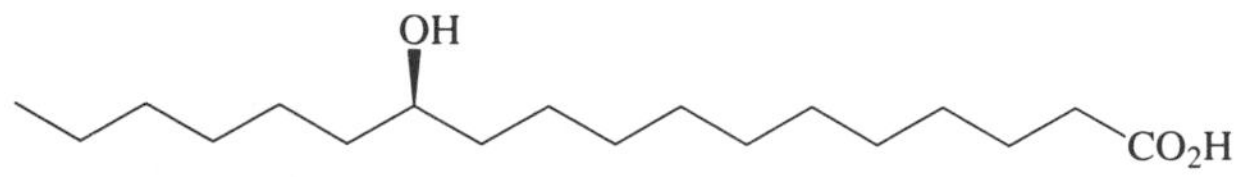

OH

CO_2H

(**17**)

Similar fibers, but usually without a twist, have been made from myristic acid in water. At a 1:1 ratio of myristic acid and sodium ions, platelets precipitate upon concentration of the solution on a microscope grid.[74] With a 2:1 excess of sodium ions, however, very long whiskers, so-called soap curds, are formed.[38] Soap fibers seem to be solid in organo gels and fluid in water.

11.4.2 Amino Acid Derivatives

The study of supramolecular assemblies made of single-chain acylamino acids started with the preparation of chiral liquid crystals.[75] *N*-Lauroyl-L-glutamic acid, for example, produced weak CD spectra in methanol ($\Theta = 6–8 \times 10^{3\circ}$ cm^2 dmol^{-1}) and water ($\Theta = 1 \times 10^{3\circ}$ cm^2 dmol^{-1}), which lost intensity on heating or addition of 8 M urea. The effects became much stronger when rigid biphenyl chromophores were integrated into the hydrophobic chains ($\Theta_{273\,nm} = 1.5 \times 10^{5\circ}$ cm^2 dmol^{-1}). Short fibers or disks were observed in electron micrographs, both for the pure L-enantiomer and the D,L-racemate.[76] Analogous but nonchiral alkylammonia amphiphiles with a urea unit in the center produced ill-defined bilayer lamellae with a sharp DSC melting point of 41°C.[77] Carbohydrate derived liquid crystals also show no defined ordering.[78]

Solid, twisted fibers with a width of a few hundred nanometers have been obtained from an amusing compound, namely (**18**), combining the chiral L-valine head group with a chiral 2-hydroxydodecyl side chain.[79] At this occasion, a "jump to conclusions" has been performed which is so typical for supramolecular chemistry. The authors stated that "...for the helical aggregates to be formed at the molecular level in the substituted amino acids it is essential that they contain both an asymmetric carbon atom and hydroxy group in the attached alkyl chain." This was in 1984, when several "essentials" for successful synkineses had already been formulated, such as rigid elements, spacers of the right size, internal charge neutralization, and so on. In 1995, one must say that there is no structural condition on the molecular scale for the formation of this or that supramolecular assembly. One has to take care of the right critical assembly concentrations (10^{-5}–10^{-2} M for micellar, $< 10^{-5}$ M for vesicular assemblies) and to provide functional groups and the right stereochemistry for strong intermolecular binding if one strives after uniform chiral structures. Everything else consists of the art of development of optimal preparation conditions and intelligent use of the chiral bilayer effect (see Section 11.7).

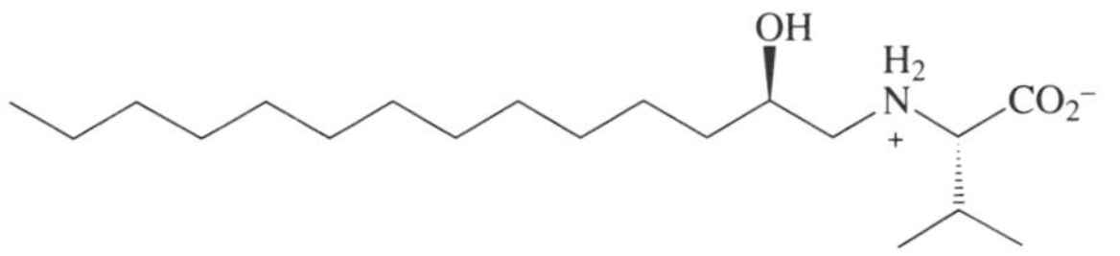

OH H_2 N + CO_2^-

(**18**)

Ribbons with 1–1.5 μm wide twists have been obtained from the biphenyl L-alanine derivative (**19**).[80] It showed a weak CD effect at 300 nm and a 45° rise angle, which corresponds to the minimum elastic energy of edge torsion and bilayer bending of a helical ribbon.[81] The diazo benzene analogue (**20**) gave straight fibrils at first, then the D-alanyl derivative slowly transformed into a right-handed helix and the L-enantiomer into a left-handed helix.[82] The racemate gave irregular fibers, but upon 4 h of incubation the appearance of left- and right-handed helices indicated some enantiomeric separation.[80]

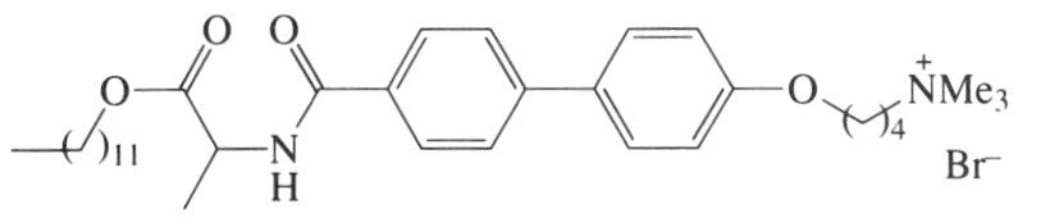

O O O ( )11 N H O ( )4 $\overset{+}{N}Me_3$ Br$^-$

(**19**)

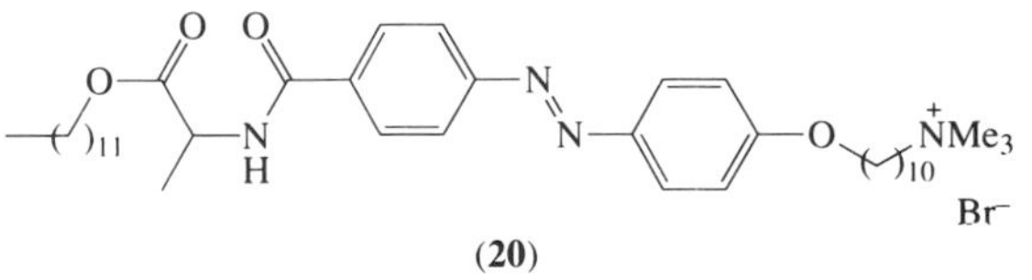

O O O ( )11 N H N N O ( )10 $\overset{+}{N}Me_3$ Br$^-$

(**20**)

Very long fibers with a uniform diameter of 12–20 nm in cryoelectron micrographs were found in aqueous gels made of *N*-myristoyl and *N*-palmitoyl-L-aspartic acid. The rheological properties of these gels corresponded to the characteristics of gelatinous linear polymer solutions and not to those of entangled micellar threads found in viscoelastic systems.[83] Furthermore, a regular helical pitch of about 65 nm was determined. It therefore seems to be clear that the fibrous aggregates are stabilized by directed binding interactions between the monomers. It is also characteristic that the corresponding glutamate amides did not form such long fibers under the given conditions, although they only contain one more methylene group (18 instead of 17) in the case of the palmitoyl derivative. Such odd–even effects are typical for supramolecular assemblies and one should always try both alternatives if one synkinetisizes supramolecular assemblies.

Another odd–even effect was demonstrated directly in absorption spectra. L-Alanine ester (**21**) was prepared with $n = 9$, 11 and 13 as the number of methylene groups and produced helical fibers with absorption maxima close to 320 nm.[26] If, however, n was even ($n = 10$, 12), a 350 nm band occurred indicating a more tilted stacking of the chromophores within the fibers (Figure 16). These assemblies thus resemble the Scheibe aggregates, which are often formed by a simple increase of monomer concentration. In the present case, it may be concluded that there was a stronger amide hydrogen bond in the even-numbered esters.

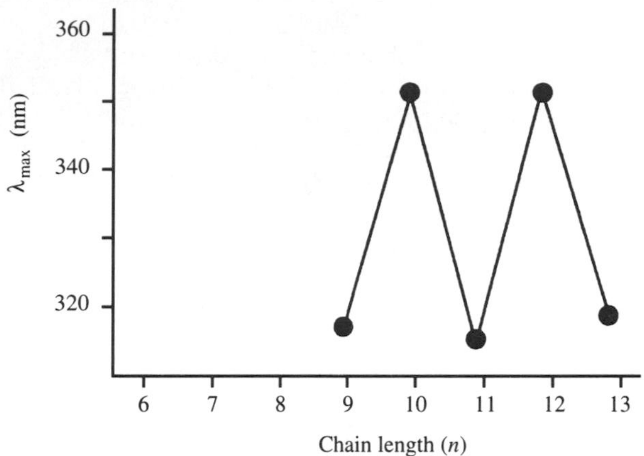

Figure 16 λ_{max} as a function of chain length of (**21**). When n is even, one observes a long wavelength J-band corresponding to a tilted stacking (see Figure 14) of azobenzene chromophores in fibers.[26] When n is odd H-aggregates or parallel stacking is found (see Figure 14).

(**21**)

Micellar fibers have also been described for the bolaamphiphile (**22**) containing an L-lysine and an amino end group connected via an amide bond. This compound was soluble at pH 5 and formed opaque gels at pH 10.5 containing single as well as clustered fibers with a uniform diameter of 2.5 nm. Two hydrogen bond chains (amide and amino-ammonium) along the rod axis and an arrangement of both head groups on two different cylinders was assumed (Figure 17). In this case, the major binding force should originate from the hydrogen bridges and not from the hydrophobic effect, because the ordered structure contains relatively few close contacts between the methylene groups.

11.4.3 Nucleotide Lipids

5′-Phosphatidylcytidine (DMPCt) (**23**) and the corresponding 2′-deoxycytidine contain two long alkyl chains and should therefore form tubules (see Section 11.5), but they do not. At first, the expected spherical vesicles were obtained by sonication at 50 °C for 45 nm, but they were transformed into circular helical strands after aging at 25 °C overnight rather than into tubules (Figure 18) at pH 8 when 0.2 M KCl was present.[85,86] The other nucleic bases in dimyristoyl-5′-phosphatidylnucleosides (DMPAd, DMPUr) usually formed less well-defined twisted ribbons.[87] The CD spectra showed in all cases strong effects between 200 nm and 300 nm and Θ values from 5 to $25 \times 10^{4\circ}$ $cm^2\,dmol^{-1}$.

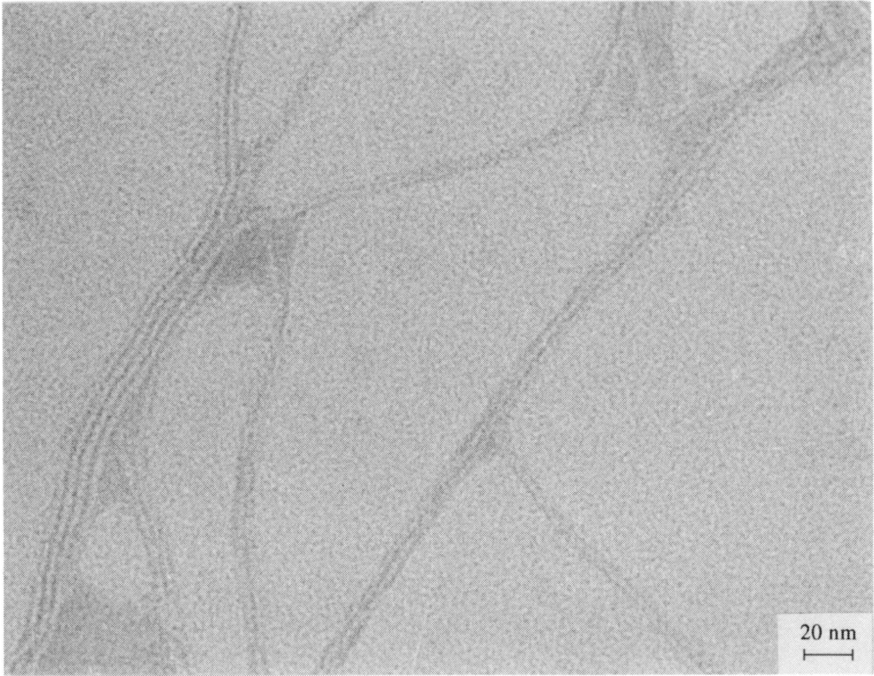

Figure 17 Monolayer lipid rods are formed from the bolaamphiphile (**22**) and are as thin as 2 nm.[84]

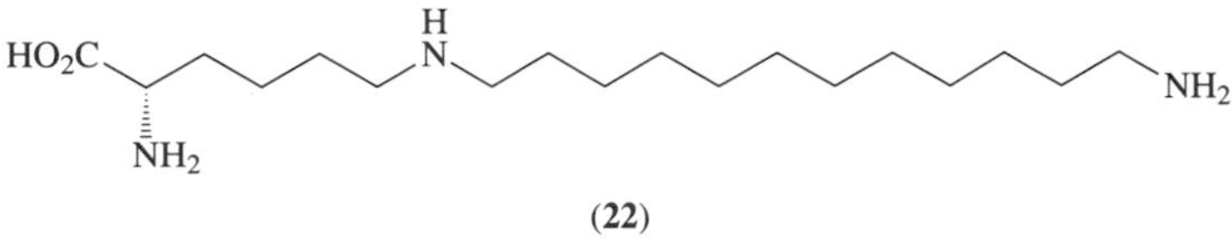

(**22**)

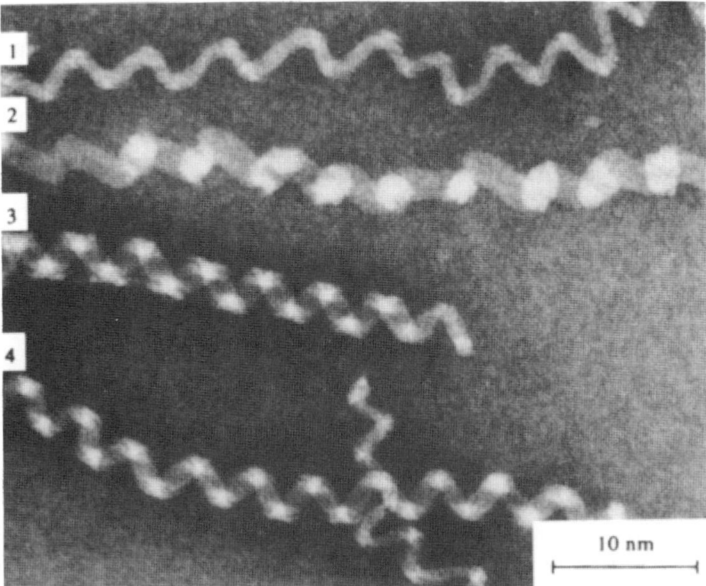

Figure 18 Double-chain amphiphiles with strongly interacting head groups often form twisted ribbons. A prominent example is the lipid nucleotide.[23] It forms vesicles upon sonification which slowly convert to the crystalline bilayer ribbons shown.[86]

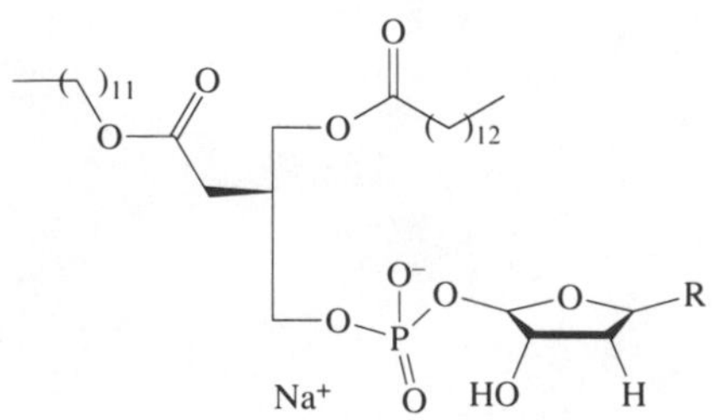

(23)

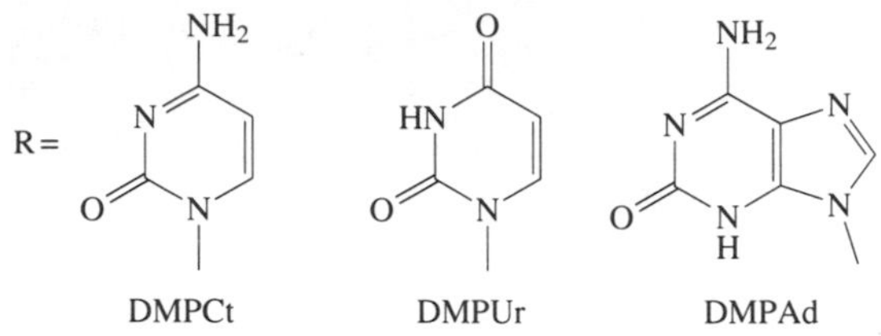

R = DMPCt DMPUr DMPAd

Since the nucleotide head group carries a negative charge on the phosphate ester the assembly structures in water also depend on pH values. DMPUr produced platelet structures in both alkaline and acidic solutions. DMPAd formed double helical structures at acidic pH which later assembled to cigar-like scrolls. In alkaline solutions, thick multihelical structures were observed, whereas DMPCt gave thin strands. They collapsed to large lamellae at acidic pH. All these observations correspond nicely to those made with the covalent homopolymers of the same nucleotides without hydrophobic side-chains.[88] A 1:1 mixture of DMPAd and DMPUr formed a new hybrid helical ribbon with a ~30 nm diameter and a helical pitch of 200 nm.[87]

11.4.4 Glyconamides

Most supramolecular assemblies are characterized by light and electron microscopies as well as melting curves obtained by DSC, and various methods measuring light scattering and macroscopic rheology measurements. All of these methods give hints on the molecular arrangements of the monomers within the fibers, but only the combination of solid-state CPMAS NMR spectroscopy and x-ray crystallography allows the determination of molecular conformations within supramolecular assemblies. This is, however, only applicable if the curved assembly structures can be obtained in a relatively dry state in order to allow spectroscopic measurements at the high spin rates used in magic angle spinning. Today, the lyophilization of intact fibers and their systematic structural characterization has only been realized with glyconamide fibers. Therefore, we discuss these fibers in more detail than the other assemblies.

11.4.4.1 Liquid crystals

Emil Fischer noted in 1911 the double melting point of some long-chain *n*-alkyl D-pyranosides.[89] This was recognized later as evidence for the formation of thermotropic liquid crystals without water. A sequence of crystal-to-crystal phase transitions often precedes another sequence of liquid crystal phases prior to the formation of a homogeneous molten liquid. The molecular clusters of ordered molecules are retained past the melting point, even though the three-dimensional order of the crystal is lost. It is usually the difference between hydrogen bonds and van der Waals cohesive forces which give rise to such thermomesophases. Thermotropic liquid crystal formation can be thought of as a two-stage melting process.

In the presence of water or other solvents, the difference in solubilites, on the other hand, give rise to lyomesophases or lyotropic liquid crystals. The molecular clusters in both phases arrange in laminar (neat), hexagonal (middle), or cubic liquid crystals. The macroscopic appearance of lyotropic liquid crystal formation is an increase in opacity of the solution, abnormal high viscosity, and changes in their dynamic properties.[90–3] Characteristically, the solubility of amhiphilic molecules increases greatly above a certain temperature, the so-called Krafft point, which has been defined as the melting point of the crystalline hydrate of the surfactant.[94]

The molecular structure of a liquid crystal cannot be determined by electron microscopy or x-ray diffraction. Liquid crystals of heptyl α-D-mannopyranoside[92] and 1-*S*-β-D-xylopyranoside,[93] for example, only show "powder diffraction" spectra indicating periodicities of 2.2 nm and 3.0 nm. These periodicities are greater than one but less than two times the respective molecular lengths. Electron micrographs of liquid crystals usually give ill-defined patterns.[94,95] Since liquid crystals of carbohydrate amphiphiles are probably not well ordered on the molecular level, they are not further considered here. For a review of carbohydrate amphiphiles in the lyotropic phase, see Jeffrey,[91a] and van Doren and Wingert.[96]

11.4.4.2 Transmission electron microscopy

The first paper which described well-defined electron micrographs of carbohydrate amphiphile fibers in water was by Pfannemüller and Welte.[97] They dissolved *N*-alkyl-D-gluconamides **(24)** in hot water ($<90\,^{\circ}C$) and cooled the solution to obtain a slightly opaque gel over a wide range of concentrations (1–50% w/v). They showed that the gel network was composed of ropes with right-handed helical twists. The diameter of these ropes was estimated to be 12.5 nm, the repeat unit 18.0 nm, with an angle of rise 35°. Crystalline needles began to grow after a short time. It was also shown that only the secondary gluconamides formed such well-defined helical structures. *N*-Methylated tertiary amides gave gels only at low temperatures and relatively high concentrations and the helices showed no regular pitch.

The same procedure with the same material was later carried out with phosphotungstate added to the aqueous solution (Figure 19). At first, a double helix of micellar strands with a diameter of 3.9 nm was characterized,[98] which was later shown to be a quadruple helix.[99] Cryoelectron microscopy (Figure 19) and direct evaluation of the projection image[100] finally yielded the 9.7 nm ribbon quadruple helix, in which each ribbon had the thickness of a molecular bilayer (Figure 20).

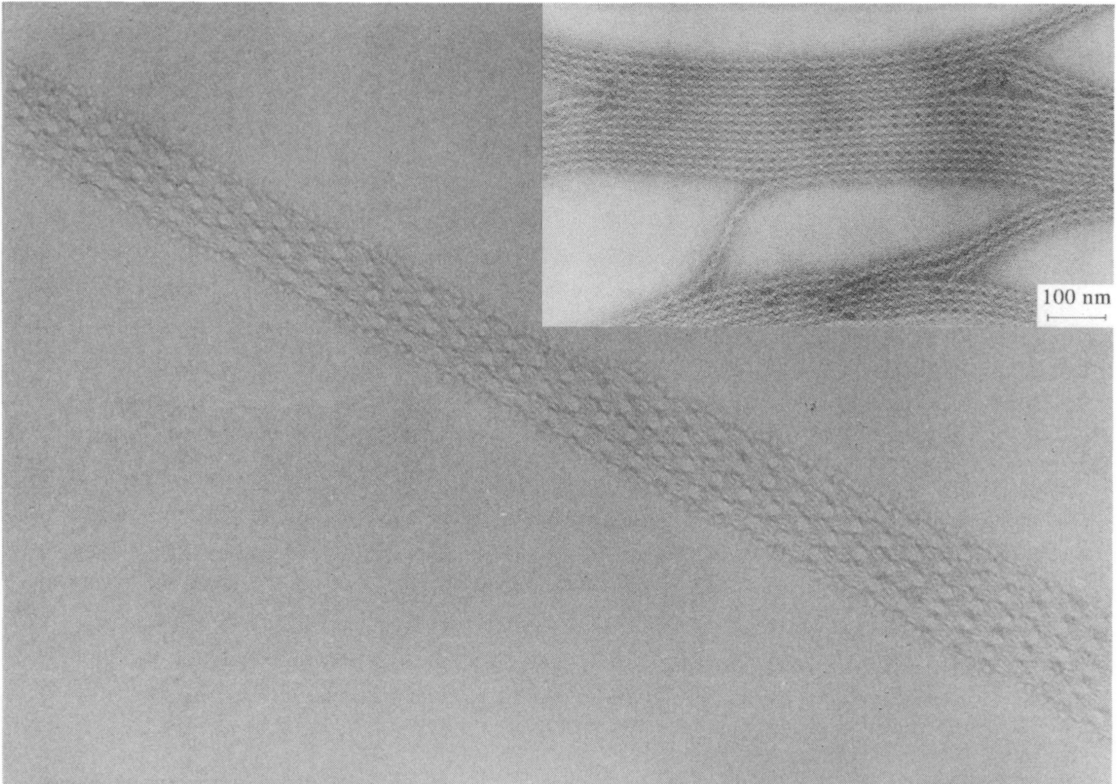

Figure 19 The central large strand corresponds to a cryoelectron micrograph of a bundle *N*-octyl-D-gluconamide fibers fixed with phosphotungstate. The whole bundle has a width of ~100 nm. The quadruple helix model at the left-hand corner was deduced from the direct computer analysis of such a bundle. The electron micrograph at the right-hand corner shows a typical dried preparation fixed by phosphotungstate.[99]

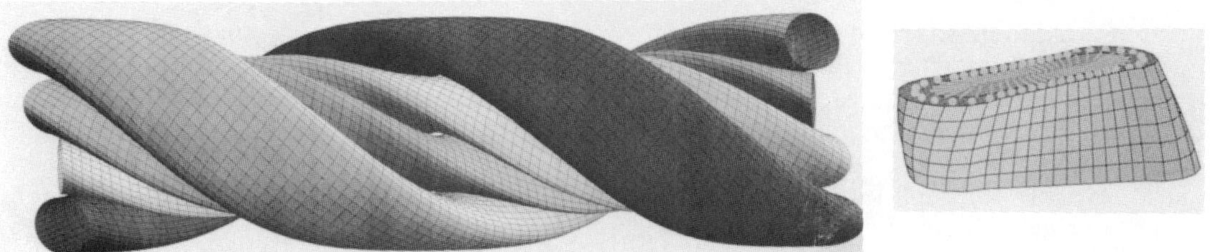

Figure 20 The quadruple helix model corresponding to the micrographs in Figure 19.

The formation of these helices can be rationalized with the assumption that spherical micelles are formed in water at high temperatures, when the amide hydrogen bonds are broken. Upon cooling, an amide hydrogen bond chain is formed in a rapid, cooperative process and the spherical micelle is converted to a chiral disk structure (Figure 20). The chirality of the monomers enforces some unidirectional deformations.[100] Similar micellar disks are observed when vesicle membranes are stiffened by the addition of large amounts of steroids.[101,102] The gluconamide disks have two hydrophobic surfaces and therefore rapidly grow to (helical) fibers. In order to reduce the surface energy of such fibers, they intertwine to form the common quadruple helix, which was also observed, for example, when planar cyclopeptides assembled in water or membranes.[103]

The diastereomeric D-mannonamides (**25**) and D-galactonamides (**26**) form rolled-up bilayer scrolls (Figure 21(a)) and twisted ribbons (Figure 21(b)),[104] respectively.

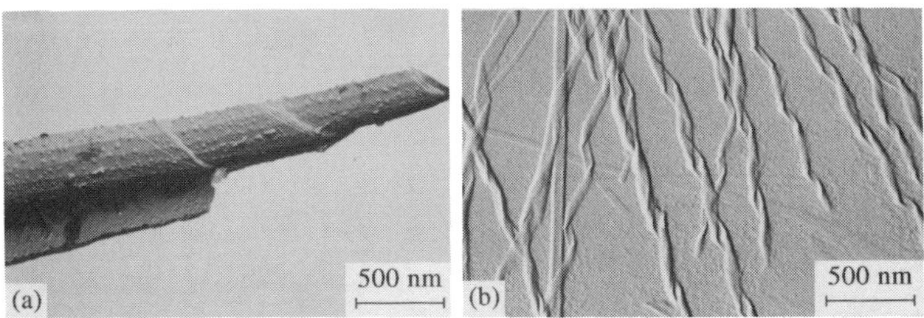

Figure 21 Electron micrographs of (a) rolled-up mannonamide (**25**) and (b) twisted galactonamide (**26**) bilayers.[104]

The remaining five diastereomeric hexonamides and the pentonamides[105] do not form well-defined membrane structures in water. They are either extremely well soluble in water, for example, altron- and idonamides,[104] or they crystallize rapidly, for example, ribonamide.[105]

The reason for the extremely different solubilities and supramolecular structures of the hexonamides can be guessed from a comparison of their molecular structures in the all-*anti*- or all-*trans* conformations. In the case of the mannon and galacton chains, there are no 1,3-*syn*-configured hydroxyl pairs in the all-*trans* conformation. This conformation should therefore be the most stable. In glucon-, gulon-, and talonamides, on the other hand, one finds one such 1,5-interaction of hydroxyl groups and one may predict a single *gauche* bent between the corresponding carbon atoms. Intermediate solubility and curved supramolecular structures are expected. In the allon- and idon, (**28**) and (**29**), chains all four hydroxyl groups are effected by 1,3-*syn* repulsion and the conformation should be irregular. High water solubilities and no supramolecular assemblies are expected. In general, these expectations are fulfilled from the first inspection. *N*-Octyl-D-galactonamide dissolves to about 16 mmol L^{-1} at 100 °C, its racemate to 10 mmol L^{-1}. D-Mannonamide is suprisingly 10 times more soluble (98 mmol L^{-1} at 100 °C). For the glucon-, gulon-, and talon-*N*-octylamides (**24**), (**27**), and (**30**), the solubilities again increase by a factor of ~100 or more (glucon: 1.6; talon 2.3 and gulon > 3 mol L^{-1}). The critical micellar concentrations are barely influenced by solubilities. They are close to 3 mmol L^{-1} in the cases of *N*-octyl-D-mannon, D-glucon, D,L-glucon, and D-gulonamides.[106]

This first approximation, however, in no way explains why gluconamides form beautifully defined helices and crystallize so easily, whereas the corresponding gulon- and talonamides give both bad fibers and three-dimensional crystals. The most important question is how to determine the molecular conformations within fibers. Is it possible to check on the purity or on eventual mixtures of conformers and if that is answered positively, can one differentiate between different

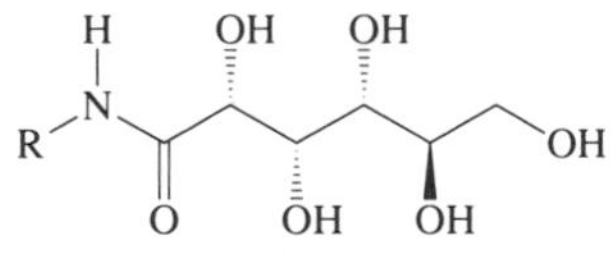

(**24**) D-Glu

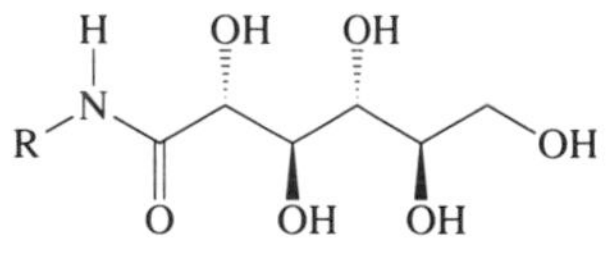

(**28**) D-All

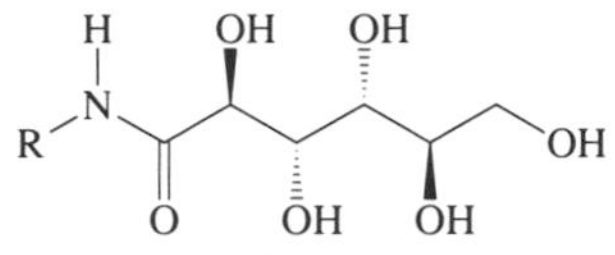

(**25**) D-Man

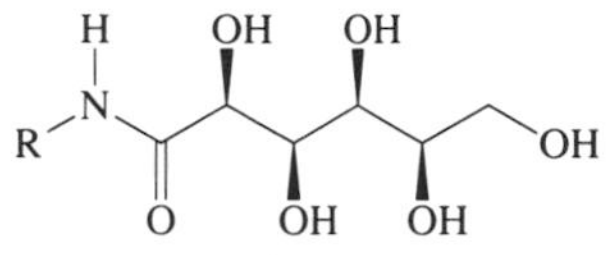

(**29**) D-Ido

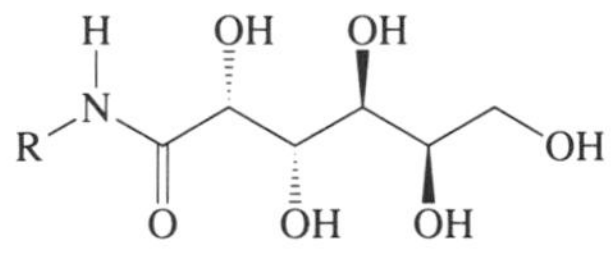

(**26**) D-Gal

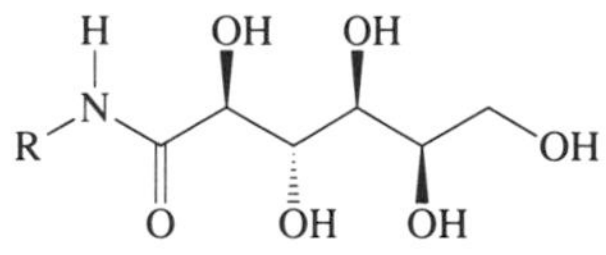

(**30**) D-Tal

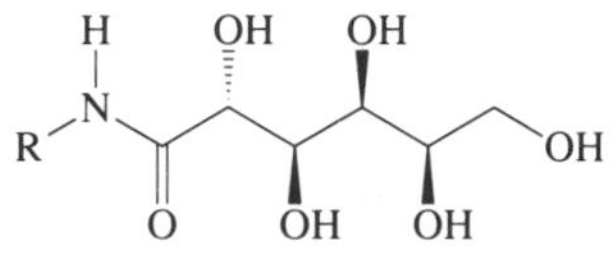

(**27**) D-Gul

(**a**) $R = C_8H_{17}$
(**b**) $R = C_{12}H_{25}$

conformers of different diastereomers? We shall now show that a combination of x-ray diffraction of single crystals, microcrystals, and lyophilized fibers together with CPMAS ^{13}C NMR and FTIR solid-state spectra of the same preparations and solution ^{1}H NMR spectra and electron micrographs provide enough information to solve this problem.

11.4.4.3 X-ray diffraction

Three three-dimensional crystal structures of glyconamide amphiphiles (D-glucon,[107] D-gulon,[108] D-talon[109]), one of a bolaamphiphile with two D-gluconamide end groups,[110] and one of a nonamphiphilic gluconamide[111] have been solved. D,L-Glucon, D-galacton, and D-mannon-amides could only be obtained in the form of microcrystalline powders.[112] In the case of *N*-octyl-gluconamide, a linear all-*trans* conformation of the carbohydrate chain was observed, a homodromic hydrogen bond cycle in which four short hydrogen bonds run unidirectionally from one crystal plane to the other and a head-to-tail arrangement of the crystal sheets. The terminal 6-OH group is fixated by hydrogen bonds within the crystal sheet and its surface is therefore formed by the hydrogen atoms of the terminal methylene group (Figure 22(a)). In the gulonamide (**27**) crystal structure, on the other hand, one also finds a linear carbohydrate chain, but a hydrophilic surface of the crystal sheets. Correspondingly, a head-to-head arrangement with interlayer hydrogen bonds occurs (Figure 22(b)). Both end groups have similar homodromic hydrogen bond cycles on the same side of the chain extension. In nonamphiphilic model compounds, for example, *N*-isopropyl-D-gluconamide, one finds the same all-*trans* glucon conformation and the same homodromic cycle.

Powder diagrams of the microcrystals of D,L-glucon, D-mannon-, and D-galacton *N*-octylamides showed that all three amphiphiles occurred in bilayers or in head-to-head arrangements of the sheets.[112]

11.4.4.4 Solid-state CPMAS ^{13}C spectra and assignment by solution spectra

Solid-state NMR spectra of supramolecular assemblies serve two purposes: they may characterize the mobility and the chemical environment of a given carbon or nitrogen atom. Deuterium labeled compounds show doublets, whose shape and linewidth characterize equilibria between *anti* and *gauche* conformers in which the particular carbon atom may participate.[106] CPMAS ^{13}C NMR

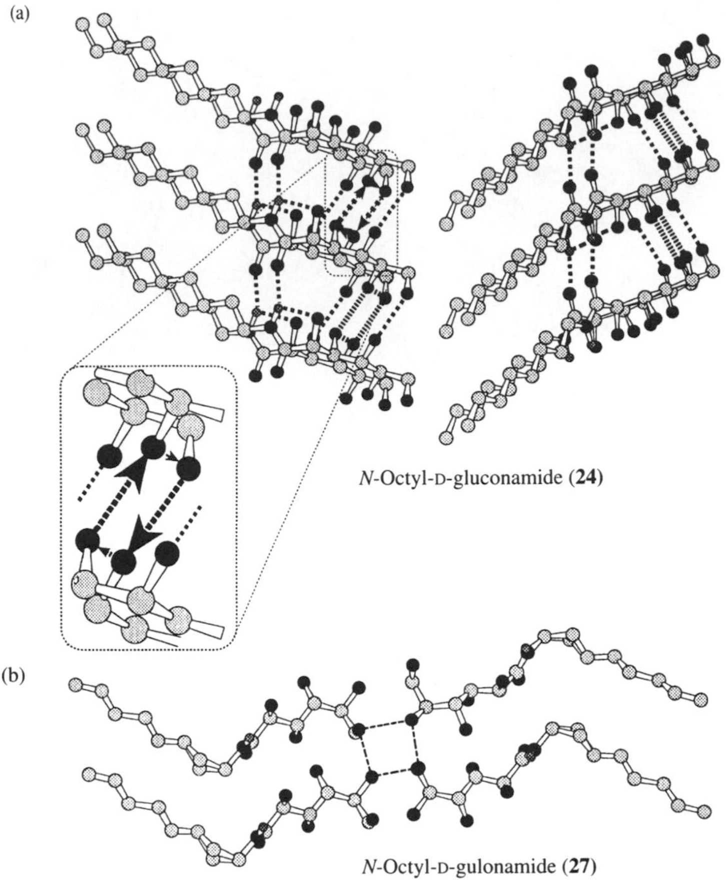

Figure 22 Crystal structures of (a) gluconamide (**24**)[107] (single layer, head-to-tail, homodromic intralayer hydrogen bond cycle) and (b) gulonamide (**27**)[108] (bilayer, tail-to-tail, interlayer hydrogen bonds).

spectra, on the other hand, yield a chemical shift for each individual carbon atom in the glycone chains. No labeling is necessary. Figure 23 shows that lyophilized fibers and three-dimensional single crystals of *N*-octyl-D-gluconamide give very different patterns for the CH and CH_2 carbon atoms of the carbohydrate part, indicating different molecular conformations. The spectrum of the D,L-crystallites is again different, but *N*-isopropyl and *N*-octyl-D-gluconamides are practically the same (not shown). The latter two compounds also show the same carbohydrate conformation in single crystal structures. Diastereomeric galacton-, mannon-, talon-, and gulonamides are again very different (not shown). In all cases the terminal methyl group signal of the oligomethylene chain is found close to 14 ppm in tail-to-tail arrangements and close to 16 ppm in head-to-tail oriented sheets (Figure 23).

The conformation of *N*-octyl-D-gluconamide is thus known from the crystal structure and the chemical shifts of the carbon atoms from the CPMAS spectrum. The ^{13}C signals now need to be assigned. This is possible by measurement of the ^{1}H NMR spectrum in DMSO solution where hydrogen bonds occur. NOE experiments, irradiation experiments, and measurement of coupling constants allow unequivocal assignments of proton signals as well as of the conformations present. When the proton signals are assigned, one uses a two-dimensional ^{13}C, ^{1}H shift-correlated spectrum in DMSO to assign ^{13}C signals in solution.[113] An example of a well-resolved ^{1}H spectrum including the OH protons and an irradiation experiment with a ^{1}H–^{13}C correlation spectrum are given in Figure 24.[114]

The solid-state conformation of *N*-octyl-D-gluconamide in crystals, the assigned ^{13}C spectrum in DMSO solution, and its conformation in this solution are now known. Comparison of the ^{13}C chemical shifts in solution and solid-state spectra then immediately show that carbon atoms in similar environments show similar shifts, whereas changes in conformations cause large shifts. The most important examples are the CPMAS solid-state ^{13}C NMR spectra for the crystals and for the fiber of D-gluconamide (**24**). The all-*trans* conformation is changed in both the quadruple helical

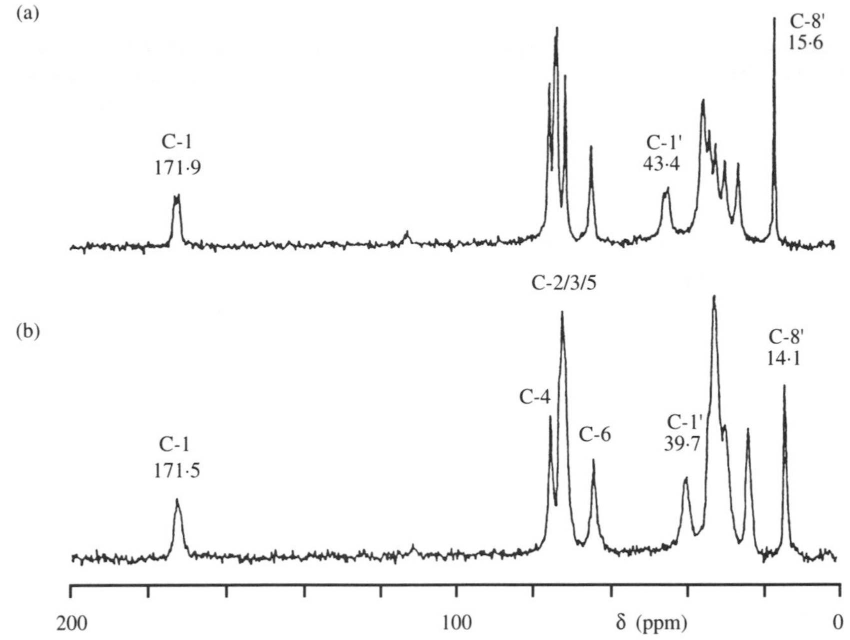

Figure 23 Two examples of solid-state CPMAS ^{13}C NMR spectra of *N*-octyl-D-gluconamide (**24**): (a) single layer crystals, (b) lyophilized bilayer fibers. Different chemical shifts correspond to different conformations. The terminal methyl group C-8′ atom is at 15.6 ppm in the polar environment of head-to-tail crystals and at 14.1 ppm in the apolar tail-to-tail arrangement.[112]

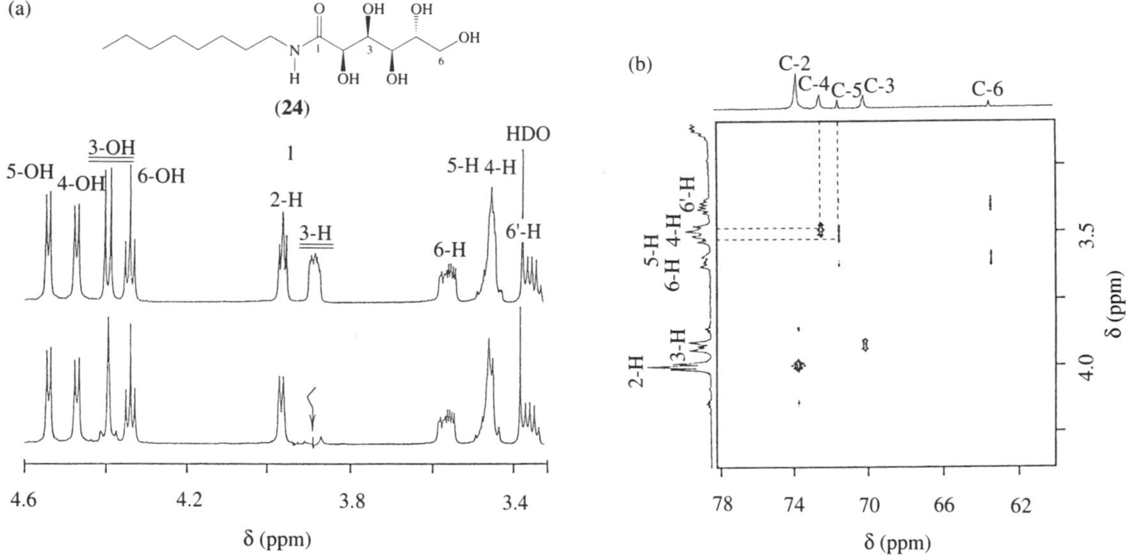

Figure 24 (a) An example of a solution NMR spectrum and a double resonance experiment with D-gluconamide (**24**) in DMSO.[114] The coupling constants give the gluconamide conformation in solution, the irradiation experiments together with ^{13}C. (b) Proton correlation spectra allow the assignment of ^{13}C signals.

fiber and in the racemic D,L-crystallites and the largest effects occur on C-2 and its neighbors. This indicates a *gauche* conformation at the C-2 carbon atom next to the amide group.[112]

It has thus been established that the conformations of gluconamide head groups are the same in highly curved fibers and in tail-to-tail crystallites ($_2G^-$), whereas they change drastically in head-to-tail crystals (P). Moreover, the bend close to the amide group obviously disturbs hydrogen bonding and favors hydrated, curved micellar structures and destabilizes planar crystal sheets in aqueous environments.

The same conclusions on conformations were reached when *gauche*- and *syn*-diaxial effects were directly evaluated to predict chemical shifts in the solid-state spectra and when CPMAS spectra of crystalline materials were compared, where the molecular conformations were known from x-ray data.[112]

11.4.4.5 FTIR spectra of alkyl chains

The conformation of the head groups in crystalline fibers has thus been established. It still remains a question whether the alkyl chains are fluid or solid and where *gauche* bonds possibly occur. In principle, this question can only be answered by deuterium labeling of each individual methylene group. In the case of glyconamide fibers, this experiment is fortunately superfluous, because FTIR spectra are very sensitive to *gauche* conformers in alkyl chains; 5% *gauche* conformation in an all-*trans* chain can easily be detected especially in the "wagging region" between 1200 cm^{-1} and 1400 cm^{-1}. However, none was found in any of the preparations of supramolecular glyconamide assemblies (Figure 25). No 2930 cm^{-1} stretching absorption for a *gauche* conformer was found either. These findings provide strong evidence for a pure all-*trans* conformation. In the more fluid vesicle membranes, *gauche* conformers have been detected even below the liquid crystal transition point.[115]

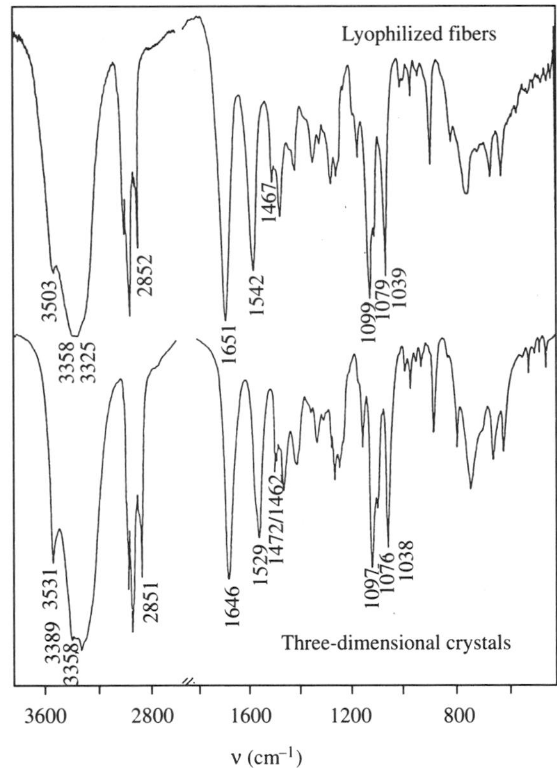

Figure 25 A comparison of the IR spectra of authentic D-gluconamide (**24**) three-dimensional crystals and lyophilized fibers shows identity in the stretching (≈2850 cm^{-1}) and wagging (1200–1400 cm^{-1}) CH_2 regions. There are no *gauche* conformations either in the crystal or in the fibers.[114]

11.4.4.6 Differential scanning calorimetry

Micellar fibers of *N*-octyl-gluconamides produce narrow, reversible melting peaks in DSC thermograms at 63 °C and 69 °C. Thermotropic rearrangements in crystals from a head-to-tail to a tail-to-tail orientation, on the other hand, produce only weak signals in both aqueous suspensions (69 °C Krafft point) and dry crystals (86 °C). The racemate crystals show no rearrangement peak (Figure 26).

DSC thermograms are in general very useful for detecting cooperative processes such as the formation of hydrogen bond chains (sharp peaks) and changes in crystal sheet orientations. The

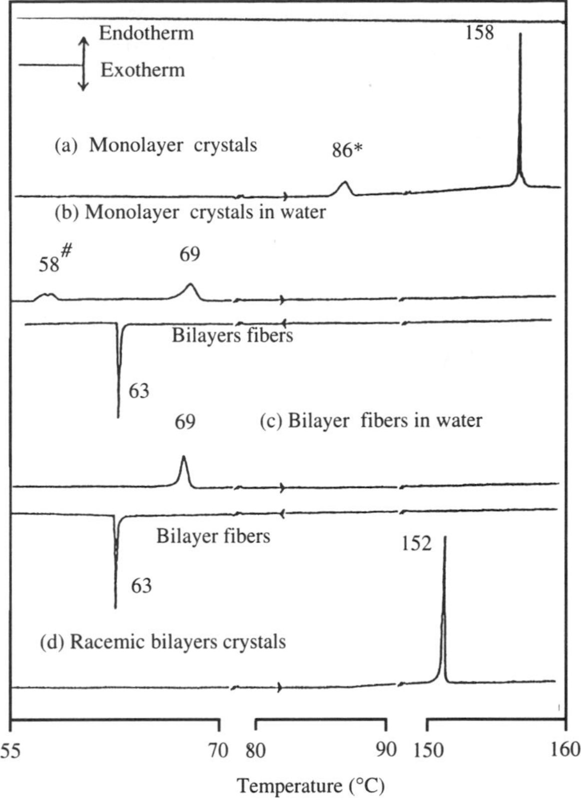

Figure 26 DSC shows the same melting and crystallization points for aqueous probes of both the D-gluconamide (**24**) quadruple helix (c) and three-dimensional crystals (b). The melting points of (d) dry bilayer and (a) monolayer crystals are also similar. DSC and other thermodynamic measurements are often insensitive to curvature and stereochemical effects and hardly discriminate between monolayer and bilayer structures.

latter are also detectable: they show an exact doubling of the replating unit thickness, when the monolayers switch to bilayers.

11.4.4.7 Glyconamide conformation

The procedure for determining molecular conformations in supramolecular assemblies can be summarized as follows: (i) collect crystal structures of your chiral head group from the literature, produce the same crystals, and run CPMAS ^{13}C NMR spectra of them. Look if general rules (γ-effect, *syn*-diaxial effect)[112] agree with the observed chemical shifts in CPMAS spectra and known crystal structure conformations; (ii) confirm with ^{1}H NMR solution spectra if you feel that this is necessary. Quite often the ^{13}C solid-state spectra of crystalline and fiber materials are partially superimposable, and the solution spectra are very similar. Large shifts of two carbon signals in otherwise very similar spectra indicate a change of conformation.

In the case of open-chain hexonamides, approximately 12 different conformations have thus been characterized.[112–14] The major general results were (i) that one cannot predict the conformations from 1,3 *syn*-diaxial repulsion effects alone. The γ-*gauche* effect (as many heavy atoms as possible are in the *gauche* and not in the *anti* position to each other)[116] is at least of the same importance; (ii) that disturbances of hydrogen bonding patterns within single crystal sheets lead to curvature and helicity.

More interesting are, of course, the detailed individual conformations. In the case of fibrous and microcrystalline galacton- and mannonamide materials, for example, it was shown that galactonamide (**26**) always occurs in the all-*anti* conformation (not shown), whereas mannonamide (**25**) has a $_4G^+$ sickle conformation (not shown). D-Gluconamide (**24**), on the other hand, changes its conformation from $_4G^+$ in solution to $_2G^-$ in lyophilized bilayer fibers (Figure 27) and to all-*anti* in monolayer crystals (Figure 27). Racemic bilayers are made of the same $_2G^-$ conformers as the fibers. Curvature is therefore not directly related to the molecular conformations of molecules, because both the planar racemate crystals as well as the quadruple helices with highly curved edges are made of the same conformers. Nevertheless, the *gauche* bend at a carbon–carbon bond adjacent to the amide hydrogen bond renders the crystal plane so flexible that highly curved edges can easily be formed by hydration. Large crystals can only be formed after rearrangement of this labile conformer to the all-*anti* form which is accompanied by sheet interconversion.

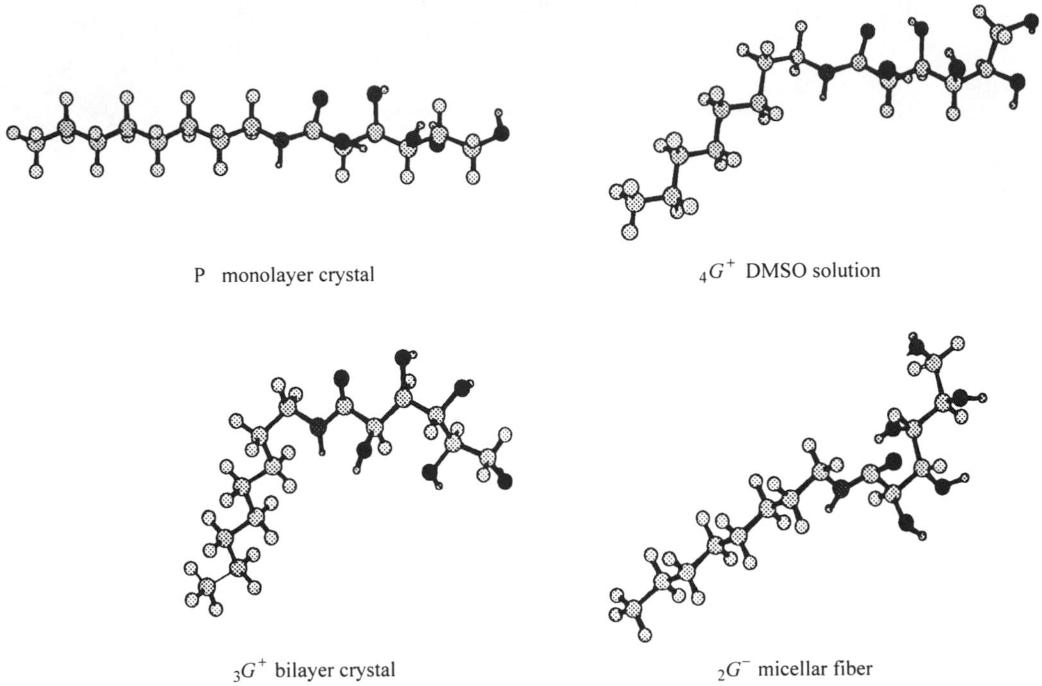

P monolayer crystal

$_4G^+$ DMSO solution

$_3G^+$ bilayer crystal

$_2G^-$ micellar fiber

Figure 27 One all-*anti* and three different *gauche* confirmations were characterized for *N*-octyl-gluconamide (**24**) in solution, crystals, and a lyophilized micellar fiber. Curvature is only found when the carbon–carbon bond next to the amide group is disturbed.[112]

11.4.5 *N*-Dodecoyl Tartaric Acid Monoamides

D- and L-tartaric acids provide cheap starting materials for the preparation of uniform monoamides with long alkyl chains. The easiest way is to start with the anhydride diacetate and to open it with long-chain alkyl amines, for example, dodecylamine. After saponification of the acetates with sodium, potassium, or ammonium hydroxides, one obtains the corresponding salts of the monoamides. They are highly soluble in water at pH 8 and form fibers at pH 5, which corresponds to half protonation. These fibers have the unique property to assemble to huge cell-like membranes of various shapes, presumably by interfiber $CO_2H \cdots O_2C$ hydrogen bridges (Figure 28).[117] In more acidic solution they precipitate, at higher pH they redissolve.

In none of the cases is it known whether the amide hydrogen bond chains run perpendicular or parallel (as in proteins, see Section 11.2.3) to the helical axis. The formation of micellar disks certainly indicates primary merry-go-round type hydrogen bonding along the disk surface. Once these disks have been assembled to fibers, however, one may well envisage a rearrangement to the form of protein helices with hydrogen bonds along the axis.

11.4.6 Porphyrins

It is now evident that long alkyl chains in supramolecular fibers usually exist in a frozen all-*trans* state. The equally rigid and hydrophobic porphyrin moiety can also act as a skeleton for amphiphilic fiber formation. This may either occur by stacking, where the electron-rich pyrrole rings repulse each other and prefer a staggered arrangement (Figure 29),[118] or a lateral assembly or a combination of both.

Figure 28 Electron micrographs of solid bilayer assemblies of tartaric acid monoamide (**5**) (negatively stained with phosphotungstate).[117] (a) Assemblies with micrometer-dimensions are formed from (b)–(d) molecular bilayer threads at pH 5.5. At pH 8, the assemblies dissolve, while at pH 3 planar platelets precipitate. Curvature of crystalline bilayers not only depends on chirality (racemic amides do not form long-lived micellar rods), but also on partial surface charge.

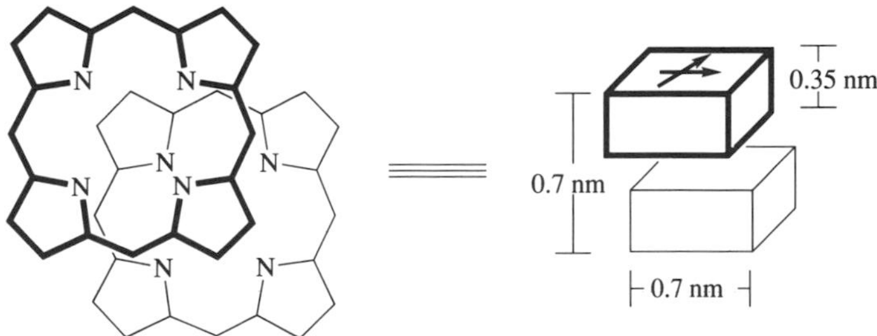

Figure 29 The porphyrin chromophore behaves like a rigid box and to date cannot be co-crystallized with solid host bilayer fibers. Amphiphilic porphyrins, however, form fibers, the stacking geometry[118] usually implies that the electron-poor central cavity attracts an electron-rich pyrrole ring of a neighboring porphyrin. A Jelley-type lateral arrangement is thus mixed with face-to-face stacking. Since the dipole of the excited state may point in both the *y*- and *x*-axes directions (see right-hand drawing), further complications of electronic (see Figure 31) and circular dichroism (see Figure 33) spectra occur.

Protoporphyrin IX (**31a**) is a typical amphiphilic porphyrin with a hydrophilic southern edge and three hydrophobic edges. Only micrometer-sized crystals are known[119a] and the crystal structure of the related hemin, an iron(III) complex, has been solved.[119] Protoporphyrin dissolves in water above pH 9 and then produces a visible spectrum with a sharp Soret band at 402 nm. At pH 5, the pK_a of the propionic acid side chains, large aggregates with a molecular mass of $>10^6$ Da and ill-defined electron microscopic structures[120] are formed.[121] These aggregates show a broad split Soret band with peaks at 350 nm and 470 nm. A much more stable and better defined porphyrin assembly was obtained when the carboxylate ions were replaced by amino groups (**31b**) and the assembly was produced at pH 6 in DMSO–water (1:1) or water.[122] Very long fibers with a typical diameter of 3 nm were observed by electron microscopy (Figure 30) and the broad Soret band shifted from 370 nm at pH 4.6 to 439 nm at pH 8.8 (Figure 31).

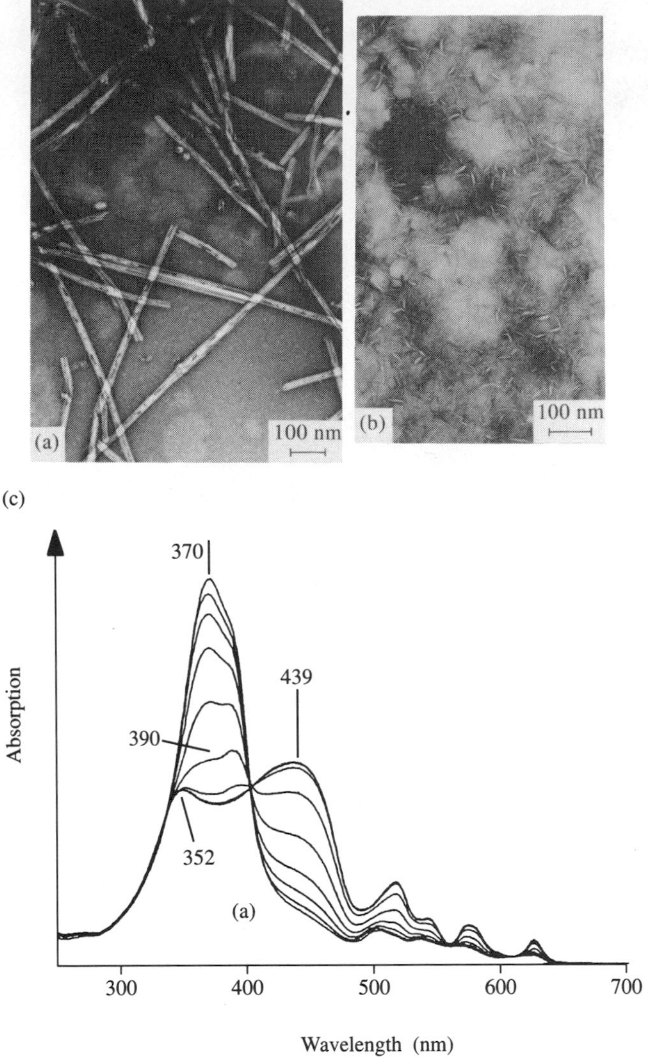

Figure 30 The diaminoporphyrin ((**31b**), see Figure 31) shows a 439 nm and 352 nm split Soret band in aqueous solution at pH 6 (c) and appears as tubules (a) in electron micrographs. At pH 9 a single Soret band at 370 nm appears (c) and short 6 nm wide ribbons are observed (b). It is thought that the pH 6 fiber contains a lateral hydrogen-bonded assembly, whereas the pH 9 species is a columnar face-to-face polymer.[122]

Amphiphilic porphyrins with chiral carbohydrate head groups produce chiral fibers with strong CD bands.[117,123] In the case of copper and zinc porphyrinates (**31c**) and (**31d**) with two gluconamide side chains similar intensities and opposite signs of CD effects were observed, which points to a different screw sense in both supramolecular assemblies[123] (Figure 31). The more complex CD spectra of the free bases (not shown) allow detailed modeling by application of the exciton theory,[121,124] and a lateral assembly usually has to be combined with the assumption of some stacking. Application of the theory is more complicated than in the polyene-type Scheibe complexes, because the porphyrin chromophore has a circular conjugation system and two transition dipoles (see Figure 29) must be considered. Tin(IV) porphyrinates contain two axial ligands which may lead to polymeric stacks, if the counterion is bivalent, or lateral assemblies only, if the counterion

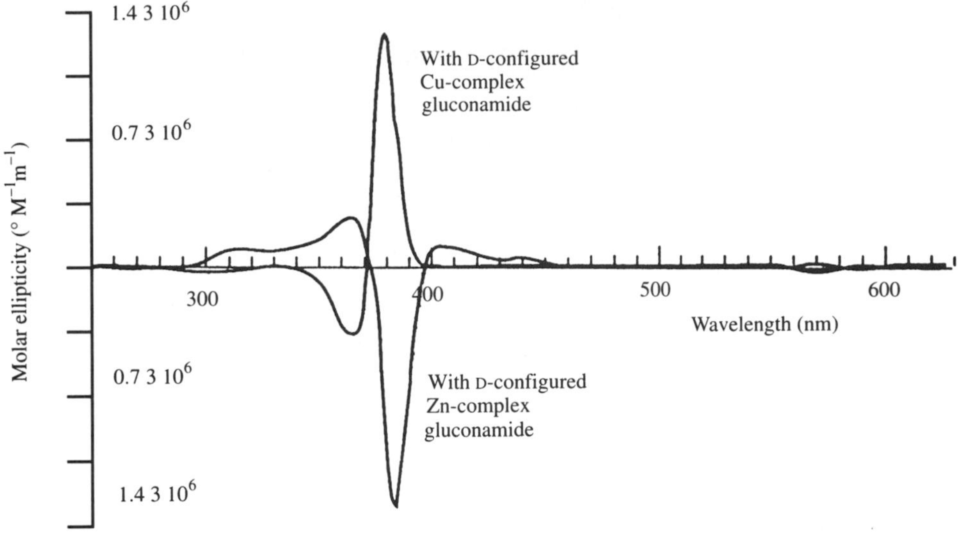

Figure 31 CD spectra signs of chiral metallo-porphyrin fibers depend on the central metal ion,[123] indicating that subtle electronic interactions between the porphyrin π-systems dictate the fiber structures.

(31)

(**a**) R^1, R^2=CH=CH$_2$;	R^3, R^4=CH_2CO_2H;	M=2H
(**b**) R^1, R^2=Et;	R^3, R^4=NH_2;	M=2H
(**c**) R^1, R^2=Et		M=Cu
(**d**) R^1, R^2=Et	R^3, R^4= –NH–C(=O)–CH(OH)–CH(OH)–CH(OH)–CH(OH)–CH$_2$OH;	M=Zn
(**e**) R^1, R^2=Et		M=$SnCl_2$

is monovalent. With a tin(IV) porphyrinate dichloride (**31e**), extremely long 6 nm fibers have been realized at pH 0, in which protonated chloride ions act as facial head groups (Figure 32).[125]

All protoporphyrin derivatives have the disadvantage that they do not fluoresce. The fibers are therefore extremely stable against photooxidation, but are also relatively inactive in charge separation processes. Only in the case of amino porphyrin fibers do photochemical reactions occur readily and porphyrin anion formation by charge separation is likely.[122,123]

Fluorescent porphyrin fibers have been obtained from the octopus porphyrin (**32**) in water.[126] In this tetraphenylporphyrin derivative, the porphyrin chromophores are so far apart from each other that π–π interactions become negligible. The porphyrin is of interest to the modeling of charge separation systems, because quinones and other hydrophobic electron-acceptor systems may be dissolved by the surrounding alkyl chains.

Bolaamphiphilic porphyrins with hydrophilic northern and southern edges are usually too water soluble to form assemblies. In the case of the type II isomer of β-tetraethyl-β-tetrapyridinyl porphyrins (**33**), however, stable monolayer leaflets are formed in water (not shown), which also strongly fluoresce. These leaflets also produce a split Soret band (470 nm and 350 nm)[127] and can be combined with redox-active metal ions on both sides of the molecular monolayer.

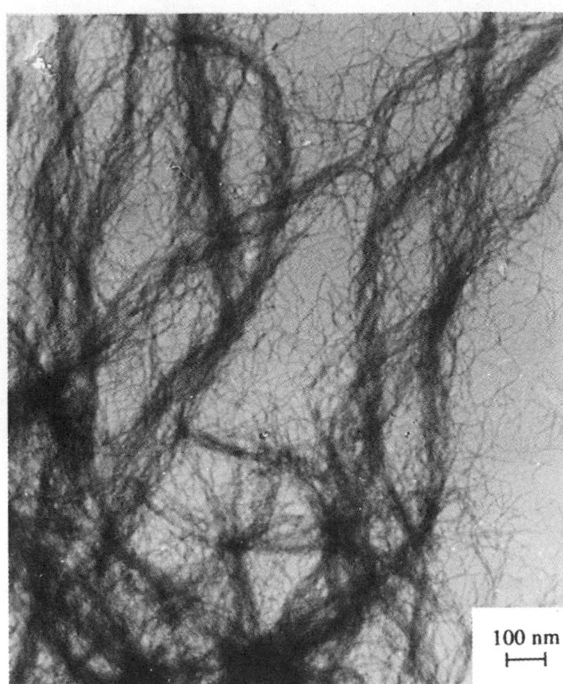

Figure 32 Tin(IV) dichloride porphyrinates with two axial ligands cannot form cofacial stacks. Nevertheless, (**31e**) produces ultrathin fibers which produce an electronic spectrum with a split Soret band. Lateral interactions between the porphyrins and hydrogen bonds between the axial chloride counterions ("facial amphiphile") are responsible for the assemblage.[125]

Short fibers have also been observed for chlorophyll derivatives.[128]

Polymeric porphyrin assemblies have emerged in the 1990s and may be useful as photoconductors, charge separation, and/or collecting systems.

11.4.7 Co-crystalline Fibers

The micellar fibers containing hydrogen-bonded chains must be considered as crystals of bimolecular thickness since all the head groups which have the same conformation are aligned in the same way and are connected by uniform and directed bonds. Such fibers cannot dissolve anything, they have the character of a solid. The admixture of other compounds will only be possible if these compounds are amphiphiles with fitting head groups and adjustable hydrophobic cores. In the case of glyconamides, it is most tempting to try diastereomeric mixtures and to check on the morphological changes of the fibers. For enantiomers, see Section 11.6.

The only systematic study applied 1:1 mixtures *N*-octyl- and *N*-dodecyl-glucon, -galacton and mannonamides in D- and L-enantiomers.[5] The pure gluconamides formed quadruple helices, and galactonamide gave twisted ribbons and mannonamide rolled up sheets (see Section 11.4.4).

The mixture of *N*-octyl- and dodecyl gluconamides has already been described as "nonordered aggregate" (see Section 11.2.1) containing irregular micellar lumps. From these lumps, however, the helical structures of pure octyl- and dodecylamides grow within a few minutes. Largely different chain lengths are obviously not allowed in co-crystalline fibers, as complete separation occurs. This is a fundamental difference from fluid assemblies such as vesicles, where domains may be formed, but never a separation in individual vesicles of compounds A and B. If enantiomeric head groups are used together with different chain lengths, one first observes the formation of racemic bilayer platelets and later the separation into right- and left-handed helices.[5] In the case of octadecyl-D-mannonamide (**25b**), one obtains P- and M-helices next to each other if one precipitates them from solutions (Figure 33).[106]

The mixture of the flexible gluconamide (**24a**) fibers with the more rigid mannonamide (**25b**) sheets gives platelets with the octylamides (Figure 34(a)) and separated fibers and scrolls with the much less soluble dodecylamides (Figure 34(b)). *N*-Octyl-D-glucon and -galactonamides do not mix at all. They separate into the quadruple helix and twisted ribbon structures shown in Section 11.4.4.

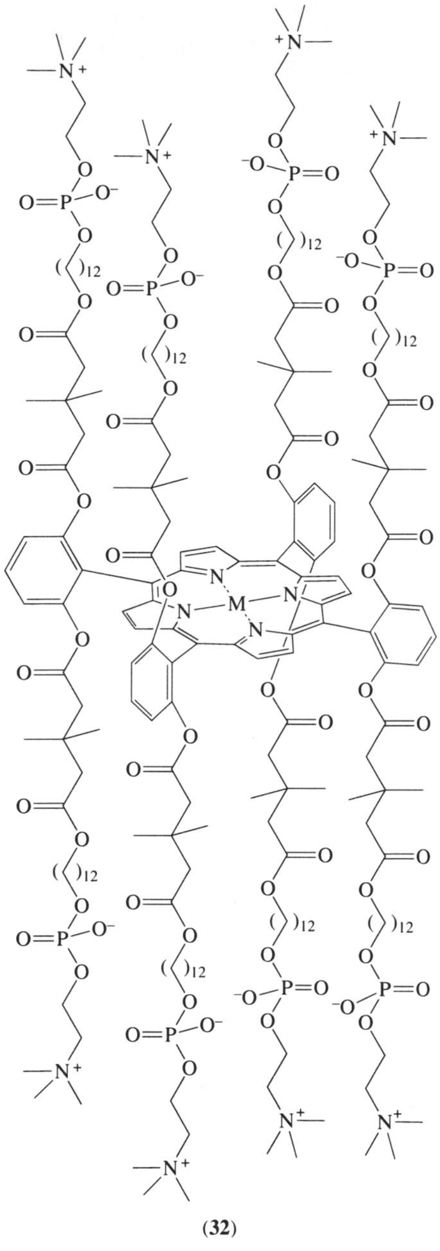

(32)

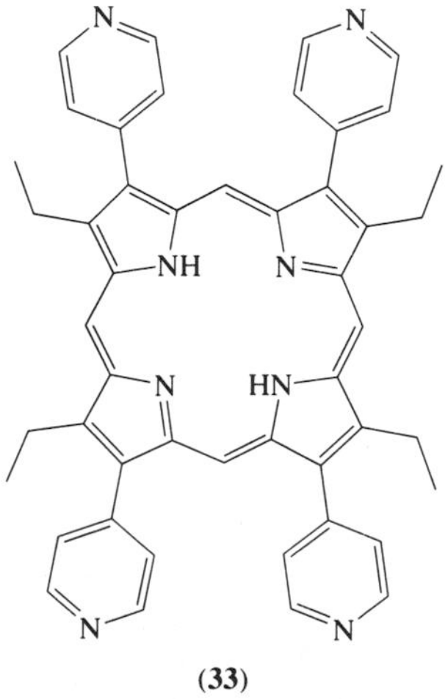

(33)

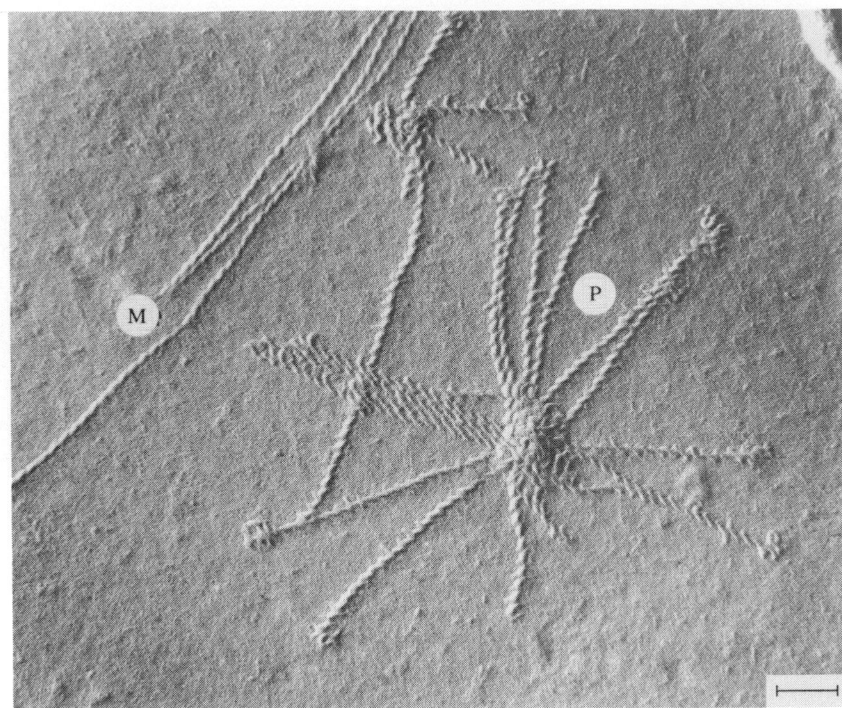

Figure 33 The electron micrograph shows the spontaneous separation of *N*-octadecyl-D-mannonamide, M- and P-helical fibers in the precipitate upon cooling of hot SDS-micellar (sodium dodecylsulfate) solutions.[106]

The general trends in these co-crystals or "alloys" of diastereomeric glyconamides seem to be that compounds with similar solubilities form co-crystals, whereas compounds with largely different solubilities separate into individual fibers, even if primarily formed co-crystals have much smaller surfaces than the individual fibers which follow as a result of the separation.

Application of the bilayer fibers in functional machineries asks for the co-crystallization of redox- and photoactive components. It has been found that terminal nitrobenzene, phenol, and pyridyl substituents in the alkyl chains are admitted in the fibers,[130] so that polymeric electron donor–acceptor assemblies become accessible. The co-crystallization of a carotenoid amphiphile, namely bixin-D-gluconamide (**34**) with *N*-octyl-D(and L)-gluconamides (**24a**) has also been achieved.[131] The slender polyene chromophore can obviously be fitted into the helices, although the methyl substituents should lead to strong perturbations. At 20 °C, a positive CD effect is observed with right-handed helices of *N*-octyl-D-gluconamide and a negative CD effect with the L-gluconamide fiber (Figure 35).

Attempts to co-crystallize the corresponding porphyrin gluconamides with the quadruple helix led to the discovery of the porphyrin fibers which are discussed in Section 11.4.5. To date, there is no example of real mixing of the porphyrin macrocycle into the crystalline or liquid crystalline phases of mono- or bilayer structures.[132] Stiff multiple protein helices are much more appropriate for entrapping the box-shaped porphyrin chromophore.[133]

11.4.8 An Arborol

A bolaamphiphilic micellar fiber was described in Section 11.4.2 (Figure 17). Its formation cannot be explained by the hydrophobic effect, because water molecules cannot be removed from the hydrophobic core in the presence of two bulky head groups at both ends of an oligomethylene chain. An even more extreme case has been realized with an "arborol" amphiphile.

An arborol has been defined as a one-directional cascade polymer possessing a central trunk attached to a single cascade sphere. The two-directional arborol (**35**) then corresponds to a bolaamphiphile with two very bulky head groups and a thin connecting oligomethylene chain. In this compound, one cannot expect much of the hydrophobic effect, because the few methylene groups will hardly be expelled from water. Nevertheless, the bolaamphiphile is not very soluble in water as a monomer and forms gels above 2 wt.%. Electron microscopy of negatively stained and

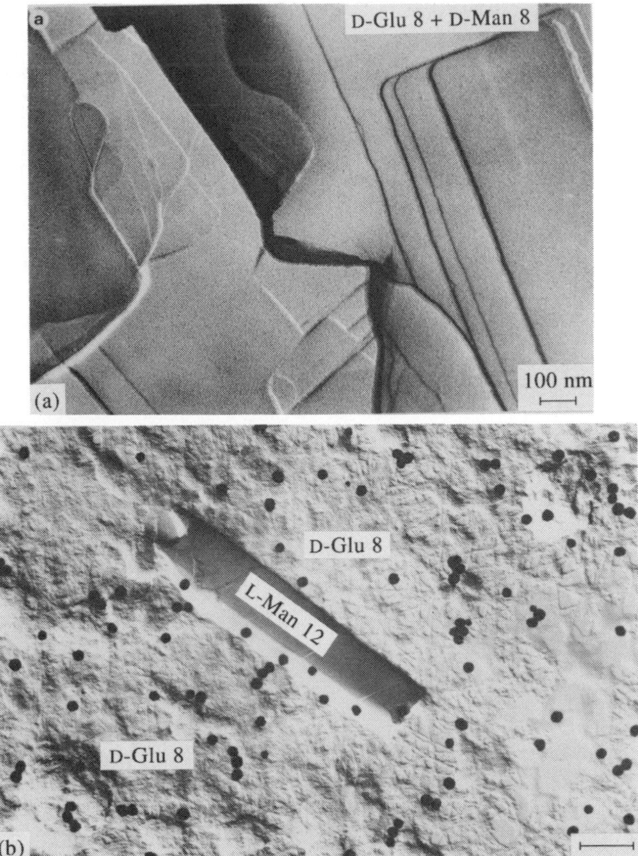

Figure 34 Different diastereomers may form perfect co-crystals, for example, *N*-octyl-D-glucon and -mannon amides (a), whereas a variation of chain lengths often leads to a complete separation, for example, of fibers with *N*-octyl-D-gluconamide (**24a**) and *N*-dodecyl-L-mannonamide (**25b**) (b). The black dots in (b) are caused by tritiated gluconamide molecules on ultrafine silver halide films. Their distribution indicates that the mannonamide scroll has not taken up any gluconamide.[5]

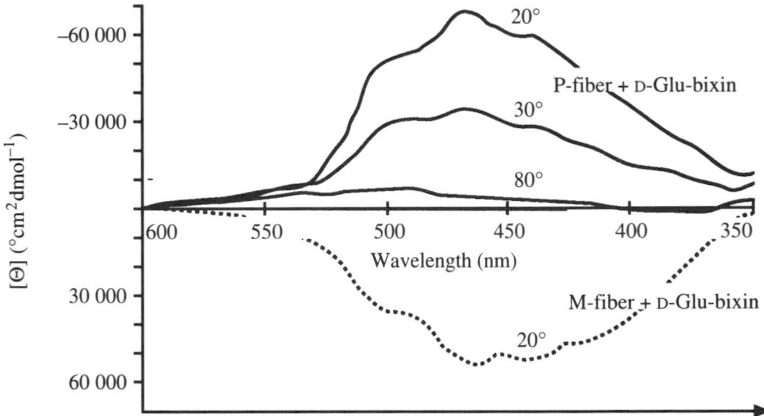

Figure 35 To date, the only dye which can be co-crystallized with solid bilayer fibers is the D-gluconamide of bixin (**34**) with an ethylenediamine linker. The sign of the CD bands is determined by the host fiber, not by the chromophore head group.[131]

dried probes showed 3.5 nm rods with typical lengths of hundreds of nanometers. In this case, it is clearly the hydrogen bonding between the terminal alcohol groups which holds the assembly together in water, since neither the amide groups nor the alkyl chains are accessible for strong intermolecular interactions. Nevertheless, small hydrophobic dyes are dissolved within the hydrophobic cavities within the fiber.[134]

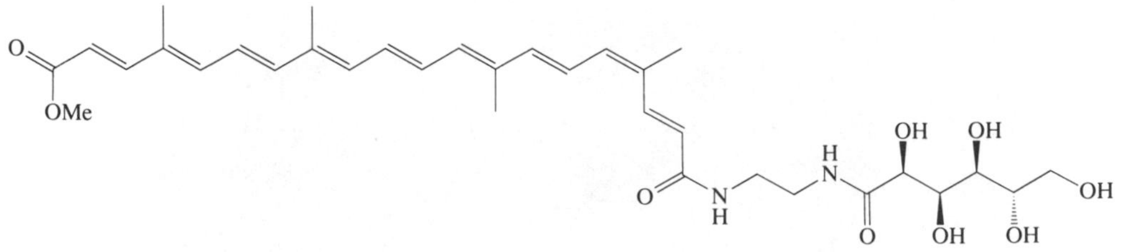

(34)

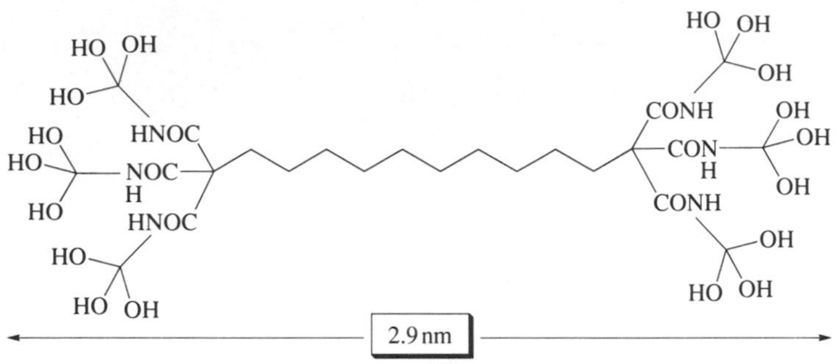

2.9 nm

(35)

11.5 SOLID VESICULAR TUBULES

In Section 11.3, synkinons for spherical micelles and vesicles were introduced. It was indicated that relatively soluble amphiphiles and bolaamphiphiles form small micellar assemblies at relatively high critical concentrations ($\gg 10^{-5}$ M), whereas vesicle synkinons are much less soluble ($\leq 10^{-5}$ M) and assemble at correspondingly low critical concentrations. There are no geometric or stereochemical rules (e.g., critical packing parameters[12]) which determine the ability of a given amphiphilic or bolaamphiphilic molecule to produce micelles or vesicles. Very often micelles are converted to vesicles if the solubility of the amphiphile changes by modification of the environmental conditions (pH, temperature, time) or if different diasteromers or enantiomers are used.

What has been established for spherical micelles and vesicles may also hold true for most of their fibrous analogues, the micellar threads, rods and ribbons and the vesicular tubules. Longer alkyl chains or two alkyl chains instead of one short one tend to favor tubules of rods. For this generalization, for which it is too early because only a few lipid tubules are known, there are, however, already some astounding exceptions in the case of bolaamphiphiles. It may very well be that cooperative binding processes between the head groups and stacking interactions between alkyne groups in the hydrophobic core also enforce tubule formation. The latter effect was introduced in Section 11.3.2.

11.5.1 Amino Acid Derivatives

About 25 different α-amino acids are commercially available as L-enantiomers, most of them also as D-enantiomers. Many of the side chains are reactive and very versatile: —OH, —NH_2, and —CO_2H groups can be used in condensation reactions. —SH is a perfect nucleophile and can be oxidatively dimerized, guanidine is an unsurpassed base and a good metal ligand, and the aromatic side chains fluoresce. In addition to these groups, one can also execute condensation reactions with the amino acid functionalities. As a result, one may introduce one, two, or three hydrophobic chains, any distribution of surface charges, strong hydrogen bridges of secondary amides, and chirality. And all of this comes in the form of a large variety of protected synthons and at a low price. This combination of properties is irresistible and many synthetic amphiphiles are therefore based on amino acids. In the case of fiber synkinesis, alanine and valine, the acidic and basic amino acids, have so far been applied. Vesicular fibers have only been made from glutamic acid, ornithine, and lysine.

The classic compound in this file is the glutamic acid diester (**3**) containing two long-chain alcohols and an ω-substituted fatty acid on the amino group. The amide is then responsible for fiber formation, and the terminal trimethylammonium group provides the water-soluble head group. If this compound is suspended in water and ultrasonicated, spherical vesicles are formed. The thermal energy of the sonication procedure is obviously sufficient to break the amide hydrogen bonds and to enforce the formation of fluid vesicle membranes. However, these vesicles are short-lived, and within a few hours flexible filaments are formed instead. They are detectable under the light microscope. Next, one observes large circular structures and within a month or so huge right-handed helices appear, which later close to tubules (Figure 36).[135]

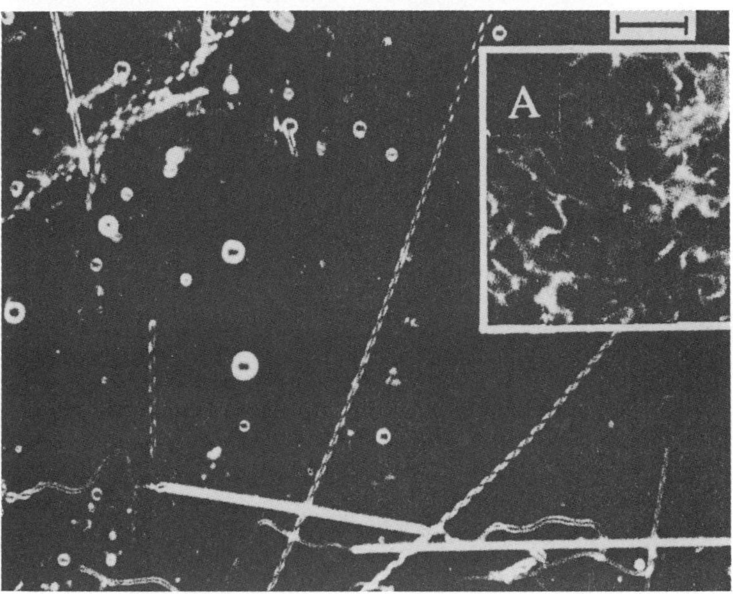

Figure 36 Triple-chain amphiphile (**7**) (Table 2) with a secondary amide link forms vesicles upon sonication. The growth of solid fibers from these vesicular solutions may take weeks, but as a result one eventually obtains beautifully regular tubules (courtesy of Professor T. Kunitake).[135]

Enantiomeric glutamic acid derivatives produced left-handed helices and the racemate gave no chiral structures. Above 40–50 °C, the amide bonds broke and vesicles were formed again. Electron micrographs only showed ill-defined twisted ribbons.[136,137]

Double- or triple-chain amphiphiles made from amino acids and containing only one secondary amide group often produce hydrogen bond chains, or the corresponding carboxyl–carboxylate and phosphoric acid–phosphate diester chains, very slowly and cooperative "melting" occurs at relatively low temperatures (0–40 °C). One often obtains preparations containing fluid and solid bilayer structures side-by-side. Fast assembly processes of uniform preparations with high melting points can usually only be expected for single-chain amphiphiles which contain one or two secondary amide groups.

N-Dodecoyl-β-alanylamide (**36**), a nonchiral single-chain amphiphile, dissolves in water above pH 7 and forms very long 40 nm cylinders at pH 5.2. These cylinders have a high tendency to form bundles and are therefore similar to the tartrate amide rods described in Section 11.4. The published electron micrographs[83] do not reveal an inner water volume, but it should certainly exist.

H_2N H N O

(**36**)

The best defined tubules, however, come from lysine and ornithine bolaamphiphiles (**37**) and (**38**) with a terminal amino group, and one or two amide links. The ornithine derivative (**38a**), for example, can be made on a 10 g scale and is easily soluble in dilute HCl at pH 2. Between pH 4 and pH 9, the compound precipitates without forming supramolecular structures. Above pH 9, the precipitate redissolves and at pH 10.5 uniform, very long tubules appear. The thickness of the membrane is ~3 nm which agrees with the molecular length of ~2.5 nm (Figure 37). The bolaamphiphiles (**37**) and (**38**) carry a large α-amino acid and a small amino head group. One may therefore expect that the ω-amino groups are mostly found at the inner surface. This, however, is very difficult to prove in the solid state. The dialkynic derivative (**38b**) of the α-ornithine, ω-amino bolaamphiphile produced shorter tubules.[138] There was much more phosphotungstate to be found at the inside of the tubules than at the outer surface. It is possible that these tubules are much more asymmetric with respect to head group distribution than the parent compound (**38a**). Similar tubules have also been obtained from the less soluble lysine derivative (**37**) with much longer alkyl chains and two amide links.[139]

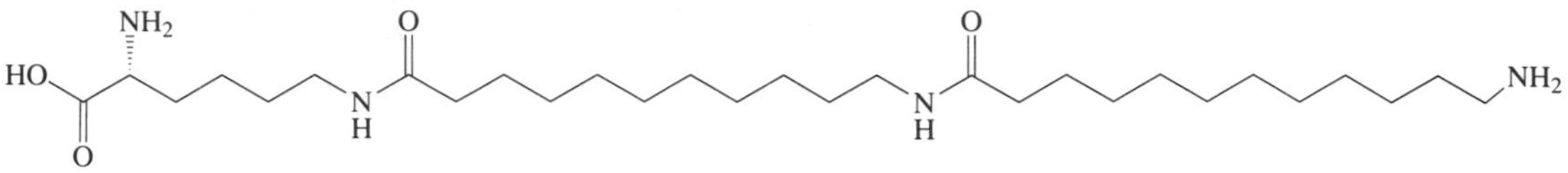

(**37**)

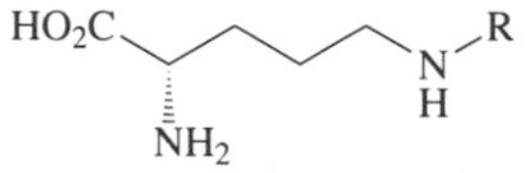

(**38**)

(**a**) $R = (CH_2)_{11}NH_2$
(**b**) $R = (CH_2)_5$—≡—≡—$(CH_2)_2NH_2$

11.5.2 Tartaric and Gluconic Acid Amides

Perfect tubules are formed from D-tartaric acid dodecylamide (**5**) (NH_4^+, Na^+, K^+) salts in water at pH 5, the pK_a value of the carboxyl group (Figure 38). These tubules can easily be prepared in a very large scale and are very long-lived in the case of ammonium and potassium salts.[117] Attempts to introduce staining materials into the central cavities failed, however, because both uranyl and phosphotungstate reacted with the charged head groups or destroyed $CO_2^- \cdots HO_2C$ hydrogen bonds. Electroneutral analogues, namely glyconamides, did not form bilayer membranes in tubular form. They have, however, a very high tendency to roll up into multilayered scrolls. If one introduces a dialkynic unit into the alkyl chain of an *N*-alkyl-D-gluconamide, beautiful bilayer tubules with very high length:diameter ratios are obtained (not shown), which are easily stained with heavy metal salts.

The dialkyne units are completely inactive to UV irradiation, if more than one methylene group separates the triple bonds from the amide group. If, however, only one methylene group is present as in (**11**), rapid and total polymerization occurs. Dialkynic mannonamide scrolls, on the other hand, always polymerize upon UV irradiation. Success depends on both ordering of the molecules and the relative reactivity of the excited dialkyne units. Electron micrographs (Figure 39) show just one big net of fibers with no detectable fiber endings.[31] The resulting polymer is insoluble in all solvents.

11.6 THE CHIRAL BILAYER EFFECT

Many fibrous assemblies, but by no means all of them, were only formed if the head groups were chiral and if pure enantiomers had been used for their preparation. The most prominent case was the *N*-octyl-D-gluconamide (**24**), in which the pure enantiomers produced helices (D-gluconamide is a left-handed helix), whereas the racemate gave platelets without any curvature. In this case, a kinetic argument was used as an explanation. Since the pure enantiomer crystallized only after rearrangement of the bilayer to head-to-tail monolayers, it may be assumed that this rearrangement is slow in water. When crystallization is slow, bilayer fibers, which correspond to very fast-growing whisker-type crystals, may be favored. Racemates, on the other hand, form racemic

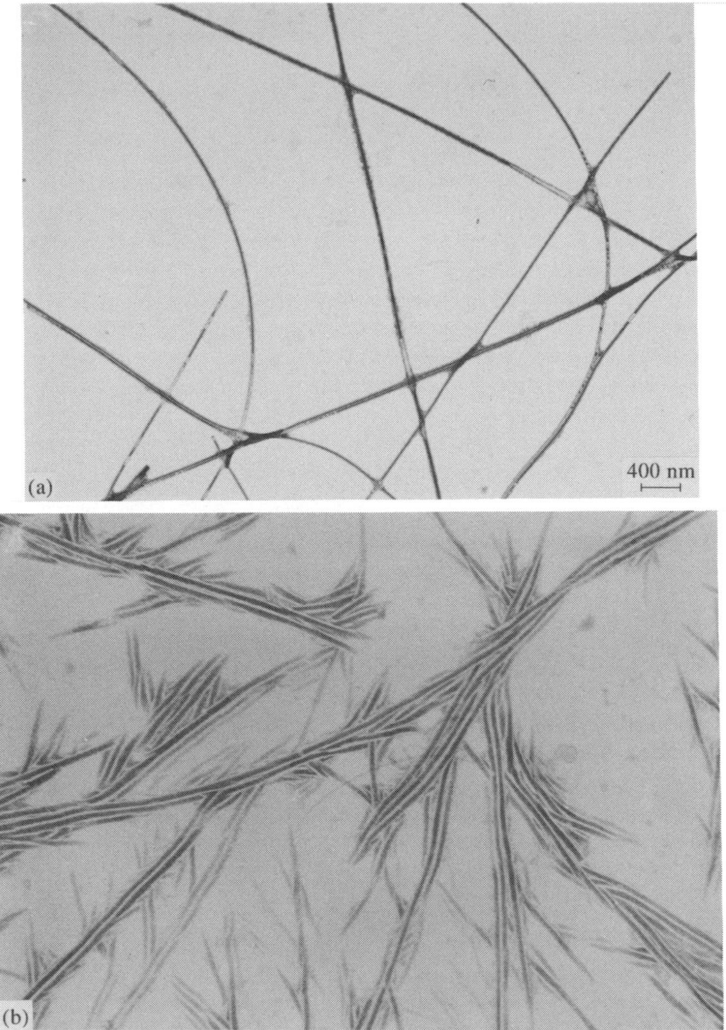

Figure 37 Fast growing tubules of an electron microscopic scale are obtained from bolaamphiphiles with an amino acid and an amino end,[139] for example, (a) (**38a**) and (b) (**38b**).

bilayers without any preference for curved structures. This effect has been called the chiral bilayer effect (Figure 40).[98] In the case of pseudoracemates with different chain lengths, racemic crystals were observed at first, which then separated into left- and right-handed helices (see Figure 37).

Similar effects have been observed in many other instances, but were often accompanied by different morphological consequences. In the case of helical 12-hydroxy-octadecanoic acid salts (**17**) in organic solvents, for example, helices were again found for the pure enantiomers and platelets for the racemate.[70–2] This behavior is quite similar to the aqueous system described above. Mannon- and galactonamides which only occasionally show chiral superstructures, form, on the other hand, similar sheets and ribbons as pure enantiomers and racemates.

Triple chain-glutamides such as (**7**) with a tetraalkylammonium head group give beautiful helices of light microscopic dimensions in water, but produce nontwisted fibers immediately when the racemate is used.[135,136] D-Tartaric acid amides give tubules, the racemate platelets. (*R*)-Eicosanol (**14**) gives tubules in organic media, the racemate only platelets.[117]

As a general result it may be stated that supramolecular helices with a uniform pitch over very long regions are only formed from chiral molecules, in which the chirality leads to a pronounced disturbance of a regular mono- or bilayer. This may possibly be caused by a *gauche* deformation of an all-*trans* chain in a region close to a structure determining amide or hydroxyl hydrogen bond chains within the hydrophobic membrane. Very thin tubules with a very high curvature also require chirality. Nonchiral or racemic amphiphiles only make tubules with diameters of the same magnitude found in vesicles (>20 nm). However, even in the case of tubules with diameters in the micrometer range, very strong chirality effects may be observed if the tubules are produced in apolar solvents.

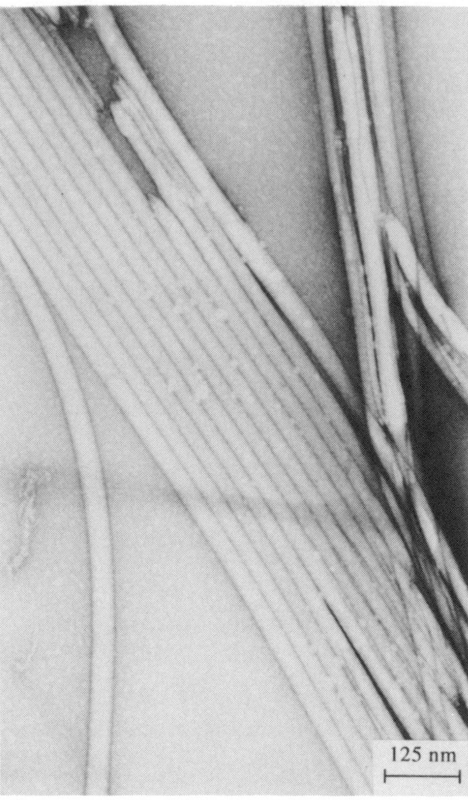

Figure 38 Charged amido amphiphiles, for example, the tartaric acid monoamide (**5**), form a variety of solid fibers when the counterion is changed. The ammonium salt of *N*-(1-dodecyl)-D-tartrate, for example, forms vesicular tubules, whereas the sodium salt tends to produces micellar rods[117] (see Figure 28).

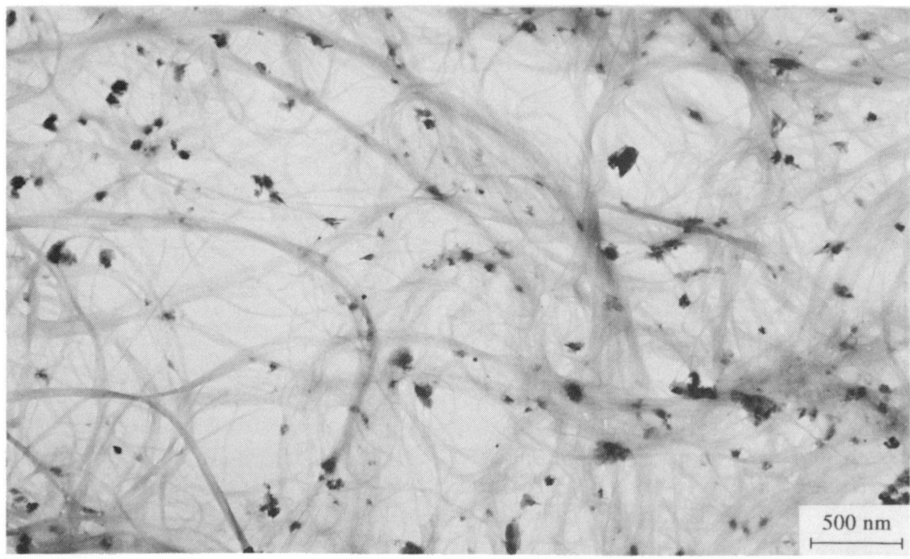

Figure 39 Micellar rods and vesicular tubules may be polymerizable if they contain dialkynic units, for example, in (**11**). The crystalline order of the fibers allows the formation of virtually endless polymers.[31]

The most important generalization, however, is that chirality is only relevant to curvature if there are binding interactions between the head groups. Fluid fibers are not dependent on chiral effects.

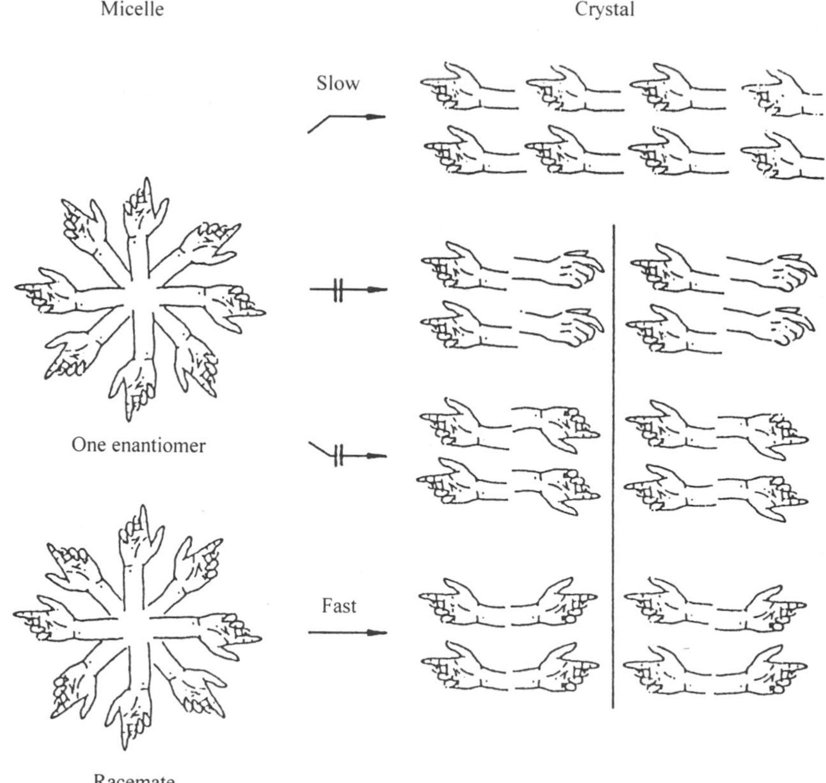

Figure 40 Long-lived solid fibers can only be obtained from chiral amphiphiles. Achiral compounds or racemates tend to form planar crystallites or crystals instead of fibers with strongly curved molecular arrangements. In the case of *N*-octyl-D-gluconamide, it has been shown that bilayer crystals are unstable, whereas monolayer crystals are stable (see Sections 11.4.4 and 11.4.5). Since rearrangement of bilayer fibers to monolayer crystals is slow, the fibrous solids are long-lived (chiral bilayer effect).[98]

11.7 SYNKINETIC RULES FOR FIBERS

Synkinetic fibers are equivalent to synthetic covalent polymers. Their advantage is the extreme simplicity and speed of preparation on a large scale, as well as the easy accessibility of complex monomers. Carbohydrate, amino acid, and porphyrin noncovalent polymers, for example, can all be made within minutes on a 100 mg scale. The preparations are uniform, chiral, and can be used to entrap water-soluble salts and polymers. Their limitation is that nothing but aqueous or organic gels has been made from these noncovalent fibers to date. They have not been processed to any other useful material. If one wishes to apply them, it has to be in the form of aqueous gels or as dry rigid rods in photophysical or photochemical devices.

Very long fluid threads can be made from charged single-chain amphiphiles of medium chain length. So far C_{14} and C_{16} chains have been most successful. The counterion should be aromatic and contain hydrogen-bonding groups. Stacking of the aromatic rings in the membrane surface presumably induces linear ordering and short hydrogen bond chains stabilize it. CTAB–salicylic acid and myristic acid–ephedrine are typical examples. Such fibers can be spun.

Solid rods have only been made from amides. Ultrathin bilayers (<4 nm) can be made from chiral acids as head groups (amino acids, gluconic, tartaric) and alkyl or porphyrin amines. Helices of light microscopic or micrometer dimensions are better obtained from chiral amide amphiphiles with two alkyl chains.

Long tubules with inner diameters of ≥ 8 nm and wall thicknesses down to 2 nm are most easily made from bolaamphiphiles with an amino head group on one end and almost any other head group at the other, provided that there is a secondary amide link in the connecting alkyl chain. The combination of strongly interacting chiral head groups and weakly interacting amino groups is probably the most potent combination. Bilayer tubules of larger sizes are available from all poorly-soluble amphiphiles, which form vesicles and contain secondary amide groups. Rearrangement of vesicles to fibers is often a very slow process. It may take many days. Fiber formation from micelles, on the other hand, occurs in the timescale of seconds or less.

11.8 REFERENCES

1. D. Voet and J. G. Voet, "Biochemistry," Wiley, New York, 1990.
2. J.-H. Fuhrhop and J. Köning, "Membranes and Molecular Assemblies: The Synkinetic Approach," The Royal Society of Chemistry, Cambridge, 1994.
3. (a) F. M. Menger, *Acc. Chem. Res.*, 1979, **12**, 111; (b) K. Raghavan, M. R. Reddy, and M. L. Berkowitz, *Langmuir*, 1992, **8**, 233.
4. J.-H. Fuhrhop and W. Helfrich, *Chem. Rev.*, 1993, **93**, 1565.
5. J.-H. Fuhrhop and C. Böttcher, *J. Am. Chem. Soc.*, 1990, **112**, 2827.
6. E. M. Landau and P. L. Luisi, *J. Am. Chem. Soc.*, 1993, **115**, 2102.
7. B. Henne, Dissertation, Freie Universität Berlin, 1992.
8. P. Meakin, in "The Fractal Approach to Heterogeneous Chemists," ed. D. Anir, Wiley, New York, 1990.
9. M. Matsushita, in "The Fractal Approach to Heterogeneous Chemistry," ed. D. Anir, Wiley, New York, 1990.
10. K. Kalyanasundaram and J. K. Thomas, *J. Phys. Chem.*, 1976, **80**, 1462.
11. G. Cevc and D. Marsh, "Phospholipid Bilayers," Wiley, New York, 1987.
12. J. N. Israelachvili, "Intermolecular and Surface Forces," Academic Press, London, 1985.
13. (a) J. H. Fendler, "Membrane Mimetic Chemistry," Wiley, New York, 1982; (b) D. D. Lasic, "Liposomes," Elsevier, Amsterdam, 1993.
14. (a) W. Helfrich, *Z. Naturforsch., Teil C*, 1973, **28**, 693; (b) M. D. Mitor, J. F. Faucon, P. Méteard, and P. Bothorel, *Adv. Supramol. Chem.*, 1992, **2**, 93.
15. J.-H. Fuhrhop and J. Mathieu, *Angew. Chem., Int. Ed. Engl.*, 1984, **23**, 100.
16. J.-H. Fuhrhop and D. Fritsch, *Acc. Chem. Res.*, 1986, **19**, 130.
17. J.-H. Fuhrhop and J. Mathieu, *J. Chem. Soc., Chem. Commun.*, 1983, 144.
18. J.-H. Fuhrhop, H.-H. David, J. Mathieu, U. Liman, H.-J. Winter, and E. Boekema, *J. Am. Chem. Soc.*, 1986, **108**, 1785.
19. K. Liang and Y. Hui, *J. Am. Chem. Soc.*, 1992, **114**, 6588.
20. J.-H. Fuhrhop, D. Fritsch, B. Tesche, and H. Schmiady, *J. Am. Chem. Soc.*, 1984, **106**, 1998.
21. G. A. Jeffrey and W. Saenger, "Hydrogen Bonding in Biological Structures," Springer, Berlin, 1991.
22. (a) A. J. Doig and D. H. Williams, *J. Am. Chem. Soc.*, 1992, **114**, 338; (b) F. M. Di Capua, S. Swaminathan, and D. L. Beveridge, *J. Am. Chem. Soc.*, 1991, **113**, 6145.
23. N. E. Zhou, B.-Y. Zhu, B. D. Sykes, and R. S. Hodges, *J. Am. Chem. Soc.*, 1992, **114**, 4320.
24. W. R. Hargreaves and D. W. Deamer, *Biochemistry*, 1978, **17**, 3759.
25. H. Ringsdorf, B. Schlarb, and J. Venzmer, *Angew. Chem., Int. Ed Engl.*, 1988, **27**, 113.
26. N. Yamada and M. Kawasaki, *J. Chem. Soc., Chem. Commun.*, 1990, 568.
27. M. Shimomura and T. Kunitake, *J. Am. Chem. Soc.*, 1982, **104**, 1757.
28. N. Nakashima, S. Asakuma, and T. Kunitake, *J. Am. Chem. Soc.*, 1985, **107**, 509.
29. J. H. Georger, A. Singh, R. R. Price, J. M. Schnur, P. Yager, and P. E. Schoen, *J. Am. Chem. Soc.*, 1987, **109**, 6169.
30. P. A. Fraenkel and D. F. O'Brien, *J. Am. Chem. Soc.*, 1991, **113**, 7436.
31. J.-H. Fuhrhop, P. Blumtritt, C. Lehmann, and P. Luger, *J. Am. Chem. Soc.*, 1991, **113**, 7437.
32. V. Enkelmann, *Adv. Polym. Sci.*, 1984, **63**, 91.
33. N. K. P. Samuel, M. Singh, K. Yamaguchi, and S. L. Regen, *J. Am. Chem. Soc.*, 1985, **107**, 42.
34. (a) R. Elbert, A. Laschewsky, and H. Ringsdorf, *J. Am. Chem. Soc.*, 1985, **107**, 4134; (b) J.-H. Fuhrhop, D. Spiroski, and P. Schnieder, *Reactive Polym.*, 1991, **15**, 215.
35. Y. Saikaigudin, T. Shikata, H. Urakami, A. Tamura, and H. Hirata, *J. Electron Microsc.*, 1987, **36**, 168.
36. T. M. Clausen, P. K. Vinson, J. R. Minter, H. T. Davis, J. Talmon, and W. G. Miller, *J. Phys. Chem.*, 1992, **96**, 474.
37. H. Hoffmann and G. Ebert, *Angew. Chem., Int. Ed. Engl.*, 1988, **27**, 902.
38. H. Hoffmann and W. Ulbricht, *Tenside Surfactants, Deterg.*, 1987, **24**, 1.
39. (a) C. Manohar, U. R. K. Rao, B. S. Valanlikar, and R. M. Iyer, *J. Chem. Soc. Chem. Commun.*, 1986, 379; (b) T. Shikata, H. Hirata, and T. Kotaka, *Langmuir*, 1989, **5**, 398.
40. J. C. Brackman and J. B. F. N. Engberts, *J. Am. Chem. Soc.*, 1990, **112**, 872.
41. D. Papahadjopoulos, W. J. Vail, K. Jacobson, and G. Poste, *Biochim. Biophys. Acta*, 1975, **394**, 483.
42. K.-C. Liu, R. M. Weis, and H. M. McConnell, *Nature*, 1982, **296**, 164.
43. I. Sakurai, T. Karvabura, A. Dakurai, A. Kegami, and T. Setoi, *Mol. Cryst. Liq. Cryst.*, 1985, **130**, 203.
44. H. Ringsdorf, B. Schlarb, and J. Venzmer, *Angew. Chem.*, 1988, **100**, 117; *Angew. Chem., Int. Ed. Engl.*, 1988, **27**, 113.
45. F. M. Menger and N. Balachander, *J. Am. Chem. Soc.*, 1992, **114**, 5863.
46. J. H. Georger, A. Singh, R. R. Price, J. M. Schnur, P. Yager, and P. E. Schoen, *J. Am. Chem. Soc.*, 1987, **109**, 6169.
47. A. Singh and J. M. Schnur, *Synth. Commun.*, 1986, **16**, 847.
48. A. S. Rudolph, J. M. Calvert, M. E. Ayers, and J. M. Schnur, *J. Am. Chem. Soc.*, 1989, **111**, 8516.
49. J. M. Schnur, R. R. Price, P. Schoen, P. Yager, J. M. Calvert, J. Georger, and A. Singh, *Thin Solid Films*, 1987, **152**, 181.
50. A. Singh, T. G. Burke, J. M. Calvert, J. H. Georger, B. Herendeen, R. R. Price, P. E. Schoen, and P. Yager, *Chem. Phys. Lipids*, 1988, **47**, 135.
51. E. M. Arnett and J. M. Gold, *J. Am. Chem. Soc.*, 1982, **104**, 636.
52. B. W. Thair and G. R. Lister, *Can. J. Bot*, 1975, **53**, 1063.
53. G. R. Lister and B. W. Thair, *Can. J. Bot.*, 1981, **59**, 640.
54. M. Riederer, *Ecol. Stud.*, 1989, **77**, 157.
55. M. S. Günthardt, *Bot. Helv.*, 1985, **95**, 5.
56. J.-H. Fuhrhop, T. Bedurke, A. Hahn, S. Grund, J. Gatzmann, and M. Riederer, *Angew. Chem., Int. Ed. Engl.*, 1994, **33**, 350.
57. S. Beckmann and H. Schühle, *Z. Naturforsch., Teil B*, 1968, **23**, 471.
58. F. Rybnikar and P. H. Geil, *Biopolymers*, 1972, **11**, 271.
59. (a) G. Scheibe, *Angew. Chem.*, 1936, **49**, 13; (b) G. Scheibe, E-Daltruzzo, O. Wörz, and J. Veiss, *Z. Phys. Chem.*, 1969, **64**, 97.
60. E. E. Jelley, *Nature (London)*, 1936, **138**, 1009.

61. B. Norden, *J. Phys. Chem.*, 1977, **81**, 151.
62. F. Katheder, *Kolloid Z.*, 1941, **93**, 28.
63. S. F. Mason, *Proc. Chem. Soc. London*, 1964, 119.
64. C. Honda and H. Hada, *Tetrahedron Lett.*, 1976, 177.
65. E. S. Emerson, M. A. Conlin, A. E. Rosenoff, K. S. Norland, H. Rodriguez, D. Chin, and G. R. Bird, *J. Phys. Chem.*, 1967, **71**, 2396.
66. F. Dietz, *Tetrahedron*, 1972, **28**, 1403.
67. A. H. Herz, *Adv. Colloid. Interface Sci.*, 1977, **8**, 237.
68. M. Kawasaki and H. Ishii, *Chem. Lett.*, 1994, 1079.
69. A. H. Herz, *Photogr. Sci. Eng.*, 1974, **18**, 323.
70. T. Tachibana, S. Kitazawa, and H. Takeno, *Bull. Chem. Soc. Jpn.*, 1970, **43**, 2418.
71. T. Tachibana and H. Kambara, *J. Am. Chem. Soc.*, 1965, **87**, 3015.
72. T. Tachibana, T. Yoshizumi, and K. Hori, *Bull. Chem. Soc. Jpn.*, 1979, **25**, 34.
73. A. L. McClellan, *J. Chem. Phys.*, 1960, **32**, 1271.
74. W. Kling and H. Mahl, *Fette Seifen Anstrichm.*, 1955, **57**, 643.
75. K. Sakamoto and M. Hatano, *Bull. Chem. Soc. Jpn*, 1980, **53**, 339.
76. T. Kunitake, N. Nakashima, and K. Morimitsu, *Chem. Lett.*, 1980, 1347.
77. T. Kunitake, N. Yamada, and N. Fukunaga, *Chem. Lett.*, 1984, 108.
78. K. Praefcke, A.-M. Levelut, B. Kohne, and A. Echert, *Liq. Cryst.*, 1989, **6**, 263.
79. H. Hidaka, M. Murata, and T. Onai, *J. Chem. Soc., Chem. Commun.*, 1984, 562.
80. T. Kunitake and N. Yamada, *J. Chem. Soc., Chem. Commun.*, 1986, 655.
81. W. Helfrich, *J. Chem. Phys.*, 1986, **85**, 1085.
82. N. Yamada, T. Sasaki, H. Murate, and T. Kunitake, *Chem. Lett.*, 1989, 205.
83. T. Imae, Y. Takahashi, and H. Muramatsu, *J. Am. Chem. Soc.*, 1992, **114**, 3414.
84. J.-H. Fuhrhop, D. Spiroski, and C. Böttcher, *J. Am. Chem. Soc.*, 1993, **115**, 1600.
85. (a) H. Yanagawa, Y. Ogawa, H. Furuta, and K. Tsuno, *Chem. Lett.*, 1988, 269; (b) H. Yanagawa, Y. Ogawa, H. Furuta, and K. Tsuno, *Chem. Lett.*, 1989, 403.
86. H. Yanagawa, Y. Ogawa, H. Furuta, and K. Tsuno, *J. Am. Chem. Soc.*, 1989, **111**, 4567.
87. Y. Hojima, Y. Ogawa, K. Tsuno, N. Hand, and H. Yanagawa, *Biochemistry*, 1992, **31**, 4757.
88. W. Saenger, "Principles of Nucleic Acid Structure," Springer, Berlin, 1984, chap. 13.
89. E. Fischer and B. Helfrich, *Liebigs Ann. Chem.*, 1911, **383**, 68.
90. C. R. Noller and W. C. Rockwell, *J. Am. Chem. Soc.*, 1938, **60**, 2076.
91. (a) G. A. Jeffrey, *Acc. Chem. Res.*, 1986, **19**, 168; (b) G. A. Jeffrey and L. M. Wingert, *Liq. Cryst.*, 1992, **12**, 179.
92. D. C. Carter, J. R. Ruble, and G. A. Jeffrey, *Carbohydr. Res.*, 1982, **102**, 59.
93. G. A. Jeffrey and S. Bhattacharjee, *Mol. Cryst. Liq. Cryst.*, 1983, **101**, 247.
94. K. Shinoda and E. Hutchinson, *J. Phys. Chem.*, 1962, **66**, 577.
95. B. Pfannemüller and W. Welte, *Liq. Cryst.*, 1986, **1**, 357.
96. H. A. van Doren and L. M. Wingert, *Recl. Trav. Chim. Pays-Bas*, 1994, **113**, 260.
97. B. Pfannemüller and W. Welte, *Chem. Phys. Lipids*, 1985, **37**, 227.
98. J.-H. Fuhrhop, P. Schnieder, J. Rosenberg, and E. Boekema, *J. Am. Chem. Soc.*, 1987, **109**, 3387.
99. J. Köning, C. Böttcher, H. Winkler, E. Zeitler, Y. Talmon, and J.-H. Fuhrhop, *J. Am. Chem. Soc.*, 1993, **115**, 693.
100. J.-H. Fuhrhop and J. Köning, "Membranes and Molecular Assemblies: The Synkinetic Approach," The Royal Society of Chemistry, Cambridge, 1994, pp. IX, 118.
101. A. D. Bangham and R. N. Horne, *J. Mol. Biol.*, 1964, **8**, 660.
102. P. Fromherz, C. Röcker, and D. Ruppel, *Faraday Discuss. Chem. Soc.*, 1986, 39.
103. (a) M. R. Ghadiri, J. R. Granja, R. A. Milligan, D. E. McRee, and N. Khazanovich, *Nature (London)*, 1993, **366**, 324; (b) M. R. Ghadiri, J. R. Granja, and L. K. Buehler, *Nature (London)*, 1994, **369**, 301.
104. J.-H. Fuhrhop, P. Schnieder, E. Boekema, and W. Helfrich, *J. Am. Chem. Soc.*, 1988, **110**, 2861.
105. J.-H. Fuhrhop, S. Svenson, C. Böttcher, R. Bach, and P. Schnieder, in "Molecular Mechanisms in Bioorganic Processes," eds. C. Bleasdale and B. Golding, Royal Society of Chemistry, Cambridge, 1990, pp. 65–82.
106. J.-H. Fuhrhop, S. Svenson, C. Böttcher, E. Rösler, and H.-M. Vieth, *J. Am. Chem. Soc.*, 1990, **112**, 4307.
107. V. Zabel, A. Müller-Fahrnow, R. Hilgenfeld, W. Saenger, B. Pfannemüller, V. Enkelmann, and W. Welte, *Chem. Phys. Lipids*, 1986, **39**, 313.
108. C. André, P. Luger, S. Svenson, and J.-H. Fuhrhop, *Carbohydr. Res.*, 1992, **230**, 31.
109. C. André, P. Luger, S. Svenson, and J.-H. Fuhrhop, *Carbohydr. Res.*, 1993, **241**, 47.
110. A. Müller-Fahrnow, W. Saenger, D. Fritsch, P. Schnieder, and J.-H. Fuhrhop, *Carbohydr. Res.*, 1993, **242**, 11.
111. N. Darbon-Meyssonnier, Y. Oddon, E. Decoster, A. A. Pavia, G. Pepe, and J. P. Reboul, *Acta Crystallogr., Sect. C*, 1985, **41**, 1324.
112. S. Svenson, B. Kirste, and J.-H. Fuhrhop, *J. Am. Chem. Soc.*, 1994, **116**, 11969.
113. S. Svenson, J. Köning, and J.-H. Fuhrhop, *J. Phys. Chem.*, 1994, **98**, 1022.
114. S. Svenson, A. Schäfer, and J.-H. Fuhrhop, *J. Chem. Soc., Perkin Trans. 2*, 1994, 1023.
115. H. H. Mantsch and R. N. McElhany, *Chem. Phys. Lipids*, 1991, **57**, 213.
116. J. K. Whitesell, T. La Cour, R. L. Lovell, J. Pojman, P. Ryan, and A. Yamada-Nosaka, *J. Am. Chem. Soc.*, 1988, **110**, 991, and references therein.
117. J.-H. Fuhrhop, C. Demoulin, J. Rosenberg, and C. Böttcher, *J. Am. Chem. Soc.*, 1990, **112**, 2827.
118. C. A. Hunter and J. K. M. Sanders, *J. Am. Chem. Soc.*, 1990, **112**, 5525.
119. (a) H. Fischer and H. Orth, "Die Chemie des Pyrrols," Akademie Verlag, Leipzig, 1937, vol. 2, p. 396; (b) R. G. Little, K. R. Dymock, and J. A. Iber, *J. Am. Chem. Soc.*, 1975, **97**, 4532.
120. J.-H. Fuhrhop, C. Demoulin, C. Böttcher, J. Köning, and U. Siggel, *J. Am. Chem. Soc.*, 1992, **114**, 4158.
121. I. Inamura and K. Uchida, *Bull. Chem. Soc. Jpn.*, 1991, **64**, 2005.
122. J.-H. Fuhrhop, U. Bindig, and U. Siggel, *J. Am. Chem. Soc.*, 1993, **115**, 11036.
123. J.-H. Fuhrhop, U. Bindig, and U. Siggel, *New. J. Chem.*, 1995, **19**, 427.
124. M. Kaska, M. A. El-Bayoumi, and W. Thodes, *J. Chim. Phys.*, 1964, **58**, 916.

125. J.-H. Fuhrhop, U. Bindig, and U. Siggel, *J. Chem. Soc., Chem. Comm.*, 1994, 1583.
126. T. Komatsu, K. Nakao, H. Nishide, and E. Tsuchida, *J. Chem. Soc., Chem. Commun.*, 1993, 728.
127. C. Endisch, C. Böttcher, and J.-H. Fuhrhop, *J. Am. Chem. Soc.*, 1995, **117**, 8273.
128. J. R. E. Fisher, V. Rosenbach-Belkin, and A. Scherz, *Biophys. J.*, 1990, **58**, 461.
129. J.-H. Fuhrhop, P. Schnieder, E. Boekema, and W. Helfrich, *J. Am. Chem. Soc.*, 1988, **110**, 2861.
130. S. Neustadt, Ph. D. Thesis, Freie Universität Berlin, 1995.
131. J.-H. Fuhrhop, M. Krull, A. Schulz and D. Möbius, *Langmuir*, 1990, **6**, 497.
132. H. Möhwald, A. Miller, W. Stich, and W. Knoll, *Thin Solid Films*, 1986, **141**, 261.
133. R. E. Dickerson and I. Geis, "Hemoglobin," Benjamin, Menlo Park, CA, 1983.
134. G. R. Newkome, G. R. Baker, S. Arai, M. J. Saunders, P. S. Russo, K. J. Theriot, C. N. Moorefield, L. E. Rogers, J. E. Müller, T. R. Lieux, M. E. Murray, B. Phillips, and L. Pascal, *J. Am. Chem. Soc.*, 1990, **112**, 8458.
135. N. Nakashima, S. Asakuma, and T. Kunitake, *J. Am. Chem. Soc.*, 1985, **107**, 509.
136. N. Nakashima, S. Asakuma, T. Kunitake, and H. Hotani, *Chem. Lett.*, 1984, 227.
137. T. Kunitake and N. Yamada, *J. Chem. Soc., Chem. Commun.*, 1986, 655.
138. S. Rocchetti, Ph. D. Thesis, Freie Universität Berlin, 1995.
139. J.-H. Fuhrhop, D. Spiroski, and C. Böttcher, *J. Am. Chem. Soc.*, 1993, **115**, 1600.

12
Control of Peptide Architecture via Self-assembly and Self-organization Processes

DAVID H. LEE and M. REZA GHADIRI
The Scripps Research Institute, La Jolla, CA, USA

12.1 INTRODUCTION

Natural polypeptides can display astonishingly large sequence-space variations.[1] Exactly how the one-dimensional sequence information of polypeptides translates into the three-dimensional folded structure of functional proteins is only partially understood. The goal of this chapter is to review recent advances in the rational design and control of peptide self-assembly and self-organization directed at the design of supramolecular structures. It begins by reviewing some pertinent studies that aim to understand as well as mimic simple naturally occurring peptide- and protein-folding motifs followed by the description of advances in the design of peptide-based supramolecular structures and biomaterials. Section 12.2 deals with self-organization of secondary structures in short peptides by describing strategies by which secondary structures were formed in peptides which would otherwise have random coil conformations. Examples of peptides which can interconvert between helical and β-strand conformations will also be discussed. Section 12.3 deals with higher order polypeptide structures and the various strategies used to assemble them. Studies directed at elucidating the features important for coiled coil assembly have provided much valuable information. Complementary to these studies are those of *de novo* protein design. While there are several good examples of artificial proteins, all are still akin to the molten globule intermediate found on the folding pathway of some proteins. It is anticipated that the information learned from studies on coiled coil systems will soon be used to increase the specificity of structure formation in the design of helical bundles. With several successful designed scaffolds in hand, a few groups have even extended those works as far as to make functional proteins and many others are now trying as well. Finally, peptide assembly on nanoscopic and macroscopic scales, and their utility as biomaterials, is discussed.

12.2 SELF-ORGANIZATION OF PEPTIDE SECONDARY STRUCTURE

What are the factors that lead to peptide self-organization and assembly? Peptides can be dissected into two closely interacting parts—the amide backbone and the side chain moieties. The amide backbone functionality endows the polymer chain with several unique properties. First, the planarity of the amide bond restricts the backbone to only two dihedral $\Phi\Psi$ angles. Second, side chain–backbone steric interactions further reduce allowed conformational states of the backbone. Finally, the amide portion of the backbone has the important feature of being able to act simultaneously as both a hydrogen bond donor and acceptor. This property allows the backbone to fold onto itself to create secondary structures such as α-helices, β-sheets, turns, and loops, forming the scaffold for side chain functional group presentation.

The amino acid side chains which confer the desired functions on a folded structure also play intimate roles in assisting peptide folding and self-organization. The role of the hydrophobic effect as the major stabilizing force of tertiary structure is well known. Hence, the presence of hydrophobic side chains in the polypeptide drives certain parts of the backbone into the core of the protein where they are sequestered from water, while hydrophilic residues are presented at the surface to provide water solubility to an otherwise hydrophobic particle. The interplay of hydrophobicity and hydrophilicity of side chains and backbone hydrogen bonding gives rise to the complex structures seen in proteins. Exactly how to control these factors to produce structurally and functionally predetermined protein-like structures is currently the central goal of protein design and engineering.

12.2.1 Conformational Switching in Peptides

Conformational changes in proteins are essential for proper function but are poorly understood phenomena. They are often invoked to explain some mechanism of enzyme function but the nature of these changes, for the most part, has not been well characterized. For instance, lysozyme was observed to change conformation while catalyzing proteolysis by atomic force microscopy,[2] and conformational changes were inferred from modifications to the membrane protein colicin.[3] In both these cases, however, detailed knowledge of the conformational change is lacking. One dramatic example of a well-characterized conformational change is the pH-induced conversion of loops into α-helices to form a long three-stranded coiled coil in hemagluttinin.[4] While this conformational change is not reversible, many others are. Understanding of this conformational switching is critical to understanding protein structure and function and will likely be important in future designs of artificial proteins.

The α-helix–β-strand transitions of poly-L-lysine were first documented long ago.[5–7] The conformation of this homopolymer is dependent on both the pH and the temperature.[7] At a pH of less than 10.5 the polymer is disordered. Above pH 10.5 and at temperatures below 30 °C the peptide adopts an α-helical structure. Elevating the temperature converts the α-helix to β-sheet conformation. The sheet size likely depends on the concentration of the solution. At low concentrations of polylysine, it is thought that the peptides form intramolecular β-sheets; at higher concentrations, both inter- and intramolecular forms were observed. No explanations have been offered as to why this transition occurs. Thus, while the polymer clearly needs to be uncharged in order to adopt a conformation which places its lysine side chains in proximity to one another, the heat-induced transition from helical to sheet structure is not understood.

Design principles learned in the synthesis of artificial proteins, however, have enabled several groups to make peptides which can switch conformation depending on their environment. One such peptide was designed by the Mutter laboratory wherein it underwent medium-induced conformational switching.[8] The peptide was amphiphilic only if it adopted a β-sheet conformation. However, the residues making up the peptide had high helical preference. In 2,2,2-trifluoroethanol

(TFE) the peptide adopted a helical conformation. In water, however, it clearly adopted a β-sheet conformation, despite the high helical preference of the composite amino acid residues. It seems that the hydrophobic interactions between the peptides overrode the helical preference of the peptide, forcing the formation of sheet aggregate structure. TFE, however, disrupts such hydrophobic interactions and promotes helix formation. A cooperative transition from α-helix to β-sheet occurred at 55% water/TFE.

Peptides which switch conformations with changes in pH were also designed by the Mutter laboratory.[9] The design here was based on amphiphilicity of the peptide in both helical and sheet conformations. This design criterion was necessary to ensure stabilization of the secondary structure through association once it was formed. An amphiphilic β-sheet was first designed. The sequence was then modified by replacing hydrophobic and hydrophilic residues that came to rest on the hydrophilic and hydrophobic faces, respectively, of an amphiphilic helix, with alanine. Alanine was felt to be indifferent as to its hydrophobicity relative to lysine, glutamic acid, and leucine, which were used to make the peptide. Four 16-residue peptides were made and all showed pH-dependent conformational transitions. Indeed, two of the peptides exhibited three conformational transitions. However, the pH values at which these conformational transitions occur could not be predicted, and so like most other examples the cause of the switching is not well understood.

Dado and Gellman designed a conformation-switching peptide which also made use of amphiphilic structures in both the helical and sheet conformations.[10] The unique feature of this design, however, is the incorporation of methionine, which can act as either a hydrophobic residue or, upon oxidation to the sulfoxide form, a hydrophilic residue. Thus, they designed an amphiphilic α-helix with methionines forming part of the hydrophobic face, particularly those residues that came to rest on the hydrophilic side of an amphiphilic β-sheet. In the reduced form the peptide was helical. Oxidation, however, converted the methionines into the more polar sulfoxide form. In this oxidation state, the only way the peptide could maintain the hydrophobic stabilization it enjoyed as an α-helix was to adopt an aggregated β-sheet conformation, forming yet another amphiphilic structure. The reversible conversion between the two structures was monitored by circular dichroism (c.d.) and IR spectroscopies.

Thus, some successful examples of designed peptides with adjustable conformations exist. Such design principles should enable interesting applications in the design of novel artificial protein constructs.

12.2.2 Metal Ion Assisted Nucleation of Secondary Structure in Short Peptides

The vast majority of stabilizing interactions in proteins are through hydrophobic and electrostatic interactions, many of them being long-range interactions. Unlike proteins, most short peptides (less than about 20 residues) do not adopt a stable conformation in solution since they do not have enough stabilizing interactions to maintain secondary structure. However, if the peptide has a propensity to form a certain structure, the loss in entropy which occurs on structure formation can often be overcome by some externally imposed conformational restraint. While templates and disulfide bridges have also been used to nucleate or constrain the structure, one of the easiest ways to induce ordered conformations in short peptides is through metal ion chelation by ligands appropriately spaced throughout its sequence.

This principle was employed in the stabilization of α-helices as shown in Figure 1. Ghadiri and Choi designed two peptides of high helical propensity with the added feature of having two metal chelators separated by three amino acid residues.[11] The rationale of this design was that chelation of a metal ion by both residues would constrain that region of the peptide to one turn of an α-helix. Since helix formation is a classical nucleation event, it was expected that the rest of the peptide should spontaneously fold once the first turn was formed. In addition, these metal-binding sites were situated near the carboxy-terminus in order to maximize the ion charge–helix dipole interaction. One peptide had histidine and cysteine as ligands (peptide A); the other peptide had two histidines, (peptide B). Titration of both peptides with various transition metals resulted in an increase in helicity, from ~50% in the absence of metal to as high as 90% upon metal ion titration, as observed by c.d. spectroscopy. Moreover, metal ion complexation was corroborated by NMR data. Interestingly, the two peptides showed metal ion selectivity. For instance, peptide A preferentially bound Cd^{2+} over Zn^{2+}. In addition, Ni^{2+} induced more helicity than Cd^{2+} in peptide B. In a separate study, Ghadiri *et al.* went on to make monomeric α-helices of exceptional stability by using the two histidines of 17-residue metal-binding peptides to form exchange-inert ruthenium(III) tetraamine bis(imidazole) complexes.[12] These metallopeptides exhibited 80% helicity at 21°C and were stable enough to be purified by ion-exchange or reversed-phase chromatography. Exclusive derivatization of the histidine residues was supported by NMR studies, thereby illustrating the chemoselectivity of the reaction. A peptide having its metal-binding site along the middle of the α-helix was as helical as its counterpart with the binding site at the carboxy-terminus. Thus, it seems that the location of the metal-binding site along the helix is not important; the critical feature of the peptide is to have two histidines separated by three residues.

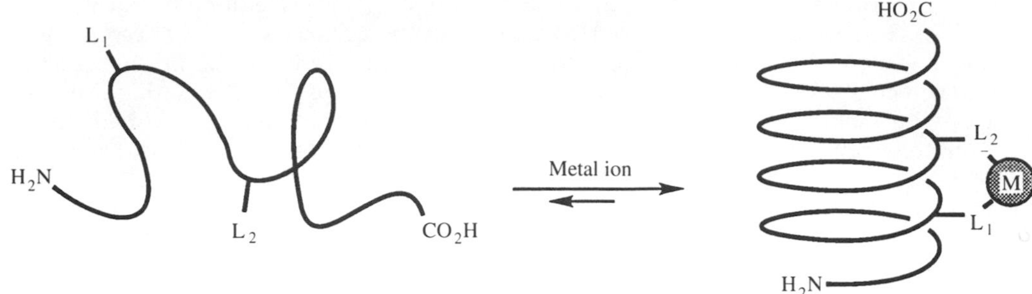

Figure 1 Schematic representation of metal ion assisted nucleation of an α-helix. Two metal-binding ligands separated by three residues can chelate a metal, thereby constraining the peptide backbone to one turn of an α-helix. Nucleation of the rest of the helix ensues.

The effects of changing the spacing between metal-binding residues as well as the length of their side chains on helical structure were illustrated in a study undertaken by the Hopkins laboratory.[13,14] Several peptides were endowed with an unnatural amino acid moiety containing an aminodiacetic acid group connected to the backbone by an alkyl chain varying in length from one to four methylene groups. These amino acid residues served as the high-affinity metal ion chelator. Thus, two such residues were incorporated into the peptide separated by either two or three residues in order to assess the effect of spacing between metal binders on helical structure. They found that the simultaneous coordination of a transition metal generally stabilized a helical structure when the aminodiacetic acid containing moieties were spaced three residues apart. Peptides having their chelating residues spaced two residues apart were generally not stabilized into a helical conformation upon metal ion binding. Not surprisingly, the metal ion specificity of each peptide was dependent on the length of the alkyl side chain.

In 1993, Imperiali and Kapoor provided an example of a metal ion complexing reverse turn.[15] Here the three-residue sequence Val-Pro-D-Ser nucleated a hairpin turn in an eight-residue peptide. Chelation of zinc by the imidazole groups of three histidines was thought to stabilize the structure further. Like the studies on the metal ion stabilized α-helices, the existence of β-structure was supported by c.d. spectroscopy. Involvement of the three histidines in zinc binding was verified by NMR spectroscopy. However, while peptides designed to be α-helices had random coil structure in the absence of the metal, the reverse turn containing peptide already had significant structure. Addition of zinc to a peptide having a flexible glycine-containing region instead of the turn did not induce formation of any ordered structure so the role of zinc clearly was not to organize the conformation of the peptide. Rather, zinc binding most likely stabilized an already prevalent turn structure.

While these secondary structures were *de novo* designed, synthetic peptides derived from proteins have been shown to be organized into ordered structures upon metal ion chelation as well. Most notable are peptides derived from calcium-binding proteins including calbindin D_{9k} and troponin C (TnC). For example, Tsuji and Kaiser synthesized a variety of peptides designed to mimic the amino-terminal calcium-binding helix–loop–helix found in calbindin D_{9k}.[16] By introducing some helix-stabilizing features such as glutamic acid–lysine salt bridges and an interhelical disulfide bridge between helices, they were able to obtain structures which bound Ca^{2+} with millimolar affinity. Calcium binding also significantly increased the helical content in several peptides. While the calcium ion may have helped to organize the structure, Tsuji and Kaiser acknowledge that the increase in helicity may also simply have been due to an increase in ionic strength.

A dramatic example of Ca^{2+}-assisted assembly of supersecondary structure was provided by Shaw *et al.*[17] They demonstrated that a synthetic 34-residue peptide representing the third Ca^{2+}-binding site (SCIII) from the carboxy-terminal domain of TnC adopted a random coil conformation but could be organized into a helix–loop–helix conformation upon addition of Ca^{2+}. Previous studies by Reid *et al.* indicated that shorter peptides were much less efficient at binding calcium ions.[18] The solution structure of the Ca^{2+}-bound form was similar to that described in the crystal structure of TnC.[19] Moreover, it indicated that this helix–loop–helix formed a symmetric homodimer in solution. Interestingly, although the homodimer contains two Ca^{2+}-binding loops, calcium titration experiments monitored by ^{1}H NMR spectroscopy showed that only one Ca^{2+} need be bound before the peptides dimerized.[20] A mechanism was proposed whereby the Ca^{2+}-binding event organizes one peptide into the helix–loop–helix structure, which in turn induces structure into a second free peptide. The cooperativity of the folding process implied by this model is reflected in the equilibrium constants obtained for initial Ca^{2+} binding, binding of the second peptide, and then binding of the second Ca^{2+} ion.

Further studies on synthetic peptides indicate that SCIII prefers to form a heterodimer with the fourth Ca^{2+}-binding domain (SCIV) rather than the homodimer.[21] Ca^{2+}-induced folding in a 1:1 mixture of SCIII and SCIV resulted in a well-dispersed NMR spectrum that is significantly different from homodimers of either SCIII or SCIV. Comparison of the NMR spectrum of the synthetic peptide mixture was very similar to a proteolytic fragment of TnC containing both domains represented by SCIII and SCIV, further supporting heterodimer formation. The Ca^{2+}-binding and folding behavior of the heterodimer was found to be very similar to that observed for the SCIII homodimer.[22]

The zinc finger, shown in Figure 2, is another naturally occurring motif that undergoes metal ion assisted folding.[23,24] The sequence can be represented by (Phe-X-Cys-X_{2-4}-Cys-X_3-Phe-X_5-Leu-X_2-His-X_{3-4}-His-X_{2-6}) where X represents nonconserved amino acids. Frankel *et al.* first showed that 30-residue peptides containing the characteristic zinc-finger sequence from *Xenopus* TFIIIA undergoes a dramatic change from a random coil structure to some other conformation upon titration with zinc.[25] A zinc finger peptide derived from yeast also showed zinc-dependent folding and a preliminary structure based on NMR data was presented.[26] A high-resolution structure was later provided by Lee *et al.*,[27] which revealed the existence of a small two-stranded antiparallel β-sheet connected to an α-helix. The α-helix provided two histidine side chains and the β-turn provided two cysteine side chains to form a tetrahedral metal-binding site. This binding site is surprisingly selective for zinc. The selectivity is attributed to the unique coordinating environment provided by the two imidazole and two thiolate ligands.[28,29] Replacement of each histidine with a "softer" thiolate ligand (in accordance with hard–soft acid–base theory) increased the preference for a softer metal like Cd^{2+}. Furthermore, selectivity of Zn^{2+} over Co^{2+} was rationalized on the basis of a higher ligand field stabilization energy for a tetrahedrally coordinated zinc than a tetrahedrally coordinated Co^{2+}.

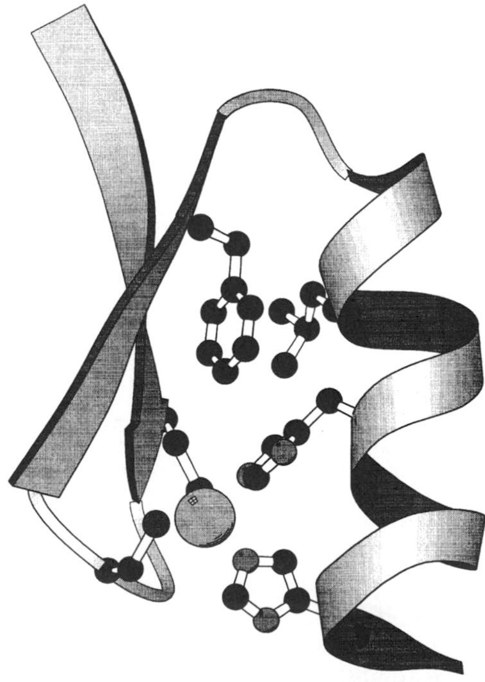

Figure 2 Ribbon diagram of a canonical zinc finger. The metal-binding and hydrophobic side chains essential for proper folding are displayed.

The enormous contribution of metal ion assisted assembly to the folding of zinc-finger peptides was demonstrated by the Berg laboratory when all nonconserved residues in a zinc-finger peptide were replaced with alanine.[30] The only characteristic features of zinc fingers that were retained were the metal-binding residues and the conserved phenylalanine and leucine. Amazingly this "minimalist" zinc-finger peptide folded into a structure very similar to a canonical zinc finger upon addition of 1–1.5 equiv. of Zn^{2+}. The presence of the α-helix and the β-turn was unambiguously determined. Resonance overlap, however, prevented determination of a high-resolution structure.

12.2.3 Nucleating Formation of Secondary Structure through Covalent Strategies

In addition to metal ion assisted assembly or other noncovalent strategies such as side chain–side chain hydrophobic interactions to form secondary structure, various groups have successfully made use of covalent linkages to accomplish this task. One popular strategy was to form covalent links between side chains of appropriately spaced residues to constrain the peptide backbone and nucleate helix formation. Examples include disulfide[31] or lactam bridges.[32]

A second strategy is to attach a template for helix formation onto the peptide. For example, Satterthwait and co-workers constrained a small piece of a peptide backbone into one turn of an α-helix by incorporating a hydrazone linkage to mimic a backbone–backbone hydrogen bond.[33,34] This hydrazone-containing segment successfully templated helix formation when attached to another peptide as determined by NMR spectroscopy.[35] These α-helices are being developed as antigen mimics for the development of vaccines.

Another helix template was made and characterized by the Kemp laboratory.[36] In this case the template consisted of a tricyclic molecule which presented two carbonyl groups spatially positioned to resemble one turn of an α-helix, as shown in Figure 3. As in Satterthwait's example, the entropic cost of forming the first turn of a helix was paid for in the synthesis of the template. One particularly interesting feature of this system is that the template slowly interconverts between a templating form, wherein the amide bond is in a *trans* conformation and a nontemplating *cis*-amide form.[37] Kemp and co-workers took advantage of this process to monitor the helix/random coil ratio of the peptide–template conjugate by NMR spectroscopy instead of fractional helicity. This approach offers results which are less affected by experimental errors and remains one of the best characterized systems to date.

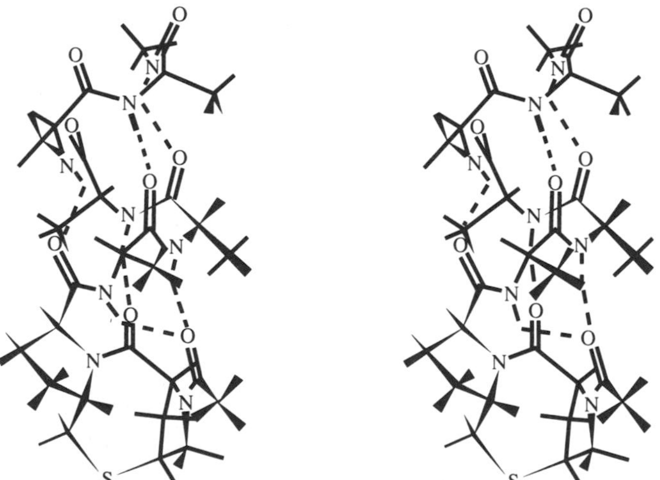

Figure 3 Stereodiagram of the covalent template–α-helix conjugate synthesized by Kemp and co-workers (reprinted with permission from *J. Am. Chem. Soc.*, 1995, **117**, 6641. Copyright 1995 American Chemical Society).

The Kemp group has also made a template which nucleated β-sheet formation.[36,38,39] The template consisted of a diacylaminoepindolidione which presented an array of amide hydrogen bond donors and acceptors in the pattern and spacing of a two-stranded antiparallel β-sheet, as shown in Figure 4. Since the multiring structure is rigid, it serves as a good template for the nucleation of short peptides into an extended conformation. Only one hydrogen bond donor and one acceptor on each side of the epindolidione was required to nucleate a β-sheet structure in a dipeptide attached to an L-proline-D-alanine containing turn. The conformation of this molecule was studied by NMR spectroscopy in dimethylsulfoxide. A later model designed by Kemp and Li used a small organic molecule to mimic a β-turn instead of a β-sheet.[40,41] Methyl-1,2-iodobenzene and 2-ethynylaniline were coupled to make a 2-aminodiphenylacetylene-2′-carboxylic acid to which two amino acids could be attached in an antiparallel orientation. Amino acids attached to this molecule were observed by NMR spectroscopy in DMSO to adopt an extended conformation typical of amino acid residues found in β-sheets. A structure of this molecule in water was not reported.

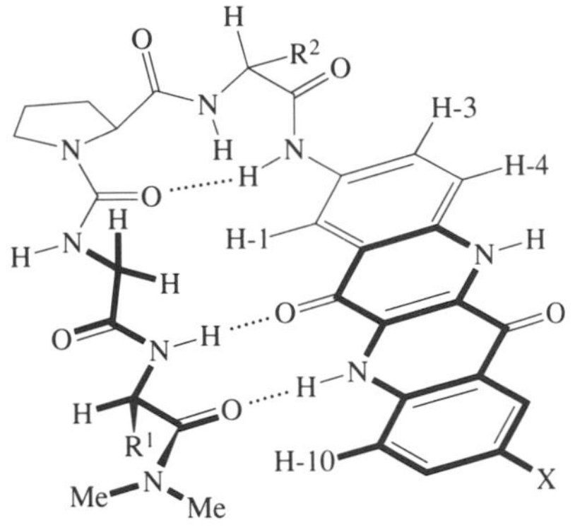

X = repeat of peptide unit

Figure 4 Epindolidione molecule used to template β-sheet formation (reprinted from *Tetrahedron Lett.*, **29**, D. S. Kemp and B. R. Bowen, Conformational analysis of peptide-functionalized diacylaminoepindolidiones. ^{1}H NMR evidence for β-sheet formation, 5081, Copyright 1988, with kind permission from Elsevier Science Ltd, The Boulevard, Langford Lane, Kidlington OX5 1GB, UK).

Kelly and co-workers have successfully used a turn mimetic to form a water-soluble β-structure from short peptides.[42] Diaz *et al.* linked two copies of the tripeptide Val-Lys-Leu in an antiparallel fashion to a dibenzofuran-based amino acid and found that this turn mimetic induced β-sheet formation. This peptide was monomeric in solution over a wide concentration range. The structure was inferred from c.d. and NMR studies carried out in aqueous buffers. Curiously, nuclear Overhauser effects (NOEs) were also observed between protons on the dibenzofuran ring and those on the side chains of the flanking leucine and valine residues.[43] This phenomenon was interpreted to mean that a hydrophobic interaction was occurring among these groups and it was felt that this interaction was important for nucleation of the β-sheet itself. Indeed, further studies were carried out where the flanking hydrophobic residues were replaced with a polar one like lysine or one of small surface area like alanine.[44] These peptide variants could not adopt any β-structure and this was attributed to the hypothesis that these residues could not undergo the hydrophobic clustering that leucine and valine could. LaBrenz and Kelly sought to extend the utility of this hydrophobic association in making a simple peptide receptor.[45] Another dibenzofuran-containing peptide was designed, only this time the attached peptides were parallel, and with a larger spacing between the two than in the original design. This larger spacing was implemented in order to accommodate another peptide. The salient feature of this third peptide was its carboxy-terminal benzamide group which was supposed to provide the initial driving force for binding through an aromatic–aromatic interaction with the dibenzofuran ring. Subsequent interpeptide hydrogen-bonding interactions would then induce the formation of a three-stranded antiparallel β-sheet. Though β-structure was evident from its c.d. spectrum, the self-assembly process did not stop at the three-stranded structure and soluble high molecular weight aggregates were formed. This problem is typical of peptides which form β-sheets and is not surprising considering that they can aggregate not only in a side-to-side manner to form an uninterrupted hydrogen-bonding network but also in a face-to-face fashion through sedition packing, which may be continuous as well.

12.3 ARTIFICIAL PROTEIN DESIGN VIA SELF-ASSEMBLING AND SELF-ORGANIZING PROCESSES

12.3.1 Controlling the Aggregation State

Despite the vast number of conformations that can be adopted by polypeptides, they fold into a limited set of ordered structures. The complexity of proteins is such that the protein-folding problem has not yet been solved. However, some general principles have been learned since 1970. For instance, the role of the hydrophobic effect is crucial to the proper folding of the peptide chain. The pioneering work of Kaiser and Kezdy have shown that this principle can be applied to control the conformation of even short peptides.[46,47] Also, some amino acids are more prevalent in certain elements of secondary structure than others. Since the late 1980s, several ways have been found to bypass the protein-folding problem. One strategy was to use a minimalist approach wherein self-assembling peptides much simpler than those found in Nature were designed. While not every peptide assembled into the desired aggregation state, the simplicity of the molecules enabled DeGrado and co-workers to learn from their mistakes and create better designs. Mutter and co-workers cleverly avoided the protein-folding problem in the synthesis of protein mimics by folding short peptides attached to a template.[48–50] Close proximity of the amphiphilic peptides induced secondary structure. Similarly, Ghadiri and co-workers,[51,52] and Sasaki and co-workers[53,54] discovered a metal-ion assisted assembly process in order to bring together peptides to form helix bundles. These peptides included unnatural amino acids in their sequence to chelate metal ions and thus induce tertiary structure through proximity effects. Finally, in 1993, a novel β-sheet-containing protein having tailored properties was created.[55] These designs and others will be discussed in the following sections.

12.3.2 Coiled Coils

The coiled coil is characterized by two amphiphilic α-helices wound about each other in a left-handed supercoil such that the hydrophobic residues appearing on one face of each helix make contact with each other, as shown in Figure 5. These hydrophobic residues occur in a regular fashion, at positions *a* and *d* of an *abcdefg* heptad sequence which results in a hydrophobic stripe on the surface of the α-helix gently winding about the helical axis. Thus, hydrophobic interactions between two helices are maximized when they coil about each other. This simple motif was first discovered in fibrous proteins of the k-m-e-f (keratin, myosin, epidermin, and fibrinogen) class[56,57] but has since been found in a large number of transcription factors as well such as GCN4, Fos, and Jun. Coiled coils in transcription factors were originally called leucine zippers because it was thought that the isobutyl side chains of the leucine residues occurring in the heptad repeat from each helix were interdigitated, much like the teeth in a closed zipper.[58] This was shown not to be the case from both NMR[59] and x-ray crystallographic studies[60] on 33-residue peptides whose sequence corresponded to the leucine zipper domain of GCN4. Rather, the hydrophobic side chains pack together in a "knobs into holes"-type fashion first proposed by Crick in 1953.[61]

The types of hydrophobic residues that occupy positions *a* and *d* are very important for determining the overall stability and even the oligomerization state of a coiled coil. For instance, the Hodges laboratory addressed the question of the relative contribution of leucine residues in positions *a* and *d* to the stability of a 35-residue coiled coil.[62] The original sequence repeated the heptapeptide Lys-Leu-Glu-Ala-Leu-Glu-Gly five times over and also had a cysteine near either the beginning or the end of the sequence in order to align the peptides in a parallel and in-register fashion when forming a homodimer. A single leucine was systematically changed to an alanine in either the *a* or *d* position and the stability of the mutant coiled coils were determined from guanidine hydrochloride denaturation studies. While the mutant coiled coils were less stable than the original sequence they found that the contribution to stability from leucines at position *a* was greater than at position *d*. Substitution of leucine with β-branched amino acids was also studied in order to assess the changes in coiled coil stability.[63] Placing isoleucine into the middle three *a* positions of the same sequence stabilized the coiled coil by 7.5 $kJ\,mol^{-1}$ relative to the native sequence; substitution at the *d* position destabilized the coiled coil by 3.8 $kJ\,mol^{-1}$. This is consistent with the observation that isoleucine is preferred over leucine in naturally occurring coiled coils.[56] Replacing leucine with valine at either *a* or *d* positions destabilized the coiled coil relative to the original sequence by 3.8 $kJ\,mol^{-1}$ and 22.2 $kJ\,mol^{-1}$, respectively. Moreover, the peptide containing valine at position *d* multimerized to a four-stranded coiled coil. Thus it seems that subtle packing interactions can have a large effect on the stability and indeed, the global fold of a small protein.

This point was further illustrated in a spectacular way by Harbury *et al.* who discovered a "switch between two-, three-, and four-stranded coiled coils in GCN4 leucine zipper mutants."[64] Sets of amino acids in a 33-residue GCN4 peptide at the *a* and/or *d* positions were changed to a particular hydrophobic amino acid, which caused the formation of two-, three- or four-stranded parallel coiled coils as determined by size-exclusion chromatography and sedimentation equilibrium

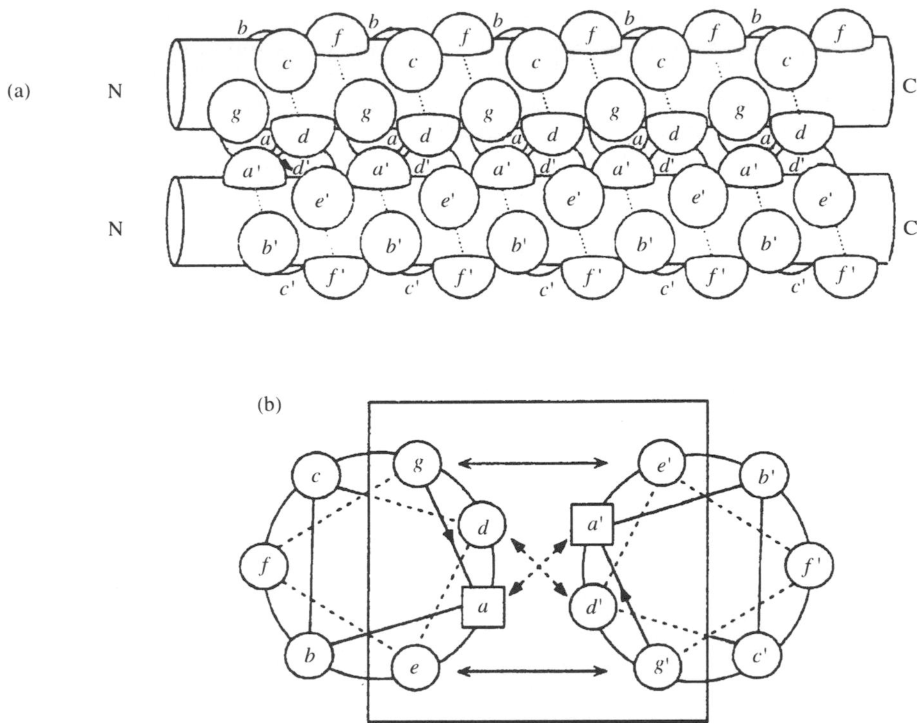

Figure 5 (a) Side and (b) top view of a dimeric parallel coiled coil. Arrows indicate interactions between residues. The position of each residue in the heptad is denoted *a* through *g*. The left-handed supertwist is omitted (reproduced by permission of Cambridge University Press from *Protein Sci.*, 1995, **4**, 237).

experiments. The orientation of the helices was determined by amino- and carboxy-terminal disulfide bond formation when the reduced peptides were placed into a redox buffer. Occupation of position *a* by isoleucine and position *d* by leucine gave rise to a dimer. Peptides containing isoleucines at both *a* and *d* positions formed a trimer. If leucine occupied position *a* and isoleucine occupied position *d* then the mutant peptide associated to form a four-helix bundle. Curiously, the LV mutant oligomerized to dimeric and unidentifiable states in this study, whereas Zhu *et al.*[63] found that their peptide containing the same core mutation formed a tetramer (see previous paragraph). The differences between the two peptides occur not only outside the hydrophobic core but in the hydrophobic core as well, at the ends. The GCN4 mutant had valine at every *d* position, whereas the peptide from the Hodges laboratory had leucine at the first and last *d* positions. Thus, there is difficulty in determining the cause of this difference in oligomerization between the two peptides.

The crystal structures of a GCN4 peptide as well as two GCN4 mutants in the trimeric and tetrameric state have been determined.[60,64,65] A "knobs into holes" type of packing at the helix interfaces was observed in all three cases where the knobs formed by the side chains of one helix fit into the holes defined by side chains of a neighboring helix. A comparison of the three structures reveals packing preferences.[65] Three modes of packing exist: (i) parallel packing, where the C-α—C-β bond of the knob side chain is directed out of the interface, (ii) perpendicular packing, where the C-α—C-β bond of the knob side chain is directed into the adjacent helix, and (iii) acute packing, which was named so because the C-α—C-β bond of the knob side chain makes an acute angle with the C-α—C-α vector defining the bottom of the hole into which the side chain fits. Leucine adopted the perpendicular mode of packing, while β-branched amino acids like isoleucine and valine adopted the parallel type of packing in the tetrameric and dimeric structures. Isoleucine, however, took up the acute mode of packing in the trimeric structure.

Assembly of two peptides having heptad repeats into an antiparallel orientation may also occur, and this would naturally change the nature of the hydrophobic side chain packing. Such an antiparallel coiled coil was found in the crystal structure of a seryl transfer RNA synthetase.[66] Here, residues at position *a* made contact not with residues at *a′* but with residues at *d′*. Residues at *d* made contact with those at *a′*. In this case the side chains were turned out of the middle in order to avoid unfavorable steric interactions. Bearing this mode of packing in mind the obvious question to ask is what is the molecular basis for the choice of helix orientation? Model studies indicate that the parallel orientation seems to be the preferred one.[67] Two very similar 33-residue peptides were synthesized, the only difference between them being the presence of a cysteine near either the carboxy- or amino-terminus. Mixing the two reduced peptides and subsequent air oxidation resulted mostly in the formation of the two parallel homodimers.

One salient feature of the GCN4 peptide which plays a directing role in the assembly of the coiled coil is the presence of the single asparagine in the hydrophobic repeat. Assembly of two GCN4 peptides in a parallel fashion placed the side chain of asparagine in the hydrophobic core where they formed hydrogen bonds to each other. O'Shea *et al.* noted that assembly of the peptide in an antiparallel fashion would result in the unfavorable interaction of a polar side chain with a hydrophobic environment.[60] This single feature not only disfavors an antiparallel orientation but also disfavors the oligomerization state other than the dimeric one. For instance, Betz *et al.* found that a peptide containing valine and leucine at positions *a* and *d*, respectively, existed as a triple-stranded coiled coil unless a single asparagine is included in the hydrophobic repeat of the peptide.[68] In this case, the peptide dimerized. Similarly, Lumb and Kim found that replacement of the asparagine with leucine in a heterodimeric coiled coil resulted in the tetramerization of the peptides.[69] Moreover, the peptides in the tetramer did not adopt any unique orientation.

There is evidence to show that electrostatic interactions direct the orientation of the helices rather than hydrophobic packing.[70] Two peptides were synthesized such that they would experience electrostatic complementarity at positions *e* and *g* if they oligomerized as an antiparallel coiled coil. Not only was spontaneous formation of the antiparallel coiled coil observed but surprisingly, it was favored over formation of the two homodimers by more than what would be expected from a simple statistical distribution, despite the fact that the two homodimers also enjoyed charge complementarity. Guanidinium hydrochloride denaturation studies, which eliminate ionic effects, showed that antiparallel coiled coils are more stable than parallel coiled coils. However, the antiparallel coiled coil suffering electrostatic repulsions between helices did not form spontaneously. This implies that electrostatic interactions play a more dominant role than hydrophobic interactions in orienting the helices. Myzska and Chaiken successfully applied these principles to the design and synthesis of an intramolecular antiparallel coiled coil.[71]

Electrostatic interactions at the *e* and *g* positions are thought to guide the formation of hetero- vs. homodimer as well. Peptides corresponding to the leucine zipper regions from the Fos and Jun gene products preferentially formed the Fos–Jun parallel heterodimer over the two possible parallel homodimers by at least 1000 fold.[72] No antiparallel coiled coil formation was reported. Helical wheel diagrams revealed some charge complementarity for the heterodimer, no charge–charge interactions between helices for the Jun–Jun homodimer, and some charge anti-complementarity for the Fos homodimer. O'Shea *et al.* assessed the role of the charged and hydrophobic residues by making GCN4-based hybrid peptides with either the Fos or Jun sequence at positions *e* and *g* or positions *a* and *d* and measuring the dependence of coiled coil stability on pH.[73] The heterodimer of the peptides having the Fos and Jun sequences at positions *e* and *g* were found to interact specifically (based on thermal denaturation experiments) and had pH-dependent stabilities similar to the native Fos–Jun heterodimer. Coiled coils formed from GCN4 hybrids having Fos and Jun sequences at positions *a* and *d* did not exhibit the same pH-dependent stability as the native heterodimer. Using electrostatic complementarity as a key design principle, several groups were able to direct the preferred assembly of heterodimer formation over homodimer.[70,74–6] All three groups synthesized peptides containing heptad repeats but also included exclusively either lysine or glutamic acid at both *e* and *g* positions. The rationale behind this design was that homodimer formation would be strongly disfavored because of electrostatic repulsion between residues of the same charge at positions *e* and *g*.[75] Several groups also cited electrostatic stabilization as a significant driving force for the formation of the heterodimer.[70,74,76] Whatever the reason, heterodimers were exclusively formed in all the studies. By themselves the peptides containing only glutamic acid or lysine had little ordered structure. However, when mixed together in a 1:1 ratio the peptides became highly helical.

While the theory that electrostatic destabilization disfavors unwanted structures is generally accepted, the issue of whether electrostatic stabilization is significant or even present is hotly debated. Many groups have cited ionic interactions as a stabilizing force in coiled coils and for good reasons. Salt bridges were observed in the crystal structures of the GCN4 peptide dimer, the mutant trimer, as well as the GCN4 mutant tetramer.[60,64,65] Sequence analysis of fibrous proteins suggests not only interhelical ionic interactions but intrahelical attractions between oppositely charged residues spaced four residues apart.[56,57,77,78] Furthermore, the Hodges laboratory carried out a series of experiments to quantify the strength of these ionic attractions.[76,79,80] For instance, the magnitude of electrostatic repulsion was evaluated by comparing the stability of coiled coils having potential interchain repulsions between glutamic acid side chains positioned at 13 and 20 (E13,20), 15 and 20 (E15,20), or 20 and 22 (E20,22), with a similar coiled coil (Native) which had glutamine instead of glutamic acid at those positions and so had no electrostatic repulsions. Compared to the Native coiled coil, the others were less stable when denatured with urea. The least stable was the coiled coil formed by E(15,20); E(13,20) and E(20,22) coiled coils were almost as stable as the Native coiled coil. This implies that the electrostatic repulsion between residues at i and $i + 5$ is felt much more than those at i and $i' + 2$ or i and $i' + 7$. The stability of these coiled coils increased greatly when the pH was changed from pH 7 to pH 3 and is attributed to the loss of interhelical ionic repulsions upon protonation of glutamic acid side chains. This increase may also be due to an increase in helical propensity as well as hydrophobicity of protonated glutamic acid side chains (see next paragraph).[81] Denaturing the coiled coils with guanidinium hydrochloride eliminated this difference in stability between coiled coils of different net charge, presumably because the high ionic strength of denaturing solutions of guanidinium hydrochloride masks any interhelical ionic interactions.

The Hodges laboratory was also able to remove the contribution of helical propensity to the stability of the coiled coil and therefore was able to quantify the strength of the electrostatic attraction between glutamic acid and lysine side chains.[81] The stability of the model coiled coil EK, having the heptad sequence KgLaGbAcLdEeKf repeated five times over, was compared with those of mutant coiled coils QK, EQ, and QQ. Coiled coil QK had glutamine instead of glutamic acid at position e, EQ had glutamine instead of lysine at position g, and QQ had glutamine at both e and g positions. A double-mutant cycle analysis of this series of coiled coils allowed for the isolation of interhelical electrostatic attractions from the contributions due to the hydrophobic effect and differences in helical propensities between glutamine and glutamic acid or lysine. Curiously EK and EQ are more stable to urea denaturation at pH 3 than at pH 7 despite the fact that there are no interhelical ionic attractions at such a low pH. This phenomenon was attributed to the greater hydrophobicity and helical preference for a protonated glutamic acid compared to an ionized glutamic acid or even glutamine. The contribution of a protonated glutamic acid residue to coiled coil stability was found to be significantly larger than that of the ionized residue ($2.7\,kJ\,mol^{-1}$) and outweighed the contribution of an interhelical salt bridge ($1.5\,kJ\,mol^{-1}$).

Several groups, however, argue that salt bridges do not stabilize protein structure at all.[75,82–4] In particular, the Kim laboratory argued that if these salt bridges did exist then a decrease in the pK_a of those glutamic acid side chains which were participating in salt-bridging interactions would be observed. Measurement of the pK_a values of glutamic acid side chains in a GCN4 peptide as well as in the designed heterodimer "Velcro" by ^{13}C NMR spectroscopy at physiological ionic strength indicated that the pK_a values of these glutamic acid residues were not decreased at all.[75,84] In fact, some pK_a values had increased in the folded state, indicating that the presence of the glutamic acid can be slightly destabilizing despite the fact that it has the potential to form favorable ionic interactions with a nearby lysine. This observation is supported by calculations on salt bridges found in nine different proteins.[83] The cause of this phenomenon is not known.

Thus there is a conundrum in the field at present. While model coiled coils designed to have ionic interactions behave as if they are present, such as the dependence of stability on ionic strength, pH titrations on glutamic acid residues observed to participate in salt bridges in the crystal structure of a GCN4 peptide would seem to indicate that these putative ionic interactions do not exist in solution.

12.3.3 Helix Bundles

Helix bundles have been the most popular target for *de novo* design, and have met with some success. This success owes much to the modular nature of the α-helix. Balaram, Karle, and coworkers made stable individual α-helices but used aminoisobutyric acid (Aib) in their peptides to strongly bias helix formation.[85–7] The rationale for such a design was that stable helices could eventually be packed together in a controlled fashion. This hierarchical strategy is somewhat reminiscent of the diffusion–collision model of protein folding. Crystal structures of several short hydrophobic Aib-containing peptides were solved in order to gain insight into peptide conformation, helix packing, and solvation. These peptides were for the most part α-helical, although some had a small amount of 3_{10} helical structure as well.[85,86] Specific interactions were not designed, so there was no surprise when the crystal structure revealed that the interhelical packing was poor, with many voids filled by disordered water molecules. While the typical Gly-Pro linker became helical when used to connect two Aib-containing helices, the linker containing ε-aminocaproic acid did not.[87] Crystal structure analysis showed that the helix–linker–helix adopted an extended rather than a bundle structure. Its behavior, however, suggested a compact structure. Further progress based on this strategy is awaited.

DeGrado and co-workers also adopted an incremental strategy to protein design. Ironically, however, successful structure formation was not achieved through separate stabilization of each helical module; the dominant stabilizing force was the hydrophobic interactions between helices. Thus the design strategy was modular in nature but the assembly process giving rise to the structure was not. An incremental strategy to designing a four-helix bundle was adopted (Figure 6) with the reasoning that such an approach would allow them to study each factor involved in the folding of the bundle separately.[88–90] First, they designed peptides that would aggregate into stable tetramers. Next, linkers were added between two helices and their sequence optimized so that two helix–turn–helix molecules dimerized to form a four-helix bundle. Finally, the two helical hairpins were tethered together with another linker. Two very similar peptides were designed to form stable amphiphilic α-helices, called α_1A and α_1B. The hydrophobic face of the helices was composed solely of leucine residues, while glutamic acid and lysine residues occupied the hydrophilic face. It was expected that the helices would pack together by association at the hydrophobic faces with the leucine side chains adopting a "ridges into grooves" type of packing. The glutamic acid and lysine residues were placed at intervals of i and $i + 4$ or $i + 3$ in order to facilitate stabilizing intrahelical and interhelical electrostatic interactions. Both peptides aggregated to a tetrameric state and so two different linkers were used to dimerize α_1B peptides into the helical hairpin α_2B. A single proline used to link the two helices gave rise to trimeric aggregates of helical hairpins. However, the three-residue linker consisting of proline and two arginines yielded the desired assemblage: a dimer of helical hairpins forming a four-helix bundle. Moreover, the presence of the three-residue linker greatly enhanced the stability of the four-helix bundle against guanidinium hydrochloride denaturation. These were the first four-helix bundles to be designed, synthesized, and characterized. Another peptide linker was later added between the two dimers to form one polypeptide which assembled into a four-helix bundle.

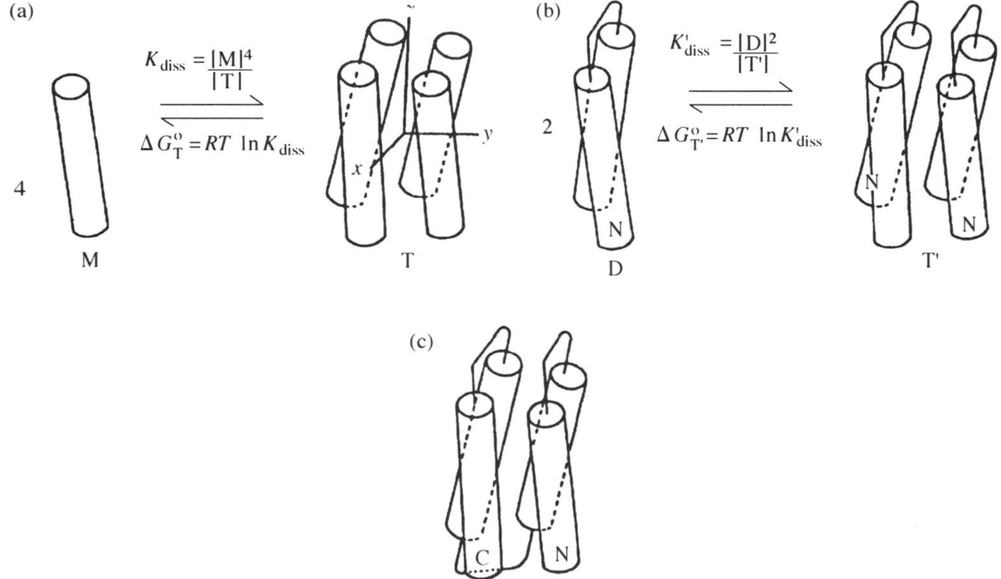

Figure 6 Minimalist approach to creating an antiparallel four-helix bundle. (a) A peptide was designed to oligomerize as a tetramer. (b) Subsequently, a linker was used to connect two helices. Dimerization produces a four-helix bundle. (c) In the last stage a third linker was added to connect the two molecules from (b). (reprinted with permission from *J. Am. Chem. Soc.*, 1987, **109**, 6751. Copyright 1987 American Chemical Society).

While peptides can assemble to form helix bundles, the *de novo* designed constructs do not show the well-defined structure of their native counterparts. Several designed proteins have been shown to bind hydrophobic dyes, which indicates that the packing in the interior of these bundles is not sufficiently tight to exclude other molecules.[90–3] The state of these artificial proteins has been likened to the molten globule intermediate observed in the folding pathway of some proteins. Spectroscopic studies on fluorescent probes within the hydrophobic core has given some insight into the structure of several bundles. For instance, a tryptophan residue was introduced into the hydrophobic core of α_4 to probe its dynamics.[94] The presence of tryptophan significantly destabilized the four-helix bundle, presumably because of the added difficulty of accommodating the large indole ring in the core. This implies that the interior was reasonably well-packed in the original design. Moreover, the tryptophan was able to adopt two different conformations which did not interconvert on the nanosecond timescale, which also supports the conclusion that the interior is reasonably well-packed. However, the change in heat capacity for the unfolding of α_4, as well as its ability to bind ANS (1-anilino-8-naphthalenesulfonate) suggests that it also has molten globule-like properties. The inclusion of hydrophobic fluorophores was also studied using fluorescence spectroscopy. The polarization decay constants measured in this study were only half that determined for α_4 and could be interpreted as rapid movement of the dyes in the protein core. This implies that these bundles were poorly packed.

Nuclear magnetic resonance studies on these designed protains can quickly give an indication of the state of their packing as well.[95–8] Proton resonances of poorly-packed side chains tend to be broad and overlapped, while the resonances of well-packed side chains are sharper and better dispersed. NMR spectroscopy was used to monitor changes in the hydrophobic interior of α_2 and α_4 variants, which were changed in order to improve packing. One design change was the substitution of some leucines with β-branched and aromatic amino acids into α_2 in order to introduce shape complementarity into the hydrophobic interface as well as the ridges-into-grooves style of packing.[98] The new protein, called α_2C, showed a thermally induced transition from a native-like state to a molten globule-like state as measured by one-dimensional ^{1}H NMR spectroscopy. The resonances in the upfield region of the spectrum were relatively well-dispersed at 285 K compared to the spectrum obtained at 300 K. A second change which resulted in a sharper and better-dispersed ^{1}H NMR spectrum is the addition of a zinc-binding site to the α_2 peptide.[95] Three histidine residues were incorporated into the hydrophilic face of the peptide such that the three imidazole side chains could converge to bind a single zinc ion. Thus, dimerization of this new peptide, called H3α_2, gave rise to a four-helix bundle with two zinc-binding sites. The downfield region of the peptide became much better dispersed with increasing concentrations of zinc acetate, up to 1 equiv. This observation, as well as the NMR spectrum, suggest that the peptide binds Zn^{2+} with its three histidines in a 1:1 complex.

In order to obtain larger multimeric states, the hydrophobic surface area on the amphiphilic peptide can be increased. In this way, Chin *et al.* successfully made two peptides which assembled into hexameric helical bundles.[99] Lutgring and Chmielewski synthesized an amphiphilic peptide reminiscent of α_1B which was supposed to assemble to form a four-helix bundle.[100] Surprisingly, the assemblage was a pentamer of peptides. One significant feature of this design was the presence of two homocysteine residues at the hydrophobic–hydrophilic interface, which provided covalent stabilization to the assemblage through disulfide cross-linking upon oxidation. However, the presence of homocysteine may have resulted in a small increase in the hydrophobic surface area such that an extra peptide could be accommodated in the bundle. Poorly understood packing effects may also play a role in dictating the oligomerization state.

12.3.4 Template-assembled Synthetic Proteins

The oligomerization state of peptides can be further controlled through covalent attachment of the assembling peptides onto a rigid template. The resulting protein-like constructs are branched molecules rather than the usual linear polypeptide chain seen in natural proteins. In addition to using amphiphilic peptides as building blocks, the success of making template-assembled synthetic proteins (TASPs) lies in the use of an appropriate template. Porphyrin derivatives have successfully served as templates[101–3] as have peptides containing β-turns, β-turn mimetics, and cyclic peptides.[48,49] In particular, the peptidic templates offer advantages in ease of synthesis (solid-phase peptide synthesis, SPPS) and more importantly, orthogonal protection of lysine residues to which the secondary structure building blocks are to be attached, allowing for the attachment of different peptides at predetermined sites in a parallel or antiparallel fashion. The Mutter laboratory called these templates regioselectively addressable functionalized templates (RAFTs).[104] Thus, mimics of a variety of protein structural motifs can be made via the TASP approach while circumventing the protein-folding problem. Using this pioneering strategy, the Mutter laboratory made TASPs which contained four-helix bundle, βαβ, β-barrel-like motifs, and even a TASP having both the four-helix bundle and a four-stranded β-barrel assembled on one template,[48,49,105] as shown in Figure 7. Size-exclusion chromatography showed that this "two-domain" molecule existed as a monomer; c.d. and IR spectroscopies support the existence of both helical and β-strand structures. Stabilization of these supersecondary structures is achieved through the packing of hydrophobic residues on amphiphilic peptides so the proper spatial arrangement of the peptides on the template is all-important.

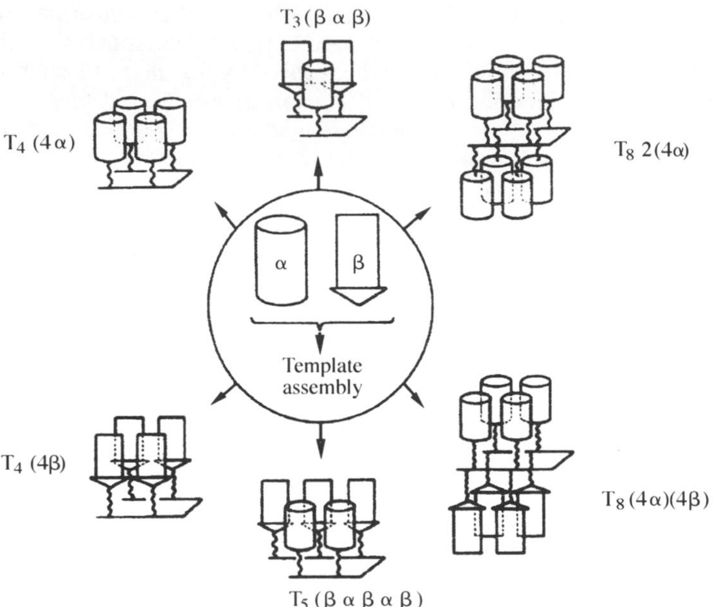

Figure 7 Schematic representation of various TASP molecules designed by the Mutter laboratory. Cylinders represent helices while arrows represent β-strands (reprinted from *Tetrahedron*, **44**, M. Mutter, K. H. Altmann, G. Tuchscherer, and S. Vuilleumier, Strategies for the *de novo* design of proteins, 771, Copyright 1988, with kind permission from Elsevier Science Ltd, The Boulevard, Langford Lane, Kidlington OX5 1GB, UK).

Rigidity of the template is also very important for organizing the peptides close enough for self-assembly to occur. For instance, some templates are only about 10 amino acid residues long and will have a restricted conformation when cyclized.[106] The peptides connecting the turns are also thought to adopt an antiparallel β-sheet structure, adding additional rigidity to the template. Moreover, the template containing a β-turn was not as good as the template containing the more rigid turn mimic aminomethyl-2-naphthoic acid (AMNA) in inducing α-helical structure. Furthermore, the template containing two AMNAs was the best template for inducing structure in a lysozyme model TASP. However, the ability of the peptides to assemble plays a role as well, for other peptides could organize into a four-helix bundle when attached to the template containing the same proline-glycine turn which could not organize the model lysozyme peptide into a four-helix bundle.

12.3.5 Metal Ion Assisted Formation of Helical Bundles

A complementary strategy to the TASP approach is metal ion assisted self-assembly, as shown in Figure 8. The strategy here is to use nonnatural metal-chelating amino acid residues to organize peptide self-assembly upon metal ion binding. Thus, the metal complex acts as the noncovalent template but this method also offers: (i) ease of synthesis, and (ii) convergent syntheses by taking advantage of chemo- and regioselectivity in metal ion complexation. Both the Ghadiri and Sasaki laboratories successfully used bipyridine derivatives at the amino-terminus of their amphiphilic peptides to complex six-coordinate transition metals thereby organizing the peptides into parallel three-helix bundles.[52,54] The binding energy gained in metal ion complexation compensated for the loss in entropy associated with organizing the peptides together. The high effective concentration of hydrophobic residues in the trimeric intermediate drove hydrophobic collapse to give the desired three-helix bundle. Small differences in the anchoring point on the bipyridyl ligands, however, were shown to affect the number of structures obtained. Lieberman and Sasaki chose to use 2,2′-bipyridine-4,4′-dicarboxylic acid, thereby attaching the bipyridyl ligand to the peptide through its C-4 position.[53,54] Use of this symmetric moiety gave rise to four stereoisomers at the metal complex center and therefore at least four closely related helix bundles: one resulting from the *fac*-Λ isomer, another from the *fac*-Δ isomer, a third from the *mer*-Λ isomer, and finally the fourth from the *mer*-Δ isomer. The helices in the *fac* isomers are estimated to be 0.9 nm apart whereas two helices are 1.4 nm apart and are both 0.9 nm away from the third. Recall also that the peptides themselves are stereogenic and so the Λ and Δ enantiomers of the *fac* and *mer* isomers become diastereomeric when the peptides are attached. Each of these four species were detected by c.d. and NMR spectroscopies and were resolved by reversed-phase HPLC. However, use of the bipyridyl moiety 5-carboxy-2,2′-bipyridine eliminated the existence of isomers.[52] Furthermore it was thought that the left-handed supercoiling topology of the three-helix bundle enforced the formation of only one diasteromer. Indeed, the Λ isomer was shown to be the preferred isomer by Ghadiri and co-workers. In this case, complex formation was not only chemo- and regioselective but diastereoselective as well.

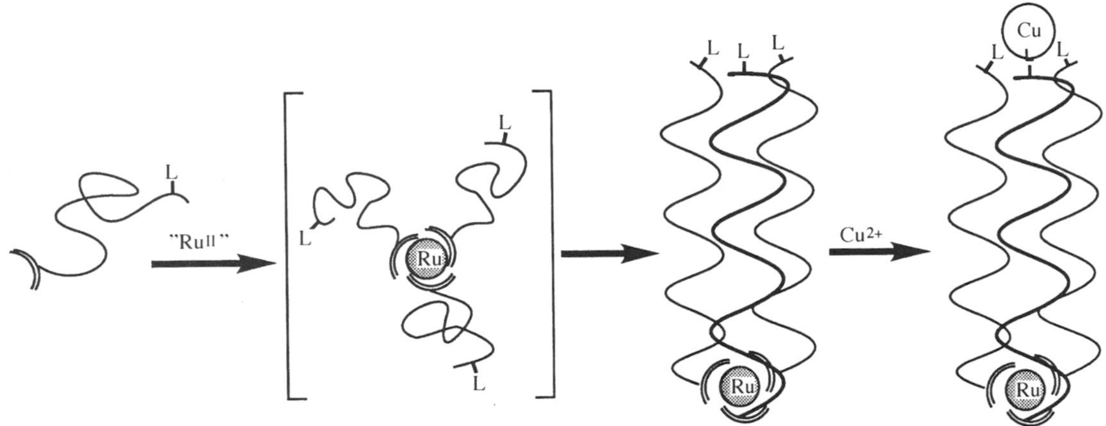

Figure 8 Schematic representation of the Ru^{II}-assisted assembly of a three-helix bundle. Metal ion complexation initially sequesters three peptide chains thereby biasing formation of a trimeric ensemble. The binding energy of metal ion complexation not only drives the formation of a trimeric intermediate but compensates for the loss of entropy associated with spatially organizing three peptides. Hydrophobic collapse ensues, thereby forming the helical structure.

Expanding on the use of the metal ion assisted self-assembly process, Ghadiri and co-workers also synthesized a pyridyl-containing amphiphilic peptide which complexed ruthenium(II) and self-assembled exclusively into a stable, parallel four-helix bundle.[51] Metal clusters composed of five or six peptides were not formed, most likely due to steric and electronic effects. The use of ruthenium gave the additional advantage of forming exchange-inert complexes. Thus, the ruthenium-containing four-helix bundle is a very stable structure. Ghadiri and Case also used ruthenium to assemble a rationally modified form of the previously mentioned three-helix bundle, wherein a metal-chelating histidine was added as the penultimate residue to the carboxy-terminus of each peptide,[107] as shown in Figure 9. Assembly into the three-helix bundle created a second composite metal-binding site composed of the three histidine side chains. This second preorganized metal-binding site had a high affinity for copper(II) with an apparent dissociation constant of $K_D = 3.0 \times 10^{-7}$ M. While the copper did not impart additional helicity to the structure, it did impart an additional 6.2 kJ mol^{-1} in stability over the ruthenium(II) metalloprotein without the bound copper. This is the only published example of a *de novo* designed heterodinuclear metalloprotein.

12.3.6 *De novo* Designed Functional Proteins

The scarcity of functional *de novo* designed proteins in the literature is an indication of how difficult it is to overcome the challenges of obtaining the desired structure and creating a preorganized active site. For instance, Stewart and co-workers published a paper describing a

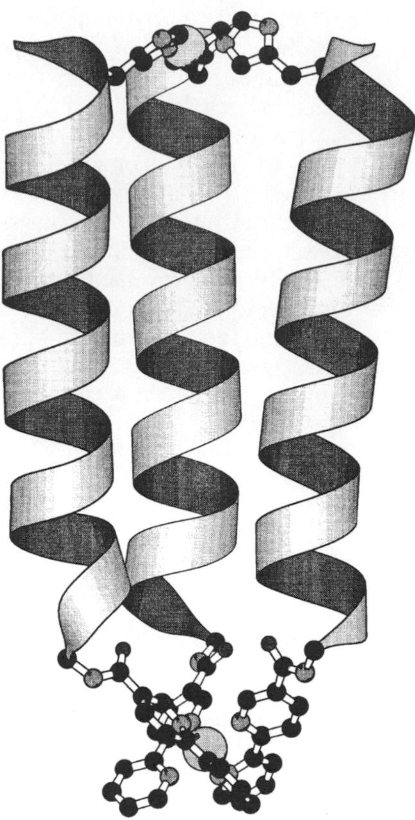

Figure 9 Ribbon diagram of a heterodinuclear three-helix bundle. A ruthenium tris(bipyridine) complex is formed at the amino terminus while the composite binding site, formed upon assembly, binds a copper(II) atom.

TASP-like four-helix bundle with catalytic residues preorganized at the amino-terminus into an "oxyanion hole"-like structure typical of serine proteases.[108] The protein called "chymohelizyme " was thought to have activity reminiscent of chymotrypsin because it could hydrolyze acetyltyrosine ethyl ester, (a chymotrypsin substrate) but not benzoyl arginine ethyl ester (a trypsin substrate). This proved not to be the case as later, the authors performed controls that were omitted from the initial work.[109] The Benner laboratory recently reported the design, synthesis, and characterization of amphiphilic α-helices that catalyze the decarboxylation of oxaloacetate.[110] The design makes use of positive charges near an amino-terminal amine to increase its reactivity towards imine formation. The Benner laboratory also speculated that the high density of positive charges imparts oxaloacetate-binding. The rate of catalysis was several orders of magnitude faster than that for catalysis by simple amines, suggesting that other structural features of the peptides helped enhance catalysis. In particular, catalytic activity was correlated with the fraction of helix formed. This observation is consistent with the grouping of the positive charges to one face of the α-helix. While the peptide was predicted to form helix bundles, a well-defined structure was not described. The postulated imine intermediate was trapped, lending support to the proposed catalytic mechanism.

A fruitful collaboration between the DeGrado and Dutton laboratories resulted in the design and synthesis of peptides which bound hemes. In their first study, helical hairpins were synthesized which aggregated into four-helix bundles.[111] The unique feature of these peptides was the presence of two histidine residues, one from each peptide, which were present in the core of the bundle. Their side chains were used to ligate the iron in the heme, forming a bis(imidazole) complex. Thus, the plane of the heme runs parallel with the helical axes of the bundle. Peptides were later designed which could each bind two hemes and then oligomerize to form a four-helix bundle containing four hemes.[112] These heme-binding four-helix bundles were to be used for studying electron transfer in the simplest possible protein structure that could be easily changed. Unfortunately, the structures of the oligomerized products were never established in the multiheme proteins so it is unknown whether the molecules dimerized in a parallel or antiparallel fashion, or both.

Other examples of four-helix bundle structures that contain porphyrins exist. These, however, differ from the previous examples in that they are TASP-like in design. Sasaki and Kaiser were the first to attach four peptides to a coproporphyrin which assembled into a four-helix bundle.[102,103] After incorporating iron(II) into the porphyrin the protein was found to have aniline hydroxylase activity. Furthermore, the four-helix bundle had enhanced activity over that of unmodified hemin, implying that the protein scaffold played a role in the catalytic cycle. They speculated that the presence of the bundle provided a binding pocket between the bottom of the bundle and one face of the porphyrin into which aniline could bind. The laboratories of DeGrado and Groves also synthesized two such four-helical bundle TASPs, one called tetraphilin, which formed proton channels in lipid bilayers,[113] and a second one which was to be used for electron-transfer studies.[114]

12.3.7 Novel β-Proteins

There have been far fewer successful designs of β-sheet proteins than α-helical ones. This lack of success can be considered to be a reflection of the general lack of knowledge of protein folding and difficulties associated with aggregate behavior of β-sheet structures. Several examples of β-sheet-containing proteins having molten globule characteristics have been reported[115,116] but are beyond the scope of this chapter from an intermolecular self-assembly standpoint. Controlling the various ways in which β-strand peptides aggregate poses a formidable challenge for those designing proteins. However, these same modes of assembly have been used in a constructive way to create a class of β-sheet-containing proteins never before seen in Nature.[55] Ghadiri and co-workers named them peptide nanotubes because rationally designed cyclic peptide monomers self-assemble to form tubular structures of nanoscale diameters. Since the diameter is determined by the number of amino acid residues, it is easily specified. A subtle but critical feature of the monomer is the chirality of the amino acid residues used. The peptide is composed of an even number of amino acids alternating in their chirality. The consequence of this design is that the cyclic peptide adopts a flat, ring-shaped β-strand conformation with all its side chains directed out of the ring, as shown in Figure 10. Intramolecularly folded structures would experience unfavorable steric interactions between the side chain of a given amino acid residue and the peptide backbone. In the flat ring conformation, the hydrogen bond donors and acceptors of the peptide backbone are presented perpendicular to the plane of the ring. The hydrogen-bonding requirements of the peptide backbone can be satisfied if the rings stack upon one another. Thus, hydrogen bonding drives the assembly process. Billions of monomers come together this way, so the synthesis of nanotubes is highly convergent. Moreover, the alternating chirality of amino acid residues results in the patterning of hydrogen bond donors and acceptors such that hydrogen bonding may occur only between homochiral residues, preferably in an antiparallel fashion. Steric interactions between side chains and the backbone of the neighboring subunit may also prevent alternative sheet registries.

The concept of cyclic peptides having alternating D–L residues which can stack was first suggested in 1974,[117] but the few reported attempts to realize the tubular ensemble in the interim were unsuccessful[118,119] and the idea was brought to fruition only in 1993.[55] The first peptide nanotube was an eight-residue peptide of the sequence *cyclo*[-(D-Ala-L-Glu-D-Ala-L-Gln)$_2$]. The salient feature is the incorporation of glutamic acid, the so-called "proton trigger," into the peptide. The presence of the glutamic acid residue enhanced the peptide's solubility in water at neutral and alkaline pH. However, acidification of the solution neutralized the negative charge from the glutamic acid sedition thereby drastically reducing its solubility. Rapid precipitation ensued, furnishing needle-shaped microcrystals of nanotube bundles.

Initial characterization of the nanotubes furnished only indirect evidence but all the data taken together made a very strong case for the existence of the proposed structure. FTIR spectroscopy indicated that hydrogen bonding between β-strands was present. Electron diffraction data provided evidence that the spacing between monomers is characteristic of the spacing observed between antiparallel β-strands, and modeling showed that conformations other than that of a flat ring could not possibly account for the observed diffraction pattern. Finally, highly magnified images of the microcrystals taken using cryoelectron microscopy revealed striations micrometers long running along the length of the crystal and having the thickness of the diameter expected for the proposed eight-residue peptide ring (Figure 11(b)). Later, nanotube-mediated ion transport across phospholipid bilayers by a related peptide exhibited channel behavior.[120,121] This phenomenon can be explained only if the peptides assembled into a tubular structure in the membrane, so this last example may be considered proof, at least for that particular peptide, that nanotubes are formed.

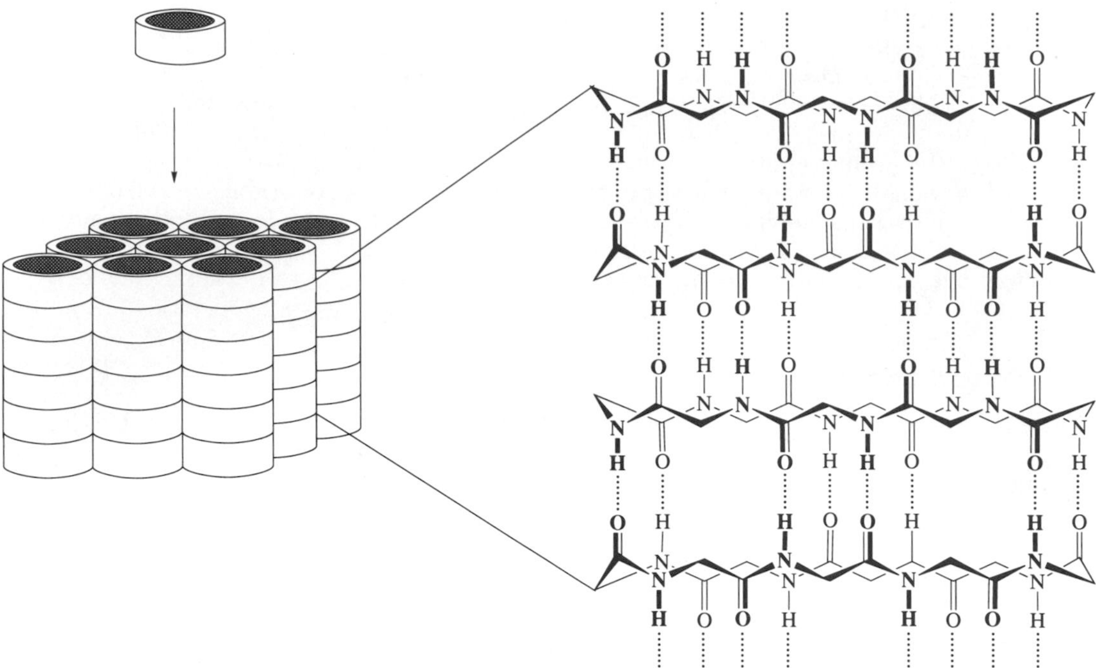

Figure 10 Assembly of cyclic peptide monomers into nanotubes. The monomers are represented as rings. Assembly along the length of the tube is mediated by backbone–backbone hydrogen bonding, while the mode of lateral assembly is determined by side chain–side chain interactions. The side chains were omitted from this diagram for clarity.

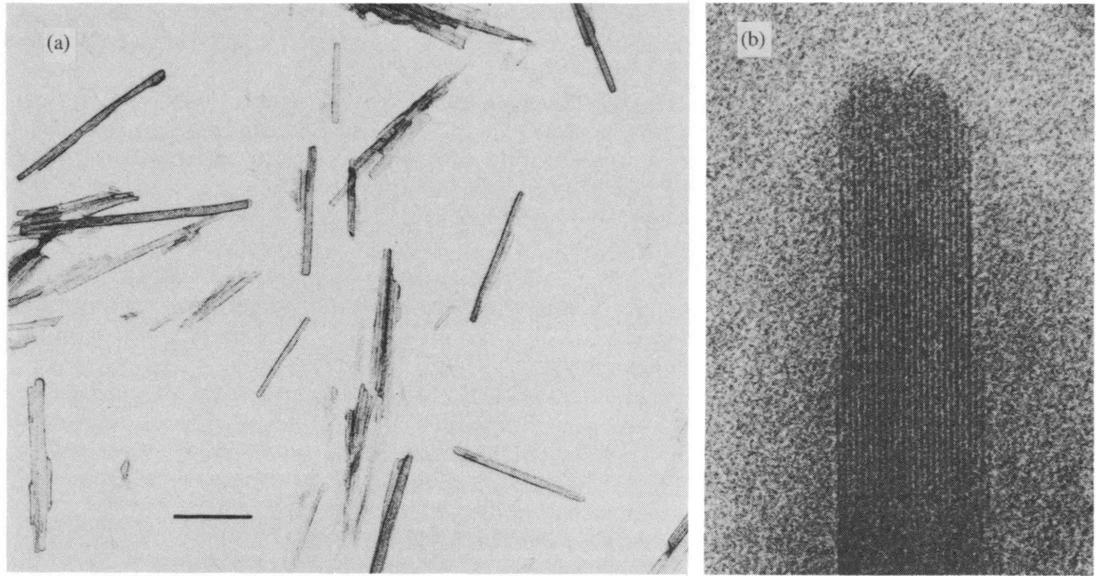

Figure 11 (a) Low magnification electron micrograph of nanotubes adsorbed on a carbon support film. (b) High magnification electron micrograph of a single nanotube bundle. Longitudinal striations are separated by 2.5 nm (reproduced by permission of VCH from *Adv. Mater.*, 1995, **7**, 675).

While nanotube formation is guided by hydrogen bonding, nanotube bundling and crystal formation is dictated by the properties of the nanotube exterior. These properties, of course, are determined by the sequence, and the problem of nanotube packing can be viewed as a problem of packing an element of protein secondary structure. A study was undertaken to probe the effects of increasing hydrophobic surface area as well as the role of hydrogen bonding on the crystal packing of nanotubes.[122] It was hoped that increasing the hydrophobic surface area of the nanotube would increase the tendency to associate nanotubes laterally when transferred to an aqueous milieu. Several peptides were synthesized and, while some nanotubes essentially packed in the predicted fashion as determined by electron diffraction data combined with molecular modeling (Figure 12), unfortunately thus far, none of the peptides provided crystals large enough for x-ray analysis. Thus, this first attempt at crystal engineering of nanotubes was limited in success but the lessons learned from the various designs should help in rationally designing peptides for a second generation of nanotubes.

Figure 12 Model of the most likely mode of packing by nanotubes composed of *cyclo*[Gln-D-Leu]$_4$. The glutamine residues form a hydrogen bond network while the side chains of D-leucine associate through hydrophobic interactions (reprinted with permission from *J. Am. Chem. Soc.*, 1996, **118**, 43. Copyright 1996 American Chemical Society).

The Lorenzi laboratory obtained x-ray crystal structures of cyclic peptides with amino acid residues of alternating chirality in the hope of demonstrating a tubular structure.[118] The peptides were hexameric and composed solely of either valine or phenylalanine. While they were ring-shaped, the hydrogen bond donors and acceptors were not positioned properly for ring stacking to occur so, unfortunately, rigorous proof of an extended tubular structure at the atomic level must still be shown. However, both the Ghadiri and Lorenzi laboratories have obtained solution and crystal structures of dimeric cylindrical peptide structures.[123,124] The extended structure was prevented from assembling through *N*-methylation of one face of the peptide. This eliminated hydrogen bond donors and prevented association through that face by virtue of unfavorable steric interactions. The dimers of the hexapeptide made by the Lorenzi laboratory and the octapeptide synthesized by the Ghadiri laboratory, which adopted the predicted flat ring-shaped β-conformation, underwent hydrogen bonding at the nonmethylated faces to form the cylinder first predicted by De Santis and co-workers. The interior of the cylinder was filled with water in both peptide dimers. From variable-temperature solution studies in chloroform Ghadiri *et al.* went on to show that the enthalpic contribution of intermolecular hydrogen bonding was the major driving force for self-assembly. Furthermore, the dimer was favored over the monomer by as much as 23.4 kJ mol^{-1}. This gain in stabilization energy would be additive as the number of assembled rings increased, thereby providing a rigorous thermodynamic argument for the formation of extended tubular structures.

This *N*-methylated system also provided a unique opportunity to study the energetic differences between parallel and antiparallel β-sheets.[125] While homodimers form antiparallel β-sheets, two enantiomeric *N*-methylated peptides must associate in a parallel fashion to achieve hydrogen bonding between homochiral residues. Since *N*-methylation prevents large-scale aggregation, the difference in stability between a two-stranded parallel and an antiparallel β-sheet can be easily measured. Note that despite the cyclic nature of the peptides, all $\phi\Psi$ angles lie well within the β-sheet regions of a Ramachandran plot. The peptide *cyclo*[-(D-Phe-L-NMe-Ala)$_4$] and its enantiomer were synthesized and the relative amounts of parallel and antiparallel dimers were measured by NMR spectroscopy in chloroform. Kobayashi *et al.* found that the antiparallel arrangement was more stable by 3.3 kJ mol^{-1} (see Figure 13). Furthermore, a C_2 symmetric peptide having the sequence *cyclo*[-(L-Phe-D-NMe-Ala-L-Leu-D-NMe-Ala)$_2$-] was also synthesized to assess the contribution of side chain interactions. The consequence of being twofold symmetric along the axis perpendicular to ring plane is that two supramolecular diastereomers are possible upon oligomerization, one with the phenylalanines and leucines forming homocross-strand pairs and the other having cross-strand formation between leucines and phenylalanines. NMR spectroscopy revealed that both diastereomers were formed in equal amounts, and so established that the side chain did not affect the structural stability of β-sheets in this model system. Thus the difference in stability is most likely attributed only to the difference intrinsic to being parallel vs. antiparallel.

12.3.8 Membrane Proteins

The current interest in signal transduction and the observation that channel proteins are often composed of transmembrane amphiphilic α-helices have made ion channels such as that shown in Figure 14 attractive targets for protein design. While an example of a *de novo* designed integral membrane protein having multiple transmembrane segments has been reported,[126] it was not designed nor shown to be functional in any way. Thus, the best models are still those employing synthetic peptides. One well-characterized example is that described by Lear *et al.*, who adopted a "minimalist" approach to the design of ion channel peptides.[127,128] Reasoning that naturally occurring sequences may be too complicated to relate structure with observed channel behavior, they chose to make a 21-residue amphiphilic peptide consisting of only leucine and serine residues. A peptide of this length would be long enough to span the thickness of a diphytanoyl phosphatidylcholine bilayer. All the serine residues were sequestered to one face if the peptide adopted an α-helical conformation. Since the sequence (Leu-Ser-Ser-Leu-Leu-Ser-Leu)$_3$ has a heptad repeat, it was expected that the helices would be able to oligomerize into helical bundles with "knobs into holes" type packing much like coiled coils. Models of this peptide in various parallel helical bundles show that it may adopt a variety of aggregation states. However, only the pentameric or higher aggregation states formed pores large enough to pass ions. All models were made with the leucine residues facing out towards the lipid, thereby forming a hydrophilic core. This peptide showed cation-selective channel activity reminiscent of the acetylcholine receptor and was estimated to have an 0.8 nm pore diameter. Furthermore, the conductance was shown to be greater at a given potential than if the potential was reversed for the *N*-acetylated, *C*-amidated peptide.[129] This observation can be interpreted to mean that the peptide inserts into the lipid bilayer in a unidirectional fashion and it was hypothesized that current flow was greater proceeding towards the amino-terminus than towards the carboxy-terminus, in accordance with the helix dipole. Finally, fluorescence data obtained for a mutant peptide containing tryptophan at various sites indicated that in the nonconducting state the peptide was associated with the lipids parallel with the plane of the bilayer and its serine side chains were thought to be associated with the hydrophilic portion of the phospholipid.[130] Application of a potential was then thought to force the helix into a transmembrane orientation through alignment of the helix dipole with the imposed electric field. Aggregation of the peptide into helical bundles then would provide a pore through the membrane.

The ambiguity in the oligomerization state and orientation of channel-forming peptides in membranes prompted the design of membrane proteins via the TASP strategy.[131,132] A review in this field was published in 1995 so this particular topic will not be discussed in detail.[133] Porphyrins have been successfully used as templates for four-helix bundles which insert into phospholipid bilayers and one even showed proton channel activity.[101,113] The vast majority of membrane-inserting TASP molecules, however, are four-helix bundles using β-turn-containing peptides as templates. The sequences of the amphiphilic peptides were typically derived from putative α-helical pore-lining segments of natural channels including melittin,[134] the acetylcholine receptor,[135] cystic fibrosis transmembrane regulator,[133] and the glycine receptor.[136] The results are surprising in that many of these TASPs behave similarly to the membrane proteins from which the peptides were derived. A particularly good example is the TASP representing the dihydropyridine (DHP)-sensitive

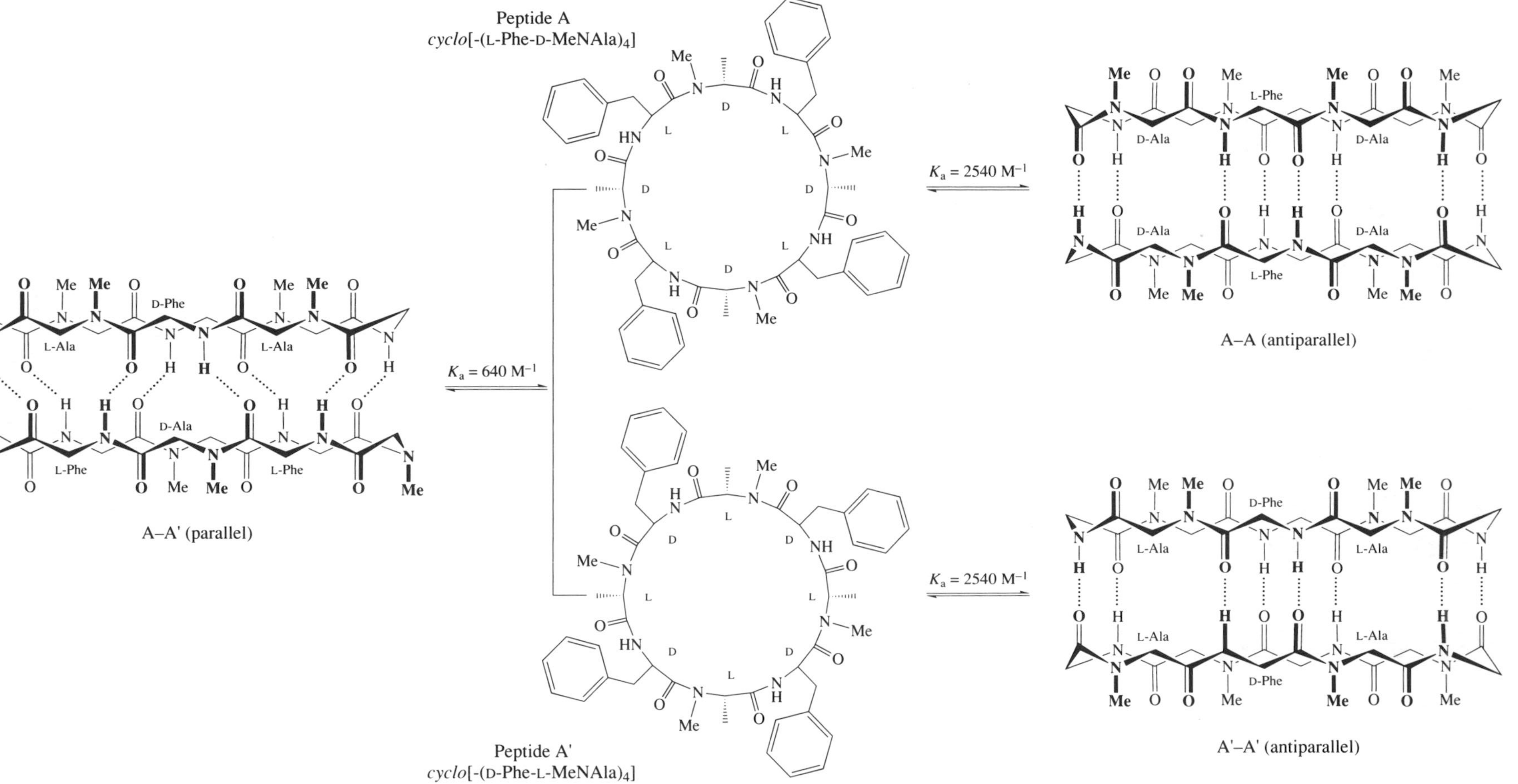

Figure 13 Schematic illustration of the system used to evaluate the relative stability of parallel and antiparallel β-sheet structures. Side chains are omitted for clarity. D and L refer to the chirality of the amino acid residues. Note that only homochiral residues make up the nearest-neighbor cross-strand pairs (reproduced by permission of *Angew. Chem., Int. Ed. Engl.*, 1995, **34**, 95).

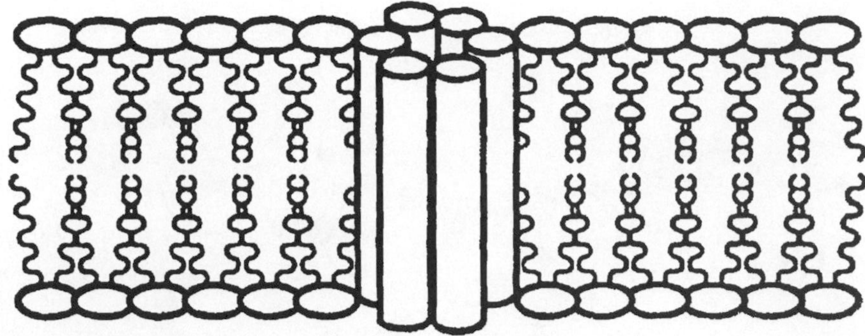

Figure 14 Schematic representation of a helix-based ion channel. Amphiphilic helices aggregate perpendicularly to the plane of the membrane to form a pore lined with polar residues (reprinted with permission from *Biochemistry*, 1992, **31**, 6608. Copyright 1992 American Chemical Society.)

calcium channel.[137] Both the natural protein and the TASP are thought to be four-helix bundles. The ion selectivity and conductance of the TASP resemble the natural protein. Its activity is also blocked by Cd^{2+}, and a variety of cation-channel blockers like veramapil and the lidocaine derivative QX-222. Most interestingly, the TASP interacts stereospecifically with DHP derivatives. Addition of the antagonist (+) BAY K8644 drastically decreases channel-opening probability and the open state lifetime. Use of the enantiomer had the opposite, agonist effect. Both enantiomers have similar effects on the natural channel. Other TASPs were shown to be sensitive to pharmacological agents as well. For instance, the channel activity of the nicotinic cholinergic receptor TASP was blocked by cation-channel inhibitors chlorpromazine and QX-222.[135] Moreover, the TASP version of the brain glycine receptor, an anion-selective channel, was sensitive to anion-channel blockers such as 9-anthracene carboxylic acid and niflumic acid but not a cation-channel blocker like QX-222.[136] Interestingly, the glycine receptor antagonist strychnine did not affect the channel activity of the TASP, presumably because it is thought to act on the ligand-binding domain of the receptor rather than on the channel domain. The fact that these compounds antagonize the function of these artificial channels implies that they adopt structures similar to ones found *in vivo*. Thus these TASPs may serve as excellent models for studying the molecular mechanisms by which inhibitors function. This conjecture must be qualified, however, with the fact that no dose response curves have been shown for any of these TASP molecules and so the observed inhibition may not be a pharmacological effect.

The pores of many channels are thought to be lined by five or even six amphiphilic helices, not four. To address this issue, the Montal laboratory synthesized a parallel bundle consisting of five helices in order to represent the nicotinic cholinergic receptor better than the original four-helix bundle.[138] The pentamer had a lower open probability than the tetrameric version but exhibited conductances to within 10% of authentic nicotinic cholinergic receptor. The minor difference was attributed to the fact that the natural channel is a heteropentamer, whereas the TASP is a homopentamer. Fortunately, orthogonal protection schemes for the template lysines are available and will in principle allow the synthesis of a heteropentamer,[104] and so further exciting developments with this model system are expected.

The latest development in designing model channels was achieved using peptide nanotubes.[120,121] Like amphiphilic helices, the sequence can be judiciously chosen to provide a hydrophobic exterior, which would cause the cyclic peptide to partition into the hydrophobic phase provided by the phospholipid bilayer. Backbone hydrogen-bonding requirements are satisfied if the peptides stack upon each other to form a tubular structure, as shown in Figure 15. Thus, both entropic and enthalpic effects bias the formation of channels in the membrane. Several noteworthy differences exist between nanotube-based channels and helical channels. The first one is the difference in structure. The cyclic β-strand-like peptide, although somewhat reminiscent of the β-helix gramicidin, represents a motif unlike any yet discovered in Nature. Second, a consequence of the structural difference in monomeric subunits as well as differences in the way that they are assembled leads to fundamental differences in channel structure. The nanotubes have a well-defined pore diameter which is determined by the number of amino acid residues in the cyclic peptide, but an ill-defined channel length which is determined by the number of peptides assembled into the tube; the opposite is true of helix-based channels because the length of such channels is dictated by the number of amino acid residues, whereas the pore diameter is dependent on the number of peptides assembling to form the pore. Third, the hydrogen-bonding network between cyclic peptides necessitates a tubular structure, while interhelical side chain packing results in a helical bundle with a gentle left-handed supertwist. Thus, helical channels are only roughly approximated by the models used to describe all channels. Fourth, because the stereochemistry of the residues in the cyclic peptide directs all side chains out of the ring, the interior of the nanotubes is relatively featureless while the lumen of helical channels is "decorated" with hydrophilic side chains, which may serve to chelate ions or constrict the channel diameter thereby adding to the ion selectivity of the channel.

Nanotubes exhibit only weak selectivity based on charge, but show selectivity based on the size of the ion. The first channel-forming nanotubes were formed from the assembly of *cyclo*[-(Trp-D-Leu)$_3$Gln-D-Leu-].[120] Hydrogen bond formation between peptides having a β-sheet conformation was observed by FTIR spectroscopy thereby supporting channel formation. This channel had a pore size of 0.75 nm and allowed passage of Na^+ and K^+ ions at rates in excess of 10^7 ions s^{-1}, roughly three times faster than conductances obtained for the naturally occurring channel gramicidin A. However, neither this nanotube nor gramicidin A could transport glucose, which requires greater than 0.9 nm pore size for passage.[121] A similar 10-residue peptide *cyclo*[Gln-(D-Leu-Trp)$_4$-D-Leu] having a pore diameter of 1.0 nm did transport glucose at rates dependent on the concentration of glucose (Figure 16). This observation further supports a channel-mediated diffusion process as opposed to a carrier-mediated event.

12.3.9 Self-assembling Peptides in Biomaterials

Peptide-based biopolymers have many desirable material properties. These can include high tensile strength, elasticity, therapeutic applications, and perhaps most importantly, the potential to have dictated molecular structure and organization.[139] Since solid-state properties such as nonlinear optics or electrical conductivity depend on these two factors, the use of peptides may provide a potentially exciting avenue for the design of such functional materials. While the field of biomaterials is still very young, the number of researchers active in this area is increasing. Much work in this field lies beyond the scope of this chapter so this section will cover only biomaterials derived from self-assembled peptides. There exist several such examples which powerfully illustrate the great promise of this area of research for producing useful substances.

One such example which was previously discussed was the self-assembling peptide nanotubes. Their ion and small-molecule channel activities are obviously very encouraging results, which provide the motivation to develop such systems into vehicles for the delivery of drugs or perhaps even nucleic acids. Their nonnatural structure as well the use of D-amino acid residues should provide significant protection against metabolic degradation, an important consideration in the development of pharmacological agents based on such design principles. The nanotube's ability to include ions and molecules also suggests many other potential uses. For instance, a nanotube could be designed with an appropriate diameter such that two molecules could be trapped within the lumen, enabling the study of individual molecules undergoing a reaction. Indeed, the organization of molecules within a cavity may lead to catalysis via a proximity effect. Thus, the capability to adjust the cavity size would be important in specifying which reaction could occur within the cavity. Perhaps the size selectivity of the nanotubes may also be harnessed to produce novel sensor devices.

Another system which has great potential for having definable surfaces with predetermined properties was developed by the Tirrell laboratory. They based their design on a peptide sequence derived from silk fibers of the moth *Bombyx mori*.[140] In designing the protein, Krejchi and coworkers consecutively repeated the sequence -[(Ala-Gly)$_3$-Glu-Gly]- 36 times. The Ala-Gly segments were predicted to form well-ordered antiparallel β-sheets, while each -Glu-Gly- unit served as a turn. The final structure was expected to be a uniform lamellar assembly of Ala-Gly-containing β-sheets with the turns making up the surface, as shown in Figure 17. This surface would be decorated with carboxylates, which perhaps could be derivatized in the future in order to make a surface of predetermined character. The gene encoding such a protein was expressed in *E. coli* and the purified protein was subsequently crystallized from a 70% solution of formic acid. FTIR and CP MAS NMR spectroscopic analyses verified that antiparallel β-sheets were indeed present in the protein crystal. Further characterization of the crystalline protein by vibrational spectroscopy revealed that the ionization state of the carboxylic acid side chains did not affect the structure.[141] However, water was observed to hydrate the polymer progressively, thereby inducing a small amount of disorder.

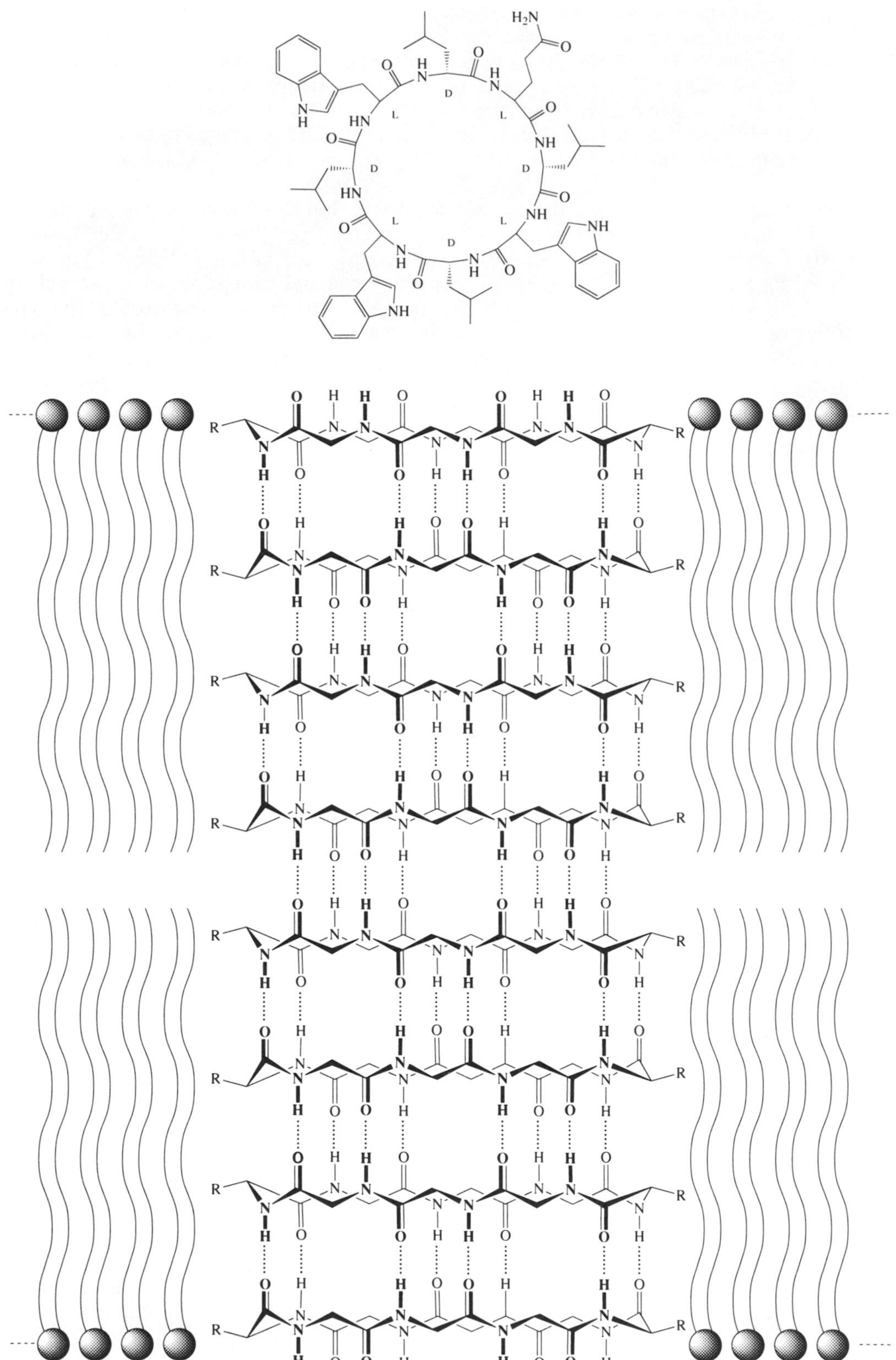

Figure 15 Schematic representation of a nanotube self-assembled in a lipid membrane. The monomer is composed mostly of hydrophobic residues. The hydrophobicity of the nanotube exterior provides the driving force for the peptide to partition into the lipid membrane. Upon dissolution in the low dielectric medium of lipid membranes, peptide subunits self-assemble to form hydrogen-bonded cylindrical channel structures. Such ensembles show remarkable transmembrane ion-transport efficiencies ($>10^7$ ions s^{-1}).

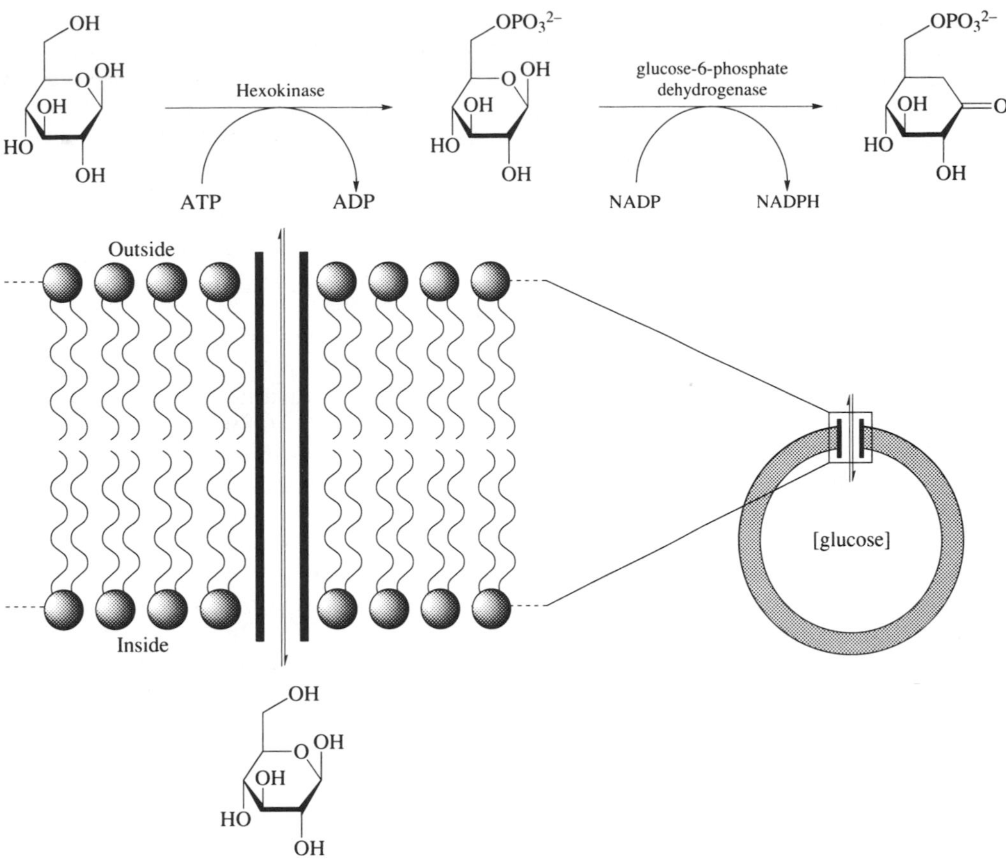

Figure 16 Schematic diagram of a nanotube membrane channel which transports glucose across lipid bilayers and the enzymatic assay used to measure glucose efflux. Formation of the channel in the membrane allows for the diffusion-driven escape of glucose from the liposome, whose rate can be monitored by measuring the production of NADPH (reprinted with permission from *J. Am. Chem. Soc.*, 1994, **116**, 10785. Copyright 1994 American Chemical Society).

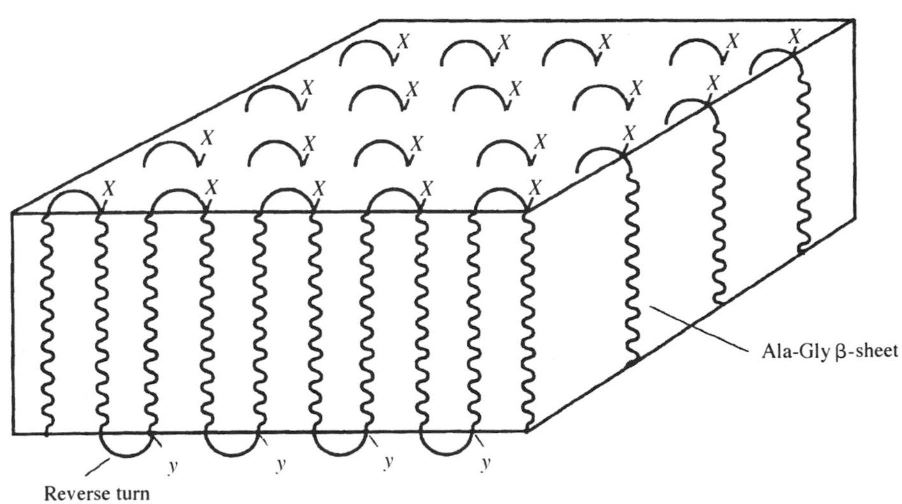

Figure 17 Schematic diagram of a lamellar protein crystal composed of β-strands oriented perpendicularly to the plane of the crystal. Turns are exposed at the surface and may be potentially derivatized to give a desired property (reproduced by permission of D. A. Tirrell from *Chem Eng. News*, 1994, **72**, 40).

Strangely, the closely related protein containing the repeat -[(Ala-Gly)$_3$-Pro-Glu-Gly]$_n$- did not form any β-structure.[142] Results of modeling studies suggest if antiparallel β-sheets were to form, the strands making up the sheets would hydrogen bond with each other in register. However, McGrath *et al.* proposed that a protein having an odd number of amino acid residues in its repeat sequence could not maintain a hydrogen-bonding network over the entire length of the sheet and that this inability prevented the formation of the desired structure.

In addition to making protein materials containing surface carboxylate groups, similar proteins have been expressed containing either a *p*-fluorophenylalanine[143] or a 3-thienylalanine[144] in each turn. Vibrational spectroscopy verified β-structure in the *p*-fluorophenylalanine-containing protein, while structural characterization beyond showing the presence of the thienylalanine moiety was not reported for the second protein. If these proteins adopt the designed lamellar structure then this system offers an exciting way of synthesizing protein-based materials with easily controlled, predefined properties.

The α-helix is another peptide motif used as components of various materials. For instance, poly(γ-benzyl-L-glutamate) which forms highly stable helices has been formed into monolayers by the LB technique and was also self-assembled onto gold.[145] Vibrational spectroscopic data suggested that the helices could be oriented in an ordered manner. A second helix-based system was designed by Whitesell and Chang[146] wherein directionally aligned α-helices were assembled into a multilamellar structure built over gold or silicon (Figure 18). A monolayer of aminotrithiol was first assembled onto a layer of gold. Subsequently, units of trialanine were polymerized onto this surface via the carboxy-terminus to furnish a third layer composed of polyalanine. Vibrational spectroscopy indicates that these polyalanine units are helical and that the helical axes are aligned perpendicular to the surface of the gold substrate. The observation of uneven chain growth illustrates that improvements in the synthesis of the peptide layers are necessary. However, the design offers a great deal of flexibility:

(i) the thickness of the layers is simply controlled by the length of the peptide chain;

(ii) the properties of the surface are dictated by the amino acid composition;

(iii) the amino-terminus is potentially available for derivatization, thereby allowing for surface engineering of the structure; and

(iv) the spacing between helices can be controlled.

The first two points are inherent in the design. The last advantage was demonstrated by forming a layer of polyphenylalanine. First, modeling showed that the interhelical spacing would be too small to accommodate the large phenylalanine side chains if the original aminotrithiol "tripod" was used. Thus the chloroacetyl chloride was reacted with aminotrithiol on gold to give a tris-α-chloroamide. Addition of more aminotrithiol furnished a new tripod with a wider base composed of three original aminotrithiols. Polymerization of the *N*-carboxyanhydride of phenylalanine onto this new tripod furnished a layer of polyphenylalanine which had high helicity.

The Urry laboratory has provided examples of cyclic peptides that, after cross-linking, can carry out mechanical work upon a change in temperature. This research was well reviewed in 1993 by Urry[147] so to summarize briefly, heating of these peptides causes an increase in their order by changing the hydration state as indicated by crystallization of the peptide. Cross-linking of these cyclic peptides does not seem to affect this property. Thus, heating strips composed of these cross-linked peptides resulted in measurable contractions which were strong enough to move weights of gram masses. Particularly interesting is the fact that the pK_a of glutamic acid and therefore the transition temperature can be rationally modified by adjusting the relative amounts of hydrophobic and acidic residues.[147,148] Moreover, Urry and co-workers were able to incorporate a nicotinamide moiety which increases in hydrophobicity upon reduction. Its reduction caused not only a shift in the transition temperature[149] but in the pK_a of ionizable residues as well.[150,151] Hence, Urry and co-workers offer these systems as models of electromechanical and electrochemical transduction in proteins.

There exist numerous other examples of self-assembling peptides which have the potential to be developed as useful biomaterials. The formation of amyloid fibrils by the scrapie form of prion proteins is well known.[152,153] Studies using the prion protein[154] as well as model peptides[155] showed that the prion protein underwent a conformational transition from a highly α-helical soluble form to a β-sheet-containing form in a scrapie protein dependent manner. The β-sheet-containing form could subsequently aggregate. This, and other evidence[152,156,157] strongly supports the notion that prion replication is a nucleic acid free event and that the mechanism of infection occurs through this scrapie-mediated conformational transition. However, the mechanism by which this transition is brought about is hotly disputed. One hypothesis was proposed wherein amyloid formation is a paracrystallization event.[158] Another hypothesis was offered

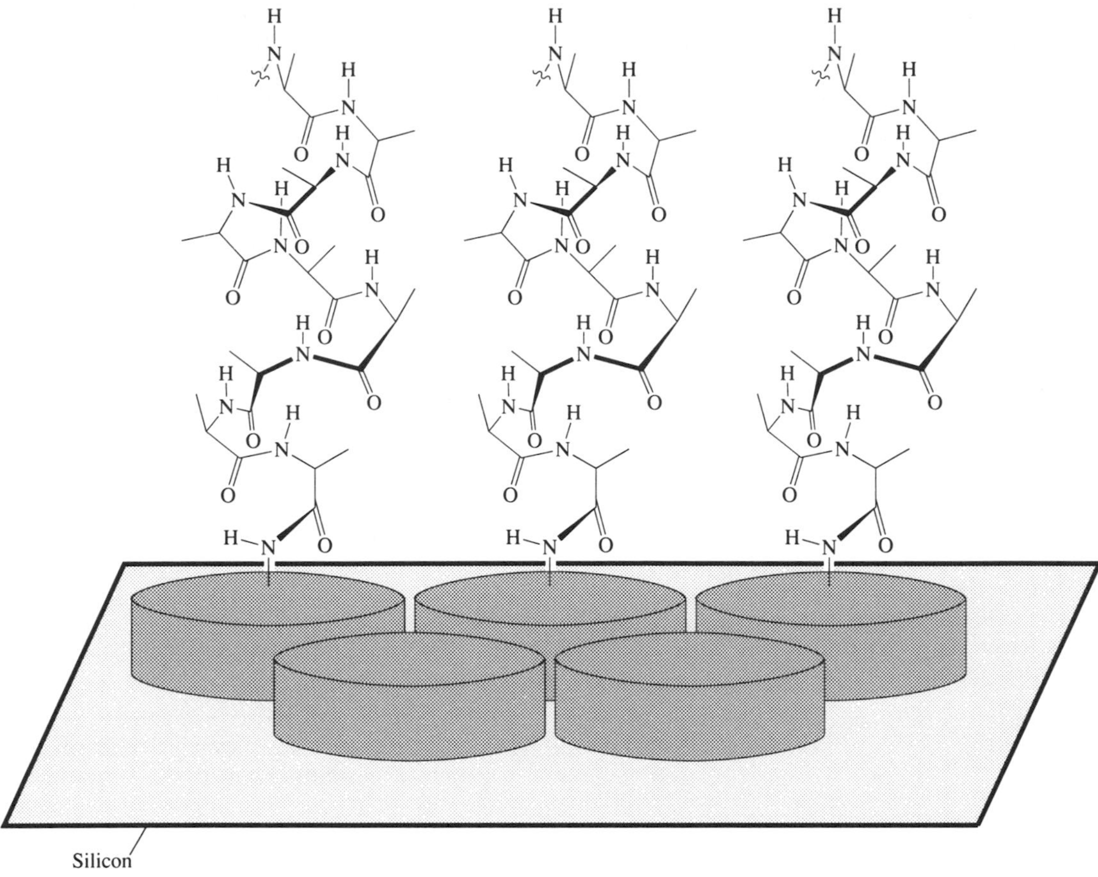

Figure 18 Schematic diagram of polyalanine α-helices adsorbed onto a layer of silicon. The covalently linked cylinders represent the aminotrithiol tripod used to anchor the peptide to the solid phase (reproduced by permission of the American Association for the Advancement of Science from *Science*, 1993, **261**, 73).

wherein the scrapie form induces the conformational change only on partially unfolded cellular structures.[159] If the mechanism of this transition could be understood, it may be possible to incorporate it into a self-assembling system to produce desired aggregates instead of amyloid fibrils.

An interesting self-assembling peptide system was serendipitously discovered by the Rich laboratory.[160,161] A 16-residue peptide which alternated glutamic acid and alanine, then lysine and alanine, was synthesized and adopted a β-structure, even at low concentration. Addition of salt caused spontaneous assembly of the peptide into an extremely stable macroscopic membrane. Examination of these membranes by electron microscopy revealed that the membranes are composed of a tangled mesh of fibers somewhat reminiscent of amyloid deposits. Its structure remained intact despite extremes of temperature, pH, and salt. Rich and co-workers hypothesized that the stability of the membranes was due largely to ionic pairing between glutamate and lysine side chains. Thus, they speculated that the associative forces between sheets alternated between hydrophobic packing of alanine side chains and salt bridging between glutamate and lysine.

Peptidomimetic systems may also be useful for making self-assembling biomaterials. Peptoids[162,163] and oligocarbamate polymers[164] have not yet been shown to be useful for making defined structures but their ease of synthesis makes these compounds suitable for materials research. In the mid-1990s, Karle and Ranganathan studied the solid-state conformation of various oxalo retropeptides.[165] They generally consisted of two peptides oriented in opposite polarity and joined at their amino-termini by an oxalo group. The presence of the oxalo group in the backbone of the retropeptide was thought to impart rigidity and an extended conformation to the peptide, as evidenced by the crystal structures of several such peptides. This structure may be due to the formation of pseudo-C_5-type hydrogen bonds between each carbonyl group and the amide proton of the adjacent peptide. Thus, the two adjacent carbonyl groups were in a *trans* conformation. Despite the common backbone oxalo group, the peptides adopted different conformations depending on their sequences. For instance, the retropeptide MeO-Aib-CO-CO-Aib-OMe crystallized into antiparallel crescent-shaped β-sheets, while MeO-Leu-CO-CO-Leu-OMe formed helices composed of intermolecularly hydrogen-bonded extended structures. Ribbon structures were formed through intermolecular hydrogen bonding by the retropeptide MeO-Ser-Leu-CO-CO-Leu-Ser-OMe. The particularly interesting feature of this molecule is the fact that the side chain of serine was also observed to take part in intermolecular hydrogen bonding and so may serve as a simple mimic of side chain–backbone interactions in proteins.

Peptide-nucleic acids (PNAs) form a unique class of peptide-like compounds with interesting self-assembling properties. They consist of poly-*N*-(2-aminoethyl)glycine which is derivatized by purines and pyrimidines.[166,167] The nucleobases furnish the molecule with the ability to form Watson–Crick base pairs with a complementary PNA molecule,[168] RNA,[169] or DNA.[170] They may also form triple helix structures, thereby making them potentially valuable tools for controlling gene transcription or mRNA translation. While the polyamide backbone is achiral, PNA assemblies can form structures having dichroic properties. The PNA-bound RNA was shown to adopt the A-form structure typical of RNA.[169] Astonishingly, duplex PNA was shown to adopt a structure with chirality when tagged with L-lysine, and formed a structure having an equal but opposite c.d. spectrum when tagged with D-lysine.[168] Thus, it seems that a chiral template can propagate chirality throughout the entire molecule. This property has broad implications for the field of biomaterials, for potentially a material can be fashioned from PNA with the desired structure and chirality given an appropriate template.

12.4 SUMMARY AND PERSPECTIVES

In summary, a thorough understanding of peptide self-assembly and self-organization is not yet attained. Factors such as hydrogen bonding and hydrophobic interactions which are involved in the formation of autonomous secondary and tertiary structures are important in the formation of large supramolecular structures, and the relatively few examples of each type is partially indicative of the difficulty in controlling aggregation states of peptides and proteins. The studies reviewed in this chapter illustrate some of the general principles currently known about self-assembly and self-organization in peptidic systems. Studies on short peptides in secondary structural motifs have verified the prediction of Zimm and Bragg that helix formation is a nucleation event[171] and provide simple models in which to study helical preferences of amino acid residues. Studies on small β-sheet systems illustrated the importance of templating and nucleation on their formation, but also revealed the added difficulties in controlling the aggregation state of β-sheet proteins. Thus, designs of most higher-ordered polypeptide structures have been all-helical proteins. The minimalist, TASP, and metal ion assisted approaches to helix bundle design have been fruitful thus far. The success of these approaches, however, must be qualified with the fact that all of these examples, while adopting the desired fold and oligomerization state, exhibit molten globule behavior and so should not be considered "native" proteins as yet. It is anticipated that results of studies on coiled coil peptides will be invaluable for improving future designs. Only a few examples of designed β-sheet-containing proteins exist. One example in which the self-assembly process is particularly well-understood is the nanotube class of β-proteins. While representing an unnatural motif, they can still serve as models for β-sheet proteins.

Knowledge gained from studying the assembly of such unnatural as well as natural peptide systems has aided in the development of peptides as biomaterials. Nanotubes, β-proteins based on silk fiber sequences, as well as helical proteins have been assembled in an organized fashion in macroscopic proportions. Moreover, these peptide systems have tunable surface properties. Together these two features make the examples attractive for further development into useful biomaterials. In particular, nanotubes have already been shown to be functional as ion channels in lipid vesicles. While efficient large-scale production has yet to be demonstrated for most peptide biomaterials (an obvious requirement for any material destined for widespread use), it is likely that such a problem will be eventually solved.

12.5 REFERENCES

1. C. Levinthal, *J. Chim. Phys.*, 1968, **65**, 44.
2. M. Radmacher, M. Fritz, H. G. Hansma, and P. K. Hansma, *Science*, 1994, **265**, 1577.
3. S. L. Slatin, X.-Q. Qui, K. S. Jakes, and A. Finkelstein, *Nature*, 1994, **371**, 158.
4. P. A. Bullough, F. M. Hughson, J. J. Skehel, and D. C. Wiley, *Nature*, 1994, **371**, 37.

5. P. K. Sarkar and P. Doty, *Proc. Natl. Acad. Sci. USA*, 1966, **55**, 981.
6. S.-Y. C. Wooley and G. Holzworth, *Biochemistry*, 1970, **9**, 3604.
7. B. Davidson and G. D. Fasman, *Biochemistry*, 1967, **6**, 1616.
8. M. Mutter and R. Hersperger, *Angew. Chem., Int. Ed. Engl.*, 1990, **29**, 185.
9. M. Mutter, R. Gassmann, U. Buttkus, and K. Altmann, *Angew. Chem., Int. Ed. Engl.*, 1991, **30**, 1514.
10. G. P. Dado and S. H. Gellman, *J. Am. Chem. Soc.*, 1993, **115**, 12609.
11. M. R. Ghadiri and C. Choi, *J. Am. Chem. Soc.*, 1990, **112**, 1630.
12. M. R. Ghadiri and A. K. Fernholz, *J. Am. Chem. Soc.*, 1990, **112**, 9633.
13. F. Ruan, Y. Chen, and P. B. Hopkins, *J. Am. Chem. Soc.*, 1990, **112**, 9403.
14. F. Ruan, Y. Chen, K. Itoh, T. Sasaki, and P. B. Hopkins, *J. Org. Chem.*, 1991, **56**, 4347.
15. B. Imperiali and T. M. Kapoor, *Tetrahedron*, 1993, **49**, 3501.
16. T. Tsuji and E. T. Kaiser, *Proteins: Structure, Function, Genetics*, 1991, **9**, 12.
17. G. S. Shaw, R. S. Hodges, and B. D. Sykes, *Science*, 1990, **249**, 280.
18. R. E. Reid, J. Gariepy, A. K. Saund, and R. S. Hodges, *J. Biol. Chem.*, 1981, **256**, 2742.
19. G. S. Shaw, R. S. Hodges, and B. D. Sykes, *Biochemistry*, 1992, **31**, 9572.
20. G. S. Shaw, L. F. Golden, R. S. Hodges, and B. D. Sykes, *J. Am. Chem. Soc.*, 1991, **113**, 5557.
21. G. S. Shaw, W. A. Findlay, P. D. Semchuk, R. S. Hodges, and B. D. Sykes, *J. Am. Chem. Soc.*, 1992, **114**, 6258.
22. G. S. Shaw, R. S. Hodges, and B. D. Sykes, *Biopolymers*, 1992, **32**, 391.
23. J. M. Berg, *Acc. Chem. Res.*, 1995, **28**, 14.
24. J. M. Berg, *Curr. Opin. Struct. Biol.*, 1993, **3**, 585.
25. A. D. Frankel, J. M. Berg, and C. O. Pabo, *Proc. Natl. Acad. Sci. USA*, 1987, **84**, 4841.
26. G. Parraga, S. J. Horvath, A. Eisen, W. E. Taylor, L. Hood, E. T. Young, and R. E. Klevit, *Science*, 1988, 1489.
27. M. S. Lee, G. P. Gippert, K. V. Soman, D. A. Case, and P. E. Wright, *Science*, 1989, **245**, 635.
28. J. M. Berg and D. L. Merkle, *J. Am. Chem. Soc.*, 1989, **111**, 3759.
29. B. A. Krizek, D. L. Merkle, and J. M. Berg, *Inorg. Chem.*, 1993, **32**, 937.
30. S. F. Michael, V. J. Kilfoil, M. H. Schmidt, B. T. Amann, and J. M. Berg, *Proc. Natl. Acad. Sci. USA*, 1992, **89**, 4796.
31. D. Y. Jackson, D. S. King, J. Chmielewski, S. Singh, and P. G. Schultz, *J. Am. Chem. Soc.*, 1991, **113**, 9391.
32. G. Osapay and J. W. Taylor, *J. Am. Chem. Soc.*, 1992, **114**, 6966.
33. A. C. Satterthwait, T. Arrhenius, R. A. Hagopian, F. Zavala, V. Nussenweig, and R. A. Lerner, *Philos. Trans. R. Soc., London Ser. B.*, 1989, **323**, 565.
34. A. C. Satterthwait, L. Chiang, T. Arrhenius, E. Cabezas, F. Zavala, H. J. Dyson, P. E. Wright, and R. A. Lerner, *Bull. World Health Org.*, 1990, **68**, 17.
35. E. Cabezas, L. Chiang, and A. C. Satterthwait, in "Peptides: Chemistry, Structure, and Biology," eds. R. S. Hodges and J. A. Smith, Escom, Leiden, 1994, p. 287.
36. D. S. Kemp, *Trends Biotech.*, 1990, **8**, 249.
37. D. S. Kemp, T. J. Allen, and S. L. Oslick, *J. Am. Chem. Soc.*, 1995, **117**, 6641.
38. D. S. Kemp and B. R. Bowen, *Tetrahedron Lett.*, 1988, **29**, 5077.
39. D. S. Kemp and B. R. Bowen, *Tetrahedron Lett.*, 1988, **29**, 5081.
40. D. S. Kemp and Z. Q. Li, *Tetrahedron Lett.*, 1995, **36**, 4179.
41. D. S. Kemp and Z. Q. Li, *Tetrahedron Lett.*, 1995, **36**, 4175.
42. H. Diaz, J. R. Espina, and J. W. Kelly, *J. Am. Chem. Soc.*, 1992, **114**, 8316.
43. H. Diaz, K. Y. Tsang, D. Choo, J. R. Espina, and J. W. Kelly, *J. Am. Chem. Soc.*, 1993, **115**, 3790.
44. K. Y. Tsang, H. Diaz, N. Graciani, and J. W. Kelly, *J. Am. Chem. Soc.*, 1994, **116**, 3988.
45. S. R. LaBrenz and J. W. Kelly, *J. Am. Chem. Soc.*, 1995, **117**, 1655.
46. J. W. Taylor and E. T. Kaiser, *Pharm. Rev.*, 1986, **38**, 291.
47. E. T. Kaiser and F. J. Kezdy, *Science*, 1984, **223**, 249.
48. M. Mutter, K.-H. Altmann, G. Tuchscherer, and S. Vuilleumier, *Tetrahedron*, 1988, **44**, 771.
49. M. Mutter and S. Vuilleumier, *Angew. Chem., Int. Ed. Engl.*, 1989, **28**, 535.
50. G. Tuchscherer, B. Domer, U. Sila, B. Kamber, and M. Mutter, *Tetrahedron*, 1993, **49**, 3559.
51. M. R. Ghadiri, C. Soares, and C. Choi, *J. Am. Chem. Soc.*, 1992, **114**, 4000.
52. M. R. Ghadiri, C. Soares, and C. Choi, *J. Am. Chem. Soc.*, 1992, **114**, 825.
53. T. Sasaki and M. Lieberman, *Tetrahedron*, 1993, **49**, 3677.
54. M. Lieberman and T. Sasaki, *J. Am. Chem. Soc.*, 1991, **113**, 1470.
55. M. R. Ghadiri, J. R. Granja, R. A. Milligan, D. E. McRee, and N. Khazanovich, *Nature*, 1993, **366**, 324.
56. C. Cohen and D. A. D. Parry, *Proteins: Structure, Function, Genetics*, 1990, **7**, 1.
57. C. Cohen and D. A. D. Parry, *Trends Biochem. Sci.*, 1986, **11**, 245.
58. W. H. Landschulz, P. F. Johnson, and S. L. McKnight, *Science*, 1988, **240**, 1759.
59. T. G. Oas, L. P. McIntosh, E. K. O'Shea, F. W. Dahlquist, and P. S. Kim, *Biochemistry*, 1990, **29**, 2891.
60. E. K. O'Shea, J. D. Klemm, P. S. Kim, and T. Alber, *Science*, 1991, **254**, 539.
61. F. H. C. Crick, *Acta Crystallogr.*, 1953, **6**, 685.
62. N. E. Zhou, C. M. Kay, and R. S. Hodges, *Biochemistry*, 1992, **31**, 5739.
63. B.-Y. Zhu, N. E. Zhou, C. M. Kay, and R. S. Hodges, *Protein Sci.*, 1993, **2**, 383.
64. P. B. Harbury, T. Zhang, P. S. Kim, and T. Alber, *Science*, 1993, **262**, 1401.
65. P. B. Harbury, P. S. Kim, and T. Alber, *Nature*, 1994, **371**, 80.
66. S. Cusack, C. Berthet-Colominas, M. Hartlein, N. Nassar, and R. Leberman, *Nature*, 1990, **347**, 249.
67. O. D. Monera, N. E. Zhou, C. M. Kay, and R. S. Hodges, *J. Biol. Chem.*, 1993, **268**, 19218.
68. S. Betz, R. Fairman, K. O'Neil, J. Lear, and W. F. DeGrado, *Philos. Trans. R. Soc., London, Ser. B*, 1995, **348**, 81.
69. K. J. Lumb and P. S. Kim, *Biochemistry*, 1995, **34**, 8642.
70. O. D. Monera, C. M. Kay, and R. S. Hodges, *Biochemistry*, 1994, **33**, 3862.
71. D. G. Myszka and I. M. Chaiken, *Biochemistry*, 1994, **33**, 2363.
72. E. K. O'Shea, R. Rutkowski, W. F. Stafford, III, and P. S. Kim, *Science*, 1989, **245**, 646.
73. E. K. O'Shea, R. Rutkowski, and P. S. Kim, *Cell*, 1992, **68**, 699.
74. T. J. Graddis, D. G. Myzska, and I. M. Chaiken, *Biochemistry*, 1993, **32**, 12664.

75. E. K. O'Shea, K. J. Lumb, and P. S. Kim, *Curr. Biol.*, 1993, **3**, 658.
76. N. E. Zhou, C. M. Kay, and R. S. Hodges, *J. Mol. Biol.*, 1994, **237**, 500.
77. A. Letai and E. Fuchs, *Proc. Natl. Acad. Sci. USA*, 1995, **92**, 92.
78. K. Beck, T. W. Dixon, J. Engel, and D. A. D. Parry, *J. Mol. Biol.*, 1993, **231**, 311.
79. W. D. Kohn, C. M. Kay, and R. S. Hodges, *Protein Sci.*, 1995, **4**, 237.
80. N. E. Zhou, C. M. Kay, and R. S. Hodges, *Protein Eng.*, 1994, **7**, 1365.
81. K. P. McGrath and D. L. Kaplan, in "Biomolecular Materials by Design," eds. M. Alper, H. Bayley, D. Kaplan, and M. Navia, Materials Research Society, Pittsburgh, PA, 1994, p. 61.
82. M. K. Gilson, *Curr. Opin. Struct. Biol.*, 1995, **5**, 216.
83. Z. S. Hendsch and B. Tidor, *Protein Sci.*, 1994, **3**, 211.
84. K. J. Lumb and P. S. Kim, *Science*, 1995, **268**, 436.
85. I. L. Karle, J. L. Flippen-Anderson, K. Uma, and P. Balaram, *Biopolymers*, 1990, **30**, 719.
86. I. L. Karle, J. L. Flippen-Anderson, K. Uma, M. Sukumar, and P. Balaram, *J. Am. Chem. Soc.*, 1990, **112**, 9350.
87. P. Balaram, *Pure Appl. Chem.*, 1992, **64**, 1061.
88. S. P. Ho and W. F. DeGrado, *J. Am. Chem. Soc.*, 1987, **109**, 6751.
89. W. F. DeGrado, *Science*, 1989, **243**, 622.
90. W. F. DeGrado, D. P. Raleigh, and T. Handel, *Curr. Opin. Struct. Biol.*, 1991, **1**, 984.
91. N. Nishino, H. Mihara, Y. Tanaka, and T. Fujimoto, *Tetrahedron Lett.*, 1992, **33**, 5767.
92. N. Nishino, H. Mihara, T. Uchida, and T. Fujimoto, *Chem. Lett.*, 1993, 53.
93. S. F. Betz, D. P. Raleigh, and W. F. DeGrado, *Curr. Opin. Struct. Biol.*, 1993, **3**, 601.
94. T. M. Handel, S. A. Williams, D. Menyhard, and W. F. DeGrado, *J. Am. Chem. Soc.*, 1993, **115**, 4457.
95. T. Handel and W. F. DeGrado, *J. Am. Chem. Soc.*, 1990, **112**, 6710.
96. T. M. Handel, S. A. Williams, and W. F. DeGrado, *Science*, 1993, **261**, 879.
97. J. J. Osterhout, *et al.*, *J. Am. Chem. Soc.*, 1992, **114**, 331.
98. D. P. Raleigh and W. F. DeGrado, *J. Am. Chem. Soc.*, 1992, **114**, 10 079.
99. T.-M. Chin, K. D. Berndt, and N.-c. C. Yang, *J. Am. Chem. Soc.*, 1992, **114**, 2279.
100. R. Lutgring and J. Chmielewski, *J. Am. Chem. Soc.*, 1994, **116**, 6451.
101. H. Mihara, N. Nishino, R. Hasegawa, and T. Fujimoto, *Chem. Lett.*, 1992, 1805.
102. T. Sasaki and E. T. Kaiser, *J. Am. Chem. Soc.*, 1989, **111**, 380.
103. T. Sasaki and E. T. Kaiser, *Biopolymers*, 1990, 79.
104. P. Dumy, I. M. Eggleston, S. Cervigni, U. Sila, X. Sun, and M. Mutter, *Tetrahedron Lett.*, 1995, **36**, 1255.
105. M. Mutter, R. Hersperger, K. Gubernator, and K. Muller, *Proteins: Structure, Function, Genetics*, 1989, **5**, 13.
106. S. Vuilleumier and M. Mutter, *Biopolymers*, 1993, **33**, 389.
107. M. R. Ghadiri and M. A. Case, *Angew. Chem., Int. Ed. Engl.*, 1993, **32**, 1594.
108. K. W. Hahn, W. A. Klis, and J. M. Stewart, *Science*, 1990, 1544.
109. M. J. Corey, E. Hallakova, K. Pugh, and J. M. Stewart, *Appl. Biochem. Biotech.*, 1994, **47**, 199.
110. K. Johnsson, R. K. Allemann, H. Widmer, and S. A. Benner, *Nature*, 1993, **365**, 530.
111. C. T. Choma, J. D. Lear, M. J. Nelson, P. L. Dutton, D. E. Robertson, and W. F. DeGrado, *J. Am. Chem. Soc.*, 1994, **116**, 856.
112. D. E. Robertson, *et al.*, *Nature*, 1994, **368**, 425.
113. K. S. Akerfeldt, R. M. Kim, D. Camac, J. T. Groves, J. D. Lear, and W. F. DeGrado, *J. Am. Chem. Soc.*, 1992, **114**, 9656.
114. C. T. Choma, K. Kaestle, K. Akerfeldt, R. M. Kim, J. T. Groves, and W. F. DeGrado, *Tetrahedron Lett.*, 1994, **35**, 6191.
115. Y. Yan and B. W. Erickson, *Protein Sci.*, 1994, **3**, 1069.
116. T. P. Quinn, N. B. Tweedy, R. W. Williams, J. S. Richardson, and D. C. Richardson, *Proc. Natl. Acad. Sci. USA*, 1994, **91**, 8747.
117. P. De Santis, S. Morosetti, and R. Rizzo, *Macromolecules*, 1974, **7**, 52.
118. V. Pavone, E. Benedetti, B. Di Blasio, A. Lombardi, C. Pedone, L. Tomasic, and G. P. Lorenzi, *Biopolymers*, 1989, **28**, 215.
119. L. Tomasic and G. P. Lorenzi, *Helv. Chim. Acta*, 1987, **70**, 1012.
120. M. R. Ghadiri, J. R. Granja, and L. K. Buehler, *Nature*, 1994, **369**, 301.
121. J. R. Granja and M. R. Ghadiri, *J. Am. Chem. Soc.*, 1994, **116**, 10 785.
122. J. D. Hartgerink and M. R. Ghadiri, manuscript in preparation.
123. X. Sun and G. P. Lorenzi, *Helv. Chim. Acta*, 1994, **77**, 1520.
124. M. R. Ghadiri, K. Kobayashi, J. R. Granja, R. K. Chadha, and D. E. McRee, *Angew. Chem., Int. Ed. Engl.*, 1995, **34**, 93.
125. K. Kobayashi, J. R. Granja, and M. R. Ghadiri, *Angew. Chem., Int. Ed. Engl.*, 1995, **34**, 95.
126. P. Whitley, I. Nilsson, and G. v. Heijne, *Nature Struct. Biol.*, 1994, **1**, 858.
127. J. D. Lear, Z. R. Wasserman, and W. F. DeGrado, *Science*, 1988, **240**, 1177.
128. K. S. Akerfeldt, J. D. Lear, Z. R. Wasserman, L. A. Chung, and W. F. DeGrado, *Acc. Chem. Res.*, 1993, **26**, 191.
129. P. K. Kienker, W. F. DeGrado, and J. D. Lear, *Proc. Natl. Acad. Sci. USA*, 1994, **91**, 4859.
130. L. A. Chung, J. D. Lear, and W. F. DeGrado, *Biochemistry*, 1992, **31**, 6608.
131. M. Montal, M. S. Montal, and J. M. Tomich, *Proc. Natl. Acad. Sci. USA*, 1990, **87**, 6929.
132. M. Montal, *FASEB J.*, 1990, **4**, 2623.
133. M. Montal, *Ann. Rev. Biophys. Biomol. Struct.*, 1995, **24**, 31.
134. M. Pawlak, U. Meseth, B. Dhanapal, M. Mutter, and H. Vogel, *Protein Sci.*, 1994, **3**, 1788.
135. M. Oblatt-Montal, L. K. Buhler, T. Iwamoto, J. M. Tomich, and M. Montal, *J. Biol. Chem.*, 1993, **268**, 14 601.
136. G. L. Reddy, T. Iwamoto, J. M. Tomich, and M. Montal, *J. Biol. Chem.*, 1993, **268**, 14 608.
137. A. Grove, J. M. Tomich, and M. Montal, *Proc. Natl. Acad. Sci. USA*, 1991, **88**, 6418.
138. M. O. Montal, T. Iwamoto, J. M. Tomich, and M. Montal, *FEBS Lett.*, 1993, **320**, 261.
139. J. G. Tirrell, M. J. Fournier, T. L. Mason, and D. A. Tirrell, *Chem. Eng. News*, 1994, **72**, 40.
140. M. T. Krejchi, E. D. T. Atkins, A. J. Waddon, M. J. Fournier, T. L. Mason, and D. A. Tirrell, *Science*, 1994, **265**, 1427.

141. C. C. Chen, M. T. Krejchi, D. A. Tirrell, and S. L. Hsu, *Macromolecules*, 1995, **28**, 1464.
142. K. P. McGrath, M. J. Fournier, T. L. Mason, and D. A. Tirrell, *J. Am. Chem. Soc.*, 1992, **114**, 727.
143. E. Yoshikawa, M. J. Fournier, T. L. Mason, and D. A. Tirrell, *Macromolecules*, 1994, **27**, 5471.
144. S. Kothakota, T. L. Mason, D. A. Tirrell, and M. J. Fournier, *J. Am. Chem. Soc.*, 1995, **117**, 536.
145. E. P. Enriquez and E. T. Samulski, in "Hierarchically Structured Materials," eds. I. A. Aksay, E. Baer, M. Sarikaya, and D. A. Tirrell, Materials Research Society, Pittsburgh, PA, 1992, p. 423.
146. J. K. Whitesell and H. K. Chang, *Science*, 1993, **261**, 73.
147. D. W. Urry, *Angew. Chem., Int. Ed. Engl.*, 1993, **32**, 819.
148. D. W. Urry, S. Peng, and T. Parker, *J. Am. Chem. Soc.*, 1993, **115**, 7509.
149. D. W. Urry, L. C. Hayes, and D. C. Gowda, *Biochem. Biophys. Res. Commun.*, 1994, **204**, 230.
150. D. W. Urry, D. C. Gowda, S. Q. Peng, and T. M. Parker, *Chem. Phys. Lett*, 1995, **239**, 67.
151. D. W. Urry, L. C. Hayes, D. C. Gowda, S. Q. Peng, and N. Jing, *Biochem. Biophys. Res. Commun.*, 1995, **210**, 1031.
152. S. B. Prusiner, *Annu. Rev. Microbiol.*, 1994, **48**, 655.
153. S. B. Prusiner, *Biochemistry*, 1992, **31**, 12 277.
154. K.-M. Pan, *et al.*, *Proc. Natl. Acad. Sci. USA*, 1993, **90**, 10 962.
155. J. Nguyen, M. A. Baldwin, F. E. Cohen, and S. B. Prusiner, *Biochemistry*, 1995, **34**, 4186.
156. R. A. Bessen, D. A. Kocisko, G. J. Raymond, S. Nandan, P. T. Lansbury, and B. Caughey, *Nature*, 1995, **375**, 698.
157. D. A. Kocisko, J. H. Come, S. A. Priola, B. Chesebro, G. J. Raymond, P. T. J. Lansbury, and B. Caughey, *Nature*, 1994, **370**, 471.
158. J. H. Come and P. T. J. Lansbury, *J. Am. Chem. Soc.*, 1994, **116**, 4109.
159. S. B. Prusiner, *Science*, 1991, **252**, 1515.
160. S. Zhang, T. Holmes, C. Lockshin, and A. Rich, *Proc. Natl. Acad. Sci. USA*, 1993, **90**, 3334.
161. S. Zhang, C. Lockshin, R. Cook, and A. Rich, *Biopolymers*, 1994, **34**, 663.
162. H. Kessler, *Angew. Chem., Int. Ed. Engl.*, 1993, **32**, 543.
163. R. J. Simon, *et al.*, *Proc. Natl. Acad. Sci. USA*, 1992, **89**, 9367.
164. C. Y. Cho, E. J. Moran, S. R. Cherry, J. C. Stephans, S. P. A. Fodor, C. L. Adams, A. Sundaram, J. W. Jacobs, and P. G. Schultz, *Science*, 1993, **261**, 1303.
165. I. L. Karle and D. Ranganathan, *Int. J. Pep. Protein Res.*, 1995, **46**, 18.
166. P. E. Nielsen, M. Egholm, R. H. Berg, and O. Buchardt, *Science*, 1991, **254**, 1497.
167. M. Egholm, O. Buchardt, P. E. Nielsen, and R. H. Berg, *J. Am. Chem. Soc.*, 1992, **114**, 1895.
168. P. Wittung, P. E. Nielsen, O. Buchardt, M. Egholm, and B. Norden, *Nature*, 1994, **368**, 561.
169. S. C. Brown, S. A. Thomson, J. M. Veal, and D. G. Davis, *Science*, 1994, **265**, 777.
170. M. Egholm, *et al.*, *Nature*, 1993, **365**, 566.
171. B. H. Zimm and J. K. Bragg, *J. Chem Phys.*, 1959, **31**, 526.

13

Self-assembled Columnar Mesophases Based on Guanine-related Molecules

GIOVANNI GOTTARELLI and GIAN PIERO SPADA
Università di Bologna, Italy
and
ANNA GARBESI
C.N.R.-I.Co.C.E.A., Bologna, Italy

13.1 INTRODUCTION

Guanine (**1**) and pterine (**2**) are heterocyclic bases present in important biological molecules: guanine is one of the bases of nucleic acids and pterine is the active moiety of folic acid (**3**).

The two molecules possess a particular sequence of groups that act as donors or acceptors of hydrogen bonds and are of fundamental importance in the self-recognition and self-assembly processes.

While the first report on the self-assembly of folates appeared in 1993,[1] the ability of guanosine derivatives to form gels has been known since the early twentieth century, after Bang,[2] reported that concentrated solutions of guanylic acid originate gels. Fifty years later, Khorana and co-workers described the amazing thermal stability of the aggregates formed by very short deoxy-guanosine oligonucleotides bearing 5′-phosphomonoester groups.[3]

An x-ray study on the fibre obtained from a gel of guanosine 3′-phosphate (Gp) (**4**) allowed Gellert *et al.*[4] to propose the tetrameric arrangement (G-quartet) depicted in Figure 1(a).

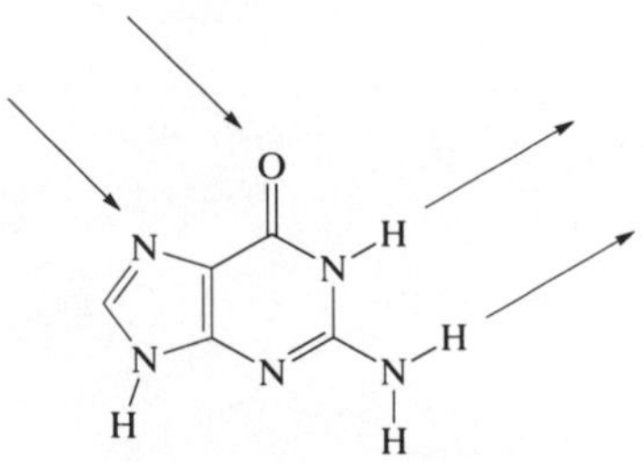

(1)

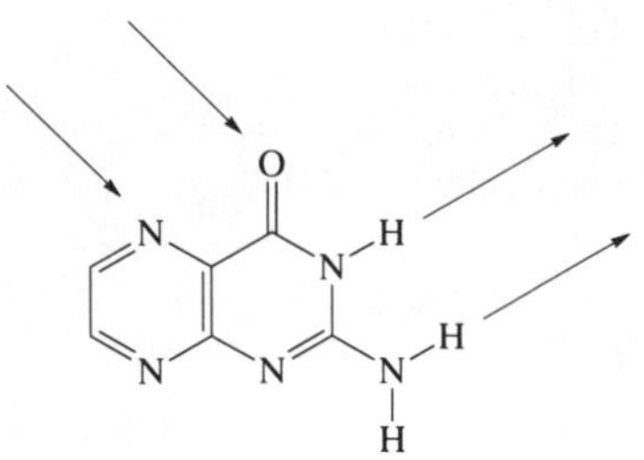

(2)

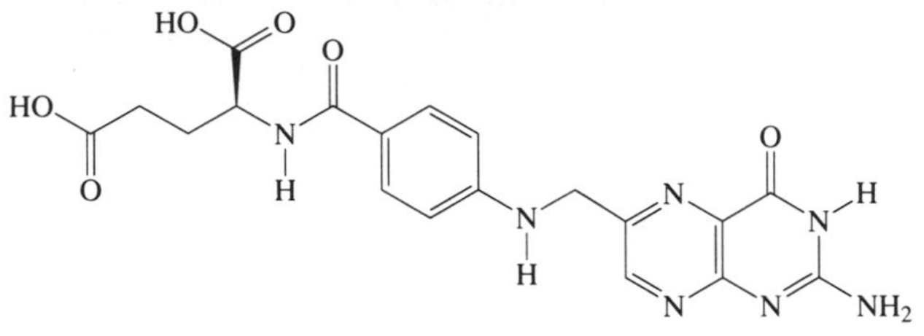

(3)

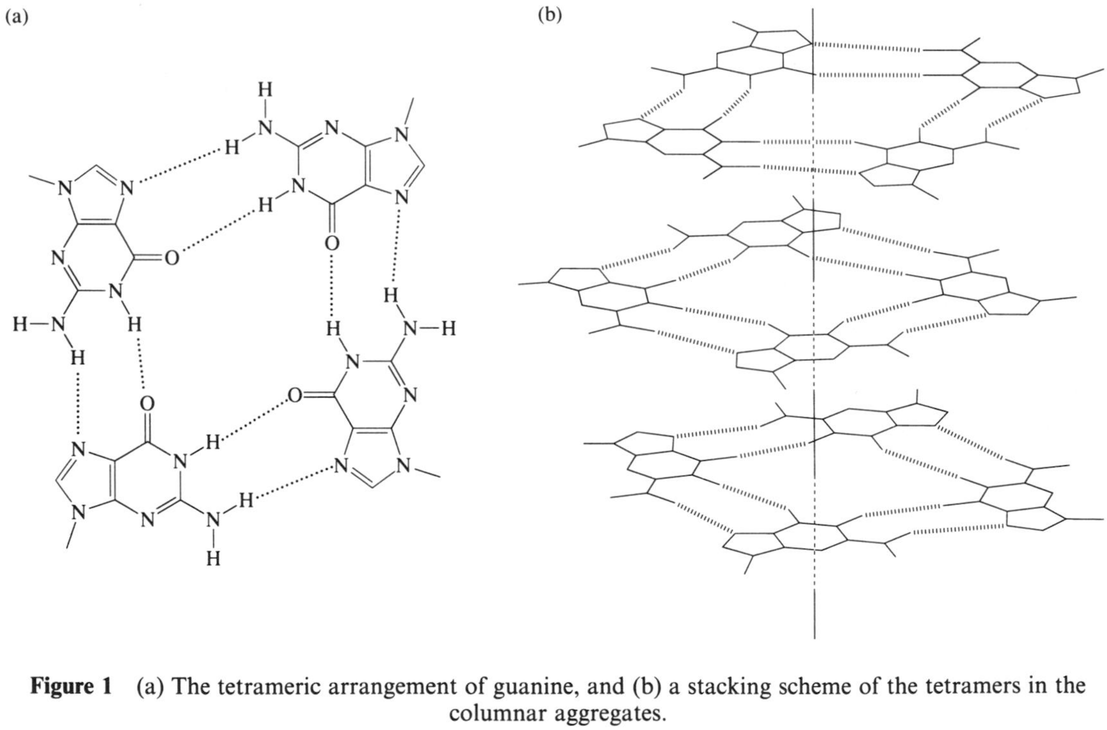

(a)

(b)

Figure 1 (a) The tetrameric arrangement of guanine, and (b) a stacking scheme of the tetramers in the columnar aggregates.

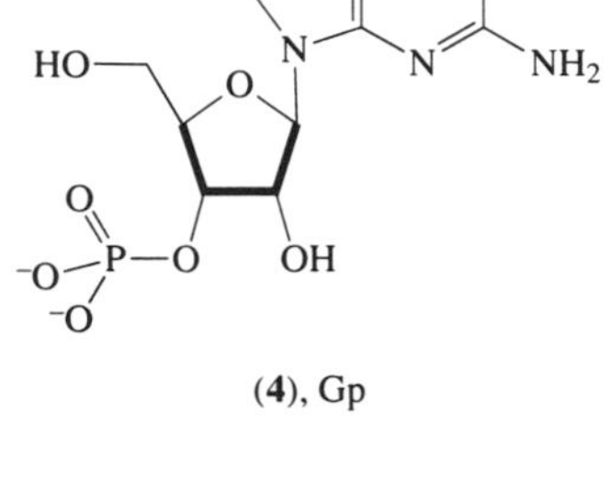

(**4**), Gp

The four molecules are connected by a special scheme of hydrogen bonds (Hoogsteen mode), as a consequence of the pattern of donor and acceptor groups present in the base. In the fibre, the guanosine tetramers are piled on top of one another, with sugar protruding at the periphery.[5,6] However, the layers do not stack in register, but are rotated with respect to each other to give an overall helical structure.

In the case of guanosine 5′-phosphate (pG) (**5**) at pH of ~5, a continuous helix seems to be formed instead; however, this change from the planar tetramers requires only small distorsions of hydrogen bonds.

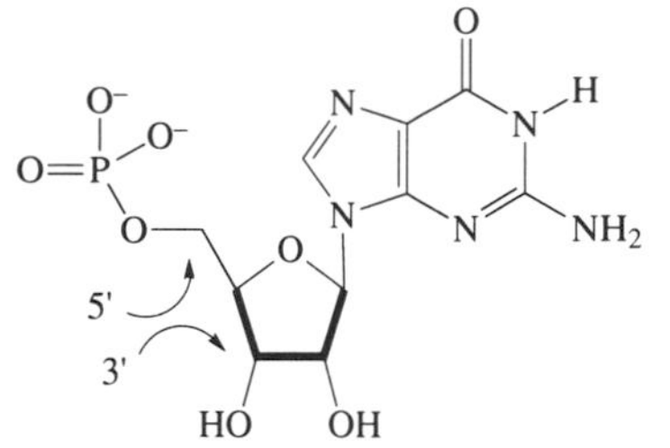

(**5**), pG

At pH values of 7–8, at which no gel formation has been observed, an assembled form composed of helically arranged stacks of planar tetramers seems to be present.[6] If the fibre is obtained in the presence of an excess of NaCl, the planar tetramers seem to form again.[5] The tetrameric arrangement was further confirmed by another fibre diffraction study of several guanosine derivatives.[7] The same tetrameric motif was found also in the self-structure formed by polyguanylic acid by two independent fibre x-ray investigations:[8,9] the assembled species is a four-stranded helix with parallel sugar phosphate filaments held together by the familiar G-quartet. The tetrameric planes are perpendicular to the helix axis and are rotated one with respect to the other at a positive angle to give a right-handed four-stranded helix (see Figure 1(b)).

An extensive investigation of gels formed by several guanosine derivatives was carried out by Guschlbauer and co-workers;[10] the presence of the tetramers was assumed also in gels formed at very low concentration. The effect of cations was also extensively investigated. The assembled forms are particularly stabilized by potassium ions, which fit the central cavity between two G-quartets perfectly and coordinate the eight carbonyl oxygens.[11,12]

The property of a deoxyguanosine derivative to form liquid crystalline phases in water was reported[13] only in 1988; subsequently, several other examples were observed in monomers and oligomers of deoxyguanosine phosphate.[14,15] The study of these phases, particularly of the more ordered hexagonal one, has given fundamental information on the structure of the assembled columnar species and was ancillary to the study of the self-assembly process in isotropic water solution.

13.2 LYOTROPIC COLUMNAR LIQUID CRYSTALS FORMED BY DEOXYGUANOSINE DERIVATIVES AND ALKALINE FOLATES

In the guanosine family, the first molecule found to give liquid crystal (LC) phases was the dinucleoside monophosphate 2′-deoxyguanylyl-(3′-5′)-2′-deoxyguanosine (d(GpG)) (**6**).[13,15] In water, this dimer forms LC phases even at relatively low concentration (2.5% w/w). Higher oligomers[14] d(GpGpG) (**7**), d(GpGpGpG) (**8**) and d(GpGpGpGpGpG) (**9**) also exhibit LC phases. Microscopic textures (see Figure 2) were indicative of the presence of a cholesteric and a hexagonal phase at lower and higher concentration, respectively.

The cholesteric phase could easily be aligned with a magnetic field to give a planar or a fingerprint texture (Figure 3) without unwinding the cholesteric helix that is oriented parallel to the applied field.

In the planar texture, the cholesteric helix axis is perpendicular to the cell window; this texture is very useful for spectroscopic measurements as it shows little or no birefringence. In the fingerprint texture, however, the helix axis is parallel to the cell windows instead and a typical striped pattern appears; from the distance between the stripes, the cholesteric pitch can be determined. This

(**6**) $n=1$, d(GpG)
(**7**) $n=2$, d(GpGpG)
(**8**) $n=3$, d(GpGpGpG)
(**9**) $n=5$, d(GpGpGpGpGpG)

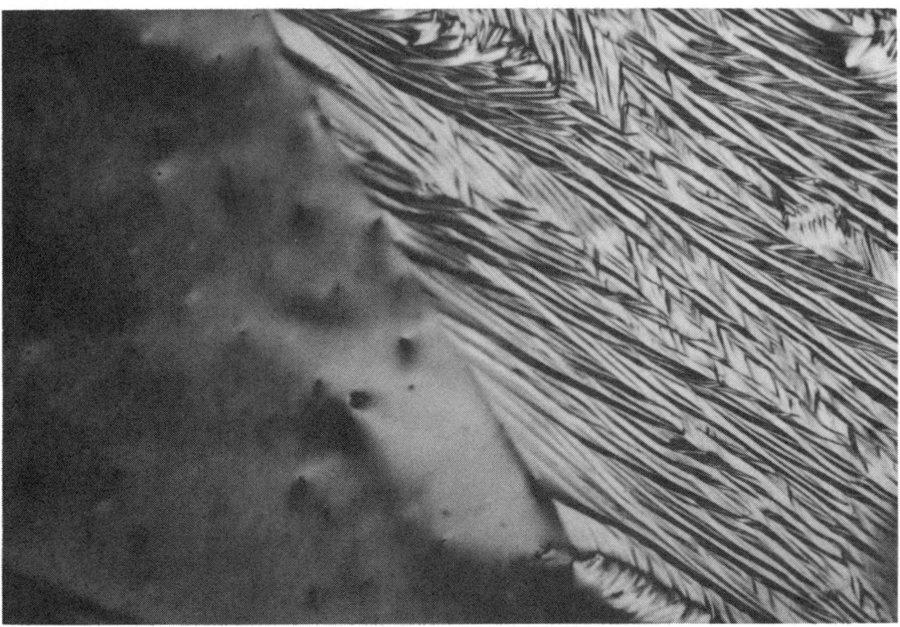

Figure 2 Microscopic textures of the cholesteric (left side) and hexagonal (right side) phases formed by a water solution d(GpGpG) at the cholesteric–hexagonal transition.

behaviour indicates that the objects composing the phase have negative diamagnetic anisotropy. The overall characteristics of the LC phases, including the magnetic behaviour, are very similar to those displayed by DNA with an average length of around 100 base pairs.[16,17] This similarity, together with what was known about the structure of the gels and the four-stranded helix of polyguanylic acid (poly(G)), indicated a probable structure for the aggregate.[13]

Low angle x-ray diffraction work[14,15] confirmed the assignments of the phases given by optical microscopy: the low concentration phases display only a broad peak, while the high concentration ones show a strong peak and few very weak peaks (up to fifth order); the ratio of their spacings indicates a two-dimensional hexagonal lattice. The hexagonal phases are of the ordered type.

In the high-angle region, both the cholesteric and hexagonal phases show a sharp peak corresponding to a periodicity of 340 pm, typical of stacked aromatic systems. From intensity data of

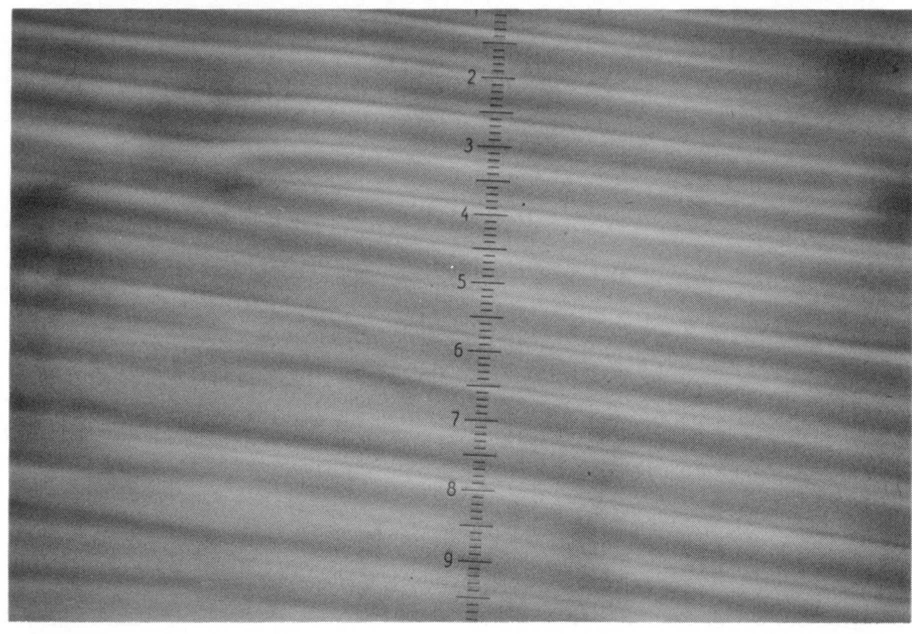

(a)

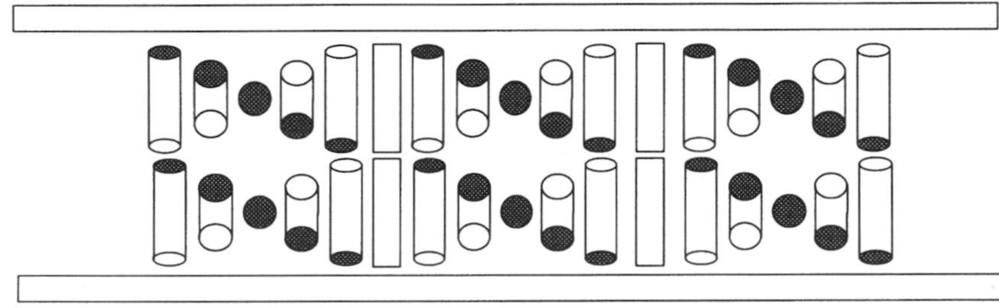

Figure 3 (a) Fingerprint and (b) planar texture of the cholesteric phase formed by a water solution of d(GpGpG), together with corresponding schemes showing the orientation of the cylindrical aggregates.

the hexagonal phase in the low-angle region, an electron density map can be calculated (Figure 4). The diameter of the section is in excellent agreement with that of the guanine tetramer; furthermore, the central region is characterized by decreased electron density corresponding to the 'hole' in the middle of the tetramer.[14] The diameter deduced from the position of the peaks is in excellent agreement with that deduced from intensity data.

From all this information, a model for the 'objects' that are the building blocks of the mesophases can be deduced (Figure 5): they are chiral columnar aggregates composed of a stacked array of equally spaced, planar or quasiplanar tetramers; specifically, the same tetramers formed by Hoogsteen-bonded guanine that characterize the gels formed by guanosine derivatives. The G-quartet planes are not stacked in register but are rotated, one with respect to the other, at a positive or negative angle to give a right-handed or a left-handed helix, respectively. As reported for poly(G),[8,9] the sugar phosphate backbones are protruding at the periphery of the structure.

In the cholesteric phase, which exists in the low concentration region of the phase diagram, the columns are relatively free to move and their surface chirality generates the cholesteric helical structure. In the hexagonal phase, which occurs at higher concentration, the free space available is much less and the columns are forced to adopt a more compact hexagonal packing in which the weak chiral forces are overpowered by the severe space limitations. In all cases, no long-range column-to-column correlation of the tetramer position exists as the rods may freely translate in a direction perpendicular to the two-dimensional hexagonal cell.

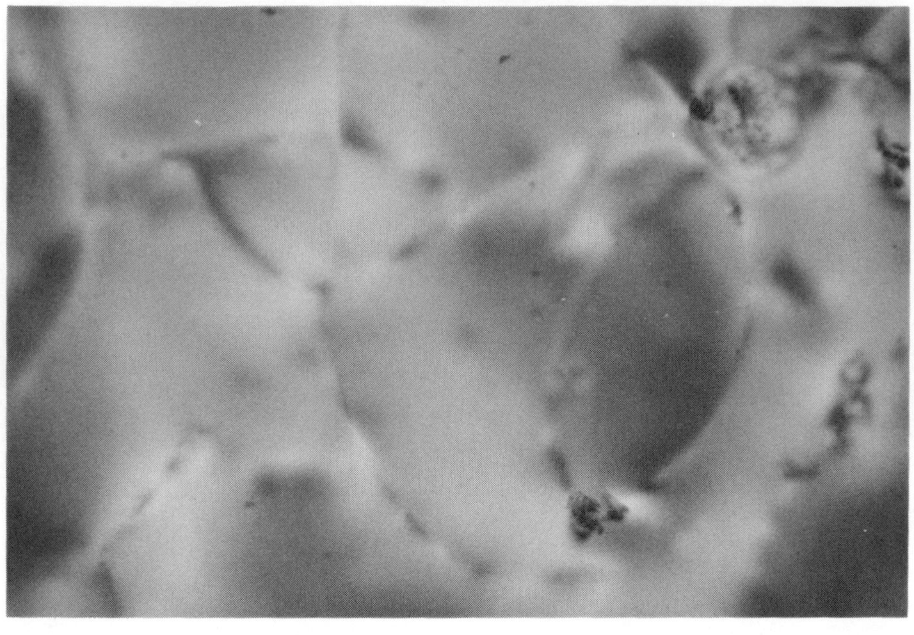

(b)

Figure 3 (continued)

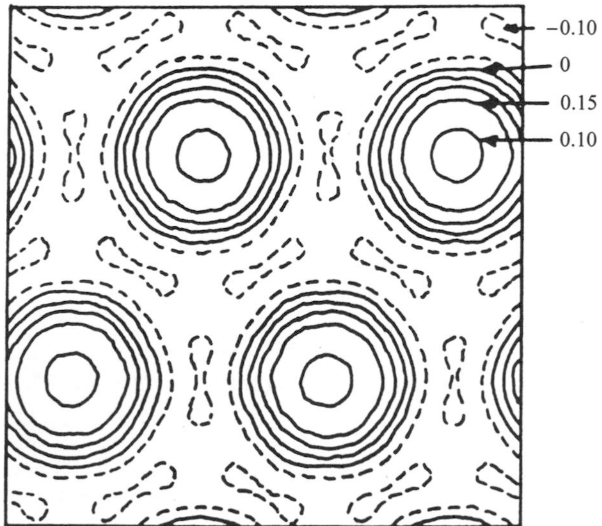

Figure 4 Electron density distribution of the hexagonal phase of d(GpGpG): the density lines are equally spaced and negative levels are dotted (reprinted with permission from *J. Am. Chem. Soc.*, 1991, **113**, 5809. Copyright 1991 American Chemical Society).

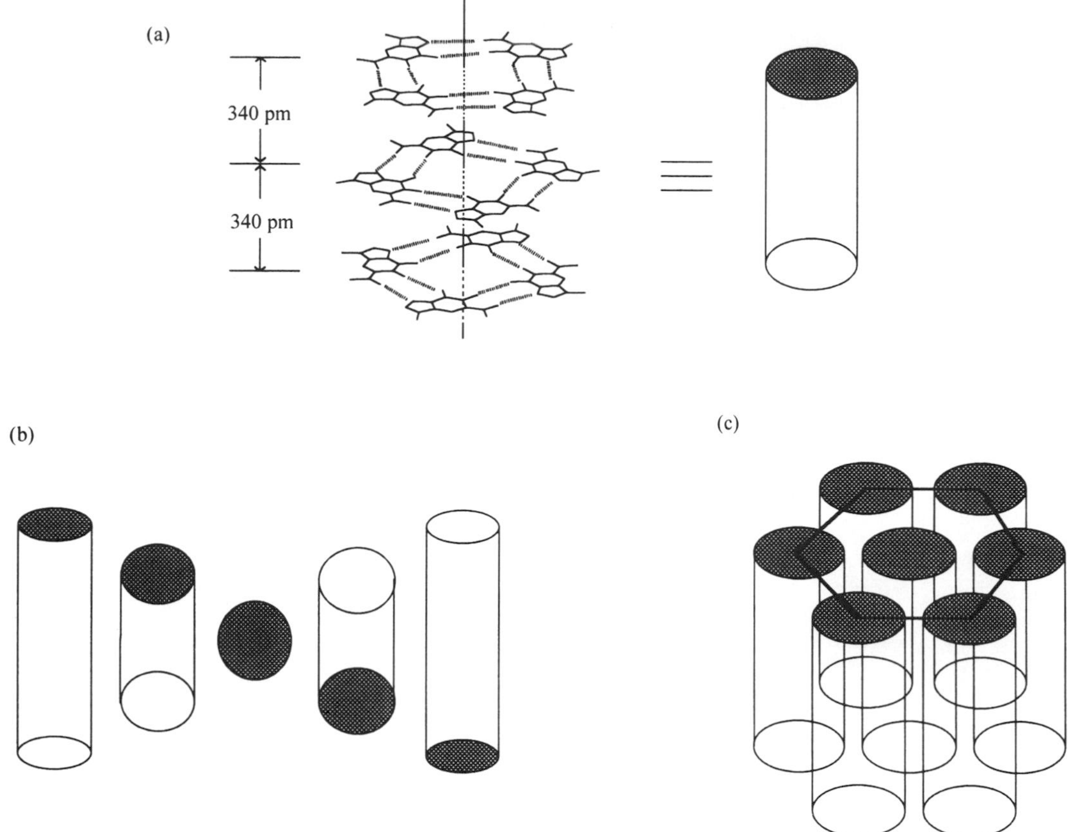

Figure 5 (a) A scheme of the columnar aggregate and their relative arrangements in the (b) cholesteric and (c) hexagonal phases.

In the case of d(GpG) (**6**), two possible self-assembly pathways leading to the columnar structure (Figure 6) can be envisaged: one is the formation of discrete aggregates by coupling four molecules of the dinucleoside phosphate followed by association of these first aggregates to give the full column; the other consists of a continuous one-step process. The importance of these possible mechanisms has been investigated in isotropic water solution in the light of circular dichroism (c.d.) melting experiments.[18]

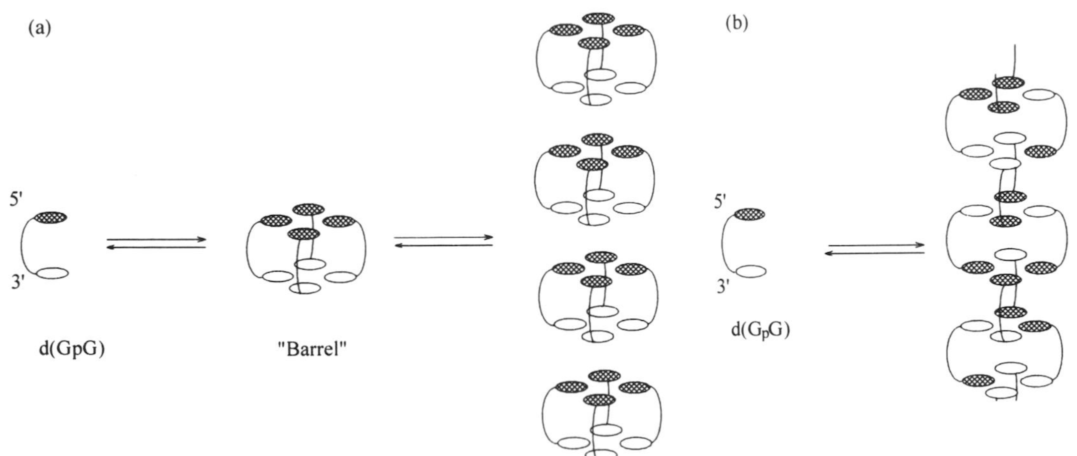

Figure 6 Models of the (a) stepwise and (b) continuous self-assembly pathway of d(GpG). Guanines are schematized as small ellipses: the hatched and open ellipses indicate the 5′ and 3′ ends, respectively (reprinted by permission of Società Chimica Italiana from *Gazz. Chim. (It.)*, 1995, **125**, 483).

Table 1 Phase sequences and transition concentrations for the guanosine derivatives investigated.

Compound		Counterion	Phase sequence									Ref.
(6)	d(GpG)	Na^+	I	2.5%	N*			18%			H	15
		K^+	I	1.5%	N*			15%			H	18
(7)	d(GpGpG)	Na^+	I	8%	N*			25%			H	14
(8)	d(GpGpGpG)	Na^+	I	13%	N*			ca. 35%			H	14
(9)	d(GpGpGpGpGpG)	Na^+	I	18–20%	N*			35%			H	14
(10)	d(pG)	NH_4^+	I	30%	N*			40%			H	14
(11)	d(Gp)	NH_4^+	I	5%	N*	35%	H_b	58%	Sq	68%	H	19
(12)	d(Bu^i-pG)	NH_4^+	I									14
(13)	d(Gp-Bu^i)	NH_4^+	I	25%	N*	38%	H_b	52%	Sq	70%	H	19
(14)	d(cGp)	NH_4^+	I	6.5%	N*			16%			H	20
(15)	d(cGpGp)	NH_4^+	I			40%					H	21
(16)	d(GpGpApGpG)	Na^+	I	12–20%	N*			35–40%			H	22

A stepwise model has already been proposed for the assembly of Gp (**4**), while pG (**5**) seems to follow a continous one-step aggregation.[10] Some of the melting data reported[3] also indicate a stepwise process.

All the investigated guanosine derivatives that show the formation of LCs, characterized by the presence of the tetrameric motif, are presented in Table 1. In the homologous series d(GpG), d(GpGpG), d(GpGpGpG) and d(GpGpGpGpGpG) (**6**)–(**9**), which possess only internucleoside phosphate groups, the critical concentration at which the cholesteric phase appears seems to follow a definite trend related to the ratio of negative charges per guanine unit. This ratio is indicative of both the electrostatic repulsive interaction and the hydrophilic/hydrophobic balance: the smaller the ratio, the easier the formation of the cholesteric phase.[14]

For the other derivatives, this ratio seems of minor importance: the example[14] of d(pG) (**10**) and its monoisobutyl ester (**12**) is typical, where the less hydrophilic molecule does not show LC phases; steric factors that affect the assembly process seem to be predominant. The comparison between d(Gp) (**11**) and its isobutyl ester (**13**) leads to the same conclusion.[19] Also, the difference in the critical concentration for the formation of the cholesteric phases of d(pG) (**10**) (30% w/w) and d(Gp) (**11**) (5% w/w) is impressive and leads to similar conclusions; on the other hand, the gel formation[10] and the fibre structure[4,7] of the two isomers are also rather different: as reported in the introduction, the former molecule seems to form a continuous helix and follows a one-step assembly process, while the latter seems to follow a stepwise aggregation and gives a fibre with strictly planar tetramers.

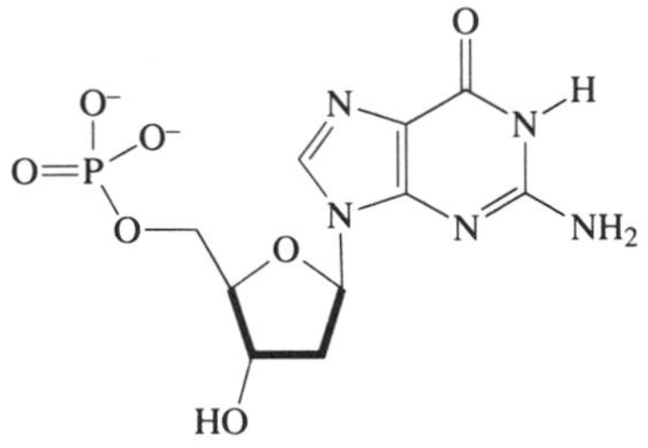

(**10**), d(pG)

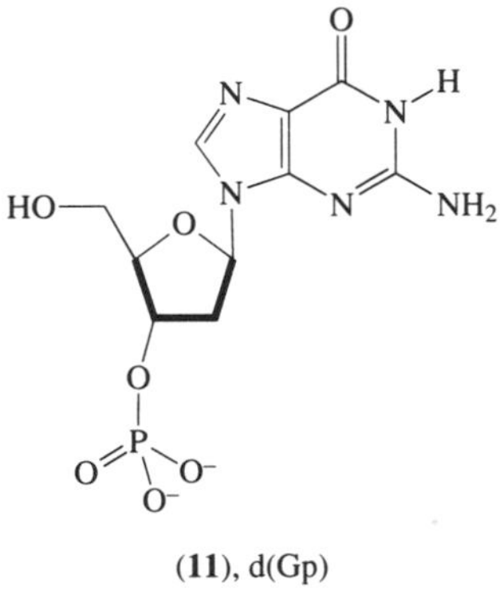

(**11**), d(Gp)

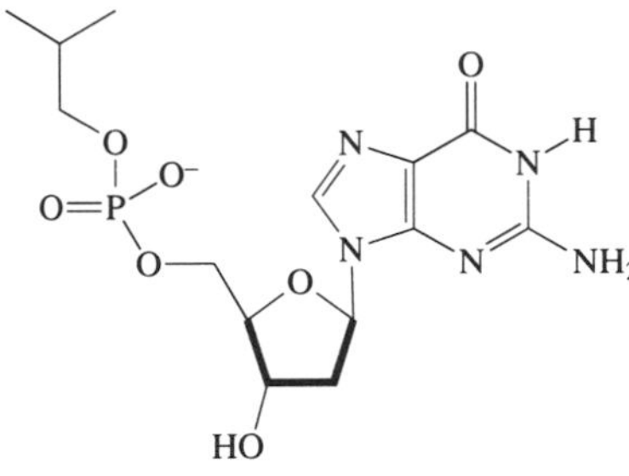

(**12**), d(Bu^i-pG)

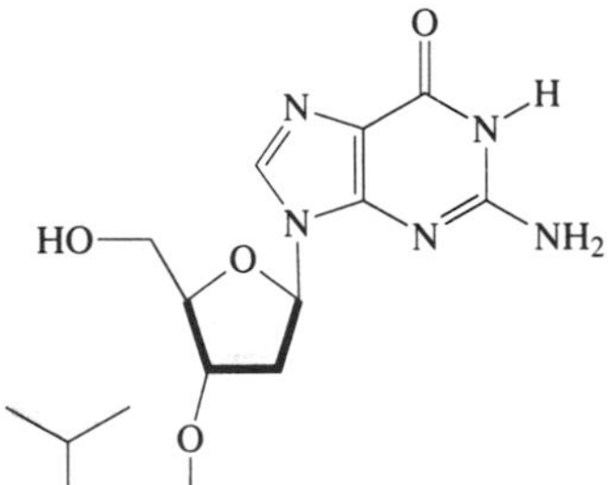

(**13**), d(Gp-Bu^i)

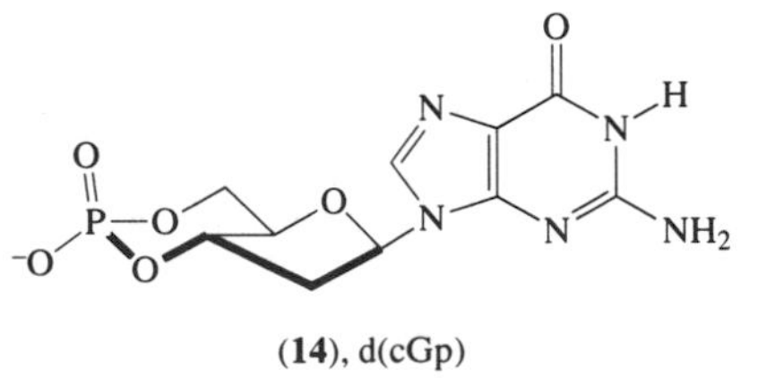

(**14**), d(cGp)

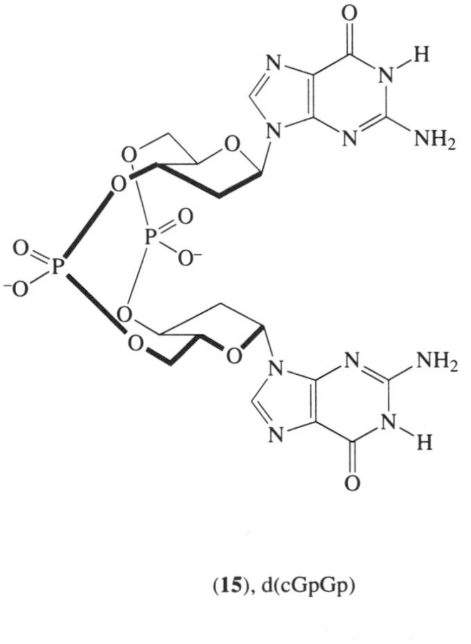

(**15**), d(cGpGp)

The cyclic derivative d(cGpGp) (**15**), in which the sugar phosphate backbone forms a twelve-membered ring from which the two guanines protrude, shows[21] LC phases: there is a very small biphasic cholesteric and hexagonal domain and a stable hexagonal phase; this hexagonal phase is similar to those formed by the other guanosine derivatives and shows the typical 340 pm spacing. No 680 pm periodicity could be detected, indicating that, in the hexagonal phase, the tetrameric planes are in contact; no cage-like cavities seem to exist with the tetramers at 680 pm distance, as postulated for the ribo derivative cGpGp in isotropic solution.[23] However, the existence of this cavity is not necessary, as suggested in Ref. 23, for including dyes within the guanosine aggregates: ethidium bromide is aggregated, probably via intercalation, in the assembled species given by d(GpG), d(GpGpG) and d(GpGpGpGpGpG).[14] This property can be used to deduce information on the chirality of the cholesteric phases (see Section 13.5).

d(GpGpApGpG) (**16**) also forms cholesteric and hexagonal phases similar to the others. The critical concentrations are not very different from those of d(GpGpGpG) and d(GpGpGpGpGpG), indicating that the presence of adenine in the central position does not inhibit the self-assembly process.[22]

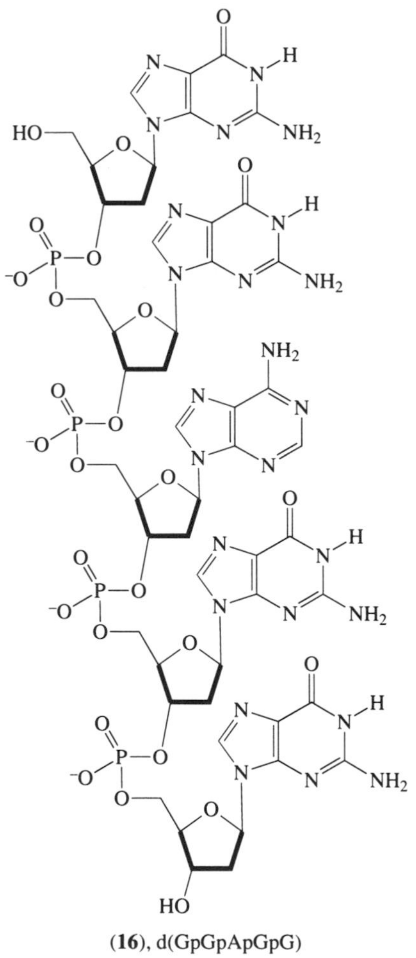

(**16**), d(GpGpApGpG)

It should be noted that the cyclic monomer phosphate d(cGp) (**14**) gives[20] both cholesteric and hexagonal phases, thus showing that the columnar aggregates are formed even in the absence of a free OH in the 3′ or 5′ position. Hydrogen bonds are therefore not essential for promoting the vertical stacking of the tetramers. Stacking interactions and binding effects from ions seem to give dominant contributions to the formation of the columns.

Considering that (i) gel formation has not been described for nucleotides other than guanosine derivatives, and (ii) several other dinucleoside phosphates (dimers were chosen because d(GpG) shows the cholesteric mesophase at the lowest concentration) containing different combinations of the bases were screened for liquid crystal formation but failed to give any sign of mesophases,[20] it seems evident that the G-quartet arrangement is unique in promoting the formation of columnar aggregates. This is likely to be due to the large surface of the tetramers, which enhances the stacking interactions, and also to the presence of the four oxygens, which allows the coordination of the binding ions.

With regard to these points, the high-resolution crystal structure of the hexanucleotide d(TG_4T) in the presence of sodium ions, reported by Laughlan *et al.*,[24] is particularly instructive. The usual G-quartets are present and connect four molecules of d(TG_4T) with parallel strands. The crystal cell is composed of two of these aggregates that stack coaxially with opposite polarity (head-to-head), forming a right-handed four-stranded helix composed only of guanine; the thymine molecules protrude out from the cylindrical structure. Sodium ions lie along the axis of the tetraplex and are between the tetrameric planes. In particular, there is a sodium ion between the two tetrameric planes connecting the two aggregates.

Alkaline folates (**3**) form, in pure water, only a hexagonal phase.[1] The overall characteristics of this phase are of the type described for the guanosine family (Figure 7): in particular, the diameter of the cylinders, calculated from diffraction experiments, is ca. 3.0 nm and corresponds to a pterine tetramer.

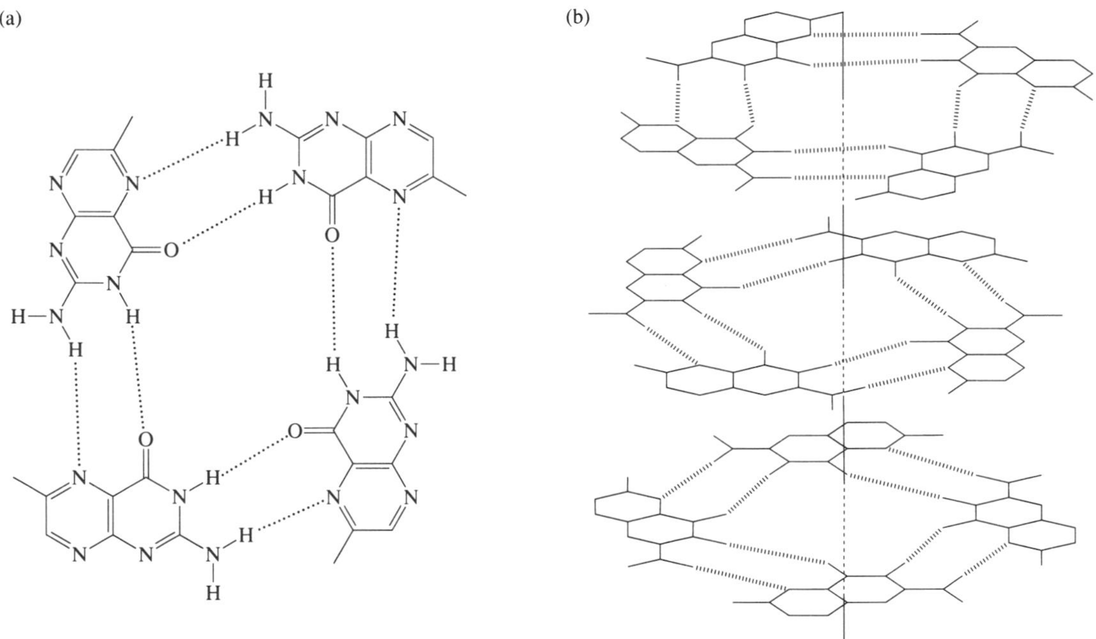

Figure 7 The tetrameric arrangement of (a) pterine and (b) a stacking scheme of the tetramers in the columnar aggregates.

The critical concentration is rather high (35% w/w). However, in the presence of 1 mol L^{-1} NaCl, at ca. 27% w/w, the solution becomes cholesteric.[25] This cholesteric can be aligned with a magnetic field to give planar or fingerprint textures, just like the majority of guanosine derivatives; this enables the determination of the pitch of the cholesteric superhelix, which is rather short (ca. 14 μm). All the data indicate that we are again dealing with a chiral columnar aggregate with pterine tetramers perpendicular to the long axis of the rod; the chiral glutamic residue protrudes from the column and gives the surface chirality which is the origin of the short pitch observed.[25]

From the data in Table 1, it can be seen that in one case, d(cGpGp) (**15**) in pure water, a stable cholesteric phase cannot be detected; the same is true for folate in water. For folate, a stable cholesteric phase exists instead in the presence of Na^+ ions. Furthermore, in the case of d(GpG), the critical concentrations change in passing from the sodium (I 2.5% N* 18% H) to the potassium (I 1.5% N* 15% H) salt. Also for d(pG), the critical concentration diminishes remarkably in passing from pure water to a 4 mol L^{-1} KCl solution.[26]

In order to explain these and other experimental data, one needs to understand the process of growth of the column and its stability as a function of concentration and ions present both in the isotropic and in the mesomorphic phases.

13.3 THE SELF-ASSEMBLY PROCESS IN ISOTROPIC SOLUTION

13.3.1 Deoxyguanosine Derivatives

13.3.1.1 SANS and SAXS investigations

Small-angle neutron scattering (SANS) is a powerful technique for studying the shape and dimensions of assembled species in isotropic solution.[27]

SANS experiments were carried out in pure D_2O for d(pG), d(GpG), d(GpGpG), d(GpGpGpG) and d(GpGpGpGpGpG) (**6**)–(**10**) at 1 and 1.5% w/w concentrations; except for d(pG) all the compounds revealed the presence of aggregates with a cylindrical shape. By modelling the cylinders with the radius of the guanosine tetramer, as deduced from x-ray work on the hexagonal phases, their length can be extracted. Furthermore, from the analysis of the scattered intensity at zero angle, one can also deduce the number of scattering particles per unit volume and hence the aggregation percentage. The results[28] are presented in Table 2.

Table 2 Aggregation percentage and length (L) of the columnar aggregates formed by homoguanylic derivatives (**6**)–(**9**) at 1 and 1.5% w/w concentration as deduced from SANS experiments.

		(6) *d(GpG)*	*(7)* *d(GpGpG)*	*(8)* *d(GpGpGpG)*	*(9)* *d(GpGpGpGpGpG)*
c = 1%	L (nm) ± 0.3	6.92	6.21	6.29	5.97
	Aggregation (%)	43.6	36.6	33.1	35.8
c = 1.5%	L (nm) ± 0.3	7.12	6.91		
	Aggregation (%)	50.2	39.0		

Source: Carsughi *et al.*[28]

The length of the aggregates (L) decreases very slowly in passing from the dimer to the hexamer and the same also seems to happen for the aggregation percentage. At 1.5% concentration, the aggregation increases by roughly 10%. All the data reported refer to freshly prepared solutions not treated thermically. We shall see that, in certain cases, the thermal history of the sample plays an important role in determining the dimensions and the fine structure of the aggregates.

The effect of thermal cycles and ions can be studied in some detail by a related technique, small-angle x-ray scattering (SAXS);[29] the geometrical parameters obtained by measurements carried out on 1% solutions in water with and without added electrolytes[18] are shown in Table 3.

Table 3 Length of the columnar aggregates formed by d(GpG) (**6**) at 1% w/w concentration as deduced from SAXS experiments.

		T	L (nm)
In water	Na salt	20 °C	7.0 ± 1.5
		60 °C	n.d.
		20 °C after cooling	n.d.
	K salt	20 °C	7.67 ± 0.69
		60 °C	n.d.
		20 °C after cooling	9.4 ± 1.1
In 1 mol L^{-1} KCl	Na salt	20 °C	9.64 ± 0.72
		60 °C	9.5 ± 1.0
		20 °C after cooling	10.7 ± 1.0
	K salt	20 °C	11.08 ± 0.88
		60 °C	11.15 ± 0.47
		20 °C after cooling	11.91 ± 0.64

Source: Bonazzi *et al.*[18]
[a]n.d., not detectable.

The different behaviour of the sodium and potassium salts of d(GpG) in pure water is remarkable: at 20 °C both salts show observable cylinders with comparable length; at 60 °C (above the melting temperature as determined by c.d.), no aggregates can be detected in either case; after cooling again to 20 °C, the potassium salt shows cylindrical aggregates again, which are, incidentally, slightly longer than those in the freshly prepared sample, while the sodium salt shows no detectable aggregates.

The effect of an added excess of ions is also remarkable. In the presence of an excess of KCl no melting is observed at 60 °C, and, after cooling, slightly longer aggregates are formed.

13.3.1.2 Circular dichroism investigations

The self-assembly process of all these chiral molecules can easily and conveniently be followed by c.d. spectroscopy. This technique has also been employed successfully in the study of guanosine gels.[30] The spectra of the isolated monomeric species can be observed in diluted solutions in pure water or, in more concentrated solutions, at higher temperature; they are usually drastically different from those of the assembled species (see Figure 8).

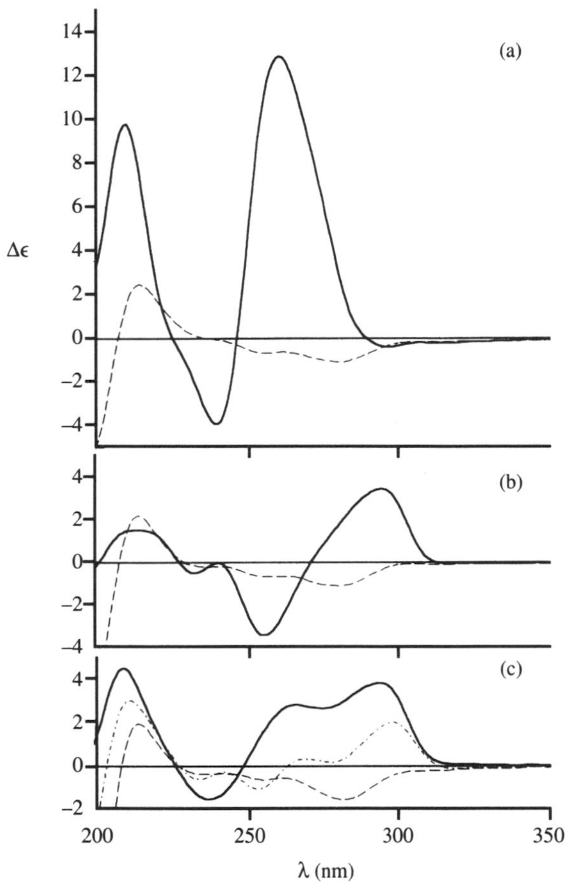

Figure 8 Circular dichroism spectra of d(GpG)Na at a 0.46% w/w concentration: (a) 0.05 mol L^{-1} KCl recorded at 20 °C (——) and 75 °C (- - - -); (b) in 0.05 mol L^{-1} RbCl recorded at 25 °C (——) and 50 °C (- - - -); (c) in 0.05 mol L^{-1} NaCl recorded at 10 °C (——) and in water recorded at 8 °C (– · – · – ·) and 40 °C (- - - -) (reprinted by permission of Società Chimica Italiana from *Gazz. Chim. (It.)*, 1995, **125**, 483).

Circular dichroism spectroscopy is very sensitive to the assembly process and, for assembled species, exciton double-signed bands are often observed corresponding to a single absorption band.[31] The spectra become similar to those of biological polymers (e.g., DNA and proteins). The technique is also very useful for measuring melting temperatures and can be more sensitive and selective than ordinary hypochromism measurements.

In the case of guanosine derivatives, several advantages are available in interpreting c.d. spectra. First, the c.d. spectra of the four-stranded helices of poly(G) and poly(dG) are known and also the structure of this unusual helix, which is similar to our columnar aggregates, is known in detail from fibre x-ray work on poly(G): the helix is right-handed and the bases are stacked perpendicularly to the helix axis.[8,9] Second, the spectroscopic properties of guanine have been extensively investigated. Consequently, it is possible to calculate with a high degree of confidence the c.d. spectra of poly(G) and of similar molecules with a relatively simple exciton treatment,[32,33] since a reliable model on which to check the calculations is available. The important part of the aggregate, which is the origin of the main features of the c.d. spectra between 230 nm and 300 nm, is the stack of the guanine tetramers. In the model used, only this pile of tetramers was considered, each tetramer being rotated with respect to the other at a fixed angle; all contributions from the sugar phosphate backbone were neglected.[33] From this approach, the chirality of the columnar aggregates can be deduced.

In the low-energy part of the spectrum, the guanine chromophore displays two well-characterized electronic transitions corresponding to an absorption maximum at ca. 250 nm and to a shoulder at ca. 280 nm. In the four-stranded helix of poly(G), the transition at ca. 250 nm gives rise to a nonsymmetric exciton couplet with a stronger positive component at ca. 260 nm and a weaker negative band at ca. 240 nm (see Figure 9). The second transition appears as a shoulder of the 260 nm band. It follows that, whenever spectra similar to this are obtained, they can be related to the presence of right-handed four-stranded structures. The correlation has both empirical and theoretical validity.

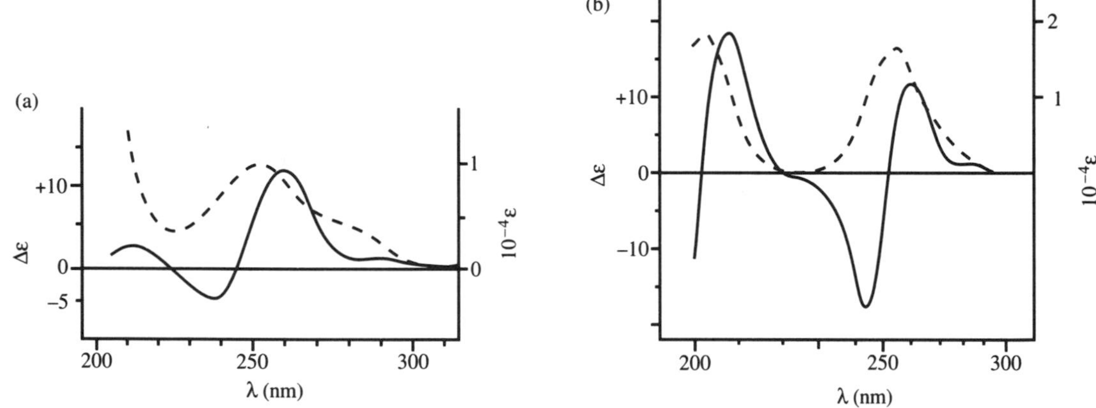

Figure 9 (a) Experimental and (b) simulated absorption (- - - -) and c.d. (——) spectra of poly(G) (reprinted by permission of Società Chimica Italiana from *Gazz. Chim. (It.)*, 1990, **120**, 101).

All monomeric derivatives have a left-handed columnar helicity, while the oligomeric compounds generate right-handed columns (see also the right-handed four-stranded helix deduced from the crystal structure of d(TG_4T)).[24] The results are summarized in Table 4.

Table 4 Columnar helicity (as deduced by the exciton c.d. corresponding to the guanine chromophore; − and + signs refer to left and right handedness, respectively) and cholesteric helicity and pitch for guanosine and pterine derviatives.

Compound		*Columnar helicity*	*p* (μm)	*Cholesteric helicity*	*Ref.*
d(Gp)	**(11)**	(−)		M	19
d(Gp-Bu^i)	**(13)**	(−)		M	19
d(pG)	**(10)**	(−)	−31	M	14
d(GpG)	**(6)**	(+)	−70	M	14
d(GpGpG)	**(7)**	(+)	+140	P	14
d(GpGpGpG)	**(8)**	(+)	+100	P	14
d(GpGpGpGpGpG)	**(9)**	(+)	+43	P	14
Sodium folate + NaCl	**(3)**		−14	M	25

Other information on the geometry of the columns can be deduced from c.d. spectra. In derivatives d(pG) (**10**), d(GpGpG) (**7**), d(GpGpGpG) (**8**), d(GpGpGpGpGpG) (**9**)[14] and d(cGp) (**14**),[21] the band shapes are similar to poly(G): the c.d. spectrum is dominated by the 260 nm band, which is positive for d(GpGpG), d(GpGpGpG) and d(GpGpGpGpGpG), and negative for d(pG) and d(cGp). However, d(GpG)[14,18] displays either a positive 260 nm band in the presence of K^+ or Sr^{2+} ions (see Figure 8, frame A) or a positive 290 nm band in the presence of Rb^+ or Cs^+ (see Figure 8, frame B); with Na^+, both 260 nm and 290 nm bands are observed (see Figure 8, frame C). Similar behaviour is shown also by d(GpGpApGpG) (**16**), since in this case again the 260 nm band becomes predominant only in the presence of an excess of K^+ ions.[22]

The 260 or 290 nm bands also characterize the c.d. spectra of the assembled species of G-rich oligodeoxynucleotides connected to telomere repeat sequences (see Section 13.6). From the stereochemical point of view, the presence of the 260 nm band seems to be associated to a rather regular structure, similar to that of poly(G), with the tetrameric planes perpendicular to the helix axis, while a dominant 290 nm band indicates consistent variations with respect to this situation.[18] The finding that K^+ ions promote the transition from a distorted to a regular structure is related to the size of this ion, which is ideal for positioning itself between two adjacent tetrameric planes coordinating all eight surrounding oxygen atoms.[6] For d(GpG), an indication of the stability of the aggregates in terms of melting temperature (T_m) was obtained by a detailed variable temperature c.d. investigation.[18] The stabilizing effect of the different ions, as deduced from melting temperatures, are in the order $Sr^{2+} \approx K^+ > Rb^+ > Na^+ > Cs^+$. Similar data were obtained from NMR analysis on d(pG).[11,12] The results are in excellent agreement with those from SAXS reported above: the conditions that favour a lengthening of the columns are the same as those which increase T_m values; in particular, the effect of K^+ ions is remarkable. We therefore have parallel behaviour of stability and length of the aggregates, both factors playing an important role in the LC formation.

13.3.2 Folates

In the case of folates, in pure D_2O no scattering aggregates can be detected by SANS even at a relatively high concentration (4% w/w). At the same concentration and with an excess of Na^+ or K^+ ions instead, the presence of scattering aggregates can be detected. The aggregates are cylindrical and, assuming a diameter of ca. 3.0 nm, as deduced from x-ray work on the hexagonal phase,[1,25] the length of the cylinders can be calculated: the aggregates are composed of a stack of ca. nine discs.[34] As data on pterine electronic transitions are not reported in the literature, c.d. spectroscopy can only be used empirically and no indication of the absolute handedness of the columnar aggregates can be deduced. Circular dichroism spectra also indicate very clearly the presence of aggregates at lower concentrations (beginning at 10^{-4} mol L^{-1}), for which no SANS signals can be measured; this points to the formation of aggregates of smaller dimensions, too small to be detected by the scattering technique: as will be reported in Section 13.4, the folate column length varies considerably with the concentration.

For the aggregated state, different spectra are obtained in pure water and in the presence of an excess of Na^+ and K^+ ions. The spectra in pure water display a single melting to give the spectrum of the free folate. The spectra obtained in the presence of salts give a clear double melting; after the first melting, the same spectrum observed in pure water is obtained and this melts again to that of the free molecule.

These results, together with NOESY and NOE data, were interpreted considering the existence of two assembled forms: one (with added ions) is more compact and has a stronger tetramer-to-tetramer interaction; the other has a different stereochemistry of the aromatic core region and a weaker tetramer-to-tetramer interaction. In both cases, the glutamic chains are rather mobile.[34] These data in isotropic solutions are in full agreement with those obtained in the mesomorphic state reported in the next section. In the case of folates, as for guanine derivatives, the polymorphism observed is related to the growth of the columnar aggregates; the factors that promote the growing process lower the transition concentrations and stabilize, in particular, the cholesteric phases.

13.4 COLUMN LENGTH AND GROWTH IN THE MESOMORPHIC STATE

Information on the column length in the mesomorphic state can be obtained by x-ray diffraction techniques. From the low-angle region, one can extract the unit cell dimension and hence the lateral distance a between the cylindrical aggregates. This distance varies as a function of the concentration c_V. A dependence of the type $a \propto c_V^{-1/2}$ is specific of a situation in which dilution occurs only laterally (normally to the cylinder elongation axis) and not in the vertical dimension, that is, the end-to-end columnar distance remains almost constant.

A dependence of the type $a \propto c_V^{-1/3}$ indicates that dilution occurs in all three dimensions.[35,36] The two situations can be interpreted in terms of length (and flexibility) of the aggregates: a $c_V^{-1/2}$ dependence is indicative of infinite (with respect to the lateral distance) cylinders, while a $c_V^{-1/3}$

dependence can be interpreted in terms of finite cylinders. More detailed information on the distance between the tetrameric discs and the length of the columns can be obtained by studying the effect of dilution and temperature on the shape and position of the high-angle peak. In order to correlate the peak width with the average number of stacked tetramers, numerical simulations of the scattering profile must be carried out to find the best fit to the experimental profile.[35] Using these approaches and considering the results for the self-assembly in the isotropic solutions (see Section 13.3), the following pictures can be obtained.

In the case of d(GpG), the dependence is of the $c_v^{-1/3}$ type in the cholesteric phase and indicates finite cylinders, while in the hexagonal phase, infinite cylinders seem to be present. The cross-over between finite and infinite objects occurs close to the N* to H transition.[36] The description that can be deduced is that the aggregate length grows rather quickly with concentration in the isotropic and cholesteric phases. In the hexagonal phase, they are so long that their length cannot be compared to the lateral distance.

In the case of derivatives d(pG), the columns seem instead to be finite also in the hexagonal phase and the length growth seems to be slow.[35] The fact that SANS is unable to detect aggregates at 1% concentration is in agreement with these results. Hence, the assembly process of d(pG) seems to be rather difficult, possibly for stereochemical reasons, and, to a lesser extent, also for hydrophobicity effects. Accordingly, the critical concentration at which the N* phase appears is much higher than for d(GpG).

Also for folates, with and without added salts, the $-1/3$ exponent indicates the formation of finite cylinders in the hexagonal phase.[25] In pure water, the distance between the discs increases as a function of the water content and temperature. The average length of the aggregates increases continuously as a function of concentration from five piled discs at the I to H transition to ca. 30 at a 60% concentration.

In NaCl, however, the stacking distance and the size of the aggregates remain almost constant: even at low concentration there are ca. 20 piled discs (see Figure 10). The stabilization effect of Na^+ ions (and, to a lesser extent, of K^+), seems similar to that described above for K^+ in the guanine systems. This behaviour is in full agreement with the result obtained by SANS in isotropic solution: at 4% concentration, no aggregates could be detected in pure water, while in NaCl, the data indicate the presence of aggregates composed of ca. nine discs.[34]

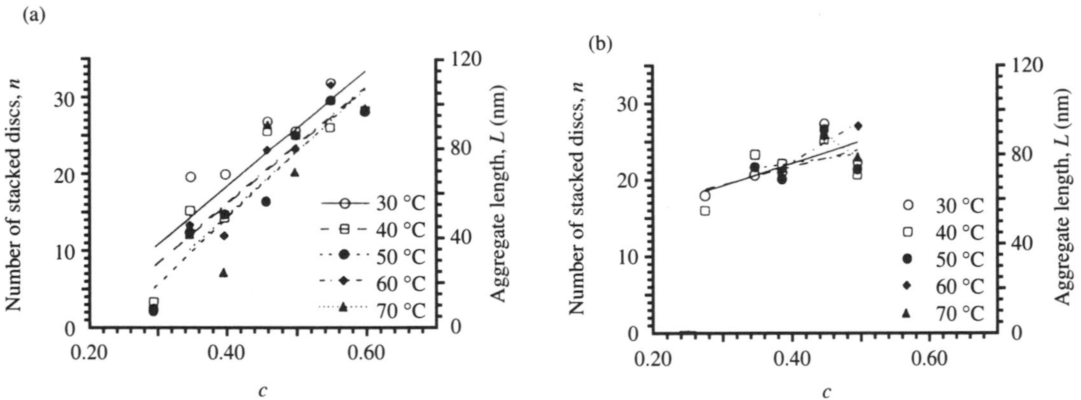

Figure 10 Average number of stacked discs as a function of folate concentration (w/w) for different temperatures: (a) disodium folate in water; (b) disodium folate in 1 mol L^{-1} NaCl. The lines are linear fits to the data to show the general trend. The right-side scale is the average columnar length calculated by multiplying the number of discs by the stacking distance (reprinted with permission from *J. Am. Chem. Soc.*, 1994, **116**, 7064. Copyright 1994 American Chemical Society).

The formation of a stable cholesteric mesophase only in the presence of NaCl is related to the fact that only in this condition are the aggregates long enough at relatively low concentration (ca. 27% w/w). Without added salts one needs a higher concentration to obtain aggregates long enough to give a mesophase and, at this concentration, there is no space available for the formation of the cholesteric helix and the more compact H phase is observed. This picture agrees very well with the theoretical prediction of phase diagrams for assembling lyotropics obtained recently by Taylor and Herzfeld.[37] The calculated phase diagrams display a nematic between the isotropic and the hexagonal phase only when the aggregation is strong; for weak aggregation the direct I to H transition is predicted.

13.5 HANDEDNESS AND PITCH OF THE CHOLESTERIC MESOPHASES

In order to obtain data on the handedness and pitch of the cholesteric columnar mesophases, one has to use the magnetic alignment described above (Section 13.2). The pitch values are obtained by measuring the spacing between the lines of the fingerprint textures. The determination of the cholesteric handedness is carried out by c.d. spectroscopy on samples with planar alignment (in order to avoid artefacts connected to strong birefringence of the sample),[38] and relies on Mauguin's model extended to the absorption region[39,40] A measurement of c.d. on a planarly aligned sample gives a signal $(A_l{-}A_r)_j$ which is correlated to the cholesteric characteristics by Equation (1):

$$(A_l - A_r)_j = \tfrac{1}{2}\, p\, \nu_j^3\, \Delta n\, (A_\parallel - A_\perp)_j / (\nu_j^2 - \nu_0^2) \qquad (1)$$

where ν_j is the frequency at which the measurement is carried out, ν_0 is the wavenumber of the selective reflection, Δn and $(A_\parallel{-}A_\perp)_j$ are the linear birefringence and the linear dichroism, respectively, of the individual columns that compose the cholesteric phase and p is the helical pitch, positive and negative for a right-handed (P) and left-handed (M) cholesteric, respectively. The frequency ν_0 is known from pitch measurements and, in all the cases investigated, is much smaller than the frequency ν_j, the linear birefringence and dichroism are both negative as guanine has electronic transitions polarized in the molecular plane, that is, perpendicularly to the cylinder axis. A positive c.d. signal in the absorption region therefore indicates a right-handed cholesteric (P), while a negative c.d. indicates a left-handed cholesteric (M). The handedness was confirmed by recording the c.d. spectra in the presence of ethidium bromide, which intercalates into four-stranded oligoguanylates.[14] The handedness and pitch values are reported in Table 4. With the exception of d(GpG) (**6**), in all the guanines investigated, right-handed columns generate right-handed cholesterics.

A model for the packing of hard helical screws, which recognizes their reciprocal chirality and spontaneously generates an overall helical structure, is shown in Figure 11. The model gives a simplified picture of the interaction, by hard-core repulsion only, of chiral columns in the cholesteric phase. In the model of Figure 11, left-handed screws pack to give a left-handed superstructure. However, the model is sensitive not only to the handedness of the screws, but also to the ratio pitch/diameter (p/d) of the individual objects: for a given handedness of the screws (i.e., right-handed) one can obtain a right-handed superstructure only if $p/d < \pi$; if $p/d > \pi$ a left-handed superstructure is generated (Figure 12).[38]

It is certainly an oversimplification to compare the cholesteric phases given by our guanine derivatives to an idealized model composed of screws, which are not charged and in which dispersive interaction does not exist. Therefore, in the screw model, the angle of rotation between screws is extremely large with respect to real cholesterics. Furthermore, the geometry of our four-stranded aggregates is complex and it is impossible precisely to evaluate the external diameter of the cylinder. Nevertheless, this simplified model, which has a critical value of the ratio p/d of the individual objects around which there is an inversion of the superhelical handedness, has some utility as it gives a qualitative indication of the effect that a process of winding or unwinding the helical aggregates might have on the cholesteric handedness.

If we consider the guanine columnar assembled species as screw-like objects (the fact that they are four-stranded helices does not affect qualitatively the results), all of them, except for that derived from d(GpG), behave as if their p/d ratio is smaller than the critical value. For d(GpG), the helical system seems to be unwound to give p/d larger than the critical value. In the homologous series d(GpG), d(GpGpG), d(GpGpGpG) and d(GpGpGpGpGpG), in which only internucleoside phosphate groups are present, the different characteristics of the cholesterics, and also the opposite sign displayed by d(GpG), can be understood: in order to construct an aggregate composed of 18 discs (which, for example, is the average length observed by SANS in isotropic solutions of ca. 1% concentration), in the case of d(GpGpGpGpGpG), three discrete tetrameric aggregates have to be connected by two sets of similar noncovalent interactions (stacking forces, ionic contributions and probably also hydrogen bonds between the free 3′ and 5′ OH groups). In the case of d(GpG) instead, nine discrete aggregates have to be connected by eight sets of noncovalent interactions (see Figure 13).

It is known from c.d. data (and x-ray work) on poly(G) and poly(dG) that the four-stranded helices are right-handed, and the same applies also to the aggregates d(GpG), d(GpGpG), d(GpGpGpG) and d(GpGpGpGpGpG) (even if in the case of d(GpG), in the absence of added

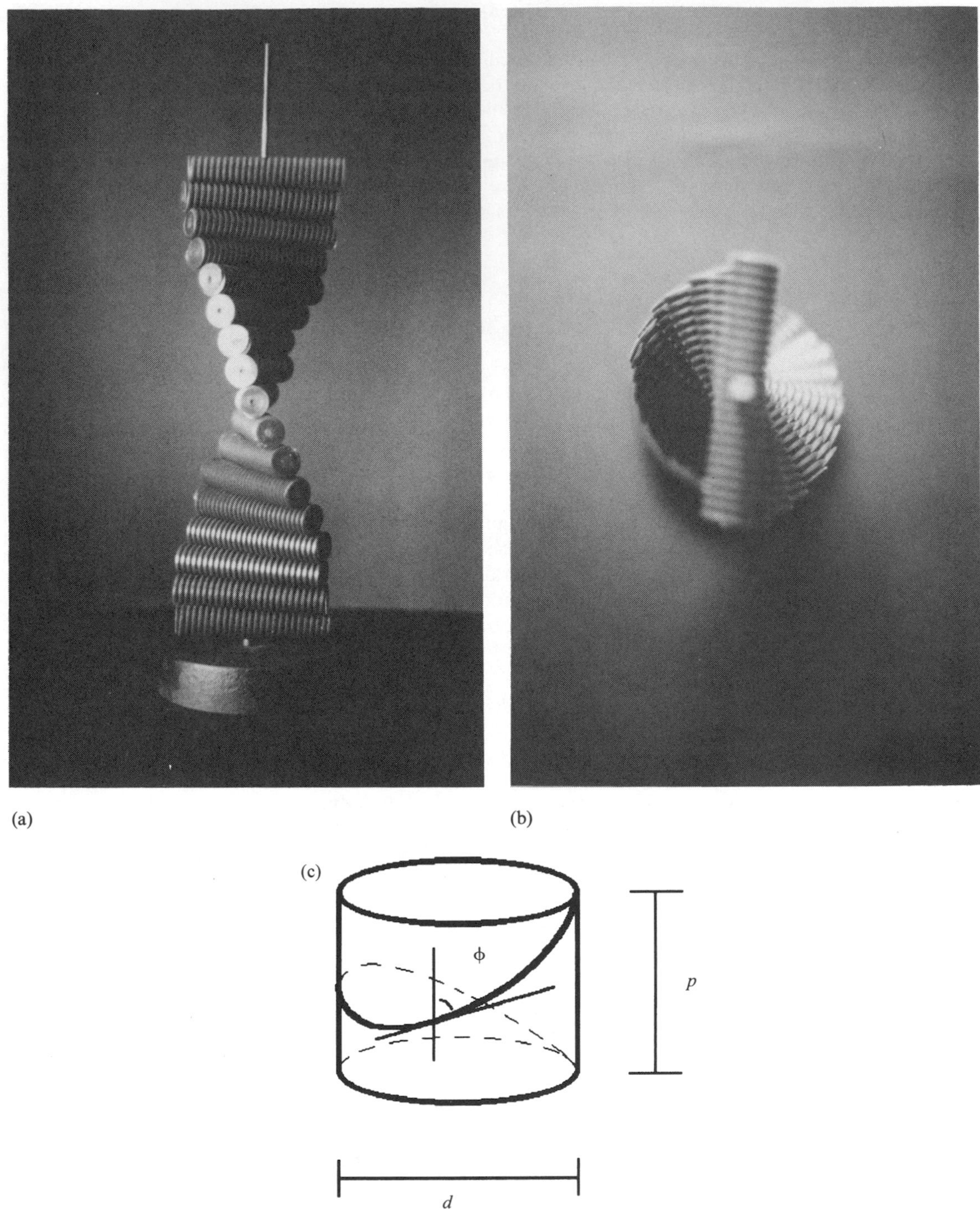

Figure 11 A simplified macroscopic model for a cholesteric: (a) lateral and (b) top view of a left-handed superhelix composed of left-handed helical screws. Helical parameters are defined in (c). Photographs display a half-pitch length of the superhelix; the helix angle ϕ of the individual screw is ca. 84° ($p/d \ll \pi$) (reprinted with permission by VCH Publishers © (1995) from 'Circular Dichroism—Principles and Applications', 1994).

K^+, the c.d. intensity is much weaker). The presence of phosphodiester linkages between the tetrameric planes is therefore connected to a right-handed structure. The contribution of noncovalent interactions to the helical handedness is not necessarily of the same sign. Let us assume that they are of the opposite sign (as for d(pG)) or maybe of the same sign, but with a smaller angle of rotation; in both cases they cause unwinding of the four-stranded helices.

Figure 12 The different handedness of the packing of right-handed helices with different pitch-to-diameter ratio: (a) right-handed helices with $p/d < \pi$ interact to give a right-handed superhelix; (b) when $p/d > \pi$, the superhelix is left-handed (reprinted with permission by VCH Publishers © (1995) from 'Circular Dichroism—Principles and Applications', 1994).

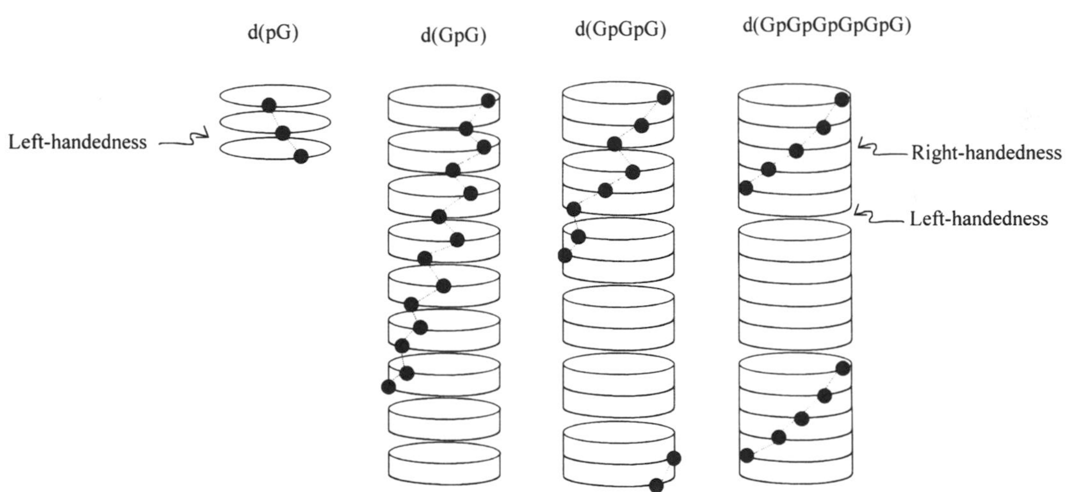

Figure 13 A representation of the chiral columnar aggregates formed by d(pG), d(GpG), d(GpGpG) and d(GpGpGpGpGpG). The filled circles represent the sugar units (for clarity, only one sugar per tetramer is shown). The screw thread is obtained by joining the filled circles: the screw pitches are in the order d(GpG) > d(GpGpG) > d(GpGpGpGpGpG), indicating parallel unwinding of the helical structure.

The unwinding effect is obviously greater for d(GpG) > d(GpGpG) > d(GpGpGpG) > d(GpGpGpGpGpG). In the case of d(GpG), it is not surprising that we have an inversion of the sense of packing as a result of the increased p/d ratio, which could be greater than the critical value. The variation of the reciprocal pitch of the cholesterics with the number of guanosine residues is in full agreement with this hypothesis (see Figure 14). The value of p^{-1} (which is a measure of the twist between the chiral columns in the cholesteric phase) varies smoothly with the number of residues as predicted by the screw model. The assumption that the columns are formed by packing of discrete objects (for which we have only some evidence from the shape of c.d. spectra for d(GpGpG), d(GpGpGpG) and d(GpGpGpGpGpG); in the case of d(GpG), a continuous aggregation mechanism could also exist in some cases),[18] is not a necessary prerequisite and a similar reasoning could apply also to a continuous assembly model.

With regard to this point, an interesting study on the self-recognition in helicate self-assembly from mixtures of ligands of different length and copper ions has been reported;[41] in this system, only "closed architectures" have been detected and justified on the basis of energy and entropy considerations. The case of discrete or continuous aggregation in the guanosine system discussed in the previous paragraph, although not strictly parallel to that reported by Kremer *et al.*,[41] has

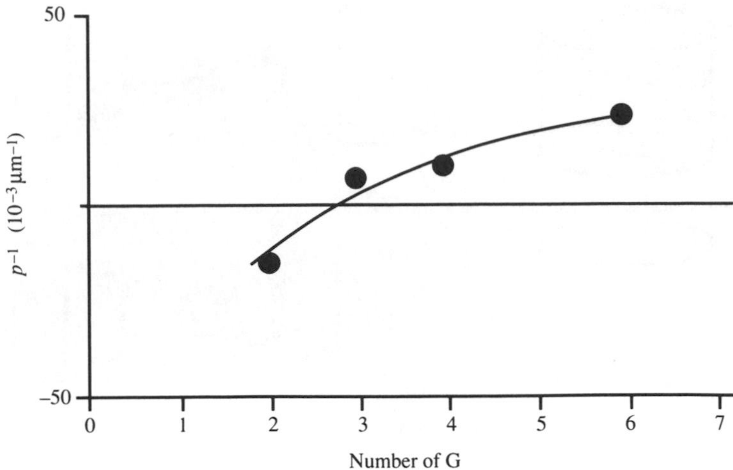

Figure 14 Reciprocal pitches of the cholesterics obtained from compounds d(GpG), d(GpGpG), d(GpGpGpG) and d(GpGpGpGpGpG). The signs − and + refer to left-handed and right-handed cholesterics, respectively (reprinted with permission from *J. Am. Chem. Soc.*, 1991, **113**, 5809. Copyright 1991 American Chemical Society).

some common points that which could support the major importance of the discrete mechanism, particularly at the onset of the process, where the formation of isolated, blunt-end species could be favoured with respect to that of species with pendant free binding sites.

13.6 BIOLOGICAL RELEVANCE OF GUANOSINE TETRAMERS

Cholesteric phases are formed *in vivo*, in animals and plants, by molecules such as DNA, cellulose and collagen, which have helical chirality. For example, in *Dinoflagellates* and bacteria whose chromatin is not organized into histones, a cholesteric organization of DNA has been found and suggested to characterize the 'resting' state of the nuclear material.[42] Needless to say, no physiological role can be envisaged for the liquid crystalline phases that result from the multistep aggregation of the deoxyguanylates just described, for the very reason that such molecules are not present *in vivo*.

However, the formation of G-quartets, which is the first step of the self-assembly process leading to liquid crystals, is not restricted to homoguanylates, but is also displayed by G-rich DNA and RNA sequences, with strings of at least three contiguous guanines (it can be noticed that GGAGG also gives G-quartets as well as liquid crystalline phases, see Section 13.2).

Such sequences are found throughout the genome, often in evolutionary and functionally conserved regions such as telomeres, gene promoters and immunoglobulin switch regions. Telomeres,[43] the ends of linear eukaryotic chromosomes, are made of tandemly repeated short sequences, with one strand containing clusters of G residues. This strand is oriented 5′ to 3′ toward the chromosomal terminus and protrudes 12–16 nucleotides beyond the complementary C-rich strand. Telomeres, whose DNA is not packaged in nucleosomes, are essential for the stability and complete replication of eukaryotic chromosomes.

The first report that single-stranded DNA oligomers, containing short guanine-rich tracts, give parallel four-stranded complexes, was published in 1988 by Sen and Gilbert,[44] based on gel electrophoresis results. Interestingly, when the complementary C-rich sequence was added to the tetrameric G-rich complex, an octameric adduct was formed, together with varying amounts of the normal Watson–Crick duplexes. One of the investigated sequences is

TAGTCCAGGCTGAGCAGGTACGGGGGAGCTGGGGTAGATGGGAATGT

The driving force for the association of the G-rich oligomers was ascribed to the formation of Hoogsteen hydrogen-bonded guanine tetrads. The authors suggested that this self-recognition of the G-rich segments of chromosomes promotes the association, in register, of the four homologous chromatids during meiosis. Later, it was reported that the association process of model sequences, in particular the telomere-type ones, strongly depends on the nature of the alkali cation present (sodium or potassium).[45]

Many telomere G-rich repeats have sequences of the general type $T_{1-4}G_4$ and $T_{2-4}AG_3$. Sundquist and Klug[46] have studied the association of an oligonucleotide consisting of a double-helical section (29 base-pairs) followed by a single-stranded overhang (12 nucleotides), which contains five repeats of the *Tetrahymena* telomeric sequence TTGGGG:

GTCGACCCGGGTTGGGGTTGGGGTTGGGGTTGGGGTTGGGG
CAGCTGGGCCCAACCCCAACCCCAACCCC

This sequence dimerizes by the formation of guanine tetrads between the intramolecular hairpins formed by the single-stranded overhang (see Figure 15). The length of the thymine tract and the nature of the cation play a dominant role in the association of model oligonucleotides, as shown by c.d. studies with $d(G_4T_nG_4)$[47] and $d(T_nG_4)_2$.[48]

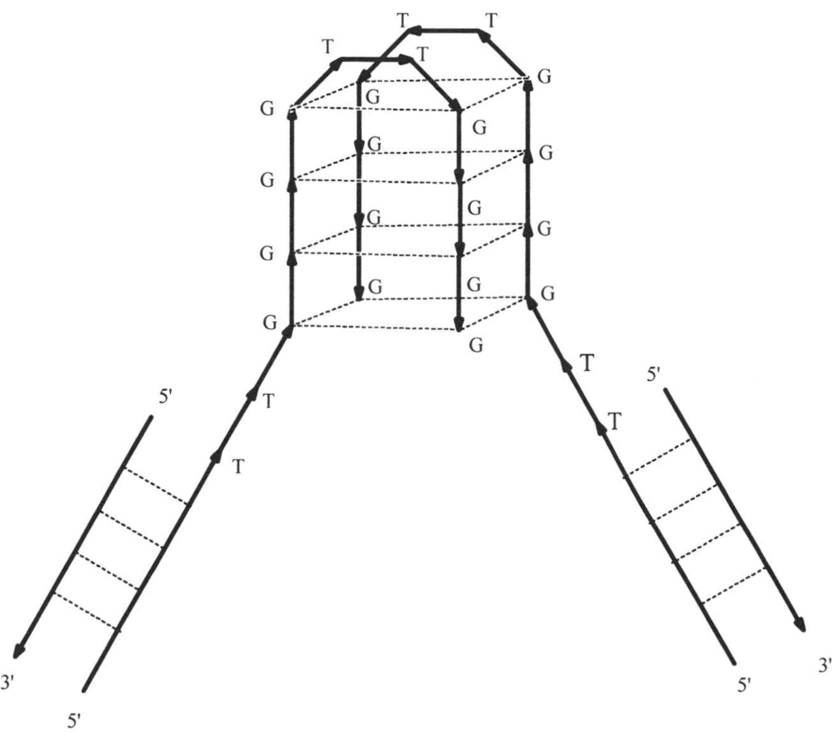

Figure 15 Structure of the hairpin dimer of the *Tetrahymena* telomeric end.

Sequences with $n = 1$, in the presence of either sodium or potassium cations, give a strong, positive c.d. signal around 260 nm, which is characteristic of the parallel-stranded tetrameric aggregates. Sequences with $n = 3, 4$ give this type of spectrum only in the presence of K^+, while in the presence of Na^+, the c.d. spectra present a positive band around 290 nm, which has been assigned to the formation of dimeric aggregates resulting from the association of folded-back hairpin structures.

The finding that the K^+ ion stabilizes the four-stranded aggregates characterized by a strong, positive, c.d. signal at 260 nm parallels our observation on short homoguanylates: as reported in Section 13.3.1.2, d(GpG) also gives a dominant 260 nm band only in the presence of an excess of K^+ ions; a dominant positive 290 nm c.d. is instead observed in the presence of an excess of Rb^+ or Na^+ ions.

In the case of d(GpG), which cannot form folded-back structures, the interpretation of the occurrence of the 260 nm or 290 nm bands must be different from that currently given for telomere-like models, which focuses on parallel (260 nm) or antiparallel hairpin (290 nm) arrangements of the interacting strands.

Our interpretation, based on exciton calculations,[18,32,33] focuses instead on the chromophoric tetrameric planes: regular structures, with planes perpendicular to the helix axis, display the 260 nm band, while distorted structures are related to the 290 nm band. The spectral changes are therefore related only to modifications of the relative geometry of the chromophores and not, in a direct way, to the orientation of the sugar phosphate backbones.

Many oligonucleotides consisting of one to several repeats of the natural telomeric sequences, have been studied by physicochemical methods, leading to a wealth of structural and thermodynamical information, whose detailed description is, however, far beyond the aim of this chapter.

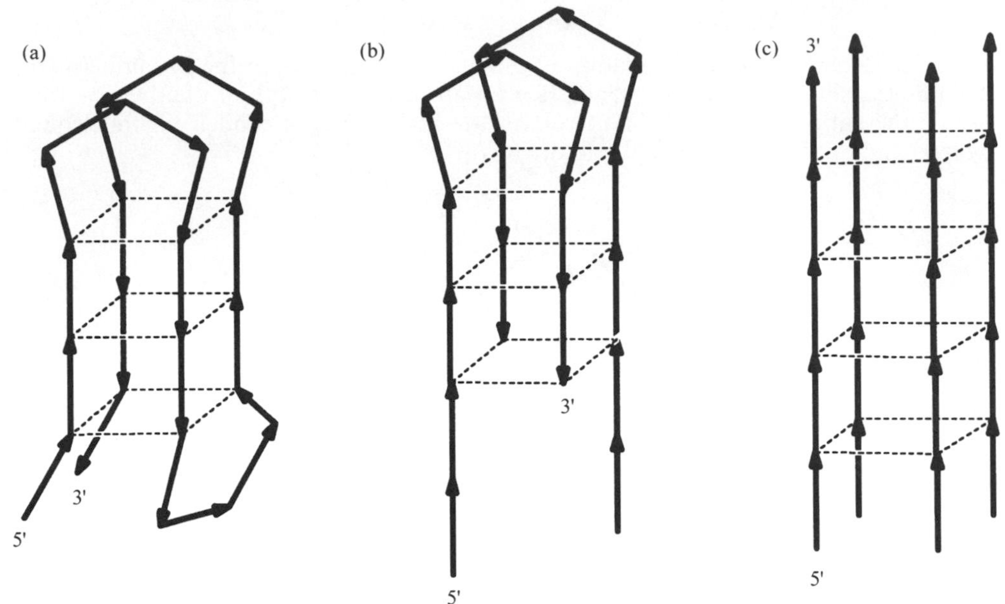

Figure 16 Basic assembly modes for G-rich strands: (a) intramolecular folded-back structure, (b) dimer of intramolecular hair-pin loops, (c) parallel-stranded tetraplex. The dashed squares represent G-quartets.

To summarize, the picture that emerges is that, depending on the salt conditions, specific sequences and DNA concentration, telomeric G-rich strands may form (i) intramolecular folded-back structures (Figure 16(a)), (ii) dimers of intramolecular hairpin loops (Figure 16(b)), or (iii) parallel-stranded tetramers (Figure 16(c)). All these structures are stabilized by the familiar G-quartet motif.

It has been suggested[49] that the same telomeric sequence will adopt one of the three possible structures, depending on the physiological ion (sodium and potassium, mainly) and fluctuations; furthermore, the different structures may be connected to different telomere functions. The very recent findings[50–3] that C-rich stretches, such as the Watson–Crick complements of the G-rich sequences in the telomeres, can self-associate to give four-stranded structures, may be taken as further, very important, evidence of the peculiar behaviour of telomeric-type DNA. However, despite the accumulation of *in vitro* experiments, the existence and biological relevance of this structural polymorphism of telomeric sequences in natural chromosomes are still to be proved. It can be foreseen that the search will be speeded up by the recent reports that telomere length may be a 'mitotic clock' for the cells of higher animals whose breakdown may contribute to the uncontrolled cell proliferation of cancer.[54–6]

Another possible role for G-rich sequences has been proposed by Sundquist and Heaphy[57] in the dimerization of HIV-1 genomic RNA. These authors have observed the spontaneous dimerization, *in vitro*, of an oligonucleotide containing the dimerization domain of genomic RNA and have shown that it depends on the presence of the sequence GGGGGAGAA; the stability of the dimer is maximal in the presence of physiological concentrations of K^+ ions. While the role of G-quartets in the life cycle of HIV-1 awaits its confirmation *in vivo*, cell culture experiments have shown that the phosphorothioate oligonucleotide $T_2G_4T_2$ is an inhibitor of HIV infection. At odds with all other phosphorothioates, the active form of this oligonucleotide is the parallel-stranded tetrameric aggregate, which binds to the retroviral envelope protein gp120 and inhibits cell-to-cell and virus-to-cell infection.[58]

Another molecule that owes its biological activity to the formation of the G-quartet motif, this time by an intramolecular process, is shown in Figure 17.[59] This oligonucleotide, selected by a directed evolution methodology,[60] is a very potent inhibitor of thrombin and has a therapeutic value in cardiovascular surgery.

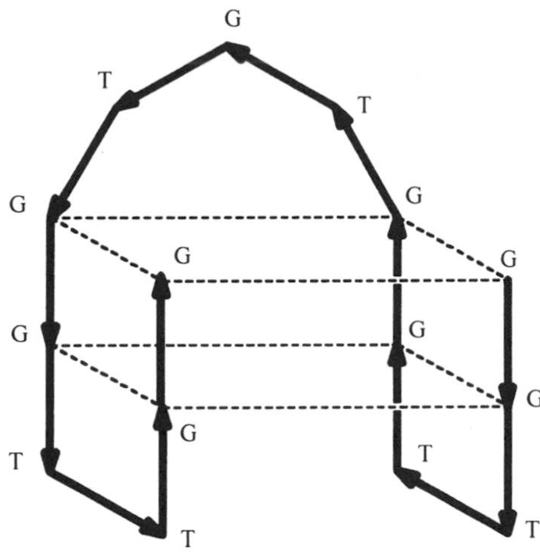

Figure 17 An oligonucleotide molecule of current interest in cardiovascular surgery.

13.7 REFERENCES

1. S. Bonazzi, M. M. De Morais, G. Gottarelli, P. Mariani and G. P. Spada, *Angew. Chem., Int. Ed. Engl.*, 1993, **32**, 248.
2. I. Bang, *Biochem. Z.*, 1910, **26**, 293.
3. R. K. Ralph, W. J. Connors and H. G. Khorana, *J. Am. Chem. Soc.*, 1962, **84**, 2266.
4. M. Gellert, M. N. Lipsett and D. R. Davies, *Proc. Natl. Acad. Sci. USA*, 1962, **48**, 1463.
5. S. B. Zimmermann, *J. Mol. Biol.*, 1976, **106**, 663.
6. W. Saenger, 'Principles of Nucleic Acid Structures', Springer, New York, 1984, p. 317.
7. P. Tsougard, J. F. Chantot and W. Guschlbauer, *Biochim. Biophys. Acta*, 1973, **308**, 9.
8. S. Arnott, R. Chandrasekaran and C. M. Martilla, *Biochem. J.*, 1974, **141**, 537.
9. S. B. Zimmermann, G. H. Cohen and D. R. Davies, *J. Mol. Biol.*, 1975, **92**, 181.
10. For a review see: W. Guschlbauer, J. F. Chantot and D. Thiele, *J. Biomol. Struct. Dyn.*, 1990, **8**, 491.
11. T. S. Pinnavaia, C. L. Marshall, C. M. Mettler, C. L. Fisk, H. T. Miles and E. D. Becker, *J. Am. Chem. Soc.*, 1978, **100**, 3625.
12. C. Detellier and P. Lazlo, *J. Am. Chem. Soc.*, 1980, **102**, 1135.
13. G. P. Spada, A. Carcuro, F. P. Colonna, A. Garbesi and G. Gottarelli, *Liq. Cryst.*, 1988, **3**, 651.
14. S. Bonazzi, M. Capobianco, M. M. De Morais, A. Garbesi, G. Gottarelli, P. Mariani, M. G. Ponzi Bossi, G. P. Spada and L. Tondelli, *J. Am. Chem. Soc.*, 1991, **113**, 5809.
15. P. Mariani, C. Mazebard, A. Garbesi and G. P. Spada, *J. Am. Chem. Soc.*, 1989, **111**, 6369.
16. F. Livolant, A. M. Levelut, J. Doucet and J. P. Benoit, *Nature (London)*, 1989, **339**, 724.
17. G. P. Spada, P. Brigidi and G. Gottarelli, *J. Chem. Soc., Chem. Commun.*, 1988, 953.
18. S. Bonazzi, G. Gottarelli, G. P. Spada, P. Mariani, S. Romanzetti, A. Garbesi and A. La Monaca, *Gazz. Chim. Ital.*, 1995, **125**, 483.
19. P. Mariani, M. M. De Morais, G. Gottarelli, G. P. Spada, H. Delacroix and L. Tondelli, *Liq. Cryst.*, 1993, **15**, 757.
20. G. P. Spada, S. Bonazzi, A. Garbesi, S. Zanella, F. Ciuchi and P. Mariani, manuscript in preparation.
21. S. Bonazzi, M. M. De Morais, A. Garbesi, G. Gottarelli, P. Mariani and G. P. Spada, *Liq. Cryst.*, 1991, **10**, 495.
22. S. Bonazzi, A. Garbesi, G. Gottarelli, P. Mariani, G. Proni and G. P. Spada, manuscript in preparation.
23. Y. C. Liaw, Y. G. Gao, H. Robinson, G. M. Sheldrick, L. A. J. M. Sliedregt, G. A. Van der Marel, J. H. Van Boom and A. H. J. Wang, *FEBS Lett.*, 1990, **264**, 223.
24. G. Laughlan, A. I. H. Murchie, D. G. Norman, M. H. Moore, P. C. E. Moody, D. M. J. Lilley and B. Luisi, *Science*, 1994, **265**, 520.
25. F. Ciuchi, G. Di Nicola, H. Franz, G. Gottarelli, P. Mariani, M. G. Ponzi Bossi and G. P. Spada, *J. Am. Chem. Soc.*, 1994, **116**, 7064.
26. L. Q. Amaral, A. Gulik, R. Itri and P. Mariani, *Phys. Rev. A.*, 1992, **46**, 3548.
27. S. H. Chen and T. H. Lin, in 'Methods in Experimental Physics', eds. R. Celotta and J. Levine, Vol. 23, 'Neutron Scattering', eds. D. L. Price and K. Skold, Part B, Academic Press, San Diego, 1987, p. 489.
28. F. Carsughi, M. Ceretti and P. Mariani, *Eur. Biophys. J.*, 1992, **21**, 155.
29. A. Guiner and G. Fournet, 'Small Angle Scattering of X-Rays', Wiley, New York, 1955.
30. J. F. Chantot, T. Haertle and W. Guschlbauer, *Biochimie*, 1974, **56**, 501.
31. S. F. Mason, 'Molecular Optical Activity and the Chiral Discrimination', Cambridge University Press, Cambridge, 1982, p. 88.
32. C. L. Cech and I. Tinoco, Jr., *Nucleic Acids Res.*, 1976, **3**, 399.
33. G. Gottarelli, P. Palmieri and G. P. Spada, *Gazz. Chim. Ital.*, 1990, **120**, 101.
34. F. Carsughi, G. Di Nicola, G. Gottarelli, P. Mariani, E. Mezzina, A. Sabatucci, G. P. Spada and S. Bonazzi, *Helv. Chim. Acta*, 1996, **79**, 220.

35. H. Franz, F. Ciuchi, G. Di Nicola, M. M. De Morais and P. Mariani, *Phys. Rev. E*, 1994, **50**, 395.
36. L. Amaral, R. Itri, P. Mariani and R. Micheletto, *Liq. Cryst.*, 1992, **12**, 913.
37. M. P. Taylor and J. Herzfeld, *Phys. Rev. A.*, 1991, **43**, 1892.
38. G. Gottarelli and G. P. Spada, in 'Circular Dichroism—Principles and Applications', eds. K. Nakanishi, N. Berova and R. W. Woody, VCH, New York, 1994, p. 108.
39. E. Sakmann and J. Voss, *Chem. Phys. Lett.*, 1972, **14**, 528.
40. R. J. Dudley, S. F. Mason and R. D. Peacock, *J. Chem. Soc., Faraday Trans. 2*, 1975, 997.
41. R. Kremer, J.-M. Lehn and A. Marquis-Rigault, *Proc. Natl. Acad. Sci. USA*, 1993, **90**, 5394.
42. F. Livolant, *Eur. J. Cell Biol.*, 1984, **33**, 300.
43. E. H. Blackburn, *Trends Biochem. Sci.*, 1991, **16**, 378.
44. D. Sen and W. Gilbert, *Nature (London)*, 1988, **334**, 364.
45. D. Sen and W. Gilbert, *Nature (London)*, 1990, **344**, 410.
46. W. I. Sundquist and A. Klug, *Nature (London)*, 1989, **342**, 825.
47. P. Balagurumoorthy, S. K. Brahmachari, D. Mohanty, M. Bansal and V. Sasisekharan, *Nucleic Acids Res.*, 1992, **20**, 4061.
48. Q. Guo, M. Lu and N. R. Kallenbach, *Biochemistry*, 1993, **32**, 3596.
49. C. C. Hardin, E. Henderson, T. Watson and J. K. Prosser, *Biochemistry*, 1991, **30**, 4460.
50. K. Gehring, J. L. Leroy and M. Gueron, *Nature (London)*, 1993, **363**, 561.
51. S. Ahmed, A. Kintanar and E. Henderson, *Nat. Struct. Biol.*, 1994, **1**, 83.
52. J. L. Leroy, M. Gueron, J. L. Mergny and C. Helene, *Nucleic Acids Res.*, 1994, **22**, 1600.
53. G. Manzini, N. Yathindra and L. E. Xodo, *Nucleic Acids Res.*, 1994, **22**, 4634.
54. C. M. Counter, H. W. Hirte, S. Bacchetti and C. B. Harley, *Proc. Natl. Acad. Sci. USA*, 1994, **91**, 2900 and references therein.
55. N. W. Kim, M. A. Piatyszek, K. R. Prowse, C. B. Harley, M. D. West, P. L. C. Ho, G. M. Coviello, W. E. Wright, S. L. Weinrich and J. W. Shay, *Science*, 1994, **266**, 2011.
56. L. Seachrist, *Science*, 1995, **268**, 29.
57. W. I. Sundquist and S. Heaphy, *Proc. Natl. Acad. Sci. USA*, 1993, **90**, 3393.
58. J. R. Wyatt, T. A. Vickers, J. L. Roberson, R. W. Buckheit, Jr., T. Klimkait, E. DeBaets, P. W. Davis, B. Rayner, J. L. Imbach and D. J. Ecker, *Proc. Natl. Acad. Sci. USA*, 1994, **91**, 1356.
59. K. Y. Wang, S. H. Krawczyk, N. Bischofberger, S. Swaminathan and P. H. Bolton, *Biochemistry*, 1993, **32**, 11285.
60. L. C. Boc, L. C. Griffin, J. A. Latham, E. H. Vermaas and J. J. Toole, *Nature (London)*, 1992, **355**, 564.

14

Layered Nanoarchitectures via Directed Assembly of Anionic and Cationic Molecules

GERO DECHER

Université Louis Pasteur and CNRS, Institut Charles Sadron, Strasbourg, France

14.1 THE LENGTH SCALE OF COMPLEX SYSTEMS

The length scale of regularities in matter common to physicists, chemists, biologists, and engineers ranges approximately from subatomic particles up to objects of several meters in size; smaller and larger objects are generally not of interest in chemistry and biology. If one looks at the levels of structural and functional complexity found in this range of length scales,[1,2] one finds atoms, molecules, and supramolecular aggregates, but then the system differentiates into materials science continuing into nanoscopic and mesoscopic systems and finally into engineering and into life sciences, continuing into prebiotic chemistry, nanobiology, cellular life and various life forms based on the accumulation of specialized cells (Figure 1).[3] Clearly, as one goes from atoms to molecules and larger assemblies, the complex properties grow with the size of the objects, but the properties of the higher level assemblies can in general not be completely predicted from the properties of its constituents. Clearly, only a limited number of the properties of molecules are truly predictable from the properties of its constituent atoms, simply because properties such as toxicity require calculation of the electronic details of living bodies or $\sim 10^{20}$ atoms.

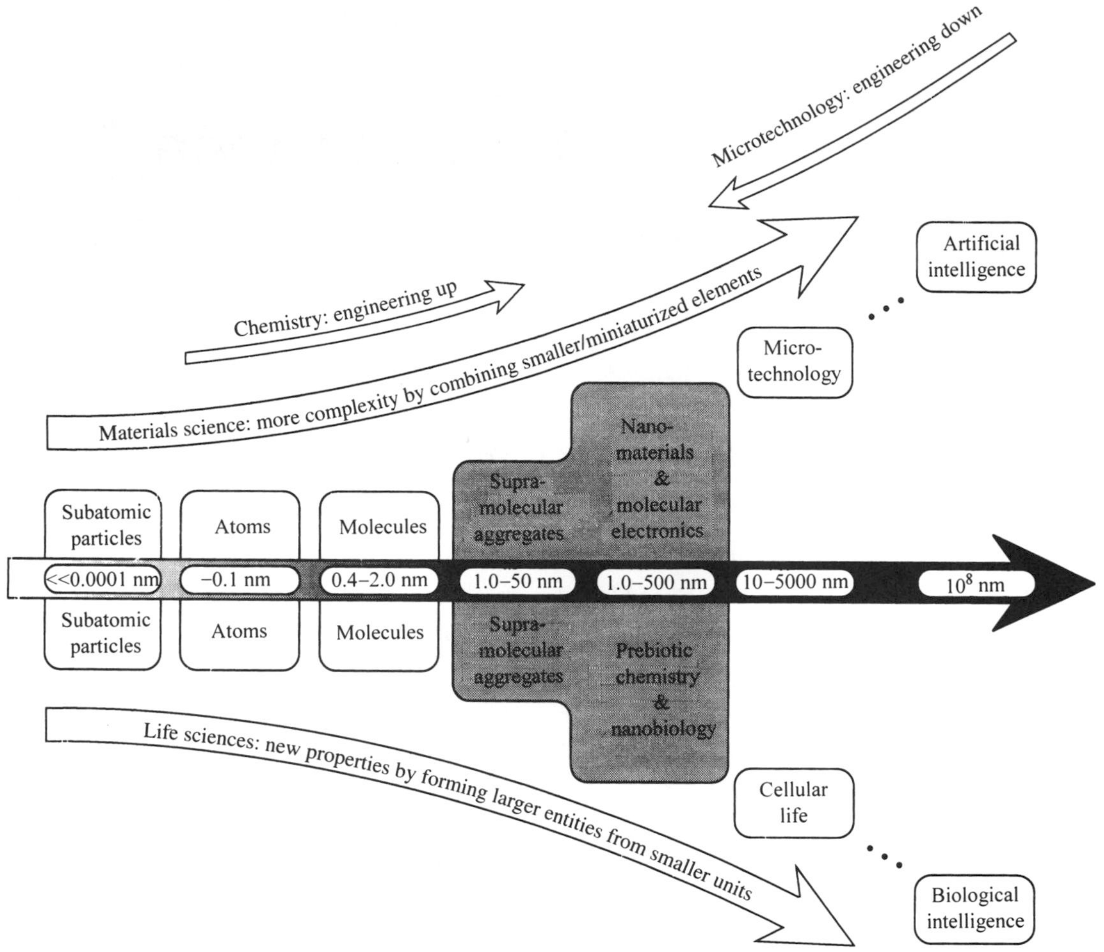

Figure 1 Complex properties can only be realized within a certain length scale of structured matter. Consequently, a higher level of complexity is reached by combination of smaller functional entities. However, the new properties of the resulting assembly cannot readily be predicted from the properties of its constituents. Whereas scientists have accumulated tremendous knowledge in manipulating matter both at the levels of atoms and molecules (length scale from 0.01 nm up to 2.0 nm) and in the macroscopic world (length scale from 0.5 µm up to 100 m), there is very little understanding of the structures and processes occurring on a length scale from 0.5 µm up to 2.0 nm. This area of supramolecular aggregation and nanostructures, which also encloses the size range of prebiotic chemistry and early subcellular life, is the length scale at which materials science and life sciences separate. Whereas new levels of complexity in biological systems are always reached by combining smaller subunits, man has also strived to make more complex systems by combination and miniaturization of existing units (silicon technology, integrated circuits). However, the length scale of nanomaterials can hardly be addressed by making things smaller, so the creation of well-defined organic nanostructures has become a challenge for materials science at the supramolecular level.

Interestingly, the structural elements in both materials science and life science consist of identical objects up to the size range of a few nanometers (atoms, molecules, and supramolecular aggregates), but nature has exploded into a vast variety of more and more complex systems by the feedback mechanism of evolution, whereas physicists, chemists, and engineers are just beginning to explore the synthesis or fabrication of objects in the nano- and mesoscopic range. Whereas engineers are fabricating smaller and smaller structures (microtechnology) in order to make more complex machines (engineering down), chemists synthesize larger molecules or supramolecular assemblies (engineering up). It is this length scale of supramolecular chemistry and beyond that will be a new challenge for synthetic work in the foreseeable future.[1] In this respect, we focus on the spatial positioning of molecules in the form of multilayer films as prepared by directed assembly.

14.2 THE ART OF STACKING MOLECULES: CONCEPTS FOR THE PREPARATION OF ONE-DIMENSIONALLY ORDERED SOLIDS

14.2.1 From "Supramolecular" to "Nanoscopic" Systems

Supramolecular chemistry is the science of intermolecular association as driven by multiple weak interactions.[4–9] At first sight, scientists in this area seem to be interested in answering the question of why one should use two or more molecules for a certain purpose instead of a single molecule, but in principle this is a question of dimensions and not of numbers. When molecules are above a certain size, parts of a molecule may interact with other parts of the same molecule, producing intra("supra")molecular aggregates (e.g., secondary and tertiary structure of proteins) and thus inter- and intramolecular interactions become indistinguishable. So the key issue in studying noncovalent interactions is the enormous diversity of organic structures and their dynamics that occur on the length scale of 1 nm and above and that are responsible for the immense complexity of living systems.[10]

Although noncovalent bonds are much weaker than covalent bonds, multiple weak bonds might add up to yield association constants of this order of magnitude, as in the case of the interaction of avidin with biotin. The interactions between different molecules are electrostatic attraction between opposite charges and dipoles, hydrogen bonding, aromatic π-stacking, charge transfer, or hydrophobic effects. However, the structure and dynamics of an aggregate are also determined by nonattractive interactions such as electrostatic repulsion between equal charges, parallel dipoles, or steric repulsion. Therefore, the unique properties of an aggregate depend on the topologies of the interacting molecules, that is, on the surface of each molecule as experienced by its partner molecule, which again is the result of the exact arrangement of attractive and repulsive groups on both partners in a given solvent.

Examples range from complex formation in solution, as in the case of two interacting molecules, to micelles, thermotropic or lyotropic liquid crystals, or even crystals whose structure results from the tailored interaction of its constituent molecules. Generally, it is very difficult to control the number of molecules in an aggregate unless the molecules are assembled in a stepwise fashion starting from a template. One example of aggregation controlled by self-organization which results from both the molecular topology and from the surface tension, is spherical micelles.

Furthermore, it is usually difficult to exploit the directionality of the intermolecular interactions since the resulting assemblies normally stay dissolved in an isotropic solution. An example of making use of directional effects is the tailoring of molecular interactions in such a way that the molecules form a noncentrosymmetric crystal which might eventually be useful for second-order nonlinear optical applications. However, this is generally not possible in the case of assemblies of nanoscopic dimensions that are too small to be handled macroscopically and whose manipulation by tools such as may be derived from scanning probe microscopes[11] or optical tweezers[12] is limited to individual molecules or objects. In this case, again, one can use a macroscopic template on which the molecules are assembled in such a way that the resulting target structures preserve the directionality of the intermolecular interaction. In contrast to self-assembly and self-organization, one is tempted to call such a process directed assembly.

14.2.2 Towards Directed Assembly: How to Control the Aggregation?

The classical example of a directed assembly procedure is the Langmuir–Blodgett technique, in which amphiphilic molecules are allowed to self-organize at the air–water interface and are subsequently transferred on to solid substrates.[13–16] Layer-by-layer deposition of multiple materials leads to multilayer films in which distance and orientation of the different compounds can in principle be controlled to such a degree that effects such as distance-dependent energy transfer between donor and acceptor dyes has been observed.[17] Despite of the obvious beauty of this approach, the realization of a given supramolecular architecture often remains a very difficult task, because the resulting multilayer films are often not equilibrium structures. Among other reasons, this is due to the fact that the interface of water, used for the assembly and self-organization of each monolayer, is no longer present in the multilayer film and other substrate-induced and intermolecular interactions come into play, usually causing structural changes.

This is different in the consecutive adsorption of molecules from solution on to solid surfaces, which has previously been employed by Merrifield[18] in solid-state peptide synthesis and later for the fabrication of multilayer structures by Sagiv and co-workers[5,19] using covalent chemistry and by Mallouk and co-workers[20–2] using ionic-covalent and coordinative covalent bonds between adjacent layers. Both strategies proved to be so successful that the concepts were later taken further by other groups,[23–5] The whole field of ultrathin organic films was reviewed in 1991 by Ulman.[26]

However, when looking at the respective structures, one immediately realizes that very high steric demands exist when synthesizing multilayer stacks of laterally close packed monolayers by using either covalent or coordinative bonds between adjacent layers. In contrast, electrostatic attraction offers a type of bonding which is much less demanding in terms of directionality of the bond and the distance between the two oppositely charged functional groups, and which additionally offers the possibility of forming multicenter bonds. It therefore seemed highly promising to employ electrostatic attraction for the formation of supramolecular multilayer architectures using directed assembly, that is, the consecutive deposition of molecular layers by subsequent adsorption of anions and cations from solution. That electrostatic attraction can in principle be used for the fabrication of multilayers had already been demonstrated as early as 1966.[27] At that time, Iler had prepared layered films by consecutively adsorbing anionic and cationic inorganic colloidal particles from aqueous solution. However, this nice idea seems not to have been followed and hard analytical data never appeared, most likely owing to the lack of appropriate equipment at that time. Additionally, there was no information available on whether the process would work with other (molecular) materials, except the relatively large and polydisperse colloids. However, as ionic compounds are easily soluble in aqueous solutions, an important side aspect of such a method would be that it would be a soft technology and that it would offer the possibility of working with biological materials.

14.2.3 Homo- or Heterobifunctional Molecules?

The layer-by-layer adsorption of molecules from solution requires that the functionality of a given surface is replaced by a different functionality after a monolayer of a certain molecule has been deposited. If the surface functional density can be maintained in every adsorption step, the formation of multilayer structures becomes possible. This concept[5,28] suggests that homo- or heterobifunctional molecules should be well suited for consecutive chemi- or physisorption from solution. If the two functional groups are separated from each other, the central spacer unit can be tailored for lateral interaction, thus giving additional stability to each monolayer. Such rodlike homobifunctional molecules are often termed bolaform in the literature.

For the fabrication of multilayers by directed assembly techniques, one also needs to consider the aspect of cross-reactivity of the two functional groups (Figure 2). First, the two functional groups must not react with each other in solution, which would lead to polymerization. In solid-state peptide synthesis this was achieved by protective groups which are cleaved off after the adsorption has taken place. This is depicted schematically in Figure 2(a). Similarly, as in the case of multilayer fabrication by consecutive chemisorption as introduced by Sagiv and co-workers,[19] a noncross-reactive end group (e.g., a double bond) is activated by transformation (e.g., hydroboration and subsequent treatment with basic H_2O_2) into a reactive group (e.g., OH) after chemisorption of the molecule to the surface via its other reactive group (e.g., trichlorosilane). Another choice would be the use of two heterobifunctional compounds, each of whose end groups are not cross-reactive, but which would react with one of the groups of the respective partner molecule, thus allowing for consecutive chemisorption of the two compounds without deprotection or activation steps. This process leads to polar films and is depicted schematically in Figure 2(b). However, the simplest case remains the use of two different homobifunctional molecules which are cross-reactive with each other; in the case of electrostatically driven adsorption, the two molecules would be boladianions and boladications (Figure 3).

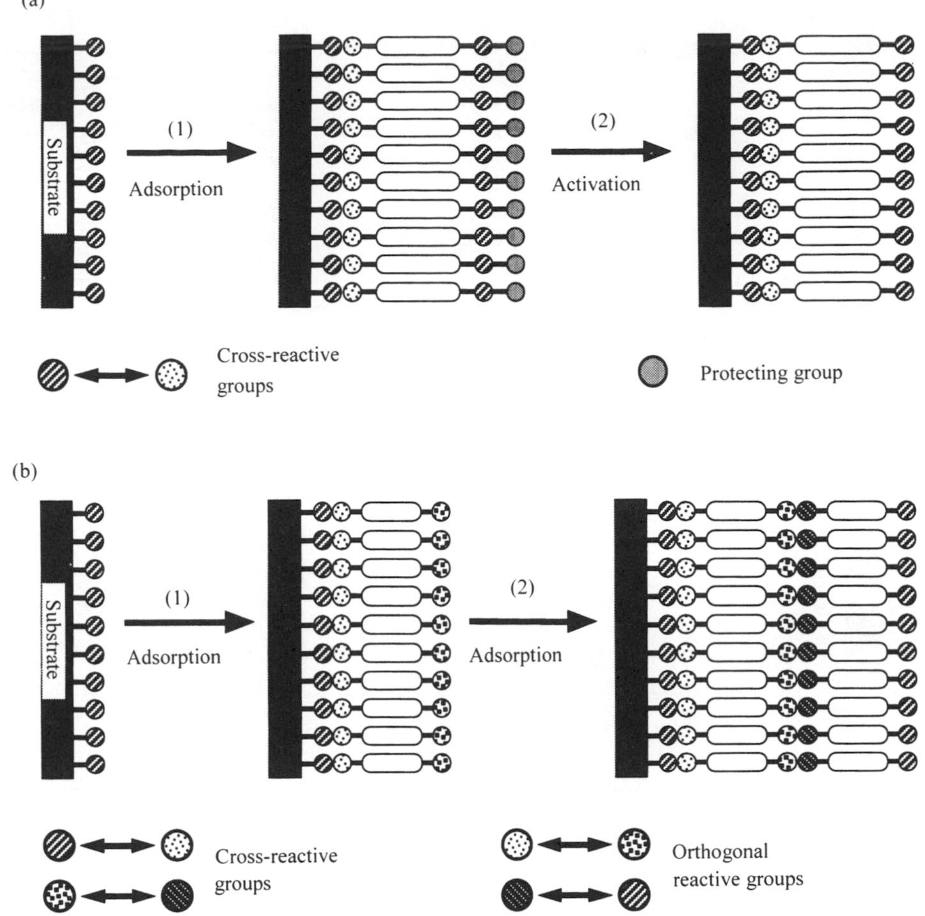

Figure 2 In the simplest case of heterobifunctional molecules, layer-by-layer surface modifications (≡ multilayer buildup) can in principle be achieved by either adsorption/activation steps using a single compound (a) or by consecutively alternating adsorption of two different compounds (b). Cross-reactive groups represent functional groups that react easily with each other and orthogonal reactive groups represent functional groups that must not react with each other.

14.2.4 Simple Nomenclature for Multilayer Film Architectures

At this point, it is convenient to introduce a simple nomenclature for the description of a certain film architecture. If a film is composed of an equal number m of layers of the two molecules A and B, this would be written as $(A/B)_m$, the total number of layers being $2m$. More complex architectures such as a superlattice structure composed of the three compounds A, B, and C in which there are n superlattice repeat units composed each of $2m + 2$ layers would be written as $[(A/B)_m A/C]_n$, the total number of layers being $2n(m + 1)$. If $m = 2$ and $n = 2$, the superlattice would be deposited on top of a single layer D, and one would write $D[(A/B)_2 A/C]_2$, which would correspond to the layer sequence D/A/B/A/B/A/C/A/B/A/B/A/C. We even use this nomenclature in an automatic film deposition device[29] in order to fabricate complex film architectures.

14.3 MULTILAYERS OF BI-, OLIGO-, AND MULTIPOLAR COMPOUNDS

14.3.1 Start with the Obvious: Bolaform Molecules

14.3.1.1 Molecular design

Of course, it would be most elegant to use only a single compound with a positive and a negative charge at each end for the deposition of multilayer assemblies. However, this is not a practicable

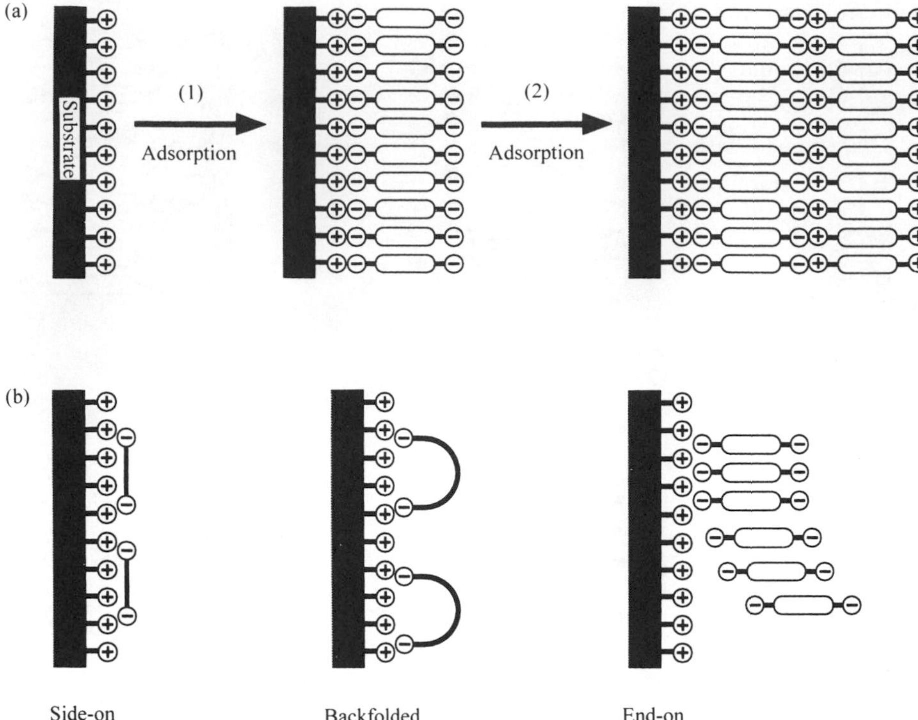

Figure 3 Layer-by-layer adsorption using two different homobifunctional molecules. The drawing shows an example of a positively charged surface being subsequently coated by a layer of a boladianion and a layer of a boladication (a). A quasiunlimited repetition of steps 1 and 2, which is the prerequisite for the formation of multilayers, can only be carried out if side-on adsorption and backfolding can be avoided and the desired end-on adsorption can be achieved in every cycle (b). It is not implied that the molecules have to be ordered in-plane and that they have to be aligned along the layer normal. A surface can also successfully be refunctionalized if the molecules are tilted and have different tilt directions. For reasons of clarity, the counterions have been omitted in this representation.

approach, for two reasons. First, α,ω-zwitterions would have a high tendency to precipitate from solution and, if not, there would consequently only be a small tendency for the adsorption of one end to an oppositely charged surface. Second, if multilayers were formed by adsorbing a single compound from solution, it would be very problematic to terminate multilayer growth after a given number of layers. Clearly, this is a major prerequisite for the fabrication of complex multilayer architectures of two or more compounds. An elegant way out of this dilemma would be the use of molecules with a permanent charge at one end and a chargeable group at the other. This would still require that the chargeable group would be neutral in the bulk of the solution, but could be charged in a second step once the molecule has been adsorbed to the surface, but such systems have not yet been described. In this respect, one should always keep in mind that the directed assembly must generally be at least a two-step process, either in the form of an adsorption and an activation step or in the form of two or more consecutive adsorption steps using two or more molecules. From this point of view, boladianions and boladications should be the simplest choice for the fabrication of multilayer assemblies.[30] This approach is depicted schematically in Figure 3(a).

This still leaves the question open of how one can achieve complete charge inversion but constant surface charge density in every adsorption step. Two immediately obvious problems that would cause a loss of surface functional density would be flat (side-on) adsorption and backfolding (Figure 3(b)). If the distance between the two end groups is small, there is very little preference for end-on adsorption of a rodlike molecule and in fact molecules such as terphenylbis(sulfonic acid) tend to do this and therefore are displaced from a positively charged surface when the sample is immersed in a solution of cationic molecules in order to deposit the following layer. The end-on adsorption can be made more favorable by increasing the lateral interaction in each monolayer and by making side-on adsorption less likely. In the case of adsorption from aqueous solution, both can be achieved by increasing the length of the rod by using a hydrophobic moiety, for example, an alkyl chain. In this case the lateral interaction is increased by the hydrophobic effect and the flat adsorption is disfavored since this would cause a long alkyl chain into contacts with many charged groups on the surface. At the same time, however, an elongation of the end-to-end distance of a bolaform molecule would also increase its tendency for backfolding, which would also retard multilayer formation. We have therefore started with the concept of using rodlike bolaform dianions and dications with a molecular length of 4–5 nm, but with a rigid central segment. Molecules of this type are easily synthesized starting from an inflexible unit such as used in smectic liquid crystals (i.e., biphenyl or stilbene) by attaching alkyl chains of about 10—CH_2— units and finally ionic groups at each end. The structure of the boladianion and boladication we have been using in our first experiments are depicted in Figure 4 along with boladiions used by other groups.

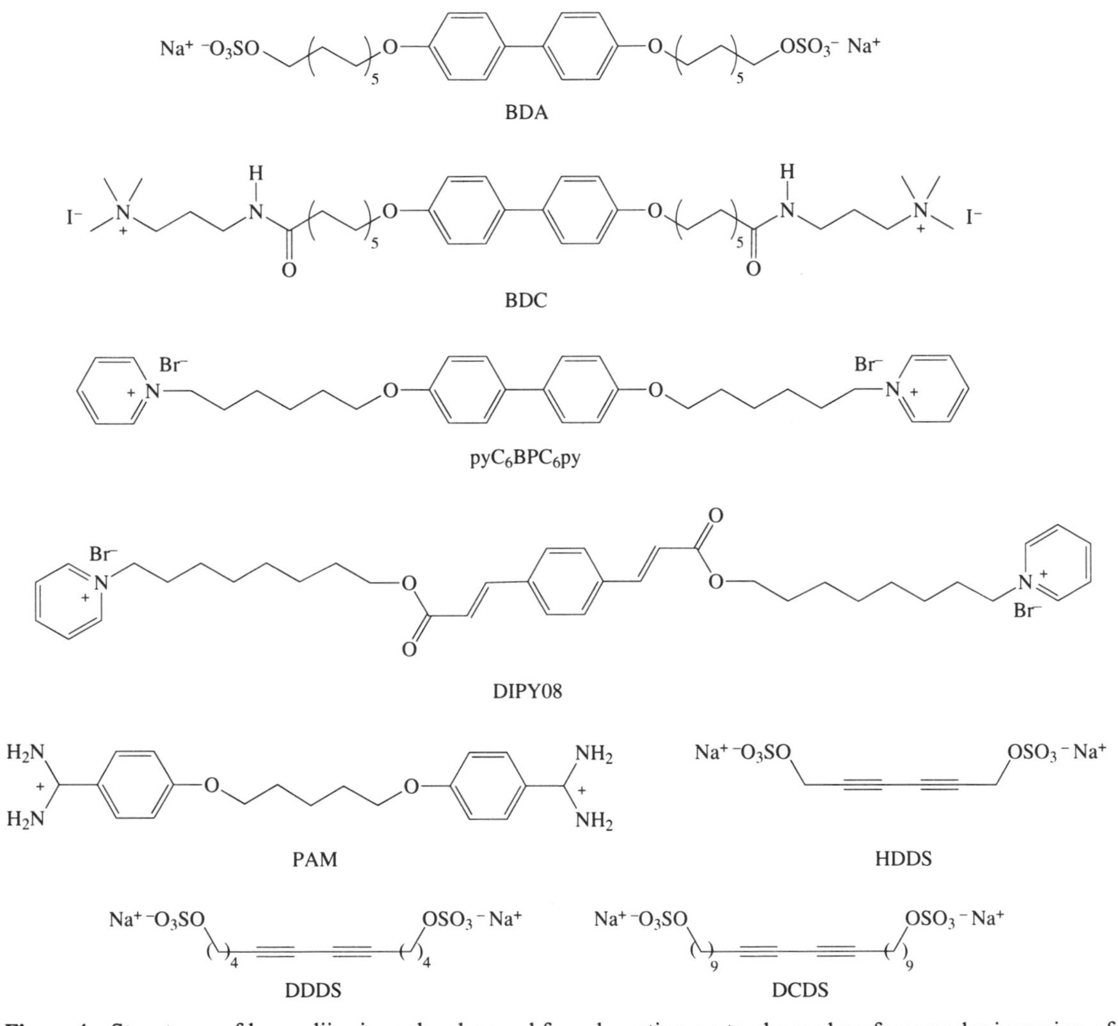

BDA

BDC

pyC_6BPC_6py

DIPY08

PAM

HDDS

DDDS

DCDS

Figure 4 Structures of homodiionic molecules used for adsorption on to charged surfaces under inversion of the respective surface charge and for the fabrication of multilayers. Counterions have been omitted in some cases; most of the abbreviations of the substances follow the original literature. The compounds are referenced as follows: BDA, BDC;[30] pyC_6BPC_6py;[31–4] DIPY08;[35] PAM;[36] HDDS, DDDS;[37] and DCDS.[37,38]

14.3.1.2 Monolayers of boladiions

If molecules such as BDA or BDC are being alternately adsorbed on to charged solid surfaces according to Figure 3, one observes a change in the optical absorbance (Figure 5(a)), which can be used to monitor the formation of the adsorbed layer in an on-line fashion (Figure 5(b)). This spectral change should be due to chromophore aggregation or orientation on adsorption. Furthermore, NMR spectra of BDA in D_2O are broadened, a typical sign of aggregation already in solution. From the fact that BDA is a homodiionic amphiphile, one might expect that it could form disk-shaped micelles in aqueous solution,[39] which would make the formation of a well-oriented monolayer upon adsorption to a surface a very likely process. This is further supported by the finding that simple amphiphiles such as sodium dodecyl sulfate are reversibly adsorbed onto aminopropylsilanized flat surfaces (e.g., functionalized silicon single crystals), thereby forming a well-oriented monolayer.[40] Figure 5(b) shows that binding and rearrangement of BDA on the solid surface of aminopropylsilanized quartz are completed after 20 min, a time span we have adopted in a general procedure for the adsorption of bolaform molecules for multilayer buildup.[30]

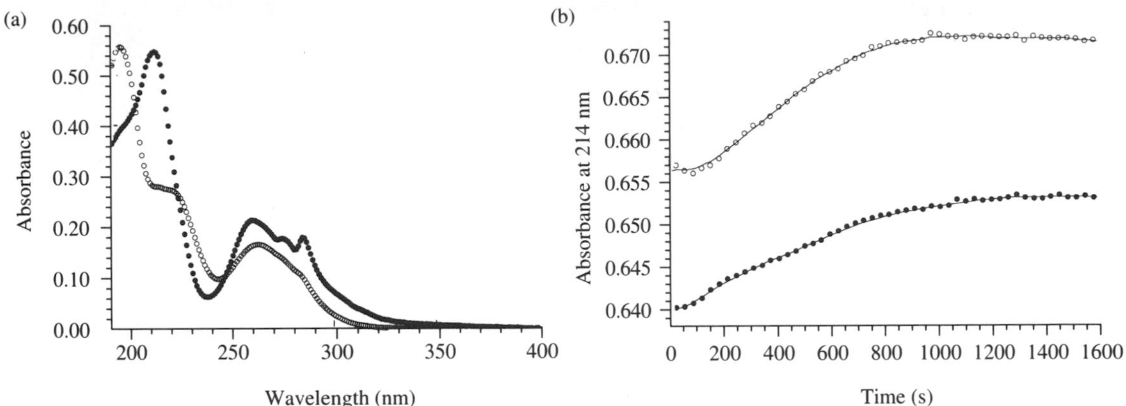

Figure 5 (a) UV spectra of an aqueous solution of the boladication BDC (concentration 9×10^{-6} mol L^{-1}) (open circles) and of a multilayer of BDC and the boladianion BDA, whose solution spectrum is almost identical with that of BDC (closed circles). The difference in the two spectra indicates a change in the chromophore aggregation or orientation upon adsorption. (b) The increase of the band at 214 nm with time reflects the amount of adsorbed BDC and of molecular rearrangements taking place after the adsorption and can be used to monitor the formation of a stable surface layer, while excess BDC is still present in solution. The graph shows for concentrations of BDC of 1.72×10^{-5} mol L^{-1} (closed circles) and 4.31×10^{-5} mol L^{-1} (open circles) that the formation of a stable layer is finished after 20 min.

14.3.1.3 Multilayers of boladiions

The fabrication of a multilayer according to Figure 3 now becomes a straightforward approach. One simply immerses a freshly aminopropylsilanized object in a solution of a boladication (e.g., BDA) for 20 min, washes it a few times with pure water, immerses the object in a solution of a boladication (e.g., BDC) for 20 min, washes it a few times with pure water, and repeats the whole process as often as desired. The whole procedure is carried out in simple beakers under regular laboratory conditions, which has recently prompted Mallouk and co-workers[41] to coin the phrase molecular beaker epitaxy (MBE).

Using BDA and BDC as the two oppositely charged bolaform homodianions, the formation of multilayers with several tens of layers was observed and verified by UV spectrophotometry (Figure 6). The linear increase in the absorbance of biphenyl chromophores with increase in the number of layers is a good indication of the directed assembly of a multilayer. In principle, the observation of a linear correlation means that the surface refunctionalization in each layer can be carried out in a consecutive fashion and is thus a reproducible process throughout the multilayer buildup.

It is readily envisioned that more complex film architectures can be easily obtained by simply immersing the film in a third or fourth solution of an appropriate boladiion at the desired layer numbers.

Multilayer formation which shows a linear increase in optical absorbance with increase in the number of adsorption cycles over several tens of layers is normally not observed in the consecutive chemisorption of heterobifunctional compounds as described above, which often leads to a deviation from linearity after the deposition of about 10 layers. Since no covalent bond formation, especially if carried out at a solid surface, has a yield of exactly 100% and since heterobifunctional compounds cannot compensate for the resulting loss of surface functional density, it is not unexpected that the continued assembly of a multilayer often becomes impossible after a few layers.

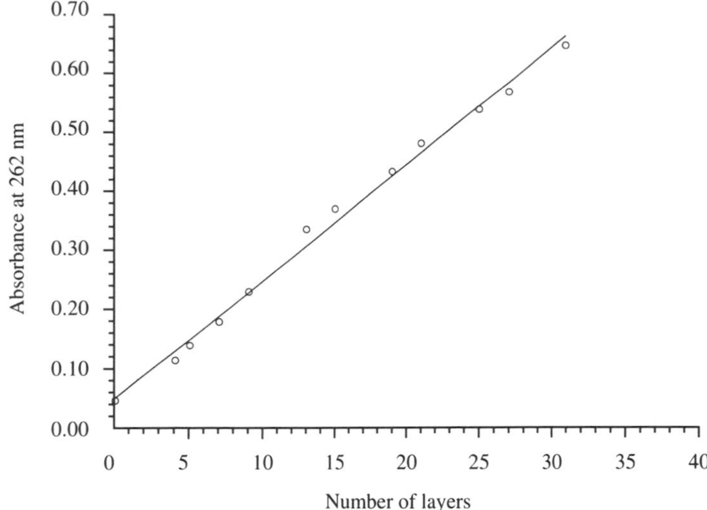

Figure 6 Buildup of a multilayer film of the boladianion BDA and the boladication BDC on the surface of aminopropylsilanized quartz, as monitored by the UV absorbance of the biphenyl chromophores at 262 nm.

Analogous results are long known from solid-state peptide synthesis[18] (a similar but simpler case, because it was not attempted to form densely packed layered structures), where the percentage of oligopeptides with the wrong sequence increases with increasing number of amino acids in the peptide chain.

Nevertheless, multilayers of boladianions and boladications are also not perfect and often show patches of different interference colors on different regions of a substrate, clearly visible to the naked eye, especially when substrates such as single-crystal silicon wafers are being used. These spots correspond to regions of slightly different thickness or refractive index which might originate from slightly different tilt angles of the molecules within the layers. They are especially pronounced on substrates which have suffered from poor silanization in the procedure of creating a charged surface. The presence of these inhomogeneities clearly indicates that the multilayer properties are not entirely controlled by the choice of materials and deposition conditions. This raises the question how multilayer buildup can be further improved.

14.3.2 The Next Step: Oligo- and Multipolar Compounds

Although electrostatically driven adsorption is sterically less demanding than chemisorption and probably leads to a strongly reduced number of defects within the layers, the use of homobifunctional implies that the number of molecules per unit area is a constant in the linear growth regime. This is due to the fact that in the ideal case the maximum number of molecules that adsorb to a surface is equal to the number of charges on that surface, if the surface charge density is equal to or smaller than the lateral packing density of the boladiions. If the surface charge density for some reason is lowered during deposition, a homobifunctional molecule has little chance of compensating for this loss. This principle feature of hetero- or homobifunctional molecules can be overcome by using oligo- or multifunctional compounds of such a shape that no matter with which part they adsorb to a surface, they still expose some number of functional groups to the solution interface and thus effectively overcompensate surface functionality. If such molecules are sufficiently large, they will also tolerate surface defects by simply spanning across them. This concept is, of course, independent of the type of bonding between the layers and we have used the sixfold reactivity of fullerene C_{60} (oligofunctional building block) and a polyamine (multifunctional building block) for the formation of highly ordered chemisorbed multilayer films.[42,43] This is similar to the idea of using branched monomers in order to fill the increasing surface area of a dendrimer in the starburst synthesis with increasing numbers of generations[44] or, in the case of chemisorbed multilayers, the use of lateral bonding[5] or polymeric "capping" layers.[25] It should be noted that the synthesis of starburst dendrimers is also a two-step process, normally using two different heterobifunctional molecules.

14.3.2.1 *Mixed boladiion–oligoion multilayers*

The incorporation of oligo-charged molecules into multilayer assemblies has meanwhile been demonstrated by Shen and co-workers.[45] They prepared alternating multilayers of a homodicationic bolaform molecule (pyC_6BPC_6py) and two different homotetraanionic disk-shaped molecules, a *meso*-tetra(4-sulfenyl)porphyrin ($tppS_4$) and the sodium salt of copper phthalocyaninetrasulfonic acid (CuTsPc); their structures are depicted in Figure 7. Interestingly, they have observed a Bragg peak in small-angle x-ray scattering, which indicates that the molecules form well-ordered layers. The fact that we have not observed any Bragg peaks in our multilayers of boladiions might indicate that oligofunctional molecules are better suited than bifunctional molecules to compensate for small defects.

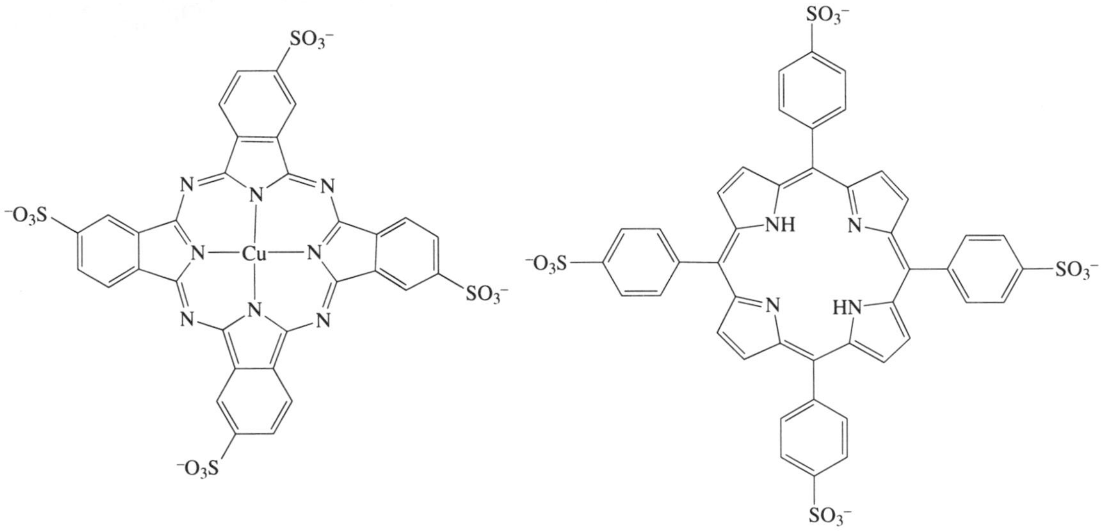

CuTsPc

$tppS_4$

Figure 7 Structures of multipolar molecules used for adsorption on to charged surfaces under inversion of the respective surface charge and for the fabrication of multilayers. Counterions have been omitted; the abbreviations of the substances follow the original literature. The compounds are referenced as follows: CuTsPc, $tppS_4$.[45]

14.3.2.2 *Mixed boladiion–polyelectrolyte multilayers*

Obviously, polyelectrolytes as multifunctional ions should be well suited for the spanning of surface defects in the case of the electrostatically driven consecutive adsorption of anions and cations. Although the adsorption of polymers in general[46] and charged macromolecules[47–52] on to solid surfaces have been subjects of considerable research, it was not clear whether their adsorption could be controlled in such a way that the surface charge could effectively be overcompensated in every adsorption cycle, which is a prerequisite for multilayer formation. That there exists a critical polyelectrolyte concentration that leads to the overcompensation of the surface charge was experimentally verified by Kunitake and co-workers[53] in 1992, although the concepts of "flat" and "loopy" conformations of polymers at interfaces, which are analogous descriptions to the charge overcompensation concept, have long been discussed in order to describe the adsorption behavior of polymers. It was also not clear if the adsorption of the second polyion on to the first layer of the oppositely charged polyion would not displace the first layer and lead to the formation of soluble nonstochiometric polyelectrolyte complexes as known from the mixing of polyanions and polycations in bulk solutions.[54]

On the way to polyelectrolyte multilayers, we first tried to replace only one of the boladiions by a polyion of the same charge;[55] the principle of this type of multilayer buildup is depicted schematically in Figure 8. As in the case of multilayers composed only of boladiions, the deposition of each layer was monitored by UV spectrophotometry and the absorbance of the phenyl chromophores of the polycation PVBDEMA at 225 nm and the biphenyl chromophores of the boladianion BDA at 262 nm were plotted vs. the number of layers of the corresponding material

(Figure 9) and a linear correlation was obtained. A mixed bola–polymer multilayer composed of 19 layers of BDA and 20 layers of PVBDEMA had a thickness of 151 nm as judged from x-ray reflectivity which is in agreement with the molecular dimensions. Interestingly, the mixed bola–polymer films always showed very homogeneous optical interference colors when viewed by the naked eye or under a microscope and patches as in the pure bola films have never been observed. We believe that this is due to the superior defect spanning capabilities of the polyelectrolytes, which made it immediately very interesting to attempt the fabrication of pure polyanion–polycation multilayers.[56]

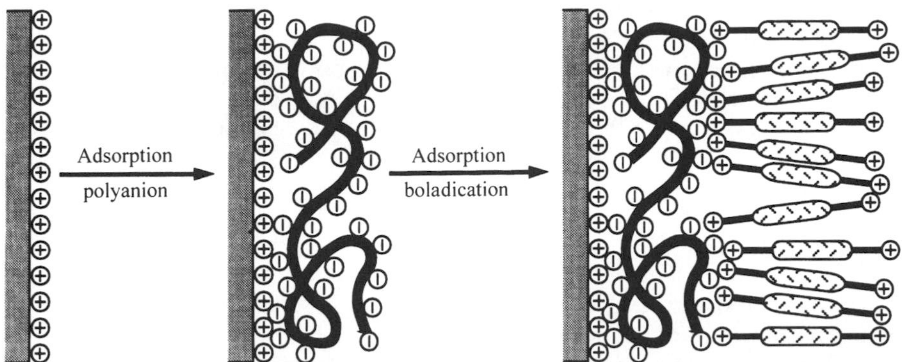

Figure 8 Layer-by-layer adsorption using one polyfunctional and one homobifunctional molecule. The drawing shows an example of a positively charged surface being subsequently coated by a layer of a polyanion and a layer of a boladication. Multilayers are simply prepared by cyclic repetition of both steps. It is not implied that the symbols used for the polyelectrolytes represent their actual structure in solution or after the adsorption. For reasons of clarity, the counterions have been omitted in this figure.

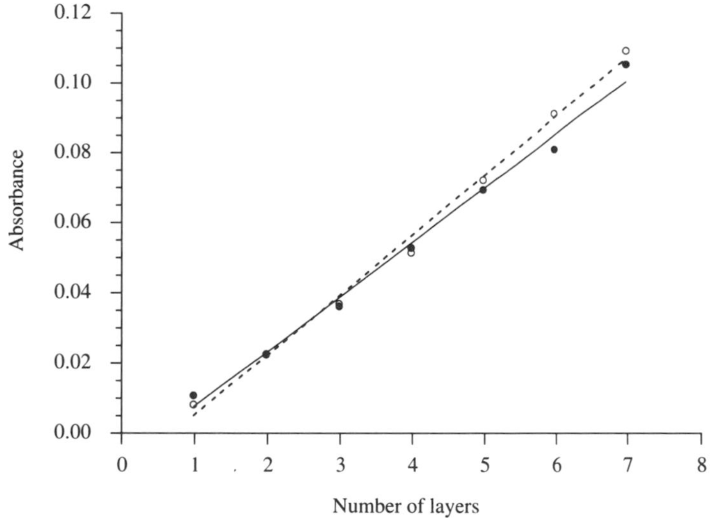

Figure 9 Buildup of a multilayer film of poly[4-vinylbenzyl-(*N*,*N*-diethyl-,*N*-methyl)ammonium iodide] (PVBDEMA) and of the anionic bipolar amphiphile BDA as monitored by UV absorbance of the phenyl chromophores at 225 nm (closed circles) and of the biphenyl chromophores at 262 nm (open circles).

Other groups have also reported on the preparation and characterization of mixed bola–polymer multilayers and many interesting results have been obtained. In 1993, Tirrell and co-workers[35] reported on multilayers composed of the boladication DIPY08 and the polyanion PAMPSA; the system also shows a linear relationship between optical absorbance and the number of deposited layers. However, since DIPY08 is a polymerizable boladianion, they showed by scanning force microscopy (AFM) that the surface of a polymerized layer of the boladiion is more ordered than the surface of an unpolymerized layer. Additionally, measurements with the surface forces apparatus (SFA) showed that a polymerized layer of DIPY08 is mechanically more stable than an unpolymerized layer and that the rigidity and integrity of the first layer are important for anchoring the whole film to the substrate.

In 1995, Tieke and co-workers[37,38] reported on the fabrication of multilayers composed of one of the three boladianions HDDS, DDDS, or DCDS and the polycation poly(allylamine hydrochloride) (PAH). They found that the boladianion DCDS with a rigid core of diacetylene can be photopolymerized in a mixed bola–polymer multilayer. Since this reaction is lattice controlled, it requires a high degree of order in the layer of the boladianion and it is interesting that polymerization does not take place in multilayers of the shorter chain analogues HDDS and DDDS with PAH. The fact that this polymerization occurs at all makes it clear that the same degree of order previously obtained in the polymerization of diacetylene fatty acid multilayers prepared by the Langmuir–Blodgett technique can also be obtained in the directed assembly of oppositely charged compounds. X-ray measurements revealed that polymerized PAH–DCDS multilayers possess a layer structure with a repeat unit of 2.7 ± 0.1 nm, but AFM studies show that the multilayers are not homogeneous films, but consist of a multitude of small separate domains of up to 300 nm in size.[37]

14.3.3 The Playground: Polyanion–Polycation Multilayers

With the already established results in mind, the layer-by-layer adsorption of oppositely charged polyelectrolytes becomes a very promising approach for the fabrication of layered organic heterostructures. There are especially two facts that would make this approach interesting: (i) the ease with which one can obtain multilayer films and (ii) the adsorption from aqueous solution which should allow the incorporation of charged biopolymers in a straightforward way. However, in the case of weak polyelectrolytes, one has to find polyanions and polycations whose pK values (or, in the case of proteins, isoelectric points) fit together in such a way that both polymers are sufficiently charged in order to adsorb irreversibly in a consecutive fashion. However, let us consider what the word "layer" means in the case of a polymer at an interface before we go into the experimental details.

14.3.3.1 What is a polymer monolayer?

In contrast to monolayers of small molecules, the expression monolayer is not exactly defined in the case of polymers, since they are polydisperse materials which adopt a multitude of Gaussian coils of different conformation in solution and the situation is even more difficult at an interface (let us omit polymers with stable unique conformations such as proteins). Clearly, there are two hypothetical cases that would represent the thinnest and the thickest possible "true" monolayer of an adsorbed polymer. These would be the situations in which all polymer chains have the same length, are completely stretched out, and either adsorbed end-on (maximum layer thickness) or perfectly parallel and flat (minimum layer thickness) in a dense packing, both of which would be analogous to a "true" monolayer as in the case of small molecules.

If only a few polymer chains are attached to a unit area and the average distance between adjacent chain segments or different chains is large in comparison with the diameter of the polymer chain, the resulting structures may be called "submonolayer." Starting from a coverage where the average distance between adjacent chain segments or different chains becomes smaller than the diameter of the polymer chain until the hypothetical situation in which all polymer chains are end-attached to the surface, completely extended, and densely packed, one could speak of a "monolayer." Hence submonolayers would always be thinner than individual chains and show a substantial free surface area, and monolayers are always thicker than individual chains and have a hypothetical maximum length corresponding to the average length of the polymer in its maximum extended conformation. As a result, a multilayer composed of random coil polymers cannot exist as a structure in which individual layers can clearly be identified as blocklike tiers, simply because the surface of a polymer monolayer cannot be atomically smooth and thus two adjacent layers will always interpenetrate to some degree. Also, one can in fact prepare a "multilayer" composed of individual "submonolayers." This is possible, as long as the previously adsorbed layer exposes some of its charged groups at the film–solution interface. Then the next polymer can bind to these charges and bridge over considerable distances. The critical surface charge density must be such that there are sufficient charges on the surface for the adsorbing polymer to bind irreversibly in such a way that it itself creates a sufficient charge density for the next polyion layer to bind. As we shall show later, all of this does not rule out that layer-by-layer adsorbed polyanion–polycation films indeed possess long-range order. It simply means that their long-range order, which is determined by the sequence of the deposited materials, is not based on a periodic correlation of single atoms, but of the average segment distribution of the polymer coils. Therefore, polyelectrolyte multilayers possess structural hierarchy[2] and represent another case (besides, e.g., lyotropic liquid crystalline phases) in which long-range (= molecular) order is present despite local (= atomic) disorder.

14.3.3.2 Preparation, manipulation, and structure of polyelectrolyte films

(i) Fabrication of polyanion–polycation multilayers

In 1992, we prepared multilayer films in which both the anionic and the cationic layers were polyelectrolytes;[56] the principle of the layer-by-layer adsorption of polyanions and polycations is depicted schematically in Figure 10. Again, as mentioned above, the drawing is strongly oversimplified and should only be taken as a representation of the most fundamental aspects of the process and the resulting structures. The chemical structures of the polyions poly(styrene sulfonate) (NaPSS), poly[4-vinylbenzyl-(*N*,*N*-diethyl-,*N*-methyl)ammonium iodide] (PVBDEMA) and poly(allylamine hydrochloride) (PAH) that we have been using in our first experiments are depicted in Figure 11 along with the polyelectrolytes meanwhile used by other groups for the preparation of polyelectrolyte-containing multilayer films.

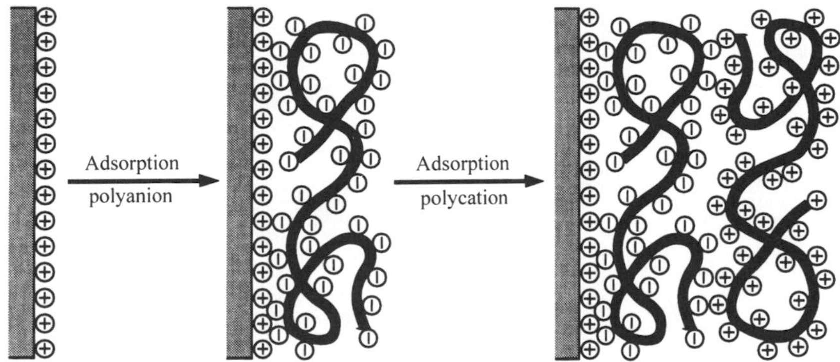

Figure 10 Oversimplified schematic diagram of the buildup of multilayer assemblies by consecutive adsorption of anionic and cationic polyelectrolytes (cyclic repetition of both steps). It is not implied that the symbols used for the polyelectrolytes represent their actual structure in solution or after the adsorption. Additionally, the interpenetration of neighboring polymer layers was neglected. For reasons of clarity, the counterions have also been omitted in this representation.

Following the general scheme of adsorption outlined in Figure 8, the following conditions are common for the preparation of polyelectrolyte multilayers: adsorption solutions contain typically 2 g of polyelectrolyte per liter of ultrapure water or an aqueous solution of an appropriate salt (e.g., NaCl). Adsorption times are generally 20 min, although adsorption times of less than 1 min have also been employed. After every adsorption step the sample is washed three times in pure water.

In this way we have prepared polyelectrolyte films composed of more than 100 layers of poly(styrene sulfonate) (NaPSS) and poly(4-vinylbenzyl-(*N*,*N*-diethyl-*N*-methyl)ammonium iodide) (PVBDEMA) on silicon single-crystal surfaces.[56] Samples such as this show bright, homogeneous interference colors when viewed with the naked eye. Both polyelectrolytes contain phenyl chromophores and we have monitored the film buildup on quartz substrates by UV spectrophotometry at 225 nm up to a total of 38 layers and verified that the absorbance is linear with the number of layers (Figure 12). Similar results were obtained for the system poly(styrene sulfonate) (NaPSS) and poly(allylamine hydrochloride) (PAH). In the latter case we additionally registered x-ray reflectivity curves in order to determine the increase in film thickness with increasing number of adsorbed polymer layers.

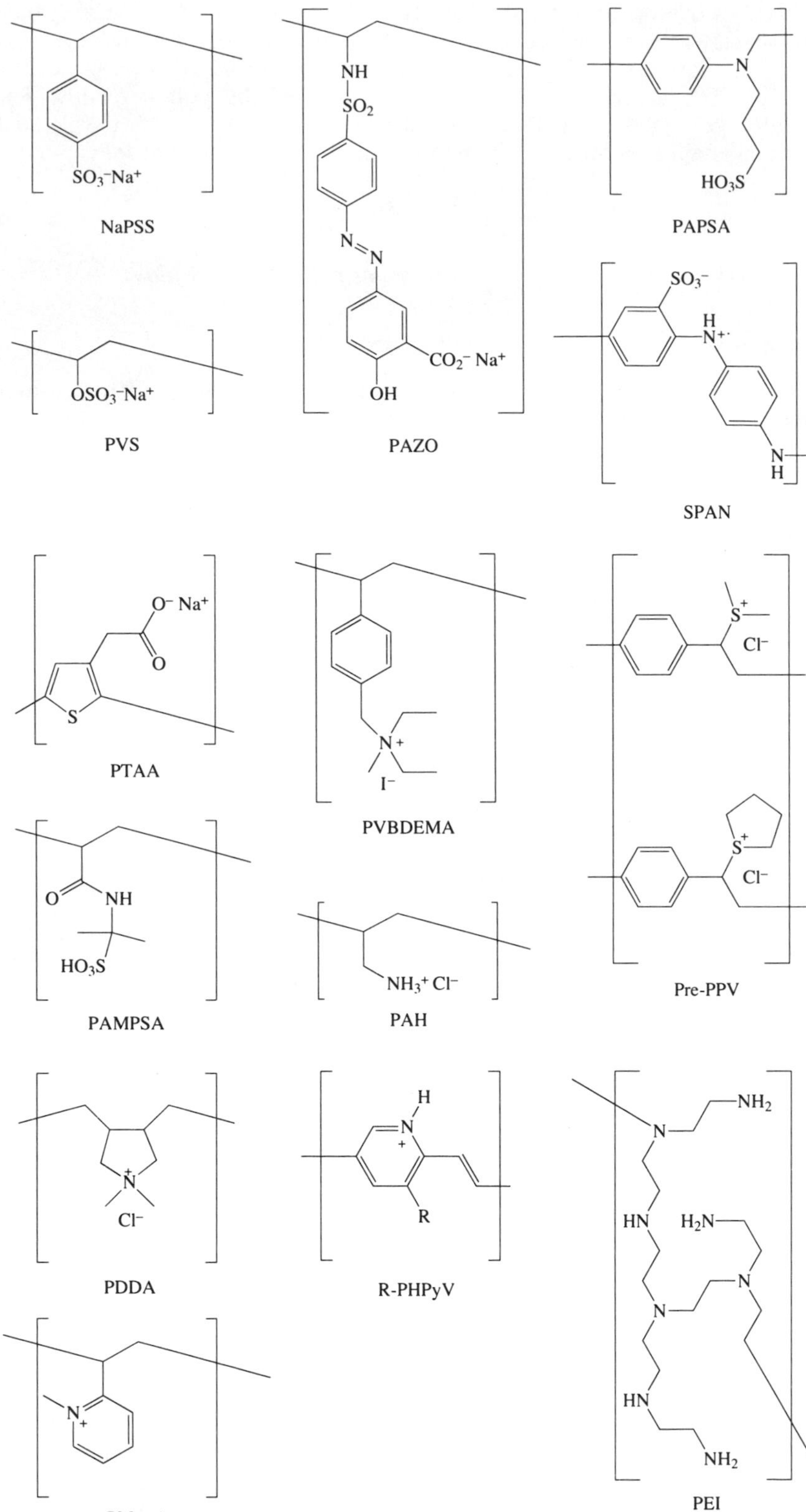

Figure 11 Structures of synthetic polyelectrolytes used for the fabrication of multilayers. There are considerably more polyelectrolytes whose adsorption on to solid surfaces has been studied previously. Counterions have been omitted in some cases; most of the abbreviations of the substances follow the original literature. The compounds are referenced as follows: NaPSS;[56–65] PVS;[66] PAZO;[59] PAPSA;[64] SPAN;[67,68] PTAA:[67] PAMPSA;[35] PVBDEMA;[55,56] PAH;[37,38,41,56–8,63,64,66,67] Pre-PPV;[61,67–9] PDDA;[64,70] PMpyA;[67] R-PHpyV;[71] and PEI.[42,64]

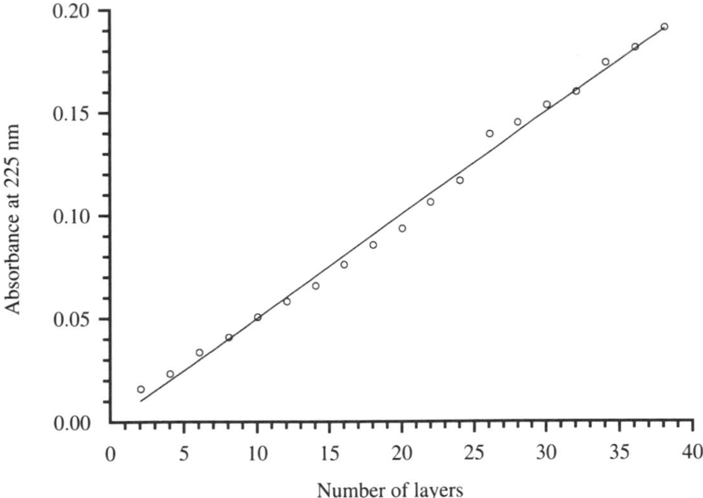

Figure 12 Buildup of a multilayer film of poly(styrene sulfonate) (NaPSS) and poly[4-vinylbenzyl-(*N*,*N*-diethyl-,*N*-methyl)ammonium iodide] (PVBDEMA) as monitored by the UV absorbance of phenyl chromophores at 225 nm.

Figure 13 shows so-called Kiessig fringes[72] obtained by x-ray reflectivity after the deposition of different numbers of layers. In this case the NaPSS was deposited from a solution containing 1.0 mol L^{-1} NaCl and the PAH was deposited from pure water. The distance between the minima in the curves in Figure 13(a) allows calculation of the respective film thicknesses, which are plotted vs. the layer numbers in Figure 13(b); more precisely, the curves in Figure 10(a) correspond to the closed triangles in Figure 13(b) and the slope of the linear fit yields the average thickness of a polyanion–polycation layer pair.[57]

(ii) Linear vs. nonlinear growth

Although all three data sets in Figure 13(b) give straight lines with good correlation coefficients for the average thickness increment of a layer pair, thus demonstrating again a reproducible refunctionalization of the surface in each adsorption step, a linear relationship with the number of layers is not always observed. Imagine the case when one starts with a surface of very low charge density, which will lead to the adsorption of very few polymer chains. However, also in this case the chains will adsorb, forming "loops," "trains," and "tails," hence it is likely that, although still at submonolayer coverage, the surface charge will increase. This will continue from layer to layer until the surface charge density becomes constant and a linear correlation of the thickness with the layer number is reached. Such behavior,[61] permanently growing or shrinking of the average thickness increment per layer pair and similar surface effects (e.g., decreasing[66] or increasing surface roughness with increasing number of layers), is indeed observed, but one can normally find conditions that lead quickly to a regular (= linear) deposition.

(iii) Salt effects: fine-tuning of the layer pair thickness

Figure 13(b) also shows that the average thickness increment per polyanion–polycation layer pair is easily fine tuned with high precision, which leads to remarkable control of the film architecture, as shown in Figure 14(a). On going from 1.0 mol L^{-1} to 2.0 mol L^{-1} salt in the solution of NaPSS, one raises the average thickness per layer pair by less than 0.5 nm from 1.77 nm to 2.26 nm; the absolute precision that can be reached by adjusting the salt concentration is less than 0.1 nm. This astonishing control of individual average layer thicknesses is simply due to the well-known changes in the coil structure of polyions in solutions of different ionic strength. At low concentrations of salt, the charged groups along the polyelectrolyte chain repel each other, the polymer adopts an extended conformation, and the chain adsorbs rather "flat." At higher concentrations of salt the repulsion of charged groups is partly screened, the polyelectrolyte

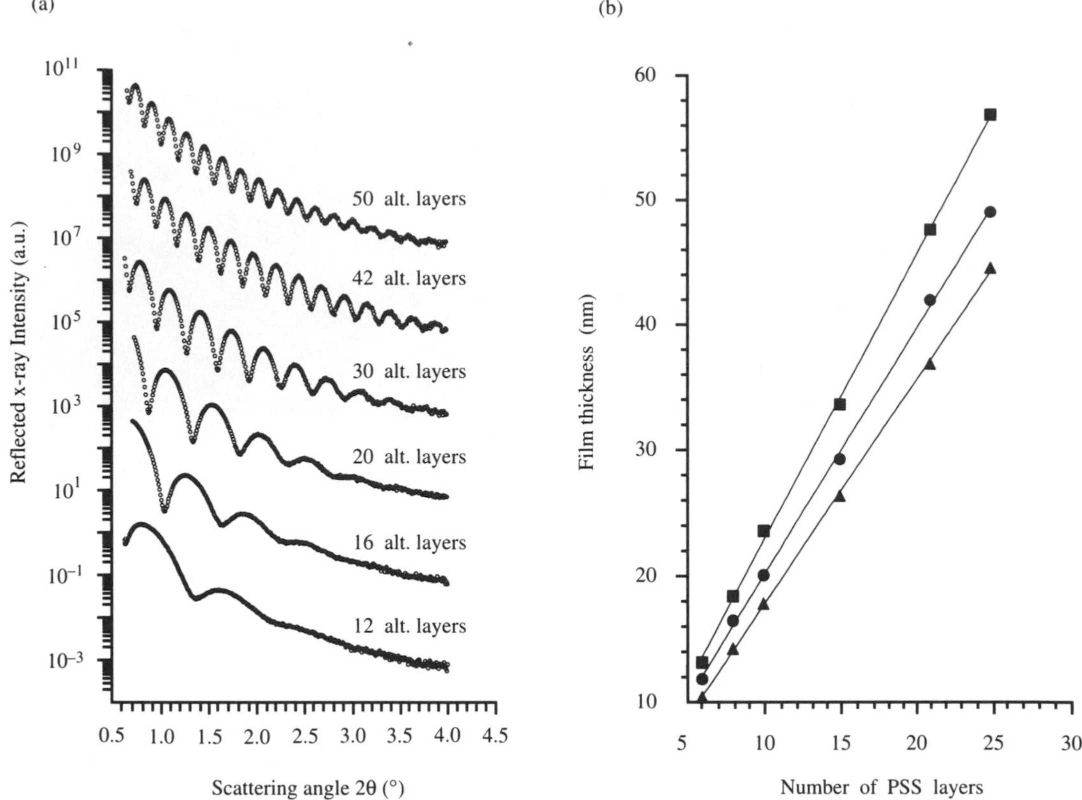

Figure 13 (a) Multilayer buildup as monitored by small-angle x-ray scattering. The multilayer is composed of poly(allylamine hydrochloride) (PAH) and sodium poly(styrene sulfonate) (NaPSS), the latter adsorbed from a solution containing 1 mol L^{-1} NaCl. Note that the curves are linearly shifted in order to permit a better comparison of the different traces. (b) Dependence of the total film thickness on the number of layers for three different concentrations of NaCl in the NaPSS solution. The triangles represent the values of the 1.0 mol L^{-1} (1.77 nm per layer pair) (thicknesses were calculated from the x-ray data shown in (a), the circles the values of the 1.5 mol L^{-1} (1.94 nm per layer pair) and the squares the values of the 2.0 mol L^{-1} NaCl solution (2.26 nm per layer pair). Errors are too small to be displayed. Note that the PAH was always adsorbed from salt-free solutions.

conformation approaches a random coil, and the chain adsorbs rather "loopy." This behavior in solution leads to the fact that adsorbed polyelectrolyte layers are thin (~0.5 nm) if the adsorption takes place in pure water and thick (up to ~3.5 nm) if adsorbed from solutions of very high ionic strength.

Figure 14(a) shows that the fine tuning of the thicknesses of individual layers can indeed be used for the construction of complex film architectures, even in a multilayer composed only of two materials, a single polyanion, and a single polycation. In this case, the polyelectrolytes NaPSS and PAH were both either adsorbed from pure water or a high concentration of salt and the resulting film thicknesses per layer were 0.81 nm and 2.18 nm, respectively. That different thicknesses per layer can be maintained in films that have been in contact with solutions containing different amounts of salt is not trivial, as seen in Figure 14(b). In this case a film composed of 12 alternating layers of PAH and NaPSS, both deposited from solutions containing 0.5 mol L^{-1} NaCl, was alternately immersed for 2 h in either pure water or a solution of 0.1 mol L^{-1} NaCl. Although there is some error in the data, one generally observes that the thickness of a polyelectrolyte film increases by up to 15% when immersed in a salt solution of high ionic strength and shrinks again to about its initial value when reimmersed in pure water. The reason that two (or more) different layer thicknesses of the same material can be maintained in a single film by adjusting the ionic strength of the adsorption solution (Figure 14(a)) is simply that the effect of swelling of a polyelectrolyte film in solutions of different ionic strength is only of the order of 10–15% (Figure 14(b)), whereas the thickness increment itself can be varied over a much wider range (from about 0.5 nm per layer up to about 3.5 nm per layer).

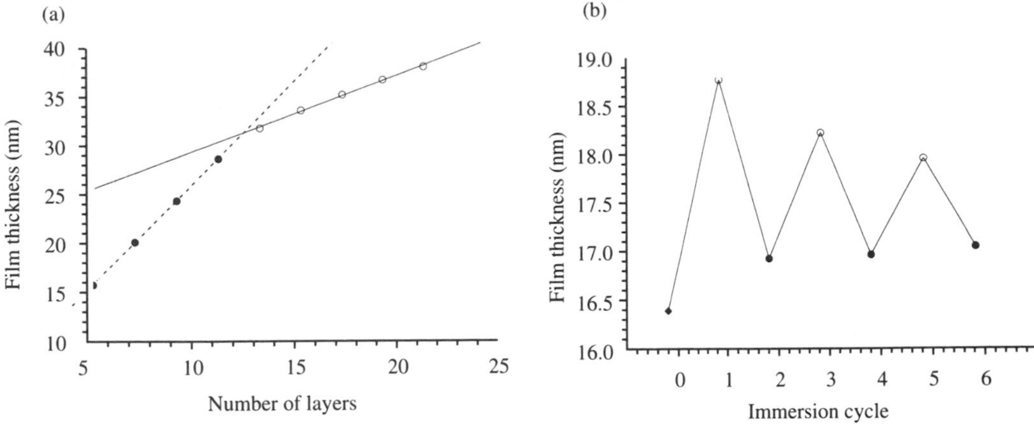

Figure 14 (a) Buildup of a two-component multilayer film that was prepared in two steps, namely adsorption with and without added salt, as monitored by x-ray scattering. The multilayer was composed of poly(allylamine hydrochloride) (PAH) and sodium poly(styrene sulfonate) (NaPSS), the latter adsorbed from a solution either containing salt (solid circles) or with no salt (open circles). Slopes: solid circles, 2.18 nm per layer; open circles, 0.81 nm per layer. Note that it is possible to obtain complex film architectures using only two compounds. (b) Postpreparation treatment of a polyelectrolyte multilayer film by 2 h immersion in pure water (solid circles) and a solution containing 0.1 mol L^{-1} NaCl (open circles). The film was composed of 12 alternating layers of poly(allylamine hydrochloride) and sodium poly(styrene sulfonate) $(PAH/NaPSS)_6$, both deposited from an aqueous solution containing 0.5 mol L^{-1} NaCl. Note that the film thicknesses after immersion in the respective solutions are identical with respect to experimental error and it is currently not clear if the effect decreases in dependence of the immersion cycle. The initial film thickness (shown by the diamond) is small because generally all films are washed in pure water after the deposition and prior to the first determination of film thickness.

(iv) Temperature effects: reversible changes of the film thickness

Although polyelectrolyte films have now been stored for several years, they are sensitive to certain environmental effects. Interestingly, films composed of PAH and NaPSS can be heated to 200 °C for more than 1 week without showing noticeable effects of deterioration, but they lose some water at elevated temperatures, which causes them to shrink.[60] The effect of temperature on film thickness and surface roughness can also be followed directly, as shown in Figure 15. Whereas the thickness of the film shrinks by approximately 15% upon heating from ambient temperature to 110 °C, the roughness of the film–air interface, as calculated from numerical fits to x-ray reflectivity data, stays constant.

This change in film thickness is due to the loss of water[60] and is reversible, at least in the case of PAH and NaPSS and at moderate temperatures. Figure 16(a) shows that polyelectrolyte films that have been heated at 110 °C for over 1 h return to their initial thickness (within experimental error) on equilibration at ambient conditions. It shows both experimental data and the corresponding numerical fits from which one obtains detailed information on the electron density profile of the films, equivalents of which are shown in Figure 16(b). Both profiles are identical, given the experimental error of the two data sets in Figure 16(a).

(v) Internal film structure: tailoring the distance between molecular sheets on the nanometer scale

Up to now we have only discussed structural details such as optical absorbance, film thickness and surface roughness of polyelectrolyte multilayers, and not their internal layer structure as discussed in Section 14.3.3.1. The problem is that x-ray reflectivity is sensitive to the electron density profile of the films such as shown in Figure 16(b). Unfortunately, the electron density differences of different polymers are only small and, as discussed above, adjacent layers penetrate into each other, thus making the small electron density gradient between the layers even more diffuse and thus difficult to observe. As a consequence, most x-ray reflectivity curves, such as those in Figures 13(a) and 16(a) or the trace at the top of Figure 17, show only the interferences of the substrate–film and film–air interfaces (Kiessig fringes) and no signals for the internal layer structure (Bragg peaks). Nevertheless, there are cases in which the internal structure of polyanion–polycation

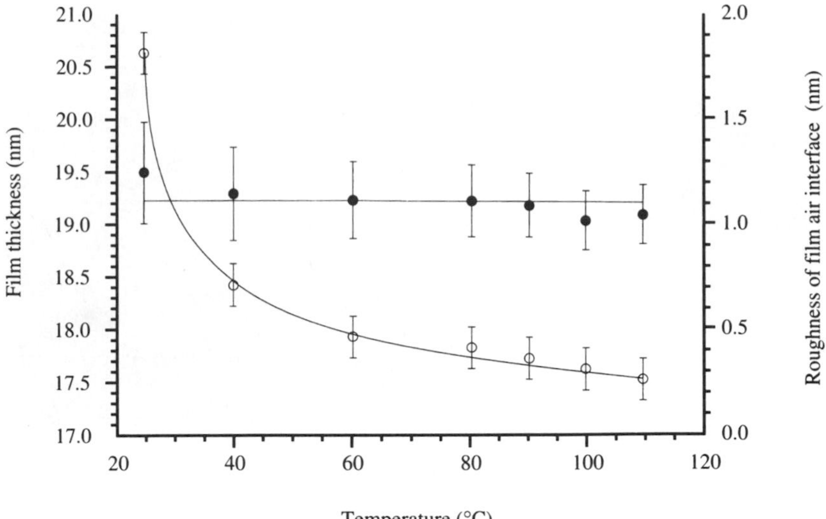

Figure 15 Thermal behavior of a typical polyelectrolyte multilayer film as monitored by x-ray reflectivity. Whereas heating clearly leads to a reduction in film thickness (left ordinate open circles), the roughness of the film–air interface remains constant (right ordinate solid circles). Typically, films prepared from salt solutions shrink by about 10% when heated from ambient temperature to 100 °C; films prepared from pure water shrink less.

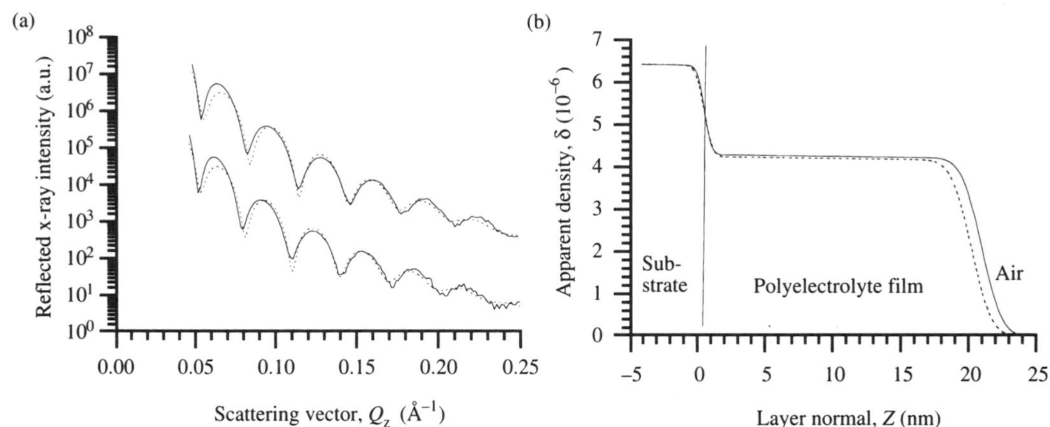

Figure 16 (a) The thermally induced changes in film thickness are reversible. On standing in the laboratory atmosphere for some time, the original film thickness is restored. The plot shows x-ray measurements (solid lines) and numerical fits to the data (dotted lines). Before the heating experiment $d = 20.6 \pm 0.5$ nm; 3 weeks after the heating experiment 19.8 ± 0.5 nm. (b) Apparent film density profiles as obtained from the numerical fits in (a). The profiles before heating (solid line) and after heating (dotted line) are identical with respect to experimental error.

multilayers has been observed by x-ray reflectivity.[59] That an internal layer structure may be present, even if x-ray reflectivity cannot resolve it, is demonstrated in Figure 17.

Both reflectivity experiments depicted in Figure 17 were performed with the same polyelectrolyte multilayer sample which was composed of 48 alternating layers of sodium poly(styrene sulfonate) (NaPSS) and poly(allylamine hydrochloride) (PAH), both deposited from solutions containing 2 mol L^{-1} NaCl. The film architecture was such that every sixth polyelectrolyte layer (= every third layer of NaPSS) was perdeuterated.[58] Due to the fact that x-ray reflectivity is sensitive to the electron density, which is identical for hydrogen and deuterium, and that neutron reflectivity is especially sensitive to the deuterium concentration, the apparent film architectures for the two techniques are different. In x-ray reflectivity the apparent layer sequence is $(NaPSS/PAH)_{24}$, yielding 120.5 ± 2.0 nm for the film thickness as derived from the Kiessig fringes, whereas in neutron reflectivity the apparent layer sequence is $((NaPSS/PAH)_2NaPSS\text{-}d_7/PAH)_8$, yielding

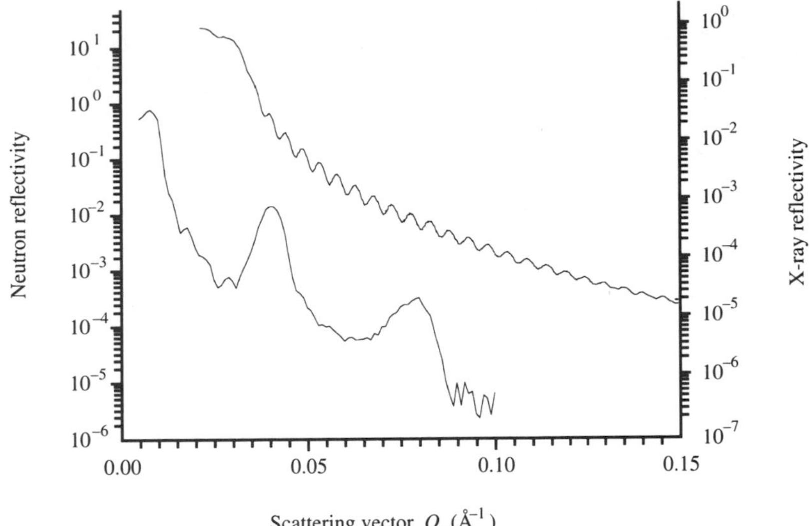

Figure 17 Structural investigation of a polyelectrolyte multilayer composed of 48 alternating layers of sodium poly(styrene sulfonate) (NaPSS) (A) and poly(allylamine hydrochloride) (PAH) (B), both deposited from solutions containing 2 mol L^{-1} NaCl. The film architecture is such that every sixth polyelectrolyte layer (= every third layer of NaPSS) was perdeuterated (C). Owing to their sensitivities to electron density, which is identical for hydrogen and deuterium (x-ray reflectivity) and deuterium concentration (neutron reflectivity), the apparent film architectures for the two techniques are different, $(A/B)_{24}$ or $(NaPSS/PAH)_{24}$, in x-ray reflectivity (top trace) yielding 120.5 ± 2.0 nm for the film thickness as derived from the Kiessig fringes and $[(A/B)_2C/B]_8$ or $[(NaPSS/PAH)_2NaPSS\text{-}d_7/PAH]_8$ in neutron reflectivity (bottom trace) yielding 15.9 ± 0.2 nm for the layer distance of the deuterated layers as derived from the Bragg peaks.

15.9 ± 0.2 nm for the layer distance of the deuterated layers as derived from the Bragg peaks. Neutron reflectivity measurements of various superlattice architectures have shown up to four Bragg peaks, indicating a well-ordered layer structure.[73] Detailed analysis shows that layer thickness and interpenetration depend to a large extent on added salt and very little on the molecular weight of the polyelectrolytes. However, we have never observed Bragg peaks in multilayers in which every second layer was deuterated. Interestingly, a film with almost the same length of the repeat unit, but in which every fourth layer is deuterated, shows a well-resolved Bragg peak.[74] This demonstrates again how much the structural details of polyelectrolyte multilayers can be influenced.

14.3.3.3 Functionality in multilayers: the incorporation of proteins, charged colloidal particles, and functional synthetic polymers

The previous sections have described how multilayers of di-, oligo-, and polyionic materials are fabricated by consecutively adsorbing anionic and cationic compounds and how parameters such as film thickness, structure, and roughness are influenced. It is easy to envisage that many other materials are easily incorporated into polyanion–polycation films. The precision with which one can adjust the average distance between two layers of functional components makes the films interesting even for the applied sciences. A tremendous advantage over other molecular deposition techniques is that no dedicated and sensitive equipment (e.g., ultrahigh vacuum apparatus or Langmuir troughs) is needed and that the adsorption is carried out from aqueous solutions, which also makes the technique environmentally attractive. Therefore, a number of interesting multilayer films containing a variety of functional synthetic and biological materials have been realized.

We have started with multilayers containing DNA[75] and later obtained films containing biological colloids such as charged virus particles.[76] The use of partially biotinylated poly(L-lysine) allows one to incorporate the protein streptavidin into multilayer films,[60,61,77] even with some lateral structure using simple photopatterning, which might be interesting for biosensing applications. Other groups have incorporated charged globular proteins such as myoglobin, hemoglobin, lysozyme, cytochrome c, glucose isomerase, glucoamylase, or glucose oxidase.[31,32,41,62,64] Even the deposition of spherical inorganic colloids[34] or inorganic platelets exfoliated from clay minerals is possible.[41,70] It is thought that the stiff platelets might reduce the amount of interpenetration of adjacent layers. A similar approach to the reduction of layer interpenetration is the incorporation of Langmuir–Blodgett layers[78] or of polymers bearing mesogenic groups.[79] Another interesting functional material for incorporation in thin transparent films is the precursor polyelectrolyte (Pre-PPV)[80] of the electroluminescent material poly(*p*-phenylenevinylene) (PPV).[81] In the case of polyelectrolyte multilayers the precursor is alternated with a polymeric counterion to fabricate the multilayer and then thermally converted into PPV, whereby one obtains transparent films with an intense yellow color whose thickness can be adjusted from about 2 nm to several tens of nanometers or more depending on the number of layers deposited.[61,67,68,71,82] Very interesting electroluminescent properties were reported for such PPV-containing multilayers. The incorporation of conducting polyelectrolytes[67,83] has yielded conductivities up to $40\,S\,cm^{-1}$ in a multilayer and $>10\,S\,cm^{-1}$ in a four-layer film. It is expected that the field will expand further in the future.

14.4 ADDENDUM

Due to the rapidly growing interest in molecular multilayers of oppositely charged materials, a considerable number of publications has appeared since submission of this chapter. In order to keep changes at this stage to an absolute minimum, recent developments are added at the end of this chapter and only introduced by a few words. The addendum is mainly intended to provide the interested reader with references to the original literature that have appeared in the meantime.

The fabrication of conducting multilayer films is now described in detail and conductivities over $300\,S\,cm^{-1}$ are reported.[84,85] Light-emitting diodes composed of naphthalene containing conjugated polymers have been obtained and characterized and even tunable emission was observed for such multilayer devices.[86] Progress towards light-emitting and photorectifying diodes was also made using poly(phenylene vinylene) in multilayer assemblies.[87] Polyelectrolyte multilayers were selectively grown on defined areas of patterned surfaces.[88] The incorporation of polypeptide dyes[89] and semiconductor nanoparticles[90] in polyelectrolyte multilayers was reported. Our group has reported on the fabrication of gold colloid/polyelectrolyte multilayers and analyzed the structure and properties of such systems.[91] Multilayer heterostructures of, for example, molybdenum oxide multinuclear complexes and polyelectrolytes were reported[92] and novel concanavalin A/polymer multilayer architectures were also fabricated by the same group.[93] Glucose oxidase and glucamylase were assembled in a single multilayer structure with the purpose of maltose sensing.[94] Most interestingly, it is even possible to obtain layered assemblies of charged amphiphiles and polyelectrolytes by consecutive adsorption from solution.[95] Finally, photoinduced charge separation was observed in multilayer thin films grown by sequential polyelectrolyte adsorption.[96]

14.5 REFERENCES

1. G. M. Whitesides, J. P. Mathias, and C. T. Seto, *Science*, 1991, **254**, 1312.
2. R. Lakes, *Nature*, 1993, **361**, 511.
3. G. Decher, J.-D. Hong, K. Lowack, Y. Lvov, and J. Schmitt, in "Self-Production of Supramolecular Structures — From Synthetic Structures to Models of Minimal Living Systems," eds. G. R. Fleischaker, S. Colonna, and P. L. Luisi, Nato ASI Series, Series C: Mathematical and Physical Sciences, Kluwer, Dordrecht, 1994, p. 267.
4. J.-M. Lehn, *Science*, 1985, **227**, 849.
5. R. Maoz, L. Netzer, J. Gun, and J. Sagiv, *J. Chim. Phys.*, 1988, **85**, 1059.
6. J.-M. Lehn, *Angew. Chem.*, 1990, **102**, 1347.
7. F. Vögtle, "Supramolecular Chemistry—An Introduction," Wiley, Chichester, 1991.
8. G. Decher and H. Ringsdorf, *Liq. Cryst.*, 1993, **13**, 57.
9. J. M. Lehn, "Supramolecular Chemistry—Concepts and Perspectives," VCH, Weinheim, 1995.
10. J. H. Fendler, "Advances in Polymer Science," Springer, Berlin, 1994, vol. 113.
11. J. A. Stroscio and D. M. Eigler, *Science*, 1991, **254**, 1319.
12. J. T. Finer, R. M. Simmons, and J. A. Spudich, *Nature*, 1994, **368**, 113.
13. K. B. Blodgett, *J. Am. Chem. Soc.*, 1934, **56**, 495.
14. K. B. Blodgett and I. Langmuir, *Phys. Rev.*, 1937, **51**, 964.
15. G. L. Gaines, "Insoluble Monolayers at Liquid–Gas Interfaces," Interscience, New York, 1966.
16. H. Kuhn and D. Möbius, *Angew. Chem.*, 1971, **83**, 672.
17. H. Kuhn, D. Möbius, and H. Bücher, in "Physical Methods of Chemistry, Part 3B," eds. A. Weissberger and P. Rossiter, Wiley, New York, 1972, p. 577.
18. R. B. Merrifield, *Science*, 1965, **150**, 178.
19. L. Netzer, R. Iscovici, and J. Sagiv, *Thin Solid Films*, 1983, **99**, 235.
20. H. Lee, L. J. Kepley, H.-G. Hong, S. Akhter, and T. E. Mallouk, *J. Phys. Chem.*, 1988, **92**, 2597.
21. G. Cao, H.-G. Hong, and T. E. Mallouk, *Acc. Chem. Res.*, 1992, **25**, 420.
22. C. M. Bell, S. W. Keller, V. M. Lynch, and T. E. Mallouk, *Mater. Chem. Phys.*, 1993, **35**, 225.
23. N. Tillman, A. Ulman, and T. L. Penner, *Langmuir*, 1989, **5**, 101.

24. H. E. Katz, G. Scheller, T. M. Putvinski, M. L. Schilling, W. L. Wilson, and C. E. D. Chidsey, *Science*, 1991, **254**, 1485.
25. A. K. Kakkar, S. Yitzchaik, S. B. Roscoe, F. Kubota, D. S. Allan, T. J. Marks, W. Lin, and G. K. Wong, *Langmuir*, 1993, **9**, 388.
26. A. Ulman, "An Introduction to Ultrathin Organic Films: From Langmuir–Blodgett to Self-Assembly," Academic Press, Boston, MA, 1991.
27. R. K. Iler, *J. Colloid Interface Sci.*, 1966, **21**, 569.
28. J. D. Swalen, D. L. Allara, J. D. Andrade, E. A. Chandross, S. Garoff, J. Israelachvili, T. J. McCarthy, R. Murray, R. F. Pease, J. F. Rabolt, K. J. Wynne, and H. Yu, *Langmuir*, 1987, **3**, 932.
29. Apparatus developed by Riegler & Kirstein GmbH Wiesbaden, Germany.
30. G. Decher and J.-D. Hong *Makromol. Chem., Macromol. Symp.*, 1991, **46**, 321.
31. W. Kong, X. Zhang, M. L. Gao, H. Zhou, W. Li, and J. C. Shen, *Macromol. Rapid Commun.*, 1994, **15**, 405.
32. W. Kong, L. P. Wang, M. L. Gao, H. Zhou, X. Zhang, W. Li, and J. C. Shen, *J. Chem. Soc., Chem. Commun.*, 1994, 1297.
33. M. Gao, X. Zhang, B. Yang, and J. Shen, *J. Chem. Soc., Chem. Commun.*, 1994, 2229.
34. M. Gao, X. Zhang, Y. Yang, B. Yang, and J. Shen, *J. Chem. Soc., Chem. Commun.*, 1994, 2777.
35. G. Mao, Y. Tsao, M. Tirrell, H. T. Davis, V. Hessel, and H. Ringsdorf, *Langmuir*, 1993, **9**, 3461.
36. B. Sellergren, A. Swietlow, T. Arnebrandt, and K. Unger, *Anal. Chem.*, 1996, **68**, 402.
37. F. Saremi, E. Maassen, B. Tieke, G. Jordan, and W. Rammensee, *Langmuir*, 1995, **11**, 1068.
38. F. Saremi and B. Tieke, *Adv. Mater.*, 1995, **7**, 378.
39. B. Lühmann and H. Finkelmann, *Colloid Polym. Sci.*, 1986, **264**, 189.
40. I. Haller, *J. Am. Chem. Soc.*, 1978, **100**, 8050.
41. S. W. Keller, H.-N. Kim, and T. E. Mallouk, *J. Am. Chem. Soc.*, 1994, **11**, 8817.
42. G. Decher, *Nachr. Chem. Tech. Lab.*, 1993, **41**, 793.
43. J. Schmitt, K. Kjær, M. Lösche, and G. Decher, manuscript in preparation.
44. D. A. Tomalia, A. M. Naylor, and W. A. Goddard, III, *Angew. Chem., Int. Ed. Engl.*, 1990, **29**, 138.
45. X. Zhang, M. Gao, X. Kong, Y. Sun, and J. Shen, *J. Chem. Soc., Chem. Commun.*, 1994, 1055.
46. G. J. Fleer, M. A. Cohen-Stuart, J. M. H. M. Scheutjens, T. Cosgrove, and B. Vincent, "Polymers at Interfaces," Chapman and Hall, London, 1993.
47. B. W. Greene, *J. Colloid Interface Sci.*, 1971, **37**, 144.
48. M. C. Cafe and I. D. Robb, *J. Colloid Interface Sci.*, 1982, **86**, 411.
49. H. A. van der Schee and J. Lyklema, *J. Phys. Chem.*, 1984, **88**, 6661.
50. T. Cosgrove, T. M. Obey, and B. Vincent, *J. Colloid Interface Sci.*, 1986, **111**, 409.
51. J. Blaakmeer, M. R. Böhmer, M. A. C. Stuart, and G. J. Fleer, *Macromolecules*, 1990, **23**, 2301.
52. H. G. M. van de Steeg, M. A. C. Stuart, A. d. Keizer, and B. H. Bijsterbosch, *Langmuir*, 1992, **8**, 2538.
53. P. Berndt, K. Kurihara, and T. Kunitake, *Langmuir*, 1992, **8**, 2486.
54. B. Philipp, H. Dautzenberg, K.-J. Linow, J. Kötz, and W. Dawydoff, *Prog. Polym. Sci.*, 1989, **14**, 91.
55. G. Decher and J.-D. Hong, *Ber. Bunsenges. Phys. Chem.*, 1991, **95**, 1430.
56. G. Decher, J.-D. Hong, and J. Schmitt, *Thin Solid Films*, 1992, **210/211**, 831.
57. G. Decher and J. Schmitt, *Prog. Colloid Polym. Sci.*, 1992, **89**, 160.
58. J. Schmitt, T. Grünewald, K. Kjær, P. Pershan, G. Decher, and M. Lösche, *Macromolecules*, 1993, **26**, 7058.
59. G. Decher, Y. Lvov, and J. Schmitt, *Thin Solid Films*, 1994, **244**, 772.
60. J.-D. Hong, K. Lowack, J. Schmitt, and G. Decher, *Prog. Colloid Polym. Sci.*, 1993, **93**, 98.
61. G. Decher, B. Lehr, K. Lowack, Y. Lvov, and J. Schmitt, *Biosensors Bioelectron.*, 1994, **9**, 677.
62. Y. Lvov, K. Ariga, and T. Kunitake, *Chem. Lett.*, 1994, 2323.
63. K. Lowack and C. A. Helm, *Macromolecules*, 1995, **28**, 2912.
64. Y. Lvov, K. Ariga, and T. Kunitake, *J. Am. Chem. Soc.*, 1995, **117**, 6117.
65. R. v. Klitzing and H. Möhwald, *Langmuir*, 1995, **11**, 3554.
66. Y. Lvov, G. Decher, and H. Möhwald, *Langmuir*, 1993, **9**, 481.
67. M. F. Rubner, *et al.*, data presented at the 205th American Chemical Society Meeting, Denver, CO, 1993, and M. Ferreira, J. H. Cheung, and M. F. Rubner, *Thin Solid Films*, 1994, **244**, 806.
68. M. Onoda and K. Yoshino, *Jpn. J. Appl. Phys.*, 1995, **34**, L260.
69. B. Lehr, Diploma Thesis, Mainz, 1993; G. Decher, B. Lehr, M. Seufert, and G. Wenz, *Supramol. Sci.*, accepted for publication.
70. E. R. Kleinfeld and G. S. Ferguson, *Science*, 1994, **265**, 370.
71. J. Tian, C.-C. Wu, M. E. Thompson, J. C. Sturm, R. A. Register, M. J. Marsella, and T. M. Swager, *Adv. Mater.*, 1995, **7**, 395.
72. H. Kiessig, *Ann. Phys.*, 1931, **10**, 769.
73. J. Schmitt, K. Kjær, M. Lösche, and G. Decher, manuscript in preparation.
74. D. Korneev, Y. Lvov, G. Decher, J. Schmitt, and S. Yaradaikin, *Physica B*, 1995, 1995, **213/214**, 954.
75. Y. Lvov, G. Decher, and G. Sukhorukov, *Macromolecules*, 1993, **26**, 5396.
76. Y. Lvov, H. Haas, G. Decher, H. Möhwald, A. Michailov, B. Mtchedlishvily, E. Morgunova, and B. Vainshtain, *Langmuir*, 1994, **10**, 4232.
77. G. Decher, F. Essler, J.-D. Hong, K. Lowack, J. Schmitt, and Y. Lvov, *Polym. Prepr.*, 1993, **34**, 745.
78. Y. Lvov, F. Essler, and G. Decher, *J. Phys. Chem.*, 1993, **97**, 13773.
79. M. Gao, X. Kong, X. Zhang, and J. Shen, *Thin Solid Films*, 1994, **244**, 815.
80. H.-H. Hörhold, M. Helbig, D. Raabe, J. Opfermann, U. Scherf, R. Stockmann, and D. Weiss, *Z. Chem.*, 1987, **27**, 126.
81. J. H. Burroughes, D. D. C. Bradley, A. R. Brown, R. N. Marks, K. McMackay, R. H. Friend, P. L. Burns, and A. B. Holmes, *Nature*, 1990, **347**, 539.
82. J. Tian, M. E. Thompson, C.-C. Wu, J. C. Sturm, R. A. Register, M. J. Marsella, and T. M. Swager, *Polym. Prepr.*, 1994, **35**, 761.
83. J. H. Cheung, A. C. Fou, and M. F. Rubner, *Thin Solid Films*, 1994, **244**, 985.

84. M. Ferreira and M. F. Rubner, *Macromolecules*, 1995, **28**, 7107.
85. A. C. Fou and M. F. Rubner, *Macromolecules*, 1995, **28**, 7115.
86. H. Hong, D. Davidov, Y. Avny, H. Chayet, E. Z. Faraggi, and R. Neumann, *Adv. Mater.*, 1995, **7**, 846.
87. A. C. Fou, O. Onitsuka, M. Ferreira, M. F. Rubner, and B. R. Hsieh, *Mater. Res. Soc. Symp. Proc.*, 1995, **369**, 575.
88. P. T. Hammond and G. M. Whitesides, *Macromolecules*, 1995, **28**, 7569.
89. T. M. Cooper, A. L. Campbell, and R. L. Crane, *Langmuir*, 1995, **11**, 2713.
90. N. A. Kotov, I. Dékány, and J. H. Fendler, *J. Phys. Chem.*, 1995, **99**, 13065.
91. J. Schmitt, G. Decher, W. J. Dressik, S. L. Brandow, R. E. Geer, R. Shashidhar, and J. M. Calvert, *Science*, 1996, submitted for publication.
92. I. Ichinose, S. Ohno, T. Kunitake, Y. Lvov, and K. Ariga, *Polym. Prepr. Jpn.*, 1995, **44**, 2544.
93. Y. Lvov, K. Ariga, I. Ichinose, and T. Kunitake, *J. Chem. Soc., Chem. Commun.*, 1995, 2313.
94. Y. Sun, X. Zhang, C. Sun, B. Wang, and J. Shen, *Macromol. Chem. Phys.*, 1996, **197**, 147.
95. I. Ichinose, K. Fujiyoshi, S. Mizuki, Y. Lvov, and T. Kunitake, *Chem. Lett.*, 1996, 257.
96. S. W. Keller, S. A. Johnson, E. S. Brigham, E. H. Yonemoto, and T. E. Mallouk, *J. Am. Chem. Soc.*, 1995, **117**, 12879.

15
Self-assembly in Biomineralization and Biomimetic Materials Chemistry

STEPHEN MANN
University of Bath, UK

15.1 INTRODUCTION

Oyster shells, corals, ivory, sea urchin spines, cuttlefish bone, limpet teeth and magnetic crystals in bacteria are just a few of the vast variety of biological minerals engineered by living organisms. Most of these biominerals are composites of inorganic and organic materials fabricated into a fascinating variety of functional shapes and forms by supramolecular processes involving assembly, templating and organization. Traditionally, scientists have been interested in the mechanical design of biominerals because many of the bioceramics, such as bone and shell, exhibit unusual toughness (as well as strength) for materials containing high proportions of an inorganic phase. For example, the mother-of- pearl (nacre) layer of seashells is a laminate of thin calcium carbonate (aragonite) crystals sandwiched between nanometre-thick sheets of biomolecules. This arrangement prevents crack propagation through the shell wall and dissipates the energy associated with crack formation within the organic layers. More recently, chemists and materials scientists have become interested in the role of the organic components of biominerals in the construction strategies giving rise to such highly organized materials. This is of immediate technological relevance because the ability to construct organized nanoscale, microscopic and bulk inorganic materials from molecular processes is of enormous importance in electronics, catalysis, magnetism, sensory devices and mechanical design. In this respect, the study of biomineralization is providing valuable insight into the scope and nature of materials chemistry at the inorganic–organic interface.[1–5]

A pivotal feature of biomineralization is the use of organic supramolecular assemblies to control the synthesis and construction of inorganic-based materials and composites. This is a very surprising, yet logical, fusion of inorganic and organic materials chemistry. It is well accepted that biological systems are replete with examples of organic supramolecular assemblies (double and triple helices, multisubunit proteins, membrane-bound reaction centres, vesicles, tubules, etc.), some of which (collagen, cellulose and chitin) extend to microscopic dimensions in the form of hierarchical structures. However, less well known is that organic architectures can be utilized to direct and control the structure, size, shape, orientation, texture and assembly of inorganic minerals. What captures the imagination is how relatively simple inorganic minerals such as $CaCO_3$, SiO_2, Fe_3O_4, and so on, can be constructed into such precise architectures compared with the limited shapes offered by purely chemical or geological processes. Moreover, the resulting materials are highly functional—not only as tough, durable and adaptive polymer–ceramic composites but as specialized sensing devices, for example, of ambient magnetic and gravitational fields.

In this chapter I will highlight the important role of organic supramolecular assemblies in the fabrication of organized inorganic-based nanomaterials and composites. An underlying theme is that the understanding of self-assembly in biomineralization directly impacts on the search for new biomimetic strategies in materials chemistry. I will consider the effect of self-assembly on biomineralization in terms of four distinct stages of construction. Each stage will be discussed first with reference to the control of specific aspects of biomineralization, and second, with regard to the potential for mimicking such processes in the fabrication of synthetic materials. I begin with a brief introduction to the types of minerals formed by organisms and their associated functions.

15.2 BIOMINERALS: TYPES AND FUNCTIONS

Of the 20–25 essential elements required by living organisms, H, C, O, Mg, Si, P, S, Ca, Mn and Fe are common constituents of biological minerals. Other essential elements such as N, F, Na, K, Cu and Zn are constituents of biominerals but less widespread. Nonessential elements, such as Ag, Au, Pb and U, are found in association with the external cell walls of bacteria, whereas Sr and Ba are accumulated and deposited as intracellular minerals in algae. The following sections briefly describe some of the main biominerals. Further details are presented elsewhere.[6–10]

15.2.1 Calcium Biominerals

The major mineralized tissues such as bone, teeth and shells are composed of calcium phosphate or carbonate minerals, respectively (Table 1), in combination with a complex organic macromolecular matrix of proteins, polysaccharides and lipids. These materials are often used for structural support but other functions are known. For example calcium carbonates are used as

Table 1 The types and functions of the main inorganic solids found in biological systems.

Mineral	*Formula*	*Organism/function*
Calcium carbonate		
Calcite[a]	$CaCO_3$	Algae/exoskeletons Trilobites/eye lens
Aragonite	$CaCO_3$	Fish/gravity device Molluscs/exoskeleton
Vaterite	$CaCO_3$	Ascidians/spicules
Amorphous	$CaCO_3 \cdot n\,H_2O$	Plants/calcium store
Calcium phosphate		
Hydroxyapatite	$Ca_{10}(PO_4)_6(OH)_2$	Vertebrates/endoskeletons, teeth, calcium store
Octa-calcium phosphate	$Ca_8H_2(PO_4)_6$	Vertebrates/precursor phase in bone Mussels/calcium store
Amorphous		Vertebrates/precursor phase in bone
Calcium oxalate		
Whewellite	$CaC_2O_4 \cdot H_2O$	Plants/calcium
Weddellite	$CaC_2O_4 \cdot 2\,H_2O$	Plants/calcium
Group IIA metal sulfates		
Gypsum	$CaSO_4$	Jellyfish larvae/gravity device
Barite	$BaSO_4$	Algae/gravity device
Celestite	$SrSO_4$	Acantharia/cellular support
Silicon dioxide		
Silica	$SiO_2 \cdot n\,H_2O$	Algae/exoskeletons
Iron oxides		
Magnetite	Fe_3O_4	Bacteria/magnetotaxis Chitons/teeth
Geothite	α-FeO_2H	Limpets/teeth
Lepidocrocite	γ-FeO_2H	Chitons (Mollusca) teeth
Ferrihydrite	$5Fe_2O_3 \cdot 9\,H_2O$	Animals and plants Iron storage proteins

[a]A range of magnesium-substituted calcites are also formed.

gravity sensors in a wide range of animals.[11] These gravity devices function in a similar way to the fluid in the semicircular canals (which detect changes in angular momentum). The crystals are sited on a membrane under which sensory cells are located. During a change in linear acceleration, the movement of the crystal mass relative to the delicate hair-like extensions of the cells results in the electrical signalling of the applied force to the brain. A further spectacular role is the use of biogenic calcite in the compound eyes of extinct trilobites. These creatures are well preserved as fossils such that the structure and organization of the corneal lenses have been determined. The eyes consist of hexagonally packed calcite single crystals. Interestingly, single crystals of calcite are renowned for their ability to doubly refract white light, suggesting that the trilobites suffered a life of continual double vision! However, studies of well-preserved fossilized material show that each crystal is aligned in the eye such that the unique (nonrefracting) c-axis is perpendicular to the surface of each lens.[12] In this orientation, the calcite lens behaves isotropically like glass, and a single well-defined image is formed. The sophisticated design of bioinorganic solid-state assemblies in this manner exemplifies the fundamental interdependence of structure, organization and function in many biomineralization processes.

Perhaps bone, more than any other biomineral, reflects the greatest distinction between an inorganic and bioinorganic solid. The structure and mechanical properties of bone are derived from the organized mineralization of hydroxyapatite (often as carbonated apatite) within a matrix of collagen fibrils, proteoglycans and many other proteins. For many purposes the calcium phosphate of bone is best thought of as a 'living mineral' since it undergoes continual growth, dissolution and remodelling. The dynamic property of bone indicates that the mineral is more than just a structural support. The mineral acts as an important calcium store in homeostasis as well as a supply of calcium in times of high demand.

The structure and organization of tooth enamel, like bone, derives from a highly complex system designed to withstand specific types of mechanical stress. Enamel is unique in that it consists of long ribbon-like hydroxyapatite crystals that make up a very high weight percentage of the mature tissue (95 wt.%; cf. bone, 65 wt.% (average)). This is achieved by depletion of the organic components (amelogenins and enamelin proteins) during tooth maturation and crystal growth.

The predominance of calcium biominerals over other group 2 metals (Table 1) can be explained by the low solubility products of the carbonates, phosphates, pyrophosphates, sulfates and oxalates and the relatively high levels of calcium in extracellular fluids (10^{-3} M). Significantly, the acid–base equilibria of many of these oxyanions (e.g., $HCO_3^- \rightarrow CO_3^{2-} + H^+$), provides a stringent means of regulating the activity products of supersaturated biological solutions. In some instances, this is so critical that enzymes such as carbonic anhydrase and alkaline phosphatase are employed. Magnesium salts are generally more soluble and no simple magnesium biominerals are known. However, magnesium has an important role in influencing the structure of both carbonate and phosphate biominerals through lattice and surface substitution reactions. Other ions, such as Na^+, NH_4^+, K^+ and F^- in particular, also have a pronounced affect on the solubility of hydroxyapatite.

15.2.2 Silica

Although most biominerals are ionic salts, the stability of Si—O—Si units in water gives rise to hydrated inorganic polymers that are moulded into elaborate shapes by many unicellular organisms (Table 1). Why some organisms utilize silica rather than calcium carbonate as a structural material is unknown. One possibility is that because the fracture and cleavage planes inherent to crystalline structures are missing, the amorphous biomineral can be subsequently moulded without loss of strength into a wide variety of elaborate architectures (Figure 1). Although only a few systems have been studied, it appears that biological silicas formed in plant cell walls are often associated with organized networks of polysaccharides.[13] Further details of biosilicification can be found elsewhere.[14]

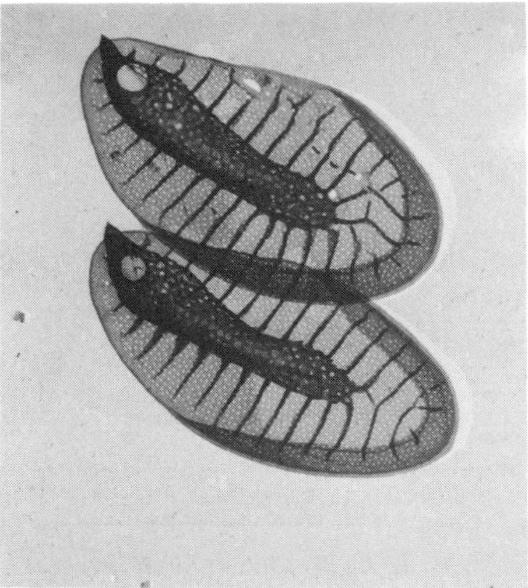

Figure 1 Isolated silica scales from the freshwater alga *Synura petersenii*. The oval-shaped scales are curved and perforated. In the intact organism, the scales overlap around the cell wall to give a mineralized shield. The size of each scale is 4.5 µm (courtesy of Dr. B. S. C. Leadbeater, University of Birmingham).

15.2.3 Transition Metals

Of the many transition metals which display a rich biocoordination chemistry, only iron, and to a lesser extent, manganese, have extensive roles in biomineralization (Table 1). The bioinorganic solid-state chemistry of these elements is dominated by the redox behaviour of the $+2/+3$ oxidation states, an affinity for O, S and OH ligands, and ease of hydrolysis in aqueous solution. Like the calcium biominerals, biological iron oxides are used to strengthen soft tissues and as storage depots (Fe^{III}, OH^- and HPO_4^{2-}). Furthermore, the magnetic properties of mixed valence phases are utilized by several types of bacteria as a means of navigation in the ambient geomagnetic field.[15] Most magnetotactic bacteria synthesize intracellular magnetite (Fe_3O_4), although recent studies have shown that species inhabiting sulfide-rich environments deposit the isomorphic mineral, greigite (Fe_3S_4) (Figure 2). In both systems, the crystals must be aligned in chains and have dimensions compatible with that of a single magnetic domain, if they are to function efficiently as biomagnetic compasses. The key component in achieving this specificity is the construction of an organized assembly of lipid molecules in the form of spherical or elongated vesicles (see Section 15.5).

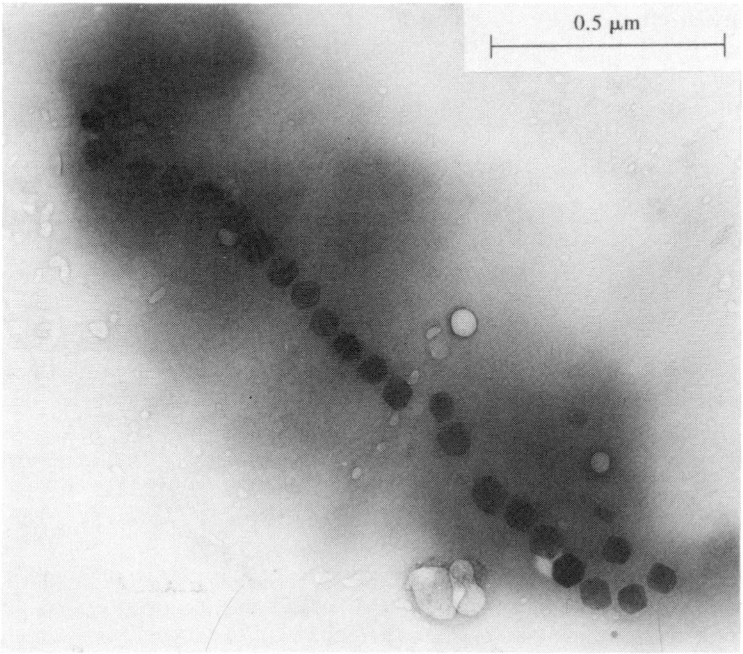

Figure 2 Electron micrograph of a magnetotactic bacterial cell containing an intracellular chain of organized ferrimagnetic greigite (Fe_3S_4) crystals.

Many elements, such as Cu, Zn and Pb, are deposited in or on the external mucopolysaccharide walls of bacterial cells in the form of metal sulfides. These processes are biologically induced, rather than biologically controlled. Interestingly, some yeasts mineralize nanometre-size intracellular CdS particles within short chelating peptides of general structure (γ-Glu-Cys)*n*-Gly.[16] As the number of CdS units per particle is small (~85), these peptide–mineral complexes can be considered as large-nuclearity clusters capped by cysteinyl thiolate ligands, analogous with proteins such as metallothionein which sequester smaller (three- and four-metal ion) clusters.

15.3 MINERALIZATION SITES

The formation of biominerals takes place in well-defined spatially delineated sites. In bacteria (Monera) and unicellular organisms (Protoctista), these sites are either epicellular (i.e., on or within the cell wall) or intracellular. Bacterial mineralization is generally associated with cell wall processes in which extruded metabolic products (ions, gases, polypeptides, electrons) coprecipitate with extraneous metal ions in the surrounding environment. The resulting biominerals are often structurally ill-defined, physically heterogeneous and spatially disorganized. Intracellular mineralization, on the other hand, is characterized by prodigious fine-tuning of the crystal chemistry. This process is uncommon in bacteria, but prevalent in algae and protozoa because intracellular compartments, usually in the form of membrane-bound vesicles, can be readily synthesized. In many cases, the vesicles and their mineral products remain in the cell, but in some organisms, such as coccolithophores, they are translocated to the external surface of the cell where they are assembled into complex microstructures. In multicellular systems, specialized cells have evolved to regulate mineralization in extracellular spaces within the organism. Secretion of biopolymeric matrices, such as collagen and chitin, into this space enables complex large-scale composite materials (e.g., bone, shell) to be fabricated. In organisms such as coralline algae, the cells secrete inorganic and organic molecules into the intercellular spaces which subsequently become mineralized.

Clearly, the intimate association of inorganic and organic phases is a hallmark of biomineralization. In many cases, the integration is at the superstructural level where mineral particles and biopolymers are organized to give composites of unusual strength and toughness. In other systems the association of inorganic and organic components is more subtle. For example, the spine of a sea urchin, with its elaborate (noncrystallographic) texture and porosity, generates a highly ordered single crystal x-ray diffraction pattern, yet it contains 0.02 wt.% of protein.[17] The intercalate has a marked effect on the mechanical properties, such that echinoderm tests and spines fracture conchoidally (like a glass) and not along the low-energy {104} cleavage planes of magnesium-doped calcite. It may turn out that many biominerals are of this form, that is, a nanocomposite of inorganic and organic constituents, with the organic macromolecules (proteins, polysaccharides, lipids) residing at the interfaces between essentially coherent crystal domain boundaries. If this is the case, then we have much to learn from biomineralization in terms of nanofabrication.

15.4 MOLECULAR TECTONICS

Currently, we know much about the ultrastructures of biominerals and how they vary in different organisms, but relatively little about the precise details concerning the molecular interactions governing their construction. The aqueous precipitation of many of the minerals listed in Table 1 is a relatively straightforward laboratory procedure, but while it may be relatively simple to grow crystals of a compound, controlling the size, shape, orientation and assembly of these crystals, as is typical of many biominerals, is a much more complex task.

In essence, the fabrication of biomineralized architectures is ultimately governed by the nature of the molecular interactions occurring at solid–liquid and solid–solid interfaces over a range of length scales. In this respect, biomineralization processes are archetypes of synthetic strategies that can be orchestrated towards the construction of organized materials. This ability to carry out 'synthesis-with-construction'—what I will call molecular tectonics[18] (Gk. *tekton*, builder)—is a hallmark of biomineralization. Moreover, this paradigm is applicable to the development of new materials chemistry in which the traditional methods of solid-state synthesis are replaced by an approach based on molecular design, construction and organized assembly.

Supramolecular chemistry is central to the development of molecular tectonics. In biomineralization, organized assemblies of organic macromolecules are utilized at various length scales. At the nanoscale, biomineralization involves the molecular construction of discrete self-assembled organic supramolecular systems (micelles, vesicles, etc.) that are used as preorganized spatial and chemical environments for controlling the formation of finely-divided inorganic materials, ~1–100 nm in size. The fabrication of consolidated biominerals, such as bone and teeth, also involves the construction of preorganized organic frameworks but the length scale is greater (micrometres) and the matrix is often polymeric (collagen, enamel tubules, etc.). Furthermore, the building of discrete or extended organic architectures in biomineralization often involves hierarchical processing in which the molecular-based construction of organic assemblies is used to provide frameworks for the synthesis of organized materials which in turn are exploited as prefabricated units in the production of higher-order complex microstructures.

The general features of the stages of construction in biomineralization can be illustrated by the formation of calcified scales (coccoliths) in the unicellular marine alga *Emiliania huxleyi*[19] (Figure 3). The constructional problem is to build individual scales consisting of 30–40 calcite crystals arranged within an oval-shaped unit. First, the deposition of each crystal takes place at specific intracellular sites and this is achieved by constructing discrete building plots around the inner ring. Thus, the first stage of biomineralization is the production and organization of a series of membrane-bound vesicles that act as delineated reaction volumes for subsequent inorganic crystallization. The second stage of construction involves the oriented nucleation of calcite crystals within these spatially discrete sites. Electron diffraction studies indicate that the crystallographic *c*-axis of each crystal is aligned radially, suggesting that a molecular blueprint is encoded within each building site. Although the details are not known, it is probable that an organic template present within the vesicles is responsible for templating the crystal unit cell. The third stage of coccolith construction involves vectorial crystal growth of the oriented calcite crystals such that the crystals adopt a complex architecture that includes a hammer-shaped extension. The overall shape is species-specific and replicated over millions of years. Although the coccolith scale is now complete, there is a fourth process in which individual scales are exocytosed and used to construct a higher-order architecture (coccosphere) (Figure 4). This networking of individual coccoliths to produce an enclosed cellular shield is of significant functional value as it provides the alga with protection, bouyancy, and a putative light-focusing device for photosynthesis.

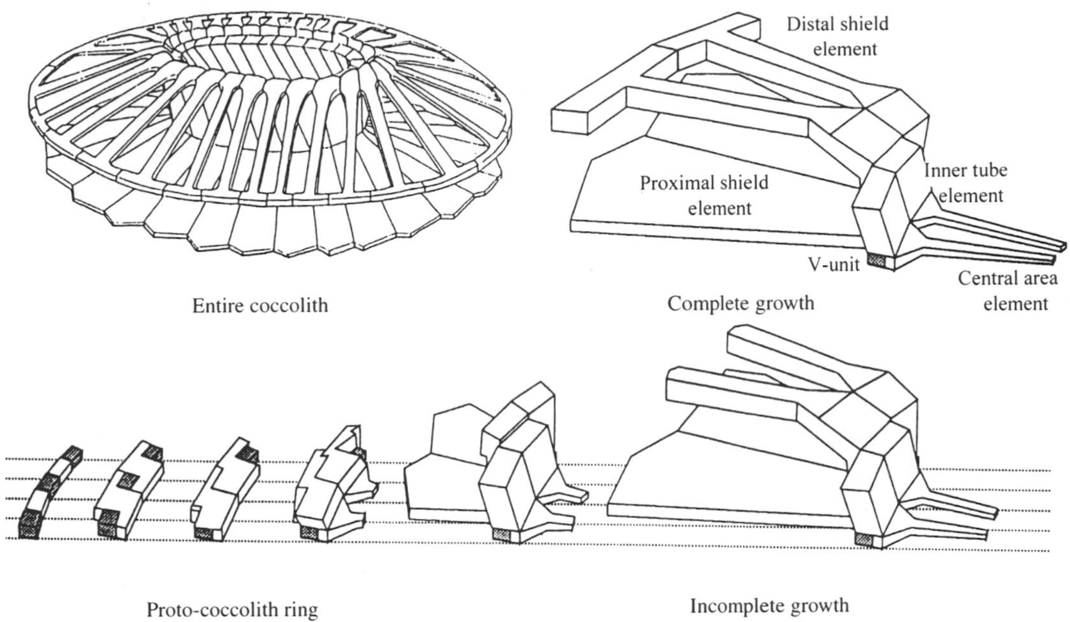

Figure 3 Constructional stages of coccolith development in the marine alga, *Emiliania huxleyi.*

The 'total synthesis' of a coccolith is clearly a very sophisticated process of molecular tectonics. The alga cell is not specialized for biomineralization (unlike, for example, bone cells), yet it contains the machinery to undertake complex materials synthesis, suggesting that many of the processes responsible for biomineralization are generic in character. Four constructional stages, involving self-assembly and organization, can be identified (Table 2). A commonplace analogy, such as house building, can be used to illustrate the underlying connections. The first stage in house construction involves the establishment of a building site. This is a pivotal stage because the size, shape and connectivity of the individual plot determines the context (preorganization) of the building programme. The next stage is to lay the foundations using a plan (blueprint) which determines the basic architecture of the dwelling. Once this is achieved, the house can now be fabricated in line with the predefined specifications of a template such that bay windows, gable roofs, and so on, are positioned in their correct locations. This requires the construction to proceed along specific directions at a given magnitude, (i.e., with vectorial regulation). If a series of building sites are established within a location, the result is not only the predetermined construction of individual houses but the higher-order organization (processing) of houses into a street. In principle, this network can be highly complex, such as in town planning in which streets are organized into housing estates, housing estates into towns, towns into cities, and so on.

Each of these biologically-derived concepts—supramolecular preorganization, interfacial molecular recognition (blueprinting), vectorial regulation and cellular processing—is inspiring new ideas in materials chemistry (Table 2).[18] For example, the preorganization of supramolecular assemblies in biomineralization can be considered as an archetype of a host–guest approach to the nanoscale synthesis of composites. Similarly, the concept of interfacial molecular recognition suggests that inorganic materials synthesis can be regulated through molecular blueprinting with organic substrates. Likewise, the crystal engineering of materials can be viewed from the perspective of vectorial regulation with concomitant concepts such as templating, directed growth and

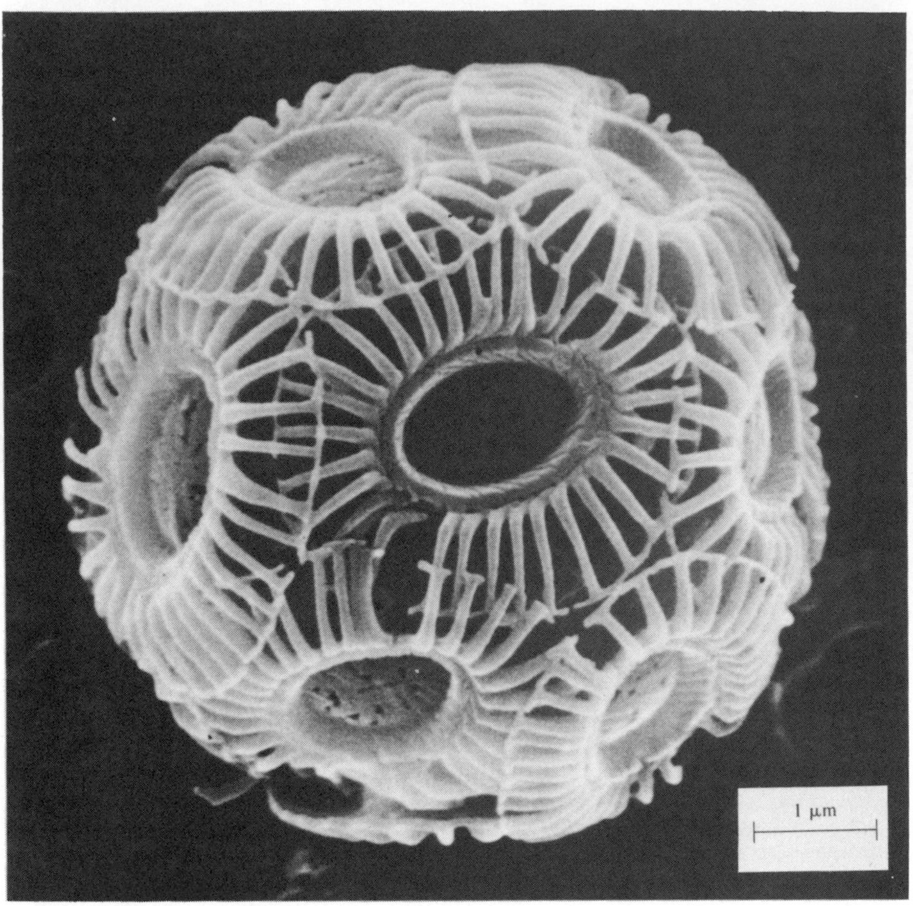

Figure 4 Scanning electron micrograph of an intact coccosphere of *Emiliania huxleyi* showing the assembly of individual coccolith scales against the cell wall (courtesy of Dr. J. R. Young, Natural History Museum).

Table 2 The four stages of construction of biominerals and associated biomimetic principles.

Building Process	*Function*	*Biological archetypes*	*Biomimetic principle*
Plot development (building site)	Delineated reaction sites	Supramolecular preorganization	Host–guest materials chemistry
Foundations (blueprint)	Site-specific inorganic nucleation	Interfacial molecular recognition	Molecular blueprinting
Assembly (bricks and mortar)	Crystal growth and termination	Vectorial regulation	Crystal engineering
Networking (town planning)	Higher-order architectures	Cellular processing	Crystal tectonics

microstructural fabrication. Finally, the cellular processing of biominerals can be used to develop methods for the controlled architectural assembly of preformed particles or crystals with tailored properties—an approach that extends all the way from molecular to crystal tectonics.

It is important to note that although these four stages will be discussed separately, it is most probable that in reality they are integrated within feedback systems. Indeed, it may turn out that the ability to fine tune both the rates of inorganic precipitation and the self-assembly of organic molecules at the same time within a reaction media—such that there is co-organization of the two phases via synergism—is an important key to the synthetic fabrication of organized organic–inorganic materials of the future.

In Section 15.5, the constructional stages of biomineralization are discussed in detail, and show how they are being used for the development of biomimetic routes in materials chemistry.

15.5 SUPRAMOLECULAR PREORGANIZATION

15.5.1 Self-assembly and Biomineralization

The initial stage in the fabrication of biominerals is the construction of an organized reaction environment prior to mineralization. This preorganized system is built by the self- and facilitated assembly of biological molecules such as phospholipids and structural proteins. A variety of structural organizations based on enclosed or extended architectures are known (Table 3). The former often involves direct self-assembly and is prevalent in intracellular biomineralization, while the extended networks are often facilitated by auxiliary molecules in the extracellular (intercellular) spaces generated by multicellular organisms.

Table 3 Supramolecular assemblies in biomineralization.

Assembly	*Constituents*	*System*
Nanoaggregates	Amelogenins	Enamel
	Ice-nucleating proteins	Bacterial cell walls
Nanocapsules	Ferritin proteins	Mammals, bacteria
Vesicles	Phospholipids	Magnetic bacteria
		Diatoms, etc.
Two-dimensional templates	S-layer proteins	Bacterial cell walls
	β-sheet glycocoproteins	Shells
Three-dimensional crystalline nets	Collagen	Bone
Three-dimensional macromolecular nets	α/β-Chitin	Limpet teeth
		Crab cuticle
	Cellulose	Plant walls
	Mucopolysaccharides	Bacterial cell walls

15.5.1.1 Biomolecular vesicles

Vesicles are ubiquitous in the cytoplasm of eukaryotic cells where they are involved in a wide range of transport, storage and secretory functions. Not only are the vesicles assembled with well-defined compositions but they may undergo a variety of specific transactions involving fusion events, notably during protein trafficking between cisterna in the Golgi stack. The molecular construction of these architectures is based on the balancing of hydrophobic–hydrophilic interactions that exist for amphiphilic molecules in aqueous environments, and in the absence of external scaffolds, vesicles generally adopt a spherical morphology. With regard to biomineralization, the enclosed intravesicular space provides a means of controlling the size and location of discrete intracellular mineral products such as silica[20] (Figure 5) and iron oxides. The composition of such membranes does not appear to be unusual, for example, the magnetosome membrane surrounding crystals of the iron oxide, magnetite, in magnetotactic bacteria has an overall composition similar to other cell membranes except for two specific proteins.[21] It is possible that these proteins could be involved in ion transport across the vesicle membrane and hence in the control of the physicochemical conditions (supersaturation, ionic strength, complexation, pH) pertaining to crystal nucleation and growth within the intravesicular space. In coccolithophores such as *E. huxleyi*, the vesicle membrane is associated with significant amounts of polysaccharide.[22] One advantage of these lipid-based architectures is that they are often labile, dynamic constructions, which can be subsequently shaped and transported during the biomineralization process.

15.5.1.2 Ferritin

Polypeptide assemblies, in comparison, are often rigid compact structures (e.g., globular or linear proteins) because of the strong intersubunit interactions acting to minimize the hydrophobic–hydrophilic interactions. The iron storage protein, ferritin, is an exception in that it consists of a perforated 8 nm diameter spherical cage formed by the self-assembly of 24 polypeptide subunits (Figure 6).[23] In the native protein, the cavity is filled with an iron oxide core consisting of no more than 4500 iron atoms. The polypeptide subunits of mammalian ferritin are of two different types (H and L) but both comprise conformations based on a bundle of four long helices, a short helix and a long loop. Self-assembly of the subunits is facilitated by the high propensity to form stable

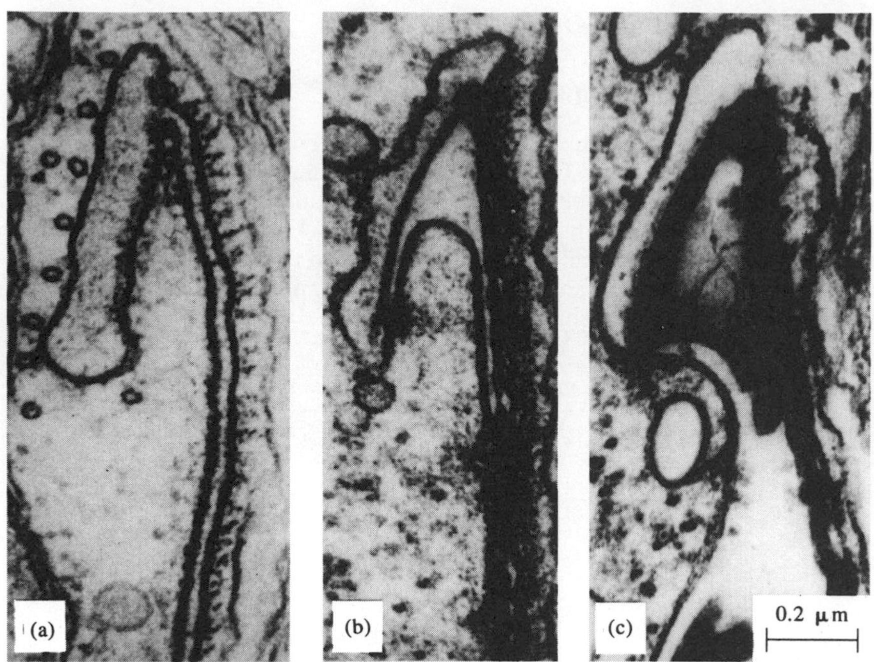

Figure 5 Thin section through part of a cell of the freshwater alga, *Synura petersenii*. This unicellular organism produces intracellular silica scales (see Figure 1). (a) Cross-section through part of a curved vesicle prior to silica deposition. Note the presence of microtubules (small circles in the micrograph) which provide an underlying framework for the shaping of the membrane vesicle. (b) Initial stage in silicification involving the secretion and construction of an organic matrix within the shaped vesicle. (c) Site-directed condensation of silica onto the preorganized organic substrate (courtesy of Dr. B. S. C. Leadbeater, University of Birmingham).

dimer intermediates through apolar interactions involving the external surfaces of the loop and one of the long helices. Additional intersubunit interactions involving hydrogen bonding, salt bridges and hydrophobic forces generate higher oligomers and stabilize the quaternary structure of the intact shell which is perforated by molecular channels at the intersubunit junctions such that metal ions and small molecules can enter and exit from the internal aqueous cavity.

Figure 6 Section through a x-ray diffraction map (α-carbon plot) of ferritin showing the micellular nature of the protein architecture. The polypetide cage is assembled and the internal cavity is then mineralized with iron oxide.

The structure of ferritin cores has been extensively studied.[24] The results indicate that the iron-containing cores of mammalian ferritins are crystalline with a unit cell based on a four-layer repeat of hexagonally close-packed oxygen atoms with variable octahedral occupancy of Fe^{III} ions. The analogous inorganic mineral is ferrihydrite ($5\,Fe_2O_3{\cdot}9\,H_2O$). High-resolution electron microscopy has been undertaken to determine the structural similarities and differences of individual ferritin cores isolated from vertebrate (human), invertebrate (limpet, *Patella vulgata*) and bacterial (*Pseudomonas aeruginosa*) sources.[25] Whereas a predominance of single crystalline cores was observed for human ferritin cores, limpet and bacterial ferritins gave very few lattice images. Limpet cores imaged with resolvable lattice fringes contained crystalline domains with dimensions in the range 3–5 nm. Bacterial ferritin cores, on the other hand, showed only incoherent fringe patterns suggesting lamella-like structures in which the ordering was extremely short range (often 1–2 nm). These differences were also reflected in the different ^{57}Fe Mössbauer spectra obtained from these samples.[24] Human ferritin cores were superparamagnetic (i.e., antiferromagnetically coupled particles of small dimensions) with a blocking temperature of $\sim$40 K. Limpet ferritin had a lower blocking temperature ($\sim$20 K) and there was also evidence for magnetic disorder. Bacterial ferritins, in contrast, were magnetically disordered down to a temperature of 4 K.

These results show unequivocally the fundamental difference in the crystallographic nature of ferritin mineral cores isolated from human, limpet and bacterial sources. Factors governing this change in crystal chemistry include the rate of nucleation and growth of the iron oxide nanoparticles within the protein cavity. These processes can be elucidated by taking advantage of the fact that ferritin can be readily demetallated by reductive dissolution of the iron core without loss of structural integrity of the protein shell. Moreover, the apoferritin molecules can be subsequently reconstituted under controlled experimental conditions involving incubation of the protein with aqueous Fe^{II} solutions.These experiments have stimulated biomimetic research (Section 15.5.2.2) as well as identifying key amino acid sites involved in biomineralization in the native protein (Section 15.6.1.1).

15.5.1.3 Biopolymeric networks

Extended polymeric networks are often associated with the construction of consolidated biominerals such as bone, enamel and shell, invertebrate teeth and siliceous plant materials. A description of the hierarchical structure of bone is well documented.[26] The primary unit is based on the nucleation of calcium phosphate in nanospaces organized within the supramolecular assembly of collagen fibrils (Figure 7).[27] Type I collagen molecules secreted by osteoblasts self-assemble into helical trimers which are subsequently stacked into microfibrils with a specific tertiary structure of periodicity 64 nm, with 40 nm gaps between the ends of the molecules. The precise hole arrangement is governed by the strong covalent cross-links that exist between the triple helical molecules when staggered by 64 nm along their long axis. Other biomolecules, such as phosphoproteins and proteoglycans, are also associated with the collagen supramolecular assembly. The mineralized microfibril is itself organized in various spatial patterns — woven bone, lamellar bone, haversian bone, and so on — such that specific mechanical properties can be induced from the same basic mineralized building unit.

Details of the assembly of other biopolymeric networks involved in biomineralization are less precise. Many shells and teeth are constructed within frameworks that may be lamellar, columnar or recticular. In each case, there appears to be two key features of the assembly (Figure 8).[9] First, a relatively inert structural frame is built from insoluble macromolecules such as β-sheet hydrophobic proteins and/or polysaccharides (chitin). Second, acidic proteins rich in aspartic acid, and often in association with sulfated polysaccharides, are assembled on the hydrophobic scaffold. The onset of mineralization then takes place at the interface between the acidic proteins and aqueous environment.

15.5.2 Host–Guest Biomimetic Nanosynthesis

Since nanoscale particles exhibit quantum-size effects in their electronic, optical and chemical properties, there has been much activity in this area of synthetic materials chemistry.[28] A complementary biomimetic strategy involves the use of organic supramolecular cages which act as spatial hosts and reactive interfaces for inorganic guest materials of nanometre dimensions (Table 4). Here, I discuss two examples: first, the use of synthetic vesicles, and second, application of the protein cage of ferritin in the synthesis of inorganic nanocomposites.

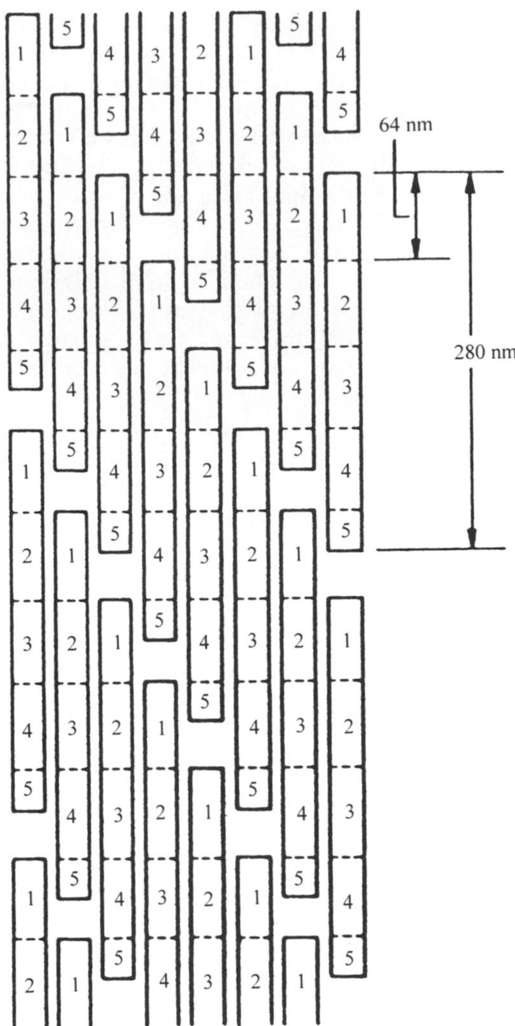

Figure 7 The revised quarter-stagger model for collagen. The planar view shows the specific overlap of collagen trimers to give a packing arrangement which includes specific interstitial sites in which calcium phosphate nucleation occurs (reprinted with permission from *Nature*, **219**, 157. Copyright 1968 Macmillan Magazines Ltd).

15.5.2.1 Phospholipid vesicles

Unilamellar phosphatidylcholine vesicles, ~30 nm in diameter, can be readily prepared by sonicating dispersions of the lipid in aqueous solution at a temperature above the gel–liquid transition point. These are versatile systems because the reaction environments can have variable diameters (1–500 nm) and the surface functional groups can be modified by molecular engineering. When formed in the presence of metal ions, the internal space contains encapsulated species which can subsequently undergo crystallization reactions with membrane permeable species such as OH^- and H_2S (Figure 9). For example, a range of nanometre-dimension inorganic materials such as Ag_2O,[29] Fe_3O_4,[30] Al_2O_3,[31] calcium phosphates,[32] and CdS[33] have been prepared. As each particle is surrounded by a 4.5 nm thick bilayer membrane, particle–particle interactions are negligible and reaction rates can be diffusion controlled. The aim of much of this work is focused on the types of nanoparticles that can be synthesized in confined vesicular environments and the control mechanisms exerted by these chemically well-defined supramolecular assemblies.

Before reaction, encapsulated cations such as Fe^{II}, Fe^{III} and Co^{II} bind strongly to the head-group phosphates of the phospholipid bilayer so that nucleation is localized at the organic surface. Reaction with OH^- results in finely divided intravesicular particles. Electron diffraction patterns arising from the particles obtained from Fe^{III} solutions had *d*-spacings corresponding to poorly ordered goethite (α-FeO_2H) while similar reactions with entrapped $Fe^{II}_{(aq.)}$ and Fe^{II}–Fe^{III} solutions gave intravesicular particles of spherulitic magnetite (Fe_3O_4) and ferrihydrite ($Fe_2O_3{\cdot}n\,H_2O$),

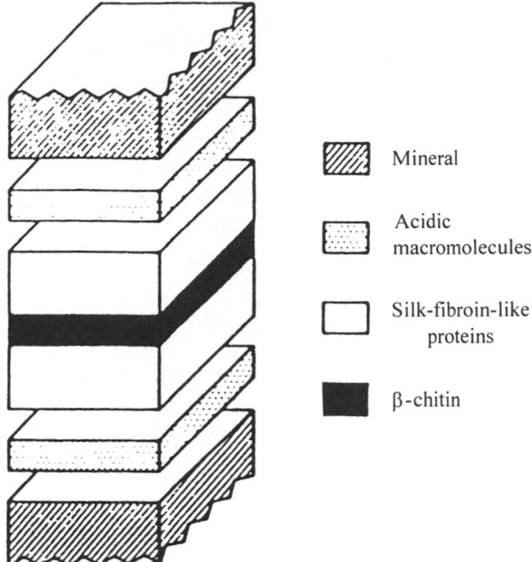

Figure 8 Schematic of the composite structure of an individual organic matrix sheet bounded on both sides by shell mineral. The organic constituents are precisely assembled prior to mineralization (reproduced by permission of the Royal Society of London from *Proc. R. Soc. London, Ser. B*, 1984, **304**, 425).

Table 4 Biomimetic approaches to the nanoscale synthesis of inorganic materials and composites.

Approach	*Strategy*	*Product*	*Systems*	*Materials*
Nanoscale synthesis	Host–guest	Clusters	Reverse micelles	CdS,
			Microemulsions	Pt, Co, metal borides
				Fe_3O_4, $CaCO_3$
		Nanoparticles	Vesicles	Pt, Ag
				CdS, ZnS
				Ag_2O, FeO_2H, Fe_3O_4, Al_2O_3
				Calcium phosphates
			Ferritin	MnO_2H, UO_3, FeS, Fe_3O_4
			LB films	CdS
			Polystyrene resin	γ-Fe_2O_3
	Ligand capping		(γ-EC)$_n$G peptides	CdS

Source: Mann.[18]

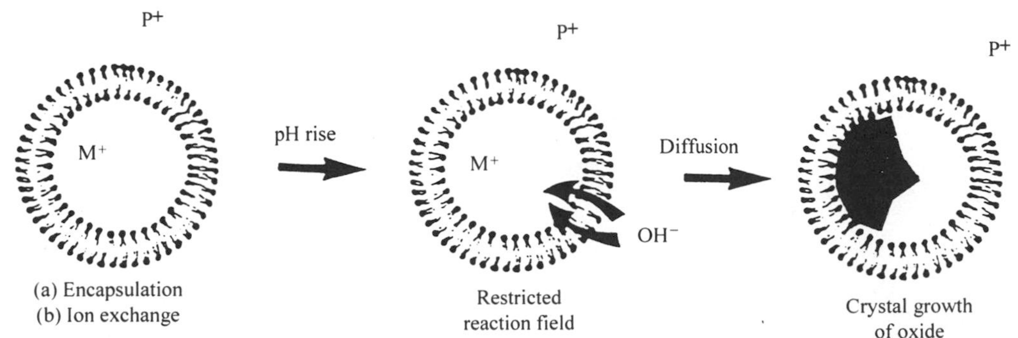

Figure 9 Use of phospholipid vesicles in the membrane-mediated precipitation of metal oxides. Cations (M^+) are encapsulated by sonication and replaced in the external phase by inert cations (P^+) by ion exchange chromatography. Increases in the extravesicular pH result in the slow OH^- influx and subsequent nucleation of the oxide on the inner membrane surface.

respectively.[30] These products are different to those formed under identical starting conditions in the absence of vesicles. For example, precipitation of Fe^{III} solutions results in extended aggregates of ferrihydrite, Fe^{II} solutions give acicular needles of lepidocrocite (γ-FeO_2H) and goethite, and irregularly shaped 10–50 nm magnetite particles are precipitated from bulk solutions containing both Fe^{II} and Fe^{III}. Thus there are distinct modifications in structure, morphology, and particle size for precipitation reactions undertaken within unilamellar vesicles. These differences can be attributed primarily to the kinetic control exerted by the vesicle membrane on the rate of OH^- diffusion into the intravesicular space although the charged organic surface may be important for stabilizing the accumulation of ionic charge and the subsequent formation of embryonic crystallites.

Similar studies involving the intravesicular precipitation of Ag_2O from encapsulated Ag^I solutions reacted with hydroxide ions resulted in single-domain nanocrystals of cubic Ag_2O (Figure 10).[29] The kinetics of these reactions was studied by light scattering which showed a linear relationship between the initial rate of precipitation and trapped Ag^I concentration, at constant hydroxide ion concentration. Thus, over the concentration range investigated the initial kinetics were first order with respect to intravesicular Ag^I concentration, $[Ag]_{in}$. However, the dependency on extravesicular pH, $[OH^-]_{out}$, was more complex; at pH values < 10 no changes in turbidity were observed, at pH 11–12 the reaction was strongly dependent on $[OH^-]_{out}$ and above pH 12, the initial rate was essentially independent of $[OH^-]_{out}$. These data suggest a two-step reaction mechanism for intravesicular Ag_2O precipitation (Equations (1) and (2)).

$$[OH^-]_{out} \rightarrow [OH^-]_{in} \quad (1)$$

$$2[OH^-]_{in} + 2[Ag^+]_{in} \rightarrow [Ag_2O]_{in} + H_2O \quad (2)$$

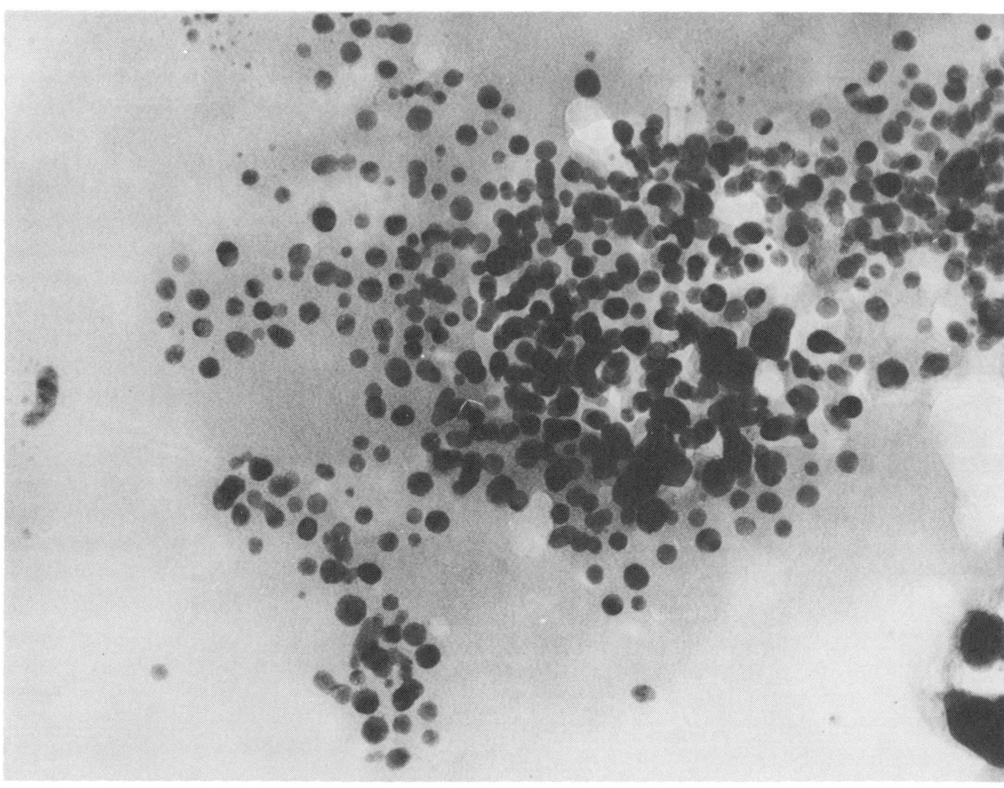

Figure 10 Electron micrograph showing discrete intravesicular crystals of Ag_2O. The average particle diameter is 10 nm.

The first step (Equation (1)) is diffusion-controlled and depends on the rate of passage of OH^- ions through the lipid membrane. At pH_{out} values below 11.0, the rate of OH^- influx is very small and supersaturation is never attained within the vesicles. At pH_{out} values above 12.0, the rate of crystal growth becomes less dependent on initial pH gradients across the membrane because the limiting rate of diffusion through the membrane is attained.

The relationship between the intra- and extravesicular pH in the above experiments has been studied by ^{31}P NMR spectroscopy.[34] Vesicles containing a mixture of NaH_2PO_4 and $NaNO_3$ were studied to determine whether the presence of the diffusable NO_3^- ion would permit the influx of hydroxide as observed for intravesicular Ag_2O formation. The intravesicular phosphate resonance, initially at −14.98 ppm, did not shift significantly until above a pH_{out} value of 11.0, after which it shifted steadily downfield due to OH^- influx. Similar experiments undertaken with phosphate in the absence of nitrate showed that a pH gradient of ~six units could be maintained across the bilayer membrane at an external pH of 12.5.

Reduction of the pH gradient across the vesicle membrane in these experiments requires the net transport of OH^- ions into the vesicles. In order to preserve electroneutrality, this can only occur if there is an equivalent net migration of anions out of the vesicles. In the experiments with vesicles containing NO_3^- the internal pH responded to a change in pH_{out} above 11.0, whereas the vesicles containing only the more highly charged $[HPO_4]^{2-}$ and $[PO_4]^{3-}$ ions showed no changes in internal pH. Thus, the presence of encapsulated diffusible anions is important for determining the role played by the membrane in controlling intravesicular pH, and hence the rate of inorganic oxide precipitation within the internal cavity of the vesicles.

The dependency of the rate at which supersaturation is maintained within the vesicles on anion diffusion rates is clearly revealed when anions of different permeabilities are substituted for OH^- and the corresponding intravesicular precipitation is followed by turbidity measurements.[29] In the case of intravesicular AgCl formation the increase in turbidity with time due to the addition of Cl^- was much slower than for Ag_2O formation indicating that diffusion of Cl^- across the lipid membrane was reduced compared with OH^-. The formation of intravesicular Ag_2S, on the other hand, occurred instantaneously after the addition of $(NH_4)_2S$ due to the diffusion of free molecular H_2S across the supramolecular membrane.

15.5.2.2 Protein cages

Lipid-based assemblies often have a limited stability with regard to aggregation and hydrolysis, although the use of polymerized vesicles has ameliorated some of these problems. An alternative approach is to use the biomolecular cage of the iron storage protein, ferritin, as a more robust environment for inorganic materials synthesis.[35] Two strategies in nanosynthesis are currently under investigation (Figure 11); (i) *in situ* chemical reaction of the native iron oxide cores of ferritin, and (ii) removal of the native iron oxide cores followed by reconstitution of the empty protein cages with metal salt solutions.

Addition of aqueous sodium sulfide under an argon atmosphere to acidic (pH = 5.4) solutions of native horse spleen ferritin containing 3000 iron atoms per molecule resulted in the formation of single-phase iron sulfide cores within the protein cage of ferritin.[36] UV–visible spectra showed the formation of a new material which exhibited absorption maxima at 426 nm and 338 nm. These spectra were similar to the visible absorption spectra of cluster compounds such as $[Fe_4S_4(SR)_4]_2^-$ and $[Fe_2S_2(SR)_4]_2^-$. Transmission electron microscopy of the transformed native ferritin showed discrete electron dense cores, of ~6.5 nm diameter, which were similar in size to the unreacted ferritin biomineral. The sulfided cores exhibited no discernible electron diffraction reflections, whereas the unreacted ferritin cores showed six lines corresponding to the mineral, ferrihydrite.

Similar observations were recorded for the sulfidation reaction of reconstituted ferritin containing ferrihydrite cores of ~500 iron atoms per molecule. The intraferritin particles were significantly smaller in diameter (< 2 nm) and amorphous. EXAFS analysis[36] showed that the iron atoms were coordinated to four sulfur atoms at 222 pm, and one iron atom at 275 pm. These parameters are characteristic of the edge-shared FeS_4 tetrahedra of cubane-type clusters such as $[Fe_4S_4(SPh)_4]^{2-}$, suggesting that the 500 iron atom sulfided cores are best described as a disordered array of edge-shared FeS_4 units.

Iron-57 Mössbauer spectra of the ferritin–iron sulfide nanocomposite showed a single quadrupole split doublet with parameters indicative of high-spin Fe^{III} in tetrahedral sites.[36] Synthesis of Fe^{III} sulfide within ferritin is surprising because most iron sulfides are considered to consist of Fe^{II} ions except for a mixed oxidation state mineral, greigite (Fe_3S_4). Spectra recorded at 4.2 K showed clear evidence of magnetic ordering (with a mean hyperfine field of approximately 26 T) and an additional component solely in the 500 iron atom cluster which was assigned to a tetrahedral iron centre with an effective mean charge of approximately +2.5.

A second strategy involving the reconstitution of demetallated ferritin has resulted in the synthesis of a range of non-native oxide materials, such as MnO_2H,[35,37] magnetite (Fe_3O_4)[38] and

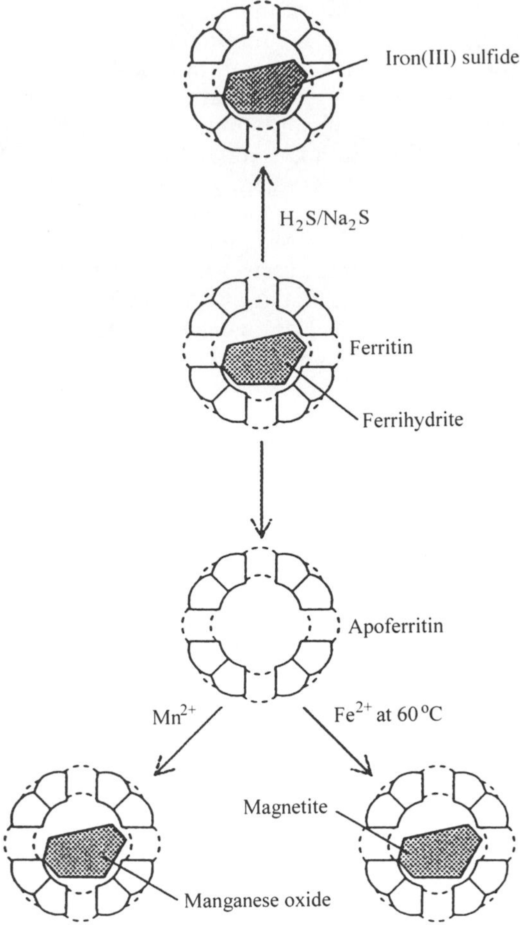

Figure 11 Use of the protein cage of ferritin in the biomimetic nanoscale synthesis of inorganic materials.

hydrated uranium oxides.[35,39] In all cases the resulting materials are nanocomposites of mineral and protein which form stable sols of discrete biocompatible particles. Simple incubation of apoferritin with Mn^{II} salt solutions results in the specific uptake of Mn^{II} by the protein and subsequent aerial oxidation and deposition of MnO_2H in the polypeptide cage. This chemistry is reminiscent of Fe^{II} oxidation. The protein-encapsulated MnO_2H cores are amorphous and structural characterization by EXAFS reveals a local Mn^{III} environment consisting of six oxygen atoms in a distorted octahedral geometry.[40] A further advantage of this approach is that metal loading of the ferritin cavity can be controlled to some degree by the stoichiometry of the reaction mixtures. Thus, nanophase cores with different particle sizes can be prepared. In addition, mixed-metal oxide cores (FeO_2H–MnO_2H) can be synthesized by sequential reconstitution of the protein cage with different metal salt-containing solutions.[37]

Whereas the iron atoms in the mineral cores of native ferritin are antiferromagnetically coupled, anaerobic reaction of apoferritin with Fe^{II} solutions at elevated temperature and pH in the presence of a stoichiometric oxidant results in a ferrimagnetic nanocomposite, called magnetoferritin.[38] This material consists of a monodisperse sol of 7 nm diameter magnetite–maghemite (Fe_3O_4–γ-Fe_2O_3) protein-encapsulated nanocrystals (Figure 12). The ^{57}Fe Mössbauer spectra were more similar to the oxidized maghemite phase than the mixed valence magnetite mineral.[41] Interestingly, the field dependent magnetization curves for magnetoferritin, measured at 300 K, showed saturation behaviour but no hysteresis, consistent with superparamagnetic properties.[42] The data could be fitted with a Langevin function with an approximate magnetic moment of 13 000 Bohr magnetons per molecule. In addition, magnetoferritin was shown to significantly increase the relaxation times for water protons.[43] Thus, coupling the magnetism of the mineral core with the biocompatibility of the polypeptide shell suggests that magnetoferritin will have potential applications as an effective contrast agent in magnetic resonance imaging of biological tissue.

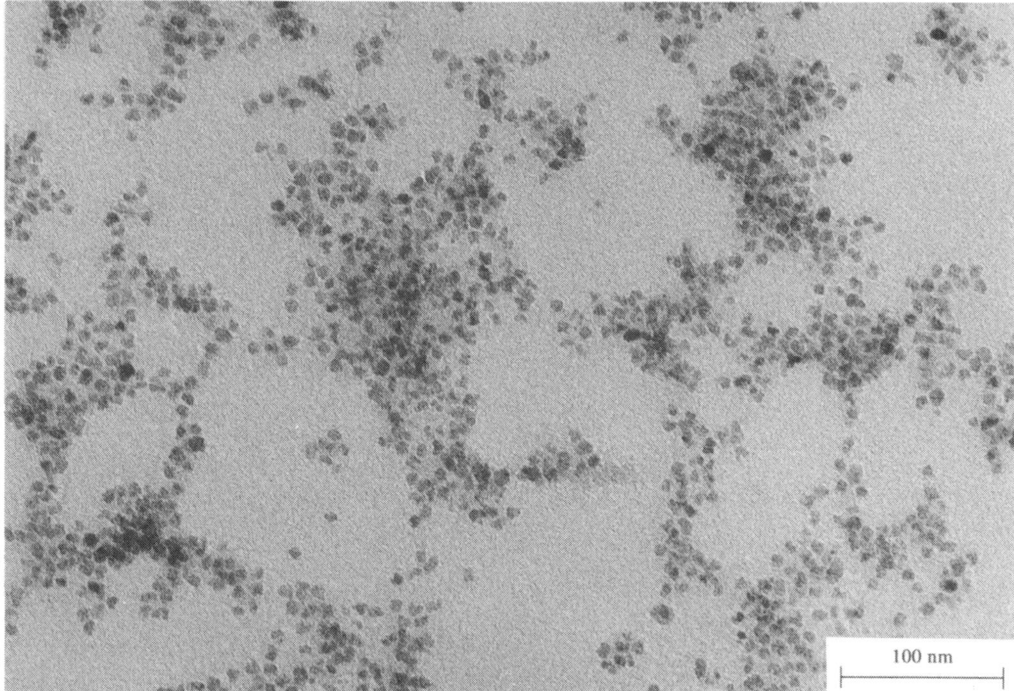

Figure 12 Ferrofluid of magnetoferritin molecules formed by the chemical synthesis of magnetite (Fe_3O_4) within the 8 nm protein cage of ferritin.

Finally, mineralization of apoferritin with uranyl oxyhydroxide by hydrolytic polymerization of UO_2^{2+} solutions results in a bioinorganic nanocomposite which may have potential applications in neutron-capture therapy.[39] The uranyl cation has been shown to enter the protein cavity and bind to specific residues.[23] At pH 8.0, slow hydrolytic condensation will displace the bound cations as the oxide develops. However, the process is not highly specific since bulk precipitation is also observed along with the intraferritin material.[35] Incubation of the protein in UO_2^{2+} solutions first at pH = 4.5 followed by an increase to 7 in the presence of phosphate buffer resulted in the formation of uranyl phosphate mostly within the polypeptide cage.[39] One possibility is that the rise in pH results in a nonequilibrium pH gradient across the protein shell which affords a kinetic preference for intraprotein nucleation over precipitation in the bulk solution.

15.6 INTERFACIAL MOLECULAR RECOGNITION

15.6.1 Molecular Blueprinting in Biomineralization

The second stage in the fabrication of biological minerals involves the controlled nucleation of inorganic clusters from aqueous solution. In this process, the first-order molecular construction of organic supramolecular systems, such as vesicles, ferritin micelles and polymeric networks, provides a framework for the second-order assembly of the inorganic phase. These preorganized architectures consist of functionalized surfaces which serve as blueprints for site-directed inorganic nucleation (Figure 5). Although specific details are not yet available, it is generally considered that the assembly of mineral nuclei is governed by electrostatic, structural and stereochemical complementarity at the inorganic–organic interface.[5,44] This aspect of molecular tectonics, in which interfacial molecular recognition facilitates the construction of nuclei, often with specific crystallographic structures and orientation, is not only a central feature of controlled biomineralization, but has important generic implications in synthetic materials chemistry.

In general, we can consider the role of an organic matrix, such as collagen or ferritin, analogous to that of an enzyme in solution, with the incipient inorganic nucleus as the corresponding substrate. However, the long-range electrostatic forces of ionic surfaces and the requirement of space symmetry indicate that factors such as lattice geometry, spatial charge distribution, hydration, defect states and surface relaxation need to be considered along with the stereochemical requirements of ion binding at the interface. Unfortunately, a description of the molecular forces operating at interfaces involving inorganic clusters and macromolecular frameworks is not currently available. However, we can state that the role of an organic surface involved in inorganic crystallization is primarily to lower the activation energy of nucleation ($\Delta G^{\ddagger}$). (There is also the possibility of influencing the collision frequencies (pre-exponential factors) of the nucleation rate.) For the most simple case,

$$\Delta G^{\ddagger} = B\sigma^3\upsilon^2/(kT\ln S)^2 \tag{3}$$

where B is a constant ($16\pi/3$ for a spherical nucleus), σ is the interfacial energy, υ is the molecular volume, k is the Boltzmann constant, T is the temperature and S is the supersaturation. Equation (3) is derived by assuming that nuclei will only develop into stable entities if the energy released through the formation of bonds in the solid state is greater than that required to maintain the newly created solid–liquid interface. No account is taken of the dependence of $\Delta G^{\ddagger}$ on the two-dimensional structure of different crystal faces or of the effect of extraneous surfaces in the medium. Clearly, both these factors influence the interfacial energy such that there may be an ensemble of nucleation profiles that are crystallographically specific and dependent on the nature of the substrate. Although unconventional, it is useful to consider the nucleation of biominerals in terms of the general ideas of transition state theory, with incipient nuclei of different structure or orientation represented as a series of activated clusters of different $\Delta G^{\ddagger}$. Consequently, their steady-state concentration and frequency with which they transform into thermodynamically stable entities will be dependent on their corresponding reaction trajectories which, in turn, are determined by the specificity of molecular recognition processes. In this way, metastable polymorphs (e.g., vaterite, ferrihydrite) and specific crystal faces can be preferentially nucleated by the stabilization of particular transition states at the matrix surface.

The specific lowering of the activation energy of nucleation reflects a requirement for structural and stereochemical complementarity between the inorganic and organic surfaces. Coordination environments in the mineral phase can be simulated by metal ion binding to appropriate ligands exposed at the organic surface. Carbonate and phosphate biominerals tend to be associated with carboxylate-rich (aspartate, glutamate) and phosphorylated (phosphoserine) proteins, respectively;[9] in both cases, the organic residues can mimic the oxyanion stereochemistry of particular crystal faces, and this may be sufficient to induce oriented nucleation. Similarly, the biological deposition of silica[45] and ice[46] is associated with hydroxy-rich macromolecules such as polysaccharides and serine–threonine-rich proteins.

15.6.1.1 Ion binding and spatial charge

The mineralization of iron oxide cores within the protein cavity of mammalian ferritin molecules indicates that nucleation and growth are specifically influenced by the nature of the polypeptide surface. Spectroscopic evidence and chemical modification studies suggest that the important sites are carboxylic residues lying close to the subunit dimer interface.[23] Work using recombinant human mutant ferritins has shown that the H-chain but not the L-chain subunit has ferroxidase activity, and crystallographic studies have identified the ferroxidase centre ligands as two neighbouring glutamates and a histidine close to the inner surface which are not present in the L-chain.[23] Although oxidation can proceed at these sites, it is unlikely that they act as unique nucleation centres since a critical cluster involving many Fe^{III} ions must be stabilized. However, the clustering of glutamate residues at the subunit dimer interface could provide the necessary charge density to act as the nucleation zone (Figure 13).[47] It is feasible that iron(III) species formed at the ferroxidase centre can readily migrate into the cavity because of the electrostatic field of neighbouring glutamate residues located on the cavity surface (Figure 14). This localized anionic patch might then serve to accumulate cationic species thereby increasing the probability of condensation to oligomeric oxo-containing species. However, the level of molecular recognition in this system is limited in length scale because of the short-range charge and polar interactions. Similar kinetic processes are envisaged to occur in lipid vesicles containing charged headgroups and intercalated proteins, as well as in ice-nucleating bacteria, where nucleation centres are established by the aggregation of specific proteins within domains of the cell wall.[46]

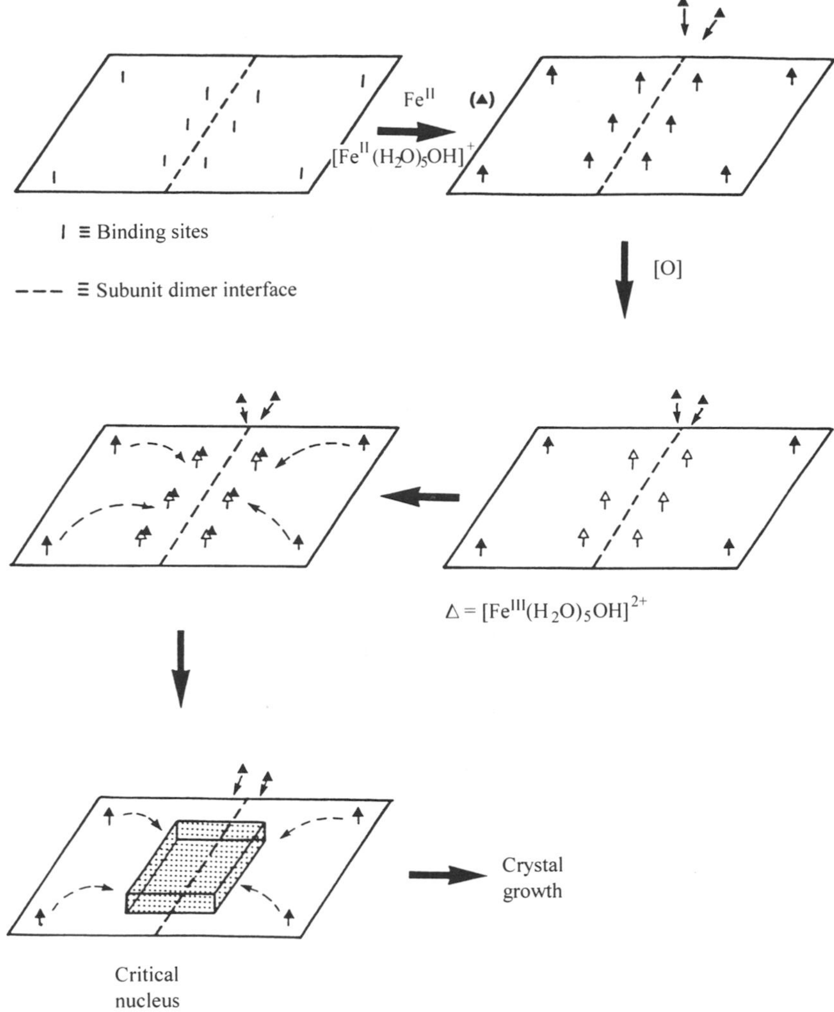

Figure 13 Possible nucleation mechanism of iron oxide in ferritin. The subunit dimer interface contains a cluster of possible binding sites for Fe^{II}. Oxidation of Fe^{II} at the ferroxidase site creates a localized concentration of Fe^{III}. This region is now activated such that it becomes the focus of additional Fe^{II} diffusion and oxidation. The activation energy of nucleation is overcome when the number of localized Fe^{III} species attains a critical size. Crystal growth continues by further oxidation of Fe^{II} within the nucleation zone.

Studies have investigated the effect of site-directed modifications in ferritin on the reconstitution of MnO_2H cores within the protein cavity.[37] With native horse spleen apoferritin, the kinetics of MnO_2H mineralization were very similar to those measured for FeO_2H formation in recombinant L-chain ferritin (rLF), suggesting no specific metallo-oxidase activity for Mn^{II}. Kinetic studies of recombinant H- and L-chain homopolymers, as well as H-chain variants containing site-directed modifications at the putative ferroxidase and nucleation centres, showed no significant differences in uptake or oxidation of Mn^{II}. This suggests that MnO_2H formation within ferritin occurs by a nonspecific pathway, at least in the initial stages of the reaction. One possibility is that the supramolecular assembly has two chemically complementary surfaces—an outer surface which through electrostatic interaction serves to inhibit the development of MnO_2H nuclei in bulk solution, and an inner cavity surface that is chemically neutral, being essentially passive toward nucleation and growth. Thus, this 'Janus' architecture facilitates nucleation within the protein cage via an indirect pathway. Moreover, this process is autocatalytic because once a nucleus is formed then it acts as a promotor of crystal growth which effectively competes against subsequent nucleation in bulk solution.

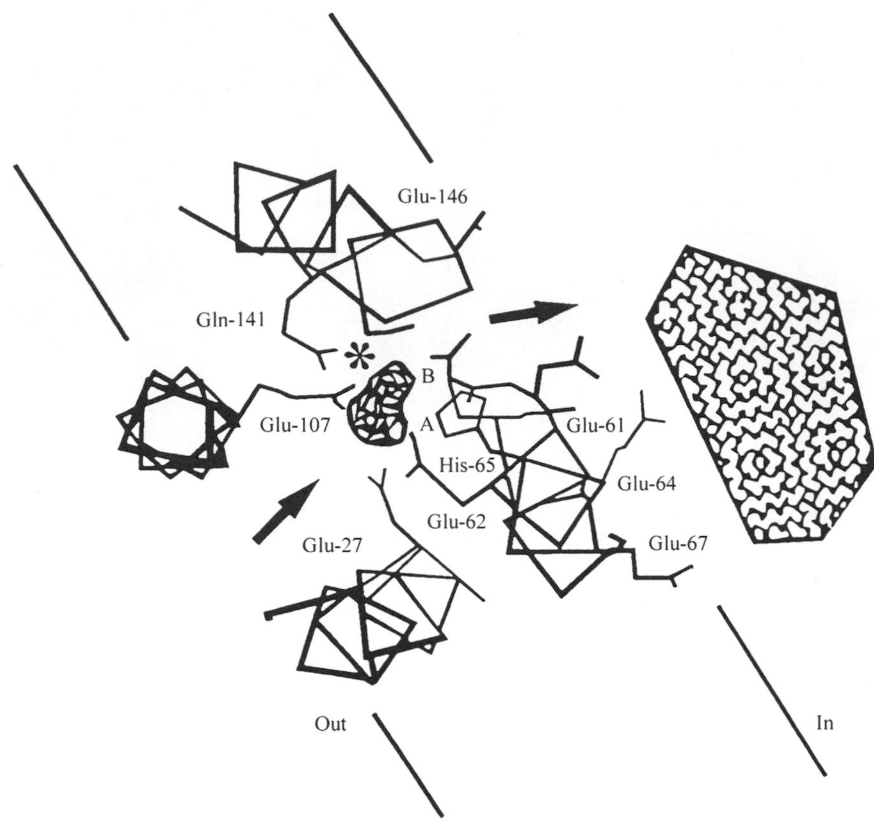

Figure 14 Molecular details of the oxidation and nucleation sites for iron oxide mineralization in ferritin. One polypeptide subunit is shown spanning the protein shell. Iron(II) species, present in the external environment, bind at the ferroxidase centre (*) where they undergo rapid oxidation. Migration of Fe^{III} into the cavity results in mineral nucleation at a site comprising three glutamate residues.

15.6.1.2 Structural and stereochemical complementarity

In extended assemblies, periodic structures such as antiparallel β-pleated sheets (shells), α-helices (fish antifreeze proteins), polysaccharide matrices, and proteolipid and phospholipid membranes, can control the assembly of nuclei along specific crystallographic directions. In this respect, the secondary, tertiary and quaternary conformations of macromolecules are key features of the blueprint required for controlled nucleation. In such cases, the long-range translational symmetry of inorganic lattices is established at the surface of the organized organic surface by regulating the spatial disposition of functional groups across the matrix surface (Figure 15).

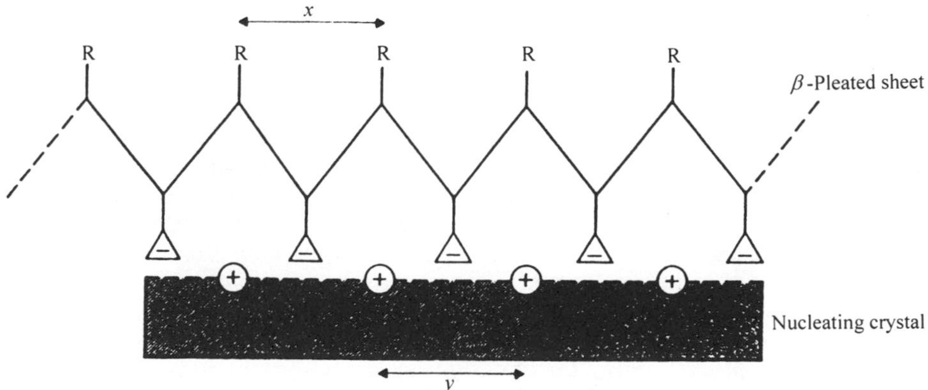

Figure 15 Geometric matching (epitaxy) in biomineralization. Cation–cation distances in one specific crystal face are commensurate with the spacing of periodic binding sites on an organized supramolecular surface.

Unfortunately, compared with ferritin, much less is known about biomineralization matrices involved in oriented nucleation. The exception is collagen which has been studied in detail.[48] However, although it is generally accepted that bone crystals are nucleated in the interstices of a crystalline assembly of collagen fibrils,[49] it remains unresolved whether nucleation is activated directly by the matrix or indirectly through noncollagenous proteins adsorbed near to or in the hole zones. In invertebrates, electron diffraction studies of partially demineralized mollusc shells have shown that in some species both the *a* and *b* axes of an antiparallel β-pleated sheet protein are aligned with the *a* and *b* crystallographic directions of aragonite ($CaCO_3$).[50] Partial amino acid sequencing of these acidic proteins[51] has indicated that there are repeated domains of poly-aspartate which could be the nucleation centres. In general, evidence is mounting that surface-binding motifs involving blocks of sequences such as $[Asp]_n$ and $[PSer]_n$ are common throughout biomineralization (Table 5).

Table 5 Amino acid motifs associated with biomineralization.

Function	*System*	*Macromolecule*	*Motif*
Structural framework	Bone/dentine	Collagen (type 1)	$[GXZ]_{338}$
	Crab cuticle	α-Chitin	β-(1,4)GlcNAc
	Mollusc shell	β-Chitin	β-(1,4)GlcNAc
	Enamel	Amelogenin	MPLPPHPGHPGYINFSYEVLTPLKWYQ (1–27)
			TDKTKREEVD (170–180)
	Plants	Cellulose	β-(1,4)Glc
Binding, nucleation, inhibition	Bone (rat)	Sialoprotein	EEEGEEEE (77–84)
			DEEEEEEEEEE (155–164)
		Osteopontin	DDDDDDDDDG (70–79)
		BAG-75	EEEEDDED/E (10–17)
	Dentine	Phosphophoryn	[DDDDDDYSDSDSSDSDD]
			[SSSSSSSS]
	Oyster shell	Glycoproteins	$[E]_{15-20}$, $[EX]_n$, $[S]_m$
	Sea urchin larval spicule	Glycoproteins	$[WVGDNQAWVENPE]_{15}$
	Bovine milk	β-Casein	ESLSSSEES (14–22)
	Cells/plasma	Ferritin	EEE (61,64,67)
	Yeast	CdS peptides	$[\gamma\text{-EC}]_n G$
	Bacteria	INA proteins	$[AGYGSTLT]_{122}$
	Polar fish	α-Helical peptides	$[AAT]_n$
	Coccoliths	Polysaccharides	$[Man]_n$ + [GalU]

Source: Mann.[10]
Residue positions given in parentheses. A = alanine, C = cysteine, D = aspartic acid, E = glutamic acid, F = phenylalanine, G = glycine, H = histidine, I = isoleucine, K = lysine, L = leucine, M = methionine, N = asparagine, P = proline, Q = glutamine, R = arginine, S = serine, T = threonine, V = valine, W = tryptophan, Y = tyrosine, X, Z = spacer residues, T = threonine O-linked to disaccharide, S = serine oxygen-linked to phosphate, BAG = bone acidic glycoprotein, INA = ice nucleating active, Man = sulfated mannose, GalU = D-galacturonic acid, Glc = glucose, GlcNAc = *N*-acetylglucosamine.

How do these surfaces regulate oriented nucleation? One possibility is that there is geometric matching (epitaxy) between the lattice spacings of ions in crystal faces and functional groups arranged across the organic surface. For example, the distances between aspartic acid residues deployed along a β-pleated sheet are similar to the calcium–calcium distances in the nucleated aragonite (001) face observed in the mollusc shell.[50] As the binding constants for calcium at carboxylate sites are not high, it has been suggested that nucleation is a cooperative process involving structurally disordered sulfate groups of flexible oligosaccharide side chains and organized motifs of carboxylate ligands within β-pleated sheet surface domains of the matrix.[4] The former provides a flux of calcium to the nucleation site, while the latter induces oriented nucleation. These suggestions are borne out by model systems in which calcite crystals were grown on sulfonated polystyrene films with or without adsorbed polyaspartic acid.[52] Rigid, highly sulfonated films induce preferential nucleation of the calcite (001) face and this is increased 10-fold in the presence of adsorbed polyaspartate in the β-sheet conformation. Adsorption of polyglutamate, which mainly adopts a random conformation, does not show this effect.

15.6.2 Biomimetic Assemblies and Oriented Inorganic Nucleation

Controlling the nucleation of inorganic materials is an important consideration in materials and colloidal chemistry. Oriented inorganic substrates such as gold and silicon are used as epitaxial surfaces in chemical vapour deposition methods, and it would be significant if similar processes could be routinely undertaken in aqueous solutions. Over the last few years, a biomimetic approach involving the use of compressed Langmuir monolayers as organized organic substrates for the oriented nucleation of inorganic materials has been developed (Table 6).[53,54] These ideas have been applied to the use of functionalized self-assembled monolayers (on gold or silica substrates) in the oriented nucleation of zeolites[55] and thin films.[56] One particular advantage of these approaches is that the functionality and packing of the supramolecular surface can be readily modified to provide a molecular blueprint for inorganic nucleation. In our studies, the effect of changing the head group identity, polarity and packing conformation of the amphiphile on the crystallization of inorganic solids has been investigated. For this work, compressed Langmuir monolayers of long-chain alkyl carboxylates ($Me(CH_2)_{17-24}CO_2H$), sulfates ($Me(CH_2)_{20}OSO_3$), amines ($Me(CH_2)_{17}NH_2$) and alcohols ($Me(CH_2)_{17}OH$) were spread on supersaturated solutions of metal salts. With the exception of aliphatic alcohols, which inhibited nucleation, the pattern of precipitation was consistent; the crystals nucleate almost exclusively at the monolayer–solution interface, the particles thus formed exhibit a unimodal size distribution and the induction time to nucleation is significantly reduced. Most importantly, in many instances, complementarity between the surface chemistry and structure of the surfactant film and crystal faces of the incipient nuclei results in the oriented nucleation of two-dimensional arrays of discrete crystals at the monolayer–solution interface.

Table 6 Biomimetic approaches to the oriented nucleation of inorganic materials.

Approach	*Strategy*	*Product*	*Systems*	*Materials*
Molecular blueprinting	Oriented nucleation	Single crystals	Langmuir monolayers	NaCl, $CaCO_3$, $BaSO_4$, PbS, $CaSO_4{\cdot}2H_2O$
			Self-assembled monolayers	Zeolites, TiO_2
			PolyAsp/polystyrene	$CaCO_3$

Source: Mann.[18]

In general, work has indicated that the inductive influence of ordered organic surfaces on inorganic nucleation can be explained by a model based on a combination of three interfacial processes[54] (Figure 16): (i) the accumulation of charged species and stabilization of specific electrostatic interactions by cation binding; (ii) the translation of geometric information through organized cation binding; and (iii) the provision of stereochemical complementarity between cations bound to the monolayer headgroups and the local coordination environments adopted by ions in the crystal structure of the incipient nuclei.

15.6.2.1 Calcium carbonate

Compressed monolayers of long-chain alkyl carboxylates induce the oriented nucleation of $CaCO_3$ in different polymorphic forms depending on the Ca^{2+} concentration of the subphase.[57] Calcite oriented with the $[1\bar{1}.0]$ axis perpendicular to the monolayer forms at subphase conditions of $[Ca^{2+}] = 10\,mM$, whereas crystals of the metastable polymorph, vaterite, nucleate on a (00.1) face at $[Ca^{2+}] = 5\,mM$. If the polarity of the monolayer is reversed (e.g., by using positively charged films of octadecylamine) then only vaterite nucleates across this concentration range. In this case, however, two distinct orientations of the vaterite crystals are observed; type I crystals oriented with the *c*-axis perpendicular to the monolayer, and type II crystals with the *a*-axis aligned perpendicular to the monolayer–solution interface.[58] The precise orientation of the crystals was deduced from morphological studies of the mature crystals and confirmed by electron diffraction investigations of immature crystals collected at different times during the growth process.[59] On the basis of this morphological and crystallographic data, and in the light of information concerning the putative packing conformation of the surfactant molecules and possible structural motifs which characterize the faces which nucleate, a molecular model was proposed to explain the interfacial relationships responsible for templated nucleation.

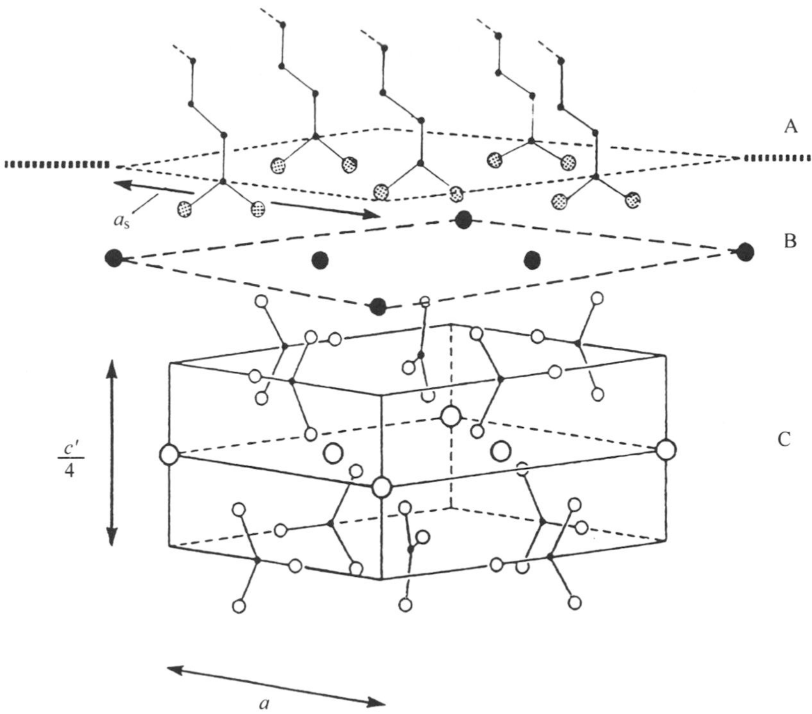

Figure 16 Schematic showing possible features of interfacial molecular recognition between an organized surfactant monolayer (A) and a specific crystal face of an incipient inorganic ($CaCO_3$) nucleus. The Stern layer (B) of bound Ca^{2+} ions is electrostatically equivalent to a specific face of the crystal. In addition, the distances between the carboxylate head groups and Ca^{2+} ions in the crystal face are similar. There is also stereochemical complementarity between the oxygen atoms of the carboxylate head groups and carbonate anions in the first layer of the nucleus (C).

First, two-dimensional crystallographic studies of amphiphilic molecules spread on the air–water interface have determined that long-chain carboxylates pack in a slightly distorted hexagonal unit cell (area per molecule, $A = a^2 \sin 60°$) with the molecules aligned vertically.[60] Second, there is evidence from compression isotherms[58] for the formation of a diffuse layer of metal ions beneath the charged surfactant surface, although recent data from x-ray and neutron reflectivity measurements of arachidate monolayers cast on a Ca^{2+} subphase suggest that the density of this layer is low with one or perhaps two water molecules screening each calcium ion and forming OH···O hydrogen bonds with the monolayer carboxylates. Under these conditions it is unlikely that the preferential accumulation of calcium ions in a diffuse, two-dimensional network beneath the monolayer would be sufficient to catalyse calcite nucleation. This is supported by the observation that oriented vaterite nucleation under carboxylate monolayers occurs at low calcium concentrations, whereas oriented calcite nucleation is observed at higher supersaturation levels where a well-defined Stern layer is present. Furthermore, addition of extraneous cations (Na^+ or Li^+), or monolayer-soluble molecules such as cholesterol, disrupt the Stern layer by competitive ion binding or structural modifications in the packing arrangement, respectively. This effect significantly reduces the number of oriented crystals formed under the monolayer.[54]

Clearly, as the oriented nucleation of vaterite under octadecylamine monolayers does not require the formation of a Ca^{2+} diffuse layer—since there is no bound Ca^{2+}—the translation of structural information other than geometric parameters must be involved at the monolayer–crystal interface. In essence, nucleation is a three- rather than two-dimensional process, and factors such as the stereochemical requirements of lattice ions must also be significant. For example, the $\{1\bar{1}.0\}$ face of calcite contains rows of carbonate anions oriented with their C_{2v} axis perpendicular to the surface, a motif which mimics the orientation of the CO_2^- (carboxylate) groups in the monolayer. Moreover, the carbonate–carbonate spacings (0.496 nm) in this face are commensurate with a possible interhead-group spacing of the monolayer (0.5 nm) suggesting a role for both geometric correspondence and stereochemical complementarity in the interfacial events mediating nucleation. Similar criteria can be invoked to explain the preferential nucleation of vaterite under carboxylate monolayers. No structural relationship exists between the (00.1) face of vaterite and the monolayer, but the coplanarity of the proposed hydrogen bonded $OH\cdots OCO_2^{2-}$ system favours the orientation of bound hydrogen carbonate ions from the subphase perpendicular to the monolayer, an orientation which mimics the crystal structure of vaterite. The selection of both the (00.1) and {11.0} faces of vaterite under amine monolayers is perhaps more difficult to rationalize, although both have carbonates aligned perpendicular to the crystal face nucleated under the organic surface. In this instance, solution conditions (pH 6.0) may favour the electrostatic binding of HCO_3^- ions orthogonal to the $—NH_3^+$ head group which would in turn facilitate some degree of stereochemical recognition between the monolayer and the nascent crystals. In summary, the results suggest that calcite nucleation is favoured by ion binding and the formation of a well-defined calcium carboxylate motif which complements the first layer of the $\{1\bar{1}.0\}$ face of the unit cell, whereas the criteria for vaterite nucleation appear to be less rigourous. Other factors such as the stoichiometry of Ca^{2+} binding at the charged organic surface or the inherent dynamical freedom of the monolayer may be significant in this case.

The requirement for stereochemical complementarity as an important determinant of oriented nucleation was confirmed in studies where the bidentate motif of carboxylate head groups was replaced with the trigonal pattern of a sulfate moiety.[61] It was predicted, and subsequently confirmed by experimentation, that a tridentate template of three oxygens favours the nucleation of calcite on a (00.1) face as it mimics the stereochemical disposition of the planar carbonates in this face. On the basis of this stereochemical complementarity, it was also predicted that a sulfate monolayer cast upon a saturated calcium bicarbonate subphase would, under suitable solution conditions ($[Ca^{2+}]$ 10 mM; $Ca^{2+}:Mg^{2+} = 1:5$), induce the nucleation of aragonite crystals with their [001] axes perpendicular to the monolayer–solution interface. This is because there is a close similarity in the crystal lattices of calcite and aragonite viewed along the *c*-axis. This prediction was confirmed by experimental observation.[61]

15.6.2.2 Barium sulfate

In all of the systems discussed above the influence of the organic template was confined solely to the events associated with nucleation. Studies with barium sulfate highlighted another aspect of template-directed crystallization since it was shown that the controlling influence of the organic surface could be extended to the growth process. Under compressed monolayers of an aliphatic long-chain sulfate, barium sulfate (barite) crystals nucleate with the (100) face parallel to the plane of the monolayer (Figure 17).[62,63] Once again, the nucleation events can be explained by the ability of the template to mimic both the geometry and stereochemistry of ions in the (100) face and translate this information to the forming crystals. The model is based on the assumption that the sulfate monolayer adopts a pseudohexagonal packing conformation, although at the present time there is no direct structural information on these monolayers. Given this caveat, a potential lattice match exists along the ⟨011⟩ and [010] directions (97% and 94%, respectively) if a distortion of 8° (the ⟨011⟩ and [010] axes subtend an angle of 128°) is present in the putative hexagonal cell of the close-packed surfactant molecules (Figure 18). This geometric correspondence could be augmented by the stereochemical complementarity between the three oxygens of the exposed sulfates of the (100) face and the alkyl sulfate headgroups pendent at the monolayer–solution interface. A close lattice match can also be proposed between the (001) face of barite and the hexagonal packing conformation of the surfactant molecules, but in this case, the oxyanions expose only two oxygens at the crystal surface compared with the tridentate motif of the sulfate headgroups.

In addition to a preferred orientation relative to the monolayer, a unique feature of the $BaSO_4$ crystals grown under the sulfate monolayers was their unusual morphology.[63] Each crystal was a thin anisometric plate distinguished by a nanodimensional texture comprising an interconnecting mosaic of some 20–40 nm diamond-shaped subunits bounded by four well-defined {011} side faces (Figure 17). Electron diffraction confirmed that despite their complex architecture each plate was a single crystal and high-resolution lattice imaging across the boundaries of adjacent subunits revealed them to be crystallographically continuous.

Substituting the sulfate head group with a phosphonate moiety did not influence the crystallographic orientation of the nucleated barium sulfate crystals,[64] a not unexpected result given the close similarity in the electrostatic, structural and stereochemical features of these anionic monolayers. The limiting area per molecule of the phosphonate monolayer (0.52 nm) is only marginally smaller than that of the sulfate monolayer (0.55 nm) and a good two-dimensional geometric match can be identified along the ⟨011⟩ and [010] axes of the (100) face assuming, (i) that the amphiphile will pack in a distorted hexagonal pattern once compressed, and (ii) that Ba^{2+} will bind in two

Figure 17 Oriented single crystal of $BaSO_4$ nucleated under compressed alkyl sulfate monolayers. The {100} face is aligned parallel to the surfactant head groups and is highly textured.

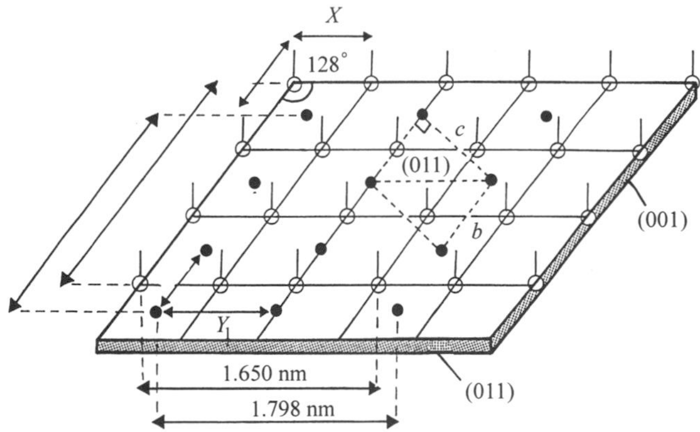

Figure 18 Schematic showing the possible binding motifs of Ba^{2+} ions (filled circles) in the (100) face of barite with the sulfate head groups of a compressed monolayer. Note that there is geometric correspondence at length scales of approximately 1.6–1.7 nm.

distinct motifs at the interface. As discussed above, there is the possibility that the interfacial oxygens of the phosphonate monolayers mirror the stereochemical disposition of sulfate anions in the (100) crystal surface with the result that nucleation is facilitated by local coordination requirements. However, a notable textural difference was observed between the barite crystals formed under phosphonate and sulfate monolayers. The former exhibited a single rhombic unit which showed considerable growth extension along the [100] axis. In mature crystals, a single thin crystalline plate was seen to extend directly from this prism. Another morphology (representing some 15% of the crystal population) was also seen in the phosphonate system; in this case the rhombic unit formed the central prominance of a 'bow-tie' structure formed by the rapid growth of the crystal along the [001] directions. The side faces defining this 'bow-tie' were again of the {011} type.

15.6.2.3 Calcium sulfate dihydrate

It was assumed that the failure of alcohol monolayers to induce the oriented nucleation of calcium carbonate or barium sulfate was related to the neutrality of the head group which limits cation binding and the formation of a complementary stereochemical motif.[54] However, polar alcohol head groups should be complementary to crystal layers containing hydroxyl ions or water molecules. This assumption has been verified by work on template-directed ice[65] and hydrated phase gypsum ($CaSO_4{\cdot}2\,H_2O$)[66] nucleation. Monolayers of amphiphilic alcohols ($C_nH_{2n+1}OH$), in contrast to the analogous acids, will nucleate ice crystals from supercooled water droplets as a result of precise structural matching between the monolayer OH groups and the layer of ice in the *ab* plane. Similar experiments with gypsum showed that whereas negatively charged head groups (sulfate and phosphate) give a 60:40 mixture of *c*-axis elongated needles aligned normal and parallel to the monolayer, respectively, octadecanol, although less effective at promoting nucleation, induced nucleation of the (010) face such that essentially 100% of the gypsum needles were oriented with the *c*-axis parallel to the monolayer surface. The (010) face lies parallel to the layers of water molecules in the unit cell and appears to be effectively stabilized by polar interactions with the hydroxyl head groups of the octadecanol monolayer. Thus, although charge and stereochemical interactions are not selective in determining the orientation of the gypsum crystals under the monolayer surface, hydrogen bonding at the crystal–monolayer interface is an important aspect of nucleation for hydrated crystals.

15.7 VECTORIAL REGULATION

15.7.1 Pattern Formation and Biomineralization

The third stage of biomineralization is associated with the assembly of the mineral phase through crystal growth and termination. In the absence of further cellular intervention, mineral nuclei will continue to grow within the confines of their supramolecular hosts according to the laws of crystallization. The resulting particles are constrained in size but exhibit normal crystallographic structures and morphologies. In many organisms, however, a third order of tectonic complexity can be established by vectorial regulation of crystallization, in which biominerals are endowed with unusual and elaborate textures and shapes.

The subtle and intricate decoration of diatom frustules, coccolith scales, acantharian skeletons, and so on, arises from the dynamic shaping of vesicles under cellular stresses. Simple elongated morphologies are probably produced by affiliation of the vesicles with cytoskeletal frameworks such as radial (tubulins) or tangential (spectrins) stress filaments and microtubules. Construction of long vesicles against a cell wall is common, for example, in the production of elongated magnetite crystals in magnetotactic bacteria[67] and silica rods in choanoflagellates.[68] In the latter case, individual vesicles are stretched out to such an extent that each associated mineral rod has a radius of curvature determined by the shape of the external cell membrane. This is essentially a static process with the physical shaping of the vesicle assembly analogous to the casting of a hard material into an artisan's mould. The siliceous architecture of diatom frustules is generated in a similar fashion except that the vesicles are no longer one-dimensional but extend into a two-dimensional network against the cell wall.[69] More complex shapes, such as the single crystal calcite elements of coccolith plates (Figure 4) are often produced under the regulation of dynamical processes that involve both radial and tangential movements. The spatial and temporal sequence of morphogenetic development is under genetic control such that the mineralized architectures are often unique and species-specific.[19]

Crystal and particle growth within shaped biological compartments is coupled to the chemical processing of these environments.[70] For instance, specific polymorphic structures are deposited, often at different locations within the same biomineralized material (e.g., calcite and aragonite in some shells, three different iron oxides ($Fe_2O_3{\cdot}n\,H_2O$, γ-FeO_2H, and Fe_3O_4) in molluscan teeth). Clearly, the regulation of physiochemical properties of these localized biological environments is an important determinant of crystal growth, aggregation and texture. Figure 19 illustrates some generalized strategies for controlling the supersaturation conditions of solutions encapsulated within preformed supramolecular assemblies.[71] In summary, regulation can be achieved by facilitated ion flux, complexation–decomplexation switches, local redox and pH modifications, and changes in local ion activities via ionic strength and vectorial water fluxes.

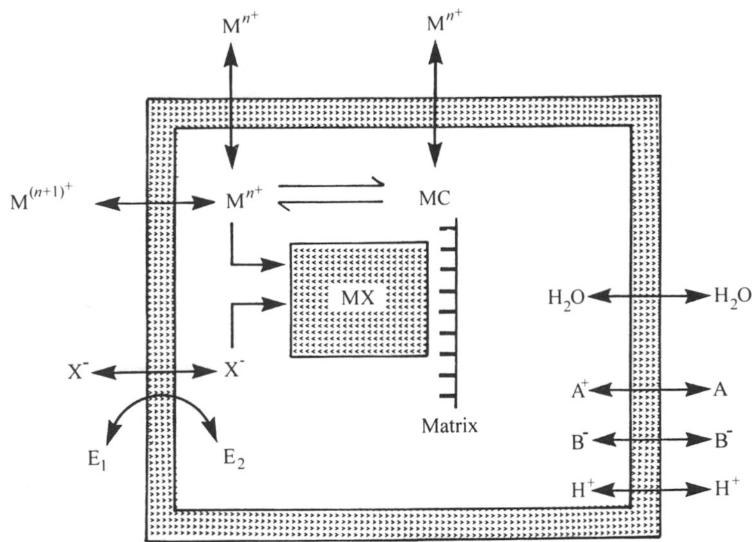

Figure 19 Generalized strategies for controlling supersaturation in biomineralization. The mechanisms can be either direct (membrane pumps, complexation (C), and enzymatic regulation (E_1,E_2)) or indirect (water, H^+ and ion fluxes).

In some systems, for example, in the formation of shaped crystals of magnetite in magnetotactic bacteria, there is also the possibility that the spatial location of ion pumps within an enclosed supramolecular assembly can provide a vectorial mechanism for shaping the crystal morphology. If ions flow into the localized compartments only at specific ports of entry, then these sites will be the initial regions of mineral growth. If these sites are now turned off and other pumps further along the membrane are switched on, then the mineral will develop along preferred directions due to vectorial flow of the ion stream. There is no definitive evidence for this mechanism although it provides an intriguing prospect for the design of synthetic systems exhibiting process control within shaped architectures.

The vectorial regulation of biomineral shapes often reflects a compromise between the constraints of crystal physics and the genetic manipulation of morphology. For example, crystalline minerals, such as calcite ($CaCO_3$), exhibit preferential growth directions and habits; this anisotropy can be utilized by organisms in the construction of elongated architectures (e.g., spicules) in which the direction of fast growth (*c*-axis) is aligned with the morphological extension (stress filaments) of the vesicle system. In other cases, the genetic patterning of vesicle morphogenesis offsets the intrinsic crystallographic symmetry to produce complex shapes (e.g., coccoliths) that bear no resemblance to the underlying crystal structure. Other organisms, such as acantharians,[72] fabricate biomineral architectures that integrate structural units based on both crystallographic and biological constraints.

15.7.2 Supramolecular Templates and Crystal Engineering

In general, inorganic crystals nucleated on organized organic supramolecular substrates such as Langmuir monolayers show no modifications in shape because the growth processes are unperturbed from those operating in bulk solution. One possibility is to induce shape in the inorganic material through templating the growth process across a preformed patterned organic surface that is suitably functionalized to induce site-specific nucleation (Table 7). For example, biomolecular templates such as bacterial fibres,[73] rhapidosomes[74] and S-layer proteins[75] can be used for this purpose. The resulting crystals are often randomly arranged, but the microscopic morphology is controlled. Another possibility is to extend the two-dimensional organization of monolayers into three dimensions by utilizing the ability of certain chiral lipids to self-assemble into high axial ratio microstructures.[76,77] The theoretical basis for some of these structures involves the generation of bilayer curvature due to packing long strings of strongly interacting chiral amphiphiles within the gel-phase bilayer.[78] Given that the microstructures ultimately result from the repeated molecular recognition of a particular side of an amphiphile by its neighbour, it is not surprising that pure lipids have been the main materials studied. However, cylindrical microstructure formation occurs in complex biologically derived mixtures of galactosylceramide (galactocerebroside) and its two subfractions, hydroxy fatty acid cerebroside (HFA-Cer) and nonhydroxy fatty acid cerebroside (NFA-Cer).[79,80] Under conditions of dehydration with 1,2-ethanediol, these cerebrosides can form stable microstructures very similar to the cochleate cylinders and helical ribbons that have been obtained with the pure synthetic and semisynthetic lipids.

Table 7 Biomimetic approaches to the crystal engineering of inorganic materials.

Approach	*Strategy*	*Product*	*Systems*	*Materials*
Crystal engineering	Templating	Shaped composites	Tubules	Cu, Ni, Al_2O_3, Fe oxides
			Bacterial fibres	Fe_2O_3, $CaCO_3$, CuCl
			Bacterial rhapidosomes	Pd
			S-layer proteins	Ta/W
	Directed growth	Textured crystals	Sea urchin proteins	$CaCO_3$
			Polyanionic peptides	$CaCO_3$
	Microstructural fabrication	Mineral–polymer composites	Polystyrene–butadiene	$CaCO_3$, CdS, Ca phosphate
			Polyvinyl chloride	TiO_2
			Polyacrylate films	Fe oxides, $BaTiO_3$
			Polysiloxanes	$CaCO_3$, $CaSiO_3$
			Polyethylene oxide	CdS
			Collagen	Ca phosphate

Source: Mann.[18]

Mineralization of these preformed high axial ratio templates results in the fabrication of fibrous organic–inorganic composites. For example, helical ribbons consisting of diacetylenic phosphatidylcholine can be coated with aluminum hydroxide or carbonate,[81] metallized with electroless nickel,[82] or mineralized with preformed colloidal silica.[83] However, many of these mineralization reactions are somewhat nonspecific with much of the inorganic component also being formed in the bulk solution. This is probably because the binding of metal ions to the lipid head groups is relatively weak due to the presence of strong interhead-group interactions between neighbouring lipid molecules in the crystalline assembly. We have observed a similar effect with rolled-up multilamellar cylinders of uncharged galactocerebrosides (HFA-Cer) but the specificity can be significantly increased by the inclusion of low levels of anionic (sulfated) galactocerebroside (S-Cer) molecules within the noncharged tubular assembly (Figure 20).[84] The resulting cylinders are open-ended multilamellar tubules with a diameter of ~120 nm and a central lumen of 10–30 nm. Incubation of the lipid suspension in ethylene glycol with freshly prepared 10 mM $FeCl_3$ in 0.10 mM HCl at room temperature under air gave mineralized tubes coated in a granular and sheet-like iron oxide (Figure 20, route A, and Figure 21). Selected area electron diffraction patterns recorded on individual tubules indicated that the inorganic coating was poorly crystalline lepidocrocite (γ-FeO_2H) with a small fraction of ferrihydrite ($Fe_2O_3{\cdot}nH_2O$). The sheet-like material was identified as γ-FeO_2H by high-resolution lattice images which clearly showed a 0.66 nm (020) *d* spacing. FTIR spectroscopy of the lipid–mineral composite suggested that the hydrocarbon and amide functionalities were largely unchanged within the molecular structure associated with mineralized tubules, while the galactosyl interface was partially modified due to the presence of iron oxyhydroxide and/or water of hydration.

Significantly, controls without iron did not cause lipid coagulation or sedimentation, and microstructures not doped with S-Cer (i.e., 100% HFA-Cer) did not develop the orange colour.[84] Thus, the sulfated sugar head groups act as metal-ion binding sites for Fe^{III} which, under appropriate solution conditions, facilitate the specific nucleation and templated growth of iron oxides from aqueous solution. The resulting material is an organic–inorganic composite consisting of discrete fibres of the mineralized supramolecular assembly, often several micrometres in length yet only 30–50 nm in cross-section.

In situ conversion of the lepidocrocite-coated tubules to a magnetic composite was carried out under N_2 at room temperature, by adding a small volume of deaerated $(NH_4)_2SO_4{\cdot}Fe^{II}SO_4{\cdot}6\,H_2O$ solution to a suspension of the Fe^{III} mineral-coated tubules to give a 1:1 molar ratio of Fe^{II}:Fe^{III} (Figure 20, route B).[84] The phase transformation reaction was then initiated by addition of 100 mM NaOH to raise the pH to 8.5. In some experiments, iodate (IO_3^-) was also added as an oxidant to produce tubules extensively mineralized with crystals of poorly-ordered goethite (α-FeO_2H). In other experiments, the reaction mixture was maintained at pH 8–9 by additional

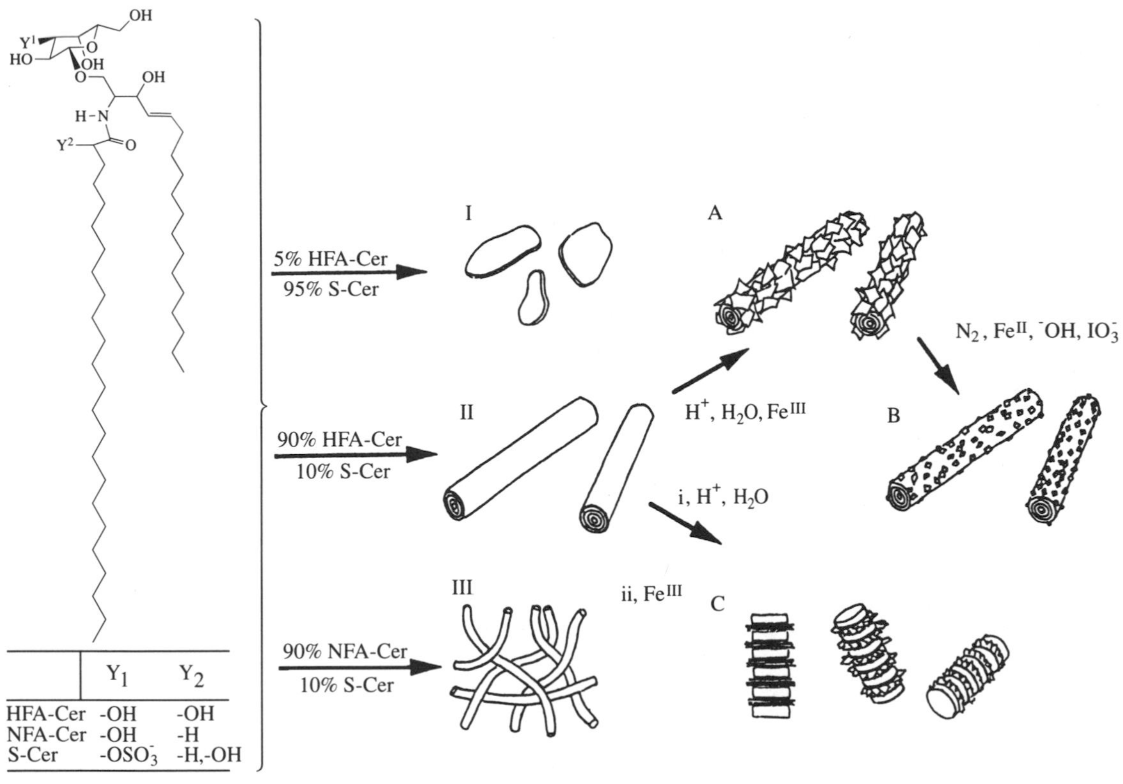

Figure 20 Generalized scheme for the fabrication of fibrous synthetic organic–inorganic composites. Galactocerebroside mixtures can produce a variety of microstructures in ethylene glycol suspensions: I, lamellar disks; II, multilamellar tubules; III, viscoelastic gels of unilamellar tubules. Various synthetic routes are available (A, B, C).

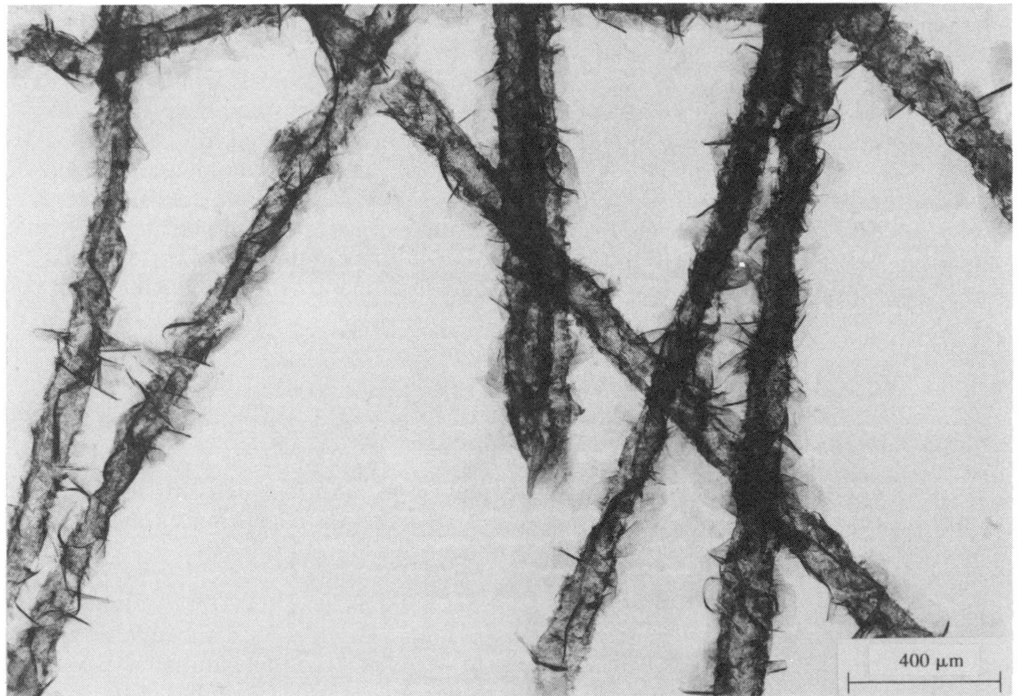

Figure 21 Lipid–mineral fibrous composites formed from the nucleation and growth of iron oxide (γ-FeO_2H) on sulfate-doped galactocerebroside tubules.

aliquots of NaOH and left for 4 d. A black tubule-containing precipitate, which was responsive to the field of a bar magnet, was isolated. Electron microscopy images showed that the lipid tubules were encrusted with clusters of fine-grained inorganic material which was identified by electron diffraction as the ferrimagnetic inverse spinel phase magnetite (Fe_3O_4).

To date, mineralization of cerebroside templates has only been achieved on the external surfaces of the microstructure. It would be very interesting if one could synthesize reproducibily tubule-based composites in which the inorganic phase resided solely within the 10–30 nm central lumen of the supramolecular assembly. The synthesis of organic-lined inorganic filaments is currently in progress.

Very different results were obtained if a S-Cer-doped lipid tubule suspension in ethylene glycol was incubated for 10 min in acid solution prior to addition of Fe^{III} (Figure 20, route C).[84] With this protocol, a S-Cer-rich phase was observed by transmission electron microscopy (TEM) to originate from partial phase separation from the parent tubules. Subsequent Fe^{III} addition caused aggregation of the segregated anionic lipid into stacks of lamellae interspaced with nanometre thin layers of poorly-crystalline iron oxyhydroxide. Presumably, Fe^{III} binds preferentially to the surfaces of the discs because of their high negative charge and the concomitant neutralization of surface charge gives rise to stack formation. The initial mineral was transformed over a period of weeks into sheet-like crystals of lepidocrocite which were aligned by confinement between the stacked lipid lamellae. The stacked disc nanocomposite was also produced directly in pure S-Cer controls but the product was less well-defined. As the discs are approximately 6.5 nm thick, the mineralized system is a nanocomposite of inorganic and organic layers. Although more structural work is required to determine the details of this association, it is possible that organized stacks of inorganic–organic materials could be of potential significance in the catalytic, photochemical and electronic application of finely divided solids.

Knowledge of the phase behaviour of galactocerebrosides[85] offers scope for the production of a range of supramolecular microstructures that are potential templates for mineralization reactions resulting in shaped hybrid composites. Thermal cycling of the NFA-Cer resulted in a viscoelastic gel composed of a significant fraction of very long unilamellar cylinders with lumen diameters of ~17 nm. Anisotropic lamellar micelle shards of NFA-Cer were also detected by negative staining. The sulfated cerebroside formed short unilamellar cylinders (lumin diameter ~44 nm) with no obvious helical substructure. The mineralization of these templates will be the subject of future work.

The above approaches have been concerned with the shaping of discrete inorganic materials through the use of preformed organic templates. Similar strategies could be explored in the formation of extended composites with hybrid microstructures. There is much activity in the synthesis and characterization of consolidated materials prepared by *in situ* inorganic precipitation within organic matrices such as functionalized polymers,[86–8] polymer gels[89] and biomolecular matrices (Table 7).[90] One of the main problems to date is the low mineral content of the resulting inorganic–organic composites and the absence of well-defined organic microstructures over these longer length scales. One possible way around this is to run the system in reverse, that is, incorporate a soluble polymer (polyethylene oxide) into a preformed mineral matrix such as the interlayers of a mica,[91] or alternatively, precipitate both mineral (calcium silicate/aluminate) and polymer (polyvinyl alcohol) simultaneously.[92,93] The ability to simultaneously tune both the rates of inorganic precipitation and the self-assembly of organic molecules will be an important step forward in the synthesis of consolidated organized materials. The recent application of surfactant self-assemblies and templating in the formation of mesoporous silicas[94] is essentially based on this principle.

15.8 CELLULAR PROCESSING

15.8.1 Higher-order Networks in Biomineralization

The final stage of biomineralization is associated with a variety of constructional processes involving larger scale cellular activity and the production of higher-order architectures with elaborate ultra- and microstructural properties. It is at this level of construction that the disparity between biological and synthetic materials is most striking.

The shaping of individual biominerals is often accompanied by their assembly into organized ultrastructures. Construction of these higher-order architectures within the cellular space can be based on either static or dynamic principles. An example of an organizational motif that is relatively rigid and spatially fixed is the linear assembly of a chain of membrane-bound magnetite crystals in magnetotactic bacteria (Figure 2). Here, the organization of individual vesicles apposed to the cell wall results in a compass needle of crystals and a permanent magnetic dipole within the organism. By contrast, silica rods formed in the protozoan *Stephanoeca diplocostata Ellis* are fabricated within intracellular vesicles but subsequently exocytosed as building blocks of an extracellular microstructure.[68] This single-celled organism (choanoflagellate) is commonly found in coastal waters around Europe and the Mediterranean. The cells comprise a colourless protoplast with a single anteriorly directed flagellum surrounded by a ring of tentacles (the collar). The protoplast is lodged in an open-ended basket-like casing (lorica) constructed of 150–180 amorphous silica strips (costal rods) (Figure 22). New costal rods are produced in advance of mitosis within long thin vesicles in the peripheral cytoplasm and then released sideways through the plasmalemma so that on cell division the juvenile, taking the supernumerary strips with it as it leaves the parent lorica, is able to assemble its own basket within 2–3 minutes.[68]

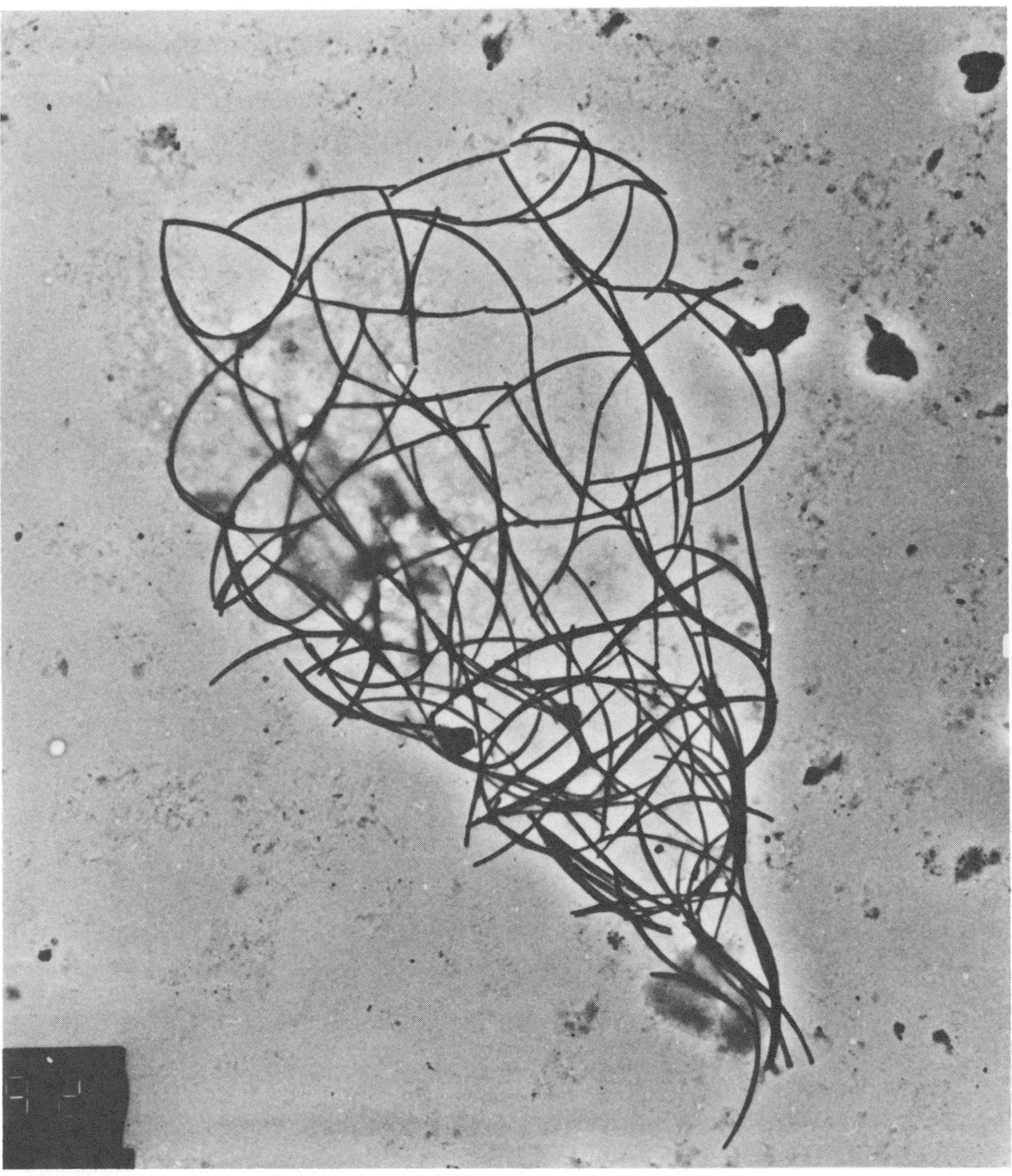

Figure 22 Intact basket (lorica) of the unicellular protozoan, *S. diplocostata*, constructed through the organized assembly of curved rods of silica. The rods are glued together during the assembly process.

The process of basket construction involves the joining together of costal rods in a manner which allows the intact lorica to be resistant to forces arising in the marine environment. The junctions between costal rods are glued in place during the construction process. The connective material is generally less electron dense than the adjacent siliceous material of the costal strips. High-resolution electron micrographs have indicated that silica and probably organic material constitute the join.[95] This suggests that the inorganic polymer has some residual flow properties which enable silica to move into the junction prior to hardening of the join. Interestingly, the silica-containing costal rods are metastable and slowly redissolve in the aqueous medium. The process of demineralization results in the formation of hollow rods rather than surface fragmentation.[95] An interesting implication of this preferential dissolution is that the rods become hollow without significant reduction in their mechanical strength (compare the use of tubular steel rods in building scaffolding). Thus the silica basket remains functional, that is, intact, even though the costal strips have undergone extensive demineralization. Only when the tubular walls become very thin does fracture and buckling occur.

Macroscopic organized architectures such as bone, shells and teeth are associated with extended extracellular matrices and their hierarchical structure produces composite materials capable of being remodelled by active cellular processing in response to mechanical stress. A feature of these systems is the slow repetitive processing of the mineralization zone. For example, the organized architecture of cuttlefish bone is constructed from the episodic deposition and processing of an extracellular polymeric (chitin) matrix.[96] Similarly, in the nacreous layer of shells, sheet-like organic assemblies are secreted periodically beyond the mineralization front, such that progressive infilling results in the construction of a highly organized lamellar architecture.[97]

While the details of the recognition and organizational processes involved in the construction of these higher-order biomineralized architectures are currently unknown, it is probable that the phenomenon is a consequence of what one might describe as cellular logistics. In particular, in unicellular organisms, such as coccoliths and choanoflagellates, it appears that the organized architectures are generated from the coupling of the (intracellular) construction process with generic processes involving the movement and organization of materials to and from the cell periphery. These processes utilize motive forces such as those generated by vesicle trafficking along cytoskeletal frameworks, flagella activity or cytoplasmic streaming. By contrast, multicellular systems possibly construct higher-order mineralized architectures by coupling biomineralization to intercellular communication concerned with the organization of supplies and functions (services) across relatively long length scales.

15.8.2 Crystal Tectonics and Organized Inorganic Materials

Controlling the organized assembly of biominerals by cellular processing of mineralized building blocks is clearly the most difficult aspect of molecular tectonics to translate into the world of materials chemistry. These strategies are based on molecular interactions but involve constructional order at much longer length scales such that the organization is best described as a process of crystal tectonics. Clearly, the sophisticated use of cellular activity in constructing higher-order architectures is difficult to reconcile with the static methods of chemical fabrication. However, there are some encouraging signs in the work reported in the 1990s (Table 8).

Table 8 Biomimetic approaches to the architectural assembly of organized inorganic materials.

Approach	*Strategy*	*Product*	*Systems*	*Materials*
Crystal tectonics	Architectural assembly	Organized materials	Monolayers/Au	CdS (preformed)
			Monolayers	Fe_3O_4 (preformed)
			Cast bilayers films	Fe_3O_4
			Hydroxyethyl cellulose	$CaCO_3$
			Polyacrylate sols	$BaSO_4$
			SiO_2/OH gel	$CaCO_3$
			Lipid stacks	FeO_2H
			Bicontinuous media	$Ca_{10}(OH)_2(PO_4)_6$

Source: Mann.[18]

One possible approach is to organize preformed CdS particles from solution by chemisorption onto self-assembled 1,6-hexanedithiol monolayers on gold surfaces.[98] A similar process has been observed with conventional monolayers in the presence of aqueous-dispersed iron oxide crystals.[99] Alternatively, preformed inorganic particles can be rendered hydrophobic by adsorbed long-chain surfactants and subsequently organized at the air–water interface as thin monolayer films.[100] Although the level of materials organization is not high, these studies suggest that organized materials can be fabricated if we can develop appropriate recognition processes between the pre-formed constituent units.

We have adopted a different approach in which the construction of inorganic architectures is directed by the use of a reaction medium with a reticulated microstructure.[101] Bicontinuous microemulsions can be readily formed from mixtures of the cationic surfactant, didodecyldimethyl ammonium bromide (DDAB), water, and a long-chain alkane. On the basis of many physical measurements (NMR spectroscopy, neutron scattering, etc.), there is a general consensus that the microstructure of this system is based on a fluctuating reticulated network of nanoscale water conduits within an oil matrix. The DDAB surfactant resides at the oil–water interface, thus stabilizing the framework from phase separation (Figure 23). The idea behind our approach was to replace the water with a supersaturated calcium phosphate solution such that crystals would slowly nucleate at the surfactant head groups and grow along the spatially delineated aqueous channels. Furthermore, the junctions between channels would provide sites for the intergrowth of individual crystals such that a replica of the bicontinuous framework would be established. Since the connectivity of the liquid network is dynamic, long-chain alkane oils with freezing points above 0 °C were used to generate immobilized (frozen) frameworks which still contained liquid water. Scanning electron micrographs of the mineralized replicas extracted after two weeks showed the presence of highly reticulated microstructures of interconnecting needle-like (0.2–1 μm in length) crystals of hydroxyapatite ($Ca_{10}(OH)_2(PO_4)_6$) (Figure 24). The majority of materials prepared were macroporous with pore diameters up to several micrometres in diameter and wall thicknesses of 50–130 nm depending on the storage time and composition of the microemulsion mixture. Modifications in the microstructure were observed with changes in composition. For example, calcium phosphates formed within microemulsions with compositions close to the edge of the bicontinuous phase domain were composed of bead-like arrangements of interconnected spherical calcium phosphates particles, consistent with the onset of spherical micelle formation prior to macroscopic phase separation.

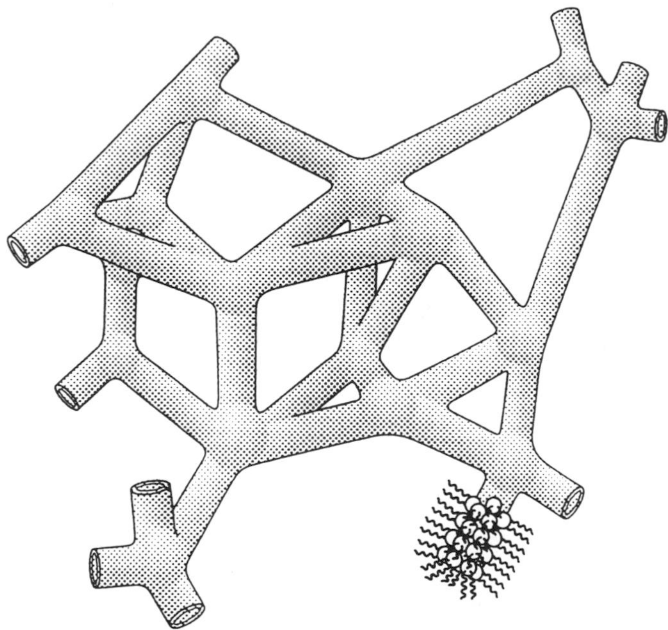

Figure 23 Schematic showing the network microstructure of water-in-oil bicontinuous microemulsions used for the fabrication of macroporous reticulated calcium phosphate.

A striking feature of the mature reticulated microstructures was that the wall diameters (~80 nm) were incommensurate with the 1 nm diameter water channels considered to be present in bicontinuous microemulsions. Microstructures commensurate with these dimensions were only observed at the very early stages of precipitation (2–6 h at +2 °C) where filamentous strings of mineral particles dominate. The results suggest that nucleation of calcium phosphate is initially

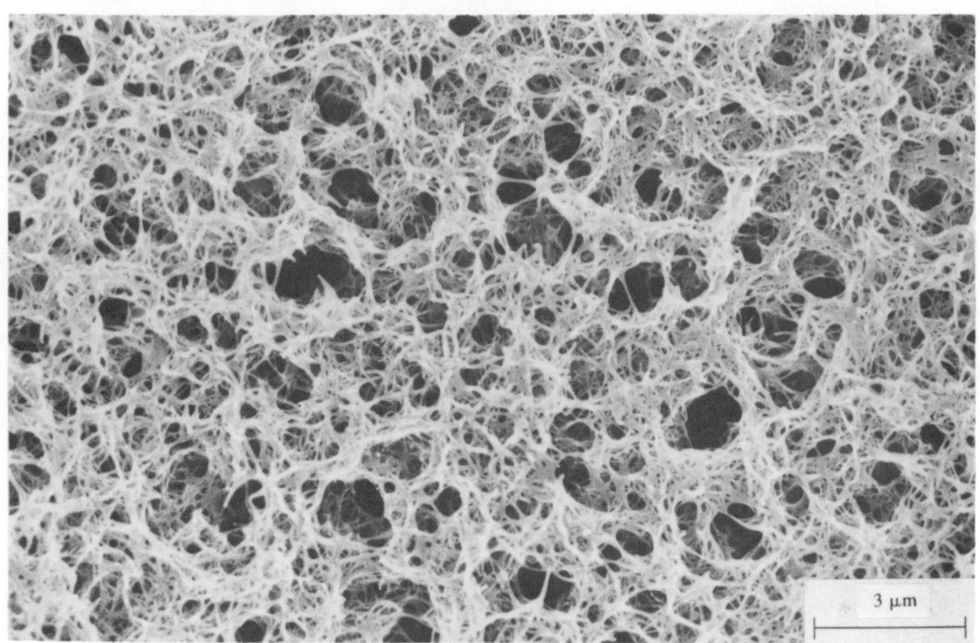

Figure 24 Scanning electron micrograph of reticulated calcium phosphate grown within a frozen bicontinuous microemulsion.

restricted to the interconnecting water conduits of the DDAB bicontinuous microemulsion but that subsequent growth within the frozen oil medium disrupts the microstructure possibly by adsorption of the surfactant molecules onto the developing crystal surfaces. Moreover, construction of the reticulated framework appears to be facilitated by the needle-like morphology of the hydroxyapatite crystals which serves to provide effective interlinking and interlocking of the architecture at relatively early stages of formation. Indeed, analogous experiments with CdS and calcite, which do not adopt acicular habits, failed to produce macroporous materials.

15.9 CONCLUSIONS

In this chapter I have attempted to indicate how the chemical fabrication of organized inorganic-based architectures in biomineralization is critically dependent on organic supramolecular assemblies. The coupling of supramolecular systems with inorganic materials chemistry represents a molecular tectonics approach to the science of chemical construction of organized architectures, and is providing new insights into biomimetic strategies in materials research.

Four stages have been identified in the synthesis-with-construction of biominerals. First, a building plot is constructed through the use of preorganized organic supramolecular assemblies. Second, the head groups of this assembly are spatially organized and chemically functionalized such that the organic surface acts as a blueprint for inorganic nucleation. In many cases, the level of interfacial molecular recognition is sufficiently high to result in oriented crystallization. Third, spatial patterning of the assembly over longer length scales controls the vectorial aspects of crystal growth so that elaborate microarchitectures are fabricated. Finally, networks of supramolecular assemblies are processed by cellular forces in the construction of higher-order architectures, often exhibiting hierarchical organization. Each of these bioconstructional stages has a synthetic counterpart in materials chemistry such that exploration of the interconnections between supramolecular assembly, templating and organization, and the self- or facilitated construction of new inorganic-based materials offers much promise in the design and fabrication of advanced products and processes.

15.10 REFERENCES

1. P. D. Calvert and S. Mann, *J. Mater. Sci.*, 1988, **23**, 3801.
2. S. Mann, D. D. Archibald, J. M. Didymus, B. R. Heywood, F. C. Meldrum and V. J. Wade, *Mater. Res. Soc. Bull.*, 1992, **17**, 32.

3. A. H. Heuer, D. J. Fink, V. J. Laraia, J. L. Arias, P. D. Calvert, K. Kendall, G. L. Messing, J. Blackwell, P. Rieke, D. H. Thompson, A. P. Wheeler, A. Veis and A. I. Caplan, *Science*, 1992, **225**, 1098.
4. L. Addadi and S. Weiner, *Angew. Chem., Int. Ed. Engl.*, 1992, **31**, 153.
5. S. Mann, D. D. Archibald, J. M. Didymus, T. Douglas, B. R. Heywood, F. C. Meldrum and N. J. Reeves, *Science*, 1993, **261**, 1286.
6. H. A. Lowenstam and S. Weiner, 'On Biomineralization', Oxford University Press, New York, 1989.
7. S. Mann, J. Webb and R. J. P. Williams (eds.), 'Biomineralization: Chemical and Biochemical Perspectives', VCH, Weinheim, 1989.
8. K. Simkiss and K. Wilbur, 'Biomineralization: Cell Biology and Mineral Deposition', Academic Press, San Diego, CA, 1989.
9. S. Weiner, *CRC Crit. Rev. Biochem.*, 1986, **20**, 365.
10. S. Mann, *J. Chem. Soc., Dalton Trans.*, 1993, 1.
11. M. D. Ross and K. G. Pote, *Philos. Trans. R. Soc. London, Ser. B*, 1984, **304**, 445.
12. K. M. Towe, *Science*, 1973, **179**, 1007.
13. C. C. Perry, S. Mann and R. J. P. Williams, *Proc. R. Soc. London, Ser. B.*, 1984, **222**, 427.
14. B. E. Volcani and T. L. Simpson, 'Silicon and Siliceous Structures in Biological Systems', Springer, Berlin, 1982.
15. S. Mann, N. H. C. Sparks and R. G. Board, *Adv. Microb. Physiol.*, 1990, **31**, 125.
16. C. T. Dameron, R. N. Reese, R. K. Mehra, A. R. Kortan, P. J. Carroll, M. L. Steigerwald, L. E. Brus and D. R. Winge, *Nature*, 1989, **338**, 596.
17. A. Berman, J. Hanson, L. Leiserowitz, T. F. Koetzle, S. Weiner and L. Addadi, *Science*, 1993, **259**, 776.
18. S. Mann, *Nature*, 1993, **365**, 499.
19. J. R. Young, J. M. Didymus, P. Bown, B. Prins and S. Mann, *Nature*, 1992, **356**, 516.
20. B. S. C. Leadbeater, *Proc. R. Soc. London, Ser.B.*, 1984, **304**, 529.
21. Y. A. Gorby, T. J. Beveridge and R. P. Blakemore, *J. Bacteriology*, 1988, **170**, 834.
22. E. W. de Vrind-de Jong, A. H. Borman, R. Thierry, P. Westbroek, M. Grüter and J. P. Kamerling, in 'Biomineralization in Lower Plants and Animals', eds. B. S. C. Leadbeater and R. Riding, Clarendon Press, Oxford, 1986, pp. 205–217.
23. P. M. Harrison, S. C. Andrews, P. J. Artymiuk, G. C. Ford, J. R. Guest, J. Hirzmann, D. M. Lawson, J. C. Livingstone, J. M. A. Smith, A. Treffry and S. J. Yewdall, *Adv. Inorg. Chem.*, 1991, **36**, 449.
24. T. G. St Pierre, S. H. Bell, D. P. E. Dickson, S. Mann, J. Webb, G. R. Moore and R. J. P. Williams, *Biochim. Biophys. Acta*, 1986, **870**, 127.
25. S. Mann, J. V. Bannister and R. J. P. Williams, *J. Mol. Biol.*, 1986, **188**, 225.
26. J. Currey, 'The Mechanical Adaptations of Bones', Princeton Scientific, NJ, 1984.
27. J. W. Smith, *Nature*, 1968, **219**, 157.
28. G. A. Ozin, *Adv. Mater.*, 1992, **4**, 612.
29. S. Mann and R. J. P. Williams, *J. Chem. Soc., Dalton Trans.*, 1983, 311.
30. S. Mann, J. P. Hannington and R. J. P. Williams, *Nature*, 1986, **324**, 565.
31. S. Bhandarkar and A. Bose, *J. Colloid. Interface Sci.*, 1990, **135**, 531.
32. B. R. Heywood and E. D. Eanes, *Calcif Tissue Int.*, 1987, **41**, 192.
33. H.-C. Youn, S. Baral and J. H. Fendler, *J. Phys. Chem.*, 1988, **92**, 6320.
34. S. Mann, M. J. Kime, R. G. Ratcliffe and R. J. P. Williams, *J. Chem. Soc., Dalton Trans*, 1983, 771.
35. F. C. Meldrum, V. J. Wade, D. L. Nimmo, B. R. Heywood and S. Mann, *Nature*, 1991, **349**, 684.
36. T. Douglas, D. P. E. Dickson, S. Betteridge, J. Charnock, C. D. Garner and S. Mann, *Science*, 1995, **269**, 54.
37. F. C. Meldrum, T. Douglas, S. Levi, P. Arosio and S. Mann, *J. Inorg. Biochem.*, 1995, **58**, 59.
38. F. C. Meldrum, B. R. Heywood and S. Mann, *Science*, 1992, **257**, 522.
39. J. F. Hainfeld, *Proc. Natl. Acad. Sci. USA*, 1992, **89**, 11064.
40. P. Mackle, J. M. Charnock, C. D. Garner, F. C. Meldrum and S. Mann, *J. Am. Chem. Soc.*, 1993, **115**, 8471.
41. Q. A. Pankhurst, S. Betteridge, D. P. E. Dickson, T. Douglas, S. Mann and R. B. Frankel, *Hyperfine Interact.*, 1994, **91,** 847.
42. J. W. M. Bulte, T. Douglas, S. Mann, R. B. Frankel, B. M. Moskovitz, R. A. Brooks, C. D. Baumgartner, J. Vymazel, M.-P. Strub and J. A. Frank, *J. Magn. Reson. Imaging*, 1994, **4**, 497.
43. J. W. M. Bulte, T. Douglas, S. Mann, R. B. Frankel, B. M. Moskovitz, R. A. Brooks, C. D. Baumgartner, J. Vymazel and J. A. Frank, *Invest. Radiol.*, 1994, **29**, S214.
44. S. Mann, *Nature*, 1988, **332**, 119.
45. D. M. Swift, and A. P. J. Wheeler, *Phycol.*, 1992, **28**, 202.
46. P. K. Wolber and G. J. Warren, *Trends Biol. Sci.*, 1989, **14**, 179.
47. V. J. Wade, S. Levi, P. Arosio, A. Treffry, P. M. Harrison and S. Mann, *J. Mol. Biol.*, 1991, **221**, 1443.
48. E. P. Katz, E. Wachtel, M. Yamauchi and G. Mechanic, *Connect. Tissue Res.*, 1989, **21**, 149.
49. W. Traub, T. Arad and S. Weiner, *Connect. Tissue Res.*, 1992, **28**, 99.
50. S. Weiner and W. Traub, *Proc. R. Soc. London, Ser. B.*, 1984, **304**, 425.
51. K. W. Rusenko, J. E. Donachy and A. P. Wheeler, in 'Surface Reactive Peptides and Polymers', eds. C. S. Sikes and A. P. Wheeler, 'ACS Symposium Series 444', American Chemistry Society, Washington, DC, 1991, pp. 107–124.
52. L. Addadi, J. Moradian, E. Shay, N. G. Maroudas and S. Weiner, *Proc. Natl. Acad. Sci. USA*, 1987, **84**, 2732.
53. S. Mann, B. R. Heywood, S. Rajam and J. D. Birchall, *Nature*, 1988, **334**, 692.
54. B. R. Heywood and S. Mann, *Adv. Mater.*, 1994, **6**, 9.
55. S. Feng and T. Bein, *Nature*, 1994, 368, 834.
56. H. Shin, R. J. Collins, M. R. de Guire, A. H. Heuer and C. N. Sukenik, *J. Mater. Res.*, 1995, **10**, 692.
57. S. Mann, B. R. Heywood, S. Rajam, J. B. A. Walker, R. J. Davey and J. D. Birchall, *Adv. Mater.*, 1990, **2**, 257.
58. S. Rajam, B. R. Heywood, J. B. A. Walker, S. Mann, R. J. Davey and J. D. Birchall, *J. Chem. Soc., Farad. Trans.*, 1991, **87**, 727.
59. B. R. Heywood, S. Rajam and S. Mann, *J. Chem. Soc., Faraday Trans.*, 1991, **87**, 735.
60. D. Jacquemain, S. Grayer-Wolf, J. Leveiller, M. Deutsch, K. Kjaer, J. Als-Nielson, M. Lahav and L. Leiserowitz, *Angew. Chem., Int. Ed. Engl.*, 1992, **31**, 130.

61. B. R. Heywood and S. Mann, *Chem. Mater.*, 1994, **6**, 311.
62. B. R. Heywood and S. Mann, *Adv. Mater.*, 1992, **4**, 278.
63. B. R. Heywood and S. Mann, *J. Am. Chem. Soc.*, 1992, **114**, 4682.
64. B. R. Heywood and S. Mann, *Langmuir*, 1992, **8**, 1492.
65. J. Majewski, L. Margulis, D. Jacquemain, F. Leveiller, C. Bohm, T. Arad, Y. Talmon, M. Lahav and L. Leiserowitz, *Science*, 1993, **261**, 899.
66. T. Douglas and S. Mann, *Mater. Sci. Eng.*, 1994, **1**, 193.
67. S. Mann, N. H. C. Sparks and R. P. Blakemore, *Proc. R. Soc. London., Ser. B*, 1987, **231**, 469.
68. B. S. C. Leadbeater, *Protoplasma*, 1979, **98**, 241.
69. C.-W. Li and B. E. Volcani, *Proc. R. Soc. London, Ser. B.*, 1984, **304**, 519.
70. S. Mann, *Struct. Bonding (Berlin)*, 1983, **54**, 125.
71. S. Mann, in 'Biomineralization: Chemical and Biochemical Perspectives', eds. S. Mann, J. Webb and R. J. P. Williams, VCH, Weinheim, 1989, pp. 35–62.
72. C. C. Perry, J. R. Wilcock and R. J. P. Williams, *Experientia*, 1988, **44**, 638.
73. N. H. Mendelson, *Science*, 1992, **258**, 1633.
74. M. Pazirandeh, S. Baral and J. R. Campbell, *Biomimetics*, 1992, **1**, 41.
75. K. Douglas, N. A. Clark and K. J. Rothschils, *Appl. Phys. Lett.*, 1990, **56**, 692.
76. P. Yager and P. E. Schoen, *Mol. Cryst. Liq. Cryst.*, 1984, **106**, 371.
77. N. Nakashima, S. Asakuma, J. M. Kim and T. Kunitake, *Chem. Lett.*, 1984, 1709.
78. W. Helfrich and J. Prost, *Phys. Rev. A.*, 1988, **38**, 3065.
79. D. D. Archibald and P. Yager, *Biochemistry*, 1992, **31**, 9045.
80. P. J. McCabe and C. Green, *Chem. Phys. Lipids*, 1977, **20**, 319.
81. J. S. Chappell and P. J. Yager, *Mater. Sci. Lett.*, 1992, **11**, 633.
82. J. M. Schnur, R. Price, P. Schoen, P. Yager, J. M. Calvert, J. Georger and A. Singh, *Thin Solid Films*, 1987, **152**, 181.
83. S. Baral and P. Schoen, *Chem. Mater.*, 1993, **5**, 145.
84. D. D. Archibald and S. Mann, *Nature*, 1993, **364**, 430.
85. D. D. Archibald and S. Mann, *Chem. Phys. Lipids*, 1994, **69**, 51.
86. E. Dalas, *J. Mater. Chem.*, 1991, **1**, 473.
87. P. D. Calvert, *Mater. Res. Soc. Bull.*, 1992, **17**, 37.
88. K. K. W. Wong, B. J. Brisdon, B. R. Heywood, A. G. W. Hodson and S. Mann, *J. Mater. Chem.*, 1994, **4**, 1387.
89. P. A. Bianconi, J. Lin and A. R. Strzelecki, *Nature*, 1991, **349**, 315.
90. O. Nakamura, D. J. Fink and A. I. Caplan, *Mater. Res. Soc. Symp.*, 1991, **218**, 275.
91. E. P. Gianellis, *J. Min. Met. Mater. Soc.*, 1992, **44**, 28.
92. P. B. Messersmith and S. I. Stupp, *Polym. Prepr.*, 1991, **32**, 536.
93. K. Kendall, J. D. Birchall and A. J. Howard, *Philos. Trans. R. Soc. London, Ser. A*, 1983, **310**, 139.
94. C. T. Kresge, M. E. Leonowicz, W. J. Roth, J. C. Vartuli and J. S. Beck, *Nature*, 1992, **359**, 710.
95. S. Mann and R. J. P. Williams, *Proc. R. Soc. London, Ser. B*, 1982, **216**, 137.
96. J. D. Birchall and N. L. Thomas, *J. Mater. Sci.*, 1983, **18**, 2081.
97. G. Bevelander and H. Nakahara, *Calcif. Tiss. Res.*, 1969, **3**, 84.
98. V. L. Colvin, A. N. Goldstein and A. P. Alivisatos, *J. Am. Chem. Soc.*, 1992, **114**, 5221.
99. X. Peng, Y. Zhang, J. Yang, B. Zou, L. Xiao and T. Li, *J. Phys. Chem.*, 1992, **96**, 3170.
100. F. C. Meldrum, N. A. Kotov and J. H. Fendler, *J. Phys. Chem.*, 1994, **98**, 4506.
101. D. Walsh, J. D. Hopwood and S. Mann, *Science*, 1994, **264**, 1576.

16

Hydrogen Bonding Control of Molecular Self-assembly: Recent Advances in Design, Synthesis, and Analysis

JOHN R. FREDERICKS and ANDREW D. HAMILTON
University of Pittsburgh, PA, USA

16.1 INTRODUCTION

16.1.1 Defining Self-assembly

Definitions of the terms "self-assembly" or "self-organization" abound in the literature. It will be helpful to begin this chapter by examining some of those definitions, looking for similarities among them.

The definition put forth by Hamilton and co-workers is "the noncovalent interaction of two or more molecular subunits to form an aggregate whose novel structure and properties are determined by the nature and positioning of the components."[1] This chapter will concentrate on hydrogen bonding as the primary noncovalent interactions used to stabilize and control aggregate formation. Other stabilizing forces such as hydrophobic interactions, charge–charge interactions, van der Waals attractions, π-stacking between aromatic systems, and metal–ligand binding are beyond the scope of this work.

Whitesides *et al.* define molecular self-assembly as "the spontaneous assembly of molecules into structured, stable, noncovalently joined aggregates."[2] Whitesides recognizes cooperativity as an important characteristic in self-assembling systems. There is a large decrease in entropy associated with bringing many molecules together to form a single aggregate. The enthalpy of the noncovalent interactions must be large enough to offset this unfavorable decrease in entropy. Individually, each noncovalent interaction may not be large enough to overcome this unfavorable entropy change, but taken together, they are able to promote aggregation. Also, the molecular components of an aggregate often will undergo a conformational change to promote aggregation. Cooperativity involves both conformational changes and multiple noncovalent interactions offsetting the decrease in entropy.

Lindsey defines self-assembly as "the spontaneous formation of higher-ordered structures."[3] Using examples from biology, he mentions advantages of self-assembly such as the need for only one or a few subunits in the assembly of large aggregates, control over the assembly of the aggregates depending upon reaction conditions, and error checking, whereby improperly placed subunits are removed in the process of building the aggregate. This last factor of error checking can be considered as a form of cooperativity: all of the subunits cooperate to form a larger, more stable aggregate. Lindsey separates self-assembly into seven different classifications, all of which contain these essential elements.

Lehn and Pfeil define self-organization as "the process by which specific components spontaneously assemble in a highly selective fashion into a well-defined, discrete supramolecular architecture."[4] For Lehn, molecules undergo self-assembly when they (i) selectively bind via molecular recognition, (ii) grow through sequential binding of the components in the proper orientation, and (iii) terminate assembling once the process has reached completion. Lehn also considers cooperativity to be an important characteristic of self-assembling systems.

This work will not limit itself to any one specific definition of self-assembly but will aim to include the following factors as characteristic of self-assembly.

(i) Self-assembling units are held together by noncovalent interactions.

(ii) The assembly of these subunits into larger aggregates is selective: the subunits bind cooperatively to form the most stable aggregate.

(iii) The aggregates can be recognized by their properties which differ from those of the individual components.

(iv) The aggregates are not infinite lattices, but rather of a definite size and molecular composition.

In general these are the factors that will be used to identify a self-assembling system, although occasionally we will venture outside these definitions. Some infinite lattices, liquid crystals, and oligomeric and polymeric chains have been designed using principles brought to light in the study of self-assembling systems, and are worth mentioning in this work.

16.1.2 Goals of Recent Work

Recent work in self-assembly can be divided into three areas: the study of aggregates which mimic natural systems, "nanoarchitecture," the design of solids and solution structures which act as molecular devices, and the study of the factors which influence self-assembly.

Much of the work in mimicking natural systems has been in designing helical structures which resemble DNA and RNA. Hamilton and co-workers have designed a supramolecular helix which is held together by hydrogen bonds,[5] and many groups have designed and characterized double and triple helices which are assembled via metal–ligand coordination. The study of self-assembling systems which are held together by hydrogen bonding may shed light on the biology of viruses.[6] The self-assembly of small peptides into larger, topologically well-defined proteins has been achieved using metal–ligand coordination.[7]

The future looks promising for the field of nanoarchitecture. Systems have been designed which act as host molecules for organic guests[8] and enclatharating agents.[9] Much progress has been made in the design of a molecule which can act as a molecular abacus,[10–12] and the groundwork has been laid for the creation of a molecular building set, systems which are fancifully but aptly named "Tinkertoys"[13] and "Molecular Meccano."[14] Hydrogen bonding,[15] π-stacking interactions[16] and metal–ligand coordination phenomena[17] have all been utilized in designing solution and solid-state structures, and liquid crystals.

Woven through all of these studies is the goal of elucidating factors which influence the structure of the self-assembled aggregate. Some recent work has shown the importance of preorganization,[18–20] the existence of cooperativity[21] and error-checking in the formation of discrete structures,[22] and the thermodynamics and kinetics of self-assembly.[23]

16.1.3 Scope and Structure of This Work

Two reviews of self-assembly were published in 1991.[2,3] It is the purpose of this chapter to review the advances made in the design and study of self-assembling systems since the publication of those reviews. Our main focus will be on aggregates which exist in solution, although some examples of solid-state structures will be given. These examples, however, will be limited to solid-state structures which were designed using principles of solution phase self-assembly. When examining the literature concerning self-assembly, interesting work on systems such as infinite crystal lattices, liquid crystals, and bilayer membranes presents itself. Unfortunately, these systems are beyond the scope of this chapter because we will mostly examine those systems which are of a finite size, whose assembly includes a "termination" step.[4]

16.1.4 Nomenclature

There are many different conventions used in the literature for naming self-assembling aggregates, which we hope will not create confusion when reading this chapter. In the interest of fairness, we generally will use the convention expressed in the original paper. For example, if a system is referred to as a "(1 + 2)" aggregrate, this means that the aggregate is composed of three subunits, one of one type and two of another. Adding the numbers in parenthesis will yield the number of subunits in the aggregate. However, the term "1·2" would refer to a complex formed between the molecule labeled "1" and the molecule labeled "2."

16.2 EXPERIMENTAL TECHNIQUES FOR THE CHARACTERIZATION OF SELF-ASSEMBLING SYSTEMS

Most standard techniques for characterizing covalently bonded molecules are used for characterizing self-assembling aggregates. These include, but are not limited to, ^{1}H NMR, ^{13}C NMR, cyclic voltammetry, fast-atom bombardment mass spectrometry (FAB-MS), x-ray crystallography, and various types of spectroscopy. However, there are solution phase techniques which yield specific information concerning aggregation. Several of these are detailed below.

16.2.1 Solubility

Often one insoluble molecule will dissolve upon the addition of another molecule, qualitatively suggesting the formation of a noncovalently bonded aggregate. The data can also provide the stoichiometry of self-assembled complexes. For example, in a paper by Whitesides and Seto, it was found that one molecule was able to solubilize up to, but not more than, one equivalent of another, lending evidence to a 1:1 stoichiometry for the complex.[18]

16.2.2 Gel Permeation Chromatography

Gel permeation chromatography (GPC) is used in polymer science to determine the molecular weight distribution of a polymer. It can also be used to determine the formula weight of an aggregate. The column is composed of a cross-linked polystyrene gel whose pores are in the range 1–10^{6} nm.[24] Molecules are separated on the basis of their hydrodynamic volume or Stokes radius.[25] Larger molecules are unable to enter the pores of the gel, and they pass through the column by way of the volume occupied only by solvent and not by gel beads. Therefore, larger molecules will elute before smaller molecules which can permeate the pores of the gel.[24]

The technique can be used to examine the kinetics of dissociation of an aggregate. If no dissociation takes place on the experimental timescale, then a single, well-defined peak will result, representing the aggregate. If some dissociation takes place, several different effects can be seen in the chromatogram. The aggregate peak may develop a tail, representing a small amount of dissociation, or one or two smaller peaks may elute after the aggregate peak, representing the individual components of the aggregate.[26] Some dissociation of self-assembling aggregates has been observed by GPC.[18]

Advantages of GPC are that it is a rapid technique and it requires less than 50 mg of sample.[25]

16.2.3 Vapor Pressure Osmometry

Vapor pressure osmometry (VPO) is a method for determining the molecular weight of a given species in solution. The instrumentation involves two thermistor beads which make up the arms of a Wheatstone bridge. A drop of pure solvent is placed on one bead, and a drop of solution containing solvent and the compound to be analyzed is placed on the other bead. Both thermistors are immersed in a chamber whose atmosphere is saturated with pure solvent vapor. Because the activity of pure solvent is less in the sample bead than in the pure solvent bead, some of the solvent vapor will condense onto the sample bead, its heat of condensation warming the sample bead. This warming changes the resistance and unbalances the bridge. The experimentally measured quantity is the amount of resistance required to rebalance the bridge.[27] The instrument is calibrated using solutions whose vapor pressure is well known. A calibration plot is created by plotting Dr/m vs. m, where Dr is the change in resistance and m is the molality. From this plot the formula weight of the molecule or aggregate is calculated.[28]

When VPO is applied to aggregation, each species in a solution is counted as one unit, whether it is an unbound molecule or a fully self-assembled complex. Different models for association can be considered and the stoichiometry and equilibrium constant of the aggregate can be determined.[29]

The advantages of VPO are that it is rapid, "fairly precise and accurate," and only requires a small amount of sample.[28] Whitesides and Seto's work suggests that the technique does have some limitations when applied to aggregation because different standards can give different calibration plots.[18]

16.2.4 ESR

ESR has been used in characterizing double and triple helicates assembled via metal–ligand coordination. Splitting in the ESR signal of metal–ligand complexes suggests that the metal centers are close enough to interact with one another. This splitting has been used as evidence for the formation of a helix, in which the metal centers would be held a short distance from one another by the ligand.[30]

16.2.5 NMR Techniques

Standard NMR techniques such as correlation spectroscopy (COSY), nuclear Overhauser enhancement (NOE) experiments, and temperature-dependent studies have all been used to characterize self-assembling complexes.

The Saunders–Hyne method has been utilized by Zimmerman and Duerr to determine the stoichiometry of a hydrogen-bonded aggregate.[31] The method uses data from an NMR dilution experiment to determine the state of aggregation. It assumes that at very low concentrations, only the monomer (or separate, unbound species if two different molecules are used) exists. At intermediate concentrations, it is assumed that there is an equilibrium between a monomer and some predominant n-mer (or between two species which can hydrogen bond with one another). By plotting the frequency of a given resonance vs. the log of the concentration, the data can be fit to different aggregation models to determine the stoichiometry of the predominant complex and the equilibrium constant.[32] An excellent review of NMR approaches to the study of aggregation states has been published.[33]

16.2.6 Mass Spectral Analysis

There are two mass spectral techniques which have been used considerably for the characterization of self-assembling aggregates. The most frequently used of these FAB-MS. In this technique, a liquid sample is dissolved in a matrix (often glycerol), and then bombarded with an atom beam. This bombardment produces gas-phase ions which can then be mass analyzed. This technique is particularly useful for aggregation studies because a liquid sample can be used, giving insight into the solution behavior of self-assembled aggregates. High molecular weights are accessible by this technique.[34]

Another ionization technique which has recently been utilized is electrospray mass spectrometry (ESMS). In this technique, a liquid sample is sprayed through a hypodermic needle which is held at a potential of a few kilovolts. This spraying produces charged droplets, which are able to give up their solute as multiply charged ions. ESMS may prove to be a more effective method of ionization than FAB-MS because ESMS has a higher ionization efficiency, and a much higher mass limit.[35]

Also, no sample preparation is required for ESMS. A liquid sample can be directly introduced without having to use an additional solvent, as in FAB-MS. One advantage to this is that the same liquid sample which has been examined by another technique, such as UV–visible spectroscopy, can be quickly introduced to the ESMS system.[36]

16.3 SELF-ASSEMBLY VIA HYDROGEN BONDING

16.3.1 Systems From a Single, Self-complementary Molecule

16.3.1.1 Dipyridone motifs

Perhaps the most economical approach to designing a self-assembling aggregate is to use only one, self-complementary molecule. This has been the approach of Wuest and Ducharme in their work with systems based on the 2-pyridone motif. Crystalline 2-pyridones are known to exist typically as hydrogen-bonded dimers (Figure 1). This type of association was also found to occur in solution, as confirmed by ^{1}H NMR and ^{13}C NMR studies.[37]

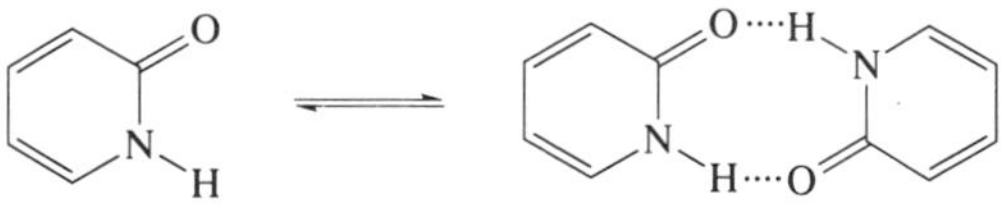

Figure 1 Hydrogen-bonded dimerization of 2-pyridones.

This motif was used to design a (1 + 1) self-assembling system in solution. Two pyridone units were linked by an alkyne group as a rigid spacer to form asymmetric and symmetric dipyridones (**1**) and (**2**), respectively. X-ray crystallographic studies showed that the asymmetric dipyridone (**1**) exists primarily as the dimer (**3**) in the solid state. Using ^{1}H NMR, VPO, and IR spectroscopy, the self-association was found to be particularly strong: the dipyridone exists as a dimer even at low concentrations (3.6×10^{-4} M), and has a free energy of formation greater than 27.2 kJ mol^{-1} at 25 °C. In contrast, crystalline symmetric dipyridone (**2**) adopts the planar polymeric motif (**4**). The choice of solvent was found to affect the ability of these pyridones to self-associate. In methanol and other hydrogen-bonding polar solvents, (**1**) and (**2**) are strongly solvated and show little aggregation, whereas they exist as self-associated forms in nonpolar solvents such as chloroform.[38]

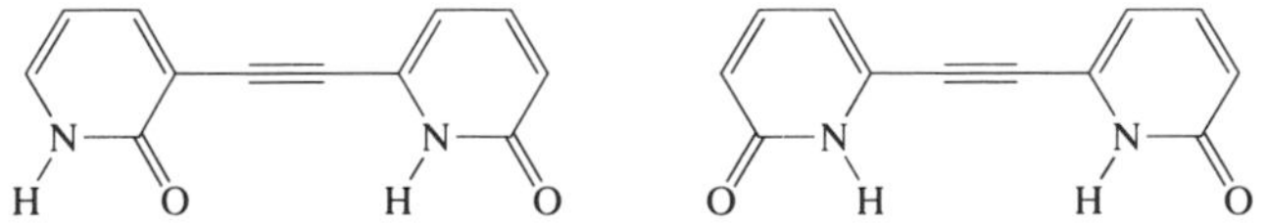

(**1**) (**2**)

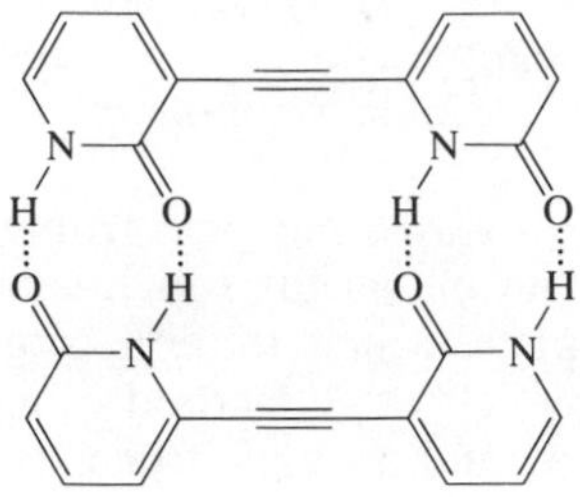

(3)

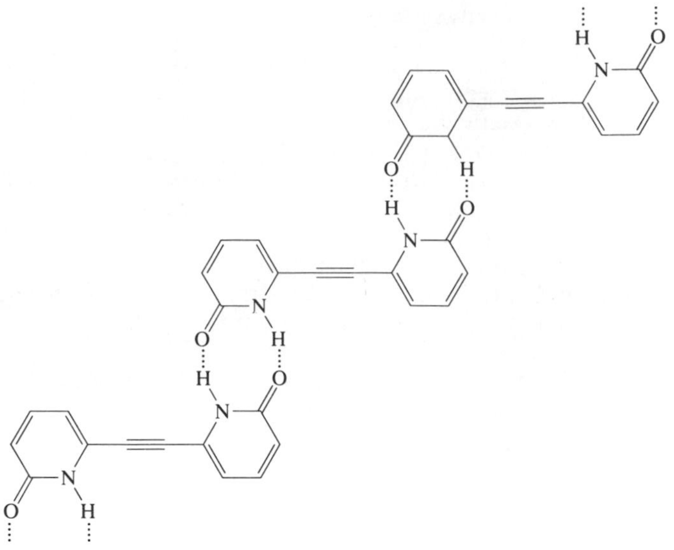

(4)

To study the role of preorganization in the formation of these aggregates flexible analogues of dipyridones (1) and (2) were synthesized. Instead of using ethyne as a rigid linker, a flexible spacer was incorporated into asymmetric dipyridone (5) and symmetric dipyridone (6). Vapor pressure osmometry indicates that asymmetric (5) dimerizes in chloroform, whereas symmetric (6) forms a linear oligomer. This behavior seems to mimic that of the alkyne dipyridones until one considers the solid-state structure of the complexes. In the crystal, dipyridone (5) does not form a discrete dimer, but takes up a linear polymeric motif. This form includes both intermolecular and intramolecular hydrogen bonds, suggesting that the dimer formed in solution may be either structure (7) or (8). Low-temperature NMR studies indicated that the dipyridone also forms a cyclic trimer. The flexible spacer provides a lower degree of preorganization than the rigid spacer, leading to the formation of several different complexes instead of one discrete self-organized unit.[19]

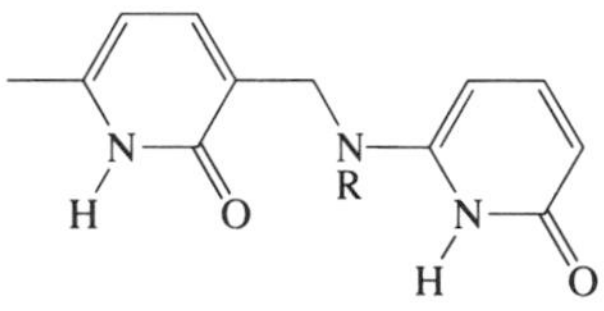

(5)

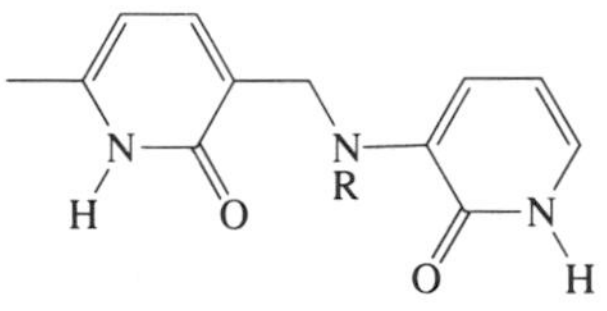

(6)

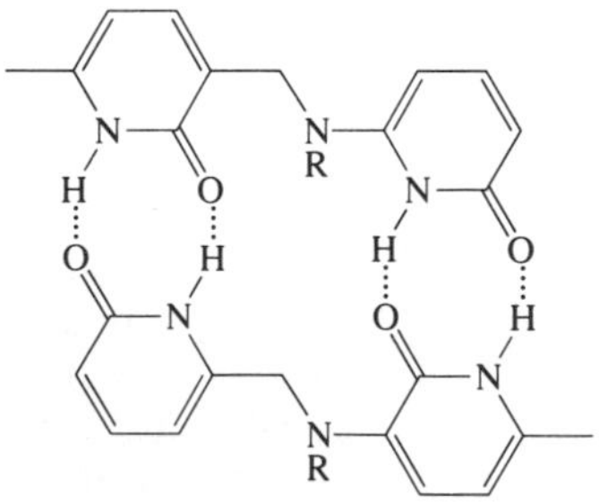

(7)

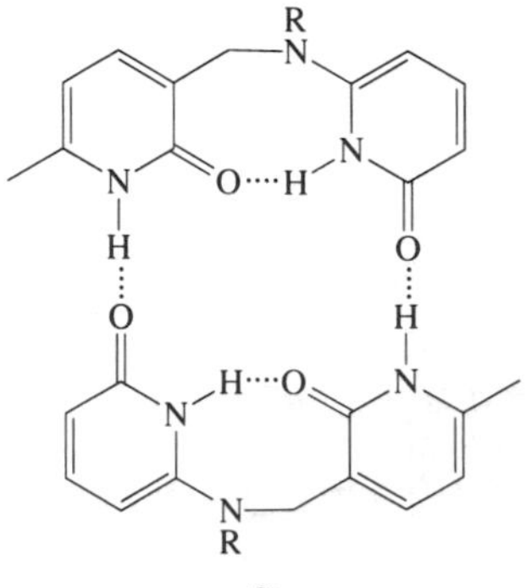

(8)

The corresponding 3,5′-ethyne linked pyridone-pyrimidone was shown by x-ray crystallography to form the extended chain structure (as in (9)) as opposed to the expected cyclic trimer (10).[39] This work established that preorganizing the hydrogen bonding groups with rigid spacers leads to greater control over the self-assembled system. The next step was to use the dipyridone motif in the design of a cubic diamondoid network. The basic unit of the structure is the rigid tetrapyridone (**11**). The name "tecton" was proposed to refer to individual units which will self-assemble into a larger aggregate. The crystallization of (**11**) into a diamondoid network (Figure 2) was achieved from a solvent mixture of butyric acid/methanol/hexane, or valeric acid/methanol/hexane. The choice of a C_4 or C_5 carboxylic acid was critical. Only acids of this size are enclathrated by the chambers of the network. Other acids, such as acetic or propionic, form bidentate hydrogen bonding complexes with the pyridone unit.[9] This diamondoid system, while not involving discrete self-organization in solution, was included here to show how the successful design of small self-assembling systems such as the dipyridones can lead to the design of large and porous solid-state structures.

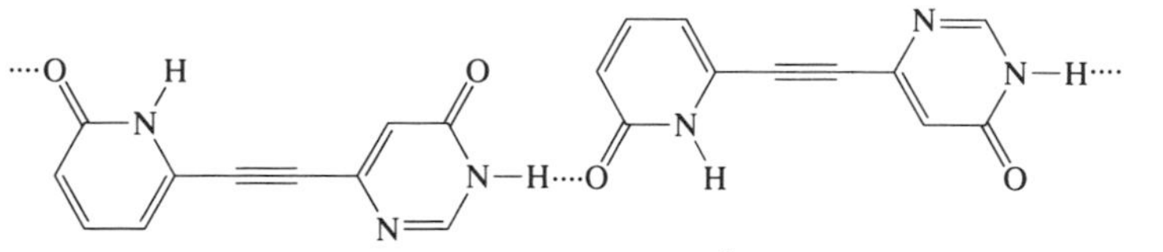

(**9**)

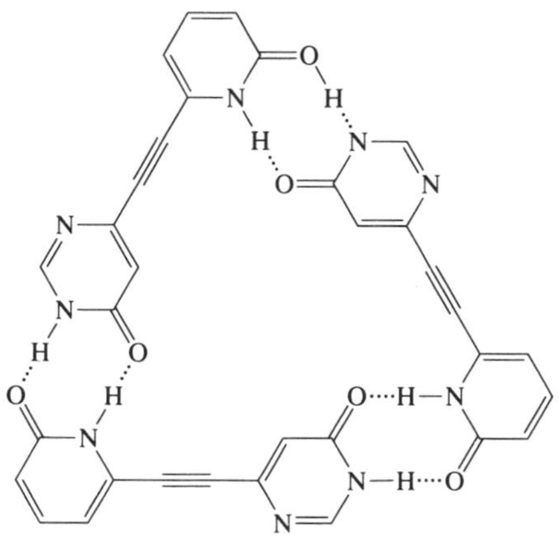

(**10**)

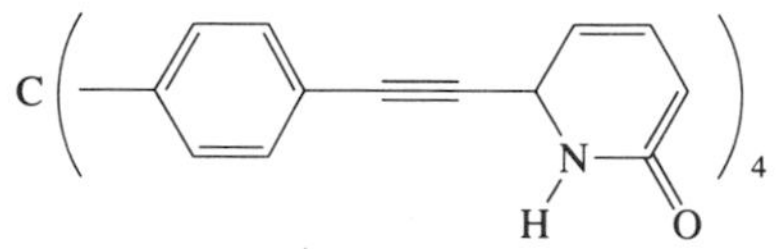

(**11**)

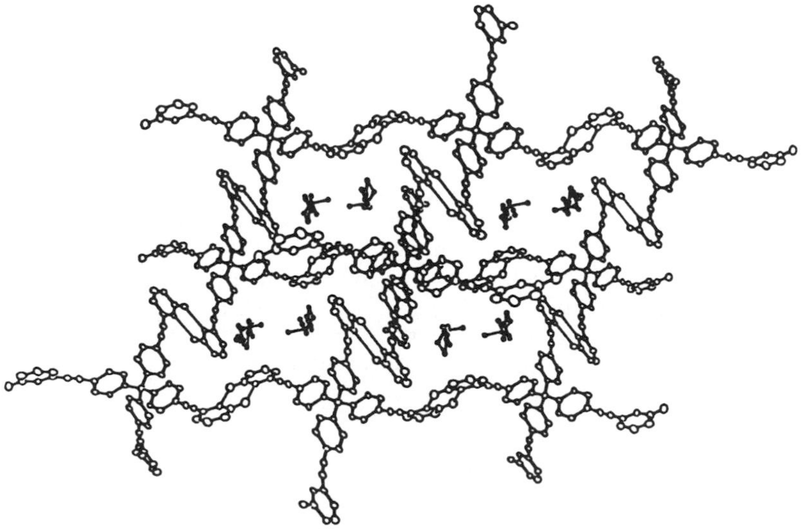

Figure 2 ORTEP drawing of the infinite diamonoid network present in crystals of 3(**11**)·2*n*-butyric acid (reproduced by permission of the American Chemical Society from *J. Am. Chem. Soc.*, 1991, **113**, 4696).

Another potential application of dipyridone self-assembly is to use the motif as a template for self-replication. Wuest and Persico synthesized rigid dipyridones **(12)** and **(13)** in an attempt to create templates for self-replication. It was hoped that the dipyridones would bring the subunits into the proper position for oxidative coupling (Figure 3). However, the strategy was not effective in orchestrating self-replication, perhaps because the dipyridones were too strongly self-associated in solution to bind to the monomeric subunits.[40]

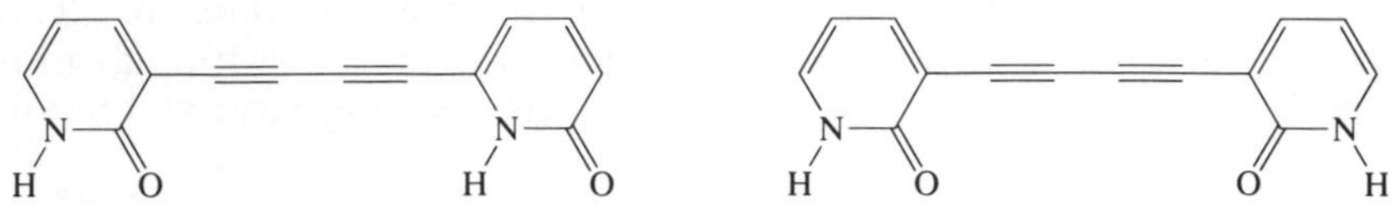

(12) **(13)**

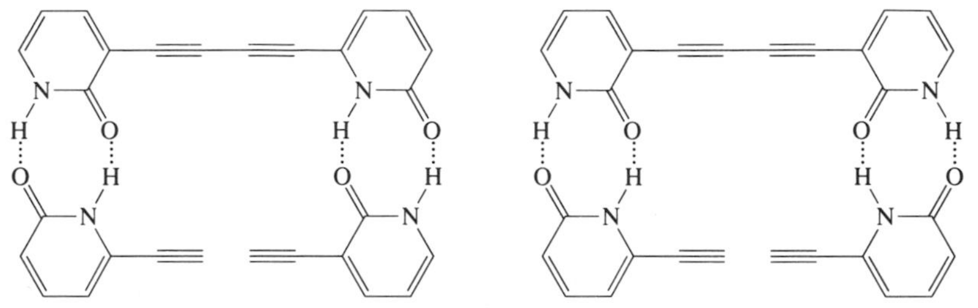

Figure 3 Attempted self-replication system using bis(pyridone) units.

The dipyridone hydrogen-bonding motif was also used by Zimmerman and Duerr in the design of a self-assembling cyclic trimer. Pyridoquinoline **(14)** was synthesized and its potential to form either oligomeric aggregates or a cyclic trimer (as shown in Figure 4) was investigated.[31] Analysis of changes in the ^{1}H NMR spectrum of **(14)** as a function of concentration using the Saunders–Hyne method gave strong evidence that it is present as the cyclic trimer in solution. The molecular weight of the aggregate determined by VPO in chloroform-*d* at 35 °C is 942 ± 80, within experimental error of the expected molecular weight of the trimer (MW = 1014). The trimer seems to be present even in polar solvents such as DMSO-d_6/$CDCl_3$. This stability relative to the linear oligomer is presumably due to the formation of two hydrogen bonds per molecule in the cyclic trimer compared to only $(2n + 2)/(n + 2)$ in the oligomer and confirms the importance of cooperativity in self-assembly.[31]

16.3.1.2 Folic acid motifs

Folic acid salts also have been shown to self-assemble into cyclic aggregates. After observing that several deoxyguanidinium derivatives form liquid crystal phases composed of chiral columnar aggregates with four molecules in each stack,[41,42] Spada and co-workers studied similar phenomena in folates. Circular dichroism spectra and x-ray crystallographic data show that the dipotassium salt of folic acid **(15)** forms the cyclic tetramer **(16)** in water, which then stacks to form a chiral cylindrical column. These columns take up a parallel hexagonal structure, forming a liquid crystal mesophase stabilized by both hydrogen bonding and π-stacking interactions (Figure 5). A related hydrogen bonding motif has been exploited by Lehn in the design of a self-assembling molecular ribbon composed of one self-complementary molecule. The crystal structure of pyrimidine **(17)** shows that it takes up a planar ribbon motif with six hydrogen bonds between each subunit and its two neighbors, as in **(18)**.[43]

16.3.1.3 Barbiturate motifs

A similar ribbon structure can be formed by dialkylbarbiturate derivatives (Figure 6). These take up a linear packing pattern with bidentate hydrogen bonds between each heterocycle.[44] As a result, the 5,5-dialkyl side chains are projected in an alternating arrangement on opposite sides of

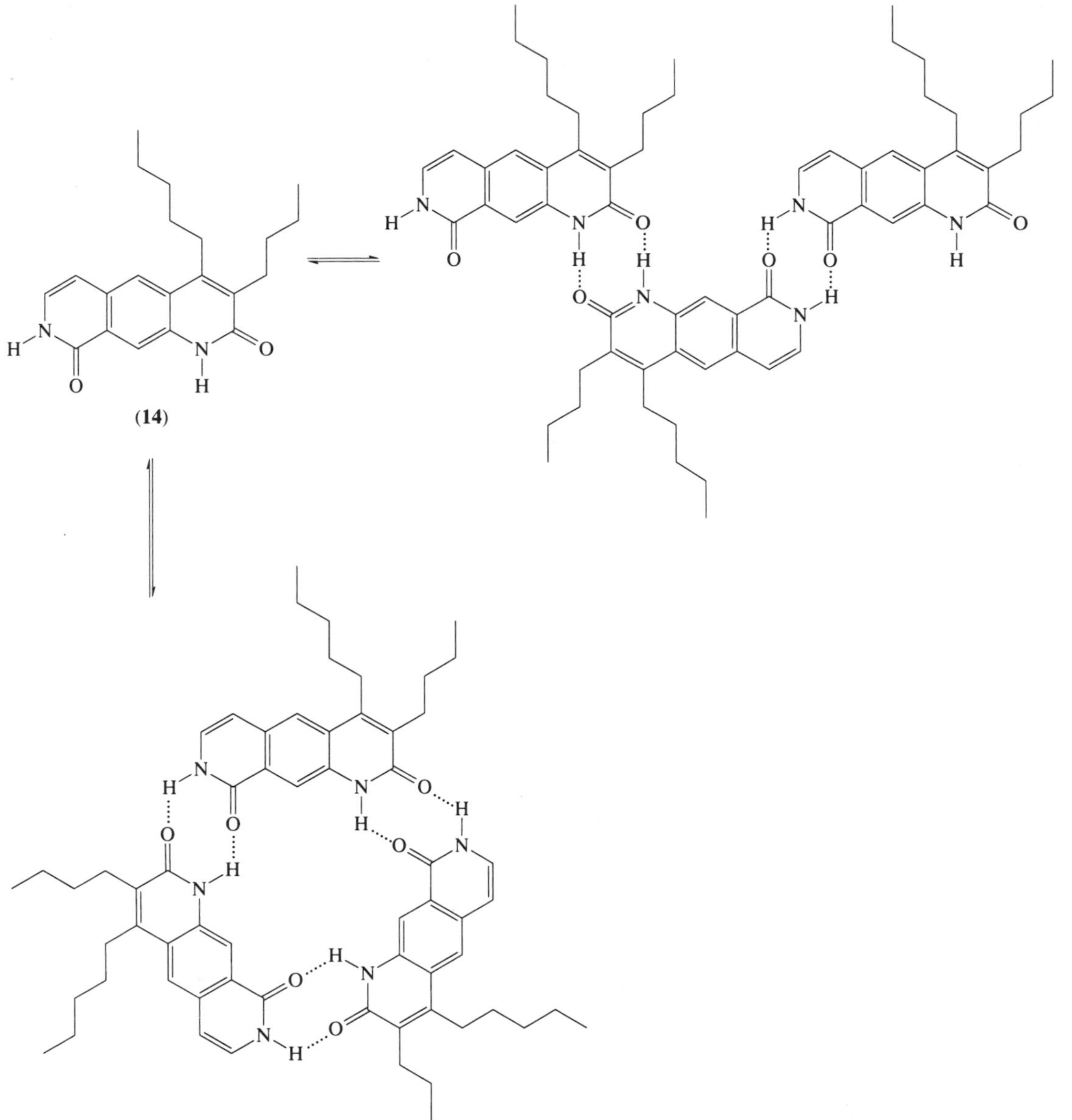

(14)

Figure 4 Self-assembling possibilities of pyridoquinoline (**14**).

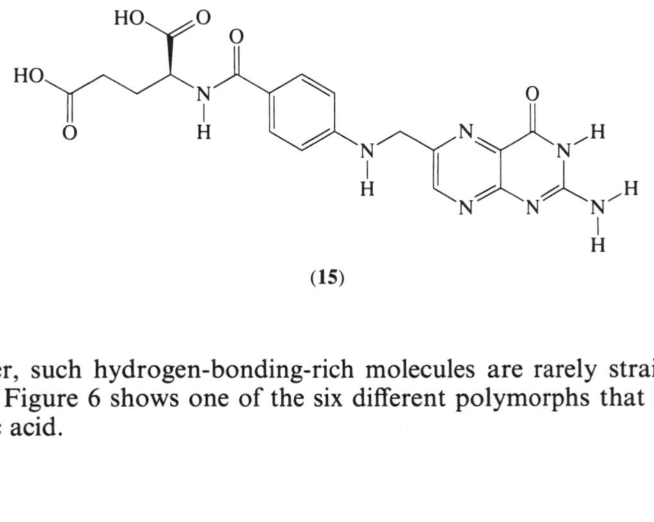

(15)

the ribbon. However, such hydrogen-bonding-rich molecules are rarely straightforward in their crystal packing and Figure 6 shows one of the six different polymorphs that have been identified for diethylbarbituric acid.

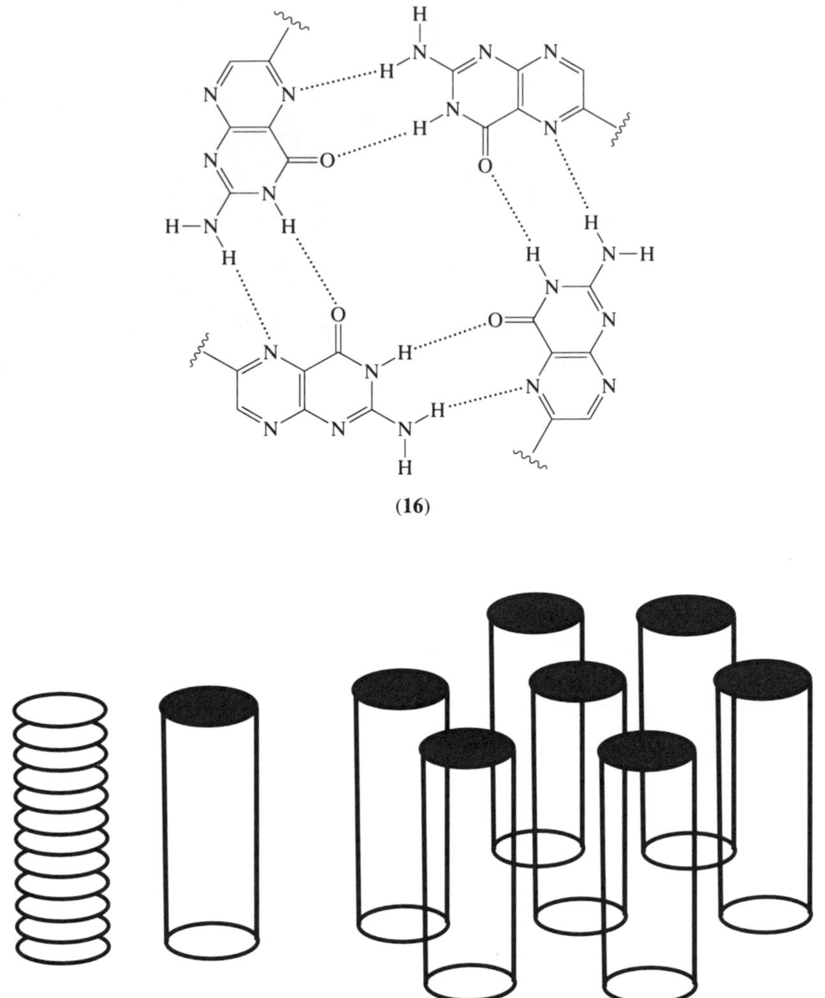

Figure 5 Structure of the proposed tetramers formed by four folic acid molecules. These tetramers may stack to form cylindrical aggregates that in the liquid crystalline phase exhibit tetragonal order.

(17)

(18)

16.3.1.4 Isophthalic acid motifs

The x-ray structure of trimesic acid provides a remarkable example of the hydrogen bond control of molecular packing. The solid-state structure is complex, composed of a concatenated arrangement of hydrogen-bonded rings.[45] However, each interlinked sheet shows a hexagonal lattice arrangement composed of two motifs, a linear ribbon, and a ring motif (Figure 7). Isophthalic acid derivatives can aggregate either in the ribbon motif (less symmetrical forms of this structure are also possible) or cyclic hexamers corresponding to the ring motif in Figure 7. Unsubstituted isophthalic acid forms the extended ribbon motif in the solid state (Figure 8) with bidentate

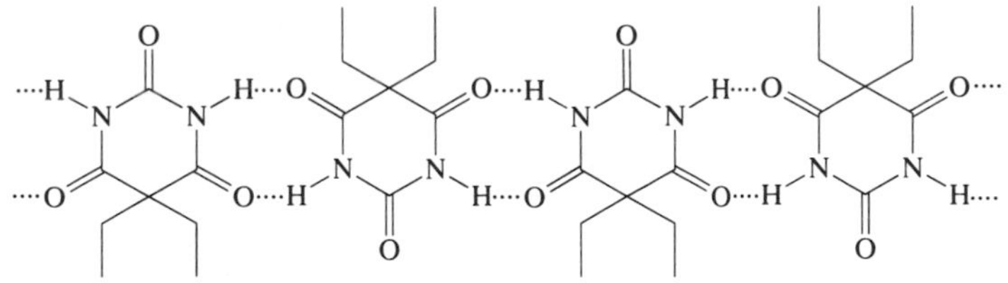

Figure 6 Ribbon packing structure formed by diethylbarbituric acid.

hydrogen bonds between the carboxylic acid groups playing a key role in organizing the structure.[46] The ribbon motif allows both optimal formation of the bidentate hydrogen bonds and efficient packing interactions between the ribbons in vertical and horizontal directions. In order to direct formation of the cyclic motif in the solid state it is necessary to disrupt the linear packing arrangement by placing a bulky substituent in the 5-position of isophthalic acid. The x-ray structure of 5-decyloxyisophthalic acid (**19**) (Figure 9) shows the formation of a hexameric aggregate corresponding to the ring motif in Figure 7. The six isophthalic acid molecules encircle a cavity that is 1.4 nm in diameter (from opposite isophthaloyl-2H sites) and is stabilized by 12 hydrogen bonds. The isophthalic core of the structure is planar with only a 4° angle between phenyl and carboxylate planes.[47]

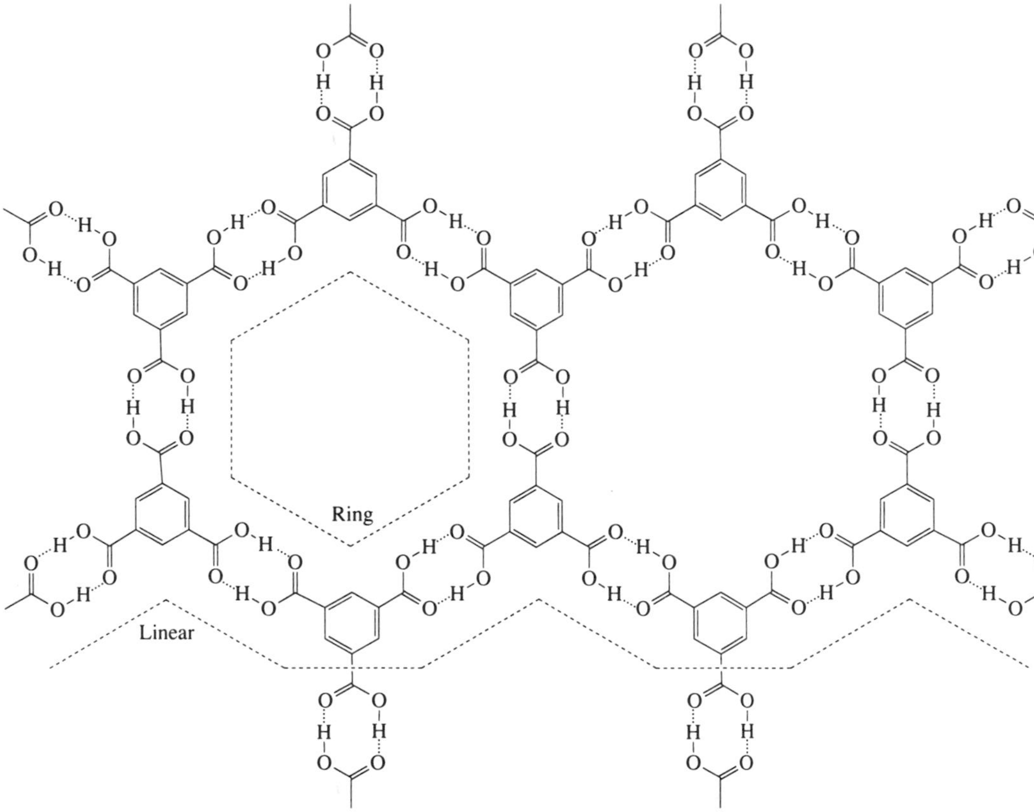

Figure 7 Extended sheet in x-ray structure of trimesic acid.

A possibly important factor in the formation of the hexameric aggregate in the solid state is the efficient interleaving of the alkyl chains within each layer, as shown in Figure 10. Furthermore, the three alkyl chains in the space between three hexamers (e.g., represented by A, B, and C in Figure 10) are positioned in the central cavity of the hexamer in the layer directly above. The size of the decyl side chain may be critical for allowing formation of the tightly packed structure. Recent work on 5-alkyloxyisophthalic acid derivatives of different lengths shows that with longer alkyl chains the extended ribbon motif is preferred.[48]

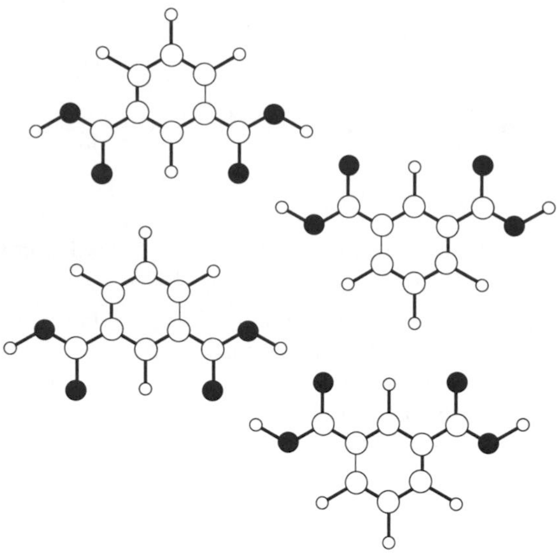

Figure 8 X-ray structure of isophthalic acid.

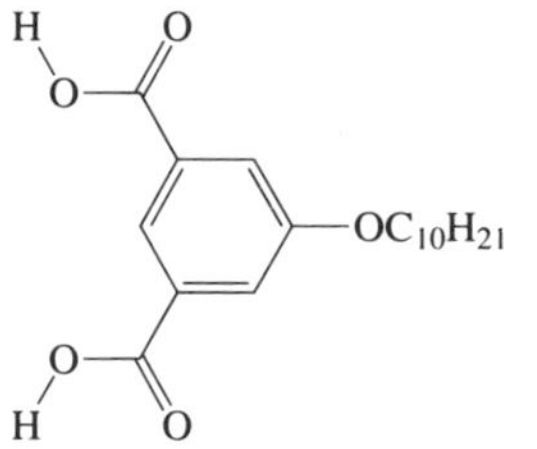

(19)

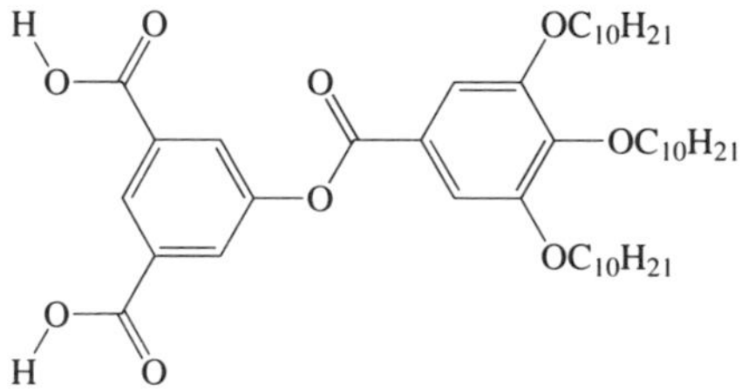

(20)

Entropically, these cyclic aggregates should also be favored in solution, because the cyclic motif allows the formation of 12 hydrogen bonds by six molecules rather than the seven that would be necessary in a linear aggregate. That the cyclic motif exists in solution was supported by VPO and ^{1}H NMR. Diacid (**20**) was used in these studies because it is soluble in both benzene and toluene. The molecular weight of the aggregate was determined by VPO and is in agreement with a cyclic aggregate at concentrations above 10 mM. The ^{1}H NMR spectrum of (**20**) in C_6D_6 showed a concentration dependence consistent with the formation of the cyclic aggregate. Between 0.1 mM and 15 mM, at 55 °C, there is a gradual shifting upfield by the isophthaloyl-2H resonance, indicating the formation of hydrogen bonds. Above 15 mM, the spectrum is that of the discrete cyclic aggregate. When the temperature is lowered to 25 °C, at 25 mM the spectrum becomes more complex, suggesting the formation of some higher order aggregate. This may involve the formation of π-stacked hexamer structures.[49] Related stacked structures have been seen in a series of covalently linked hexa(phenylacetylene) macrocycles.[50]

16.3.1.5 Lactam motifs

Most of the self-complementary motifs discussed above lead to planar aggregates with little possibility of interaction with other guest molecules. In an elegant series of experiments, Rebek and co-workers introduced a tetralactam subunit (**21**) based on diphenyl glycoluril.[51] The disposition of hydrogen bonding groups in (**21**) leads to self-complementary formation of a dimeric aggregate, (**22**). The spherical structure is stabilized by eight hydrogen bonds and the two subunits follow the shape and the seams of a tennis ball. VPO measurements on a $CHCl_3$ solution of (**21**) gave a molecular weight of 1500–2000, within reasonable agreement with the calculated molecular weight of the dimer (1429.58). Similarly, EI, FAB, and laser desorption mass spectrometry all gave significant peaks at 1429. The solution stability of the aggregate was assessed by ^{1}H NMR. In

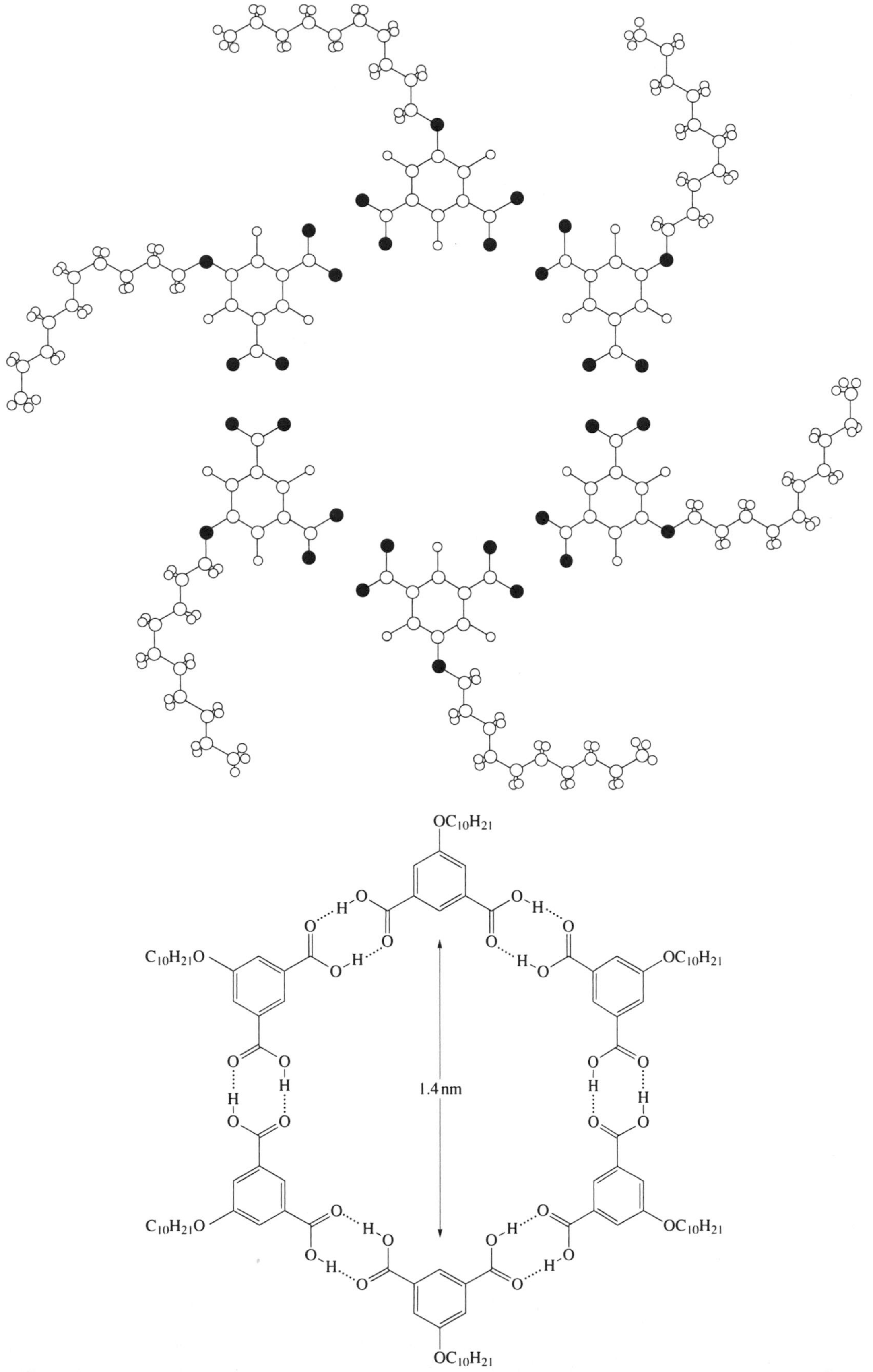

Figure 9 X-ray and schematic structure of the hexameric aggregate of 5-decyloxyisophthalic acid (**19**).

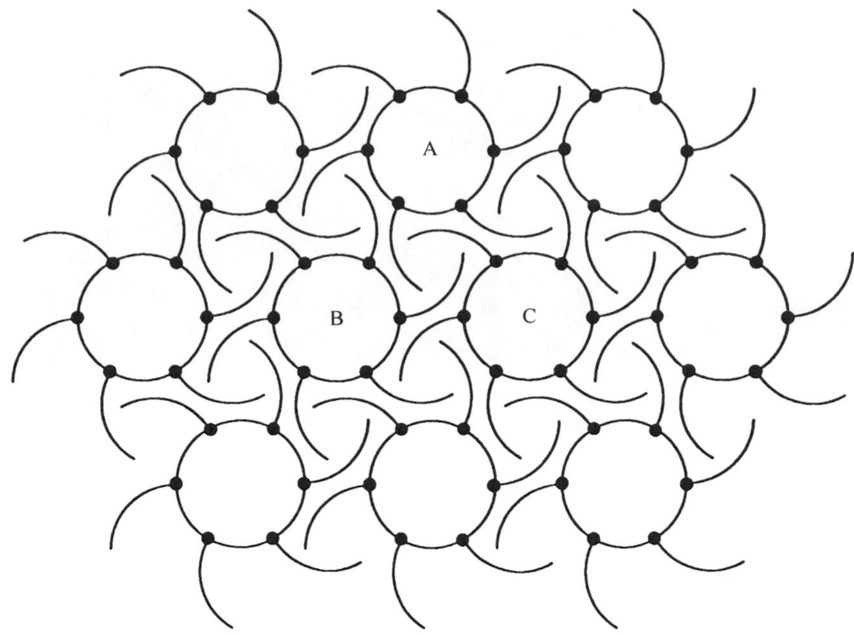

Figure 10 Packing arrangement for cyclic hexamer of 5-decyloxyisophthalic acid.

C_6D_6 and $CDCl_3$, large downfield shifts of the NH protons were observed consistent with extensive hydrogen-bonding aggregation. The association was also intact in 20% DMSO–$CDCl_3$ but was disrupted by strong solvation of the hydrogen-bonding groups in pure DMSO.

(21)

(22)

The inner dimensions of the cavity in **(22)** are 0.050–0.055 nm^3 and are large enough to encapsulate small organic molecules. In $CDCl_3$, two forms of the aggregate are seen by NMR, one empty and the other occupied by a solvent molecule. A particularly good fit is seen with methane which has an estimated molecular volume of 0.028 nm^3. In $CDCl_3$ solution, bound and unbound CH_4 slowly exchange as seen by the presence of two distinct ^1H NMR resonances for the guest at –0.91 and 0.23 ppm, respectively. The strong binding of methane ($K_a = 300\,M^{-1}$) is largely enthalpy driven ($\Delta H = -37.67\,kJ\,mol^{-1}$) with a large unfavorable binding entropy ($\Delta S = -83.72\,kJ\,mol^{-1}$), presumably due to the restriction of guest motion within the confines of the cavity.[52]

Lehn and co-workers introduced a simple bicyclic bis(lactam) derivative that can form bidentate hydrogen bonds with its neighbors, **(23)**. In its racemic form **(23)** would be expected to form a zig-zag alternating aggregate of type **(24)** while the enaniomerically pure bis(lactam) should form cyclic hexamers of type **(25)**. The bis(lactam) was prepared and resolved and its solid-state structures determined by x-ray crystallography.[53]

Zig-zag structure **(24)** with racemic **(23)** was confirmed; however, enantiomerically pure **(23)** did not give **(25)** but rather a cylic tetramer with a more extensive network of hydrogen bonds in the crystal lattice.

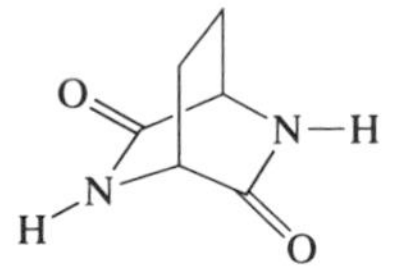

(23)

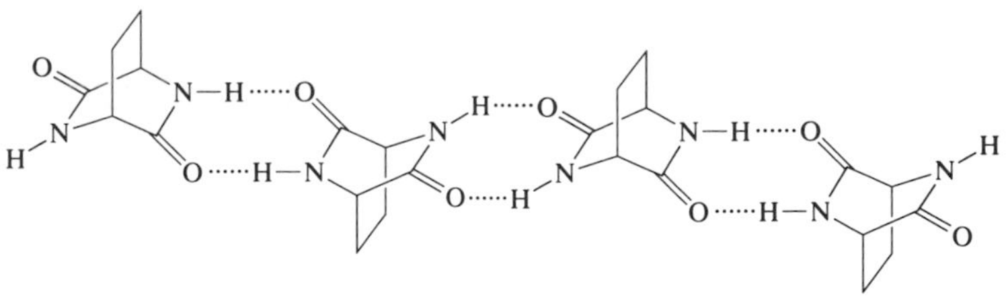

(24)

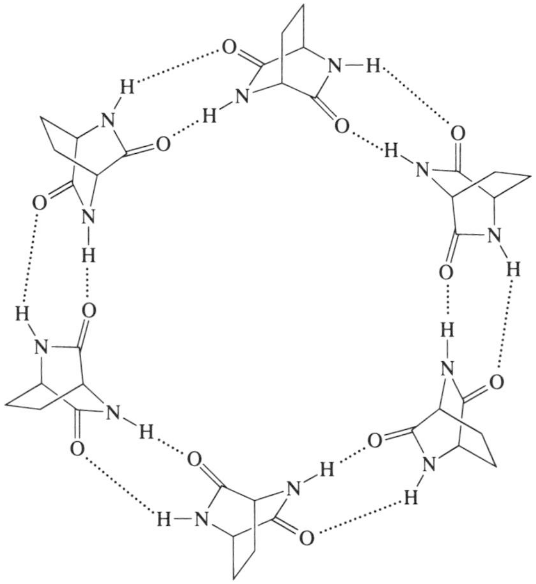

(25)

16.3.1.6 Cyclic peptide motifs

Ghadiri and co-workers have exploited the ability of cyclic peptides to interact by an extensive network of hydrogen bonds to form well-defined self-assembled structures.[54] The incorporation of alternating D- and L-amino acids into a cyclic octamer structure (e.g., (**26**)) leads to a low-energy, ring-shaped conformation in which all the α-substituents project out from the central cavity. As a result, the amide NH and CO groups in the main chain lie perpendicular to the plane of the ring and in a position to hydrogen bond to neighboring molecules in an extended stacked structure, as in (**27**). The presence of two ionizable glutamic acid groups in the ring provides a switch mechanism for self-association. In basic solution the dianionic rings are mutually repulsive; however, on acidification a spontaneous aggregation occurs. Support for the formation of tubular structures of type (**27**) in the solid state comes from transmission electron microscopy which shows longitudinal striations in the crystals corresponding to a 1.49 nm distance between the central axis of each nanotube. The electron diffraction patterns also show a 0.47 nm axial periodicity as would be expected for a structure made up of stacked peptide rings hydrogen-bonded in the manner of an antiparallel β-sheet. The internal diameter of the nanotubes is 0.7–0.8 nm. Further support for the structure comes from FTIR spectroscopy which shows shifts in the amide bands consistent with tight backbone–backbone hydrogen bonding and similar to those of gramicidin, a natural alternating D-, L-oligopeptide that also forms channel structures.[55] A closely related design strategy has been used by other workers for the formation of peptide based tubes.[56]

The tubular structure of these cyclic peptides clearly suggested their possible function as channels for the transport of ions or molecules across membranes. Modifying the sequence of the cyclic octamer to (D-Val, L-Trp, D-Val, L-Gln)$_2$ increased its hydrophobicity and solubility in phospholipid bilayer membranes. These rings presumably form tubular structures of at least eight stacked subunits creating a pore 0.7–0.8 nm in diameter from one side of the membrane to the other. Time-resolved fluorescence measurements showed that very fast ion transport took place when the

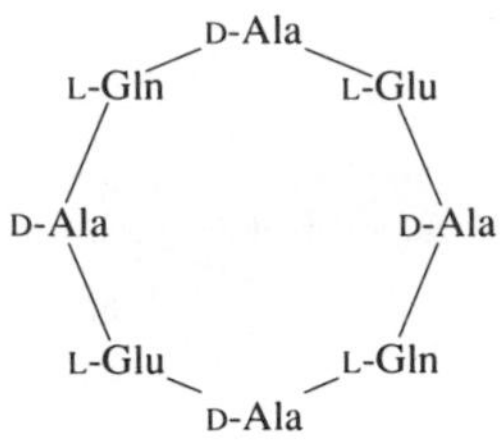

(26)

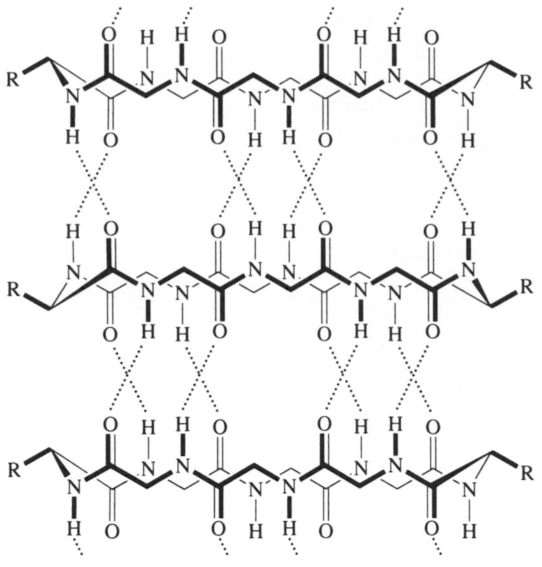

(27)

cyclic peptides were added to a vesicle containing entrapped fluorescein. Rates as high as 10^7 ions s^{-1} were measured and compared favorably with natural ion channel forming molecules such as gramicidin and amphotericin B.[57]

The strategy is also applicable to larger ring structures. Cyclic dodecapeptides of the sequence (D-Ala, L-Glu, D-Ala, L-Gln)$_3$ form tubular structures closely related to (**27**) but with a larger (1.3 nm) internal diameter.[58] Hydrophobic modifications of the larger ring structures lead to the formation of channels that have sufficiently large internal cavities to allow the efficient transport of glucose across membranes.[59]

16.3.2 Systems Which Utilize Two (or More) Unique Complementary Molecules

16.3.2.1 Cyanuric acid/melamine motifs

Whitesides and co-workers have utilized the hydrogen-bonding pattern of melamine (M) and cyanuric acid (CA) or dialkyl barbiturates to design a number of novel solid-state structures and solution-phase aggregates (Figure 11). The CA·M motif has the advantage that the heterocyclic core faciliatates bidirectional hydrogen bonding ideal for ribbon and cyclic motifs. The core is rigid and is therefore not dependent on the conformation of a flexible linker as in other systems. Furthermore, the simple trigonal shapes of the molecules require the consideration of only a few orientations when designing an aggregate.[60]

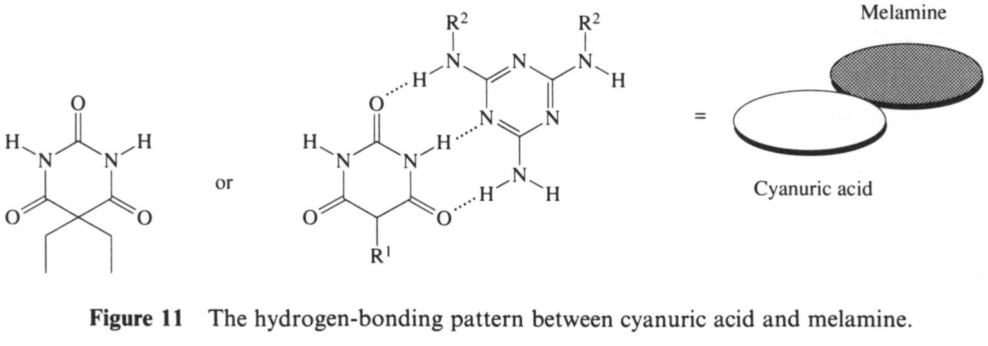

Figure 11 The hydrogen-bonding pattern between cyanuric acid and melamine.

The CA·M motif has been utilized to design three types of solid-state structure; "linear" tapes, "crinkled" tapes, and cyclic "rosettes." One strategy for structural control involves varying the substitution on the *para* position of a phenyl substituent of the melamine derivative (R_2 in Figure 11). When the *para* substituent is small (e.g., F, Cl, Br, I, Me), the simple linear tape motif is observed (Figure 12(a)). As the substituents increase in size (e.g., CO_2Me), unfavorable non-bonding interactions that would occur in the linear tape structure are relieved by formation of the crinkled tape form (Figure 12(b)). When substituents are very bulky (e.g., Bu^t), the crystal structure of the aggregate reveals a rosette motif (Figure 12).[61,62] This strategy of peripheral crowding represents an important approach to directing the aggregate structure both in solution and in the solid state.

Polar substituents (such as CO_2H, CN, and $CONH_2$) do not form tape structures, perhaps because of competition for the hydrogen-bonding sites.[62] Other tapes have been designed by varying substitution in the *meta* position of the phenyl ring.[63] Substitution on the cyanuric acid, including the use of various dialkylbarbiturates, may also be used to produce crinkled tape structures.[64]

The "rosette" form of the CA·M motif has proved useful in designing a large class of self-assembling aggregates in solution. Two types of preorganization have been used to promote "rosette" structures: covalent linking, and peripheral crowding. The first of these involved linking three melamine units through a rigid spacer to form the rigid $hubM_3$, **(28)**.[65] Interaction of one equivalent of these units with three equivalents of a cyanuric acid derivative RCA, **(29)** gave the (1 + 3) complex $hubM_3(RCA)_3$, **(30)** (see Figure 13). Evidence for the formation of these complexes was obtained from solubility data, ^1H NMR, ^{13}C NMR, UV spectroscopy, GPC, and VPO.

The molecular weight determined by VPO was 15–35% higher for $hubM_3(RCA)_3$ depending on the standard used. An explanation for the higher molecular weight may be that some intercomplex hydrogen bonding is occurring. Solvent effects on the formation of the complexes were uncovered by the NMR studies: the compounds are better defined in *o*-dichlorobenzene than in $CDCl_3$, presumably because the $CDCl_3$ can compete for the hydrogen-bonding sites. Similar results to those in $CDCl_3$ were seen in polar solvent mixtures such as $CHCl_3/MeOH$ and $CHCl_3/DMSO$. The role of preorganization in the formation of these complexes was studied by creating three flexible analogues of $hubM_3$. In competition experiments, it was found that in an equimolar mixture of $hubM_3$ and its more flexible analogue ($flexM_3$), cyanuric acid derivatives only complexed with $hubM_3$; no $flexM_3(RCA)_3$ was detected.

This hydrogen-bonding motif has been extended to design (2 + 3) complexes. In these complexes, two units of $hubM_3$ were allowed to react with bis(cyanuric acid) units containing either benzene or furan spacers ($benzCA_2$ or $furanCA_2$) between the hydrogen-bonding groups. Characterization of these structures, which correspond to two units of **(30)** linked through the cyanuric acid groups, was performed using solubility data, UV spectroscopy, ^1H NMR, GPC, and VPO. All tests, with the exception of VPO, confirm the existence of the (2 + 3) aggregates. Again, the molecular weight determined using VPO gave molecular weight values that varied between 10–20% above or below the expected value for $(hubM_3)_2(benzCA_2)_3$ depending on the choice of standard.[66]

To study the thermodynamics of self-assembly, a (1 + 1) system was designed based on the interaction of linked tris(melamine) and linked tris(cyanuric acid).[67] The resulting aggregate corresponds to **(30)** but with the three cyanuric acids linked through a similar hub. This study was performed on a (1 + 1) aggregate and not a larger system to simplify the analysis. The exchange reaction of the $hubCA_3$ between two different tris(melamine) derivatives was studied with ^1H NMR because it was not possible to directly measure the equilibrium binding constants. From variable temperature studies activation parameters of $\Delta H^\dagger = 100.5 \pm 8.4\,kJ\,mol^{-1}$, $-T\Delta S^\dagger = 8.4 \pm 16.7\,kJ\,mol^{-1}$, and $\Delta G^\dagger = 108.8 \pm 25.1\,kJ\,mol^{-1}$ were measured. The activation enthalpy $\Delta H^\dagger$ presumably reflects the breaking of 18 hydrogen bonds. The enthalpy for each individual hydrogen bond can be calculated as $5.44 \pm 0.41\,kJ\,mol^{-1}$, a value similar to literature precedents. The value of $T\Delta S^\dagger$ near zero is unexpected for the dissociation of one highly structured particle into two flexible particles. One way to rationalize this is to consider the entropy of solvation in the transition state. Dissociation of the complex should have a large negative entropy ($-T\Delta S^\dagger$), but this may be offset by the decrease in entropy of the solvent molecules as they hydrogen bond to the free subunits in the transition state.

The strategy of preorganizing by covalent linkages allows the design of more complex structures. The main building block of these compounds contains two melamine units linked through a xylyl group and then to a hub unit. This subunit was allowed to react with bis(cyanuric) acid

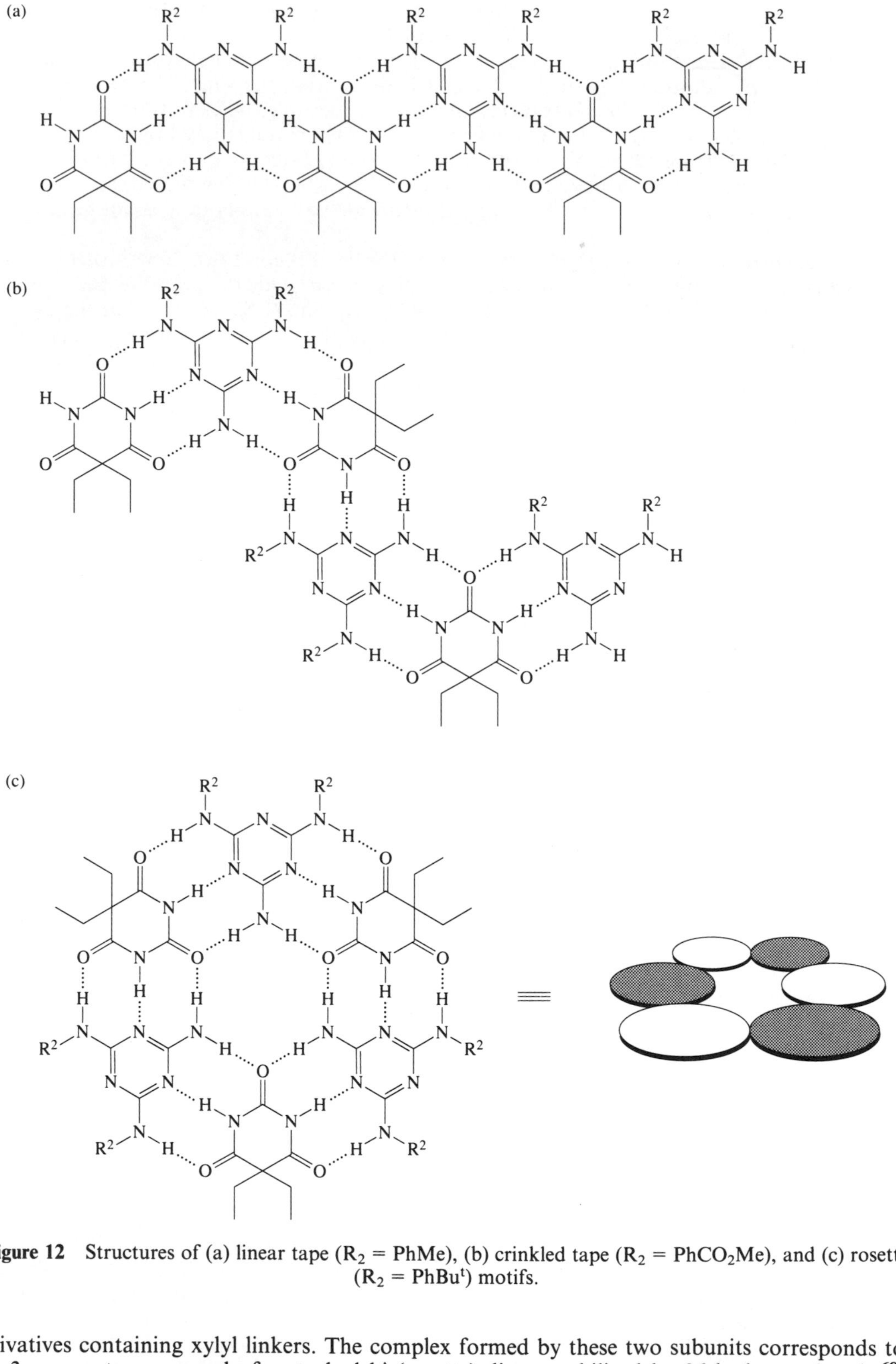

(a)

(b)

(c)

Figure 12 Structures of (a) linear tape (R_2 = PhMe), (b) crinkled tape (R_2 = $PhCO_2Me$), and (c) rosette (R_2 = $PhBu^t$) motifs.

derivatives containing xylyl linkers. The complex formed by these two subunits corresponds to a 1 + 3 aggregate composed of a stacked bis(rosette) dimer stabilized by 36 hydrogen bonds.[68] A related more complex aggregate is formed by a subunit containing three melamine groups linked to each other and to a hub in a linear arrangement. Addition of nine equivalents of cyanuric acid leads to an aggregate composed of 10 particles stabilized by 54 hydrogen bonds. The structure of this aggregate corresponds to three stacked rosette motifs and is confirmed by solubility, ^{1}H NMR, VPO, and GPC measurements.[69]

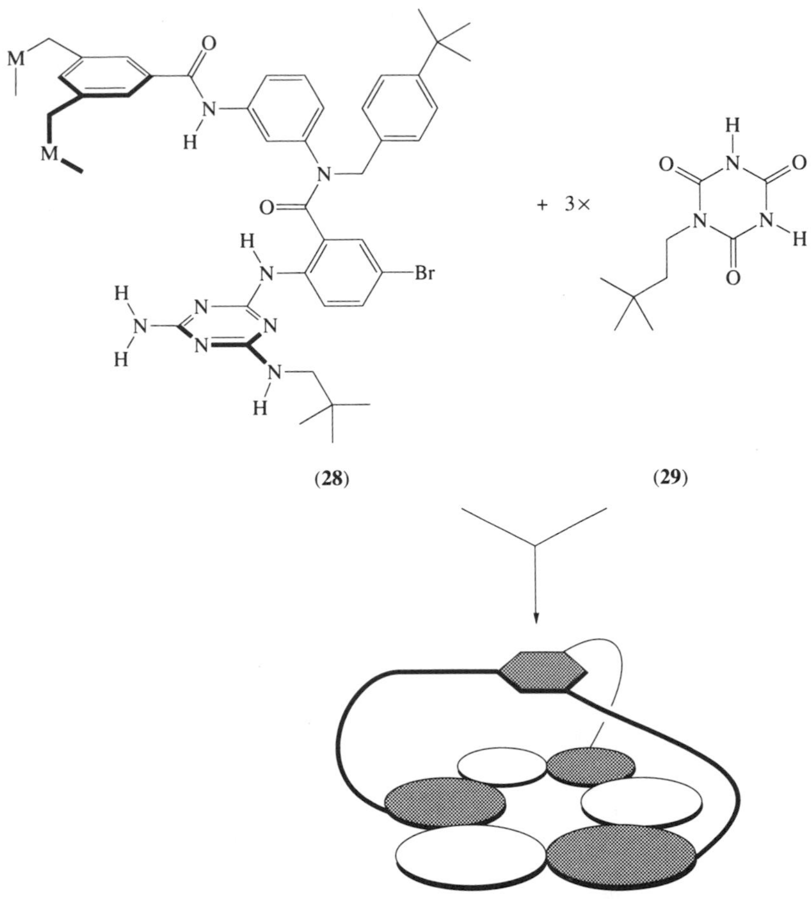

(28) (29)

(30)

Figure 13 Self-assembly of hubM_3 RCA to give a 1 + 3 supramolecular aggregate.

A second way to preorganize cyanuric acid and melamine units so that they form rosette structures involves peripheral crowding. Bulky substituents on the units should force the molecules to take up a cyclic motif, as in Figure 12(c). This strategy, combined with preorganization via covalent spacers, has allowed the creation of a new class of self-assembled aggregates.

Evidence for the formation of the simple rosette aggregate in solution can be found in the solubility data, since insoluble barbital and melamine components dissolve in chloroform in each other's presence. The ^{1}H NMR spectra of the melamine derivative show changes in the chemical shifts corresponding to hydrogen bonding upon addition of the cyanuric acid derivative. At room temperature, the absence of separate peaks associated with "free" and "bound" components suggests that the exchange reaction is fast on the NMR timescale. Cooling the mixture can slow the exchange, so that at 226 K, >90% of the components are present as the $CA_3{\cdot}M_3$ rosette. The molecular weight of the complex as determined by VPO is in agreement with the formation of the $CA_3{\cdot}M_3$ aggregate, although when the concentrations of the component molecules are less than 4 mM, the unassociated components and larger hydrogen-bonded oligomers appear to exist alongside the rosette. The aggregate could not be observed by GPC as the complex dissociated during elution. This reveals that this aggregate is less stable than the other hydrogen-bonded aggregates mentioned above.[70]

Rosettes have been designed using either peripheral crowding or covalent linkers to preorganize the molecules for assembly. Both of these strategies have been combined to design stable bis(rosettes) in solution. Figure 14 shows the self-assembly of bis(rosettes) composed of nine particles and stabilized by 36 hydrogen bonds. A variety of substitituents were placed on the bis(melamine) to assess the role of peripheral crowding in the formation of the aggregates. Stable aggregates were formed in all the experiments, as judged by ^{1}H NMR spectroscopy. The resonances of the isocyanurate protons (between δ 13 and 17 ppm) proved particularly useful in confirming the presence of the aggregates. Other regions of the spectra often overlap with other resonances. There are two possible conformations for the bis(rosette), an "eclipsed" conformation in which melamine units lie one over the other (Figure 15(b)), and a "staggered" conformation (Figure 15(a)). The observation of two resonances for the isocyanurate protons in some of these aggregates suggests that the staggered conformation is preferred, presumably to avoid steric crowding.[60]

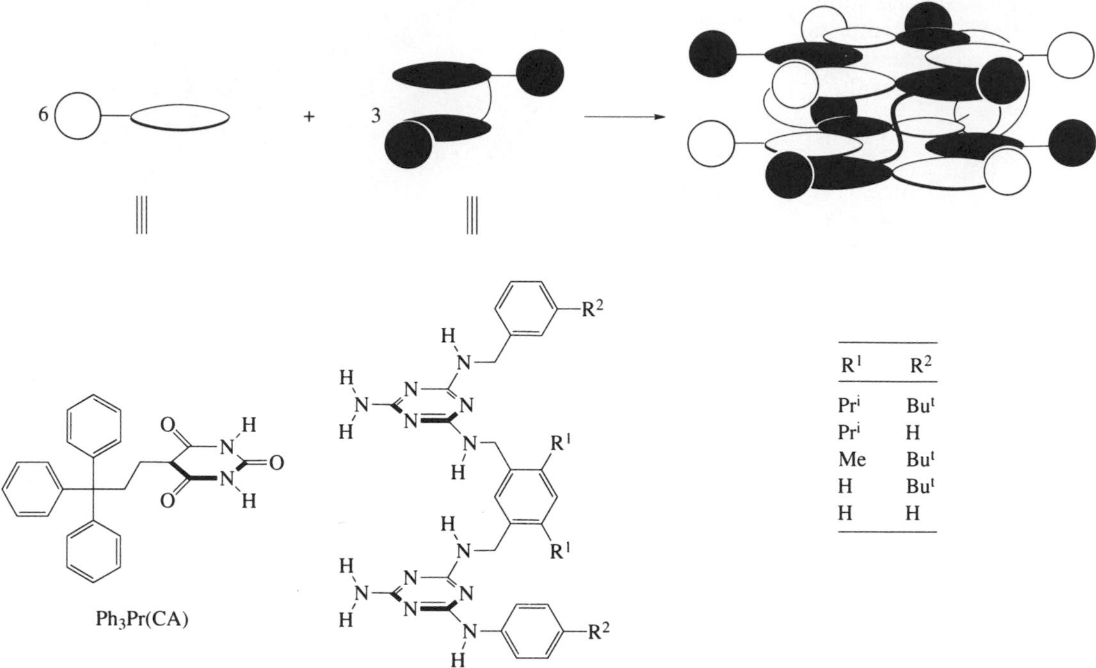

R^1	R^2
Pr^i	Bu^t
Pr^i	H
Me	Bu^t
H	Bu^t
H	H

Figure 14 Peripheral crowding as a route to the stabilization of "stacked rosette" aggregates.

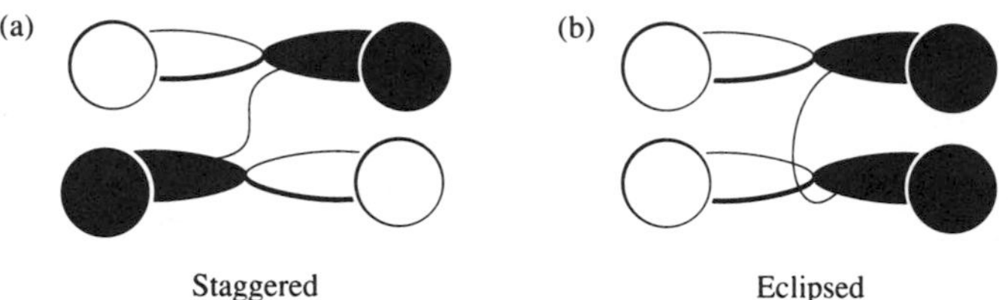

Figure 15 Two conformations available to the "stacked rosette" aggregates: (a) staggered, (b) eclipsed.

Solubility and VPO data provided further evidence for the formation of these aggregates. GPC traces show sharp peaks with little tailing, as compared to the single $CA_3 \cdot M_3$ rosette. Aggregates consisting of a bis(isocyanurate) and a melamine derivative also were designed and characterized using solubility, ^1H NMR, GPC, and VPO. NMR data suggest that these aggregates take up a staggered conformation to avoid steric crowding. Separate resonances were observed for the "free" and "bound" components, indicating an exchange reaction which is slow on the NMR timescale. In comparision, the exchange for the single rosette structure was fast on the NMR timescale. When the *para-t*-butylphenyl substituent on the melamine was replaced with less sterically demanding substituents, no aggregates were observed, pointing to the increased stability induced by peripheral crowding.

16.3.2.2 Acylaminopyridine–carboxylic acid motifs

The bidentate hydrogen-bonding interaction between 2-acylaminopyridine and carboxylic acids (Figure 16) provides a strong complex that can be used, like the cyanuric acid–melamine unit above, as a modular component in the design of self-assembling structures in solution and in the solid state.[71] For example, bis(2-acylaminopyridines) (**31**)–(**34**) of increasing spacer size were synthesized. Each of these complexes can take up two conformations, *syn* or *anti*, with respect to the amidopyridine receptor moeities. If the receptor takes up the *syn* conformation, then it should form a 1:1 host–guest complex with a dicarboxylic acid of complementary length, as in (**35**). If the receptor takes the *anti* conformation, then it should only form oligomeric chains, as in (**36**). For example, receptor (**31**) and adipic acid are complementary in length and form a 1:1 complex of type (**35**) in the solid state. Evidence for the persistence of the 1:1 complex in solution was seen in the ^1H NMR spectrum in $CDCl_3$. However, when crystals are grown from a 1:1 mixture of (**33**) and the noncomplementary diacid (e.g., 1,12-dodecanedicarboxylic acid), an oligomeric aggregate of type (**36**) is the result.[72]

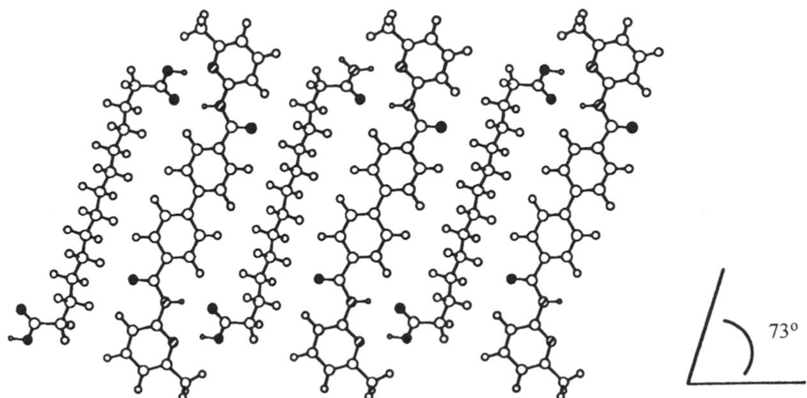

Figure 16 X-ray structure of the aggregate between (**33**) and dodecanedicarboxylic acid.

(**31**)

(**32**)

(**33**)

(**34**)

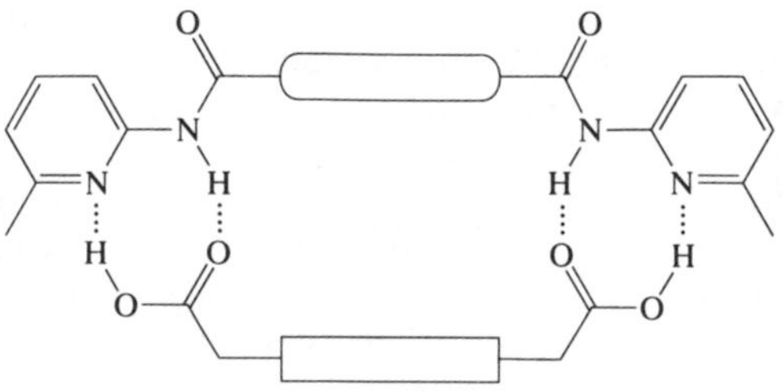

(35)

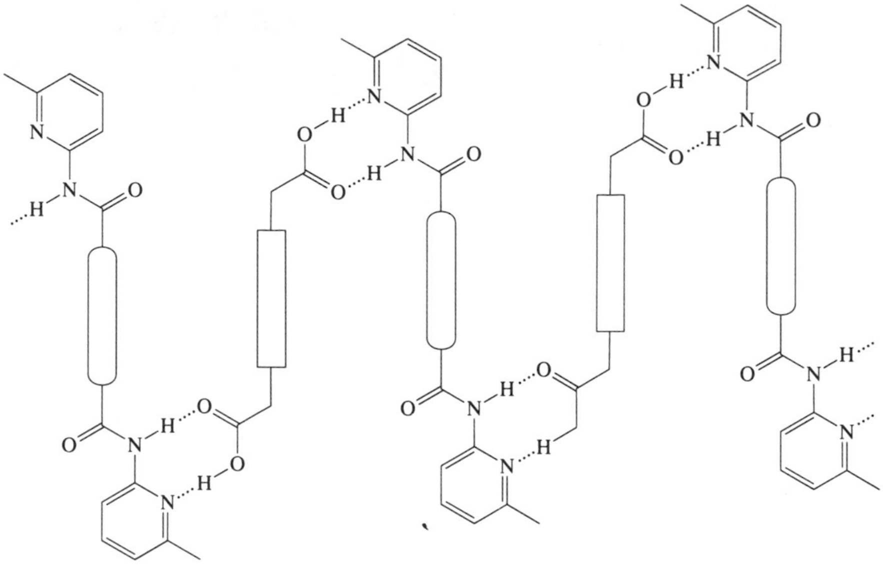

(36)

These latter structures represent elongated molecular sheets whose dimensions are imposed by the hydrogen-bonding network and the relative size match of the alternating components. Figure 16 shows the x-ray structure of the complex formed between (**33**) and 1,12-dodecanedicarboxylic acid. An almost flat sheet structure is taken up with a 73° slip or tilt angle between the polymethylene chains and the horizontal (defined by a line drawn through the pyridine nitrogen atoms). This hydrogen-bonding motif is dominant and is retained despite changes in the size of the molecular components. Indeed, the extent of the slip angle can be varied in a systematic and predictable way by changing the size matching of the diamide and diacid. Shortening the diamide spacer from biphenyl to naphthyl, (**32**) (~0.22 nm shorter), leads to a decrease in the slip angle to 60° (Figure 17). Decreasing the length of the diacid with respect to the receptor increases the slip angle. Crystals grown from a 1:1 mixture of naphthyl-(**32**) and 1,8-octanedicarboxylic acid has two slip angles due to a slight asymmetry in the crystal structure (Figure 18). These two slip angles, 89.0° and 77.6°, are larger than the slip angle for the complex between (**32**) and 1,12-dodecanedicarboxylic acid. The ability to control the structure of the 1:1 complex and the slip angle offers the possibility of designing crystals with specific electrical or optical properties.

A receptor similar to (**31**) was used in the design of a hydrogen-bonded helix in the solid state. The diacid used was pimelic acid (heptanedioc acid), and the diaminopyridine receptor was (**38**), which is shown in its *syn–syn* conformation. In this conformation, its binding cavity is too small to form 1:1 complexes with pimelic acid. The carboxylic acid groups on two different molecules bind to (**38**) with one aminopyridine unit directed above and the other below the plane of the binding cavity. This binding structure propagates through the crystal to form the helix (**39**). The existence of helix (**39**) in the solid state was confirmed by x-ray crystallography as shown in Figure 19. The receptor is not strictly planar in the crystal; one acylaminopyridine ring is 20.5° above and the other 9.4° below the plane of the isophthalate spacer. Each pimelic acid molecule remains in its preferred all-*trans* conformation and forms two pairs of bidentate hydrogen bonds to different receptors (average hydrogen bond lengths are 210 pm for NH—O and 180 pm for OH—N).

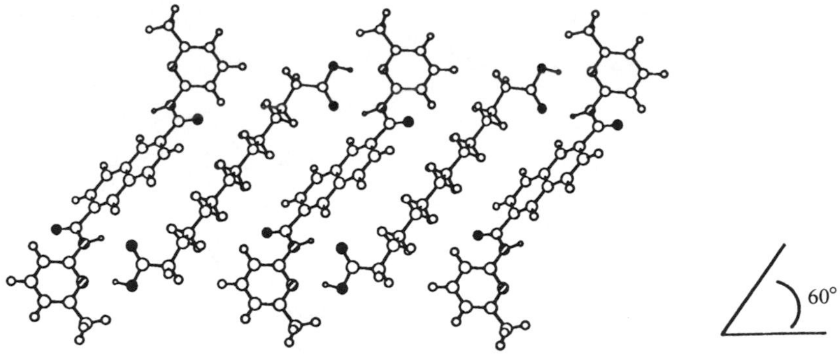

Figure 17 X-ray structure of the aggregate between (**32**) and dodecanedicarboxylic acid.

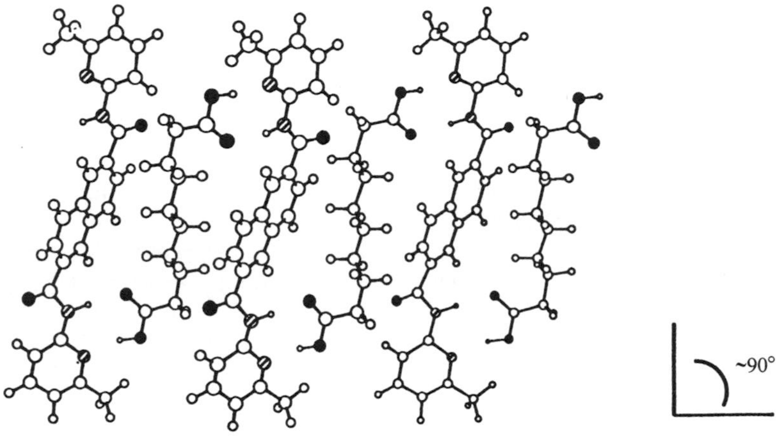

Figure 18 X-ray structure of the aggregate between (**32**) and octanedicarboxylic acid.

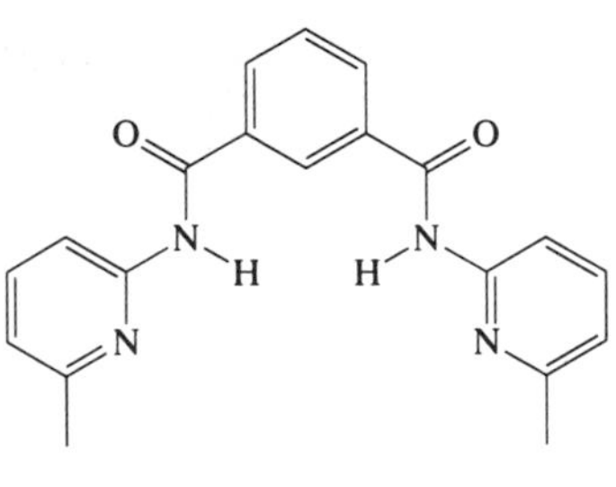

(**38**)

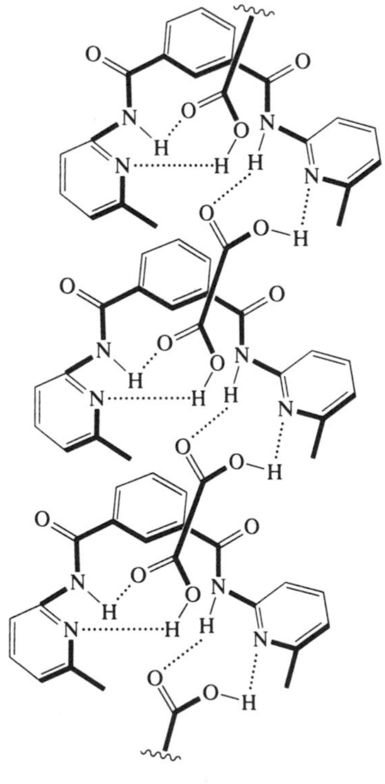

(**39**)

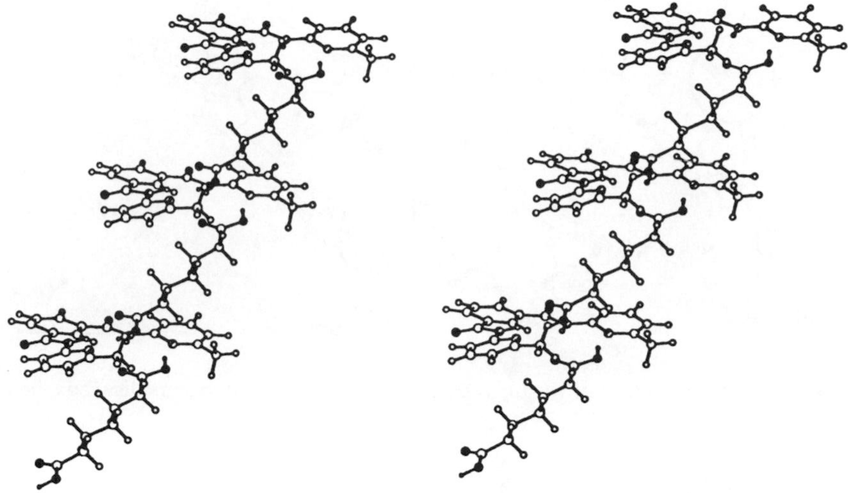

Figure 19 Stereoview of a single strand of the helix from the x-ray structure of the aggregate between (**38**) and pimelic acid.

The length of the diacid affects the stability of the helix in the solid state. When crystals were grown from a 1:1 mixture of (**38**) and glutaric acid, the structure was not a helix. The bis(acylaminopyridine) took up a *syn,anti* conformation and the aggregate took up a planar ribbon structure (Figure 20).[73] The diacid–2-acylaminopyridine motif was also used in the design of a cyclic (2 + 2) aggregate (as in Figure 21) in solution and the solid state. Scheme 1 shows how diacid (**40**) was allowed to react with (**38**). Due to its size and conformational restriction, the diacid would be unable to bind to both aminopyridine sites of (**38**). The diacid has two low-energy conformations, *syn* and *anti*, each of which should produce a different complex with (**38**). A 1:1 interaction with the *syn* isomer should lead to the complex (**41**), which can further dimerize to the (2 + 2) complex, (**42**). A 1:1 interaction with the *anti* isomer should lead to complex (**43**), whose hydrogen-bonding sites are disposed in different directions so that only polymeric aggregates can form.

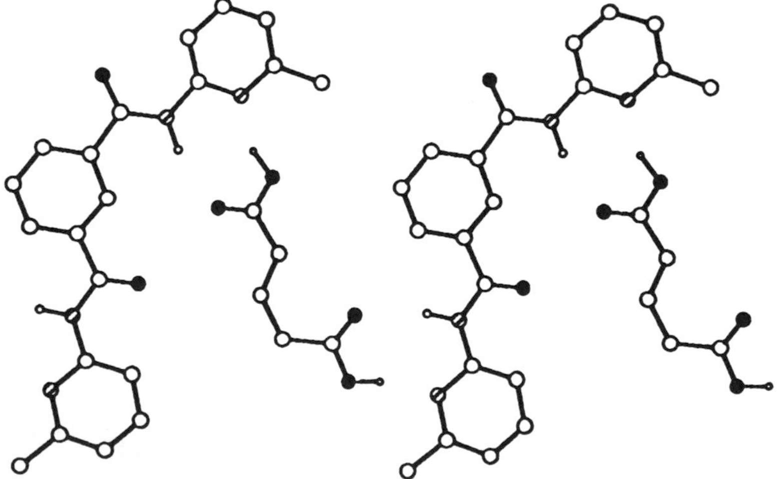

Figure 20 X-ray structure of the aggregate between (**38**) and glutaric acid.

Crystallization of (**38**) and (**40**) (R = H) produced the (2 + 2) aggregate (**42**), whose crystal structure is shown in Figure 22. The overall shape of the complex is that of a figure eight stabilized by the formation of eight hydrogen bonds. No hydrogen bonds were observed between adjacent aggregates. By allowing the more soluble biphenyl diacid (**40**) (R = $CO_2C_{10}H_{21}$) to react with (**38**), the existence of a (2 + 2) complex in solution could be confirmed. Characteristics of the complex

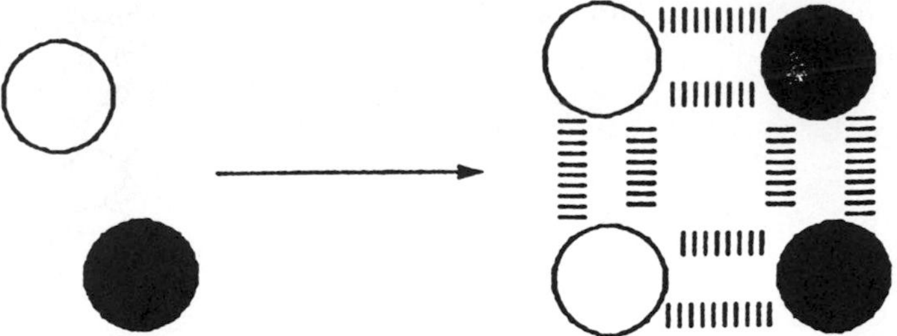

Figure 21 Schematic design for 2 + 2 aggregation.

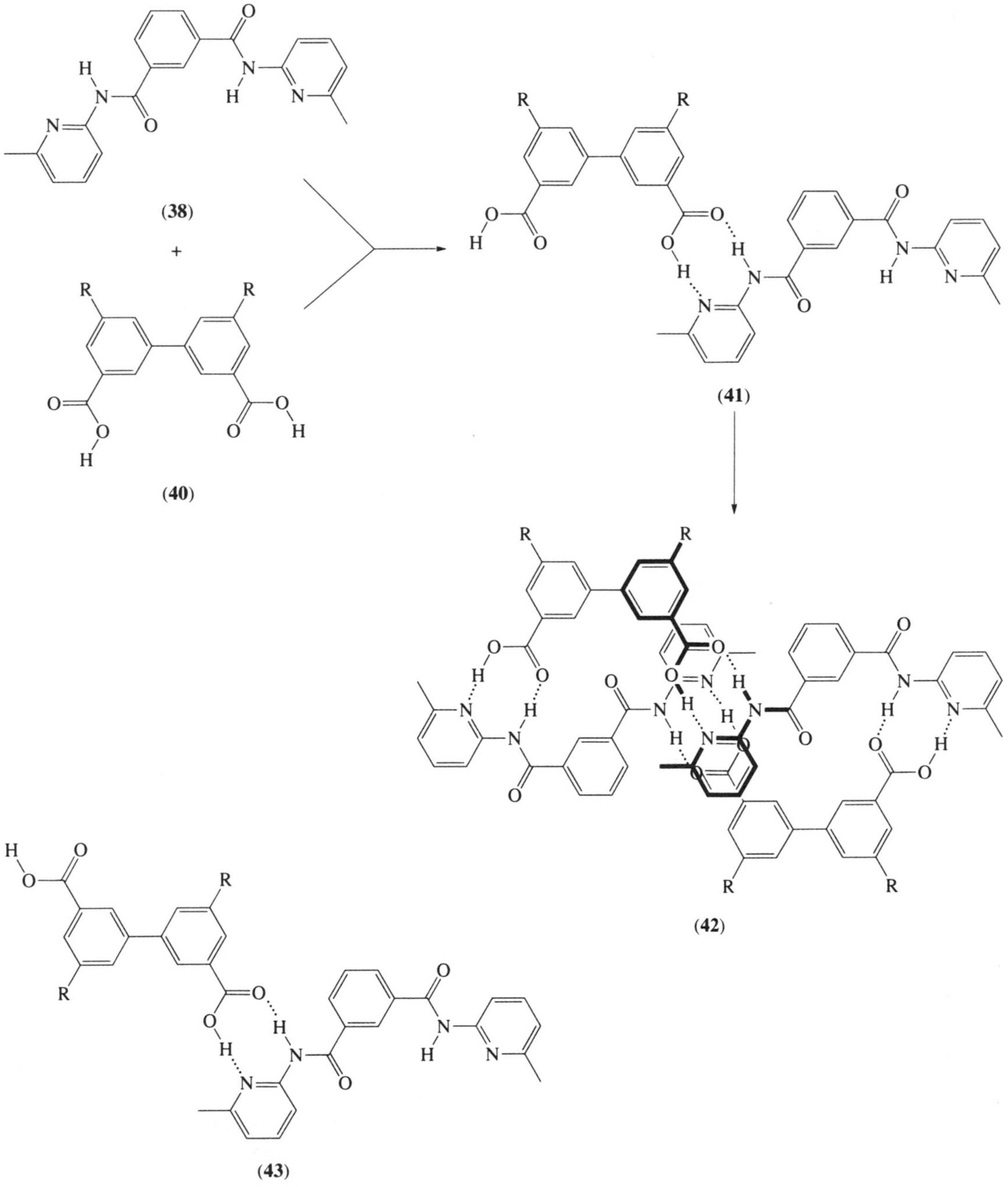

(**38**)

+

(**40**)

(**41**)

(**42**)

(**43**)

Scheme 1

were determined using ^{1}H NMR, GPC, and VPO. The molecular weight determined with VPO was 1400 ± 100 amu. This is lower than that calculated for the complex, presumably due to the existence of smaller aggregates in solution. Analyzing the VPO data by the method of Schrier[29] shows that > 80% of the species in solution are the (2 + 2) complex.[74]

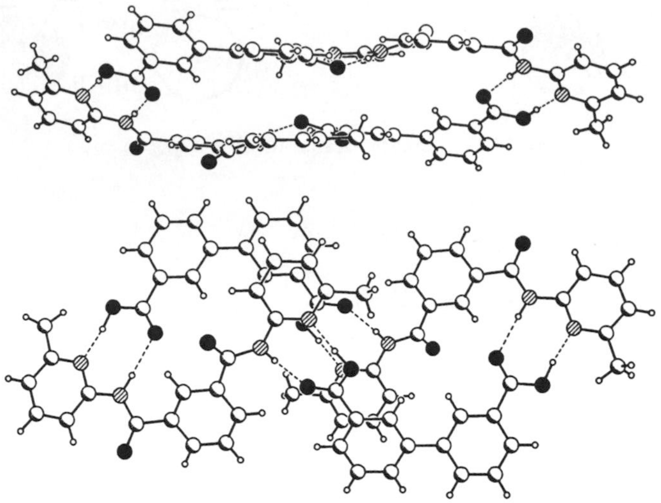

Figure 22 X-ray structure of the 2 + 2 aggregate between (**38**) and biphenyl-3,3′-dicarboxylic acid (**40**).

16.3.2.3 Diaminopyridine-based motifs

Of special interest in the field of host–guest chemistry is the ability to link redox or photoactive chromophores to simple hydrogen-bonding subunits. Hamilton and co-workers designed (1 + 1) and (1 + 2) self-assembling complexes based on the hexahydrogen-bonding complementarity between barbiturate derivatives and two 2,6-diaminopyridine units. The diaminopyridine units were linked through an isophthalate spacer. Reaction of the porphyrin compound (**44**) with (**45**) produced the (1 + 1) complex (**46**). The structure of (**46**) was confirmed by ^{1}H NMR, and the distance between chromophore centers was estimated to be about 2.3 nm.

The extent of communication between the chromophores was assessed by fluorescence emmision spectroscopy. Titrating porphyrin (**44**) with (**45**) led to quenching of the dansyl fluorescence at 528 nm. Evidence that this quenching is due to porphyrin (**44**) comes from competition experiments where (**45**) is titrated with both the porphyrin and excess barbital. A smaller decrease in the fluorescence was observed, and the spectra could be subtracted to give a curve which corresponds to the binding curve between (**44**) and (**45**). The association constant (K_a) was calculated to be 1.0×10^6 M^{-1}.[1]

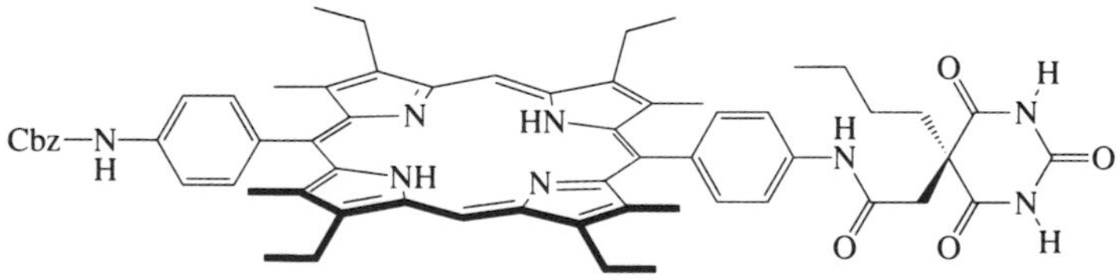

Cbz—N—H N HN NH N H N O O O H N O N H

(**44**)

Another approach investigated by the Hamilton group involved the design of a molecule which could act as either a convergent receptor or a self-assembling subunit depending on the conditions. This switching function can be controlled by the directionality of the hydrogen-bonding groups. For 1 + 1 recognition the binding groups are usually directed inwards (or *endo*) to converge on a central cleft or cavity (Figure 23(a)). In contrast, positioning the hydrogen-bonding groups in an outwards (or *exo*) direction can lead to a self-assembly of the same components into an alternating polymeric complex (Figure 23(b)).

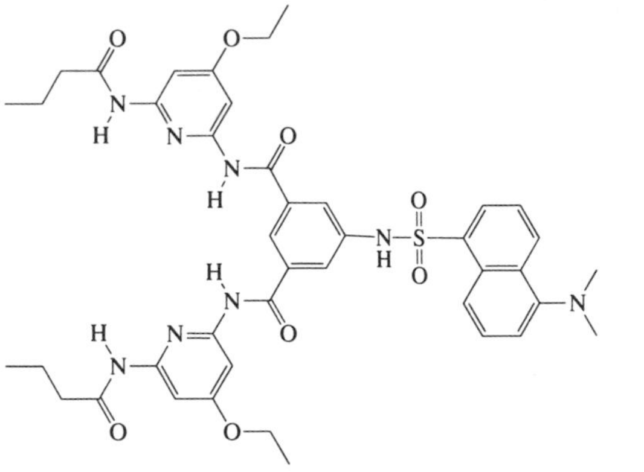

(45)

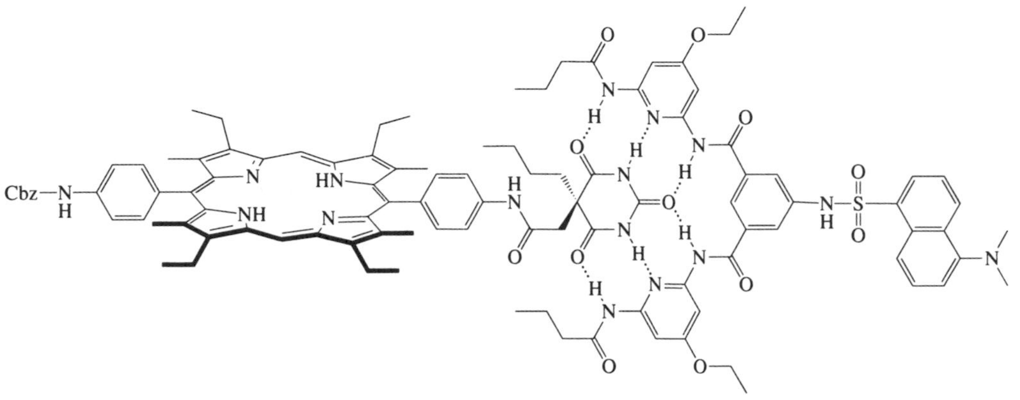

(46)

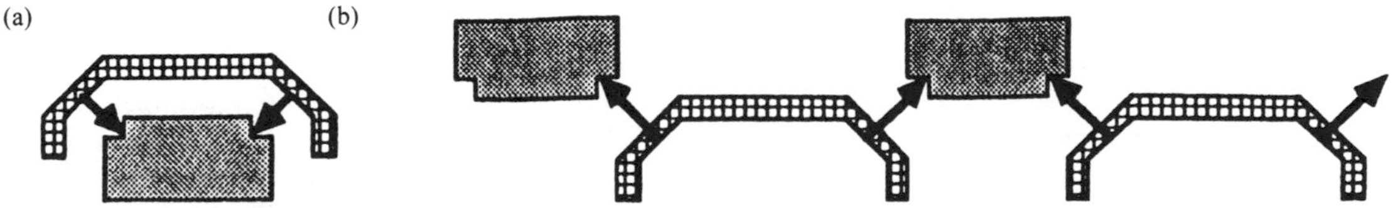

(a) (b)

Figure 23 Switching of recognition functions from (a) *endo* or 1 + 1 binding to (b) *exo* or polymeric aggregation.

A molecule that can switch between structures as a function of environment is 2,6-dibutyraminopyridine (**47**).[75] It is shown in its *endo* conformation, capable of making multiple hydrogen bonds to a single substrate, such as an imide or carboxylic acid.[76] This conformation is the expected conformation of the free and substrate bound receptor owing to favorable electrostatic interaction between the pyridine-nitrogen and amide NH groups. However, protonation of (**47**) with an acid such as HCl introduces a repulsive interaction, (**48**), causing the receptor to take up an *exo* conformation. Rotation about the pyridine–amide bonds now allows the receptor to hydrogen bond to two substrate molecules as shown in (**49**).

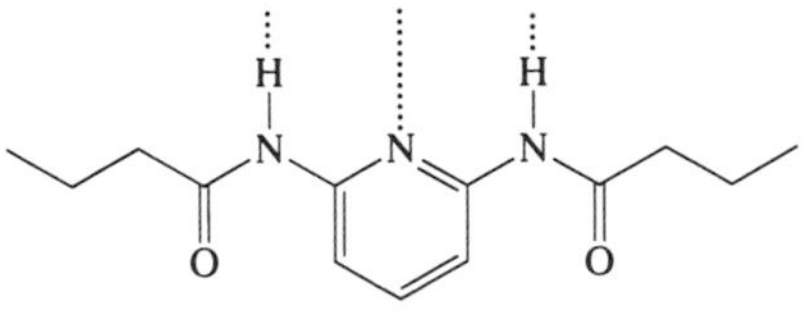

(47)

One equivalent of the proton donor bis(4-nitrophenyl) hydrogen phosphate was allowed to react with (**47**) in CH_2Cl_2. The absorption spectra showed a decrease in the absorption at 292 nm and

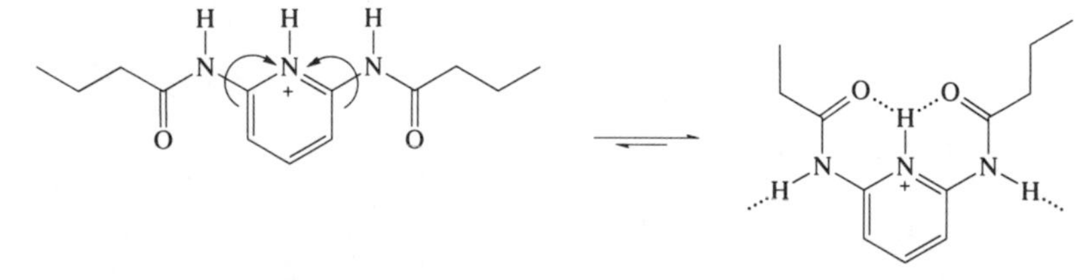

(48) (49)

the formation of a new maximum at 325 nm, corresponding to protonation of the pyridine ring. The x-ray structure of the resulting complex shows that **(47)** has taken up the *exo* conformation and formed a hydrogen-bonded polymeric aggregate to the phosphodiester as in Figure 24. A similar strategy was extended to design asymmetrical crystals from (R)-(–)-1,1′-binaphthyl-2,2′-diyl hydrogen phosphate and **(47)**.

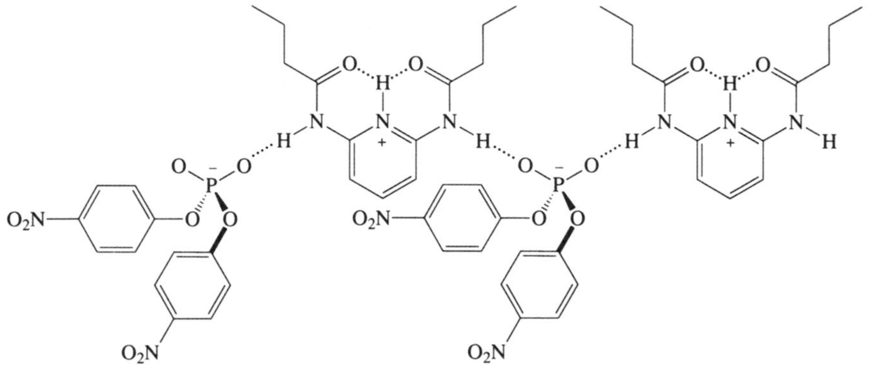

Figure 24 Self-assembly of 2,6-dibutyramidopyridine with bis(4-nitrophenyl) hydrogen phosphate.

16.3.2.4 Other hydrogen-bonding motifs

A self-assembled, bis(porphyrin) cage was designed by Lehn and co-workers (Scheme 2). The binding motif used was that between the 5-alkyluracil groups attached to the porphyrin ring (Scheme 2) and two alkyltriaminopyrimidine (TAP) units. The aggregate was characterized using ^{1}H NMR, electrospray mass spectrometry, and VPO, although a small amount of open-ended chain structures may also be present. Luminescence measurements also supported the structure. The uncomplexed porphyrin showed fluorescence quenching with increased concentration due to π-stacking. The self-assembled complex, shown in Figure 25, did not show this quenching even up to concentrations of 5 mmol dm^{-3}. When 0.1% HCl was added, the emission reverted to that of the unassembled, protonated porphyrin. It was found that the zinc complex binds 4,4′-bipyridine more strongly than zinc tetraphenylporphyrin or zinc octethylporphyrin, suggesting possible use of these types of aggregates as binding agents.[77]

Sessler has exploited the strong hydrogen-bonding interaction between nucleotide bases to prepare a series of self-assembled multichromophore aggregates.[78] For example, metalloporphyrins linked to an aryl guanosine derivative associate with metal-free porphyrins linked to an arylcytosine group. The resulting complex **(50)** shows strong energy transfer between the two chromophores.

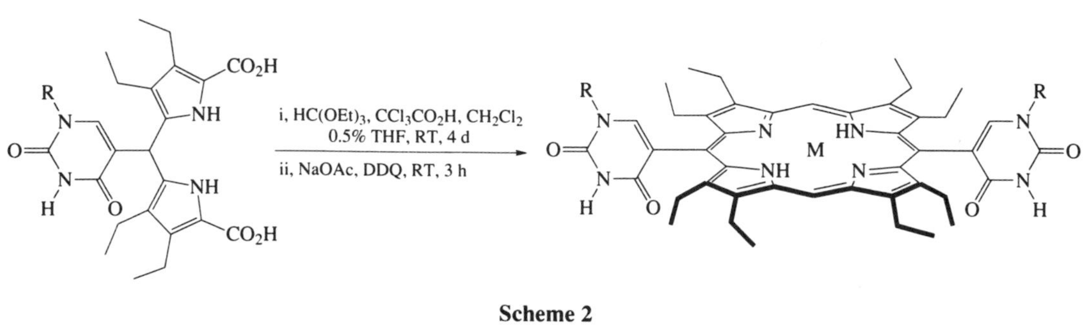

i, HC(OEt)$_3$, CCl$_3$CO$_2$H, CH$_2$Cl$_2$ 0.5% THF, RT, 4 d
ii, NaOAc, DDQ, RT, 3 h

Scheme 2

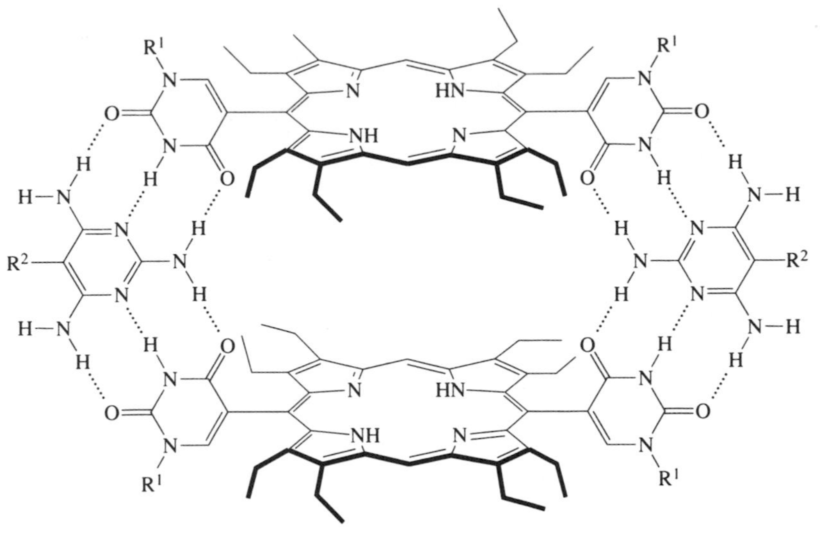

R^1 = decyl, R^2 = octyl or decyl

Figure 25 Supramolecular cage-like structure formed by the self-assembly of two porphyrins containing 5-alkyluracil recognition groups and two TAP units.

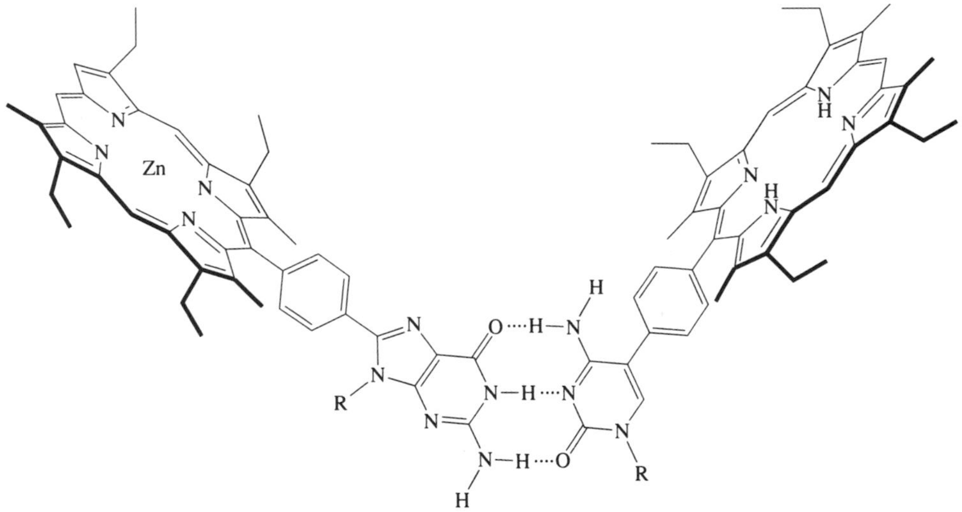

(**50**)

16.4 REFERENCES

1. P. Tecila, R. P. Dixon, G. Slobodkin, D. S. Alavi, D. H. Waldeck, and A. D. Hamilton, *J. Am. Chem. Soc.*, 1990, **112**, 9408.
2. G. M. Whitesides, J. P. Mathias, and C. T. Seto, *Science*, 1991, **254**, 1312.
3. J. S. Lindsey, *New J. Chem.*, 1991, **15**, 153.
4. A. Pfeil and J. M. Lehn, *J. Chem. Soc., Chem. Commun.*, 1992, 838.
5. S. J. Geib, C. Vicent, E. Fan, and A. D. Hamilton, *Angew. Chem., Int. Ed. Engl.*, 1993, **32**, 119.
6. G. Stubbs and K. Namba, *Science*, 1986, **231**, 1401.
7. M. R. Ghadiri, C. Soares, and C. Choi, *J. Am. Chem. Soc.*, 1992, **114**, 825.
8. S. Ruttimann, G. Bernardinelli, and A. F. Williams, *Angew. Chem., Int. Ed. Engl.*, 1993, **32**, 392.
9. M. Simard, D. Su, and J. D. Wuest, *J. Am. Chem. Soc.*, 1991, **113**, 4696.
10. P. R. Ashton, R. A. Bissell, N. Spencer, J. F. Stoddart, and M. S. Tolley, *Synlett*, 1992, 914.
11. P. R. Ashton, R. A. Bissell, R. Gorski, D. Philp, N. Spencer, J. F. Stoddart, and M. S. Tolley, *Synlett*, 1992, 919.
12. P. R. Ashton, R. A. Bissell, N. Spencer, J. F. Stoddart, and M. S. Tolley, *Synlett*, 1992, 923.
13. P. Kaszynski, A. C. Friedli, and J. Michl, *J. Am. Chem. Soc.*, 1992, **114**, 601.

14. P. L. Anelli, P. R. Ashton, R. Ballardini, V. Balzani, M. Delgado, M. T. Gandolfi, T. T. Goodnow, A. E. Kaifer, D. Philp, M. Pietraszkiewicz, L. Prodi, M. V. Reddington, A. M. Z. Slawin, N. Spencer, J. F. Stoddart, C. Vicent, and D. S. Williams, *J. Am. Chem. Soc.*, 1992, **114**, 193.
15. S. Bonazzi, M. M. De Morais, G. Gottarelli, P. Mariani, and G. P. Spada, *Angew. Chem., Int. Ed. Engl.*, 1993, **32**, 248.
16. P. R. Ashton, M. Grognuz, A. M. Z. Slawin, J. F. Stoddart, and D. F. Williams, *Tetrahedron Lett.*, 1991, **32**, 6235.
17. P. Baxter, J. M. Lehn, A. DeCian, and J. Fischer, *Angew. Chem., Int. Ed. Engl.*, 1993, **32**, 69.
18. C. T. Seto and G. M. Whitesides, *J. Am. Chem. Soc.*, 1993, **115**, 905.
19. M. V. Gallant, M. T. P. Viet, and J. D. Wuest, *J. Org. Chem.*, 1991, **56**, 2284.
20. S. Ruttimann, C. Piguet, G. Bernardinelli, B. Bocquet, and A. F. Williams, *J. Am. Chem. Soc.*, 1992, **114**, 4230.
21. J. Rebek, Jr., *Acc. Chem. Res.*, 1990, **23**, 399.
22. R. Kramer, J. M. Lehn, and A. Marquis-Rigault, *Proc. Natl. Acad. Sci. USA*, 1993, **90**, 5394.
23. C. T. Seto and G. M. Whitesides, *J. Am. Chem. Soc.*, 1993, **115**, 1330.
24. J. Cazes, *J. Chem. Educ.*, 1966, **43**, A567.
25. F. J. Stevens, *Biochemistry*, 1986, **25**, 981.
26. F. J. Stevens, *Biophys. J.*, 1989, **55**, 1155.
27. D. E. Burge, *J. Phys. Chem.*, 1963, **67**, 2590.
28. T. N. Solie, *Methods Enzymol.*, 1972, **26**, 50.
29. E. E. Schrier, *J. Chem. Educ.*, 1968, **45**, 176.
30. E. C. Constable, M. D. Ward, and D. A. Tocher, *J. Chem. Soc., Dalton. Trans.*, 1991, 1675.
31. S. C. Zimmerman and B. F. Duerr, *J. Org. Chem.*, 1992, **57**, 2215.
32. M. Saunders and J. B. Hyne, *J. Chem. Phys.*, 1958, **29**, 1319.
33. J. C. Davis and K. K. Deb, *Adv. Magn. Res.*, 1970, **4**, 201.
34. M. Barber, R. S. Bordoli, G. J. Elliot, D. Sedgwick, and A. Tyler, *Anal. Chem.*, 1982, **54**, 645A.
35. M. Mann, *Org. Mass Spectrosc.*, 1990, **25**, 575.
36. E. Leize, A. Van Dorsselaer, R. Kramer, and J. M. Lehn, *J. Chem. Soc., Chem. Commun.*, 1993, 990.
37. Y. Ducharme and J. D. Wuest, *J. Org. Chem.*, 1988, **53**, 5787.
38. M. Gallant, M. T. P. Viet, and J. D. Wuest, *J. Am. Chem. Soc.*, 1991, **113**, 721.
39. E. Boucher, M. Simard, and J. D. Wuest, *J. Org. Chem.*, 1995, **60**, 1408.
40. F. Persico and J. D. Wuest, *J. Org. Chem.*, 1993, **58**, 95.
41. P. Mariani, C. Mazabard, A. Garbesi, and G. P. Spada, *J. Am. Chem. Soc.*, 1989, **111**, 6369.
42. S. Bonazzi, M. Capobianco, M. M. De Morais, A. Garbesi, G. Gottarelli, P. Mariani, M. G. P. Bossi, G. P. Spada, and L. Tondelli, *J. Am. Chem. Soc.*, 1991, **113**, 5809.
43. J. M. Lehn, M. Mascal, A. DeCian, and J. Fischer, *J. Chem. Soc., Perkin Trans. 2*, 1992, 461.
44. J. M. Lehn, M. Mascal, A. DeCian, and J. Fischer, *J. Chem. Soc., Chem. Commun.*, 1990, 479. B. Craven and E. A. Vizzini, *Acta Crystallogr., Sect. B*, 1971, **27**, 1917.
45. D. J. Duchamp and R. E. Marsh, *Acta Crystallogr., Sect. B.*, 1969, **25**, 5.
46. R. Alcala and S. Martinez-Carrera, *Acta Crystallogr.*, Sect. B, 1972, **28**, 1671.
47. J. Yang, J. L. Marendaz, S. J. Geib, and A. D. Hamilton, *Tetrahedron Lett.*, 1994, 3665.
48. V. Valiyaveettil, G. Enkelmann, and K. Mullen, *J. Chem. Soc., Chem. Commun.*, 1994, 2097.
49. J. Yang, J. L. Marandez, S. J. Geib, and A. D. Hamilton, submitted for publication.
50. J. Zhang, J. S. Moore, Z. Xu, and R. A. Aguirre, *J. Am. Chem. Soc.*, 1992, **114**, 2273.
51. R. Wyler, J. de Mendoza, and J. Rebek, Jr., *Angew. Chem., Int. Ed. Engl.*, 1993, **32**, 1699.
52. N. Branda, R. Wyler, and J. Rebek, Jr., *Science*, 1994, **263**, 1267.
53. M. J. Brienne, J. Gabard, M. Leclercq, J. M. Lehn, M. Cesario, C. Pascard, M. Cheve, and C. Dutruc-Rosset, *Tetrahedron Lett.*, 1994, 8157.
54. M. R. Ghadiri, J. R. Granja, R. A. Milligan, D. E. McRee, and N. Khazanovich, *Nature*, 1993, **366**, 324.
55. B. A. Wallace and K. Ravikumar, *Science*, 1988, **241**, 182.
56. X. Sun and G. P. Lorenzi, *Helv. Chim. Acta*, 1994, **77**, 1520.
57. M. R. Ghadiri, J. R. Granja, and L. K. Buehler, *Nature*, 1994, **369**, 301.
58. N. Khazanovich, J. R. Granja, D. E. McRee, R. A. Milligan, and M. R. Ghadiri, *J. Am. Chem. Soc.*, 1994, **116**, 6011.
59. J. R. Granja and M. R. Ghadiri, *J. Am. Chem. Soc.*, 1994, **116**, 10785.
60. J. P. Mathias, E. E. Simanek, and G. M. Whitesides, *J. Am. Chem. Soc.*, 1994, **116**, 4326.
61. J. A. Zerkowski, C. T. Seto, and G. M. Whitesides, *J. Am. Chem. Soc.*, 1992, **114**, 5473.
62. J. A. Zerkowski, J. C. MacDonald, C. T. Seto, D. A. Wierda, and G. M. Whitesides, *J. Am. Chem. Soc.*, 1994, **116**, 2382.
63. J. A. Zerkowski, J. P. Mathias, and G. M. Whitesides, *J. Am. Chem. Soc.*, 1994, **116**, 4305.
64. J. A. Zerkowski and G. M. Whitesides, *J. Am. Chem. Soc.*, 1994, **116**, 4298.
65. C. T. Seto and G. M. Whitesides, *J. Am. Chem. Soc.*, 1990, **112**, 6409.
66. C. T. Seto, J. P. Mathias, and G. M. Whitesides, *J. Am. Chem. Soc.*, 1993, **115**, 1321.
67. C. T. Seto and G. M. Whitesides, *J. Am. Chem. Soc.*, 1993, **115**, 1330.
68. J. P. Mathias, C. T. Seto, and G. M. Whitesides, *J. Am. Chem. Soc.*, 1994, **116**, 1725.
69. J. P. Mathias, C. T. Seto, and G. M. Whitesides, *Angew. Chem., Int. Ed. Engl.*, 1993, **32**, 1767.
70. J. P. Mathias, E. E. Simanek, J. A. Zerkowski, C. T. Seto, and G. M. Whitesides, *J. Am. Chem. Soc.*, 1994, **116**, 4316.
71. F. Garcia-Tellado, S. Goswami, S. K. Chang, S. J. Geib, and A. D. Hamilton, *J. Am. Chem. Soc.*, 1990, **112**, 7393.
72. F. Garcia-Tellado, S. J. Geib, S. Goswami, and A. D. Hamilton *J. Am. Chem. Soc.*, 1991, **113**, 9265.
73. E. Fan, C. Vicent, S. J. Geib, and A. D. Hamilton, *Chem. Mater.*, 1994, **6**, 1113.
74. J. Yang, E. Fan, S. J. Geib, and A. D. Hamilton, *J. Am. Chem. Soc.*, 1993, **115**, 5314.
75. S. J. Geib, S. C. Hirst, C. Vicent, and A. D. Hamilton, *J. Chem. Soc., Chem. Commun.*, 1991, 1283.
76. A. D. Hamilton and D. Van Engen, *J. Am. Chem. Soc.*, 1987, **109**, 5035.
77. C. M. Drain, R. Fischer, E. G. Nolen, and J. M. Lehn, *J. Chem. Soc., Chem. Commun.*, 1993, 243–5.
78. J. L. Sessler, B. Wang, and A. Harriman, *J. Am. Chem. Soc.*, 1995, **117**, 704, and references therein.

17

Cyanuric Acid and Melamine: A Platform for the Construction of Soluble Aggregates and Crystalline Materials

ERIC E. SIMANEK, XINHUA LI, INSUNG S. CHOI, and GEORGE M. WHITESIDES
Harvard University, Cambridge, MA, USA

17.1 INTRODUCTION

This chapter summarizes efforts to construct soluble aggregates and crystalline materials utilizing the noncovalent interactions (principally hydrogen bonds) between derivatives of cyanuric acid (CA) and melamine (M). The discussion is divided into two parts: soluble aggregates and crystalline materials.

Organic chemistry is turning from the synthesis of covalent molecules to the assembly of noncovalent aggregates. This shift in interest reflects both a high degree of mastery of covalent synthesis, and realization of the importance of noncovalent interactions. The dissection of natural systems—exemplified by Mrksich and Dervan's examination of *ds*DNA·distamycin derivatives,[1] Schreiber's study of FK-506·FKBP,[2] and Williams *et al.*'s inquiries into vancomycin·D-Ala-D-Ala[3] —has begun to clarify their structural and thermodynamic foundations. The specificity of these naturally occurring systems is not matched by design-based efforts of the type pioneered by Pedersen,[4] Cram,[5] and Lehn.[6] These efforts have, nonetheless, produced increasing numbers of aggregates of different classes.

The design of noncovalent aggregates is now a central challenge for organic chemistry. By keeping the number of particles and/or the number of different types of noncovalent interactions to a manageable number, it is beginning to be possible to develop general principles of self-assembly. While several classes of designed assemblies have been explored, four main classes have emerged as foci for current interest.

17.1.1 Ion Receptors

Aggregates such as Ca^{2+}·ionomycin, and related complexes of alkali and alkaline earth cations with polyether antibiotics, offer a natural paradigm for the recognition of alkali metals.[7] Similarly, the complexation of Fe^{III} by siderophiles has suggested structural bases for the selective recognition of transition metal ions. Figure 1 shows some representative assemblies that recognize monoatomic and polyatomic ions.

17.1.2 Small Molecule Receptors

There are many natural paradigms for these receptor–target assemblies; the recognition of D-Ala-D-Ala vancomycin is one of the best understood, both theoretically and experimentally.[3] The fundamental stimulus for this work is the need to understand the recognition of small molecules by enzymes and receptors.[8] The recognition of other small molecules, such as barbituric acid, imidazole, and molecules of biological interest (including the amino acids and cholesterol) is being examined by many groups. This interest has, however, broadened to encompass recognition in nonaqueous solution, and to include classes of aggregates for which there are no precedents in nature. Figure 2 shows representative examples of receptors that recognize small organic molecules. The relative ease with which calixarenes, cryptands, and cyclodextrins can be synthesized has led to active communities engaged almost exclusively in their study.

17.1.3 Molecular "Machines"

The work of Stoddart and co-workers (Equation (1)) and Shanzer[9a] and co-workers (Equation (2))[9b] exemplifies the concept of molecular machines as primitive switches that show "off" and

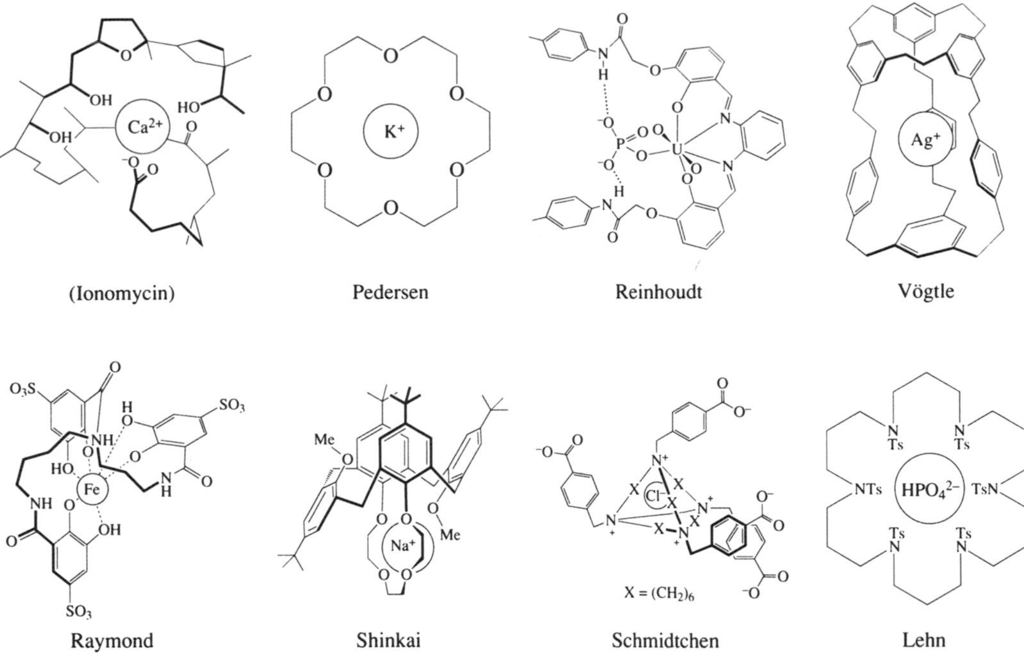

Figure 1 Recognition of ions in solution.

"on" states as a function of a physical stimulus such as redox or pH. Current molecules and aggregates are molecules that exist in different states under different conditions, but show no switching function. The gap between structure and useful function in these materials remains large, and the classes of the materials ultimately developed in this area will probably reflect the requirements for "utility" as defined by information and sensor technologies, rather than biological precedent.

$+e^-$ $-e^-$ (1)

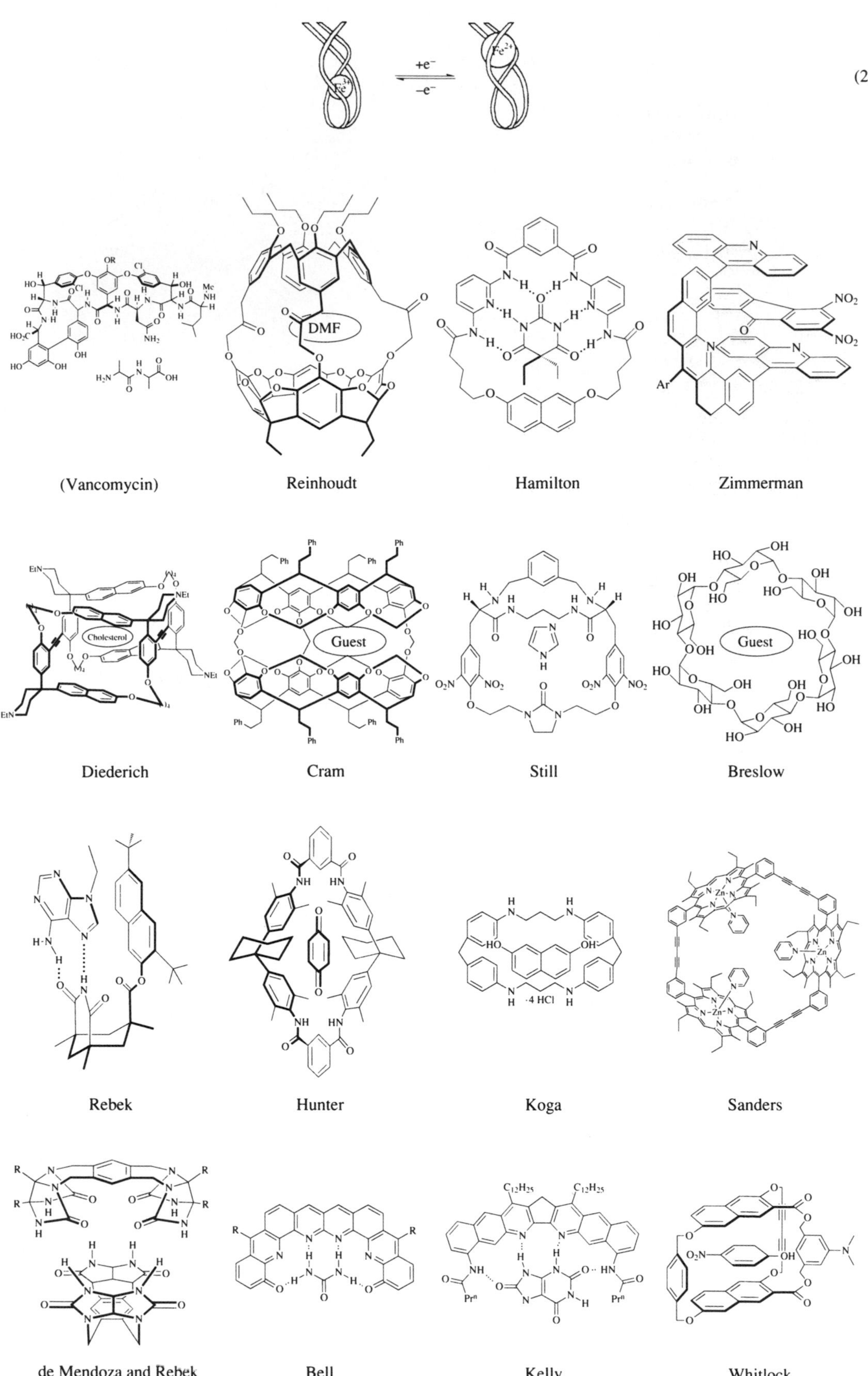

Figure 2 The recognition of small molecules.

171.4 Multiparticle Aggregates

Many important recognition events involve multiple particles (either connected or independent): formation of molecular crystals; folding of proteins; assembly of multiprotein structures such as ribosomes and viral capsids; formation of lipid bilayers and liposomes.[10] These large assemblies can serve as structural elements in materials and catalysts, and as molecular scaffolds that orient and position functional components. tRNA is one example of a molecular scaffold: holding recognition and reaction domains at a specific distance in specific orientations is accomplished by a combination of hydrogen bonding and other interactions present on an oligonucleotide backbone. Enzymes provide another sophisticated example in which many recognition and catalytic elements are positioned by a self-assembled structure. Figure 3 shows examples of synthetic scaffolds.

17.2 OUR WORK

Our own efforts in noncovalent synthesis and molecular self-assembly center on the construction of multiparticle aggregates. We have used derivatives of CA and M to investigate self-assembly both in solution and the solid state. (We refer to any derivative of isocyanuric acid or barbituric acid as CA.) The system offers, as one of its assets, a strong conceptual complementarity between soluble aggregates and crystalline solids. While the majority of this chapter is dedicated to the description of soluble aggregates, efforts in the solid state have suggested important concepts—especially those based on steric constraints to shape (peripheral crowding)—that are also directly applicable in soluble assemblies.

17.3 CA·M

The CA·M lattice is shown in Figure 4. We have identified three different motifs that are useful as scaffolds for the construction of soluble aggregates and crystalline materials: the rosette, and the linear and crinkled tapes. The construction of soluble scaffolds requires a simple, soluble, and discrete system: our efforts focus on the rosette. The study of crystalline materials benefits by

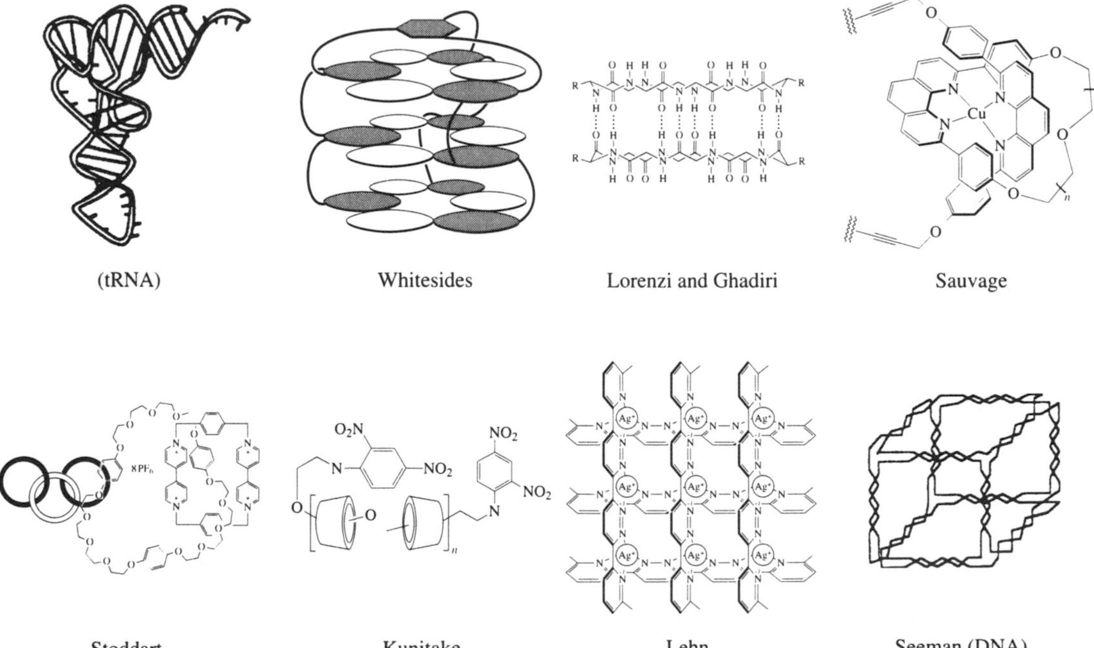

(tRNA) Whitesides Lorenzi and Ghadiri Sauvage

Stoddart Kunitake Lehn Seeman (DNA)

Figure 3 Scaffolds based on noncovalent interactions.

reducing the dimensionality of the problem: learning the rules governing the alignment of quasi-one-dimensional tapes in the solid state should be easier than understanding the packing of "zero-dimensional" molecules with no constraints on their orientation. The solubility of extended tape structures is low and they are, for that reason, good choices for work with solids, but poor choices for soluble systems.

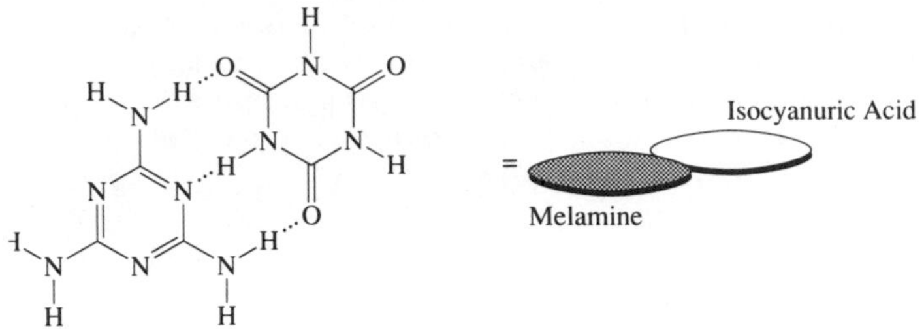

Isocyanuric Acid = Melamine

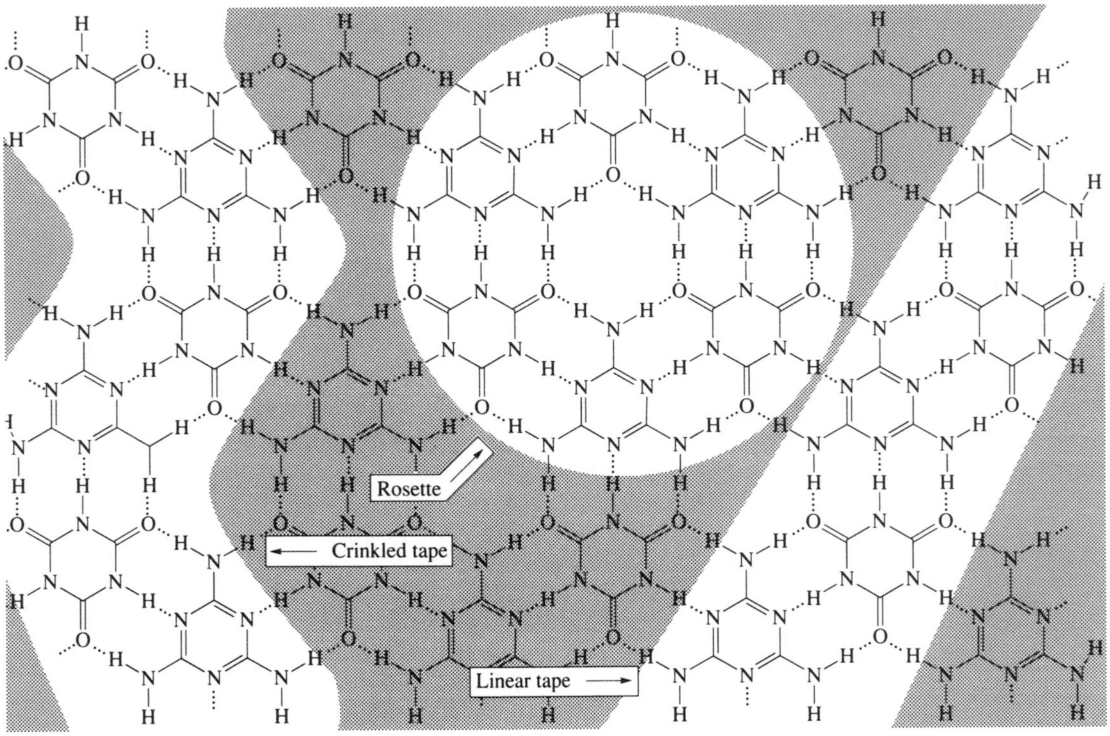

Rosette

Crinkled tape

Linear tape

Figure 4 The lattice of isocyanuric acid and melamine. Three motifs are highlighted: two extended tape motifs and the discrete rosette. We represent melamines as darkened disks and isocyanuric acids as white

In addition to affording motifs for the design and construction of soluble aggregates or crystalline solids, CA·M offers additional advantages: the relatively high stability reflects the network of hydrogen bonds between molecules of CA and M; the ease of characterization due to the symmetries of the individual molecules and resulting assemblies; the straightforward synthesis of derivatives of CA and M. We comment on each of these advantages separately.

17.3.1 Hydrogen Bonds

The assembly of an aggregate in solution requires that the enthalpic gain resulting from the formation of new bonds exceeds the entropic costs of bringing molecules together. The enthalpy of aggregation can be influenced in two ways: by increasing the strength of the interaction (van der Waals forces < hydrogen bonds < metal chelates < covalent bonds), and by increasing the number of bonds being formed. The assembly of a multiparticle aggregate must occur reversibly if it is to lead to a structurally well-defined system. The importance of reversibility stems from the fact that irreversible association of components in solution will usually result in a number of different aggregates, and kinetic assembly will not normally cleanly generate the most thermodynamically stable aggregate. For the target (equilibrium) structure to form, the components of the assembly must be able to reorganize repeatedly and with small expense in energy: if the enthalpy of the interactions is too high, the favored product can be made kinetically inaccessible.

We (and many others) choose to use hydrogen bonds for our studies of noncovalent assembly. Each hydrogen bond in a structure based on the CA·M lattice contributes 4.186–12.56 kJ mol^{-1} to the enthalpy of aggregation in chloroform solution. This value is comparable to RT = 2.51 kJ mol^{-1} at 300 K, and equilibrium among aggregates is usually rapid. Although this value is relatively small compared to charge–charge interactions in nonpolar solvents, and to some metal–chelate interactions, the total enthalpy of aggregation can be "tuned" by increasing or decreasing the total number of hydrogen bonds. Hydrogen bonds offer additional advantages: they involve distinct donor and acceptor groups, and they are directional: they thus offer more possibility for structural design than forces that are symmetric and nondirectional.[11]

17.3.2 Symmetries

Molecules of CA and M are effectively C_2 symmetric with respect to the axis connecting the two sets of interacting hydrogen bonds. Functionalizing the third edge of a CA usually does not influence this symmetry. Functionalizing molecules of M can decrease this symmetry: it introduces a potential for isomerism in some of these aggregates (Figure 5). While this isomerism can complicate structural characterization, it can also provide important information about the rigidity of the framework that organizes the hydrogen-bonded components. Isomerism due to the linker groups on M does not change the positions at which CA is recognized, that is, isomerism does not affect shape, and therefore, probably does not affect function. We discuss isomerism in more detail in Section 17.5.2.8.

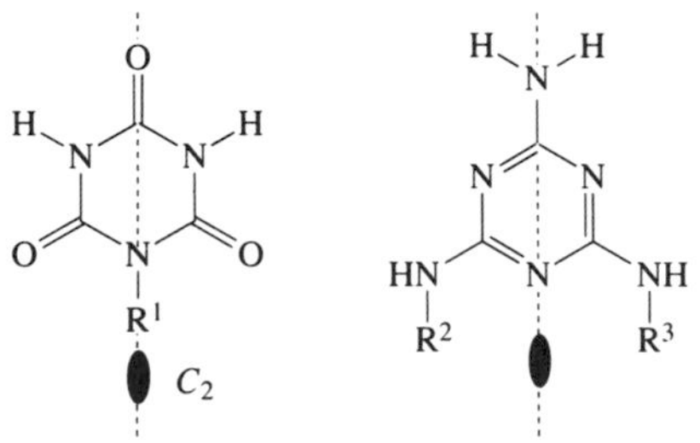

If $R^2 = R^3 \Rightarrow C_2$

$R^2 \neq R^3 \Rightarrow C_1$

Figure 5 CA are C_2 symmetric unless R[1] is chiral. For M, unless $R^2 = R^3$, there is no C_2 symmetry.

17.3.3 Synthesis

The synthesis of the linker groups used for tethering melamines or cyanuric acids is not difficult; details are available in the original sources.[12] The efficient generation of derivatives of CA and M is also relatively straightforward, but is sufficiently challenging that it deserves some comment. The methods used for the "synthesis" of the noncovalent aggregates that form when derivatives of CA and M are mixed also requires description.

Two types of molecules related to CA are used in these studies. Diethylbarbital (and other derivatives of barbituric acid) come from the condensation product of diethyl dialkylmalonate and urea (Equation (3)). Neohexylisocyanuric acid and other *N*-substituted derivatives of CA are synthesized in two steps from the appropriate amine (Equation (4)). Reaction of an alkylamine with nitrobiuret yields an isolable intermediate—an alkylbiuret—which is cyclized with a carbonyl source such as diethyl carbonate and carbonyl diimidazole. Yields are often high (60%), but depend strongly on the amine used. Many derivatives of CA are commercially available.

Cyanuric chloride reacts selectively with amines and generates trisubstituted melamines (Equation (5)). The first equivalent of amine (typically an unreactive aniline derivative) is added at 0 °C. After 10 min, a second amine (such as NH_3) is added. After stirring at room temperature for 90 min, the final amine (often a reactive alkyl or benzyl amine) is added. Refluxing these reactants for 8 h yields the trisubstituted melamine in ~80% yield.

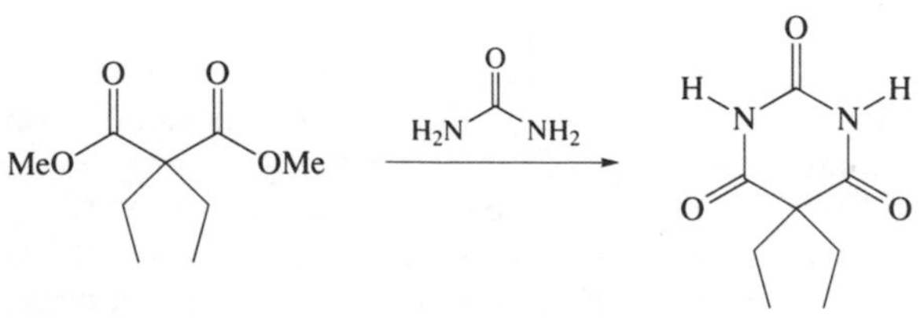

$$\text{MeO(O=)C–C(Et)_2–C(=O)OMe} \xrightarrow{H_2N\text{–}C(=O)\text{–}NH_2} \text{5,5-diethylbarbituric acid} \tag{3}$$

yields ~80%

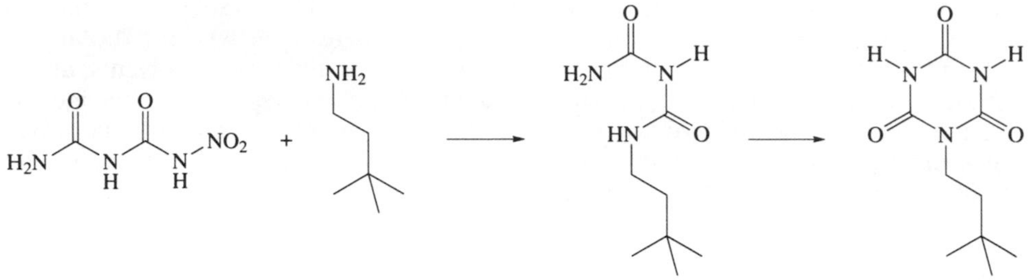

$$H_2N\text{–}C(=O)\text{–}NH\text{–}C(=O)\text{–}NH\text{–}NO_2 + NH_2CH_2CH_2C(CH_3)_3 \longrightarrow \text{biuret derivative} \longrightarrow \text{N-substituted cyanuric acid} \tag{4}$$

yields 50–80%

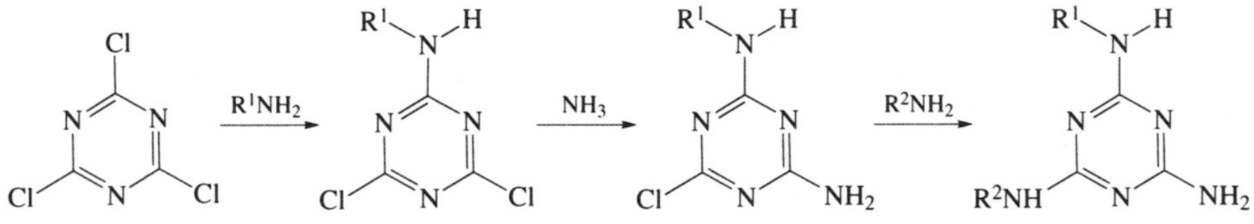

$$\text{cyanuric chloride} \xrightarrow{R^1NH_2} \xrightarrow{NH_3} \xrightarrow{R^2NH_2} \tag{5}$$

Yields ~80%

The ease of synthesis of an aggregate from its components depends partially on their solubility in chloroform: most molecules of CA are only slightly soluble in organic solvents. When both components are soluble, simply mixing them generates the aggregate. Heating the components at reflux in chloroform for ~30 s yields aggregates for all but those containing the least soluble components. In these final cases (i.e., the bis(rosette) that will be introduced later), approximately 5% methanol is added to the chloroform to increase solubility, and the mixture of components is refluxed. The solution is evaporated to dryness. The observation that the resulting powder is completely soluble in chloroform is an indication that aggregation is occurring.

17.4 ORGANIZATION OF THIS CHAPTER

For simplicity, we divide this review into two sections—soluble aggregates and crystalline tapes. We discuss similarities between these systems as they arise.

17.5 SOLUBLE AGGREGATES

We divide our discussion of soluble aggregates (shown pictorially in Figure 6) into four parts. Design describes two strategies for noncovalent synthesis: preorganization and peripheral crowding. Characterization outlines the methods we employ to assign structure; evaluating isomerism is discussed in this section. Computation summarizes our initial efforts to rationalize the relative stabilities of a series of simple aggregates. Theory sketches our efforts to arrive at a conceptual framework and structure-based parameters for comparing stabilities of aggregates, in order to make it possible to design new structures.

17.5.1 Design

We have applied two strategies to the design of noncovalent assemblies based on the CA·M lattice: preorganization and peripheral crowding. Cram's rule of preorganization qualitatively highlights the entropic aspects of self-assembly.[13] The rule states that the assembly occurring with

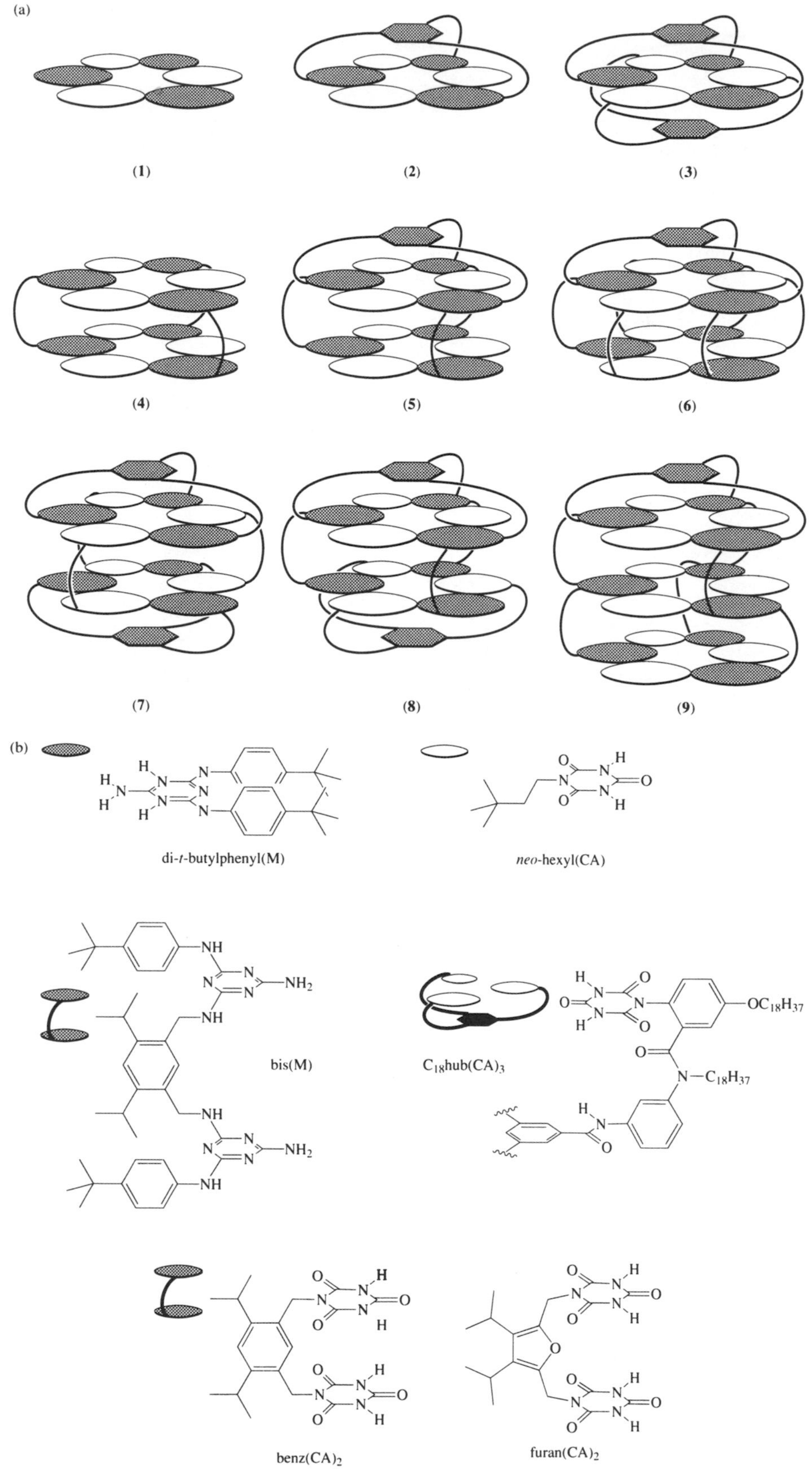

Figure 6 (a) Nine structurally similar classes of aggregates have been explored. All are based on the rosette motif; atomic detail is provided in (b) and (c) (source: Whitesides *et al.*[10]).

(c)

Figure 6 (continued)

the smallest loss of entropy is most favorable. Preorganization is most efficiently discussed in terms of the entropies affected: translational, rotational, and conformational. Peripheral crowding describes a strategy for design that favors one recognition motif over a second by designing enthalpic costs into the two through steric overlaps. This strategy is related, conceptually, to the use of the *gauche* effect in influencing conformational equilibria in covalent organic molecules.

17.5.1.1 The effect of preorganization on the translational entropy (ΔS_{trans}) of assembly

If N particles assemble, the magnitude of ΔS_{trans} is proportional to N–1 (Equation (6)).[14]

$$\Delta S_{trans} = (N-1)\left[\frac{5}{2}R + R\ln\frac{V}{N}\left(\frac{2\pi kT}{h^2}\right)^{3/2}\left(\frac{M_{\rm P}}{M_{\rm ass}}\right)^{3/2}\right] = (N-1)C_{trans} \tag{6}$$

We expect (and observe experimentally) that (**1**), comprising 6 molecules joined by 18 hydrogen bonds, will be less stable than (**3**), comprising 2 molecules joined by 18 hydrogen bonds. Linking the recognition domains (such as M in hub$(M)_3$ and CA in hub$(CA)_3$ by tethers—and thereby reducing ΔS_{trans}—has yielded a series of stable aggregates (Figure 6(a)).

17.5.1.2 The effect of preorganization on rotational entropy (ΔS_{rot})

The entropic cost associated with loss of rotational entropy on assembly, like that of translational entropy, is proportional to the number of particles (Equation (7)).[14] Reducing the number of particles will also reduce the magnitude of the entropic cost of aggregation.

$$\Delta S_{rot} = (N-1)\left[R\ln\left(\frac{8\pi^2 ekT}{h^2}\right)\left(\frac{I_{\rm P}}{I_{\rm ass}}\right)\right] = (N-1)C_{rot} \tag{7}$$

17.5.1.3 The effect of preorganization on conformational entropy (ΔS_{conf})

Most of our efforts are invested in understanding and minimizing the costs of the reduction of conformational entropy on assembly. These costs come from requiring freely rotating single bonds to adopt a single conformation (or a limited set of conformations) upon aggregation. Ideally, we want the geometries of the preassembled components to be similar to those adopted in the aggregate. Hub$(M)_3$·3*neo*-hexyl(CA), (**2**)—the model system for most of the aggregates generated to date—incorporates attempts that we envision to lead to preorganization: rigid linking groups; curved tethering arms.

The majority of bonds in hub$(M)_3$ are rigid: aromatic rings and amides. By contrast, flex$(M)_3$ contains polymethylene chains linked by esters. We have not quantitatively explored the relative stabilities of aggregates derived from these compounds; stability correlates qualitatively with rigidity. We comment on this correlation further in Section 17.5.3.

All of the tethered molecules were designed to place the recognition element in the appropriate orientation. Much of this positioning is accomplished by the substitution of aromatic groups. We hypothesized that *m*-phenylenediamine and *o*-anthranilate of hub$(M)_3$ might cooperate in forming a 180° turn that oriented the melamine groups beneath the central core, that is, into a geometry that allows them to form networks of hydrogen bonds with molecules of CA. We believe that the *m*-xylenyl linker of hub$(MM)_3$ (of (**5**)) offers the same type of preorganization.

17.5.1.4 Peripheral crowding

Peripheral crowding is a strategy for favoring a mono- or bis(rosette) (**1**) or (**4**) in solution rather than a tape by increasing the size of the groups on the periphery (Figure 7). Unfavorable steric interactions between these substituents maximize the distance between them; this distance increases as the three motifs are ordered: linear tape < crinkled tape < rosette. Tables 1 and 2 show the dependence of stability of bis(rosettes) on the size of the peripheral groups.

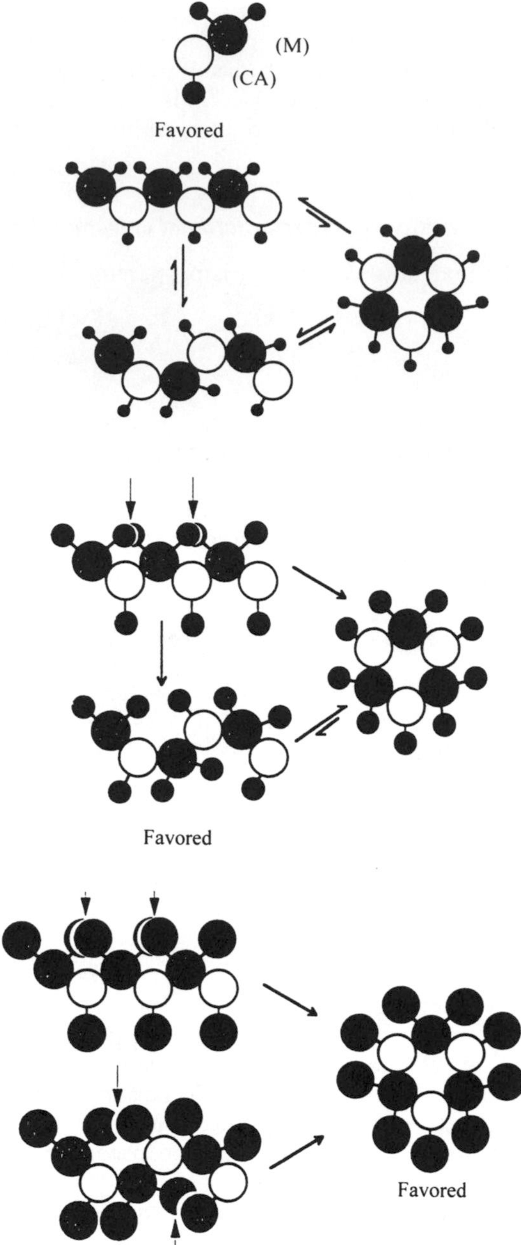

Figure 7 Peripheral crowding describes the preference for the indicated motif based on the size of the peripheral groups (pg: indicated as blackened disks). When the pg are small, we observe linear tapes in the solid state. As pg get larger, steric overlap (arrows) between them result in the formation of crinkled tapes over linear tapes. Rosettes are favored over both tape motifs when the pg become large.

17.5.2 Characterization

We assign structures to soluble aggregates based on the summation of inferences from a number of techniques; no single technique is completely convincing in these systems, and we have not, with one exception, been able to obtain diffraction-quality crystals for x-ray structure analysis.[15] Experiments that use NMR spectroscopy to follow the titration of the M component by addition of CA have established the relative stoichiometries of components; differences in solubility between aggregates and individual molecules of CA and M are a useful indicator of aggregation. Molecular weights can be determined by vapor phase osmometry (VPO), electrospray-ionization mass spectrometry (ESI-MS), and gel permeation chromatography (GPC). The shape of the GPC

Table 1 Increasing the size of the peripheral groups favors rosette formation. Aggregates which are stable indefinitely are marked with (++). Unsuccessful formation of an aggregate is indicated with (–).

	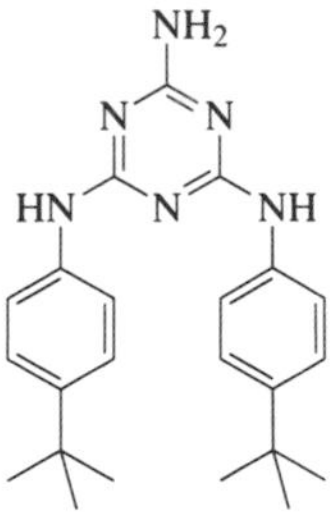	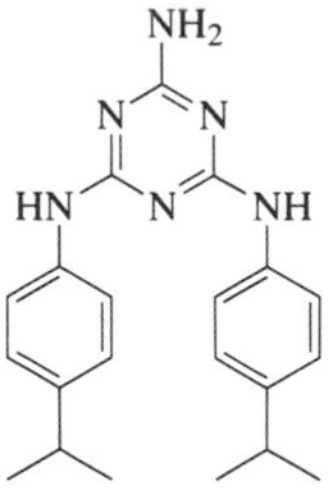	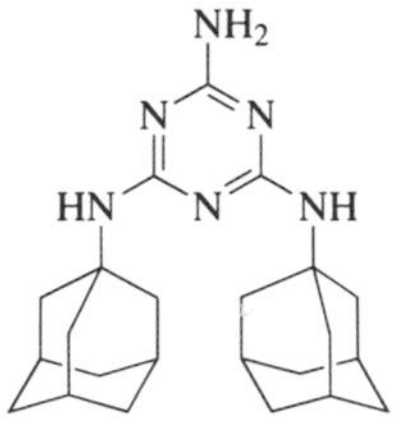	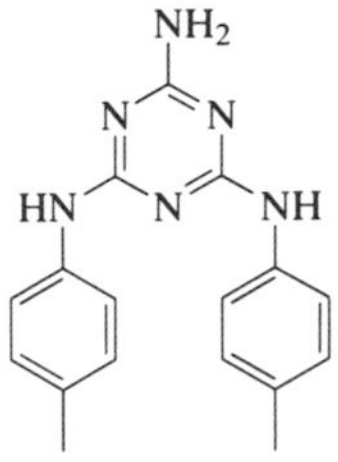
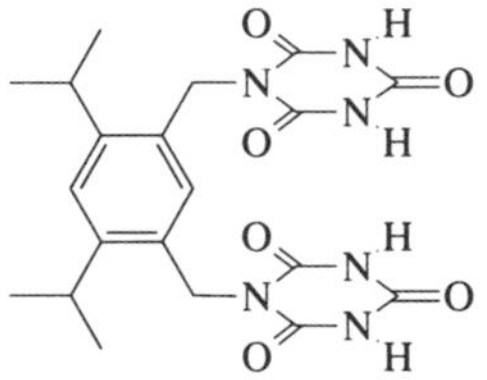	++	–	–	–
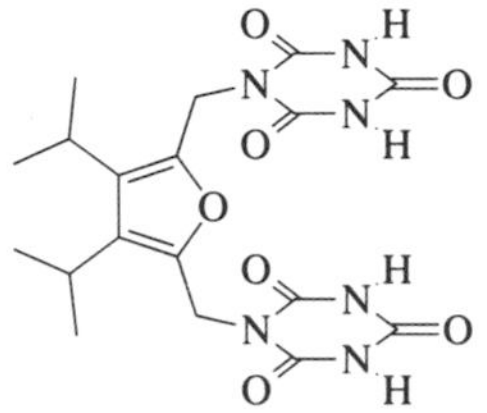	++	–	–	–

trace and methanol titrations provide another type of estimate of the stabilities of these aggregates. The majority of detail about structure (number and symmetries of isomers) comes from ^{1}H NMR spectroscopy; ^{13}C NMR spectroscopy is also useful.

17.5.2.1 Stoichiometry

Proton-NMR spectroscopy can be used to monitor titration experiments that yield the ratio of CA:M. Visual inspection corroborates the result. To a solution of the soluble M component, the CA component (which is insoluble by itself) is added. Assembly is cooperative in most cases—sharp lines (those of the final aggregate) appear superimposed on the broadened spectrum of the melamine; there is no spectroscopic evidence for intermediate complexes with different stoichiometries. The intensities of the lines increase until some integral stoichiometry is reached (for (**2**), lines increase in intensity until three equivalents of CA have been added); beyond this point, CA is no longer soluble and a suspension of insoluble material can be observed.

17.5.2.2 The determination of molecular weight: vapor phase osmometry

Molecular weights can be estimated using VPO—a technique based on the difference in vapor pressures of a pure solvent and solutions containing aggregates. Known amounts of aggregate (or standard) are dissolved in chloroform and applied in small volumes (300 μL) to a thermistor. Pure chloroform is applied to a second thermistor. The two thermistors are kept at the same temperature by passing current through heating elements; the voltage is monitored. This voltage is recorded for a series of concentrations (5–100 mM), and the values are extrapolated to infinite dilution. The results are compared to values obtained using standards of known values of MW. The result is a number-averaged MW. The value for MW is typically accurate to within only 20% of the expected value, and depends strongly on the standard used as a reference of MW. Many standards are used in an effort to check for (and control) the nonidealities of the system.

Table 2 Increasing the size of the peripheral groups favors rosette formation. Aggregates which are stable indefinitely are marked with (++). Aggregates that formed, but were stable in solution for days, are marked with (+). Unsuccessful formation of an aggregate is indicated with (–).

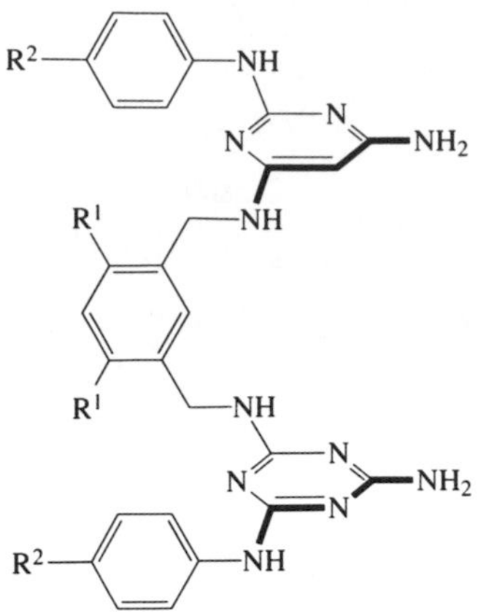	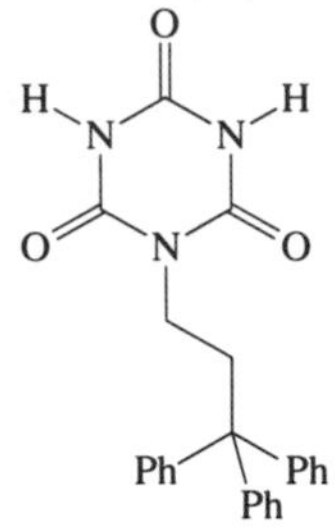	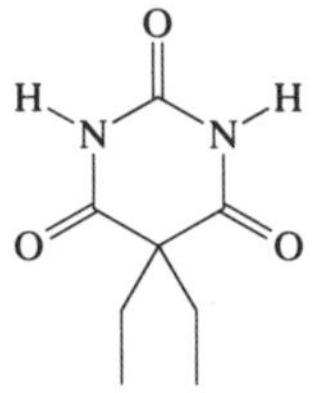	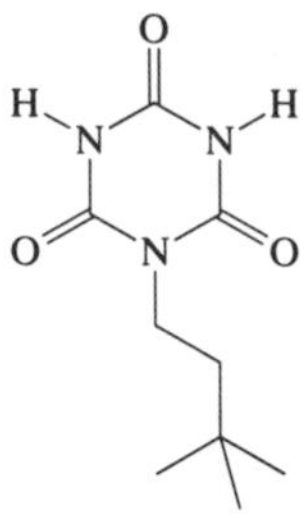	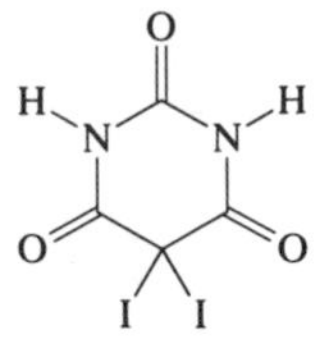
Increasing steric bulk of R^2				
$R^2=H$, $R^1=Pr^i$	++	++	–	–
$R^2=Bu^t$, $R^1=Pr^i$	++	++	++	+
Increasing preorganization (size of R^1)				
$R^2=Bu^t$, $R^1=Me$	++	+	–	–
$R^2=Bu^t$, $R^1=H$	+	+	–	–
$R^2=Me$, $R^1=H$	+	–	–	–

17.5.2.3 The determination of molecular weight: gel permeation chromatography

GPC is a size exclusion technique. Solutions of aggregates are injected onto a column containing semiporous polystyrene beads having a narrow distribution of pore sizes. Molecules (or aggregates) small enough to diffuse into the stationary liquid in the pores move through the column more slowly than larger molecules that are sterically excluded from the pores. Figure 8 shows the correlation between retention time and molecular weight (which correlates with size). The trend observed with the aggregates is similar to that seen with soluble polystyrene, although the slopes of the lines are significantly different. We show two dashed lines on the graph corresponding to aggregates comprising one and two rosettes. While the difference between these lines suggest that it may ultimately, with greater understanding of GPC, be possible to distinguish between these and other structural motifs for CA·M-based aggregates based on their shape, it is not currently possible to draw conclusions (based on the limited numbers of aggregates explored) beyond molecular weight. Even in this limited sphere, GPC is not useful in estimating molecular weight unless compounds that are structurally closely related are being compared. We use GPC primarily to estimate relative stabilities of aggregates (see below), rather than their molecular weights.

17.5.2.4 The determination of molecular weight: electrospray ionization mass spectrometry

Molecular ions corresponding to the assembled aggregates associated with one or more equivalents of Cl^- have been recorded using ESI-MS for many of our more stable aggregates. Ionization is accomplished from solutions of the aggregate in chloroform containing $Ph_4P^+Cl^-$ as a soluble source of Cl^-.[16] In general, the values of MW obtained from ESI-MS are those expected for the hypothesized structure.

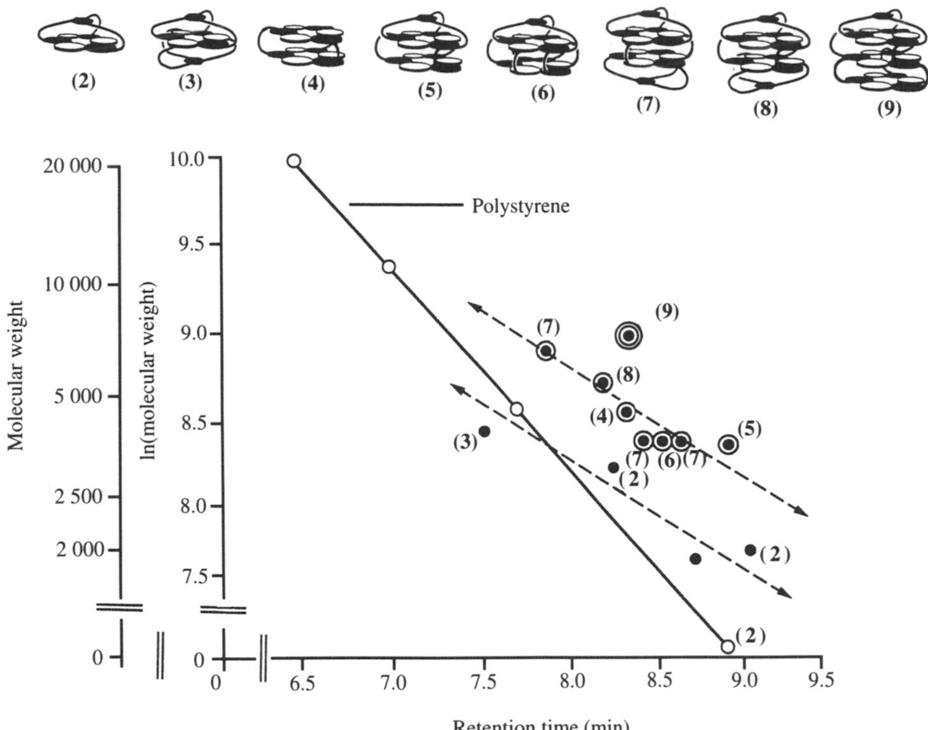

Figure 8 The retention time from gel permeation chromatograms of **(2)**–**(9)** correlate with molecular weight. The dotted lines organize similar classes of aggregates: those comprising single rosettes (solid dot) are distributed around the lower line; aggregates comprising two rosettes (ringed dot) are distributed around the higher line. Aggregate **(9)** (with three rosettes) is shown with a double-ring dot. Both lines appear to have similar slopes, suggesting a similar shape-dependent behavior.

Table 3 compares the MW determined by a variety of techniques. While ESI-MS is by far the most accurate, GPC and VPO still offer interesting and useful information about stabilities and nonideal behaviors.

17.5.2.5 Relative stability: methanol titrations

One of the aggregates based on CA·M **(1)** must be at concentrations above 4 mM to be observable by NMR spectroscopy—at concentrations below 4mM, the NMR spectrum appears as a superposition of the spectra of the individual components. The remaining aggregates exhibit concentration-independent stability: they are observable by NMR spectroscopy up to dilutions (< 0.05 mM) at which covalent molecules are not observed. The relative stability of these aggregates can be determined by measuring the lifetime of the aggregate in solutions containing methanol-d_4.[17] The most stable aggregate slowly dissociates over 3 d in 20% methanol (v:v): ~5 M CD_3OD in 10 M $CDCl_3$. The remaining aggregates can be ordered by their lifetime in solutions containing 5% or 10% methanol in chloroform. The results of this ranking are shown in Table 4 and will be discussed in greater detail later.

17.5.2.6 Relative stability: analysis of the GPC trace

The extent of tailing of the GPC trace yields information about the stability of the aggregate. We consider any aggregate that elutes from the column to be stable; aggregates that dissociate and precipitate on the column are unstable. Figure 9 shows the GPC traces of representative aggregates. Of the aggregates that elute, we believe that the extent of tailing of the GPC trace correlates with the stability of the aggregate: more stable aggregates elute as sharp peaks; less stable aggregates tail. Tailing results when the partial dissociation of CA molecules provides a soluble aggregate of slightly different shape and mass than the original aggregate (and often, a longer elution

Table 3 Calculated and experimental molecular weights for a series of aggregates.

Aggregate		*Calculated*	*ESI-MS*	*VPO*
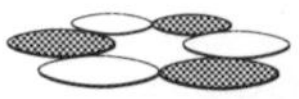	**(1)**	1724.2	a	1700
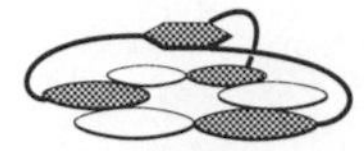	**(2a)**	2732.7	2732.7	3200
	**(2b)**	1961.2	3922.6[b]	4000
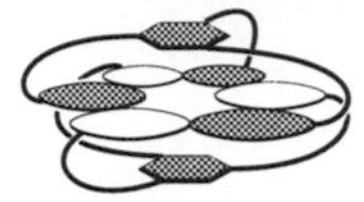	**(3)**	4100.6	4100.2	6500
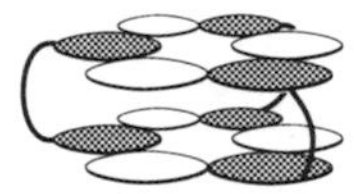	**(4)**	3635.5	a	3700
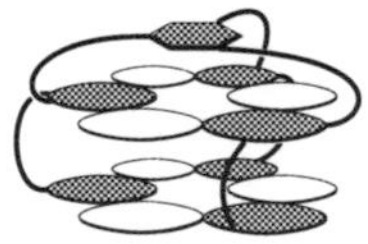	**(5)**	4142.6	4142.6	4200
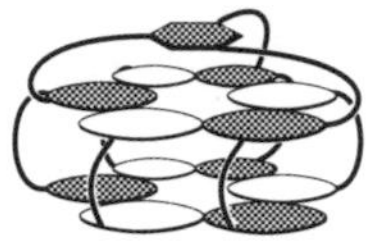	**(6)**	4196.4	4196.1	4500
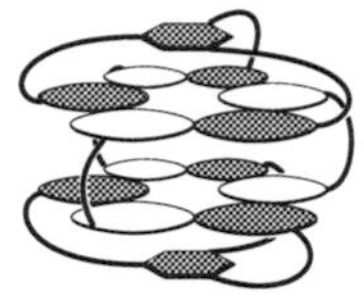	**(7)**	5519.5	5519.1	5400
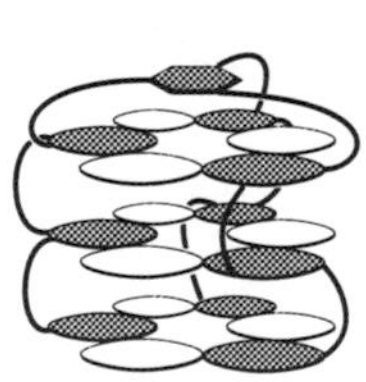	**(9)**	6440.6	6433.8	6500

[a]Molecular ions were not observed for these aggregates. [b]No molecular ion corresponding to the hypothesized structure was observed, only an ion corresponding to the dimer.

time). If the dissociation of components yields insoluble by-products (dissociation triggers complete and irreversible dissociation and precipitation of all components onto the matrix), the eluting peak may appear uncharacteristically sharp. The unexpected sharpness of **(4)** may result from this situation: both components are insoluble in chloroform, unless organized into a bis(rosette).

17.5.2.7 ^{1}H and ^{13}C NMR spectroscopies

NMR spectroscopies have yielded the majority of structural information about these aggregates. The ^{1}H and ^{13}C NMR spectra of unassociated components—especially the polymelamines—show very broad peaks due to nonspecific aggregation. Few lines are visible in the ^{13}C NMR

Table 4 Aggregates (**1**)–(**9**) are arranged in order of decreasing stability based on the percentage of methanol at which the aggregate survives. HB is the number of hydrogen bonds. *N* is the number of particles.

	Percentage	*HB/(N-1)*	*HB*	*N*
	20(3)	18	18	2
	20	12	36	4
	20	9	36	5
	10	9	36	5
	5	6	18	4
Stability increases	5	6	36	7
	5	6	54	10
	< 5	4	18	6
	< 5	4.5	36	6
	0	3.8	18	6

spectra. Upon formation of the aggregate, sharp lines appear (Figure 10) in the ^{1}H NMR spectrum. Lines due to the N—H protons of the CA groups also appear at δ16–13; this spectral region contains no other signals, and the pattern of lines for the N—H protons are highly diagnostic for the number of aggregates present in solution, and the symmetries of these aggregates.

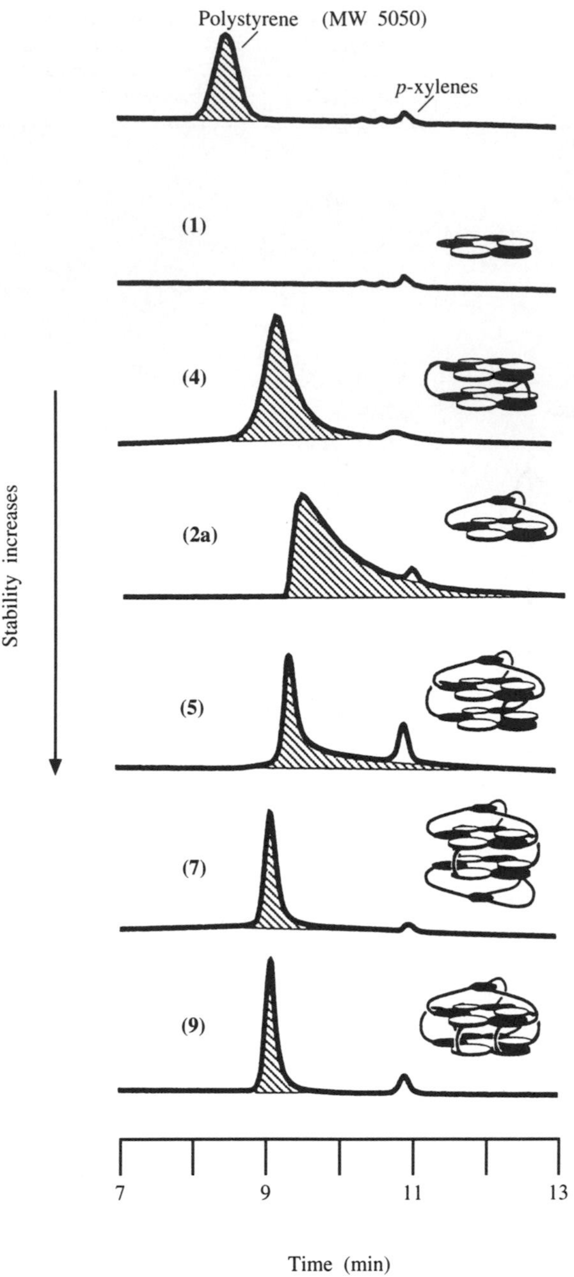

Figure 9 GPC traces for selected aggregates arranged in order of increasing stability of the aggregate.

17.5.2.8 Isomerism

As aggregates get larger, conformational isomers become more important. The NMR spectra also become increasingly complex, and the δ10–0 region almost impossible to interpret. While we envisage (and observe) only two isomers of hub(M)$_3$·3CA, a structure based on a monorosette, structures based on bis(rosettes), such as hub(MM)$_3$·6CA, can exist as 16 isomers. Understanding these ^{1}H NMR spectra and correlating their characteristics with conformational isomers of the rosettes provides a powerful method for analyzing the structures of these aggregates.

The region of the ^{1}H NMR spectrum at δ16–13 is diagnostic for the number of isomers, and symmetries of the isomers of hub(M)$_3$·3CA.[18] The hydrogen-bonded imide (N—H···O) groups appear in this region. The symmetric isomer shows two lines as a result of the unsymmetric substitution of the adjacent melamine group. The asymmetric isomer shows six lines (Figure 11). Ratios of isomers depend on the structure of the CA used, but are not significantly influenced by

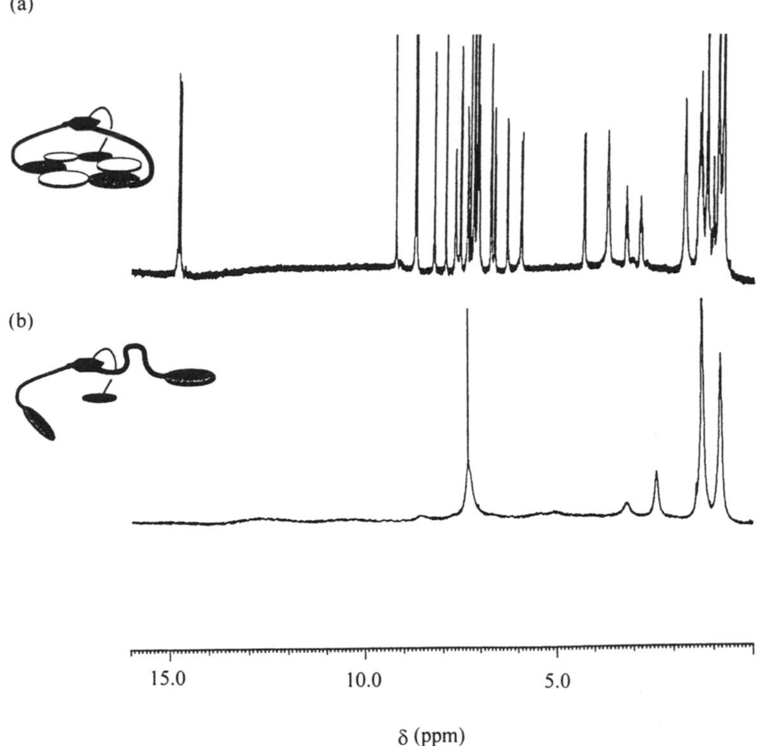

Figure 10 The [1]N NMR spectrum of (a) hub(M)$_3$·3CA and (b) hub(M)$_3$. The imide region (δ16–13) is uncrowded and diagnostic for the number and symmetries of aggregates present. For a complete assignment of peaks, see Simanek *et al.*[18]

solvent. The energy barrier to interconversion of these isomers at 328 K is ~58.6 kJ mol^{-1}. This value is similar to the barrier of exchange of CA molecules between different pairs of adjacent melamine groups.

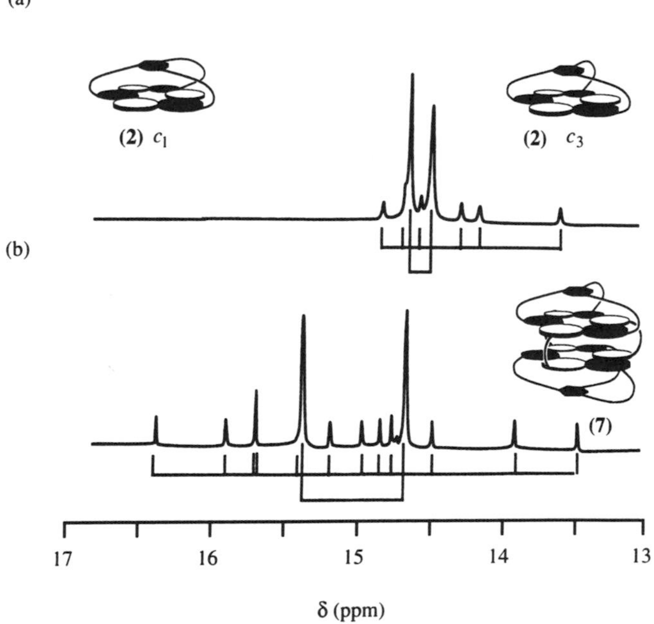

Figure 11 The imide regions of two aggregates reveals the number and symmetries of the isomers. Each spectrum shows two different aggregates. Assignment of (a) is straightforward—only two isomers exist. Assignment of (b) is less straightforward—12 isomers can exist.

More complicated aggregates behave similarly. The aggregate 2 hub$(M)_3$·3 bis(CA) shows either two or five isomers, depending on the derivative of bis(CA) that is used: at least 12 isomers are theoretically possible for this aggregate. The spectra of **(2)** and **(7)** are shown in Figure 11. We do not currently understand why we observe only two isomers with benz$(CA)_2$ and at least five with furan$(CA)_2$. The fact that isomers are observed, however, provides additional information for the assignment of structure.

17.5.3 Computation

Utilizing hub$(M)_3$·3CA and related structures, we have explored the use of computational models[19] for determining relative stabilities. Estimating relative entropies and enthalpies for these structures computationally is not currently practical, because these calculations would require enormous amounts of time. Ideally, the method for estimating the values of entropy and enthalpy —or arriving at some stability factor that can be used to compare two aggregates—should be accessible to members of the experimental community, fast, and general enough to be applied to a large number of related structures. The role of solvent must be considered in many cases.

17.5.3.1 Starting point for the calculations and minimization strategy

The rosette was constructed by placing three M and three CA rings on a common plane in a rosette at a distance of 180 pm (measured from the N—H to O of any hydrogen bond donor–acceptor pair). The spokes were attached to preserve the C_3 symmetry observed in the ^{1}H NMR spectrum. All torsional angles were rotated to preserve this symmetry and to maximize the distance between the rosette and central hub. Energetically unfavorable interactions were minimized by a 1000 step conjugate gradient energy-minimizing algorithm. Additional details are available in the original source.[19] From this starting point, minimizations (described above) could be carried out for any of the aggregates with a structure shown by **(2)**.

We have explored a surrogate for these values—a value that we refer to as "DP" (distortion from planarity). To calculate DP, we determine the minimized structures of aggregates built using valence geometries from QUANTA 3.3 using the CHARMm 22 force field. (The role of solvent in these calculations is discussed in Section 17.5.3.2) A plane is fit through the rosette portion of the minimized aggregate so that the sum of squared distances from that plane is minimized. The distance of all nonhydrogen atoms of the M and CA groups from the plane is measured and DP is calculated using Equation (8).

$$\mathrm{DP} = \left[\frac{\left(\sum_{i=1}^{N}(\Delta r_i)^2\right)}{N}\right]^{1/2} \tag{8}$$

where Δr is the distance of an atom from the plane and N is the number of atoms in the rosette. Large values of DP correspond to rosettes that are more bowl-shaped; small values of DP correspond to rosettes that are more planar (Figure 12). The values of DP correlate with experimentally determined stabilities for hub$(M)_3$·3CA > triaz$(M)_3$·3CA and Ad$(M)_3$·3CA > flex$(M)_3$·3CA: the larger the value of DP, the lower the stability of the aggregate.

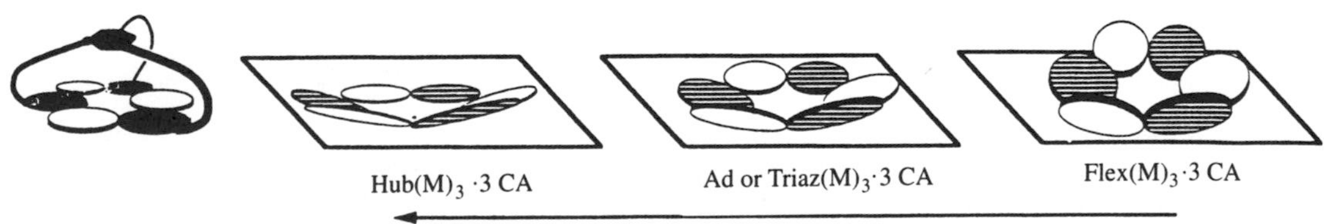

Figure 12 The stability of aggregates (determined experimentally) can be correlated to the computational parameter, DP, that describes the distortion of the heavy atoms of CA and M from a mean plane. The most stable aggregate (left) shows the most planar rosette after minimization (small DP). The least stable aggregate (right) shows the most bowled rosette after minimization (large DP). Both triaz$(M)_3$ and ad$(M)_3$ yield comparable values of DP and are of comparable stability. The bowling of the rosettes is shown crudely: for a more rigorous diagram, please see the original source (Ref. 19).

17.5.3.2 *The role of solvent*

Initial simulations carried out for (**2**) in vacuum resulted in a structure that was inconsistent with the ^{1}H NMR spectra: symmetry was lost due to strong interactions between the three amides of the central hub and portions of the rosette. The aggregate collapsed upon itself. Simulations done in chloroform, with one molecule of chloroform placed within the central cavity of the aggregate (between the rosette and hub portions), resulted in structures more consistent with the ^{1}H NMR spectrum—and more similar to the starting structure than those without solvent. For the timescale of the experiment (80 ps), no significant diffusion of molecules of chloroform away from the aggregate was observed. Of all the molecules of chloroform in the simulation, four interacted most strongly: the caged chloroform and one between each of the three spokes. The trend in value of DP for an abbreviated (and faster) minimization strategy including only these four chloroform molecules showed the same trend in stability.

17.5.4 Theory

One goal of this project has been to develop an intuitive sense for the design of aggregates and their stabilities. This strategy requires an understanding of the balance between enthalpic gains resulting from the formation of hydrogen bonds, and entropic costs from the aggregation of independent particles. Ideally we would like to be able to estimate ΔG, or a surrogate of it. To accomplish this goal, we require simple estimates of ΔH and ΔS.

17.5.4.1 *The number of hydrogen bonds (HB) is proportional to the enthalpy of aggregation (ΔH)*

The total enthalpy of assembly can be described as the sum of hydrogen bonds and van der Waals forces (Equation (9): no ionic or charge–charge interactions exist in these aggregates). Since recognition occurs in organic solvents, we believe that the contribution of van der Waals interactions to the stability of an aggregate is minimal—that is, organic molecules show little preference for contact with molecules of solvent or molecules of solute. If all the hydrogen bonds have equal energy, or are identical in the aggregates compared (i.e., while the hydrogen bonds of a rosette are likely to be different in strength, the contribution of an entire rosette should be constant), the total enthalpy of aggregation should be proportional to the number of hydrogen bonds (HB).

$$\Delta H = \Delta H_{\mathrm{HB}} + \Delta H_{\mathrm{vdW}} = \mathrm{HB} \tag{9}$$

17.5.4.2 *The number of particles is proportional to the entropy of aggregation*

Equations (6) and (7) describe the rotational and translational entropies of aggregation. In both cases, the total entropy is proportional to $(N-1)$ Estimating the contribution of conformation entropy is difficult. If the molecules that are being compared have similar connectivities, then this component can be ignored. While this approximation limits the generality of any subsequent model, models that rationalize stabilities have emerged.

$$\Delta S = (N - 1) \tag{10}$$

Estimating ΔG using Equations (9) and (10) yields Equations (11)–(13). This expression is of limited use because the constants c_1 and c_2 cannot be easily evaluated and may be system dependent. If we choose to evaluate these expressions as described by Equation (13), a situation suggesting a melting process is implied: this melting can be examined experimentally by titrations of solutions of aggregates with methanol.

$$\Delta G = \Delta H - T\Delta S \tag{11}$$

$$\Delta G = c_1\mathrm{HB} - c_2 T\Delta S \tag{12}$$

$$(c_2/c_1)T = \mathrm{HB}/(N - 1) \tag{13}$$

While evaluating the constant terms and solving Equation (12) rigorously would be preferred, HB/($N-1$) is qualitatively useful: it seems to correlate with stability. Values for HB/($N-1$) are given in Table 4 for a number of aggregates.

17.6 CRYSTALLINE SOLIDS

Our discussion of crystalline solids based on CA·M is divided into three parts. Section 17.6.1 describes our motivations for this study. Section 17.6.2 introduces our studies of crystalline materials and discusses representative crystal structures. Section 17.6.3 summarizes the conclusions we can so far draw from our efforts with this system.

17.6.1 Crystal Engineering

The phrase "crystal engineering" describes both the effort to rationalize the relationship between the structure of molecules and the structures of the crystals that form from them, and the attempt to use this information to design molecules that will pack in desired crystal structures. The strategy that we have adopted—a strategy that has been used by many others including Etter and co-workers,[20] Leiserowitz and co-workers,[21] Taylor and Kennard,[22] Jeffrey and Saenger,[23] Bishop and Dance,[23] Hsu and Craven,[23] and Shimizu *et al.*[23]—is to use directional interactions such as hydrogen bonds to impose constraints on the packing arrangement.

Many enthalpic constraints have been identified and used successfully: repulsive van der Waals forces prevent energetically unfavorable close contacts; attractive van der Waals interactions tend to bring molecules into arrangements that minimize the free volume of the crystal; specific interactions—hydrogen bonds,[24] charge–charge,[25] dipole–induced dipole,[26] and charge transfer[27]—orient molecules in specific arrangements. The magnitude of the enthalpic gain from engineering these forces into a crystalline architecture must exceed the entropic costs associated with that specific packing arrangement: vibrational motion and conformational freedom of the molecules influences the packing arrangement in unpredictable and nonintuitive ways.

Although there is some progress both experimentally and theoretically,[28] prediction of a crystal structure based on the structures of the component molecule(s) is impossible. In an effort to simplify the problem, there has been a substantial focus on the design and development of specific architectures based on directional interactions: tapes,[29] sheets,[30] and diamondoid[31] structures have been investigated most thoroughly. These motifs limit the range of possible arrangements of the molecules in space and begin to reduce the dimensionality of the problem—that is, the ability to predict how two-dimensional sheets stack to form a three-dimensional crystal would represent a substantial contribution to this field.

Our work has focused on tapes comprising 1:1 cocrystals of CA and M: we have observed both linear and crinkled tapes derived from the CA·M lattice. Here we will summarize only the broad features of this work, and will emphasize information relevant to the idea that the CA·M lattice can be considered a scaffold for the design in the solid state as well as design in solution. We now believe that the problem in rationalization posed by this system, although simpler than that for an arbitrary organic molecule without constraints, is still too complex to be solved at the current state of the science in this area. The problem of polymorphism attributable (at least in part) to rotational isomers around the *N*-phenyl bond and the ethyl groups of diethylbarbituric acid is a particular problem that has proved intractable, although it may be possible to use molecules with substituents chosen to eliminate the conformational isomerism that results from these groups.

17.6.2 Motifs and Structure

Three motifs can be obtained from the cocrystallization of CA and M: rosettes and linear or crinkled tapes.[32–5] These motifs arise from different connectivities of CA and M. The pattern of connectivity is highly dependent on the size of the groups on the periphery: peripheral crowding is an effective strategy for design in the solid state.

17.6.2.1 Linear tapes: small peripheral groups

The majority of crystalline tapes that we have obtained are linear. When the peripheral groups are small (Figure 13), linear tapes appear to be favored over crinkled tapes and rosettes, although for no obvious reason. We have obtained tapes from two related series of molecules: the cocrystallization of 5,5-diethylbarbituric acid with any of seven *N*, *N'*-bis(*p*-X-phenyl)melamines (X = H, F, Cl, Br, I, Me, and CF_3) or any of four *N*,*N'*-bis(*m*-X-phenyl)melamines (X = F, I, Me, and CF_3).[33] Four additional complexes also form linear tapes (Figure 14).[34] These last four were pursued to assess opportunities for secondary interactions (R—I···NC—R), to observe what happened in the absence of peripheral groups, and to determine the propensity of imperfect molecules to form tapes.

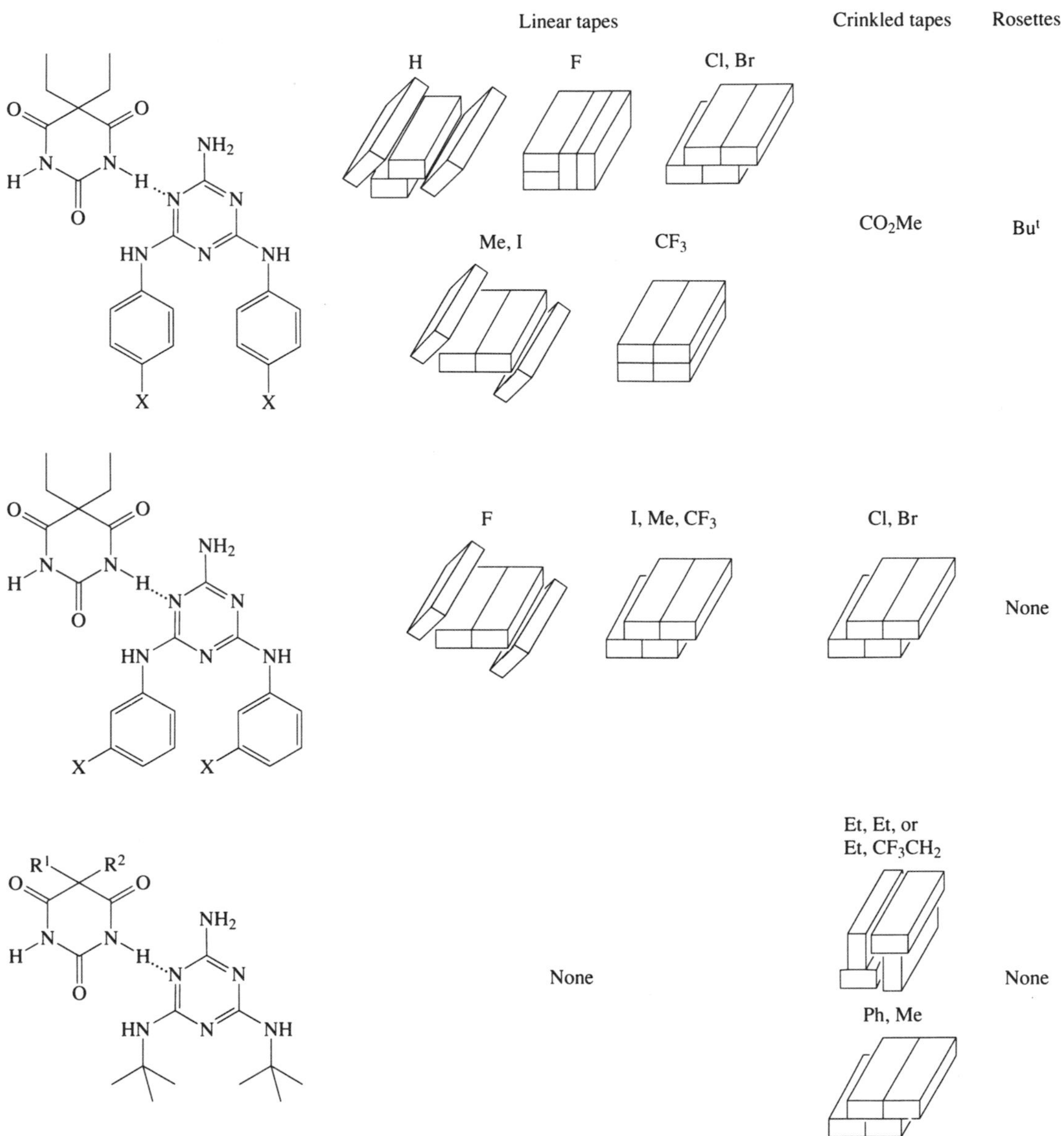

Figure 13 Linear tapes, crinkled tapes, and rosettes are classified based on the structures of the molecules comprising them. In all cases, tapes pack with their long axes (the hydrogen-bonded axis) parallel (long edge of the block). The block diagrams serve to communicate crudely the three-dimensional packing of these tapes. Similar arrangements do not imply polymorphism: the crystal structures are less similar than these diagrams imply.

With one exception, all tapes lie with their long axes parallel. The packing of linear tapes shows no simple trend with the size of the groups on the periphery. Figures 13 and 14 attempt to represent some of this diversity. Translations in all three dimensions make these structures even less similar: only one isomorphous pair exists (Br and Cl). Generalizations about the role of

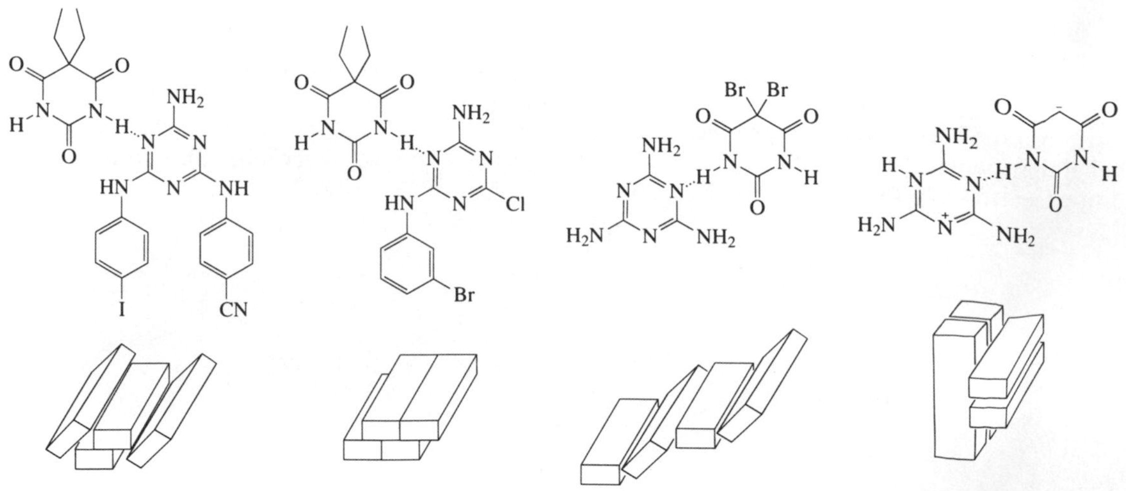

Figure 14 Linear tapes resulted from the crystallization of the molecules shown. In one instance, the tape axes did not align parallel.

peripheral groups can be made, but the complexity of these molecules and polymorphism (discussed later) complicate these issues.

17.6.2.2 Crinkled tapes: medium-sized peripheral groups

When the size of the substituents on the melamines is increased, crinkled tapes are observed: 5,5-diethylbarbituric acid and *N,N′*-bis(4-X-phenyl)melamines (X is CO_2-Me) yield crinkled tapes, while X < CF_3 yields linear tapes (Figure 13).[35] Crinkled tapes are obtained from 1:1 complexes between *N,N′*-di(*t*-butyl)melamine and 5,5-disubstituted barbituric acids (diethyl, dimethyl, diphenyl, ethyltrifluoroethyl). While the crinkling of one tape may effect the packing of its neighbor, no general trends emerge about the relative structure of these tapes.

17.6.2.3 Rosette: large peripheral groups

If the size of the peripheral group is further increased (as shown in Figure 12), the infinite crinkled tape structure is replaced by the rosette. We have been able to obtain only one crystal structure of a rosette: the 1:1 complex of *N,N′*-bis(4-*t*-butyl phenyl)melamine and 5,5-diethyl barbital. The crystal structure is shown in Figure 15.[15]

17.6.3 Implications for Crystal Engineering

While the synthetic accessibility to a large number of substituted melamines and barbiturates makes this system attractive, many issues suggest that CA·M is not the system of choice for these studies.

17.6.3.1 Cocrystallization is difficult

The tapes based on cyanuric acid (barbital) and melamine lattices are difficult to crystallize. Growing crystals of the quality and size suitable for single-crystal x-ray diffraction requires patience. Crystal growth is favored kinetically along the tape axis (in the direction that leads to new hydrogen-bond formation): crystals obtained are usually needles that are too thin to be used for single-crystal x-ray diffraction study.

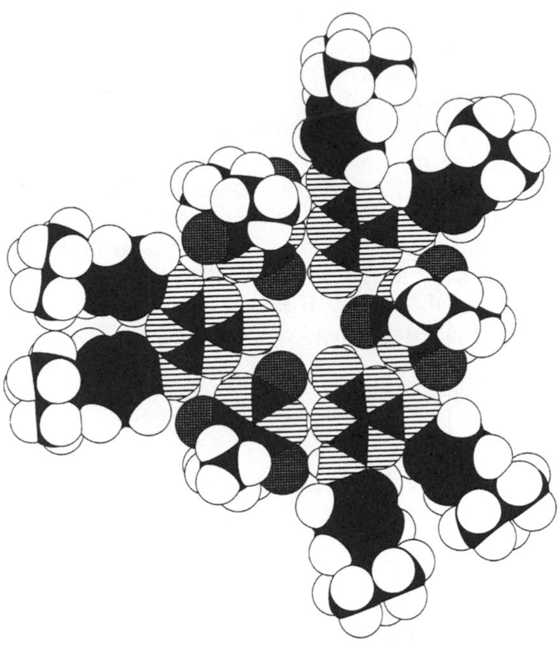

Figure 15 The crystal structure of the rosette. Carbon atoms are black. Hydrogen atoms are white. Nitrogen atoms are striped. Oxygen atoms are hatched.

17.6.3.2 Conformational complexity and polymorphism

The number of ways that molecules of CA and M can arrange their peripheral groups while maintaining the same tape motif leads to the existence of polymorphs. Orientations of the ethyl groups of diethylbarbital and free rotation along the *N*-phenyl bond is probably responsible for polymorphism in the CA·M system.

17.6.3.3 The utility of powder diffraction

One of the key concerns in efforts to develop a rational understanding of the organic solid state is the unpredictable, but substantial, occurrence of polymorphism. The crystal structure obtained in any single-crystal study could be a true minimum energy structure, or a local minimum, or a metastable phase formed kinetically under the conditions of a specific crystallization. To survey for polymorphism, we have relied on x-ray powder diffraction (XPD). XPD performed on powders of mixtures of CA and M from different solvents (such as acetone, MeCN, THF–$CHCl_3$, and MeOH) has led us to believe that a majority of the reported structures are minimum energy structures. Using XPD we were, however, able to determine that the tape from *N,N'*-bis(*P*-bromophenyl)melamine does crystallize in two different structures. By comparison of the XPD, patterns calculated based on the single-crystal coordinates with the experimental XPD patterns from the powders obtained from different solvents suggest that one of the polymorphs is isomorphous to the X = Cl complex, the other is related to the X = Me complex.

17.7 CONCLUSIONS

CA·M is a useful platform for the construction of soluble aggregates and crystalline materials, although its use for exploring the latter is limited by the low solubility of the components, by the difficulties of cocrystallization, and by the complexities of polymorphism and structural isomerism. The greatest benefit from efforts with this system is the observation that important principles which emerge from the solid state also seem relevant to solution-based design. The most important of these is the value of steric effects (peripheral crowding) in excluding structures: if one wishes a particular structure not to form, it may be possible to exclude it by arranging substituents in a way that make it sterically unfavorable This strong conceptual complementarity is a strong motivation to us, and others,[36] for pursuing this system.

Exploration of the CA·M system provides trials of techniques of characterization useful for the study of noncovalent systems. While these techniques are not new to the community of organic chemists, their application to this class of compounds has suggested both their strengths and weaknesses. GPC, VPO, and ESI-MS will undoubtedly prove useful to others in the community. Additional techniques are being surveyed: analysis of limited regions of the ^{1}H NMR spectra is proving to be both useful and general.

The most significant contribution of this work to date is the generation of a data set: we have constructed more than 20 different aggregates based on the rosette. With this data set we can begin exploring the use of computation and theory for more efficient design. The use of computation to rationalize stabilities and to aid in design remains in its early stages, although the surrogate for stability, DP, appears to be general to many of these aggregates based on monorosettes. This set of data sharpens intuition about the balance between enthalpy and entropy in these systems. A useful result is rule-of-thumb: increased stability correlates with higher values of $HB/(N-1)$.

ACKNOWLEDGMENTS

This work was supported by the NSF grant CHE-91-2233 1. EES gratefully acknowledges Eli Lilly for a predoctoral fellowship (1995). Much of the work reported on aggregates in solution was carried out by Drs. Christopher Seto, John Mathias, and Dana Gordon. Dr. Donovan Chin performed the computational work. Mathai Mammen contributed to the section on theory. Dr. Mohammed I. M. Wazeer assisted with NMR analysis of isomerism. Work in the solid state was done by Drs. Jonathon Zerkowski and John MacDonald.

17.8 REFERENCES

1. M. Mrksich and P. B. Dervan, *J. Am. Chem. Soc.*, 1995, **117**, 3325.
2. S. L. Schreiber, *Science*, 1991, **251**, 283.
3. D. H. Williams, M. S. Searle, J. P. Mackay, U. Gerhard, and R. A. Maplestone, *Proc. Natl. Acad. Sci. USA*, 1993, **90**, 1172.
4. C. J. Pedersen, *Angew. Chem., Int. Ed. Eng.*, 1988, **27**, 1021 and references cited therein.
5. D. J. Cram, *Angew. Chem., Int. Ed. Eng.*, 1988, **27**, 1009.
6. J. M. Lehn, *Angew. Chem., Int. Ed. Eng.*, 1988, **27**, 89.
7. References for these ion receptors are identified by the investigator as denoted in the Figure. (Ionomycin) B. K. Toeplitz, A. I. Cohen, P. T. Funke, W. L. Parker, and J. Z. Gougoutas, *J. Am. Chem. Soc.*, 1979, **101**, 3344; (Lehn) B. Dietrich, M. W. Hosseini, J. M. Lehn, and R. B. Sessions, *J. Am. Chem. Soc.*, 1981, **103**, 1282; (Pederson) See Ref. 4; (Raymond) T. D. Y. Chung and K. N. Raymond, *J. Am. Chem. Soc.*, 1993, **115**, 6765; (Reinhoudt) D. M. Rudkevich, W. Verboom, Z. Brzozka, M. C. Palys, W. P. R. V. Stauthamer, G. J. van Hummel, S. M. Franken, S. Harkema, J. F. J. Engbersen, and D. N. Reinhoudt, *J. Am. Chem. Soc.*, 1994, **116**, 4341; (Schmidtchen) F. P. Schmidtchen, *J. Am. Chem. Soc.*, 1986, **108**, 8249 and references therein; (Shinkai) H. Yamamoto and S. Shinkai, *Chem. Lett.*, 1994, 1115; (Vogtle) J. Gross, G. Harder, F. Vogtle, H. Stephan, and K. Gloe, *Angew. Chem., Int. Ed. Engl.*, 1995, **34**, 481.
8. References for these small molecule receptors are identified by the investigator as denoted in the Figure. (Bell) T. W. Bell and J. Liu, *J. Am. Chem. Soc.*, 1988, **110**, 3673; (Breslow) J. M. Desper and R. Breslow, *J. Am. Chem. Soc.*, 1994, **116**, 12081; (Cram) C. N. Eid, Jr., C. B. Knobler, D. A. Gronbeck, and D. J. Cram, *J. Am. Chem. Soc.*, 1994, **116**, 8506; (Diederich) B. R. Peterson, T. Morasini-Denti, and F. Diederich, *Chem. Biol.*, 1995, **2**, 139; (Hamilton) S.-K. Chang, D. Van Engen, E. Fan, and A. D. Hamilton, *J. Am. Chem. Soc.*, 1991, **113**, 7640; (Hunter) C. A. Hunter, *J. Chem. Soc., Chem. Commun.* 1991, 749; (Kelly) T. R. Kelly and M. P. Maguire, *J. Am. Chem. Soc.*, 1987, **109**, 6549; (Koga) K. Odashima, A. Itai, Y. Iitaka, and K. Koga, *J. Am. Chem. Soc.*, 1980, **102**, 2504; (de Mendoza and Rebek) R. Wyler, J. de Mendoza, and J. Rebek, *Angew. Chem., Int. Ed. Eng.*, 1993, **32**, 1699; (Rebek) J. Rebek, K. Williams, K. Parris, P. Ballester, and K. S. Jeong, *Angew. Chem., Int. Ed. Eng.*, 1987, **26**, 1244; (Still) K. T. Chapman and W. C. Still, *J. Am. Chem. Soc.*, 1989, **111**, 3075; S.-S. Yoon and W. C. Still, *J. Am. Chem. Soc.*, 1993, **115**, 823; (Reinhoudt) P. Timmerman, W. Verboom, F. C. J. M. van Veggel, W. P. van Hoorn, and D. N. Reinhoudt, *Angew. Chem. Int. Ed. Engl.*, 1994, **33**, 2345; (Whitlock) J. E. Cochran, T. J. Parrott, B. J. Whitlock and H. W. Whitlock, *J. Am. Chem. Soc.*, 1992, **114**, 2269; (Zimmerman) S. C. Zimmerman, C. M. Vanzyl, and G. S. Hamilton, *J. Am. Chem. Soc.*, 1989, **111**, 1373.
9. References for these molecular machines are identified by Equations. (a) Equation (1) R. A. Bissell, E. Cordova, A. E. Kaifer, and J. F. Stoddart, *Nature*, 1994, **369**, 133; (b) Equation (2) L. Zelikovich, J. Libman, and A. Shanzer, *Nature* 1995, **374**, 790.
10. References for these scaffolds are identified by the investigator as denoted in the Figure. (Ghadiri) M. R. Ghadiri, K. Kobayashi, J. R. Granja, R. K. Chadha, and D. E. McRee, *Angew. Chem., Int. Ed. Engl.*, 1995, **34**, 93 and references therein; (Lehn) P. N. W. Baxter, J. M. Lehn, J. Fischer, and M. T. Youinou, *Angew. Chem., Int. Ed. Engl.*, 1994, **33**, 2284; (Sauvage) C. O. Dietrich-Buchecker and J. P. Sauvage, *Chem. Rev.*, 1987, **87**, 798; (Stoddart) D. B. Amabilino, P. R. Ashton, A. S. Reder, N. Spencer, and J. F. Stoddart, *Angew. Chem., Int. Ed. Engl.*, 1994, **33**, 1286; (Seeman) J. Chen and N. C. Seeman, *Nature*, 1991, **350**, 631; (Whitesides) G. M. Whitesides, E. E. Simanek, J. P. Mathias, C. T. Seto, D. Chin, M. Mammen, and D. M. Gordon, *Acc. Chem. Res.*, 1995, **28**, 37.
11. R. Taylor and O. Kennard, *Acc. Chem. Res.*, 1984, **17**, 320 and references therein.

12. Aggregate (**1**): J. P. Mathias, E. E. Simanek, J. A. Zerkowski, C. T. Seto, and G. M. Whitesides, *J. Am. Chem. Soc.*, 1994, **116**, 4316; Aggregate (**2**): C. T. Seto and G. M. Whitesides, *J. Am. Chem. Soc.*, 1993, **115**, 905; Aggregate (**3**) C. T. Seto and G. M. Whitesides, *J. Am. Chem. Soc.*, 1993, **115**, 1330; Aggregate (**4**): J. P. Mathias, E. E. Simanek, and G. M. Whitesides, *J. Am. Chem. Soc.*, 1994, **116**, 4326; Aggregates (**5**) and (**6**): J. P. Mathias, C. T. Seto, E. E. Simanek, and G. M. Whitesides, *J. Am. Chem. Soc.*, 1994, **116**, 1725; Aggregate (**7**): C. T. Seto and G. M. Whitesides, *J. Am. Chem. Soc.*, 1991, **113**, 712; C. T. Seto, J. P. Mathias, and G. M. Whitesides, *J. Am. Chem. Soc.*, 1993, **115**, 1321; Aggregate (**8**): reported with (**5**) and (**6**); Aggregate (**9**): J. P. Mathias, E. E. Simanek, C. T. Seto and G. M. Whitesides, *Angew. Chem., Int. Ed. Eng.*, 1993, **32**, 1766.
13. R. C. Helgeson, B. J. Selle, I. Goldberg, C. B. Knobler and D. J. Cram, *J. Am. Chem. Soc.*, 1993, **115**, 11 506.
14. D. A. McQuarrie, "Statistical Thermodynamics," Harper & Row, New York, 1973, chaps. 5 and 6.
15. J. A. Zerkowski, C. T. Seto, and G. M. Whitesides, *J. Am. Chem. Soc.* 1992, **114**, 5473.
16. X. Cheng, Q. Gao, R. D. Smith, E. E. Simanek, M. Mammen and G. M. Whitesides, *Rapid Commun. Mass Spectrom.*, 1995, **9**, 312; For an alternative strategy see: K. C. Russell, E. Leize, A. V. Dorsselaer, and J. M. Lehn, *Angew. Chem., Int. Ed. Engl.*, 1995, **34**, 209.
17. Determining stability by doing titrations with methanol focuses on the kinetic stability of these aggregates. We have not thoroughly examined all the aggregates to determine whether this trend in stability is identical to the desired trend in thermodynamic stability. We are currently pursuing these studies.
18. E. E. Simanek, M. I. M. Wazeer, J. P. Mathias, and G. M. Whitesides, *J. Org. Chem.*, 1994, **59**, 4904.
19. Full details of our efforts in applying computation to these systems can be found in: D. N. Chin, D. M. Gordon, and G. M. Whitesides, *J. Am. Chem. Soc.*, 1994, **116**, 12033; X. Li, D. N. Chin, and G. M. Whitesides, *J. Org. Chem.*, 1996, **61**, 1779.
20. M. C. Etter, K. S. Hung, G. M. Frankenbach, and D. A. Adsmond, in "Materials for Nonlinear Optics: Chemical Perspectives," eds. S. R. Marder, J. E. Sohn, and G. D. Stucky, American Chemical Society, Washington, DC, 1991, vol. 455, p. 446; M. C. Etter, G. M. Frankenbach, and D. A. Adsmond, *Mol. Cryst. Liq. Cryst.*, 1990, **187**, 25; M. C. Etter, *Acc. Chem. Res.*, 1990, **23**, 120.
21. L. Leiserowitz and G. M. J. Schmidt, *J. Chem. Soc. (A)*, 1969, 2372; L. Leiserowitz and F. Nader, *Acta Crystallogr., Sect. B*, 1977, **33**, 2719; L. Leiserowitz and M. Tuval, *Acta Crystallogr., Sect. B*, 1978, **34**, 1230; S. Weinstein, L. Leiserowitz, and E. Gil-Av, *J. Am. Chem. Soc.*, 1980, **102**, 2768.
22. R. Taylor and O. Kennard, *J. Am. Chem. Soc.*, 1982, **104**, 5063.
23. G. A. Jeffrey and W. Saenger, "Hydrogen Bonding in Biological Structures," Springer, Berlin, 1991; R. Bishop and I. G. Dance, in "Inclusion Compounds," eds. J. L. Atwood, D. D. MacNicol and J. E. D. Davies, Academic Press, London, 1991, vol. IV; I. Hsu and B. M. Craven, *Acta Crystallogr., Sect. B*, 1974, **30**, 974; N. Shimizu, S. Nishigaki, Y. Nakai, and K. Osaki, *Acta. Crystallogr., Sect. B*, 1982, **38**, 2309.
24. C. B. Aakeroy and K. R. Seddon, *Chem. Soc. Rev.*, 1993, 397; M. C. Etter, *J. Phys. Chem.*, 1991, **95**, 4601; G. R. Desiraju, "Crystal Engineering: The Design of Organic Solids," Elsevier, New York, 1989; M. C. Etter, *J. Phys. Chem.*, 1991, **95**, 4601.
25. U. Zimmerman, G. Schnitzler, N. Karl, E. Umbach, and R. Dudde, *Thin Solid Films*, 1989, **175**, 85; E. A. Sudbeck and M. C. Etter, *Chem. Mater.*, 1994, **6**, 1192; M. C. Grossel, P. B. Hitchcock, K. R. Seddon, T. Welton, and S. C. Weston, *Chem. Mater.*, 1994, **6**, 1106.
26. D. S. Reddy, K. Panneerselvam, T. Pilati, and G. R. Desiraju, *J. Chem. Soc., Chem. Commun.*, 1993, 661.
27. P. J. Fagan, M. D. Ward, and J. C. Calabrese, *J. Am. Chem. Soc.*, 1989, **111**, 1698; M. D. Ward, P. J. Fagan, J. C. Calabrese, and D. C. Johnson, *J. Am. Chem. Soc.*, 1989, **111**, 1719.
28. J. Perlstein, *J. Am. Chem. Soc.*, 1994, **116**, 455; J. Perlstein, *Chem. Mater.*, 1994, **6**, 319; A. Gavezzotti, *J. Am. Chem. Soc.*, 1991, **113**, 4622.
29. J. C. MacDonald and G. M. Whitesides, *Chem. Rev.*, 1994, **94**, 2383; M. W. Hosseini, T. Ruppert, P. Shaeffer, A. Decian, N. Kyritsakas, and J. Fischer, *J. Chem. Soc. Chem. Commun.*, 1994, 2135; E. Fan, J. Yang, S. B. Geib, T. C. Stoner, M. D. Hopkins, and A. D. Hamilton, *J. Chem. Soc. Chem. Commun.*, 1994, 1251.
30. D. S. Reddy, B. S. Goud, K. Panneerselvam, and G. R. Desiraju, *J. Chem. Soc., Chem. Commun.*, 1993, 663. Y.-L Chang, M.-A. West, F. W. Fowler, and J. W. Lauher, *J. Am. Chem. Soc.*, 1993, **115**, 5991; M. D. Hollingsworth, M. E. Brown, B. D. Santarsiero, J. C. Huffman, and C. R. Goss, *Chem. Mater.*, 1994, **6**, 1227. V. A. Russel, M. C. Etter, and M. D. Ward, *J. Am. Chem. Soc.*, 1994, **116**, 1941; K. D. M. Harris and M. D. Hollingsworth, *Nature*, 1989, **341**, 19.
31. M. Simard, D. Su, and J. D. Wuest, *J. Am. Chem. Soc.*, 1991, **113**, 4696; X. Wang, M. Simard, and J. D. Wuest, *J. Am. Chem. Soc.*, 1990, **116**, 12119; O. Ermer, *J. Am. Chem. Soc.*, 1988, **110**, 3747; O. Ermer and A. Eling, *J. Chem. Soc. Perkin Trans, 2*, 1994, 925; D. S. Reddy, D. C. Craig, A. D. Rae, and G. R. Desiraju, *J. Chem. Soc., Chem. Commun.*, 1993, 1737; M. J. Zaworotko, *Chem. Soc. Rev.*, 1994, **23**, 283.
32. J.-M. Lehn, M. Mascal, A. DeCian, and J. Fischer, *J. Chem. Soc., Chem. Commun.*, 1990, 479.
33. (Linear) J. A. Zerkowski, J. C. MacDonald, C. T. Seto, D. A. Wierda, and G. M. Whitesides, *J. Am. Chem. Soc.*, 1994, **116**, 2382; J. A. Zerkowski and G. M. Whitesides, *J. Am. Chem. Soc.*, 1994, **116**, 4298; (Linear and crinkled) J. A. Zerkowski, J. P. Mathias, and G. M. Whitesides, *J. Am. Chem. Soc.*, 1994, **116**, 4305.
34. J. A. Zerkowski, J. C. MacDonald, and G. M. Whitesides, *Chem. Mater.*, 1994, **6**, 1250.
35. J. A. Zerkowski, C. T. Seto, and G. M. Whitesides, *J. Am. Chem. Soc.*, 1990, **112**, 9025. See also Refs. 16 and 34.
36. See for example: K. Matesharei and D. C. Myles, *J. Am. Chem. Soc.*, 1994, **116**, 7413, N. Kimizuka, T. Kawasaki, K. Hirata, and T. Kunitake, *J. Am. Chem. Soc.*, 1995, **117**, 6360.

Author Index

This Author Index comprises an alphabetical listing of the names of the authors cited in the text and the references listed at the end of each chapter in this volume.

Each entry consists of the author's name, followed by a list of numbers, for example

Templeton, J. L., 366, 385[233] (350, 366), 387[370] (363)

For each name, the page numbers for the citation in the reference list are given, followed by the reference number in superscript and the page number(s) in parantheses of where that reference is cited in the text. Where a name is referred to in text only, the page number of the citation appears with no superscript number. References cited in both the text and in the tables are included.

Although much effort has gone into eliminating inaccuracies resulting from the use of different combinations of initials by the same author, the use by some journals of only one initial, and different spellings of the same name as a result of the transliteration processes, the accuracy of some entries may have been affected by these factors.

Subject Index

J. NEWTON
David John Services Ltd., Slough, UK

This Subject Index contains individual entries to the text pages of Volume 9. The index covers general types of supramolecular compound, specific supramolecular compounds, general and specific supramolecular compounds where their synthesis or use involves supramolecular compounds, types of reaction (insertion, oxidative addition, etc.), spectroscopic techniques (NMR, IR, etc.), and topics involving supramolecular compounds.

The index entries are presented in letter-by-letter alphabetical sequence. Compounds are normally indexed under the parent compound name, with the substituent component separated by a comma of inversion. An entry with a prefix appears after the same entry without any attachments, and in alphanumerical sequence. For example, "paracyclophane," "[2.2]-paracyclophane," and "2,9-diketo-[2.2]-paracyclophane" will appear as:

Paracyclophane
[2.2]-Paracyclophane
[2.2]-Paracyclophane, 2,9-diketo-

Because authors may have approached similar topics from different viewpoints, index entries to those topics may not always appear under the same headings. Both synonyms and alternatives should therefore be considered to obtain all the entries on a particular topic. Commonly used synonyms include alkyne/acetylene, compound/complex, preparation/synthesis, etc.

WITHDRAWAL